Selected Titles in This Series

29 **Fritz Gesztesy, Helge Holden, Jürgen Jost, Sylvie Paycha, Michael Röckner, and Sergio Scarlatti, Editors,** Stochastic processes, physics and geometry: New interplays. II: A volume in honor of Sergio Albeverio

28 **Fritz Gesztesy, Helge Holden, Jürgen Jost, Sylvie Paycha, Michael Röckner, and Sergio Scarlatti, Editors,** Stochastic processes, physics and geometry: New interplays. I: A volume in honor of Sergio Albeverio

27 **Michel Théra, Editor,** Constructive, experimental, and nonlinear analysis

26 **Luis G. Gorostiza and B. Gail Ivanoff, Editors,** Stochastic models

25 **V. Jurdjevic and R. W. Sharpe, Editors,** Geometric control and non-holomorphic mechanics

24 **Idun Reiten, Sverre O. Smalø, and Øyvind Solberg, Editors,** Algebras and modules II

23 **Idun Reiten, Sverre O. Smalø, and Øyvind Solberg, Editors,** Algebras and modules I

22 **Vlastimil Dlab and László Márki, Editors,** Trends in ring theory

21 **S. W. Drury and M. Ram Murty, Editors,** Harmonic analysis and number theory: Papers in honour of Carl S. Herz

20 **J. Borwein, P. Borwein, L. Jörgenson, and R. Corless, Editors,** Organic mathematics

19 **Raymundo Bautista, Roberto Martínez-Villa, and José Antonio de la Peña, Editors,** Representation theory of algebras and related topics

18 **Raymundo Bautista, Roberto Martínez-Villa, and José Antonio de la Peña, Editors,** Representation theory of algebras

17 **V. Kumar Murty, Editor,** Seminar on Fermat's Last Theorem

16 **Bruce N. Allison and Gerald H. Cliff, Editors,** Representations of groups

15 **Karl Dilcher, Editor,** Number theory

14 **Vlastimil Dlab and Helmut Lenzing, Editors,** Representations of algebras

13 **R. A. G. Seely, Editor,** Category theory 1991

12 **A. Nicas and W. F. Shadwick, Editors,** Differential geometry, global analysis, and topology

11 **H. Tachikawa and V. Dlab, Editors,** Representations of finite dimensional algebras

10 **A. Bialynicki-Birula, J. Carrell, P. Russell, and D. Snow, Editors,** Group actions and invariant theory

9 **Joel S. Feldman and Lon M. Rosen, Editors,** Mathematical quantum field theory and related topics

8 **F. V. Atkinson, W. F. Langford, and A. B. Mingarelli, Editors,** Oscillation, bifurcation and chaos

7 **H. Kisilevsky and J. Labute, Editors,** Number theory

6 **J. Carrell, A. V. Geramita, and P. Russell, Editors,** Proceedings of the 1984 Vancouver Conference in algebraic geometry

5 **D. J. Britten, F. W. Lemire, and R. V. Moody, Editors,** Lie algebras and related topics

4 **I. Hambleton and C. R. Riehm, Editors,** Quadratic and Hermitian forms

3 **Zeev Ditzian, Amram Meir, Sherman D. Riemenschneider, and Ambikeshwar Sharma, Editors,** Second Edmonton conference on approximation theory

2 **Richard M. Kane, Stanley O. Kochman, Paul S. Selick, and Vector P. Snaith, Editors,** Current trends in algebraic topology (Parts 1 and 2)

1 **Carl Herz and R. Rigelhof, Editors,** 1980 Seminar on harmonic analysis

STOCHASTIC PROCESSES, PHYSICS AND GEOMETRY: NEW INTERPLAYS. II

A VOLUME IN HONOR OF SERGIO ALBEVERIO

This volume is dedicated to **Sergio Albeverio**
on the occasion of his **60th birthday**
by his students, colleagues, and friends

Canadian Mathematical Society
Société mathématique du Canada

CONFERENCE PROCEEDINGS • Volume 29

STOCHASTIC PROCESSES, PHYSICS AND GEOMETRY: NEW INTERPLAYS. II

A VOLUME IN HONOR OF SERGIO ALBEVERIO

Proceedings of the Conference on Infinite
Dimensional (Stochastic) Analysis and Quantum Physics
Max Planck Institute for Mathematics in the Sciences, Leipzig
January 18–22, 1999

Fritz Gesztesy
Helge Holden
Jürgen Jost
Sylvie Paycha
Michael Röckner
Sergio Scarlatti
Editors

Published by the American Mathematical Society
for the Canadian Mathematical Society
PROVIDENCE, RHODE ISLAND

CMS CONFERENCE PROCEEDINGS
Series Editors
Anthony V. Geramita and Niky Kamran
Volume 29

Accepted for publication by the Canadian Mathematical Society
Published by the American Mathematical Society

2000 *Mathematics Subject Classification.* Primary 58J20, 58J65, 60G15, 60G20, 60H07, 60H10, 60H15, 60H30, 60H40, 60J60, 60J65, 82B31, 82B44.

Library of Congress Cataloging-in-Publication Data

Conference on Infinite Dimensional (Stochastic) Analysis and Quantum Physics (1999 : Leipzig, Germany)

Stochastic processes, physics and geometry : new interplays : a volume in honor of Sergio Albeverio : proceedings of the Conference on Infinite Dimensional (Stochastic) Analysis and Quantum Physics, Max Planck Institute for Mathematics in the Sciences, Leipzig, January 18–22, 1999 / Fritz Gesztesy... [et al.], editors.

p. cm. — (Conference proceedings / Canadian Mathematical Society, ISSN 0731-1036 ; v. 28–29)

Includes bibliographical references.

ISBN 0-8218-1959-3 (v. 1 : alk. paper) — ISBN 0-8218-1960-7 (v. 2 : alk. paper)

1. Stochastic analysis—Congresses. 2. Mathematical physics—Congresses. 3. Quantum theory—Congresses. I. Albeverio, Sergio. II. Gesztesy, Fritz, 1953– III. Title. IV. Conference proceedings (Canadian Mathematical Society) ; v. 28–29.

QC20.7.S8 .C66 1999
530.15′923—dc21 00-046237

---

10 9 8 7 6 5 4 3 2 1 05 04 03 02 01 00

Contents

Preface

In order to celebrate Sergio Albeverio's 60th birthday that took place on January 17, 1999, an international conference was organized in Leipzig during January 18–22, 1999. The theme of the conference, "Infinite Dimensional (Stochastic) Analysis and Quantum Physics" was chosen to reflect the wide range of Sergio's scientific interests, without exhausting it, however. The number of Sergio's friends, colleagues, collaborators, and students that attended the conference, almost 250 participants altogether, exceeded the most optimistic estimates of the organizers.

The venue of the conference was the Alte Handelsbörse, a beautiful rococo style building at the Naschmarkt in Leipzig, behind the Renaissance style Altes Rathaus (Old Town Hall) in the city center. The atmosphere of the interior architecture and decoration provided a unique background for the various lectures.

The lecturers presented surveys in some fields of mathematics and physics covered by the conference theme as well as recent advances in those areas. Many lecturers emphasized the scientific impact of Sergio's contributions and expressed both their respect for him as a scientist and their admiration for him as a human being. One afternoon was reserved for presentations concerning his scientific contributions and his personal influence, with an opening address by Philippe Blanchard describing the life and personality of Sergio, included in this volume, and lectures by Helge Holden, Werner Kirsch, and Michael Röckner that covered some of the directions of Sergio's research, as experienced by those speakers in many years of intense and stimulating collaboration with Sergio.

A lecture by Helge Holden paid special tribute to Raphael Høegh-Krohn, a close friend of Sergio's, and to the intense collaboration between them, that was cut short by Raphael's untimely death in 1988.

The conference succeeded in presenting a vivid picture of Sergio Albeverio as a scientist and a human being, of the enormous range of scientific interests that he manages to cover, of his many profound insights and results, of his devotion to science, of his outstanding diligence and productivity, and of his kind and generous personality that leaves a deep impression on everyone associated with him. Contrary to most scientists, Sergio's scientific productivity has not declined with age. At age sixty one, his current list of publications runs at well over 400 items, with numerous papers presently in preparation. Given the breadth of his interests, and the sheer volume of his productivity, a comprehensive overview of his scientific accomplishments is not possible in a mere preface. It is hoped that the conference lectures and the large number of contributions to both volumes, can offer a glimpse at Sergio's encyclopedic view of science. Sergio himself was the most avid listener at every lecture, taking a deep interest in the results and ideas presented, asking many profound questions, mapping out further ideas and perspectives, and being deeply moved by the personal touches and memories recalled in many of the talks.

The conference also succeeded in describing the state-of-the art in current research at the interface between mathematics and physics. The present two-volume collection of the conference proceedings covers a broad range of fields and provides a detailed overview of many recent developments in the areas of stochastic processes, mathematical physics, and geometry, especially highlighting their deep interplay. More specifically, they contain contributions written by internationally recognized experts in the fields of stochastic analysis, linear and nonlinear (stochastic) partial differential equations, mathematics of finance, infinite dimensional analysis, functional analysis, nonstandard analysis, fractals, commutative and noncommutative probability theory, integrable systems, quantum and statistical mechanics, geometric quantization, neural networks, and applications to various other fields, including biology.

While the vast majority of these contributions are high-level research papers, a few reviews on subjects of general interest are also included. The contributions were selected with the intention of introducing the reader to advanced topics, emphasize various interdisciplinary connections, and stress possible future research directions. Volume I contains invited contributions; contributed papers are collected in Volume II.

The organizing committee of the conference, Ph. Blanchard, E. B. Dynkin, J. Fröhlich, J. Jost, and M. Röckner, would like to take this opportunity to express their gratitude to the Max Planck Society (MPG) and the German Science Foundation (DFG) for the financial support through the Max-Planck-Institut für Mathematik in den Naturwissenschaften, Leipzig, respectively, the Sonderforschungsbereich "Diskrete Strukturen in der Mathematik", Bielefeld, and the DFG-Schwerpunkt "Interagierende stochastische Systeme hoher Komplexität".

We owe the success of the conference to Birgit Kruschwitz and Walter Hoh, whose hard work and dedication to details, in spite of an ever increasing number of participants that required numerous adjustments, ensured that everything ran smoothly.

We are grateful to N. Kamran and C. Thivierge for valuable advice throughout the preparations of both volumes. In addition, we thank all authors for their contributions and the referees for their assistance.

The Editors
May 2000

Canadian Mathematical Society
Conference Proceedings
Volume **29**, 2000

Valuation of bonds and options under floating interest rate

V. Adamyan and B. Pavlov

Dedicated to Professor Sergio Albeverio on the occasion of his sixtieth birthday.

ABSTRACT. Evolution operators, generators of which contain a numerical parameter forming a Markov process, are considered in connection with problems of financial mathematics. Under certain conditions exact and explicit expressions for the values of the evolution operators averaged over trajectories of the process and for the corresponding variances are derived. Obtained results are applied for the valuation of some financial products that take into account the floating interest rates.

1. Introduction: Averaged evolution versus stochastic calculus.

The standard Black-Scholes analysis, [**3, 4, 14**] usually is based on the hypothesis, that the asset price follows a lognormal random walk and that the corresponding risk-free interest rate $r \equiv r_0$ and asset volatility are assumed to be known functions of time. The Black-Scholes algorithm for option pricing is derived under the assumption that there are no arbitrage possibilities, which means that all risk-free portfolios earn the same return and the trading of the underlying asset goes continuously so that one can sell or buy any share of the asset immediately, if it is necessary. For the fixed interest rate $r = r_0$ these assumption give the following standardized form of the Black-Scholes equation for an European call option value normalized with respect to exercise price E in the logarithmic scale of prices $S = Ee^x$ and reversed time $\tau : t = T - \frac{2\tau}{\sigma^2}$:

$$\frac{\partial v}{\partial \tau} = \frac{\partial^2 v}{\partial x^2} + \left(\frac{2r}{\sigma^2} - 1\right)\frac{\partial v}{\partial x} - \frac{2r}{\sigma^2}v$$

with the initial condition

$$v(x, 0) = \max(e^x - 1, 0)$$

1991 *Mathematics Subject Classification.* 47D03, 47D07, 91B28.

Our research was supported by the Marsden Fund of the Royal Society of New Zealand, grant 3368152. V. Adamyan was also supported by the Civilian Research and Development Foundation of USA and the Government of Ukraine in the framework of joint project, Award UM 1 - 298. V. Adamyan is grateful to the Department of Mathematics of the University of Auckland for hospitality during his visit to New Zealand in 1998.The authors are grateful to the referee for careful reading of the text and suggested improvements.

and exponential asymptotics as $x \to \infty$. Here σ is the volatility of the underlying asset's price and r is the interest rate, see for instance [**14**]. The analysis of Bond pricing requires a random interest rate, see for instance [**6, 10**]. In this case the interest rate is usually assumed to be governed by some stochastic differential equation defining the local increment of interest rate in terms of time and the absolute value of the interest rate, e.g.

$$dr = w(r,t)dX + u(r,t)dt.$$

For special choices of the parameters u, w the equation for a Bond price is similar to Black-Scholes equation and was solved by different authors, see [**9**] via a stochastic calculus approach.

In this note we study the problem of **option pricing** under a floating interest rate using remarkable properties of the averaged evolution first noticed by R. Griego and R. Hersh, see [**8**]. For the mathematically similar problem of the description of quantum evolution with Markov background these properties were rediscovered and intensely used by physicists for investigation of Mössbauer Scattering in disperse media, see [**2**]. The rigorous analysis of the models of quantum systems with Markov background was done in [**12**] and [**5**]. In both papers [**12, 5**] the investigation of the averaged dynamics was successfully reduced to the study of the spectral properties of a well-defined nonselfadjoint (dissipative) operator - the generator of the semigroup formed by the resolving operator for the Schroedinger equation with random Hamiltonian averaged over all trajectories of the stochastic process.

Our work, presented here, is an attempt to demonstrate that the same approach may be applied to problems concerning pricing of some financial products under *random* parameters in stochastic differential equations for underlying assets. In Section 2 we consider the Cauchy problem for the equation

$$\begin{cases} \frac{\partial}{\partial t}\psi = -H(t)\psi, \ t > 0, \\ \psi(0) = \psi_0 \end{cases} \tag{1.1}$$

in the Hilbert space $\mathcal{K}$ with densely defined closed operator function $H(t)$, consequent values of which form some stochastic process. Following the main idea of [**8**] for the problem (1.1) we obtain under certain assumptions explicit formulae for the values of the quadratic form and for the matrix elements of resolving operator averaged over all trajectories of the process and for the corresponding variance at any moment of time t. In Section 3 the general formulae so obtained are used in the simplest case $\dim \mathcal{K} = 1$ for the calculation (in the frameworks of a model) of present values of zero-coupon bonds under the assumption that the spot values of interest rates form a Markov process. In Section 4 under the same assumption the exact expression for the analogue of the Black-Scholes formulae for averaged values of European options and corresponding variances are presented.

2. Operators of Averaged Evolution

If the variable interest rate r is a bounded measurable function of time, then the capital gain from the investment of x_0 dollars under this variable interest rate can be calculated via solution of the differential equation

$$\frac{dx}{dt} = r(t)x,$$

$$x(0) = x_0,$$

which gives

$$x(T) - x_0 = x_0 \left[e^{\int_0^T r(t)dt} - 1 \right].$$

We assume now that $r(t)$ is a Markov process governed by the Fokker-Planck equation for transition probabilities w:

$$\frac{dw}{dt} + Mw = 0, \tag{2.1}$$

where M is a bounded operator acting in a Hilbert space $\mathcal{L}$ of functions $f(r)$ on the corresponding stochastic space - the set $\mathcal{E}$ of all values of $r(t)$. The operator M is assumed to satisfy the condition that the Markov property for transition probabilities holds. For the purpose of pricing financial derivatives under a floating interest rate it is enough to consider the case, where $\mathcal{E}$ is a finite or countable set. Therefore we will assume that $\mathcal{L}$ is realized as $\mathbf{C}_N, N = \mathrm{card}\mathcal{E} < \infty$, with the euclidean norm or as the space $\mathbf{l}_2$ in countable case, respectively. We will assume also that the finite or infinite matrix M has the form

$$M = \lambda(I - W), \tag{2.2}$$

where $\lambda > 0$ and W is a stochastic matrix. Actually (2.2) means that possible changes of $r(t)$ are only instantaneous random jumps between points of $\mathcal{E}$ with the matrix of the transition probabilities W, which may happen at random moments of time distributed according to the Poisson law, that is, the probability of the occurance of n such moments in the interval $(0, t)$ is equal to

$$e^{-\lambda t} \frac{(\lambda t)^n}{n!}.$$

In other words (2.2) means that $r(t)$ is a pseudo-Poisson process [**7**]. For such a process the result of the averaging of the present values of financial derivatives over all anticipated trajectories $r(t), t > 0$, which begin at $r(0) = r_{in}$, may be calculated in general on the basis of the following special statement, Theorem 1 below, which is a result of application of the general ideas of the theory of quantum evolution with Markov background [**8, 2, 12, 5, 11**] to our special case.

Let us consider operators $H(r), r \in \mathcal{E}$, which acts in a Hilbert space $\mathcal{K}$ and

- are closed and have the same dense domain $\mathcal{D}$ for each $r \in \mathcal{E}$;
- $$\Re(H(r)\varphi, \varphi) \geq \gamma \|\varphi\|^2, \varphi \in \mathcal{D}, \tag{2.3}$$

  where $\gamma > -\infty$ is also independent of r;
- there exists $C > 0$ and $\alpha > 0$ such that the evolution operators

  $$Q(t; r) = \exp(-tH(r)), t > 0, \tag{2.4}$$

  satisfy the following uniform estimate for arbitrary $r, r' \in \mathcal{E}$ and $t \in (0, T), 0 < T < \infty$:

  $$\|Q(t; r) - Q(t; r')\| \leq Ct^{\alpha}. \tag{2.5}$$

The transformation of the Cauchy data $\psi(0) \in \mathcal{K}$ of the equation

$$\frac{\partial}{\partial t}\psi = -H(r(t))\psi, \tag{2.6}$$

along the selected trajectory $r(t)$ is formally produced by the following time-ordered product or $\mathbf{T}$ - product

$$\psi(t) = U(t;r)\psi(0) = \mathbf{T}\exp\left(-\int_0^t H(r(s))ds\right)\psi(0) :=$$

$$\lim_{\max \Delta_j \to 0} e^{-\Delta_n H[r(t_n')]} e^{-\Delta_{n-1} H[r(t_{n-1}')]} \ldots e^{-\Delta_1 H[r(t_1')]}, \tag{2.7}$$

where $\Delta_j = t_j - t_{j-1}, t_j' \in [t_{j-1}, t_j]; 0 = t_0 < t_1 < \ldots < t_n = t$.

Under our assumptions $r(t)$, and hence $H(r(t))$, with the probability one are piecewise constant functions having only a finite number of jumps on each interval $[0, t]$.

Let $t_1 < t_2 < \ldots < t_s$ be the set of all possible jumps on the time interval $(0, t)$ for the fixed trajectory $r(t)$. The evolution operator along this trajectory of the Markov process

$$U(t;r) = Q(t - t_s; r(t_s + 0)) \cdot Q(t_s - t_{s-1}; r(t_{s-1} + 0)) \cdot \ldots \cdot Q(t_1; r(0)) \tag{2.8}$$

generally does not form a semigroup. Nevertheless averaging over trajectories generates a semigroup of bounded operators which may be used to calculate the evolution operator (2.8) averaged over all trajectories starting and finishing at the fixed points $r(0) = r_{in}$ and $r(t) = r_{out}$:

$$\langle U(t;r) \rangle \mid_{r_{out}, r_{in}}$$

as described below.

Let $\text{diag}_r H$ be the linear operator on the tensor product $\mathcal{K} \times \mathcal{L}$ of the Hilbert spaces $\mathcal{K}$ and $\mathcal{L}$, which acts on vectors of the form $x \otimes f(r)$, where $x \in \mathcal{D}, f(\cdot) \in \mathcal{L}$, as follows

$$\text{diag}_r H\,(x \otimes f(\cdot))\,(r) = H(r)x \otimes f(r).$$

It is easy to see that $\Re\,\text{diag}_r H$ is a densely defined closed operator in $\mathcal{K} \times \mathcal{L}$ and $\Re\,\text{diag}_r H \geq \alpha > -\infty$. For M defined in (2) $\widehat{M}$ be the bounded operator on $\mathcal{K} \times \mathcal{L}$ defined as $I_{\mathcal{K}} \times M$, where $I_{\mathcal{K}}$ is the identity operator on $\mathcal{K}$, $I_{\mathcal{K}} e = e,\ e \in \mathcal{K}$. Under our assumptions each vector $\mathbf{h} \in \mathcal{K} \times \mathcal{L}$ can be represented as vector functions $h(r), r \in \mathcal{E}$, taking values in $\mathcal{K}$ such that

$$\|\mathbf{h}\|^2 = \sum_{r \in \mathcal{E}} \|h(r)\|_{\mathcal{K}}^2 .$$

We denote by $P_r, r \in \mathcal{E}$, operators from $\mathcal{K} \times \mathcal{L}$ onto $\mathcal{K}$, which acts on the vector $\mathbf{h}$ in $\mathcal{K} \times \mathcal{L}$ as

$$P_r \mathbf{h} = h(r).$$

Observe that $(P_r^* x)(r') = x \otimes \delta_{r,r'}, x \in \mathcal{K}$.

THEOREM 2.1. *Let the operators $H(r)$ and the process $r(t)$ satisfy the above assumptions. Then the evolution operator (2.8) averaged over all trajectories of the process with fixed ends $r(0) = r_{in}, r(t) = r_{out}$, is given by the expression*

$$\langle U(t;r) \rangle \mid_{r_{out}, r_{in}} = P_{r_{out}} \mathbf{G}(t) P_{r_{in}}^*, \tag{2.9}$$

where $\mathbf{G}(t), t > 0$, is the semigroup of bounded operators in $\mathcal{K} \times \mathcal{L}$ generated by $\text{diag}_r H + \widehat{M}$, i.e.

$$\mathbf{G}(t) = \exp\left(-t\left(\text{diag}_r H + \widehat{M}\right)\right), t > 0. \tag{2.10}$$

Proof. Let us consider the evolution operator $U(t;r)$ for the equation (2.8) on a fixed trajectory $r(t)$ which has s jumps in $(0,t)$. This operator we approximately represent in the form of a **T**-product, see (8) above:

$$\begin{aligned} U_n(t;r) &= \prod_{k=1}^{n} \exp\left(-\Delta_n H(r_k+0)\right), \\ \Delta_n &= \frac{t}{n},\ r_k = r((k-1)\Delta_n + 0). \end{aligned} \tag{2.11}$$

According to (2.3), (2.5) and (2.6, 2.7) for the uniform error of this approximation we have the estimate

$$\|U(t;r) - U_n(t;r)\| \leq \frac{s\ Ct^{\alpha} e^{-\gamma t}}{n^{\alpha}}. \tag{2.12}$$

To calculate the average of the evolution over the set of all trajectories with the fixed ends r_{in}, r_{out} we form first the **T**-product (2.11) over all trajectories passing the prescribed "gates" r_k, $k = 1, ..., n-1$ at the fixed exit times at $t_k = k\Delta_n$. By (2.1) the transition probability along such a trajectory is given by the product

$$\begin{aligned} w(r_1, ..., r_{n-1} \quad &| \quad r_{in}(=r_0), r_{out}(=r_n)) = \prod_{k=1}^{n} G(r_k,\, r_{k-1}, \Delta_n), \\ G(r_k,\, r_{k-1}, \Delta_n) \quad &= \quad (\exp(-M\Delta_n))_{|r_k,\, r_{k-1}}. \end{aligned} \tag{2.13}$$

Multiplying the evolution operator $U(t;r)_{|r_{out}, r_{in}}$ along a certain trajectory by the corresponding transition probability (2.13) and forming the sum of the obtained expressions over all trajectories passing at the times t_k the prescribed "gates" $r_k, 1 \leq k \leq N-1$, r_0, r_n fixed , we obtain $P_{r_{out}} \langle U(t;r)\rangle P^*_{r_{in}}$. Applying the same procedure to $U_n(t;r)$ yields the expression

$$P_{r_{out}} \langle U_n(t;r)\rangle P^*_{r_{in}} = \sum_{r_k \in \mathcal{E}} \prod_{k=1}^{n} G(r_k,\, r_{k-1}, \Delta_n) \exp\left(-\Delta_n H(r_{k-1})\right),$$

where first the product is formed over the fixed trajectory r_0, r_1, $\ldots r_n$, and then the summation is taken over all trajectories with fixed beginning $r_0 = r_{in}$ and fixed end $r_n = r_{out}$:

$$P_{r_{out}} \langle U_n(t;r)\rangle = P_{r_{out}} \prod_{k=1}^{n} \left(\exp(-M\Delta_n) \exp\left(-\mathrm{diag}_r H\Delta_n\right)\right) P^*_{r_{in}} =$$

$$P_{r_{out}} \left(\exp(-\widehat{M}\frac{t}{n}) \exp\left(-\mathrm{diag}_r H \frac{t}{n}\right)\right)^n P^*_{r_{in}}. \tag{2.14}$$

By our assumptions the probability of a trajectory $r(t)$ having exactly s jumps on the interval $(0,t)$ is

$$e^{-\lambda t} \frac{(\lambda t)^s}{s!}.$$

Therefore by (2.12)

$$\begin{aligned} &\|\langle U(t;r)\rangle\, |_{r_{out}, r_{in}} - \langle U(t;r)\rangle_n\, |_{r_{out}, r_{in}}\| \leq \\ &\leq \langle \|U(t;r)\, |_{r_{out}, r_{in}} - U(t;r)_n\, |_{r_{out}, r_{in}}\|\rangle \leq \end{aligned}$$

$$\leq \langle \|U(t;r) - U(t;r)_n\| \rangle \leq \frac{\lambda t^{1+\alpha}}{n^\alpha} e^{-\gamma t}. \tag{2.15}$$

Applying to (2.14) the Trotter formula [**13**], which is valid under our assumptions, we obtain the following formula for the strong limit of the products (2.14)

(2.16)

$$s - \lim_{n\to\infty} \left(\exp(-\widehat{M}\frac{t}{n}) \exp\left(-\mathrm{diag}_r H \frac{t}{n}\right)\right)^n = \exp\left(-t \left(\mathrm{diag}_r H + \widehat{M}\right)\right),$$

Combining (2.16) and (2.15) yields the assertion of the theorem.

Corollary *Let the operators $H(r)$ and the Markov process $r(t)$ satisfy the above assumptions and $\psi_{in}(t), t > 0$, be a random vector function defined by the expression*

$$\psi_{in}(t) := U(t;r)\psi_{in} = \mathbf{T}\exp\left(-\int_0^t H(r(s))ds\right)\psi_{in},\ \psi_{in} \in \mathcal{K}.$$

Then the values of the functionals $\Psi_{out,in}(t) := (\psi_{in}(t), \psi_{out}), \psi_{out} \in \mathcal{K}$ averaged over all trajectories $r(t)$ beginning at r_{in} or ending at r_{out} are equal respectively to

$$\langle \Psi_{out,in}(t)\rangle_{r_{in}} = \sum_{r_{out}} (P_{r_{out}} \mathbf{G}(t) P^*_{r_{in}} \psi_{in}, \psi_{out})_{\mathcal{K}} \tag{2.17}$$

or

$$\langle \Psi_{out,in}(t)\rangle_{r_{out}} = \sum_{r_{in}} (P_{r_{out}} \mathbf{G}(t) P^*_{r_{in}} \psi_{in}, \psi_{out})_{\mathcal{K}},. \tag{2.18}$$

We use the avaraging techniques to compute the variance of the random evolution. We assume in addition now that

- the linear operator

  $$\mathrm{diag}_r H \times I \times I + I \times \mathrm{diag}_r H \times I$$

  on the tensor product $\mathcal{K}\times\mathcal{K}\times\mathcal{L}$, which is defined on the linear set $\mathcal{D}\times\mathcal{D}\times\mathcal{L}$ and acts on vectors of the form $x_1 \otimes x_2 \otimes f(r)$, where $x_1, x_2 \in \mathcal{D}, f(\cdot) \in \mathcal{L}$, as follows

  $$\begin{aligned} &(\mathrm{diag}_r H \times I \times I + I \times \mathrm{diag}_r H \times I))\,(x_1 \otimes x_2 \otimes f(\cdot))\,(r) \\ = \;& H(r)x_1 \otimes x_2 \otimes f(r) + x_1 \otimes H(r)x_2 \otimes f(r) \end{aligned}$$

  is closable and its closure $\mathrm{diag}_r W$ has the property

  $$\Re\mathrm{diag}_r W \geq 2\gamma > -\infty.$$

- the semigroups generated by all the operators $H(r), \mathrm{diag}_r H, \mathrm{diag}_r W$ are *real* i.e. the congugation operator J,

  $$\mathrm{J}^2 = I,\ (\mathrm{J}x, y) = \overline{(x, y)}$$

  defined on the whole space $\mathcal{K}$, commutes with operators each of the above semigroups in the corresponding Hilbert spaces.

Let $\widehat{\widehat{M}}$ be the bounded operator in $\mathcal{K}\times\mathcal{L}$ defined by $I_{\mathcal{K}} \times I_{\mathcal{K}} \times M$. Observe that M and $\widehat{\widehat{M}}$ are real operators. As above, we represent each vector $\mathbf{h} \in \mathcal{K}\times\mathcal{K}\times\mathcal{L}$ as vector functions $h_{\times 2}(r), r \in \mathcal{E}$,with values in $\mathcal{K}\times\mathcal{K}$ and denote by $P_r^{\times 2}, r \in \mathcal{E}$, operators from $\mathcal{K}\times\mathcal{K}\times\mathcal{L}$ onto $\mathcal{K}$, which act on $\mathbf{h} \in \mathcal{K}\times\mathcal{K}\times\mathcal{L}$ by

$$P_r^{\times 2}\mathbf{h} = h_{\times 2}(r).$$

Applying the same arguments as in the proof of Theorem 2.1 we obtain in addition to (2.17) a tool for the calculation of the variance

$$\mathrm{Var}[\Psi_{out,in}(t)] := \left\langle |\Psi_{out,in}(t)|^2 \right\rangle_{r_{in}} - \left| \langle \Psi_{out,in}(t) \rangle_{r_{in}} \right|^2 . \tag{2.19}$$

THEOREM 2.2. *Let the operators $H(r)$ and the Markov process $r(t)$ satisfy the all above assumptions. Then the value of the functional $|\Psi_{out,in}(t)|^2$ averaged over all trajectories $r(t)$ beginning at r_{in} or ending at r_{out} equals, respectively*

$$\left\langle |\Psi_{out,in}(t)|^2 \right\rangle_{r_{in}} = \sum_{r_{out}} (P^{\times 2}_{r_{out}} \mathbf{G}_{\times 2}(t) P^{\times 2 *}_{rin} \psi_{in} \otimes \mathrm{J}\psi_{in}, \psi_{out} \otimes \mathrm{J}\psi_{out})_{\mathcal{K}\times\mathcal{K}} \tag{2.20}$$

and

$$\left\langle |\Psi_{out,in}(t)|^2 \right\rangle_{r_{out}} = \sum_{r_{in}} (P^{\times 2}_{r_{out}} \mathbf{G}_{\times 2}(t) P^{\times 2 *}_{r_{in}} \psi_{in} \otimes \mathrm{J}\psi_{in}, \psi_{out} \otimes \mathrm{J}\psi_{out})_{\mathcal{K}\times\mathcal{K}}, \tag{2.21}$$

where $\mathbf{G}_{\times 2}(t), t > 0$, is the semigroup of bounded operators on $\mathcal{K}\times\mathcal{K}\times\mathcal{L}$ generated by $\mathrm{diag}_r W + \widehat{\widehat{M}}$, i.e.

$$\mathbf{G}_{\times 2}(t) = \exp\left(-t \left(\mathrm{diag}_r W + \widehat{\widehat{M}} \right) \right), t > 0. \tag{2.22}$$

Proof. Under our assumptions for a given trajectory $r(t)$ with arbitrary initial point r_{in} the operator $U(t;r)$ is real. Therefore

$$|\Psi_{out,in}(t)|^2 = (U(t;r)\psi_{in}, \psi_{out})\overline{(U(t;r)\psi_{in}, \psi_{out})} =$$

$$= (U(t;r)\psi_{in}, \psi_{out})(U(t;r)\mathrm{J}\psi_{in}, \mathrm{J}\psi_{out}) =$$

$$= (U_{\times 2}(t;r)(\psi_{in} \otimes \mathrm{J}\psi_{in}), \psi_{out} \otimes \mathrm{J}\psi_{out})_{\mathcal{K}\times\mathcal{K}} , \tag{2.23}$$

where $U_{\times 2}(t;r) = U(t;r) \otimes U(t;r)$ is the resolving operator for the Cauchy problem for the equation

$$\frac{\partial}{\partial t}\psi = \overline{(H(r(t)) \times I + I \times H(r(t)))\psi}$$

in $\mathcal{K}\times\mathcal{K}$. Now using (2.23), and repeating word for word the reasoning given in the proof of Theorem 2.1 we obtain (2.20) with $\mathbf{G}_{\times 2}(t)$ defined by (2.22).

3. Bond valuation

Recall that a pure discount bond, or zero coupon bond, makes no intermediate payments between its issue date and its maturity date T. It promises only to pay a certain amount at its maturity, which is called its par value, or face value. The difference at any intermediate t between the par value Z and the lower selling price $V(t,T)$ is the bond discount. If the interest rate varies deterministically, then

$$V(t,T) = \exp\left(- \int_t^T r(s)ds \right) Z. \tag{3.1}$$

If the values of the interest rate $r(s)$ form a Markov process governed by (2.1), $r(t) = r_{in}$ and there are no arbitrage opportunities, then

$$V(t,T) = \left\langle \exp\left(- \int_t^T r(s)ds \right) Z \right\rangle_{r_{in}}, \tag{3.2}$$

where, as before, the brackets with subscript r_{in} denote the ensemble averaging over all trajectories of the process beginning from r_{in}.

Let us assume that $r(s)$ takes only $2N+1$ or $2N$ equally spaced values

$$r_m = r_0 + m\delta,\ m = 0, \pm 1, ..., \pm N,\ \delta > 0, \tag{3.3}$$

or

$$r_m = r_0 + m\delta,\ m = \pm\frac{1}{2}, ..., \pm\frac{2N+1}{2},\ \delta > 0, \tag{3.4}$$

respectively. Then for the problem of the calculation of $V(t,T)$ the operator M in (2.1) can be considered as a $(2N+1) \times (2N+1)$ or $2N \times 2N$ matrix satisfying the Markov property and the conservation law for transition probabilities. If the state space $\mathcal{K}$ for each value r_m is one-dimensional then the operator $H(r_m)$ is simply the multiplication by r_m and hence $\text{diag}_r H$ is the diagonal matrix $r_0 I + R$ with elements $r_m \delta_{m,m'}$, where $\delta_{m,m'}$ is the Kronecker symbol and I is the identity matrix. Applying Theorem 2.1 and Corollary 2 we obtain an expression for the averaged selling price of the Bond in the form:

$$V(t,T) = \sum_m \left(e^{-(R+M)(T-t)} \right)_{m,m_{in}} \cdot e^{-r_0(T-t)} Z. \tag{3.5}$$

with the initial value of the interest rate $r(0) = r_0 + m_{in}\delta$, defined by the stochastic state m_{in}.

From now on we shall consider the special case, where for r taking $N' = 2N+1$ or $N' = 2N$ values (3.3) or (3.4), respectively we have

$$M = \lambda(I - P), P = \frac{1}{N'} \begin{pmatrix} 1 & 1 & ... & 1 \\ 1 & 1 & ... & 1 \\ ... & ... & ... & ... \\ 1 & 1 & ... & 1 \end{pmatrix}, \lambda > 0. \tag{3.6}$$

Here I is the identity matrix; P is a rank one matrix, which represents projection onto the one-dimensional subspace of column vectors with the all elements equal to each other. Observe that such a choice of M has the property that

$$e^{-(T-t)M} = P + (I - P)e^{-\lambda(T-t)},$$

$$\lim_{T\to\infty} e^{-(T-t)M} = P = \frac{1}{N'} \begin{pmatrix} 1, & ... & 1 \\ ... & ... & ... \\ 1, & ... & 1 \end{pmatrix}.$$

Using the notation

$$\mathbf{e} = \frac{1}{\sqrt{N'}} \begin{pmatrix} 1 \\ ... \\ 1 \end{pmatrix}.$$

we obtain the representation for P of the form

$$P = (*, \mathbf{e})_{\mathbf{C}_{N'}} \mathbf{e}.$$

Consequently, for the generator of the averaged evolution, we have

$$R + M = \lambda I + R - \lambda(.,\mathbf{e})_{\mathbf{C}_{N'}}\mathbf{e}.$$

Based on the Theorem 2.1 we calculate now the averaged evolution of the investment via the Riesz integral of the resolvent of $R+M$. Let us consider the resolvent of the "nonperturbed" diagonal operator R :

$$G_w = (R-w)^{-1}.$$

Then the resolvent of $R+M$ is given by the Krein formula

$$(R+M-w)^{-1} = G_{w-\lambda} + \lambda\Theta(w-\lambda)^{-1}(*, G_{\overline{w}-\lambda}\mathbf{e})_{\mathbf{C}_{N'}} G_{w-\lambda}\mathbf{e}$$

involving the perturbation denominator (called also perturbation determinant in general matrix case when P is a multi-dimensional operator):

$$\Theta(w) = 1 - \lambda(G_w\mathbf{e},\mathbf{e}) = 1 - \frac{\lambda}{N'}\sum_m \frac{1}{m\delta - w}.$$

For $N' = 2N+1$ and $N' = 2N$ we have respectively

$$\begin{aligned} \Theta_{2N+1}(w) &= 1 + \frac{\lambda}{2N+1}\left(\frac{1}{w} + \sum_{n=1}^{N}\frac{2w}{w^2 - n^2\delta^2}\right), \\ \Theta_{2N}(w) &= 1 + \frac{\lambda}{N}\sum_{n=1}^{N}\frac{w}{w^2-(n-\frac{1}{2})^2\delta^2}. \end{aligned}$$

Considering the Riesz Integral over a simple contour Λ encircling all eigenvalues of the operators R and $R+M$ we may represent the evolution operator as

$$e^{-(R+M)(T-t)} = -\frac{1}{2\pi i}\oint_\Lambda e^{-(T-t)w}(R+M-w)^{-1}dw =$$

$$-\frac{1}{2\pi i}\oint_\Lambda e^{-(T-t)w}\left(G_{w-\lambda} + \lambda[\Theta(w-\lambda)]^{-1}(*, G_{\overline{w}-\lambda}\mathbf{e})_{\mathbf{C}_{N'}} G_{w-\lambda}\mathbf{e}\right)dw. \tag{3.7}$$

Since

$$\begin{aligned} &\sum_m \left(G_{w-\lambda} + \lambda[\Theta(w-\lambda)]^{-1}(*, G_{\overline{w}-\lambda}\mathbf{e})_{\mathbf{C}_{N'}} G_{w-\lambda}\mathbf{e}\right)_{m,m_{in}} \\ = \ &[\Theta(w-\lambda)]^{-1}\frac{1}{m_{in}\delta + \lambda - w}, \end{aligned}$$

then for the present value of the interest rate $r_0 + m_i\delta$ from (3.5) and (3.7)we obtain the sum of residues:

$$V(t,T) = e^{-(T-t)(r_0+\lambda)}\sum_{w_n}\frac{1}{\Theta'(w_n)(w_n - m_{in}\delta)}e^{-(T-t)w_n}, \tag{3.8}$$

where w_n runs over the set of all zeros of the perturbation denominator $\Theta(w)$.

In the simplest case when the interest rate $r(t)$ randomly jumps between two values $r_0 \pm \frac{\delta}{2}$, applying (3.8) yields

$$V(t,T) = e^{-(r_0+\frac{\lambda}{2})(T-t)}\cosh\frac{1}{2}\sqrt{\lambda^2+\delta^2}(T-t)Z+$$

$$e^{-(r_0+\frac{\lambda}{2})(T-t)}\frac{\lambda\pm\delta}{\sqrt{\lambda^2+\delta^2}}\sinh\frac{1}{2}\sqrt{\lambda^2+\delta^2}(T-t)Z. \tag{3.9}$$

Since in the case of bond valuation $\dim \mathcal{K} = \dim(\mathcal{K} \times K) = 1$ Theorem 2.2 appears to be trivial for the calculation of the mean-square deviation of the bond price from its expected value

$$D_{m_{in}}(V) := \left[\left\langle \exp\left(-2\int_t^T r(s)ds\right)\right\rangle_{r_{in}} - V^2(t,T)\right]^{\frac{1}{2}}. \tag{3.10}$$

Indeed, by (3.10) the average of the distribution of squares of the bond prices may be found by the same formulas (3.8), (3.9), in which only r_0 and δ must be doubled.

For the model (3.9) under the condition $\delta \ll \lambda$, $\lambda(T-t) \ll 1$ we have

$$\frac{D_{m_{in}}(V)}{V(T,t)} \approx O\left(\sqrt{\lambda\delta^2}(T-t)^{\frac{3}{2}}\right).$$

Note that the above obtained expressions may be used for the specification of time values of a series of payments, future values etc. which takes into account random jumps in interest rates.

4. The averaged formula for the Black-Scholes option pricing model

From the formal point of view the calculation of averaged values of financial derivatives in the framework of the Black-Scholes model is reduced to the determination of the averaged resolving operator of the Cauchy problem for the equation

$$\frac{\partial v}{\partial \tau} = \frac{\partial^2 v}{\partial x^2} + (q(\tau) - 1)\frac{\partial v}{\partial x} - q(\tau)v,\; q(\tau) = \frac{2r(\tau)}{\sigma^2}, -\infty < x < \infty, \tag{4.1}$$

Here $r(\tau) \geq 0, 0 < \tau < \infty$, is the stochastic process with transition probabilities governed by the equation

$$\frac{dw}{d\tau} = -\frac{2}{\sigma^2}Mw,\; \tau = \frac{\sigma^2}{2}(T-t). \tag{4.2}$$

Under our assumptions the set of values of $r(\tau)$ is finite or countable. Equation (4.1) has the form (1.1) with densely defined closed operator

$$H_{BS}(r) = -\frac{\partial^2}{\partial x^2} - (q(\tau) - 1)\frac{\partial}{\partial x} + q(\tau)$$

in the Hilbert space $\mathcal{K} = \mathbf{L}^2(-\infty, \infty)$, which is defined on the r-independent linear set of all absolutely continuous functions with absolutely continuous first derivative and square integrable second derivative.

Proposition *The operators $H_{BS}(r)$ satisfy conditions (2.3) and (2.5) with $\alpha = \frac{1}{2}$.*

Proof The standard unitary Fourier transform in $\mathbf{L}^2(-\infty,\infty)$

$$(\mathcal{F}f)(k) := \frac{1}{\sqrt{2\pi}}\int_{-\infty}^{\infty} e^{-ikx}f(x)dx,\; f \in \mathbf{L}^2(-\infty,\infty),$$

converts $H_{BS}(r)$ into the multiplication operator by the function

$$\widehat{H}_{BS}(r;k) := k^2 + q - ik(q-1).$$

We have (2.3) since

$$\Re \widehat{H}_{BS}(r;k) = k^2 + q \geq 0.$$

The image of the semigroup of operators $Q(t;r)$ generated by $H_{BS}(r)$ under the Fourier transformation is the semigroup of the multiplication operators

$$\widehat{Q}(t;r)(k) = \exp\left(-\widehat{H}_{BS}(r;k)t\right),\ t > 0.$$

Taking into account that for different values r_1, r_2 satisfying the condition

$$\left|e^{-r_1 t} - e^{-r_2 t}\right| \leq t$$

we have the inequality

$$\begin{aligned} 4e^{-2k^2 t}\sin^2\left(\frac{r_1 - r_2}{2}kt\right) &= \frac{1}{2}t(r_1 - r_2)^2\left(2k^2 t e^{-2k^2 t}\right)\frac{4\sin^2\left(\frac{r_1 - r_2}{2}kt\right)}{(r_1 - r_2)^2 k^2 t^2} \\ &\leq \frac{1}{2e}(r_1 - r_2)^2 t. \end{aligned}$$

Hence we obtain for any $f \in \mathbf{L}^2(-\infty,\infty)$, the estimate

$$\begin{aligned} &\int_{-\infty}^{\infty} \left|Q(t;r_1)(k) - Q(t;r_2)(k)\right|^2 \left|f(k)\right|^2 dk \\ =& \int_{-\infty}^{\infty} \left|e^{-(ik+1)r_1 t} - e^{-(ik+1)r_2 t}\right|^2 e^{-2k^2 t}\left|f(k)\right|^2 dk \\ \leq& \left|e^{-r_1 t} - e^{-r_2 t}\right|^2 \|f\|^2 + 4\int_{-\infty}^{\infty} e^{-2k^2 t}\sin^2\left(\frac{r_1 - r_2}{2}kt\right)\left|f(k)\right|^2 dk \\ \leq& \left(t^2 + \frac{1}{2e}t\right)(r_1 - r_2)^2 \|f\|^2. \end{aligned}$$

Therefore

$$\|Q(t;r_1) - Q(t;r_2)\|_{t\to 0} = O(\sqrt{t}).$$

Due to Proposition above the operators $H_{BS}(r)$ satisfy all the conditions of Theorem 2.1. Hence the averaged resolving operator for the Cauchy problem for the equation (4.1) can be calculated via Theorem 2.1 and Corollary 2 using as a tool the related semigroup $\mathbf{G}(t)$ in the space $\mathbf{L}^2(-\infty,\infty) \times \mathcal{L}$, which may be considered as the Hilbert space $\mathbf{L}^2(-\infty,\infty;\mathcal{L})$ of vector functions $f(x)$ on the real axis with values in $\mathcal{L}$ and norm

$$\|f\|^2 = \int_{-\infty}^{\infty} \|f(x)\|_{\mathcal{L}}^2\,dx.$$

Under the unitary Fourier Transform the semigroup $\mathbf{G}(t)$ is converted into the semigroup of operators acting as "multiplication" by the operator functions

$$\widehat{\mathbf{G}}(t)(k) = \exp\left(-(\widehat{H}_{BS}(r;k) + \frac{2}{\sigma^2}M)\tau\right).$$

We assume now that $r(\tau)$ takes only a finite number of different equally spaced values as in (3.3), (3.4) and M has the form (3.6) as in Section 3 and keep the notations r_0, r_m introduced there:

$$\begin{aligned} q_0(k) &= \frac{2r_0}{\sigma^2} - ik(\frac{2r_0}{\sigma^2} - 1) + k^2; \\ \epsilon(k) &= (1 - ik)\frac{2\delta}{\sigma^2}, \ \eta = \frac{2\lambda}{\sigma^2}. \end{aligned}$$

Then for each $k \in (-\infty, \infty)$ we see that

$$\operatorname{diag}_r \widehat{H}(r; k) = q_0(k) \cdot I + R(k)$$

is the diagonal matrix with elements $(q_0(k) + \epsilon(k)m)\delta_{m,m'}$. With this notations

$$\widehat{\mathbf{G}}(t)(k) = e^{-q_0(k)\tau} \exp\left(-(R(k) + \eta M)\tau\right). \tag{4.3}$$

If M is defined by (3.6), then introducing as above the function

$$\Theta(z, k) = 1 - \eta((R(k) - z)^{-1}\mathbf{e}, \mathbf{e}) = 1 - \frac{\eta}{N'} \sum_m \frac{1}{m\epsilon(k) - z}$$

and repeating the arguments that resulted in (3.8) we obtain the following expression for the Fourier transform of the resolving operator of the Black - Scholes equation (4.1) averaged over all trajectories of r with $r(t) = r_{in}$. For the problem with given data at time $T > t$ this coincides with multiplication by the function

$$G(k, T - t) = \exp\left(-\frac{\sigma^2}{2} q_0(k)(T - t)\right) \times$$

$$\sum_{w_n} \frac{1}{\Theta'(w_{n;r}, k)(w_{n;k} - m_{in}\epsilon(k))} e^{-(T-t)w_{n;k}}, \tag{4.4}$$

where $w_{n;k}$ runs over the set of all zeros of $\Theta(w, k)$.

Observe that the factor

$$G_0(k, T - t) = \exp\left(-\frac{\sigma^2}{2} q_0(k)(T - t)\right)$$

in (4.4) is just the Fourier transform of the Green function for the Black-Scholes equation with constant interest rate r_0 equal to the simple mean of those over the stochastic space $\mathcal{E}$.

Let us introduce the Fourier transform of (4.4)

$$D(x, T - t; r_{in}) := \sum_{w_n} \int_{-\infty}^{\infty} \frac{dk}{2\pi} \frac{1}{\Theta'(w_{n;r}, k)(w_{n;k} - m_{in}\epsilon(k))} e^{-(T-t)w_{n;k} + ikx}$$

and denote by $C(S, T - t; E, r_0, \sigma)$ the price of the simple European call option on some asset with present value $S = Ee^x$ calculated via the Black-Scholes formula[**3**] for given strike price E, volatility σ and the interest rate r_0. The following assertion is an interpretation of the expression (4.4).

THEOREM 4.1. *If the random process $r(\tau)$ takes only a finite number of different equally spaced values and the matrix of transition probabilities M for the stochastic process has form (3.6). Then the present value of the simple European* *call option averaged over all trajectories of the process $r(t')$ begining at $t' = t$ from r_{in} is given by the formula*

$$\langle C(S,T-t;E,r,\sigma)\rangle_{r_{in}} = \int_0^\infty \frac{dS'}{S'} D(\ln\frac{S}{S'}, T-t; r_{in}) C(S', T-t; E, r_0, \sigma).$$

Now on the basis of Theorem 2.2 let us calculate $\langle C^2(S,T-t;E,r,\sigma)\rangle_{r_{in}}$. The role of the operator W appearing in this theorem now is played the well-defined closed normal operator

$$H_{\times 2BS}(r) = -\frac{\partial^2}{\partial x_1^2} - \frac{\partial^2}{\partial x_2^2} - (q(\tau)-1)\left(\frac{\partial}{\partial x_1} + \frac{\partial}{\partial x_2}\right) + 2q(\tau)$$

in the Hilbert space

$$\mathcal{K} = \mathbf{L}^2(-\infty,\infty) \otimes \mathbf{L}^2(-\infty,\infty) = \mathbf{L}^2(\mathbf{E}_2).$$

Using the two-dimensional Fourier transformation we can prove as above that the operators $H_{\times 2BS}(r)$ satisfy conditions (2.3), (2.5) with $\alpha = \frac{1}{2}$. Therefore the averaged resolving operator for the Cauchy problem for the equation

$$\frac{\partial v}{\partial \tau} = H_{\times 2BS}(r)v, \ -\infty < x_1, x_2 < \infty,$$

can also be calculated by Theorem 2.1 and Corollary 2 . The image of resulting semigroup $\mathbf{G}_{\times 2}(t)$ under the two-dimensional Fourier transformation acts as a multiplication operator in $\mathbf{L}^2(\mathbf{E}_2) \otimes \mathcal{L}$ by the matrix function

$$\widehat{\mathbf{G}}_{\times 2}(t)(k) = e^{-q_0(k_1)\tau} e^{-q_0(k_2)\tau} \exp\left(-(R(k_1+k_2) + \eta M)\tau\right).$$

This yields the following assertion:

THEOREM 4.2. *Let the random process $r(\tau)$ be such as in Theorem 4.1. Then the square of the present value of the simple European call option averaged over all trajectories of the process $r(t')$ beginning from $t' = t$ at r_{in} is given by the formula*

$$\langle C^2(S,T-t;E,r,\sigma)\rangle_{r_{in}} = \int_0^\infty \frac{dS'}{S'} D(\ln\frac{S}{S'}, T-t; r_{in}) C^2(S', T-t; E, r_0, \sigma).$$

Acknowledgments. It is a great pleasure to thank the referee for a very careful reading of our manuscript and for numerous suggestions concerning improvements of our presentation.

References

[1] S. Albeverio, *Non linearity and disorder in classical and quantum systems, deterministic and stochastic*, in: Proc. El Escorial Summer School, 1993, Ed. L. Vazquez et al.

[2] G. Belozerskij, Y. Pavljuhin, *Mössbauer effect in disperse media*, Zhurnal eksperimentalnoj i teoreticheskoj fisiki**70**,(1976), 717, in russian.

[3] F.Black, M.Scholes, *The pricing of options and corporate Liabilities*, Journ. Pol. Econ., **81** (1973), 637–659.

[4] S.C. Blank, C.A. Carter, B.H. Schmiesing, *Futures and Option Markets*, Prentice-Hall, 1991.

[5] S. Cheremshantsev, *Theory of Scattering by a Brownian particle*, Proceedings of Steklov Institute of Acad.Sci.USSR,**184**, (1990).

[6] J.C. Cox, J. Ingersol, S. Ross, *A theory of the term structure of interest rates*, Econometrica, **53**, 1985, 385–467.

[7] W. Feller, *An Introduction to probability theory and its Application*, Vol. II, John Willey & Sons, Inc., 1966

[8] R. Griego, R. Hersh, *Random evolutions, Markov chains and systems of partial differential equations*, Proc. Nat. Acad. USA,**62**,(1969), 305–308

[9] J. Hull, *Options, Futures and other Derivative Securities*, second edition, Prentice Hall, 1993.

[10] J. Hull, A. White , *Pricing interest rate derivative Securities*, Rev. Fin. Stud. **3**, (1990) 573–592.

[11] B. Pavlov, *Quantum Dynamics on Markov Background and Irreversibility*,in the book : Nonlinear Dynamics, Chaotic and Complex Systems, ed.I.Infeld, R, Zelazny and A. Galkowsky, Cambridge University Press,1997, 190–205.

[12] B. Pavlov, A. Ryshkov *Scattering on a random point*, in the book: Lecture notes in Physics, **324**, Springer-verlag, 1989

[13] E. Trotter, *Approximation of semi-groups of operators*, Pacific J. Math., **8**(1958), 887–919

[14] P. Wilmott, S. Howison, J. Dewynne, *The Mathematics of Financial Derivatives*, Cambridge University Press, 1995.

Department of Theoretical Physics, Odessa State University, Odessa 65026, Ukraine
E-mail address: vadamjan@m-vox.odessa.ua

Department of Mathematics, The University of Auckland Private Bag 92019 Auckland , New Zealand; International Solvay Institute for Physics, Brussels, Belgium
E-mail address: pavlov@scitec.auckland.ac.nz

Canadian Mathematical Society
Conference Proceedings
Volume **29**, 2000

The Large Time Behaviour in Quantum Field Theory and Quantum Chaos

I.Ya. Aref'eva and I.V. Volovich

Dedicated to Sergio Albeverio on the occasion of his 60th birthday.

ABSTRACT. An exact general formula for the matrix elements of the evolution operator in quantum theory is established. The formula ("ABC-formula") has the form $< U(t) >= \exp(At + B + C(t))$. The constants A and B and the decreasing function $C(t)$ are computed in perturbation theory. The ABC-formula is valid for a general class of Hamiltonians used in statistical physics and quantum field theory. The formula includes the higher order corrections to the well known Weisskopf-Wigner approximation and to the stochastic (van Hove) limit which are widely used in considerations of problems of radiation, decay, quantum decoherence, derivation of master and kinetic equations etc. The function $C(t)$ admits an interpretation as an analogue of the autocorrelation function describing quantum chaos for the quantum baker's map.

1. Introduction

Scattering theory is one of the very old and developed topics in quantum theory. The study of the large time behaviour of the evolution operator in statistical physics and quantum field theory is the subject of numerous investigations. The basic object to study in quantum field theory is the scattering matrix, see in particular [**1**]–[**6**] and references therein.

However, there are many important problems in quantum field theory where we are interested in the large but not infinite time and where the standard S-matrix description is not very convenient or even not applicable. These include processes with unstable particles [**7, 8**], atom-photon interactions [**9**], elementary particles in "semidressed states" with non-equilibrium proper fields [**10**], electroweak baryogenesis and phase transitions in the early Universe and in high-energy collisions [**11**], quantum optics [**12**], quantum decoherence (see for example [**12, 13**]) etc. In the consideration of such processes we are interested in the time regime smaller than the "infinite" time when the S-matrix description becomes applicable. The

1991 *Mathematics Subject Classification.* 81T10, 82C10, 81Q50.

This work is supported in part by INTAS 96-0698, by RFFI-99-01-00166 (I.Ya.A.) and by RFFI-99-01-00105 (I.V.V.) and by grants for the leading scientific schools 96-15-96208 (I.Ya.A.) and 96-15-96131 (I.V.V.).

consideration of such processes belongs to non-equilibrium quantum field theory, see [**14**] for more discussions.

Various approximate methods of consideration of time evolution for classical and quantum systems have been developed by Bogoliubov [**15**], Weisskopf and Wigner (see [**8**]), van Hove [**16**], Prigogine [**17**] and many others, see [**18, 19**]. The purpose of this paper is to obtain an exact general result about the large time behaviour of certain matrix elements of the evolution operator

If H is a generic self-adjoint operator in a Hilbert space and we are interested in the study of the corresponding evolution operator e^{-itH} then we can say not much interesting about the behaviour of its matrix elements

$$(\psi, e^{-itH}\psi) = \int e^{-it\sigma} d\rho(\sigma)$$

We have to restrict ourself to the Hamiltonians of some special forms if we want to obtain useful results. In this paper we consider a general class of Hamiltonians used in solid state physics and quantum field theory

$$H = H_0 + \lambda V, \tag{1.1}$$

where H_0 is a free Hamiltonian, V describes an interaction and λ is the coupling constant. We shall study the evolution operator

$$U(t) = e^{itH_0}e^{-itH}. \tag{1.2}$$

The main result of this paper is the following exact formula (we call it the *ABC*-formula) which is valid for arbitrary time t

$$\langle U(t)\rangle = e^{At+B+C(t)}. \tag{1.3}$$

Here A and B are constants for which a representation in perturbation theory will be given and $C(t)$ is a function which under rather general assumptions can be represented for large time t as

$$C(t) = \frac{f(t)}{t^\alpha}. \tag{1.4}$$

Here $f(t)$ is a bounded function and the exponent α depends on the model and on the dimension of space ($\alpha = 3/2$ for the physical 3-dimensional space). The function $C(t)$ admits a representation which has the form similar to the autocorrelation function describibg quantum chaos, see below.

The well known Weisskopf-Wigner approximation [**8**] is given by

$$\langle U(t)\rangle \simeq e^{at}, \tag{1.5}$$

where a is a constant. It corresponds to the particular case of the *ABC* - formula (1.3).

Expectation value in (1.3) is taken over the vacuum vector. For the case of one-particle states we obtain

$$\langle p|U(t)|p'\rangle = e^{iA(p)t+B(p)+C(t,p)}\delta(p-p'). \tag{1.6}$$

The formulae (1.3) and (1.6) have a very general character. We prove (1.3) in Section 3 for a wide class of Hamiltonians. The class of considered Hamiltonians includes the Bose and Fermi gases, phonon self-interaction and electron-phonon interaction, quantum electrodynamics in external fields etc.

We derive the main formula (1.3) by using the theory of perturbation of spectra and renormalized wave operators [**5, 6**]. This method can be used only under some restrictions to the form of the Hamiltonian when one has not decay. Another method based on the direct examination of perturbation theory which can be used also in the case of decay will be considered in another publication.

2. Notations and auxiliary results

2.1. Hamiltonians. We consider Hamiltonians of the form (1.1) where H_0 is a free Hamiltonian

$$H_0 = \sum_i \int \omega_i(k) a_i^*(k) a_i(k) d^d k \tag{2.1}$$

and V is the sum of Wick monomials. Creation and annihilation operators $a_i^*(k)$, $a_i(k)$ describe particles or quasiparticles and they satisfy the commutation or anti-commutation relations

$$[a_i(k), a_j^*(k')]_\pm = \delta_{ij}\delta(k-k') \tag{2.2}$$

Here $k, k^{'} \in R^d$ and $i, j = 1, .., N$ label the finite number of different types of (quasi)particles. Creation and annihilation operators have the standard realization in the Fock space with the vacuum vector which will be denoted $|0\rangle$ or Φ_0 . Examples of one-particle energy $\omega_i(k)$ include the relativistic ($\omega(k) = (k^2 + m^2)^{1/2}$) and non-relativistic ($\omega(k) = k^2/2 - \omega_0$) laws, the Bogoliubov spectrum ($\omega(k) = (bk^4 + k^2 v(k))^{1/2}$), the Fermi quasiparticle spectrum ($\omega(k) = |k^2/2m - \mu|$) etc.

We consider two different types of Wick polynomials. The first type describes an interaction in the case when there is not the translation invariance

$$V = \sum_{I,J,i,j} \int v(p_1, i_1 \dots p_I, i_I | q_1, j_1 \dots q_J, j_J) \prod_{l=1}^{I} a_{i_l}^*(p_l) dp_l \prod_{r=1}^{J} a_{j_r}(q_r) dq_r \tag{2.3}$$

were $v(p_1, i_1, \dots p_I, i_I | q_1, \dots q_J, j_J)$ are some test functions.

The second type is described by the translation invariant Hamiltonian

$$V = \sum_{I,J} V_{I,J} = \sum_{I,J,i,j} \int \hat{v}(p_1, i_1, \dots p_I, i_I | q_1, j_1 \dots q_J, j_J) \tag{2.4}$$

$$\delta\left(\sum_l^I p_l - \sum_r^J q_r\right) \prod_{l=1}^{I} a_{i_l}^*(p_l) dp_l \prod_{r=1}^{J} a_{j_r}(q_r) dq_r$$

Clearly the delta function causes the trouble and there are singular terms in (2.4). Namely, $V_{I,0}\phi$ does not belong to the Fock space unless the vector $\phi = 0$. This singularity is called the volume singularity. To give a meaning to the Hamiltonian with interaction (2.4) one has to introduce a volume cut-off, then perform the vacuum renormalization and vacuum dressing and only after that remove the cut-off. This procedure defines the Hamiltonian in a new space (see **[5, 20]** for details). To avoid this difficulty in this paper we will assume that for translation invariant interaction there are no pure creation and annihilation terms.

2.2. Friedrichs diagrams. We will study the evolution operator

$$U(t) = e^{itH_0}e^{-it(H_0+\lambda V)}. \tag{2.5}$$

In perturbation theory the evolution operator (2.5) has the representation

$$U(t) = 1 - i\lambda\int_0^t V(t_1)dt_1 + (-i\lambda)^2\int_0^t dt_1\int_0^{t_1} dt_2 V(t_1)V(t_2) + \ldots \tag{2.6}$$

where

$$V(t) = e^{itH_0}Ve^{-itH_0}$$

We will use for $V_{I,J}$ the Friedrichs diagram representation [**3**]. The corresponding diagram has one vertex and I lines going from the vertex to the left and J lines going to the right. The first I lines represent creation operators and the last J lines represent annihilation operators. In what follows we will use the Wick theorem and also the following notions. The line of the graph is called the internal if it connects two vertices of the graph. A graph is the connected graph if all its vertices are connected by a set of internal lines otherwise it is called the disconnected one. A connected graph is called the one-particle reducible (1PR) if after the removing a line it becomes disconnected. A connected graph is called the one-particle irreducible (1PI) if after the removing any line it is still connected.

The following "linked cluster theorem" [**5**] will be used:

$$U(t) =: e^{U_c(t)} : \tag{2.7}$$

where

$$U_c(t) = \sum_{n=1}^{\infty}(-i\lambda)^n\int_0^t dt_1 \ldots \int_0^{t_{n-1}} dt_n\,(V(t_1)\ldots V(t_n))_c$$

Here :: means the Wick normal ordering and the index c in U_c indicates that one takes only the connected diagrams.

Below for simplicity of notations we consider interactions with the only one type of particles, but the main results are valid for arbitrary number of types of particles.

3. Evolution operator for non-translation invariant Hamiltonians

3.1. Second order. To get an insight to the problem it is very instructive to start with the consideration of matrix elements of the evolution operator in the second order of perturbation theory. For the vacuum matrix element of the evolution operator we obtain from (2.7) the representation:

$$\langle 0|U(t)|0\rangle = e^{\mathcal{E}(t)} \tag{3.1}$$

where

$$\mathcal{E}(t) = \langle 0|(-i\lambda\int_0^t dt_1 V(t) + (-i\lambda)^2\int_0^t dt_1\int_0^{t_1} dt_2 V(t_1)V(t_2) + \ldots)_c|0\rangle. \tag{3.2}$$

Here the symbol $(\ldots)_c$ means that we keep only connected diagrams.

Representation (3.1), (3.2) permits us to calculate the leading terms of the asymptotic behaviour of the matrix elements of the evolution operator for large time t as well the corrections to the leading terms. In fact we will show that $\mathcal{E}(t)$ has the following form

$$\mathcal{E}(t) = At + B + C(t) \tag{3.3}$$

where one has the perturbative expansions

(3.4)
$$A = \lambda^2 A_2 + \lambda^3 A_3 + ..., \; B = \lambda^2 B_2 + \lambda^3 B_3 + ..., \; C(t) = \lambda^2 C_2(t) + \lambda^3 C_3(t) + ...$$

and $C_n(t)$ vanishes for large t.

Let us find explicitly these terms in the second order of perturbation theory for the Hamiltonian

$$H = H_0 + \lambda V, \tag{3.5}$$

where

$$H_0 = \int \omega(p) a^*(p) a(p) dp \tag{3.6}$$

and the interaction has the form

$$V = \int (v(p_1, ..., p_n) a^*(p_1) \ldots a^*(p_n) + h.c.). dp_1 ... dp_n \tag{3.7}$$

Here $\omega(p)$ is a positive smooth function, for example $\omega(p) = \sqrt{p^2 + m^2}, m > 0$ and $v(p_1, ..., p_n)$ is a test function. For this interaction the first term in (3.2) is identically zero. The second term in (3.2) equals to

$$\mathcal{E}^{(2)}(t) = (-i\lambda)^2 \int dp_1 ... dp_n |v(p_1, ..., p_n)|^2 \int_0^t dt_1 \int_0^{t_1} dt_2 e^{it_1 E_1 + it_2 E_2} \tag{3.8}$$

where

$$E_2 = -E_1 = E(p_1, ..., p_n) = \sum_{i=1}^n \omega(p_i) \tag{3.9}$$

By using the equality

$$\int_0^t dt_1 \int_0^{t_1} dt_2 e^{-it_1 E + it_2 E} = -\frac{i}{E} t + \frac{1}{E^2} - \frac{1}{E^2} e^{-itE} \tag{3.10}$$

we get

$$\mathcal{E}^{(2)}(t) = (-i\lambda)^2 \int dp_1 ... dp_n |v(p_1, ..., p_n)|^2 (-\frac{i}{E} t + \frac{1}{E^2} - \frac{1}{E^2} e^{-itE}) \tag{3.11}$$

Therefore we obtain the expression of the form (3.3)

$$\mathcal{E}(t) = \lambda^2 A_2 t + \lambda^2 B_2 + \lambda^2 C_2(t) + ... \tag{3.12}$$

where

$$A_2 = i \int \frac{|v(p_1, ..., p_n)|^2}{E(p_1, ..., p_n)} dp_1 ... dp_n, \tag{3.13}$$

$$B_2 = -\int \frac{|v(p_1, ..., p_n)|^2}{E(p_1, ..., p_n)^2} dp_1 ... dp_n, \tag{3.14}$$

$$C_2(t) = \int \frac{|v(p_1, ..., p_n)|^2}{E(p_1, ..., p_n)^2} e^{-itE(p_1, ..., p_n)} dp_1 ... dp_n \tag{3.15}$$

We have obtained the following

THEOREM 3.1. *The vacuum expectation value of the evolution operator for the Hamiltonian (3.5) in the second order of perturbation theory has the form*

$$< U(t) >= e^{\lambda^2 A_2 t + \lambda^2 B_2 + \lambda^2 C_2(t)} \tag{3.16}$$

where A_2, B_2 and $C_2(t)$ are given by (3.13),(3.14) and (3.15).

REMARK. By using the stationary phase method one can prove that the function $C_2(t)$ vanishes as $t \to \infty$ (see below).

3.2. Decay. We have proved theorem 3.1 under the assumption $\omega(p) > 0$. However the obtained formula (3.16) is valid in the more general case when one has the decay. In this case formula (3.16) still is true but in the expressions (3.13)-(3.15) one has to substitute $E \to E - i0$. Let us consider the important case when

$$\omega(p) = \frac{p^2}{2} - \omega_0, \ \omega_0 > 0 \tag{3.17}$$

Instead of (3.10) we will use now the identity

$$\int_0^t dt_1 \int_0^{t_1} dt_2 e^{i(t_2 - t_1)E} = t \int_0^t (1 - \frac{\sigma}{t}) e^{-i\sigma E} d\sigma. \tag{3.18}$$

We have

$$\mathcal{E}^{(2)}(t) = (-i\lambda)^2 \int dp_1 ... dp_n |v(p_1, ..., p_n)|^2 \int_0^t dt_1 \int_0^{t_1} dt_2 e^{i(t_2 - t_1)E(p_1,...,p_n)} \tag{3.19}$$

$$= -\lambda^2 t \int_0^t d\sigma (1 - \frac{\sigma}{t}) \int dp_1 ... dp_n |v(p_1, ..., p_n)|^2 e^{-i\sigma E(p_1,...,p_n)}$$

$$= \lambda^2 t A_2(t) + \lambda^2 B_2(t)$$

where

$$A_2(t) = - \int_0^t d\sigma F(\sigma), \ B_2(t) = \int_0^t d\sigma \sigma F(\sigma) \tag{3.20}$$

and

$$F(\sigma) = \int dp_1 ... dp_n |v(p_1, ..., p_n)|^2 e^{-i\sigma E(p_1,...,p_n)}. \tag{3.21}$$

By using the stationary phase method we obtain the following asymptotic behaviour of the function $F(\sigma)$ as $\sigma \to \infty$:

$$F(\sigma) = (\frac{2\pi i}{\sigma})^{dn/2} e^{in\sigma\omega_0} |v(0)|^2 [1 + o(\frac{1}{\sigma})].$$

Therefore for $dn \geq 3$ there exist the limits

$$\lim_{t \to \infty} A_2(t) = A_2 = - \int_0^\infty d\sigma F(\sigma) \tag{3.22}$$

$$\lim_{t \to \infty} B_2(t) = B_2 = \int_0^\infty d\sigma \sigma F(\sigma)$$

because there exists the limit

$$\lim_{t \to \infty} \int_1^t e^{i\sigma\omega_0} \frac{d\sigma}{\sigma^{1/2}}.$$

Moreover one has

$$A_2(t) = -\int_0^\infty d\sigma F(\sigma) + o(\frac{1}{t^2}),\ B_2(t) = \int_0^\infty d\sigma\sigma F(\sigma) + o(\frac{1}{t}).$$

If $dn \geq 5$ one gets also

(3.23) $$A_2 = i\int dp_1...dp_n \frac{|v(p_1,...,p_n)|^2}{E(p_1,...,p_n) - i0}$$

(3.24) $$B_2 = -\int dp_1...dp_n \frac{|v(p_1,...,p_n)|^2}{(E(p_1,...,p_n) - i0)^2}.$$

Indeed one has

$$A_2 = -\int_0^\infty d\sigma F(\sigma) = -\lim_{\epsilon\to 0}\int_0^\infty d\sigma F(\sigma)e^{-\sigma\epsilon}.$$

This is true due to the Lebesgue theorem since $|F(\sigma)e^{-\sigma\epsilon}| \leq |F(\sigma)|$ and $F(\sigma) \in L_1(R_+)$ (L_1 is the space of absolute integrable functions) if $nd \geq 3$. Substituting in the above formula the representation (3.21) and changing the order of integrations (we can do this due to the Fubini theorem since for positive ϵ the function $|v(p_1,...,p_n)|^2 e^{-i\sigma(E(p_1,...,p_n)-i\epsilon)}$ belongs to the space $L_1(R_+ \times R^{nd})$ of absolute integrable functions), we can perform the integration over σ explicitly

$$A_2 = \lim_{\epsilon\to 0}(i)\int dp_1...dp_n \frac{|v(p_1,...,p_n)|^2}{E(p_1,...,p_n) - i\epsilon} = i\int dp_1...dp_n \frac{|v(p_1,...,p_n)|^2}{E(p_1,...,p_n) - i0}.$$

The same calculation is true for B_2 with the more strong assumption : $dn \geq 5$,

$$B_2 = \int_0^\infty d\sigma\sigma F(\sigma) = \lim_{\epsilon\to 0}\int_0^\infty d\sigma\sigma \int dp_1...dp_n |v(p_1,...,p_n)|^2 e^{-i\sigma(E(p_1,...,p_n)-i\epsilon)}$$

$$= -\lim_{\epsilon\to 0}\int dp_1...dp_n \frac{|v(p_1,...,p_n)|^2}{(E(p_1,...,p_n) - i\epsilon)^2} = -\int dp_1...dp_n \frac{|v(p_1,...,p_n)|^2}{(E(p_1,...,p_n) - i0)^2}.$$

We have proved the following theorem.

THEOREM 3.2. *The asymptotic behaviour as $t \to \infty$ of the vacuum expectation value of the evolution operator for the Hamiltonian (3.5) with the dispersion low (3.17) in the second order of perturbation theory is*

(3.25) $$< U(t) >= e^{\lambda^2 A_2 t + \lambda^2 B_2 + \lambda^2 o(1/t)}$$

where A_2 and B_2 are given by (3.22) (or (3.23) and (3.24)). After the rescaling $t \to t/\lambda^2$ one gets the λ^2 corrections to the stochastic limit

(3.26) $$< U(t/\lambda^2) >= e^{A_2 t + \lambda^2 B_2 + \lambda^2 o(\lambda^2/t)}.$$

3.3. Example. We discuss here the evolution operator for the simple explicitly solvable model described by the Hamiltonian

(3.27) $$H = \int \omega(k)a^+(k)a(k)d^dk + \lambda\int (a(k)\overline{v}(k) + a^*(k)v(k))d^dk.$$

We will see that the vacuum expectation value of the evolution operator has the form obtained in theorems 3.1 and 3.2. Under assumptions

(3.28)

$$A = \lambda^2 A_2 = i\lambda^2 \int \frac{|v(k)|^2}{\omega(k)}d^dk < \infty, \qquad B = \lambda^2 B_2 = -\lambda^2 \int \frac{|v(k)|^2}{\omega^2(k)}d^dk < \infty\ ,$$

one has the following

PROPOSITION 3.1. *The vacuum expectation value of the evolution operator* $U(t) = e^{itH_0}\, e^{-itH}$ *for the model (3.27) is*

$$\langle U(t)\rangle = \exp\left[At + B + C(t)\right] \tag{3.29}$$

where A *and* B *are given by (3.28) and* $C(t)$ *is*

$$C(t) = \lambda^2 \int dk \frac{|v(k)|^2}{\omega^2(k)}\, e^{-i\omega(k)t}$$

PROOF. It follows from the known explicit solution of the model. □

From Proposition 3.1 we obtain

PROPOSITION 3.2. *The asymptotic behaviour of the expectation value (3.29) for* $t \to \infty$ *has the form*

$$\langle U(t)\rangle = \exp\left[At + B + \lambda^2 \left(\frac{1}{t}\right)^{\frac{d}{2}} (2i\pi)^{\frac{d}{2}} \frac{|v(k_0)|^2}{\omega^2(k_0)}\, e^{-i\omega(k_0)t} + ...\right]. \tag{3.30}$$

where k_0 *is a critical point,* $\nabla\omega(k_0) = 0$ *and we assume there is only one nondegenerate critical point.*

PROOF. It follows immediately from (3.29) by using the stationary phase method. □

REMARK. If $\omega(k) = k^2 - \omega_0$, $\omega_0 > 0$ then one has the decay. We can not use in this case the diagonalization of the Hamiltonian (3.27) but the formula (3.1) still is true. We have

$$< U(t) >= \exp[-\lambda^2 \int_0^t dt_1 \int_0^{t_1} dt_2 \int d^d k |v(k)|^2 e^{i(t_2-t_1)\omega(k)}]$$
$$= \exp[\lambda^2 t A_2(t) + \lambda^2 B_2(t)]$$

where

$$A_2(t) = -\int_0^t d\sigma F(\sigma),\ B_2(t) = \int_0^t d\sigma\sigma F(\sigma)$$

and

$$F(\sigma) = \int d^d k |v(k)|^2 e^{-i\sigma\omega(k)}$$

Again we obtain the ABC-formula.

3.4. Wave operators and the main formula. In this section we show how the spectral theory and renormalized wave operators can be used for the derivation of the main formula. In particular explicit expressions for the parameters A, B and $C(t)$ will be obtained.

We consider the Hamiltonian (2.3) for one type of particles with $\omega(p) = \sqrt{p^2 + m^2}$, $m > 0$ in the space $R^d, d > 2$. We will work with the formal perturbation series for the evolution operator. In fact if the interaction (2.3) includes only fermionic operators or it is linear in bosonic operators then one can prove that the series are absolutely convergent. The following operator plays the crucial role in the scattering theory

$$T =: \exp \sum_{n=1}^{\infty} (-\lambda)^n \Gamma(V...\Gamma(V)...)_L : \tag{3.31}$$

Here Γ means the Friedrichs Γ - operation

$$\Gamma(V) = \lim_{\epsilon \to +0} (-i) \int_0^\infty e^{-\epsilon t} e^{itH_0} V e^{-itH_0} dt \tag{3.32}$$

and $()_L$ means that only connected non-vacuum diagrams are included. The operator T is equal in fact to the non-vacuum part of the conjugate wave operator:

$$T = \lim_{\epsilon \to 0} \lim_{t \to \infty} U_\epsilon^*(t) / < U_\epsilon^*(t) >$$

One has the following relations [**5**]:

$$HT = T(H_0 + E_0), \tag{3.33}$$

$$E_0 = \sum_{n=1}^{\infty} (-\lambda)^{n+1} < V\Gamma(V...\Gamma(V)...)_c >,$$

$$T^*T = TT^* = Z^{-1},$$

$$Z^{-1} = ||T\Phi_0||^2$$

We will use these relations to derive the main formula for matrix elements of the evolution operator and in particular to compute corrections to the stochastic limit.

From (3.33) it follows

$$H = T(H_0 + E_0)T^* e^B$$

where

$$B = \ln Z \tag{3.34}$$

Therefore one has

$$U(t) = e^{itH_0} e^{-itH} = e^{itH_0} T e^{-it(H_0+E_0)} T^* e^B = e^{At+B} e^{itH_0} T e^{-itH_0} T^* \tag{3.35}$$

where

$$A = -iE_0 \tag{3.36}$$

By taking the expectation value of the equality (3.35) we obtain

$$< \psi, U(t)\psi >= e^{At+B+C(t)} \tag{3.37}$$

where

$$e^{C(t)} =< \psi, e^{itH_0} T e^{-itH_0} T^* \psi >$$

If ψ is the vacuum vector then one can prove that $C(t) \to 0$ as $t \to \infty$ and we obtain the main formula (1.3). If ψ is a non-vacuum vector then the asymptotic behaviour of $C(t)$ is more complicated. We have proved the following theorem.

THEOREM 3.3. *If the Hamiltonian satisfies the indicated above assumptions then there exists the following representation for the vacuum expectation value of the evolution operator*

$$< \Phi_0, U(t)\Phi_0 >= e^{At+B+C(t)}$$

where constants A and B are given by (3.36) and (3.34) and $C(t)$ is defined by

$$e^{C(t)} =< \Phi_0, T(t)T^*\Phi_0 >, \ T(t) = e^{itH_0} T e^{-itH_0} \tag{3.38}$$

Here the weak limit of $T(t)$ as $t \to \infty$ is equal to 1 and $\lim_{t\to\infty} C(t) = 0$.

This theorem also shows the physical meaning of constants A, B and the function $C(t)$.

4. One particle matrix elements of the evolution operator for translation invariant Hamiltonian

In this section we study the asymptotic behaviour of one particle matrix elements of the evolution operator $< p|U(t)|p' >$ for translation invariant Hamiltonian (2.4) without vacuum polarization, i.e. when $V_{I0} = V_{0J} = 0$, and we assume also $V_{1,1} = 0$. We will prove that the following representation is true

$$< p|U(t)|p' >= \exp\{A(p, \lambda)t + B(p, \lambda) + C(t, p, \lambda)\}\delta(p - p') \tag{4.1}$$

where $A(p, \lambda)$, $B(p, \lambda)$ and $C(t, p, \lambda)$ are formal series in λ

$$A(p, \lambda) = \sum_{n=1}^{\infty} \lambda^n A_n(p), \ B(p, \lambda) = \sum_{n=1}^{\infty} \lambda^n B_n(p), \ C(t, p, \lambda) = \sum_{n=1}^{\infty} \lambda^n C_n(t, p) \tag{4.2}$$

and functions $C_n(t, p)$ vanish as $t \to \infty$.

We will show how the spectral theory and renormalized wave operators can be used for the derivation of the main formula. In particular, explicit recursive relations for the parameters $A_n(p)$, $B_n(p)$ and $C_n(t, p)$ will be obtained. Note that for this derivation we have to assume that there is no decay.

The intertwining operator T is defined as the solution of the following equation

$$HT = T(H_0 + M), \tag{4.3}$$

Here M has the form

$$M = \int M(p, \lambda)a^*(p)a(p)dp \tag{4.4}$$

The operator T plays the crucial role in the scattering theory. Its singular part defines the renormalized wave operators. Taking T in the following form

$$T =: \exp \Gamma(Q) : \tag{4.5}$$

one gets [**6**] equations to derive Q and M

$$Q + V \perp T - (V \perp T)_{1,1} - \Gamma(Q) - \circ - M = 0, \tag{4.6}$$

$$M = (V \perp T)_{1,1} \tag{4.7}$$

Here the symbol $\perp$ means that in the operator expansion of $A \perp B$ all connected parts of B are paired with A. If B is connected then $A \perp B = A - \circ - B = (AB)_c$. For a special form of the interaction, when $V_{0I} = V_{I0} = V_{1,1} = 0$ we can write $M = (V \perp T)_{1,1} = (V\Gamma(Q))_{1,1}$.

One can construct [**3, 5, 6**] in perturbation theory an operator Z of the form $Z =: \exp z_{1,1}$: such that

$$TT^* Z = 1 \tag{4.8}$$

By using (4.3) and (4.8) one gets

$$U(t) = e^{itH_0}e^{-itH} = e^{-itM}e^{it(H_0+M)}Te^{-it(H_0+M)}T^* Z \tag{4.9}$$

By taking the one particle expectation value of the equality (4.9) we obtain (4.1), where

$$A(p, \lambda) = -iM(p, \lambda),$$
$$e^{B(p,\lambda)}\delta(p - p') =< p|Z|p' >,$$

(4.10) $$e^{C(t,p,\lambda)}\delta(p-p') =< p|e^{it(H_0+M)}Te^{-it(H_0+M)}T^*|p' >$$

One can prove that $C(t,p,\lambda)$ vanishes as $t \to \infty$. We have proved the following

THEOREM 4.1. *For translation invariant Hamiltonians without vacuum polarization the one particle matrix elements of the evolution operator are given by the formula (4.1) where functions $A(p,\lambda)$ and $B(p,\lambda)$ are solutions of equations (4.5)-(4.7) and (4.8). Function $C(t,p,\lambda)$ is defined by (4.10) and it vanishes as $t \to \infty$.*

Now we compute explicitly the first few terms in the expansion of M and Q in the power series in λ

(4.11) $$M = \sum_{n=1}^{\infty} \lambda^n M_n, \ Q = \sum_{n=1}^{\infty} \lambda^n Q_n,$$

We have the recursive relations (4.6), (4.7) to define M_n and Q_n. Let us compute explicitly the first terms solving these equations. We obtain

(4.12) $$Q_1 = -V$$

(4.13) $$M_2 = -(V\Gamma V)_{1,1}$$

(4.14) $$Q_2 = (V\Gamma V)_c - (V\Gamma V)_{1,1}$$

By using thes expressions we get

(4.15) $$Q_3 = -(V\Gamma_r(V\Gamma(V)))_c - \frac{1}{2}V \perp \ :\Gamma(V)^2: + \Gamma V - \circ - M_2$$

(4.16) $$M_4 = -(V\Gamma(V\Gamma_r(V\Gamma(V))))_{1,1} + (V\Gamma^2(V)M_2)_{1,1}$$

or

(4.17) $$M_4 = -(V\Gamma(V\Gamma_r(V\Gamma(V))))_{1,1} - (V\Gamma^2(V)(V\Gamma(V))_{1,1})_{1,1}$$

Here we use the notation

(4.18) $$\Gamma_r(Q) = \Gamma(Q - Q_{1,1})$$

5. Quantum Chaos

The representations (3.38) and (4.10) for the function $C(t)$ are similar in fact to the representation for the autocorrelation function describing quantum chaos for the quantum baker's map. The quantum baker's map is a simple model invented to study quantum chaos, see for example [**21, 22, 23, 24, 25**] and references therein. The evolution operator for the quantum baker's map has the form $U(t) = B^t$ where B is a unitary matrix and t is an integer number. The autocorrelation function $F(t) =< \psi|U(t)|\psi >$ where ψ is a coherent vector, admits a representation of the form [**23, 22**]

$$F(t) = a(t)\sum e^{W(\beta,p,q,t)}$$

where $a(t)$ falls off exponentially for large t and W is quadratic polynomial in q, p and β. Here p and q are parameters in the coherent vectors, β is a string in the symbolic dynamics representation and the sum runs over all the strings. This form is similar to the representations (3.38) and (4.10) for the function $C(t)$, compare also with the explicit formula (3.15) for $C_2(t)$. It would be very interesting to study further this analogy between the functions $F(t)$ and $C(t)$. In particular one has to investigate the sensitivity of the function $C(t)$ to the initial data, i.e. in the one particle case the dependence of the function $C(t, p, \lambda)$ on p. Also one has to investigate the analogous function for coherent states. To this end one can use the perturbation theory as well the semiclassical expansion. It would be interesting to study a relation with the wave chaos [**26**] and also with the spectral and holographic properties of quantum chaos, see [**27**].

6. Conclusion

We have obtained in this paper the explicit representations (1.3) for the vacuum and one-particle matrix elements of the evolution operator. By using these representations we have computed the corrections to the known results for the large time exponential behaviour of these matrix elements. This opens the way for further investigations of the large time behaviour in quantum theory. In particular the problems of quantum decoherence and decay and also the relation with quantum chaos require the further study.

References

[1] S. Albeverio , D. Bolle , F. Gesztesy , R. Høegh-Krohn and L. Streit, *Low-energy parameters in nonrelativistic scattering theory*, Ann. Phys., New York, **148** (1983), 308–337.
[2] N. N. Bogoliubov and D. V. Shirkov, *Introduction to the theory of quantum fields*, Nauka, Moscow, 1973.
[3] K. O. Friedrichs, *Perturbation of Spectra in Hilbert Space*, AMS, Providence, 1965
[4] L. D. Faddeev, *Separation of selfinteraction effects and scattring in the perturbation theory*, Dokladi AN USSR, **152** (1963), 573–578
[5] K. Hepp, *Theorie de la renormalisation*, Springer, New York, 1969.
[6] I. Ya. Aref'eva, *Renormalized waves operators, I*, Teor. Mat. Fis. **14** (1973), 3–17.
[7] J. Schwinger, *Field theory of unstable particles*, Ann. Phys., New York, **9** (1960), 169-193
[8] M. L. Goldberger and K. M. Watson, *Collision theory*, John Wiley & Sons, Inc, New York-London-Sydney, 1964.
[9] C. Cohen-Tannoudji, J. Dupont-Roc and G. Grinberg, *Atom-Photon Interactions, Basic Processes and Applications*, John Wiley & Sons, Inc., 1992.
[10] E. L. Feinberg, *A particle with non-equilibrium proper field*, in: Problems of theoretical physics, Memorial volume to Igor E. Tamm, Nauka, Moscow, 1972, pp. 134–149.
[11] V. A. Rubakov and M. E. Shaposhnikov, *Elektroweak baryon number non-conservation in the early Universe and in high-enegry collisions*, Uspehi Fisicheskih nauk, **166** (1996), 493–538.
[12] D. F. Walls and G. J. Milburn, *Quantum Optics*, Springer-Verlag, New York, 1994.
[13] I. V. Volovich, *Models of Quantum Computers and Decoherence Problem*, e-Print archive http://xxx.lanl.gov/ quant-ph /9902055; Proc. of the International Conference on Quantum Information, Meijo Univ. Nagoya, 4-8 Nov. 1997.
[14] L. Accardi, I. Ya. Aref'eva and I. V. Volovich, *Non-Equilibrium Quantum Field Theory and Entangled Commutation Relations*, e-Print archive http://xxx.lanl.gov/hep-th/9905035.
[15] N.N. Bogoliubov, *Problems of dynamical theory in statistical physics*, Gostehizdat, Moscow, 1946
[16] L. van Hove, *Quantum mechanical perturbations giving rise to a transport equation*, Physica, **21** (1955), 517–540.
[17] I. Prigogine, *Non-equilibrium statistical mechanics*, Pergamon, 1963.
[18] L. Accardi, S. V. Kozyrev, I. V. Volovich, *Dynamics of dissipative two–state systems in the stochastic approximation*, Phys. Rev. **A 57** (1997), 1111-1127; e-Print archive http://xxx.lanl.gov/quant-ph/9706021
[19] L. Accardi, Y. G. Lu and I. V. Volovich, *Quantum theory and its stochastic limit*, Springer, (in press)
[20] I. Ya. Aref'eva, Teor. Mat. Fis. *Renormalized waves operators, II*, **15** (1973), 207-215.
[21] N. L. Balazs and and A. Voros, *The quantum Baker's Transformation*, Ann. Phys., New York, **190** (1989), 1–31.

[22] F. Toscano, R. O. Vallejos and M. Saraceno, *Boundary contributions to the semiclassical traces of the baker's map*, e-Print archive http://xxx.lanl.gov/chao-dyn/9703005.
[23] L. Kaplan and E. J. Heller, *Overcoming the Wall in the Semiclassical Baker's Map*, e-Print archive http://xxx.lanl.gov/chao-dyn/9809008.
[24] A. N. Soklakov and R. Schack, *Classical limit in terms of symbolic dymanics for the quantum baker's map* , e-Print archive http://xxx.lanl.gov/quant-ph/9908040
[25] K. Inoue, M. Ohya and I. V. Volovich, *Semiclassical Properties and Chaos Degree for the Quantum Baker's Map* (in preparation).
[26] S. Albeverio and P. Seba, *Wave chaos in quantum systems with point interaction*, Preprint BIBOS-447-90, 1991, 13pp.
[27] I. Ya. Aref'eva, P. B. Medvedev, O. A. Rytchkov and I. V. Volovich, *Chaos in M(atrix) Theory* , Chaos, Solitons and Fractals, **10** (1999), 213–223.

Steklov Mathematical Institute, Russian Academy of Sciences, Gubkin St.8, GSP-1, 117966, Moscow, Russia

E-mail address: `arefeva@mi.ras.ru`

E-mail address: `volovich@mi.ras.ru`

Canadian Mathematical Society
Conference Proceedings
Volume **29**, 2000

Correlation Effects in the Trapping Problem: General Approach and Rigorous Results

L. V. Bogachev, S. A. Molchanov, Yu. A. Makhnovskii, and A. M. Berezhkovskii

Dedicated to Professor Sergio Albeverio on the occasion of his sixtieth birthday.

ABSTRACT. The problem of Brownian survival among randomly located traps is considered with emphasis on the role of trap correlations. We proceed from the general representation of the survival probability as the expected value of the emptiness probability function applied to the Wiener sausage. Using the definition of (pure) trap attraction vs. repulsion in terms of the emptiness probability function, we prove the physical conjecture about the trapping slowdown or acceleration, according to the "sign" of correlations. Two specific models are studied along this line, in which the emptiness probability can be found explicitly; in particular, the long-time survival asymptotics is derived. A remarkable correlation effect of the survival probability dependence on the trap size in one dimension is also discussed.

1. Introduction

The problem of Brownian particle's survival among randomly located killing traps, usually referred to as the *trapping problem*, is well known in the general theory of transport processes in disordered media (see, e.g., [**16**] and the bibliography therein). Of primary interest is the time dependence of the survival probability $P(t)$. The case of *noncorrelated* trapping medium, where the traps are distributed in space according to a Poisson law, has been studied rather extensively (see [**17**] for a survey of analytic, approximation and simulation methods and results, and a recent monograph [**27**] for a rigorous mathematical treatment). Although the exact solution is available only in one dimension [**2**], two important general features of the trapping kinetics have been revealed in the noncorrelated situation:

1. At small and intermediate times, the decay of the particle species basically follows the conventional theory dated back to Smoluchowski [**25**]. In particular, an exponential decrease of the survival probability $P_{\rm nc}(t)$ is predicted in dimensions $d \geq 3$. (Here and below, the subscript "nc" refers to the case of noncorrelated

1991 *Mathematics Subject Classification.* 60J70, 60J65, 60G55.

This work was supported in part by DFG Grant # 436 RUS 113/534, RFBR Grant # 99-01-00298, and NSF Grant # DMS-9971592.

traps.) The conventional approach is based on the mean-field approximation by neglecting the many-body effects due to "competition" between the traps (for an up-to-date account of this theory, see [**4**], [**20**]).

2. At long times ($t \to \infty$) the mean-field approach breaks down, since the behavior of the $P_{\rm nc}(t)$ is governed by the particle's untypical survival in untypically big "holes" in the Poisson cloud. This leads to a fractional-exponent form of the survival tail [**2**], [**22**]. More precisely, Donsker and Varadhan [**12**] proved that

$$\lim_{t\to\infty} t^{-d/(d+2)} \log P_{\rm nc}(t) = -\gamma_d c^{2/(d+2)}, \tag{1.1}$$

where c is the intensity of the Poisson law (i.e., the trap concentration) and $\gamma_d > 0$ is an explicit dimension-dependent constant (see also [**27**, § 4.5B]).

The problem of allowing for *trap correlations* has attracted considerable interest in the last 15 years (see [**3**], [**6**]–[**9**], [**19**], [**23**], [**28**], [**29**] and further references therein). In particular, it was argued, mainly on physical grounds, that trap *attraction* (*repulsion*) induces the *slowdown* (respectively, *acceleration*) of the trapping as compared to the case of noncorrelated traps[1] [**6**]–[**8**], [**23**], [**28**]. This conjecture was also supported by the calculation of the long-time survival tail in some models [**1**], [**9**], [**19**], [**26**], [**29**], which is of mathematical interest in its own right. We would like to point out a result proved by Sznitman [**26**] (see also an earlier paper by Kayser and Hubbard [**19**]) stating that for Gibbsian trap distribution (with finite-range interaction potential Φ), the asymptotical behavior of the survival probability $P(t)$ follows the law (1.1), but with concentration c replaced by the quantity βp, where β is the inverse temperature and p is the pressure [**24**, § 3.4].[2]

In the present contribution, we pursue a systematic approach to this problem. Analysis of the qualitative influence of correlations is of our main concern, complemented whenever possible by the calculation of the survival asymptotics. We proceed from the general representation of the survival probability as the expectation of a certain Brownian functional, in which the trapping medium enters through the *emptiness probability function* applied to the corresponding Wiener sausage (Sect. 2). On the other hand, a "phenomenological" definition of pure attraction vs. repulsion of traps can also be given in terms of the emptiness probabilities (Sect. 3), which essentially amounts to distinguishing between positive vs. negative correlations in the distribution of the trap point process.[3] Using this link, we are able to prove the aforementioned general conjecture about the qualitative role of trap correlations.

We consider along this line two concrete models where the distribution of traps is determined by a Poisson cluster point process (Sect. 4) or by a one-dimensional renewal process (Sects. 5, 6). These models are exactly solvable in that the emptiness probability function can be found in a closed form, which paves the way for studying the survival probability in some detail. It is worth pointing out that in the Poisson cluster model, correlations prove to be of attractive type, regardless of the intra-cluster interaction. The renewal model, specified by the choice of the gamma distribution for distances between the adjacent traps, exhibits both types

---

[1] In such a comparison, it is implied that the trap concentration c is fixed.

[2] In the Gibbsian framework, the noncorrelated case corresponds to the ideal gas with null interaction, $\Phi \equiv 0$. The pressure $p_{\rm nc}$ is easily identified as c/β.

[3] Of course, in general the qualitative character of the trap "interaction" may not be universal, that is, *purely* attractive or repulsive.

of correlations, depending on the distribution parameter. The long-time survival asymptotics is derived for these models as well. We also discuss some novel correlation effects arising in one dimension, in particular a remarkable dependence of the trapping rate upon the trap size (Sect. 6).

2. Survival probability

Let us briefly recall the setting of the trapping problem (see [**1**] for more details). Consider a standard Brownian motion $(B_t,\ t \geq 0)$ on $\mathbf{R}^d$ $(d \geq 1)$ starting from the origin, and let P_0 and E_0 stand for the corresponding Wiener measure and expectation, respectively. The trapping medium is constituted by random configuration of spherical traps of radius b centered at the points of a background random point process $Z = \{Z_i\}$. In other words, the trapping set is of the form $\cup_i\{Z_i + V_b\}$, where V_b is the closed ball of radius b centered at the origin (note that, generally speaking, the traps may overlap). The traps are supposed to be perfectly absorbing, which means that the particle is terminated at the first contact with a trap.

The law of the process Z denoted by $\mathbb{P}$ is assumed to be independent of B_t, and therefore the trapping process is governed by the product probability measure $\mathbb{P} \otimes P_0$. We also suppose that Z is *simple* (i.e., without multiple points) and *stationary* (i.e., $\mathbb{P}$ is invariant under space shifts). For a Borel subset $S \in \mathcal{B}(\mathbf{R}^d)$, let N_S stand for the number of points $Z_i \in S$. We assume that the expectation $\mathbb{E}[N_S]$ is finite for all bounded $S \in \mathcal{B}(\mathbf{R}^d)$. Thanks to stationarity, $\mathbb{E}[N_S]$ is proportional to the Lebesgue volume of S,

$$\mathbb{E}[N_S] = c|S|,$$

where the constant c is called the *trap concentration*.

In this work, we are concerned with the *survival probability* $P(t)$ defined as

$$P(t) := \mathbb{P} \otimes P_0\{T > t\} = \mathbb{E}[P_0\{T > t|Z\}], \tag{2.1}$$

where $T := \min\{t \geq 0 \colon B_t \in \cup_i(Z_i + V_b)\}$ is the particle's lifetime. Note that generally $P(0) \leq 1$, since the origin where the particle is born may be covered by a trap. Hence, somewhat more natural is the *normalized survival probability*

$$\tilde{P}(t) := \mathbb{P} \otimes P_0\{T > t | T > 0\} = \frac{P(t)}{P(0)}.$$

The following basic representation of the function $P(t)$ is obtained from its definition (2.1) via interchanging the order of averages (see [**1**], [**8**], [**9**]):

$$P(t) = E_0[g(S_b(t))]. \tag{2.2}$$

Here $S_b(t) := \cup_{0 \leq s \leq t}\{B_s + V_b\}$ is the b-tubular neighborhood of a Brownian path on time interval $[0, t]$ (referred to as the *Wiener sausage*, due to M. Kac [**18**]), and $g(\cdot)$ called the *emptiness probability function*[4] is a set function defined as

$$g(S) := \mathbb{P}\{N_S = 0\}, \qquad S \in \mathcal{B}(\mathbf{R}^d). \tag{2.3}$$

In particular, $P(0) = E_0[g(S_b(0))] = g(V_b)$, and therefore

$$\tilde{P}(t) = \frac{E_0[g(S_b(t))]}{g(V_b)}. \tag{2.4}$$

[4]In the general theory of point processes, where $g(\cdot)$ is also called the *avoidance function*, it is proved that this function completely characterizes the process distribution (see [**11**, § 7.3]).

In the case of noncorrelated traps where $Z = \{Z_i\}$ is a Poisson point process (with intensity c), the emptiness probability is explicitly given by

$$g_{\mathrm{nc}}(S) = e^{-c|S|}, \qquad S \in \mathcal{B}(\mathbf{R}^d). \tag{2.5}$$

As a result, equation (2.2) takes the form

$$P_{\mathrm{nc}}(t) = E_0[\exp(-c|S_b(t)|)]. \tag{2.6}$$

We would like to remark here that the Smoluchowski approximation amounts to carrying the expectation sign in (2.6) through the exponent:

$$P_{\mathrm{nc}}(t) \approx \exp(-cE_0|S_b(t)|), \tag{2.7}$$

which in fact yields a lower bound for $P_{\mathrm{nc}}(t)$ [**4**]. Evaluation of the expected volume of the Wiener sausage and related topics have been addressed in many papers (see, e.g., [**5**]). The corresponding expression is of particularly simple form for $d = 1$ and $d = 3$:

$$E_0|S_b(t)| = \begin{cases} \dfrac{4}{\sqrt{2\pi}}\, t^{1/2} + 2b, & d = 1, \\ 2\pi bt + 4\sqrt{2\pi}\, b^2 t^{1/2} + \dfrac{4\pi b^3}{3}, & d = 3. \end{cases} \tag{2.8}$$

In one dimension, the general formula (2.2) for the survival probability $P(t)$ reads as follows:

$$P(t) = \int_0^\infty g(\ell + 2b) f(\ell; t)\, d\ell, \tag{2.9}$$

where $g(\ell)$ is the probability that an interval of length ℓ is free from points $\{Z_i\}$,[5] and $f(\ell; t)$ is the probability density of the range of a Brownian motion on $[0, t]$:

$$R(t) := \max_{0 \le s \le t} B_s - \min_{0 \le s \le t} B_s, \tag{2.10}$$

that is the "volume" of the one-dimensional Wiener sausage (with $b = 0$). It is well known [**5**], [**13**] that the function $f(\ell; t)$ can be expressed in either of the two forms:

$$f(\ell; t) = \frac{8}{\sqrt{2\pi t}} \sum_{k=1}^{\infty} (-1)^{k-1} k^2 \exp\left(-\frac{k^2 \ell^2}{2t}\right) \tag{2.11}$$

$$= \frac{8t}{\ell^3} \sum_{k=1}^{\infty} \left[\frac{(2k-1)^2 \pi^2 t}{\ell^2} - 1\right] \exp\left(-\frac{(2k-1)^2 \pi^2 t}{2\ell^2}\right). \tag{2.12}$$

Accordingly, the normalized survival probability $\tilde{P}(t)$ is represented as

$$\tilde{P}(t) = \int_0^\infty \tilde{g}(\ell)\, f(\ell; t)\, d\ell, \tag{2.13}$$

where the function

$$\tilde{g}(\ell) := \frac{g(\ell + 2b)}{g(2b)} \tag{2.14}$$

can be viewed as an effective emptiness probability (a detailed discussion of formulas (2.13), (2.14) is deferred to Sect. 6).

[5]Thanks to stationarity of the point process Z, this probability does not depend on the position of the interval.

For noncorrelated traps, one has $\tilde{g}_{\rm nc}(\ell) = g_{\rm nc}(\ell) = e^{-c\ell}$ (see equations (2.5), (2.14)). Using representation (2.11), from (2.13) one then arrives, after some transformations, to the known Balagurov–Vaks expression [**2**]

$$\tilde{P}_{\rm nc}(t) = \frac{4}{\pi^2}\int_0^\infty \exp\left(-\frac{\pi^2 c^2 t}{2x^2}\right)\frac{x}{\sinh x}\,dx. \tag{2.15}$$

As would be expected, at short times ($t \to 0$) the survival probability (2.15) is reduced to the Smoluchowski-type form (cf. (2.7), (2.8))

$$\tilde{P}_{\rm nc}(t) = 1 - \frac{4c}{\sqrt{2\pi}}\,t^{1/2} + O(t). \tag{2.16}$$

It should be noted that this result is not restricted by the explicit formula (2.15) and can be derived directly from the basic equation (2.13). Indeed, substituting $\tilde{g}_{\rm nc}(\ell) = 1 - c\ell + O(\ell^2)$ into (2.13) and using expression (2.12) for the probability density $f(\ell;t)$ of the Brownian range $R(t)$ (see (2.10)), we obtain

$$\begin{aligned}\tilde{P}_{\rm nc}(t) &= \int_0^\infty f(\ell;t)\,d\ell - c\int_0^\infty \ell f(\ell;t)\,d\ell + O(1)\int_0^\infty \ell^2 f(\ell;t)\,d\ell \\ &= 1 - c\mathbb{E}[R(t)] + O(t).\end{aligned}$$

In view of (2.8) (with $d = 1$ and $b = 0$), this yields (2.16).

The long term asymptotics of $\tilde{P}_{\rm nc}(t)$ (as $t \to \infty$) can be derived via the Laplace method from (2.15) or, alternatively, directly from (2.9) using (2.12) [**2**]:

$$\tilde{P}_{\rm nc}(t) \sim 8\left(\frac{2c^2 t}{3\pi}\right)^{1/2}\exp\left(-\frac{3}{2}\left(\pi^2 c^2 t\right)^{1/3}\right). \tag{2.17}$$

As compared to the general Donsker–Varadhan result (1.1) (for $d = 1$), this is a more precise formula since it identifies the pre-exponential factor.

3. Taking account of trap correlations

Representations (2.2) and (2.4) show that the emptiness probability function $g(\cdot)$ is the sole characteristic of the trap distribution that enters the survival probability. In particular, the influence of trap correlations on the particle's survival is determined by the difference between the corresponding emptiness probability functions, $g(\cdot)$ and $g_{\rm nc}(\cdot)$.

On the other hand, the emptiness probabilities furnish a convenient tool to characterize the trap interaction (at least on a qualitative level). We give the following basic definition (cf. [**1**]):

DEFINITION 3.1. A random point process $\{Z_i\}$ is called (*purely*) *attractive* or *repulsive* if for any disjoint (bounded) Borel sets $S_1, S_2 \in \mathcal{B}(\mathbf{R}^d)$, the emptiness probability function $g(\cdot)$ defined in (2.3) satisfies respectively

$$g(S_1 \cup S_2) \ge g(S_1)\cdot g(S_2) \qquad \text{(attraction)} \tag{3.1}$$

$$g(S_1 \cup S_2) \le g(S_1)\cdot g(S_2) \qquad \text{(repulsion)} \tag{3.2}$$

and, moreover, these inequalities do not degenerate to identity. Furthermore, we say that attraction or repulsion is *strict* if the equality sign in (3.1), (3.2) is only possible with $|S_i| = 0$, $i = 1, 2$, so that both sides amount to 1.

REMARK. The requirement for S_1, S_2 to be disjoint can be relaxed to the condition that $|S_1 \cap S_2| = 0$.

In fact, the above definition is just a suitable reformulation of the property for a random point process to have positive or negative correlations. Indeed, introducing the indicator random variables $\delta_S := I\{N_S \geq 1\}$, so that $\mathbb{E}[\delta_S] = 1 - g(S)$, it is straightforward to check that the condition $\mathrm{Cov}(\delta_{S_1}, \delta_{S_2}) \geq 0$ (or ≤ 0) amounts to inequality (3.1) (or (3.2), respectively).

Note that in the noncorrelated case, due to (2.5) we have identically

$$g_{\mathrm{nc}}(S_1 \cup S_2) \equiv g_{\mathrm{nc}}(S_1) \cdot g_{\mathrm{nc}}(S_2), \qquad S_1 \cap S_2 = \emptyset,$$

and therefore the deviation from the equality is an evidence of some dependence. To get more insight why the specific inequality sign in (3.1) and (3.2) is relevant to a common notion of "attraction" and "repulsion", let us introduce the *conditional emptiness probability* $g(S|x_0)$, conditioned on the presence of a "probe" trap center at point $x_0 \in S$:

$$g(S|x_0) := \mathbb{P}\{\delta_{S\setminus\{x_0\}} = 0 | \delta_{\{x_0\}} = 1\}.$$

Obviously, in the Poisson case the two emptiness probabilities coincide, $g_{\mathrm{nc}}(S|x_0) \equiv g_{\mathrm{nc}}(S)$. Moreover, it is easy to check that each of the hypotheses (3.1), (3.2) implies for all S and $x_0 \in S$ respectively

$$g(S|x_0) \leq g(S) \text{ (attraction) or } g(S|x_0) \geq g(S) \text{ (repulsion)}, \tag{3.3}$$

which is in agreement with an intuitive understanding of the qualitative types of "interaction" [**1**], [**9**].

The following theorem provides a rigorous basis for the physical conjecture mentioned in Sect. 1.

THEOREM 3.1. *Suppose that the random point process $\{Z_i\}$ is purely attractive or repulsive, in the sense of Definition* 3.1, *and let c be its concentration. Then the emptiness probability $g(S)$ for all (bounded) Borel sets S satisfies respectively*

$$g(S) \geq g_{\mathrm{nc}}(S) \text{ (attraction) or } g(S) \leq g_{\mathrm{nc}}(S) \text{ (repulsion)}, \tag{3.4}$$

where $g_{\mathrm{nc}}(S) = \exp(-c|S|)$ is the emptiness probability of the Poisson point process with intensity c. In the case of strict attraction and repulsion, the inequalities (3.4) *are strict provided* $|S| > 0$.

PROOF. We give an outline of the proof. By induction, inequalities (3.1), (3.2) are readily extended to finite unions of disjoint sets. Therefore, via an appropriate approximation of S we obtain accordingly $g(S) \gtrless \prod_i g(\Delta S_i)$. It remains to make use of Korolyuk's theorem (see [**11**, § 7.2]), which implies that for a "small" set ΔS one can write $g(\Delta S) = 1 - c|\Delta S| + o(|\Delta S|)$. □

Applying Theorem 3.1 to (2.2), we arrive at our main claim in this section.

THEOREM 3.2 (Correlation Conjecture). *Suppose that the trap ensemble $\{Z_i\}$ is purely attractive or repulsive, according to Definition* 3.1, *and let c denote its concentration. Then the survival probability $P(t)$ for each $t \geq 0$ satisfies*

$$P(t) \geq P_{\mathrm{nc}}(t) \text{ (attraction) or } P(t) \leq P_{\mathrm{nc}}(t) \text{ (repulsion)},$$

where $P_{\mathrm{nc}}(t)$ corresponds to the Poisson trap ensemble with intensity c. The same is valid for the normalized survival probabilities, that is, for all $t \geq 0$

$$\tilde{P}(t) \geq \tilde{P}_{\mathrm{nc}}(t) \text{ (attraction) or } \tilde{P}(t) \leq \tilde{P}_{\mathrm{nc}}(t) \text{ (repulsion)}.$$

In case attraction or repulsion be strict, so are the above inequalities for all $t > 0$.

PROOF. To prove the statement for the normalized probabilities, being less evident, we use Definition 3.1 and Theorem 3.1 to write

$$\frac{g(S_b(t))}{g(V_b)} \gtrless g(S_b(t) \setminus V_b) \gtrless \exp(-c|S_b(t) \setminus V_b|) = \frac{\exp(-c|S_b(t)|)}{\exp(-c|V_b|)} = \frac{g_{\mathrm{nc}}(S_b(t))}{g_{\mathrm{nc}}(V_b)},$$

whence in view of (2.4) the result follows. □

REMARK. Theorems 3.1, 3.2 can be extended to the nonstationary case where the expected number of points is determined by a certain (non-constant) concentration function $c(x)$ via the formula $\mathbb{E}[N_S] = \int_S c(x)\,dx$.

4. Poisson cluster model

We now proceed to studying a few specific models. In this section, we analyze the trapping problem in a locally inhomogeneous trapping environment, in which correlations are displayed in the presence of *trap clusters* [**1**], [**3**], [**9**], [**21**]. More precisely, suppose that each trap is attached to a certain group (cluster), to the effect that the process $\{Z_i\}$ can be represented in the form $\{X_i + Y_j^{(i)},\ j = 1, \dots, \nu_i\}$, where X_i denotes the cluster "center" and a random aggregate of random vectors $\mathbf{Y}^{(i)} := \{Y_1^{(i)}, \dots, Y_{\nu_i}^{(i)}\}$ determines the positions of ν_i traps belonging to the cluster, relative to its center.

Let us consider the simplest model of such kind, the *Poisson cluster point process* [**11**, § 8.3], in which $\{X_i\}$ is a Poisson point process (with intensity c_0) and the aggregates $\{\mathbf{Y}^{(i)}\}$ are assumed to be independent both of each other and of $\{X_i\}$, and identically distributed. The total trap concentration for such an ensemble is $c = c_0\bar{\nu}$, where $\bar{\nu} := \mathbb{E}[\nu]$ is the expected number of traps in a cluster. For $\nu \equiv 1$, our model obviously reduces to the noncorrelated one, since the Poisson process is invariant under independent identically distributed random shifts of its points (see [**11**, § 8.2]).

THEOREM 4.1 ([**1**, Proposition 3.10]; cf. [**11**, § 8.3]). *For the Poisson cluster process, the emptiness probability $g(S)$ has the form*

$$g(S) = \exp\left(-c_0 \int_{\mathbf{R}^d} \mathbb{P}\Big(\cup_{j=1}^{\nu} \{x + Y_j \in S\}\Big)\,dx\right), \tag{4.1}$$

where $\{Y_1, \dots, Y_\nu\} =: \mathbf{Y}$ represents the random structure of a generic cluster.

The survival probability $P(t)$ can now be obtained by substituting expression (4.1) into the general formula (2.2). However, via interchanging the order of the integration and expectation in the exponent, one can rewrite the equation for $P(t)$ in a more suggestive way (cf. (2.6)).

THEOREM 4.2 ([**1**, p. 173], [**9**]). *The survival probability $P(t)$ for the Poisson cluster trapping model can be represented as*

$$P(t) = E_0\left[\exp(-c_0\mathbb{E}|S_b^*(\mathbf{Y};t)|)\right], \tag{4.2}$$

where $S_b^*(\mathbf{Y};t) := \cup_{j=1}^{\nu}(S_b(t) - Y_j)$, $S_b(t)$ being the Wiener sausage.

That is to say, $S_b^*(\mathbf{Y};t)$ is a random region in $\mathbf{R}^d$ defined as the union of ν identical copies of the Wiener sausage $S_b(t)$ shifted at $-Y_1, \dots, -Y_\nu$. We call such an object the *bunch of Wiener sausages*, or simply the *Wiener bunch.* This concept is useful in the trapping context [**3**], [**21**] and is of mathematical interest in itself.

Let us now cite some known general results for the Poisson cluster model. We assume henceforth that the cluster point process under consideration is proper, that is, $1 < \bar{\nu} < \infty$.

THEOREM 4.3 ([**1**, Proposition 4.1]). *The Poisson cluster process is purely attractive, regardless of the inner structure of the clusters.*

In accordance with Theorem 3.2, this leads to the trapping slowdown.

THEOREM 4.4 ([**1**, Propositions 5.2, 5.2′]). *The survival probability* (4.2) *for all* $t > 0$ *satisfies the inequality* $P(t) > P_{\rm nc}(t)$. *Moreover, the same inequality holds for the normalized survival probabilities*: $\tilde{P}(t) > \tilde{P}_{\rm nc}(t)$.

Note that the inequalities in this theorem are strict, although attraction itself may not be strict (cf. Theorem 4.3) and therefore the last statement of Theorem 3.2 cannot be directly applied.

Physical discussion of the trapping slowdown predicted by Theorem 4.4 can be found in [**3**], [**21**], where the "transparency" of individual clusters is shown to be an important characteristic responsible for the manifestation of the effect at short and intermediate times.

The behavior of the survival probability $P(t)$ as $t \to \infty$ is described by the following theorem ([**1**, Proposition 5.3]; see also [**3**], [**9**]).

THEOREM 4.5. *The long-time asymptotics of the survival probability* $P(t)$ *in the Poisson cluster model is of the form*

$$\lim_{t\to\infty} t^{-d/(d+2)} \log P(t) = -\gamma_d c_0^{2/(d+2)}, \tag{4.3}$$

where γ_d *is the same constant as in* (1.1).

Thus, the form of the long-time survival tail in the Poisson cluster model reproduces that in the noncorrelated case, but with parameter c replaced by c_0. Expressing this in a picturesque way: at long times the Brownian particle "feels" the concentration of clusters, c_0, rather than the total trap concentration, c. Note also that the limit (4.3) does not depend on the inner structure of clusters.

The remainder of this section is devoted to a particular one-dimensional version of the Poisson cluster model which illuminates the role of the clusters' overlapping. Assume that each cluster is a segment of length L containing $n \geq 2$ *pointwise* traps ($b = 0$). More specifically, let two traps be attached to the segment ends and the remaining $n - 2$ traps be distributed over the segment completely at random, that is, uniformly and independently of each other.

Adapting the general formula (4.1) to this particular case, we obtain

THEOREM 4.6. *The emptiness probability function in this model is of the form*

$$g(\ell) = \exp\left\{-\rho\left[1 + \frac{\ell}{L} - \left(1 - \frac{\ell}{L}\right)_+^{n-1}\right]\right\},$$

where $u_+ := \max(u, 0)$ *and* $\rho := c_0 L$.

Let Λ_k be the random subset of the line $\mathbf{R}$ consisting of the points "covered" by exactly k clusters ($k = 0, 1, \dots$). Obviously, once the Brownian particle has been born in the Λ_k, it will henceforth stay there until being trapped. Let us point out that via cutting out those intervals on $\mathbf{R}$ which are not in Λ_k and gluing together the remaining segments, one can realize Λ_k as an *effective k-layer trapping medium* in which the distribution of the new traps is induced by the cutting procedure (cf. [**10**]). In the next theorem, we find the basic characteristics of Λ_k.

THEOREM 4.7. *The probability π_k for a given point to occur in the k-layer medium Λ_k follows the Poisson distribution with parameter $\rho = c_0 L$:*

$$\pi_k = \frac{\rho^k}{k!}\, e^{-\rho}, \qquad k = 0, 1, 2, \dots \tag{4.4}$$

The conditional emptiness probability $g_k(\ell)$, conditioned on Λ_k, equals

$$g_k(\ell) = \left(1 - \frac{\ell}{L}\right)_+^{k(n-1)} e^{-\rho\ell/L}, \qquad k = 0, 1, 2, \dots, \tag{4.5}$$

where expression u_+^0 is understood to be equal to 1. Finally, the inner concentration of traps in Λ_k is given by

$$c_k = c_0 \left(1 + \frac{k(n-1)}{\rho}\right), \qquad k = 0, 1, 2, \dots \tag{4.6}$$

PROOF. In order for a given point to occur in Λ_k, its two-sided $(L/2)$-vicinity must contain exactly k cluster centers, which has the probability (4.4).

Furthermore, it is easy to verify that the probability for an interval I of length ℓ to lie in Λ_k is given by the expression

$$e^{-2c_0\ell} \cdot \frac{c_0^k (L-\ell)_+^k}{k!}\, e^{-c_0(L-\ell)}.$$

To ensure the emptiness of I, this has to be multiplied by $(1 - \ell/L)_+^{k(n-2)}$, that is the probability that none of the $k(n-2)$ inner traps belonging to the k covering clusters ("layers") occurs in I. Conditioning on Λ_k, which amounts to dividing by π_k, then yields (4.5).

Equation (4.6) follows from the observation that due to each of the k layers, the "ground" concentration c_0 is increased by $(n-1)/L = c_0(n-1)/\rho$ traps per unit length. To check (4.6) more formally, we find the probability distribution function of a random spacing $\Delta \subset \Lambda_k$ between two adjacent traps, which can be done similarly to the above calculation, and then compute the concentration c_k as the reciprocal of the conditional expectation $\mathbb{E}[\Delta \,|\, \Delta \subset \Lambda_k]$. We omit the details. □

Note that for $k = 0$, one has $g_0(\ell) \equiv e^{-c_0\ell}$, which is actually due to the fact that the 0-layer medium Λ_0 is noncorrelated (cf. Sect. 2 and also Sect. 6 below). Using (4.5) and (4.6), one can also verify that for each $k \geq 1$ and all $\ell > 0$

$$g_k(\ell) < e^{-c_k\ell}.$$

That is to say, the emptiness probability inside Λ_k appears to be smaller than that in a Poisson medium with the same concentration c_k. In view of Theorem 3.1, this is an evidence of the inner trap *repulsion* in each Λ_k, $k \geq 1$. On the other hand, the clusterized medium on the whole exhibits trap *attraction* (cf. Theorem 4.3).

5. One-dimensional renewal model

In the remaining two sections we assume that $Z = \{Z_i\}$ is a stationary *renewal process* on the line[6] $\mathbf{R}$, so that random spacings $\{\Delta_i\}$ between adjacent points are independent and identically distributed. In this section we consider the case of pointwise traps ($b = 0$). Let $\varphi(x)$, $x > 0$, be a common probability density of Δ_i's. Due to the classical renewal theorem (see, e.g., [**14**, § XI.1]), the concentration c is equal to the reciprocal mean spacing, $1/\mathbb{E}[\Delta]$:

$$c = \left(\int_0^\infty x\,\varphi(x)\,dx\right)^{-1}. \tag{5.1}$$

Furthermore, the emptiness probability function $g(\ell)$ is explicitly given by the expression (see [**14**, § XI.4])

$$g(\ell) = c\int_0^\infty x\,\varphi(x+\ell)\,dx. \tag{5.2}$$

Let us now take $\varphi(x)$ in the form of the gamma density, first proposed in the trapping context in [**6**]:

$$\varphi_\alpha(x) = \frac{\alpha c}{\Gamma(\alpha)}(\alpha c x)^{\alpha-1}e^{-\alpha c x}, \qquad x > 0, \tag{5.3}$$

where $\alpha > 0$ and $\Gamma(\alpha) = \int_0^\infty x^{\alpha-1}e^{-x}\,dx$ is the gamma function. (The parameter c here plays the role of the concentration, since $\varphi_\alpha(x)$ satisfies the relation (5.1).) Equation (5.2) then takes the form

$$g_\alpha(\ell) = \frac{e^{-\alpha c\ell}}{\Gamma(\alpha+1)}\int_0^\infty x(x+\alpha c\ell)^{\alpha-1}e^{-x}\,dx. \tag{5.4}$$

Note that for $\alpha = 1$ the process Z reduces to a Poisson one. For $\alpha \neq 1$, the behavior of the density $\varphi_\alpha(x)$ at zero suggests that $\alpha < 1$ and $\alpha > 1$ correspond to trap attraction and repulsion, respectively. This is made precise by the following theorem which can be proved by a direct analysis of equation (5.4).

THEOREM 5.1. *The gamma ensemble governed by the density* (5.3) *is strictly attractive or repulsive, in the sense of Definition* 3.1, *according as* $\alpha < 1$ *or* $\alpha > 1$. *Therefore, by Theorem* 3.1 *one has* $g_\alpha(\ell) \gtrless g_1(\ell) \equiv g_{\rm nc}(\ell)$ *whenever* $\alpha \lessgtr 1$. *Moreover,* $g_\alpha(\ell)$ *is strictly decreasing as a function of* α.

This amounts to saying that the deviation of the emptiness probability $g_\alpha(\ell)$ from its noncorrelated counterpart $g_1(\ell) \equiv g_{\rm nc}(\ell)$ becomes bigger as $|1-\alpha|$ grows, while the sign of $1-\alpha$ determines the "direction" of correlations. Note that in the limit $\alpha \uparrow \infty$, corresponding to the infinitely strong repulsion, from (5.3) we have $\varphi_\alpha(\cdot) \to \delta_{1/c}(\cdot)$, which means the formation of a regular lattice of traps on the line $\mathbf{R}$ with period c^{-1}. Hence, the limiting emptiness probability $g_\infty(\ell) = (1-c\ell)\,\mathbf{1}_{\{c\ell\le 1\}}$ provides the minimal possible value of $g_\alpha(\ell)$, and thus the lattice appears to be the most unfriendly environment for the surviving particle.

The next result following from Theorem 5.1 specifies the general Theorem 3.2.

---

[6] Such a process can be directly constructed on the whole line $\mathbf{R}$ (see [**11**, § 12.3]), without referring to the limit regime arising for the renewal process on a half-line as the starting point is moved to $-\infty$.

THEOREM 5.2. *For each $t > 0$ the survival probability $P_\alpha(t)$ is a strictly decreasing function of α. In particular, for all $t > 0$ one has $P_\alpha(t) \gtrless P_{\rm nc}(t)$ whenever $\alpha \lessgtr 1$. Moreover, the same is valid for the normalized survival probability $\tilde{P}_\alpha(t)$.*

The long-time survival asymptotics can be found from equations (2.13), (5.4) using representation (2.12) (cf. (2.17)):

THEOREM 5.3. *As $t \to \infty$,*

$$\tilde{P}_\alpha(t) \sim 8\sqrt{\frac{2}{3}}\,\frac{\pi^{-3/2}}{\Gamma(\alpha+1)}\left(\pi^2\alpha^2c^2t\right)^{(2\alpha+1)/6}\exp\left(-\frac{3}{2}\left(\pi^2\alpha^2c^2t\right)^{1/3}\right). \tag{5.5}$$

REMARK. Comparing (5.5) with the result for Gibbsian traps mentioned in Sect. 1 suggests that the quantity αc here is related to the "pressure" as $\alpha c = \beta p$, which can be viewed as the equation of state. In fact, the gamma ensemble can be embedded into a Gibbsian framework by picking the nearest-neighbor interaction potential $\Phi(x) = (1-\alpha)\log(cx)$ (see [**15**]). Note that for $\alpha < 1$ (attractive case) this potential appears to be non-stable in the sense of Ruelle [**24**, § 3.2]. However, the 'nearest-neighbor' constraint prevents the system from the collapse.

6. Role of the trap size in one dimension

In this section, we analyze the dependence of the trapping rate upon the trap size b in one dimension. Contrary to the common opinion that the (normalized) survival probability is independent of b, we show that such a dependence does arise provided the traps are correlated (for the physical discussion, see [**10**]).

Recalling (2.13), we have[7]

$$\tilde{P}(t;b) = \int_0^\infty \tilde{g}(\ell;b)f(\ell;t)\,d\ell, \tag{6.1}$$

where $f(\ell;t)$ is specified in (2.11)–(2.12) and (cf. (5.2))

$$\tilde{g}(\ell;b) = \frac{g(\ell+2b)}{g(2b)} \tag{6.2}$$

is a new emptiness probability function determined according to (5.2) by the conditional probability density

$$\tilde{\varphi}(\ell;b) = \frac{\varphi(\ell+2b)}{\int_0^\infty \varphi(x+2b)\,dx}. \tag{6.3}$$

Such a conditioning can also be realized as passing from the original ensemble of spacings $\{\Delta_i\}$ between the adjacent trap centers $\{Z_i\}$ to an effective ensemble of *void* intervals $\{\tilde{\Delta}_i\}$ which do not intersect with any trap $[Z_i - b, Z_i + b]$.[8] It is not difficult to check that the random variables $\{\tilde{\Delta}_i\}$ are again independent and identically distributed, so that the ensemble $\{\tilde{Z}_i\}$ of effective pointwise traps generated by $\{\tilde{\Delta}_i\}$ is a renewal process as well.

As mentioned in Sect. 2, in the noncorrelated case one has $\tilde{g}_{\rm nc}(\ell;b) \equiv \tilde{g}_{\rm nc}(\ell;0)$ $(= e^{-c\ell})$, so that actually the survival probability does not depend on b:

$$\tilde{P}_{\rm nc}(t;b) \equiv \tilde{P}_{\rm nc}(t;0).$$

[7]We introduce b in the notation in order to emphasize the presence of this parameter.

[8]In terminology of Sect. 4, the void intervals $\{\tilde{\Delta}_i\}$ constitute the 0-layer medium Λ_0.

Formulas (6.1) and (6.2) indicate that in general such a dependence does take place. Moreover, physical arguments suggest **[10]** that the effective trap concentration

$$\tilde{c}(b) = \left(\int_0^\infty x\,\tilde{\varphi}(x;b)\,dx\right)^{-1} \tag{6.4}$$

is likely to be smaller (attraction) or greater (repulsion) than the original concentration $c \equiv \tilde{c}(0)$. Accordingly, one can expect that nonzero trap radius b either promotes or inhibits the Brownian survival, depending on the qualitative type of trap interaction. For short times, $t \to 0$, this is justified by the Smoluchowski-type formula

$$\tilde{P}(t;b) = 1 - \frac{4\tilde{c}(b)}{\sqrt{2\pi}}\,t^{1/2} + o\big(t^{1/2}\big),$$

which can be derived from (6.1) in much the same way as (2.16); in so doing, one only needs to apply Korolyuk's theorem (mentioned in the proof of Theorem 3.1) to write $\tilde{g}(\ell;b) = 1 - \tilde{c}(b)\ell + o(\ell)$ for $\ell \to 0$.

The following general theorem makes these considerations precise.

THEOREM 6.1. *Suppose that the renewal (one-dimensional) point process* $\{Z_i\}$ *is purely attractive or repulsive, according to Definition* 3.1. *Then for all* $t \ge 0$

$$\tilde{P}(t;b) \ge \tilde{P}(t;0) \quad (\textit{attraction}) \quad \textit{or} \quad \tilde{P}(t;b) \le \tilde{P}(t;0) \quad (\textit{repulsion}).$$

Moreover, the deviation of the effective concentration is as predicted:

$$\tilde{c}(b) \le c \quad (\textit{attraction}) \quad \textit{or} \quad \tilde{c}(b) \ge c \quad (\textit{repulsion}).$$

PROOF. To be definite, consider the case of attraction. Using (6.2) and applying condition (3.1) of Definition 3.1, we can write

$$\tilde{g}(\ell;b) = \frac{g(\ell+2b)}{g(2b)} \ge g(\ell) = \tilde{g}(\ell;0),$$

whence the first assertion of the theorem immediately follows due to (6.1). In view of (6.4) and (6.3), in order to prove the second assertion we have to check that

$$\int_0^\infty \varphi(x+2b)\,dx \le c\int_0^\infty x\,\varphi(x+2b)\,dx. \tag{6.5}$$

According to (5.2), the right-hand side of (6.5) coincides with $g(2b)$, whereas the left-hand side is equal to

$$\mathbb{P}\{\Delta > 2b\} = \mathbb{P}\{N_{(0,2b]} = 0 \,|\, \delta_{\{0\}} = 1\} = g(2b|0).$$

Thus, inequality (6.5) amounts to $g(2b|0) \le g(2b)$. But from (3.3) it follows that the last inequality is true, and we are done. □

For the gamma model introduced above, the b-dependence can be studied in more detail. Substituting (5.4) into (6.2), we find the emptiness probability

$$\tilde{g}_\alpha(\ell;b) = \frac{\int_0^\infty x(x+\alpha c\ell+2\alpha cb)^{\alpha-1}e^{-x}\,dx}{\int_0^\infty x(x+2\alpha cb)^{\alpha-1}e^{-x}\,dx}\,e^{-\alpha c\ell}, \tag{6.6}$$

whereas the trap concentration is given by (see (6.4), (6.3))

$$\tilde{c}_\alpha(b) = c\left(1 - 2bc\left[1-(2\alpha bc)^{\alpha-1}\frac{e^{-2\alpha bc}}{\Gamma(\alpha,2\alpha bc)}\right]\right)^{-1}, \tag{6.7}$$

where $\Gamma(\alpha,z) = \int_z^\infty x^{\alpha-1}e^{-x}\,dx$ is the incomplete gamma function.

The following two results are proved similarly to Theorems 5.1, 5.2 using the explicit expressions (6.6), (6.7).

THEOREM 6.2 (cf. Theorem 5.1). *The emptiness probability* $g_\alpha(\ell; b)$ *is strictly decreasing in* α. *For a fixed* $\alpha \neq 1$, *both* $\tilde{g}_\alpha(\ell; b)$ *and* $1/\tilde{c}_\alpha(b)$ *monotonically increase or decrease as functions of* b, *depending on whether* $\alpha < 1$ *or* $\alpha > 1$, *respectively.*

THEOREM 6.3 (cf. Theorem 5.2). *For a fixed* $\alpha \neq 1$, *the probability* $\tilde{P}_\alpha(t; b)$ *is a strictly increasing or decreasing function of* b, *whenever* $\alpha < 1$ *or* $\alpha > 1$, *respectively.*

Combined with Theorem 5.2, this means that the effect of the trapping slowdown ($\alpha < 1$) or acceleration ($\alpha > 1$) is enhanced as the trap size b grows. This result may prove useful in experimental detecting of trap correlations in real one-dimensional systems.

References

[1] S. Albeverio and L. V. Bogachev, *Brownian survival in a clusterized trapping medium*, Rev. Math. Phys. **10** (1998), 147–189.

[2] B. Ya. Balagurov and V. G. Vaks, *Random walks of a particle on lattices with traps*, Zh. Eksper. Teoret. Fiz. **65** (1973), 1939–1945 (Russian). English translation: Soviet Phys. JETP **38** (1974), 968–970.

[3] A. M. Berezhkovskii, Yu. A. Makhnovskii, L. V. Bogachev, and S. A. Molchanov, *Brownian-particle trapping by clusters of traps*, Phys. Rev. E **47** (1993), 4564–4567.

[4] A. M. Berezhkovskii, Yu. A. Makhnovskii, and R. A. Suris, *Kinetics of diffusion-controlled reactions*, Chem. Phys. **137** (1989), 41–49.

[5] ———, *Wiener sausage volume moments*, J. Statist. Phys. **57** (1989), 333–346.

[6] A. M. Berezhkovskii, Yu. A. Makhnovskii, R. A. Suris, L. V. Bogachev, and S. A. Molchanov, *Trap correlation influence on Brownian particle death. One-dimensional case*, Phys. Lett. A **161** (1991), 114–117.

[7] ———, *Diffusion-limited reactions with correlated traps*, Chem. Phys. Lett. **193** (1992), 211–214.

[8] ———, *Trap correlation influence on diffusion-limited process rate*, Phys. Rev. A **45** (1992), 6119–6122.

[9] L. V. Bogachev and Yu. A. Makhnovskii, *Brownian motion with absorption in a clusterized random medium*, Dokl. Akad. Nauk **340** (1995), 300–302 (Russian). English translation: Russian Acad. Sci. Dokl. Math. **51** (1995), 51–53.

[10] L. V. Bogachev, Yu. A. Makhnovskii, and A. M. Berezhkovskii, *Trapping rate dependence on the trap size in one dimension*, Phys. Rev. E **52** (1995), 6900–6903.

[11] D. J. Daley and D. Vere-Jones, *An introduction to the theory of point processes*, Springer, New York, 1988.

[12] M. D. Donsker and S. R. S. Varadhan, *Asymptotics for the Wiener sausage*, Comm. Pure Appl. Math. **28** (1975), 525–565.

[13] W. Feller, *The asymptotic distribution of the range of sums of independent random variables*, Ann. Math. Statist. **22** (1951), 427–432.

[14] ———, *An introduction to probability theory and its applications*, Vol. 2, 2nd ed., Wiley, New York, 1971.

[15] F. Gürsey, *Classical statistical mechanics of a rectilinear assembly*, Proc. Cambridge Philos. Soc. **46** (1950), 182–194.

[16] S. Havlin and D. Ben-Avraham, *Diffusion in disordered media*, Adv. in Phys. **36** (1987), 695–798.

[17] F. den Hollander and G. H. Weiss, *Aspects of trapping in transport processes*, Contemporary Problems in Statistical Physics, G. H. Weiss, ed., SIAM, Philadelphia, PA, 1994, pp. 147–203.

[18] M. Kac, *Probabilistic methods in some problems of scattering theory*, Rocky Mountain J. Math. **4** (1974), 511–537.

[19] R. F. Kayser and J. B. Hubbard, *Reaction diffusion in a medium containing a random distribution of nonoverlapping traps*, J. Chem. Phys. **80** (1984), 1127–1130.

[20] E. Kotomin and V. Kuzovkov, *Modern aspects of diffusion-controlled reactions*, Elsevier, Amsterdam, 1996.
[21] Yu. A. Makhnovskii, L. V. Bogachev, and A. M. Berezhkovskii, *The effect of trap grouping on the decay kinetics of Brownian particles*, Chem. Phys. Rep. **14** (1995), 710–728.
[22] A. A. Ovchinnikov and Ya. B. Zeldovich, *Role of density fluctuations in bimolecular reaction kinetics*, Chem. Phys. **28** (1978), 215–218.
[23] P. M. Richards, *Diffusion to nonoverlapping or spatially correlated traps*, Phys. Rev. B **35** (1987), 248–256.
[24] D. Ruelle, *Statistical mechanics: Rigorous results*, Benjamin, Reading, Mass., 1969.
[25] M. von Smoluchowski, *Versuch einer mathematischen Theorie der Koagulationskinetik kolloider Lösungen*, Z. Phys. Chem. **29** (1917), 129–168.
[26] A. S. Sznitman, *Brownian survival among Gibbsian traps*, Ann. Probab. **21** (1993), 490–508.
[27] ———, *Brownian motion, obstacles and random media*, Springer, Berlin, 1998.
[28] S. Torquato, *Concentration dependence of diffusion-controlled reactions among static reactive sinks*, J. Chem. Phys. **85** (1986), 7178–7179.
[29] G. H. Weiss and S. Havlin, *Trapping of random walks on the line*, J. Statist. Phys. **37** (1984), 17–25.

Faculty of Mechanics and Mathematics, Moscow State University, Moscow 119899, Russia
E-mail address: `lvb@mech.math.msu.su`

Department of Mathematics, University of North Carolina at Charlotte, Charlotte, NC 28223, USA
E-mail address: `smolchan@uncc.edu`

Topchiev Institute of Petrochemical Synthesis, Russian Academy of Sciences, Moscow 117912, Russia
E-mail address: `yuam@tips-gw.ips.ac.ru`

Karpov Institute of Physical Chemistry, Moscow 103064, Russia
E-mail address: `berezh@cc.nifhi.ac.ru`

Canadian Mathematical Society
Conference Proceedings
Volume **29**, 2000

Boundary Reduction of Spectral Invariants – Results and Puzzles

Bernhelm Booss–Bavnbek

This report is dedicated to Sergio Albeverio on the occasion of his 60th birthday.

ABSTRACT. We summarize concepts of infinite–dimensional analysis which arise from the study of spectral invariants of finite–dimensional manifolds: index, spectral flow, and determinant. We investigate the conceptional changes and computational advantages in infinite–dimensional analysis which can be achieved when 'adding' one dimension by evolving the available information to a bounding manifold, or when 'subtracting' by reducing the information to a hypersurface. The situation typically arises when we consider (elliptic) operators of Dirac type and families of such operators over a compact smooth Riemannian manifold with boundary or over a closed partitioned manifold. We apply suitable boundary conditions and discuss three types of results regarding the involved function spaces, operators, and spectral invariants: reduction to the boundary, correction formulas for changing the boundary condition, and pasting results. We emphasize the joint features of the different approaches, but also focus on some puzzling differences between them.

Introduction

In this survey we present and analyze various cases of boundary reduction of spectral invariants. The cases are related to the index, the spectral flow, and the ζ–function regularized determinant of Dirac operators, resp. families of Dirac operators over a compact manifold with boundary.

We address three types of results under the general heading of 'boundary reduction': *genuine boundary reduction* where a global invariant can be recovered from expressions defined on the boundary; *correction formulas* where the difference between two spectral invariants can be localized at the boundary; and *pasting formulas* where a global invariant on a partitioned manifold can be split into components or localized along the splitting hypersurface.

In Section 1 we give general restriction results for Dirac operators and spaces of sections and distributions.

1991 *Mathematics Subject Classification.* 58G25, 58G20.

Key words and phrases. Cauchy data spaces, elliptic boundary problems, families of Dirac operators, Fredholm determinant, index, Lagrangian subspaces, Maslov index, quadratic functionals, spectral flow, symplectic functional analysis, ζ–function regularized determinant.

In Sections 2-3 we treat the index and the spectral flow in two almost complementary set-ups, distinguished by their assumptions about the differentiability and naturalness of the Cauchy data spaces. Since the index and the spectral flow are both topological invariants, the main results clearly have quite a lot in common. At some points one feels tempted to put up a categorial framework and a 'functor' to describe the boundary reduction in terms of a homology theory. However, the delicacy and the variability of some of the results speak against it.

In Section 4 we recall the three basic concepts of the determinant as used in quantum field theories: the quadratic functional, the Fredholm determinant, and the ζ–determinant. We present our view of the recent Scott–Wojciechowski Formula, which relates the Fredholm determinant and the ζ–determinant over manifolds with boundary. It is interesting that many of the basic concepts discussed in the first sections also apply to the study of the determinant which is not a topological invariant but much 'finer'.

Acknowledgments. *I would like to thank S. Albeverio (Bonn), E. Balslev and A. Venkov (both Aarhus), K. Furutani and N. Otsuki (both Tokyo), M. Lesch (Tucson), G. Morchio and F. Strocchi (both Pisa), S.G. Scott (London), I.M. Singer (M.I.T.) and K.P. Wojciechowski (Indianapolis) for sharing some of their ideas and calculations with me and for valuable suggestions and comments which have entered in this report.*

1. Boundary Reduction of Function Spaces and Distributions

1.1. The Tangential Operator. Let M be a compact smooth Riemannian manifold with boundary Σ. We shall assume that M does not contain a connected component which is closed (i.e. with empty boundary). Let

$$\mathrm{A} : C^\infty(M;S) \longrightarrow C^\infty(M;S)$$

be an operator of Dirac type acting on sections of a Hermitian bundle S of Clifford modules over M, i.e. $\mathrm{A} = \mathbf{c} \circ \nabla$ where $\mathbf{c}$ denotes the Clifford multiplication and ∇ is a connection for S which is compatible with $\mathbf{c}$ (i.e. $\nabla \mathbf{c} = 0$). To begin with, we assume that all metric structures of M and S are product in a collar neighbourhood $\mathcal{N}$ of the boundary. Then

$$\mathrm{A}\,|_{\mathcal{N}} = \sigma\left(\frac{\partial}{\partial\tau} + \mathcal{B}\right), \tag{1.1}$$

where τ denotes the (inward) normal coordinate, σ denotes the Clifford multiplication with $d\tau$ and $\mathcal{B}$ denotes the canonically associated Dirac operator over Σ, called the *tangential operator*. Here the point of the product structure is that then σ and $\mathcal{B}$ do not depend on the normal variable. We note that σ is unitary with $\sigma^2 = -\mathrm{Id}$ and $\sigma\mathcal{B} = -\mathcal{B}\sigma$.

1.2. General Restriction Results. We consider the sth order Sobolev spaces $H^s(M;S)$ and $H^s(\Sigma;S|_\Sigma)$ ($s \in \mathbf{R}$). There is not one single space, one single s to pick up as a canonical choice for solving the system of Dirac equations or determining the spectral quantities: from the point of view of physics, e.g., one would be mainly interested in smooth sections and in the smooth category it is no problem to restrict a section to the boundary. We shall write

$$\gamma_\infty : C^\infty(M;S) \longrightarrow C^\infty(\Sigma;S|_\Sigma).$$

We recall the Green–Stokes Theorem as model for *all* boundary reduction formulas. It takes the following form in our context.

LEMMA 1.1. (Green's Formula). *All (compatible) Dirac operators are symmetric with*

$$(\mathrm{A}\, f, g) - (f, \mathrm{A}\, g) = -\int_\Sigma \langle \sigma\gamma_\infty f, \gamma_\infty g\rangle \, \mathrm{dvol}_\Sigma$$

for any $f, g \in C^\infty(M; S)$.

From the point of view of analysis one is mainly interested in L_2–sections or even in distributional sections. But γ_∞ extends to a bounded map

$$\gamma_s : H^s(M;S) \longrightarrow H^{s-\frac{1}{2}}(\Sigma; S|_\Sigma) \tag{1.2}$$

only for $s > \frac{1}{2}$. For general s we can prove the restriction theorem below by a Poisson operator type argument for elliptic operators of first order. First we define:

DEFINITION 1.2. For any real s we distinguish the *null spaces*

$$\ker(\mathrm{A}, s) := \{f \in H^s(M;S) \mid \mathrm{A}\, f = 0\}$$

and the corresponding *Cauchy data* (or *Hardy*) *spaces*

$$\Lambda(\mathrm{A}, s) := \overline{\gamma_\infty\{f \in C^\infty(M;S) \mid \mathrm{A}\, f = 0 \text{ in } M \setminus \Sigma\}}^{H^{s-\frac{1}{2}}(\Sigma;S|\Sigma)}.$$

The null spaces consist of sections which are distributional for negative s; by elliptic regularity they are smooth in the interior; and by a Riesz operator argument they can be shown to possess a trace over the boundary in $H^{s-\frac{1}{2}}(\Sigma; S|_\Sigma)$. More precisely, we have the following *General Restriction Theorem*:

THEOREM 1.3. ([**11**]). (a) *Let* $s, t \in \mathbf{R}$, $f \in H^s(M;S)$, $\mathrm{A}\, f \in H^t(M;S)$, *and* $t > -\frac{1}{2}$. *Then the trace of* f *on* Σ *is well-defined in* $H^{s-\frac{1}{2}}(\Sigma; S|_\Sigma)$ *for any real* s.
(b) *For any real* s *the mapping*

$$\mathcal{K} := r_+ \widetilde{\mathrm{A}}^{-1} \gamma_\infty^* \sigma : C^\infty(\Sigma; S|_\Sigma) \longrightarrow C^\infty(M;S)$$

extends to a continuous map $\mathcal{K}^{(s)} : H^{s-1/2}(\Sigma; S|_\Sigma) \to H^s(M;S)$ *with* range $\mathcal{K}^{(s)} = \ker(\mathrm{A}, s)$.
(c) *In fact, the restriction*

$$\mathcal{K}^{(s)}|_{\Lambda(\mathrm{A},s)} : \Lambda(\mathrm{A}, s) \longrightarrow \ker(\mathrm{A}, s)$$

is an homeomorphism (relative to the respective Sobolev norms), and we have $\Lambda(\mathrm{A}, s) = \gamma_s(\ker(\mathrm{A}, s))$ *for all real* s.
(d) *The Cauchy data space* $\Lambda(\mathrm{A}, \frac{1}{2})$ *is a Lagrangian subspace of the Hilbert space* $L_2(\Sigma; S|_\Sigma)$ *equipped with the symplectic form* $\omega(\varphi, \psi) := (\sigma\varphi, \psi)$.

In the preceding theorem, $\widetilde{\mathrm{A}}$ denotes the invertible double of A defined on the closed double $\widetilde{M} = -M \cup_\Sigma M$, where the bundles are glued by σ. We denote the restriction operator by $r_+ : H^s(\widetilde{M}; \widetilde{S}) \to H^s(M;S)$ and the dual of γ_∞ in the distributional sense by γ_∞^*. The composition $\mathcal{P}(\mathrm{A}) := \gamma_\infty \circ \mathcal{K}$ is called the *(Szegö–) Calderón projection.* It is a pseudo–differential projection. Its extension $\mathcal{P}(\mathrm{A})^{(s)}$ to the sth Sobolev space over Σ has the corresponding Cauchy data space $\Lambda(\mathrm{A}, s + \frac{1}{2})$ as its range.

2. Boundary Reduction of the Index

Until now we have discussed only the restriction of the section spaces to the boundary, and in particular the restriction of the null spaces. These are all infinite–dimensional spaces. To obtain finite–dimensional null spaces we must impose suitable boundary conditions. First we restrict ourselves to the Grassmannian $\mathcal{G}r(\mathrm{A})$ of all pseudo–differential projections which differ from the Calderón projection $\mathcal{P}(\mathrm{A})$ by an operator of order -1. It has countably many connected components; two projections P_1, P_2 belong to the same component, if and only if the *virtual codimension*

$$\mathbf{i}(P_2, P_1) := \operatorname{index}\{P_2 P_1 : \operatorname{range} P_1 \to \operatorname{range} P_2\} \tag{2.1}$$

of P_1 in P_2 vanishes; the higher homotopy groups of each connected component are given by Bott periodicity.

2.1. Boundary Reduction Formulas. Let $P \in \mathcal{G}r(\mathrm{A})$. We consider the extension

$$\mathrm{A}_P : \operatorname{dom}(\mathrm{A}_P) \longrightarrow L_2(M; S) \tag{2.2}$$

of A defined by the domain

$$\operatorname{dom}(\mathrm{A}_P) := \{f \in H^1(M; S) \mid P^{(0)}\gamma_1(f) = 0\}. \tag{2.3}$$

It is a closed operator in $L_2(M; S)$ with finite–dimensional kernel and cokernel. That A_P is a closed L_2 realization can be deduced from the explicit description of a left parametrix for A by

$$(r^+(\widetilde{\mathrm{A}})^{-1}e^+)\,\mathrm{A} = \mathrm{Id} - \mathcal{K}\gamma_\infty, \tag{2.4}$$

which is a direct consequence of the Calderón construction. We have an explicit formula for the adjoint operator

$$(\mathrm{A}_P)^* = \mathrm{A}_{\sigma(\mathrm{Id}-P)\sigma^*}, \tag{2.5}$$

and *Seeley's Boundary Reduction Formula for the Index* ([**40**], see also [**11**]):

PROPOSITION 2.1.

$$\operatorname{index} \mathrm{A}_P = \operatorname{index}\{P\mathcal{P}(\mathrm{A}) : \Lambda(\mathrm{A}, \tfrac{1}{2}) \to \operatorname{range}(P^{(0)})\}.$$

To prove the Proposition, one begins with the 1-1 relation between the subspace $\ker \mathrm{A}_P$ of $H^1(M; S)$ and the space $\ker\{P\mathcal{P}(\mathrm{A}) : \Lambda(\mathrm{A}, \tfrac{1}{2}) \longrightarrow \operatorname{range}(P^{(0)})\}$ which is contained in the kernel of the elliptic pseudo–differential operator of order zero (the 'fan') $(\mathrm{Id} - \mathcal{P}(\mathrm{A})) + \mathcal{P}(\mathrm{A})P^*P\mathcal{P}(\mathrm{A})$ and therefore only consists of smooth sections. So, the reduction to the fan is the true boundary reduction, even though the fan does not appear in the final formulation of Proposition 2.1.

Note that the bundle S of Clifford modules splits naturally into $S = S^- \oplus S^+$ on even–dimensional manifolds. Correspondingly, the Dirac operator splits

$$\mathrm{A} = \begin{pmatrix} 0 & \mathrm{A}^- \\ \mathrm{A}^+ & 0 \end{pmatrix} \tag{2.6}$$

with the chiral Dirac operators A^+ and $\mathrm{A}^- = (\mathrm{A}^+)^*$. Also the Calderón operator and the Grassmannian split, and all the preceding results remain valid in the chiral setting. Notice, however, that the Lagrangian property of the Cauchy data space (Theorem 1.3d) has to be replaced by the *chiral twisting property*

$$\sigma(\Lambda(A^+,\tfrac{1}{2})) = \Lambda(A^-,\tfrac{1}{2})^\perp . \tag{2.7}$$

NOTE. In the rest of this section, we shall not always distinguish between the *total* and the *chiral* Dirac operator since the index formulas presented are valid in both cases.

A more specific boundary reduction formula can be obtained by choosing the *Atiyah–Patodi–Singer boundary condition* defined by the spectral projection $\Pi_{\geq}(\mathcal{B})$ of $L_2(\Sigma; S|_\Sigma)$ onto the space spanned by the eigensections of the boundary Dirac operator $\mathcal{B}$ corresponding to the non–negative eigenvalues. It is a pseudo–differential operator with the same principal symbol as the Calderón projection $\mathcal{P}(\mathrm{A})$. It can be shown that the difference $\mathcal{P}(\Sigma) - \Pi_{\geq}(\mathcal{B})$ is a smoothing operator ([**34**], see also [**20**], [**28**], and [**47**] for related, partially also adiabatic results). The Atiyah–Patodi–Singer Index Theorem ([**2**]) gives

$$\operatorname{index} \mathrm{A}_{\Pi_{\geq}} + \tfrac{1}{2}(\eta_{\mathcal{B}}(0) + \dim\ker\mathcal{B}) = \int_M \alpha(x)\,dx. \tag{2.8}$$

Here $\alpha(x)$ denotes the locally defined *index density* of A, and

$$\eta_{\mathcal{B}}(z) := \sum_{\lambda\in\operatorname{spec}\mathcal{B}\setminus\{0\}} \operatorname{sign}\lambda|\lambda|^{-z} = \frac{1}{\Gamma(\frac{z+1}{2})}\int_0^\infty t^{\frac{z-1}{2}}\operatorname{Tr}(\mathcal{B}e^{-t\mathcal{B}^2})\,dt \tag{2.9}$$

denotes the η*-function* of $\mathcal{B}$. It is (i) well defined through absolute convergence for $\Re(z)$ large; (ii) it extends to a meromorphic function in the complex plane with isolated simple poles; (iii) its residues are given by a local formula; and (iv) it has a finite value at $z = 0$ (see e.g. Gilkey [**19**]).

Equation (2.8) has the celebre Gauss–Bonnet Theorem as its model. It separates the contributions to the index from the whole manifold and from the structure on the boundary. A special feature is that it expresses the spectral quantities on the left side in terms of a classical integral on the right side.

2.2. Boundary Correction Formulas. Since the index of the operator $\mathrm{A}_{\mathcal{P}(\mathrm{A})}$ vanishes by definition, we may read Proposition 2.1 as providing a boundary correction formula for the difference index A_P – index $\mathrm{A}_{\mathcal{P}(\mathrm{A})}$. We have, however, also true *boundary correction formulas* for arbitrary changes of the boundary conditions. If P_1, $P_2 \in \mathcal{G}r(\mathrm{A})$, we obtain a version of the classical *Agronovič–Dynin formula*:

$$\operatorname{index}(\mathrm{A}_{P_1}) - \operatorname{index}(\mathrm{A}_{P_2}) = \mathbf{i}(P_1, P_2). \tag{2.10}$$

On odd–dimensional manifolds, the chiral splitting of the bundle $S|_\Sigma$ over the (even–dimensional) boundary provides a splitting $\mathcal{B} = \begin{pmatrix} 0 & B^- \\ B^+ & 0 \end{pmatrix}$ and two local boundary conditions induced by the respective chiral projections $\Pi_\pm$. The corresponding boundary correction formula becomes

$$\operatorname{index} \mathrm{A}_{\Pi_-} - \operatorname{index} \mathrm{A}_{\Pi_+} = \operatorname{index} B^+ \tag{2.11}$$

which immediately leads to the *Cobordism Theorem*:

THEOREM 2.2. (Atiyah, Singer [**3**]) *The index of a (chiral) Dirac operator* $B^+ : C^\infty(\Sigma; S^+) \longrightarrow C^\infty(\Sigma; S^-)$ *over a closed even–dimensional manifold* Σ *vanishes, if the couple* (Σ, S^+) *is a 'boundary', i.e. if there exists a manifold* M *with boundary* Σ *and a bundle of Clifford modules over* M *with compatible connection such that* B^+ *becomes the chiral component of the induced tangential Dirac operator.*

PROOF. The result follows from (2.11), since index $\mathrm{A}_{\Pi_\pm}$ vanishes by Green's formula (Lemma 1.1). □

2.3. Pasting Formulas. A third type of reduction results to be discussed are pasting formulas. Let $X = M_1 \cup M_2$ be an even–dimensional closed partitioned manifold with $\partial M_1 = \partial M_2 = M_1 \cap M_2 = \Sigma$. As always in this report, we assume that no connected component of M_1 or M_2 is closed. Let A be a Dirac operator over X and let A_j denote its restriction to M_j. By combining the boundary correction formula (2.10) and the Atiyah–Patodi–Singer index theorem (2.8) with Seeley's index formula for closed manifolds

$$\operatorname{index} \mathrm{A} = \int_X \alpha(x)\, dx\,,$$

we obtain a non–additivity formula for splitting the index over partitioned manifolds. The formula is valid for the total and (on even–dimensional manifolds) for the chiral Dirac operator.

THEOREM 2.3. *Let* P_i *be projections belonging to* $\mathcal{G}r(\mathrm{A}_j)$, $j = 1, 2$. *Then*

$$\operatorname{index} \mathrm{A} = \operatorname{index}(\mathrm{A}_1)_{P_1} + \operatorname{index}(\mathrm{A}_2)_{P_2} - \mathbf{i}(P_2, \mathrm{Id} - P_1).$$

It is immediate that $\mathbf{i}(P_2, \mathrm{Id} - P_1) = \operatorname{index}(\sigma(\partial_u + \mathcal{B}); P_2, P_1)$, where the last operator is on the cylinder $[0,1] \times \Sigma$ with boundary condition P_2 at $\tau = 0$ and P_1 at $\tau = 1$. The preceding pasting formula provides an analytic explanation for the combinatorial additivity of the Euler characteristic of compact $2k$–dimensional manifolds and the *Novikov additivity* of the signature of compact $4k$–dimensional manifolds because, in both cases, the projections are the complementary spectral projections. Hence the correction term vanishes on the separating hypersurface Σ. It may seem strange that the precise additivity can be obtained topologically just by applying the two related cohomology sequences (see [**4**], p. 588), whereas the analytical proof is rather delicate. But that is typical with spectral invariants where the analytical gains are highest when the topology can *not* make headway.

The *Bojarski Conjecture* is another pasting formula for the index. It relates the 'quantum' quantity index (coming from the multiplicity of the 0–eigenvalues) with a 'classical' quantity (coming from the Lagrangian geometry of the Cauchy data spaces) and was suggested in [**5**], proved in [**11**], and generalized in [**16**].

PROPOSITION 2.4. *Let* X *be a partitioned manifold as before and let* $\mathcal{P}(\mathrm{A}_j)$ *and* $\Lambda(\mathrm{A}_j, \frac{1}{2})$ *denote the corresponding Calderón projections and* L_2 *closures of the Cauchy data spaces,* $j = 1, 2$. *Then*

$$\operatorname{index} \mathrm{A} = \mathbf{i}(\mathrm{Id} - \mathcal{P}(\mathrm{A}_2), \mathcal{P}(\mathrm{A}_1)) = \operatorname{index}(\Lambda(\mathrm{A}_1, \tfrac{1}{2}), \Lambda(\mathrm{A}_2, \tfrac{1}{2})).$$

Recall that

$$\begin{aligned}\operatorname{index}(\Lambda(\mathrm{A}_1, \tfrac{1}{2}), \Lambda(\mathrm{A}_2, \tfrac{1}{2})) :=\ & \dim(\Lambda(\mathrm{A}_1, \tfrac{1}{2}) \cap \Lambda(\mathrm{A}_2, \tfrac{1}{2})) \\ & - \dim(L_2(\Sigma; S|_\Sigma)/(\Lambda(\mathrm{A}_1, \tfrac{1}{2}) + \Lambda(\mathrm{A}_2, \tfrac{1}{2})),\end{aligned}$$

and that pairs of closed subspaces for which the two dimensions in the preceding definition are finite are called *Fredholm pairs* of subspaces.

The proof depends on the unique continuation property for Dirac operators and the Lagrangian property of the Cauchy data spaces (Theorem 1.3d), more precisely, the chiral twisting property (2.7).

3. Boundary Reduction of the Spectral Flow

A particular clarity of the concept of boundary reduction is required and can be obtained when discussing the spectral flow $\mathbf{sf}\{\mathrm{A}_{t,D}\}$ of a continuous family of (from now on always *total*) Dirac operators $\{\mathrm{A}_t\}$ with the same principal symbol and the same domain D. Roughly speaking, the *spectral flow* is the difference between the number of eigenvalues, which change the sign from $-$ to $+$ as t goes from 0 to 1, and the number of eigenvalues which change the sign from $+$ to $-$. It can be defined in a satisfactory, purely functional analytical way, following a suggestion by J. Phillips (see [**6**] and [**31**]; see also [**30**] which defines the spectral flow for curves of a broad class of unbounded self–adjoint Fredholm operators with *varying* domain).

3.1. Boundary Reduction of Global Sections, Revisited. We give a systematic presentation of the boundary reduction of the solution spaces, inspired by M. Krein's construction of the maximal space of boundary values for closed symmetric operators (see [**6**], [**7**]). In all this section we stay in the real category and do not assume product structure near Σ unless otherwise stated.

We denote by A_0 the restriction of the (total, compatible) Dirac operator A to the space $C_0^\infty(M;S)$ of smooth sections with support in the interior of M. As mentioned above, there is no natural choice of the order of a Sobolev space for the boundary reduction. Therefore, a systematic treatment of the boundary reduction may begin with the minimal closed extension $\mathrm{A}_{\min} := \overline{\mathrm{A}_0}$ and the adjoint $\mathrm{A}_{\max} := (\mathrm{A}_0)^*$ of A_0. Then $\mathrm{A}_{\max}$ is the maximal closed extension. We have

$$D_{\min} := \operatorname{dom}(\mathrm{A}_{\min}) = \overline{C_0^\infty(M;S)}^{\mathcal{G}} = \overline{C_0^\infty(M;S)}^{H^1(M;S)}$$

and

$$\begin{aligned} D_{\max} &:= \operatorname{dom}(\mathrm{A}_{\max}) \\ &= \{u \in L_2(M;S) \mid \mathrm{A}\,u \in L_2(M;S) \text{ in the sense of distributions}\}. \end{aligned}$$

Here, the superscript $\mathcal{G}$ means the closure in the graph norm which coincides with the 1st Sobolev norm on $C_0^\infty(M;S)$. We form the space $\boldsymbol{\beta}$ of *natural boundary values* with the *natural trace map* γ in the following way:

$$\begin{array}{ccc} D_{\max} & \xrightarrow{\gamma} & D_{\max}/D_{\min} =: \boldsymbol{\beta} \\ x & \mapsto & \gamma(x) = [x] := x + D_{\min}\,. \end{array}$$

The space $\boldsymbol{\beta}$ becomes a symplectic Hilbert space with the scalar product induced by the graph norm

$$(x,y)_{\mathcal{G}} := (x,y) + (\mathrm{A}\,x, \mathrm{A}\,y) \tag{3.1}$$

and the symplectic form given by Green's form

$$\omega([x],[y]) := (\mathrm{A}\,x, y) - (x, \mathrm{A}\,y) \qquad \text{for } [x],[y] \in \boldsymbol{\beta}. \tag{3.2}$$

One shows easily that ω is non–degenerate.

We define the *natural Cauchy data space* $\Lambda(\mathrm{A}) := \gamma(\ker \mathrm{A}_{\max})$ as a Lagrangian subspace of $\boldsymbol{\beta}$.

By Theorem 1.3a and, alternatively and in greater generality, by Hörmander [**21**] (Theorem 2.2.1 and the Estimate (2.2.8), p. 194), the space $\boldsymbol{\beta}$ is naturally embedded in the distribution space $H^{-\frac{1}{2}}(\Sigma; S|_\Sigma)$. If the metrics are product close to Σ, we can give a more precise description of the embedding, namely as a *graded* space of distributions. Let $\{\varphi_k, \lambda_k\}$ be a spectral resolution of $L_2(\Sigma)$ by eigensections of $\mathcal{B}$. (Here and in the following we do not mention the bundle S). For simplicity, we assume $\ker \mathcal{B} = \{0\}$. Then $\mathcal{B}\varphi_k = \lambda_k \varphi_k$ for all $k \in \mathbf{Z} \setminus \{0\}$, and $\lambda_{-k} = -\lambda_k$, $\sigma(\varphi_k) = \varphi_{-k}$, and $\sigma(\varphi_{-k}) = -\varphi_k$ for $k > 0$. We have ([**7**], Proposition 7.15, see also [**12**] for a more general setting)

$$(3.3) \quad \boldsymbol{\beta} = \boldsymbol{\beta}_- \oplus \boldsymbol{\beta}_+ \quad \text{with}$$

$$\boldsymbol{\beta}_- := \overline{[\{\varphi_k\}_{k<0}]}^{H^{\frac{1}{2}}(\Sigma)} \text{ and } \boldsymbol{\beta}_+ := \overline{[\{\varphi_k\}_{k>0}]}^{H^{-\frac{1}{2}}(\Sigma)}.$$

Then $\boldsymbol{\beta}_-$ and $\boldsymbol{\beta}_+$ are Lagrangian and transversal subspaces of $\boldsymbol{\beta}$. Let us define two Lagrangian and transversal subspaces $L_\pm$ of $L_2(\Sigma)$ in a similar way, namely by the closure in $L_2(\Sigma)$ of the linear span of the eigensections with negative, resp. with positive eigenvalue. We have that L_+ is dense in $\boldsymbol{\beta}_+$, and $\boldsymbol{\beta}_-$ is dense in L_-. This anti–symmetric relation may explain some of the well–observed delicacies of dealing with spectral invariants of continuous families of Dirac operators.

Moreover, we have $\gamma(D_{\mathrm{aps}}) = \boldsymbol{\beta}_-$, where

$$(3.4) \qquad D_{\mathrm{aps}} := \{f \in H^1(M) \mid \Pi_>(f|_\Sigma) = 0\}$$

denotes the domain corresponding to the Atiyah–Patodi–Singer boundary condition. Note that a series $\sum_{k<0} c_k \varphi_k$ may converge to an element $\varphi \in L_2(\Sigma)$ without converging in $H^{\frac{1}{2}}(\Sigma)$. So, such $\varphi \in L_-$ can not appear as the trace at the boundary of any $f \in D_{\max}$.

For all domains D with $D_{\min} \subset D \subset D_{\max}$ and $\gamma(D)$ Lagrangian in $\boldsymbol{\beta}$, we have that the extension $\mathrm{A}_D := \mathrm{A}_{\max}|_D$ is self–adjoint. It becomes a Fredholm operator, if and only if the pair $(\gamma(D), \Lambda(\mathrm{A}))$ of Lagrangian subspaces of $\boldsymbol{\beta}$ becomes a Fredholm pair. In particular, $(\boldsymbol{\beta}_-, \Lambda(\mathrm{A}))$ is a Fredholm pair. In the product case, a fairly general symplectic 'Criss–Cross' Reduction Theorem ([**8**], Theorem 1.1) implies:

PROPOSITION 3.1. *The $L_2(\Sigma)$ part $\Lambda(\mathrm{A}) \cap L_2(\Sigma)$ of the natural Cauchy data space $\Lambda(\mathrm{A})$ is closed in $L_2(\Sigma)$. Actually, it is a Lagrangian subspace of $L_2(\Sigma)$ and it forms a Fredholm pair with the component L_-.*

3.2. Spectral Flow and Maslov Index. Inspired by the surgery operations, so successful in topology, one may consider the spectral flow of a family of Dirac operators with the same principal symbol on a partitioned manifold and want a pasting formula for it. To do that, one replaces the spectral flow of a continuous 1–parameter family of self–adjoint Fredholm operators, which is a 'quantum' quantity just like the index, by the *Maslov index* of a corresponding path of Lagrangian Fredholm pairs, which is a 'quasi–classical' quantity. The idea is due to Floer and was worked out subsequently by Yoshida in dimension 3, by Nicolaescu in all odd dimensions, and pushed further by Cappell, Lee and Miller, Daniel and Kirk and many other authors. For a survey, see [**6**], [**7**], [**17**].

First, we give a general result - without assuming differentiability of the path, invertibility at the ends, regular crossings, or a product structure near Σ. On a compact manifold M with boundary Σ, we consider a continuous family of Dirac operators $\{\mathrm{A}_t\}_{t\in[0,1]}$ (induced by a continuous family $\{\nabla_t\}_{t\in[0,1]}$ of connections). We can fix the space $\boldsymbol{\beta}$ for the family and derive the continuity of the corresponding family $\{\Lambda(\mathrm{A}_t\}$ of natural Cauchy data spaces from the weak unique continuation property

$$\ker \mathrm{A}_{t,\max} \cap D_{\min} = \{0\} \quad \text{for all } t \in [0,1]$$

and the existence of a Fredholm extension with domain D. We obtain the *General Boundary Reduction Formula* for the spectral flow ([**6**]), which gives a family version of the Bojarski conjecture (our Proposition 2.4):

THEOREM 3.2. (a) *The spectral flow of the family* $\{\mathrm{A}_{t,D}\}$ *is well defined under the preceding assumptions.*
(b) *The family* $\{\Lambda(\mathrm{A}_t) := \gamma(\ker \mathrm{A}_{t,\max})\}$ *is a continuous curve of Lagrangian subspaces of* $\boldsymbol{\beta}$ *which all make Fredholm pairs with* $\gamma(D)$.
(c) *The Maslov index* $\mathbf{mas}\left(\{\Lambda(\mathrm{A}_t)\}, \gamma(D)\right)$ *is well defined and we have*

$$\mathbf{sf}\,\{\mathrm{A}_{t,D}\} = \mathbf{mas}\left(\{\Lambda(\mathrm{A}_t)\}, \gamma(D)\right). \tag{3.5}$$

We have two corollaries for the spectral flow on closed manifolds with fixed hypersurface. The first corollary treats the case of a separating hypersurface, the second the case of a non–separating hypersurface. Both cases can be reduced to the situation of the preceding theorem by cutting the manifold along Σ. Then we get a manifold with two isometric boundary components in both cases (see [**7**]).

For product structures near Σ, one can apply a sharper version of Proposition 3.1 providing also for the continuity of the L_2 parts of a continuous family of natural Cauchy data spaces and the invariance of the Maslov index under taking the L_2–intersections. Then one obtains an L_2–version of the preceding Theorem which gives a new proof and a slight generalization of the Yoshida–Nicolaescu Formula (for details see [**8**], Section 3).

3.3. Correction Formula for the Spectral Flow. Let D, D' with $D_{\min} \subset D, D' \subset D_{\max}$ be two domains such that both $\{\mathrm{A}_{t,D}\}$ and $\{\mathrm{A}_{t,D'}\}$ become families of self–adjoint Fredholm operators. We assume that D and D' differ only by finite dimension, more precisely, that

$$\dim \gamma(D)/\gamma(D) \cap \gamma(D') = \dim \gamma(D')/\gamma(D) \cap \gamma(D') < +\infty\,. \tag{3.6}$$

Then we find from Theorem 3.2 (for details see [**7**], Theorem 6.5):

$$\begin{aligned} \mathbf{sf}\Big(\{\mathrm{A}_{t,D}\}\Big) - \mathbf{sf}\Big(\{\mathrm{A}_{t,D'}\}\Big) \\ = \mathbf{mas}\Big(\{\Lambda(\mathrm{A}_t)\}, \gamma(D')\Big) - \mathbf{mas}\Big(\{\Lambda(\mathrm{A}_t)\}, \gamma(D)\Big). \end{aligned} \tag{3.7}$$

On the right side of (3.7), the difference of the Maslov indices does not depend on the curve $\{\Lambda(\mathrm{A}_t)\}$, but only on the endpoints and is also called the *Hörmander index* of the four determining Lagrangian subspaces. The assumption (3.6) is rather restrictive. The pair of domains, for instance, defined by the Atiyah–Patodi–Singer projection and the Calderón projection, may not always satisfy that condition. For the present proof, however, it seems indispensable.

4. Boundary Reduction of the Determinant

The understanding of the determinant for Dirac operators is in rapid development. A formula, recently obtained by Scott and Wojciechowski, shows that the concept of boundary reduction is crucial for illuminating the relations between the competing concepts of the Fredholm determinant and the ζ–function regularized determinant. This section presents the Scott–Wojciechowski formula and puts it under the perspective of boundary reduction.

4.1. Three Determinant Concepts. Let us begin with the most simple integral of statistical mechanics, the *partition function* which is the model for all quadratic functionals:

$$Z(\beta) := \int_\Gamma e^{-\beta(Tx,x)}\, dx\,. \tag{4.1}$$

To begin with, let $\dim \Gamma = d < \infty$ and β real with $\beta > 0$ and assume that T is a strictly positive, symmetric endomorphism. In suitable coordinates we evaluate the Gaussian integrals and find

$$Z(\beta) = \pi^{d/2} \cdot \beta^{-d/2} \cdot (\det T)^{-\frac{1}{2}}\,.$$

Two fundamental problems arise when we try to take a Dirac operator for T and all sections in a bundle S over a compact manifold M for Γ according to the Feynman recipes in the Matthews and Salam program ([**25**], [**26**]. What if T is not > 0? And what if $d = +\infty$ (i.e. if M is not a finite set of points)? To get around the first problem, we reproduce a calculation made by Adams and Sen, [**1**]:

We decompose $\Gamma = \Gamma_+ \times \Gamma_-$ and $T = T_+ \oplus T_-$ with $T_+, -T_-$ strictly positive in $\Gamma_\pm$ and $\dim \Gamma_\pm = d_\pm$. Formally, we obtain by a suitable path in the complex plane approaching $\beta = 1$:

$$Z(1) = \pi^{\zeta/2}\, e^{\pm i\frac{\pi}{4}(\zeta-\eta)}\, (\det|T|)^{-\frac{1}{2}}\,, \tag{4.2}$$

with $\zeta := d_+ + d_-$ and $\eta := d_+ - d_-$.

We shall not discuss the various stochastic approaches to evaluate the integral when $d = +\infty$, but present two other concepts of the determinant.

From the point of view of functional analysis, the only natural concept is the *Fredholm determinant* of bounded operators acting on a separable Hilbert space of the form e^α or, more generally, $\mathrm{Id} + \alpha$ where α is of trace class. We recall the formulas

$$det_{Fr}\, e^\alpha = e^{\mathrm{Tr}\,\alpha} \quad \text{and} \quad det_{Fr}(\mathrm{Id} + \alpha) = \sum_{k=0}^{\infty} \mathrm{Tr}\, \wedge^k \alpha\,. \tag{4.3}$$

The Fredholm determinant is notable for obeying the product rule, in difference to other generalizations of the determinant to infinite dimensions where the error of the product rule leads to new invariants, see e.g. [**22**].

Clearly, the parametrix (or Green's function) of a Dirac operator leads to operators for which the Fredholm determinant can be defined, but the relevant information about the spectrum of the Dirac operator does not seem sufficiently maintained. Note also that Quillen and Segal's construction of the *determinant line bundle* is based on the concept of the Fredholm determinant, though without leading to *numbers* when the bundle is non–trivial.

A third concept is the *ζ-function regularized determinant*, based on Ray and Singer's observation that, formally,

$$\det T = \prod \lambda_j = \exp\{\sum \ln \lambda_j e^{-z\ln\lambda_j}|_{z=0}\} = e^{-\frac{d}{dz}\zeta_T(z)|_{z=0}},$$

where $\zeta_T(z) := \sum_{j=1}^{\infty} \lambda_j^{-z} = \frac{1}{\Gamma(z)}\int_0^\infty t^{z-1}\operatorname{Tr} e^{-tT}\,dt$. By a result of Seeley, for a *positive definite self-adjoint elliptic* operator T of second order, acting on sections of a Hermitian vector bundle over a closed manifold M of dimension m, the function $\zeta_T(z)$ is holomorphic for $\Re(z)$ sufficiently large and can be extended meromorphically to the whole complex plane with $z = 0$ no pole.

The preceding definition does not apply immediately to the Dirac operator A which has infinitely many positive λ_j and negative eigenvalues $-\mu_j$. As an example, consider the operator $\mathrm{A}_a := -i\frac{d}{dx} + a : C^\infty(S^1) \to C^\infty(S^1)$ with $\mathrm{A}_a\,\varphi_k = k\varphi_k + a\varphi_k$ where $\varphi_k(x) = e^{ikx}$. It follows that $\operatorname{spec}\mathrm{A}_a = \{k+a\}_{k\in\mathbf{Z}}$.

Choosing the branch $(-1)^{-z} = e^{-i\pi z}$, we find

$$\begin{aligned}\zeta_{\mathrm{A}}(z) - \sum \lambda_j^{-z} + \sum (-1)^{-z}\mu_j^{-z} \\ = \frac{1}{2}\left\{\zeta_{\mathrm{A}^2}(\frac{z}{2}) + \eta_{\mathrm{A}}(z)\right\} + \frac{1}{2}e^{-i\pi z}\left\{\zeta_{\mathrm{A}^2}(\frac{z}{2}) - \eta_{\mathrm{A}}(z)\right\},\end{aligned}$$

where $\eta_{\mathrm{A}}(z)$ is defined as in (2.9). Thus:

$$\zeta'_{\mathrm{A}}(0) = \frac{1}{2}\zeta'_{\mathrm{A}^2}(0) - \frac{i\pi}{2}\{\zeta_{\mathrm{A}^2}(0) - \eta_{\mathrm{A}}(0)\}$$

and

$$\det\nolimits_\zeta \mathrm{A} = e^{-\zeta'_{\mathrm{A}}(0)} = e^{\frac{i\pi}{2}\{\zeta_{\mathrm{A}^2}(0) - \eta_{\mathrm{A}}(0)\}} \cdot e^{-\frac{1}{2}\zeta'_{\mathrm{A}^2}(0)}. \tag{4.4}$$

REMARK 4.1. **(a)** The three spectral invariants which enter in the preceding formula are of very different character. The first invariant, $\zeta_{\mathrm{A}^2}(0)$, is the most stable of the three. It is given by $\int_M \alpha(x)dx$, where $\alpha(x)$ denotes the index density which is a certain coefficient in the heat kernel expansion and is locally expressed by the coefficients of A. In particular, $\zeta_{\mathrm{A}^2}(0)$ vanishes on a closed odd–dimensional manifold. The second invariant, $\eta_{\mathrm{A}}(0)$, is not given by an integral, not by a local formula. It depends, however, only on finitely many terms of the symbol of the resolvent $(\mathrm{A} - \lambda)^{-1}$ and will not change when one changes or removes a finite number of eigenvalues. Its derivative is local. The third invariant, $\zeta'_{\mathrm{A}^2}(0)$, is the most delicate of the three: even small changes of the eigenvalues will change the ζ'–invariant and hence the determinant and make it capable of detecting small *anomalies*.

Keeping in mind that the Atiyah–Singer Index Theorem is given by local invariants, and that the η–invariant (and R–torsion) are the next level, since their derivatives are local, we meet a question repeatedly put forward by Gelfand: "what comes next?" To this, Singer remarked in personal communication ([**43**]): "Just as η arises in boundary value problems for smooth boundaries, I think the next level will come from corner contributions when the boundary has corners."

(b) It should be noted that for Dirac operators in even dimensions also a vanishing determinant can provide some insight by replacing the operator A on the finite–dimensional kernel by the rotation $M(m, \theta_m) := me^{i\gamma_5\theta_m}$ with suitable 'mass' m and 'angle' θ_m and chiral switch $\gamma_5 := \begin{pmatrix} 1 & 0 \\ 0 & -1\end{pmatrix}$. Then the standardized determinant takes the form $\det'_\zeta \cdot m^{n_+ + n_-} e^{i(n_+ - n_-)\theta_m}$, where $\det'_\zeta$ denotes the rest–determinant after removing the zero–eigenvalues and $n_\pm$ denotes the dimension of the kernel of the chiral Dirac operators (see [**9**] and [**46**], based on the manifold–*with*–boundary case discussed in [**27**]).

(c) Our choice of the branch $e^{-i\pi z}$ for $(-1)^{-z}$ does not coincide with Singer's original choice $e^{i\pi z}$ in [**42**] which gives an extra term in the Scott–Wojciechowski Formula (our Theorem 4.2, see [**38**], Theorem 7.1), unless the argument for the Fredholm determinant is also suitably reversed.

4.2. The Scott–Wojciechowski Formula. In a recent paper by Wojciechowski it was shown that the ζ–regularized determinant can also be defined for certain self–adjoint Fredholm extensions of the Dirac operator on a compact manifold with boundary, namely when the domain is defined by a projection belonging to the *smooth, self–adjoint* Grassmannian

$$\mathcal{G}r^*_\infty(\mathrm{A}) = \{P \in \mathcal{G}r(\mathrm{A}) \mid P \text{ is self–adjoint, } P - \mathcal{P}(\mathrm{A}) \text{ is smoothing and range}(P^{(0)}) \text{ is Lagrangian in } L_2(\Sigma; S|_\Sigma)\}. \tag{4.5}$$

We refer to [**45**] for the details of the delicate estimates needed for establishing the three involved invariants in that case.

Scott and Wojciechowski have now established a *boundary correction formula* which relates the ζ–determinant and the Fredholm determinant (see [**37**], [**38**]).

THEOREM 4.2. (a) *Let* A *be a Dirac operator over an odd–dimensional compact manifold* M *with boundary* Σ *and let* $P \in \mathcal{G}r^*_\infty(\mathrm{A})$. *Then the range of* $\mathcal{P}(\mathrm{A})$ *and the range of* P *can be written as the graphs of unitary, elliptic operators of order 0,* K*, resp.* T *which differ from the operator* $(B^+B^-)^{-1/2}B^+ : C^\infty(\Sigma; S^+|_\Sigma) \to C^\infty(\Sigma; S^-|_\Sigma)$ *by a smoothing operator. Moreover,*

$$\det_\zeta \mathrm{A}_P = \det_\zeta \mathrm{A}_{\mathcal{P}(\mathrm{A})} \cdot \det_{Fr} \tfrac{1}{2}(\mathrm{Id} + KT^{-1}) . \tag{4.6}$$

(b) *If we pick the alternative sign for the phase of the determinant and define*

$$\det_\zeta \mathrm{A}_P = e^{\frac{i\pi}{2}\{\eta_{\mathrm{A}}(0) - \zeta_{\mathrm{A}^2}(0)\}} \cdot e^{-\frac{1}{2}\zeta'_{\mathrm{A}^2}(0)} ,$$

we obtain

$$\det_\zeta \mathrm{A}_P = \det_\zeta \mathrm{A}_{\mathcal{P}(\mathrm{A})} \cdot \det_{Fr} \tfrac{1}{2}(\mathrm{Id} + KT^{-1}) \cdot \det_{Fr}(g_P) , \tag{4.7}$$

where g_P *denotes the unique unitary elliptic pseudo–differential operator of order zero acting on* $C^\infty(\Sigma; S^-|_\Sigma)$ *which differs from the identity by a smoothing operator, such that*

$$P = \begin{pmatrix} \mathrm{Id} & 0 \\ 0 & g_P \end{pmatrix} \mathcal{P}(\mathrm{A}) \begin{pmatrix} \mathrm{Id} & 0 \\ 0 & g_P^{-1} \end{pmatrix} .$$

In geometric terms, the key to understanding the preceding Theorem is that the determinant line bundle, parametrized by the projections belonging to the smooth self–adjoint Grassmannian, is *trivial* so that one can attribute complex numbers (up to a multiple) to the canonical determinant section. This may explain why earlier attempts to relate the concept of the ζ–determinant with the Fredholm determinant (see for instance [**15**]) had to be content with discussing the metric of the determinant bundle in terms of the ζ–determinant, and why the break–through in understanding the mutual relation required a concept of boundary reduction.

4.3. A 1–Dimensional Toy Model. Some basic ideas of the proof of the Scott–Wojciechowski Formula can be best understood by analysing a simple example on the interval $M = [0, 2\pi]$. We follow **[10]** which gives a setting consistent with the Scott–Wojciechowski Formula, though with the alternative sign choice in the definition of the ζ–determinant. For a review of other and more general approaches we refer to **[23]** (see also **[13]** and **[34]**).

Let T be a unitary $N \times N$–matrix and $\mathrm{A} := -i\frac{d}{dx} \oplus \cdots \oplus -i\frac{d}{dx}$ be acting on

$$\operatorname{dom} \mathrm{A}_T := \{f \in H^1([0, 2\pi]; \mathbf{C}^N) \mid f(2\pi) = Tf(0)\}.$$

The natural space $\boldsymbol{\beta}$ of boundary values is now finite–dimensional and the Grassmannian of boundary conditions of Atiyah–Patodi–Singer type is not really defined. Anyhow, we can let $-\mathrm{Id} \in U(N)$ play the role of the Calderón projection. We find $\operatorname{spec} \mathrm{A}_{-\mathrm{Id}} = \{\frac{2k+1}{2}\}_{k \in \mathbf{Z}}$. By the Hurwitz ζ–function we find $\det_\zeta \mathrm{A}_{-\mathrm{Id}} = 2^N$. Then the Scott–Wojciechowski Formula would imply:

PROPOSITION 4.3.

$$\det{}_\zeta \mathrm{A}_T = 2^N \det \frac{\mathrm{Id}_{\mathbf{C}^N} - T^{-1}}{2}. \tag{4.8}$$

This is precisely what we obtained in **[10]**.

PROOF. The proof of (4.8) in **[10]** uses a variational argument. Let $T_r := e^{ir\alpha}T$ be such a variation with α a self–adjoint $N \times N$–matrix. For the phase and the modulus of the boundary term (the right side) in (4.8) we find at once

$$\Im\Big(\frac{d}{dr} \ln 2^N \det \frac{\mathrm{Id}_{\mathbf{C}^N} - T_r^{-1}}{2}\Big) = -\frac{\operatorname{tr}\alpha}{2} \quad \text{and} \tag{4.9}$$

$$\Re\big(\cdot\big) = -\frac{i}{2} \operatorname{tr}\Big(\alpha(\mathrm{Id} - T)^{-1}(\mathrm{Id} + T)\Big).$$

The variation of the phase and modulus of the ζ–determinant is more delicate. First we fix the domain under the variation by replacing the operator A_{T_r} by the unitarily equivalent operator $\mathrm{A}_r := (U_r \mathrm{A} U_r^{-1})_T$, where $U_r(x) := T^{-1}e^{ir\chi(x)\alpha}T$ with a smooth cut–off function χ being constant equal 1 near 0 and constant equal 0 near 2π. Then $\zeta_{\mathrm{A}_r^2}(0)$ is constantly vanishing. By Duhamel's principle we can replace the heat kernel for $\chi'(x)e^{-\varepsilon \mathrm{A}_0^2}$ by the standard heat kernel and we obtain

$$\frac{\pi}{2}\frac{d}{dr}\eta_{\mathrm{A}_r}(0)|_{r=0} = \frac{\pi}{2}\frac{2}{\sqrt{\pi}} \lim_{\varepsilon \to 0} \sqrt{\varepsilon} \operatorname{Tr} \dot{\mathrm{A}}_0 e^{-\varepsilon \mathrm{A}_0^2} = -\frac{\operatorname{tr}\alpha}{2} \tag{4.10}$$

and, plugging in the standard integral kernel for A^{-1},

$$\begin{aligned} \frac{d}{dr}\Big(-\frac{1}{2}\zeta'_{\mathrm{A}_r^2}(0)\Big)|_{r=0} &= \lim_{\varepsilon \to 0} \operatorname{Tr} \dot{\mathrm{A}}_0 \, \mathrm{A}_0^{-1} \, e^{-\varepsilon \mathrm{A}_0^2} \\ &= -\frac{i}{2} \operatorname{tr}\Big(\alpha(\mathrm{Id} - T)^{-1}(\mathrm{Id} + T)\Big). \end{aligned} \tag{4.11}$$

□

Comparing the middle terms in equations (4.10) and (4.11) we see that the variation of the ζ'–term is much more delicate than the variation of the η–invariant.

We may rest assured that Proposition 4.3 will not lead to new number theory insights since the relevant integrals of one variable have probably all been checked before in analytic number theory. We may, however, exploit the simple relation

$\zeta_{-\frac{d^2}{dx^2}|S^1}(z) = 2\zeta_{Riem}(2z)$ to develop an approach which in higher dimensions, applied to Theorem 4.2 for suitable symmetric spaces, may lead to new insight in the Dirichlet ζ–function. The following calculation, suggested by Wojciechowski ([**44**]), shall serve not only as a pilot for future, hopefully more relevant number theoretical calculations, but also to check the validity and in particular the signs of the reduction formula of Proposition 4.3.

We consider the case $N = 1$ and a path $\{T_r := e^{2\pi i r}\}$ close to $r = \frac{1}{2}$. We find

$$\operatorname{spec} \mathrm{A}_{T_r} = \operatorname{spec}(-i\frac{d}{dx} + r)_{\mathrm{Id}} = \operatorname{spec}(-i\frac{d}{dx} + r)|_{S^1} = \{k + r\}_{k \in \mathbf{Z}}\,.$$

Note that $\dot{\mathrm{A}}_r = 1$. We find

$$\begin{aligned}(4.12)\quad \frac{d}{dr}(\ln \det{}_\zeta \mathrm{A}_r^2)|_{r=\frac{1}{2}} &= \frac{d}{dr}\Big(-\int_0^\infty \frac{1}{t}\operatorname{Tr} e^{-t\,\mathrm{A}_r^2}\,dt\Big)|_{r=\frac{1}{2}}\\ &= 2\int_0^\infty \operatorname{Tr} \dot{\mathrm{A}}_{\frac{1}{2}}\,\mathrm{A}_{\frac{1}{2}}\, e^{-t\,\mathrm{A}_{\frac{1}{2}}^2}\,dt\\ &= 2\frac{1}{\Gamma(\frac{z+1}{2})}\int_0^\infty t^{\frac{z-1}{2}}\operatorname{Tr} \mathrm{A}_{\frac{1}{2}}\, e^{-t\,\mathrm{A}_{\frac{1}{2}}^2}\,dt|_{z=1} = 2\eta_{\mathrm{A}_{\frac{1}{2}}}(1).\end{aligned}$$

We expand like in [**24**], but at $r = \frac{1}{2}$, and get for the right side of (4.12)

$$\begin{aligned}\eta_{\mathrm{A}_{\frac{1}{2}}}(1) &= \frac{1}{r} - 2\sum_{n=0}^\infty r^{2n+1}\zeta_{Riem}(2(n+1))\\ &= 2\Big(1 - \sum_{n=0}^\infty 2^{-2n-1}\zeta_{Riem}(2(n+1))\Big).\end{aligned}$$

By elementary summation, the difference vanishes in the preceding equation in the parentheses on the right side. Alternatively, we could determine it by using our boundary reduction formula for the determinant of Proposition 4.3. We recall from [**19**] that $\eta_{\mathrm{A}_{T_r}}(0) = \operatorname{sign} r - 2r$ and find

$$\det{}_\zeta \mathrm{A}_{T_r} = 2^1\frac{1 - e^{-2\pi i r}}{2} = e^{\frac{i\pi}{2}(1-2r)} \cdot 2\sin \pi r.$$

So, $\ln(2\sin \pi r) = -\zeta'_{\mathrm{A}_{T_r}^2}(0)$ and we get for the left side of (4.12)

$$\frac{d}{dr}(\ln \det{}_\zeta \mathrm{A}_{T_r}^2)|_{r=\frac{1}{2}} = \frac{1}{2\sin \pi r}2\cos \pi r|_{r=\frac{1}{2}} = 0$$

in nice agreement with our previous completely elementary result.

REMARK 4.4. As seen in (4.11), the variation of the modulus of the ζ–determinant contains a truly global term and can not be localized near the boundary. This may explain why the proof method, which was successful in the 1–dimensional case, can not literally be imitated in higher dimensions. It works with the phase as demonstrated in [**36**], but not with the modulus. In [**38**], the authors get around that problem by varying a quotient of determinants. That idea has been applied before by Forman [**18**] for a boundary reduction of the determinant of the Laplacian with local boundary conditions.

Then the main ingredients in the Scott–Wojciechowski proof are first to assume that A_P is invertible or, equivalently, that the boundary integral operator $P\mathcal{P}(\mathrm{A})$:

$\Lambda(\mathrm{A}, \frac{1}{2}) \to \operatorname{range} P^{(0)}$ is invertible. Then they determine a parametrix for A_P and find

$$\mathrm{A}_P^{-1} = \mathrm{A}^{-1} - \mathcal{K} \circ (P\mathcal{P}(\mathrm{A}))^{-1} \circ P \circ \gamma \circ \mathrm{A}^{-1},$$

where A^{-1} denotes the parametrix for A introduced in (2.4). Then they show that the difference $\mathrm{A}_{P_1}^{-1} - \mathrm{A}_{P_2}^{-1}$ is a smoothing operator and they prove the variation formula

$$\frac{d}{dr}\Big(\ln \det{}_\zeta \mathrm{A}_{P_1,r} - \ln \det{}_\zeta \mathrm{A}_{P_2,r}\Big)|_{r=0} = \operatorname{Tr} \dot{\mathrm{A}}_0(\mathrm{A}_{P_1}^{-1} - \mathrm{A}_{P_2}^{-1}),$$

from which the boundary reduction follows.

References

[1] D.H. Adams and S. Sen, "Partition Function of a quadratic functional and semiclassical approximation for Witten's 3-manifold invariant", Preprint, Dublin 1995.

[2] M.F. Atiyah, V.K. Patodi, and I.M. Singer, Spectral asymmetry and Riemannian geometry. I, *Math. Proc. Cambridge Phil. Soc.* **77** (1975), 43–69.

[3] M.F. Atiyah and I.M. Singer, The index of elliptic operators on compact manifolds, *Bull. Amer. Math. Soc.* **69** (1963), 422–433.

[4] —, —: The index of elliptic operators. III, *Ann. of Math.* **87** (1968), 564–604.

[5] B. Bojarski, The abstract linear conjugation problem and Fredholm pairs of subspaces, *in* "In Memoriam I.N. Vekua", Tbilisi Univ, Tbilisi, 1979, pp. 45–60 (Russian).

[6] B. Booss–Bavnbek and K. Furutani, The Maslov index – a functional analytical definition and the spectral flow formula, *Tokyo J. Math.* **21** (1998), 1–34.

[7] —, —, Symplectic functional analysis and spectral invariants, in: B. Booss-Bavnbek, K.P. Wojciechowski (eds.), "Geometric Aspects of Partial Differential Equations", Amer. Math. Soc. Series *Contemporary Mathematics*, vol. 242, Providence, R.I., 1999, pp. 53–83.

[8] B. Booss–Bavnbek, K. Furutani, and N. Otsuki, "Criss–Cross Reduction of the Maslov Index and a Proof of the Yoshida–Nicolaescu Theorem", Tekster fra IMFUFA no. 373, Roskilde, 1999 (*preprint*).

[9] B. Booss–Bavnbek, G. Morchio, F. Strocchi, and K.P. Wojciechowski, Grassmannian and chiral anomaly, *J. Geom. Phys.* **22** (1997), 219–244.

[10] B. Booss–Bavnbek, S.G. Scott, and K.P. Wojciechowski, The ζ-determinant and $\mathcal{C}$-determinant on the Grassmannian in dimension one, *Letters in Math. Phys.* **45** (1998), 353–362.

[11] B. Booss–Bavnbek and K.P. Wojciechowski, "Elliptic Boundary Problems for Dirac Operators", Birkhäuser, Boston, 1993.

[12] J. Brüning and M. Lesch, Spectral theory of boundary value problems for Dirac type operators, in: B. Booss-Bavnbek, K.P. Wojciechowski (eds.), "Geometric Aspects of Partial Differential Equations", Amer. Math. Soc. Series *Contemporary Mathematics*, vol. 242, Providence, R.I., 1999, pp. 203–215.

[13] S.E. Cappell, R. Lee, and E.Y. Miller, On the Maslov index, *Comm. Pure Appl. Math.* **47** (1994), 121–186.

[14] —, —, —, Selfadjoint elliptic operators and manifold decompositions Part I: Low eigenmodes and stretching, *Comm. Pure Appl. Math.* **49** (1996), 825–866. Part II: Spectral flow and Maslov index, *Comm. Pure Appl. Math.* **49** (1996), 869–909. Part III: Determinant line bundles and Lagrangian intersection, *Comm. Pure Appl. Math.* **52** (1999), 543–611.

[15] X. Dai and D.S. Freed, η–invariants and determinant lines, *J. Math. Phys.* **35** *(10)* (1994), 5155–5194.

[16] X. Dai and W. Zhang, Splitting of the family index, *Comm. Math. Phys.* **182** (1996), 303–318.

[17] M. Daniel and P. Kirk, with an appendix by K.P. Wojciechowski, "A general splitting formula for the spectral flow", Bloomington preprint, 1999.

[18] R. Forman, Functional determinants and geometry, *Invent. Math.* **88** (1987), 447–493.

[19] P.B. Gilkey, "Invariance Theory, the Heat Equation, and the Atiyah–Singer Index Theory" (Second Edition), CRC Press, Boca Raton, Florida, 1995.

[20] G. GRUBB, Trace expansions for pseudodifferential boundary problems for Dirac–type operators and more general systems, *Ark. Mat.* **37** (1999), 45–86.

[21] L. HÖRMANDER, Pseudo–differential operators and non–elliptic boundary problems, *Ann. of Math.* **83** (1966), 129–209.

[22] M. KONTSEVITCH AND S. VISHIK, Geometry of determinants of elliptic operators. In: Gindikin, S. (ed.) et al., *Functional Analysis on the eve of the 21st century'*. Vol. I. In honor of the eightieth birthday of I. M. Gelfand. Boston. Birkhäuser, Prog. Math. 131, 1993, pp. 173–197.

[23] M. LESCH AND J. TOLKSDORF, On the determinant of one–dimensional elliptic boundary value problems, *Comm. Math. Phys.* **193** (1998), 643–660.

[24] M. LESCH AND K.P. WOJCIECHOWSKI, On the η–invariant of generalized Atiyah–Patodi–Singer problems, *Illinois J. Math.* **40** (1996), 30–46.

[25] P.T. MATTHEWS AND A. SALAM, The Green's function of quantized field, *Nuovo Cimento Series 9* **12** (1954), 563–565.

[26] —, —, Propagators of quantized field, *Nuovo Cimento Series 11* **2** (1955), 120–134.

[27] G. MORCHIO AND F. STROCCHI, Boundary terms, long range effects, and chiral symmetry breaking, in: Mitter, H., and Schweifer, W. (eds.), *Fields and Particles*, Proceedings Schladming, Austria, 1990, Springer–Verlag, Berlin–Heidelberg–New York, pp. 171–214.

[28] L. NICOLAESCU, The Maslov index, the spectral flow, and decomposition of manifolds, *Duke Math. J.* **80** (1995), 485–533.

[29] —, "Generalized Symplectic Geometries and the Index of Families of Elliptic Problems", Mem. Amer. Math. Soc. vol. 609, Providence, 1997.

[30] —, "On the Space of Fredholm Self–adjoint Operators", Note, University of Notre Dame, 1999 (*multiplied*).

[31] J. PHILLIPS, Self–adjoint Fredholm operators and spectral flow, *Canad. Math. Bull.* **39** (1996), 460–467.

[32] D.G. QUILLEN, Determinants of Cauchy-Riemann operators over a Riemann surface, *Funkcionalnyi Analiz i ego Prilozhenya* **19** (1985), 37–41.

[33] D. RAY AND I.M. SINGER, R–torsion and the Laplacian on Riemannian manifolds, *Adv. Math.* **7** (1971), 145–210.

[34] S.G. SCOTT, Determinants of Dirac boundary value problems over odd–dimensional manifolds, *Comm. Math. Phys.* **173** (1995), 43–76.

[35] —, Splitting the curvature of the determinant line bundle, *Proc. Am. Math. Soc.*, to appear.

[36] S.G. SCOTT AND K.P. WOJCIECHOWSKI, Determinants, Grassmannians and elliptic boundary problems for the Dirac operator, *Letters in Math. Phys.* **40** (1997), 135–145.

[37] —, —, ζ-determinant and the Quillen determinant on the Grassmannian of elliptic self-adjoint boundary conditions, *C. R. Acad. Sci., Serie I,* **328** (1999), 139–144.

[38] —, —, The ζ-determinant and Quillen determinant for a Dirac operator on a manifold with boundary, *GAFA* (1999), (*in print*).

[39] R.T. SEELEY, Complex powers of an elliptic operator, AMS *Proc. Symp. Pure Math. X.* AMS Providence, 1997, 288–307.

[40] —: Topics in pseudo–differential operators, in: *CIME Conference on Pseudo–Differential Operators (Stresa 1968)*. Ed. Cremonese, Rome, 1969, pp. 167–305.

[41] G.B. SEGAL, "The definition of conformal field theory", Oxford preprint, 1990.

[42] I.M. SINGER, Families of Dirac operators with applications to physics, *Asterisque, hors série*, 1985, 323–340.

[43] —: *Personal Communication*, 1999.

[44] K.P. WOJCIECHOWSKI, *Personal Communication*, 1998.

[45] —: The ζ-determinant and the additivity of the η-invariant on the smooth, self-adjoint Grassmannian, *Comm. Math. Phys.* **201** (1999), 423–444.

[46] K.P. WOJCIECHOWSKI, S.G. SCOTT, G. MORCHIO, AND B. BOOSS–BAVNBEK, Determinants, manifolds with boundary and Dirac operators. In: V. Dietrich et al. (eds.), *Clifford Algebras and their applications in mathematical physics*, Kluwer Academic Publishers, 1998, pp. 423–432.

[47] T. YOSHIDA, Floer homology and splittings of manifolds, *Ann. of Math.* **134** (1991), 277–323.

INSTITUT FOR MATEMATIK OG FYSIK, ROSKILDE UNIVERSITY, DK-4000 ROSKILDE, DENMARK
E-mail address: `booss@mmf.ruc.dk`

Canadian Mathematical Society
Conference Proceedings
Volume **29**, 2000

Loop Group Factorization of Biorthogonal Wavelet Bases

Silviu Borac
Eigenval Research Inc., Toronto

and Ruedi Seiler
Technical University of Berlin, Fachbereich Mathematik

ABSTRACT. We present a factorization theorem for certain polynomial loops in the matrix group $GL(2, \mathbb{C})$. The theorem leads to the construction of an algorithm for the factorization of pairs of biorthogonal filters with finite impulse response in simple terms, resulting in a reduction of the complexity of the corresponding wavelet transform. We show that the theorem can be extended to coefficients that do not belong to a field and illustrate this using the ring of dyadic integers.

1. Introduction

Biorthogonal wavelets give rise to perfect reconstruction filters which are useful for audio and image data compression. We propose a method for factorization of the corresponding polyphase matrix in factors of low computational complexity. The result is close but different from the factorization proposed by Daubechies and Sweldens [**2**].

Factorization of filters, as a method for diminishing computational complexity, has been discussed by several authors, in particular by Nguyen and Vaidyanathan [**10**], Vetterli [**15**] and more recently by Daubechies and Sweldens [**2**].

Loop groups are interesting examples of infinite dimensional groups and have arisen in physics in the context of two-dimensional integrable models and string theory [**4**] [**11**]. Factorization is an important standard tool. Recently it has found particularly nice applications in the theory of two-dimensional surfaces [**3**].

2000 *Mathematics Subject Classification.* Primary 42C40, 22E67, 43–04, 43A80, 94A11.
e-mail: silviu@eigenval.com
e-mail: seiler@math.tu-berlin.de

The main result of this paper is a simple algorithm for the decomposition of biorthogonal filters with finite impulse response into elementary filters of the following 3 types, where a, b, c, z denote complex and n, m integer numbers:

$$\begin{array}{ll}
\text{Type 1a} & h(z) = 1 \\
 & g(z) = 1 + a/z \\
\text{Type 1b} & h(z) = 1 + a/z \\
 & g(z) = 1 \\
\text{Type 2} & h(z) = z^{-2n} \\
 & g(z) = z^{-(2m+1)} \\
\text{Type 3a} & h(z) = a + b/z \\
 & g(z) = c \\
\text{Type 3b} & h(z) = a \\
 & g(z) = b + c/z
\end{array}$$

Type 1 and Type 3 come in conjugate pairs. Type 3 arises in a given filter decomposition at most once.

Factorisation is based on the observation that biorthogonal filters with finite impulse response are in one to one correspondence with polynomial loops in $GL(2,\mathbb{C})$. The polynomial loop group $L_{pol}GL(2,\mathbb{C})$ is graded according to the winding number of the determinant map. The shift acts as a ladder operator.

Factorisation gives an exhaustive parametrisation of all biorthogonal FIR filters. An algebraic characterisation of filter pairs leading to reagular wavelets is, however, still missing.

Factorization of polyphase matrices in the context of wavelet analysis has been presented for the orthogonal case by Holschneider and Pinkall [**8**] and most recently for the biorthogonal case by Daubechies and Sweldens [**2**]. The approach taken here is different from the one in [**2**] in that the polyphase matrix is interpreted as a loop in $GL(2,\mathbb{C})$, whereas in the [**2**] the filter pair $(h(z), g(z))$ is decomposed using the Euclidean algorithm. The factorisation presented here is of course closely related to the lifting concept introduced by Sweldens[**12**]; furthermore it is of the "product form" in the sense of Vetterli and Le Gall [**16**].The algorithm by Holschneider and Pinkall [**8**] has been implemented by Tanriver [**13**] and applied to the factorization of Daubechies wavelets.

The article is structured as follows:
In a first section the connection of biorthogonal filters and loops in $GL(2,\mathbb{C})$ is presented. The next section presents our method of factorizing loops followed by its extension to the ring of dyadic integers. In the last section we give an estimate on computational complexity and show some examples and the pseudo code for the factorization algorithm.

2. From Biorthogonal Filters to Loops

A multiresolution analysis of $L^2(\mathbb{R})$ [**1**] consists of a a sequence of closed spaces V_j satisfying:

$$\cdots \subset V_{-2} \subset V_{-1} \subset V_0 \subset V_1 \subset V_2 \subset \cdots$$
$$\overline{\bigcup_{j\in\mathbb{Z}} V_j} = L^2(\mathbb{R})$$
$$\bigcap_{j\in\mathbb{Z}} V_j = \{0\} \tag{2.1}$$

For every j we have a decomposition of V_j into an average space and a detail space:

$$V_j = V_{j-1} \oplus W_{j-1}, \quad j \in \mathbb{Z} \tag{2.2}$$

The multiresolution analysis is built using two basic functions: a scaling function ϕ and a wavelet ψ. The integer translates of the 2^j-dilation of ϕ form an L^2-basis [1] for $V_j = \langle\langle \phi\left(2^j x - m\right), \; m \in \mathbb{Z}\rangle\rangle$. Similarly, we have $W_j = \langle\langle \psi\left(2^j x - m\right), \; m \in \mathbb{Z}\rangle\rangle$. Furthermore $L^2(\mathbb{R}) = \langle\langle \psi\left(2^j x - m\right), \; m \in \mathbb{Z}, j \in \mathbb{Z}\rangle\rangle$. Due to eq.(2.1) the functions ψ and ϕ satisfy two-scale refinement relations of the form:

$$V_{-1} \ni \phi\left(\frac{x}{2}\right) = \sqrt{2}\sum h_k \phi\left(x-k\right)$$
$$\psi\left(\frac{x}{2}\right) = \sqrt{2}\sum g_k \phi\left(x-k\right) \tag{2.3}$$

The dual basis is given in terms of the so called dual scaling function $\tilde{\phi}$ and dual wavelet $\tilde{\psi}$. They generate another multiresolution analysis of $L^2(\mathbb{R})$:

$$\cdots \subset \widetilde{V}_{-2} \subset \widetilde{V}_{-1} \subset \widetilde{V_0} \subset \widetilde{V_1} \subset \widetilde{V_2} \subset \cdots$$
$$\overline{\bigcup_{j\in\mathbb{Z}} \widetilde{V_j}} = L^2\left(\mathbb{R}\right) \quad \bigcap_{j\in\mathbb{Z}} \widetilde{V_j} = \{0\}$$
$$\widetilde{V_j} = \widetilde{V}_{j-1} \oplus \widetilde{W}_{j-1} \tag{2.4}$$

The functions $\phi\,,\ \psi\,,\ \tilde{\phi}\,,\ \tilde{\psi}$ satisfy the following orthogonality relations in $L^2(\mathbb{R})$:

$$\langle \tilde{\phi}\,,\ \psi(\cdot - n)\rangle = \langle \tilde{\psi}\,,\ \phi(\cdot - n)\rangle = 0 \quad n \in \mathbb{Z}$$
$$\langle \tilde{\phi}\,,\ \phi(\cdot - n)\rangle = \langle \tilde{\psi}\,,\ \psi(\cdot - n)\rangle = \delta_{n,0} \tag{2.5}$$

The dual scaling function and wavelet satisfy refinement relations similar to (2.3) with filter coefficients $\tilde{h}_k$ and $\tilde{g}_k$, respectively.

A function $f \in V_0 \subset L^2\left(\mathbb{R}\right)$ can be decomposed in the basis $\{\phi(\cdot - n)\,,\ n \in \mathbb{Z}\}$ as follows:

$$f(x) = \sum_{j\in\mathbb{Z}} f_n\,\phi(x-n) \quad \text{with} \quad f_n = \langle f(x)\,,\ \tilde{\phi}(x-n)\rangle \tag{2.6}$$

where $\langle \;\;, \;\;\rangle$ denotes the scalar product in L^2. The sequence of coefficients $\{f_n\}$ is in $l^2\left(\mathbb{Z}\right)$. To this sequence corresponds, via the discrete Fourier transform, a

[1]Here, an L^2-basis is what is sometimes called an unconditional L^2-basis.

square integrable function on the circle $\mathbb{T}$:

$$f(z) = \sum_{n\in\mathbb{Z}} f_n z^{-n}, \quad z \in \mathbb{T} = \{z \in \mathbb{C}, \mid z \mid = 1\} \tag{2.7}$$

We have the sequence of bijections:

$$\begin{array}{rcccc} L^2(\mathbb{R}) \supset V_0 = \langle\langle \phi(\cdot - n) \mid n \in \mathbb{Z}\rangle\rangle & \longrightarrow & l_2(\mathbb{Z}) & \longrightarrow & L^2(\mathbb{T}) \\ f(x) = \sum f_n \phi(x-n) & \longmapsto & \{f_n\}_{n\in\mathbb{Z}} & \longmapsto & f(z) = \sum f_n z^{-n} \end{array}$$

To the filter sequences h_k, g_k, $\tilde{h}_k$, $\tilde{g}_k$ correspond the functions $h(z)$, $g(z)$, $\tilde{h}(z)$, $\tilde{g}(z)$.

There is another representation, called the polyphase representation, that is very convenient for expressing wavelet transforms because it diagonalizes the double shift $\{f_n\} \longmapsto \{f_{n-2}\}$ (see below). It is based on the generalized Fourier transformation:

$$\begin{array}{rcl} l_2(\mathbb{Z}) & \longrightarrow & L^2(\mathbb{T}, \mathbb{C}^2) \\ \{f_n\}_{n\in\mathbb{Z}} & \longmapsto & \mathbf{f}(z) = \sum_n z^{-n}\mathbf{f_n} = \sum_n z^{-n}\binom{f_{2n}}{f_{2n+1}} \end{array} \tag{2.8}$$

where $L^2(\mathbb{T}, \mathbb{C}^2)$ is the set of column vectors with two components that are complex functions on the circle, of finite L^2 norm with respect to the product $\langle f, g\rangle = \sum_n \mathbf{f}_n^* \mathbf{g}_n$, where $\mathbf{f}_n^* = (\overline{f_{2n}}, \overline{f_{2n+1}})$.

The polyphase representation $\mathbf{f}(z)$ of the sequence $\{f_n\}$ is related to the shift operator S, $Sf_n := f_{n-1}$, through the formula

$$f(z) = \sum_n z^{-n}(S^{-2n}f)_m, \qquad (m = 0, 1) \tag{2.9}$$

$$= \sum_n z^{-n} f_{2n+m} \tag{2.10}$$

We call (as in [**2**]) the first component of $\mathbf{f}(z) \in \mathbb{C}$ the "even" part and the second component the "odd" part: $\mathbf{f}(z) = \begin{pmatrix} f_e(z) \\ f_o(z) \end{pmatrix}$.

The relation between the $L^2(\mathbb{T})$ representation and the $L^2(\mathbb{T}, \mathbb{C}^2)$ is given by:

$$f(z) = (1\,,\, z^{-1})\,\mathbf{f}(z^2) \tag{2.11}$$

and

$$\mathbf{f}(z^2) = \frac{1}{2}\begin{pmatrix} 1 & 1 \\ z & -z \end{pmatrix}\begin{pmatrix} f(z) \\ f(-z) \end{pmatrix} \tag{2.12}$$

In $L^2(\mathbb{T}^1, \mathbb{C}^2)$ the shift operator becomes the following 2×2 matrix:

$$(S\mathbf{f}) = \begin{pmatrix} 0 & z^{-1} \\ 1 & 0 \end{pmatrix}\mathbf{f}(z) \tag{2.13}$$

The square of the shift operator is diagonal:

$$(S^2\mathbf{f})(z) = \begin{pmatrix} z^{-1} & 0 \\ 0 & z^{-1} \end{pmatrix}\mathbf{f}(z) \tag{2.14}$$

The shift operator is, up to a constant permutation matrix Π, the same as the even and odd translation operator T_e, T_o defined by:

$$T_e = \begin{pmatrix} z^{-1} & 0 \\ 0 & 1 \end{pmatrix}, T_o = \begin{pmatrix} 1 & 0 \\ 0 & z^{-1} \end{pmatrix}, \Pi = \begin{pmatrix} 0 & 1 \\ 1 & 0 \end{pmatrix} \tag{2.15}$$

$$T_e = S\Pi, \quad T_o = \Pi S \tag{2.16}$$

The forward wavelet transform uses $\tilde{h}$ and $\tilde{g}$ as analysis filters while the inverse wavelet transform uses h and g as synthesis filters. [**1**]

The conditions for perfect reconstruction are:

$$h(z)\overline{\tilde{h}(z)} + g(z)\overline{\tilde{g}(z)} = 2 \qquad h(z)\overline{\tilde{h}(-z)} + g(z)\overline{\tilde{g}(-z)} = 0 \tag{2.17}$$

They can be expressed in terms of the polyphase matrices:

$$P(z) = \begin{pmatrix} h_e(z) & g_e(z) \\ h_o(z) & g_o(z) \end{pmatrix}, \quad \tilde{P}(z) = \begin{pmatrix} \tilde{h}_e(z) & \tilde{g}_e(z) \\ \tilde{h}_o(z) & \tilde{g}_o(z) \end{pmatrix} \tag{2.18}$$

Perfect reconstruction is achieved when:

$$P(z)\tilde{P}^*(z) = 1 \tag{2.19}$$

The equation above expresses $\tilde{P}$ in terms of P and vice versa. Each polyphase matrix is a loop in the group of invertible two by two matrices $GL(2,\mathbb{C})$. In the case of an orthogonal wavelet, $P(z)$ is equal to $\tilde{P}(z)$. So the corresponding loop is in the group of unitary matrices $U(2)$. For the special case of filters with real coefficients, equation (2.19) becomes:

$$P(z)\tilde{P}^t(z^{-1}) = 1 \tag{2.20}$$

In this work we shall consider only the case of finite impulse filters (FIR). In this case, the entries in both matrices $P(z)$ and $\tilde{P}(z)$ are Laurent polynomials. This implies that $\det P(z)$ is also a Laurent polynomial and, by the perfect reconstruction condition (2.19), it is in fact a monomial:

$$\det P(z) = \lambda z^n \tag{2.21}$$

The problem of finding biorthogonal filters with finite impulse response is thus equivalent to finding polynomial loops in $GL(2,\mathbb{C})$. Hence classification of polynomial loops in $GL(2,\mathbb{C})$ is tantamount to classification of finite impulse filters. This is the subject of the following section.

3. Factorization of Loops

A loop group LG is the group of maps form the circle $\mathbb{T}$ into a group G. The group composition law of LG is given by the pointwise composition in G. [**11**]. LG contains the subgroup of polynomial loops $L_{pol}G$. P belongs to $L_{pol}G$, if P and its inverse in LG are polynomials in z with coefficients in G.

As mentioned in the previous section, each FIR biorthogonal filter can be identified with a polynomial loop in $GL(2,\mathbb{C})$ For several reasons it is appropriate to consider only loops $P(z)$ which are normalized by $|\det P(z)| = 1, \quad z \in \mathbb{T}$. First, a loop which corresponds to a wavelet, has to satisfy a minimal regularity condition[**1**]. In terms of $P(z)$ this is

$$P(z=1) = \frac{1}{\sqrt{2}}\begin{pmatrix} 1 & 1 \\ 1 & -1 \end{pmatrix} \tag{3.1}$$

Hence, for the class of loops arising from wavelets, $\det P(z=1) = 1$. Second, in the orthogonal case, this condition is automatic. Third, and most importantly, the shift operators S and T_e, T_o belong to this class. Fourth, each finite impulse filter is of this type up to a non vanishing factor $\lambda \in \mathbb{C}$. For the purpose of classification of filters, it suffices to consider the following restricted class of loops:

DEFINITION 3.1. *Let X be the group of Laurent polynomial loops $P : \mathbb{T} \to GL(2, \mathbb{C})$ such that the determinant is a circle map $\det P : \mathbb{T} \to \mathbb{T}$. We refer to elements in X as polynomial loops.*

X is clearly a subgroup of $LGL(2\mathbb{C})$ and its elements are in one to one correspondence with normalized polyphase matrices of biorthogonal filters with finite impulse response. Due to the following proposition it is graded:

PROPOSITION 3.1. *Let $P \in X$. Then there exist $\lambda \in \mathbb{T}$ and $n \in \mathbb{Z}$ such that $\det P(z) = \lambda z^{-n}$.*

PROOF. Since $P(z)$ is a Laurent polynomial loop we have

$$\det P(z) = \sum_{n=n_s}^{n_e} a_n z^{-n}.$$

The leading term of the Laurent polynomial $\mid \det P(z) \mid^2 = 1$ is $\sum a_{n_s} \overline{a_{n_e}} z^{-n_s+n_e}$ which implies $n_s = n_e$ and $\mid a_{n_s} \mid = 1$. □

We get the first structural result about X:

PROPOSITION 3.2. *The loop group X is graded. The grading is compatible with the multiplication:*

$$\begin{aligned} X = \quad & \bigcup_n X_n, \quad X_n = \quad \{P \in X \mid \det P(z) = \lambda z^{-n}, \lambda \in T\} \\ m : \quad & X_{n_1} \times X_{n_2} \quad \longrightarrow X_{n_1+n_2} \end{aligned}$$

The operator S is a ladder operator for the grading: $S : X_n \longrightarrow X_{n+1}$, $S^{-1} : X_{n+1} \longrightarrow X_n$.

As a consequence, for any loop $A \in X$, there is an $n \in \mathbb{Z}$ such that $S^n A$ is a loop with a determinant of modulus one.

It turns out that every polynomial loop is a product of elementary loops of three different types which we call upper triangular, lower triangular and diagonal loops.

DEFINITION 3.2. *Let $k, m, n \in \mathbb{Z}$ and $a \in \mathbb{R}$. Then,*

$$U_+(k, a) = T_e^k \begin{pmatrix} 1 & a \\ 0 & 1 \end{pmatrix} T_e^{-k}$$

and

$$U_-(k,a) = T_o^k \begin{pmatrix} 1 & 0 \\ a & 1 \end{pmatrix} T_o^{-k}$$

are special loops in $SL(2,\mathbb{C})$ referred to as upper and lower triangular loops, respectively. Furthermore

$$D(m,n) = \begin{pmatrix} z^m & 0 \\ 0 & z^n \end{pmatrix}$$

is an elementary diagonal loop. It is an element of X_{m+n}.

The following factorization of polynomial loops holds [**2**]

THEOREM 3.1. *Consider a polynomial loop $P \in X$. There exist finitely many upper and lower triangular loops, an elementary diagonal loop and a matrix B with $|\det B| = 1$ such that:*

$P = U_{\sigma(1)} \cdots U_{\sigma(N)} DB.$

$\sigma(1) \ldots \sigma(N)$ denotes an appropriate choice of signs.

Before we prove the theorem let us add some remarks and a corollary:

REMARK 3.1. The above theorem implies the decomposition of loops using lifting in the sense of Daubechies and Sweldens [**2**], since:

$$U_+(k_1,a_1) \cdots U_+(k_s,a_s) = \begin{pmatrix} 1 & \sum_1^s a_i z^{-k_i} \\ 0 & 1 \end{pmatrix} \tag{3.2}$$

$$U_-(k_1,a_1) \cdots U_-(k_s,a_s) = \begin{pmatrix} 1 & 0 \\ \sum_1^s a_i z^{-k_i} & 1 \end{pmatrix} \tag{3.3}$$

REMARK 3.2. If $P(z)$ is a loop in $SL(2,\mathbb{C})$ the factorization involves a diagonal matrix of the type $D(\lambda = 1; m, n = -m)$. Hence the factorization is in terms of loops in $SL(2,\mathbb{C})$.

REMARK 3.3. The above factorization is not of the Birkhoff type, where a loop is factorized into a term entire in z, a diagonal - monomial and a third term which is entire in $1/z$. It is, however, possible to derive a Birkhoff factorization from the theorem above, since in the statement of the theorem T can be replaced by S and the identity:

$$SAS^{-1} = S^{-1}AS \tag{3.4}$$

holds for any 2×2 matrix A. Hence all entire terms can be shuffled to the left and the terms entire in z^{-1} to the right.

REMARK 3.4. For a dense set in $SL(2,\mathbb{C})$ B can be factorized into an upper and a lower triangular matrix.

The proof of Theorem 3.1 is based upon the following technical result:

LEMMA 3.1. *Consider $P(z) \in X_m$ and denote by $P_e(z)$ and $P_o(z)$ the $\mathbb{C}^2$-valued polynomials: $P_e(z) = \sum_{n_{e,s}}^{n_{e,f}} P_{e,n} z^{-n}$ and $P_o(z) = \sum_{n_{o,s}}^{n_{o,f}} P_{o,n} z^{-n}$ such that $P(z) = \begin{pmatrix} P_e(z) \\ P_o(z) \end{pmatrix}$ and all four vectors $P_{e,n_{e,s}}, P_{e,n_{e,f}}, P_{o,n_{e,s}}, P_{o,n_{o,f}}$ are non-vanishing vectors in $\mathbb{C}^2$. By construction the determinant of P is of the form:*

$$det P(z) = \Delta_s z^{-(n_{e,s}+n_{o,s})} + \cdots + \Delta_f z^{(n_{e,f}+n_{o,f})} \tag{3.5}$$

with $\Delta_s = det\begin{pmatrix} P_{e,n_{e,s}} \\ P_{o,n_{o,s}} \end{pmatrix}$ and $\Delta_f = det\begin{pmatrix} P_{e,n_{e,f}} \\ P_{o,n_{o,f}} \end{pmatrix}$. Now, either of the three cases holds (exclusively):

Case 1: $\Delta_s = \Delta_f = 0$
Case 2: $(|\Delta_s| = 1 \text{ and } \Delta_f = 0) \quad or \quad (\Delta_s = 0 \text{ and } |\Delta_f| = 1)$
Case 3: $n_{e,s} = n_{e,f}, n_{o,s} = n_{o,f}, n_{e,s} + n_{o,s} = m$

PROOF. 1. $\det P(z)$ is a Laurent polynomial in z, where Δ_s and Δ_f are the coefficients with the smallest and largest power of z^{-1}, respectively.
2. If $(n_{e,s} + n_{o,s}) < m < (n_{e,f} + n_{o,f})$ both coefficients have to vanish.
3. If $(n_{e,s} + n_{o,s}) = m < (n_{e,f} + n_{o,f})$ we find $|\Delta_s| = 1$ and $\Delta_f = 0$.
4. If $(n_{e,s} + n_{o,s}) = m = (n_{e,f} + n_{o,f})$ we argue as follows: $n_{e,s}$ and $n_{o,s}$ as well as $n_{e,f}$ and $n_{o,f}$ are symmetrically arranged with respect to m. On the other hand, $n_{e,s} \leq n_{e,f}$ and $n_{os,} \leq n_{o,f}$. This is only consistent with $n_{e,s} = n_{e,f}$ and $n_{o,s} = n_{o,f}$.

To prove the theorem, consider a loop $P \in X_m$. Assume that we are not in Case 3 of the previous Lemma. Then at least one of the two determinants Δ_s and Δ_f vanishes. Say $\Delta_s = 0$. In this case we work at the start of the filter. Then either $n_{e,s} \geq n_{o,s}$ or $n_{e,s} \leq n_{o,s}$ If $n_{e,s} \geq n_{o,s}$, apply a lower left shift $T_-^{(n_{e,s}-n_{o,s})}$ by multiplying from the left. The even and the odd parts $P_e(n_{e,s})$ and $P_o(n_{o,s})$ of the new filter at $n_{e,s}$ - again denoted by the same symbol - are both not zero. Multiplying by an appropriate upper triangular matrix of the type $\begin{pmatrix} 1 & a \\ 0 & 1 \end{pmatrix}$ one can annihilate the even part of the filter at the position $n_{e,s}$. A shift to the right by $T_-^{-(n_{e,s}-n_{o,s})}$ produces a loop which has $n_{s,e}$ reduced by one. If $\Delta_f = 0$, we operate on the final end of the filter analogously. If $n_{e,s} \leq n_{o,s}$ we apply an upper left shift $T_+^{(n_{o,s}-n_{e,s})}$. Again, the new filter has $P_e(n_{e,s}) \neq 0$ and $P_o(n_{o,s}) \neq 0$. A lower triangular matrix annihilates the odd part. Multiplication by $T_-^{(n_{o,s}-n_{e,s})}$ produces a filter where $n_{os,}$ is reduced by one. If we encounter case 3 of the previous Lemma, we get one single term of the diagonal type times a matrix with a determinant of absolute value one. □

A consequence of Theorem 3.1 is that every loop in X can be written as a product of shifts, i.e. loops given by the shift operator and of constant loops. More precisely we have:

COROLLARY 3.1. *Consider the two subgroups of X:*

$$\begin{aligned} &1. \quad \mathcal{C} = \{C(z) = C_0 \in SL(2,\mathbb{C})\} \\ (3.6) \qquad &2. \quad \mathcal{S} = \{S^m \mid m \in \mathbb{Z}\} \end{aligned}$$

Then X is generated algebraically by $\mathcal{C}$ and $\mathcal{S}$, i.e. each $P \in X$ is of the form:

$$P = \lambda S^m C_1 S^{n_1} C_2 S^{n_2} \dots C_l S^{n_l} C_{l+1} \tag{3.7}$$

where $n_i \in \{-1, 0, +1\}$, $l \in \mathbb{N}$, $m \in \mathbb{Z}$, and $\lambda \in \mathbb{T}$; all factors S are to the left and the factors S^{-1} to the right.

REMARK 3.5. If $P \in LU(2) \subset X$ the C matrices can be chosen from $U(2)$. This corresponds to orthogonal filters. [**8**]

3.1. Factorization in the ring of dyadic integers. Let us consider $D = \{a2^k \| a \in \mathbb{Z},\ k \in \mathbb{Z},\ (a,2) \text{ coprime}\}$ the ring of dyadic numbers. We would like to extend our factorization approach to D. We are indebted to Wim Sweldens for inquiring about such a possibility.

We have to show that for all $a_0, b_0 \in D$ there exist (L_i, U_i), $i = 1, \dots, n$ such that $\begin{pmatrix} a_0 \\ b_0 \end{pmatrix} = \prod_{i=1}^n L_i U_i \begin{pmatrix} 0 \\ b_n \end{pmatrix}$ and (L_i', U_i') such that $\begin{pmatrix} a_0 \\ b_0 \end{pmatrix} = \prod_{i=1}^n L_i' U_i' \begin{pmatrix} a_n' \\ 0 \end{pmatrix}$, where L_i, L_i', U_i, U_i' are lower (upper) triangular matrices with entries in D.

The units [**9, 6**] of the ring D are $U = \{\pm 2^k \mid k \in \mathbb{Z}\}$. It is immediate that D is an integral domain. In fact, D is an Euclidean domain [**9, 6**] as shown by the following theorem. (Note: a domain D is called Euclidean if there exists a map $\delta : a \to \delta(a)$ of D into $\mathbb{N}$ such that if $a, b \neq 0 \in D$ there exist $q, r \in D$ such that $a = bq + r$, where $\delta(r) < \delta(b)$.)

THEOREM 3.2. *D is an Euclidean domain with Euclidean function $\delta(a2^k) = \mid a \mid$ (with $(a,2)$ coprime).*

PROOF. We do not just show that the statement is true but also choose the (q, r) pair that minimizes $\delta(r)$. Assume in the following that gcd $(a, b) = 1$. (If it is not 1 everything will be multiplied by $d = \gcd(a, b)$.) Find $m \in \mathbb{N}$ in $2^m a = bq + r$ that minimizes $\mid r \mid$ by evaluating the *signed* residues modulo b of $2^m a$:$\mid r \mid = \min_m(\min\{2^m a \bmod b, b - (2^m a \bmod b)\})$. There are at most $\Phi(b)$ values of m to be checked, [**5**] where ϕ is Euler's function.

The division of $a\,2^k$ by $b\,2^l$ can now be written as:

$$a2^k = b\,2^l(q2^{-l+k-m}) + r(2^{k-m}) \tag{3.8}$$

It is clear that $\delta(r\,2^{k-m}) < \delta(b\,2^l)$ and therefore D is an Euclidean domain. □

As immediate consequences of the theorem are that D has the unique factorization property and D is a principal ideal domain. More important for our problem is that we have an Euclidean algorithm.

Using the Euclidean algorithm a column vector $\begin{pmatrix} a \\ b \end{pmatrix}$ can be written

$$\begin{pmatrix} a \\ b \end{pmatrix} = \prod_{i=0}^{n-1} \begin{pmatrix} q_i & 1 \\ 1 & 0 \end{pmatrix} \begin{pmatrix} a_n \\ 0 \end{pmatrix} \tag{3.9}$$

and

$$\begin{pmatrix} a \\ b \end{pmatrix} = \prod_{i=0}^{n-1} \begin{pmatrix} 0 & 1 \\ 1 & q_i \end{pmatrix} \begin{pmatrix} 0 \\ a_n \end{pmatrix} \tag{3.10}$$

with the choice for (q, r) as in the proof of the theorem, the Euclidean algorithm will terminate in a minimum number of steps.

Starting from 3.9 and 3.10 we can formulate the $\begin{pmatrix} a \\ b \end{pmatrix}$ reduction in terms of L and U matrices by inserting pairs of permutation matrices $\begin{pmatrix} 0 & 1 \\ 1 & 0 \end{pmatrix}$ as in [**2**].

One can thus reduce the column vector $\begin{pmatrix} a \\ b \end{pmatrix}$ by using only lower and upper triangular matrices as follows:

1. to $\begin{pmatrix} a_n \\ 0 \end{pmatrix}$ by using 3.9 when n is even,
2. to $\begin{pmatrix} 0 \\ a_n \end{pmatrix}$ by using 3.9 when n is odd,
3. to $\begin{pmatrix} 0 \\ b_n \end{pmatrix}$ by using 3.10 when n' is even,
4. to $\begin{pmatrix} b_n \\ 0 \end{pmatrix}$ by using 3.10 when n' is odd.

Since n is in general different from n' neither of these tranformations achieves the desired reduction. One can change the parity of n or n' at the expense of adding a step in the Euclidean algorithm since $\begin{pmatrix} q & 1 \\ 1 & 0 \end{pmatrix} = \begin{pmatrix} q-1 & 1 \\ 1 & 0 \end{pmatrix} \begin{pmatrix} 1 & 1 \\ 1 & 0 \end{pmatrix}$.

4. Computational Aspects and Examples

Factorization of filters is of great importance for reducing computational complexity of transformations. After computing a bound for complexity, we apply the algorithm to some examples. The algorithm is finally presented in pseudocode form.

4.1. Complexity. Computational complexity of the factorized wavelet transform is essentially determined by the number of triangular matrices in the factorization of the loop $P(z)$. This is because the upper and lower shifts affect only the indexing.

Each term in the factorization adds to the total complexity $C(P)$ according to the following rule:

$$C(triangular\,matrix) = 2 \qquad \text{(one multiplication and two sums)} \tag{4.1}$$

$$C(B) = 6 \qquad \text{(four multiplications and one sum)} \tag{4.2}$$

$$C(shift) = C(D) = 0 \qquad \text{(reindexing)} \tag{4.3}$$

A rough estimate leads to the bound

$$C(P) \leq 2(n_{e,f} - n_{o,f}) + 2(n_{o,f} - n_{o,s}) + 6 \tag{4.4}$$

The first two terms count the complexities related to the application of the triangular matrices, while the last term arises from the factor B.

In order to get an estimate in terms of the lengths of the filters h and g, it is more convenient to apply the factorization algorithm to the transposed loop $P(z)^t$, since now the upper row contains the filter h and the lower one the filter g. Denote by $l(h)$ and $l(g)$ the two smallest even numbers larger or equal to the lengths of the two filters. In the same manner as above, we get the estimate

$$C(P) \leq l(h) + l(g) + 2$$

4.2. Examples. We apply the algorithm to three examples.

The first example has the following polyphase matrix:

$$\begin{pmatrix} z^{-1} + 3 + z & z^{-2} + 5z^{-1} + 5 + z \\ 1 + z & z^{-1} + 3 + z \end{pmatrix}$$

The factorization is as follows:

$$\begin{pmatrix} 1 & 1 \\ 0 & 1 \end{pmatrix} \begin{pmatrix} 1 & z^{-1} \\ 0 & 1 \end{pmatrix} \begin{pmatrix} 1 & 0 \\ z & 1 \end{pmatrix} \begin{pmatrix} 1 & 0 \\ 1 & 1 \end{pmatrix} \begin{pmatrix} 1 & z^{-1} \\ 0 & 1 \end{pmatrix} \begin{pmatrix} 1 & 1 \\ 0 & 1 \end{pmatrix} =$$

$$= \begin{pmatrix} 1 & 1 + z^{-1} \\ 0 & 1 \end{pmatrix} \begin{pmatrix} 1 & 0 \\ 1 + z & 1 \end{pmatrix} \begin{pmatrix} 1 & 1 + z^{-1} \\ 0 & 1 \end{pmatrix}$$

The next example is the cubic B-spline. The polyphase matrix is defined by:

$$\begin{pmatrix} \frac{1}{8}(z^{-1} + 6 + z) & \frac{1}{32}(-3z^{-2} - 5z^{-1} - 5 - 3z) \\ \frac{1}{2}(1 + z) & \frac{1}{8}(-3z^{-1} + 10 - 3z) \end{pmatrix}$$

The factorisation of the cubic B-spline is:

$$\begin{pmatrix} 1 & \frac{1}{4} \\ 0 & 1 \end{pmatrix} \begin{pmatrix} 1 & \frac{1}{4}z^{-1} \\ 0 & 1 \end{pmatrix} \begin{pmatrix} 1 & 0 \\ z & 1 \end{pmatrix} \begin{pmatrix} 1 & 0 \\ 1 & 1 \end{pmatrix} \begin{pmatrix} 1 & -\frac{3}{16}z^{-1} \\ 0 & 1 \end{pmatrix} \begin{pmatrix} \frac{1}{2} & -\frac{3}{8} \\ 0 & 2 \end{pmatrix} \tag{4.5}$$

The last example is one by Vetterli and Le Gall. As the two authors note, it can not be factorized into simple pseudo-unitary terms. [2] It is, however, factorizable in the sense of Theorem 3.1.

The defining polyphase matrix is:

$$\begin{pmatrix} z^{-2}+3z^{-1}+2+z & -\frac{1}{14}z^{-1}-\frac{5}{14}+\frac{2}{7}z+\frac{1}{14}z^2 \\ z^{-2}+2z^{-1}+3+z & -\frac{1}{14}z^{-1}-\frac{2}{7}+\frac{5}{14}z+\frac{1}{14}z^2 \end{pmatrix}$$

The factorisation of this filter is:

$$\begin{pmatrix} 1 & 1 \\ 0 & 1 \end{pmatrix}\begin{pmatrix} 1 & 0 \\ -z & 1 \end{pmatrix}\begin{pmatrix} 1 & 0 \\ z^{-1} & 1 \end{pmatrix}\begin{pmatrix} 1 & 0 \\ -4 & 1 \end{pmatrix}\begin{pmatrix} 1 & \frac{1}{7}z \\ 0 & 1 \end{pmatrix}\times$$
$$\times\begin{pmatrix} 1 & \frac{1}{7} \\ 0 & 1 \end{pmatrix}\begin{pmatrix} 1 & 0 \\ -\frac{7}{2}z^{-1} & 1 \end{pmatrix}\begin{pmatrix} -2 & 0 \\ 0 & -\frac{1}{2} \end{pmatrix}=$$
$$=\begin{pmatrix} 1 & 1 \\ 0 & 1 \end{pmatrix}\begin{pmatrix} 1 & 0 \\ -z+z^{-1}-4 & 1 \end{pmatrix}\begin{pmatrix} 1 & \frac{1}{7}(1+z) \\ 0 & 1 \end{pmatrix}\begin{pmatrix} 1 & 0 \\ -\frac{7}{2}z^{-1} & 1 \end{pmatrix}\begin{pmatrix} -2 & 0 \\ 0 & -\frac{1}{2} \end{pmatrix}$$

4.3. Pseudocode. $P(z) = \begin{pmatrix} \sum_{n_{e,s}}^{n_{e,f}} P_{e,n} z^{-n} \\ \sum_{n_{o,s}}^{n_{e,f}} P_{o,n} z^{-n} \end{pmatrix}$

`factorList` $= \{\}$

`assert:` $\det P(z) = \mu z^{-m}$

$\lambda = \mu^{1/2}$

$P(z) = (-1)^m \lambda^{-1} T_+^{-m} P(z)$

`AppendToFactorList` $\{\lambda I, T_+^m\}$

`while` $(n_{e,f} - n_{e,s}) > 0$ `or` $(n_{o,f} - n_{o,s}) > 0$

`(* reduce the first row until it becomes shorter than the second *)`

  `while` $(n_{e,f} - n_{e,s}) \geq (n_{o,f} - n_{o,s})$ `and` $(n_{e,f} - n_{e,s}) > 0$

    `if` det $\begin{pmatrix} P_{e,n_{e,s}} \\ P_{o,n_{o,s}} \end{pmatrix} = 0$ `then`

      `ReduceSUpper()`

    `else`

[2]Pseudo-unitary means, that the matrix is a unitary matrix up to a scaling in the directions of the coordinate lines.

ReduceFUpper()

end while

(* reduce the second row until it becomes shorter than the first *)

while $(n_{o,f} - n_{o,s}) \geq (n_{e,f} - n_{e,s})$ and $(n_{o,f} - n_{o,s}) > 0$

if $\det \begin{pmatrix} P_{e,n_{e,s}} \\ P_{o,n_{o,s}} \end{pmatrix} = 0$ then

ReduceSLower()

else

ReduceFLower()

end while

end while

AppendToFactorList $\{P(z)\}$

begin subroutine ReduceSUpper()

if $h_{o,n_{o,s}} \neq 0$ then

$$U = \begin{pmatrix} 1 & -\frac{h_{e,n_{e,s}}}{h_{o,n_{o,s}}} \\ 0 & 1 \end{pmatrix}$$

else

$$U = \begin{pmatrix} 1 & -\frac{g_{e,n_{e,s}}}{g_{o,n_{o,s}}} \\ 0 & 1 \end{pmatrix}$$

$$U^{-1} = \begin{pmatrix} 1 & -U_{12} \\ 0 & 1 \end{pmatrix}$$

AppendToFactorList $(\{T_+^{(n_{o,s}-n_{e,s})}\, U^{-1}\, T_+^{-(n_{o,s}-n_{e,s})}\})$

$$P(z) = T_+^{-(n_{o,s}-n_{e,s})}\, U\, T_+^{(n_{o,s}-n_{e,s})}\, P(z)$$

end ReduceSUpper()

begin subroutine ReduceSLower()

if $h_{e,n_{e,s}} \neq 0$ then

$$L = \begin{pmatrix} 1 & 0 \\ -\frac{h_{o,n_{o,s}}}{h_{e,n_{e,s}}} & 1 \end{pmatrix}$$

`else`

$$L = \begin{pmatrix} 1 & 0 \\ -\frac{g_{o,n_{o,s}}}{g_{o,n_{o,s}}} & 1 \end{pmatrix}$$

$$L^{-1} = \begin{pmatrix} 1 & 0 \\ -L_{21} & 1 \end{pmatrix}$$

`AppendToFactorList` $(\{T_-^{(n_{e,s}-n_{o,s})} \, L^{-1} \, T_-^{-(n_{e,s}-n_{o,s})}\})$

$P(z) = T_-^{-(n_{e,s}-n_{o,s})} \, L \, T_-^{(n_{e,s}-n_{o,s})} \, P(z)$

`end ReduceSLower()`

`begin subroutine ReduceFUpper()`

`if` $h_{o,n_{o,f}} \neq 0$ `then`

$$U = \begin{pmatrix} 1 & -\frac{h_{e,n_{e,f}}}{h_{o,n_{o,f}}} \\ 0 & 1 \end{pmatrix}$$

`else`

$$U = \begin{pmatrix} 1 & -\frac{g_{e,n_{e,f}}}{g_{o,n_{o,f}}} \\ 0 & 1 \end{pmatrix}$$

$$U^{-1} = \begin{pmatrix} 1 & -U_{12} \\ 0 & 1 \end{pmatrix}$$

`AppendToFactorList` $(\{T_+^{(n_{o,f}-n_{e,f})} \, U^{-1} \, T_+^{-(n_{o,f}-n_{e,f})}\})$

$P(z) = T_+^{-(n_{o,f}-n_{e,f})} \, U \, T_+^{(n_{o,f}-n_{e,f})} \, P(z)$

`end ReduceFUpper()`

`begin subroutine ReduceELower()`

`if` $h_{e,n_{e,f}} \neq 0$ `then`

$$U = \begin{pmatrix} 1 & 0 \\ -\frac{h_{o,n_{o,f}}}{h_{e,n_{e,f}}} & 1 \end{pmatrix}$$

```
else
```

$$L = \begin{pmatrix} 1 & 0 \\ -\frac{g_{o,n_{o,f}}}{g_{e,n_{e,f}}} & 1 \end{pmatrix}$$

$$L^{-1} = \begin{pmatrix} 1 & 0 \\ -U_{21} & 1 \end{pmatrix}$$

`AppendToFactorList` $(\{T_-^{(n_{e,f}-n_{o,f})} L^{-1} T_-^{-(n_{e,f}-n_{o,f})}\})$

$$P(z) = T_-^{-(n_{e,f}-n_{o,f})} L T_-^{(n_{e,f}-n_{o,f})} P(z)$$

```
end ReduceFLower()
```

References

[1] I. Daubechies. Ten Lectures on Wavelets. *CBMS-NSF Regional Conf. Series in Appl. Math.*, 61, 1992.

[2] I. Daubechies and W. Sweldens. Factoring Wavelet Transform into Lifting Steps. *Technical Report*, 1996, 1996.

[3] F. Dorfmeister, J. , Pedit and H. Wu. Weierstrass type reprsentation of harmonic maps into symmetric spaces. *Communications in Analysis and Geometry*, to appear.

[4] L.D. Faddeev and L.A. Takhtajan. *Hamiltonian Methods in the Theory of Solitons.* Springer Verlag, 1987.

[5] G.H. Hardy and E.M. Wright. *An Introduction to the Theory of Numbers.* Oxford University Press, 1979.

[6] B. Hartley. *Rings Modules and Linear Algebra.* Chapman and Hall Ltd., 1970.

[7] M. Holschneider. *Wavelets: An Analysis Tool.* Clarendon Press Oxford, 1995.

[8] M. Holschneider and U. Pinkall. Quadrature Mirror Filters and Loop Groups. *SFB-Preprint, TU-Berlin*, 1993.

[9] N. Jacobson. *Basic Algebra I.* W.H.Freeman, 1974.

[10] T.Y. Nguyen and P.P. Vaidyanathan. Perfect reconstruction QMF structures which yield linear, phase FIR analysis filters. In *Proceedings of the Intenational Symposium on Circuits Systems*, pages 297–300. 1988.

[11] A. Pressley and G. Segal. *Loop Groups.* Oxford Math. Monographs. Clarendon Press, 1986.

[12] W. Sweldens. The Lifting Scheme: A Construction of Second Generation Wavelets. *Technical Report*, 1995:6, 1995.

[13] L. Tanriver. Factorisation of daubechies wavelet. *Masters Thesis, Technische Universität Berlin*, 1997.

[14] P. P. Vaidyanathan. *Theory and Design of M–Channel Maximally Decimated Quadrature Mirror Filters with Arbitrary M, having Perfect Reconstruction Property.* IEEE Trans. Acoust. Speech Signal Process, 35(2):476–492 1987.

[15] M. Vetterli. Filter banks allowing perfect reconstruction. *Signal Process*, 10:219–244, 1986.

[16] M. Vetterli and D. Le Gall. Perfect reconstruction FIR filter banks: Some properties and factorizations. *IEEE Trans. Acoust. Speech Signal Process*, 37:1057–1071, 1989.

Canadian Mathematical Society
Conference Proceedings
Volume **29**, 2000

Weyl functions and singular continuous spectra of self – adjoint extensions

J. F. Brasche, M. M. Malamud, and H. Neidhardt

Dedicated with great pleasure to Sergio Albeverio on the occasion of his 60th birthday.

ABSTRACT. Let A be the orthogonal sum of infinitely many pairwise unitarily equivalent symmetric operators with strictly positive deficiency numbers and gap J. For every perfect set F there exists a self – adjoint extension $\tilde{A}$ of A such that

$$\sigma_{sc}(\tilde{A}) \cap J = F \cap J.$$

1. Introduction

In his celebrated work [**K**] M. Krein has, in particular, given first results on the spectra of self – adjoint extensions of symmetric operators with finite deficiency indices. Recently S. Albeverio, J. F. Brasche, H. Neidhardt and J. Weidmann have created an inverse spectral theory for self – adjoint extensions ([**ABN**], [**B95**] – [**BNW**]) of symmetric operators with infinite deficiency indices. As in other areas, e. g. the theory of rank – one – perturbations (cf., e. g., [**AK**], [**dRJLS**] – [**GS**], [**G**],[**L**] – [**SW**]) and the theory of Schrödinger operators, one has obtained many affirmative results on the absolutely continuous spectra and the eigenvalue distributions but only little is known about the singular continuous spectra.
For a subset F of $\mathbb{R}$ there exists a self – adjoint operator R such that the singular continuous spectrum of R equals F,

$$\sigma_{sc}(R) = F,$$

if and only if F is perfect, i. e. F is closed and has no isolated points (cf. [**B95**], Remark 6, or [**dRS**]). In [**ABN**] one has conjectured that for every symmetric operator A with infinite deficiency indices and gap J and every perfect set F there exists a self – adjoint extension $\tilde{A}$ of A such that

$$\sigma_{sc}(A) \cap J = F \cap J.$$

In the present note we shall prove that this conjecture is true at least under the additional hypothesis that A equals the orthogonal sum of infinitely many pairwise unitarily equivalent operators.

1991 *Mathematics Subject Classification.* Primary 47A10, 47A60; Secondary 47B25.

In order to achieve this result we shall combine methods from the mentioned articles [**ABN**], [**B95**] – [**BNW**] with results by V. A. Derkach and M. M. Malamud ([**DM91**], [**DM95**]) on boundary triplets and Weyl functions of symmetric operators with infinite deficiency indices. In the present note we shall concentrate on singular continuous spectra as sets. In forthcoming work we shall show that the new tools in the inverse spectral theory of self – adjoint extensions can also be used in order to obtain detailed information about spectral measures.

Notation

Let A be a symmetric operator in the Hilbert space H. $\mathcal{N}_z := \ker(A^* - z)$ denote its deficiency subspaces and $n_\pm(A) := \dim\ker(A^* \mp i)$ its deficiency numbers.

The open interval (α, β) is a gap of the symmetric operator A if for all $f \in D(A)$

$$\begin{array}{rcll} \| (A - \frac{\alpha+\beta}{2})f \| & \geq & \frac{\beta-\alpha}{2} \| f \|, & \text{if} \quad -\infty < \alpha < \beta < \infty, \\ (Af, f) & \geq & \beta \| f \|^2, & \text{if} \quad \alpha = -\infty, \\ (Af, f) & \leq & \alpha \| f \|^2, & \text{if} \quad \beta = \infty. \end{array}$$

$[H_1, H_2]$ denotes the set of bounded linear mappings from the Hilbert space H_1 to the Hilbert space H_2. $[H] := [H, H]$.

Let R be a self – adjoint operator in the Hilbert space H and E_R be its projection – operator valued measure (or the spectral measure, the resolution of identity). We shall denote the distribution function of the measure E_R also by E_R. For every Borel set F in $\mathbb{R}$ we put

$$R_F := R_{| D(R)\cap \operatorname{ran}(E_R(F))}.$$

Let us recall that $\operatorname{ran}(E_R(F))$ is a closed subspace of H and reduces R. Thus R_F can and will be regarded as an operator in the Hilbert space $\operatorname{ran}(E_R(F))$.

For every $f \in H$ the unique measure with distribution function $(E_R(\cdot)f, f)$ is called the spectral measure of f (with respect to R). The space of all f with absolutely continuous, singular continuous and pure point spectral measures is called the absolutely continuous, singular continous and pure point spectral subspace of R and denoted by $H^{ac}(R)$, $H^{sc}(R)$ and $H^{pp}(R)$, respectively. Let us recall that each of these spaces reduces the self – adjoint operator R and

$$R^{ac} := R_{|D(R)\cap H^{ac}(R)},\ R^{sc} := R_{|D(R)\cap H^{sc}(R)},\ R^{pp} := R_{|D(R)\cap H^{pp}(R)},$$

denotes the absolutely continuous part of R, the singular continuous part of R and the pure point part of R, respectively. $\sigma_p(R)$, $\sigma(R)$ and $\rho(R)$ denotes the set of eigenvalues of R, the spectrum of R and the resolvent set of R, respectively.

$$\begin{array}{ll} \sigma_{ac}(R) := \sigma(R^{ac}), & \sigma_{sc}(R) := \sigma(R^{sc}) \\ \sigma_{pp}(R) := \sigma(R^{pp}), & \sigma_c(R) := \sigma_{ac}(R) \cup \sigma_{sc}(R) \end{array}$$

denotes the absolutely continuous spectrum, the singular continuous spectrum, the point spectrum and the continuous spectrum of the self – adjoint operator R, respectively. Note that the set $\sigma_p(R)$ of eigenvalues of R is a dense subset of the point spectrum $\sigma_{pp}(R)$ of the self – adjoint operator R, $\overline{\sigma_p(R)} = \sigma_{pp}(R)$.

Let us recall that for every Borel subset F of G the orthogonal projections $E_R(F)$ and $E_{R_G}(F)$ coincide on $\operatorname{ran}E_R(G)$ provided that G is open. In particular, for open G we have

$$\sigma_{ac}(R) \cap G = \sigma_{ac}(R_G) \cap G, \quad \sigma_{sc}(R) \cap G = \sigma_{sc}(R_G) \cap G$$

and within G the operators R and R_G have the same eigenvalues with the same eigenspaces.

2. Preliminaries

In this subsection we shall collect some of the tools used for the solution of the problem mentioned in the introduction. We refer to [**DM91**], [**DM95**], [**GG**] and [**Ko**] for the proofs.

2.1. Boundary triplets and self – adjoint relations. In what follows A will always denote a closed symmetric operator in the separable Hilbert space H and we shall suppose that A has equal deficiency indices $n_+(A) = n_-(A)$. Our approach to the inverse spectral theory of self – adjoint extensions is based on the concept of boundary triplets, introduced by A. N. Kochubei ([**Ko**]), which may be considered as an abstract version of the second Green's formula.

DEFINITION 1. *A triple $\Pi = \{\mathcal{H}, \Gamma_0, \Gamma_1\}$ consisting of an auxiliary Hilbert space $\mathcal{H}$ and linear mappings*

$$\Gamma_i : D(A^*) \longrightarrow \mathcal{H}, \quad i = 0, 1,$$

is called a boundary triplet for the adjoint A^* of A if the following two conditions are satisfied:
1) (second Green's formula)

$$(A^*f, g) - (f, A^*g) = (\Gamma_1 f, \Gamma_0 g) - (\Gamma_0 f, \Gamma_1 g) \quad \forall f, g \in D(A^*).$$

2) the mapping

$$\Gamma : D(A^*) \longrightarrow \mathcal{H} \oplus \mathcal{H},$$
$$\Gamma f := \{\Gamma_0 f, \Gamma_1 f\} \quad \forall f \in D(A^*),$$

is surjective.

REMARK 2. *(i) The boundary triplet of the adjoint A^* of the symmetric operator A is not unique. Actually this lack of uniqueness is an advantage! In the same way as in analytic geometry a good choice of the coordinate system may make the equations simple, a good choice of a boundary triplet can be very helpful for the solution of different kinds of direct and inverse spectral problems.*
(ii) It is easily seen that for each self – adjoint extension $\tilde{A}$ of A there exists a boundary triplet $\{\mathcal{H}, \Gamma_0, \Gamma_1\}$ such that

$$D(\tilde{A}) = \ker(\Gamma_0).$$

Once a boundary triplet for A^* is fixed one can give explicitly a 1 -1 - correspondence between the set of self – adjoint extensions of A and the set of self – adjoint linear relations in $\mathcal{H}$.
First we recall the definition of self – adjoint relations:

DEFINITION 3. *(i) A closed linear relation Θ in $\mathcal{H}$ is a closed subspace Θ of $\mathcal{H} \oplus \mathcal{H}$.*
(ii) The closed linear relation Θ is symmetric if

$$(g_1, f_2) - (f_1, g_2) = 0 \quad \forall \{f_1, g_1\}, \{f_2, g_2\} \in \Theta.$$

(iii) The closed linear relation Θ is self – adjoint if it is maximal symmetric, i.e., Θ is symmetric and there does not exist a closed symmetric relation $\tilde{\Theta}$ which is a proper superset of Θ.

EXAMPLE 4. *(i) Let B be a self – adjoint operator in $\mathcal{H}$, not necessarily bounded. Then the graph $G(B)$ of B is a self – adjoint relation in $\mathcal{H}$.*
(ii)

$$\Theta_0 := \{0\} \times \mathcal{H}, \quad \Theta_1 := \mathcal{H} \times \{0\},$$

are self – adjoint relations in $\mathcal{H}$. Obviously Θ_0 is not the graph of an operator.

DEFINITION 5. *Let $\{\mathcal{H}, \Gamma_0, \Gamma_1\}$ be a boundary triplet for A^*.*
(i) For every self – adjoint relation Θ in $\mathcal{H}$ we put

$$\begin{aligned} D_\Theta &:= \{f \in D(A^*) | \{\Gamma_0 f, \Gamma_1 f\} \in \Theta\}, \\ A_\Theta &:= A^*_{\restriction D_\Theta}. \end{aligned}$$

(ii) We put

$$A_B := A_\Theta$$

if

$$\Theta = G(B)$$

for some self – adjoint operator B.

The following proposition describes the mentioned $1-1$ – correspondence between the set of self – adjoint extensions of A and the set of self – adjoint linear relations in $\mathcal{H}$.

PROPOSITION 6. *Let $\{\mathcal{H}, \Gamma_0, \Gamma_1\}$ be a boundary triplet for A^*. Then for every self – adjoint relation Θ in $\mathcal{H}$ the operator A_Θ, given by Definition 5 above, is a self – adjoint extension of A.*
The mapping

$$\Theta \mapsto A_\Theta$$

from the set of self – adjoint relations in $\mathcal{H}$ onto the set of self – adjoint extensions of A is bijective.

The self – adjoint extensions $\tilde{A}$ of A preserving the gap (α, β) of A (i. e. $(\alpha, \beta) \subset \rho(\tilde{A})$) play an important role for the investigation of the "spectral properties within the gap (α, β)" of the other self – adjoint extensions of A. A self - adjoint extension preserving a gap (α, β) always exists (we refer to [**BN94**] for a simple characterization of all self – adjoint extensions preserving a gap) and therefore the following holds true:

REMARK 7. *Suppose that the symmetric operator A has a gap (α, β). Then we can choose a self – adjoint extension A_0 of A preserving the gap (α, β). By Remark 2, there exists a boundary triplet $\{\mathcal{H}, \Gamma_0, \Gamma_1\}$ of A^* such that*

$$A_0 = A^*_{\restriction \ker \Gamma_0}.$$

2.2. Weyl functions. It is well known that Weyl functions, introduced by Weyl, are an important tool in the direct and inverse spectral theory of singular Sturm – Liouville operators. The concept of the Weyl function has been generalized (see [**DM91**], [**DM95**]) and now it is also possible to define the Weyl function of a symmetric operator A with infinite deficiency indices $n_+(A) = n_-(A)$. In this subsection we shall recall some known facts about Weyl functions.

DEFINITION 8. *Let* $\{\mathcal{H},\Gamma_0,\Gamma_1\}$ *be a boundary triplet for the operator* A^* *and* A_0 *the self – adjoint extension of* A *such that* $D(A_0)=\ker\Gamma_0$. *The Weyl function of* A *corresponding to the boundary triplet* $\{\mathcal{H},\Gamma_0,\Gamma_1\}$ *is the unique mapping*

$$M:\rho(A_0)\longrightarrow[\mathcal{H}]$$

satisfying

$$\Gamma_1 f_z = M(z)\Gamma_0 f_z \quad \forall f_z\in\mathcal{N}_z(A) \quad \forall z\in\rho(A_0).$$

It is well known that the above implicit definition of the Weyl function is correct. Moreover the function M is continuous, the operator $M(\lambda)$ is self – adjoint for every $\lambda\in\mathbb{R}\cap\rho(A_0)$ and on every gap (α,β) of A_0 the function $(M(\cdot)f,f)$ is strictly increasing for every $f\neq 0$.
The Weyl function immediately provides some information about the "spectral properties" of the self – adjoint extensions of the symmetric operator A.

PROPOSITION 9. *Suppose that* $\{\mathcal{H},\Gamma_0,\Gamma_1\}$ *is a boundary triplet for* A^*, A_0 *the self – adjoint extension of* A *such that* $D(A_0)=\ker\Gamma_0$, M *is the corresponding Weyl function,* Θ *a self – adjoint relation in* $\mathcal{H}$ *and* $\lambda\in\rho(A_0)$. *Then the following holds:*
(i) $\lambda\in\rho(A_\Theta)$ *if and only if* $0\in\rho(\Theta-M(\lambda))$.
(ii) $\lambda\in\sigma_i(A_\Theta)$ *if and only if* $0\in\sigma_i(\Theta-M(\lambda))$ $(i=p,c)$.

In forthcoming work we shall give more detailed results on the relation between the Weyl function and the spectral measures of self – adjoint extensions.

2.3. Singular continuous spectra. In this subsection we shall prove the result mentioned in the introduction. Actually we have the following stronger theorem.

THEOREM 10. *Let* A_{01} *be a symmetric operator in a separable Hilbert space* H_0. *Suppose that* A *has a gap* $J=(\alpha,\beta)$ *and strictly positive deficiency numbers. Suppose also that* $A_{0k}(k\in\mathbb{N})$ *acts in* H_k *and is unitarily equivalent to* A_{01} *and* R *is an arbitrary self – adjoint operator in a separable Hilbert space. Then there exists a self – adjoint extension* $\tilde{A}$ *of the operator*

$$A:=\oplus_{k=1}^{\infty}A_{0k} \qquad in \qquad H:=\oplus_{k=1}^{\infty}H_k$$

such that

$$\tilde{A}_J^{ac}\simeq R_J^{ac},\quad \tilde{A}_J^{pp}\simeq R_J^{pp},\quad \sigma_{sc}(\tilde{A})\cap J=\sigma_{sc}(R)\cap J.$$

I. First we shall prove the following:
1. It suffices to show that for every singular continuous positive Radon measure μ such that the topological support $S(\mu)$ of μ is a compact subset of the gap J and $|S(\mu)|=0$ there exists a self – adjoint extension $\tilde{A}$ of A with the following properties:
(i) $\sigma(\tilde{A})\cap J=S(\mu)$.
(ii) $\tilde{A}$ has no eigenvalues in J
The topological support $S(\mu)$ is the complement of the largest open set G such that $\mu(G)=0$ and $|F|:=\lambda(F)$ denotes the Lebesgue measure of the Borel set F.
2. Without loss of generality we may assume that both deficiency numbers of A_{01} equal one.

ad 1. A can be decomposed as $A = \bigoplus_{n=0}^{\infty} A_n$ where for every n the operator A_n is the orthogonal sum of suitably chosen infinitely many of the operators A_{0k}. Let

$$F := \{E \in \sigma_{sc}(R) | \quad \forall \varepsilon > 0 : |\sigma_{sc}(R) \cap (E - \varepsilon, E + \varepsilon)| > 0\},$$

$$G := \sigma_{sc}(R) \setminus F. \tag{1}$$

Obviously F is the topological support of the absolutely continuous measure $\chi_F \lambda$. Thus, by [**B00**], Theorem 24, there exists a self – adjoint extension $\tilde{A}_0$ of A_0 such that

$$\tilde{A}_{0\,J}^{ac} \simeq R_J^{ac}, \quad \tilde{A}_{0\,J}^{pp} = R_J^{pp}, \quad \sigma_{sc}(\tilde{A}_0) \cap J = F \cap J. \tag{2}$$

Since $\mathbb{R}$ with the Euclidean metric is a separable metric space also $\sigma_{sc}(R)$ with the Euclidean metric is separable. To distinguish between "open in $\mathbb{R}$" and "open in the metric space $\sigma_{sc}(R)$" sets which are open in the metric space $\sigma_{sc}(R)$ will be called "relatively open". Of course a set G_1 is relatively open if and only if $G_1 = G_2 \cap \sigma_{sc}(R)$ for some open subset G_2 of $\mathbb{R}$. Obviously $G \cap J$ is relatively open.

Since $\sigma_{sc}(R)$ is a separable metric space it is also a Lindelöf space, i. e. every relatively open cover has a countable subcover.

For every $E \in G \cap J$ there exists an $\varepsilon_E > 0$ such that

$$\sigma_{sc}(R) \cap [E - \varepsilon_E, E + \varepsilon_E] \subset G \cap J$$

and

$$|\sigma_{sc}(R) \cap [E - \varepsilon_E, E + \varepsilon_E]| = 0.$$

Since $\sigma_{sc}(R)$ is the singular continuous spectrum of a self – adjoint operator it is perfect. Thus we can choose ε_E such that in addition also $\sigma_{sc}(R) \cap [E - \varepsilon_E, E + \varepsilon_E]$ is perfect. Since $G \cap J$ is covered by the relatively open sets $\sigma_{sc}(R) \cap (E - \varepsilon_E, E + \varepsilon_E)$, $E \in G \cap J$, the set $G \cap J$ is also covered by suitably chosen countably many of these sets. Thus $J \cap G$ can be represented as

$$G \cap J = \bigcup_{n \in \mathbb{N}} C_n \tag{3}$$

where for every $n \in \mathbb{N}$ the set C_n is compact and perfect and $|C_n| = 0$.

Thus it suffices to prove that for every $n \in \mathbb{N}$ the operator A_n has a self – adjoint extension $\tilde{A}_n$ such that

$$\sigma_{pp}(\tilde{A}_n) \cap J = \emptyset \tag{4}$$

and

$$\sigma(\tilde{A}_n) \cap J = C_n.$$

In fact since $|C_n| = 0$ it follows that

$$\sigma_{ac}(\tilde{A}_n) \cap J = \emptyset, \quad \sigma_{sc}(\tilde{A}_n) \cap J = C_n, \tag{5}$$

for every $n \in \mathbb{N}$. Thus the operator $\tilde{A} := \bigoplus_{n=0}^{\infty} \tilde{A}_n$ is a self – adjoint extension of A and, by (1) – (5), it has the required properties.

Since for every perfect set C there exists a singular continuous positive Radon measure μ such that $S(\mu) = C$ (cf. [**B95**], Remark 6 or [**dRS**]) we have proved 1. above.

ad 2. Let T be a self – adjoint extension of A_{01} preserving the gap J. By von Neumann's theory there exists a unitary transformation

$$W : \ker(A_{01}^* + i) \longrightarrow \ker(A_{01}^* - i)$$

such that

$$T = A_{01}^{W}$$

in the sense of the second von Neumann formula, i. e.

$$D(T) = \{f + g + Wg | f \in D(\bar{A}_{01}), g \in \ker(A_{01}^* + i)\},$$
$$Th = A_{01}^* h \quad \forall h \in D(T).$$

Here $\bar{B}$ denotes the closure of the operator B.
We choose any vector $e \in \ker(A_{01}^* + i)$, $e \neq 0$, and denote by W_1 the restriction of W to the set of vectors in $\ker(A_{01}^* + i)$ which are orthogonal to e. Then obviously

$$T_1 := A_{01}^{W_1}$$

is a symmetric extension of A_{01}. Since it is a restriction of T the gap J of A_{01} and T is also a gap of T_1.
Moreover both deficiency numbers of T_1 equal 1. In fact, by von Neumanns second formula, we have

(6) $$\operatorname{ran}(T_1 - i) = \{h - ig | \, h \in \operatorname{ran}(A_{01} - i), g \in \ker(A_{01}^* + i), (g, e) = 0\}.$$

Since $e \in \ker(A_{01}^* + i) = \operatorname{ran}(\bar{A}_{01} - i)^{\perp}$ it follows that $e \in \operatorname{ran}(T_1 - i)^{\perp} = \ker(T_1^* + i)$. Conversely let $h \in \operatorname{ran}(T_1 - i)^{\perp} \subset \ker(A_{01}^* + i)$. By (6), $(h, g) = 0$ for all $g \in \ker(A_{01}^* + i)$ satisfying $(g, e) = 0$. Thus h belongs to the space spanned by e. Thus we have shown that $\ker(T_1^* + i)$ equals the one – dimensional space spanned by e. By using that $(g, e) = 0$ if and only if $(Wg, We) = 0$ we can show in an analogous way that $\ker(T_1^* - i)$ equals the one – dimensional space spanned by We.
By the hypothesis of the theorem there exist unitary operators U_k from H_k to H_0 such that $A_{0k} = U_k^* A_{01} U_k \quad \forall k \in \mathbb{N}$. Let $k \in \mathbb{N}$.
Let $\lambda \in \rho(A_{01})$. Since $A_{0k} = U_k^* A_{01} U_k$ we have $U_k A_{0k}^* = A_{01}^* U_k$ and consequently $A_{01}^* U_k f_\lambda^{(k)} = U_k A_k^* f_\lambda^{(k)} = \lambda U_k f_\lambda^{(k)} \quad \forall f_\lambda^{(k)} \in \mathcal{N}_\lambda(A_{0k})$, i. e. $U_k f_\lambda^{(k)} \in \mathcal{N}_\lambda(A_{01}) \, \forall f_\lambda^{(k)} \in \mathcal{N}_\lambda(A_{0k})$. In the same way one obtains $U_k^* f_\lambda^{(1)} \in \mathcal{N}_\lambda(A_{0k}) \, \forall f_\lambda^{(1)} \in \mathcal{N}_\lambda(A_{01})$. Thus U_k maps $\mathcal{N}_\lambda(A_{0k})$ unitarily onto $\mathcal{N}_\lambda(A_{01})$.
Let

$$D(W_k) := \{U_k^* f | \, f \in D(W_1)\},$$
$$W_k h := U_k^* W_1 U_k h \quad \forall h \in D(W_k).$$

Since W_1 is a partial isometry from $\ker(A_{01}^* + i)$ onto $\ker(A_{01}^* - i)$ and U_k maps $\mathcal{N}_\lambda(A_{01})$ unitarily onto $\mathcal{N}_\lambda(A_{0k})$ the operator W_k is a partial isometry from $\ker(A_{0k}^* + i)$ onto $\ker(A_{0k}^* - i)$. It is easily verified that

$$T_k := A_{0k}^{W_k} = U_k^* T_1 U_k.$$

Thus, by replacing A_{0k} by T_k if necessary, we may assume that both deficiency numbers of A_{01} equal 1.
II. By Remark 7, there exists a self – adjoint extension T of A_{01} and a boundary triplet $\Pi_1 = \{\mathcal{H}_1, \Gamma_0^1, \Gamma_1^1\}$ of A_{01}^* such that $ker \Gamma_0^1 = D(T)$ and J is a gap of T. Let $m_1 : \rho(T) \longrightarrow \mathbb{C}$ be the corresponding Weyl function, i.e

$$\Gamma_1^1 f_z = m_1(z) \Gamma_0^1 f_z \quad \forall f_z \in \mathcal{N}_z(A_{01}) \quad \forall z \in \rho(T).$$

Since $J = (\alpha, \beta)$ is a gap of T the Weyl function $m := m_1$ is continuous and strictly increasing within the gap J.
We define the boundary triplet $\Pi_k = \{\mathcal{H}_k, \Gamma_0^k, \Gamma_1^k\}$ by setting

(7) $$\mathcal{H}_k := \mathcal{H}_1, \qquad \Gamma_i^k := \Gamma_i^1 U_k, \qquad (i = 1, 2), \quad \forall k \in \mathbb{N}.$$

Let us calculate the Weyl function m_k corresponding to the boundary triplet $\Pi_k = \{\mathcal{H}, \Gamma_0^k, \Gamma_1^k\}$. In accordance with Definition 8 and taking (7) into account one derives

$$\Gamma_1^1 U_k f_\lambda^{(k)} = m_k(\lambda)\Gamma_0^1 U_k f_\lambda^{(k)} \qquad \forall f_\lambda^{(k)} \in \mathcal{N}_\lambda(A_{0k}). \tag{8}$$

Setting $g_\lambda^{(k)} := U_k f_\lambda^{(k)}$ and observing that $U_k \mathcal{N}_\lambda(A_{0k}) = \mathcal{N}_\lambda(A_{01})$ one rewrites (8) as

$$\Gamma_1^1 g_\lambda = m_k(\lambda)\Gamma_0^1 g_\lambda \qquad \forall g_\lambda \in \mathcal{N}_\lambda(A_{01}). \tag{9}$$

Thus

$$m_k = m \quad \forall k \in \mathbb{N}. \tag{10}$$

Let $\Pi := \{\mathcal{H}, \Gamma_0, \Gamma_1\}$ where

$$\mathcal{H} := \oplus_{k=1}^\infty \mathcal{H}_k, \quad \Gamma_i(\oplus_{k=1}^\infty f_k) := \oplus_{k=1}^\infty \Gamma_i^k f_k \qquad (i = 0, 1). \tag{11}$$

It is easily verified that Π is a boundary triplet for the operator $A^* = \oplus_{k=1}^\infty A_{0k}^*$. Combining (10) and (11) one concludes that the Weyl function M of A corresponding to the boundary triplet Π satisfies

$$M(\lambda)f = m(\lambda)f \quad \forall f \in \mathcal{H}. \tag{12}$$

Let $\mu \neq 0$ be a singular continuous positive Radon measure on $\mathbb{R}$ such that the topological support $C := S(\mu)$ of μ is a compact subset of the gap $J = (\alpha, \beta)$ and $|C| = 0$. Since $\mathcal{H}$ is an infinite dimensional separable Hilbert space there exists a unitary mapping V from $\mathcal{H}$ onto $L_\mu^2 := L^2(C, \mu)$.
We may introduce a new boundary triplet $\tilde{\Pi} = \{L_\mu^2, \tilde{\Gamma}_0, \tilde{\Gamma}_1\}$ with $\tilde{\Gamma}_i := V\Gamma_i (i = 1, 2)$. It is clear that the Weyl function $\tilde{M}$ of A^* corresponding to $\tilde{\Pi}$ coincides with $m(\lambda)I_{L_\mu^2}$. Indeed,

$$\tilde{M}(\lambda) = VM(\lambda)V^* = Vm(\lambda)I_{\mathcal{H}}V^* = m(\lambda)VV^* = m(\lambda)I_{L_\mu^2}. \tag{13}$$

The relation (13) allows us to identify in the sequel $\mathcal{H}$ with L_μ^2 and $\tilde{M}(\lambda) = m(\lambda)I_{L_\mu^2}$ with $M(\lambda)$.
We denote by B the operator in L_μ^2 of multiplication by the function m. Recall that J is a gap of T and m the Weyl function of A_{01} corresponding to a boundary triplet $\{\mathcal{H}_1, \Gamma_0^1, \Gamma_1^1\}$ such that $D(T) = \ker(\Gamma_0^1)$. Thus on J the function m is continuous and strictly increasing and $m(\lambda) \in \mathbb{R}$ for all $\lambda \in J$. Thus the operator B is bounded and self – adjoint and

$$D(B) = L_\mu^2,$$
$$Bf(x) = m(x)f(x) \quad \mu\text{–a.e.} \quad \forall f \in L_\mu^2.$$

Since m is continuous, strictly increasing and real – valued on J and the measure μ is continuous we have that

$$\sigma(B) = \{m(\lambda) |\, \lambda \in C = S(\mu)\}, \quad \sigma_p(B) = \emptyset.$$

Thus

$$\begin{aligned} 0 &\in \sigma(B - M(\lambda)) && \forall \lambda \in S(\mu), \\ 0 &\in \rho(B - M(\lambda)) && \forall \lambda \in J \setminus S(\mu), \\ 0 &\notin \sigma_p(B - M(\lambda)) && \forall \lambda \in J, \end{aligned}$$

and, by Proposition 9, the self – adjoint extension A_B of A, introduced in Definition 5, satisfies

$$\sigma(A_B) \cap J = S(\mu), \quad \sigma_p(A_B) \cap J = \emptyset.$$

References

[ABN] S. Albeverio, J. F. Brasche, H. Neidhardt: On inverse spectral theory for self – adjoint extensions; J. Funct. Anal. 154 (1998), 130 – 173.

[AK] S. Albeverio, P. Kurasov: Rank one perturbations, approximations and self – adjoint extensions; J. Funct. Anal. 148 (1997), 152 – 169.

[B95] J. F. Brasche: On spectra of self – adjoint extensions and generalized Schrödinger operators. Habilitationsschrift. Ruhr – Universität Bochum, 1995.

[B00] J. F. Brasche: On inverse spectral theory for self–adjoint extensions: Nowhere dense singular continuous spectra and Hausdorff dimension of spectra; to appear in J. Op. Theory 43, 2000.

[BN96] J. F. Brasche, H. Neidhardt: On the singular continuous spectrum of self–adjoint extensions; Math. Zeitschr. 222 (1996), 533–542.

[BN95] J. F. Brasche, H. Neidhardt: On the absolutely continuous spectrum of self–adjoint extensions; J. Funct. Analysis 131, (1995), 364–385.

[BNW] J. F. Brasche, H. Neidhardt, J. Weidmann: On the point spectrum of self–adjoint extensions; Math.Zeitschr. 214 (1993), 343–355.

[BN94] J. F. Brasche, H. Neidhardt: Some Remarks on Krein's extension Theory; Math. Nachr. 165 (1994), 159–181.

[DM91] V. A. Derkach, M. M. Malamud: Generalized resolvents and the boundary value problems for Hermitean operators with gap; J. Funct. Anal. 95 (1991), 1 – 95.

[DM95] V. A. Derkach, M. M. Malamud: The extension theory of Hermitean operators and the moment problem. J. Math. Sc. 73,2 (1995), 141 – 242.

[dRJLS] R. del Rio, S. Jitomirskaya, Y. Last, B. Simon: Operators with singular continuous spectrum IV. Hausdorff dimensions, rank one perturbations, and localization; J. D'Analyse Mathematique 69, (1996), 153 – 200.

[DMS] R. del Rio, N. Makarov, B. Simon: Operators with singular continuous spectrum II. Comm. Math. Phys.164 (1994), 59 – 67.

[dRS] R. del Rio, B. Simon: Point spectrum and mixed spectral types for rank one perturbations; Proc. Amer. Math. Soc. 125 (1997), 3593 – 3599.

[GS] F. Gesztesy, B. Simon: Rank one perturbations at infinite coupling; J. Funct. Anal. 128 (1995), 242 – 252.

[GG] M. L. Gorbachuk, V. I. Gorbachuk: Boundary Value Problems for Operator – Differential Equations; Naukova Dumka, Kiev, 1984. Engl. transl.: Soviet Series 48, Dordrecht, Kluwer Acad. Publ. 1991.

[G] A. Gordon: Pure point spectrum under 1 – parameter perturbations and instability of Anderson localization; Comm. Math. Phys. 164 (1994), 489 – 505.

[Ko] A. N. Kochubei: Extensions of symmetric operators and of symmetric binary relations, Mat Zametki 17, 41 – 48 (1975) (in Russian).

[K] M. G. Krein: Theory of self – adjoint extensions of semi – bounded Hermitean transformations and its applications; Mat. Sbornik 2062, 431 – 495 (1947) (in Russian).

[L] Y. Last: Quantum dynamics and decompositions of singular continuous spectra. J. Funct. Anal. 142, 406–445 (1996).

[S] B. Simon: Operators with singular continuous spectrum: I. General operators. Annals of Math. 141, 131–145 (1995).

[SW] B. Simon, T. Wolff: Singular continuous spectrum under rank one perturbations and localization for random Hamiltonians Comm. Pure Appl. Math. 39 (1986), 75 – 90.

Institute of Applied Mathematics, University of Bonn, Wegelerstr. 10, 53115 Bonn, Germany
E-mail address: brasche@riemann.iam.uni-bonn.de

Departement of Mathematics, Universitetskaya str. 24, Donetsk 340055, Ukraine
E-mail address: mmm@univ.donetsk.ua

Institute of Mathematics, University of Potsdam, Am Neuen Palais 10, 14469 Potsdam, Germany
E-mail address: neidhard@rz.uni-potsdam.de

Canadian Mathematical Society
Conference Proceedings
Volume **29**, 2000

Irregular Spectral Asymptotics

Jochen Brüning

Dedicated to Sergio Albeverio

Abstract. In this note we derive the resolvent expansion for elliptic operators with irregular singularities in the coefficients, like Laplace type operators on manifolds with metric horns. The main technical tool is a generalization of the "Singular Asymptotics Lemma" of Brüning and Seeley.

1. Introduction

The resolvent trace expansion for elliptic operators is a well established tool with many important applications in Geometric Analysis, see e.g. [Gi] and its bibliography. It is known for a long time in the case of elliptic differential operators on compact manifolds [MP], [S1] and differential boundary value problems [S2], [Gre]. Recently, this expansion has been extended fully to pseudodifferential elliptic operators and boundary value problems [GruS], [Gru].

The method has been developed further to cover spaces with singularities, notably of conic type, including families of cones [Ch1], [BS2], [BS3]. Here, operators of Laplace type are considered which are most interesting in applications. Typically, near a conic singularity such an operator takes the form

$$\Delta^\circ = -\left(\frac{\partial}{\partial x}\right)^2 + x^{-2}A(x), \quad x \in (0, \varepsilon_0], \tag{1.1}$$

where A is a family of Laplace type operators on the base of the cone (a compact smooth manifold), and A varies smoothly with $x \in [0, \varepsilon_0]$. The next step towards more complicated (isolated) singularities will introduce operators of the form

$$\Delta^\circ = -\left(\frac{\partial}{\partial x}\right)^2 + x^{-2\alpha_0}A(x), \quad x \in (0, \varepsilon_0], \tag{1.2}$$

where $\alpha_0 \geq 1$ and A is no longer smooth but admits an asymptotic expansion in powers of x, as $x \to 0$. This comprises the case of "metric horns" which refers to Riemannian manifolds with metric singularities of the form $dx^2 + x^{2\alpha_0}g_N$ on $(0, \varepsilon_0] \times N$, with (N, g_N) a compact Riemannian manifold; in this case we have

$$A(x) = x^{-2\alpha_0}A_0 + x^{-\alpha_0 - 1}A_1 + x^{-2}A_2. \tag{1.3}$$

Various aspects of this class of singular spaces have been studied, like Hodge theory [Ch2] and index theorems [LPe]. In [B1], we have announced the full asymptotic expansion of the heat kernel for the Laplacian (Thm 1.2) and have used it to derive a Signature Theorem (Thm.1.3). The proof of the expansion result was based on an extension of the Singular Asymptotics Lemma (SAL) (Thm.2.2) which was introduced in [BS1] (cf. also the "push forward lemma" of Melrose [M]), a very convenient tool to produce asymptotic expansions in singular situations which admit scaling. Callias [Ca] has given a version of the SAL not requiring smoothness up to $x = 0$.

In this note we supply the proofs of [B1, Thms.1.2 and 2.2] and, at the same time, provide substantial generalizations. The main observation concerns the structure of the SAL. Recall that it deals with a function, $\sigma(x, \xi)$, of two variables $x, \xi > 0$, and derives an asymptotic expansion of the function

$$I(\sigma; z) := \int_0^\infty \sigma(x, xz)dx, \tag{1.4}$$

as $z \to \infty$. To do so one assumes a differentiable asymptotic expansion of σ as $\xi \to \infty$, with coefficients in $\mathcal{S}(\mathbb{R}_+)$ (as functions of x), and smoothness in $x \geq 0$, in order to use the Taylor expansion at $x = 0$; both expansions are matched by a special "integrability condition". The emphasis on smoothness at $x = 0$ links the SAL in this form closely to the special structure of the Taylor expansion, whereas the expansion at ∞ is unrestricted.

The main generalization we introduce here is that we drop the differentiability assumptions and put the expansions at 0 and at ∞ on an equal footing. This leads us to different remainder estimates which replace the "Integrability Condition" in [BS1]. It seems that they are more easily verified, but for the time being we do not see the precise relationship between both types of remainder estimates. The price we have to pay for now is the exclusion of "small eigenvalues" which can be easily treated, however, by ode methods. It should be noted that our expansion includes a new term which is not present if we assume smoothness at $x = 0$.

In Section 2, we prove Theorem 1, the generalization of the SAL, which was announced (in a more restrictive form) in [B1, Thm.1.2]. In Section 3, we use this result to deduce Theorem 2, as announced in [B1, Thm.2.2]. Thanks are due to the referee who pointed out a gap in the proof of Theorem 1.

2. The Generalized Singular Asymptotics Lemma

We begin with some terminology which is essentially taken from [L, Ch. II]. By Γ we denote a discrete closed subset of $\mathbb{C}$ with

$$\underline{\gamma} := \inf \text{Re } \gamma > -\infty, \tag{2.1}$$

and

$$\Gamma_N := \{\gamma \in \Gamma; \text{Re } \gamma \leq N\} \tag{2.2}$$

finite for all $N \in N$, and equipped with a "weight function" $\nu : \Gamma \to \mathbb{Z}_+$. In addition, we write $\Gamma_N^* := \Gamma_N \cap \{\text{Re } z > 0\}$. Then we consider the following types of asymptotic expansions.

(1) For $f \in L^1_{\text{loc}}(0,1]$, we want the expansion

$$f(x) = \sum_{\substack{\alpha\in\Gamma_N \\ k\le\nu(\alpha)}} f^0_{\alpha k} x^\alpha \log^k x + R^0_N(f;x), \tag{2.3}$$

with certain numbers $f^0_{\alpha k} \in \mathbb{C}$ and remainder estimate

$$|R^0_N(f;x)| \le C_{N\delta} x^{N-\delta}, \tag{2.4}$$

uniformly in $x \in (0,1]$, for every $\delta > 0$. This class of functions we denote by $\mathcal{F}^0_{\Gamma,\nu}$.

(2) For $f \in L^1_{\text{loc}}[1,\infty)$, we want the expansion

$$f(x) = \sum_{\substack{\beta\in\Gamma_N \\ l\le\nu(\beta)}} f^\infty_{\beta l} x^{-\beta} \log^l x + R^\infty_N(f;x), \tag{2.5}$$

with certain numbers $f^\infty_{\beta l} \in \mathbb{C}$ and remainder estimate

$$|R^\infty_N(f;x)| \le C_{N,\delta} x^{\delta-N}, \tag{2.6}$$

uniformly in $x \in [1,\infty)$, for every $\delta > 0$. This class of functions we denote by $\mathcal{F}^\infty_{\Gamma,\nu}$. Then we put

$$\mathcal{F}_{\Gamma,\nu} := \{f \in L^1_{\text{loc}}(0,\infty); f_0 := f \mid (0,1] \in \mathcal{F}^0_{\Gamma,\nu}, f_\infty := f \mid [1,\infty) \in \mathcal{F}^\infty_{\Gamma,\nu}\}; \tag{2.7}$$

this is the function space we are interested in. We can define the Mellin transform on $\mathcal{F}_{\Gamma,\nu}$ by

$$\mathcal{M}f(z) := \mathcal{M}f_0(z) + \mathcal{M}f_1(z). \tag{2.8}$$

$\mathcal{M}f$ is meromorphic in $\mathbb{C}$, hence we can define the "regularized integral" by

$$⨍_0^\infty f(x)dx := \text{Res}_0 \mathcal{M}f(1), \tag{2.9}$$

where "Res_0" denotes the constant term in the Laurent expansion. This is the appropriate notion for our purposes. We also employ the notion of "two-variable asymptotic expansion", associated with Γ and ν, as explained e.g. in [B2]. By this we mean an asymptotic expansion for functions, f, in $L^1_{\text{loc}}((0,1]\times(0,1])$ of the form

$$f(\varepsilon,\omega) = \sum_{\substack{\alpha,\beta\in\Gamma_N \\ k\le\nu(\alpha),l\le\nu(\beta)}} f^{kl}_{\alpha\beta} \varepsilon^\alpha \log^k \varepsilon\ \omega^\beta \log^l \omega + R_N(f;\varepsilon,\omega), \tag{2.10}$$

with certain $f^{kl}_{\alpha\beta} \in \mathbb{C}$ and remainder estimate

$$|R_N(f;\varepsilon,\omega)| \le C_{N\delta}(\varepsilon+\omega)^{N-\delta}, \tag{2.11}$$

uniformly in $0 < \varepsilon,\omega \le 1$, for every $\delta > 0$. We will mainly use the fact that (2.10) determines the coefficients $f^{kl}_{\alpha\beta}$ uniquely. We are interested in the asymptotic expansion of expressions of the form

$$I(\sigma;z) := ⨍_0^\infty \sigma(x,xz)dx, \quad 1 \le z \to \infty, \tag{2.12}$$

where σ is in $C((0,\infty)\times(0,\infty),\mathbb{C})$. To ensure the existence of the integral and its asymptotic expansion, we introduce two assumptions. (A_0) For any $C_0>0$ and $0<x\le C_0,\ \xi\in(0,\infty)$, $N\in\mathbb{N}$, we have the expansion

$$\sigma(x,\xi)=\sum_{\substack{\alpha\in\Gamma_N\\ k\le\nu(\alpha)}}\sigma^0_{\alpha k}(\xi)x^\alpha\log^k x+R^0_N(x,\xi),$$

where $\sigma^0_{\alpha k}\in\mathcal{F}_{\Gamma,\nu}$ and, for $0<\delta\le1$ and certain numbers $\alpha^0_0\le0,\alpha^0_\infty\ge0$,

$$|R^0_N(x,\xi)|\le C(N,\delta,C_0)x^{N-\delta}\xi^{\alpha^0_0}(1+\xi)^{\alpha^0_\infty}.$$

(A_∞) For any $C_0>0$ and $\xi\ge1/C_0$, $x\in(0,\infty)$, $N\in\mathbb{N}$ we have the expansion

$$\sigma(x,\xi)=\sum_{\substack{\beta\in\Gamma_N\\ l\le\nu(\beta)}}\sigma^\infty_{\beta l}(x)\xi^{-\beta}\log^l\xi+R^\infty_N(x,\xi),$$

where $\sigma^\infty_{\beta l}\in\mathcal{F}_{\Gamma,\nu}$ and, for $0<\delta\le1$ and certain numbers $\alpha^\infty_0\le0$, $\alpha^\infty_\infty\ge0$,

$$|R^\infty_N(x,\xi)|\le C(N,\delta,C_0)\xi^{\delta-N}x^{\alpha^\infty_0}(1+x)^{\alpha^\infty_\infty}.$$

We point out the following consequence of (A_0), (A_∞): as $x\to0$, $\xi\to\infty$ we obtain the expansion, from (A_0),

$$\sigma(x,\xi)\sim\sum_{\substack{\alpha,\beta\in\Gamma\\ k\le\nu(\alpha),l\le\nu(\beta)}}[\sigma^0_{\alpha k}]^\infty_{\beta l}\ x^\alpha\log^k x\ \xi^{-\beta}\log^l\xi,$$

and from (A_∞),

$$\sigma(x,\xi)\sim\sum_{\substack{\alpha,\beta\in\Gamma\\ k\le\nu(\alpha),l\le\nu(\beta)}}[\sigma^\infty_{\beta l}]^0_{\alpha k}\ x^\alpha\log^k x\ \xi^{-\beta}\log^l\xi.$$

By uniqueness of the expansion coefficients, we conclude

$$[\sigma^0_{\alpha k}]^\infty_{\beta l}=[\sigma^\infty_{\beta l}]^0_{\alpha k}=:\sigma_{\alpha k,\beta l}. \tag{2.13}$$

Now we can formulate the first result of this paper.

THEOREM 2.1. *Let $\sigma\in C((0,\infty)\times(0,\infty),\mathbb{C})$. Under the assumptions (A_0) and (A_∞) we have the following asymptotic expansion, as $1\le z\to\infty$:*

$$I(\sigma;z)\sim\sum_{\substack{\alpha\in\Gamma\\ k\le\nu(\alpha)}}\fint_0^\infty(x/z)^\alpha\log^k(x/z)\sigma^0_{\alpha k}(x)dx/z \tag{2.14}$$

$$+\sum_{\substack{\beta\in\Gamma\\ l\le\nu(\beta)}}\fint_0^\infty(xz)^{-\beta}\log^l(xz)\sigma^\infty_{\beta l}(x)dx \tag{2.15}$$

(2.16)

$$+\sum_{\substack{\alpha\in\Gamma\\ k\le\nu(\alpha)\\ 0\le k''\le k,0\le k'\le\nu(\alpha+1)}}z^{-\alpha-1}\log^{k+k'+1}z\binom{k}{k''}(-1)^{k-k'+1}(k+k'+1)^{-1}[\sigma^0_{\alpha k}]^0_{-\alpha-1,k'}$$

$$(2.17)\qquad + \sum_{\substack{\beta\in\Gamma\\ k\leq\nu(\beta+1),l\leq\nu(\beta)\\ 0\leq k'\leq k}} z^{-\beta}\log^{k+l+1} z\binom{k}{k'}(-1)^{k-k'}(k'+l+1)^{-1}\sigma_{\beta l;\beta-1,k}.$$

In particular, $I(\sigma;z)\in\mathcal{F}^{\infty}_{\Gamma,\tilde{\nu}}$ *with* $\tilde{\nu}(\alpha)=\nu(\alpha)+\nu(\alpha+1)+1$.

REMARKS. Comparing with [BS1] we see that the terms (Ia) and (Ib) there - which equal (2.14) and (2.15) here - are now seen in a more symmetric way; indeed, our condition (A_0) replaces the Taylor expansion used before. Also, the fairly subtle integrability condition [BS1, (1.2b)] and the remainder estimate [BS1, (1.2a)] are now replaced by the rather straightforward estimates in (A_0) and (A_∞). They are enough for the application we have in mind here but it is not clear at the moment how they relate to the previous assumptions in general. In particular, it is not obvious how they imply the vanishing of (2.16). The systematic use of the regularized integral throughout also adds some transparency, as pointed out in [L]. In particular, we see that the "singular" contributions (2.17) - which corresponds to [BS1, (Ic)] - and the new contribution (2.16) arise simply from the formula for the change of variables [L, Lemma 2.1.4]; this was pointed out in [L,Ch.II].

PROOF. We clearly need to split the integral into a part where x is small, $x\leq C_0$, and one where x is large, $xz\geq 1/C_0$, in order to use the different asymptotics (A_0), (A_∞). This will be the case if we split at $x=1/C_0z$ where the choice of C_0 is arbitrary. We take advantage of this by introducing two new expansion variables:

$$(2.18)\qquad \varepsilon\in(0,1] \text{ and } w:=\varepsilon z\geq 1.$$

Then we write

$$\begin{aligned} I(\sigma;z) &= \,⨍_0^\infty(\phi(xw)+(1-\phi(xw)))\sigma(x,xz)dx \\ (2.19)\qquad &=: I_0(\sigma;\varepsilon,w)+I_\infty(\sigma;\varepsilon,w), \end{aligned}$$

with $0\leq\phi\leq 1,\phi\in C_0^\infty(-2,2)$ and $\phi\mid[-1,1]=1$. We will show that I_0 and I_∞ admit asymptotic expansions as $\varepsilon+w^{-1}\to 0$, which will imply the theorem. Beginning with I_0, we obtain from (A_0) and [L, Lemma 2.1.4.], since $x\leq 2$ on the support of the integrand,

$$\begin{aligned} I_0(\sigma;\varepsilon,w) &= ⨍_0^\infty \phi(xw)\sigma(x,xz)dx \\ &= ⨍_0^\infty \phi(xw)\left[\sum_{\substack{\alpha\in\Gamma_N\\ k\leq\nu(\alpha)}} x^\alpha\log^k x\ \sigma^0_{\alpha k}(xz)+R^0_N(x,xz)\right]dx \\ (2.20)\qquad &=: \sum_{\substack{\alpha\in\Gamma_N\\ k\leq\nu(\alpha)}}\{⨍_0^\infty \phi(x\varepsilon)(x/z)^\alpha\log^k(x/z)\sigma^0_{\alpha k}(x)dx/z \end{aligned}$$

(2.21)
$$+\sum_{\substack{\alpha\in\Gamma_N\\ k\le\nu(\alpha)\\ 0\le k''\le k, 0\le k'\le\nu(\alpha+1)}} z^{-\alpha-1}\log^{k+k'+1} z\binom{k}{k''}(-1)^{k-k'+1}(k+k'+1)^{-1}[\sigma^0_{\alpha k}]^0_{-\alpha-1,k'}$$

(2.22)
$$+\tilde{R}^0_N(\sigma;\varepsilon,w).$$

Now, (2.21) already gives the first singular contribution, (2.16). (2.20) we rewrite as

(2.23)
$$\sum_{\substack{\alpha\in\Gamma_N\\ k\le\nu(\alpha)}}[\fint_0^\infty (x/z)^\alpha\log^k(x/z)\sigma^0_{\alpha k}(x)dx/z$$

(2.24)
$$-\fint_0^\infty(1-\phi(x\varepsilon))(x/z)^\alpha\log^k(x/z)\sigma^0_{\alpha k}(x)dx/z],$$

such that (2.23) gives (2.14), while (2.24) will be dealt with later. Using (A_0) again, we estimate the remainder (2.24), with $\delta=1/2$:

$$|\tilde{R}^0_N(\sigma;\varepsilon,w)| \le C_N\int_0^\infty \phi(xw)x^{N-1/2}(xz)^{\alpha^0_0}(1+xz)^{\alpha^0_\infty}dx$$

(2.25)
$$\le C_N\ z^{\alpha^0_0+\alpha^0_\infty}\ w^{-N-1/2-\alpha^0_0}.$$

I_∞ is now treated analogously. Since $xz\ge\varepsilon^{-1}\ge1$ on the support of the integrand, we can use (A_∞) to get

$$I_\infty(\sigma;\varepsilon,w) = \fint_0^\infty(1-\phi(xw))\left[\sum_{\substack{\beta\in\Gamma_N\\ l\le\nu(\beta)}}\sigma^\infty_{\beta l}(x)(xz)^{-\beta}\log^l xz+R^\infty_N(x,xz)\right]dx$$

(2.26)
$$=: \sum_{\substack{\beta\in\Gamma_N\\ l\le\nu(\beta)}}\{\fint_0^\infty(xz)^{-\beta}\log^l xz\ \ \sigma^\infty_{\beta l}(x)dx$$

(2.27)
$$-\fint_0^\infty\phi(xw)(xz)^{-\beta}\log^l xz\ \ \sigma^\infty_{\beta l}(x)dx\}$$

(2.28)
$$+\ \tilde{R}^\infty_N(\sigma;\varepsilon,w).$$

Again, (2.26) gives the desired term (2.15); and the remainder we estimate using (A_∞) and, again, $\delta=1/2$:

$$\begin{aligned}
|\tilde{R}^\infty_N(\sigma;\varepsilon,w)| &\le C_N\int_{w^{-1}}^\infty (xz)^{1/2-N}x^{\alpha^\infty_0}(1+x)^{\alpha^\infty_\infty}dx\\
&= C_N\int_{\varepsilon^{-1}}^\infty z^{-1-\alpha^\infty_0}x^{1/2-N+\alpha^\infty_0}(1+x/z)^{\alpha^\infty_\infty}dx\\
&\le C_N\ z^{-1-\alpha^\infty_0}\ \varepsilon^{N-3/2-\alpha^\infty_0-\alpha^\infty_\infty}
\end{aligned}$$

(2.29)
$$\le C_N\ z^{-\alpha^\infty_0}\ \varepsilon^{N-(3/2+\alpha^\infty_0+\alpha^\infty_\infty)}.$$

It remains to deal with the terms (2.27) and (2.24). In the integrand of (2.24) we have $x \geq \varepsilon^{-1}$, so we plug in the expansion of $\sigma^0_{\alpha k}$ as $x \to \infty$; the resulting remainder of order $L > N$ can be estimated by

$$C_{LN} \int_{\varepsilon^{-1}}^{\infty} z^{1/2-\alpha} \; x^{\alpha-L-1/2} dx = C_{LN} \; z^{1/2-\alpha} \; \varepsilon^{L-\alpha-3/2} \leq C_{LN} \; z^{1/2-\underline{\gamma}} \; \varepsilon^{L-N-3/2}.$$

Using [L, II (1.12)], this leads to the expansion

$$\begin{aligned} & - \sum_{\substack{\alpha\in\Gamma_N \\ k\leq\nu(\alpha)}} ⨍_0^{\infty} (1-\phi(x\varepsilon))(x/z)^{\alpha} \log^k(x/z)\sigma^0_{\alpha k}(x)dx/z \\ &= \sum_{\substack{\alpha\in\Gamma_N,\beta\in\Gamma_L \\ k\leq\nu(\alpha),l\leq\nu(\beta)}} [\sigma^0_{\alpha k}]^{\infty}_{\beta l} ⨍_0^{\infty} \phi(x\varepsilon)(x/z)^{\alpha} \log^k(x/z) x^{-\beta} \log^l x \; dx/z \\ &\quad + \; O_{NL}(z^{1/2-\underline{\gamma}}\varepsilon^{L-N-3/2}). \end{aligned} \tag{2.30}$$

In the integrand of (2.27) we have $x \leq 2w^{-1} \leq 2$, so we can use the expansion of $\sigma^{\infty}_{\beta l}$ as $x \to 0$. Then we obtain quite similarly:

$$\begin{aligned} & - \sum_{\substack{\beta\in\Gamma_N \\ l\leq\nu(\beta)}} ⨍_0^{\infty} \phi(xw)(xz)^{-\beta} \; \log^l(xz)\sigma^{\infty}_{\beta l}(x)dx \\ &= - \sum_{\substack{\beta\in\Gamma_N,\alpha\in\Gamma_L \\ l\leq\nu(\beta),k\leq\nu(\alpha)}} [\sigma^{\infty}_{\beta l}]^0_{\alpha k} ⨍_0^{\infty} \phi(x\varepsilon z)(xz)^{-\beta} \log^l(xz) x^{\alpha} \log^k x dx \\ &\quad + \; O_{NL}(z^{1/2-\underline{\gamma}} w^{1/2-(L-N)}). \end{aligned} \tag{2.31}$$

But now we see from (2.13) and [L, Lemma 2.1.4 and II(2.12)] again that the sums in (2.30) and (2.31) with $\alpha, \beta \in \Gamma_N$ add up to the term

$$\sum_{\substack{\beta\in\Gamma_N \\ k\in\nu(\beta+1),l\leq\nu(\beta)}} \sigma_{\beta l;\beta-1,k} \; z^{-\beta} \; \log^{k+l+1} z \sum_{k'=0}^{k} (-1)^{k-k'} \binom{k}{k'} (k'+l+1)^{-1}, \tag{2.32}$$

which coincides with the second singular contribution, (2.17). It now remains to estimate the terms in (2.30) with $N < \beta \leq L$, and in (2.31) with $N < \alpha \leq L$. For a term from (2.30), we have $\alpha \leq N < \beta$. We estimate with [L,Ch.2,(1.6)]

$$\begin{aligned} & |⨍_0^{\infty} \phi(x\varepsilon)(x/z)^{\alpha} \log^k(x/z) x^{-\beta} \log^l x \; dx/z| \\ \leq \; & C_{NL} z^{-\alpha-1} \varepsilon^{\beta-\alpha-1}(1+|\log^k \varepsilon|) \log^k z \\ \leq \; & C_{NL} w^{-1} z^{-\alpha+1/2} \varepsilon^{(\beta-\alpha)/2}. \end{aligned}$$

If $\alpha \geq N/2$, we can estimate this by

$$C_{NL} z^{-(N-1)/2} \leq C_{NL} \; \varepsilon^{(N-1)/2},$$

and for $\alpha < N/2$, implying $\beta - \alpha > N/2$, by

$$C_{NL} z^{1/2-\underline{\gamma}} \; \varepsilon^{N/4},$$

hence altogether, with $D_1 := \max\{0,\, 1/2 - \underline{\gamma}\}$ and for $N \geq 2$:

$$
\begin{aligned}
&|\int_0^\infty \phi(x\varepsilon)(x/z)^\alpha \log^k(x/z) x^{-\beta} \log^l x \; dx/z| \\
(2.33) \qquad &\leq \; C_{NL} \; z^{D_1} \varepsilon^{N/4}.
\end{aligned}
$$

A similar reasoning gives for a term in (2.31), with $N \geq 2$ and $\beta \leq N < \alpha$

$$
\begin{aligned}
&|\int_0^\infty \phi(xw)(xz)^{-\beta} \log^l(xz) x^\alpha \log^k x \; dx| \\
(2.34) \qquad &\leq \; C_{NL}(z^{D_1} w^{-N/4} + \varepsilon^{(N-1)/2}).
\end{aligned}
$$

Putting now $L = 2N$ and collecting all the remainder terms, (2.25), (2.29), (2.30), (2.31), (2.33), and (2.34), we see that there is a problem only if one of the quantities

$$D_1, \; D_2 := \alpha_0^0 + \alpha_\infty^0, \; D_3 := -\alpha_0^\infty, \; D_4 := -\underline{\gamma},$$

is positive. Thus, we put

$$D := \sum_{i=1}^{4} |D_i|$$

and consider the expansion of $\tilde{I}(\sigma; z) := z^{-D} I(\sigma; z)$ resulting from our considerations. Then all remainder terms are small, so we obtain an asymptotic expansion of $\tilde{I}$ hence also of I. The proof is complete. □

3. Irregular Singular Equations

We want to apply Theorem 1 to obtain the resolvent expansion of certain elliptic equations with irregular singularities. The prototypical example is given by the differential operator

$$\Delta^\circ := -\left(\frac{\partial}{\partial x}\right)^2 + \frac{a}{x^{2\alpha_0}}, \; x > 0, \tag{3.1}$$

with domain $C_0^2(0,\infty)$, here $a > 0$ and $\alpha_0 \geq 1$. Δ° is symmetric in $L^2(\mathbb{R}_+)$ and bounded from below, hence we can consider its Friedrichs extension, Δ, which is self-adjoint and bounded from below with the same lower bound. The expansion of the resolvent trace (or, almost equivalently, of the heat trace) for Δ has been announced by Callias [Ca]. In [B1], we have announced a different proof, in the spirit of 'regular singular analysis', and have applied it to derive the Signature Theorem for metric horns. Here we supply the missing details, in a more general and probably also more accessible form then studied before.

In order to apply the results to singular spaces like metric horns, we generalize (3.1) to an operator valued setting. Thus we consider a Hilbert space, H, a dense subspace, H_1, and a family of self-adjoint operators in H, $A(x)$, defined for $x > 0$, with common domain H_1. We assume in addition the following properties

$$(0,\infty) \ni x \mapsto A(x) \in \mathcal{L}(H_1, H) \text{ is smooth,} \tag{3.2}$$

$$A(x) \geq a > 0 \text{ for all } x > 0. \tag{3.3}$$

The smoothness of A in $\mathbb{R}_+$ - which holds in the conic case - is replaced by assuming an asymptotic expansion

(3.4) $$A(x) = A_0 + \sum_{\gamma \in \Gamma^*_{N-1}} A_\gamma x^\gamma + R_N(x),$$

where all A_γ are defined and symmetric on H_1, and

(3.5) $$|R_N(x)| \leq C_N x^{\mu_n} A_0^{1-\delta_N},$$

for some $\delta_N \in (0,1]$ and some sequence $\mu_N \nearrow \infty$ (note that A_0 is self-adjoint with domain H, and $A_0 \geq a$). We also assume, with $\mathcal{C}_p$ the Schatten-von Neumann class of order p, that

(3.6) $$A_0^{-1} \in \mathcal{C}_p(H)$$

for some $p > 0$. Then we can consider the analogue of (3.1),

(3.7) $$\triangle^\circ := -\left(\frac{\partial}{\partial x}\right)^2 + \frac{A(x)}{x^{2\alpha_0}},$$

with domain $C_0^2((0,\infty), H_1)$, and also its Friedrichs extension, $\triangle$, in the Hilbert space $\mathcal{H} := L^2(\mathbb{R}_+, H)$. Because $\mathbb{R}_+$ is not compact, the resolvent may not be trace class. To remedy this, we choose $\phi \in C_0^\infty(-2,2)$ with $\phi = 1$ in a neighbourhood of 0 and consider for $z \geq 1$ and $l \in \mathbb{N}$, l sufficiently large, the function

(3.8) $$r(z) := \mathrm{tr}_{\mathcal{H}}[\phi(\triangle + z^2)^{-l}].$$

We will show that r is well defined and admits an asymptotic expansion in powers of z and $\log z$, as $z \to \infty$; this will be a consequence of Theorem 1. To put r into the framework of the SAL we use the natural scaling, as in [BS 2,3], and the Trace Lemma from [BS 2]. Starting with the latter, we have the following result.

LEMMA 3.1 *For $l \geq p + 1/2$, $(\triangle + z^2)^{-l}$ has an operator kernel, $(\triangle + z^2)^{-l}(x,y)$, on $\mathbb{R}_+$ such that*

(3.9) $$(\triangle + z^2)^{-l}(x,x) \in \mathcal{C}_1(H), \quad x > 0.$$

Moreover,

(3.10) $$r(z) = \int_0^\infty \phi(x)\mathrm{tr}_H[(\triangle + z^2)^{-l}(x,x)]dx.$$

PROOF. We write $\mathcal{S} := C_0^\infty(\mathbb{R}_+)$ and note that, by construction,

(3.11) $$\psi\mathcal{D}(\triangle) \subset \mathcal{D}(\triangle) \quad \text{for } \psi \in \mathcal{S}.$$

We also write $\psi_2 > \psi_1$, for $\psi_1, \psi_2 \in \mathcal{S}$, if $\psi_2 = 1$ in a neighbourhood of supp ψ_1. Then, by the Trace Lemma, it is enough to prove that

(3.12) $$\psi_2(\triangle + z^2)^{-l}\psi_1 \in \mathcal{C}_1(\mathcal{H}),$$

for all such ψ_1 and ψ_2 in $\mathcal{S}$ if l is large enough. Thus, we fix $\psi_i \in \mathcal{S}$, $1 \leq i \leq 3$, with $\psi_3 > \psi_2 > \psi_1$ and supp $\psi_3 \subset [0,T)$, for some $T > 0$. We want to prove that

(3.13) $$\psi_2(\triangle + z^2)^{-l}\ \psi_1 \in \mathcal{C}_{(p+1/2)/l}(\mathcal{H}),$$

with p from (3.6). To show this for $l = 1$, we denote by $\triangle_T$ the Friedrichs extension of $\triangle^0$ in $L^2([0,T],H)$ with domain $C_0^2((0,T),H_1)$. Then we obtain from the max-min principle and (3.4), (3.5) the estimate

$$\begin{aligned} \| (\triangle_T + C_1)^{-1} \|^{p+1/2}_{\mathcal{C}_{p+1/2}(\mathcal{H}_T)} &\leq C_2 \sum_{i,j\geq 1} \left(\frac{4\pi^2}{T^2} i^2 + \lambda_j\right)^{-(p+1/2)} \\ &\leq C_2 \sum_{j\geq 1} \int_0^\infty \left(\frac{4\pi^2}{T^2} x^2 + \lambda_j\right)^{-(p+1/2)} dx \\ &\leq C_3 \| A_0^{-1} \|_{\mathcal{C}_p(H)}, \end{aligned} \tag{3.14}$$

where we have written $\mathcal{H}_T = L^2((0,T),H)$, $(\lambda_j)_{j\geq 1} = \operatorname{spec} \mathrm{A}_0$, and the constants C_i depend only on A_0 and p. Now we find by standard arguments

$$\begin{aligned} \psi_2(\triangle + z^2)^{-1}\psi_1 &= \psi_2(\triangle_T + z^2)^{-1}\psi_1 \\ &+ \psi_2(\triangle_T + z^2)^{-1}[\triangle_T, \psi_3](1-\psi_2)(\triangle + z^2)^{-1}\psi_1. \end{aligned} \tag{3.15}$$

Next we observe the easy a priori estimate

$$\| [\triangle_{(T)}, \psi] u \|_{\mathcal{H}_{(T)}} \leq C(\psi)((\triangle_{(T)} u, u) + \| u \|^2_{\mathcal{H}_{(T)}}), \tag{3.16}$$

which holds for any $\psi \in \mathcal{S}_{(T)}$ and $u \in C_0^2((0,\infty),H_1)$, $(C_0^2((0,T),H_1))$, where $\mathcal{S}_T = C_0^\infty[0,T]$. But then it is readily seen that $(\triangle_{(T)} + I, \mathcal{S}_{(T)})$ satisfy the axioms (Co1) through (Co5) in [B3, Sec.4]; in particular, from (Co2) we see that

$$(\triangle_T + I)^{-1/2}[\triangle_T, \psi_3] \text{ is bounded in } \mathcal{H}_T. \tag{3.17}$$

Thus, we derive from (3.14), (3.15), and (3.17) that

$$\psi_1(\triangle + z^2)^{-1}\psi_1 \in \mathcal{C}_{2p+1}(\mathcal{H}),$$

hence

$$\psi_1(\triangle + z^2)^{-1/2} \in \mathcal{C}_{4p+2}(\mathcal{H}).$$

But then it follows from [B3, Lemma 4.1] that

$$(1-\psi_2)(\triangle + z^2)^{-1}\psi_1 \in \mathcal{C}_q(\mathcal{H}), \text{ for all } q > 0,$$

which gives, with (3.15), (3.13) for $l = 1$. If we have proved this for some $l \geq 1$, then we write

$$\begin{aligned} \psi_2(\triangle + z^2)^{-l-1}\psi_1 &= \psi_2(\triangle + z^2)^{-l}\psi_3(\triangle + z^2)^{-1}\psi_1 \\ &+ \psi_2(\triangle + z^2)^{-l}(1-\psi_3)(\triangle + z^2)^{-1}\psi_1. \end{aligned}$$

Using [B3, Lemma 4.1] again and the Hölder inequality for $\mathcal{C}_p$-norms, we complete the induction; hence the assertion. □ The scaling mentioned above is effected

through the unitary family

$$U_\varepsilon u(X) := \varepsilon^{1/2} u(\varepsilon x), \ \varepsilon > 0, \ u \in \mathcal{H}. \tag{3.18}$$

Clearly, with $\alpha_0 =: 1 + \beta_0$,

$$\begin{aligned} U_\varepsilon \triangle^\circ U_\varepsilon^\star &=: \varepsilon^{-2}\left(-\partial_x^2 + \varepsilon^{-2\beta_0}\frac{A(\varepsilon x)}{x^{2\alpha_0}}\right) \\ &= \varepsilon^{-2}\triangle_\varepsilon^\circ. \end{aligned} \tag{3.19}$$

It follows that for the Friedrichs extension, $\triangle_\varepsilon$, we have

$$\varepsilon^2 U_\varepsilon \triangle U_\varepsilon^* = \triangle_\varepsilon. \tag{3.20}$$

It follows also that

$$\varepsilon^{-2l} U_\varepsilon (\triangle + z^2)^{-l} U_\varepsilon^* = (\triangle_\varepsilon + (\varepsilon z)^2)^{-l}, \tag{3.21}$$

and for the kernels

$$\varepsilon^{1-2l}(\triangle + z^2)^{-l}(\varepsilon x, \varepsilon y) = (\triangle_\varepsilon + (\varepsilon z)^2)^{-l}(x,y). \tag{3.22}$$

Thus, we obtain with Lemma 3.1 (and $x = y = 1, \varepsilon = y$)

$$r(z) = \int_0^\infty y^{2l-1}\phi(y)\mathrm{tr}_H[(\triangle_y + (yz)^2)^{-l}(1,1)]dy. \tag{3.23}$$

In the conic case (when $\alpha_0 = 1$) we have applied the SAL directly to (3.9), cf. [BS2, Sec.7]. To be able to do this also in this case, we rewrite our operator as

$$\triangle_y + (yz)^2 = -\left(\frac{\partial}{\partial x}\right)^2 + y^{-2\beta_0}\left(\frac{A(yx)}{x^{2\alpha_0}} + (yz^{1/2\alpha_0})^{2\alpha_0}\right). \tag{3.24}$$

This suggests that we introduce $\tilde{z} := z^{1/2\alpha_0}$ and write

$$\tilde{r}(\tilde{z}) = r(z) =: \int_0^\infty \sigma(y, y\tilde{z})dy, \tag{3.25}$$

where

$$\sigma(y,\zeta) = y^{2l-1}\phi(y)\mathrm{tr}_H[(-(\frac{\partial}{\partial x})^2 + y^{-2\beta_0}(\frac{A(yx)}{x^{2\alpha_0}} + \zeta^{2\alpha_0}))^{-l}(1,1)]. \tag{3.26}$$

In what follows, it is convenient to write

$$\tilde{A}(x) := \frac{A(x)}{x^{2\alpha_0}}, \tag{3.27}$$

such that

$$y^{-2\beta_0}\frac{A(yx)}{x^{2\alpha_0}} = y^2\tilde{A}(yx) =: \tilde{A}_y(x). \tag{3.28}$$

We also write

$$P(y,\xi;\zeta) := (\xi^2 + y^2\tilde{A}(y) + \zeta^{2\alpha_0}/y^{2\beta_0})^{-1}. \tag{3.29}$$

Now, for any fixed $y > 0$, we have to determine the resolvent kernel expansion of the elliptic operator with operator coefficients, (3.24), at a *regular* point, in order to verify the conditions of Theorem 1. To do so, we have to add some assumptions on resolvent expansions for the operator family $\tilde{A}$. First of all, we assume now that the expansion (3.4) can be differentiated any number of times, i.e. that for $j \in \mathbb{Z}_+$ we have

$$A^{(j)}(x) = \sum_{\gamma\in\Gamma_N^*} A_\gamma j!\binom{\gamma}{j}x^{\gamma-j} + R_{Nj}(x),$$

where, with μ_N and δ_N as above,

$$|R_{Nj}(x)| \le C_N x^{\mu_N - j} A_0^{1-\delta_N}. \tag{3.30}$$

Moreover, we will have to consider expressions of the form

$$I_\alpha(A, B_1, \dots, B_{j-1}; x, \eta)$$

$$:= \mathrm{tr}_H[(A(x)+\eta^2)^{-\alpha_1}B_1(A(x)+\eta^2)^{-\alpha_2}B_2\cdots B_{j-1}(A(x)+\eta^2)^{-\alpha_j}]. \tag{3.31}$$

Here, $\alpha \in \mathbb{N}^j$ is a multiindex satisfying

$$|\alpha| \geq j+p, \tag{3.32}$$

and the operators B_i are closed with domain containing H_1; in addition, we assume that

$$A_0^{-1}B_i \in \mathcal{L}(H), \tag{3.33}$$

and that there is a locally convex space, $\mathcal{B}$, such that

$$B_i \in \mathcal{B}, \ \ 1 \leq i \leq j. \tag{3.34}$$

We observe that (3.30), (3.32), and (3.33) imply that (3.31) is, indeed, well defined. If we now add the assumption that

$$A_\gamma \in \mathcal{B} \ \text{and} \ R_N(x) \in \mathcal{B}, \tag{3.35}$$

for all $\gamma \in \Gamma^*$, $x > 0$, $N \in \mathbb{N}$, then we see from (3.4,5) that, for x sufficiently small, the Neumann series

$$\begin{aligned}(A(x)+\eta^2)^{-1} &= (A_0+\eta^2)^{-1}(I+R_1(x)(A_0+\eta^2)^{-1})^{-1}\\ &= \sum_{j\geq 0}(-1)^j(A_0+\eta^2)^{-1}(R_1(x)(A_0+\eta^2)^{-1})^j\end{aligned} \tag{3.36}$$

is asymptotic in $\mathcal{C}_1(H)$ as $\eta \to \infty$, in the sense that the trace norm of a suitable remainder decays faster than any given power of $(1+\eta)^{-1}$. Hence the η-expansion of I_α is reduced to the expansion of

$$I_\alpha^0(B_1,\cdots,B_{j-1};\eta) :=$$

$$\mathrm{tr}_H[(A_0+\eta^2)^{-\alpha_1}B_1(A_0+\eta^2)^{-\alpha_2}B_2\ldots B_{j-1}(A_0+\eta^2)^{-\alpha_j}], \tag{3.37}$$

for $\alpha \in \mathbb{N}^j$, $|\alpha| \geq j+p$, and $B_i \in \mathcal{B}$ (possibly also depending on x). Thus, we are lead to the following assumption:

There is an asymptotic expansion of the form

$$\begin{aligned}I_\alpha^0(B_1,\ldots,B_{j-1};\eta) &= \sum_{\substack{\gamma\in\Gamma_N\\ m\leq\nu(\gamma)}} I_{\alpha,\gamma m}^0(B_1,\ldots,B_{j-1})\eta^\gamma\log^m\eta\\ &+ R_N(B_1,\ldots,B_{j-1};\eta),\end{aligned} \tag{3.38}$$

where the multilinear forms $I_{\alpha,\gamma m}^0$ and $\eta^{\mu_N}R_N(B_1,\ldots,B_{j-1};\eta)$ are bounded on bounded sets in $\mathcal{B}$, uniformly in $\eta \geq 0$.

Note that in view of (3.3) we can in fact allow $\eta \geq 0$ for the resolvent estimates. With these preparations we can turn to the expansion theorem we are after.
We begin with the following lemma which follows e.g. from the material in [BS2, Sec.2]; a detailed treatment will be given in a forthcoming publication.

LEMMA 3.2 *For $l \geq p$, $j \geq 1$, and $\alpha \in \mathbb{N}_+^j$, $\beta \in \mathbb{Z}_+^{j-1}$ there are homogeneous polynomials $p_{\alpha\beta}$ in $\mathbb{R}$, of degree $d_{\alpha\beta}$, such that for $y, \zeta > 0$*

$$\begin{aligned}\sigma(y,\zeta) &= \sum_{\substack{1\leq j\leq N\\ \alpha,\beta\in\mathcal{M}_j}} y^{2l-1}\phi(y)\int_{\mathbb{R}} p_{\alpha\beta}(\xi)\mathrm{tr}_H[P(y,\xi;\zeta)^{\alpha_1}y^{2+\beta_1}\tilde{A}^{(\beta_1)}(y)..\\ &\quad ..\ y^{2+\beta_{j-1}}\tilde{A}^{(\beta_{j-1})}(y)P(y,\xi;\zeta)^{\alpha_j}]d\xi\\ &\quad + R_N(y,\zeta). \end{aligned} \tag{3.39}$$

Here, the index set is

(3.40)
$$\mathcal{M}_j := \{(\alpha,\beta)\in\mathbb{N}_+^j\times\mathbb{Z}_+^{j-1}; |\alpha|\geq l+j-1, d_{\alpha\beta}-2|\alpha| = -2(l+j-1)-|\beta|\},$$

and we have the uniform estimate

$$|R_N(y,\zeta)| \leq C_N(\|\tilde{A}(y)^{-1}A_{2\alpha}\|_H)(a/2+\zeta/y^{2\beta})^{-N}, \tag{3.41}$$

with some continuous function $C_N : \mathbb{R}_+ \to \mathbb{R}_+$. Now we are ready for the proof of our main result. THEOREM 2. *Assume that the operator family, $A(x)$, satisfies the assumptions (3.2) through (3.6) and (3.30), (3.35), and (3.38). Then the function σ introduced in (3.26) satisfies the assumptions of Theorem 1. This implies a complete asymptotic expansion for*

$$r(z) = \mathrm{tr}_{\mathcal{H}}[\phi(\triangle+z^2)^{-l}] \text{ as } z\to\infty,$$

whenever $l \geq p+1/2$. In particular, $r \in \mathcal{F}^{\infty}_{\Gamma,\tilde{\nu}}$, with $\tilde{\nu}$ as in Theorem 1.

PROOF. We have to analyze the expansion of σ given in Lemma 3.2. Note first that, since supp $\phi \subset [-2,2]$, we derive from (3.41) the remainder estimate

$$|R_N(y,\zeta)| \leq C_N(C_0)y^{2N\beta_0}(a/2+\zeta)^{-N}, \tag{3.42}$$

uniformly in $\zeta > 0$. This estimate is good for (A_0) and (A_∞) since $\sigma(y,\zeta) = 0$ for $y > 2$. Hence it only remains to verify that the individual terms in the sum (3.39) satisfy the assumptions (A_0) and (A_∞). We begin with (A_0). Using (3.3), and (3.40), we obtain for a typical term in (3.20) the form

$$\begin{aligned} & y^{2l-1+2\beta_0|\alpha|}\phi(y)\int_{-\infty}^{+\infty} p_{\alpha\beta}(\xi)\tilde{I}_\alpha(y,y^{\beta_0}\xi,\zeta)d\xi \\ &= y^{2l-1+\beta_0(2|\alpha|-d_{\alpha\beta}-1)}\phi(y)\int_{-\infty}^{+\infty} p_{\alpha\beta}(\xi)\tilde{I}_\alpha(y,\xi,\zeta)d\xi, \end{aligned} \tag{3.43}$$

where

$$\tilde{I}_\alpha(y,\tilde{\xi},\zeta)$$
$$:= I_\alpha(A(y)-a/2, y^{2+\beta_1}\tilde{A}^{(\beta_1)},\ldots,y^{2+\beta_{j-1}}\tilde{A}^{(\beta_{j-1})}(y);y,(\tilde{\xi}^2+\zeta^{2\alpha_0}+a/2)^{1/2}).$$

Now, we observe that, from (3.27) and (3.30),

$$y^{2+i}\tilde{A}^{(i)}(y) = O_i(y^{-2\beta_0}), \tag{3.44}$$

and admits a full asymptotic expansion with coefficients in $\mathcal{B}$. Plugging this into (3.43) and using (3.36), we obtain an asymptotic expansion as $y \to 0+$, with typical term

$$y^\delta \phi(y) \int_{\mathbb{R}} p(\xi) I_\alpha^0(B_1, \dots, B_{k-1}; (\xi^2 + \zeta^{2\alpha_0} + a/2)^{1/2}) d\xi, \tag{3.45}$$

for some δ and some polynomial p, and with A_0 replaced by $A_0 - a/2$. Using (3.38) we see that each term of the form (3.45) is in $\mathcal{F}_{\Gamma,\nu}$, as required. For (A_∞), we use (3.36) and (3.38) first, as $\zeta \to \infty$, and plug in the asymptotic expansions, as $y \to 0+$, for the operators (3.28) in each term. This completes the proof. □
We remark in conclusion that that condition (3.5) is unnecessarily restrictive, even

though it suffices to deal with Laplace type operators in metric horns, as explained in the Introduction. It is fairly obvious how to modify the assumptions on the functions I_α in (3.38) in such a way that they are satisfied if $A(y)$ is a family of elliptic differential operators on a compact manifold, and still imply Theorem 2. It would be desirable, however, to find a more general and less clumsy set of conditions also implying this result. Let us finally spell out the result for the "metric horns".
COROLLARY 3. *Denote by $\triangle$ the Friedrichs extension of the Laplacian on forms*

on a manifold with a metric horn, as in (1.3). Then we have for $l > m/2$

$$\operatorname{tr}(\triangle + z^2)^{-l} \in \mathcal{F}^\infty_{\Gamma,\nu},$$

where, with $\alpha_0 =: 1 + \beta_0$,

$$\Gamma := \{k + l\beta_0; k, l \in \mathbb{Z}, k \geq 2l - m, l \geq 0\},$$

and $\nu(\alpha) \equiv 1$. PROOF: Since the conic case has been dealt with in [Ch1] and

[BS2], we assume $\beta_0 > 0$. In view of (1.3), Theorem 2 applies if we can prove that $A(x) \geq a > 0$ for x near 0. The explicit form of the coefficients in (1.3), given in [B1, (2.4)] shows that this holds if $\ker A_2 = 0$. Otherwise, the full operator function $A(x)$ is reduced by $\ker A_2 \cap \ker A_0$ which splits the problem as a sum of a "small eigenvalue part" and a "large eigenvalue part". To the latter, Theorem 2 applies, and the former is dealt with by standard ode methods. □

References

[B1] J. Brüning: The signature theorem for manifolds with metric horns. Proceedings "Equations aux derivées partielles", St.Jean de Monts, 1996

[B2] J. Brüning: On the asymptotic expansion of some integrals. Arch. Math. **42** (1984), 253-259

[BS1] J. Brüning and R.T. Seeley: Regular singular asymptotics. Advances in Math. **58** (1985), 133-148

[BS2] J. Brüning and R.T. Seeley: The resolvent expansion for regular singular operators. J. Funct. Anal. **73** (1987), 369-429

[BS3] J. Brüning and R.T. Seeley: The expansion of the resolvent near a singular stratum of conic type. J. Funct. Anal. **95** (1991), 255-290

[Ca] C.J. Callias: On the existence of small-time heat expansions for operators with irregular singularities in the coefficients. Math. Res. Lett. **2** (1995), 129-146

[Ch1] J. Cheeger: Spectral geometry of singular Riemannian spaces. J. Diff. Geom. **18** (1983), 575-657

[Ch2] J. Cheeger: On the Hodge theory of Riemannian pseudomanifolds. In: Proc. Sympos. Pure Math. **36** (1980), 21-45

[Gi] P. Gilkey: Invariance theory, the heat equation, and the Atiyah-Singer index theorem. Boca Raton 1995

[Gre] P. Greiner: An asymptotic expansion for the heat equation. Arch. Rat. Mech. Anal. **41** (1971), 163-218

[Gru] G. Grubb: A weakly polyhomogeneous calculus for pseudodifferential boundary problems. University of Copenhagen, Institute for Mathematical Sciences, Preprint **13** (1999)

[GruS] G. Grubb and R.T. Seeley: Weakly parametric pseudodifferential operators and Atiyah-Patodi-Singer boundary problems. Inventiones Math. **121** (1995), 481-529

[L] M. Lesch: Operators of Fuchs type, conical singularities, and asymptotic methods. Teubner-Texte zur Mathematik **136**, Stuttgart, Leipzig 1997

[LPe] M. Lesch and N. Peyerimhoff: On index formulas for manifolds with metric horns. Commun. Partial Differ. Equations **23** (1998), 649-684

[M] R.B. Melrose: Calculus of conormal distributions on manifolds with corners. Internat. Math. Res. Notices **3** (1992), 51-61

[MPl] S. Minakshisundaram and Å. Pleijel: Some properties of the eigenfunctions of the Laplace operator on Riemannian manifolds. Can. J. Math. **1** (1949), 242-256

[S1] R.T. Seeley: Complex powers of an elliptic operator. In: Proceedings Sympos. Pure Math. **10** (1968), 308-315

[S2] R.T. Seeley: The resolvent of an elliptic boundary value problem. Amer. J. Math. **91** (1969), 963-983

Canadian Mathematical Society
Conference Proceedings
Volume **29**, 2000

The Lagrangian Gauge Problem in Classical Mechanics and Group Cohomology

L. Cattaneo and U. Cattaneo

Ga sarà bé quaicoss da nüm / che la pas di selvi e dal lagh / insegolàda da vus e da riciàm / la tegnarà da cünt / par quand la carezzarà / la memoria di altri.
da F. GRIGNOLA, *I noss compleanni*

A Sergio in occasione del suo 60mo compleanno.

ABSTRACT. A cohomology class of differentiable 1-cocycles of a second countable Lie group G with values in the quotient by the constants of a suitable topological vector space of real-valued functions is associated univocally with every equivalence class of G-gauge functions defined on a G-homogeneous manifold. In addition, an equivalence class of differentiable 2-cocycles of G with values in **R** is associated with this class too. An extension group of G which leaves strictly invariant particular extensions of G-gauge invariant Lagrangians is determined. The construction of a representative of every equivalence class of G-gauge functions is exemplified in two simple cases.

1. Introduction

In Lagrangian mechanics, the motion of a system is described in its configuration space, a differentiable manifold Q on which operates a group of diffeomorphisms. We have a Lagrangian dynamical system whenever a (real-valued) function $\mathcal{L}$, the Lagrangian of the system, is defined on the tangent bundle T(Q) of Q. However, this Lagrangian is not uniquely determined by the dynamics of the system: the addition to $\mathcal{L}$ of a total derivative, with respect to an evolution parameter, of a "gauge function" defined on Q does not modify the Euler-Lagrange equations of motion.

When the Lagrangian is G-gauge invariant, i.e., invariant up to gauges depending on a group G, then G is a symmetry group of the system since the equations of motion are invariant under the action on Q of any $g \in$ G. If, in addition, G is a Lie group and Q a G-homogeneous manifold, we obtain G-gauge functions, namely, C^∞-mappings of G into a suitable topological vector space of functions on Q. These mappings are 1-cocycles on G up to constants. It is possible to associate

1991 *Mathematics Subject Classification.* Primary 70H33, 57T10; Secondary 70D10, 22E41.

with every G-gauge function a real-valued 2-cocycle on G^2 in such a way that, given these 2-cocycles, we can reconstruct the G-gauge functions.

The algebraic association of 2-cocycles with G-gauge functions defined on a G-homogeneous space was done by Lévy-Leblond [**LL**], without using explicitly group cohomology and overlooking on purpose topological questions; in particular, without proving that the constructed functions are differentiable. In the present paper, we complete the work of Lévy-Leblond and we show that, in every equivalence class of G-gauge invariant Lagrangians on T(Q), there exists an extended Lagrangian which is left invariant under the action of a central extension group E of G by $\mathbf{R}^n$, where n is the dimension of the vector space of equivalence classes of admissible 2-cocycles. This means that, in every equivalence class, we can find a Lagrangian $\mathfrak{L}$ on T(Q) and a Lagrangian $\widetilde{\mathfrak{L}}$ on $T(\tilde{Q})$, with $\tilde{Q} \supset Q$, so that $\widetilde{\mathfrak{L}} | T(Q) = \mathfrak{L}$ and $\widetilde{\mathfrak{L}}$ is E-invariant.

In Section 2, after some preliminaries on differentiable G-modules and group cohomology, we prove that, when G is a second countable Lie group and Q a G-homogeneous manifold, the action of G on G-gauge invariant Lagrangians on T(Q) singles out particular real-valued functions on Q of class C^∞, namely, G-gauge functions. We define the vector space of equivalence classes of G-gauge functions and, with every one of these classes, we associate univocally a cohomology class of differentiable 1-cocycles of G with values in the quotient by the constants of a suitable topological vector space of real-valued functions, as well as a proper equivalence class of differentiable 2-cocycles of G with values in $\mathbf{R}$.

For each admissible 2-cocycle, we define in Section 3 an extended Lagrangian which is (strictly) invariant under a central extension group of G by $\mathbf{R}$. Moreover, we find a central extension group of G, independent on the particular 2-cocycles, which leaves invariant an extended Lagrangian in every equivalence class of G-gauge invariant Lagrangians.

In Section 4, the construction of the G-gauge functions is exemplified in the well-known cases of the free 1-particle system in the 4-dimensional space-time under the action of the restricted Poincaré and Galilei groups.

Only a brief sketch of the proofs will be given in this paper. The details, and a generalized version of the theory, will be published later.

In the following,

- every group G, which is not a priori Abelian, is written multiplicatively and its neutral element denoted by e_G; the additive notation and the neutral element 0 are used for Abelian groups;
- every Lie group is tacitly understood to be *real* and *finite-dimensional*;
- we denote by $[f]$ the equivalence class of any mapping f modulo a given equivalence relation, left unexpressed if there is no possibility of misunderstanding;
- the Borel structures considered are always those generated by the closed sets;
- the symbol $\cong$ stays for a vector space isomorphism.

We remember that a Polish group is a second countable topological group which is metrically topologically complete.

2. G-gauge invariance and group cohomology

2.1. Differentiable G-modules and cohomology. Let V be a locally convex topological vector space and Ψ a continuous linear representation of a Lie group G on V, i.e., a linear representation Ψ of G on V such that the mapping $(g, v) \mapsto \Psi(g)v$ of $G \times V$ into V is continuous. Equipped with the representation Ψ, the space V becomes a topological G-module V_Ψ, which we write simply V if $\Psi(g) = \mathrm{Id}_V$ (the identity mapping on V) for all $g \in G$. Let V_∞ be the vector subspace of all differentiable vectors for Ψ, i.e., the vector subspace of all $v \in V$ such that the mapping $g \mapsto \Psi(g)v$ of G into V is of class C^∞. If $V = V_\infty$, the continuous representation Ψ and the topological G-module V_Ψ are said to be *differentiable.*

We denote by $\mathcal{C}^k(Q, \mathbf{R})$ the topological vector space of all k-times continuously differentiable real-valued functions on a (finite-dimensional) C^∞-manifold Q $(1 \leqslant k \leqslant +\infty)$ with the topology of uniform convergence of the functions and of their derivatives of order $\leqslant k$ on compact subsets. It is well-known that $\mathcal{C}^k(Q, \mathbf{R})$ is a Fréchet space and is separable, as may be shown by adapting the proof for k-times continuously differentiable real-valued functions on an open subset of $\mathbf{R}^n$ ([**DI**], 17.1.2). It follows that the quotient topological vector space $\mathcal{C}^k(Q, \mathbf{R})/\iota(\mathbf{R})$, where ι is the natural injection of $\mathbf{R}$ into $\mathcal{C}^k(Q, \mathbf{R})$ sending $r \in \mathbf{R}$ into the constant function with the value r, is also a separable Fréchet space because $\iota(\mathbf{R})$ is closed in $\mathcal{C}^k(Q, \mathbf{R})$. We use the symbol $\mathcal{C}^k_{/\mathbf{R}}(Q, \mathbf{R})$ for this space.

We consider the representation Φ of G on $\mathcal{C}^k(Q, \mathbf{R})$ $(1 \leqslant k \leqslant +\infty)$ defined by

$$(\Phi(g)f)(q) = f(g^{-1} \cdot q) \tag{2.1}$$

for all $q \in Q$, where the dot "$\cdot$" denotes the left action of G on the local coordinates of Q. This representation is differentiable since the mapping $g \mapsto \Phi(g)f = f(g^{-1} \cdot .)$ of G into the barelled space $\mathcal{C}^k(Q, \mathbf{R})$ is of class C^1 for all $f \in \mathcal{C}^k(Q, \mathbf{R})$ ([**WA**], 4.1.1 and Proposition 4.4.1.7); therefore $\mathcal{C}^k(Q, \mathbf{R})_\Phi$ $(1 \leqslant k \leqslant +\infty)$ is a separable differentiable Fréchet G-module. The quotient representation $\tilde{\Phi}$ of G on $\mathcal{C}^k_{/\mathbf{R}}(Q, \mathbf{R})$, obtained from Φ by passing to the quotient, is differentiable because the mapping $g \mapsto \tilde{\Phi}(g)[f] = [\Phi(g)f]$ is of class C^1 for all $f \in \mathcal{C}^k(Q, \mathbf{R})$ and so $\mathcal{C}^k_{/\mathbf{R}}(Q, \mathbf{R})_{\tilde{\Phi}}$ $(1 \leqslant k \leqslant +\infty)$ is also a separable differentiable Fréchet G-module.

In the following, we will use the just considered differentiable G-modules only for $k = 3$, but our results remain valid for $1 \leqslant k \leqslant +\infty$. We shall often identify $\mathbf{R}$ and $\iota(\mathbf{R})$ via ι.

Let G be a Lie group and V_Ψ a differentiable Fréchet G-module. We consider the cochain complex $(C_\infty(G, V_\Psi), \delta)$ of the differentiable cohomology of G with values in V_Ψ, or, as we shall say, of the differentiable cohomology of V_Ψ. The cochain space of this complex is the $\mathbf{Z}$-graded vector space $C_\infty(G, V_\Psi)$ whose homogeneous elements of degree $n > 0$ are the normalized C^∞-mappings f of G^n into V, while $C^0_\infty(G, V_\Psi) = V$ and $C^n_\infty(G, V_\Psi) = \{0\}$, when $n < 0$. Here and in the following, we say that a mapping f of G^n into V is "normalized" when $f(g_1, \cdots, g_n) = 0$ whenever $g_i = e_G$ for some $i \in \{1, \cdots, n\}$. The coboundary operator δ is a $\mathbf{Z}$-graded linear mapping of degree 1 such that $\delta \circ \delta = 0$ and is defined by

$$\begin{aligned} \delta f(g_1, \cdots, g_{n+1}) &= (-1)^{n+1} f(g_1, \cdots, g_n) + \Psi(g_1) f(g_2, \cdots, g_{n+1}) \\ &+ \sum_{i=1}^{n} (-1)^i f(g_1, \cdots, \hat{g}_i, g_i g_{i+1}, g_{i+2}, \cdots, g_{n+1}) \end{aligned}$$

with $f \in C^n_\infty(G, V_\Psi)$ and $n \geqslant 0$, where $\hat{g}_i$ means that g_i is omitted. In particular, we shall need the operator δ for $n=0$, $n=1$, and $n=2$:

$$\delta f(g) = -f + \Psi(g) f,$$
$$\delta f(g, g') = f(g) + \Psi(g) f(g') - f(gg'),$$

$$\delta f(g, g', g'') = -f(g, g') + \Psi(g) f(g', g'') - f(gg', g'') + f(g, g'g'')$$

for all g, g', g'' in G. From the cochain complex $(C_\infty(G, V_\Psi), \delta)$ we get the **Z**-graded vector spaces

$$Z_\infty(G, V_\Psi) = \operatorname{Ker} \delta, \ \ B_\infty(G, V_\Psi) = \operatorname{Im} \delta, \ \ H_\infty(G, V_\Psi) = Z_\infty(G, V_\Psi)/B_\infty(G, V_\Psi),$$

which are, respectively, the vector spaces of differentiable cocycles, coboundaries, and cohomology of V_Ψ. If, instead of C^∞-mappings, we consider mappings which are only continuous (respectively, Borel) and a G-module V_Ψ topological but not necessarily differentiable, we have the cochain complex $(C_c(G, V_\Psi), \delta)$ of the continuous (respectively, $(C_b(G, V_\Psi), \delta)$ of the Borel) cohomology of V_Ψ and the relative vector spaces of cocycles, coboundaries, and cohomology. If we forget the topology, we have the cochain complex $(C(G, V_\Psi), \delta)$ of the cohomology of V_Ψ and the vector spaces of cocycles, coboundaries, and cohomology. We denote by δ the coboundary operators of all these cochain complexes since their algebraic expressions are the same.

2.2. G-gauge invariant Lagrangians. A *Lagrangian dynamical system on* Q is an ordered pair $(Q, \mathfrak{L}_0)$, where Q is the extended configuration space and $\mathfrak{L}_0$ is the Lagrangian of the system. The space Q is the direct product manifold $Q_0 \times \mathbf{R}$, with Q_0 a C^∞-manifold (the configuration space), and $\mathfrak{L}_0$ is a real-valued function on $T(Q_0) \times \mathbf{R}$ of class C^2, with $T(Q_0)$ the tangent bundle of Q_0. A mapping $\gamma : \mathbf{R} \to Q$ of class C^1 is a motion in $(Q, \mathfrak{L}_0)$ if γ is an extremal of the functional

$$\gamma \mapsto \int_I \mathfrak{L}_0(\dot{\gamma}) dt\,,$$

where I is a compact time interval and $\dot{\gamma}(t) \in T_{\gamma(t)}(Q_0)$, the tangent space to Q_0 at $\gamma(t)$. The equations of motion of the system are the Euler-Lagrange equations of this variational problem.

The Lagrangian $\mathfrak{L}_0$ of a dynamical system on Q is not uniquely determined by the equations of motion: every Lagrangian $\widehat{\mathfrak{L}}_0$ given by

$$\widehat{\mathfrak{L}}_0\left(q^0, \frac{dq^0}{dt}, t\right) = \mathfrak{L}_0\left(q^0, \frac{dq^0}{dt}, t\right) + \frac{df}{dt}(q^0, t)$$

in the local coordinates $\left(q^0, \dfrac{dq^0}{dt}, t\right)$ of $T(Q_0) \times \mathbf{R}$, with f a real-valued function on $Q_0 \times \mathbf{R}$ of class C^3, determines the same equations of motion as $\mathfrak{L}_0$ and is said to be *gauge equivalent to* $\mathfrak{L}_0$. From now on, instead of $(Q, \mathfrak{L}_0)$, we will consider the ordered pair $(Q, \mathfrak{L})$, where the Lagrangian $\mathfrak{L}$ on $T(Q)$ is the real-valued function of class C^2 given, in local coordinates $(q, \dot{q})$ with $q = (q^0, t)$, by

$$\mathfrak{L}(q, \dot{q}) = \mathfrak{L}_0\left(q^0, \frac{\dot{q}^0}{\dot{t}}, t\right) \dot{t},$$

the dot denoting derivation with respect to an evolution parameter τ. Therefore, two Lagrangians $\mathfrak{L}$ and $\mathfrak{L}'$ on T(Q) are said to be *gauge equivalent* if there exists a real-valued function f on Q of class C^3 such that

$$\mathfrak{L}'(q,\dot{q}) = \mathfrak{L}(q,\dot{q}) + \frac{df}{d\tau}(q)$$

for all $(q,\dot{q}) \in T(Q)$. Here and in the following, $(q,\dot{q}) \in T(Q)$ means that $(q,\dot{q})$ are local coordinates of T(Q).

Let G be a Lie group, M a (finite-dimensional) C^∞-manifold, and $(g,m) \mapsto g{\cdot}m$ a C^∞-transitive left action of G on M. There exists a closed subgroup K of G such that M and the homogeneous manifold $Q = G/K$ are diffeomorphic. From now on we identify M and Q. Let $(Q,\mathfrak{L})$ be a Lagrangian dynamical system; for each $g \in G$, we define a Lagrangian $\mathfrak{L}_g$ on T(Q) by

$$\mathfrak{L}_g(q,\dot{q}) = \mathfrak{L}(g^{-1}\cdot q, g^{-1}\cdot \dot{q}),$$

where $g \cdot \dot{q}$ stays for $(g\cdot q)^{\cdot}$. The Lie group G is a *symmetry group of* $(Q,\mathfrak{L})$, and $\mathfrak{L}$ is a G-*invariant Lagrangian*, if $\mathfrak{L}_g = \mathfrak{L}$ for all $g \in G$; the Lie group G is a *gauge symmetry group of* $(Q,\mathfrak{L})$, and $\mathfrak{L}$ is a G-*gauge invariant Lagrangian*, if $\mathfrak{L}_g$ and $\mathfrak{L}$ are gauge equivalent for all $g \in G$. This last assertion means that there exists a mapping f of G into the real-valued functions on Q of class C^3 such that

$$\mathfrak{L}_g(q,\dot{q}) = \mathfrak{L}(q,\dot{q}) + \frac{df(g)}{d\tau}(q) \tag{2.2}$$

for all $(q,\dot{q}) \in T(Q)$ and all $g \in G$: the equations of motion determined by $\mathfrak{L}$ and $\mathfrak{L}_g$ the same. We have then

$$\mathfrak{L}_{gg'}(q,\dot{q}) = \mathfrak{L}(q,\dot{q}) + \frac{df(gg')}{d\tau}(q)$$

for all g, g' in G and all $(q,\dot{q}) \in T(Q)$. On the other hand, the associativity of the group law implies

$$\mathfrak{L}_{gg'}(q,\dot{q}) = \mathfrak{L}_g(q,\dot{q}) + \frac{d\Phi(g)f(g')}{d\tau}(q) = \mathfrak{L}(q,\dot{q}) + \frac{d}{d\tau}(f(g) + \Phi(g)f(g'))(q),$$

applying first g' and then g. It follows that

$$f(g) + \Phi(g)f(g') - f(gg') = \xi(g,g') \tag{2.3}$$

for all g, g' in G, where ξ is a mapping of G^2 into $\mathbf{R}$. By using the coboundary operator defined in Subsection 2.1, we may write (2.3) as

$$\delta f = \xi\,; \tag{2.4}$$

this shows that $\xi \in Z^2(G,\mathbf{R})$. The mapping f is uniquely defined by (2.2) only up to a mapping χ of G into $\mathbf{R}$ and can be chosen to be of class C^∞. In fact, we have the following results.

LEMMA 2.1. *Let* G *be a second countable Lie group and* f *a Borel mapping of* G *into a separable differentiable Fréchet* G*-module* V_Ψ *satisfying*

$$f(g) + \Psi(g)f(g') - f(gg') = \xi(g,g')$$

for all g, g' *in* G, *with* ξ *a* C^∞*-mapping of* G^2 *into* V. *Then* f *is of class* C^∞.

The continuity of f is proven using a well-known argument of Banach ([**BA**], Ch. I, Théorème 4) and the C^∞-differentiability follows on account of a regularization formula of Blanc ([**BL**], 5).

LEMMA 2.2. *Let* G *be a second countable Lie group and* V_Ψ *a separable differentiable Fréchet* G*-module. Then*

$$H^n_b(G, V_\Psi) = H^n_c(G, V_\Psi) = H^n_\infty(G, V_\Psi)$$

if $n = 0$ *or* $n = 1$ *and*

$$H^2_b(G, V_\Psi) \cong H^2_c(G, V_\Psi) \cong H^2_\infty(G, V_\Psi).$$

More precisely, $Z^n_b(G, V_\Psi) = Z^n_c(G, V_\Psi) = Z^n_\infty(G, V_\Psi)$ *if* $n = 0$ *or* $n = 1$ *and, for each* $f \in Z^2_b(G, V_\Psi)$*, there exist* $f' \in Z^2_c(G, V_\Psi)$*,* $h \in C^1_b(G, V_\Psi)$*,* $f'' \in Z^2_\infty(G, V_\Psi)$*,* $h' \in C^1_c(G, V_\Psi)$ *such that* $f' = f + \delta h$ *and* $f'' = f' + \delta h'$*.*

The result is obvious for $n = 0$ and $n = 1$. In order to prove the vector space isomorphism $H^2_b(G, V_\Psi) \cong H^2_c(G, V_\Psi)$, we notice that Ψ and every $f \in Z^2_b(G, V_\Psi)$ determine a topological extension of G by V, namely, an ordered pair (E, ρ), where the extension group E is topological and ρ is an open continuous group homomorphism of E onto G with $\operatorname{Ker} \rho = V$. This means that such a pair exists and we can find a Borel mapping $\sigma : G \to E$, with $\rho(\sigma(g)) = g$ for all $g \in G$ and $\sigma(e_G) = e_E$, which satisfies $f(g, g') = \sigma(g)\sigma(g')\sigma(gg')^{-1}$ and $\Psi(g)v = \sigma(g)v\sigma(g)^{-1}$ for all g, g' in G and all $v \in V$ ([**BW**], Theorem). We get $f' \in Z^2_c(G, V_\Psi)$ cohomologous to f by a result of Michael ([**MI**], Corollary 7.3). The vector space isomorphism $H^2_c(G, V_\Psi) \cong H^2_\infty(G, V_\Psi)$ is proven in Theorem 5.1 of Ref. [**HM**] and in Théorème 5.2 of Ref. [**BL**].

PROPOSITION 2.3. *Let* G *be a second countable Lie group,* K *a closed subgroup of* G*,* Q *the homogeneous manifold* G / K*,* $\mathcal{L}$ *a Lagrangian on* T(Q)*, and* f *a mapping of* G *into the real-valued functions on* Q *of class* C^3 *satisfying* (2.2) *and* (2.3)*. There exists a mapping* χ *of* G *into* **R** *such that* $\bar{f} = f + \chi$ *is normalized and of class* C^∞*.*

By integration of (2.2) with respect to τ on a compact interval, we have that $g \mapsto f(g) - f(g)(q_0)$, where $q_0 = K \in Q$, is a continuous mapping of G into the topological vector space of all real-valued functions on Q of class C^3 equipped with the topology of pointwise convergence, hence a Borel mapping of G into $\mathcal{C}^3(Q, \mathbf{R})$ ([**HJ**], Ch. III, § 2, Corollary 6). We find the required χ on account of LEMMA 2.2 and LEMMA 2.1.

Now the following definition makes sense.

DEFINITION 2.4. Let G be a second countable Lie group, K a closed subgroup of G and Q the homogeneous manifold G / K. A G-*gauge function on* Q *defining* ξ is a normalized C^∞-mapping f of G into $\mathcal{C}^3(Q, \mathbf{R})$ satisfying (2.3).

We denote by $\mathcal{G}(G, \mathcal{C}^3(Q, \mathbf{R}))$ the vector space of all G-gauge functions on Q and say that a G-gauge function f on Q is a G-*gauge function of a Lagrangian* $\mathcal{L}$ if f satisfies (2.2).

A mapping f is a G-gauge function of a Lagrangian $\mathcal{L}$ if and only if the mapping f' given by

$$f'(g) = f(g) - h + \Phi(g)h + \chi(g), \tag{2.5}$$

where Φ is defined by (2.1), $h \in \mathcal{C}^3(\mathrm{Q}, \mathbf{R})$, and $\chi \in \mathrm{C}^1_\infty(\mathrm{G}, \mathbf{R})$, is a G-gauge function of the gauge equivalent Lagrangian $\mathcal{L}'$ given by

$$\mathcal{L}'(q, \dot{q}) = \mathcal{L}(q, \dot{q}) + \frac{dh}{d\tau}(q).$$

In fact, we have (2.2) if and only if

$$\mathcal{L}'_g(q, \dot{q}) = \mathcal{L}(q, \dot{q}) + \frac{df(g)}{d\tau}(q) + \frac{d\Phi(g)h}{d\tau}(q),$$

hence if and only if

$$\mathcal{L}'_g(q, \dot{q}) = \mathcal{L}'(q, \dot{q}) + \frac{df'(g)}{d\tau}(q)$$

for all $g \in \mathrm{G}$ and all $(q, \dot{q}) \in \mathrm{T(Q)}$. This justifies the following definition.

DEFINITION 2.5. Let G, K, and Q be as in DEFINITION 2.4. Two G-gauge functions f, f' on Q are said to be *equivalent* if there exist $h \in \mathcal{C}^3(\mathrm{Q}, \mathbf{R})$ and $\chi \in \mathrm{C}^1_\infty(\mathrm{G}, \mathbf{R})$ satisfying (2.5).

Two G-gauge invariant Lagrangians on T(Q) are said to be *equivalent* if their G-gauge functions are equivalent.

As (2.5) determines an equivalence relation $\mathcal{R}$ in $\mathcal{G}(\mathrm{G}, \mathcal{C}^3(\mathrm{Q}, \mathbf{R}))$, we may consider the vector space $\mathcal{G}_{\mathrm{cl}}(\mathrm{G}, \mathcal{C}^3(\mathrm{Q}, \mathbf{R}))$ of all equivalence classes of G-gauge functions on Q modulo $\mathcal{R}$, i.e., we may put

$$\mathcal{G}_{\mathrm{cl}}(\mathrm{G}, \mathcal{C}^3(\mathrm{Q}, \mathbf{R})) = \mathcal{G}(\mathrm{G}, \mathcal{C}^3(\mathrm{Q}, \mathbf{R}))/\mathcal{G}_0(\mathrm{G}, \mathcal{C}^3(\mathrm{Q}, \mathbf{R})),$$

where $\mathcal{G}_0(\mathrm{G}, \mathcal{C}^3(\mathrm{Q}, \mathbf{R}))$ is the vector space of all G-gauge functions on Q equivalent to the G-gauge function $g \mapsto 0$. If $f \in \mathcal{G}_0(\mathrm{G}, \mathcal{C}^3(\mathrm{Q}, \mathbf{R}))$, there exist $h \in \mathcal{C}^3(\mathrm{Q}, \mathbf{R})$ and $\chi \in \mathrm{C}^1_\infty(\mathrm{G}, \mathbf{R})$ such that

$$f(g) = -h + \Phi(g)h + \chi(g)$$

for all $g \in \mathrm{G}$.

2.3. G-gauge functions and cocycles. The relation (2.3) defines a 1-cocycle up to constants and so we are naturally led to the following proposition which reduces the study of G-gauge functions to a problem in group cohomology.

PROPOSITION 2.6. *If* G *is a second countable Lie group,* K *a closed subgroup of* G, Q *the homogeneous manifold* G / K, *then*

$$\mathcal{G}_{\mathrm{cl}}(\mathrm{G}, \mathcal{C}^3(\mathrm{Q}, \mathbf{R})) \cong \mathrm{H}^1_\infty(\mathrm{G}, \mathcal{C}^3_{/\mathbf{R}}(\mathrm{Q}, \mathbf{R})_{\tilde{\Phi}}),$$

where $\tilde{\Phi}$ *is the quotient of the representation* Φ *defined by* (2.1).

Let $\mathrm{C}^n_\infty(\mathrm{G}, \mathrm{K}, \mathbf{R})$ be the vector subspace of all $\xi \in \mathrm{C}^n_\infty(\mathrm{G}, \mathbf{R})$ such that $\xi(kg_1, g_2, \cdots, g_n) = \xi(g_1, g_2, \cdots, g_n)$ for all $k \in \mathrm{K}$ and all $g_1, \cdots, g_n$ in G when $n > 0$, and $\mathrm{C}^n_\infty(\mathrm{G}, \mathrm{K}, \mathbf{R}) = \mathrm{C}^n_\infty(\mathrm{G}, \mathbf{R})$ when $n \leqslant 0$. Since $\delta\xi \in \mathrm{C}^{n+1}_\infty(\mathrm{G}, \mathrm{K}, \mathbf{R})$ whenever $\xi \in \mathrm{C}^n_\infty(\mathrm{G}, \mathrm{K}, \mathbf{R})$, we get the cochain complex $(\mathrm{C}_\infty(\mathrm{G}, \mathrm{K}, \mathbf{R}), \delta)$ of the differentiable relative cohomology of G modulo K with values in $\mathbf{R}$ and the related

$\mathbf{Z}$-graded vector spaces $Z_\infty(G, K, \mathbf{R})$, $B_\infty(G, K, \mathbf{R})$, and $H_\infty(G, K, \mathbf{R})$ as in Subsection 2.1. Notice that $\xi \in Z^2_\infty(G, K, \mathbf{R})$ if and only if $\xi | K \times G = 0$.

We will show in the following proposition that it is possible to associate univocally with every element of $H^1_\infty(G, \mathcal{C}^3_{/\mathbf{R}}(Q, \mathbf{R})_{\tilde{\Phi}})$, hence with every equivalence class of G-gauge functions, an element of $H^2_\infty(G, K, \mathbf{R})$. Obviously, it is by far easier to compute $H^2_\infty(G, K, \mathbf{R})$ than $H^1_\infty(G, \mathcal{C}^3_{/\mathbf{R}}(Q, \mathbf{R})_{\tilde{\Phi}})$.

PROPOSITION 2.7. *Let* G, K, Q, *and* $\tilde{\Phi}$ *be as in* PROPOSITION 2.6. *Then*

$$H^1_b(G, \mathcal{C}^3_{/\mathbf{R}}(Q, \mathbf{R})_{\tilde{\Phi}}) = H^1_c(G, \mathcal{C}^3_{/\mathbf{R}}(Q, \mathbf{R})_{\tilde{\Phi}}) = H^1_\infty(G, \mathcal{C}^3_{/\mathbf{R}}(Q, \mathbf{R})_{\tilde{\Phi}}) \cong H^2_\infty(G, K, \mathbf{R}).$$

We have $H^1_b(G, \mathcal{C}^3_{/\mathbf{R}}(Q, \mathbf{R})_{\tilde{\Phi}}) = H^1_c(G, \mathcal{C}^3_{/\mathbf{R}}(Q, \mathbf{R})_{\tilde{\Phi}}) = H^1_\infty(G, \mathcal{C}^3_{/\mathbf{R}}(Q, \mathbf{R})_{\tilde{\Phi}})$ by LEMMA 2.2. The vector space isomorphism $H^1_\infty(G, \mathcal{C}^3_{/\mathbf{R}}(Q, \mathbf{R})_{\tilde{\Phi}}) \cong H^2_\infty(G, K, \mathbf{R})$ is defined via the relation

$$f(g)(g'K) = \xi((g'K)^{-1}, g) \qquad (g, g' \text{ in } G),$$

which follows from (2.4) by choosing $f \in C^1_\infty(G, \mathcal{C}^3(Q, \mathbf{R})_{\tilde{\Phi}})$ such that $f(g)(q_0) = 0$ for all $g \in G$, whence $\xi \in Z^2_\infty(G, K, \mathbf{R})$.

REMARK 2.8. If ξ and ξ' are two elements of $Z^2_\infty(G, K, \mathbf{R})$ which are cohomologous modulo $B^2_\infty(G, \mathbf{R})$ with $\xi' = \xi + \delta\nu$, the mapping $\nu | K$ is a 1-dimensional continuous representation of K. Then ξ and ξ' are cohomologous modulo $B^2_\infty(G, K, \mathbf{R})$ with $\xi' = \xi + \delta\mu$ if and only if $\nu | K$ may be extended to a 1-dimensional continuous representation of G. In fact, the extension $\nu - \mu$ of $\nu | K$ is a 1-dimensional continuous representation of G which defines μ when we know ν ([**LL**], Proposition 4). In particular, we have $H^2_\infty(G, K, \mathbf{R}) \cong H^2_\infty(G, \mathbf{R})$ if the group K has no nontrivial 1-dimensional continuous representations, for instance if K is simple (of dimension > 1).

REMARK 2.9. In Lévy-Leblond's paper [**LL**] the Lie group G operates on Q with a right action, whence the 2-cocycles defined by the G-gauge functions vanish on $G \times K$, not on $K \times G$. However, there is a vector space isomorphism of $Z^2_\infty(G, K, \mathbf{R})$ onto the vector space of Lévy-Leblond's 2-cocycles: for every $\xi \in Z^2_\infty(G, K, \mathbf{R})$, we can define $\bar{\xi} \in Z^2_\infty(G, \mathbf{R})$ such that $\bar{\xi} | G \times K = 0$ by

$$\bar{\xi}(g, g') = \xi(g'^{-1}, g^{-1}) \tag{2.6}$$

and then we have $\bar{\xi} = -\xi + \delta\nu$ with $\nu \in C^1_\infty(G, \mathbf{R})$ given by $\nu(g) = \xi(g^{-1}, g)$.

It follows from PROPOSITION 2.7 that, for each $\xi \in Z^2_\infty(G, K, \mathbf{R})$, a G-gauge function f in the equivalence class univocally associated with $[\xi] (\operatorname{mod} B^2_\infty(G, K, \mathbf{R}))$ is given by

$$f(g)(g'K) = \xi((g'K)^{-1}, g) = \bar{\xi}(g^{-1}, g'K) \qquad (g \in G), \tag{2.7}$$

where $\bar{\xi}$ is defined by (2.6). In addition, the real-valued function $f(g)$ is of class C^∞ for all $g \in G$. We say that a Lagrangian $\mathcal{L}$ on $T(Q)$ which satisfies (2.2) with f given by (2.7) is an *admissible Lagrangian for* ξ.

3. Invariant extended Lagrangians

Let G be a second countable Lie group and V_Ψ a separable Fréchet G-module. If $f \in Z^2_c(G, V_\Psi)$, then the set of all ordered pairs $(v, g) \in V \times G$ equipped with the composition law defined by

$$(v,g)(v',g') = (v + \Psi(g)v' + f(g,g'), gg')$$

and endowed with the product topology is a Polish group $V_\Psi f\, G$ ([**CA**], Proposition 3.1 and § 4), which is a second countable Lie group if V is finite-dimensional ([**IW**], Theorem 7). The ordered pair $(V_\Psi f\, G, \mathrm{pr}_2)$ is a topological extension of G by V (cf. LEMMA 2.2) and is a topological central extension if $\Psi(g) = \mathrm{Id}_V$ for all $g \in G$ (since, in this case, V is in the center of $V_\Psi f\, G$).

Let G be a gauge symmetry group of a Lagrangian dynamical system $(Q, \mathcal{L})$ defining a 2-cocycle ξ. It was shown by Marmo, Morandi, Simoni, and Sudarshan [**MM**] that a suitable central extension of the Lie algebra of G is the Lie algebra of a symmetry group $\tilde{G}_{[\xi]}$ of a Lagrangian dynamical system $(\tilde{Q}, \tilde{\mathcal{L}})$: $\tilde{\mathcal{L}}$ is $\tilde{G}_{[\xi]}$-invariant and not only $\tilde{G}_{[\xi]}$-gauge invariant. Here, $\tilde{\mathcal{L}}$ is an extension of $\mathcal{L}$ defined on the tangent bundle of $Q \times \mathbf{R}$. The following proposition shows that it is possible to do the same at the level of Lie groups.

PROPOSITION 3.1. *Let* G, K, Q *be as in* PROPOSITION 2.6, $\xi \in Z^2_\infty(G, K, \mathbf{R})$ *the 2-cocycle defined by a G-gauge function f of a Lagrangian $\mathcal{L}$ on* T(Q), $\bar{\xi}$ *the 2-cocycle given by* (2.6), *and* $\tilde{G}_{[\xi]}$ *the Lie group* $\mathbf{R}\bar{\xi}\,G$. *Then the Lagrangian* $\tilde{\mathcal{L}}$ *on* $T(\tilde{Q})$, *defined by* $\tilde{\mathcal{L}}(\tilde{q}, \dot{\tilde{q}}) = \mathcal{L}(q, \dot{q}) - \dot{s}$ *for all* $\tilde{q} = (s, q)$ *in the homogeneous manifold* $\tilde{Q} = \tilde{G}_{[\xi]}/(\{0\}, K)$, *is* $\tilde{G}_{[\xi]}$*-invariant.*

We have

$$(r,g)^{-1} \cdot \dot{s} = \frac{d}{d\tau}(s - r - \bar{\xi}(g, g^{-1}) + f(g))(q)$$

for all $(r, g) \in \tilde{G}_{[\xi]}$, whence the $\tilde{G}_{[\xi]}$-invariance of the Lagrangian $\tilde{\mathcal{L}}$.

The Lie group $\tilde{G}_{[\xi]}$ of PROPOSITION 3.1 depends on the cohomology class $[\xi](\mathrm{mod}\, B^2_\infty(G, K, \mathbf{R}))$ of the 2-cocycle ξ: we get different groups for different equivalence classes of G-gauge functions. Proceeding in analogy with what may be done in the theory of continuous unitary projective representations [**CT**], it is possible to define a unique Lie group $\tilde{G}$, depending only on G and not on the Lagrangians, such that we may always find a Lagrangian $\mathcal{L}$ in every equivalence class and an extension $\tilde{\mathcal{L}}$ of $\mathcal{L}$ which is $\tilde{G}$-invariant. We need, in general, many ausiliary coordinates which are the components $s_1, \ldots, s_n$ of a vector $\boldsymbol{s} \in \mathbf{R}^n$, where $n = \dim H^2_\infty(G, K, \mathbf{R})$.

With G and K as above, let λ be a linear section of $Z^2_\infty(G, K, \mathbf{R})$ fibered by $B^2_\infty(G, K, \mathbf{R})$, namely, a linear mapping of $H^2_\infty(G, K, \mathbf{R})$ into $Z^2_\infty(G, K, \mathbf{R})$ such that $[\lambda([\xi])] = [\xi]$ for all $[\xi] \in H^2_\infty(G, K, \mathbf{R})$. Moreover, let $\kappa : [\xi] \mapsto \boldsymbol{x}_{[\xi]}$ be a vector space isomorphism of $H^2_\infty(G, K, \mathbf{R})$ onto $\mathbf{R}^n$, where $n = \dim H^2_\infty(G, K, \mathbf{R})$. The mapping $\mathbf{ev}$ of G^2 into $\mathbf{R}^n$ defined by

$$\langle \mathbf{ev}(g, g') \,|\, \boldsymbol{x}_{[\xi]} \rangle = \lambda([\xi])(g, g'), \tag{3.1}$$

with $\langle \cdot \,|\, \cdot \rangle$ the canonical scalar product on $\mathbf{R}^n$, is an element of $Z^2_c(G, \mathbf{R}^n)$ satisfying $\mathbf{ev}|\, K \times G = 0$ and we call it the *evaluation 2-cocycle determined by* λ *and* κ.

DEFINITION 3.2. Let G be a second countable Lie group and K a closed subgroup of G. A *Lagrangian* G*-group* is an extension group $\mathbf{R}^n \overline{\mathbf{ev}}\, G$ of G by $\mathbf{R}^n$, where

$n = \dim \mathrm{H}^2_\infty(\mathrm{G},\mathrm{K},\mathbf{R})$ and the continuous 2-cocycle $\overline{\mathbf{ev}}$ such that $\overline{\mathbf{ev}}|\,\mathrm{G}\times\mathrm{K} = 0$ is defined by

$$\overline{\mathbf{ev}}(g,g') = \mathbf{ev}(g'^{-1},g^{-1}), \tag{3.2}$$

with $\mathbf{ev}$ the evaluation 2-cocycle given by (3.1).

The following proposition may be proven along the lines of the proof of PROPOSITION 3.1.

PROPOSITION 3.3. *Let* G, K, Q *be as in* PROPOSITION 2.6, λ *a linear section of* $\mathrm{Z}^2_\infty(\mathrm{G},\mathrm{K},\mathbf{R})$ *fibered by* $\mathrm{B}^2_\infty(\mathrm{G},\mathrm{K},\mathbf{R})$, $\kappa : [\xi] \mapsto \boldsymbol{x}_{[\xi]}$ *a vector space isomorphism of* $\mathrm{H}^2_\infty(\mathrm{G},\mathrm{K},\mathbf{R})$ *onto* $\mathbf{R}^n$, *and* $\tilde{\mathrm{G}}$ *the Lagrangian* G*-group* $\mathbf{R}^n\overline{\mathbf{ev}}\,\mathrm{G}$, *where* $n = \dim \mathrm{H}^2_\infty(\mathrm{G},\mathrm{K},\mathbf{R})$ *and* $\overline{\mathbf{ev}}$ *is defined by* (3.2) *with* $\mathbf{ev}$ *the evaluation* 2*-cocycle determined by* λ *and* κ. *If* $\lambda([\xi])$ *is the* 2*-cocycle defined by a* G*-gauge function* f *of a Lagrangian* $\mathfrak{L}$ *on* T(Q), *then the Lagrangian* $\tilde{\mathfrak{L}}_{[\xi]}$ *on* $\mathrm{T}(\tilde{\mathrm{Q}})$, *given by*

$$\tilde{\mathfrak{L}}_{[\xi]}(\tilde{q},\dot{\tilde{q}}) = \mathfrak{L}(q,\dot{q}) - \langle \dot{\boldsymbol{s}} \,|\, \boldsymbol{x}_{[\xi]}\rangle$$

for all $\tilde{q} = (\boldsymbol{s},q)$ *in the homogeneous manifold* $\tilde{\mathrm{Q}} = \tilde{\mathrm{G}}/(\{0\},\mathrm{K})$, *is* $\tilde{\mathrm{G}}$*-invariant.*

REMARK 3.4. The vector $\boldsymbol{s} \in \mathbf{R}^n$ of PROPOSITION 3.3, as well as the coordinate s of PROPOSITION 3.1, are devoid of any dynamical significance: the Euler-Lagrange equations of the systems $(\tilde{\mathrm{Q}},\tilde{\mathfrak{L}})$ and $(\tilde{\mathrm{Q}},\tilde{\mathfrak{L}}_{[\xi]})$ are the same as those of the system $(\mathrm{Q},\mathfrak{L})$. Moreover, no other integral of motion is added by Noether's theorem to those of $(\mathrm{Q},\mathfrak{L})$. In fact, let $([\xi_1],\dots,[\xi_n])$ be an orthonormal basis of $\mathrm{H}^2_\infty(\mathrm{G},\mathrm{K},\mathbf{R})$ and $[\xi] = \sum_{i=1}^{n}\mu_i[\xi_i]$. We choose κ such that $(\boldsymbol{x}_{[\xi_1]},\dots,\boldsymbol{x}_{[\xi_n]})$ is the canonical basis of $\mathbf{R}^n$ and then

$$\langle \dot{\boldsymbol{s}} \,|\, \boldsymbol{x}_{[\xi]}\rangle = \sum_{i=1}^{n}\mu_i\dot{s}_i,$$

where $\dot{s}_i = \langle \dot{\boldsymbol{s}} \,|\, \boldsymbol{x}_{[\xi_i]}\rangle$. By Noether's theorem, the integrals of motion associated with $\boldsymbol{s}$ are the real parameters $\mu_1,\dots,\mu_n$, obviously constant. However, the μ_i's are essential parameters of the Lagrangian $\mathfrak{L}$ and deserve a physical interpretation (cf. Subsection 4.2).

As the entries of the Hessian matrix

$$\left(\frac{\partial^2\tilde{\mathfrak{L}}}{\partial\dot{\tilde{q}}_i\partial\dot{\tilde{q}}_j}\right)$$

vanish whenever one of the local coordinates $\tilde{q}_i,\tilde{q}_j$ is a component of $\boldsymbol{s}$, the Lagrangian $\tilde{\mathfrak{L}}_{[\xi]}$ has constraints: it is not possible to obtain all the accelerations $\ddot{\tilde{q}}_i$ as functions of the $\tilde{q}_i$ and $\dot{\tilde{q}}_i$ from the Euler-Lagrange equations [**MS**].

We will now consider two well-known examples: the free 1-particle system in the 4-dimensional space-time under the action of the restricted Poincaré and Galilei groups.

4. Examples

Let G be a second countable Lie group, K a closed subgroup of G, Q the homogeneous manifold G / K, $\xi \in \mathrm{Z}^2_\infty(\mathrm{G},\mathrm{K},\mathbf{R})$, and $\mathfrak{L}$ an admissible Lagrangian for ξ. For each $g \in \mathrm{G}$, we have

$$\mathcal{L}_g(q,\dot{q}) = \mathcal{L}(q,\dot{q}) + \frac{d\bar{\xi}(g^{-1},\cdot)}{d\tau}(q), \tag{4.1}$$

where $\bar{\xi}$ is obtained from ξ via (2.6). This relation may be seen as an equation satisfied by the admissible Lagrangians for ξ and the problem of its solution was tackled by Houard [**HO**].

We need the following result. Let E be a topological extension group of G by a second countable compact Lie group F; then ([**GU**], Chap. III, Corollaire 5.4)

$$\mathrm{H}^2_{\mathrm{c}}(\mathrm{E},\mathbf{R}) \cong \mathrm{H}^2_{\mathrm{c}}(\mathrm{G},\mathbf{R}). \tag{4.2}$$

4.1. The restricted Poincaré group. Let G be the restricted Poincaré group $\mathbf{P}^{\uparrow}_{+}$ (the neutral component of the Poincaré group $\mathbf{P}$), K the restricted Lorentz group $\mathbf{L}^{\uparrow}_{+}$, and Q the Minkowski space-time $\mathbf{M}$. The action of every $(a,\Lambda) \in \mathbf{P}^{\uparrow}_{+}$ on $\mathbf{M}$, with $a \in \mathbf{R}^4$ and $\Lambda \in \mathbf{L}^{\uparrow}_{+}$, is

$$((a,\Lambda),(t,\boldsymbol{x})) \mapsto \Lambda(t,\boldsymbol{x}) + a \qquad ((t,\boldsymbol{x}) \in \mathbf{M})$$

and the group law of $\mathbf{P}^{\uparrow}_{+}$ is given by $(a,\Lambda)(a',\Lambda') = (a+\Lambda a', \Lambda\Lambda')$.
Since $\mathrm{H}^2_{\infty}(\mathbf{ISL}(2,\mathbf{C}),\mathbf{R}) = \{0\}$ ([**BR**], (6.17)), where $\mathbf{ISL}(2,\mathbf{C})$ is the inhomogeneous $\mathbf{SL}(2,\mathbf{C})$ (the universal covering group of $\mathbf{P}^{\uparrow}_{+}$), we have

$$\mathrm{H}^2_{\infty}(\mathbf{P}^{\uparrow}_{+},\mathbf{R}) = \mathrm{H}^2_{\infty}(\mathbf{ISL}(2,\mathbf{C}),\mathbf{R}) = \{0\}$$

by (4.2). On the other hand, the simple group $\mathbf{L}^{\uparrow}_{+}$ has no nontrivial 1-dimensional continuous representations; therefore $\mathrm{H}^2_{\infty}(\mathbf{P}^{\uparrow}_{+},\mathbf{L}^{\uparrow}_{+},\mathbf{R}) = \{0\}$ by REMARK 2.8 and there does not exist any nontrivial equivalence class of $\mathbf{P}^{\uparrow}_{+}$-gauge functions:

$$\mathcal{G}_{\mathrm{cl}}(\mathbf{P}^{\uparrow}_{+},\mathcal{C}^3(\mathbf{M},\mathbf{R})) = \{0\}.$$

It follows that every $\mathbf{P}^{\uparrow}_{+}$-gauge invariant Lagrangian is gauge equivalent to a $\mathbf{P}^{\uparrow}_{+}$-invariant Lagrangian $\mathcal{L}$ and, from (4.1) with $\xi = 0$, we get

$$\mathcal{L}((t,\boldsymbol{x}),(\dot{t},\dot{\boldsymbol{x}})) = -m\sqrt{\dot{t}^2 - \dot{\boldsymbol{x}}^2},$$

where m is an arbitrary real number physically interpretable as the rest mass of the particle when strictly positive ([**LL**], § 4a).

4.2. The restricted Galilei group. Let G be the restricted Galilei group $\mathbf{G}^{\uparrow}_{+}$ (the neutral component of the Galilei group $\mathbf{G}$), K the restricted homogeneous Galilei group, and Q the Newton space-time. The action of every element $(b,\boldsymbol{a},\boldsymbol{v},R)$ of $\mathbf{G}^{\uparrow}_{+}$, with $(b,\boldsymbol{a})$ a space-time translation, $\boldsymbol{v}$ a Galilean boost, and R a space rotation, is

$$((b,\boldsymbol{a},\boldsymbol{v},R)(t,\boldsymbol{x})) \mapsto (t+b, R\boldsymbol{x}+t\boldsymbol{v}+\boldsymbol{a}) \qquad ((t,\boldsymbol{x}) \in \mathrm{Q})$$

and the group law is given by

$$(b,\boldsymbol{a},\boldsymbol{v},R)(b',\boldsymbol{a}',\boldsymbol{v}',R') = (b+b', \boldsymbol{a}+R\boldsymbol{a}'+b'\boldsymbol{v}, \boldsymbol{v}+R\boldsymbol{v}', RR').$$

Since $\mathrm{H}^2_{\infty}(\overline{\mathbf{G}}^{\uparrow}_{+},\mathbf{R}) \cong \mathbf{R}$ ([**BR**], § 6f), where $\overline{\mathbf{G}}^{\uparrow}_{+}$ is the universal covering group of $\mathbf{G}^{\uparrow}_{+}$, we have $\mathrm{H}^2_{\infty}(\mathbf{G}^{\uparrow}_{+},\mathbf{R}) \cong \mathrm{H}^2_{\infty}(\overline{\mathbf{G}}^{\uparrow}_{+},\mathbf{R}) \cong \mathbf{R}$ by (4.2). The subgroup K has no nontrivial continuous 1-dimensional representations, since there are no nontrivial normal subgroups of dimension $\geqslant 5$ in K. Therefore $\mathrm{H}^2_{\infty}(\mathbf{G}^{\uparrow}_{+},\mathrm{K},\mathbf{R}) \cong \mathbf{R}$ and there exists a 1-parameter family of equivalence classes of $\mathbf{G}_+^\uparrow$-gauge functions: $\mathcal{G}_{\mathrm{cl}}(\mathbf{G}_+^\uparrow, \mathcal{C}^3(\mathrm{Q},\mathbf{R})) \cong \mathbf{R}$. We have

$$\mathrm{H}^2_\infty(\mathbf{G}_+^\uparrow, \mathrm{K}, \mathbf{R}) = \{m[\xi_0] \mid m \in \mathbf{R}\},$$

where ξ_0 is defined by

$$\xi_0((b,\boldsymbol{a},\boldsymbol{v},R),(b',\boldsymbol{a}',\boldsymbol{v}',R')) = \langle \boldsymbol{a} - b\boldsymbol{v} \mid R\boldsymbol{v}'\rangle - \frac{1}{2}b\boldsymbol{v}'^2,$$

satisfies $\xi_0|\,\mathrm{K}\times\mathbf{G}_+^\uparrow = 0$, and is cohomologous to Bargmann's 2-cocycle $\tilde{\xi}_0$ ([**BR**], (6.28)) given by

$$\tilde{\xi}_0((b,\boldsymbol{a},\boldsymbol{v},R),(b',\boldsymbol{a}',\boldsymbol{v}',R')) = \frac{1}{2}\left(\langle \boldsymbol{a} + b'\boldsymbol{v} \mid R\boldsymbol{v}'\rangle - \langle \boldsymbol{v} \mid R\boldsymbol{a}'\rangle\right),$$

since $\xi_0 = \tilde{\xi}_0 + \delta\nu$ with $\nu((b,\boldsymbol{a},\boldsymbol{v},R)) = \frac{1}{2}\langle \boldsymbol{v} \mid b\boldsymbol{v} - \boldsymbol{a}\rangle$. A representative f_m of every equivalence class of $\mathbf{G}_+^\uparrow$-gauge functions is defined by

$$f_m((b,\boldsymbol{a},\boldsymbol{v},R))(t,\boldsymbol{x}) = m\left(\frac{1}{2}t\boldsymbol{v}^2 - \langle \boldsymbol{v} \mid \boldsymbol{x}\rangle\right) \qquad (m \in \mathbf{R})$$

for all $(t,\boldsymbol{x}) \in \mathrm{Q}$ because, by (2.7),

$$\begin{aligned} f_m((b,\boldsymbol{a},\boldsymbol{v},R))(t,\boldsymbol{x}) &= f_m((b,\boldsymbol{a},\boldsymbol{v},R))((t,\boldsymbol{x},\mathbf{0},I_3)\cdot(0,\mathbf{0})) \\ &= m\xi_0((-t,-\boldsymbol{x},\mathbf{0},I_3),(b,\boldsymbol{a},\boldsymbol{v},R)). \end{aligned}$$

It follows that every $\mathbf{G}_+^\uparrow$-gauge invariant Lagrangian is gauge equivalent to a $\mathbf{G}_+^\uparrow$-gauge invariant Lagrangian $\mathcal{L}$ with a $\mathbf{G}_+^\uparrow$-gauge function defining $m\xi_0$ $(m \in \mathbf{R})$ and, from (4.1) with $\xi = m\xi_0$, we obtain

$$\mathcal{L}((t,\boldsymbol{x}),(\dot{t},\dot{\boldsymbol{x}})) = \frac{1}{2}m\frac{\dot{\boldsymbol{x}}^2}{\dot{t}},$$

where m is an arbitrary real number which is physically interpreted as the mass of the particle when strictly positive ([**LL**], § 4a).

REMARK 4.1. We stress the different role of the real parameter m in the Poincaré and Galilei cases. In the first, m is introduced as a proportionality constant when we determine the Lagrangian via (4.1) with $\xi = 0$, while in the second it is already present in the 2-cocycle $\xi = m\xi_0$ and therefore it leads, in quantum mechanics, to the mass superselection rule.

References

[BA] S. Banach, *Théorie des operations linéaires*, Monografje Matematyczne Tom I, Warszawa, 1932

[BR] V. Bargmann, *On unitary ray representations of continuous groups*, Ann. Math. **59** (1954), 1-46

[BL] Ph. Blanc, *Sur la cohomologie continue des groupes localement compacts*, Ann. Sci. Éc. Norm. Supér. **12** (1979), 137-168

[BW] L. G. Brown, *Extensions of topological groups*, Pac. J. Math. **39** (1971), 71-78

[CA] L. Calabi, *Sur les extensions des groupes topologiques*, Ann. Mat. Pura Appl. **32** (1951), 295-370

[CT] U. Cattaneo, *Splitting and representation groups for Polish groups*, J. Math. Phys. **19** (1978), 452-460

[DI] J. Dieudonné, *Treatise on Analysis*. III, Academic Press, New York, 1972

[GU] A. Guichardet, *Cohomologie des groupes topologiques et des algèbres de Lie*, CEDIC, Paris, 1980

[HJ] J. Hoffmann-Jørgensen, *The Theory of Analytic Spaces*, Various Publication Series No. 10, Aarhus Universitet, Aarhus, 1970

[HM] G. Hochschild, G. D. Mostow, *Cohomology of Lie groups*, Ill. J. Math. **6** (1962), 367-401

[HO] J.-C. Houard, *On invariance groups and Lagrangian theories*, J. Math. Phys. **18** (1977), 502-516

[IW] K. Iwasawa, *On some types of topological groups*, Ann. Math. **50** (1949), 507-558

[LL] J.-M. Lévy-Leblond, *Group-theoretical foundations of classical mechanics: the Lagrangian gauge problem*, Commun. Math. Phys. **12** (1969), 64-79

[MI] E. Michael, *Convex structures and continuous selections*, Can. J. Math. **11** (1959), 556-575

[MM] G. Marmo, G. Morandi, A. Simoni, E. C. G. Sudarshan, *Quasi-invariance and central extensions*, Phys. Rev. D **37** (1988), 2196-2205

[MS] G. Marmo, N. Mukunda, J. Samuel, *Dynamics and symmetry for constrained systems: a geometrical analysis*, Riv. Nuovo Cim. **6** (1983), 1-62

[WA] G. Warner, *Harmonic Analysis on Semi-Simple Lie Groups*. I, Springer-Verlag, Berlin, 1972

Institut für theoretische Physik, ETH Zürich
E-mail address: `laura@itp.phys.ethz.ch`

CH-6673 Maggia
E-mail address: `cattaneo@lilo.lic.ti-edu.ch`

Canadian Mathematical Society
Conference Proceedings
Volume **29**, 2000

Perron Theory for Positive Maps and Semigroups on von Neumann algebras.

Fabio Cipriani

Dedicated to Sergio Albeverio on the occasion of his 60th birthday.

ABSTRACT. In this work we generalize the Perron Theorem on the uniqueness of the eigenvector corresponding to highest eigenvalue of a positive matrix to positive maps A on Hilbert spaces of Standard Forms of general von Neumann algebras. In particular we show that the uniqueness of a strictly positive eigenvector corresponding to $||A||$, the ergodicity of A and the indecomposability of A are equivalent properties.
We then apply the result to prove the ergodicity of the Quantum Ornstein–Uhlenbeck semigroups.

1. Introduction.

The aim of this paper is to generalize a well known result due to O. Perron [P], concerning the uniqueness of the eigenvector corresponding to the eigenvalue $||A||$ of an $n \times n$ matrix A with positive entries.
This result has been generalized by L. Gross [Gr] to the infinite dimensional cases of a positivity preserving, selfadjoint operator A acting on the Hilbert space $L^2(X, m)$ over a finite measure space (X, m). In the same paper, a noncommutative version of the theorem has been proved for a positivity preserving, selfadjoint operator A acting on the Hilbert space $L^2(\mathcal{M}, \tau)$ over a Type II_1 von Neumann algebras $\mathcal{M}$ equipped with a faithful, normal, finite trace τ.
As L. Gross showed this is the right framework in which consider the problem of the uniqueness of the Ground State for physical hamiltonians describing fermions systems in Quantum Field Theory (see also [F]).
However, since, on one hand, von Neumann algebras need not to admit faithful, normal traces (as in the case of type III von Neumann algebras) and, on the other hand, the operator A considered need not to be selfadjoint with respect to the trace, even if this is avaible, a further generalization seems to be needed. This is precisely

1991 *Mathematics Subject Classification.* 47A35, 46L57, 58F19, 47C15, 47B65.
Key words and phrases. von Neumann algebras, Standard Forms, Positive Maps, Ornstein–Uhlenbeck semigroups.
The work has been supported by MURST–Italy grant No. 9801262841–02.

the situation one has to face when dealing with models from the Noncommutative Geometry of A. Connes [Con1] (Anosov foliation, non unimodular topological groups), convergence to the equilibrium in Quantum Statistical Mechanics [OZ], dissipative evolution in Quantum Open Systems [D].
The theme of positive maps on ordered vector spaces is an old subject which grew out since the early days of Ergodic Theory. A state of the art up to the 1950 can be found in the paper of M.G. Krein and M.A. Rutman [KR], while for a more recent account see [BaRo]. More in the spirit of the present paper, we like to cite, in addition the works of L. Gross [Gr], W. Faris [F], the paper of S. Albeverio and R. Hoegh–Krohn [AHK] on the extension to von Neumann algebras of the Frobenius Theory on the cyclic structure of the eigenvalues of positive matrices.
The paper is organized as follows. In Section 2 we recall the main properties of Standard Forms of von Neumann algebras. In Section 3 we introduce a notion of indecomposability for maps and we prove its equivalence with ergodicity. Section 4 contains the generalized Perron Theorem for positive maps and positive semigroups. In the final Section 5 we show the ergodicity and the indecomposability of the Ornstein–Uhlenbeck semigroups.

In a companion work we will approach the ergodicity of Markovian semigroups from the point of view of the Noncommutative Dirichlet forms associated to their generators, in the spirit of [Cip].

2. Standard Forms of von Neumann Algebras.

The structure of *Standard Form* of a von Neumann algebra $\mathcal{M}$, we are going to recall (see [A], [Con2], [H]), generalizes that of the abelian von Neumann algebra $L^\infty(X,m)$ acting on the Hilbert space $L^2(X,m)$ ordered by the cone $L^2_+(X,m)$.
A *Standard Form* $(\mathcal{M},\mathcal{H},\mathcal{P})$ of a von Neumann algebra $\mathcal{M}$ consists of an Hilbert space $\mathcal{H}$ on which $\mathcal{M}$ is acting and a subset $\mathcal{P}\subset\mathcal{H}$ such that

$$\{\xi\in\mathcal{H}:(\xi,\eta)\geq 0\,\forall\eta\in\mathcal{P}\}=\mathcal{P}. \tag{2.1}$$

Property (2.1), called the *selfduality* of $\mathcal{P}$, implies that $\mathcal{P}$ is a closed, convex cone in $\mathcal{H}$ and that $\mathcal{H}$ is the complexification of the real subspace $\mathcal{H}^J:=\{\xi\in\mathcal{H}:(\xi,\eta)\in\mathbb{R}\,\forall\eta\in\mathcal{P}\}$, whose elements are called J–real: $\mathcal{H}=\mathcal{H}^J\bigoplus i\mathcal{H}^J$. Let J be the antiunitary involution on $\mathcal{H}$, defined by $J(\xi+i\eta):=\xi-i\eta$ for all $\xi,\eta\in\mathcal{H}^J$, that preserves $\mathcal{P}$ and $\mathcal{H}^J$.
In a standard form these objects are subject to the following requirements:

- $J\mathcal{M}J=\mathcal{M}'$;
- $JxJ=x^*$, $\forall x\in\mathcal{M}\bigcap\mathcal{M}'$ (the center of $\mathcal{M}$);
- $J\xi=\xi$, $\forall\xi\in\mathcal{P}$;
- $xJxJ(\mathcal{P})\subseteq\mathcal{P}$, $\forall x\in\mathcal{M}$,

where $\mathcal{M}'$ is the commutant of $\mathcal{M}$ in $\mathcal{H}$: $\mathcal{M}':=\{y\in B(\mathcal{H}):xy=yx\,\forall x\in\mathcal{M}\}$.
The cone $\mathcal{P}$, whose element are called *positive*, defines on $\mathcal{H}^J$ a structure of ordered Hilbert space. An element $\xi\in\mathcal{H}^J$ can be uniquely decomposed as a difference $\xi=\xi_+-\xi_-$, of two orthogonal, positive elements, called the *positive* and *negative* part of ξ: $\xi_+,\xi_-\in\mathcal{P}$, $(\xi_+,\xi_-)=0$. The positive part ξ_+ coincides with the projection of $\xi\in\mathcal{H}^J$ onto the closed, convex set $\mathcal{P}$ while the negative ξ_- part is defined to be the difference $\xi-\xi_+$. This is called the *polar* or *Jordan* decomposition.
The *support* $S^{\mathcal{M}}(\xi)$ of a vector $\xi\in\mathcal{P}$ is the smallest projection in $\mathcal{M}$ containing ξ in its range. Two positive vectors are then orthogonal if and only if they have

orthogonal supports. A positive vector $\xi \in \mathcal{P}$ is said to be *strictly positive* if its support is the identity of $\mathcal{M}$. This happens if and only if ξ is cyclic and sepataring for $\mathcal{M}$.
The conjugation J induces an antilinear isomorphism $j : \mathcal{M} \to \mathcal{M}'$ defined by $j(x) := Jx^*J$ for all $x \in \mathcal{M}$. If $e \in \mathrm{Proj}(\mathcal{M})$ is a projection in $\mathcal{M}$, $j(e)$ is a projection in the commutant $\mathcal{M}'$ so that

$$P_e := ej(e) \tag{2.2}$$

is a projection in $B(\mathcal{H})$. The usefulness of the notion of standard form derive from its essential uniqueness [H]: two standard forms $(\mathcal{M}, \mathcal{H}_i, \mathcal{P}_i)$ $i = 1, 2$ of the same von Neumann algebra $\mathcal{M}$ are always isomorphic in the sense that there exists a unitary operator U from $\mathcal{H}_1$ onto $\mathcal{H}_2$ such that $\mathcal{P}_2 = U(\mathcal{P}_1)$.

We recall that a standard form of a von Neumann algebra $\mathcal{M}$ can always be constructed, using the Tomita–Takesaki Theory, in the GNS–semicyclic representation space $L^2(\mathcal{M}, \Phi)$, associated to a faithful, normal, semifinite weight Φ on $\mathcal{M}$.
In Section 5 we will make use of the Hilbert–Schmidt standard form $(L^2(h), L^2_+(h))$ of the Type I algebra $L^\infty(h)$ of bounded operators on a fixed Hilbert space h, where $L^2(h)$ (resp. $L^2_+(h)$) is the class of Hilbert–Schmidt (resp. positive Hilbert–Schmidt) operators on h.
As a final example we recall that if G is unimodular, locally compact group endowed with its Haar measure, the Hilbert space $L^2(G)$, ordered by the cone $\mathcal{P}(G)$ of positive definite functions on G, gives a standard form for the left von Neumann algebra $\lambda(G)''$ generated by the left regular representation λ of G.
Finally, we recall that J-L. Sauvageot [Sau] has shown that for general von Neumann algebras, a standard form can always be seen as a generalized Hilbert–Schmidt class.

3. Ergodicity and Indecomposability of Positive Maps.

Let us consider a fixed standard form $(\mathcal{M}, \mathcal{H}, \mathcal{P})$ of the von Neumann algebra $\mathcal{M}$. Recall that a *face* $F \subseteq \mathcal{P}$ of the cone $\mathcal{P}$ is an hereditary subcone in the sense that given $\xi \in \mathcal{P}$ and $\eta \in F$, one has $\xi \in F$ whenever $\eta - \xi \in \mathcal{P}$.
In the rest of the paper we are concerned with the following classes of maps:

DEFINITION 3.1. A map bounded map $A : \mathcal{H} \to \mathcal{H}$ is said to be

- *positive* or *positivity preserving* if it leaves globally invariant the cone $\mathcal{P}$: $A(\mathcal{P}) \subseteq \mathcal{P}$;
- *ergodic* if it is positive and if for all $\xi, \eta \in \mathcal{P}$, $\xi, \eta \neq 0$ there exist $n \in \mathbb{N}$ such that $(\xi, A^n\eta) > 0$;
- *indecomposable* or *irreducible* if it is positive and if it leaves globally invariant no proper closed, face F of $\mathcal{P}$: $A(F) \subseteq F$ implies $F = \{0\}$ or $F = \mathcal{P}$.

LEMMA 3.2. *Let $e \in \mathrm{Proj}(\mathcal{M})$ be a projection in $\mathcal{M}$. Then: i) $e \neq 0$ if and only if $P_e \neq 0$; ii) if $e \neq 0$ there exist a positive vector $\xi \in \mathcal{P}$ in the range of P_e.*

PROOF. Suppose $e \in \mathrm{Proj}(\mathcal{M})$ and $P_e = 0$. Then take a cyclic and separating vector $\xi_0 \in \mathcal{P}$. Denote by S_{ξ_0} the *sharp operator* $S_{\xi_0} = J_{\xi_0}\Delta^{1/2}_{\xi_0}$ of the Tomita–Takesaki theory [Tak]. We have $J = J_{\xi_0}$ because ξ_0 is cyclic and separating in $\mathcal{P}$. Moreover

$$0 = \|P_e\xi_0\|^2 = (\xi_0; P_e\xi_0) = (\xi_0; eJeJ\xi_0) =$$

$$= (e\xi_0; Je\xi_0) = (e\xi_0; Je^*\xi_0) = (e\xi_0; JS_{\xi_0}(e\xi_0)) = (e\xi_0; \Delta_{\xi_0}^{1/2} e\xi_0) =$$

$$= \|\Delta_{\xi_0}^{1/4} e\xi_0\|^2.$$

The injectivity of the map $x \mapsto \Delta_{\xi_0}^{1/4} x\xi_0$ from $\mathcal{M}$ into $\mathcal{H}$ (see [BR]) implies $e = 0$.
ii) Let $e \in \mathrm{Proj}(\mathcal{M})$ be a non zero projection. Since, by i), $P_e \neq 0$ and P_e commutes with J, there exists a nontrivial J-real vector ξ in the range of P_e: $\xi \in \mathcal{H}^J = \mathcal{P} - \mathcal{P}$, $P_e\xi = \xi$. Let $\xi = \xi_+ - \xi_-$ the polar decomposition of ξ. Since P_e is positivity preserving (Lemma 2.9. in [Con2]) then $P_e\xi_0 \in \mathcal{P}$. But since $\xi \neq 0$ at least one among $P_e\xi_+$ and $P_e\xi_-$ is different from zero. □

Our first result relates the indecomposability of a map A with its commutation properties with the projections P_e, $e \in \mathrm{Proj}(\mathcal{M})$.

THEOREM 3.3. *A positive map $A : \mathcal{H} \to \mathcal{H}$ is indecomposable if and only if it commmutes with no proper projection P_e with $e \in \mathrm{Proj}(\mathcal{M})$.*

PROOF. Suppose A is indecomposable. Let $e \in Proj(\mathcal{M})$, $e \neq 0, I$ and suppose that $P_eA = AP_e$. Let us consider the closed face $F_e \equiv P_e(\mathcal{P})$ (see Theorem 4.2. [Con2]). For each $\xi \in \mathcal{P}$, since A is positivity preserving, we have that $AP_e\xi = P_eA\xi \in P_e(\mathcal{P}) \equiv F_e$. The first part follows since $e \neq 0, I$ implies that F_e is non trivial. The converse is also true because if F is a closed face of $\mathcal{P}$, then there exists a projection $e \in \mathrm{Proj}(\mathcal{M})$ in $\mathcal{M}$ such that $F = P_e(\mathcal{P})$ (Theorem 4.2. [Con2]) and F is non trivial if and only if $e \neq 0, I$. Now if A preserves F, taking ξ in the range of P_e and noticing that $P_e\xi_\pm \in \mathcal{P}$ implies $AP_e\xi_\pm \in F = P_e(\mathcal{P})$, we have:

$$P_eA\xi = P_e(AP_e\xi_+ - AP_e\xi_-) = AP_e\xi_+ - AP_e\xi_- = A\xi$$

and this concludes the proof. □

Next theorem shows the equivalence between ergodicity and indecomposability.
In the rest of the paper the range of a projection $e \in \mathrm{Proj}(\mathcal{M})$ will be denoted by R_e.

THEOREM 3.4. *A positive map $A : \mathcal{H} \to \mathcal{H}$ is indecomposable if and only if it is ergodic.*

PROOF. Let us suppose that A is not indecomposable. Then by Theorem 3.3 there exists a non trivial projection $e \in \mathrm{Proj}(\mathcal{M})$ such that $AP_e = P_eA$. Since e is not trivial so are P_e and $P_{e^\perp}$, and by Lemma 3.2 *ii)* there exist non zero positive vectors $\xi, \eta \in \mathcal{P}$ in the range of P_e and $P_{e^\perp}$ respectively: $P_e\xi = \xi$, $P_{e^\perp}\eta = \eta$. Then we have, for each $n > 0$:

$$(A^n\xi; \eta) = (A^nP_e\xi; \eta) = (P_eA^n\xi; \eta) = (A^n\xi; P_e\eta) = (A^n\xi; P_eP_{e^\perp}\eta) = 0.$$

Thus A cannot be ergodic. Suppose now that A is not ergodic. Then there exist positive vectors $\xi, \eta \in \mathcal{P}$ such that $(A^n\xi; \eta) = 0$ $\forall n > 0$. Let e_n be the projection in $\mathcal{M}$ whose range is

$$\ker S^{\mathcal{M}}(A^n\xi) = R^\perp_{S^{\mathcal{M}}(A^n\xi)}.$$

Since $A^n\xi$ and η are positive orthogonal support:

$$S^{\mathcal{M}}(A^n\xi) \cdot S^{\mathcal{M}}(\eta) = 0.$$

Then $e_n \geq S^{\mathcal{M}}(\eta) \neq 0$ $\forall n \geq 0$ and also $e \equiv \bigwedge_{n\geq 1} e_n \geq S^{\mathcal{M}}(\eta) \neq 0$. Again, since e is a non zero projection in $\mathcal{M}$, there exists a non zero vector $\zeta \in \mathcal{P}$ in the range of P_e, $P_e\zeta = \zeta$, and moreover $\forall n \geq 0$:

$$(A^n\xi; A\zeta) = (A^{n+1}\xi; \zeta) = (A^{n+1}\xi; P_e\zeta) = (P_eA^{n+1}\xi; \zeta) = (j(e)eA^{n+1}\xi; \zeta) = 0$$

since $A^{n+1}\xi \in \ker e_n = R^{\perp}e_n = \left(\ker S^{\mathcal{M}}(A^n\xi)\right)^{\perp} = R_{S^{\mathcal{M}}(A^n\xi)}$. So we have $S^{\mathcal{M}}(A^n\xi) \cdot S^{\mathcal{M}}(A\xi) = 0$, $A\xi \in R_{e_n}$ for any $n \geq 0$ and $A\xi \in R_e$. Finally

$$P_eA\xi = j(e)eA\xi = j(e)A\xi = JeJA\xi = (A\xi \in \mathcal{P}) = JeA\zeta = JA\zeta = A\zeta.$$

Applying Lemma 3.2 *ii)* we have $P_eA = AP_e$ and, by Theorem 3.3, A cannot be indecomposable. □

4. A generalized Perron theorem and its application to positive semigroups.

The following theorem contains the Perron uniqueness type result for indecomposable, symmetric, positive maps mentioned in the introduction. It was proved by L. Gross in [Gr] in case of a regular gage space. A version of it was proved by W. Faris [F] with indecomposability replaced by ergodicity.

THEOREM 4.1. *Let $A : \mathcal{H} \to \mathcal{H}$ be a symmetric, positive map and suppose that $\|A\|$ is an eigenvalue of A. The following properties are then equivalent:*
i) $\|A\|$ has multiplicity one and there exists an associated eigenvector which is strictly positive (i.e. cyclic and separating for $\mathcal{M}$);
ii) A is indecomposable;
iii) A is ergodic.

PROOF. We have already proved that ii) and iii) are equivalent. Let us show that ii) implies i). Since $\mathcal{H}$ splits as $\mathcal{H}^J + i\mathcal{H}^J$ and A commutes with J, the eigenspace of $\|A\|$ is spanned by J-real eigenvectors. Let $\xi \in \mathcal{H}^J$ be a J-real eigenvector of A corresponding to $\|A\|$. Given the polar decomposition of ξ, $\xi = \xi_+ - \xi_-$, we want to prove firstly that ξ_+ and ξ_- are eigenvectors corresponding to $\|A\|$. We have, for $|\xi| = \xi_+ + \xi_-$:

$$\begin{aligned} \|A\|(|\xi|;|\xi|) &= \|A\|(\xi;\xi) \\ &= (A\xi_+;\xi_+) - (A\xi_-;\xi_+) - (A\xi_+;\xi_-) + (A\xi_+;\xi_-) \\ &\leq (A\xi_+;\xi_+) + (A\xi_-;\xi_+) + (A\xi_+;\xi_-) + (A\xi_+;\xi_-) \\ &= (A|\xi|;|\xi|) \leq \|A\|(|\xi|;|\xi|). \end{aligned} \tag{4.1}$$

Then

$$\begin{aligned} \|A\|(|\xi|;|\xi|) &= (A|\xi|;|\xi|) \\ ((\|A\| - A)|\xi|;|\xi|) &= \|(\|A\| - A)^{1/2}|\xi|\|^2 = 0 \\ (\|A\| - A)^{1/2}|\xi| &= 0 \\ (\|A\| - A)|\xi| &= 0 \\ A|\xi| &= \|A\||\xi|. \end{aligned} \tag{4.2}$$

Then $|\xi|, \xi_\pm$ are eigenvectors of A corresponding to $\|A\|$. We have shown that each eigenvector of A corresponding to $\|A\|$ splits as

$$\xi = (\xi_+^r - \xi_-^r) + i(\xi_+^i - \xi_-^i)$$

where $\xi_\pm^\tau \in \mathcal{P}$, $\tau = i, r$, are mutually orthogonal and are all eigenvectors corresponding to $\|A\|$. Consider now a positive, non zero eigenvector ξ corresponding to $\|A\|$and let e be the projection $S^{\mathcal{M}}(\xi)^\perp$. Then $e \in \mathrm{Proj}(\mathcal{M})$ and $e\xi = 0$. We want to show that A commutes with P_e. Since both these operators are positivity preserving, it is sufficient to show that they commute on $\mathcal{P}$. We have:

$$\begin{aligned}(\xi; AP_e\eta) = (A\xi; P_e\eta) = \|A\|(A\xi; P_e\eta) = \\ \|A\|(P_e\xi;\eta) = \|A\|(\jmath(e)e\xi;\eta) = 0.\end{aligned} \tag{4.3}$$

Since $AP_e\eta \in \mathcal{P}$ (A and P_e are positivity preserving) then $AP_e\eta \in R_e$, i.e. $eAP_e\eta = AP_e\eta$ and also $P_e(AP_e\eta) = AP_e\eta$. Then A takes the range of P_e into itself and then A and P_e commute. The indecomposability of A then implies $e = I_{\mathcal{H}}$, i.e. $S^{\mathcal{M}}(\xi) = I_{\mathcal{M}}$ so that $||A||$ has multiplicity one.
The proof that i) implies iii) is contained in [F Theorem 1,2] (where it is shown to hold in the more general context of ordered Hilbert spaces). □

To apply Theorem 4.1 to semigroups we have to adapt Definition 3.1.

DEFINITION 4.2. A symmetric, strongly continuous, positive semigroup $\{T_t : t > 0\}$ on $\mathcal{H}$ is called

- *ergodic* if, for each $\xi, \eta \in \mathcal{P}$, $\xi, \eta \neq 0$, there exists $t > 0$ such that $(\xi; T_t\eta) > 0$;
- *indecomposable* if there exists $t > 0$ such that T_t is indecomposable.

Let H be the selfadjoint generator of $\{T_t : t > 0\}$: $T_t = e^{-tH}$. An eigenvector of H corresponding to the lowest eigenvalue $\inf\sigma(H)$ will be called a *ground state* for H (or $\{T_t : t > 0\}$).

In applications the operator H may represent the hamiltonian of a physical system, and this justifies the name of *ground state* given to an eigenvector with these properties.

THEOREM 4.3. *Uniqueness of the Ground State. Let $\{T_t : t > 0\}$ be a symmetric, strongly continuous, positive semigroup. Assume that* $\inf\sigma(H)$ *is an eigenvalue of the generator H of $\{T_t : t > 0\}$ (i.e. there exists a ground state). Then the following conditions are equivalent:*
i) zero has multiplicity one and there exists a strictly positive ground state;
ii) $\{T_t : t > 0\}$ is indecomposable;
iii) $\{T_t : t > 0\}$ is ergodic.

PROOF. The equivalence of i) and ii) follows from Theorem 3.4. The equivalence of i) and iii) (or ii)) follows applying Theorem 4.1 to $A := T_{t_0}$ where $t_0 > 0$ is chosen in such a way that T_{t_0} is indecomposable (or ergodic). Notice in fact that, by the Spectral Theorem, a ground state for $\{T_t : t > 0\}$ is an eigenvector of A associated to the eigenvalue $\|A\| = e^{-t_0 \inf\sigma(H)}$). □

REMARK 4.4. Versions of the above theorem were proved by L. Gross [Gr] and W. Faris [F] in the special case of a regular gage space and in the abstract case of a Hilbert space with a Hilbert cone, respectively. In [Gr] L. Gross applied the theorem to prove the uniqueness of the ground state of fermion systems. There, a hamiltonian is given on the antisymmetric Fock-Hilbert space, of some one-particle

Hilbert space. The above mentioned version of the theorem is applied to a new operator, which is unitarily equivalent to the previous one, and is defined on the space $L^2(\mathcal{C})$ over the Clifford algebra $\mathcal{C}$.

5. Ergodicity and Indecomposability of the Quantum Ornstein–Uhlenbeck Semigroups.

In this section we apply the results obtained above to prove ergodocity and indecomposability for a class of semigroups, constructed in [CFL], describing dissipative evolutions in mathematical models of maser and lasers and also in the weak–coupling models of quantum open systems.

Let h be the sequence space $l^2(\mathbb{N})$ and consider the Type I von Neumann algebra $L^\infty(h)$ of all bounded operators on h. This algebra acts by left composition on the Hilbert space $L^2(h)$ of all Hilbert–Schmidt operators on h. The cone $L^2_+(h)$ of positive, Hilbert–Schmidt operators is selfdual and $(L^\infty(h), L^2(h), L^2_+(h))$ is a standard form for $L^\infty(h)$. The conjugation J is then the operation of taking the adjoint $J\xi = \xi^*$ of an Hilbert–Schmidt operator $\xi \in L^2(h)$.

In [CFL] a class of Markovian, in particular Positivity Preserving, semigroups, called the Ornstein–Uhlenbeck semigroups, is constructed using the theory of Noncommutative Dirichlet forms developed in [Cip].
To describe these forms let $\{e_n : n \in \mathbb{N}\}$ be the standard Hilbert basis of $l^2(\mathbb{N})$ and denote by A, A^* and N the *annihilation*, *creation* and *number* operators on h respectively:

- $Ae_0 = 0$, $Ae_n = \sqrt{n}e_{n-1}$ for $n \geq 1$,
- $A^*e_n = \sqrt{n+1}e_{n+1}$ for $n \in \mathbb{N}$;
- $Ne_n = ne_n$ for $n \in \mathbb{N}$.

We recall that these operators are unbounded and closed on suitable domains we do not need to specify here. For a fixed closed operator X on h, let us denote by $(L_X, D(L_X))$ (resp. $(R_X, D(R_X))$) the left (resp. right) closed multiplication operator by X on $L^2(h)$ on its maximal domain of definition. Then, for each $\mu > \lambda > 0$ fixed, the quadratic form $(\mathcal{E}_\nu, \mathcal{F})$ defined by

$$\mathcal{F} := D(L_{\sqrt{N}}) \bigcap D(R_{\sqrt{N}}) \subset L^2(h), \tag{5.1}$$

$$\mathcal{E}_\nu[\xi] := \frac{1}{2}\{||(\mu L_A - \lambda R_A)\xi||^2 + ||(\mu L_A - \lambda R_A)\xi^*||^2\} \qquad \xi \in \mathcal{F}, \tag{5.2}$$

is closed and Markovian (in the sense of [Cip]) with respect to the positive vector

$$\xi_\nu := (1-\nu)^{1/2} \sum_{n\in\mathbb{N}} \nu^{n/2} e_n \otimes e_n, \tag{5.3}$$

where $\nu := \lambda^2/\mu^2$. The selfadjoint operator $(H_\nu, D(H_\nu))$ on $L^2(h)$ associated to $(\mathcal{E}_\nu, \mathcal{F})$ is the closure of the operator

$$H_\nu\xi = \frac{1}{2}(\lambda^2 + \mu^2)(N\xi + \xi N) - \lambda\mu(A\xi A^* + A^*\xi A) + \lambda^2\xi, \tag{5.4}$$

defined for ξ in the linear span $C_{00}(h)$ of $\{e_n \otimes e_m : n, m \in \mathbb{N}\}$. By the general theory (see [Cip]), $(H, D(H))$ generates a Markovian (hence contractive and positivity preserving) semigroup on $L^2(h)$: $T_t = e^{-tH_\nu}, t > 0$. Finally, associated to $\{T_t : t > 0\}$ there exists also a weakly*–continuous Markovian semigroup on the von

Neumann algebra $L^\infty(h)$ whose generator is the closure of the following operator $\mathcal{L}_\nu$ of Lindblad type:

(5.5)

$$\mathcal{L}_\nu x = -\frac{\mu^2}{2}(A^*Ax - 2A^*xA + xA^*A) - \frac{\lambda^2}{2}(AA^*x - 2AxA^* + xAA^*) \quad x \in C_{00}(h).$$

We are now in the position to state the announced result concerning the ergodic properties of this family of semigroups.

THEOREM 5.1. *The Ornstein–Uhlenbeck semigroups are ergodic and indecomposable.*

PROOF. Given the expression (5.2) of the Dirichlet form $(\mathcal{E}_\nu, \mathcal{F})$, a lengthy but not difficult computation reveals that the vector ξ_ν in (5.3) is the only one on which $\mathcal{E}_\nu$ vanishes: $\mathcal{E}_\nu[\xi_\nu] = 0$. By the Spectral Theorem $H_\nu \xi_\nu = 0$ so that $\xi_n u$ is the unique ground state of the Ornstein–Uhlenbeck semigroup. A direct application of Theorem 4.3 concludes the proof of this theorem. □

REFERENCES

[AHK] S. Albeverio, R. Høegh–Krohn, Frobenius Theory for Positive Maps of von Neumann algebras, *Commun. Math. Phys.* **64** (1978), 83-94.

[A] H. Araki, Some properties of modular conjugation operator of von Neumann algebras and noncommutative Radon–Nikodym theorem with chain rule, *Pacific J. Math.* **50** (1974), 309-354.

[BaRo] C.J.K. Batty, D.W. Robinson, Positive one–parameter semigroups on ordered Banach spaces, *Acta Appl. Math.* **2** (1984), 221-296.

[BR] O. Bratteli, D.W. Robinson, "Operator Algebras and Quantum Statistical Mechanics 1", Springer Verlag, New York/Berlin, 1987.

[Cip] F. Cipriani, Dirichlet forms and Markovian semigroups on Standard Forms of von Neumann algebras, *J. Functional Analysis* **147** (1997), 259-300.

[CFL] F. Cipriani, F. Fagnola, J.M. Lindsay, Spectral Analysis and Feller Property for Quantum Ornstein–Uhlenbeck semigroups, to appear Commun. Math. Phys.1999.

[Con1] A. Connes, "Noncommutative Geometry", Academic Press, New York San Francisco London, 1994.

[Con2] A. Connes, Caracterization des espaces vectoriel ordonneés sous–jacents aux algébre de von Neumann, *Ann. Inst. Fourier Grenoble* **24** (1974), 121-155.

[D] E.B. Davies, "Quantum Theory of Open Systems", Academic Press, New York San Francisco London, 1976.

[F] W. Faris, Invarianr cones and uniqueness of the ground state for Fermion systems, *J. Math. Phys.* **13** (1972), 1285-1290.

[Gr] L. Gross, Existence and Uniqueness of Physical Ground States, *J. Functional Analysis* **10** (1972), 52-109.

[H] U. Haagerup, The standard form of a von Neumann algebra, *Math. Scand.* **37** (1975), 271-283.

[KR] M.G. Krein, M.A. Rutman, Linear operators leaving invariant a cone in a Banach space, *Amer. Math. Soc. Transl.* **10** (1950), 199-325.

[OZ] R. Olkiewiczb, Zegarlinski, Hypercontractivity in Noncommutative L_p Spaces, *J. Funct. Anal.* **161** (1999), 246-285.

[P] O. Perron, Zur Theorie der Matrices, *Math. Ann.* **64** (1908), 248-263.

[Sau] J-L. Sauvageot, Sur le produit tensoriel relatif d'espaces de Hilbert, *J. Oper. Theory* **9** (1983), 237-252.

Dipartimento di Matematica, Politecnico di Milano, piazza Leonardo da Vinci 32, 20133 Milano, Italy.

E-mail address: `fabcip@mate.polimi.it`

Canadian Mathematical Society
Conference Proceedings
Volume **29**, 2000

Some remarks on the method of sums

Ph. Clément, G. Gripenberg, and V. Högnäs

ABSTRACT. This note presents a reasonably short but self-contained proof of the method of sums due to DaPrato and Grisvard. Explicit constants for the regularity estimates are given.

1. Introduction

In this note we consider the method of sums of operators, devised by DaPrato and Grisvard. The method of sums gives conditions under which the problem $Ay + By = x$ can be solved. Here A and B are linear operators mapping, respectively, $\mathcal{D}(A)$ and $\mathcal{D}(B)$ into X, where X is a Banach space and $x \in X$ is given. In general, only the existence of a mild solution can be guaranteed, but if this solution y belongs to either $\mathcal{D}(A)$ or $\mathcal{D}(B)$, then it is a strict solution. In particular, if x belongs to a certain interpolation space, then one has a strict solution y. Moreover, then Ay and By belong to the same interpolation space, i.e., one has maximal regularity.

Our purpose is twofold. First, the aim is to present a brief but concise and self-contained proof of several previously known results scattered in the literature. Second, our aim is to make explicit the constants occurring in the estimates for the various interpolation norms of Au and Bu. In addition, we extend the method to give some regularity results for the case where neither A nor B is invertible, but then the existence of a strict solution is assumed.

We make very little claim as to originality; most of the results that we present can, in one form or another, be found in [**3**]–[**6**]. See also [**1**] and [**2**].

We begin by defining the class of operators considered. If X is a Banach space, then we denote the norm in X by $\|\cdot\|$ (or $\|\cdot\|_X$) and we let $\|\cdot\|$ denote the norm of bounded linear operators on X as well.

DEFINITION 1. *Let X be a (complex) Banach space. A linear operator $L : \mathcal{D}(L) \subset X \to X$ is nonnegative if $(-\infty, 0) \subset \rho(L)$ (the resolvent set of L) and*

$$\sup_{t>0} \|t(L + tI)^{-1}\| < \infty.$$

1991 *Mathematics Subject Classification.* 47A99, 47A10, 46B70.
Key words and phrases. Method of sums, maximal regularity, interpolation spaces.

If L is a nonnegative operator on X, then

$$\phi_L \stackrel{\text{def}}{=} \sup\{\,\phi \in [0,\pi] \mid \sup_{\substack{|\arg(\lambda)|\le\phi \\ \lambda\neq 0}} \|\lambda(L+\lambda I)^{-1}\| < \infty\,\},$$

and

$$M(L,\phi) \stackrel{\text{def}}{=} \sup_{\substack{|\arg(\lambda)|=\phi \\ \lambda\neq 0}} \|\lambda(L+\lambda I)^{-1}\|.$$

In Definition 1 we, of course, take $\|(L+\lambda I)^{-1}\| = \infty$ if $-\lambda$ does not belong to the resolvent set of L, i.e., if $L+\lambda I$ is not invertible. Observe also that if L is a nonnegative operator, then $\phi_L \geq \arcsin(1/M(L,0))$. One usually says that $\omega_L \stackrel{\text{def}}{=} \pi - \phi_L$ is the spectral angle of L.

DEFINITION 2. *Let X be a (complex) Banach space and let L be a nonnegative operator on X. If $\gamma \in (0,1)$ and $p \in [1,\infty]$, then*

$$\mathcal{D}_L(\gamma,p) \stackrel{\text{def}}{=} \{\,x \in X \mid [x]_{\mathcal{D}_L(\gamma,p)} < \infty\,\},$$

where

$$[x]_{\mathcal{D}_L(\gamma,p)} = \begin{cases} \left(\int_0^\infty \left(t^\gamma \|L(L+tI)^{-1}x\|\right)^p \frac{dt}{t}\right)^{\frac{1}{p}}, & 1 \le p < \infty, \\ \sup_{t>0} t^\gamma \|L(L+tI)^{-1}x\|, & p = \infty. \end{cases}$$

Moreover,

$$\mathcal{D}_L(\gamma,\infty_0) \stackrel{\text{def}}{=} \{\,x \in \mathcal{D}_L(\gamma,\infty) \mid \lim_{t\to\infty} t^\gamma \|L(L+tI)^{-1}x\| = 0\,\},$$

with $[\cdot]_{\mathcal{D}_L(\gamma,\infty_0)} = [\cdot]_{\mathcal{D}_L(\gamma,\infty)}$.

It is easy to see that $[\cdot]_{\mathcal{D}_L(\gamma,p)}$ is (at least) a seminorm. Note that for notational convenience we write $\mathcal{D}_L(\gamma,\infty_0) = \mathcal{D}_L(\gamma)$. The interpolation spaces between X and $\mathcal{D}(L)$, defined by, e.g., the K-method, are denoted by $(X,\mathcal{D}(L))_{\gamma,p}$ where $0<\gamma\le 1$ and $p \in [1,\infty]\cup\{\infty_0\}$, (where again $(X,\mathcal{D}(L))_{\gamma,\infty_0} = (X,\mathcal{D}(L))_\gamma$); see [**7**, Chap. 1.2] or the proof of Proposition 3 below.

For completeness we state (and prove) the following well-known facts:

PROPOSITION 3. *Let X be a (complex) Banach space and let L be a nonnegative operator on X with domain $\mathcal{D}(L)$. Let the norm in $\mathcal{D}(L)$ be either $\|\underline{x}\|_{\mathcal{D}(L)} = \|L\underline{x}\| + \|\underline{x}\|$ or $\|\underline{x}\|_{\mathcal{D}(L)} = \|L\underline{x}\|$ (if L is invertible). Suppose that $\gamma \in (0,1)$ and $p \in [0,\infty]\cup\{\infty_0\}$. Then $\mathcal{D}_L(\gamma,p) = (X,\mathcal{D}(L))_{\gamma,p}$ and for each $x \in X$,*

$$\frac{1}{1+M(L,0)}[x]_{\mathcal{D}_L(\gamma,p)} \le \|x\|_{(X,\mathcal{D}(L))_{\gamma,p}}$$

$$\le 2[x]_{\mathcal{D}_L(\gamma,p)} + \begin{cases} 0, & \text{if } \|\underline{x}\|_{\mathcal{D}(L)} = \|L\underline{x}\|, \\ M(L,0)^{1-\gamma}\left(p\gamma(1-\gamma)\right)^{-\frac{1}{p}}\|x\|, & \text{if } \|\underline{x}\|_{\mathcal{D}(L)} = \|L\underline{x}\| + \|\underline{x}\|. \end{cases}$$

Next we state a theorem due to DaPato and Grisvard on the method of sums.

THEOREM 4. *Let X be a (complex) Banach space and assume that*

(i) *A and B are two linear operators on X with domains $\mathcal{D}(A)$ and $\mathcal{D}(B)$, respectively, and there are numbers α and β in the resolvent sets $\rho(A)$ and $\rho(B)$ of A and B, respectively, such that*

$$(A-\alpha I)^{-1}(B-\beta I)^{-1} = (B-\beta I)^{-1}(A-\alpha I)^{-1}.$$

(ii) *A and B are nonnegative operators on X and*

$$\phi_A + \phi_B > \pi \text{ or, equivalently, } \omega_A + \omega_B < \pi.$$

(iii) $0 \in \rho(A) \cup \rho(B)$, *i.e., at least one of the operators A and B is invertible.*

Then the following statements hold true:

(a) *There is a bounded linear operator* $S : X \to X$ *such that*

$$S = \frac{1}{2\pi \mathrm{i}} \int_\gamma (A + zI)^{-1}(B - zI)^{-1}\,\mathrm{d}z,$$

where γ *is a path in* $\rho(-A) \cap \rho(B)$ *separating the spectra of* $-A$ *and* B, *and*

$$S + BA^{-1}S = A^{-1} \quad \text{if } A \text{ is invertible,}$$
$$AB^{-1}S + S = B^{-1} \quad \text{if } B \text{ is invertible.}$$

(b) *If* $y \in \mathcal{D}(A) \cap \mathcal{D}(B)$, *then* $S(Ay + By) = y$.

(c) *If* $Sx \in \mathcal{D}(A) \cup \mathcal{D}(B)$ *for some* $x \in X$, *then* $Sx \in \mathcal{D}(A) \cap \mathcal{D}(B)$ *and* $ASx + BSx = x$.

(d) *The operator* $A + B$ *with domain* $\mathcal{D}(A) \cap \mathcal{D}(B)$ *is closable in X and if* $\mathcal{D}(A) + \mathcal{D}(B)$ *is dense in X, then* $S = (\overline{A + B})^{-1}$.

(e) *If* $x \in \mathcal{D}_A(\gamma, p)$ *for some* $\gamma \in (0, 1)$ *and* $p \in [1, \infty] \cup \{\infty_0\}$, *then* $Sx \in \mathcal{D}(A) \cap \mathcal{D}(B)$, $ASx \in \mathcal{D}_A(\gamma, p) \cap \mathcal{D}_B(\gamma, p)$ *and* $BSx \in \mathcal{D}_A(\gamma, p)$. *Moreover*

$$[ASx]_{\mathcal{D}_A(\gamma,p)} \le c_1 [x]_{\mathcal{D}_A(\gamma,p)},$$
$$[BSx]_{\mathcal{D}_A(\gamma,p)} \le (1 + c_1)[x]_{\mathcal{D}_A(\gamma,p)},$$
$$[ASx]_{\mathcal{D}_B(\gamma,p)} \le c_2 [x]_{\mathcal{D}_A(\gamma,p)},$$

where

$$\text{(1)} \quad \begin{aligned} c_1 &= \frac{1}{\pi} M(B, \pi - \theta)\Big(1 + 2\sin(\tfrac{\theta}{2}) M(A, \theta)\Big) \int_0^\infty \frac{s^{\gamma - 1}}{|s - \mathrm{e}^{\mathrm{i}\theta}|}\,\mathrm{d}s, \\ c_2 &= \frac{1}{\pi} M(B, \pi - \theta)\Big(1 + 2\sin(\tfrac{\theta}{2}) M(A, \theta)\Big) \int_0^\infty \frac{s^{\gamma - 1}}{|s + \mathrm{e}^{\mathrm{i}\theta}|}\,\mathrm{d}s, \end{aligned}$$

and $\theta \in (\pi - \phi_B, \phi_A)$.

The statements (a)–(c) have, in the form stated here, previously been formulated in [**2**, Thm. 3.3 and Prop. 3.4]. Related results can be found in [**4**]–[**6**]. For (d) and for the claims $ASx \in \mathcal{D}_A(\gamma, p)$ and $BSx \in \mathcal{D}_A(\gamma, p)$ in (e), see [**6**, Thm. 2.7, p. 315], where however $\overline{\mathcal{D}(A)} = \overline{\mathcal{D}(B)} = X$ is assumed, and [**4**, Thm. 3.7, p. 324 and Thm. 3.11, p. 328]. In [**3**] a cross-regularity result ($ASx \in \mathcal{D}_B(\gamma, p)$) is proved for the case where both $-A$ and $-B$ generate bounded semigroups.

In the case where neither A nor B is invertible, we have the following result:

COROLLARY 5. *Let X be (complex) Banach space and suppose that assumptions* (i) *and* (ii) *of Theorem 4 hold true. If* $x \in \mathcal{D}_A(\gamma, p)$ *for some* $\gamma \in (0, 1)$ *and* $p \in [1, \infty] \cup \{\infty_0\}$ *and if* $y \in \mathcal{D}(A) \cap \mathcal{D}(B)$ *is a solution to the equation* $Ay + By = x$, *then* $Ay \in \mathcal{D}_A(\gamma, p) \cap \mathcal{D}_B(\gamma, p)$ *and* $By \in \mathcal{D}_A(\gamma, p)$. *Moreover*

$$[Ay]_{\mathcal{D}_A(\gamma,p)} \le c_1 [x]_{\mathcal{D}_A(\gamma,p)},$$
$$[By]_{\mathcal{D}_A(\gamma,p)} \le (1 + c_1)[x]_{\mathcal{D}_A(\gamma,p)},$$
$$[Ay]_{\mathcal{D}_B(\gamma,p)} \le c_2 [x]_{\mathcal{D}_A(\gamma,p)},$$

where c_1 *and* c_2 *are as in* (1).

We shall repeatedly make use of the following lemma, and for completeness we give a proof below.

LEMMA 6. *Let X be a (complex) Banach space and let assumption* (i) *of Theorem* 4 *hold true. Then*

(a) *If $x \in \mathcal{D}(A) \cap \mathcal{D}(B)$, $Ax \in \mathcal{D}(B)$ and $Bx \in \mathcal{D}(A)$, then $ABx = BAx$.*
(b) *$A^m(A-\mu I)^{-1}B^n(B-\nu I)^{-1} = B^n(B-\nu I)^{-1}A^m(A-\mu I)^{-1}$ for all $\mu \in \rho(A)$ and $\nu \in \rho(B)$ and all m, $n \in \{0,1\}$.*

2. Proofs

PROOF OF PROPOSITION 3. Recall that if X and Y are two Banach spaces with norms $\|\cdot\|_X$ and $\|\cdot\|_Y$, respectively, and if $Y \subset X$, then one defines $K(\tau,x) = \inf_{\substack{a+b=x \\ a\in X, b\in Y}} \Big(\|a\|_X + \tau\|b\|_Y\Big)$, where $x \in X$ and $\tau > 0$, and if $p \in [1,\infty]$, then $(X,Y)_{\gamma,p} \stackrel{\text{def}}{=} \{x \in X \mid \|x\|_{(X,Y)_{\gamma,p}} < \infty\}$ where

$$\|x\|_{(X,Y)_{\gamma,p}} \stackrel{\text{def}}{=} \begin{cases} \left(\int_0^\infty \left(\tau^{-\gamma}K(\tau,x)\right)^p \frac{d\tau}{\tau}\right)^{\frac{1}{p}}, & 1 \le p < \infty, \\ \sup_{\tau>0} \tau^{-\gamma}K(\tau,x), & p = \infty. \end{cases}$$

Moreover, $(X,Y)_{\gamma,\infty 0} \stackrel{\text{def}}{=} \{x \in (X,Y)_{\gamma,\infty} \mid \lim_{\tau\downarrow 0} \tau^{-\gamma}K(\tau,x) = 0\}$, with norm $\|\cdot\|_{(X,Y)_{\gamma,\infty 0}} = \|\cdot\|_{(X,Y)_{\gamma,\infty}}$.

First suppose that $x \in (X,\mathcal{D}(L))_{\gamma,p}$ and that $\tau > 0$. If $\epsilon > 0$ there are $a \in X$ and $b \in \mathcal{D}(L)$ such that $x = a+b$, $\|a\|_X \le (1+\epsilon)K(\tau,x)$ and $\tau\|Lb\|_X \le \tau\|b\|_{\mathcal{D}(L)} \le (1+\epsilon)K(\tau,x)$. If $t = \frac{1}{\tau}$ we get

$$\begin{aligned} \|L(L+tI)^{-1}x\|_X &\le \|L(L+tI)^{-1}a\|_X + \|(L+tI)^{-1}Lb\|_X \le \|a\|_X \\ &+ \|L(L+tI)^{-1}a\|_X + \|t(L+tI)^{-1}\tau Lb\|_X \le (1+M(L,0))(1+\epsilon)K(\tau,x). \end{aligned}$$

This inequality shows that $x \in \mathcal{D}_L(\gamma,p)$. Since $\epsilon > 0$ is arbitrary, a change of variables in the integral shows that $[x]_{\mathcal{D}_L(\gamma,p)} \le (1+M(L,0))\|x\|_{(x,\mathcal{D}(L))_{\gamma,p}}$.

Next suppose that $x \in \mathcal{D}_L(\gamma,p)$ and first assume that the norm in $\mathcal{D}(L)$ is $\|\underline{x}\|_{\mathcal{D}(L)} = \|L\underline{x}\|$. If $\tau > 0$ is given we take $t = \frac{1}{\tau}$, $b = t(L+tI)^{-1}x$ and $a = x - b$. Then

$$K(\tau,x) \le \|L(L+tI)^{-1}x\| + \frac{1}{t}\|t(L+tI)^{-1}Lx\| = 2\|L(L+tI)^{-1}x\|.$$

Thus we conclude that $x \in (X,\mathcal{D}(L))_{\gamma,p}$ and that $\|x\|_{(x,\mathcal{D}(L))_{\gamma,p}} \le 2[x]_{\mathcal{D}_L(\gamma,p)}$.

Finally we consider the case where the norm in $\mathcal{D}(L)$ is $\|\underline{x}\|_{\mathcal{D}(L)} = \|L\underline{x}\|_X + \|x\|_X$. By the same choice of a and b as above we get

$$\begin{aligned} K(\tau,x) &\le 2\|L(L+tI)^{-1}x\| + \tau\|t(L+tI)^{-1}x\| \\ &\le 2\|L(L+tI)^{-1}x\| + \tau M(L,0)\|x\|. \end{aligned}$$

Since $K(\tau,x) \le \|x\|$ we get $K(\tau,x) \le 2\|L(L+tI)^{-1}x\| + \min\{\tau M(L,0),1\}\|x\|$. This shows that $x \in (X,\mathcal{D}(L))_{\gamma,p}$ and a calculation gives $\|x\|_{(x,\mathcal{D}(L))_{\gamma,p}} \le 2[x]_{\mathcal{D}_L(\gamma,p)} + M(L,0)^{1-\gamma}(p\gamma(1-\gamma))^{-\frac{1}{p}}\|x\|$. □

PROOF OF LEMMA 6. (a) First let us assume that A and B are invertible and $A^{-1}B^{-1} = B^{-1}A^{-1}$. Then

$$A^{-1}B^{-1}(ABx - BAx) = B^{-1}A^{-1}ABx - A^{-1}B^{-1}BAx = 0,$$

and we get the claim since $A^{-1}B^{-1}$ is an injection. Since $\mathcal{D}(A-\alpha I)=\mathcal{D}(A)$ and $\mathcal{D}(B-\beta I)=\mathcal{D}(B)$ we have

$$ABx - BAx = (A-\alpha I)(B-\beta I)x - (B-\beta I)(A-\alpha I)x,$$

and we can use the calculation above with A replaced by $A-\alpha I$ and B replaced by $B-\beta I$ to get the claim.

(b) We use case (a) and we have only to observe that $((A-\mu I)^{-1}-\frac{1}{\alpha-\mu}I)^{-1} = -(\alpha-\mu)^2(A-\alpha I)^{-1}-(\alpha-\mu)I$ and $((B-\nu I)^{-1}-\frac{1}{\beta-\nu}I)^{-1} = -(\beta-\nu)^2(B-\beta I)^{-1}-(\beta-\nu)I$ so that the assumptions of case (a) are satisfied with A replaced by $(A-\mu I)^{-1}$ and B replaced by $(B-\nu I)^{-1}$. Thus we get the desired claim when $m=n=0$. If m or $n=1$ we have only to use the facts that $A(A-\mu I)^{-1} = I+\mu(A-\mu I)^{-1}$ and $B(B-\nu I)^{-1}=I+\nu(B-\nu I)^{-1}$ and the case already proved. □

PROOF OF THEOREM 4. Since $\phi_A+\phi_B>\pi$ we can choose a number $\theta\in(\pi-\phi_B,\phi_A)$. Let $r>0$ and let γ_r be a path in $\mathbb{C}$ with range consisting of the rays $\rho e^{\pm i\theta}$ with $\rho\geq r$ and the part of the circle re^{it} with $|t|\leq\theta$ if B is invertible and $|\pi-t|\leq\pi-\theta$ if A is invertible. We can choose r so small that the range of γ_r lies in the intersection of the resolvent sets of $-A$ and B and we take the direction of γ_r to be such that the imaginary part increases on the rays.

Our choice of of θ implies that we have the following estimates for $|\arg(z)|=\theta$:

$$\begin{aligned} &\|(A+zI)^{-1}\|\leq|z|^{-1}M(A,\theta),\\ &\|(B-zI)^{-1}\|\leq|z|^{-1}M(B,\pi-\theta). \end{aligned} \tag{2}$$

Since $(A+zI)^{-1}$ and $(B-zI)^{-1}$ are continuous on the range of γ_r we see that if we define the operator S by

$$S=\frac{1}{2\pi i}\int_{\gamma_r}(A+zI)^{-1}(B-zI)^{-1}\,dz, \tag{3}$$

then the integral converges absolutely, and S is a well-defined bounded operator.

Suppose now that A is invertible. Because $A^{-1}(A+zI)^{-1}=\frac{1}{z}A^{-1}-\frac{1}{z}(A+zI)^{-1}$ we get

$$A^{-1}S = A^{-1}\frac{1}{2\pi i}\int_{\gamma_r}\frac{1}{z}(B-zI)^{-1}\,dz - \frac{1}{2\pi i}\int_{\gamma_r}\frac{1}{z}(A+zI)^{-1}(B-zI)^{-1}\,dz.$$

By "closing" the curve γ_r through infinity with increasing argument we see by Cauchy's theorem that

$$\frac{1}{2\pi i}\int_{\gamma_r}\frac{1}{z}(B-zI)^{-1}\,dz=0.$$

Hence we conclude that

$$A^{-1}S=-\frac{1}{2\pi i}\int_{\gamma_r}\frac{1}{z}(A+zI)^{-1}(B-zI)^{-1}\,dz.$$

Next we note that

$$BA^{-1}S=-\frac{1}{2\pi i}\int_{\gamma_r}\frac{1}{z}B(B-zI)^{-1}(A+zI)^{-1}\,dz,$$

where the fact that the integral on the right-hand side converges absolutely implies that $A^{-1}S$ maps X into $\mathcal{D}(B)$. Finally, because $\frac{1}{z}B(B-zI)^{-1}=(B-zI)^{-1}+\frac{1}{z}I$

we get

$$BA^{-1}S = -\frac{1}{2\pi \mathrm{i}}\int_{\gamma_r}(B-zI)^{-1}(A+zI)^{-1}\,\mathrm{d}z - \frac{1}{2\pi \mathrm{i}}\int_{\gamma_r}\frac{1}{z}(A+zI)^{-1}\,\mathrm{d}z,$$

and by Cauchy's theorem, when we "close" the curve at infinity through decreasing argument, we have

$$\frac{1}{2\pi \mathrm{i}}\int_{\gamma_r}\frac{1}{z}(A+zI)^{-1}\,\mathrm{d}z = -A^{-1}.$$

Thus we have obtained the formula

$$S + BA^{-1}S = A^{-1}, \tag{4}$$

which is what we wanted to prove. In order to treat the case where B is invertible it suffices to observe that interchanging A and B is equivalent to changing the variable in the integral defining S.

We proceed to the proof of (b). Because

$$\begin{aligned}(A+zI)^{-1} &= \frac{1}{z}\big(I - A(A+zI)^{-1}\big),\\ (B-zI)^{-1} &= \frac{1}{z}\big(B(B-zI)^{-1} - I\big),\end{aligned} \qquad z \in \rho(-A)\cap\rho(B).$$

we have by Lemma 6,

$$\begin{aligned}&(A+zI)^{-1}(B-zI)^{-1}(Ay+By)\\ &\quad = (B-zI)^{-1}(A+zI)^{-1}Ay + (A+zI)^{-1}(B-zI)^{-1}By\\ &\qquad = \frac{1}{z}B(B-zI)^{-1}A(A+zI)^{-1}y - \frac{1}{z}A(A+zI)^{-1}y\\ &\qquad\quad + \frac{1}{z}B(B-zI)^{-1}y - \frac{1}{z}A(A+zI)^{-1}B(B-zI)^{-1}y\\ &\qquad = \frac{1}{z}B(B-zI)^{-1}y - \frac{1}{z}A(A+zI)^{-1}y.\end{aligned}$$

By the definition of S we therefore get that

$$S(Ay+By) = \frac{1}{2\pi \mathrm{i}}\int_{\gamma_r}\frac{1}{z}(B-zI)^{-1}By\,\mathrm{d}z - \frac{1}{2\pi \mathrm{i}}\int_{\gamma_r}\frac{1}{z}(A+zI)^{-1}Ay\,\mathrm{d}z.$$

If, for example A is invertible, then we can complete the path γ_r at infinity with increasing argument and the first integral becomes 0 by Cauchy's theorem. In the second integral we complete the path γ_r at infinity with decreasing argument and the integral is seen to be $-y$ by Cauchy's formula. Thus we get $S(Ay+By) = y$ as claimed.

Next we prove claim (c) and again we may without loss of generality assume that A is invertible. First suppose that $Sx \in \mathcal{D}(A)$. We have by (4) and Lemma 6

$$\begin{aligned}(B+I)^{-1}BA^{-1}Sx &= (B+I)^{-1}A^{-1}x - (B+I)^{-1}A^{-1}ASx\\ &= A^{-1}(B+I)^{-1}\big(x - ASx\big).\end{aligned}$$

On the other hand we have, again by Lemma 6,

$$(B+I)^{-1}BA^{-1}Sx = \big(I-(B+I)^{-1}\big)A^{-1}Sx = A^{-1}\big(I-(B+I)^{-1}\big)Sx.$$

Combining the two previous results, we see because A^{-1} is an injection, that

$$Sx = (B+I)^{-1}\big(Sx + x - ASx\big).$$

It follows that $Sx \in \mathcal{D}(B)$.

Next suppose that $Sx \in \mathcal{D}(B)$. Since $(A^{-1}+I)^{-1} = A(A+I)^{-1} = I-(A+I)^{-1}$, we see that the assumptions of Lemma 6 are satisfied with A replaced by A^{-1}. Since $\mathcal{D}(A^{-1}) = X$ and $Sx \in \mathcal{D}(B)$ we therefore conclude that

$$BA^{-1}Sx = A^{-1}BSx,$$

and by (4) we then have

$$Sx = A^{-1}x - A^{-1}BSx,$$

and it follows that $Sx \in \mathcal{D}(A)$ and in addition that

$$ASx + BSx = x.$$

For the proof of claim (e) we no longer make the assumption that A is invertible, only that A or B is invertible. Since $x \in \mathcal{D}_A(\gamma, p)$ we know that $x \in \mathcal{D}_A(\gamma, \infty)$ which implies that

$$\sup_{t>0} t^{\gamma} \|A(A+tI)^{-1}x\| = [x]_{\mathcal{D}_A(\gamma,\infty)} < \infty. \tag{5}$$

Because

$$A(A+se^{\pm i\theta}I)^{-1} - A(A+sI)^{-1} = (e^{\mp i\theta}-1)se^{\pm i\theta}(A+se^{\pm i\theta}I)^{-1}A(A+sI)^{-1},$$

we have

$$\|A(A+zI)^{-1}x\| \le \Big(1+2\sin(\tfrac{\theta}{2})M(A,\theta)\Big)\|A(A+|z|I)^{-1}x\|, \quad |\arg(z)| = \theta. \tag{6}$$

An immediate consequence is that $Sx \in \mathcal{D}(A)$ with

$$ASx = \frac{1}{2\pi i}\int_{\gamma_r} A(A+zI)^{-1}(B-zI)^{-1}x\,dz \tag{7}$$

because the integral converges absolutely by Lemma 6, (5) and (6). By claim (c) we know that $Sx \in \mathcal{D}(B)$ as well.

Now let $t > r$ be arbitrary. Because

$$A(A+tI)^{-1}A(A+zI)^{-1} = \frac{t}{t-z}A(A+tI)^{-1} - \frac{z}{t-z}A(A+zI)^{-1}, \tag{8}$$

we have by (7)

$$\begin{aligned} A(A+tI)^{-1}ASx = A(A+tI)^{-1}&\frac{1}{2\pi i}\int_{\gamma_r} \frac{t}{t-z}(B-zI)^{-1}x\,dz \\ &- \frac{1}{2\pi i}\int_{\gamma_r} \frac{z}{t-z}A(A+zI)^{-1}(B-zI)^{-1}x\,dz. \end{aligned}$$

When we "close" the path γ_r at infinity by increasing argument, we see that the first integral is 0 by Cauchy's theorem and we get from Lemma 6 that

$$A(A+tI)^{-1}ASx = -\frac{1}{2\pi i}\int_{\gamma_r} \frac{1}{t-z}z(B-zI)^{-1}A(A+zI)^{-1}x\,dz. \tag{9}$$

In this integral we may let $r \downarrow 0$ without changing the value of the integral, because the function we integrate is analytic and the integral over a part of the circle with radius r goes to 0 by the assumption that $\pi - \phi_B < \theta < \phi_A$, the definition of γ_r and by the assumption that A or B is invertible.

Thus we have by (5), (6), and (9)

$$
\begin{aligned}
(10)\quad t^{\gamma}\|A(A+tI)^{-1}ASx\| &\le c_3\int_0^{\infty}\frac{t^{\gamma}}{|t-se^{i\theta}|}\|A(A+sI)^{-1}x\|\,\mathrm{d}s\\
&= c_3\int_0^{\infty}\frac{(\frac{t}{s})^{\gamma}}{|\frac{t}{s}-\mathrm{e}^{\mathrm{i}\theta}|}s^{\gamma}\|A(A+sI)^{-1}x\|\,\frac{\mathrm{d}s}{s},
\end{aligned}
$$

where

$$
(11)\qquad c_3=\frac{1}{\pi}M(B,\pi-\theta)\Big(1+2\sin(\tfrac{\theta}{2})M(A,\theta)\Big).
$$

Let $f(\tau)\stackrel{\text{def}}{=}\mathrm{e}^{\tau\gamma}\|A(A+\mathrm{e}^{\tau}I)^{-1}x\|$, $g(\tau)\stackrel{\text{def}}{=}\mathrm{e}^{\tau\gamma}\|A(A+\mathrm{e}^{\tau}I)^{-1}ASx\|$, and $h(\tau)\stackrel{\text{def}}{=}\mathrm{e}^{\tau\gamma}/|\mathrm{e}^{\tau}-\mathrm{e}^{\mathrm{i}\theta}|$ where $\tau\in\mathbb{R}$. By changing variables ($s=\mathrm{e}^{\sigma}$) in the integral in (10) we conclude that

$$
(12)\qquad g(\tau)\le c_3\int_{-\infty}^{\infty}h(\tau-\sigma)f(\sigma)\,\mathrm{d}\sigma.
$$

Since convolution with an integrable function is a bounded mapping from $L^p(\mathbb{R})$, $1\le p\le\infty$, into itself and because a change of variable shows that $\|f\|_{L^p(\mathbb{R})}=[x]_{\mathcal{D}_A(\gamma,p)}$ and $\|g\|_{L^p(\mathbb{R})}=[ASx]_{\mathcal{D}_A(\gamma,p)}$, we conclude after another change of variables that

$$
[ASx]_{\mathcal{D}_A(\gamma,p)}\le c_3\int_0^{\infty}\frac{s^{\gamma-1}}{|s-\mathrm{e}^{\mathrm{i}\theta}|}\,\mathrm{d}s[x]_{\mathcal{D}_A(\gamma,p)}.
$$

Because convolution with an integrable function is a bounded mapping from the space of bounded functions converging to 0 at $+\infty$ into itself, the claim for the case $p=\infty_0$ follows as well.

Since $x\in\mathcal{D}_A(\gamma,p)$ and $BSx=x-ASx$ we see that $BSx\in\mathcal{D}_A(\gamma,p)$.

Finally we observe that if we instead of (8) use the equation

$$
B(B+tI)^{-1}(B-zI)^{-1}=\frac{t}{t+z}(B+tI)^{-1}+\frac{z}{t+z}(B-zI)^{-1},
$$

in (7), then we get

$$
\begin{aligned}
B(B+tI)^{-1}ASx &= (B+tI)^{-1}\frac{1}{2\pi\mathrm{i}}\int_{\gamma_r}\frac{t}{t+z}A(A+zI)^{-1}x\,\mathrm{d}z\\
&\quad+\frac{1}{2\pi\mathrm{i}}\int_{\gamma_r}\frac{z}{t+z}A(A+zI)^{-1}(B-zI)^{-1}x\,\mathrm{d}z.
\end{aligned}
$$

When we "close" the path at infinity with decreasing argument and use the fact that $t>r$, we see that the first integral is 0 and we conclude that we have instead of (9)

$$
B(B+tI)^{-1}ASx=\frac{1}{2\pi\mathrm{i}}\int_{\gamma_r}\frac{1}{t+z}z(B-zI)^{-1}A(A+zI)^{-1}x\,\mathrm{d}z.
$$

We see that the right-hand side of this equation only differs from the right-hand side of (9) by two minus signs and it follows that we get

$$
t^{\gamma}\|B(B+tI)^{-1}ASx\|\le c_3\int_0^{\infty}\frac{(\frac{t}{s})^{\gamma}}{|\frac{t}{s}+\mathrm{e}^{\mathrm{i}\theta}|}s^{\gamma}\|A(A+sI)^{-1}x\|\,\frac{\mathrm{d}s}{s}.
$$

Proceeding in the same way as above we conclude that

$$
[ASx]_{\mathcal{D}_B(\gamma,p)}\le c_3\int_0^{\infty}\frac{s^{\gamma-1}}{|s+\mathrm{e}^{\mathrm{i}\theta}|}\,\mathrm{d}s[x]_{\mathcal{D}_A(\gamma,p)}.
$$

It is also clear that if $x \in \mathcal{D}_A(\gamma,\infty_0)$ then $ASx \in \mathcal{D}_B(\gamma,\infty_0)$.

Finally we prove (d). First suppose that $\{y_n\}_{n=1}^{\infty} \subset \mathcal{D}(A)\cap\mathcal{D}(B)$ is such that $\lim_{n\to\infty} y_n = y$ and $\lim_{n\to\infty}(Ay_n + By_n) = x$. Then it follows from (b) and the continuity of S that $Sx = y$. If $y = 0$ it follows from (c) that $x = ASx + BSx = 0$ and we conclude that $A+B$ is closable. The general case (where we do not assume that $y = 0$) implies that $S(\overline{A+B})y = y$ for $y \in \mathcal{D}(\overline{A+B})$.

If $\mathcal{D}(A)+\mathcal{D}(B)$ is dense in X and $x \in X$ then there are sequences $\{a_n\}_{n=1}^{\infty} \subset \mathcal{D}(A)$ and $\{b_n\}_{n=1}^{\infty} \subset \mathcal{D}(B)$ such that $\lim_{n\to\infty}(a_n+b_n) = x$. Because clearly $\mathcal{D}(A) \subset \mathcal{D}_A(\frac{1}{2},\infty)$ and $\mathcal{D}(B) \subset \mathcal{D}_B(\frac{1}{2},\infty)$ we know by (c) and (e) (where we also interchange A and B) that Sa_n and $Sb_n \in \mathcal{D}(A+B)$ and $(A+B)S(a_n+b_n) = a_n+b_n \to x$ as $n\to\infty$. Because S is continuous we have $\lim_{n\to\infty} S(a_n+b_n) = Sx$, and so $(\overline{A+B})Sx = x$. □

Proof of Corollary 5. Let $\epsilon > 0$ be arbitrary and define $B_\epsilon = B+\epsilon I$. Since B_ϵ is invertible, we can apply Theorem 4, (and we can choose θ independent of ϵ). Let S_ϵ be the operator that exists according to Theorem 4.(a). Since $Ay+By+\epsilon y = x+\epsilon y$ we see from Theorem 4.(b) that $y = S_\epsilon(x+\epsilon y)$. Thus we conclude by Theorem 4.(e) that $Ay \in \mathcal{D}_A(\gamma,p)$ with

$$[Ay]_{\mathcal{D}_A(\gamma,p)} \le c_4\frac{1}{\pi}M(B_\epsilon,\pi-\theta)\Big(1+2\sin(\tfrac{\theta}{2})M(A,\theta)\Big)[x+\epsilon y]_{\mathcal{D}_A(\gamma,p)},$$

where $c_4 = \int_0^\infty \frac{s^{\gamma-1}}{|s-e^{i\theta}|}\,ds$. Since $y \in \mathcal{D}(A)$ we have $y \in \mathcal{D}_A(\gamma,p)$ and $[\epsilon y]_{\mathcal{D}_A(\gamma,p)} \to 0$ when $\epsilon \downarrow 0$. It is also clear that $\lim_{\epsilon\downarrow 0} M(B_\epsilon,\pi-\theta) = M(B,\pi-\theta)$ and we get the desired inequality for $[Ay]_{\mathcal{D}_A(\gamma,p)}$. Since $By = x - Ay$ we get the claim about By as well.

By Theorem 4 we also know that $Ay \in \mathcal{D}_{B_\epsilon}(\theta,p)$ and

$$[Ay]_{\mathcal{D}_{B_\epsilon}(\gamma,p)} \le c_5\frac{1}{\pi}M(B_\epsilon,\pi-\theta)\Big(1+2\sin(\tfrac{\theta}{2})M(A,\theta)\Big)[x+\epsilon y]_{\mathcal{D}_A(\gamma,p)},$$

where $c_5 = \int_0^\infty \frac{s^{\gamma-1}}{|s+e^{i\theta}|}\,ds$. Because we have $t^\gamma\Big(B_\epsilon(B_\epsilon+tI)^{-1} - B(B+tI)^{-1}\Big) = \epsilon^\gamma\Big((t+\epsilon)(B+(t+\epsilon)I)^{-1}\Big)\Big(t(B+tI)^{-1}\Big)(\tfrac{t}{\epsilon})^\gamma(I+\tfrac{t}{\epsilon})^{-1}$, it follows that

$$\Big|[x]_{\mathcal{D}_{B_\epsilon}(\gamma,p)} - [x]_{\mathcal{D}_B(\gamma,p)}\Big| \le \epsilon^\gamma M(B,0)^2[x]_{\mathcal{D}_I(\gamma,p)},$$

and we see that $\mathcal{D}_B(\theta,p) = \mathcal{D}_{B_\epsilon}(\theta,p)$ and that $\lim_{\epsilon\downarrow 0}[Ay]_{\mathcal{D}_{B_\epsilon}(\gamma,p)} = [Ay]_{\mathcal{D}_B(\gamma,p)}$. This completes the proof.

□

References

1. S. Bu, Ph. Clément, and S. Guerre-Delabrière, *Regularity of pairs of positive operators*, Illinois J. Math. **42** (1998), 357–370.
2. Ph. Clement, *On the method of sums of operators*, Lecture notes, Besançon Summer School 1998.
3. G. Da Prato, *Analisi superiore*, Lecture notes, Course SNS Pisa 1983–1984.
4. G. Da Prato and P. Grisvard, *Sommes d'opérateurs linéaires et équations différentielles opérationnelles*, J. Math Pures Appl. **54** (1975), 305–387.
5. P. Grisvard, *Espaces d'interpolation et équations opérationnelles*, C. R. Acad. Sci. Paris **260** (1965), 1536–1538.
6. P. Grisvard, *Équations différentielles abstraites*, Ann. Sci. École Norm. Sup.(4) **2** (1969), 311–395.

7. A. Lunardi, *Analytic semigroups and optimal regularity in parabolic problems*, Birkhäuser, Basel, 1995.

Faculty of Technical Mathematics, and Informatics, Delft University of Technology, P.O. Box 5031, 2600 GA Delft, The Netherlands
E-mail address: clement@twi.tudelft.nl

Institute of Mathematics, Helsinki University of Technology, P.O. Box 1100, FIN-02015 HUT, Finland
E-mail address: gustaf.gripenberg@hut.fi, www.math.hut.fi/~ggripenb

E-mail address: viking.hognas@hut.fi

Canadian Mathematical Society
Conference Proceedings
Volume **29**, 2000

Stochastic processes and the spectrum of Feller operators

M. Demuth

This paper is written in honor of the sixtieth birthday of Sergio Albeverio. Although it covers only a very small part of his active knowledge in several fields of mathematical physics, it is dedicated to him in appreciation of his very creative scientific power.

ABSTRACT. The fundamentals of stochastic spectral analysis are described. Feller operators are introduced as generators of Feller semigroups. The stochastic representation for the kernels of these semigroups is discussed. Giving a very short proof of the CDS^2-trace class criterion it is shown that the absolutely continuous spectrum is not affected by singular perturbations due to obstacles of finite capacity.

1. Feller operators and stochastic processes

Let

$$\{(\Omega, \mathcal{F}, P_x);\ (X(t);\ t \geq 0);\ (E, \mathcal{E})\}$$

be a strong Markov process. $(\Omega, \mathcal{F}, P_x)$ is a measure space with the σ-field $\mathcal{F}$, $(E, \mathcal{E})$ is the state space, where E is assumed to be a second countable, locally compact Hausdorff space. This process is related to a semigroup by the distribution

$$P_x\{X(t) \in B\} = \int_B p_0(t, x, y)dm(y)$$

with $t > 0, B$ a Borel set in E, m a Radon measure on $\mathcal{E}$. Here we have assumed implicitly the existence of the transition density function $p_0 : \mathbb{R}_+ \times E \times E \to \mathbb{R}_+$. The associated semigroup is given by

$$(P(t)f)(x) := E_x\{f(X(t))\} = \int_E p_0(t, x, y)f(y)dm(y), \tag{1}$$

where the right hand side can be defined as long as the integral is finite. That is true for instance for bounded functions f.

The processes allowed here are classified by the basic assumptions of stochastic spectral analysis (BASSA) for the density function p_0 (see[DvC]):

1991 *Mathematics Subject Classification.* Primary 47A10, 47D06; Secondary 60J35.

ASSUMPTION 1. $p_0 : \mathbb{R}_+ \times E \times E \to \mathbb{R}_+$ is assumed to be continuous in all three variables and symmetric in x and y. It has to satisfy the Chapman-Kolmogorov identity. For the total mass we assume

$$\int_E p_0(t,x,y)dm(y) \leq 1, \tag{2}$$

and we assume the L^1-L^∞smoothing property

$$\sup_{x,y\in E} p_0(t,x,y) =: c_t < \infty \tag{3}$$

for $t > 0$.

ASSUMPTION 2. Let C_∞ denote the set of continuous functions vanishing at infinity. Then we assume the Feller property

$$P(t)C_\infty \subseteq C_\infty. \tag{4}$$

Moreover, for $f \in C_\infty(E)$ we suppose

$$\lim_{t\to 0}[P(t)f](x) = f(x) \tag{5}$$

for all $x \in E$.

Because of the Feller property (see (4)) the process is called a *Feller process*. The assumptions imply that the semigroup $\{P(t), t \geq 0\}$ is positivity preserving, is a contraction in $L^p(E,m)$, $1 \leq p \leq \infty$, and is strongly continuous for $1 \leq p < \infty$.

DEFINITION 3. The L^2-generator of the semigroup $\{P(t), t \geq 0\}$ is called the free Feller generator and is denoted by K_0 i. e.

$$\begin{aligned} (e^{-tK_0}f)(x) &= E_x\{f(X(t))\} \\ &= \int_E p_0(t,x,y)f(y)dm(y) \end{aligned} \tag{6}$$

for all $f \in L^2(E,m)$ and *a.a.* $x \in E$.

It is obvious that K_0 is selfadjoint and positive such that its spectrum lies in $\mathbb{R}_+$, i. e.

$$\sigma(K_0) \subseteq [0,\infty). \tag{7}$$

By definition the operator e^{-tK_0} is an integral operator for any $t > 0$. In accordance to the following theory we denote its kernel sometimes by $(e^{-tK_0})(x,y)$, i. e.

$$(e^{-tK_0})(x,y) := p_0(t,x,y). \tag{8}$$

In order to introduce the stochastic representation of the integral kernel for perturbed Feller generators we define the pinned measure

$$E_x^{y,t}(A) := E_x\{p_0(t-s, X(s), y)1_A\} \tag{9}$$

where 1_A is the indicator function of a set A taken from the field $\mathcal{F}_s$, $s < t$. The left hand side in (9) does not depend on s, because the process: $s \to p_0(t-s, X(s), y)$ is a martingale on the interval $[0,t)$.

The objective of the article is to study the spectral behaviour if the free Feller generator K_0 is perturbed. We allow two kind of perturbations:

Regular perturbations of K_0 are given by potential functions $V = V_+ - V_- : E \to \mathbb{R}$ satisfying the socalled Kato-Feller conditions:

$$\limsup_{t\downarrow 0} \sup_x \int_0^t ds \int_E dm(y)\; p_0(s,x,y)V_-(y) = 0 \tag{10}$$

$$\limsup_{t\downarrow 0} \sup_x \int_0^t ds \int_B dm(y)\; p_0(s,x,y)V_+(y) = 0 \tag{11}$$

Here B is an arbitrary compact set in E. V is called a Kato-Feller potential. The form sum $K_0\dot{+}V$ generates again a Feller semigroup in $L^2(E)$, and the Feynman-Kac-formula holds:

$$(e^{-t(K_0\dot{+}V)}f)(x) = E_x\{e^{-\int_0^t V(X(u))du} f(X(t))\} \tag{12}$$

for all $f \in L^2(E,m)$ and $a.a.\ x \in E$. For any $t > 0$ the operators $e^{-t(K_0\dot{+}V)}$ are integral operators with the kernels

$$(e^{-t(K_0\dot{+}V)})(x,y) := E_x^{y,t}\{e^{-\int_0^t V(X(u))du}\}. \tag{13}$$

The other kind of perturbations of K_0 are called singular or perturbations due to obstacles. Let Γ be a closed set in E having a positive measure. Assume that $V(x) = \infty$ for $x \in \Gamma$. Set

$$\begin{aligned} S_\Gamma &:= \text{meas}\{u : u \le t, X(u) \in \Gamma\} \\ &= \int_0^t 1_\Gamma(X(u))du. \end{aligned} \tag{14}$$

S_Γ is called the spending time of $X(.)$ in Γ. For trajectories with $S_\Gamma > 0$ the expectation in (12) vanishes, because the potential is assumed to be infinitely high over Γ.

Let

$$P_\Gamma := \inf\{s > 0 : \int_0^s 1_\Gamma(X(u))du > 0\} \tag{15}$$

be the penetration time of Γ. Then the family of operators

$$E_x\{f(X(t));\ P_\Gamma > t\} \tag{16}$$

restricted to $L^2(\Sigma, m)$, where $\Sigma := E\backslash\Gamma$, provides a strongly continuous semigroup in this Hilbert space. Its generator is denoted by $(K_0)_\Sigma$. One can find conditions for K_0 and Γ such that $(K_0)_\Sigma$ corresponds to the free Feller generator K_0 with Dirichlet boundary conditions on the boundary of Γ. The associated semigroup consists of integral operators in $L^2(\Sigma, m)$ with the kernels

$$E_x^{y,t}\{P_\Gamma > t\} =: (e^{-t(K_0)_\Sigma})(x,y). \tag{17}$$

Let

$$H_\Gamma := \inf\{s, s > 0,\ X(s) \in \Gamma\}$$

be the first hitting time of Γ. Let $[\Gamma]^r$ be the set of H_Γ-regular points for Γ. If $[\Gamma]^r$ coincides with the H_Γ-regular points of the interior of Γ we have $H_\Gamma = S_\Gamma$ P_x - almost surely for almost all $x \in E$. In that case

$$(e^{-t(K_0)_\Sigma})(x,y) = E_x^{y,t}\{H_\Gamma > t\}. \tag{18}$$

Both perturbations can be combined. Let V be a Kato-Feller potential. Le Γ be a singularity region, i. e. a closed set in E with positive measure and for which $[\Gamma]^r = [\mathrm{int}(\Gamma)]^r$. Then

$$E_x\{e^{-\int_0^t V(X(u))du} f(X(t)), H_\Gamma > t\} \tag{19}$$

forms a strongly continuous Feller semigroup in $L^2(\Sigma, m)$. Its generator is denoted by $(K_0 \dot{+} V)_\Sigma$. The kernels of the associated operators are

$$E_x^{y,t}\{e^{-\int_0^t V(X(u))du}, H_\Gamma > t\} =: (e^{-t(K_0 \dot{+} V)_\Sigma})(x,y) \tag{20}$$

for $a.a.$ $x, y \in \Sigma$.

The operators K_0, $(K_0)_\Sigma$, $K_0 \dot{+} V$, $(K_0 \dot{+} V)_\Sigma$ generate semigroups having the Feller property (see (4)). For that they are called Feller operators, $(K_0)_\Sigma$ is a singularly perturbed Feller operator, and $K_0 + V$ is a regularly perturbed Feller operator.

The theory on Feller operators described above is a very short summary of the first three chapters in the book by Demuth, van Casteren [DvC]. The interested reader can find the details there.

2. The absolutely continuous spectrum of Feller operators

In stochastic spectral analysis one studies the behaviour of the spectra of Feller operators if they are perturbed regularly or singularly in the sense as described in the first section. Here we will restrict ourselves to changes of $\sigma_{ac}(K_0 \dot{+} V)$, i. e. of the absolutely continuous spectrum of $K_0 \dot{+} V$, if we impose additionally Dirichlet boundary conditions.

From time-independent scattering theory (see [D]) we know that $\sigma_{ac}(K_0 \dot{+} V)$ remains stable if for some $t > 0$

$$e^{-t(K_0 \dot{+} V)_\Sigma}(e^{-t(K_0 \dot{+} V)_\Sigma} J - J\, e^{-t(K_0 \dot{+} V)})e^{-t(K_0 \dot{+} V)} \in \mathcal{B}_{\text{trace}} \tag{21}$$

and if

$$e^{-t(K_0 \dot{+} V)_\Sigma} J - J\, e^{-t(K_0 \dot{+} V)} \in \mathcal{B}_{\text{comp}} \tag{22}$$

where $\mathcal{B}_{\text{trace}}$, $\mathcal{B}_{\text{comp}}$ denotes the set of trace class and compact operators. respectively. The operator J is given by $Jf := f/_\Sigma$.

For the trace class condition in (21) one can use the CDS^2-trace class criterion for products of integral operators given by van Casteren, Demuth, Stollmann, Stolz [CDSS]. It says: If A and B are integral operators in $L^2(E, m)$ with measurable kernels, then the integral operator AB is trace class if

$$\int_E \|A(.,x)\|_{L^2} \, \|B(x,.)\|_{L^2} \, dm(x) < \infty. \tag{23}$$

The proof in [CDSS] can be abbreviated. Let M_φ be the multiplication operator with a function φ, specified in (26). Then $AB \in \mathcal{B}_{\text{trace}}$ if AM_φ and $M_{\frac{1}{\varphi}}B$ are Hilbert-Schmidt operators, i. e. if both

$$\begin{aligned} (24) \qquad & \int dm(x) \int dm(y)|A(x,y)|^2|\varphi(y)|^2 \\ = & \int dm(y)\|A(.,y)\|_{L^2}^2|\varphi(y)|^2 \end{aligned}$$

and

$$(25) \qquad \int dm(y)\|B(y,.)\|_{L^2}^2 \, \frac{1}{|\varphi(y)|^2}$$

are finite. These conditions are satisfied symmetrically if we choose

$$(26) \qquad |\varphi(y)|^2 = \frac{\|B(y,.)\|_{L^2}}{\|A(.,y)\|_{L^2}} \, .$$

The CDS^2-trace class criterion can be used for the sandwiched semigroup difference as follows. Take $t = 2$ and

$$(27) \qquad A \quad = e^{-(K_0 \dot{+} V)_\Sigma}$$

$$(28) \qquad B \quad = e^{-(K_0 \dot{+} V)_\Sigma} D \; e^{-2(K_0 \dot{+} V)_\Sigma}$$

with

$$(29) \qquad D \; = e^{-2(K_0 + V)_\Sigma} J - J \; e^{-2(K_0 \dot{+} V)}$$

Then we know (see [DvC], Chapter 2 A)

$$\begin{aligned} (30) \qquad & A(x,y) \\ = \; & |(e^{-(K_0 \dot{+} V)_\Sigma})(x,y)| \\ = \; & E_x^{y,1}\{e^{-\int_0^1 V(X(u))du}, \; H_\Gamma > 1\} \\ \leq \; & E_x^{y,1}\{e^{-\int_0^1 V(X(u))du}\} \\ \leq \; & c \cdot [\sup_{x,y} p_0(\frac{1}{2}, x, y)]^{1/2} \; [p_0(1,x,y)]^{1/2} \\ \leq \; & c \cdot p_0(1,x,y)^{1/2}. \end{aligned}$$

We will not distinguish between different constants.

Hence

$$(31) \qquad \|A(.,x)\|_{L^2} \leq c \sqrt{\int_E p_0(1,x,y)dm(y)}$$

is finite.

For B in (28) we have

(32)
$$\begin{aligned}&\int \|B(x,.)\|_{L^2}\, dm(x)\\ &= \int dm(x)\sqrt{\int dm(y)\, |B(x,y)|^2}\\ &\le c\int dm(x)\int dm(u)\int dm(v)(e^{-(K_0+V)_\Sigma})(x,v)\, |D(v,u)|\\ &\le c\int dm(u)\int dm(v)\, E_u^{v,2}\{e^{-\int_0^2 V(X(\sigma))d\sigma}, H_\Gamma < 2\}\\ &= c\int dm(u) E_u\{e^{-\int_0^2 V(X(\sigma))d\sigma}, H_\Gamma < 2\}.\end{aligned}$$

Hence as a first information we obtained the sandwiched semigroup in (21) to be a trace class operator if the function

(33)
$$E_x\{e^{-\int_0^2 V(X(\sigma))d\sigma},\ H_\Gamma < 2\}$$

is in $L^1(E,m)$.

The consideration of the condition in (22) is easier. Take $t=4$, then

(34)
$$e^{-2(K_0\dot{+}V)_\Sigma} D + D\, e^{-2(K_0\dot{+}V)}$$

is a Hilbert-Schmidt operator if the function

(35)
$$E_x\{e^{-\int_0^2 V(X(\sigma))d\sigma},\ H_\Gamma < 2\}$$

is in $L^2(E,m)$.

THEOREM 4. *Let K_0 be a free Feller operator. Let V be a Kato-Feller potential. Let Γ be a singularity region as described for (18). Then we have the stability of the absolutely continuous spectra, i. e.*

(36)
$$\sigma_{ac}(K_0\dot{+}V) = \sigma_{ac}((K_0\dot{+}V)_\Sigma)$$

if the function

(37)
$$\tilde{v}_{\Gamma,V}(x) := E_x\{e^{-\int_0^2 V(X(\sigma))d\sigma},\ H_\Gamma < 2\}$$

is in $L^1(E,m)\cap L^2(E,m)$.

The function $\tilde{v}_{\Gamma,V}(.)$ is related to the equilibrium potential of Γ defined as

(38)
$$v_\Gamma(x) := E_x\{e^{-H_\Gamma},\ H_\Gamma < \infty\}.$$

For instance

(39)
$$(\tilde{v}_{\Gamma,V}(x))^2 \le E_x\{e^{-2\int_0^2 V(X(\sigma))d\sigma}\}\cdot E_x\{H_\Gamma < 2\}$$

Because V is a Kato-Feller potential,

(40)
$$\sup_{x\in E} E_x\{e^{-2\int_0^t V(X(\sigma))d\sigma}\} \le c\, e^{ct}$$

for any $t \geq 0$. Hence

$$(41) \qquad (\tilde{v}_{\Gamma,V}(x))^2 \leq c \cdot E_x\{e^{H_\Gamma} e^{-H_\Gamma}, H_\Gamma < 2\} \leq c \cdot e^2 \cdot v_\Gamma(x).$$

That implies condition (22) to be valid if $v_\Gamma(.) \in L^1(E,m)$.

On the other hand for potentials bounded from below by $V(x) \geq -a,\ a > 0$, for all $X \in E$,

$$(42) \qquad \tilde{v}_{\Gamma,V}(x) \leq c \cdot v_\Gamma(x)$$

is obvious. Another possibility is to decompose $\tilde{v}_{\Gamma,V}$ by

$$(43) \qquad \tilde{v}_{\Gamma,V}(x) \leq 2E_x\{|e^{-\int_0^2 V(X(\sigma))d\sigma} - 1|\} + E_x\{H_\Gamma < 2\}.$$

From (42) and (43) it follows that the trace class condition in (21) is given if $v_\Gamma(.) \in L^1(E,m)$ together with some additional conditions on the potential function $V(.)$.

The L^1-norm of the equilibrium potential determines the capacity of the set Γ. The capacity is defined by

$$(44) \qquad \text{cap}(\Gamma) := \inf \quad \{\|K_0^{1/2} f\|^2 + \|f\|^2;\ f \in \text{dom}\ (K_0^{1/2}),\ f \geq 1_U, U open, U \supseteq \Gamma\}.$$

This infinum is taken by the equilibrium potential and it holds

$$(45) \qquad \text{cap}(\Gamma) = \int_E v_\Gamma(x)dm(x)$$

(see Fukushima [F], p. 62). Thus we can formulate the following result.

COROLLARY 5. *Let K_0 be a free Feller operator. Let V be a Kato-Feller potential which is bounded from below or for which$E_x\{|1 - e^{-\int_0^2 V(X(\sigma))d\sigma}|\}$ is in L^1 (this is manily an L^1-condition for the potential). Let Γ be a singularity region as described for (18) with finite capacity.*

Then we obtain the invariance of the absolutely continuous spectra, i. e.

$$\sigma_{ac}(K_0 \dot{+} V) = \sigma_{ac}((K_0 \dot{+} V)_\Sigma).$$

References

[CDSS] J. v. Casteren, M. Demuth, P. Stollmann, G. Stolz; Trace norm estimates for products of integral operators and diffusion semigroups; Integral Equat. and Op. Theory 23 (1995), 145 - 153.

[D] M. Demuth; Integral conditions for the asymptotic completeness of two-space scattering systems; Helv. Phys. Acta 71 (1998), 117 - 132.

[DvC] M. Demuth, Jan van Casteren; Stochastic Spectral Theory for Selfadjoint Feller operators; to appear in Birkhäuser Verlag, 2000.

[F] M. Fukushima; Dirichlet Forms and Markov Processes; North Holland, Amsterdam 1980.

REMARK. The author wants to thank M. Korey for the translation of the following acknowledgement.

Acknowledgement

More then twenty years ago I met Sergio Albeverio at a conference on mathematical physics in Torun. At that time, he was one of the last scientific globetrotters. Later he assumed his first position in Bielefeld.

Many colleagues know, or have directly experienced themselves, how Sergio is always interested in strengthening scientific research and supporting individual researchers. During the era of the Iron Curtain and during the time the Berlin Wall still stood, he kept in contact with me. In the 1980's this helped me maintain both, my motivation for and the pleasure of doing mathematics. I should like to use the occasion of his 60th birthday to thank him sincerely and heartily for his support in many matters.

Our co-operation took a new dimension after 1990. We initiated and pursued a number of projects, both jointly and with other colleagues. In this context I should like to mention B.-W. Schulze's proposal to found a new Max-Planck-Institute of Mathematics in Leipzig. Sergio Albeverio made an important contribution to this proposal (in addition to B.-W. Schulze, R. Seiler, and myself). I am pleased that our collaboration on the editorial board of the series 'Advances in Partial Differential Equations' (Birkhäuser) will be continuing in the coming years.

The personal and scientific contacts I have had with Sergio are among the most positive experiences of my life.

INSTITUTE OF MATHEMATICS, TU CLAUSTHAL, ERZSTRASSE 1, 38678 CLAUSTHAL-ZELLERFELD, GERMANY

E-mail address: demuth@math.tu-clausthal.de

Canadian Mathematical Society
Conference Proceedings
Volume **29**, 2000

The energy representation has no non-zero fixed vectors

Bruce K. Driver† and Brian C. Hall*

ABSTRACT. We consider the "energy representation" W of the group $\mathcal{G}$ of smooth mappings of a Riemannian manifold M into a compact Lie group G. Our main result is that if $W(g) f = f$ for all $g \in \mathcal{G}$, then $f = 0$. In the language of quantum field theory this says that there are no "states." Our result follows from the irreducibility of the energy representation whenever the irreducibility theorems of Ismagilov, Gelfand–Graev–Veršik, Albeverio–Høegh-Krohn–Testard, or Wallach apply. Our result, however, applies in general, even in cases where the energy representation is known to be reducible.

We work in the more general context of the "Gaussian regular representation" of the Euclidean group of a real separable Hilbert space. We show that if a function is invariant under the action of any subgroup of the Euclidean group that has unbounded orbits, then this function must be identically zero. Our result about the energy representation is a special case.

1. Introduction and statement of result

Let M be a compact Riemannian manifold without boundary, with dimension at least one. Let G (the "structure group") be a compact connected Lie group, and let $\mathcal{G}$ (the "gauge group") be the set of smooth mappings of M into G, forming a group under pointwise multiplication. Let $\mathfrak{g}$ be the Lie algebra of G, and fix on $\mathfrak{g}$ an Ad-G-invariant inner product. In addition, fix a positive function ω on M.

We consider the real Hilbert space H of square-integrable, $\mathfrak{g}$-valued 1-forms on M. So an element A of H assigns to a point $m \in M$ an element of $T_m^*M \otimes \mathfrak{g} \cong \operatorname{Hom}(T_mM, \mathfrak{g})$. The inner product on H is given explicitly by

$$\langle A, B\rangle = \int_M \operatorname{tr}\left(A(m)^* B(m)\right) \omega(m)\, dm, \quad A, B \in H, \tag{1}$$

where $A(m)^*$ denotes the adjoint of $A(m) : T_mM \to \mathfrak{g}$ relative the given inner products on T_mM and on $\mathfrak{g}$, and where dm is the Riemannian volume measure on M. If $\dim M \neq 2$, the function ω can be absorbed into the definition of the metric; if $\dim M = 2$, the inner product on H depends only on the conformal class of the metric on M.

2000 *Mathematics Subject Classification.* Primary 22E65, 60H40, 81T13; Secondary 58Z05, 81S30.

†This research was partially supported by NSF Grants DMS 96-12651 and DMS 99-71036.
*Supported by an NSF Postdoctoral Fellowship at Univ. of California, San Diego.

For $g \in \mathcal{G}$ and $A \in H$ we define the action of g on A by

$$(g \cdot A)(v_m) = \mathrm{Ad}_{g(m)} A(v_m) - dg(v_m)\, g(m)^{-1}, \tag{2}$$

where v is a vector field on M. The action of $\mathcal{G}$ on H is isometric, the first term being an invertible isometric linear transformation applied to A and the second term being a translation.

We now consider the Gaussian measure ρ associated to H, given formally by the expression

$$d\rho(A) = \frac{1}{Z} e^{-|A|^2/2} \mathcal{D}A,$$

where $|A|^2 := \langle A, A \rangle$, $\mathcal{D}A$ is the fictitious Lebesgue measure on H, and Z is supposed to be a normalizing constant. Rigorously, ρ is defined not on H itself, but on some extension B of H, in this case a suitable space of distributional ($\mathfrak{g}$-valued) 1-forms on M. The measure ρ may be characterized, for example, by the condition that

$$\int_B e^{i\phi(A)}\, d\rho(A) = \exp\left\{-\frac{1}{2}|\phi|_{H^*}^2\right\}$$

for all continuous linear functionals ϕ on B. Here $|\phi|_{H^*}$ is the norm of ϕ as a linear functional on $H \subset B$. The action of $\mathcal{G}$ on H extends continuously to an action of $\mathcal{G}$ on B, and this action leaves the measure ρ quasi-invariant.

Finally, we consider a unitary representation of $\mathcal{G}$ acting in $L^2(B, \rho)$ given by

$$W(g) f(A) = e^{-\left|dgg^{-1}\right|^2/4} e^{-\left\langle A, dgg^{-1}\right\rangle/2} f\left(g^{-1} \cdot A\right). \tag{3}$$

The "multiplier" in front of $f\left(g^{-1} \cdot A\right)$ is the square root of the Radon-Nikodym derivative $d\rho\left(g^{-1} \cdot A\right)/d\rho(A)$. More details on the definition of $W(g)$ may be found in the next section; in particular, $W(g) = U(Ad_g, -dg\, g^{-1})$ in the notation of Definition 8 below.

We may now state the main result of this paper.

THEOREM 1. *If $f \in L^2(B, \rho)$ satisfies*

$$W(g) f = f$$

for all $g \in \mathcal{G}$, then $f = 0$.

Proof. This is an immediate consequence of Theorem 14 and Proposition 2. □

Thus in the language of quantum field theory, there are no "states." This result is expected in view of the non-compactness of the gauge orbits. However, we know of no proof of this in general in the literature.

The representation W is unitarily equivalent to the "energy representation" introduced by Ismagilov [I] in the case $G = SU(2)$ and by Gelfand, Graev, and Veršik [**GGV1**] in general. See Section 4 for details. If G is semisimple and $\dim M \geq 3$, or under additional assumptions on ω if G is semisimple and $\dim M = 2$, the energy representation is irreducible [**I, GGV1, GGV2, AKT, Wa**]. Thus, in these cases, W is also irreducible and Theorem 1 follows immediately. On the other hand, if $\dim M = 1$ or if G is commutative, then the energy representation is reducible; in any case it is desirable to have a simple direct proof of Theorem 1, which is much weaker than irreducibility.

Theorem 1 can doubtless be proved in many ways, for example, using the same techniques as in the proofs of irreducibility. Our proof will make use of coherent states and the Segal–Bargmann transform. The only properties of $\mathcal{G}$ that we will use are that $\mathcal{G}$ acts isometrically on H and the following. (See Theorem 14 below.)

PROPOSITION 2. *For any $A \in H$, the orbit of A under the action of $\mathcal{G}$ is unbounded in H.*

Proof. Since the inner product on $\mathfrak{g}$ is Ad-G-invariant, the norm of the first term in (2) is equal to the norm of A and hence independent of g. Meanwhile, the second term in (2) is independent of A. It follows that if any one $\mathcal{G}$-orbit is unbounded, then all the others are as well. Thus to show that the orbits of $\mathcal{G}$ are unbounded, we must merely show that $\sup_g \left|g^{-1}dg\right|_H = \infty$,which is evident, say, by considering g of the form $g(x) = e^{f(x)\xi}$ with $\xi \in \mathfrak{g}$ and $f \in C^\infty(M)$. $\square$

The authors thank Nolan Wallach for valuable discussions concerning the energy representation and thank Stephen Sontz for comments on and corrections to an earlier version of this paper.

2. The regular representation and Segal–Bargmann transform

Having noted that the representation W arises from an isometric action of $\mathcal{G}$ on H, and having recorded Proposition 2, we no longer need to concern ourselves with the manifold M or with the precise nature of the gauge action in (2). We will prove a general result (Theorem 14) about subgroups of the isometry group of a Hilbert space H; Theorem 1 is an immediate consequence.

2.1. Finite dimensions. Let H be a real Hilbert space, which in this subsection we take to have finite dimension d. Let $\mathcal{E}(H)$ be the Euclidean group, that is, the set of transformations of the form

$$x \to Rx + h,$$

where R is an orthogonal linear transformation and h is an element of H. We denote this transformation as (R,h) and observe that the composition of Euclidean transformations takes the form $(R,h) \circ (S,k) = (RS, Rk+h)$. Let dx be the Lebesgue measure on H, normalized so that a unit cube has volume one. We now define a unitary representation $\tilde{U}$ of $\mathcal{E}(H)$ acting on $L^2(H, dx)$ by

$$\tilde{U}(R,h) f(x) = f((R,h)^{-1} x) = f\left(R^{-1}(x-h)\right).$$

In preparation for the passage to the infinite-dimensional limit, we make a unitary change from Lebesgue measure to Gaussian measure on H. So define the measure ρ on H to be

$$d\rho(x) = (2\pi)^{-d/2} e^{-|x|^2/2} dx,$$

where $d = \dim H$; and consider the unitary map $\Phi : L^2(H, dx) \to L^2(H, \rho)$ given by

$$\Phi f(x) = (2\pi)^{d/4} e^{|x|^2/4} f(x). \tag{4}$$

Then define a unitary representation of $\mathcal{E}(H)$ acting on $L^2(H,\rho)$ by

$$U(R,h) = \Phi \tilde{U}(R,h) \Phi^{-1}.$$

We will call $\tilde{U}$ the regular representation of $\mathcal{E}(H)$ and U the Gaussian regular representation. Computing U explicitly leads to the following definition.

DEFINITION 3. *The* ***Gaussian regular representation*** *is the unitary representation U of $\mathcal{E}(H)$ acting on $L^2(H,\rho)$ given by*

$$U(R,h)f(x) = e^{-|h|^2/4}e^{\langle h,x\rangle/2}f\left(R^{-1}(x-h)\right).$$

We next turn to the Segal–Bargmann transform [**B, S1, S2, S3**], which converts the Gaussian regular representation to the "holomorphic regular representation." Let $H_{\mathbb{C}} = H + iH$ be the complexification of H and let μ be the measure on $H_{\mathbb{C}}$ given by

$$d\mu(z) = \pi^{-d}e^{-|z|^2}\,dz,$$

where $|x+iy|^2 = |x|^2 + |y|^2$ and dz denotes $2d$-dimensional Lebesgue measure on $H_{\mathbb{C}}$.

DEFINITION 4. *Let $\mathcal{H}L^2(H_{\mathbb{C}},\mu)$ denote the space of entire holomorphic functions on $H_{\mathbb{C}}$ which are square-integrable with respect to μ.*

It is easily shown that $\mathcal{H}L^2(H_{\mathbb{C}},\mu)$ is a closed subspace of $L^2(H_{\mathbb{C}},\mu)$ and therefore a Hilbert space. We will now define a unitary representation of $\mathcal{E}(H)$ on $\mathcal{H}L^2(H_{\mathbb{C}},\mu)$. Let (z,w) denote the complex *bilinear* (not Hermitian) extension of the inner product on H to $H_{\mathbb{C}}$. Given $R \in O(H)$ we consider the unique complex-linear extension of R to $H_{\mathbb{C}}$, also denoted R.

DEFINITION 5. *The* ***holomorphic regular representation*** *of $\mathcal{E}(H)$ is the unitary representation V of $\mathcal{E}(H)$ on $\mathcal{H}L^2(H_{\mathbb{C}},\mu)$ by*

$$V(R,h)F(z) = e^{-|h|^2/8}e^{(h,z)/2}F\left(R^{-1}(z-h/2)\right).$$

It straightforward computation to check directly that V is unitary and a representation. This result also is a consequence of Proposition 7 below.

THEOREM 6 (Segal–Bargmann transform). *For $f \in \mathcal{H}L^2(H_{\mathbb{C}},\mu)$, let Sf be the function on $H_{\mathbb{C}}$ given by*

$$Sf(z) = e^{-(z,z)/2}\int_H f(x)\,e^{(z,x)}\,d\rho(x).$$

Then the integral that defines Sf is absolutely convergent and depends holomorphically on z. Moreover, the map $f \to Sf$ is a unitary map of $L^2(H,\rho)$ onto $\mathcal{H}L^2(H_{\mathbb{C}},\mu)$.

This is well known and will not be proved here. See [**S3, BSZ**], or [**B**], which considers $S\Phi$, where Φ is the change of measure (4). Note that for $z \in H$, $e^{-(z,z)/2}e^{(z,x)}$ is the Radon–Nikodym derivative $d\rho(z-x)/d\rho(x)$. So Sf is the analytic continuation to $H_{\mathbb{C}}$ of the function

$$Sf(z) = \int_H f(x)\,d\rho(z-x) = \int_H f(z-x)\,d\rho(x), \quad z \in H.$$

PROPOSITION 7. *The Segal–Bargmann transform intertwines the Gaussian regular representation and the holomorphic regular representation. That is, for $(R,h) \in \mathcal{E}(H)$, $V(R,h) = SU(R,h)S^{-1}$.*

The proof of this result is an explicit computation which is carried out in the proof of Proposition 13 below in the infinite-dimensional case.

2.2. Infinite dimensions. Now let (H, B, ρ) be an abstract Wiener space [**G**]. This means that B is a real, separable Banach space; H is an infinite-dimensional, real, separable Hilbert space continuously embedded into B with dense image; and ρ is a probability measure on B such that

$$\int_B e^{i\phi(x)}\, d\rho\,(x) = \exp\left\{-\frac{1}{2}\,|\phi|_{H^*}^2\right\}$$

for all $\phi \in B^* \subset H^*$. In this the infinite-dimensional case, the Gaussian measure ρ does not exist on H, but must live on some larger space B; in particular, $\rho\,(H) = 0$.

As in the linear case, the Euclidean group $\mathcal{E}\,(H)$ is the set of transformations of H of the form $x \to Rx + h$, with $h \in H$ and R an invertible, isometric linear transformation on H. We want $\mathcal{E}\,(H)$ to act on $L^2\,(B, \rho)$ as in the finite-dimensional case. This requires some care, since now elements of $\mathcal{E}\,(H)$ are transformations of H, but our functions are defined on B.

First, there is a dense subspace of H, denoted B^*, such that for all $h \in B^*$, $\langle h, \cdot\rangle$ extends continuously from H to B. Given $h \in H$, choose $h_n \in X$ with $h_n \to h$. It is well known that $\langle h_n, \cdot\rangle$ converges in $L^2\,(B, \rho)$ to a limit $\langle h, \cdot\rangle \in L^2(H, \rho)$, independent of the choice of approximating sequence. It is easily shown that the map $h \to \langle h, \cdot\rangle$ is linear from H to $L^2\,(B, \rho)$, and that for all $h, k \in H$, $\langle h, x + k\rangle = \langle h, x\rangle + \langle h, k\rangle$ for ρ-a.e. x. The expression $\langle h, x + k\rangle$ is well defined for ρ - a.e. x because of the Cameron–Martin theorem, which asserts that

$$d\rho(x + h) = e^{-\langle h,x\rangle - |h|^2/2} d\rho(x) \qquad \forall\, h \in H. \tag{5}$$

Now we define an element f of $L^2\,(B, \rho)$ to be a **cylinder function** if there is a finite orthonormal set $\{h_1, \cdots, h_n\}$ in H and a measurable function ϕ on $\mathbb{R}^n$ such that

$$f\,(x) = \phi\,(\langle h_1, x\rangle\,, \cdots, \langle h_n, x\rangle)\,. \tag{6}$$

Suppose that R is an isometric invertible linear transformation of H and that f is a cylinder function given by (6). Then we *define* $f \circ R^{-1}$ to be the cylinder function given by

$$f \circ R^{-1}\,(x) = \phi\,(\langle Rh_1, x\rangle\,, \cdots, \langle Rh_n, x\rangle)\,. \tag{7}$$

Although a given cylinder function f can be represented in form (6) in many different ways, it is not hard to see that $f \circ R^{-1}$ is independent of the representation. It is straightforward to verify that the map $f \to f \circ R^{-1}$ is isometric on $L^2\,(B, \rho)$ and that $\left(f \circ R^{-1}\right) \circ R = f$. Hence this map extends to a unitary map of $L^2\,(B, \rho)$ to itself, which we still denote as $f \to f \circ R^{-1}$. If R^{-1} happens to have a continuous extension from H to B, then for all $f \in L^2\,(B, \rho)\,,\ f \circ R^{-1}$ will coincide with the composition of f with this extension.

Finally, by the Cameron–Martin theorem, the map $T_h : L^2\,(B, \rho) \to L^2\,(B, \rho)$ given by

$$T_h f\,(x) = e^{-|h|^2/4} e^{\langle h,x\rangle/2} f\,(x - h)$$

is unitary for all $h \in H$.

So we make the following definition.

DEFINITION 8. *The **Gaussian regular representation** of $\mathcal{E}\,(H)$ on $L^2\,(B, \mu)$ is given by*

$$U\,(R, h)\, f = T_h\left(f \circ R^{-1}\right),$$

that is,

$$U(R,h)\,f(x) = e^{-|h|^2/4} e^{\langle h,x\rangle/2} f \circ R^{-1}(x-h).$$

We turn now to the infinite-dimensional version of the Segal–Bargmann transform. While this may be defined in terms of a complex abstract Wiener space as in [**DH, HS**], we follow [**S3**] and consider a Hilbert space of holomorphic functions on $H_{\mathbb{C}}$, where we use the standard notion of holomorphicity on a Banach space [**HP**].

DEFINITION 9. *A function $F : H_{\mathbb{C}} \to \mathbb{C}$ is **holomorphic** if F is locally bounded and the restriction of F to every finite-dimensional subspace of $H_{\mathbb{C}}$ is holomorphic.*

DEFINITION 10. *The **Segal-Bargmann space** over $H_{\mathbb{C}}$, denoted $\mathcal{H}L^2(H_{\mathbb{C}},\mu)$, is the set of holomorphic functions F on $H_{\mathbb{C}}$ such that*

$$\sup_X \int_X |F(z)|^2 \, d\mu_X(z) < \infty,$$

where the supremum is over finite-dimensional subspaces X of $H_{\mathbb{C}}$, and where μ_X is the Gaussian probability measure on X given by $d\mu_X(z) = \pi^{-\dim X} e^{-|z|^2}\, dz$. For $F \in \mathcal{H}L^2(H_{\mathbb{C}},\mu)$ we define $\|F\|$ to be the square root of the above supremum.

It is a consequence of the Krée Skeleton Theorem [**K**] that this defines a Hilbertian norm on $\mathcal{H}L^2(H_{\mathbb{C}},\mu)$ and that $\mathcal{H}L^2(H_{\mathbb{C}},\mu)$ is complete in this norm. Thus $\mathcal{H}L^2(H_{\mathbb{C}},\mu)$ is a complex Hilbert space. We remark that if $F : H_{\mathbb{C}} \to \mathbb{C}$ is holomorphic on each finite-dimensional subspace and the supremum in the above definition is finite, then F is automatically locally bounded. (See (11) below.)

Note that $\mathcal{H}L^2(H_{\mathbb{C}},\mu)$ is an abuse of notation, in that there does *not* exist a measure μ on $H_{\mathbb{C}}$ such that $\|F\|$ is the L^2 norm of F with respect to μ. We may regard μ either as a cylinder-set measure on $H_{\mathbb{C}}$ or as a honest measure on a suitable extension $B_{\mathbb{C}}$ of $H_{\mathbb{C}}$–see [**G**].

Since our holomorphic functions are truly functions on $H_{\mathbb{C}}$ (not on some Banach space containing $H_{\mathbb{C}}$), there is no trouble in extending the definition of the holomorphic regular representation to infinite dimensions, and in verifying that it is unitary.

DEFINITION 11. *The **holomorphic regular representation** of $\mathcal{E}(H)$ is the unitary representation V of $\mathcal{E}(H)$ on $\mathcal{H}L^2(H_{\mathbb{C}},\mu)$ by*

$$V(R,h)\,F(z) = e^{-|h|^2/8} e^{(h,z)/2} F\left(R^{-1}(z-h/2)\right).$$

If we interpret (z,x) as $\langle \mathrm{Re} z, x\rangle + i\,\langle \mathrm{Im} z, x\rangle$ as an ρ-almost everywhere defined function on B, then the definition of the Segal–Bargmann transform carries over without change from the finite-dimensional case.

THEOREM 12 (Segal–Bargmann transform). *For all $f \in L^2(B,\rho)$, define a function Sf on $H_{\mathbb{C}}$ by*

$$Sf(z) = e^{-(z,z)/2} \int_B e^{(z,x)} f(x)\, d\rho(x).$$

Then the integral that defines Sf is absolutely convergent and depends holomorphically on z. Moreover, S is a unitary map of $L^2(B,\rho)$ onto $\mathcal{H}L^2(H_{\mathbb{C}},\mu)$.

For a proof see [**S3**].

PROPOSITION 13. *The Segal–Bargmann transform intertwines the Gaussian regular representation and the holomorphic regular representation. That is, for* $(R,h) \in \mathcal{E}(H)$, $V(R,h) = SU(R,h)S^{-1}$.

Proof. Note that any $(R,h) \in \mathcal{E}(H)$ may be factored as $(I,h)(R,0)$. So it suffices to consider rotations and translations separately. Note that by inspection, $U(R,h) = U(I,h)U(R,0)$ and similarly for V, as must be the case, since U and V are representations.

It is easily seen that a holomorphic function on $H_{\mathbb{C}}$ is determined by its values on H. So it suffices to show that $SU(R,h)f(z) = V(R,h)Sf(z)$ for all $f \in L^2(B,\rho)$ and for all $z \in H$. We begin by computing on a cylinder function f of the form (6). For any fixed $z \in H$ we have

$$\begin{aligned} Sf(z) &= \int_B \phi(\langle h_1, z-x\rangle, \cdots, \langle h_n, z-x\rangle)\, d\rho(x) \\ &= \int_B \phi(\langle h_1, z\rangle - \langle h_1, x\rangle, \cdots, \langle h_n, z\rangle - \langle h_n, x\rangle)\, d\rho(x) \\ &= \int_{\mathbb{R}^n} \phi(\langle h_1, z\rangle - x_1, \cdots, \langle h_n, z\rangle - x_n)\, e^{-(x_1^2+\cdots+x_n^2)/2} \frac{dx_1\cdots dx_n}{(2\pi)^{n/2}} \\ &= (S\phi)(\langle h_1, z\rangle, \cdots, \langle h_n, z\rangle). \end{aligned}$$

Here $S\phi$ means the Segal–Bargmann transform of ϕ defined on the finite-dimensional Hilbert space $\mathbb{C}^n$. The third equality is because the joint distribution of $\{\langle h_i, x\rangle\}$ is the given measure on $\mathbb{R}^n$.

But now recall the definition (7) of $f \circ R^{-1}$ on cylinder functions. Since R is isometric, $\{Rh_1, \cdots, Rh_n\}$ is again an orthonormal set, and so by the previous paragraph we have

$$\begin{aligned} S\left(f \circ R^{-1}\right)(z) &= (S\phi)(\langle Rh_1, z\rangle, \cdots, \langle Rh_n, z\rangle) \\ &= (S\phi)\left(\langle h_1, R^{-1}z\rangle, \cdots, \langle h_n, R^{-1}z\rangle\right) \\ &= Sf\left(R^{-1}z\right). \end{aligned}$$

Thus $S\left(f \circ R^{-1}\right) = Sf \circ R^{-1}$ on cylinder functions, and therefore for all functions.

Meanwhile, for $z \in H_{\mathbb{C}}$ we have, using the Cameron–Martin theorem,

$$\begin{aligned} SU(I,h)f(z) &= e^{-(z,z)/2} \int_B e^{(z,x)} e^{-|h|^2/4} e^{\langle h,x\rangle/2} f(x-h)\, d\rho(x) \\ &= e^{-|h|^2/4} e^{-(z,z)/2} \int_B e^{(z,x+h)} e^{\langle h,x+h\rangle/2} f(x)\, e^{-\langle h,x\rangle - |h|^2/2} d\rho(x) \\ &= e^{-|h|^2/4} e^{-(z,z)/2} e^{(z,h)} \int_B e^{(z-h/2,x)} f(x)\, d\rho(x) \\ &= e^{-|h|^2/4} e^{-(z,z)/2} e^{(z,h)} e^{(z-h/2,z-h/2)/2} Sf(z-h/2) \\ &= e^{-|h|^2/8} e^{(h,z)/2} Sf(z-h/2) = V(I,h)Sf(z). \end{aligned}$$

Combining the results of this paragraph and the previous one gives the theorem.
□

3. The gauge-fixed subspace is trivial

We now present an abstract theorem about subgroups $\mathcal{G}$ of $\mathcal{E}(H)$. Now, $\mathcal{G}$ is a set of transformations of the form (R_g, h_g), $g \in \mathcal{G}$. An orbit of $\mathcal{G}$ is thus of the form

$\{R_g x + h_g\}$. Since $|R_g x| = |x|$, we see that if any one orbit is unbounded then all others are as well.

THEOREM 14. *Suppose $\mathcal{G}$ is a subgroup of $\mathcal{E}(H)$ with the property that one (and hence every) orbit of $\mathcal{G}$ is unbounded in H. Suppose $f \in L^2(B, \rho)$ is such that*

$$U(R,h)\,f = f$$

for all $(R,h) \in \mathcal{G}$. Then $f = 0$.

Proof. In light of Proposition 13 it is equivalent to assume that $F := Sf$ satisfies $V(R,h)\,F = F$ for all $(R,h) \in \mathcal{G}$. The hypothesis on $\mathcal{G}$ implies that there is a sequence $\{(R_n, h_n)\}$ in $\mathcal{G}$ with $|h_n| \to \infty$. Then the assumption on F says that for all $z \in H_{\mathbb{C}}$,

$$F(z) = e^{-|h_n|^2/8} e^{(h_n,z)/2} F\left((R_n, h_n/2)^{-1} z\right).$$

Fixing some $w \in H_{\mathbb{C}}$ and choosing $z = z_n := (R_n, h_n/2)\,w = R_n w + h_n/2$ we get

$$F(z_n) = e^{-|h_n|^2/8} e^{(h_n,z_n)/2} F(w).$$

Doing a little algebra, this implies that

$$\frac{|F(z_n)|^2}{e^{|z_n|^2}} = \frac{|F(w)|^2}{e^{|w|^2}}. \tag{8}$$

Now, it is well known (e.g. [**B, BSZ**]) that the pointwise evaluation map

$$F \to F(z)$$

is a continuous linear functional on $\mathcal{H}L^2(H_{\mathbb{C}}, \mu)$ for all $z \in H_{\mathbb{C}}$. So there is a unique $\chi_z \in \mathcal{H}L^2(H_{\mathbb{C}}, \mu)$ such that

$$F(z) = \langle \chi_z, F \rangle \tag{9}$$

for all $F \in \mathcal{H}L^2(H_{\mathbb{C}}, \mu)$. (The inner product is complex linear on the right.) The χ_z's are called **coherent states** and are given explicitly by

$$\chi_z(w) = e^{(\bar{z},w)}. \tag{10}$$

As a consequence of (9) and (10) we have

$$\langle \chi_w, \chi_z \rangle = e^{(\bar{z},w)}.$$

In particular $\|\chi_z\|^2 = \exp|z|^2$, and so by (9)

$$|F(z)|^2 \le \|F\|^2 e^{|z|^2}. \tag{11}$$

We now introduce the normalized coherent states

$$\tilde{\chi}_z := \frac{\chi_z}{\|\chi_z\|}$$

and observe that

$$\begin{aligned} |\langle \tilde{\chi}_w, \tilde{\chi}_z \rangle| &= e^{\mathrm{Re}(\bar{z},w)} e^{-|w|^2/2} e^{-|z|^2/2} \\ &= e^{-|z-w|^2/2}. \end{aligned} \tag{12}$$

This shows that $\tilde{\chi}_w$ and $\tilde{\chi}_z$ are nearly orthogonal whenever the distance between w and z is large. Equation (8), re-stated in terms of coherent states, says

$$|\langle \tilde{\chi}_{z_n}, F \rangle|^2 = |\langle \tilde{\chi}_w, F \rangle|^2 = |F(w)|^2 e^{-|w|^2}. \tag{13}$$

We are assuming that $|h_n| \to \infty$ and so that $|z_n| \to \infty$. The idea is now to pass to a subsequence on which $|z_n - z_m|$ is large for all $n \neq m$. Then the $\tilde{\chi}_{z_n}$'s will be essentially orthonormal, and so (13) will tell us that

$$\|F\|^2 \gtrsim \sum_{n=1}^{\infty} |\langle \tilde{\chi}_{z_n}, F\rangle|^2 = \infty$$

unless $\langle \tilde{\chi}_w, F\rangle = F(w)\exp(-|w|^2) = 0$. Since this argument holds for all w, we see that $F \equiv 0$.

To be more precise, suppose that we are given a positive integer N. Since $\lim z_n = \infty$ we may choose $\{w_k\}_{k=1}^N \subset \{z_n\}_{n=1}^\infty$ such that $|w_n - w_m|$ is large enough that (by (12))

$$|\langle \tilde{\chi}_{w_n}, \tilde{\chi}_{w_m}\rangle - \delta_{m,n}| \leq \frac{1}{4N} \ \forall\ m, n \leq N.$$

Let $C_{n,m} := \langle \tilde{\chi}_{w_n}, \tilde{\chi}_{w_m}\rangle$ be the Gram matrix of $\{\tilde{\chi}_{w_n}\}_{n=1}^N$, $A = I - C$ and $D = C^{-1} - I$. Then A is an $N \times N$ matrix whose Hilbert–Schmidt norm, and therefore it's operator norm $\|A\|$, is bounded by $1/4$. Using the geometric series expansion for C^{-1}, we conclude that

$$\|D\| = \|C^{-1} - I\| \leq \frac{1}{4}(1 - 1/4)^{-1} = 1/3.$$

By the previous considerations and the fact that the norm-squared of the projection of F onto the span of $\{\tilde{\chi}_{w_n}\}_{n=1}^N$ is given by $\sum_{m,n=1}^N C_{n,m}^{-1}\overline{\langle \tilde{\chi}_{w_n}, F\rangle}\langle \tilde{\chi}_{w_m}, F\rangle$, we find that

$$\begin{aligned}
\|F\|^2 &\geq \sum_{m,n=1}^N C_{n,m}^{-1}\overline{\langle \tilde{\chi}_{w_n}, F\rangle}\langle \tilde{\chi}_{w_m}, F\rangle \\
&= \sum_{m=1}^N |\langle \tilde{\chi}_{w_m}, F\rangle|^2 + \sum_{m,n=1}^N D_{n,m}\overline{\langle \tilde{\chi}_{w_n}, F\rangle}\langle \tilde{\chi}_{w_m}, F\rangle \\
&\geq \sum_{m=1}^N |\langle \tilde{\chi}_{w_m}, F\rangle|^2 - \|D\| \sum_{m=1}^N |\langle \tilde{\chi}_{w_m}, F\rangle|^2 \\
&= \frac{2}{3}N|F(w)|^2 e^{-|w|^2}.
\end{aligned}$$

Since $\|F\|^2 < \infty$ and N is arbitrary in the above equation, it follows that $|F(w)| = 0$. This finishes the proof since w was arbitrary as well. □

COROLLARY 15. *If $F \in \mathcal{H}L^2(H_\mathbb{C}, \mu)$ then*

$$\lim_{|z|\to\infty} |F(z)|^2 e^{-|z|^2} = 0.$$

This result is reasonable, since in order for F to be in $\mathcal{H}L^2(H_\mathbb{C}, \mu)$, the integral of $|F(z)|^2 e^{-|z|^2}$ over every finite-dimensional subspace of $H_\mathbb{C}$ must be finite. Nevertheless, there is no obvious way to prove this directly from the definitions.

Proof. If the corollary were not true, then there would exist $\varepsilon > 0$ and a sequence $\{z_n\}$ with $|z_n| \to \infty$ such that

$$\frac{|F(z_n)|^2}{e^{|z_n|^2}} = |\langle \tilde{\chi}_{z_n}, F\rangle|^2 \geq \varepsilon,$$

and so the above argument would show that $\|F\| = \infty$. □

Remark. It would be possible to prove Theorem 14 directly in $L^2(B,\rho)$, by considering the coherent states $\psi_z \in L^2(B,\rho)$ defined as $\psi_z = S^{-1}(\chi_z)$, which may be computed explicitly as

$$\psi_z(x) = e^{(\bar{z},x)-(\bar{z},\bar{z})/2}.$$

However, it seems that the most natural "home" for the coherent states is the Segal–Bargmann space, especially in light of the natural interpretation of the χ_z's as representing the pointwise evaluation maps.

4. Comparison with the Energy Representation

4.1. Energy Representation. Recall the notation of Section 1. We now introduce a 1-parameter family ρ_t of Gaussian measures on B, characterized by the condition that

$$\int_B e^{i\phi(A)}\, d\rho_t(A) = \exp\left\{-\frac{t}{2}|\phi|^2_{H^*}\right\}$$

for all $\phi \in B^* \subset H^*$. Here t is an arbitrary positive number. Thus the measure ρ of the previous sections is the same as ρ_1.

We consider the "energy representation" E_t, a unitary representation of $\mathcal{G}$ acting on $L^2(B,\rho_t)$, given by the formula

$$E_t(g)f(A) = e^{i\langle A, dg\, g^{-1}\rangle} f\left(\mathrm{Ad}_{g^{-1}}(A)\right) \tag{14}$$

as introduced in [**I**] in the case $G = SU(2)$ and in [**GGV1**] in general. (Although the right hand side of (14) appears to be independent of t, the Hilbert space in which f lives depends on t.) Note that for g smooth, the adjoint action of $\mathcal{G}$ on H extends continuously to B. Equation (14) is the form of the energy representation used in [**GGV1**, Sect. 5] and in [**Wa**]. A unitarily equivalent form is used in some of the other references. Note also that parameter t could be absorbed into the density ω in (1), or into the choice of inner product on $\mathfrak{g}$.

The main theorem of this section is the following.

THEOREM 16. *The representations W and $E_{1/4}$ are unitarily equivalent.*

The proof of this theorem is an easy consequence of Proposition 18 below with $r = 1$. The intertwining operator between W and $E_{1/4}$ is a variant of the Fourier transform called the Fourier–Wiener transform, which we now introduce. The details of the proof are given at the end of the paper.

4.2. The Fourier–Wiener Transform. The Fourier–Wiener transform was introduced by Cameron and Martin [**C, CM1, CM2**]–see especially Theorem 1 of [**CM2**]. See also [**BSZ**]. The Fourier–Wiener transform is given by essentially the same formula as the Segal–Bargmann transform–see [**GM**, Sect. 4]. In fact, the unitarity of the Fourier–Wiener transform can be viewed as a limiting case of the unitarity of the Segal–Bargmann transform [**H**, Sect. 2].

For the sake of motivation, first assume that $\dim(H) < \infty$. For any $r > 0$ and $f \in L^2(H,\lambda)$ let

$$\mathcal{F}_r f(x) = \left(\frac{r}{2\pi}\right)^{d/2} \int_H e^{-ir\langle x,y\rangle} f(y)d\lambda(y) \tag{15}$$

where the integral is (as usual in the theory of the Fourier transform) to be interpreted as an improper Lebesgue integral. Since $r^{d/2}d\lambda$ is Lebesgue measure on H relative to the inner product $r\langle\cdot,\cdot\rangle$, it follows $\mathcal{F}_r : L^2(H, r^{d/2}\lambda) \to L^2(H, r^{d/2}\lambda)$ is unitary for all $r > 0$. This easily implies that $\mathcal{F}_r : L^2(H,\lambda) \to L^2(H,\lambda)$ is unitary for all $r > 0$. We now define the Fourier–Wiener transform $\mathcal{W}_r$ so that the following diagram commutes

$$\begin{array}{ccccc} & L^2(H,\lambda) & \overset{\mathcal{F}_r}{\longrightarrow} & L^2(H,\lambda) & \\ \Phi_1 & \downarrow & & \downarrow & \Phi_{1/4r^2} \\ & L^2(H,\rho) & \underset{\mathcal{W}_r}{\longrightarrow} & L^2(H,\rho_{1/4r^2}) & \end{array}$$

where

$$\Phi_t f(x) = \left(\frac{d\rho_t(x)}{dx}\right)^{-1/2} f(x) = (2\pi t)^{d/4} e^{|x|^2/4t} f(x).$$

A simple computation shows that

$$\mathcal{W}_r f(x) = e^{r^2|x|^2} \int_H e^{-ir\sqrt{2}\langle x,y\rangle} f(\sqrt{2}y) d\rho(y). \tag{16}$$

The choice of the target space $L^2(H,\rho_{1/4r^2})$ is necessary for the passage to infinite-dimensional H. Specifically, the target space is chosen so that the following property will hold. Suppose $H = H_1 \oplus H_2$ is an orthogonal direct sum decomposition, and let P_1 be the projection onto the first factor. Then our definition of $\mathcal{W}_r$ guarantees that

$$\mathcal{W}_r^H [f \circ P_1] = \left[\mathcal{W}_r^{H_1} f\right] \circ P_1, \tag{17}$$

where $\mathcal{W}_r^H$ is computed on H and $\mathcal{W}_r^{H_1}$ is computed on H_1. This consistency condition is necessary in the definition of $\mathcal{W}_r$ when $\dim H = \infty$, and is easily verified from (16) once one checks that $\mathcal{W}_r$ of the constant function $\mathbf{1}$ is $\mathbf{1}$.

With this as background we are now in a position to define the Fourier–Wiener transform in general.

DEFINITION 17 (Fourier–Wiener Transform). *The Fourier–Wiener transform is the unique unitary map* $\mathcal{W}_r : L^2(B,\rho) \to L^2(B,\rho_{1/4r^2})$ *defined on cylinder functions as follows. If* $f \in L^2(B,\rho)$ *is of the form*

$$f(x) = \phi(\langle h_1, x\rangle, \cdots, \langle h_n, x\rangle) \tag{18}$$

for some finite orthonormal set $\{h_1, \cdots, h_n\}$ *in* H, *then we set*

$$\mathcal{W}_r f(x) = \left(\mathcal{W}_r^{\mathbb{R}^n}\phi\right)(\langle h_1, x\rangle, \cdots, \langle h_n, x\rangle), \tag{19}$$

where $\mathcal{W}_r^{\mathbb{R}^n}$ *is computed on* $\mathbb{R}^n$.

Note that in the formula for $\mathcal{W}_r f$, $\langle h_i, x\rangle$ is to be computed by a limiting procedure in $L^2(B,\rho_{1/4r^2})$ and so technically does not mean the same thing as in the formula for f, where it is computed in $L^2(B,\rho)$. Although a given cylinder function f can be represented as in (18) in many different ways, it follows from (17) that $\mathcal{W}_r f$ is independent of the representation. Also note that $\mathcal{W}_r$ is an isometric transform of cylinder functions in $L^2(B,\rho)$ onto cylinder functions in $L^2(B,\rho_{1/4r^2})$ and that $\mathcal{W}_r$ is formally given by (16).

PROPOSITION 18. *For $f \in L^2(H,\rho)$ and $(R,h) \in \mathcal{E}(H)$,*

$$\mathcal{W}_r U(R,h) f(x) = \exp(-ir\langle h, x\rangle)\, (\mathcal{W}_r f) \circ R^{-1}(x)\,. \tag{20}$$

Proof. If f is a cylinder function, then using (19) we reason precisely as in the proof of Proposition 13 to show that $\mathcal{W}_r\left(f \circ R^{-1}\right) = \mathcal{W}_r f \circ R^{-1}$. By continuity, this holds for all $f \in L^2(B,\rho)$.

To consider the effect of a translation $U(I,h) = T_h$, we consider a cylinder function f as in (18). By enlarging the set $\{h_1, \cdots, h_n\}$ if necessary we may write $h = \sum_{i=1}^n a_i h_i$ with $a_i = \langle h, h_i\rangle \in \mathbb{R}$. Let $a = (a_1, a_2, \ldots, a_n) \in \mathbb{R}^n$, then

$$T_h f(x) = e^{-|h|^2/4} e^{\langle h,x\rangle/2} f(x-h) = \psi(\langle h_1, x\rangle, \cdots, \langle h_n, x\rangle),$$

where

$$\psi(u) := e^{-|h|^2/4} e^{a\cdot u/2} \phi(u-a)$$

for $u \in \mathbb{R}^n$. Using the definition of $\mathcal{W}_r^{\mathbb{R}^n}$, the change of variables $u \to u + a/\sqrt{2}$ and the fact that $a \cdot a = |h|^2$, we find that

$$\begin{aligned}
\left(\mathcal{W}_r^{\mathbb{R}^n}\psi\right)(v) &= e^{r^2|v|^2} \int_H e^{-ir\sqrt{2}v\cdot u} e^{-|h|^2/4} e^{a\cdot\sqrt{2}u/2} \phi\left(\sqrt{2}u - a\right) d\rho(u) \\
&= e^{r^2|v|^2} \int_H e^{-ir\sqrt{2}v\cdot(u+a/\sqrt{2})} e^{-|h|^2/4} e^{a\cdot\sqrt{2}(u+a/\sqrt{2})/2} \times \\
&\qquad \phi(\sqrt{2}u) e^{-u\cdot a/\sqrt{2} - |a|^2/4} d\rho(u) \\
&= e^{r^2|v|^2} \int_H e^{-ir\sqrt{2}v\cdot(u+a/\sqrt{2})} \phi(\sqrt{2}u) d\rho(u) \\
&= e^{-irv\cdot a} \left(\mathcal{W}_r^{\mathbb{R}^n}\phi\right)(v),
\end{aligned}$$

where ρ is now the standard normal distribution on $\mathbb{R}^n$. Since $\sum_{i=1}^n \langle h_i, x\rangle a_i = \langle h, x\rangle$, evaluating this last expression at $v = (\langle h_1, x\rangle, \ldots, \langle h_n, x\rangle)$ implies that

$$\mathcal{W}_r T_h f(x) = \left(\mathcal{W}_r^{\mathbb{R}^n}\psi\right)(\langle h_1, x\rangle, \ldots, \langle h_n, x\rangle) = e^{-ir\langle h,x\rangle} \mathcal{W}_r f(x)\,.$$

□

Proof of Theorem 16. As noted prior to Theorem 1, $W(g) = U(Ad_g, -dg\, g^{-1})$. Hence by Proposition 18 and the definition of $E_{1/4}$ in (14),

$$\begin{aligned}
\mathcal{W}_1 W(g) f(A) &= \mathcal{W}_1 U(Ad_g, -dg\, g^{-1}) f(A) \\
&= \exp(i\langle -dg\, g^{-1}, A\rangle)\, (\mathcal{W}_1 f) \circ Ad_g^{-1}(A) \\
&= E_{1/4}(g) \mathcal{W}_1 f(A).
\end{aligned}$$

Note that in order to make the formulas come out right we need to use $\mathcal{W}_r$ with $r = 1$, and that $\mathcal{W}_1$ maps the Hilbert space on which $W(g)$ is defined to Hilbert space on which $E_{1/4}$ is defined – see Definition 17.

References

[AKT] S. Albeverio, R. Høegh-Krohn, and D. Testard, Irreducibility and reducibility for the energy representation of the group of mappings of a Riemannian manifold into a compact semisimple Lie group, *J. Funct. Anal.* **41** (1981), 378-396.

[BSZ] J. Baez, I. Segal, and Z. Zhou, "Introduction to Algebraic and Constructive Quantum Field Theory," Princeton Univ. Press, Princeton, NJ, 1992.

[B] V. Bargmann, On a Hilbert space of analytic functions and an associated integral transform, Part I, *Comm. Pure Appl. Math.* **14** (1961), 187-214.

[C] R. H. Cameron, Some examples of Fourier–Wiener transforms of analytic functionals, *Duke J. Math.* **12** (1945), 485-488.

[CM1] R. H. Cameron and W. T. Martin, Fourier–Wiener transforms of analytic functionals, *Duke J. Math.* **12** (1945), 489-507.

[CM2] R. H. Cameron and W. T. Martin, Fourier–Wiener transforms of analytic functionals, *Duke J. Math.* **14** (1947), 99-107.

[DH] B. Driver and B. Hall, Yang–Mills theory and the Segal–Bargmann transform, *Comm. Math. Phys.* **201** (1999), 249–290.

[GGV1] I. Gelfand, M. Graev, and A. Veršik, Representations of the group of smooth mappings of a manifold X into compact Lie group, *Compositio Math.* **35** (1977), 299-334.

[GGV2] I. Gelfand, M. Graev, and A. Veršik, Representations of the group of functions taking values in a compact Lie group, *Compositio Math.* **42** (1980/81), 217-243.

[G] L. Gross, Abstract Wiener spaces, *in* "Proceedings of the Fifth Berkeley Symposium on Mathematical Statistics and Probablility," Vol. II, Univ. of California Press, 1967.

[GM] L. Gross and P. Malliavin, Hall's transform and the Segal–Bargmann map, *in* "Itô's Stochastic Calculus and Probability Theory" (M. Fukushima, N. Ikeda, H. Kunita, and S. Watanabe, Eds.), Springer-Verlag, New York/Berlin, 1996, pp. 73-116.

[H] B. Hall, A new form of the Segal–Bargmann transform for Lie groups of compact type, *Canadian J. Math.* **51** (1999), 816-834.

[HS] B. Hall and A. Sengupta, The Segal–Bargmann transform for path-groups, *J. Funct. Anal.* **152** (1998), 220-254.

[HP] E. Hille and R. Phillips, "Functional Analysis and Semigroups," Amer. Math. Soc., Providence, RI, 1957.

[I] R. S. Ismagilov, Unitary representations of the group $C_0^\infty(X, G)$, $G = \mathrm{SU}_2$ (Russian), *Mat. Sb. (N.S.)* **100** (1976), 117–131. English translation: *Math. U.S.S.R. Sb.* **142** (1976), 105-117.

[K] P. Krée, Solutions faibles d'équations aux dérivées fonctionnelles. II. (French) Séminaires Pierre Lelong (Analyse) (année 1973–1974), pp. 16–47. Lecture Notes in Math., Vol. 474, Springer-Verlag, New York/Berlin, 1975.

[S1] I. Segal, Mathematical problems of relativistic physics, Chap. VI, *in* "Proceedings of the Summer Seminar, Boulder, Colorado, 1960, Vol. II." (M. Kac, Ed.). Lectures in Applied Mathematics, American Math. Soc., Providence, Rhode Island, 1963.

[S2] I. Segal, Mathematical characterization of the physical vacuum for a linear Bose–Einstein field, *Illinois J. Math.* **6** (1962), 500-523.

[S3] I. Segal, The complex wave representation of the free Boson field, *in* "Topics in functional analysis: Essays dedicated to M.G. Krein on the occasion of his 70th birthday" (I. Gohberg and M. Kac, Eds). Advances in Mathematics Supplementary Studies, Vol. 3, pp. 321-343. Academic Press, New York, 1978.

[Wa] N. Wallach, On the irreducibility and inequivalence of unitary representations of gauge groups, *Compositio Math.* **64** (1987), 3-29.

Department of Mathematics-0112, University of California at San Diego, La Jolla, CA 92093-0112 USA

E-mail address: `driver@math.ucsd.edu`

University of Notre Dame, Department of Mathematics, Notre Dame, IN 46556 USA

E-mail address: `bhall@nd.edu`

Canadian Mathematical Society
Conference Proceedings
Volume **29**, 2000

Weakly Projective Representations, Quaternions, and Monopoles

G.G. Emch * and A.Z. Jadczyk *†

ABSTRACT. It is shown that the quaternionic Hilbert space formulation of quantum mechanics allows a quantization, based on a generalized system of imprimitivity, that leads to a less singular description of the motion of a quantum particle in the field of a magnetic monopole. The corresponding Hamilton operator is linked to the theory of projective representations in the weakened form proposed by Adler.

To Sergio, who consistently brought to light the beauty hidden in singularities.

1. INTRODUCTION

In the present paper several ideas are brought together. While symmetries are one of the most powerful tools in theoretical physics, there are few, if any, exact symmetries in Nature. Hans Ekstein [**1**] addressed this problem by introducing the concept of "presymmetry"– a pre–dynamical symmetry, that is being broken by dynamics and yet is evidenced in the algebra structure. Adler [**2**] introduced the concept of a "weak projective representation "(WPR) and analyzed it within the framework of quaternionic quantum mechanics (see also [**3, 4**] for the epistemological controversy which arose around this concept). In a recent note Adler and Emch [**5**] revisited the basic concepts of strong and weak projective representations from the point of view of Wigner's theorem [**6**] and the axiomatic formulation of quaternionic quantum mechanics extensively analyzed in [**7**]. Almost concommitantly, following the original ideas of Ekstein [**1**], the concept of a generalized imprimitivity system (GIS) was introduced. Specifically, in [**8**] the Stone – von Neumann theorem was enhanced so as to apply also to GIS's, and in [**9**], as an illustration, a GIS corresponding to a charged quantum particle in the field of Dirac's magnetic monopole was explicitly constructed.

To weave these threads together, we note that the generalized systems of imprimitivity involve operator–valued multipliers as do the weakly projective representations; we show that WPR's indeed arise naturally from GIS's and that they

1991 *Mathematics Subject Classification.* 20C25 . 20C35 ;
1992 *Physics and Astronomy classification:* 03.65.Fd , 03.65.Bz .
A.J. benefited from the financial support of the Kosciuszko Foundation.

correspond to symmetries that are only partially broken, with a remaining presymmetry (in the sense of Ekstein) holding only for an Abelian subalgebra of the algebra of all observables. Our discussion of these concepts and their relations is supported by a specific model: the quantum evolution of a charged particle in the field of a magnetic monopole. The half-spin properties [**10**] – by the very nature of the Clifford algebra [**11**] of $(\mathbb{E}^3; -1, -1, -1)$ – naturally involve the structure of the field $\mathbb{H}$ of the real quaternions [**11**]; see also [**12**] and references in both of these texts. This was part of our motivation for treating the problem here in terms of quaternionic Hilbert spaces.

In order not to distract from the main purposes of the paper, we collect in an Appendix some of the basic mathematical facts and notation pertaining to quaternionic quantum mechanics.

2. THE MOTIVATING MODEL

Our model describes quantum kinematics and dynamics of a charged particle in the field of a magnetic monopole. The model is realized in a space of square integrable sections of a Hermitian *quaternionic* line bundle over $\mathbb{R}^3 \setminus \{0\}$. The basic properties and notations relative to the field of quaternions $\mathbb{H}$ and the quaternionic Hilbert space $\mathcal{H}_{\mathbb{H}} = \mathcal{L}^2(\mathbb{R}^3, d^3x; \mathbb{H})$ are reviewed in the Appendix. From a measure–theoretical point of view the Hilbert spaces $\mathcal{L}^2(\mathbb{R}^3, d^3x; \mathbb{H})$ and $\mathcal{L}^2(\mathbb{R}^3 \setminus \{0\}, d^3x; \mathbb{H})$ are naturally isomorphic and we will not make any distinction between them until section 4, where the differential geometric aspects of the construction will be discussed.

The position operators are defined, as usual, by $[X_i\, \psi](x) = x_i\, \psi(x)$; we denote by $\{E(\Delta) | \Delta \subset \mathbb{R}^3\}$ their spectral family (see Appendix). Our model is spherically symmetric, with the rotation generators M_i given by $M_i = \epsilon_{ijk} x_j \partial_k - \frac{1}{2}\hat{e}_i$, where ϵ_{ijk} is the totally antisymmetric tensor with $\epsilon_{ijk} = 1$ for ijk any cyclic permutation of the indices 123 – so that, e.g. $\epsilon_{ijk} a_j\, b_k = (\mathbf{a} \times \mathbf{b})_i$ – e_1, e_2, e_3 are the three standard quaternion imaginary units, and $\hat{e}_i$ is defined in (A.10).

For every $0 \neq \mathbf{x} \in \mathbb{R}^3$ let $j(\mathbf{x})$ be the imaginary unit quaternion

$$j(\mathbf{x}) = \frac{\mathbf{e} \cdot \mathbf{x}}{\|\mathbf{x}\|} \quad . \tag{2.1}$$

The linear operator $J = \hat{j}$, defined by:

$$(J\psi)(\mathbf{x}) = j(\mathbf{x})\psi(\mathbf{x}) \tag{2.2}$$

is unitary and anti–hermitian, i.e.: $J^*J = I = JJ^*$ and $J^* = -J$; and thus $J^2 = -I$. Moreover J is invariant under rotations and commutes with the position operators.

For every direction $\mathbf{u} \in S^2 = \{\mathbf{u} \in \mathbb{R}^3 \mid \|\mathbf{u}\| = 1\}$, the anti-hermitian operator

$$\nabla_{\mathbf{u}} = \mathbf{u} \cdot \boldsymbol{\partial} + \frac{1}{2}\, \frac{\mathbf{e} \cdot [\mathbf{u} \times \mathbf{x}]}{\|\mathbf{x}\|^2} \tag{2.3}$$

generates a one–parameter unitary group $\{U_{\mathbf{u}}(s) \mid s \in \mathbb{R}\}$ which satisfies, for all $s \in \mathbb{R}$ and all Borel subsets $\Delta \subset \mathbb{R}^3$:

$$U_{\mathbf{u}}(s)\, E(\Delta)\, U_{\mathbf{u}}(-s) = E(\Delta - s\mathbf{u}) \, ; \tag{2.4}$$

or, infinitesimally:

$$[\nabla_i, x_j] = \delta_{ij} \quad . \tag{2.5}$$

Thus $\nabla_{\mathbf{u}}$ generates translations in the direction $\mathbf{u}$ of the position variables. Moreover, we have $[M_i, \nabla_j] = -\epsilon_{ijk}\nabla_k$, so that $\boldsymbol{\nabla}$ transforms as a vector under rotations.

The unitary evolution defined by

$$U(t) = \exp(-JHt) \quad \text{where} \quad H = -\frac{1}{2m}\boldsymbol{\nabla}^2 \quad \text{and} \quad \boldsymbol{\nabla}^2 = \sum_{i=1}^{3}(\nabla_i)^2 \tag{2.6}$$

gives the evolution equations for the position operator $\mathbf{X}$, namely :

$$\dot{X}_i = -\frac{1}{m} J\,\nabla_i \tag{2.7}$$

and

$$\ddot{X}_i = \frac{1}{2m}\,\epsilon_{ijk}(\dot{X}_j\,B_k + B_j\,\dot{X}_k) \tag{2.8}$$

with

$$[B_i\psi](\mathbf{x}) = \frac{1}{2}\,\frac{x_i}{\|\mathbf{x}\|^3}\,\psi(\mathbf{x}) \quad , \tag{2.9}$$

which correspond to the motion of a charged particle in the field of a magnetic monopole.

The translation generators do not commute:

$$[\,\nabla_i\,,\,\nabla_j\,] = -\frac{1}{2}\,\epsilon_{ijk}\frac{x^k}{\|\mathbf{x}\|^3}\,J \tag{2.10}$$

which will be seen to imply that the unitary operators $\{U(\mathbf{a}) \mid \mathbf{a} \in \mathbb{R}^3\}$ defined by $U(s\,\mathbf{u}) = U_{\mathbf{u}}(s)$ for all $s \in \mathbb{R}$ and $\mathbf{u} \in S^2$, i.e. for all $s\,\mathbf{u} \in \mathbb{R}^3$, are only a WPR of the translation group in the sense of Adler.

Note also, for future usage, that the following "splitting"relations are satisfied:

$$0 = [X_i, J] = [\nabla_{\mathbf{u}}, J] = [H, J]\,. \tag{2.11}$$

3. DETAILS OF THE CONSTRUCTION

The canonical quantization is given by the system of imprimitivity where

$$V(\mathbf{a})E(\Delta)V(-\mathbf{a}) = E(\Delta - \mathbf{a}) \quad \text{with} \quad [V(\mathbf{a})\psi](\mathbf{x}) = \psi(\mathbf{x} - \mathbf{a}) \quad , \tag{3.1}$$

where $\{V(\mathbf{a}) \mid \mathbf{a} \in \mathbb{R}^3\}$ is a continuous unitary representation with generators ∂_i. These generators correspond to covariant derivatives of the flat connection. In the presence of an external magnetic field: a vector potential enters into the connection form; covariant derivatives cease to commute; parallel transport becomes path dependent; translational symmetry is partially broken; and an operator–valued multiplier corresponding to an integral curvature enters into the group composition formula. One of the principal aims of the present paper is to draw attention to the clarifying role played by the quaternions; we skip therefore any further heuristic motivation of the construction.

We define, for every $\mathbf{a} \in \mathbb{R}^3$ and for all $\mathbf{x} \in \mathbb{R}^3$ not colinear with $\mathbf{a}$:

$$w(\mathbf{a};\mathbf{x}) = \frac{1}{\sqrt{2}}\left(\sqrt{1+\frac{\|\mathbf{x}\|^2+\mathbf{a}\cdot\mathbf{x}}{\|\mathbf{x}\|\|\mathbf{x}+\mathbf{a}\|}} + j(\mathbf{x}\times\mathbf{a})\sqrt{1-\frac{\|\mathbf{x}\|^2-\mathbf{a}\cdot\mathbf{x}}{\|\mathbf{x}\|\,\|\mathbf{x}+\mathbf{a}\|}}\right), \tag{3.2}$$

and let $W(\mathbf{a})$ denote – see (A.11) – the bounded linear operator $\hat{w}(\mathbf{a};\cdot)$, that is:

$$(W(\mathbf{a})\psi)(\mathbf{x}) = w(\mathbf{a};\mathbf{x})\,\psi(\mathbf{x}) \quad \text{a.e.} \tag{3.3}$$

SCHOLIUM 3.1.

1.
$$w(\mathbf{a};\mathbf{x})\,w(\mathbf{a};\mathbf{x})^\star = 1 \quad a.e.\,,$$
and thus $W(\mathbf{a})$ are unitary operators. They commute with the position observables.
2. *$w(\mathbf{a};\mathbf{x})$ satisfy the cocycle relations*
$$w(t\mathbf{a},\mathbf{x}+s\mathbf{a})\,w(s\mathbf{a},\mathbf{x})\,w(s\mathbf{a},\mathbf{x}) = w((s+t)\mathbf{a},\mathbf{x}), \quad a.e..$$
3. *With $U(\mathbf{a})$ defined for every $\mathbf{a} \in \mathbb{R}^3$ by $U(\mathbf{a}) = V(\mathbf{a})W(\mathbf{a})$, and with $U_{\mathbf{u}}(s)$ defined for each $\mathbf{u} \in S^2$ and $s \in \mathbb{R}$ by $U_{\mathbf{u}}(s) = U(s\mathbf{u})$, $\{U_{\mathbf{u}}(s)|s \in \mathbb{R}\}$ is a continuous unitary group representation of $\mathbb{R}$ whereas $\{U(\mathbf{a})|a \in \mathbb{R}^3\}$ will only be a weak projective representation – see below.*

4. *For every direction $\mathbf{u}$, the infinitesimal generator of $U_{\mathbf{u}}(t)$ is the operator $\nabla_{\mathbf{u}}$ defined in (2.3).*

5. *$U(\mathbf{a})$ satisfy the imprimitivity relations (2.4); and thus, for all $\mathbf{a},\mathbf{b} \in \mathbb{R}^3$:*
$$U(\mathbf{a})U(\mathbf{b}) = U(\mathbf{a}+\mathbf{b})M(\mathbf{a},\mathbf{b})$$
with
$$(M(\mathbf{a},\mathbf{b})\psi)(\mathbf{x}) = m(\mathbf{a},\mathbf{b};\mathbf{x})\psi(\mathbf{x})$$
and
$$m(\mathbf{a},\mathbf{b};\mathbf{x}) = w(\mathbf{a}+\mathbf{b};\mathbf{x})^\star w(\mathbf{a};\mathbf{x}+\mathbf{b})w(\mathbf{b};\mathbf{x}) \in \mathbb{H}\,.$$

REMARKS.

a: *$M(\mathbf{a},\mathbf{b})$ commutes with $E(\Delta)$ for all Borel subsets $\Delta \subseteq \mathbb{R}^3$.*

b: *In Adler's notation* [**1**]
$$M(\mathbf{a},\mathbf{b}) = \int d^3x\ |\mathbf{x} > m(\mathbf{a},\mathbf{b};\mathbf{x}) < \mathbf{x}|\quad .$$

c: *With $\Phi(\mathbf{a},\mathbf{b};\mathbf{x})$ denoting the flux of the monopole magnetic field through the flat triangular surface spanned by the vertices $(\mathbf{x},\mathbf{x}+\mathbf{a},\mathbf{x}+\mathbf{a}+\mathbf{b})$:*
$$m(\mathbf{a},\mathbf{b};\mathbf{x}) = \exp(J\Phi(\mathbf{a},\mathbf{b};\mathbf{x}))\,.$$

d: *The cocycle formula for $M(\mathbf{a},\mathbf{b})$ expressing the associativity of the operator product $(U(\mathbf{a})U(\mathbf{b}))U(\mathbf{c}) = U(\mathbf{a})(U(\mathbf{b})U(\mathbf{c}))$ can be interpreted here as stating that the flux through the closed tetrahedron spanned by the edges $(\mathbf{x},\mathbf{x}+\mathbf{a},\mathbf{x}+\mathbf{a}+\mathbf{b},\mathbf{x}+\mathbf{a}+\mathbf{b}+\mathbf{c})$ is an integer multiple of 2π which is indeed satisfied by the magnetic field of the monopole - see (2.9).*

4. DISCUSSION

The magnetic monopole model presented here in a *quaternionic* Hilbert space $\mathcal{H}_{\mathbb{H}}$, admits a commuting antiunitary involution J and thus can be reduced to a *complex* Hilbert space model in $\mathcal{H}_\omega$. The phenomenon of a "weak projective representation", in the sense implied by Adler, here for the translation group, shows up in both the quaternionic space and in the complex reduction. This is because the "twisted translations" $U(\mathbf{a})$ commute with J. A differential geometric interpretation of the construction is helpful in order to understand at a deeper level what is really going on here. The Hilbert space $\mathcal{H}_{\mathbb{H}} = \mathcal{L}^2(\mathbb{R}^3, d^3x; \mathbb{H})$ can be considered as a Hilbert space of square integrable sections of a trivial Hermitian complex line bundle F over $\mathbb{R}^3 \backslash \{\mathbf{0}\}$. Removing the origin results in no measure–theoretical consequences; this removal however does have differential geometric sequels. Our operators $\nabla_\mathbf{u}$ define a Hermitian connection in F. The curvature two–form Ω, with values in the Lie algebra $su(2)$, is given by the formula

$$\Omega^r = -\frac{1}{2}\epsilon_{ijk}\frac{x^k x^r}{\|\mathbf{x}\|^4}\, dx^i \wedge dx^j \,. \tag{4.1}$$

The fact that the operator J defined by (2.1-2.2) commutes with $\nabla_\mathbf{u}$ can be interpreted as stating that the map $\mathbf{x} \longrightarrow j(\mathbf{x})$ is a parallel section of the bundle of quaternionic right-linear endomorphisms of F. The formula (A.12) defining $\mathcal{H}_\omega$ describes, de facto, a construction of a Hermitian complex subbundle F_ω of F which reduces the connection $\boldsymbol{\nabla}$. The complex Hilbert space $\mathcal{H}_\omega$ consists of square–integrable sections of the bundle F_ω. Because J is invariant under rotation, it follows that the rotation group acts covariantly on F_ω and unitarily on $\mathcal{H}_\omega$ and is a two-valued representation of $SO(3)$ corresponding to spin one–half. At first sight, it appeared somewhat surprising to have spin one–half in a Hilbert space of complex, one-component, functions; [**10**]. To tame this peculiarity, note that the bundle F_ω is non–trivial; [**11**], [**12**] and the original literature cited there. This bundle admits no continuous, nowhere zero, sections – it carries a spin one–half "kink". To see that the bundle is nontrivial we compute the simplest topological invariant, that is its first Chern class. In our case it is the integral of the curvature two–form κ, with now:

$$\kappa = -\frac{1}{2}\epsilon_{ijk}\frac{x^k}{\|\mathbf{x}\|^3} dx^i \wedge dx^j \,, \tag{4.2}$$

over the sphere S^2 - the result is 2π which proves that the bundle is non–trivial.

While $\nabla_\mathbf{u}$ (and thus H) have a simple explicit form, as globally defined differential operators on a dense domain of differentiable functions in the *quaternionic* $\mathcal{H}_{\mathbb{H}}$, which is built out of sections of a trivial vector bundle over $\mathbb{R}^3 \setminus \{\mathbf{0}\}$ – their restriction to the *complex* $\mathcal{H}_\omega$ cannot be so written; this is due to the fact that $\mathcal{H}_\omega$ is defined in terms of sections of a non–trivial subbundle over $\mathbb{R}^3 \setminus \{\mathbf{0}\}$. If we were to force an explicit expression for the covariant derivative in the *reduced* bundle, a string–like singularity would have to appear – a one–point singularity on each sphere of the constant radius $0 < r \in \mathbb{R}$.

Hence there is, in this model, a definite advantage to working with the quaternionic Hilbert space $\mathcal{H}_{\mathbb{H}}$. *To bring this special mathematical feature to light was the first purpose of this paper.*

Working with singularity–free formulation does not depend by itself on the quaternionic structure – we could use as well a $\mathbb{C}^2$–bundle; nevertheless, the full gauge freedom of the theory is manifest only from within a quaternionic perspective.

Let us, finally, comment upon the relations between the present work and the GIS (generalized system of imprimitivity) studied in [**8, 9**]. To define a GIS we need an action of a group G on a space X. In the most regular case, G is a Lie group acting differentiably on a manifold X. A GIS is then defined by the relations:

$$U(g)E(\Delta)U(g)^{\star} = E(\Delta g) \tag{4.3}$$

$$U(g)U(h) = U(gh)M(g,h) \tag{4.4}$$

$$M(g,h) = \int_X dE(x)\; m(g,h;x) \quad , \tag{4.5}$$

where $g :\longrightarrow U(g)$ is a continuous map from G into unitary operators acting on the Hilbert space $\mathcal{H}$, and $m(g,h;x)$ commute with the spectral measure. In the example discussed in the present paper X is the three–dimensional Euclidean space $\mathbf{E}^3$, G is its translation group, and $m(g,h;x)$ are quaternion–valued. It is seen that a GIS always gives rise to a WPR in the sense of Adler. It is however to be remarked that the very concept of a GIS (and also of WPR) is not bound to the field over which the Hilbert space is constructed: The concept applies to real, complex or quaternionic Hilbert spaces as well.

The idea of a "presymmetry" – that is of a symmetry group which is partially broken by the dynamics, but yet still corresponds to a full symmetry group on an Abelian subalgebra – is quite naturally supported by the GIS framework; in contrast, the a priori mathematically more general concept of WPR leaves open the choice of the sub–algebra necessary to the physical interpretation of the group of (pre–)symmetries. *The formulation in terms of GIS seems therefore to help resolve physically the unresolved choice latent in the WPR formulation. Drawing attention to this physical interpretation was the second aim of this paper.*

A. Appendix: Quaternionic notations

The field $\mathbb{H}$ of the (real) quaternions is obtained upon equiping the 4–dimensional real vector space

$$\mathbb{H} = \{\, q = \sum_{\mu=0}^{3} q^{\mu} e_{\mu} \mid a^{\mu} \in \mathbb{R}\,\} \tag{A.1}$$

with the non–commutative multiplication it inherits from

$$e_o\, q = q\, e_o \;\forall\; q \in \mathbb{H}\,;\; e_i\, e_j = -\delta_{ij} e_0 + \epsilon_{ijk}\, e_k\,, \quad i,j,k = 1,2,3. \tag{A.2}$$

$\mathbb{H}$ is equiped with the involution

$$q = \sum_{\mu}^{3} a^{\mu} e_{\mu} \rightarrow q^{*} = \sum_{\mu}^{3} a^{\mu} e_{\mu}^{*} \quad \text{with} \quad e_o^{*} = e_o \quad \text{and} \quad e_i^{*} = -e_i \quad . \tag{A.3}$$

Note that $SU(2,\mathbb{C})$ is isomorphic to $\{ q \in \mathbb{H} \mid q^* q = e_0 \}$, with the isomorphism given by the identification

$$e_o = \begin{pmatrix} 1 & 0 \\ 0 & 1 \end{pmatrix} \quad e_1 = \begin{pmatrix} 0 & -i \\ -i & 0 \end{pmatrix} \quad e_2 = \begin{pmatrix} 0 & -1 \\ 1 & 0 \end{pmatrix} \quad e_3 = \begin{pmatrix} -i & 0 \\ 0 & i \end{pmatrix}$$

i.e.

(A.4) $$e_o = I \quad \text{and} \quad e_k = -i\sigma_k$$

where the σ_k are the three Pauli matrices. For any such quaternion, the map

(A.5) $$\alpha_\omega : q \in \mathbb{H} \mapsto \omega^* \, q \, \omega \in \mathbb{H}$$

is an automorphism of $\mathbb{H}$, and it is essential that every automorphism of $\mathbb{H}$ can in fact be implemented in this manner. In particular, if ω is an imaginary unit, i.e.

(A.6) $$\omega^* = -\omega \qquad \text{and} \qquad \omega^* \omega = e_o \, ,$$

then

(A.7) $$\alpha_\omega[q] = q \quad \text{iff} \quad q \in \mathbb{C}_\omega = \{ u \, e_o + v \, \omega \mid u, v \in \mathbb{R} \} \quad .$$

Note that $\mathbb{C}_\omega$ inherits from $\mathbb{H}$, the structure of the field $\mathbb{C}$ of the complex numbers.

We will henceforth use the notations $1 = e_o$, and $e = (e_1, e_2, e_3)$, and for $x \in \mathbb{R}^3$, $x \cdot e = \sum_{i=1}^3 x^i e_i$. Note that $q^* q = \|q\|^2$ defines the quaternion norm, and that $(x \cdot e)^* (x \cdot e) = \|x\|^2 = \sum_{i=1}^3 (x^i)^2$.

The Hilbert space $\mathcal{H}_\mathbb{H} = \mathcal{L}^2(\mathbb{R}^3, d^3x; \mathbb{H})$ is the space of "functions" $\psi : \mathbb{R}^3 \mapsto \mathbb{H}$, square–integrable with respect to Lebesgue measure d^3x. Its vector space structure is defined with multiplication by scalars written from the right:

(A.8) $$[\psi q](x) = \psi(x) \, q \, ,$$

and the scalar product is given by:

(A.9) $$(\varphi, \psi) = \int_{\mathbb{R}^3} d^3x \; \varphi(x)^* \, \psi(x) \quad .$$

It is linear in its *second* factor, and skew adjoint; hence $(\varphi q_1, \psi q_2) = q_1^* \, (\varphi, \psi) \, q_2$.

The linear operators $A : \mathcal{H}_\mathbb{H} \to \mathcal{H}_\mathbb{H}$ are denoted with left action, so that $A(\psi q) = (A\psi) q = A\psi q$. The adjoint is defined, as usual, by $(\varphi, A^* \psi) = (A\varphi, \psi)$ $\forall \, \varphi, \psi \in \mathcal{H}_\mathbb{H}$.

Let E be the spectral family $[E(\Delta)\psi](x) = \psi(x) \chi_\Delta(x)$ where Δ runs over all Borel subsets of $\mathbb{R}^3$, and χ_Δ is the indicator function of Δ.

We denote by $\hat{e}_i$ the linear anti-hermitian on $\mathcal{H}_\mathbb{H}$ defined by left quaternion multiplication

(A.10) $$(\hat{e}_i)\psi(x) = e_i \psi(x) \, .$$

More generally, for each bounded measurable function $f : \mathbb{R}^3 \to \mathbb{H}$ let $\hat{f}$ denote the bounded linear operator on $\mathcal{H}_\mathbb{H}$ defined by

(A.11) $$(\hat{f}\psi)(x) = f(x)\psi(x) \quad \text{a.e.}$$

The (real) commutant of E consists then exactly of the operators of the form $\hat{f}$.

For any unitary and anti–hermitian operator J and any fixed imaginary unit ω, let

(A.12) $$\mathcal{H}_\omega = \{ \psi \in \mathcal{H}_\mathbb{H} \mid J \, \psi = \psi \, \omega \} \quad .$$

Note that $\mathcal{H}_\omega$ inherits from $\mathcal{H}_\mathbb{H}$ the structure of a complex Hilbert space over the copy $\mathbb{C}_\omega$ – see $(A.7)$ – of the field of complex numbers. Specifically, $\varphi, \psi \in \mathcal{H}_\omega$ and $z \in \mathbb{C}_\omega$ imply $\varphi + \psi \in \mathcal{H}_\omega$, $(\varphi, \psi) \in \mathbb{C}_\omega$, and $\psi z \in \mathcal{H}_\omega$. Furthermore, for every $\psi \in \mathcal{H}_\mathbb{H}$, and every imaginary unit $\tilde{\omega}$ such that $\tilde{\omega}\,\omega = -\omega\,\tilde{\omega}$, there exists a unique pair

$$\text{(A.13)} \qquad \psi_1, \psi_2 \in \mathcal{H}_\omega \quad \text{such that} \quad \psi = \psi_1 + \psi_2\,\tilde{\omega} \quad ;$$

specifically

$$\text{(A.14)} \qquad \psi_1 = \frac{1}{2}(\psi - J\,\psi\,\omega) \quad \text{and} \quad \psi_2 = -\frac{1}{2}(\psi + J\,\psi\,\omega)\,\tilde{\omega} \quad .$$

Note that as vectors in $\mathcal{H}_\mathbb{H}$, ψ_1 and ψ_2 are mutually orthogonal. Therefore the "splitting" (or "dimension–doubling")

$$\text{(A.15)} \qquad \psi \in \mathcal{H}_\mathbb{H} \mapsto \begin{pmatrix} \psi_1 \\ \psi_2 \end{pmatrix} \in \mathcal{H}_\omega \oplus \mathcal{H}_\omega$$

is a bijective isometry.

Acknowledgements. The authors thank S.L. Adler for several conversations rekindling our interest in quaternionic quantum mechanics; the model discussed in this paper provides a specific application of the formalism. One of us (A.J) wants to record his appreciation to the Mathematics Department of the University of Florida, and in particular to J. Klauder, for their hospitality.

References

[1] H. Ekstein, *Presymmetry*, Phys. Rev. **153** (1967) 1397–1402; **184** (1969) 1315–1337.

[2] S.L. Adler, *Quaternionic Quantum Mechanics and Quantum Fields*, Oxford Univ. Press, New York, 1995; S.L. Adler, *Projective Group Representations in Quaternionic Hilbert Space*, J. Math. Phys. **37** (1996) 2352–2360.

[3] G.G. Emch, *Comments on a Recent Paper by S. Adler*, J. Math. Phys. **37** (1996) 6582–6585.

[4] S.L. Adler, *Response to the Comments by G. Emch*, J. Math. Phys. **37** (1996) 6586–6589.

[5] S.L. Adler and G.G. Emch, *A Rejoinder on quaternionic projective representations*, J. Math. Phys. **38** (1997) 4758–4762.

[6] G. Emch and C. Piron, *Note sur les symétries en théorie quantique*, Helv. Phys. Acta **35** (1962) 542-543; *Symmetry in Quantum Theory*, J. Math. Phys. **4** (1963) 469–473.

[7] G.G. Emch, *Mécanique quantique quaternionienne et Relativité restreinte*, Helv. Phys. Acta **36** (1963) 739–769; 770–788; *Representations of the Lorentz Group in Quaternionic Quantum Mechanics* (Lorentz Group Symposium, Summer 1964) *in* Lectures in Theoretical Physics, Vol. VIIa, W.E. Brittin, ed., U.Colorado Press, Boulder CO, 1964 (pp. 1–36).

[8] A. Jadczyk, *On a theorem of Mackey, Stone and von Neumann for projective imprimitivity systems*, Techn. Rep. IFT, Uni. Wroclaw, no. 328 (1975) 1-13.

[9] A. Jadczyk, *Magnetic charge quantization and generalized imprimitivity systems*, Int. J. Theor. Phys. **14** (1975) 183–192.

[10] A.S. Goldhaber, *Role of Spin in the Monopole Problem*, Phys.Rev. **140** (1965) 1407–1414.

[11] L.C. Biederharn and J.D. Louck, *The Racah–Wigner Algebra in Quantum Theory*, Encyclopedia of Mathematics and its Applications, Addidon–Wesley, Reading, MA.

[12] N.P. Landman, *Mathematical Topics between Classical and Quantum Mechanics*, Springer Monographs in Mathematics, Springer, New York, 1998.

[13] I.R. Porteous, *Clifford Algebras and the Classical Groups*, Cambridge Univ. Press, Cambridge, 1995.

* Dept. of Mathematics, Univ. of Florida, Gainesville, FL 32611-8105,
† On leave from: Institute of Theoretical Physics, University of Wroclaw, now with: Constellation Technology Corporation, Largo, FL 33777-1406
E-mail address: gge@math.ufl.edu

Canadian Mathematical Society
Conference Proceedings
Volume **29**, 2000

Point interactions in a tube

Pavel Exner

ABSTRACT. We discuss discuss spectral and scattering properties of a particle confined to a straight Dirichlet tube in $\mathbb{R}^3$ with a family of point interactions.

Point interactions belong to the list of problems to which Sergio Albeverio made a significant contribution. This topic combines a practical importance as a source of numerous solvable models with an aesthetic appeal as the monograph [**1**] witnesses. At the same time it is far to be closed; despite the extensive and thorough character of the mentioned treatise new questions still arise.

One of them concerns point interactions in tubular regions which represent a natural model for a "quantum wire" with impurities. The simplest situation when the tube is a straight planar strip was investigated in [**4**]; we refer to this paper for a detailed motivation and bibliography. In the present paper we present a brief discussion of a straight tube in $\mathbb{R}^3$ with a family of point interactions.

1. The one-center case

Let $\Omega := \mathbb{R} \times M$ where $M \subset \mathbb{R}^2$ is a closed compact set; we suppose that it is pathwise connected and ∂M has the segment property [**5**]. The free Hamiltonian is the corresponding Dirichlet Laplacian, $H_0 = -\Delta_D^\Omega$ with the domain $W_0^{2,2}(\Omega)$. It can be expressed by means of the one-dimensional Laplacian and $-\Delta_D^M$. The last named operator has a purely discrete spectrum; we denote by χ_n, ν_n its eigenfuctions and eigenvalues, respectively. For any $z \in \mathbb{C} \setminus [\nu_0, \infty)$ the free resolvent is an integral operator with the kernel

$$(1.1)\quad G_0(\vec{x}_1, \vec{x}_2; z) \equiv (H_0 - z)^{-1}(\vec{x}_1, \vec{x}_2) = \frac{i}{2} \sum_{n=0}^{\infty} \frac{e^{ik_n(z)|x_1 - x_2|}}{k_n(z)} \chi_n(\vec{y}_1)\chi_n(\vec{y}_2)\,,$$

where $\vec{x}_j = (x_j, \vec{y}_j)$ and $k_n(z) := \sqrt{z - \nu_n}$, which is defined and smooth except at $\vec{x}_1 = \vec{x}_2$. It is a multivalued function of z with cuts $[\nu_n, \infty)$, $n = 0, 1, \dots$.

Suppose now that a point interaction is situated at $\vec{a} = (a, \vec{b}) \in M^o$. We define it as in [**1**], i.e. as a self-adjoint extension of the operator $-\Delta_D^\Omega \upharpoonright C_0^\infty(\Omega \setminus \{\vec{a}\})$. We employ generalized boundary values

$$(1.2)\qquad L_0(\psi, \vec{a}) := \lim_{\vec{x} \to \vec{a}} \psi(\vec{x})|\vec{x} - \vec{a}|\,, \qquad L_1(\psi, \vec{a}) := \lim_{\vec{x} \to \vec{a}} \left[\psi(\vec{x}) - \frac{L_0(\psi, \vec{a})}{|\vec{x} - \vec{a}|} \right]\,;$$

2000 *Mathematics Subject Classification.* Primary 58J50; Secondary 35P05, 81V99.

The research has been partially supported by GAAS under the contract A1048801.

then the extension in question is specified by the boundary condition

$$L_1(\psi,\vec{a}) + 4\pi\alpha L_0(\psi,\vec{a}) \;=\; 0 \tag{1.3}$$

for a given $\alpha \in \mathbb{R}$. We shall denote it $H(\alpha,\vec{a})$; the case $\alpha = \infty$, i.e. $L_0(\psi,\vec{a}) = 0$, corresponds to the free Hamiltonian H_0.

The resolvent of $H(\alpha,\vec{a})$ is obtained by Krein's formula. Mimicking the argument of [**4**] we get

$$(H(\alpha,\vec{a})-z)^{-1}(\vec{x}_1,\vec{x}_2) \;=\; G_0(\vec{x}_1,\vec{x}_2;z) + \frac{G_0(\vec{x}_1,\vec{a};z)G_0(\vec{a},\vec{x}_2;z)}{\alpha-\xi(\vec{a};z)}\,, \tag{1.4}$$

where $\xi(\vec{a};z)$ is the regularized Green's function at $\vec{a}$,

$$\xi(\vec{a};z) \;=\; \lim_{u\to 0}\left[\frac{i}{2}\sum_{n=0}^{\infty}\frac{e^{ik_n u}}{k_n}|\chi_n(\vec{b})|^2 - \frac{1}{4\pi u}\right]. \tag{1.5}$$

The existence of the limit follows from the kernel behaviour at the singularity [**7**]. However, we also need a prescription how to compute it and this differs from the two-dimensional case. We use semiclassical properties of the above series terms [**5**]. The transverse eigenvalues behave as $\nu_n \approx 4\pi|M|^{-1}n$ for $n\to\infty$ so $k_n = 2i\sqrt{\pi}|M|^{-1/2}\sqrt{n}+\mathcal{O}(1)$. On the other hand, the probability densities $|\chi_n|^2$ are rapidly oscillating functions. Since M supports no potential the mean value of these oscillations equals the constant $|M|^{-1}$; assuming that $|\chi_n(\vec{b})|^2$ oscillates around this value as $n\to\infty$ we can assess the divergence rate of the first series. Next we use the identity

$$\frac{1}{4\pi u} \;=\; \beta\int_0^\infty \frac{e^{-\gamma u\sqrt{s}}}{\sqrt{s}}\,ds$$

with $\gamma := 2\sqrt{\pi}|M|^{-1/2}$ and $\beta^{-1} := 4\sqrt{\pi|M|}$ and write the r.h.s. as a sum of the integrals over $(n,n+1)$ obtaining

$$\xi(\vec{a};z) \;=\; \lim_{u\to 0}\left[\sum_{n=0}^{\infty}\frac{e^{-\kappa_n u}}{2\kappa_n}|\chi_n(\vec{b})|^2 + \frac{e^{-\gamma u\sqrt{n+1}}-e^{-\gamma u\sqrt{n}}}{4\pi u}\right],$$

where $\kappa_n(z) := -ik_n(z) = \sqrt{\nu_n - z}$. In combination with the preceding argument, it is easy to check that the summand has a uniform bound of order $o(n^{-3/2})$. Hence the limit can be interchanged with the sum and

$$\xi(\vec{a};z) \;=\; \sum_{n=0}^{\infty}\left[\frac{|\chi_n(\vec{b})|^2}{2\kappa_n(z)} + \frac{\sqrt{n}-\sqrt{n+1}}{2\sqrt{\pi|M|}}\right]. \tag{1.6}$$

REMARKS 1.1. (i) We do not give here details of the oscillation argument. Notice that the conclusions made below can be obtained even without it, up to an additive renormalization of the coupling constant, since by the Weyl formula and the uniform boundedness of the χ_n's the difference $\xi(\vec{a};z)-\xi(\vec{a};z_0)$ is given by a convergent series.
(ii) The scaling behaviour for $\Omega^\sigma = \mathbb{R}\times M^\sigma$ with $M^\sigma := \sigma M$, $\sigma>0$ is more complicated than in the two-dimensional case. We have $\xi(\vec{a^\sigma};z\sigma^{-2}) = \sigma^{-1}\xi(\vec{a};z)$, so the singularities of the resolvent kernel are related by

$$\epsilon^\sigma(\alpha^\sigma,\vec{a^\sigma}) \;=\; \sigma^{-2}\epsilon(\alpha,\vec{a})\,,\quad \alpha^\sigma := \sigma^{-1}\alpha\,. \tag{1.7}$$

PROPOSITION 1.2. *The operator $H(\alpha,\vec{a})$ has for any $\alpha \in \mathbb{R}$ a single eigenvalue $\epsilon(\alpha,\vec{a}) \in (-\infty,\nu_0)$. The corresponding eigenfunction is*

$$\psi(\vec{x};\alpha,\vec{a}) = \sum_{n=0}^{\infty} \frac{e^{-\kappa_n(\epsilon)|x-a|}}{2\kappa_n(\epsilon)} \chi_n(\vec{y})\chi_n(\vec{b}). \tag{1.8}$$

The function $\epsilon(\cdot,\vec{a})$ is strictly increasing and behaves as

$$\epsilon(\alpha,\vec{a}) = \nu_0 - \left(\frac{|\chi_0(\vec{b})|^2}{2\alpha}\right)^2 + \mathcal{O}(\alpha^{-3}), \tag{1.9}$$

in the limit of weak coupling, $\alpha \to +\infty$. Moreover, there are no eigenvalues embedded in $\sigma_c(H(\alpha,\vec{a})) = [\nu_0,\infty)$.

PROOF: Due to (1.6), $\xi(\vec{a};\cdot)$ is strictly increasing with $\operatorname{Ran}\xi = \mathbb{R}$ and $\xi(\vec{a};z) = \frac{1}{2}|\chi_0(\vec{b})|^2(\nu_0-z)^{-1/2} + \mathcal{O}(1)$ as $z \to \nu_0-$. The non-normalized eigenfunction (1.8) is given by the residue term in (1.4). To check the absence of embedded eigenvalues we have to show that $\xi(\vec{a};z) = \alpha$ has no solutions on $[\nu_0,\infty)$. Away of the thresholds, this follows from

$$\operatorname{Im}\xi(\vec{a};z) = \sum_{\{n:\,\nu_n<z\}} \frac{|\chi_n(\vec{b})|^2}{2\sqrt{\nu_n-z}} > 0. \tag{1.10}$$

If $|\chi_n(\vec{b})|^2 \neq 0$, the resolvent kernel has a finite limit as z approaches ν_n, otherwise it has the same singularity as $G_0(\vec{x}_1,\vec{a};\cdot)$ there, so in neither case it has a pole. ∎

Mimicking the argument of [**4**] we also get

PROPOSITION 1.3. *The on-shell S-matrix at energy $z = k^2$ is a $2N_{\text{open}} \times 2N_{\text{open}}$ unitary matrix with elementary blocks*

$$S_{nm} = \sqrt{\frac{k_m}{k_n}} \begin{pmatrix} t_{nm} & r_{nm} \\ \tilde{r}_{nm} & \tilde{t}_{nm} \end{pmatrix}, \quad n,m = 1,\dots,N_{\text{open}}, \tag{1.11}$$

where $N_{\text{open}} :=$ card$\{\nu_n : \nu_n < z\}$, the tilded quantities are obtained by switching sign of the longitudinal component of $\vec{a}$, $a \mapsto -a$, and

$$r_{nm}e^{-ik_m a} = (t_{nm}-\delta_{nm})e^{ik_m a} = \frac{i}{2k_m}\,\frac{e^{ik_n a}}{\alpha-\xi(\vec{a};z)}\,\chi_n(\vec{b})\chi_m(\vec{b}). \tag{1.12}$$

2. Finite number of perturbations

Denote $\vec{a} := \{a_1,\dots a_N\}$, where $\vec{a}_j = (a_j,\vec{b}_j)$, and $\alpha := \{\alpha_1,\dots,\alpha_N\}$, $j = 1,\dots,N$. The Hamiltonian $(H(\alpha,\vec{a})$ with N point interactions is defined as the self-adjoint extension of the operator $-\Delta_D^M \restriction C_0^\infty(\Omega\setminus\{\vec{a}\})$ specified by the boundary conditions

$$L_1(\psi,\vec{a}_j) + 4\pi\alpha_j L_0(\psi,\vec{a}_j) = 0, \quad j = 1,\dots,N. \tag{2.1}$$

The resolvent is again found by means of the Krein formula:

$$(H(\alpha,\vec{a})-z)^{-1}(\vec{x}_1,\vec{x}_2) = G_0(\vec{x}_1,\vec{x}_2;z) + \sum_{j,k=1}^{N} \lambda_{jk}(\alpha,\vec{a};z)G_0(\vec{x}_1,\vec{a};z)G_0(\vec{a},\vec{x}_2;z), \tag{2.2}$$

where $\lambda(\alpha,\vec{a};z) = \Lambda(\alpha,\vec{a};z)^{-1}$ with

$$\Lambda_{jj} = \alpha_j - \xi(\vec{a}_j;z), \quad \Lambda_{jk} = -G_0(\vec{a}_j,\vec{a}_k;z) \quad \text{for} \quad j \neq k, \tag{2.3}$$

where $\xi(\vec{a}_j;z)$ is given by (1.5), (1.6). With these prerequisites we can derive spectral properties of our point-interaction Hamiltonian.

THEOREM 2.1. *(a) The spectrum of $H(\alpha,\vec{a})$ consists for any $\alpha\in\mathbb{R}^N$ of the absolutely continuous part $[\nu_0,\infty)$ and eigenvalues $\epsilon_1<\epsilon_2\le\cdots\le\epsilon_m<\nu_0$ with $1\le m\le N$, given by the condition*

$$\det\Lambda(\alpha,\vec{a},z)\;=\;0\,. \tag{2.4}$$

The corresponding eigenfunctions are $\psi(\vec{x})=\sum_{j=1}^N d_jG_0(\vec{x},\vec{a}_j;z)$, where $d\in\mathbb{R}^N$ solves $\sum_{m=1}^N\Lambda(z)_{jm}d_m=0$. The ground-state eigenfunction is positive.
(b) $z>\nu_0$ cannot be an eigenvalue corresponding to an eigenvector from the subspace $\bigoplus_{\{n:\,\nu_n<z\}}L^2(\mathbb{R})\otimes\{\chi_n\}$. On the other hand, $H(\alpha,\vec{a})$ can have embedded eigenvalues if the family $\{\Omega,\,\vec{a},\,\alpha\}$ has a suitable symmetry.
(c) In the weak coupling limit, $|A|:=\min_{1\le j\le N}\alpha_j\to\infty$, there is a single eigenvalue which behaves as

$$\epsilon(\alpha,\vec{a})\;=\;\nu_0-\left(\frac{\left(\sum_{j=1}^N\chi_0(\vec{b}_j)\right)^2}{2\sum_{j=1}^N\alpha_j\chi_0(\vec{b}_j)}\right)^2+\mathcal{O}\left(|A|^{-3}\right)\,. \tag{2.5}$$

PROOF: (a) A finite-rank perturbation in the resolvent preserves $\sigma_{ac}(H_0)=[\nu_0,\infty)$. The discrete spectrum is determined by poles of the resolvent coming from the coefficients λ_{jk} in (2.2). This yields (2.4); the eigenfunctions are obtained in the same way as in [**1, 4**]. The next question concerns the existence of solutions to (2.4). If $z\to-\infty$ the matrix can be written as $\xi(\vec{a};z)\tilde{\Lambda}(\alpha,\vec{a},z)$, where $\tilde{\Lambda}\to-I$, hence all eigenvalues of $\Lambda(\alpha,\vec{a},z)$ tend to $+\infty$. On the other hand, for $z\to\nu_0-$ we have

$$\Lambda(\alpha,\vec{a},z)\;=\;-\frac{1}{2\sqrt{\nu_0-z}}\,M_1\,+\mathcal{O}(1)\,,$$

where $M_1:=\left(\chi_0(\vec{b}_j)\chi_0(\vec{b}_m)\right)_{j,m=1}^N$. This matrix has, in particular, an eigenvector $(\chi_0(\vec{b}_1),\dots,\chi_0(\vec{b}_N))$ corresponding to the *positive* eigenvalue $\sum_{j=1}^N\chi_0(\vec{b}_j)^2$, and therefore at least one of the eigenvalues of $\Lambda(\alpha,\vec{a},z)$ tends to $-\infty$ as $z\to\nu_0-$. Using the continuity we see that there is an eigenvalue which crosses zero, i.e. $H(\alpha,\vec{a})$ has at least one eigenvalue. By a straightforward differentiation we find

$$\frac{d}{dz}\Lambda(z)_{jm}\;=\;-\sum_{n=0}^{\infty}\frac{e^{-|a_j-a_m|\sqrt{\nu_n-z}}}{4(\nu_n-z)^{3/2}}\,\left(1+|a_j-a_m|\sqrt{\nu_n-z}\right)\,\chi_0(\vec{b}_j)\chi_0(\vec{b}_m)\,.$$

The matrix function $\Lambda(\cdot)$ is monotonous if for any $c\in\mathbb{C}^N$ the quantity $\frac{d}{dz}(c,\Lambda(z)c)$ has a definite sign (is non-positive in our case). This is true provided the function $f:\,f(x)=e^{-\kappa|x|}(1+\kappa|x|)$ is of positive type for any $\kappa>0$, which follows from the identity

$$(1+\kappa|x|)\,e^{-\kappa|x|}\;=\;\frac{2\kappa^3}{\pi}\int_{\mathbb{R}}\frac{e^{ipx}}{(p^2+\kappa^2)^2}\,dp$$

and Bochner's theorem [**5**, Sec.IX.2]. In fact, since the measure in the last integral is pointwise positive, $\frac{d}{dz}\Lambda(z)$ is even strictly positive; it means that all the eigenvalues of $\Lambda(\alpha,\vec{a};z)$ are decreasing functions of z and $H(\alpha,\vec{a})$ has at most N eigenvalues.

To check that the ground state is non-degenerate, we have to demonstrate that the lowest eigenvalue of $\Lambda(z)$ is simple for any $z\in(-\infty,\nu_0)$, which is equivalent to the claim that the matrix semigroup $\{ e^{-t\Lambda(z)} \,:\, t \ge 0 \}$ is positivity preserving [**5**, Sec.XIII.12]. The last property is ensured if all the non-diagonal elements of $\Lambda(z)$ are negative; we have $\Lambda(z)_{jm} = -G_0(\vec{a}_j, \vec{a}_m; z)$ by (2.3) so the desired result follows from the positivity of the free-resolvent kernel. The coefficients may be therefore chosen of the same sign for the ground state; in fact, as strictly positive because $d_{j_0} = 0$ would mean that the eigenfunction is smooth at $\vec{x} = \vec{a}_{j_0}$ so the corresponding interaction is absent, $\alpha_{j_0} = \infty$.

(b) Suppose now that $H\varphi = z\varphi$ for some $z > \nu_0$. We adapt again the argument from [**1**, Sec.II.1] and pick an arbitrary $z' \in \rho(H)$; then there is a vector $\psi_0 \in D(H_0)$ which allows us to write

$$\varphi = \psi_0 + \sum_{j=1}^{N} d_j G_0(\cdot, \vec{a}_j; z') \,. \tag{2.6}$$

Furthermore, we expand ψ_0 as a series, $\psi_0(\vec{x}) = \sum_{n=0}^{\infty} g_n(x)\chi_n(\vec{y})$ with the coefficients $g_n \in L^2(\mathbb{R})$. Using the identity $(H_0 - z)\psi_0 = (z - z')\sum_{j=1}^{N} d_j G_0(\cdot, \vec{a}_j; z')$ and the fact that $\{\chi_n\}$ is an orthonormal basis in $L^2(M)$, we obtain a system of equations; by the Fourier-Plancherel operator it is transformed into

$$(p^2 - z + \nu_n)\hat{g}_n(p) = \frac{z - z'}{2\pi} \sum_{j=1}^{N} d_j \chi_n(\vec{b}_j)\, \frac{e^{-ipa_j}}{p^2 - z' + \nu_n} \,. \tag{2.7}$$

If $g_n \in L^2$ the same has to be true for $\hat{g}_n$; this is impossible if $z > \nu_n$ and the r.h.s. of (2.7) is nonzero at $\pm p_n$, where $p_n := \sqrt{\nu_n - z}$, since $\hat{g}_n^2$ would have then a non-integrable singularity. If $N > 1$, it might happen that the r.h.s. of (2.7) is not zero identically. However, if the a_j are mutually different, $\sum_{j=1}^{N} d_j \chi_n(\vec{b}_j)\, e^{\mp ipa_j} = 0$ implies $d_j = 0$ by linear independence. On the other hand, if some of them coincide we find $\sum_j d_j \chi_n(\vec{b}_j) = 0$ where the index runs through the values with the same longitudinal coordinate a_j, and therefore $\hat{g}_n = 0$ again.

The condition $\nu_n < z$ in the above argument is crucial; the operator $H(\alpha, \vec{a})$ can have embedded eigenvalues with eigenfunctions in the orthogonal complement of the mentioned subspace if $N > 1$. Examples can be constructed as in [**4**] (or other similar systems – cf.[**3**]) using M with a symmetry: one has to choose a family of weak enough point interactions with the same symmetry.

(c) We denote $A := \mathrm{diag}(\alpha_1, \dots, \alpha_N)$, and use the decomposition

$$\Lambda(z) = \left(A - \tilde{\Gamma}(z)\right) \left[I - \left(A - \tilde{\Gamma}(z)\right)^{-1} \frac{M_1}{2\sqrt{\nu_0 - z}} \right],$$

where $\tilde{\Gamma}(z)$ is a remainder independent of α, whose norm is bounded as $z \to \nu_0-$, and M_1 is the matrix defined above. The first factor is regular for $|A|$ large enough. Since M_1 is rank one we have to solve the equation

$$\eta \sum_{j,k=1}^{N} \chi_0(\vec{b}_j)\alpha_j \left(I - \tilde{\Gamma}(z)A^{-1}\right)_{jk} \chi_0(\vec{b}_k) - \sum_{j=1}^{N} \chi_0(\vec{b}_j)^2 = 0$$

with $\eta := 2\sqrt{\nu_0 - z}$; then (2.5) follows by the implicit-function theorem. ∎

REMARKS 2.2. (i) We get also the weak-coupling asymptotics for the eigenfunction:

$$\psi(x;\alpha,\vec{a}) \approx \chi_0(\vec{y})\,\frac{2\sum_{j=1}^N \alpha_j\chi_0(\vec{b}_j)^2}{\left(\sum_{j=1}^N \chi_0(\vec{b}_j)^2\right)^2}\sum_{j=1}^N e^{-\sqrt{\nu_0-\epsilon}|x-a_j|}\chi_0(\vec{b}_j)^2$$

$$+\sum_{n=1}^\infty \chi_n(\vec{b}_j)\sum_{j=1}^N \frac{e^{-\sqrt{\nu_n-\nu_0}|x-a_j|}}{\sqrt{\nu_n-\nu_0}}\,\chi_n(\vec{b}_j)\chi_0(\vec{b}_j)\,.$$

The leading term is a product of $\chi_1(\vec{y})$ with a linear combination of the eigenfunctions of one-dimensional point interactions placed at a_j, $j=1,\dots,N$.
(ii) The scattering problem can be treated as in the one-center case. Existence and completeness of wave operators follow from the Kato–Birman theory [**5**]. The reflection and transmission amplitudes from the n-th to the m-th channel are

$$r_{nm}(z) = \frac{i}{2}\sum_{j,k=1}^N (\Lambda(z)^{-1})_{jk}\frac{\chi_m(\vec{b}_j)\chi_n(\vec{b}_k)}{k_m(z)}\,e^{i(k_m a_j+k_n a_k)}\,,$$

$$t_{nm}(z) = \delta_{nm}+\frac{i}{2}\sum_{j,k=1}^N (\Lambda(z)^{-1})_{jk}\frac{\chi_m(\vec{b}_j)\chi_n(\vec{b}_k)}{k_m(z)}\,e^{-i(k_m a_j-k_n a_k)} \tag{2.8}$$

and the unitarity condition now reads

$$\sum_{\{m:\,0\le\nu_m<z\}} k_m(t_{nm}\bar{t}_{sm}+r_{nm}\bar{r}_{sm}) = \delta_{ns}k_n\,,$$

$$\sum_{\{m:\,0\le\nu_m<z\}} k_m\left(\tilde{t}_{nm}\bar{r}_{sm}+\tilde{r}_{nm}\bar{t}_{sm}\right) = 0\,, \tag{2.9}$$

because S_{nm} is given again by (1.11), where the tilded quantities are obtained by mirror transformation, $a_j\to -a_j$.

3. The periodic case

In the infinite-center case we restrict ourselves to the periodic situation, i.e. we suppose that the set $\{\alpha,\vec{a}\}_{per}=\{[\alpha_j,\vec{a}_j]:\ j=1,2,\dots\}$ in Ω is countably infinite and has a periodic pattern with a period $\ell>0$ and N perturbations in each cell, which we denote again as $\{\alpha,\vec{a}\}$. Following the Floquet-Bloch decomposition, we find the unitary operator $U:\ L^2(\Omega)\to L^2(\mathcal{B},(\ell/2\pi)d\theta;L^2(\hat{\Omega}))$, where

$$\hat{\Omega} := [0,\ell)\times M\,,\qquad \mathcal{B} := \left[-\frac{\pi}{\ell},\frac{\pi}{\ell}\right)\times M\,; \tag{3.1}$$

the x-projections of these sets are the Wigner-Seitz cell of the underlying one-dimensional lattice and the corresponding Brillouin zone, respectively. By means of U, the operator $H(\alpha,\vec{a})$ is unitarily equivalent to

$$U\,H(\{\alpha,\vec{a}\}_{per})\,U^{-1} = \frac{\ell}{2\pi}\int_{|\theta\ell|\le\pi}^{\oplus} H(\alpha,\vec{a};\theta)\,d\theta\,, \tag{3.2}$$

where $H(\alpha,\vec{a};\theta)$ is the point-interaction Hamiltonian on $L^2(\hat{\Omega})$, i.e. the Laplacian satisfying (2.1) at the points $\vec{a}_j$, Dirichlet b.c. for $x\in[0,\ell)$, $\vec{y}\in\partial M$, and

$$\psi(\ell-,\vec{y}) = e^{i\theta\ell}\psi(0+,\vec{y})\,,\qquad \frac{\partial\psi}{\partial x}(\ell-,\vec{y}) = e^{i\theta\ell}\frac{\partial\psi}{\partial x}(0+,\vec{y}) \tag{3.3}$$

for $\vec{y} \in M$. The resolvent of $H(\alpha, \vec{a}; \theta)$ can be derived by modifying the argument of [**4**, Sec.5]. The "free" eigenvalues

$$\epsilon_{mn}(\theta) := \left(\frac{2\pi m}{\ell} + \theta\right)^2 + \nu_n\,, \quad m \in \mathcal{Z}\,,\ n = 0, 2, \dots\,, \tag{3.4}$$

correspond to the eigenfunctions $\eta_m^\theta \otimes \chi_n$, where χ_n are as above and $\eta_m^\theta(x) := \ell^{-1/2}\, e^{i(2\pi m + \theta\ell)x/\ell}$, $m \in \mathcal{Z}$. Moreover, the free resolvent kernel is in analogy with [**4**] obtained by a partial summation of the appropriate double series and equals

$$\begin{aligned} G_0(\vec{x}_1, \vec{x}_2; \theta; z) \ &= \ \sum_{n=0}^{\infty} \frac{\sinh((\ell - |x_1 - x_2|)\sqrt{\nu_n - z}) + e^{2i\eta\theta\ell}\sinh(|x_1 - x_2|\sqrt{\nu_n - z})}{\cosh(\ell\sqrt{\nu_n - z}) - \cos(\theta\ell)} \\ &\quad \times \frac{\chi_n(\vec{y}_1)\chi_n(\vec{y}_2)}{2\sqrt{\nu_n - z}}\,, \end{aligned} \tag{3.5}$$

where $\eta := \mathrm{sgn}(x_1 - x_2)$. The full kernel is then expressed by a formula analogous to (2.2) with

$$\lambda(\alpha, \vec{a}, \theta; z) \ = \ \Lambda(\alpha, \vec{a}, \theta; z)^{-1}\,, \tag{3.6}$$

where $\Lambda_{jr} = -G_0(\vec{a}_j, \vec{a}_r)$ for $j \neq r$, while the diagonal elements are given by

$$\Lambda_{jj} \ = \ \alpha_j \ - \ \frac{1}{2}\sum_{n=0}^{\infty}\left(\frac{\sinh(\ell\sqrt{\nu_n - z})}{\cosh(\ell\sqrt{\nu_n - z}) - \cos\theta\ell}\,\frac{\chi_n(\vec{y})^2}{\sqrt{\nu_n - z}} + \frac{\sqrt{n} - \sqrt{n+1}}{\sqrt{\pi|M|}}\right). \tag{3.7}$$

We may also write the last formula as $\Lambda_{jj}(\alpha, \vec{a}, \theta; z) = \alpha_j - \xi(\vec{a}_j, \theta; z)$, where the function ξ is for $z \in \mathbb{R}$ more explicitly given by

$$\begin{aligned} \xi(\vec{a}_j, \theta; z) \ &= \ \frac{1}{2}\sum_{\{n:\, \nu_n \le z\}}\left(\frac{\sin(\ell\sqrt{z - \nu_n})}{\cos(\ell\sqrt{z - \nu_n}) - \cos\theta\ell}\,\frac{\chi_n(\vec{b}_j)^2}{\sqrt{z - \nu_n}} + \frac{\sqrt{n} - \sqrt{n+1}}{\sqrt{\pi|M|}}\right) \\ &\quad + \ \frac{1}{2}\sum_{\{n:\, \nu_n > z\}}\left(\frac{\sinh(\ell\sqrt{\nu_n - z})}{\cosh(\ell\sqrt{\nu_n - z}) - \cos\theta\ell}\,\frac{\chi_n(\vec{b}_j)^2}{\sqrt{\nu_n - z}} + \frac{\sqrt{n} - \sqrt{n+1}}{\sqrt{\pi|M|}}\right). \end{aligned} \tag{3.8}$$

It is defined everywhere except at

$$\mathcal{E}(\vec{a}, \theta) \ := \ \{\, \epsilon_{mn}(\theta) \in \mathcal{E}(\theta) \ :\ \chi_n(\vec{b}_j) \neq 0 \,\}\,, \tag{3.9}$$

where $\mathcal{E}(\theta)$ is the eigenvalue set (3.4). Of course, the r.h.s. of (3.9) makes no sense if $z = \nu_n$ and $\chi_n(\vec{b}_j) \neq 0$, but applying general results on self-adjoint extensions to the one-center case, we can establish *a posteriori* that ξ can be defined there by continuity. Moreover, we find that the function $\xi(\vec{a}_j, \theta; \cdot)$ is monotonously increasing between any pair of neighboring singularities. Then we have the following result.

PROPOSITION 3.1. *For a given $\alpha \in \mathbb{R}^N$ the operator $H(\alpha, \vec{a}; \theta)$ has N eigenvalues $\epsilon_{mnj}(\alpha, \vec{a}; \theta)$, $j = 1, \dots, N$, in any gap of the set (3.9) determined by*

$$\det \Lambda(\alpha, \vec{a}, \theta; z) \ = \ 0\,. \tag{3.10}$$

The corresponding eigenfunctions are $\psi(\vec{x}) = \sum_{j=1}^N d_j G_0(\vec{x}, \vec{a}_j, \theta; z)$ where the d_j's are determined by $\Lambda(\alpha, \vec{a}, \theta; z)$ as in Theorem 2.1.

The spectrum of the original Hamiltonian $H(\{\alpha, \vec{a}\}_{per})$ consist then of bands,

$$\sigma\left(H(\{\alpha, \vec{a}\}_{per})\right) \ = \ \bigcup_{mnj} \{\, \epsilon_{mnj}(\alpha, \vec{a}; \theta) \ :\ \theta \in \mathcal{B} \,\}\,. \tag{3.11}$$

One is interested, of course, in its absolute continuity and existence of gaps. We restrict ourselves to the simplest nontrivial situation.

EXAMPLE 3.2. Let $N = 1$, i.e. let each cell contain a single point interaction. The condition (3.10) then simplifies to

$$\xi(\vec{a},\theta;z) = \alpha\,. \tag{3.12}$$

The left hand side is monotonously increasing between its singularities, i.e. the points of $\mathcal{E}(\vec{a},\theta)$. This means that for fixed α, θ there is a sequence $\{\epsilon_r(\alpha,\vec{a},\theta)\}_{r=0}^{\infty}$ arranged in the ascending order; each of them depends, in fact, only on the transverse component $\vec{b}$ of the vector $\vec{a}$. The lowest one satisfies

$$\epsilon_0(\alpha,\vec{a},\theta) < \nu_0+\theta^2$$

and between each two neighboring points of $\mathcal{E}(\vec{a},\theta)$ there is just one of the other eigenvalues. It is also clear that any of $\epsilon_r(\alpha,\vec{a},\theta)$ is continuous with respect to the parameters and $\epsilon_r(\cdot,\vec{a},\theta)$ is increasing for fixed $\vec{b}$ and θ. Concerning the θ-dependence, the implicit-function theorem tells us that

$$\frac{\partial\epsilon_r(\alpha,\vec{a},\theta)}{\partial\theta} = -\left.\frac{\partial\xi(\vec{a},\theta;z)}{\partial\theta}\left(\frac{\partial\xi(\vec{a},\theta;z)}{\partial z}\right)^{-1}\right|_{(\epsilon_r,\theta)}$$

whenever the denominator is nonzero. Away of the thresholds, $z=\nu_n$, and the points of $\mathcal{E}(\theta)$, a straightforward differentiation shows that $\xi(\vec{a},\cdot;\cdot)$ is analytic in both variables. By (3.8) the numerator is not identically zero; hence the derivative $\partial\epsilon_r(\alpha,\vec{a},\theta)/\partial\theta$ may be zero at some points but never in an interval and *the spectrum of $H(\{\alpha,\vec{a}\}_{per})$ is absolutely continuous* [**5**, Sec.XIII.16].

Let us turn now to the question about the number of gaps. Below $z=\nu_0$ the spectrum may be estimated by means of extrema of the function ξ which yield θ-independent bounds: we have $\xi(\vec{a},\theta;z)\le\xi_+(\vec{a},z)$ where

$$\xi_+(\vec{a},z) := \max_{|\theta\ell|\le\pi}\xi(\vec{a},\theta;z) = \frac12\sum_{n=0}^{\infty}\left(\frac{\chi_0(\vec{b})^2}{\sqrt{\nu_n-z}}\coth\left(\frac{\ell}{2}\sqrt{\nu_n-z}\right)+\frac{\sqrt{n}-\sqrt{n+1}}{\sqrt{\pi|M|}}\right)$$

and a similar formula for the minimum, $\xi_-(\vec{a},z)$, with coth replaced by tanh. Both functions are continuously increasing and tend to $-\infty$ as $z\to-\infty$. On the other hand, $\xi_+(\vec{a},\cdot)$ diverges as $z\to\nu_0-$ while the lower bound $\xi_-(\vec{a},\cdot)$ has a finite limit. This shows, in particular, that the spectral condition (3.12) has no solution for any θ in a left neighborhood of ν_0 provided

$$\begin{aligned}\alpha < \xi_-(\vec{a},\nu_0-) &= \frac{\ell}{4}\chi_0(\vec{b})^2-\frac{1}{2\sqrt{\pi|M|}}\\ &+\frac12\sum_{n=1}^{\infty}\left(\frac{\chi_0(\vec{b})^2}{\sqrt{\nu_n-\nu_0}}\tanh\left(\frac{\ell}{2}\sqrt{\nu_n-\nu_0}\right)+\frac{\sqrt{n}-\sqrt{n+1}}{\sqrt{\pi|M|}}\right);\end{aligned} \tag{3.13}$$

in other words, that a gap exists. The condition (3.13) is satisfied for a strong enough coupling if the point-interaction spacing is kept fixed. On the other hand, inspecting the right-hand-side we see that the gap exists also for any fixed α and the spacing ℓ large enough. In this respect the spectrum is similar to that of a straight polymer in $\mathbb{R}^3$ described in [**1**, Sec.III.1]. However, for our "coated polymer" a much stronger result is valid: we shall show that under a suitable choice of parameters it can have *any finite number* of gaps.

To this end, we consider $z \in (\nu_0+\varepsilon, \nu_1-\varepsilon)$ for a fixed $\varepsilon>0$ and $\ell \gg \sqrt{|M|}$, and rewrite the right hand side of the relation (3.8) as

$$\xi(\vec{a},\theta;z) \;=\; \xi_0(\vec{a},\theta;z)+\eta(\vec{a},\theta;z)\,,$$

where

$$\xi_0(\vec{a},\theta;z) \;:=\; \frac{\sin(\ell\sqrt{z-\nu_0})}{\cos(\ell\sqrt{z-\nu_0})-\cos\theta\ell}\,\frac{\chi_0(\vec{b})^2}{2\sqrt{z-\nu_0}}$$

and $\eta(\vec{a},\theta;z)$ is the rest. The latter is monotonously increasing with respect to z and its derivative is bounded everywhere below the second threshold, in particular, in the chosen interval of energies. Moreover, $\eta(\vec{a},\theta;z)$ is bounded from above by

$$\eta_+(\vec{a},z) \;:=\; -\,\frac{1}{2\sqrt{\pi|M|}}+\frac{1}{2}\sum_{n=1}^{\infty}\left(\frac{\chi_0(\vec{b})^2}{\sqrt{\nu_n-z}}\,\coth\left(\frac{\ell}{2}\sqrt{\nu_n-z}\right)+\frac{\sqrt{n}-\sqrt{n+1}}{\sqrt{\pi|M|}}\right)$$

and the corresponding minimum, $\eta_-(\vec{a},z)$, is obtained when coth in the last expression is replaced by tanh. These estimates shrink as ℓ becomes large: using the inequality $\coth u-\tanh u<5\,e^{-2u}$ for $2u\ge 1$, we find

$$\eta_+(\vec{a},z)-\eta_-(\vec{a},z) \;<\; \frac{5}{2}\,\frac{\chi_1(\vec{b})^2}{\sqrt{\nu_1-z}}\,e^{-\ell\sqrt{\nu_1-z}}+\frac{5}{2}\sum_{n=2}^{\infty}\frac{\chi_n(\vec{b})^2}{\sqrt{\nu_n-z}}\,e^{-\ell\sqrt{\nu_n-z}}\,.$$

The series can be estimated by an integral, which yields for a fixed $z\in(\nu_0+\varepsilon,\nu_1-\varepsilon)$ the behavior

(3.14) $$\eta_+(\vec{a},z)-\eta_-(\vec{a},z) \;=\; \mathcal{O}\left(\sqrt{|M|}\,\ell^{-1}\right).$$

On the other hand, the function $g_\theta(u):=\sin u(\cos u-\cos\theta\ell)^{-1}$ is increasing between any two zeros of its denominator. In the intervals, where it is positive, it can be estimated from below by the appropriate branch of $\tan\left(\frac{u}{2}+\pi m\right)$; when it is negative, we have a similar estimate from above with tan replaced by $-\cot$. Hence independently of θ we have either

$$\xi_0(\vec{a},\theta;z) \;\ge\; \frac{\chi_n(\vec{b})^2}{2\sqrt{z-\nu_0}}\,\tan\left(\frac{\pi}{2}\left\{\frac{\ell}{\pi}\sqrt{z-\nu_0}\right\}\right)$$

or

$$\xi_0(\vec{a},\theta;z) \;\le\; -\,\frac{\chi_n(\vec{b})^2}{2\sqrt{z-\nu_0}}\,\cot\left(\frac{\pi}{2}\left\{\frac{\ell}{\pi}\sqrt{z-\nu_0}\right\}\right),$$

where $\{\cdot\}$ denotes the fractional part. Putting the estimates together we see that the oscillating part dominates, so for sufficiently large $|\alpha|$ there are gaps having $\nu_0+\left(\frac{\pi m}{\ell}\right)^2$ as one endpoint provided it belongs to $(\nu_0+\varepsilon,\nu_1-\varepsilon)$. In addition, $\tan u+\cot u\ge 2$ which means that the gap between the lower and the upper bound to $\xi(\vec{a},\theta;z)$ never closes within the chosen interval if $\ell/\sqrt{|M|}$ is large enough; we infer that *for any* $\alpha\in\mathbb{R}$ the operator $H(\{\alpha,\vec{a}\}_{per})$ can have *an arbitrary finite number of gaps* in its spectrum provided the spacing of the point-interaction array is large enough.

Conclusions of the example can be extended to a finite number of point perturbations per cell. In a similar way one can treat a toroidal tube supporting point interactions and threaded by a magnetic flux, etc.

A more difficult question concerns the finiteness of the number of open gaps. Recall that the gap number depends strongly on the dimension of a periodic system: for one-dimensional systems it is generically infinite [**5**, Sec.XIII.16], [**1**, Sec.III.2], while for higher dimensions it is finite by the *Bethe-Sommerfeld conjecture*. The latter is known to be true, in particular, for periodic potentials or lattices of point interactions in $\mathbb{R}^3$ – see [**2, 6**]. The tube boundary makes things more complicated, but one still expects that at high energies gaps will close due to overlapping of contributions from different transverse modes. Notice in this connection that in the example we have been looking for gaps in the energy interval where transport is possible in the lowest transverse mode only. Nevertheless, it is not easy to demonstrate that no gaps remain open above a certain energy.

In a similar way, the "mixed dimensionality" of waveguide systems inspires other questions such as existence of a mobility edge in tubes with random point interactions, etc. Generally speaking, solvable models whose genealogy can be traced back to the treatise [**1**] will represent for long a useful laboratory for the spectral theory.

References

[1] S. Albeverio, F. Gesztesy, R. Høegh-Krohn, H. Holden: *Solvable Models in Quantum Mechanics*, Springer, Heidelberg 1988.

[2] S. Albeverio, V.A. Geyler: The band structure of the general periodic Schrödinger operator with point interactions, *Commun. Math. Phys.* (1999), to appear

[3] D.V. Evans, M. Levitin, D. Vassiliev: Existence theorems for trapped modes, *J. Fluid. Mech.* **261** (1994), 21–31.

[4] P. Exner, R. Gawlista, P. Šeba, M. Tater: Point interactions in a strip, *Ann. Phys.* **252** (1996), 133–179.

[5] M. Reed, B. Simon: *Methods of Modern Mathematical Physics, II. Fourier Analysis. Self-Adjointness, III. Scattering Theory, IV. Analysis of Operators*, Academic Press, New York 1975–1979.

[6] M.M. Skriganov: The spectrum band structure of the three–dimensional Schrödinger operator with a periodic potential, *Invent. Math.* **80** (1985), 107–121.

[7] E.C. Titchmarch: *Eigenfunction Expansions Associated with Second–Order Differential Equations*, vol.II, Clarendon Press, Oxford 1958.

Nuclear Physics Institute, Academy of Sciences, 25068 Řež near Prague; Doppler Institute, Czech Technical University, Břehová 7, 11519 Prague, Czech Republic

E-mail address: exner@ujf.cas.cz

Canadian Mathematical Society
Conference Proceedings
Volume **29**, 2000

Asymptotic Properties of the Maximal Subinterval of a Poisson Process

Ruzong Fan and Kenneth Lange

It is our pleasure to dedicate this paper to Dr. Sergio Albeverio on the occasion of his 60th birthday in appreciation for his encouragement and support of Ruzong Fan in the study of stochastic analysis.

ABSTRACT. The random points of a Poisson process with unit intensity divide the interval $[0, t]$ into adjacent subintervals. Let R_t denote the length of the largest subinterval. By analogy with the discrete Erdős-Rényi law for the longest head or tail run in coin tossing, $R_t/\ln t$ tends almost surely to 1. The current paper shows that $R_t - \ln t$ tends in distribution to an extreme value distribution and that all moments of $R_t - \ln t$ converge to their conjectured limits. Accurate asymptotic expansions are provided for the mean and variance of $R_t - \ln t$.

1. Introduction

Consider a Poisson process N_t on $[0, \infty)$ with unit intensity. The N_t random points on the interval $[0, t]$ define $N_t + 1$ subintervals counting the subinterval between the origin and the first random point and the subinterval between the last random point and t. If R_t is the length of the largest of these subintervals, then the almost sure limit

$$\lim_{t\to\infty} \frac{R_t}{\ln t} = 1 \tag{1.1}$$

provides a continuous analog of the Erdős-Rényi law for the longest run of heads or tails in coin tossing [**10**]. Here the exponential waiting times between adjacent random points of the Poisson process replace the geometric waiting times for the durations of head and tail runs. This analog of the Erdős-Rényi law has been perfected and generalized by many authors [**3, 4, 5, 7, 8, 19**]. It holds not only for Poisson processes but also for renewal processes. Deheuvels [**7**] has even proved

1991 *Mathematics Subject Classification.* Primary 60E05, 60E15, 60F05, 60F15, 60G55, 60G70; Secondary 92D20.

Key Words: Poisson process, extreme value distribution, geometric probability, harmonic sums, Erdős-Rényi law.

The second author is supported in part by USPHS grant GM53275.

the almost sure extension

$$\begin{aligned}\liminf_{t\to\infty} \frac{R_t - \ln t}{\ln\ln t} &= 0 \\ \limsup_{t\to\infty} \frac{R_t - \ln t}{\ln\ln t} &= 1.\end{aligned}$$

For a Poisson process, the lengths of the $N_t + 1$ subintervals of $[0, t]$ are nearly independent and exponentially distributed for large t. It is therefore natural to conjecture that the centered random variables $R_t - \ln t$ converge to the extreme value distribution $\exp(-e^{-u})$. In this paper, we prove:

THEOREM 1.1. *Given the above assumptions, $R_t - \ln t$ satisfies*

$$\lim_{t\to\infty} \Pr(R_t - \ln t \le u) = \exp(-e^{-u}) \tag{1.2}$$

for all real numbers u and

$$\begin{aligned}\lim_{t\to\infty} \mathrm{E}[(R_t - \ln t)^k] &= \int_{-\infty}^{\infty} u^k e^{-e^{-u}} e^{-u} du \\ &= \int_0^{\infty} (-\ln v)^k v^{s-1} e^{-v} dv\Big|_{s=1} \\ &= (-1)^k \Gamma^{(k)}(1)\end{aligned} \tag{1.3}$$

for all integers $k > 0$.

Section 2 is devoted to demonstrating the convergence in distribution asserted in equation (1.2). Sections 3 and 4 take up the much harder task of proving that all moments converge as stated in relation (1.3). The major stumbling block in establishing (1.3) is the lack of closed-form expressions for the moments $\mathrm{E}(R_t^k)$. Fortunately, asymptotic expansions suffice for our theoretical arguments. In the case of the mean and variance, we show in Section 5 that

$$\begin{aligned}\mathrm{E}(R_t) &= \ln t + \gamma + O(e^{-t}) \\ &= m(t) + O(e^{-t}) \\ \mathrm{Var}(R_t) &= \frac{\pi^2}{6} - \frac{2\ln t}{t} - \frac{2\gamma}{t} - \frac{2\ln t}{t^2} - \frac{2(\gamma-1)}{t^2} + O\Big(\frac{\ln t}{t^3}\Big) \\ &= v(t) + O\Big(\frac{\ln t}{t^3}\Big)\end{aligned} \tag{1.4}$$

as t approaches ∞, where $\gamma \approx .5772$ is Euler's constant [**14**]. The errors $\mathrm{E}(R_t) - m(t)$ and $\mathrm{Var}(R_t) - v(t)$ in approximating the mean and variance are plotted in Figure 1 based on our subsequent equation (4.1).

Some of the more important applications of the Erdős-Rényi law involve DNA sequence matching [**1, 2, 20**]. The calculation of p-values for global and local DNA alignments involve nontrivial extensions of the underlying theory. It would be of considerable interest to explore Poisson process analogs of these more complicated discrete problems that go beyond Chen-Stein approximation. Possibly the current paper will point in useful directions. Certainly, some of our proofs diverge sharply from existing proofs in the discrete setting.

FIGURE 1. Plot of $\mathrm{E}(R_t) - m(t)$ as a solid line and $\mathrm{Var}(R_t) - v(t)$ as a dotted line.

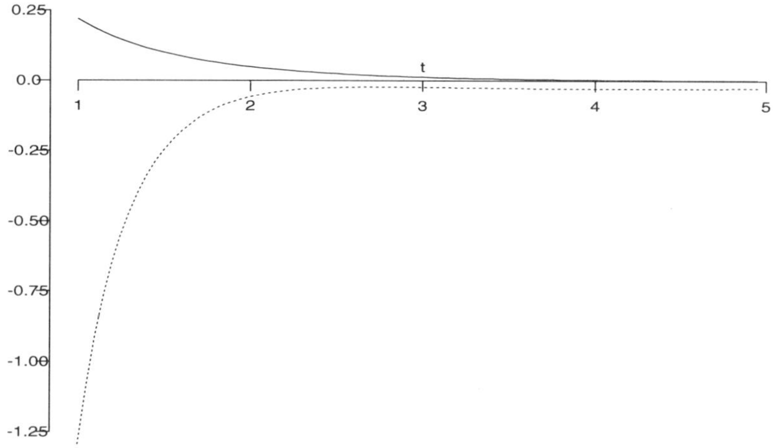

2. Convergence in Distribution

We will give two different proofs that $R_t - \ln t$ converges in distribution. Both rely on introducing a sequence of independent, exponentially distributed random variables X_n with unit means. If we put $S_n = \sum_{i=n}^{n} X_i$, then as explained by Pyke [**18**], R_t has the same distribution as the product tM_{N_t}, where $M_n = S_{n+1}^{-1} \max_{1 \le i \le n+1} X_i$. Because

$$\Pr\left(\max_{1 \le i \le n+1} X_i \le x \right) = (1 - e^{-x})^{n+1},$$

the traditional arguments of Levy [**17**] and Darling [**6**] show that

$$\lim_{n \to \infty} \Pr(nM_n - \ln n \le u) = \exp(-e^{-u})$$

for all real u. To exploit the convergence in distribution of $M_n - \ln n$, we write

$$R_t - \ln t =_d \frac{t}{N_t}\Big(N_t M_{N_t} - \ln N_t\Big) + \frac{t}{N_t} \ln \frac{N_t}{t} + \frac{t - N_t}{\sqrt{t}} \frac{t}{N_t} \frac{\ln t}{\sqrt{t}}.$$

In view of the almost sure convergence of N_t/t to 1 and the convergence in distribution of $(N_t - t)/\sqrt{t}$ to a standard normal distribution, $R_t - \ln t$ converges in distribution to the same extreme value distribution as the sequence $nM_n - \ln n$.

For our second proof, we let X_n be the nth interarrival time in the Poisson process. If we fix N_t temporarily at $n > 0$, and consider the exchangeable gaps $U_i = X_i$ for $1 \le i \le n$ and $U_{n+1} = t - S_n$, then $R_t = \max_{1 \le i \le n+1} U_i$. One can calculate the conditional distribution of R_t via the sieve (inclusion-exclusion) formula

$$\Pr(R_t > u \mid N_t = n) \quad = \quad \sum_{j=1}^{n+1} (-1)^{j-1} \binom{n+1}{j} \Pr(\min_{1 \le i \le j} U_i > u) \tag{2.1}$$

explained in [**11**]. Because the random gaps U_i can be equally well generated by randomly scattering n points uniformly on $[0, t]$, we can use de Finetti's identity

$$\Pr(U_1 > u_1, \ldots, U_{n+1} > u_{n+1}) \quad = \quad \frac{1}{t^n}(t - u_1 - \cdots - u_{n+1})_+^n$$

from the theory of geometrical probability [**12**]. This identity gives the right-tail probability

$$\Pr(\min_{1 \le i \le j} U_i > u) \quad = \quad \frac{1}{t^n}(t - ju)_+^n. \tag{2.2}$$

Note here the convention

$$(x)_+^0 \quad = \quad \begin{cases} 0 & x \le 0 \\ 1 & x > 0 \end{cases}$$

when $n = 0$.

Because N_t is Poisson distributed with mean t, we can express the right-tail probability $1 - F_{R_t}(u)$ of R_t as

$$\begin{aligned}
1 - F_{R_t}(u) &= \sum_{n=0}^{\infty} e^{-t} \frac{t^n}{n!} \sum_{j=1}^{n+1} (-1)^{j-1} \binom{n+1}{j} \frac{1}{t^n}(t - ju)_+^n \\
&= \sum_{j=1}^{\infty} \frac{(-1)^{j-1}}{j!} \sum_{n=j-1}^{\infty} e^{-t} \frac{(t-ju)_+^n}{(n+1-j)!}(n+1-j+j) \\
&= \sum_{j=1}^{\infty} \frac{(-1)^{j-1}}{j!}(t-ju)_+^j \sum_{n=j}^{\infty} e^{-t} \frac{(t-ju)_+^{n-j}}{(n-j)!} \\
&\quad + \sum_{j=1}^{\infty} \frac{(-1)^{j-1}}{(j-1)!}(t-ju)_+^{j-1} \sum_{n=j-1}^{\infty} e^{-t} \frac{(t-ju)_+^{n+1-j}}{(n+1-j)!} \\
&= \sum_{j=1}^{\infty} \frac{(-1)^{j-1} t^j}{j!} \left(1 - \frac{ju}{t}\right)_+^j e^{-t+(t-ju)_+} \\
&\quad + \frac{1}{t} \sum_{j=1}^{\infty} \frac{(-1)^{j-1} t^j}{(j-1)!} \left(1 - \frac{ju}{t}\right)_+^{j-1} e^{-t+(t-ju)_+} \\
&= \sum_{j=1}^{\infty} \frac{(-1)^{j-1} t^j e^{-ju}}{j!} \left(1 - \frac{ju}{t}\right)_+^j \\
&\quad + \frac{1}{t} \sum_{j=1}^{\infty} \frac{(-1)^{j-1} t^j e^{-ju}}{(j-1)!} \left(1 - \frac{ju}{t}\right)_+^{j-1}.
\end{aligned} \tag{2.3}$$

If we substitute $u + \ln t$ for u in equation (2.3) and let t tend to ∞, then we find that

$$\lim_{t \to \infty} [1 - F_{R_t}(u + \ln t)] \quad = \quad \sum_{j=1}^{\infty} \frac{(-1)^{j-1}}{j!} e^{-ju} \quad = \quad 1 - e^{-e^{-u}}, \tag{2.4}$$

provided the indicated interchanges of limit and summation are valid. In other words, $R_t - \ln t$ has its specified extreme-value distribution in the limit. The interchange of limit and summation in equation (2.4) hold because the series

$$\sum_{j=0}^{\infty} \frac{(-1)^j e^{-ju}}{j!} \left[1 - \frac{j(u+\ln t)}{t}\right]_+^j$$

is dominated term by term by the absolutely convergent series $\sum_{j=0}^{\infty} \frac{e^{-ju}}{j!}$. This completes our second proof of convergence in distribution.

Before leaving this topic, we make two comments. First, the formula

$$\Pr(R_t > u \mid N_t = n) \quad = \quad \sum_{j=1}^{n+1} (-1)^{j-1} \binom{n+1}{j} \frac{1}{t^n} (t - ju)_+^n$$

used implicitly in equation (2.3) is the well known Whitworth formula highlighted by Devroye [**9, 21**]. Second, the distribution function $F_{R_t}(u)$ has a point mass of size e^{-t} at $u = t$. The rest of the mass of $F_{R_t}(u)$ is determined by a density function $f_{R_t}(u)$. To calculate $f_{R_t}(u)$, we use the fact

$$\frac{d}{du}\left(1 - \frac{u}{t}\right)_+^0 \quad = \quad -\delta_t(u)$$

in calculating the formal derivative

$$\begin{aligned} \frac{d}{du} F_{R_t}(u) &= \delta_t(u) e^{-t} + \sum_{j=1}^{\infty} \frac{(-1)^{j-1}}{(j-1)!} e^{-ju} t^j \Big[\left(1 - \frac{ju}{t}\right)_+^j \\ &\quad + \frac{2j}{t}\left(1 - \frac{ju}{t}\right)_+^{j-1} + \frac{j(j-1)}{t^2}\left(1 - \frac{ju}{t}\right)_+^{j-2}\Big] \\ &= \delta_t(u) e^{-t} + f_{R_t}(u). \end{aligned}$$

Note here that the term $j(j-1)t^{-2}(1 - jut^{-1})_+^{j-2}$ is taken as 0 when $j = 1$. If we condition on $N_t > 0$, we can interpret $(1 - e^{-t})^{-1} f_{R_t}(u)$ as the conditional density of R_t. Straightforward calculations now give

$$\begin{aligned} \lim_{t\to\infty} f_{R_t}(u + \ln t) &= \sum_{j=1}^{\infty} \frac{(-1)^{j-1}}{(j-1)!} e^{-ju} \\ &= e^{-e^{-u}} e^{-u}. \end{aligned}$$

This interesting recovery of the extreme value density is a bonus of our second, less intuitive proof of convergence in distribution.

3. Moments of R_t

To investigate the moments of R_t, we exploit the sieve formula

$$\mathrm{E}(R_t^k \mid N_t = n) \quad = \quad \sum_{j=1}^{n+1} (-1)^{j-1} \binom{n+1}{j} \mathrm{E}[(\min_{1 \le i \le j} U_i)^k]$$

derived by multiplying equation (2.1) by ku^{k-1} and integrating over the interval $[0, \infty)$ [**12**]. In view of equation (2.2), we can write

$$\begin{aligned} \mathrm{E}[(\min_{1\leq i\leq j} U_i)^k] &= \frac{1}{t^n}\int_0^t ku^{k-1}(t-ju)_+^n du \\ &= \frac{k!t^k}{(n+1)\cdots(n+k)j^k} \end{aligned}$$

after reduction to a beta distribution integral via the change of variables $tv = ju$. It follows that

$$\mathrm{E}(R_t^k \mid N_t = n) = \frac{k!t^k}{(n+1)\cdots(n+k)}\sum_{j=1}^{n+1}(-1)^{j-1}\binom{n+1}{j}\frac{1}{j^k}. \tag{3.1}$$

The sieve formula (3.1) is inconvenient for asymptotic approximation. To make further progress, we define

$$h_n^{(k)} = \sum_{j=1}^{n}(-1)^{j-1}\binom{n}{j}\frac{1}{j^k}. \tag{3.2}$$

Clearly $h_n^{(0)} = -(1-1)^n + 1 = 1$. For $k > 0$ and $n > 1$, Pascal's triangle implies that

$$\begin{aligned} h_n^{(k)} &= \sum_{j=1}^{n}(-1)^{j-1}\left[\binom{n-1}{j} + \binom{n-1}{j-1}\right]\frac{1}{j^k} \\ &= h_{n-1}^{(k)} + \frac{1}{n}\sum_{j=1}^{n}(-1)^{j-1}\binom{n}{j}\frac{1}{j^{k-1}} \qquad (3.3)\\ &= h_{n-1}^{(k)} + \frac{1}{n}h_n^{(k-1)}. \end{aligned}$$

Iteration of this recurrence yields

$$\begin{aligned} h_n^{(k)} &= \sum_{j=1}^{n}\frac{1}{j}h_j^{(k-1)} \\ &= \sum_{i_k=1}^{n}\sum_{i_{k-1}=1}^{i_k}\cdots\sum_{i_1=1}^{i_2}\frac{1}{\prod_{j=1}^{k} i_j}. \qquad (3.4) \end{aligned}$$

This is revealing but still inconvenient.

Formula (3.4) can be reduced to a remarkable formula involving the harmonic sums $H_n^{(j)} = \sum_{i=1}^{n}\frac{1}{i^j}$. Because simple asymptotic expansions are available for these sums, our final substitute for equation (3.1) will be more pliable. The first four cases of formula (3.5) below are

$$\begin{aligned} h_n^{(1)} &= H_n^{(1)} \\ h_n^{(2)} &= \frac{1}{2}[H_n^{(1)}]^2 + \frac{1}{2}H_n^{(2)} \\ h_n^{(3)} &= \frac{1}{6}[H_n^{(1)}]^3 + \frac{1}{2}H_n^{(1)}H_n^{(2)} + \frac{1}{3}H_n^{(3)} \\ h_n^{(4)} &= \frac{1}{24}[H_n^{(1)}]^4 + \frac{1}{4}[H_n^{(1)}]^2H_n^{(2)} + \frac{1}{8}[H_n^{(2)}]^2 + \frac{1}{3}H_n^{(1)}H_n^{(3)} + \frac{1}{4}H_n^{(4)}. \end{aligned}$$

In general, if S_j is the set $\{\mathbf{i} = (i_1, \ldots, i_j) : \sum_{k=1}^{j} k i_k = j\}$ of j-tuples of nonnegative integers, then

$$h_n^{(j)} = \sum_{\mathbf{i}\in S_j} \frac{1}{\prod_{k=1}^{j} i_k! \prod_{k=1}^{j} k^{i_k}} \prod_{k=1}^{j} [H_n^{(k)}]^{i_k}. \tag{3.5}$$

Formula (3.5) is a special case of a more general formula proved in [**16**]; namely, if $x_1, \ldots, x_n$ are n numbers and $G_n^{(j)} = \sum_{i=1}^{n} x_i^j$ is the sum of their jth powers, then

$$\sum_{1\le i_1\le i_2\le\cdots\le i_j\le n} \prod_{k=1}^{j} x_{i_k} = \sum_{\mathbf{i}\in S_j} \frac{1}{\prod_{k=1}^{j} i_k! \prod_{k=1}^{j} k^{i_k}} \prod_{k=1}^{j} [G_n^{(k)}]^{i_k}.$$

The harmonic sums $H_n^{(k)}$ have Euler-MacLaurin expansions of the form

$$\begin{aligned} H_n^{(1)} &= \ln n + \gamma + \frac{1}{2n} - \frac{\theta_1}{12n^2} \\ H_n^{(k)} &= \zeta(k) - \frac{1}{(k-1)n^{k-1}} + \frac{1}{2n^k} - \frac{k\theta_k}{12n^{k+1}}, \quad k > 1, \end{aligned}$$

where $|\theta_k| \le 1$ and $\zeta(k)$ is Riemann's zeta function [**13, 22**].

4. Boundedness of the Moments of $R_t - \ln t$

We now undertake a more detailed examination of the moments of $R_t - \ln t$ with the ultimate goal of showing that $\mathrm{E}[(R_t - \ln t)^k]$ is bounded in t for each positive integer k. Boundedness together with convergence in distribution proves the limit relation (1.3) as noted in [**12**]. Our point of departure is the formula

$$\begin{aligned} \mathrm{E}(R_t^k) &= k!t^k \sum_{n=0}^{\infty} e^{-t}\frac{t^n}{n!}\frac{1}{(n+1)\cdots(n+k)} h_{n+1}^{(k)} \\ &= k!e^{-t} \sum_{n=0}^{\infty} \frac{t^{n+k}}{(n+k)!} h_{n+1}^{(k)} \qquad (4.1) \\ &= k!e^{-t}\Big\{\sum_{n=0}^{\infty}\frac{t^n}{n!}h_{n+1}^{(k)} - \sum_{n=k}^{\infty}\frac{t^n}{n!}[h_{n+1}^{(k)} - h_{n-k+1}^{(k)}] - \sum_{n=0}^{k-1}\frac{t^n}{n!}h_{n+1}^{(k)}\Big\} \end{aligned}$$

derived from equations (3.1) and (3.2).

The last sum

$$k!e^{-t}\sum_{n=0}^{k-1}\frac{t^n}{n!}h_{n+1}^{(k)} = O(t^{k-1}e^{-t})$$

on the far right of equation (4.1) is exponentially small in t and can be neglected. To bound the middle sum on the far right of equation (4.1), we first observe that equation (3.4) and elementary calculus imply

$$h_n^{(j)} \le \Big(\sum_{i=1}^{n}\frac{1}{i}\Big)^j \le (1+\ln n)^j$$

for all j. Therefore, the recurrence relation (3.3) gives

$$\begin{aligned} h_{n+1}^{(k)} - h_{n-k+1}^{(k)} &= \sum_{i=n-k+1}^{n} [h_{i+1}^{(k)} - h_i^{(k)}] \\ &= \sum_{i=n-k+1}^{n} \frac{1}{i+1} h_{i+1}^{(k-1)} \\ &\leq \frac{k}{n-k+2}[1+\ln(n+1)]^{k-1} \\ &\leq \frac{k(k+1)}{2(n+1)}[1+\ln(n+1)]^{k-1} \end{aligned}$$

for $1 \leq k \leq n$. Substituting this inequality into the middle sum of equation (4.1) and applying the Cauchy-Schwarz inequality produce

$$\begin{aligned} & k!e^{-t}\sum_{n=k}^{\infty}\frac{t^n}{n!}\left[h_{n+1}^{(k)} - h_{n-k+1}^{(k)}\right] \\ \leq\ & \frac{k!e^{-t}}{t}\sum_{n=k}^{\infty}\frac{t^{n+1}}{n!}\frac{k(k+1)}{2(n+1)}[1+\ln(n+2)]^{k-1} \\ \leq\ & \frac{k(k+1)!}{2t}\mathrm{E}\{[1+\ln(N_t+1)]^{k-1}\} \\ \leq\ & \frac{k(k+1)!}{2t}\sqrt{\mathrm{E}(N_t+1)}\sqrt{\mathrm{E}\left\{\frac{[1+\ln(N_t+1)]^{2(k-1)}}{N_t+1}\right\}} \\ =\ & \frac{k(k+1)!\sqrt{t+1}}{2t}\sqrt{\mathrm{E}\left\{\frac{[1+\ln(N_t+1)]^{2(k-1)}}{N_t+1}\right\}} \\ =\ & O(t^{-\frac{1}{2}}). \end{aligned}$$

Here we let N_t denote a generic Poisson random variable with mean t and use the boundedness of the function $[1+\ln(n+1)]^{2(k-1)}/(n+1)$ on the nonnegative integers.

Replacing the middle and last terms of representation (4.1) by their asymptotic bounds, we now write

$$\mathrm{E}(R_t^k) = k!e^{-t}\sum_{n=0}^{\infty}\frac{t^n}{n!}h_{n+1}^{(k)} + O(t^{-\frac{1}{2}}).$$

It follows that $R_t - \ln t$ satisfies the moment equality

$$\begin{aligned} \mathrm{E}[(R_t - \ln t)^k] &= \sum_{j=0}^{k}\binom{k}{j}\mathrm{E}(R_t^j)(-\ln t)^{k-j} \\ &= e^{-t}\sum_{n=0}^{\infty}\frac{t^n}{n!}\sum_{j=0}^{k}\binom{k}{j}j!h_{n+1}^{(j)}(-\ln t)^{k-j} + O(t^{-\frac{1}{2}}[\ln t]^k). \qquad (4.2) \end{aligned}$$

We can safely omit the error term in equation (4.2) and focus on showing that the sum in the last line is bounded.

With this end in mind, let us fix n and define variables $m_j(n) = j!h_{n+1}^{(j)}$ and $c_j(n) = (j-1)!H_{n+1}^{(j)}$, which are analogous to the j-th moment and j-th cumulant of some random variable. Formula (3.5) can be re-expressed as

$$m_j(n) = \sum_{\mathbf{i}\in S_j} \frac{j!}{\prod_{k=1}^{j} i_k!} \prod_{k=1}^{j} \Big[\frac{c_k(n)}{k!}\Big]^{i_k}, \tag{4.3}$$

which is the classical formula for the j-th moment of a random variable in terms of its cumulants [**15**]. From this perspective, we also have the formal power series identity

$$\sum_{k=1}^{\infty} c_k(n)\frac{s^k}{k!} = \ln\Big[\sum_{j=0}^{\infty} m_j(n)\frac{s^j}{j!}\Big].$$

Subtracting $s\ln t$ from both sides of this last identity yields

$$\begin{aligned}\sum_{k=1}^{\infty} d_k(n)\frac{s^k}{k!} &= [c_1(n)-\ln t]s + \sum_{k=2}^{\infty} c_k(n)\frac{s^k}{k!} \\ &= \ln\Big[\sum_{j=0}^{\infty} m_j(n)\frac{s^j}{j!}e^{-s\ln t}\Big] \\ &= \ln\Big[\sum_{k=0}^{\infty}\frac{s^k}{k!}\sum_{j=0}^{k}\binom{k}{j} m_j(n)(-\ln t)^{k-j}\Big]\end{aligned}$$

for $d_1(n) = c_1(n) - \ln t$ and $d_k(n) = c_k(n)$ for $k > 1$. Treating the $d_k(n)$ as cumulants, the analog of equation (4.3) is

$$\sum_{j=0}^{k}\binom{k}{j} m_j(n)(-\ln t)^{k-j} = \sum_{\mathbf{i}\in S_k}\frac{k!}{\prod_{j=1}^{k} i_j!}\prod_{j=1}^{k}\Big[\frac{d_j(n)}{j!}\Big]^{i_j}.$$

Substituting this expression in equation (4.2) reduces our problem to one of proving the boundedness of the sums

$$\begin{aligned}& e^{-t}\sum_{n=0}^{\infty}\frac{t^n}{n!}\sum_{j=0}^{k}\binom{k}{j} j!h_{n+1}^{(j)}(-\ln t)^{k-j} \\ (4.4)\qquad &= \sum_{\mathbf{i}\in S_k}\frac{k!}{\prod_{j=1}^{k} i_j!}e^{-t}\sum_{n=0}^{\infty}\frac{t^n}{n!}\prod_{j=1}^{k}\Big[\frac{d_j(n)}{j!}\Big]^{i_j} \\ &= \sum_{\mathbf{i}\in S_k}\frac{k!}{\prod_{j=1}^{k} i_j!}\mathrm{E}\Big\{\prod_{j=1}^{k}\Big[\frac{d_j(N_t)}{j!}\Big]^{i_j}\Big\}\end{aligned}$$

involving expectations relative to a generic Poisson random variable N_t with mean t.

For $j > 1$, the random variable $d_j(N_t) = c_j(N_t) = (j-1)!H_{N_t+1}^{(j)}$ is bounded because the harmonic sum $\sum_{i=1}^{\infty}\frac{1}{i^j}$ converges. However, the random variable $d_1(N_t) = H_{N_t+1}^{(1)} - \ln t$ is more troublesome. If we express

$$\begin{aligned} d_1(N_t)^k &= [H_{N_t+1}^{(1)} - \ln(N_t+1) + \ln(N_t+1) - \ln t]^k \\ &= \sum_{j=0}^{k}\binom{k}{j}[H_{N_t+1}^{(1)} - \ln(N_t+1)]^j[\ln(N_t+1)-\ln t]^{k-j},\end{aligned}$$

then in view of the boundedness of the function $H_n - \ln n$, it suffices to prove that all moments of $Y_t = \ln(N_t+1) - \ln t$ are bounded in t. In so doing, we first observe that the kth moment of Y_t is $O[\sqrt{t}(\ln t)^k]$. Indeed, the Cauchy-Schwarz inequality and the boundedness of the function $[\ln(n+1)]^{2j}/(n+1)$ on the nonnegative integers imply

$$\begin{aligned}
\mathrm{E}(|Y_t|^k) &= \mathrm{E}\left|\sum_{j=0}^{k}\binom{k}{j}[\ln(N_t+1)]^j(-\ln t)^{k-j}\right| \\
&\le \sum_{j=0}^{k}\binom{k}{j}(\ln t)^{k-j}\mathrm{E}\left\{\sqrt{N_t+1}\frac{[\ln(N_t+1)]^j}{\sqrt{N_t+1}}\right\} \\
&\le \sum_{j=0}^{k}\binom{k}{j}(\ln t)^{k-j}\sqrt{\mathrm{E}(N_t+1)}\sqrt{\mathrm{E}\left\{\frac{[\ln(N_t+1)]^{2j}}{N_t+1}\right\}} \\
&\le \sum_{j=0}^{k}\binom{k}{j}(\ln t)^{k-j}\sqrt{t+1}\sqrt{\mathrm{E}\left\{\frac{[\ln(N_t+1)]^{2j}}{N_t+1}\right\}} \\
&= O[\sqrt{t}(\ln t)^k].
\end{aligned}$$

If we now let $\delta(t) = \sqrt{\mathrm{E}(Y_t^{2k})} = O[t^{1/4}(\ln t)^k]$, then the Cauchy-Schwarz and Chebychev inequalities yield

$$\begin{aligned}
\mathrm{E}(|Y_t|^k) &\le \mathrm{E}\Big(|Y_t^k|1_{\{|N_t-t|\le\delta(t)\sqrt{t}\}}\Big) + \mathrm{E}\Big(|Y_t^k|1_{\{|N_t-t|>\delta(t)\sqrt{t}\}}\Big) \\
&\le \sup_{|n-t|\le\delta(t)\sqrt{t}} |\ln(n+1) - \ln t|^k + \sqrt{\mathrm{E}(Y_t^{2k})}\frac{1}{\delta(t)\sqrt{t}}\sqrt{\mathrm{Var}(N_t)} \\
&\le \left\{\max\left[\ln\Big(\frac{t}{t-\delta(t)\sqrt{t}+1}\Big), \ln\Big(\frac{t+1+\delta(t)\sqrt{t}}{t}\Big)\right]\right\}^k + 1,
\end{aligned}$$

which is bounded in t. This completes our proof that the moments of $R_t - \ln t$ are bounded.

5. Approximation of $\mathrm{E}(R_t)$ and $\mathrm{Var}(R_t)$

We now turn to the derivation of the asymptotic expansions in equation (1.4). In this regard, it is convenient to employ the identity

$$\frac{1}{k!}\int_0^t u^k e^{-u}du = e^{-t}\sum_{n=k+1}^{\infty}\frac{t^n}{n!} \tag{5.1}$$

involving the distribution of the waiting time until the appearance of random point $k+1$ in a Poisson process with unit intensity. Multiplying the identity (5.1) by $t^{-1}e^t$ and integrating yield the further identity

$$\frac{1}{k!}\int_0^s \frac{e^t}{t}\int_0^t u^k e^{-u}du\,dt = \int_0^s \frac{1}{t}\sum_{n=k+1}^{\infty}\frac{t^n}{n!}dt = \sum_{n=k+1}^{\infty}\frac{s^n}{nn!}. \tag{5.2}$$

Equations (3.4), (4.1), and (5.1) lead to

$$
\begin{aligned}
\mathrm{E}(R_t) &= e^{-t}\sum_{n=0}^{\infty}\frac{t^{n+1}}{(n+1)!}\sum_{i=1}^{n+1}\frac{1}{i} \\
&= \sum_{i=1}^{\infty}\frac{1}{i}e^{-t}\sum_{n=i}^{\infty}\frac{t^n}{n!} \\
&= \sum_{i=1}^{\infty}\frac{1}{i!}\int_0^t u^{i-1}e^{-u}du \\
&= \int_0^t \frac{1-e^{-u}}{u}du \\
&= \ln t + \int_0^1 \frac{1-e^{-u}}{u}du - \int_1^\infty \frac{e^{-u}}{u}du + \int_t^\infty \frac{e^{-u}}{u}du \\
&= \ln t - \int_0^\infty \ln u e^{-u}du + \int_t^\infty \frac{e^{-u}}{u}du \\
&= \ln t + \gamma + O(e^{-t})
\end{aligned}
$$

where $\gamma = -\Gamma'(1)$ is Euler's constant [**14, 16**]. In like manner, equations (3.4), (4.1), (5.1), and (5.2) yield the second moment expression

$$
\begin{aligned}
\mathrm{E}(R_t^2) &= 2e^{-t}\sum_{n=0}^{\infty}\frac{t^{n+2}}{(n+2)!}\sum_{i=1}^{n+1}\frac{1}{i}\sum_{j=1}^{i}\frac{1}{j} \\
&= 2e^{-t}\sum_{i=1}^{\infty}\frac{1}{i}\sum_{j=1}^{i}\frac{1}{j}\sum_{n=i-1}^{\infty}\frac{t^{n+2}}{(n+2)!} \\
&= 2\int_0^t e^{-u}\sum_{i=1}^{\infty}\frac{u^i}{ii!}\sum_{j=1}^{i}\frac{1}{j}du \\
&= 2\int_0^t e^{-u}\sum_{j=1}^{\infty}\frac{1}{j}\sum_{i=j}^{\infty}\frac{u^i}{ii!}du \\
&= 2\int_0^t e^{-u}\int_0^u \frac{e^v}{v}\int_0^v \sum_{j=1}^{\infty}\frac{w^{j-1}}{j!}e^{-w}dw\,dv\,du \qquad (5.3) \\
&= 2\int_0^t e^{-u}\int_0^u \frac{e^v}{v}\int_0^v \frac{1-e^{-w}}{w}dw\,dv\,du \\
&= 2\int_0^t \frac{e^v}{v}\int_0^v \frac{1-e^{-w}}{w}\int_v^t e^{-u}du\,dw\,dv \\
&= 2\int_0^t \frac{1-e^{v-t}}{v}\int_0^v \frac{1-e^{-w}}{w}dw\,dv.
\end{aligned}
$$

Repeated integration by parts gives the reduction

$$
\begin{aligned}
& 2\int_0^t \frac{1}{v}\int_0^v \frac{1-e^{-w}}{w}\,dw\,dv \\
&= 2\ln v \int_0^v \frac{1-e^{-w}}{w}\,dw\Big|_0^t - 2\int_0^t \ln v \frac{1-e^{-v}}{v}\,dv \\
&= 2\ln t\mathrm{E}(R_t) - (\ln v)^2(1-e^{-v})\big|_0^t + \int_0^t (\ln v)^2 e^{-v}\,dv \qquad (5.4)\\
&= 2\ln t\mathrm{E}(R_t) - (\ln t)^2(1-e^{-t}) + \Gamma''(1) - \int_t^\infty (\ln v)^2 e^{-v}\,dv \\
&= 2\ln t\mathrm{E}(R_t) - (\ln t)^2 + \frac{\pi^2}{6} + \gamma^2 + O\Big[(\ln t)^2 e^{-t}\Big] \\
&= (\ln t)^2 + 2\gamma\ln t + \frac{\pi^2}{6} + \gamma^2 + O\Big[(\ln t)^2 e^{-t}\Big],
\end{aligned}
$$

where $\Gamma''(1) = \int_0^\infty (\ln v)^2 e^{-v}dv = \frac{\pi^2}{6} + \gamma^2$ **[14, 16]**. Similarly,

$$
\begin{aligned}
& 2\int_0^t \frac{e^{v-t}}{v}\int_0^v \frac{1-e^{-w}}{w}\,dw\,dv \\
&= 2e^{-t}\int_1^t \frac{e^v}{v}\int_0^v \frac{1-e^{-w}}{w}\,dw\,dv + 2e^{-t}\int_0^1 \frac{e^v}{v}\int_0^v \frac{1-e^{-w}}{w}\,dw\,dv \\
&= 2e^{-t}\int_1^t \frac{e^v}{v}[\ln v + \gamma + O(e^{-v})]dv + O(e^{-t}) \qquad (5.5)\\
&= 2e^{-t}\int_1^t \frac{e^v}{v}\ln v\,dv + 2\gamma e^{-t}\int_1^t \frac{e^v}{v}dv + O(\ln t\, e^{-t}).
\end{aligned}
$$

Substituting the integration-by-parts expansions

$$
\begin{aligned}
e^{-t}\int_1^t \frac{e^v}{v}\ln v\,dv &= \frac{\ln t}{t} + \frac{\ln t}{t^2} - \frac{1}{t^2} + O\Big(\frac{\ln t}{t^3}\Big) \\
e^{-t}\int_1^t \frac{e^v}{v}dv &= \frac{1}{t} + \frac{1}{t^2} + O\Big(\frac{1}{t^3}\Big)
\end{aligned}
$$

in equation (5.5) yields

$$
\begin{aligned}
& 2\int_0^t \frac{e^{v-t}}{v}\int_0^v \frac{1-e^{-w}}{w}\,dw\,dv \\
&= \frac{2\ln t}{t} + \frac{2\gamma}{t} + \frac{2\ln t}{t^2} + \frac{2(\gamma-1)}{t^2} + O\Big(\frac{\ln t}{t^3}\Big). \qquad (5.6)
\end{aligned}
$$

Finally, combining the mean expansion in equation (1.4) with equations (5.3), (5.4), and (5.6) gives the variance expansion in equation (1.4).

Acknowledgment: We thank one referee for providing us with key references, suggesting the shorter proof of convergence in distribution, correcting a theoretical error in our first draft. Moreover, the other referee kindly suggests us to change the original title to the current better one.

References

[1] Arratia, R., Gordon, L., Waterman, M.S. (1990). The Erdős-Rényi law in distribution, for coin tossing and sequence matching. *Ann. Stat.* **18**:539–570.

[2] Arratia, R., Waterman, M.S. (1989). The Erdős-Rényi strong law for pattern matching with a given proportion of mismatches. *Ann. Prob.* **17**:1152–1169.

[3] Auer, P. and Hornik, K. (1994). On the number of points of a homogeneous Poisson process. *J. Mult. Anal.* **48**:115–156.

[4] Auer, P., Hornik, K. and Révész, P. (1991). Some limit theorems for the homogeneous Poisson process. *Statist. and Probab. Letters* **12**:91–96.

[5] Bacro, J.-N., Deheuvels, P. and Steinebach, J. (1987). Exact convergence rates in Erdős-Rényi-type theorems for renewal processes. *Ann. Inst. H. Poincaré Probab. Statist.* **23**:195–207.

[6] Darling, D. A. (1952). On a test for homogeneity and extreme values. *Ann. Math. Statist.* **23**:450–456.

[7] Deheuvels, P. (1985). On the Erdős-Rényi theorem for random fields and sequences and its relationships with the theory of runs and spacings. *Z. Wahrscheinlichkeit. verw. Gebiete* **70**:91–115.

[8] Deheuvels, P. and Steinebach, J. (1989). Sharp rates for increments of renewal processes. *Ann. Probab.* **17**:700–722.

[9] Devroye, L. (1981). Laws of the iterated logarithm for order statistics of uniform spacings. *Ann. Probab.* **9**:860–867.

[10] Erdős, P., Rényi, A. (1970). On a new law of large numbers. *J. Anal. Math.* **22**:103–111.

[11] Feller, W. (1968). *An Introduction to Probability Theory and Its Applications, Vol. 1*, 3rd ed. Wiley, New York.

[12] Feller, W. (1971). *An Introduction to Probability Theory and its Applications, Vol. 2*, 2nd ed. Wiley, New York.

[13] Graham, R.L., Knuth, D.E., Patashnik, O. (1988). *Concrete Mathematics: A Foundation for Computer Science.* Addison-Wesley, Reading, MA.

[14] Hochstadt, H. (1986). *The Functions of Mathematical Physics*, Dover, New York.

[15] Kendall, M.G., Stuart, A. (1969). *The Advanced Theory of Statistics, Vol. 1*, 3rd ed. Hafner Publishing, New York.

[16] Knuth, D. (1997). *The Art of Computer Programming, 1: Fundamental Algorithms*, 3rd ed. Addison-Wesley, Reading MA.

[17] Lévy, P. (1939). Sur la division d'un segment par des points choisis au hasard. *C. R. Acad. Sci. Paris* **208**:147–149.

[18] Pyke, R. (1965). Spacings. *J. Roy. Statist. Soc.* **27**:395-436.

[19] Steinebach, J. (1982). Improved Erdős-Rényi and strong approximation laws for increments of renewal processes. *Ann. Probab.* **14**:547–559.

[20] Waterman, M.S. (1995). *Introduction to Computational Biology: Maps, Sequences, and Genomes.* Chapman & Hall, London.

[21] Whitworth, W. A. (1897). *Choice and Chance.* Cambridge University Press.

[22] Wilf, H.S. (1978). *Mathematics for the Physical Sciences.* Dover, New York.

Center for Biostatistics in AIDS Research, Harvard School of Public Health, 651 Huntington Avenue, Boston, MA 02115

Current address: Department of Health Evaluation Sciences, H173, College of Medicine, The Pennsylvania State University, 500 University Drive, Hershey, PA 17033-0850

E-mail address: `rfan@hes.hmc.psghs.edu`

Departments of Biomathematics and Human Genetics, School of Medicine, University of California, Los Angeles, CA 90095

E-mail address: `klange@ucla.edu`

Canadian Mathematical Society
Conference Proceedings
Volume **29**, 2000

Inclusions of second quantization algebras

Franca Figliolini and Daniele Guido

1. Introduction.

In this note we study inclusions of second quantization algebras, namely inclusions $\mathcal{R}(M_0) \subset \mathcal{R}(M_1)$ of von Neumann algebras on the Fock space $e^{\mathcal{H}}$ ($\mathcal{H}$ is a separable complex Hilbert space) generated by the Weyl unitaries with test functions in the closed, real linear subspaces M_0, M_1 of $\mathcal{H}$. More precisely we concentrate our attention to the case where both $\mathcal{R}(M_0)$ and $\mathcal{R}(M_1)$ are standard w.r.t. the vacuum vector $e^0 \in e^{\mathcal{H}}$, since in this case the tower and tunnel associated with the inclusion (and the corresponding relative commutants) can themselves be realized as second quantization algebras on the same space:

$$\cdots \subset \mathcal{R}(M_{-1}) \subset \mathcal{R}(M_0) \subset \mathcal{R}(M_1) \subset \mathcal{R}(M_2) \ldots$$

First we show that the class of irreducible inclusions of standard second quantization algebras is non empty, and that they are depth two inclusions, namely $\mathcal{R}(M_0)' \cap \mathcal{R}(M_3)$ is a factor. Then we prove that, when $M_0 \subset M_1$ is a (not necessarily irreducible) inclusion of standard spaces with finite codimension n, $\mathcal{R}(M)$ is isomorphic to the cross product of $\mathcal{R}(N)$ with $\mathbb{R}^n$. On the contrary, when the codimension is infinite, we show that the inclusion may be non regular (see subsection 4.1).

Second quantization algebras and their inclusions occur when studying algebras of local observables for the free fields. Inclusions of local observable algebras are in general neither irreducible nor come from a finite codimension inclusion of real vector spaces. In [**9**] however, local algebras for conformal field theories on the real line are studied, and it is shown that the inclusion of the real vector space corresponding to a bounded interval for the $n+p$-th derivative of the current algebra into the real vector space for the same interval and the n-th derivative theory has codimension p (and is irreducible when $p = 1$). We show in Theorem 4.1 that the corresponding inclusion of second quantization algebras is given by a cross product for any $n \geq 0$, $p > 0$.

Our analysis was also motivated by results concerning depth two inclusions of von Neumann algebras. It is well known that, analyzing the Jones' tower associated with the inclusion of a von Neumann algebra $\mathcal{N}$ into the cross product $\mathcal{M}$ of the same algebra with an outer action of locally compact group, the family of

---

2000 *Mathematics Subject Classification.* Primary 46Lxx, 81Txx.

The second named author is supported in part by MURST and EU.

relative commutants presents some characteristic features, namely the first relative commutant is trivial, the second is abelian, and the third is a type I factor. The study of inclusions with these properties, started in [**10**], was recently enriched by a pair of papers [**5, 6**] where these properties, together with the existence of an operator valued weight from $\mathcal{M}_1$ to $\mathcal{M}_0$, where shown to characterize the inclusions given by cross products with locally compact groups. We show that inclusions of second quantization algebras may produce examples of irreducible depth two inclusions with type III third relative commutant. By a result in [**5**], these inclusions are not regular, since in that case the third relative commutant would be type I, hence do not correspond to a crossed–product with a locally compact group.

On the other hand, our examples satisfy many of the features of a crossed–product, the locally compact group being replaced by an infinite-dimensional vector space, thus furnishing examples in order to develop a theory of non-locally compact cross products.

2. Preliminaries

In this section $\mathcal{H}$ will be a separable Hilbert space and $e^{\mathcal{H}}$ the *symmetric Fock space* over it, i.e. the space

$$e^{\mathcal{H}} = \bigoplus_{n=0}^{\infty} \mathcal{H}^{\otimes s n}$$

where $\mathcal{H}^{\otimes s n}$ is the subspace of the $n-$th tensor product of $\mathcal{H}$ which is pointwise invariant under the natural action of the permutation group.

The set of *coherent* vectors in $e^{\mathcal{H}}$ consists of the vectors

$$e^h = \bigoplus_{n=0}^{\infty} \frac{h^{\otimes n}}{\sqrt{n!}} \quad .$$

This set turns out to be total in $e^{\mathcal{H}}$ (see e.g. [**8**], p.32).

There are two important classes of operators acting on $e^{\mathcal{H}}$:
Second quantization operators

$$e^A = \bigoplus_{n=0}^{\infty} A^{\otimes n} \ ,$$

where A is a densely defined, closed operator on $\mathcal{H}$, and
Weyl unitaries, which are the range of the map

$$h \to W(h)$$

from $\mathcal{H}$ to the unitaries on $e^{\mathcal{H}}$ defined by

$$W(h)e^0 = \exp\left(-\frac{1}{4}\|h\|^2\right) e^{\frac{i}{\sqrt{2}}h}, \quad h \in \mathcal{H} \tag{2.1}$$

$$W(h)W(k) = \exp\left(-\frac{i}{2}Im(h,k)\right) W(h+k) \quad h,k \in \mathcal{H} \tag{2.2}$$

The vector e^0 is called *vacuum* and the relations in the last equality are called Canonical Commutation Relations. Via the preceding equalities $W(h)$ becomes a well defined, isometric and invertible (with inverse $W(-h)$) operator on the dense set spanned by coherent vectors, and hence it extends to a unitary on $e^{\mathcal{H}}$. Weyl unitaries generate the so-called *second quantization algebras.* With each closed real linear subspace $K \subset \mathcal{H}$ (in the following we shall write $K \leq_{\mathbb{R}} \mathcal{H}$), a von Neumann algebra $\mathcal{R}(K)$ is associated, defined by

$$\mathcal{R}(K) = \{W(h), \quad h \in K\}''.$$

In the following Proposition we recall the basic facts about these algebras [**1, 4**]:

PROPOSITION 2.1. *Let us consider the map $K \to \mathcal{R}(K)$ where $K \leq_{\mathbb{R}} \mathcal{H}$. Then (i)The mentioned map is an isomorphism of complemented nets where "$\wedge$" is the intersection, "$\vee$" is the generated real subspace, resp. von Neuman algebra, and the complementation "$'$" is the polar space w.r.t. $Im(\cdot,\cdot)$, resp. the commutant.*
(ii) The map preserve the standard property, namely the vacuum vector is standard, i.e. cyclic and separating, for $\mathcal{R}(K)$ if and only if K is standard, i.e. $K \cap iK = 0$ and $K + iK$ is dense. Moreover the Tomita operator S associated with a standard algebra $\mathcal{R}(K)$ is the second quantization of the Tomita operator s on K defined by

$$s : h + ik \to h - ik \qquad h, k \in K$$

and also $J = e^{j}$, $\Delta = e^{\delta}$, where $S = J\Delta^{1/2}$, $s = j\delta^{1/2}$.

As a consequence of the preceding proposition, the properties of the second quantization algebras can be studied in terms of the generating one-particle subspaces, and in the following we shall attribute to a real linear subspace K of $\mathcal{H}$ all the properties of its second quantization algebra, for instance we will say that K is a standard type III subspace if $\mathcal{R}(K)$ is, etc. We shall also say that $M' = \{x \in \mathcal{H} : Im(x, y) = 0 \; \forall x \in M\}$ is the commutant of M. In particular, if $M_0 \subset M_1$ is an inclusion of standard subspaces of $\mathcal{H}$, the tower and the tunnel [**11, 3**] generated by the inclusion $\mathcal{M}_0 \equiv \mathcal{R}(M_0) \subset \mathcal{M}_1 \equiv \mathcal{R}(M_1)$ can be spatially realized as

$$\mathcal{M}_{i+1} = J_i \mathcal{M}'_{i-1} J_i \qquad i \geq 1$$
$$\mathcal{M}_{i-1} = J_i \mathcal{M}'_{i+1} J_i \qquad i \leq 0,$$

where J_i is the Tomita conjugation for $\{\mathcal{M}_i, e^0\}$. Therefore they are the second quantization of the tower, resp. tunnel, of standard subspaces:

$$M_{i+1} = j_i M'_{i-1} \qquad i \geq 1$$
$$M_{i-1} = j_i M'_{i+1} \qquad i \leq 0.$$

3. Inclusions of standard subspaces

DEFINITION 3.1. We shall say that a pair (E, F), $E, F \leq_{\mathbb{R}} \mathcal{H}$, is a standard pair if $E \wedge F = \{0\}$ and $E \vee F = \mathcal{H}$. A standard pair will be called strongly standard if, moreover, $E + F = \mathcal{H}$. If (E, F) is a standard pair, we define the operator $s_{E,F}$ as follows:

$$\begin{array}{rccc} s_{E,F} : & E + F & \to & E + F \\ & e + f & \mapsto & e - f \end{array}$$

Let us observe that N is standard if and only if (N, iN) is a standard pair. Also, N is a factor if and only if (N, N') is a standard pair. Indeed, if N is a factor then $N \cap N' = \{0\}$, and, taking the commutant, $N \vee N' = (N \cap N')' = \mathcal{H}$.

LEMMA 3.2. *Let (E, F) be a standard pair. Then:*
(i) $s_{E,F}$ is closed
(ii) $s_{E,F}$ is bounded iff (E, F) is a strongly standard pair.

(iii) The pair (E, F) can be properly extended by a standard pair $(\tilde{E}, \tilde{F})$ (i.e. $E \subseteq \tilde{E}$, $F \subseteq \tilde{F}$ where at least one inclusion is proper) iff (E, F) is not strongly standard.

PROOF. (*i*) Since the graph norm of $e+f$ w.r.t. $s_{E,F}$ is $(\|e+f\|^2 + \|s_{E,F}(e+f)\|^2)^{1/2} = \sqrt{2}(\|e\|^2 + \|f\|^2)^{1/2}$, the result follows.

(*ii*) Immediate by the closed graph theorem.

(*iii*) Suppose (E, F) is strongly standard and let $(\tilde{E}, \tilde{F})$ be a standard extension. For any $y \in \tilde{E}$, $y = e + f$, hence $f = y - e \in F \cap \tilde{E} \subset \tilde{F} \cap \tilde{E}$, which implies $f = 0$, i.e. $y \in E$, hence $\tilde{E} = E$. □

LEMMA 3.3. *Let $N \leq_{\mathbb{R}} \mathcal{H}$ be a standard subspace. Then*
(i) There exists a standard subspace M such that $N \subset M$ if and only if (N, iN) is not strongly standard.
(ii) If $y \in (N + iN)^c$, then $M = \{x + \lambda y : x \in N, \lambda \in \mathbb{R}\}$ is standard (but not strongly standard).

PROOF. If (N, iN) is strongly standard and $M \supset N$, then (M, iM) cannot be standard by Lemma 3.2, (*iii*).

Now let (N, iN) be not strongly standard, $y \in (N + iN)^c$ and $M = \{x + \lambda y : x \in N, \lambda \in \mathbb{R}\}$. If $z \in M \cap iM$ then $z = n_1 + \lambda_1 y = in_2 + i\lambda_2 y$ for some $n_1, n_2 \in N$, $\lambda_1, \lambda_2 \in \mathbb{R}$. However λ_1 and λ_2 should be zero, otherwise $y = (i\lambda_2 - \lambda_1)^{-1}(n_1 - in_2) \in N + iN$, hence $z \in N \cap iN = \{0\}$. Finally, if M is standard then (N, iN) is not strongly standard hence s_N is unbounded. Since $s_N \subset s_M$, s_M is unbounded, hence M is not strongly standard. □

LEMMA 3.4. *Let N be a factor subspace of $\mathcal{H}$. Then*
(i) There exists a subspace M such that $N \subset M$ is irreducible an irreducible if and only if (N, N') is not strongly standard.
(ii) If $y \in (N + N')^c$, and $M = \{x + \lambda y : x \in N, \lambda \in \mathbb{R}\}$, then $N' \cap M = \{0\}$ (and (M, M') is not strongly standard).

PROOF. If (N, N') is strongly standard and $M \supset N$, then (M, N') cannot be standard by Lemma 3.2, (*iii*).

Now let (N, N') be not strongly standard, $y \in (N + N')^c$ and $M = \{x + \lambda y : x \in N, \lambda \in \mathbb{R}\}$. If $z \in M \cap N'$ then $z = n + \lambda y = n'$ for some $n \in N$, $n' \in N'$, $\lambda \in \mathbb{R}$. However λ should be zero, otherwise $y = \lambda^{-1}(n' - n) \in N + N'$, hence $z \in N \cap N' = \{0\}$. Finally, observe that (M, M') is properly extended by (M, N'), hence $s_(M, M')$ cannot be bounded, which means that (M, M') is not strongly standard. □

PROPOSITION 3.5. *Let N be a standard subspace of $\mathcal{H}$. Then*
(i) There exists a standard subspace $M \supset N$ if and only if $0 \in \sigma(\delta_N)$.
(ii) Suppose N is a factor. There exists a subspace M such that $N \subset M$ is an irreducible inclusion if and only if $1 \in \sigma(\delta_N)$.
(iii) Suppose N is a factor. There exists a standard subspace M such that $N \subset M$ is an irreducible inclusion if and only if $\{0, 1\} \subset \sigma(\delta_N)$.
(iv) If $N \subset M$ is an irreducible inclusion of standard spaces,then N and M are type III_1 factors.

PROOF. (*i*) By Lemmas 3.2, (*ii*), 3.3, (*i*), there exists a standard subspace $M \supset N$ *iff* s_N, hence $\delta_N^{1/2}$, is unbounded. Since $j_N \delta_N^{1/2} j_N = \delta_N^{-1/2}$, this is equivalent to $0 \in \sigma(\delta_N^{1/2})$, which in turn corresponds to $0 \in \sigma(\delta_N)$ by spectral mapping.

(ii) Let Θ be the selfadjoint operator defined by $\cos\Theta = \frac{|\delta-I|}{\delta+I}$, $\delta = \delta_N$, and let $s_{N,N'} = ud^{1/2}$ be the polar decomposition of $s_{N,N'}$. By $s_{N,N'}^2 \subset I$ one obtains $u^2 = 1$ and $udu = d^{-1}$, and by **[7]**, Proposition 3.2, one gets $\sin\Theta = \frac{|d-I|}{d+I}$. This implies that there exists a subspace M such that $N \subset M$ is irreducible $\iff$ $s_{N,N'}$ is unbounded, $\iff$ $0 \in \sigma(d)$, $\iff$ $\pi/2 \in \sigma(\Theta)$, $\iff$ $1 \in \sigma(\delta)$.

(iii) By the relations among δ, d and Θ one gets that $x \in \mathcal{D}(\delta^{1/2})^c \iff \chi_{[0,\varepsilon]}(\Theta)x \in \mathcal{D}(\delta^{1/2})^c$ and $y \in \mathcal{D}(d^{1/2})^c \iff \chi_{[\pi/2-\varepsilon,\pi/2]}(\Theta)y \in \mathcal{D}(d^{1/2})^c$, therefore if $x \in \mathcal{D}(\delta^{1/2})^c$ and $y \in \mathcal{D}(d^{1/2})^c$, then $z = \chi_{[0,\pi/4]}(\Theta)x + \chi_{[\pi/4,\pi/2]}(\Theta)y \in (\mathcal{D}(\delta^{1/2}) \cup \mathcal{D}(d^{1/2}))^c$, hence $M = \{x + \lambda z : x \in N,\ \lambda \in \mathbb{R}\}$ has the required properties by Lemmas 3.3, 3.4.

(iv) As shown before, N is a factor if and only if $1 \in \sigma(\delta_N)$. By **[7]**, Proposition 4.5, this implies that N is an injective type III_1 factor. Since $M \subset j_M N' j_M$ is an irreducible inclusion, the same reasoning applies to M. □

REMARK 3.6. If the inclusion of standard spaces $M_0 \subset M_1$ is irreducible, all the algebras of the Jones tower (and tunnel) are standard type III_1 factors.

We have proved that the set of irreducible incusions of standard factors in $\mathcal{H}$ is not empty. More precisely we have shown how to costruct irreducible inclusions of standard factors with codimension 1. Iterating the process and taking direct sums one can construct irreducible inclusions of standard factors with any codimension. In the following we study the structure of such inclusions.

LEMMA 3.7. *If M is standard in $\mathcal{H}$, $(M, Ker(j+I))$ is a strongly standard pair, where j is the modular conjugation of M.*

PROOF. This proof closely follows the proof of Proposition 3.2 in **[7]**. The spaces $M, Ker(j+I)$ are both reduced by the spectral projections relative to $|\log\delta|$, where δ is the modular operator relative to M, hence it is sufficient to study the problem on the fiber of the representation of $\mathcal{H}$ as direct integral w.r.t. $|\log\delta|$. Then, as in **[7]**, Prop. 3.2, it is sufficient to study the case in which K is a standard subspace of $\mathbb{C}^2$, hence K can be seen as generated by the vectors

$$y^+ = \begin{pmatrix} \cos\frac{\vartheta}{2} \\ \sin\frac{\vartheta}{2} \end{pmatrix} \qquad y^- = \begin{pmatrix} i\cos\frac{\vartheta}{2} \\ -i\sin\frac{\vartheta}{2} \end{pmatrix}$$

and j is represented by the matrix

$$\begin{pmatrix} 0 & C \\ C & 0 \end{pmatrix}$$

where C is the complex conjugation. Therefore

$$Ker(j+I) = \{\begin{pmatrix} a \\ -\overline{a} \end{pmatrix}, \quad a \in \mathbb{R}\}$$

and it is easy to see that

$$|Re(h,k)| \leq \frac{\sqrt{2}}{2}, \quad h \in M,\ k \in Ker(j+I),\ \|h\| = \|k\| = 1,$$

which implies

$$\frac{\|h+k\|^2}{\|h\|^2+\|k\|^2} = \frac{\|h\|^2+\|k\|^2+2Re(h,k)}{\|h\|^2+\|k\|^2} \geq \frac{\|h\|^2+\|k\|^2-\sqrt{2}\|h\|\|k\|}{\|h\|^2+\|k\|^2} \geq \frac{\sqrt{2}-1}{\sqrt{2}},$$

namely the graph norm of $s_{M,Ker(j+I)}$ is equivalent to the Hilbert norm, and the strongly standard property follows by Lemma 3.2, (ii). □

If $M_0 \subset M_1$ are standard subspaces and j_0, j_1 are the modular conjugations, the tower and the tunnel of standard subspaces are inductively defined by the equations:

$$M_{k+1} = j_k M'_{k-1}, \qquad k \in \mathbb{Z}.$$

We shall use the following notations:

$$A_{k,l} = M'_k \cap M_l \qquad k \leq l \in \mathbb{Z} \tag{3.1}$$

$$B_k = M_{k+1} \cap Ker(j_k + I) \qquad k \in \mathbb{Z} \tag{3.2}$$

Now we can prove some structural results about inclusions of standard subspaces.

PROPOSITION 3.8. *Let $M_0 \subset M_1$ be an inclusion of standard subspaces. Then the following holds:*

$$(i) \quad M_{k+p} = M_k + \sum_{j=0}^{p-1} B_{k+j} \qquad k \in \mathbb{Z},\ p \geq 0$$

$$(ii) \quad A_{k-1,k+p} = A_{k-1,k} + \sum_{j=0}^{p-1} B_{k+j} \qquad k \in \mathbb{Z},\ p \geq 0$$

$$(iii) \quad B_k \quad \text{is commutative} \qquad k \in \mathbb{Z}$$

$$(iv) \quad \sum_{j=1}^{2p} B_{k+j} \quad \text{is a factor} \qquad k \in \mathbb{Z},\ p > 0$$

$$(v) \quad \mathcal{Z}\left(\sum_{j=1}^{2p+1} B_{k+j}\right) = \{(I + \sum_{j=1}^{p}(-1)^i j_{k+2i})x : x \in B_{k+1}\} \qquad k \in \mathbb{Z},\ p > 0,$$

where $\mathcal{Z}(P) = P \cap P'$. In particular, if the inclusion $M_0 \subset M_1$ is irreducible,
(vi) $\quad A_{k-1,k+1} = B_k$ is abelian $\forall k \in \mathbb{Z}$
(vii) $\quad A_{k-1,k+2} = B_k + B_{k+1}$ is a factor $\forall k \in \mathbb{Z}$

PROOF. By Lemma 3.7 and eq. (3.2), any vector in M_{k+p} can be (uniquely) decomposed in sum of a vector in M_{k+p-1} and a vector in B_{k+j-1}. Iterating this argument we get (i).

Now we note that, by the global j_k-invariance of B_k,

$$B_k \subseteq M_{k+1} \cap j_k M_{k+1} = M_{k+1} \cap M'_{k-1} = A_{k-1,k+1}.$$

In particular, B_{k+j} commutes with M_{k-1} for each $j \geq 0$. Therefore, applying the decomposition (i) to a vector in $A_{k-1,k+p}$, we get (ii).

Let us take $x, y \in B_k$. Since B_k is j_k–antiinvariant,

$$Im(x, y) = Im(x, -j_k y) = Im(y, -j_k x) = Im(y, x) = -Im(x, y),$$

which implies (iii).

Now observe that

$$B_{k+1} \cap B'_k = (B_{k+1} \cap M'_k) \cap B'_k = B_{k+1} \cap (M_k \vee B_k)' =$$
$$= B_{k+1} \cap M'_{k+1} = (M_{k+1} \cap M'_{k+1}) \cap Ker(j_{k+1} + I)$$

Since the elements in the center of M_{k+1} are j_{k+1} invariant (see Remark 1.7, [**7**]), it turns out that $B_{k+1} \cap B'_k = 0$.

Then notice that $x \in \mathcal{Z}(\sum_{j=1}^p B_{k+j})$ *iff* $x \in \sum_{j=1}^p B_{k+j}$ and

(3.3) $$Im(x, y) = 0 \ \forall y \in B_{k+j}, \qquad 1 \le j \le p.$$

Since any $x \in \sum_{j=1}^p B_{k+j}$ may be uniquely written as $x = \sum_1^p b_{k+j}$, $b_{k+j} \in B_{k+j}$, and B_m commutes with B_n if $|m - n| \ge 2$, eq. (3.3) for $j = 1$ implies $b_{k+2} = 0$. Then eq. 3.3 for $j = 3$ implies $b_{k+4} = 0$ and, iterating, $b_{k+2j} = 0$ for any $1 \le j \le p/2$. Analogously we show that b_{k+p-1} vanishes, and therefore $b_{k+p-2j-1} = 0$ for $j \le p/2 - 1$. If p is even this shows that $x = 0$, i.e. (iv).

When $p = 2m + 1$ is odd, we proved that $x = \sum_0^m b_{k+2j+1}$. Now, taking $y \in B_{k+2j}$, $1 \le j \le m$, and making use of the j-anti-invariance of B, we have

$$\begin{aligned} 0 = Im(x, y) &= Im(b_{k+2j-1} + b_{k+2j+1}, y) \\ &= Im(b_{k+2j-1}, y) - Im(b_{k+2j+1}, j_{k+2j} y) \\ &= Im(b_{k+2j-1}, y) + Im(j_{k+2j} b_{k+2j+1}, y) \\ &= Im(b_{k+2j-1} + j_{k+2j} b_{k+2j+1}, y), \end{aligned}$$

hence $b_{k+2j-1} + j_{k+2j} b_{k+2j+1} = 0$, and this implies (v).

Finally we observe that the inclusion $M_0 \subset M_1$ is irreducible *iff* $A_{0,1}$, and therefore $A_{k,k+1}$, $\forall k \in \mathbb{Z}$, are trivial. Then (vi) follows from (ii) and (iii) and (vii) follows from (ii) and (iv). □

4. Inclusions of second quantization algebras

We first recall that an inclusion of von Neumann algebras $\mathcal{M}_0 \subset \mathcal{M}_1$ is said of depth 2 if the von Neumann algebra $\mathcal{M}'_0 \cap \mathcal{M}_3$ is a factor.

THEOREM 4.1. *Let $M_0 \subset M_1$ be an inclusion of standard subspaces.*
(i) If the codimension of M_0 in M_1 is finite and equal to n

$$\mathcal{R}(M_1) = \mathcal{R}(M_0) \times_\alpha \mathbb{R}^n$$

where $\mathbb{R}^n$ is identified with B_0 and $\alpha_h(\cdot) = \mathrm{ad}\, W(h)$, $h \in B_0$.
(ii) If $M_0 \subset M_1$ is irreducible then $\mathcal{R}(M_0) \subset \mathcal{R}(M_1)$ is of depth two.

PROOF. (i) First we note that, by eq. 2.2,

(4.1) $$\mathrm{ad}\, W(h)\,(W(k)) = e^{-iIm(h,k)} W(h + k)$$

i.e. each closed commutative subspace of $\mathcal{H}$ considered as an additive group acts on any second quantization algebra via $\mathrm{ad}\, W(\cdot)$.

Then we only have to check that the hypotheses of Landstad Theorem [**12**] are satisfied. By Proposition 2.1 and Proposition 3.8 (*i*) we get

$$\mathcal{R}(M_1) = \mathcal{R}(M_0) \vee \{W(h) : h \in B_0\}''$$

i.e. $\mathcal{R}(M_1)$ is generated by $\mathcal{R}(M_0)$ and a unitary representation of B_0 which is strongly continuous by the finite-dimensionality of B_0.

Now we claim that the action

$$\beta : h \in B_1 \to \mathrm{ad}\, W(h) \in Aut(R(M_1))$$

is dual to the action α of B_0 on M_0.

On the one hand

$$\mathcal{R}(M_1)^\beta = \mathcal{R}(M_1) \cap \{W(h) : h \in B_1\}' = \mathcal{R}(M_1 \cap B_1') = \mathcal{R}(M_0)$$

where the last equality follows from Proposition 3.8 (*i*), (*iv*). On the other hand, by eq. (4.1), B_1 acts as the dual group on $\mathcal{R}(B_0)$ *iff* the pairing

$$Im(\cdot,\cdot) : B_0 \times B_1 \to \mathbb{R}$$

is a duality between real Hilbert spaces. The factoriality of $B_0 \vee B_1$ implies that the pairing is non degenerate, the finite-dimensionality of B_0 and B_1 implies that it is continuous, and the thesis follows.

(*ii*) follows immediately by Proposition 3.8 □

4.1. Non regular inclusions.

THEOREM 4.2. *There exists an irreducible inclusion $M_0 \subset M_1$ of standard subspaces with infinite codimension for which the third relative commutant is of type III.*

PROOF. Let $M_0^n \subset M_1^n$ be a sequence of irreducible inclusions of standard subspaces such that the codimension of M_0^n in M_1^n is one, and let M_0, resp. M_1 the direct sum of the M_0^n, resp. M_1^n in $\oplus_{n=1}^{\infty}\mathcal{H}$.

As it is well known, direct sums of complex orthogonal real subspaces give rise to tensor products at the second quantization level, therefore the decomposition $A_{0,3} = \oplus_{n=1}^{\infty} A_{0,3}^n$ gives rise to an ITPFI decomposition. Since the codimension of M_0^n in M_1^n is one, $A_{0,3}^n$ is two-dimensional and the angle of $A_{0,3}^n$ with $iA_{0,3}^n$ is a number θ_n.

Therefore, by [**2**] and [**7**] Proposition 2.7, the type of $A_{0,3}$ depends only on the sequence θ_n. Choosing isomorphic inclusions $M_0^n \subset M_1^n$ one gets a constant sequence θ_n hence the factor has type III_λ, with $\lambda = \tan^2\theta/2$. □

Let us recall that an inclusion $\mathcal{M}_0 \subset \mathcal{M}_1$ is called *regular* if the space of intertwiners $T : yJ_1xJ_1T = TyJ_2xJ_2$ for any $x \in \mathcal{M}_1$, $y \in \mathcal{M}_0$ is non zero. It was shown in [**5**] that for any depth two regular inclusion the third relative commutant is of type I, and that for any crossed–product inclusion regularity holds. This shows that the inclusions of Theorem 4.2 are not regular, hence cannot come from a crossed–product with a locally compact group. Indeed they constitute an example of a kind of crossed–product with the group given by an infinite dimensional Hilbert space with the additive structure.

We conclude this paper observing that, for the described inclusions, the only visible invariant is the type of the third relative commutant. Since these inclusions are, in a sense, topologically trivial, it seems reasonable to conjecture that the third relative comutant is actually a complete invariant.

ACKNOWLEDGEMENT . The first draft of this work was completed when the authors where visiting the mathematics department of the Penn State University. We thank the people in the department for the warm hospitality and Adrian Ocneanu for having suggested this line of research.

References

[1] H. Araki. *A lattice of von Neumann algebras associated with the quantum field theory of a free Bose field*", J. Math. Phys. **4** (1963), 1343–1362.

[2] H. Araki, E.J. Woods. *A classification of factors*, Publ. RIMS Kyoto Univ. **A 3** (1968), 51–130.

[3] M. Pimsner, S. Popa. *Iterating the basic construction.* Trans. Amer. Math. Soc. 310 (1988), no. 1, 127–133

[4] J.P. Eckmann, K. Osterwalder. *An application of Tomita's theory of modular Hilbert algebras: Duality for free Bose fields*, J. Funct. Analysis **13** (1973), 1–22.

[5] M. Enock, R. Nest. *Irreducible inclusions of factors, multiplicative unitaries, and Kac algebras.* J. Funct. Anal. **137** (1996), 466–543.

[6] M. Enock. *Inclusions irriductibles de facteurs et unitaires multiplicatifs. II.* J. Funct. Anal. **154** (1998), 67–109.

[7] F. Figliolini, D. Guido. *On the Type of Second Quantization Factors*, Journal of Operator Theory, **31** (1994), 229–252.

[8] A. Guichardet. *Tensor products of C*-algebras (infinite)*, Aarhus Univ., Lecture notes series, 13 (1969).

[9] D. Guido, R. Longo, H.-W. Wiesbrock. *Extensions of conformal nets and superselection structures*, Commun. Math. Phys. **192** (1998), 217-244.

[10] R. Herman, A. Ocneanu. *Index theory and Galois theory for infinite index inclusions of factors.* C. R. Acad. Sci. Paris Ser. I Math. **309** (1989), 923–927.

[11] V. Jones. *Index for subfactors*" Invent. Math. **2** (1983) 1–25.

[12] M. Landstad. *Duality theory for covariant systems.* Trans. Amer. Math. Soc. **248** (1979), 223–267.

D. GUIDO, DIP. DI MATEMATICA, UNIVERSITÀ DELLA BASILICATA, I–10085 POTENZA, ITALY.
E-mail address: guido@unibas.it

Canadian Mathematical Society
Conference Proceedings
Volume **29**, 2000

Weak Solutions to Stochastic Differential Equations with Discontinuous Drift in Hilbert Space

L. Gawarecki and V. Mandrekar

ABSTRACT. Professor Sergio Albeverio has been interested in solutions of infinite dimensional stochastic differential equations with values in $C([0,T],\mathbf{R}^{\mathbf{Z}^d})$ motivated by applications to physics ([**1**], [**2**]). Leha and Ritter in [**6**] have studied the existence problem in $C([0,T],H^w)$. Our purpose here is to extend the latter work by weakening the assumptions on the potential, using techniques developed in our earlier work. More precisely, we study the problem of the existence of weak solutions for SDE's with discontinuous drift in a Hilbert space H. The discontinuity is modeled by a countable family of real valued functions. The solution has finite dimensional Galerkin approximation and is realized in the space of continuous functions with values in H, where H is endowed with its weak topology. This work extends a result in [**6**] and also shows that under the assumptions in [**6**] both, the Galerkin approximation and the infinite dimensional approximation of [**6**] produce solutions with identical laws

1. Introduction

In our work [**3**], it was shown that with the usual continuity and growth assumptions on the coefficients of a stochastic differential equation (SDE) in an infinite dimensional real separable Hilbert space H, the solution can be obtained in a larger Hilbert space H_{-1}, such that the embedding $H \hookrightarrow H_{-1}$ is compact. As a consequence, we derived the result in [**5**], and the work in [**7**]. In some sense, the space H_{-1} was arbitrary, however needed due to a (deterministic) example of Godunov [**4**]. This forced us to extend continuously the coefficients of the SDE to H_{-1}. If one looks at H as a space isomorphic to l_2 (with the canonical isomorphism $h \to (h,e_k)_H$, $\{e_k\}_{k=1}^{\infty}$ an ONB in H), then the natural choice of a larger space is $\mathbf{R}^{\infty}$, so that the embedding $l_2 \hookrightarrow \mathbf{R}^{\infty}$ is compact.

We exploit this idea to obtain solutions in $C([0,T],H^w)$ (H^w denotes H with its weak topology) essentially using the technique of compact embedding in [**3**]. As a consequence of this approach we are able to give an extension of an interesting result due to Leha and Ritter, [**6**], on equations with discontinuous drift. Under the Lipschitz assumption of [**6**] we relate our solution to that obtained by Leha and Ritter. The advantage of our construction is that our weak solution has

1991 *Mathematics Subject Classification.* 60H10.

finite dimensional Galerkin approximation as opposed to the infinite dimensional approximation given in [**6**].

In a subsequent paper we shall study the solutions in $C([0,T], \mathbf{R}^{\mathbf{Z}^d})$ using ideas in [**2**] and our technique in [**3**] concerning existence of weak solution in the dual of a nuclear space. We first review notation and some facts from [**3**].

Let $(\Omega, \mathcal{F}, (\mathcal{F}_t)_{t\in[0,T]}, P)$ be a filtered probability space, K a real separable Hilbert space, and Q a non-negative trace-class operator on K. Let W_t be a K-valued Brownian motion with covariance Q. Consider the following SDE:

$$\left\{ \begin{array}{l} dX(t) = F(t, X(t))dt + B(t, X(t))dW(t) \\ X(0) = x \in H \quad (deterministic). \end{array} \right. \tag{1.1}$$

We assume that $F : [0,T] \times H \to H$, $B : [0,T] \times H \to L(K, H) := L$, the space of continuous linear operators. Consider the following conditions:

(A1) F and B are jointly continuous.

(A2) There exists a constant ℓ, such that $\forall x \in H$

$$\|F(t,x)\|_H + \|B(t,x)\|_L \leq \ell(1 + \|x\|_H).$$

Note, that if we define $\tilde{F} : [0,T] \times C([0,T], H) \to H$, and $\tilde{B} : [0,T] \times C([0,T], H) \to L$ by $\tilde{F}(t,x) = F(t, x(t))$, and $\tilde{B}(t,x) = B(t, x(t))$, then we are in the framework of [**3**].

Let $X(t)$ be a solution (weak or strong) to the equation (1.1). Lemma 5 and Lemma 6 in [**3**] state that under conditions (A1) and (A2), the following inequalities hold:

$$\begin{aligned} E \sup_{0\leq t\leq T} \|X(t)\|_H^{2p} &\leq C, \ p \geq 1 \\ E\|X(t+h) - X(t)\|_H^4 &\leq \quad Ch^2, \end{aligned} \tag{1.2}$$

with the constant C depending only on ℓ and T.

Similar to Corollary 1 in [**3**], we have from (1.2), that for every $\varepsilon > 0$,

(i) There exists M_ε, satisfying $\sup_n \mu_n \left(\sup_{0\leq t\leq T}\|x(t)\|_H > M_\varepsilon\right) \leq \varepsilon$

(ii) $\lim_{\delta\to 0} \sup_n \mu_n \left(\sup_{|t-s|<\delta}\|x(t) - x(s)\|_H > \varepsilon\right) = 0.$

Here, $\mu_n = Law\,(X_n)$, and $\{X_n\}_{n=1}^\infty$ is a sequence of $C([0,T], H)$–valued processes, defined possibly on different probability spaces, satisfying the conditions in (1.2), with a common constant C independent of n.

Let us denote by $j : H \hookrightarrow \mathbf{R}^\infty$ the map $j(h) = \{(h, e_k)\}_{k=1}^\infty$. Then the embedding j is compact, and $\rho_{\mathbf{R}^\infty}(j(h_1), j(h_2)) = \sum_{k=1}^{\infty} \frac{1}{2^k} \frac{|(h_1, e_k)_H - (h_2, e_k)_H|}{1 + |(h_1, e_k)_H - (h_2, e_k)_H|} \leq C\,\|h_1 - h_2\|_H$. Hence, as in Corollary 2 in [**3**], we get that if $\{X_n\}_{n=1}^\infty$ satisfy conditions (i) and (ii) above, then the sequence of measures $\mu_n \circ j^{-1}$ is relatively weakly compact on $C([0,T], \mathbf{R}^\infty)$.

2. Existence of Solutions for SDE's with discontinuous drift

An interesting existence result, Theorem 2.4 in [**6**], for SDE's with discontinuous drift in Hilbert space, has been obtained in [**6**]. The authors have presented a direct construction of a compact set in $C(\mathbf{R}_+, \mathbf{R}^\infty)$, on which, up to ε, the laws of the approximate solutions were concentrated. Here, we apply the technique discussed in the Introduction, to extend this result to non-Lipschitz coefficients and we show, that in the case of Lipschitz coefficients our approximation yields the same weak solution as the one in [**6**]. For $x \in H$ we denote $x^k = (x, e_k)_H$, $\{e_k\}_{k=1}^{\infty}$ is an ONB in H, and H^τ, the space H endowed with the topology induced by $\mathbf{R}^\infty$.

THEOREM 2.1. *Suppose that H is a real separable Hilbert space. Let $F : H \to H$, and assume that $F^k(x) := (F(x), e_k)_H : H^\tau \to \mathbf{R}$ and $q_k : \mathbf{R} \to \mathbf{R}$, $k = 1, 2, \ldots$, are continuous, and the following conditions hold true:*
(B1) $(x, F(x))_H \leq \ell(1 + \|x\|_H^2),\ x \in H,$
(B2) $u \cdot q_k(u) \leq \ell(\gamma_k + u^2),\ u \in \mathbf{R}.$
Here, trace $Q = \sum_{k=1}^{\infty} \gamma_k$, where Q is the covariance of an H-valued Brownian motion W. Then, there exists a weak, weakly continuous solution to the equation

$$X(t) = x + \int_0^t (F(X(s) + q(X(s)))\, ds + W(t)$$

in the following sense. There exists an H-valued Brownian motion $\tilde{W}$ with covariance Q and a weakly continuous process $X(\cdot) \in C([0,T], H^w)$, such that for every $k = 1, 2, \ldots$

$$X^k(t) = x^k + \int_0^t (F^k(X(s)) + q_k(X^k(s)))\, ds + \tilde{W}^k(t). \tag{2.1}$$

Proof. Consider the following equation:

$$X_n(t) = P_n X + \int_0^t (P_n F(P_n X_n(s)) + q^n(P_n X_n(s)))\, ds + W^n(t) \tag{2.2}$$

Here, P_n is a projection of H onto $span\{e_1, \ldots, e_n\}$, $q^n : \mathbf{R}^n \to \mathbf{R}^n$, $q^n(y) = (q_1(y^1), \ldots, q_n(y^n))$, and $W^n(t)$ is a Brownian motion with covariance $Q_n = P_n\, Q\, P_n$.

Equation (2.2) can be considered in $\mathbf{R}^n$ identified with a finite dimensional subspace of H, and it is well known that it has a weak solution $X_n(\cdot)$ in $C([0,T], \mathbf{R}^n)$. Denote $F_n = P_n\, F\, P_n$. The process

$$y_n(t) = x_n(t) - P_n x - \int_0^t (F_n(x_n(s)) + q^n(x_n(s)))\, ds \tag{2.3}$$

(here, the subscript n in y_n, x_n indicates the dimension of $\mathbf{R}^n$), is an n–dimensional Brownian motion with respect to the canonical filtration on $C([0,T], \mathbf{R}^n)$ and the measure $\mu_n = Law(X_n)$.

For $h \in H$, the Meyer process of the real-valued martingale $(y_n(t), P_n u)_{\mathbf{R}^n}$ is given by

$$< (y_n(\cdot), P_n u)_{\mathbf{R}^n} >_t = t\, Q\,(P_n u, P_n u) := t\, Q_n(u, u).$$

We will be in a position to apply the argument involving compact embedding, discussed in the Introduction if we prove that the conditions in (1.2) are fulfilled.

The estimate $E \sup_t \|X_n(t)\|_{\mathbf{R}^n}^{2p} \leq C$, $p \geq 1$, is established in Lemma (2.1) in [**6**], by applying Itô's formula to the function $x \mapsto \|x\|_{\mathbf{R}^n}^{2p}$.

We denote $G_n(x) = F_n(x) + q^n(x)$, and identify $\mathbf{R}^n$ with an n-dimensional subspace of H. Using Itô's formula for the function $x \to \|x\|_{\mathbf{R}^n}^4$ we obtain, that for a process $\xi(t)$, which is a solution to (2.2),

$$\begin{aligned}\|\xi(t+h)-\xi(t)\|_{\mathbf{R}^n}^4 &= 4\int_t^{t+h}\|\xi(u)-\xi(t)\|_{\mathbf{R}^n}^2(\xi(u)-\xi(t),G_n(\xi(u)))_{\mathbf{R}^n}\,du\\ &\quad+4\,trQ_n\int_t^{t+h}\|\xi(u)-\xi(t)\|_{\mathbf{R}^n}^2\,du\\ &\quad+4\int_t^{t+h}\|\xi(u)-\xi(t)\|_{\mathbf{R}^n}^2(\xi(u)-\xi(t),dW^n(u)).\end{aligned}$$

This yields the following estimate for the fourth moment:

$$\begin{aligned}&E\|\xi(t+h)-\xi(t)\|_H^4\\ &\le C_1\int_t^{t+h}E\|\xi(u)-\xi(t)\|_{\mathbf{R}^n}^2(1+|(\xi(u)-\xi(t),G_n(\xi(u))_{\mathbf{R}^n}|du\\ &\le C_2\left(\int_t^{t+h}E\|\xi(u)-\xi(t)\|_{\mathbf{R}^n}^2du\right.\\ &\quad\left.+\int_t^{t+h}E\|\xi(u)-\xi(t)\|_{\mathbf{R}^n}^3\|G_n(\xi(u))\|_{\mathbf{R}^n}du\right)\\ &\le C_2\left(\int_t^{t+h}E\|\xi(u)-\xi(t)\|_{\mathbf{R}^n}^2du\right.\\ &\quad\left.+\left(\int_t^{t+h}E\|\xi(u)-\xi(t)\|_{\mathbf{R}^n}^4du\right)^{3/4}\left(\int_t^{t+h}E\|G_n(\xi(u))\|_{\mathbf{R}^n}^4du\right)^{1/4}\right)\\ &\le C_3\left(\left(\int_t^{t+h}E\|\xi(u)-\xi(t)\|_{\mathbf{R}^n}^4du\right)^{1/2}h^{1/2}\right.\\ &\quad\left.+\left(\int_t^{t+h}E\|\xi(u)-\xi(t)\|_{\mathbf{R}^n}^4du\right)^{3/4}h^{1/4}\right)\le C_4\,h\end{aligned}\tag{2.4}$$

where we have used the fact that

$$\begin{aligned}&\int_t^{t+h}E\|G_n(\xi(u))\|_{\mathbf{R}^n}^4du\\ &\le C_4\left(\int_t^{t+h}E\|\xi_n(u)\|_{\mathbf{R}^n}^4du+\int_t^{t+h}E\|W^n(u)\|_{\mathbf{R}^n}^4du\right)\le C\end{aligned}\tag{2.5}$$

independent of t, h, and n. Substituting repeatedly the result in (2.4), we arrive at the desired inequality.

Now, as in the Introduction, the natural embedding $J\hookrightarrow\mathbf{R}^\infty$ is compact, so that the measures $\mu_n=P\circ X_n^{-1}$ can be assumed to weakly converge to a measure μ on $C([0,T],\mathbf{R}^\infty)$. Moreover, using the ideas in [**6**] and conditions in (1.2), we observe that the function $C([0,T],\mathbf{R}^\infty)\ni x\mapsto\sup_{0\le t\le T}\sum_{k=1}^\infty(x^k(t))^2\in\mathbf{R}$ is lower semicontinuous, and hence

$$\begin{aligned} E_\mu \left(\sup_{0\le t\le T} \sum_{k=1}^{\infty} (x^k(t))^2 \right) &\le \liminf_n E_{\mu_n} \left(\sup_{0\le t\le T} \sum_{k=1}^{\infty} (x^k(t))^2 \right) \\ &= \liminf_n E_P \left(\sup_{0\le t\le T} \|X_n(t)\|_H^2 \right) < \infty. \end{aligned}$$

Thus $P(\sup_{0\le t\le T} \sum_{k=1}^{\infty} (X^k(t))^2 < \infty) = 1$. The limit $\lim_{n\to\infty} X_n = X$ can be constructed, so that a.s. $X(t) \in H$ for each $t \in [0,T]$ and the measure μ is concentrated on $C([0,T], H^w)$.

Using the Skorokhod theorem, by changing the underlying probability space, we can assume that $X_n \to X$, P–a.s. as $C([0,T], \mathbf{R}^\infty)$ random variables, with $X_n(t)$, $X(t) \in H$, $t \in [0,T]$, P–a.s. We note that the random variables $(X_n^k(t))^2$ and $\int_0^t [F_n^k(X_n(s)) + q_k(X_n^k(s))]^2\, ds$ are P–uniformly integrable due to (1.2) and (2.5). Moreover, for $n \ge k$, $F_n^k(X_n(s)) = F^k(X_n(s))$, and

$$\begin{aligned} E_P \int_0^t |F_n^k(X_n(s) - F^k(X(s))|^2\, ds &\le C \left(E_P \int_0^t |F^k(X_n(s)) - F^k(P_n X(s))|^2\, ds \right. \\ &\left. + E_P \int_0^t |F^k(P_n X(s)) - F^k(X(s))|^2\, ds \right). \end{aligned}$$

The continuity of F^k on H^τ implies that both terms converge to zero, and consequently, that the sequence of Brownian motions

$$Y_n^k(t) = X_n^k(t) - x^k - \int_0^t [F_n^k(X_n(s)) + q_k(X_n^k(s))]\, ds$$

converges in $L_2(dP)$ to the process

$$Y^k(t) = X^k(t) - x^k - \int_0^t [F^k(X(s)) + q_k(X^k(s))]\, ds,$$

which is a Brownian motion. We define an H–valued Brownian motion

$$\tilde{W}(t) = \sum_{k=1}^{\infty} Y^k(t) e_k.$$

The processes $X(t)$, $\tilde{W}(t)$ satisfy (2.1). □

For the reader's convenience, we recall the result on weak solutions, Theorem 2.4, [**6**]. For a finite subset V of the set of positive integers, $q^V : H \to H$ is defined by

$$(q^V(y), e_k)_H := q_k^V(y) = \begin{cases} q_k(y_k) & k \in V \\ 0 & k \notin V \end{cases} \tag{2.6}$$

It is well known that under conditions (B1), (B2), and a local Lipschitz condition on the coefficients F and q^V, there exists a unique strong solution to the equation

$$\xi^V(t) = x + \int_0^t \left(F(\xi^V(s)) + q^V(\xi^V(s)) \right) ds + W(t).$$

The solution $\xi^V(\cdot) \in C([0,T], H)$. Theorem 2.4 in [**6**] states that under conditions (B1), (B2), and a local Lipschitz condition on F and q^V, with the Lipschitz constant independent on V, there exists a Brownian motion $\tilde{W}(\mathrm{t})$, equivalent to $W(t)$, and a weakly continuous process $\xi(t)$, $\xi(\cdot) \in C([0,T], H^w)$, constructed as a weak $\lim_{n\to\infty} \xi^{V_n}$, such that for every k,

$$\xi^k(t) = x^k + \int_0^t \left(F^k(\xi(s)) + q_k(\xi^k(s))\right) ds + \tilde{W}^k(t).$$

We now show that, under a global Lipschitz condition on H, the above solution $\xi(t)$ and the solution $X(t)$, obtained in Theorem 2.1, using the compact embedding argument, coincide.

COROLLARY 2.2. *Assume that the functions $F + q^V : H \to H$ satisfy Lipschitz conditions (global) for every V, a finite subset of positive integers, with the Lipschitz constants possibly varying with V. Assume that conditions (B1) and (B2) hold true. Then the weak solution ξ of Theorem 2.4 in* [**6**] *and the weak solution X constructed in Theorem 2.1, using the compact embedding argument, have the same law.*

Proof. We first note that under the Lipschitz condition, both approximating sequences ξ_n^V, of Theorem 2.4 in [**6**] and X_n of Theorem 2.1, can be constructed as strong solutions on the same probability space (see [**8**]). Let $F_m = P_m \circ F \circ P_m$, then for $m \geq n$ we have

$$\begin{aligned} X_{m,n}(t) &= x_m + \int_0^t \left(F_m(X_{m,n}(s)) + q^n(X_{m,n}(s))\right) ds + W^m(t) \\ \xi^{V_n}(t) &= x + \int_0^t \left(F(\xi^{V_n}(s)) + q^{V_n}(\xi^{V_n}(s))\right) ds + W(t) \end{aligned}$$

The laws $Law(\xi^V)$ are tight. For a sequence $V_n = \{1, 2, ..., n\}$, there is a subsequence V_{n_k}, such that $Law(\xi^{V_{n_k}}) \to Law(\xi)$. Therefore, for simplicity, we assume in Theorem 2.4 that $V_n = \{1, ..., n\}$, and $\xi^{V_n} \to \xi$ weakly. Denote $Y_m(t) = P_m \xi^{V_n}(t)$, then

$$Y_m(t) = x_m + \int_0^t \left(P_m F\left(\xi^{V_n}(s)\right) + q^n\left(Y_m(s)\right)\right) ds + W^m(t).$$

We obtain

$$\begin{aligned} E\|X_{m,n}(t) - \xi^{V_n}(t)\|_H^2 &\leq 2(E\|X_{m,n}(t) - Y_m(t)\|_H^2 + E\|Y_m(t) - \xi^{V_n}(t)\|_H^2) \\ &\leq 2\,E\|\int_0^t \left(F_m(X_{m,n}(s)) - P_m F(\xi^{V_n}(s))\right) + \left(q^n(X_{m,n}(s)) - q^n(Y_m(s))\right) ds\|_H^2 \\ &\quad +2E\|P_m\xi^{V_n}(t) - \xi^{V_n}(t)\|_H^2 \\ &\leq 4\left(E\int_0^t \|F(X_{m,n}(s)) - F(\xi^{V_n}(s))\|_H^2 ds\right. \\ &\quad \left. +E\int_0^t \|q^n(X_{m,n}(s)) - q^n(Y_m(s))\|_H^2 ds\right) + 2\,E\|P_m\xi^{V_n}(t) - \xi^{V_n}(t)\|_H^2. \end{aligned}$$

Using the Lipschitz condition we get

$$\begin{aligned} &E\|X_{m,n}(t) - \xi^{V_n}(t)\|_H^2 \\ &\qquad \leq C(n)\,E\int_0^t \|X_{m,n}(s) - \xi^{V_n}(s)\|_H^2 ds + 2\,E\|P_m\xi^{V_n}(t) - \xi^{V_n}(t)\|_H^2. \end{aligned}$$

The second term converges to zero as $m \to \infty$, thus it can be majorized by some $\varepsilon_{m,n} \to 0$. Using Gronwall's Lemma, we obtain that $E\|X_{m,n}(t) - \xi^{V_n}(t)\|_H^2 \leq \varepsilon_{m,n} e^{C(n)t}$, with the expression on the right hand side converging to zero as $m \to \infty$.

We conclude that for each n, $X_{m,n}(t) \to \xi^{V_n}(t)$ in $L_2(\Omega)$. But $\xi^{V_n} \to \xi$ and, for some subsequence, $X_{m_k,n_k} \to X$ weakly. Thus $Law(X) = Law(\xi)$ on $C([0,T], \mathbf{R}^\infty)$ and consequently the laws coincide on $C([0,T], H^w)$. □

EXAMPLE 2.3. Unbounded Spin Systems.

Consider an unbounded spin system of statistical mechanics with an energy function H defined on the space of configurations with finite support in $\mathbf{R}^\infty$ into $\mathbf{R}^\infty$. Typically

$$H^k(x) = \phi_k(x^k) - \sum_{l \neq k} J_{k,l} x^k x^l$$

with the functions $\phi_k : \mathbf{R} \to \mathbf{R}$. The coefficients of a SDE are defined as

$$\begin{aligned} F^k(x) &:= \frac{\partial}{\partial x^k} \sum_{l \neq k} J_{k,l} x^k x^l = \sum_{l \neq k} J_{k,l} x^l \\ q^k(u) &:= -\phi_k'(u). \end{aligned}$$

Theorem 2.1 relaxes the Lipschitz assumption on the functions $q^k(u)$. Under the condition $\sup_k \sum_{l \neq k} |J_{k,l}| < \infty$, the growth and Lipschitz conditions on $F(x)$ are satisfied. Our results show that there exists a weak solution only under the assumption that $q^k(u) : \mathbf{R} \to \mathbf{R}$ satisfy growth condition and are continuous functions. It should be noted that even if $J_{k,l} = 0$, for all k, l, the growth condition is necessary for the existence of a solution without explosion, and continuity occurs in the proof of the Peano theorem.

References

[1] S. Albeverio, R. Høegh-Krohn. Homogeneous Random Fields and Statistical Mechanics. *J. Funct. Anal.* **19**, 242 – 272 (1975).

[2] S. Albeverio, Yu. G. Kondratiev, M. Röckner, T.V. Tsikalenko. Glauber Dynamics for Quantum Lattice Systems. *Preprint* (1999).

[3] L. Gawarecki, V. Mandrekar, P. Richard. Existence of Weak Solutions for Stochastic Differential Equations and Martingale Solutions for Stochastic Semilinear Equations. *Random Operators and Stochastic Equations* **7**, No. 3, 215 – 240 (1999).

[4] A.N. Godunov. On Peano's theorem in Banach spaces, *Functional Anal. Appl.* **9**, 53–55 (1975).

[5] G. Kallianpur, I. Mitoma, R.L. Wolpert. Diffusion equations in dual of nuclear spaces, *Stoch. Stoch. Reports* **29**, 285–329 (1990).

[6] G. Leha, G. Ritter. On Solutions to Stochastic Differential Equations with Discontinuous Drift in Hilbert Space *Math. Ann.* **270**, 109 – 123 (1985).

[7] A.V. Skorokhod. Personal communication.

[8] M. Yor. Existence et unicité de diffusions á valeurs dans un espace de Hilbert, *Ann. Inst. Henri Poincaré* B **10**, 55–88 (1974).

KETTERING UNIVERSITY, DEPARTMENT OF SCIENCE AND MATHEMATICS, 1700 W. THIRD AVE., FLINT, MI 48504, U.S.A.

E-mail address: `lgawarec@kettering.edu`

MICHIGAN STATE UNIVERSITY, DEPARTMENT OF STATISTICS AND PROBABILITY, MICHIGAN STATE UNIVERSITY, WELLS HALL, EAST LANSING, MI 48840, U.S.A.

E-mail address: `mandrekar@stt.msu.edu`

Canadian Mathematical Society
Conference Proceedings
Volume **29**, 2000

Monotonicity and Concavity Properties of The Spectral Shift Function

Fritz Gesztesy, Konstantin A. Makarov, and Alexander K. Motovilov

Dedicated with great pleasure to Sergio Albeverio on the occasion of his 60th birthday.

ABSTRACT. Let H_0 and $V(s)$ be self-adjoint, V, V' continuously differentiable in trace norm with $V''(s) \geq 0$ for $s \in (s_1, s_2)$, and denote by $\{E_{H(s)}(\lambda)\}_{\lambda \in \mathbb{R}}$ the family of spectral projections of $H(s) = H_0 + V(s)$. Then we prove for given $\mu \in \mathbb{R}$, that $s \longmapsto \operatorname{tr}\left(V'(s)E_{H(s)}((-\infty, \mu))\right)$ is a nonincreasing function with respect to s, extending a result of Birman and Solomyak. Moreover, denoting by $\zeta(\mu, s) = \int_{-\infty}^{\mu} d\lambda\, \xi(\lambda, H_0, H(s))$ the integrated spectral shift function for the pair $(H_0, H(s))$, we prove concavity of $\zeta(\mu, s)$ with respect to s, extending previous results by Geisler, Kostrykin, and Schrader. Our proofs employ operator-valued Herglotz functions and establish the latter as an effective tool in this context.

1. Introduction and principal results

In the following $\mathcal{H}$ denotes a complex separable Hilbert space with scalar product $(\cdot, \cdot)_{\mathcal{H}}$ (linear in the second factor) and norm $\|\cdot\|_{\mathcal{H}}$, $\mathcal{B}(\mathcal{H})$ represents the Banach space of bounded linear operators defined on $\mathcal{H}$, $\mathcal{B}_p(\mathcal{H})$, $p \geq 1$ the standard Schatten-von Neumann ideals of $\mathcal{B}(\mathcal{H})$ (cf., e.g., [**18**], [**32**]) and $\mathbb{C}_+$ (resp., $\mathbb{C}_-$) the open complex upper (resp., lower) half-plane. Moreover, real and imaginary parts of a bounded operator $T \in \mathcal{B}(\mathcal{H})$ are defined as usual by $\operatorname{Re}(T) = (T + T^*)/2$, $\operatorname{Im}(T) = (T - T^*)/(2i)$.

The spectral shift function $\xi(\lambda, H_0, H)$ associated with a pair of self-adjoint operators (H_0, H), $H = H_0 + V$, $\operatorname{dom}(H_0) = \operatorname{dom}(H)$, where

$$V = V^* \in \mathcal{B}_1(\mathcal{H}), \tag{1.1}$$

is one of the fundamental spectral characteristics in the perturbation theory of self-adjoint operators. It is well-known (see [**23**], [**24**], [**25**], [**26**], [**27**]) that for a wide function class $\mathfrak{K}(H_0, H)$, the Lifshits-Krein trace formula holds, that is,

$$\operatorname{tr}(\varphi(H) - \varphi(H_0)) = \int_{\mathbb{R}} d\lambda\, \varphi'(\lambda)\, \xi(\lambda, H_0, H)\,, \quad \varphi \in \mathfrak{K}(H_0, H). \tag{1.2}$$

1991 *Mathematics Subject Classification.* Primary 47B44, 47A10; Secondary 47A20, 47A40.

In the case of trace class perturbations (1.1), the spectral shift function is integrable, that is,

$$\xi(\cdot, H_0, H) \in L^1(\mathbb{R}), \tag{1.3}$$

and the following relations hold

$$\|\xi(\cdot, H_0, H)\|_{L^1(\mathbb{R})} \leq \|V\|_{\mathcal{B}_1(\mathcal{H})}, \tag{1.4}$$

$$\int_{\mathbb{R}} d\lambda\, \xi(\lambda, H_0, H) = \operatorname{tr}(V). \tag{1.5}$$

The precise characterization of the class

$$\mathfrak{K} = \bigcap_{H_0, H} \mathfrak{K}(H_0, H) \tag{1.6}$$

of all those φ for which (1.2) holds for any pair of self-adjoint operators H_0 and $H = H_0 + V$ with a trace class difference (1.1), is still unknown. In particular, there are functions $\varphi \in C_0^1(\mathbb{R})$ for which (1.2) fails (cf. [**11**], [**30**]). Necessary conditions very close to sufficient ones for φ belonging to the class $\mathfrak{K}$ have been found by Peller [**28**], [**29**]. Here we only note that (1.3) and $(\varphi(H) - \varphi(H_0)) \in \mathcal{B}_1(\mathcal{H})$ hold, and (1.2) is valid, if φ' is the Fourier transform of a finite Borel measure,

$$\varphi'(\lambda) = \int_{\mathbb{R}} d\nu(t)\, e^{-it\lambda}, \quad \varphi \in C^1(\mathbb{R}), \quad \int_{\mathbb{R}} d|\nu(t)| < \infty. \tag{1.7}$$

We denote the function class (1.7) by $\mathcal{W}_1(\mathbb{R})$.

Different representations for the spectral shift function and their interrelationships can be found in [**4**], for further information we refer to [**2**, Ch. 19], [**9**], [**10**], [**16**], [**34**, Ch. 8] and the references therein).

In the present short note we will focus on two particular results: one, a monotonicity result obtained by Birman and Solomyak [**8**], the other, a concavity result obtained by Geisler, Kostrykin, and Schrader [**12**], [**21**]. We also present some extensions and new proofs that we hope might give additional insights into the subject.

We start by recalling pertinent results discovered by Birman and Solomyak [**8**] in connection with the spectral averaging formula (providing a representation for the spectral shift function via an integral over the coupling constant) and a monotonicity result of a certain trace with respect to the coupling constant parameter.

THEOREM 1.1 ([**8**]). *Let H_0 and V be self-adjoint in $\mathcal{H}$, $V \in \mathcal{B}_1(\mathcal{H})$, and define*

$$H_s = H_0 + sV, \quad \operatorname{dom}(H_s) = \operatorname{dom}(H_0), \ s \in \mathbb{R}, \tag{1.8}$$

with $\{E_{H_s}(\lambda)\}_{\lambda\in\mathbb{R}}$ the family of orthogonal spectral projections of H_s. Moreover, denote by $\xi(\cdot, H_0, H_1)$ the spectral shift function for the pair (H_0, H_1). Then for any Borel set $\Delta \subset \mathbb{R}$,

$$\int_{\Delta} d\lambda\, \xi(\lambda, H_0, H_1) = \int_0^1 ds\, \operatorname{tr}(V E_{H_s}(\Delta)). \tag{1.9}$$

In the same paper [**8**], Birman and Solomyak proved another remarkable statement concerning the monotonicity of the integrand in the right-hand side of (1.9) with respect to s for semi-infinite intervals $\Delta = (-\infty, \lambda)$, $\lambda \in \mathbb{R}$.

THEOREM 1.2 ([**8**]). *Assume the hypotheses in Theorem 1.1. Given $\mu \in \mathbb{R}$, the function*

$$s \longmapsto \operatorname{tr}\big(VE_{H_s}((-\infty,\mu))\big), \quad s \in \mathbb{R}, \tag{1.10}$$

is a nonincreasing function with respect to $s \in \mathbb{R}$.

The spectral averaging formula (1.9) combined with the monotonicity result (1.10) is a convenient tool for producing estimates for the spectral shift function [**8**]. For instance,

$$\operatorname{tr}\big(VE_{H_1}((-\infty,\mu))\big) \leq \int_{-\infty}^{\mu} d\lambda\, \xi(\lambda, H_0, H_1) \leq \operatorname{tr}\big(VE_{H_0}((-\infty,\mu))\big). \tag{1.11}$$

In particular, passing to the limit $\mu \to \infty$ in (1.11) one obtains (1.5) (see [**8**] for more details).

Another application of the pair of results (1.9) and (1.10) leads to the proof of concavity properties of the integrated spectral shift function with respect to the coupling constant, originally discovered in the case of Schrödinger operators by Geisler, Kostrykin, and Schrader [**12**] and extended by Kostrykin [**21**], [**22**] to the general case presented next.

THEOREM 1.3 ([**21**], [**22**]). *Let $\xi(\cdot, H_0, H_s)$ be the spectral shift function in Theorem 1.1. Given $\mu \in \mathbb{R}$, the integrated spectral shift function*

$$\zeta_s(\mu) = \int_{-\infty}^{\mu} d\lambda\, \xi(\lambda, H_0, H_s), \quad s \in \mathbb{R} \tag{1.12}$$

is a concave function with respect to the coupling constant $s \in \mathbb{R}$. More precisely, for any $s, t \in \mathbb{R}$, and for all $\alpha \in [0,1]$, the following inequality

$$\zeta_{\alpha s+(1-\alpha)t}(\mu) \geq \alpha\, \zeta_s(\mu) + (1-\alpha)\, \zeta_t(\mu) \tag{1.13}$$

holds. Moreover, $\zeta_s(\mu)$ is subadditive with respect to $s \in (0,\infty)$ in the sense that for any $s, t \geq 0$,

$$\zeta_{s+t}(\mu) \leq \zeta_s(\mu) + \zeta_t(\mu). \tag{1.14}$$

While Theorem 1.3 focuses on a linear coupling constant dependence in $H_s = H_0 + sV$, Kostrykin [**21**] also discusses the case of a nonlinear dependence on s for operators of the form $H(s) = H_0 + V(s)$:

THEOREM 1.4 ([**21**], [**22**]). *Suppose $f : \mathbb{R} \to \mathbb{R}$ is an nonincreasing function of bounded variation and $\{V(s)\}_{s\in\mathbb{R}} \in \mathcal{B}_1(\mathcal{H})$ is operator concave (i.e., $V(\alpha s + (1-\alpha)t) \geq \alpha V(s) + (1-\alpha)V(t)$ for all $\alpha \in [0,1]$, $s, t \in \mathbb{R}$). Then*

$$s \longmapsto g(V(s)) = \int_{\mathbb{R}} d\lambda\, f(\lambda)\xi(\lambda, H_0, H_0 + V(s)), \tag{1.15}$$

is concave in $s \in \mathbb{R}$. More precisely, for all $0 \leq \alpha \leq 1$ and all $s, t \in \mathbb{R}$, the following inequality

$$g(V(\alpha s + (1-\alpha)t)) \geq \alpha g(V(s)) + (1-\alpha) g(V(t)) \tag{1.16}$$

holds.

Actually, Kostrykin considered the general case of relative trace class perturbations in [**21**] but we omit further details in this note.

REMARK 1.5. The results of Theorems 1.1 and 1.2 in [**8**] have been obtained using the approach of Stieltjes' double operator integrals [**5**]–[**7**]. Birman and Solomyak treated the case $V(s) = sV$, $V \in \mathcal{B}_1(\mathcal{H})$, that is, they discussed the case of a linear dependence of the perturbation $V(s)$ with respect to the coupling constant parameter s. The general case of a nonlinear dependence $V(s)$ of s, assuming $V'(s) \geq 0$, in the context of the spectral averaging result (1.9) has recently been treated by Simon [**33**]. In Theorem 1.7 below we cite the most recent result of this type obtained in [**16**].

It is convenient to introduce the following hypothesis.

HYPOTHESIS 1.6. *Let H_0 be a self-adjoint operator in $\mathcal{H}$ with domain* $\operatorname{dom}(H_0)$, *and assume $\{V(s)\}_{s\in\Omega} \subset \mathcal{B}_1(\mathcal{H})$ to be a family of self-adjoint trace class operators in $\mathcal{H}$, where $\Omega \subseteq \mathbb{R}$ denotes an open interval with $0 \in \Omega$. Moreover, suppose that $V(s)$ is continuously differentiable in $\mathcal{B}_1(\mathcal{H})$-norm with respect to $s \in \Omega$. For convenience (and without loss of generality) we may assume that $V(0) = 0$ in the following.*

In the rest of the paper we will frequently use the notation $(s_1, s_2) \Subset \Omega$ to denote an open interval that is strictly contained in the interval $\Omega = (a, b)$ (i.e., $a < s_1 < s_2 < b$).

THEOREM 1.7 ([**16**]). *Assume Hypothesis 1.6 and $0 \in (s_1, s_2) \Subset \Omega$. Let*

$$H(s) = H_0 + V(s), \quad \operatorname{dom}(H(s)) = \operatorname{dom}(H_0), \quad s \in (s_1, s_2), \tag{1.17}$$

with $\{E_{H(s)}(\lambda)\}_{\lambda\in\mathbb{R}}$ the family of orthogonal spectral projections of $H(s)$ and denote by $\xi(\cdot, H_0, H(s))$ the spectral shift function for the pair $(H_0, H(s))$. Then for any Borel set $\Delta \subset \mathbb{R}$ the following spectral averaging formula holds

$$\int_\Delta d\lambda\, (\xi(\lambda, H_0, H(s_2)) - \xi(\lambda, H_0, H(s_1))) = \int_{s_1}^{s_2} ds\, \operatorname{tr}(V'(s) E_{H(s)}(\Delta)). \tag{1.18}$$

The principal new result of the present note is an extension of the monotonicity result, Theorem 1.2, to the case of a nonlinear dependence of $V(s)$ on s. In particular, we provide a new strategy of proof for such results, which appears to be interesting in itself.

THEOREM 1.8. *Assume Hypothesis 1.6 and $0 \in (s_1, s_2) \Subset \Omega$. Suppose in addition, that the derivative $V'(s)$ is continuously differentiable in $\mathcal{B}_1(\mathcal{H})$-norm with respect to $s \in (s_1, s_2)$ and that $V(s)$ is concave in the sense that*

$$0 \geq V''(s) \in \mathcal{B}_1(\mathcal{H}), \quad s \in (s_1, s_2). \tag{1.19}$$

Then, given $\mu \in \mathbb{R}$, the function

$$s \longmapsto \operatorname{tr}\big(V'(s) E_{H(s)}((-\infty, \mu))\big), \quad s \in (s_1, s_2), \tag{1.20}$$

is a nonincreasing function with respect to $s \in (s_1, s_2)$.

Combining Theorems 1.7 and 1.8 one obtains the following result.

COROLLARY 1.9. *Suppose the hypotheses of Theorem 1.8. Then, for given $\mu \in \mathbb{R}$, the integrated spectral shift function*

$$\zeta(\mu, s) = \int_{-\infty}^{\mu} d\lambda\, \xi(\lambda, H_0, H(s)) \tag{1.21}$$

is concave in $s \in (s_1, s_2)$. *More precisely, for all* $0 \le \alpha \le 1$ *and all* $s, t \in (s_1, s_2)$, *the following inequality*

$$\zeta(\mu, \alpha s + (1-\alpha)t) \ge \alpha\, \zeta(\mu, s) + (1-\alpha)\, \zeta(\mu, t) \tag{1.22}$$

holds. Moreover, $\zeta(\mu, s)$ *is subadditive with respect to* $s \in [0, s_2)$ *in the sense that for any* $s, t \ge 0$, $s + t \in [0, s_2)$,

$$\zeta(\mu, s+t) \le \zeta(\mu, s) + \zeta(\mu, t). \tag{1.23}$$

We emphasize that Corollary 1.9 is a special case of Kostrykin's Theorem 1.4.

As explained in [**3**], [**8**], and [**9**], the original proofs of Theorems 1.1 and 1.2 in [**8**] were motivated by a real analysis approach to the spectral shift function in contrast to M. Krein's complex analytic treatment. In this note we return to complex analytic proofs in the spirit of M. Krein and provide a proof of the monotonicity result, Theorem 1.8, based on operator-valued Herglotz function techniques. For various recent applications of this formalism we refer to [**13**], [**14**], [**15**], [**16**], and [**17**].

2. A property of operator-valued Herglotz functions

We recall that $f : \mathbb{C}_+ \to \mathbb{C}$ is called a *Herglotz function* if it is analytic and $f(\mathbb{C}_+) \subseteq \mathbb{C}_+$. In this case we extend f to $\mathbb{C}_-$ in the usual manner, that is, by $f(\overline{z}) = \overline{f(z)}$, $z \in \mathbb{C}_+$.

The principal purpose of this section is to obtain some generalizations of the following elementary result.

LEMMA 2.1. *Let* P *and* Q *be two rational Herglotz functions vanishing at infinity and let* Γ *be a closed clockwise oriented Jordan contour encircling some of the poles of* P *and* Q *starting from the left (and without any poles of* P *and* Q *on* Γ*). Then*

$$\frac{1}{2\pi i} \oint_\Gamma dz\, P(z) Q(z) \ge 0. \tag{2.1}$$

PROOF. By the hypotheses on P and Q we may write

$$P(z) = \sum_{j \in J_1} A_j (p_j - z)^{-1}, \quad A_j \ge 0,\ p_j \in \mathbb{R},\ j \in J_1, \tag{2.2}$$

$$Q(z) = \sum_{\ell \in J_2} B_\ell (q_\ell - z)^{-1}, \quad B_\ell \ge 0,\ q_\ell \in \mathbb{R},\ \ell \in J_2, \tag{2.3}$$

$$\tag{2.4}$$

with J_1, J_2 finite index sets. Next one decomposes P and Q with respect to their poles located in the interior and exterior of the bounded domain encircled by Γ,

$$\begin{aligned} P(z) &= P_{\rm int}(z) + P_{\rm ext}(z) \\ &= \sum_{j \in J_{1,\rm int}} A_j (p_j - z)^{-1} + \sum_{j \in J_{1,\rm ext}} A_j (p_j - z)^{-1}, \end{aligned} \tag{2.5}$$

$$\begin{aligned} Q(z) &= Q_{\rm int}(z) + Q_{\rm ext}(z) \\ &= \sum_{\ell \in J_{2,\rm int}} B_\ell (q_\ell - z)^{-1} + \sum_{\ell \in J_{2,\rm ext}} B_\ell (q_\ell - z)^{-1}, \end{aligned} \tag{2.6}$$

$$J_k = J_{k,\rm int} \cup J_{k,\rm ext}, \quad k = 1, 2. \tag{2.7}$$

Then straightforward residue computations yield

$$
\begin{aligned}
&\frac{1}{2\pi i}\oint_\Gamma dz\, P(z)Q(z) \\
&= \frac{1}{2\pi i}\oint_\Gamma dz\, P_{\text{int}}(z)Q_{\text{int}}(z) + \frac{1}{2\pi i}\oint_\Gamma dz\, P_{\text{int}}(z)Q_{\text{ext}}(z) \\
&\quad + \frac{1}{2\pi i}\oint_\Gamma dz\, P_{\text{ext}}(z)Q_{\text{int}}(z) + \frac{1}{2\pi i}\oint_\Gamma dz\, P_{\text{ext}}(z)Q_{\text{ext}}(z) \qquad (2.8)\\
&= \frac{1}{2\pi i}\oint_\Gamma dz \sum_{j\in J_{1,\text{int}}}\sum_{\ell\in J_{2,\text{int}}} A_jB_\ell(p_j-z)^{-1}(q_\ell-z)^{-1} \\
&\quad + \sum_{j\in J_{1,\text{int}}} A_jQ_{\text{ext}}(p_j) + \sum_{\ell\in J_{2,\text{int}}} B_\ell P_{\text{ext}}(q_\ell) \\
&= \frac{1}{2\pi i}\oint_\Gamma dz \sideset{}{'}\sum_{j\in J_{1,\text{int}},\ell\in J_{2,\text{int}}} A_jB_\ell(q_\ell-p_j)^{-1}\big((q_\ell-z)^{-1}-(p_j-z)^{-1}\big) \\
&\quad + \sum_{j\in J_{1,\text{int}}} A_jQ_{\text{ext}}(p_j) + \sum_{\ell\in J_{2,\text{int}}} B_\ell P_{\text{ext}}(q_\ell) \qquad (2.9)\\
&= \sum_{j\in J_{1,\text{int}}} A_jQ_{\text{ext}}(p_j) + \sum_{\ell\in J_{2,\text{int}}} B_\ell P_{\text{ext}}(q_\ell) \\
&= \sum_{j\in J_{1,\text{int}}}\sum_{\ell\in J_{2,\text{ext}}} A_jB_\ell(q_\ell-p_j)^{-1} + \sum_{j\in J_{1,\text{ext}}}\sum_{\ell\in J_{2,\text{int}}} A_jB_\ell(p_j-q_\ell)^{-1} \geq 0. \qquad (2.10)
\end{aligned}
$$

Here the last integral in (2.8) vanishes since the integrand is analytic inside Γ and the first integral in (2.9) vanishes by symmetry. Moreover, we used the symbol $\sum'$ to indicate summation only over those j and ℓ with $p_j \neq q_\ell$, since only first-order poles contribute in this calculation. □

Next we turn to operator-valued extensions of the concept of Herglotz functions.

DEFINITION 2.2. $M : \mathbb{C}_+ \to \mathcal{B}(\mathcal{H})$ is called an *operator-valued Herglotz function* if M is analytic on $\mathbb{C}_+$ and $\operatorname{Im}(M(z)) \geq 0$ for all $z \in \mathbb{C}_+$.

Any operator-valued Herglotz function admits a canonical representation, which can be considered a generalization of the dilation theory of maximal dissipative operators to the case of operator-valued Herglotz functions. In the following, however, we will focus on Herglotz functions of the resolvent-type

$$M(z) = K^*(L-z)^{-1}K, \quad z \in \mathbb{C}_+, \tag{2.11}$$

where K is a bounded operator between the Hilbert spaces $\mathcal{K}$ and $\mathcal{H}$, $K \in \mathcal{B}(\mathcal{H},\mathcal{K})$, $\mathcal{K} \supseteq \mathcal{H}$, while L is assumed to be a self-adjoint operator in $\mathcal{K}$, bounded from below, and with a gap in its spectrum as described in Theorem 2.3 below.

THEOREM 2.3. *Let $M_j : \mathbb{C}_+ \to \mathcal{B}_2(\mathcal{H})$, $j = 1,2$, be operator-valued Herglotz functions (taking values in the space of the Hilbert-Schmidt operators) admitting the representations*

$$M_j(z) = K_j(L_j-z)^{-1}K_j^*, \tag{2.12}$$

where K_j, $j = 1,2$, acts between Hilbert spaces $\mathcal{K}_j$ and $\mathcal{H}$, $\mathcal{K}_j \supseteq \mathcal{H}$, L_j is a self-adjoint operator in $\mathcal{K}_j$ bounded from below. Assume $K_j \in \mathcal{B}_4(\mathcal{K}_j,\mathcal{H})$, $j = 1,2$, and

$\mathcal{D}$ is a domain in the complex plane and (a,b) an open interval such that

$$a < \min_{j=1,2} \inf\big(\operatorname{spec}(L_j)\big) \tag{2.13}$$

and

$$\big\{\operatorname{spec}(L_1) \cup \operatorname{spec}(L_2)\big\} \cap \mathcal{D} \subset (a,b) \subset \mathcal{D}. \tag{2.14}$$

In addition, assume that Γ is a closed oriented Jordan contour in $\mathcal{D}$ encircling the interval $[a,b]$ in the clockwise direction, and φ an analytic function on $\mathcal{D}$, nonnegative and nonincreasing on (a,b), that is,

$$\varphi\big|_{(a,b)} \geq 0, \tag{2.15}$$

and

$$\varphi'\big|_{(a,b)} \leq 0. \tag{2.16}$$

Then

$$\frac{1}{2\pi i} \oint_\Gamma dz\, \varphi(z) \operatorname{tr}\big(M_1(z)M_2(z)\big) \geq 0. \tag{2.17}$$

PROOF. Given $n \in \mathbb{N}$, introduce the partition $a = t_0 < t_1 < t_2 < \ldots < t_{n-1} < t_n = b$ of the closed interval $[a,b]$,

$$t_k = a + k\frac{b-a}{n}, \quad k = 1, 2, \ldots, n, \tag{2.18}$$

and denote by $\chi^{(n)}(\lambda)$ the piecewise continuous function

$$\chi^{(n)}(\lambda) = \sum_{k=1}^{n} t_k\, \chi_{[t_{k-1},t_k)}(\lambda), \quad \lambda \in \mathbb{R}, \tag{2.19}$$

where $\chi_\Delta(\cdot)$ is the characteristic function of the set $\Delta \subset \mathbb{R}$.

In the Hilbert space $\mathcal{K}_j$ introduce the (possibly unbounded) operators

$$L_j^{(n)} = \int_{(a,b)} \chi^{(n)}(\lambda)\, dE_{L_j}(\lambda) + \int_{(b,\infty)} \lambda\, dE_{L_j}(\lambda), \tag{2.20}$$
$$\operatorname{dom}(L_j^{(n)}) = \operatorname{dom}(L_j), \quad n \in \mathbb{N},\ j = 1, 2.$$

We note, that by definition (2.20), the spectrum of L_j, $j = 1, 2$, in the interval (a,b) consists of finitely many eigenvalues (possibly of infinite multiplicity).

By (2.14) one infers that the sequence of operators $\{L_j^{(n)}\}_{n=1}^\infty$ converges in norm resolvent sense to L_j in the Hilbert space $\mathcal{K}_j$, $j = 1, 2$. This convergence, in turn, combined with the hypothesis $K_j \in \mathcal{B}_4(\mathcal{K}_j, \mathcal{H})$, $j = 1, 2$, implies the convergence

$$\lim_{n\to\infty} \frac{1}{2\pi i} \oint_\Gamma dz\, \varphi(z) \operatorname{tr}\big(M_1^{(n)}(z)M_2^{(n)}(z)\big) = \frac{1}{2\pi i} \oint_\Gamma dz\, \varphi(z) \operatorname{tr}\big(M_1(z)M_2(z)\big), \tag{2.21}$$

where in obvious notation

$$M_j^{(n)}(z) = K_j(L_j^{(n)} - z)^{-1} K_j^*, \quad j = 1, 2. \tag{2.22}$$

Thus, in order to prove (2.17) it suffices to check that every term on the left-hand side of (2.21) is nonnegative.

In the following it is useful to decompose the Herglotz functions $M_j^{(n)}(z)$ in the form

$$M_j^{(n)}(z) = N_j^{(n)}(z) + \widetilde{N}_j^{(n)}(z), \quad j = 1, 2, \tag{2.23}$$

where

$$N_j^{(n)}(z) = K_j E_{L_j}\big((a,b)\big)\,(L_j^{(n)} - z)^{-1} E_{L_j}\big((a,b)\big)\,K_j^*, \tag{2.24}$$

$$\widetilde{N}_j^{(n)}(z) = K_j\, E_{L_j}\big([b,\infty)\big)\,(L_j^{(n)} - z)^{-1} E_{L_j}\big([b,\infty)\big)\,K_j^*, \quad j=1,2 \tag{2.25}$$

are Herglotz functions associated with the spectral subspaces $E_{L_j}\big((a,b)\big)\,\mathcal{K}_j$ and $E_{L_j}\big([b,\infty)\big)\,\mathcal{K}_j$ of L_j, $j=1,2$.

According to the decomposition (2.20), the Herglotz functions $N_j^{(n)}(z)$ are rational operator-valued functions of the form

$$N_j^{(n)}(z) = \sum_{k=1}^{n} \frac{Q_j^{(n),k}}{t_k - z}, \quad j=1,2, \tag{2.26}$$

where

$$Q_j^{(n),k} = K_j E_{L_j}\big([t_{k-1}, t_k)\big)\,K_j^* \ge 0, \quad k=1,\dots,n,\ j=1,2. \tag{2.27}$$

Given $n \in \mathbb{N}$, one decomposes the integrals on the left-hand side of (2.21) as a sum of four terms

$$\begin{aligned}
\frac{1}{2\pi i}\oint_\Gamma dz\,\varphi(z)\,\operatorname{tr}\big(M_1^{(n)}(z)M_2^{(n)}(z)\big) &= \frac{1}{2\pi i}\oint_\Gamma dz\,\varphi(z)\,\operatorname{tr}\big(N_1^{(n)}(z)N_2^{(n)}(z)\big)\\
&\quad+\frac{1}{2\pi i}\oint_\Gamma dz\,\varphi(z)\,\operatorname{tr}\big(\widetilde{N}_1^{(n)}(z)\widetilde{N}_2^{(n)}(z)\big)\\
&\quad+\frac{1}{2\pi i}\oint_\Gamma dz\,\varphi(z)\,\operatorname{tr}\big(\widetilde{N}_1^{(n)}(z)N_2^{(n)}(z)\big)\\
&\quad+\frac{1}{2\pi i}\oint_\Gamma dz\,\varphi(z)\,\operatorname{tr}\big(N_1^{(n)}(z)\widetilde{N}_2^{(n)}(z)\big)\\
&= J_1 + J_2 + J_3 + J_4.
\end{aligned} \tag{2.28}$$

According to (2.26), the first integral J_1 in (2.28) can be represented as follows

$$\begin{aligned}
J_1 &= \frac{1}{2\pi i}\sum_{k,m}^{n}\oint_\Gamma dz\,\frac{\varphi(z)}{(t_k - z)(t_m - z)}\,\operatorname{tr}(Q_1^{(n),k}Q_2^{(n),m})\\
&= \frac{1}{2\pi i}\sum_{\substack{k,\,m\\ k\neq m}}^{n}\oint_\Gamma dz\,\frac{\varphi(z)}{(t_k - z)(t_m - z)}\,\operatorname{tr}(Q_1^{(n),k}Q_2^{(n),m})\\
&\quad+\frac{1}{2\pi i}\sum_{k=1}^{n}\oint_\Gamma dz\,\frac{\varphi(z)}{(t_k - z)^2}\,\operatorname{tr}(Q_1^{(n),k}Q_2^{(n),k}).
\end{aligned} \tag{2.29}$$

Applying the residue theorem to the integrals on the right-hand side of (2.29) one infers

$$\begin{aligned}
J_1 &= \sum_{\substack{k,\,m\\ k\neq m}}^{n}\left(-\frac{\varphi(t_k)-\varphi(t_m)}{t_k - t_m}\right)\operatorname{tr}(Q_1^{(n),k}Q_2^{(n),m})\\
&\quad-\sum_{k=1}^{n}\varphi'(t_k)\,\operatorname{tr}(Q_1^{(n),k}Q_2^{(n),k}) \ge 0,
\end{aligned} \tag{2.30}$$

since φ is a nonincreasing differentiable function and since the inequalities

$$\operatorname{tr}(Q_1^{(n),k}Q_2^{(n),m}) \ge 0, \qquad k,m=1,\dots,n \tag{2.31}$$

hold (due to the fact that the operators $Q_j^{(n),k}$, $k = 1,2,\dots,n$, $j = 1,2$, are nonnegative by (2.27)).

The integral J_2 vanishes,

$$J_2 = 0, \tag{2.32}$$

since $\operatorname{tr}\big(\widetilde{N}_1^{(n)}(z)\widetilde{N}_2^{(n)}(z)\big)$ is holomorphic in $\mathcal{D}$.

The remaining integrals J_3 and J_4 can also be evaluated by the residue theorem and one obtains

$$J_3 = \sum_{k=1}^{n} \varphi(t_k)\ \operatorname{tr}(\widetilde{N}_1^{(n)}(t_k)Q_2^{(n),k}), \tag{2.33}$$

$$J_4 = \sum_{k=1}^{n} \varphi(t_k)\ \operatorname{tr}(Q_1^{(n),k}\widetilde{N}_2^{(n)}(t_k)). \tag{2.34}$$

We recall that $\varphi(t_k) \geq 0$, $k = 1,\dots,n$, by (2.15). At the same time

$$\widetilde{N}_j^{(n)}(t_k) \geq 0, \quad k = 1,2,\dots,n,\ n \in \mathbb{N},\ j = 1,2, \tag{2.35}$$

by (2.20) and (2.25), while

$$Q_j^{(n),k} \geq 0, \quad k = 1,2,\dots,n,\ n \in \mathbb{N},\ j = 1,2, \tag{2.36}$$

by (2.27). Hence,

$$\operatorname{tr}(\widetilde{N}_1^{(n)}(t_k)Q_2^{(n),k}) \geq 0 \text{ and } \operatorname{tr}(Q_1^{(n),k}\widetilde{N}_2^{(n)}(t_k)) \geq 0, \quad k = 1,\dots,n, \tag{2.37}$$

and thus, $J_3 \geq 0$ and $J_4 \geq 0$. Together with (2.30) and (2.32) (combined with (2.21)) this proves (2.17). □

REMARK 2.4. (i) Conditions $K_j \in \mathcal{B}_4(\mathcal{K}_j,\mathcal{H})$, $j = 1,2$, in Theorem 2.3 can be relaxed. In fact, it suffices to require that

$$M_1(z)M_2(z) \in \mathcal{B}_1(\mathcal{H}), \quad z \in \mathbb{C}_+. \tag{2.38}$$

(ii) Hypotheses (2.13) and (2.14) concerning the semiboundedness of L_j, $j = 1,2$, and the existence of a gap in their spectra in Theorem 2.3, are also unnecessarily stringent. In fact, it suffices to assume that the holomorphy domains of the Herglotz functions $M_j(z)$ $j = 1,2$ given by (2.12) include the set $(-\infty,\alpha)\cup(\beta,\gamma)$ for some $a < \alpha$ and $b < \beta < \gamma$.
(iii) If the operators L_j are bounded and

$$\operatorname{spec}(L_j) \subset (a,b), \quad j = 1,2, \tag{2.39}$$

then the terms (2.25) in the decomposition (2.23) vanish and hence the integrals (2.33) and (2.34) vanish too. This means that under assumption (2.39), condition (2.15) is redundant.

3. Monotonicity, Concavity, and Subadditivity

Throughout this section we assume Hypothesis 1.6 and recall that

$$H(s) = H_0 + V(s), \quad \operatorname{dom}(H(s)) = \operatorname{dom}(H_0), \ s \in \Omega. \tag{3.1}$$

First, we treat the case of bounded H_0, $H_0 \in \mathcal{B}(\mathcal{H})$, and study differential and monotonicity properties of the function

$$s \longmapsto \operatorname{tr}\big(V'(s)\,\varphi(H(s))\big), \quad s \in \Omega, \tag{3.2}$$

where φ is analytic on a domain $\mathcal{D}$ containing the spectra of the family (3.1)

$$\bigcup_{s\in\Omega} (\operatorname{spec}(H(s))) \subset \mathcal{D}. \tag{3.3}$$

Next we introduce the following additional hypothesis, which is motivated in part by Remark 2.4 (iii).

HYPOTHESIS 3.1. *Let H be a bounded self-adjoint operator, $\mathcal{D}$ a domain of the complex plane, $(a,b) \subset \mathbb{R}$ an open interval, and Γ a closed clockwise oriented Jordan contour in $\mathcal{D}$ encircling the interval $[a,b]$ such that*

$$\operatorname{spec}(H) \subset (a,b) \subset \mathcal{D}. \tag{3.4}$$

The following remark shows that Hypothesis 3.1 is stable under small (compact) perturbations of H.

REMARK 3.2. Suppose that the collection $\{H(s_0), \mathcal{D}, (a,b), \Gamma\}$ satisfies Hypothesis 3.1 for some $s_0 \in (s_1, s_2)$. By perturbation arguments, one infers the existence of a neighborhood S of s_0 such that $\{H(s), \mathcal{D}, (a,b), \Gamma\}$ also satisfy Hypothesis 3.1 for $s \in S$. Moreover, if φ is analytic on $\mathcal{D}$, then the (bounded) operators $\varphi(H(s))$, $s \in S$, are well-defined by the Riesz integral

$$\varphi(H(s)) = \frac{1}{2\pi i} \oint_\Gamma dz\, \varphi(z)(H(s)-z)^{-1}, \quad s \in S. \tag{3.5}$$

LEMMA 3.3. *Assume Hypothesis 1.6 and let S be the neighborhood of s_0 in Remark 3.2. Suppose in addition that $V'(s)$ is continuously differentiable in $\mathcal{B}_1(\mathcal{H})$-norm for $s \in S$ and that the collection $\{H(s_0), \mathcal{D}, (a,b), \Gamma\}$ satisfies Hypothesis 3.1 for some $s_0 \in (s_1, s_2)$. If φ is an analytic function on $\mathcal{D}$, then*

$$s \longmapsto \operatorname{tr}\big(V'(s)\, \varphi(H(s))\big), \quad s \in S, \tag{3.6}$$

is differentiable on S and

$$\begin{aligned} \frac{d}{ds} \operatorname{tr}\big(V'(s)\, \varphi(H(s))\big) &= \operatorname{tr}\big(V''(s)\, \varphi(H(s))\big) \\ &\quad - \frac{1}{2\pi i} \oint_\Gamma dz\, \varphi(z)\, \operatorname{tr}\big[V'(s)(H(s)-z)^{-1}\big]^2. \end{aligned} \tag{3.7}$$

PROOF. By Remark 3.2, the operators $\varphi(H(s))$, $s \in S$, are well-defined (cf. (3.5)). In particular,

$$\begin{aligned} \frac{\varphi(H(s)) - \varphi(H(s_0))}{s - s_0} &= \frac{1}{2\pi i} \oint_\Gamma dz\, \varphi(z) \frac{(H(s)-z)^{-1} - (H(s_0)-z)^{-1}}{s-s_0} \\ &= -\frac{1}{2\pi i} \oint_\Gamma dz\, \varphi(z)(H(s)-z)^{-1} \frac{V(s)-V(s_0)}{s-s_0} (H(s_0)-z)^{-1}. \end{aligned} \tag{3.8}$$

Since $\operatorname*{n\text{-}lim}_{s\to s_0} H(s) = H(s_0)$, one infers

$$\operatorname*{n\text{-}lim}_{s\to s_0} (H(s)-z)^{-1} = (H(s_0)-z)^{-1}, \tag{3.9}$$

uniformly with respect to $z \in \Gamma$. Thus, combining (3.9) and (3.8), one concludes that $s \mapsto \varphi(H(s))$ is differentiable with respect to $s \in S$ in $\mathcal{B}(\mathcal{H})$-topology. Along with the existence of a continuous $V''(s)$ on S, this yields the differentiability of the function in (3.6). Equation (3.7) then follows by a straightforward computation using (3.5). □

LEMMA 3.4. *Under the assumptions of Lemma 3.3 suppose in addition that* $V(s)$ *is concave with respect to* $s \in S$ *in the sense that*

$$0 \geq V''(s) \in \mathcal{B}_1(\mathcal{H}), \quad s \in S, \tag{3.10}$$

and that the function φ *is nonnegative and nonincreasing on* (a, b). *Then the function*

$$s \longmapsto \operatorname{tr}\big(V'(s)\, \varphi(H(s))\big), \quad s \in S, \tag{3.11}$$

is differentiable and nonincreasing on S.

PROOF. Since φ is nonnegative, $0 \leq \varphi(H(s)) \in \mathcal{B}(\mathcal{H})$, $s \in S$ and hence

$$\operatorname{tr}\big(V''(s)\, \varphi(H(s))\big) \leq 0, \quad s \in S, \tag{3.12}$$

by (3.10). Applying Theorem 2.3 and Remark 2.4 (i) to the operator-valued Herglotz functions $M_1(z) = V'(s)(H(s) - z)^{-1}V'(s)$ and $M_2(z) = (H(s) - z)^{-1}$, one obtains the inequality

$$\frac{1}{2\pi i} \oint_\Gamma dz\, \varphi(z) \operatorname{tr}\big[V'(s)(H(s) - z)^{-1}\big]^2 \geq 0, \quad s \in S. \tag{3.13}$$

Combining (3.12) and (3.13) one infers

$$\frac{d}{ds} \operatorname{tr}\big(V'(s)\, \varphi(H(s))\big) \leq 0 \tag{3.14}$$

by Lemma 3.3, proving the assertion. □

In the general case of unbounded operators H_0 one can prove the following result.

THEOREM 3.5. *Assume Hypothesis 1.6. Suppose in addition that* $V'(s)$ *is continuously differentiable in* $\mathcal{B}_1(\mathcal{H})$*-norm with respect to* $s \in (s_1, s_2)$ *and that* $V(s)$ *is concave in the sense that*

$$0 \geq V''(s) \in \mathcal{B}_1(\mathcal{H}), \quad s \in (s_1, s_2). \tag{3.15}$$

Let φ *be a bounded nonnegative and nonincreasing real-analytic function on* $\mathbb{R}$ *admitting the analytic continuation to a domain* $\mathcal{D}$ *of the complex plane* $\mathbb{C}$ *containing the real axis* $\mathbb{R}$. *Then the function*

$$s \longmapsto \operatorname{tr}\big(V'(s)\, \varphi(H(s))\big), \quad s \in (s_1, s_2), \tag{3.16}$$

is nonincreasing with respect to $s \in (s_1, s_2)$.

PROOF. Introducing the sequence $\{P_n\}_{n\in\mathbb{N}}$ of spectral projections of H_0,

$$P_n = E_{H_0}((-n, n)), \quad n \in \mathbb{N}, \tag{3.17}$$

one concludes that for fixed $s \in \mathbb{R}$ the bounded operators given by

$$H^{(n)}(s) = P_n H_0 P_n + V(s), \quad \operatorname{dom}(H^{(n)}(s)) = \mathcal{H}, \tag{3.18}$$

converge to $H(s)$ in the strong resolvent sense and therefore,

$$\operatorname*{s-lim}_{n\to\infty} \varphi(H^{(n)}(s)) = \varphi(H(s)), \quad s \in (s_1, s_2), \tag{3.19}$$

by Theorem VIII.20 in [**31**].

Since by hypothesis φ is a bounded function, the family of operators $\varphi(H^{(n)}(s))$ is uniformly bounded with respect to n and hence by Theorem 1 in [**19**],

$$\lim_{n\to\infty} \operatorname{tr}\big(V'(s)\, \varphi(H^{(n)}(s)\big) = \operatorname{tr}\big(V'(s)\, \varphi(H(s))\big), \quad s \in (s_1, s_2). \tag{3.20}$$

Given $n \in \mathbb{N}$ and $s_0 \in (s_1, s_2)$, one can always find a closed oriented Jordan contour Γ in $\mathcal{D}$, and a bounded interval $(a,b) \subset \mathbb{R}$ such that the collection $\{H^{(n)}(s_0), \mathcal{D}, (a,b), \Gamma\}$ satisfies Hypothesis 3.1. Since by hypothesis φ is a nonnegative nonincreasing function on (a,b) and (3.15) holds, one concludes by Lemma 3.4 that the function

$$s \longmapsto \operatorname{tr}\big(V'(s)\,\varphi(H^{(n)}(s))\big) \tag{3.21}$$

is nonincreasing in some neighborhood of s_0 and hence on the whole interval (s_1, s_2) since $s_0 \in (s_1, s_2)$ was arbitrary. Therefore, the function

$$s \longmapsto \operatorname{tr}\big(V'(s)\,\varphi(H(s))\big), \quad s \in (s_1, s_2), \tag{3.22}$$

is also nonincreasing on (s_1, s_2) as a pointwise limit (3.20) of the nonincreasing functions in (3.21). □

Now we are able to prove Theorem 1.8, which is an extension of the monotonicity result, Theorem 1.2, of Birman and Solomyak [**8**].

Proof of Theorem 1.8. Given $\mu \in \mathbb{R}$, introducing the real-analytic function

$$\varphi_{\mu,\varepsilon}(\lambda) = \frac{1}{2} - \frac{1}{\pi}\arctan\left(\frac{\lambda-\mu}{\varepsilon} + \frac{1}{\sqrt{\varepsilon}}\right), \quad \varepsilon > 0, \tag{3.23}$$

one concludes that

$$\lim_{\varepsilon\downarrow 0} \varphi_{\mu,\varepsilon}(\lambda) = \chi_{(-\infty,\mu)}(\lambda), \tag{3.24}$$

where $\chi_\Delta(\cdot)$ denotes the characteristic function of the set Δ.

Since

$$\sup_{\varepsilon>0} \|\varphi_{\mu,\varepsilon}\|_{L^\infty(\mathbb{R})} < \infty, \tag{3.25}$$

(3.24) implies the strong convergence

$$\operatorname*{s-lim}_{\varepsilon\downarrow 0} \varphi_{\mu,\varepsilon}(H(s)) = E_{H(s)}((-\infty,\mu)) \tag{3.26}$$

by Theorem VIII.5 in [**31**]. Combining (3.24)–(3.26) with Theorem 1 in [**19**], one infers

$$\lim_{\varepsilon\downarrow 0} \operatorname{tr}\big(V'(s)\,\varphi_{\mu,\varepsilon}(H(s))\big) = \operatorname{tr}\big(V'(s)E_{H(s)}((-\infty,\mu))\big). \tag{3.27}$$

By Theorem 1.8, the function

$$s \longmapsto \operatorname{tr}\big(V'(s)\,\varphi_{\mu,\varepsilon}(H(s))\big), \quad s \in (s_1, s_2), \tag{3.28}$$

is nonincreasing, proving the assertion, since the pointwise limit of nonincreasing functions is nonincreasing. □

Next we prove Corollary 1.9.

Proof of Corollary 1.9. By Theorem 1.7

$$\int_{-\infty}^{\mu} d\lambda\, \xi(\lambda, H_0, H(s)) = \int_0^s dt\, \operatorname{tr}\big(V'(t)E_{H(t)}((-\infty,\mu))\big), \quad s \in (s_1, s_2), \ \mu \in \mathbb{R}. \tag{3.29}$$

By Theorem 1.8 the integrand on the right-hand side of (3.29) is a nonincreasing function of t and hence the left-hand side of (3.29) is a concave function of s. Thus (1.22) holds.

In order to prove (1.23) one notes that $\zeta(\mu, 0+) = 0$ and that a necessary and sufficient condition for a measurable concave function $f(t)$ to be subadditive on $(0, \infty)$ is that $f(0+) \geq 0$ (see, e.g., [**20**, Theorem 7.2.5]). □

In the case of semibounded operators the following statement might be useful. We recall that $\mathcal{W}_1(\mathbb{R})$ denotes the function class of φ with φ' the Fourier transform of a finite (complex) Borel measure (cf. (1.7)).

THEOREM 3.6. *Let H_0 be a self-adjoint operator in $\mathcal{H}$, bounded from below, and assume the hypotheses of Theorem 1.8. Denote by Λ the smallest semi-infinite interval containing the spectra of the family $H(s)$, $s \in (s_1, s_2)$,*

$$\Lambda = \Big[\inf_{s \in (s_1, s_2)} \operatorname{spec}(H(s)), \infty \Big). \tag{3.30}$$

Let $\varphi \in \mathcal{W}_1(\mathbb{R}) \cap C^2(\mathbb{R})$ be concave on Λ in the sense that

$$\varphi''\big|_\Lambda \leq 0, \tag{3.31}$$

and

$$\varphi'(\lambda) = o(1) \text{ as } \lambda \to +\infty. \tag{3.32}$$

Then the function

$$s \longmapsto \operatorname{tr}\big[\varphi(H(s)) - \varphi(H_0)\big], \quad s \in (s_1, s_2), \tag{3.33}$$

is concave in the sense that for any $s, t \in (s_1, s_2)$ and for any $0 \leq \alpha \leq 1$,

$$\begin{aligned} &\operatorname{tr}\big[\varphi(H(\alpha s + (1-\alpha)t)) - \varphi(H_0)\big] \\ &\quad \geq \alpha \operatorname{tr}\big[\varphi(H(s)) - \varphi(H_0)\big] + (1-\alpha) \operatorname{tr}\big[\varphi(H(t)) - \varphi(H_0)\big]. \end{aligned} \tag{3.34}$$

In particular,

$$\big[\varphi(H(\alpha s + (1-\alpha)t)) - \alpha\, \varphi(H(s)) - (1-\alpha)\, \varphi(H_t)\big] \in \mathcal{B}_1(\mathcal{H}) \tag{3.35}$$

and

$$\operatorname{tr}\big[\varphi(H(\alpha s + (1-\alpha)t)) - \alpha\, \varphi(H(s)) - (1-\alpha)\, \varphi(H(s))\big] \geq 0. \tag{3.36}$$

PROOF. First, one observes that by (1.4) the integrated spectral shift function $\zeta(\lambda, s)$ given by (1.21) is uniformly bounded, that is,

$$|\zeta(\lambda, s)| \leq \|V(s)\|_{\mathcal{B}_1(\mathcal{H})}, \quad \lambda \in \mathbb{R}, \ s \in (s_1, s_2). \tag{3.37}$$

Moreover, since H_0 is semibounded, one concludes that $\mathbb{R} \backslash \Lambda \neq \emptyset$ by definition (3.30) of the set Λ and hence for all $s \in (s_1, s_2)$,

$$\zeta(\lambda, s) = 0, \quad \lambda \in \mathbb{R} \backslash \Lambda. \tag{3.38}$$

Thus, using (3.32) and (3.38), one infers

$$\lim_{\lambda \to \pm\infty} \varphi'(\lambda)\, \zeta(\lambda, s) = 0, \quad s \in (s_1, s_2). \tag{3.39}$$

Next, combining (3.38) and (3.39), an integration by parts in the trace formula (1.2) yields

$$\operatorname{tr}\big[\varphi(H(s)) - \varphi(H_0)\big] = -\int_\Lambda d\lambda\, \varphi''(\lambda)\, \zeta(\lambda, s). \tag{3.40}$$

Given $\lambda \in \mathbb{R}$, the integrated spectral shift function $\zeta(\lambda, s)$ is concave with respect to $s \in (s_1, s_2)$ by Corollary 1.9 and hence the left-hand side of (3.40) is also a concave function of s by (3.31) (as a weighted mean of concave functions with a positive weight). □

REMARK 3.7. (i) If the measure ν in representation (1.7) is absolutely continuous, then condition (3.32) holds automatically by the Riemann–Lebesgue Lemma. (ii) If φ is convex on Λ, that is,

$$\varphi''\big|_{\Lambda} \geq 0, \tag{3.41}$$

then the function given by (3.33) is convex.

EXAMPLE 3.8. Under assumptions of Theorem 3.6, choosing $\varphi \in \mathcal{W}_1(\mathbb{R})$ as

$$\varphi(\lambda) = \exp\big(-\lambda t\big), \quad \lambda \in \Lambda,\ t \geq 0, \tag{3.42}$$

one concludes that for any $t > 0$,

$$s \longrightarrow \operatorname{tr}\big[\exp\big(-tH(s)\big) - \exp\big(-tH_0\big)\big] \tag{3.43}$$

is a convex function of $s \in (s_1, s_2)$.

Acknowledgments. We are indebted to Vadim Kostrykin for kindly making available to us reference [**21**] prior to its publication.

References

[1] N. Aronszajn and W. F. Donoghue, *On exponential representations of analytic functions in the upper half-plane with positive imaginary part,* J. d'Anal. Math. **5**, 321–388 (1956-57).

[2] H. Baumgärtel and M. Wollenberg, *Mathematical Scattering Theory,* Birkhäuser, Basel, 1983.

[3] M. Sh. Birman, *Spectral shift function and double operator integrals,* in Linear and Complex Analysis Problem Book 3, V. P. Havin and N. K. Nikolski (eds.), Lecture Notes in Mathematics **1573**, Springer, Berlin, 1994, pp. 272–273.

[4] M. Sh. Birman and A. B. Pushnitski, *Spectral shift function, amazing and multifaceted,* Integr. Eqs. Operator Th. **30**, 191–199 (1998).

[5] M. Sh. Birman and M. Z. Solomyak, *Double Stieltjes operator integrals,* Problems of Mathematical Physics. No. 1, Spectral Theory and Wave Processes, Izdat. Leningrad. Univ., Leningrad, 1966, pp. 33–67 (Russian); English transl., Topics in Mathematical Physics, Vol. 1, Consultants Bureau, Plenum Publ., New York, 1967, pp. 25–54.

[6] M. Sh. Birman and M. Z. Solomyak, *Double Stieltjes operator integrals. II,* Problems of Mathematical Physics. No. 2, Spectral Theory, Diffraction Problems, Izdat. Leningrad. Univ., Leningrad, 1967, pp. 26–60 (Russian); English transl., Topics in Mathematical Physics, Vol. 2, Consultants Bureau, Plenum Publ., New York, 1968, pp. 19–46.

[7] M. Sh. Birman and M. Z. Solomyak, *Double Stieltjes operator integrals. III,* Problems of Mathematical Physics. No. 6, Theory of functions. Spectral theory. Wave propagation, Izdat. Leningrad. Univ., Leningrad, 1973, pp. 27–53 (Russian).

[8] M. Sh. Birman and M. Z. Solomyak, *Remarks on the spectral shift function,* Boundary value problems of mathematical physics and related questions in the theory of functions, 6. Zap. Naučn. Sem. Leningrad. Otdel. Mat. Inst. Steklov. (LOMI) **27**, 33–46 (1972) (Russian); English transl., J. Sov. Math. **3**, 408–419 (1975).

[9] M. Sh. Birman and D. R. Yafaev, *The spectral shift function. The work of M. G. Krein and its further development,* Algebra i Analiz **4**, No. 5, 1–44 (1992) (Russian); English transl., St. Petersburg Math. J. **4**, 833–870 (1993).

[10] M. Sh. Birman and D. R. Yafaev, *Spectral properties of the scattering matrix,* Algebra i Analiz **4**, No. 6, 1–27 (1992) (Russian); English transl., St. Petersburg Math. J. **4**, 1055–1079 (1993).

[11] Yu. B. Farforovskaya, *An estimate of the norm $\|f(B) - f(A)\|$ for self-adjoint operators A and B*, Zap. Naučn. Sem. Leningrad. Otdel. Mat. Inst. Steklov. (LOMI) **56**, 143–162 (1976) (Russian); English transl., J. Sov. Math. **14**, 1133–1149 (1980).

[12] R. Geisler, V. Kostrykin, and R. Schrader, *Concavity properties of Krein's spectral shift function,* Rev. Math. Phys. **7**, 161–181 (1995).

[13] F. Gesztesy, N. J. Kalton, K. A. Makarov, and E. Tsekanovskii, *Some Applications of Operator-Valued Herglotz Functions*, Operator Theory: Advances and Applications, Birkhäuser, to appear.

[14] F. Gesztesy and K. A. Makarov, *Some applications of the spectral shift operator*, in *Operator Theory and its Applications*, A. G. Ramm, P. N. Shivakumar, and A. V.Strauss (eds.), Fields Institute Communication Series, Vol. 25, Amer. Math. Soc., Providence RI, 2000.

[15] F. Gesztesy and K. A. Makarov, *The Ξ operator and its relation to Krein's spectral shift function*, J. d'Anal. Math., to appear.

[16] F. Gesztesy, K. A. Makarov, and S. N. Naboko, *The spectral shift operator*, in *Mathematical Results in Quantum Mechanics*, J. Dittrich, P. Exner, and M. Tater (eds.), Operator Theory: Advances and Applications **108**, Birkhäuser, Basel, 1999, pp. 59–90.

[17] F. Gesztesy and E. Tsekanovskii, *On matrix-valued Herglotz functions,* Math. Nachr., to appear.

[18] I. C. Gohberg and M. G. Krein, *Introduction to the Theory of Linear Nonself-adjoint Operators,* Amer. Math. Soc., Providence, RI, 1969.

[19] H. R. Grümm, *Two theorems about $\mathcal{C}_p$,* Rep. Math. Phys. **4**, 211–215 (1973).

[20] E. Hille and R. S. Phillips, *Functional Analysis and Semi-Groups*, rev. ed., Amer. Math. Soc., Providence, RI, 1957.

[21] V. Kostrykin, *The Spectral Shift Function and Its Applications to Random Schrödinger Operators*, Habilitationsschrift, RWTH Aachen, 1999.

[22] V. Kostrykin, *Concavity of eigenvalues sums and the spectral shift function*, preprint, 1999, to appear in J. Funct. Anal.

[23] M. G. Krein, *On the trace formula in perturbation theory,* Mat. Sbornik N.S. **33(75)**, 597–626 (1953) (Russian).

[24] M. G. Krein, *Perturbation determinants and a formula for the traces of unitary and self-adjoint operators,* Dokl. Akad. Nauk SSSR **144**, 268–271 (1962) (Russian); English transl., Sov. Math. Dokl. **3**, 707–710 (1962).

[25] M. G. Krein, *On certain new studies in the perturbation theory for self-adjoint operators,* in *M. G. Krein, Topics in Differential and Integral Equations and Operator Theory,* I. Gohberg (ed.), Birkhäuser, Basel, 1983, pp. 107–172.

[26] M. G. Krein, *On perturbation determinants and a trace formula for certain classes of pairs of operators,* J. Operator Theory **17**, 129–187 (1987) (Russian); English transl., Amer. Math. Soc. Trans. (2) **145**, 39–84 (1989).

[27] I. M. Lifshits, *On a problem of perturbation theory,* Uspekhi Mat. Nauk **7**, No. 1, 171–180 (1952) (Russian).

[28] V. V. Peller, *Hankel operators in the perturbation theory of unitary and selfajoint operators,* Funktsional. Anal. i Prilozhen. **19**, 37–51 (1985) (Russian); English transl., Funct. Anal. Appl. **19**, 111–123 (1985).

[29] V. V. Peller, *Hankel operators in the perturbation theory of unbounded self-adjoint operators,* in *Analysis and Partial Differential Equations: A Collection of Papers Dedicated to Mischa Cotlar*, C. Sadosky (ed.), Marcel Dekker, New York, 1990, pp. 529–544.

[30] V. V. Peller, Commentary to H. Widom's article *When are differentiable functions differentiable?*, in *Linear and Complex Analysis Problem Book 3*, Part I, V. P. Havin and N. K. Nikolski (eds.), Lecture Notes in Math. **1573**, Springer, Berlin, 1994, pp. 269–271.

[31] M. Reed and B. Simon, *Methods of Modern Mathematical Physics I, Functional Analysis*, 2nd. ed., Academic Press, New York, 1980.

[32] B. Simon, *Trace Ideals and their Applications,* Cambridge University Press, Cambridge, 1979.

[33] B. Simon, *Spectral averaging and the Krein spectral shift,* Proc. Amer. Math. Soc. **126**, 1409–1413 (1998).

[34] D. R. Yafaev, *Mathematical Scattering Theory,* Amer. Math. Soc., Providence, RI, 1992.

Department of Mathematics, University of Missouri, Columbia, MO 65211, USA
E-mail address: `fritz@math.missouri.edu`
URL: `http://www.math.missouri.edu/people/fgesztesy.html`

Department of Mathematics, University of Missouri, Columbia, MO 65211, USA
E-mail address: `makarov@azure.math.missouri.edu`
URL: `http://www.math.missouri.edu/people/kmakarov.html`

Physikalisches Institut, Universität Bonn, D-53115 Bonn, Germany
E-mail address: `motovilov@physik.uni-bonn.de`

On leave of absence from the Laboratory of Theoretical Physics, JINR, 141980 Dubna, Russia
E-mail address: `motovilv@thsun1.jinr.ru`

Canadian Mathematical Society
Conference Proceedings
Volume **29**, 2000

Zero-range perturbations of the Schrödinger operator with a saddle-point potential

V. A. Geyler and V. A Margulis

ABSTRACT. Point perturbations of the Schrödinger operator with a saddle-point potential are considered. The complete spectral analysis of the perturbed operator are performed. The resonances of the Breit–Wigner type and the transmission coefficients are found. Some physical applications are discussed.

1. Introduction

The Schrödinger operator with the saddle-point potential

$$V_{\mathrm{sp}}(x,y,z) = \frac{m_*}{2}(\omega_1^2x^2 + \omega_2^2y^2 - \omega_3^2z^2) \tag{1}$$

(here $\omega_i > 0$ and m_* is the mass of the considered particle) is of considerable interest at least for two distinct areas of theoretical physics.

Firstly, the potential (1) is of frequent use in the ballistic conductance theory of narrow constrictions in quantum wires (see, e.g. [**25**] and references therein). In this case the potential V_{sp} simulates the confining potential for a charged particle in the constriction with an elliptic cross-section; namely, $(\hbar/\omega_i m_*)^{1/2}$ (where $i = 1, 2$) are the semiaxes of the cross-section and $(\hbar/\omega_3 m_*)^{1/2}$ is the length of the constriction.

Secondly, the Hamiltonian

$$H_{\mathrm{sp}} = \frac{1}{2m_*}\widehat{\mathbf{p}}^2 + V_{\mathrm{sp}} \tag{2}$$

arises in the spectral theory of coupled systems in spaces of nontrivial curvature. Schrödinger was the first to consider the quantum-mechanical spectral problem for the hydrogen atom in a space of constant positive curvature [**24**]. He has shown that the discrete spectrum of the problem has the form

$$E_n = E_n^0 + \frac{\hbar^2 R}{12m_*}(n^2 - 1), \tag{3}$$

where E_n^0 are the energy levels of the atom in Euclidean space, n is the principal quantum number ($n = 1, 2 \ldots$), and R is the scalar curvature. For a deutonium (i.e. weakly coupled two-particle system) modelled with the help of the Fermi pseudopotential [**5**], the energy levels in a space of constant curvature (both positive and negative) are found in [**3**]. The shift of the binding energy for the hydrogen atom $\Delta E = (E_2 - E_1) - (E_2^0 - E_1^0)$ caused by the curvature R is equal to $\Delta E = \hbar^2 R/4m_*$.

2000 *Mathematics Subject Classification.* Primary 81U15, 47N50.

For realistic cosmological models of Friedmann's type this shift is of order 10^{-30}eV, and hence is negligibly small in relation to the Lamb shift ($\sim 10^{-6}$eV). For the deutonium the shift ΔE is of the order of $\hbar^2 R/12m_*$, therefore, it is negligible, too. (Curiously, that in the space of constant *negative* curvature R this shift is equal to zero for all R, $R < 0$ [**3**]). The situation changes if one considers excited states of a hydrogenlike atom in free fall at the surface of a spherical object of mass M and radius r_0. In this case the gravitational tidal shift may be of order of the Lamb shift (e.g. if the object is a black hole with an appropriate Schwarzschild radius [**22**]). In the nonrelativistic limit ($v/c \ll 1$, not the limit of vanishing curvature), the Schrödinger operator containing an effective classical geodesic deviation potential describing the tidal interaction of the atom with the gravitational field, to first order to the Riemann tensor, reads [**20**]

$$H = \frac{1}{2m_*}\widehat{\mathbf{p}}^2 + \frac{m_* c^2}{2} R_{0l0m} x^l x^m + V_c(r), \tag{4}$$

where V_c is the confining potential of the atom, R_{0l0m} are the components of the Riemann tensor, and x^l is the position of the electron in the nucleus-centered reference frame. At least in the spherically symmetric case, H is a perturbation of a Hamiltonian of the form (2). Namely, assuming that the z-axis is directed radially, one obtains $R_{0101} = R_{0202} = GM/c^2r_0^3$, $R_{0303} = -2GM/c^2r_0^3$, and $R_{0l0m} = 0$ if $l \neq m$ [**19**]; hence we have in this case $\omega_1^2 = \omega_2^2 = c^2 R_{0101}$, $\omega_3^2 = -c^2 R_{0303}$.

In both cases, of great importance is the presence of a uniform magnetic field directed along the z-axis. For instance, one can change confinement parameters of a quantum constriction by variation of the field strength [**14**]. On the other hand, the effect of a magnetic field on the energy shift may exceed considerably the gravitational effect [**22**]. Therefore, we shall include the vector potential of a magnetic field in the Hamiltonian (2).

In the first case, the perturbation of H_{sp} by a zero-range potential (or, in other words, by a point potential) may be considered as a model of an impurity in the quantum constriction; in the second one this perturbation is a model of a deutonium in a spherically symmetric curved space. Note that zero-range impurities in the *two-dimensional* quantum constrictions have been studied earlier at the physical level of rigour in [**11**], [**13**], [**18**].

Here we present the spectral analysis of the saddle-point Hamiltonian perturbed by a point potential. The problem is complicated by the fact that the unperturbed operator is not semibounded, therefore we can not use some standard statements concerning the perturbations of rank one (see, e.g. [**10**], [**16**]). Moreover, the condition from [**26**] for the existence of the Schrödinger simigroup is not satisfied. However, using some ideas from [**2**] we prove that the spectrum of the considered problem is purely absolutely continuous and coincides with the whole real line $\mathbf{R}$. In addition, for a deep zero-range well we find a point near the real axis which is similar to the Breit–Wigner resonance; in particular, the transmission coefficients through the point potential have a maximum at this point. The position of the resonance allows to estimate the energy shift in the well caused by the saddle-point potential; our result is in good agreement with those of [**7**] and [**22**].

Dedication. Discussions with S. Albeverio have played a crucial role when preparing this paper; ideas of the book [**2**] have been used over the article. We are happy to dedicate our paper to Sergio Albeverio on the occasion of his 60th birthday.

2. Some properties of unperturbed Hamiltonian

The main object of the paper is a point perturbation of the following Schrödinger operator

$$H^0 = \frac{1}{2m_*}\left(\widehat{\mathbf{p}} - \frac{e}{c}\mathbf{A}\right)^2 + \frac{m_*}{2}\left(\omega_1^2 x^2 + \omega_2^2 y^2 - \omega_3^2 z^2\right). \tag{5}$$

Here $\widehat{\mathbf{p}} = -i\hbar\nabla$, and $\mathbf{A} = \frac{1}{2}\mathbf{B}\times\mathbf{r}$ is the vector potential of a uniform magnetic field $\mathbf{B}$ directed along the z-axis: $\mathbf{B} = B\mathbf{e}_z$. In this paper we consider only the rotationally symmetric case $\omega_1 = \omega_2$, To simplify the notation, we shall use below a system of units in which $m_* = 1/2$, $e = c = \hbar = 1$ and denote ω_3 by ω.

This section contains some auxiliary results concerning the operator H^0. First note that H^0 admits the separation of variables

$$H^0 = H_1^0 \otimes I_1 + I_2 \otimes H_2^0, \tag{6}$$

where H_1^0 is an operator in $L^2(\mathbf{R}^2)$

$$H_1^0 = (i\nabla + \mathbf{A})^2 + \frac{\omega_1^2}{4}(x^2 + y^2) \tag{7}$$

(the Hamiltonian of the two-dimensional oscillator in a uniform magnetic field), and H_1^0 is an operator in $L^2(\mathbf{R})$

$$H_2^0 = -\frac{d^2}{dz^2} - \frac{\omega^2}{4}z^2 \tag{8}$$

(the Hamiltonian of the inverse harmonic oscillator). The operators H_1^0 and H_2^0 are well-defined as the closures of the corresponding differential operators on $C_0^\infty(\mathbf{R}^\nu)$, $\nu = 2, 1$.

Denote $\omega_c = 2|B|$ (the cyclotron frequency), $\Omega = (4\omega_1^2 + \omega_c^2)^{1/2}$. It is easy to show that the operator H_1^0 has a pure point spectrum consisting of eigenvalues

$$E_{mn} = \frac{\omega_c}{2}m + \frac{\Omega}{2}(2n + |m| + 1), \tag{9}$$

where $m = 0, \pm1, \pm2, \ldots$, $n = 0, 1, 2 \ldots$. In the polar coordinates (ρ, φ), the normalized eigenfunction Ψ_{mn} of the level E_{mn} has the following form:

$$\Psi_{mn}(\rho,\varphi) = \frac{c_{mn}}{\sqrt{2\pi}} e^{im\varphi}\rho^{|m|}\exp\left(-\frac{\Omega\rho^2}{2}\right)L_n^{|m|}(\Omega\rho^2), \tag{10}$$

where

$$c_{mn} = \left[\frac{\Omega^{|m|+1}n!}{2^{|m|+1}(n+|m|)!}\right]^{1/2}$$

is the normalizing constant and $L_n^{|m|}$ is the Laguerre polynomial. The spectrum $\sigma(H_2^0)$ coincides with $\mathbf{R}$, it is purely absolutely continuous and two-fold degenerate. Namely, the generalized eigensubspace of H_2^0 associated with a point $E_\lambda^+ = E_\lambda^- = \omega\lambda$ from $\sigma(H_2^0)$ is two-dimensional; as a basis we choose the pair of functions $\Psi_\lambda^\pm(z)$,

$$\Psi_\lambda^\pm(z) = \left[2\pi\omega^{1/2}(1 + e^{-2\pi\lambda})\right]^{-1/2}\mathbf{E}(-\lambda, \pm\omega^{\mathbf{1/2}}\mathbf{z}), \tag{11}$$

where $\mathbf{E}(\lambda, \mathbf{x})$ is the modified parabolical cylinder function [**1**]. The functions $\Psi_\lambda^\pm$ are normalized by the conditions

(12) $$\int_{-\infty}^{\infty} \overline{\Psi_\lambda^j(z)}\Psi_\mu^k(z)\, dz = \delta_{jk}\delta(\lambda - \mu)\,,$$

(13) $$\int_{-\infty}^{\infty} \left[\overline{\Psi_\lambda^+(z)}\Psi_\lambda^+(u) + \overline{\Psi_\lambda^-(z)}\Psi_\lambda^-(u)\right] d\lambda = \delta(z-u)\,,$$

where $j, k \in \{+, -\}$. Rigorously, conditions (12) and (13) mean that the mapping $\mathcal{F}_2$ defined on the functions φ from $C_0^\infty(\mathbf{R})$ by the rule

(14) $$\mathcal{F}_2\varphi = (\psi_+, \psi_-), \qquad \psi_\pm(\lambda) = \int_{-\infty}^{\infty} \overline{\Psi_\lambda^\pm(z)}\varphi(z)\, dz\,,$$

may be extended to a Hilbert isomorphism of the spaces $L^2(\mathbf{R})$ and $L^2(\mathbf{R})\otimes\mathbf{C}^2$. As a result we obtain that the spectrum of H^0 is purely absolutely continuous and coincides with the whole real axis $\mathbf{R}$. A complete set of the generalized eigenfunctions for H^0 consists of the functions having the form

(15) $$\Psi_{mn\lambda}^\pm(\rho, \varphi, z) = \Psi_{mn}(\rho, \varphi)\Psi_\lambda^\pm(z),$$

($m \in \mathbf{Z}$, $n \in \mathbf{N}$, $\lambda \in \mathbf{R}$); $\Psi_{mn\lambda}^\pm$ corresponds to the point $E_{mn\lambda}^\pm = E_{mn} + E_\lambda^\pm$ of the spectrum.

We list some useful properties of the operator H^0 in the following theorem which is an immediate consequence of Theorem XIV.6.2 from [**8**] and Theorem III.5.1 from [**6**].

Theorem A. 1) *Each function from the domain $\mathcal{D}(H^0)$ is continuous.* 2) *The resolvent $R^0(\zeta) = (H^0 - \zeta)^{-1}$, $\operatorname{Im}\zeta \neq 0$, has an integral kernel (the Green function) $G^0(\mathbf{r}, \mathbf{r}'; \zeta)$ with the properties:*

(2a) *for every $\mathbf{r} \in \mathbf{R}^3$ and $\zeta \in \mathbf{C}\setminus\sigma(H^0)$*

$$\int_{\mathbf{R}^3} |G(\mathbf{r}, \mathbf{r}'; \zeta)|^2\, d\mathbf{r}' < \infty$$

(*i.e. $R^0(\zeta)$ is a Carleman operator*);

(2b) *there is a continuous function $\widetilde{G}^0(\mathbf{r}, \mathbf{r}'; \zeta)$ defined on $\mathbf{R}^3 \times \mathbf{R}^3 \times (\mathbf{C}\setminus\mathbf{R})$ such that*

$$G^0(\mathbf{r}, \mathbf{r}'; \zeta) = \frac{1}{4\pi|\mathbf{r} - \mathbf{r}'|} + \widetilde{G}^0(\mathbf{r}, \mathbf{r}'; \zeta);$$

moreover, $\widetilde{G}^0(\mathbf{r}, \mathbf{r}'; \zeta)$ is an analytic function with respect to the variable ζ.

Denote by $\mathcal{F}$ the Fourier transform that is the eigenfunction expansion for the operator H^0. The operator $\mathcal{F}$ is a Hilbert isomorphism of the spaces $L^2(\mathbf{R}^3)$ and $l^2(\mathbf{Z}\times\mathbf{N})\otimes L^2(\mathbf{R})\otimes\mathbf{C}^2$; namely, if $f \in L^2(\mathbf{R}^3)$, then $\mathcal{F}f \equiv \tilde{f}$ is a pair $(\tilde{f}^+, \tilde{f}^-)$ of functions $\tilde{f}^\pm(m, n, \lambda)$, where $m \in \mathbf{Z}$, $n \in \mathbf{N}$, $\lambda \in \mathbf{R}$. For a function $f \in C_0^\infty(\mathbf{R}^3)$ its image $(\tilde{f}^+, \tilde{f}^-)$ is defined by the rule:

(16) $$\tilde{f}^\pm(m, n, \lambda) = \int_{\mathbf{R}^3} \overline{\Psi_{mn\lambda}^\pm(\mathbf{r})}\, f(\mathbf{r})\, d\mathbf{r}\,.$$

Although the operator H^0 is not semibounded, the space $\mathcal{H}_{-2}(H^0)$ [**10**], [**16**] is well-defined; the statement (2a) of Theorem A means that $\delta(x) \in \mathcal{H}_{-2}(H^0)$. Therefore, the following assertion is valid:

Lemma 1. *Let* $g_\zeta(\mathbf{r}) = G(\mathbf{r}, \mathbf{0}; \zeta)$, *where* $\mathrm{Im}\zeta \neq 0$. *Then*

$$\tilde{g}_\zeta^{\pm}(m, n, \lambda) = \left(E^{\pm}_{mn\lambda} - \zeta\right)^{-1} \Psi^{\pm}_{mn\lambda}(\mathbf{0}).$$

3. Construction of point perturbations for H^0

In this section we define point perturbations of H^0 using the so-called "restriction– extension procedure" [**2**], [**21**]. Let S be a symmetric operator which is the restriction of H^0 to the domain

$$\mathcal{D}(S) = \{f \in \mathcal{D}(H^0) : f(\mathbf{0}) = 0\}. \tag{17}$$

By definition, the point perturbation of H^0 is a self-adjoint extension of S that differs from H^0. Because the resolvent $R^0(\zeta)$ is a Carleman operator, we can describe all such extensions by means of the Krein resolvent formula [**12**]. Namely, every self-adjoint extension of S except H^0 has the Green function $G(\mathbf{r}, \mathbf{r}'; \zeta)$ of the form

$$G(\mathbf{r}, \mathbf{r}'; \zeta) = G^0(\mathbf{r}, \mathbf{r}'; \zeta) - [Q(\zeta) + \alpha]^{-1} G^0(\mathbf{r}, \mathbf{0}; \zeta) G^0(\mathbf{0}, \mathbf{r}'; \zeta), \tag{18}$$

where $Q(\zeta)$ is the Krein $\mathcal{Q}$-function [**17**] of the pair (H^0, S) and α is a real number. Conversely, for every $\alpha \in \mathbf{R}$ Eq. (18) defines a self-adjoint extension of S which differs from H^0 (i.e., defines a point perturbation of H^0). We denote the operator having the Green function (18) as H_α and its resolvent and Green function as $R_\alpha(\zeta)$ and G_α respectively. The $\mathcal{Q}$-function is defined up to an additive constant [**17**]; by virtue of statement (2b) of Theorem A, we can fix this constant defining $Q(\zeta)$ with the help of the equality

$$Q(\zeta) = \lim_{\mathbf{r} \to \mathbf{0}} \left[G^0(\mathbf{r}, \mathbf{0}; \zeta) - \frac{1}{4\pi |\mathbf{r}|} \right]. \tag{19}$$

In this case the extension parameter α is related to the scattering length l_s of the corresponding point potential as follows [**5**]:

$$\alpha = \frac{m_*}{2\pi\hbar^2} l_s^{-1}. \tag{20}$$

We can obtain an explicit form for $Q(\zeta)$ using a series expansion for G^0. It is easy to show by a direct computation that the Green function $G_2^0(z, z'; \zeta)$ for the operator H_2^0 has the form

$$G_2^0(z, z'; \zeta) = \frac{e^{i\pi/4} \Gamma\left(\frac{1}{2} - i\zeta\right)}{\sqrt{2\pi\omega}} \times$$

$$U(-i\zeta/\omega,\, e^{-i\pi/4}\omega^{1/2} \max(z, z')) U(-i\zeta/\omega,\, e^{-i\pi/4}\omega^{1/2} \max(-z, -z')). \tag{21}$$

Here $\mathrm{Im}\,\zeta > 0$, and $U(a, x)$ is the Weber function [**1**]. Using arguments similar to those in Chapter XV of [**27**] we can obtain the following series expansion for G^0:

$$G^0(\mathbf{r}, \mathbf{r}'; \zeta) \equiv G^0(\rho, \varphi, z; \rho', \varphi', z'; \zeta) =$$

$$\sum_{m=-\infty}^{\infty}\sum_{n=0}^{\infty}\Psi_{mn}(\rho,\varphi)\overline{\Psi_{mn}(\rho',\varphi')}G_2^0(z,z';\zeta-E_{mn}) \tag{22}$$

(the expansion converges for every $(\mathbf{r},\mathbf{r}')$ with $\mathbf{r}\neq\mathbf{r}'$). Further consider an auxiliary self-adjoint operator L in $L^2(\mathbf{R}^3)$,

$$L=(i\nabla+\mathbf{A})^2+\frac{\omega_1^2}{4}(x^2+y^2)\,, \tag{23}$$

with the Green function $K(\mathbf{r},\mathbf{r}';\zeta)$. It is known [**15**] that

$$\lim_{\mathbf{r}\to\mathbf{0}}\left[K(\mathbf{r},\mathbf{0};\zeta)-\frac{1}{4\pi|\mathbf{r}|}\right]=\frac{\Omega^{1/2}}{8\pi}\mathrm{Z}\left(\frac{1}{2},\frac{1}{2}-\frac{\zeta}{\Omega}\right), \tag{24}$$

where Z is the generalized Riemann ζ-function (or the Hurwitz ζ-function), see [**4**]. On the other hand, $K(\mathbf{r},\mathbf{r}';\zeta)$ admits the following series expansion

$$K(\mathbf{r},\mathbf{r}';\zeta)\equiv K(\rho,\varphi,z;\rho',\varphi',z';\zeta)=$$

$$\sum_{m=-\infty}^{\infty}\sum_{n=0}^{\infty}\Psi_{mn}(\rho,\varphi)\overline{\Psi_{mn}(\rho',\varphi')}G_3^0(z,z';\zeta-E_{mn})\,, \tag{25}$$

where G_3^0 is the Green function of the one-dimensional free Hamiltonian $-d^2/dz^2$:

$$G_3^0(z,z';\zeta)=\frac{\exp\left(-\sqrt{-\zeta}\,|z-z'|\right)}{2\sqrt{-\zeta}}\,. \tag{26}$$

To find now the limit (19) it is sufficient to find the limit

$$\lim_{\rho\to 0}\left[G(\rho,0,0;0,0,0;\zeta)-K(\rho,0,0;0,0,0;\zeta)\right]. \tag{27}$$

As a result, after some calculations we get the required expression:

(28)

$$Q(\zeta)=\frac{\Omega}{8\pi\omega^{1/2}}\left\{\left(\frac{\omega}{\Omega}\right)^{1/2}\mathrm{Z}\left(\frac{1}{2},\frac{1}{2}-\frac{\zeta}{\Omega}\right)+\sum_{n=0}^{\infty}\left[\frac{e^{i\pi/4}}{\sqrt{2}}\frac{\Gamma\left(\frac{1}{4}+\frac{i\varepsilon_n(\zeta)}{2}\right)}{\Gamma\left(\frac{3}{4}+\frac{i\varepsilon_n(\zeta)}{2}\right)}-\frac{1}{\varepsilon_n^{1/2}(\zeta)}\right]\right\},$$

where $\varepsilon_n(\zeta)=\omega^{-1}[\Omega(n+1/2)-\zeta]$. By virtue of Eq. (6.1.47) from [**1**], we have

$$\frac{e^{i\pi/4}}{\sqrt{2}}\frac{\Gamma\left(\frac{1}{4}+\frac{i\varepsilon_n(\zeta)}{2}\right)}{\Gamma\left(\frac{3}{4}+\frac{i\varepsilon_n(\zeta)}{2}\right)}=\varepsilon_n^{-1/2}(\zeta)+\frac{1}{16}\varepsilon_n^{-5/2}(\zeta)+O\left(|\varepsilon_n(\zeta)|^{-7/2}\right) \tag{29}$$

locally uniformly with respect to ζ, $\operatorname{Im}\zeta\geq 0$, $\zeta\neq\Omega(n+1/2)$. Hence the following lemma is valid

Lemma 2. *For every $E\in\mathbf{R}$, $E\neq\Omega(n+1/2)$, $n\in\mathbf{N}$, there exists a limit*

$$\lim_{\zeta\to E,\ \operatorname{Im}\zeta>0}Q(\zeta). \tag{30}$$

We denote the limit (30) by $Q(E)$.

Lemma 3. *The function Q may be continuously extended on the whole real axis $\mathbf{R}$. The function obtained by extension obeys the condition $\operatorname{Im}Q(E)>0$ for all $E\in\mathbf{R}$.*

Proof. One can obtain from Eq. (28) that

$$Q(E) = \frac{\Omega^{1/2}}{8\pi} Z\left(\frac{1}{2}, \frac{1}{2} - \frac{E}{\Omega}\right) +$$

$$\frac{\Omega}{8\pi\omega^{1/2}} \left\{ \sum_{n=0}^{\infty} \left[\frac{1}{\sqrt{2}} \left| \frac{\Gamma\left(\frac{1}{4} + \frac{i\varepsilon_n(E)}{2}\right)}{\Gamma\left(\frac{3}{4} + \frac{i\varepsilon_n(E)}{2}\right)} \right| (1 + \exp(-2\pi\varepsilon_n(E)))^{-1/2} - \frac{1}{\varepsilon_n^{1/2}(E)} \right] + \right.$$

$$\left. \frac{i}{\sqrt{2}} \sum_{n=0}^{\infty} \left| \frac{\Gamma\left(\frac{1}{4} + \frac{i\varepsilon_n(E)}{2}\right)}{\Gamma\left(\frac{3}{4} + \frac{i\varepsilon_n(E)}{2}\right)} \right| (1 + \exp(2\pi\varepsilon_n(E)))^{-1/2} \right\}, \tag{31}$$

if $E \in \mathbf{R}$ and $E \neq \Omega(n + 1/2)$, $n \in \mathbf{N}$. Using the shift formula for the Hurwitz function

$$Z(s, x) = Z(s, m + x) + \sum_{n=0}^{m-1} (n + x)^{-s}, \tag{32}$$

(Eq. (1.10.2) from [**4**]) we show that the singularities in the two first terms of Formula (31) are cancelled. Moreover, the third term is equal to $i \operatorname{Im} Q(E)$ ∎

4. Spectral properties of point perturbations

This section contains the main results of the paper. We start with the following theorem.

Theorem 1. *For every $\alpha \in \mathbf{R}$ the spectrum of H_α is purely absolutely continuous and coincides with the whole real axis* $\mathbf{R}$.

Proof. First note that $\sigma_{\rm ess}(H_\alpha) = \sigma_{\rm ess}(H^0)$ by virtue of the Weyl theorem ([**23**], Theorem XIII.14). Therefore, $\sigma(H_\alpha) = \mathbf{R}$.

Further we show that H_α has no eigenvalues. Suppose that this is not the case, and let E be an eigenvalue of H_α with an eigenfunction ψ. Fix a number ζ with $\operatorname{Im}\zeta > 0$; then $R_\alpha(\zeta)\psi = (E - \zeta)^{-1}\psi$. Using Eq. (18) and notations from Lemma 1, we get

$$R^0(\zeta)\psi - [Q(\zeta) + \alpha]^{-1} \langle g_{\bar\zeta} \,|\, \psi \rangle g_\zeta = (E - \zeta)^{-1}\psi. \tag{33}$$

Let $\varphi = R^0(\zeta)\psi$; then Eq. (33) reads

$$(E - \zeta)\varphi - (E - \zeta)[Q(\zeta) + \alpha]^{-1} \langle g_{\bar\zeta} \,|\, (H^0 - \zeta)\varphi \rangle g_\zeta = (H^0 - \zeta)\varphi, \tag{34}$$

or

$$(H^0 - E)\varphi = c g_\zeta, \tag{35}$$

where $c = (\zeta - E)[Q(\zeta) + \alpha]^{-1}\varphi(\mathbf{0})$. Because H^0 has no eigenvalues, $c \neq 0$. Using Lemma 1, we get

$$\tilde\varphi^\pm(m, n, \lambda) = c(E^\pm_{m,n,\lambda} - E)^{-1}(E^\pm_{mn\lambda} - \zeta)^{-1}\Psi^\pm(m, n, \lambda)(\mathbf{0}). \tag{36}$$

Since $\varphi \in L^2(\mathbf{R}^3)$, the function

$$\xi(\lambda) \equiv \tilde{\varphi}^+(0,0,\lambda) = \frac{c\Omega^{1/2}}{2\pi(4\omega)^{1/4}}(1+e^{-2\pi\lambda})^{-1/2}\frac{4\mathbf{E}(-\lambda,\mathbf{0})}{(\Omega+2\omega\lambda-2\zeta)(\Omega+2\omega\lambda-2E)}$$

must belong to $L^2(\mathbf{R})$. On the other hand, the function $\lambda \mapsto \mathbf{E}(-\lambda, \mathbf{0})$ is continuous and does not vanish [**1**]; therefore $\xi \notin L^2(\mathbf{R})$. This contradiction proves the absence of eigenvalues.

It remains to prove that $\sigma_{\text{sing}}(H_\alpha) = \emptyset$. Denote by M the set of all functions f from $L^2(\mathbf{R}^3)$ having the form

$$f(x,y,z) = \Psi_{mn}(x,y)\psi(z), \tag{37}$$

where $m \in \mathbf{Z}$, $n \in \mathbf{N}$ and ψ runs through $C_0^\infty(\mathbf{R})$; it is clear that the linear hull of M is dense in $L^2(\mathbf{R}^3)$. Using Eq. (22), we show that if f has the form (37), then

$$\langle f \,|\, [R^0(\zeta) - R_\alpha(\zeta)]f \rangle = \frac{|\Psi_{mn}(0)|^2}{Q(\zeta)+\alpha}\left|\int_{-\infty}^{\infty} G_2^0(0,z;E_{mn}-\zeta)\psi(z)\,dz\right|^2. \tag{38}$$

It follows from Lemma 3 that the mapping $\zeta \mapsto \langle f \,|\, [R^0(\zeta)-R_\alpha(\zeta)]f \rangle$ is continuous in the closed half-plane $\operatorname{Im}\zeta \geq 0$. Since the spectrum of H^0 is purely absolutely continuous, we can apply Theorem XIII.19 from [**23**] to complete the proof ∎

Because the saddle-point potential V_{sp} tends at infinity to $-\infty$ very rapidly, the standard scattering theory is not applicable to the couple (H^0, H_α). Moreover, in our case the set $\mathbf{C} \setminus \sigma(H^0)$ of regular values of H^0 is not connected, and the dependence of $G^0(\mathbf{r},\mathbf{r}';\zeta)$ on the spectral parameter ζ has no singularities of the square root type; therefore, there is no analogue of the nonphysical sheet here. In this connection an alternate version of the time-independent scattering theory was proposed in [**9**]. The approach of the cited article is based on the following observation. The propagation states (11) have the asymptotics of the form $A(z)\exp\left(+i\omega z^2/4\right) + B(z)\exp\left(-i\omega z^2/4\right)$ as $z \to \pm\infty$ with slowly varying $A(z)$ and $B(z)$. Writing $P = (1/i)(d/dz)$ (the momentum operator) and noting that $Pe^{\pm iz^2} = \pm 2ze^{\pm iz^2}$, we conclude that the current associated with term proportional to $\exp\left(+i\omega z^2/4\right)$ is directed away from the origin, and the current associated with term proportional to $\exp\left(-i\omega z^2/4\right)$ is directed toward the origin. Thus we can associate the former with the outgoing current, and the latter with the incoming current. Bearing this in mind, we obtain an explicit form for the transmission coefficients. Namely, let $T_{m_0n_0\to m_1n_1}(E)$ be the partial transmission coefficient from the mode associated with the transverse eigenfunction $\Psi_{m_0n_0}$ to the mode associated with $\Psi_{m_1n_1}$. Then

$$T_{m_0n_0\to m_1n_1}(E) = \delta_{m_0m_1}\frac{\exp\left(\pi\Omega(n_0-n_1)/(2\omega)\right)}{1+\exp\left(-2\pi\lambda_{0n_0}(E)\right)}\times$$

$$\left|\delta_{n_0n_1} - \delta_{0m_0}\frac{\Omega e^{i\pi/4}2^{i\Omega(n_1-n_0)/2\omega}}{2^{7/2}[Q(E)+\alpha]}\frac{\Gamma\left(\frac{1}{4}-\frac{i\lambda_{0n_1}(E)}{2}\right)}{\Gamma\left(\frac{3}{4}-\frac{i\lambda_{0n_0}(E)}{2}\right)}\right|^2, \tag{39}$$

where

$$\lambda_{mn}(E) = \frac{E-E_{mn}}{\omega}. \tag{40}$$

Formula (31) shows that $\operatorname{Im} Q(E) \to 0$ as $E \to -\infty$. Therefore, the equation

$$\text{Re } Q(E) + \alpha = 0 \tag{41}$$

determines a point of a maximum for coefficients $T_{0n\to 0n}(E)$ if α is sufficiently large. We can find an asymptotic solution to equation (41). Namely, for $E \ll -1$ we can replace Eq. (41) by the equation

$$\frac{\Omega^{1/2}}{8\pi}\left[\mathrm{Z}\left(\frac{1}{2}, \frac{1}{2} - \frac{E}{\Omega}\right) + \left(\frac{\omega}{4\Omega}\right)^2 \mathrm{Z}\left(\frac{5}{2}, \frac{1}{2} - \frac{E}{\Omega}\right)\right] = -\alpha \tag{42}$$

(see Eq. (31)). The following asymptotics are known [**4**], [**7**]:

$$\mathrm{Z}\left(\frac{1}{2}, x\right) = -2x^{1/2} + \frac{1}{2}x^{-1/2} + \frac{1}{24}x^{-3/2} + O(x^{-7/2}), \tag{43}$$

$$\mathrm{Z}\left(\frac{5}{2}, x\right) = \frac{2}{3}x^{-3/2} + O(x^{-5/2}). \tag{44}$$

Substituting (43) and (44) in Eq. (42), we get

$$E_0 = -l_s^{-2} + \frac{1}{24}l_s^2\left(\omega_1^2 + \omega_2^2 - \omega_3^2 + \frac{\omega_c^2}{2}\right) + O(l_s^4). \tag{45}$$

Note that $-l_s^{-2}$ is the ground state of the point perturbation of the free particle Hamiltonian $-\Delta$. Therefore, Eq. (45) gives the energy shift caused by the saddle-point potential V_{sp} and the magnetic field $\mathbf{B}$. For the spherically symmetric gravitational field the correction term in Eq. (45) vanishes if $\mathbf{B} = \mathbf{0}$ and we obtain an analogue of a result from [**22**] (it was proven in the cited paper that the energy shift for the Coulomb potential in the gravitational field of a spherically symmetric object vanishes in the first order of the perturbation theory). In the limit $\omega_i \to 0$ we recover a result of [**7**]; namely, in this case the correction term in (45) gives the diamagnetic shift of the energy in a zero-range well.

Acknowledgments. The paper was partially supported by grants of RFBR and Russian Ministry of Education. One of us (V.A.G.) is grateful to the Sonderforschungsbereich 256 for the generous support during his stay at Bonn University where a part of this paper was performed. We are thankful to the referee for useful remarks.

References

[1] Abramowitz, A., Stegun, I. A. eds. *Handbook of Mathematical Functions.* Dover, New York (1972).

[2] Albeverio, S., Gesztesy, F., Høegh-Krohn, R., Holden, H. *Solvable Models in Quantum Mechanics.* Springer-Verlag, Berlin (1988).

[3] Albeverio, S., Geyler, V. A., Margulis, V. A. Paper in preparation.

[4] Bateman, H., Erdely, A. *Higher Transcendental Functions*, vol. I, II. McGraw-Hill, New York (1953).

[5] Baz', A. I., Zeldovich, Y. B., Perelomov, A. M. *Scattering, Reactions and Decay in Nonrelativistic Quantum Mechanics.* Israel Program for Sci. Translations, Jerusalem (1969).

[6] Berezansky, Yu. M. *Expansion in Eigenfunctions of Self-Adjoint Operators.* AMS, Providence, Rhode Island (1968).

[7] Demkov, Yu. N., Ostrovskii, V. N. *Zero-range Potentials and their Applications in Atomic Physics.* Plenum, New York (1988).
[8] Dunford, N., Schwartz, T. C. *Linear Operators. Part II: Spectral Theory.* Interscience, New York (1963).
[9] Fertig, H. A., Halperin, B. I. *Phys. Rev.* **B 36** (1987), 7969-7976.
[10] Gesztesy, F., Simon, B. *J. Funct. Anal.* **128** (1995), 245-252.
[11] Geyler, V. A. In: *Proc. II Int. Conf. Nanometer Scale Sci. Technol. Moscow, August* 2-6, 1993. Part C (1994), 834-842.
[12] Geyler, V. A., Margulis, V. A., Chuchaev, I. I. *Siberian Math. J.* **36** (1995), 714-726.
[13] Geyler, V. A., Margulis, V. A., Chuchaev, I. I. *JETP Lett.* **58** (1993), 648-652.
[14] Geyler, V. A., Margulis, V. A., Filina, L. I. *J. Exp. Theor. Phys.* **86** (1998), 751-762.
[15] Geyler, V. A., Popov, I. Yu. Z. Phys. **B93** (1994), 437-439.
[16] Kiselev, A., Simon, B. *J. Func. Anal.* **130** (1995), 345-356.
[17] Krein, M. G., Langer, H. K. Funct. Anal. and its Appl. **5** (1971), 217-228.
[18] Levinson, Y. B., Levin, M. I., Sukhorukov, E. V. *Phys. Rev.* **B 45** (1992), 11936-11943.
[19] Misner, C. W., Thorne, K. S., Wheler, J. A. *Gravitation.* Freeman, San-Francisco (1973).
[20] Parker, L. *Phys. Rev. Lett.* **44** (1980), 1559-1562.
[21] Pavlov, B. S.: *Russ. Math. Surv.* **42**, No. 6 (1987), 127-168.
[22] Pinto, F. *Phys. Rev. Lett.* **70** (1993), 3839-3842.
[23] Reed, M., Simon, B. *Methods of Modern Mathematical Physics.* Vol. IV: *Analysis of Operators.* Acad. Press, New York (1978).
[24] Schrödinger, E. *Proc. Roy. Irish Acad.* **46** (1940), 9-37.
[25] Scherbakov, A. G., Bogachek, E. N., Landman, U. *Phys. Rev.* **B57** (1998), 6655-6661.
[26] Simon, B. *A Feynmann–Kac formula for unbounded semigroups.* To appear in Proc. Int. Conf. "Infinite Dimensional (Stochastic) Analysis and Quantum Physics", Leipzig, 1999.
[27] Titchmarsh, E. C. *Eigenfunction Expansions Associated with Second–Order Differential Equations*, Vol.II. Clarendon Press, Oxford (1958).

Department of Mathematical Analysis, Mordovian State University, Saransk, Russia

E-mail address: `geyler@mrsu.ru`

Department of Theoretical Physics and Astronomy, Mordovian State University, Saransk, Russia

E-mail address: `margulis@mrsu.ru`

Canadian Mathematical Society
Conference Proceedings
Volume **29**, 2000

4D Gauge-like Quantum Fields from Rectangular Systems of Stochastic Partial Differential Equations

R. Gielerak and P. Ługiewicz

Dedicated with admiration to Prof. Sergio Albeverio on the occasion of his 60th birthday.

ABSTRACT. Rectangular systems of covariant stochastic differential equations are analysed and covariant schemes for solving them are presented. This leads in particular to an interesting class of indefinite quantum field theory models with local gauge group symmetry and nontrivial scattering operators. In particular, new models of a nonlinear electromagnetism on $4D$ space-time are presented.

1. Introduction

The fundamental discovery of Nelson [**14, 15**] (cf. references therein for historical background) that from homogeneous, Markov fields (obeying certain regularity conditions) it is possible to reconstruct quantum field structures fulfilling all Wightman axioms [**12**], became one of the most important achievements of the Euclidean Quantum Field Theory. In particular, Nelson's ideas gave a powerful input for dynamical developments of the Constructive Quantum Field Theory [**11, 17**]. In the case of the free scalar (massive $m > 0$), field the corresponding generalized Markov random field μ (that is a corresponding probability Borel cylindric measure on the space of tempered distributions $\mathcal{S}'(\mathbf{R}^{d+1})$ with $d \geq 1$) can be identified with the law of the homogeneous solution of the following stochastic differential equation of Itô type:

$$d\xi_t = -\sqrt{m^2 - \triangle}\xi_t dt + dB_t$$

where dB_t is the $\mathcal{S}'(\mathbf{R}^d)$- valued canonical Brownian motion (version of) [**18**].

In the pioneering papers [**1, 2, 3**], a discussion concerning the covariant stochastic partial differential equation

$$\partial\phi = H \tag{1.1.1}$$

where ∂ is the quaternionic Cauchy-Riemann operator (on the 4-dimensional Euclidean space-time) and H is a noise of zero order obeying certain regularity and covariance conditions, has been initiated and oriented into the discussion of nonlinear electrodynamics models with nonlinearity caused by the presence of the

2000 *Mathematics Subject Classification.* Primary 60H15, 81T10, 81T05.

nonzero Poisson part in H. A number of generalizations [**4, 5, 6, 16, 8**] have appeared which were, however, limited to the situation where the covariant decomposition of the corresponding Laplace operator can be given in terms of a product of first order square covariant operators with validity limited, therefore, to particular dimensions $d+1 = 2, 4, 8$ (see [**7, 16**]). A systematic approach to covariant systems of stochastic partial differential equations generalizing the framework of [**1, 2, 3**], [**4, 5, 6, 8, 16**] has been initiated in [**7**] and developed in several directions in [**4, 9, 10**]. As a product of this activity, a variety of new quantum field theory models obeying all Morchio-Strocchi axioms for indefinite metric quantum field theory (IMQFT) of gauge type [**5, 13**] and with nontrivial scattering operators was obtained.

The main aim of the present contribution is to announce some new results on the possibility to obtain a large class of IMQFT models obeying the Morchio - Strocchi system of axioms and additionally obeying a large gauge-like group symmetry and describing nontrivial scattering processes on the basis of the covariant analysis of the rectangular problems like (1.1.1). In particular, a new models of nonlinear electrodynamics based on the vector potential and field strength will be presented. More details on this topic can be found in [**9**].

2. Covariant solutions of rectangular systems of SPDEs.

Let μ_0^N be a (regular) noise (of the 0-order) on the space $\mathcal{S}'(\mathbf{R}^4)\otimes\mathbf{R}^{n_\tau}$, where τ is a real (orthogonal) representation of the rotation group $SO(4)$ and n_τ is the (finite) dimension of τ. The noise $d\mu_0^N$ is completely characterized by:

$$\text{(2.2.1)} \qquad \int_{\mathcal{S}'(\mathbf{R}^4)\otimes\mathbf{R}^{n_\tau}} \mathrm{e}^{i(\eta,f)}d\mu_0^N(\eta) = \exp\{-\frac{1}{2}\int_{\mathbf{R}^4} dx(f(x),f(x))\}\times$$
$$\times\exp\{\int_{\mathbf{R}^4} dx \int_{\mathbf{R}^{n_\tau}} d\lambda(\alpha)[\mathrm{e}^{i(\alpha,f(x))}-1]\}$$

where we assume that the so-called Levy measure $d\lambda$ is some Borel measure on $\mathbf{R}^{n_\tau}$ such that:

(i) $d\lambda$ has all moments, (ii) $d\lambda$ is τ- invariant.

Let τ' be another representation of the group $SO(4)$ in $\mathbf{R}^{n_{\tau'}}$. The action τ (τ' resp.) of the group $SO(4)$ can be naturally lifted to the action T_τ ($T_{\tau'}$ resp.) of $SO(4)$ in the space $\mathcal{S}'(\mathbf{R}^4)\otimes\mathbf{R}^{n_\tau}$ ($\mathcal{S}'(\mathbf{R}^4)\otimes\mathbf{R}^{n_{\tau'}}$ resp.). Recall that a first order (for simplicity only) differential operator $\mathcal{D}=\sum_{\mu=0}^3 B_\mu\partial_\mu+\mathcal{M}$, where $B_\mu, \mathcal{M}\in\mathrm{Hom}(\mathbf{R}^{n_\tau},\mathbf{R}^{n_{\tau'}})$, is called (τ,τ')-covariant **iff** the following diagram

$$\text{(2.2.2)} \qquad \begin{array}{ccc} \mathcal{S}'(\mathbf{R}^4)\otimes\mathbf{R}^{n_\tau} & \xrightarrow{\mathcal{D}} & \mathcal{S}'(\mathbf{R}^4)\otimes\mathbf{R}^{n_{\tau'}} \\ T_\tau\big\downarrow & & \big\downarrow T_{\tau'} \\ \mathcal{S}'(\mathbf{R}^4)\otimes\mathbf{R}^{n_\tau} & \xrightarrow{\mathcal{D}} & \mathcal{S}'(\mathbf{R}^4)\otimes\mathbf{R}^{n_{\tau'}} \end{array}$$

commutes. Let us denote by $Cov(\tau,\tau')$ the set of all such (τ,τ')- covariant operators. The complete description of these sets is given in Appendix.

Let us denote by $\mathcal{D}^T$ the adjoint of $\mathcal{D}$ in the canonical pairing
$_{\mathcal{S}'(\mathbf{R}^4)\otimes\mathbf{R}^{n_{\tau'}}}<\mathcal{D}(\cdot),\cdot>_{\mathcal{S}(\mathbf{R}^4)\otimes\mathbf{R}^{n_\tau}} = {}_{\mathcal{S}'(\mathbf{R}^4)\otimes\mathbf{R}^{n_{\tau'}}}<\cdot,\mathcal{D}^T(\cdot)>_{\mathcal{S}(\mathbf{R}^4)\otimes\mathbf{R}^{n_\tau}}$.

We shall assume from now on that $n_\tau > n_{\tau'}$. The equation

$$\text{(2.2.3)} \qquad \mathcal{D}\phi=\eta$$

where η is the noise as above, is called t-regular iff there exists:

(i) a (minimal) nuclear space $\mathcal{F}_\eta \supset \mathcal{S}(\mathbf{R}^4)\otimes\mathbf{R}^{n_{\tau'}}$;

(ii) a continuous (bijective) mapping $\mathcal{G} : \mathcal{S}(\mathbf{R}^4)\otimes\mathbf{R}^{n_\tau} \longrightarrow \mathcal{F}_\eta$ such that $\mathcal{G}\circ\mathcal{D}^T = \mathrm{id}_{\mathcal{F}_\eta}$;

(iii) the noise η can be extended continuously to $\mathcal{F}_\eta$.

If the problem (2.2.3) is t-regular we can define a new random field ϕ by giving its characteristic functional $\Gamma_\phi(f) = \Gamma_\eta(\mathcal{G}f)$, where Γ_η is given by the r.h.s. of (2.2.1). The random field ϕ given by Γ_ϕ will be called t-regular solution of (t-regular) equation (2.2.3). Informally, for a t-regular solution of (2.2.3) we have $(\phi,\mathcal{D}^T f)\sim(\eta,f)$, where $\sim$ means the equality in law.

We shall assume for the sequel that $\ker\mathcal{D}^T = \{0\}$ and we denote $\mathcal{H}\cong\mathrm{Im}\mathcal{D}^T$. Moreover we assume $\mathcal{F}_\phi/\ker\mathcal{D}\cong\mathcal{F}_\eta$ where $\mathcal{F}_\phi$ (and $\mathcal{F}_\eta$ resp.) are (nuclear) indexing spaces for the solution ϕ of (2.2.3) (and for the noise η resp.). Assume that there exists a continuous left inverse $E_\mathcal{H}$ of $\mathcal{D}^T$, i.e. $E_\mathcal{H} : \mathcal{H}\longrightarrow\mathcal{F}_\eta$ and such that $E_\mathcal{H}\circ\mathcal{D}^T = \mathrm{id}_{\mathcal{F}_\eta}$. Let Φ be a generalized random field (indexed by $\mathcal{H}$) defined by $\mathbf{E}e^{i(\Phi,f)} = \mathbf{E}e^{i(\eta,E_\mathcal{H}f)}$ for all $f\in\mathcal{H}$ and $f=\mathcal{D}^T g$ with $g\in\mathcal{F}_\eta$. In other words the field Φ is a solution of the quadratic system of equations:

$$< \Phi, f >=< \phi,\mathcal{D}^T g >=< \eta, g >$$

the meaning of which was extensively discussed in [**7**]. It follows from the covariancy of $\mathcal{D}$ that $T_\tau : \mathcal{H}\longrightarrow\mathcal{H}$ and after defining $T^\tau_\mathcal{H}\equiv T_\tau|_\mathcal{H}$ we have

$$< \Phi, T^\tau_\mathcal{H} f >=< \Phi, T_\tau\mathcal{D}^T g >=< \Phi,\mathcal{D}^T T^{\tau'} g >=$$
$$=< \eta, T^{\tau'} g >=< \eta, g >=< \Phi, f >$$

and, therefore, the field Φ is $T^\tau_\mathcal{H}$- covariant.

Now we assume:

$\mathbf{h}_1$) *there exists continuous and (τ,τ')-covariant extension $E^*_\mathcal{H}$ of $E_\mathcal{H}$ to the whole space $\mathcal{F}_\phi$.*

The existence of the desired extension $E^*_\mathcal{H}$ is equivalent to the existence of T_τ-invariant (topological) subspace $\mathcal{H}^c\subset\mathcal{F}_\phi$ complementary to $\mathcal{H}$ (see Proposition 2.10 in [**9**]).

Many extensions $E^*_\mathcal{H}$ of $E_\mathcal{H}$ obeying $\mathbf{h}_1$ may exist and, therefore, the introduced now generalized random field $\phi^{E^*_\mathcal{H}}$ may depend on a particular choice of such $E^*_\mathcal{H}$. The field $\phi^{E^*_\mathcal{H}}$ is defined by the following equality $\mathbf{E}e^{(\phi^{E^*_\mathcal{H}},f)} = \mathbf{E}e^{(\eta,E^*_\mathcal{H}f)}$. The field $\phi^{E^*_\mathcal{H}}$ can be seen as a kind of extension of the field Φ to the bigger index space $\mathcal{F}_\phi$. It follows easily that $\phi^{E^*_\mathcal{H}}$ is T_τ- covariant random field and it follows from the very definition of $\phi^{E^*_\mathcal{H}}$ that $<\phi^{E^*_\mathcal{H}}, f+\ker E^*_\mathcal{H}>\sim<\phi^{E^*_\mathcal{H}}, f>$ and, moreover, $<\phi^{E^*_\mathcal{H}}, f>\sim<\Phi, f>$ for any $f\in\mathcal{H}$.

Thus the property $<\phi^{E^*_\mathcal{H}}, f>\sim 0$ for $f\in\ker E^*_\mathcal{H}$ can be used to conclude that the distribution $\mu_{\phi^{E^*_\mathcal{H}}}$ of the field $\phi^{E^*_\mathcal{H}}$ is supported on the set $(\ker E^*_\mathcal{H})^\circ\subset\mathcal{F}'_\phi$ where $(\cdot)^\circ$ means the set of functionals in $\mathcal{F}'_\phi$ which annihilate the set $(\cdot)\subset\mathcal{F}_\phi$. This indicates that the redundant degrees of freedom of the field $\phi^{E^*_\mathcal{H}}$ in (2.2.3) lead to certain kind of gauge-invariance of $\mu_{\phi^{E^*_\mathcal{H}}}$. Thus the interesting point (illustrated explicitly in the next section) of the whole discussion is the ability to provide examples of nongaussian random fields with gauge-type symmetry, the feature which deserves further studies.

From the probabilistic point of view the field $\phi^{E^*_\mathcal{H}}$ is identical with the field Φ and, therefore, the problem of covariant solution of (2.2.3) has been reduced to the better known quadratic problems [**1, 2**].

Now we shall consider the more general situation: ker$\mathcal{D}^T \neq \{0\}$ in $\mathcal{F}_\eta$. To deal with this case we define $\mathcal{F}_\eta^* = \mathcal{F}_\eta/\text{ker}\mathcal{D}^T$. Because of continuity of $\mathcal{D}^T$ the space $\mathcal{F}_\eta^*$ is again a nuclear space and it follows from the covariancy of $\mathcal{D}$ that ker$\mathcal{D}^T$ is $T_{\tau'}$-invariant and, therefore, we can lift the representation T_τ to the space $\mathcal{F}_\eta^*$, getting new representation denoted by $T_{\tau'}^*$. We define $\mathcal{D}_*^T : \mathcal{F}_\eta^* \longrightarrow \mathcal{F}_\phi$ which is a $(T_\tau, T_{\tau'}^*)$-covariant operator. For all $f \in$ker$\mathcal{D}^T$, one can define a new noise η_* on $\mathcal{F}_\eta^*$ by the formula $\mathbf{E}e^{i(\eta_*, f_*)} = \mathbf{E}e^{i(\eta, f)}$ for all $f \in \mathcal{F}_\eta$ (this is possible only if the noise η fulfils appropriate conservation law, see below). The (well defined !) noise η_* is continuous on $\mathcal{F}_\eta^*$ (in the sense of characteristic functional). Now, instead of (2.2.3) we consider the following equation

$$\mathcal{D}_*\phi = \eta_*$$

where ker$\mathcal{D}_*^T = \{0\}$ and we take $\mathcal{H} = \overline{\text{Im}\mathcal{D}_*^T}$, i.e. we arrive at the previous situation.

By the very construction it follows that $< \eta, f > \sim 0$ for all $f \in$ker$\mathcal{D}^T$ and this should be interpreted as some conservation law of the noise η.

The outlined programme for constructing covariant solutions of rectangular systems (2.2.3) has been applied in [**9**] for a construction of new examples of non-trivial (from the physical point of view) indefinite metric quantum field theories of Morchio-Strocchi type [**13**] (i.e. quantum field theories of qauge-type and with the mild infrared singularities of the confining type).

3. New models of nonlinear $4D$-electrodynamics.

As a particular application of the general framework presented above, we shall give two versions of nonlinear electrodynamics. The first one is based upon the field potential A_μ and the second one on the stochastic Maxwell equation (in Euclidean time) for the field strength tensor $F_{\mu\nu}$. The systematic analysis of the models introduced will be presented elsewhere, see also [**9**].

3.1. The models of nonlinear electrodynamical potential. Let us consider the pair $\tau = (0, 2)$ and $\tau' = (1, 2)$ of orthogonal representation of the group $SO(4)$ (see Appendix for the introduced labelling of representations of the $SO(4)$). If $\mathcal{D} \in Cov(\tau, \tau')$ $\mathcal{D}$ has the form (without any loss of generality; see Appendix):

$$\mathcal{D}(\partial) = \begin{pmatrix} -\partial_1 & \partial_0 & -\partial_3 & \partial_2 \\ -\partial_2 & \partial_3 & \partial_0 & -\partial_1 \\ -\partial_3 & -\partial_2 & \partial_1 & \partial_0 \end{pmatrix}. \tag{3.3.1}$$

The following simple identities hold:

$$\mathcal{D}(\partial)\mathcal{D}^T(-\partial) = -\triangle \cdot \mathbf{1}_{3\times 3} \equiv \begin{pmatrix} -\triangle & 0 & 0 \\ 0 & -\triangle & 0 \\ 0 & 0 & -\triangle \end{pmatrix} \tag{3.3.2}$$

$$\mathcal{D}^T(-\partial)\mathcal{D}(\partial) = \{-\triangle \cdot \mathbf{1}_{4\times 4} + (\partial_\mu \partial_\nu)\} \tag{3.3.3}$$

where $\mathcal{D}^T(-\partial)$ is the transpose of $\mathcal{D}$ in the pairing $< \cdot, \cdot >$; $\triangle$ is the (four- dimensional) Laplace operator and $\mathbf{1}_{d\times d}$ denotes the unit matrix on the space $\mathbf{R}^d$.

We shall build the appropriate sequence of spaces $\mathcal{F}_\eta$, $\mathcal{H}$, $\mathcal{F}_\phi$ illustrating the general scheme outlined above (one can find more elaborated details in [**9**]).

The space (the Fourier transform of $\mathcal{F}_\eta$):

$$\hat{\mathcal{F}}_\eta = \{\hat{f} = (\hat{f}_1, \hat{f}_2, \hat{f}_3) \mid \mathcal{D}^T(ip)\hat{f} \in \hat{\mathcal{S}}(\mathbf{R}^4) \otimes \mathbf{R}^4\}$$

equipped with the projective topology induced by the mapping $\mathcal{D}^T(ip) : \hat{\mathcal{F}}_\eta \to \hat{\mathcal{S}}(\mathbf{R}^4) \otimes \mathbf{R}^4$ becomes a nuclear space and, moreover, $\hat{\mathcal{F}}_\eta \subset (L^1 \cap L^2)(\mathbf{R}^4) \otimes \mathbf{C}^3$. As a conclusion we get $\mathcal{F}_\eta \subset L^2(\mathbf{R}^4) \otimes \mathbf{R}^3$ and, therefore, the class of admissible noise for the problem (2.2.3) is determined by the requirement of L^2-extendability (for an account of sufficient conditions see [**8, 9**]). Then the extended to L^2-characteristic functional of the admissible noise becomes continuous in the topology of $\mathcal{F}_\eta$.

The $(0,2)$-invariant space $\mathcal{F}_\phi$ topologically included in $\mathcal{S}(\mathbf{R}^4) \otimes \mathbf{R}^4$ is defined then as

$$\mathcal{F}_\phi = \{A \in \mathcal{S}(\mathbf{R}^4) \otimes \mathbf{R}^4 \mid (\delta_{\mu\nu} - \frac{\partial_\mu \partial_\nu}{\triangle})A^\nu \in \mathcal{S}(\mathbf{R}^4) \text{ for } \mu \in \{0,1,2,3\}\}.$$

In the induced topology, the space $\mathcal{F}_\phi$ becomes nuclear, σ-hilbertian as a closed subspace of $\mathcal{S}(\mathbf{R}^4) \otimes \mathbf{R}^4$. The closed subspace $\mathcal{H}$ of $\mathcal{F}_\phi$ is defined as: $\mathcal{H} = \{H \in \mathcal{S}(\mathbf{R}^4) \otimes \mathbf{R}^4 \mid \partial_\mu H^\mu = 0\}$ and is again $(0,2)$-invariant and nuclear (moreover, the particular choice of elements $A \in \mathcal{H}$ one can interpret as a kind of "Lorentz gauge" fixing).

The principal Green function $E_\mathcal{H} : \mathcal{H} \to \mathcal{F}_\eta$ is continuous, $((0,2),(1,2))$- covariant and we denote it as: $E_\mathcal{H} = -\triangle^{-1} * D(\partial)|_\mathcal{H}$ where the kernel $\triangle^{-1}(x) = -\frac{1}{4\pi^2}\frac{1}{|x|^2}$ is given as the Fourier transform of the (tempered) distribution $-\frac{1}{p^2}$. The operator $E^*_\mathcal{H}$, i.e. a covariant and continuous extension of $E_\mathcal{H}$, is defined in the present example as $E^*_\mathcal{H} = -\triangle^{-1} * D(\partial)|_{\mathcal{F}_\phi}$. The corresponding projection Π is given explicitly by $\Pi = \{\delta_{\mu\nu} - \frac{\partial_\mu\partial_\nu}{\triangle}\}|_{\mathcal{F}_\phi}$ with $\ker\Pi = \ker E^*_\mathcal{H}$ where $\ker\Pi = \{A \in \mathcal{F}_\phi \mid A_\mu = \partial_\mu \Lambda \text{ with } \Lambda \in \mathcal{G}\}$ with the set of "gauge transformations" $\mathcal{G}$ given by: $\mathcal{G} = \{\Lambda \mid \Lambda = \triangle^{-1} * (\partial_\mu A^\mu) \text{ for some } A \in \mathcal{F}_\phi\}$. Finally, $\mathcal{F}_\phi = \mathcal{H} \oplus_t ker\Pi$ and the index space for the solution Φ of SPDE (with "physical meaning") is $\mathcal{F}_\phi / ker\Pi \cong \mathcal{H}$.

As an application of the general framework developed in [**9**] and the programme of section 2 the following theorem can be proved.

THEOREM 3.1. *1) The problem $\mathcal{D}\phi = \eta$ with $\mathcal{D}$ given by (3.3.1) is t- regular.*
2) There exists a system $\{\mathcal{W}_n\}$ of $\hat{\tau}_M$-covariant (where $\hat{\tau}_M$ is suitable continuation of τ to representation of $\mathcal{P}^\uparrow_+$) tempered distributions such that

(i) *the moments $\{S_n\}$ of the (covariant) solutions of (2.2.3) have the Laplace -Fourier property, i.e. for any n, $x^0_1 < ... < x^0_n$:*

$$S_{\tau_1\ldots\tau_n}(x^0_1, \mathbf{x}_1; ...; x^0_n, \mathbf{x}_n) =$$

$$\int_{\mathbf{R}_+ \times \mathbf{R}^{3(n-1)}} e^{-\sum_{k=1}^{n-1}(x^0_{k+1}-x^0_k)p^0_k} e^{i\sum_{k=1}^{n-1}(\mathbf{x}_{k+1}-\mathbf{x}_k)\mathbf{p}_k} \mathcal{T}^\tau_{\tau_1\ldots\tau_n}(p_1,...,p_{n-1}) \otimes_{k=1}^{n-1} d^4p_k$$

with originals $\mathcal{T}^\tau_n$ being appropriate linear combinations of Fourier transforms $\hat{W}_n$ (with support in a product of cones of nonnegative energies) of (difference) Wightman distributions.

(ii) *the system $\{\mathcal{W}_n\}$ obeys the locality axiom, the spectral condition in the weak form and moreover the Hilbert Space Structure Condition (HSSC) (of [**13**]) is valid, i.e. there exists a system of continuous Schwartz norms $\{p_n\}$ such that*

$$|\mathcal{W}_{n+m}(\underline{f}^* \otimes \underline{g})| \le p_n(f)p_m(g)$$

(iii) *the system* $\{\mathcal{W}_n\}$ *is* $\tilde{\mathcal{G}}_R$*-invariant i.e. for any* $h_k \in \tilde{\mathcal{G}}_R$, $f_k \in \mathcal{S}(\mathbf{R}^4) \otimes \mathbf{R}^4$ *(where* $k \leq n$*):*

$$\mathcal{W}_n((f_1 + h_1) \otimes ... \otimes (f_k + h_k) \otimes f_{k+1} ... \otimes f_n) = \mathcal{W}_n(f_1 \otimes ... \otimes f_n)$$

where $\tilde{\mathcal{G}}_R = \{h = (h^\mu)_{\mu=0,1,2,3} \mid h^\mu = \partial^\mu f$ *for* $f \in \mathcal{S}(\mathbf{R}^4)\}$.

3.2. The stochastic Maxwell equation. A general form of covariant operator $\mathcal{D}$ for representations $\tau = (1,2) \oplus (-1,2)$ and $\tau' = (0,2)$ is given by (see Appendix):

$$\mathcal{D}(\partial) = \alpha \begin{pmatrix} -\partial_1 & -\partial_2 & -\partial_3 & \beta\partial_1 & \beta\partial_2 & \beta\partial_3 \\ \partial_0 & -\partial_3 & \partial_2 & -\beta\partial_0 & -\beta\partial_3 & \beta\partial_2 \\ \partial_3 & \partial_0 & -\partial_1 & \beta\partial_3 & -\beta\partial_0 & -\beta\partial_1 \\ -\partial_2 & \partial_1 & \partial_0 & -\beta\partial_2 & \beta\partial_1 & -\beta\partial_0 \end{pmatrix} \tag{3.3.4}$$

where $\alpha, \beta \in \mathbf{R}$. We fix their values as $\alpha = \frac{1}{\sqrt{2}}$ and $\beta = 1$. We shall consider the following stochastic equation of Maxwell type:

$$\mathcal{D}F = J \tag{3.3.5}$$

where F corresponds to a skew-symmetric tensor $F_{\mu\nu}$ and J is a random current provided by suitable noise. One can easy verify that

$$\mathcal{D}(\partial) = -(\mathbf{1}_{6\times 6} + \Omega)\mathrm{rot}(\partial) \text{ where } \Omega = \begin{pmatrix} \mathbf{0} & \mathbf{1}_{3\times 3} \\ -\mathbf{1}_{3\times 3} & \mathbf{0} \end{pmatrix}$$

and $\mathrm{rot}(\partial) = d$ is an exterior derivative operator acting on 1- forms. Applying the similarity transformation governed by the matrix $S = \frac{1}{\sqrt{2}}(\mathbf{1}_{6\times 6} + \Omega) = \sqrt{2}((\mathbf{1}_{6\times 6} + \Omega)^{-1})^T$ to the equation (3.3.5) we obtain $\mathrm{rot}^T(\partial)[\mathbf{E}, \mathbf{B}] = J$ where $E_k = F_{0k}$ and $B_k = -\frac{1}{2}\varepsilon_{kij}F_{ij}$ for $k = 1, 2, 3$ ($\mathrm{rot}^T(\partial) = \delta$ is co-derivative when acting on 2-forms). We list the basic properties of $\mathrm{rot}(\partial)$ and $\mathrm{rot}^T(-\partial)$:

$$\mathrm{rot}^T(-\partial)\mathrm{rot}(\partial) = \{-\triangle\delta_{\mu\nu} + \partial_\mu\partial_\nu\} \tag{3.3.6}$$

and

$$\mathrm{rot}(\partial)\mathrm{rot}^T(-\partial) = \begin{pmatrix} -\partial_0^2\mathbf{1}_{3\times 3} - (\partial_k\partial_l) & -(\partial_0\varepsilon_{kjl}\partial_j) \\ (\partial_0\varepsilon_{kjl}\partial_j) & -\mathbf{1}_{3\times 3}\triangle_3 + (\partial_k\partial_l) \end{pmatrix}. \tag{3.3.7}$$

The index space $\mathcal{F}_\eta \subset L^2(\mathbf{R}^4) \otimes \mathbf{R}^4$ for the current J is defined as (Fourier transform):

$\hat{\mathcal{F}}_\eta = \{\hat{j} \in L^1(\mathbf{R}^4) \otimes \mathbf{C}^6 \cap L^2(\mathbf{R}^4) \otimes \mathbf{C}^6 \mid \mathrm{rot}(-ip)\hat{j} \in \hat{\mathcal{S}}(\mathbf{R}^4) \otimes \mathbf{R}^6\}$ with the nuclear topology induced by the map $\mathrm{rot}(-ip) : \hat{\mathcal{F}}_\eta \to \hat{\mathcal{S}}(\mathbf{R}^4) \otimes \mathbf{R}^6$. Let a continuous idempotent $\pi : \mathcal{F}_\eta \to \mathcal{F}_\eta$ be given by $\pi(j)_\mu = (\delta_{\mu\nu} - \frac{\partial_\mu\partial_\nu}{\triangle})j^\nu$. Then $\mathrm{ker}\pi = \mathrm{ker}(\mathrm{rot}(\partial))$ i.e. $\mathrm{ker}\pi$ is closed. Defining $\mathcal{F}_\eta^* = \mathcal{F}_\eta / ker\pi$ one can prove that $\mathcal{F}_\eta^* \cong \{\mathbf{j} \in \mathcal{F}_\eta \mid \partial_\mu \mathbf{j}^\mu = 0\}$.

The important point of the present construction is appearing now, namely a proper choice of the noise J. For the present exposition we take the noise J described by the characteristic functional (continuous on $\mathcal{F}_\eta$)

$$\Gamma_J(j) = \exp\{-\int_{\mathbf{R}^4}[\int_{\mathbf{R}^4\setminus\{0\}}(e^{i\alpha^\mu(\delta_{\mu\nu} - \frac{\partial_\mu\partial_\nu}{\triangle})j^\nu(x)} - 1)d\lambda(\alpha)]dx\}$$

(for any $j \in \mathcal{F}_\eta$) which from probabilistic point of view is equivalent to the Poisson noise

$$\Gamma_{J_*}(\mathcal{J}) = \exp\{-\int_{\mathbf{R}^4}[\int_{\mathbf{R}^4\setminus\{0\}}(e^{i\alpha^\mu \mathbf{j}_\nu(x)} - 1)d\lambda(\alpha)]dx\}$$

for any $\mathcal{J} \in \mathcal{F}_\eta^*$ and $\mathbf{j} \in \mathcal{J}$ obeying $\partial_\mu \mathbf{j}^\mu = 0$ ("conservation law").

According to (3.3.6) and the fact that kerπ =ker(rot(∂)) we obtain

$$\text{rot}^T(-\partial)\text{rot}(\partial) = -\triangle \mathbf{1}_{4\times 4} \tag{3.3.8}$$

where rot(∂) and the right-hand side of (3.3.8) are determined on $\mathcal{F}_\eta^*$. Let $\Pi = -\triangle^{-1} * \text{rot}(\partial)\text{rot}^T(-\partial)$. We define the $((1,2)\oplus(-1,2))$-invariant space $\mathcal{F}_\phi$ (the topological subspace of $\mathcal{S}(\mathbf{R}^4)\otimes\mathbf{R}^6$):

$$\mathcal{F}_\phi = \{\begin{pmatrix}\mathbf{E}\\ \mathbf{B}\end{pmatrix} \in \mathcal{S}(\mathbf{R}^4)\otimes\mathbf{R}^6 \mid \Pi\begin{pmatrix}\mathbf{E}\\ \mathbf{B}\end{pmatrix} \in \mathcal{S}(\mathbf{R}^4)\otimes\mathbf{R}^6\}$$

and closed subspace $\mathcal{H}\subset\mathcal{F}_\phi$

$$\mathcal{H} = \{\begin{pmatrix}\mathbf{E}\\ \mathbf{B}\end{pmatrix} \in \mathcal{S}(\mathbf{R}^4)\otimes\mathbf{R}^6 \mid \partial_0\mathbf{B} = -\nabla\times\mathbf{E} \text{ and } \nabla\cdot\mathbf{B} = 0\}$$

i.e. $\mathcal{H}$ is determined by the Gauss law and the Faraday law (or equivalently by Maxwell equation $dF = 0$).

Finally we introduce the principial Green function $E_\mathcal{H} = -\triangle^{-1} * \text{rot}^T(-\partial)|_\mathcal{H}$ (see (3.3.8)) with values in the space $\mathcal{F}_\eta^*$, i.e. $E_\mathcal{H}(\mathbf{E},\mathbf{B}) = [-\triangle^{-1}*\text{rot}^T(-\partial)(\mathbf{E},\mathbf{B})]_{ker\pi}$ for any $(\mathbf{E},\mathbf{B})\in\mathcal{H}$ ($[\cdot]_{ker\pi}$ the abstract class of $\mathcal{F}_\eta^*$). Making use of the projection π (acting in $\mathcal{F}_\eta$) we obtain

$$E_\mathcal{H}\begin{pmatrix}\mathbf{E}\\ \mathbf{B}\end{pmatrix}|_{\substack{\partial_0\mathbf{B}=-\nabla\times\mathbf{E}\\ \nabla\cdot\mathbf{B}=0}} = \pi(-\triangle^{-1} * \text{rot}^T(-\partial))\begin{pmatrix}\mathbf{E}\\ \mathbf{B}\end{pmatrix} =$$

$$= -\triangle^{-1} * \begin{pmatrix}\nabla\cdot\mathbf{E}\\ -\partial_0\mathbf{E} - \nabla\times\mathbf{B}\end{pmatrix}|_{\substack{\partial_0\mathbf{B}=-\nabla\times\mathbf{E}\\ \nabla\cdot\mathbf{B}=0}}$$

In similar way one can find the extension $E_\mathcal{H}^* = -\triangle^{-1} * \text{rot}^T(-\partial)|_{\mathcal{F}_\phi}$.

One can explicitly verify that $h\in\mathcal{H}$ iff $\Pi h = h$ and also kerΠ =ker$E_\mathcal{H}^*$. We have $\mathcal{F}_\phi = \mathcal{H}\oplus_t ker\Pi$.

Results analogous to Thm 3.1 can be formulated and proven, thus new relativistic and local models of self-interacting electrodynamics obeying additionally the spectrum condition (in the weak form) and HSSC is constructed. The models presented clearly deserve more studies but this will be presented elsewhere.

4. Appendix: The variety of $Cov(\tau,\sigma)$ for the group $SO(4)$

Let $SU(2)_1$ be the subgroup of the group Spin(4) generated by rotations in (i,j)- planes, $i,j\in\{1,2,3\}$. For any irreducible unitary representation τ of the group Spin(4) let $\tau|_1 \equiv \tau|_{SU(2)_1}$ and let $\tau|_1 = \oplus_{j\in\sigma_1(\tau)}\mathbf{D}_j$ be the spectral decomposition of $\tau|_1$ where $\sigma_1(\tau)$ is the spectrum $\tau|_1$. We define $|l_0| \equiv$min$\{j\in\sigma_1(\tau)\}$ and $|l_1| \equiv$max$\{j\in\sigma_1(\tau)\}+1$. Then it is well known that l_0 and l_1 must be either integer or half-integer. $\sigma_1(\tau) = \{|l_0|,...,|l_1|-1\}$ and moreover the multiplicity of $\mathbf{D}_j$ with $j\in\sigma_1(\tau)$ is exactly equal to the one; the pair (l_0,l_1) uniquely determines the representation τ if by convention $(l_0,l_1)\cong(-l_0,-l_1)$. In particular: $\dim(l_0,l_1) = l_1^2 - l_0^2$ and $\overline{\tau} = \overline{(l_0,l_1)} = (l_0,-l_1)$ where the overline means the complex conjugation.

PROPOSITION 4.1. *Let* $\tau = \bigoplus_{\alpha=1}^P(l_0^\alpha,l_1^\alpha)$, $\sigma = \bigoplus_{\alpha'=1}^Q(l_0^{\alpha'},l_1^{\alpha'})$ *be unitary representations of the group Spin*(4) *in Aut*$(\mathbf{C}^{dim\tau})$*, respectively, in Aut*$(\mathbf{C}^{dim\sigma})$*. For* $\mathcal{D}\in Cov(\tau,\sigma)$ *we denote* $B_k^{\alpha\alpha'}$ *the corresponding blocks of the matrices* B_k *(written in the canonical bases) defining* $\mathcal{D}$*. Then:*

(1) $\mathcal{D}^{\alpha\alpha'} \neq 0$ *iff the following coupling rules are fulfilled:*

(4.4.1) $$(l_0^{\alpha'}, l_1^{\alpha'}) = (l_0^{\alpha} \pm 1, l_1^{\alpha})$$

or

(4.4.2) $$(l_0^{\alpha'}, l_1^{\alpha'}) = (l_0^{\alpha}, l_1^{\alpha} \pm 1);$$

(2) if the pair (α, α') *fulfils (4.4.1),(4.4.2) then for* $l \in \{|l_0^\alpha|, ..., |l_1^\alpha| - 1\}$, $l' \in \{|l_0^{\alpha'}|, ..., |l_1^{\alpha'}| - 1\}$ *we have*

(4.4.3) $$(B_0^{\alpha\alpha'})_{lm_l, l'm'_{l'}} = i b_l^{\alpha\alpha'} \delta_{ll'} \delta_{m_l m'_{l'}}$$

where:

(4.4.4) $$b_l^{\alpha\alpha'} = \begin{cases} c^{\alpha\alpha'} \sqrt{(l + l_0 + 1)(l - l_0)} & \text{if } (l_0^{\alpha'}, l_1^{\alpha'}) = (l_0^{\alpha} + 1, l_1^{\alpha}) \\ c^{\alpha\alpha'} \sqrt{(l + l_1 + 1)(l - l_1)} & \text{if } (l_0^{\alpha'}, l_1^{\alpha'}) = (l_0^{\alpha}, l_1^{\alpha} + 1) \end{cases}$$

(4.4.5) $$b_l^{\alpha'\alpha} = \begin{cases} c^{\alpha\alpha'} \sqrt{(l + l_0 + 1)(l - l_0)} & \text{if } (l_0^{\alpha'}, l_1^{\alpha'}) = (l_0^{\alpha} - 1, l_1^{\alpha}) \\ c^{\alpha\alpha'} \sqrt{(l + l_1 + 1)(l - l_1)} & \text{if } (l_0^{\alpha'}, l_1^{\alpha'}) = (l_0^{\alpha}, l_1^{\alpha} - 1) \end{cases}$$

and where $b_l^{\alpha'\alpha}, b_l^{\alpha\alpha'} \in \mathbf{C}$ *are arbitrary;*

(3) if the coupling rules (4.4.1), (4.4.2) are fulfilled the nonvanishing matrix elements of $B_k^{\alpha\alpha'}$ *for* $k = 1, 2, 3$ *are listed below:*

(4.4.6)
$$(B_1)^{\alpha\alpha'}_{lm;l-1,m-1} = -i(B_2)^{\alpha\alpha'}_{lm;l-1,m-1} = \frac{i}{2}\sqrt{(l+m)(l+m-1)}[C_l^{\alpha} c_{l-1}^{\alpha\alpha'} - C_l^{\alpha'} c_l^{\alpha\alpha'}];$$

(4.4.7)
$$(B_1)^{\alpha\alpha'}_{lm;l-1,m+1} = i(B_2)^{\alpha\alpha'}_{lm;l-1,m+1} = -\frac{i}{2}\sqrt{(l-m)(l-m-1)}[C_l^{\alpha} c_{l-1}^{\alpha\alpha'} - C_l^{\alpha'} c_l^{\alpha\alpha'}];$$

(4.4.8) $$(B_1)^{\alpha\alpha'}_{lm;l,m-1} = -i(B_2)^{\alpha\alpha'}_{lm;l,m-1} = \frac{i}{2}\sqrt{(l+m)(l-m+1)} c_l^{\alpha\alpha'} [A_l^{\alpha} - A_l^{\alpha'}];$$

(4.4.9) $$(B_1)^{\alpha\alpha'}_{lm;l,m+1} = i(B_2)^{\alpha\alpha'}_{lm;l,m+1} = \frac{i}{2}\sqrt{(l-m)(l+m+1)} c_l^{\alpha\alpha'} [A_l^{\alpha} - A_l^{\alpha'}];$$

(4.4.10) $$\begin{aligned} &(B_1)^{\alpha\alpha'}_{lm;l+1,m-1} = -i(B_2)^{\alpha\alpha'}_{lm;l+1,m-1} = \\ &= \frac{i}{2}\sqrt{(l-m+1)(l-m+2)}[C_{l+1}^{\alpha} c_{l+1}^{\alpha\alpha'} - C_{l+1}^{\alpha'} c_l^{\alpha\alpha'}]; \end{aligned}$$

(4.4.11) $$\begin{aligned} &(B_1)^{\alpha\alpha'}_{lm;l+1,m+1} = i(B_2)^{\alpha\alpha'}_{lm;l+1,m+1} = \\ &= -\frac{i}{2}\sqrt{(l+m+2)(l+m+2)}[C_{l+1}^{\alpha} c_{l+1}^{\alpha\alpha'} - C_{l+1}^{\alpha'} c_l^{\alpha\alpha'}]; \end{aligned}$$

(4.4.12) $$(B_3)^{\alpha\alpha'}_{lm;l-1,m} = i\sqrt{l^2 - m^2}[C_l^{\alpha} c_{l-1}^{\alpha\alpha'} - C_l^{\alpha'} c_l^{\alpha\alpha'}];$$

(4.4.13) $$(B_3)^{\alpha\alpha'}_{lm;lm} = -imc_l^{\alpha\alpha'} [A_l^{\alpha} - A_l^{\alpha'}];$$

(4.4.14) $$(B_3)^{\alpha\alpha'}_{lm;l+1,m} = -i\sqrt{(l+1)^2 - m^2}[C_{l+1}^{\alpha} c_{l+1}^{\alpha\alpha'} - C_{l+1}^{\alpha'} c_l^{\alpha\alpha'}];$$

the numbers A_l^α and C_l^α are given by

$$A_l^\alpha = \frac{il_0l_1}{l(l+1)}, \quad C_l^\alpha = \frac{i}{l}\sqrt{\frac{(l^2-l_0^2)(l^2-l_1^2)}{4l^2-1}} \tag{4.4.15}$$

for $l \in \{|l_0|, ..., |l_1|-1\}$ if $\alpha = (l_0, l_1)$.

References

[1] S.Albeverio, R.Høegh-Krohn: *Euclidean Markov Fields and Relativistic Quantum Fields from Stochastic Partial Differential Equations in Four Dimensions*; Phys. Lett. B 177 (1986), 175–179

[2] S.Albeverio, H.Holden, R.Høegh-Krohn, T.Kolsrud: *Representation and Construction of Multiplicative Noise*; J. Funct. Anal. 87 (1989), 250–272

[3] S.Albeverio, K.Iwata, T.Kolsrud: *Random Fields as Solutions of the Inhomogeneous Quaternionic Cauchy-Riemann Equation I. Invariance and Analytic Continuation*; Commun. Math. Phys. 132 (1990), 555–580

[4] S. Albeverio, H. Gottschalk, J-L. Wu: *Convoluted Generalized White Noise, Schwinger Functions and Their Analytic Continuation to Wightman Functions*; Rev. Math. Phys **8**,763 - 817 (1996)

[5] S. Albeverio, J-L. Wu: *Models of Local Relativistic Quantum Fields with Indefinite Metric (in All Dimensions)*; Comm. Math. Phys. 184, 509-531 (1997)

[6] S. Albeverio, H. Gottschalk, J-L. Wu: *Nontrivial scattering amplitudes for some local relativistic quantum field models with indefinite metric* Phys. Lett. **B** 405 (1997), 243-248

[7] C. Becker, R. Gielerak, P. Ługiewicz: *Covariant SPDEs and Quantum Field Structures*; J. Phys. A: Math.Gen. **31** 231 - 258 (1998)

[8] C. Becker: *Wilson Loops in Two-Dimensional Space-Time Regarded as Generalized White Noise*; J. Funct. Analysis **134**, 321 - 349 (1995)

[9] R. Gielerak, P Ługiewicz: *4D Quantum Field Theory Models from Covariant Stochastic Partial Differential Equations I.Generalities* to appear in Rev. Math. Phys.;

[10] R. Gielerak, P. Ługiewicz: *From Stochastic Differential Equations to Quantum Field Theory*; Rep. Math. Phys. vol. **44** (1999) 1-11

[11] J.Glimm, A.Jaffe: *Quantum Physics: A Functional Integral Point of View*; Second Ed. Springer 1987

[12] R. Jost: *The General Theory of Quantized Fields*; Lectures in Applied Mathematics, American Mathematical Society, Providence, Rhode Island 1965

[13] G. Morchio, F. Strocchi: *Infrared Singularities, Vacuum Structure and Pure Phases in Local Quantum Field Theory*; Ann. Inst. H. Poincare **A 33**, 251-282 (1980)

[14] E. Nelson: The Construction of Quantum Field Theory from Markov Fields; J.Funct.Anal. **11** (1973) 97-112

[15] E. Nelson: The Free Markov Field; J.Funct.Anal. **12** (1973) 211-217

[16] E.P. Osipov: Octavic Markov cosurfaces and relativistic quantum fields in eight-dimensional space; Preprint PPH No **19**(161) Nowosybirsk (1988)

[17] K. Osterwalder, R. Schrader: *Axioms for Euclidean Green's functions I*; Comm.Math.Phys. **31** (1973), 83-112;
Axioms for Euclidean Green's functions II; Comm.Math.Phys. **42** (1975), 281-305

[18] G. Da Prato, J. Zabczyk : *Stochastic equations in infinite dimensions*; Cambridge Univ. Press 1992

Current address: Technical University of Zielona Gora, Dept. of Mechanics., 65-246 Zielona Gora, Poland

E-mail address: `gielerak@ift.uni.wroc.pl`

Current address: University of Wrocław, Institute of Theoretical Physics, 50-204 Wrocław, Poland

E-mail address: `piol@ift.uni.wroc.pl`

Canadian Mathematical Society
Conference Proceedings
Volume **29**, 2000

Generalized Configuration Spaces for Quantum Systems

Gerald A. Goldin and Ugo Moschella

ABSTRACT. We consider some spaces of infinite point configurations that are not necessarily locally finite. Making use of the σ-field of cylinder sets in spaces of infinite sequences, we demonstrate the measurability of sets characterized by a fixed number N of accumulation points. Unitary representations of diffeomorphism groups, describing quantum systems having infinitely many degrees of freedom, may be constructed on such generalized configuration spaces by projecting from infinite sequence spaces.

1. Introduction

Let X be a smooth manifold, equipped with the σ-field of Borel sets generated by its topology. For now we assume that X has all the "nice" topological properties needed. In general, a *configuration* is just a Borel subset of X (without any ordering or parameterization being specified). Spaces of configurations, both pointlike and extended, have been considered in the literature. In statistical physics, it is common to work with infinite but locally finite point configurations, which describe a wide variety of classical and quantum systems. Configurations that are embeddings of lower-dimensional manifolds in X, i.e. extended objects with some smoothness properties, occur in the classical and quantum dynamics of fluids, strings, membranes, and so forth.

We think it is interesting, and important, to explore situations for which the appropriate configurations are pointlike but not necessarily locally finite. The points in such configurations can cluster, allowing for fractals or even more complicated objects. Furthermore this permits pointlike approximations to embedded manifolds. We are also motivated in this study by the investigation of unitary representations of *Diff*(X), the group of compactly-supported diffeomorphisms of X under composition. *Diff*(X) acts naturally not only on X, but on configurations in X. When a configuration space Γ is endowed with a measure μ having the technical property of *quasi-invariance* under the action of diffeomorphisms [this means that the class of measure zero sets is, as a class, invariant under *Diff*(X)], then there exist unitary representations of *Diff*(X) in the Hilbert space of μ-square-integrable functions on Γ. It has long been known that such representations describe a wide variety of nonrelativistic quantum systems [**1, 2, 3, 4**]; they are relevant also to quantum

1991 *Mathematics Subject Classification.* 28C20, 81R10.

gravity [**5**]. Major contributions to this subject in quantum statistical physics have been made recently for the locally finite point configuration spaces, suggesting to us the value of a still more general theory [**6, 7, 8**].

In the case $X = \mathbb{R}$, we have elsewhere constructed certain measures on spaces of *sequences* of elements of X, quasi-invariant under diffeomorphisms [**9, 10, 11**]. Our construction makes use of self-similar random processes. The sequences can have accumulation points, and their self-similarity is an important feature in obtaining quasi-invariance of the measure. It is natural, then, to ask whether we can project from sequence spaces (ordered) to configuration spaces (unordered) in a way that respects the action of diffeomorphisms, preserves the quasi-invariance of measures, and allows the number of accumulation points in configuration space to be measured. An affirmative answer lets us use the whole machinery developed in standard probability theory and the theory of stochastic processes to generate interesting quasi-invariant measures on our new, enlarged configuration spaces. A different class of quasi-invariant measures on configurations with one accumulation point is due to Ismagilov [**12**], and a further example of interest is the relation between quasi-invariant measures on indexed and nonindexed random Cantor sets [**13, 14, 15**].

In this paper we propose definitions and address some preliminary questions concerning measurability for configuration spaces with accumulation points. The main, encouraging result is the measurability of sets having precisely N accumulation points with respect to the natural σ-field. Thus, in an important sense, it is possible to consider the accumulation points as collective coordinates that partially describe the classical or quantum statistical physics.

In Section 2, we define some general sequence and configuration spaces. Section 3 addresses measures on these spaces, and the issue of quasi-invariance under *Diff*(X). In Section 4, we demonstrate the measurability of various interesting subsets of our sequence spaces, and indicate how the whole construction for sequence spaces projects to the enlarged configuration spaces.

2. Sequence and configuration spaces

Let X be a C^∞-manifold, which we take to be connected, locally compact, σ-compact, Hausdorff and second countable. Let X^∞ be the space of all infinite sequences $\omega = (\mathbf{x}_1, \mathbf{x}_2, \dots)$ of elements of X, which we identify with the infinite Cartesian product

$$X^\infty = \prod_{i=1}^{\infty} X_i\,, \quad X_i = X \ \ (\forall\, i). \tag{1}$$

We write $(\mathbf{x}_1(\omega), \mathbf{x}_2(\omega), \dots)$ when we intend to emphasize the dependence of the jth entry on ω. The space X^∞ can be endowed with the *weak topology,* defined to be the weakest topology for which all the natural projections (or coordinate functions) $\pi_j : X^\infty \to X$ given by $\pi_j(\omega) = \mathbf{x}_j(\omega)$ are continuous. Let us also introduce the space of n-tuples of elements of X,

$$X^n = \prod_{i=1}^{n} X_i, \quad X_i = X \ \ (\forall\, i). \tag{2}$$

By convention $X^{n=0}$ contains one element, the empty sequence ( ). Now we define the set

$$S(X) = X^F \cup X^\infty = \left(\bigcup_{n=0}^{\infty} X^n \right) \cup X^\infty \tag{3}$$

whose elements are finite or infinite sequences $(\mathbf{x}_i)$; here the superscript F stands for "finite."

We shall introduce a cup above the letter denoting each set, to indicate the restriction that no two points in a (finite or infinite) sequence coincide—i.e., to remove the "diagonal" from the space. Thus

$$\breve{X}^F = \{\omega \in X^F : \mathbf{x}_j(\omega) \neq \mathbf{x}_k(\omega) \text{ for all } j \neq k\};$$

$$\breve{X}^\infty = \{\omega \in X^\infty : \mathbf{x}_j(\omega) \neq \mathbf{x}_k(\omega) \text{ for all } j \neq k\}; \text{ and} \tag{4}$$

$$\breve{S}(X) = \breve{X}^F \cup \breve{X}^\infty.$$

A point $\mathbf{x} \in X$ is an accumulation point of the infinite sequence $(\mathbf{x}_i)$ if for any neighborhood $\mathcal{U}$ of $\mathbf{x}$, the set $\mathcal{U} \setminus \{\mathbf{x}\}$ contains infinitely many points of $(\mathbf{x}_i)$. Evidently the set of accumulation points of the infinite sequence $(\mathbf{x}_i)$ may be finite, countably infinite, or uncountably infinite. For $N = 0, 1, 2, \ldots$, let us define

$$X_N^\infty = \{\omega \in X^\infty : (\mathbf{x}_i(\omega)) \text{ has exactly } N \text{ accumulation points in } X\}, \tag{5}$$

$$\breve{X}_N^\infty = \breve{X}^\infty \cap X_N^\infty .$$

In particular $\breve{X}_0^\infty$ is the set of locally finite sequences without repeated entries, which was denoted by $\tilde{X}^\infty$ in Ref. [**2**].

We now introduce some configuration spaces over the manifold X. The largest configuration space to be considered here is

$$\Sigma_X = \{\gamma \subset X : \gamma \text{ is finite or countably infinite}\}. \tag{6}$$

For $\gamma \in \Sigma_X$, we shall write $|\gamma| \in \mathbb{N}_{\text{ext}} = \{0\} \cup \mathbb{N} \cup \{\aleph_0\}$, where $|\gamma|$ denotes the cardinality of γ.

Let p be the map which assigns to a given sequence $(\mathbf{x}_i) \in S(X)$ the corresponding (unordered) set of points $\{\mathbf{x}_i\}$; we write $p(\omega) = \gamma$. The image of X^∞ under p is all of Σ_X; the possibility of repeated entries in elements of X^∞ permits the corresponding configurations to be finite or infinite. For this reason it is often natural to consider the restriction of p to sequences without repeated elements. Thus the image of $\breve{S}(X)$ under p is Σ_X. We define the following useful subsets of Σ_X:

$$\Gamma_X = \{\gamma \in \Sigma_X : \gamma \cap K \text{ is finite for every compact } K \subset X\}; \tag{7}$$

$$\Gamma_X^{(n)} = \{\gamma \in \Gamma_X : |\gamma| = n\}, \tag{8}$$

$$\Gamma_X^F = \bigcup_{n=0}^{\infty} \Gamma_X^{(n)};$$

$$\Gamma_X^{(\infty)} = \{\gamma \in \Gamma_X : |\gamma| = \aleph_0\}. \tag{9}$$

Then $\Gamma_X = \Gamma_X^{(n)} \cup \Gamma_X^{(\infty)}$. We have $p(X^F) = \Gamma_X^F$, and $p(\check{X}_0^\infty) = \Gamma_X^{(\infty)}$. Finally, define

$$\Sigma_X^{(\infty)} = \{\gamma \in \Sigma_X : |\gamma| = \aleph_0\}, \tag{10}$$

so that $p(\check{X}^\infty) = \Sigma_X^{(\infty)}$. Then $\Gamma_X^{(\infty)} \subset \Sigma_X^{(\infty)}$, and $\Sigma_X = \Gamma_X^{(n)} \cup \Sigma_X^{(\infty)}$. Note that accumulation points in X depend on the set $\{\mathbf{x}_i\}$, and not actually on the sequence $(\mathbf{x}_i)$; any sequence in $p^{-1}(\gamma)$ has the same accumulation points. Thus we can characterize the subsets $\Sigma_{X,N}^{(\infty)}$ of $\Sigma_X^{(\infty)}$, consisting of configurations having exactly N accumulation points in X, by the condition $p(X_N^\infty) = \Sigma_{X,N}^{(\infty)}$.

For $\phi \in \mathit{Diff}(X)$, the natural action on $S(X)$ is given by $\phi[(\mathbf{x}_i)] = (\phi(\mathbf{x}_i))$. The natural action on Σ_X, $\phi[\{\mathbf{x}_j\}] = \{\phi(\mathbf{x}_i)\}$, satisfies $\phi[p(\mathbf{x}_i)] = p(\phi[(\mathbf{x}_i)])$. Supposing, then, that we have a measure on $S(X)$—or some subset thereof—quasi-invariant under diffeomorphisms of X (see below). Such a measure is associated with a quantum theory of distinguishable particles, possibly having accumulation points. Our goal is to be able to project this measure to Σ_X—or a corresponding subset thereof—ensuring that sufficiently many sets of configurations are measurable, and preserving the property of quasi-invariance. Then we obtain the related quantum theory of indistinguishable particles.

Let $\mathcal{D}(X)$ be the additive group of real-valued C^∞ functions on X having compact support. For $f_1, f_2 \in \mathcal{D}(X)$ and $\phi_1, \phi_2 \in \mathit{Diff}(X)$, the group law $(f_1, \phi_1) \cdot (f_2, \phi_2) = (f_1 + f_2 \circ \phi_1, \phi_2 \circ \phi_1)$ defines the natural semidirect product of $\mathcal{D}(X)$ with $\mathit{Diff}(X)$. Unitary representations of this semidirect product group are also of interest in quantum mechanics. We remark that for each configuration $\gamma \in \Gamma_X$, there exists a functional on $\mathcal{D}(X)$ given by the sum of evaluation functionals,

$$< \gamma, f > = \sum_{\mathbf{x}_i \in \gamma} f(\mathbf{x}_i). \tag{11}$$

For this reason, measures on Γ_X quasi-invariant under $\mathit{Diff}(X)$ also permit construction of unitary representations of the semidirect product. However the functional (11) is not defined for general elements of Σ_X.

3. Measures on sequence and configuration spaces

We next look at the construction of cylinder set measures on X^∞, with an eye toward some particular measures that are supported by spaces of sequences with accumulation points in X. Our goal is to be able to project these measures to $\Sigma_X^{(\infty)}$. Denote by $\mathcal{O}(X)$ the family of open subsets of X, and by $\mathcal{B}(X)$ the corresponding family of Borel subsets of X. Let $\mathcal{O}_c(X)$ and $\mathcal{B}_c(X)$ be the respective subfamilies of sets that have compact closure in X. Define $\mathcal{O}(X^n)$, $\mathcal{B}(X^n)$, etc., analogously. Let $\mathcal{B}_0(X^\infty)$ denote the field of cylinder sets in X^∞, of the form

$$\{\omega \in X^\infty : (\mathbf{x}_{j_1}(\omega), \ldots, \mathbf{x}_{j_n}(\omega)) \in H\} \tag{12}$$

for $H \in \mathcal{B}(X^n)$. The σ-field generated by $\mathcal{B}_0(X^\infty)$ is denoted by $\mathcal{B}(X^\infty)$. It is equivalently generated by rank one cylinder sets, of the form $\{\omega \in X^\infty : \mathbf{x}_j(\omega) \in B\}$, for $B \in \mathcal{B}(X)$. Together the $\mathcal{B}(X^n)$, $n = 1, 2, 3, \ldots$, and $\mathcal{B}(X^\infty)$ give us a σ-field $\mathcal{B}[S(X)]$ on $S(X)$.

The most straightforward way to construct a countably additive measure μ on X^∞ [with the σ-field $\mathcal{B}(X^\infty)$] is to specify a compatible family of measures $\mu_{j_1, \ldots, j_n}$ on finite-dimensional spaces, by providing compatible values $\mu_{j_1, \ldots, j_n}(H)$ on the cylinders sets H. The existence of the corresponding measure μ is then assured by Kolmogorov's theorem. Thus let the locally Euclidean manifold X be equipped with a (local) Lebesgue measure ν, quasi-invariant under diffeomorphisms. For each natural number k write a conditional (Borel) probability measure μ_k on X, depending measurably on the $k-1$ real parameters $\mathbf{x}_1, \ldots, \mathbf{x}_{k-1}$ (the first $k-1$ entries in a sequence ω), and absolutely continuous with respect to ν:

$$d\mu_k(\mathbf{x}_k) = f_k(\mathbf{x}_k \,|\, \mathbf{x}_1, \ldots, \mathbf{x}_{k-1})\, d\nu(\mathbf{x}_k). \tag{13}$$

The f_k are conditional probability density functions. The joint probability measure for the first j entries is then given by

$$\mu^{(j)} = \prod_{k=1}^{j} \mu_k. \tag{14}$$

The $\mu^{(j)}, j \in \mathbb{N}$, provide a compatible family of probability measures, so there is a unique measure μ on X^∞ associated with them. Under the action of $\mathit{Diff}(X)$ on X^j, the Radon-Nikodym derivative of the transformed measure $\mu_\phi^{(j)}$ with respect to $\mu^{(j)}$ is given (where it exists) by

$$\frac{d\mu_\phi^{(j)}}{d\mu^{(j)}} = \prod_{k=1}^{j} \frac{d\mu_{k,\phi}}{d\mu_k} \tag{15}$$

with

$$\frac{d\mu_{k,\phi}}{d\mu_k}(\omega) = u_{k,\phi}(\omega) = \frac{f_k(\phi(\mathbf{x}_k) \,|\, \phi(\mathbf{x}_1), \ldots, \phi(\mathbf{x}_{k-1}))}{f_k(\mathbf{x}_k \,|\, \mathbf{x}_1, \ldots, \mathbf{x}_{k-1})} \mathcal{J}_\phi(\mathbf{x}_k), \tag{16}$$

where $\mathcal{J}_\phi(\mathbf{x}_k) = [d\nu_\phi/d\nu](\mathbf{x}_k)$. The quasi-invariance of $\mu^{(j)}$ is assured if (15) is almost everywhere defined and positive. Supposing that to be so, quasi-invariance of μ further depends on the behavior of the infinite product

$$\frac{d\mu_\phi}{d\mu}(\omega) = \prod_{k=1}^{\infty} u_{k,\phi}(\omega). \tag{17}$$

In general, a measure constructed this way will not be quasi-invariant under diffeomorphisms; some special properties are needed. An idea that leads to a whole class of quasi-invariant measures is to let the probability distribution of the jth point *scale* according to the outcomes of previous choices, establishing a self-similar random process. Our earlier results include interesting measures of this sort, where sets of sequences having one or more accumulation points are of positive measure.

To conclude this section, we introduce the needed σ-field of subsets of Σ_X, making use of the projection map p. Define $\mathcal{B}_{\Sigma(X)}$ to be the largest σ-field for which p is a measurable function with respect to the σ-field $\mathcal{B}[S(X)]$. The results of the next section then ensure that the various subsets of Σ_X introduced in Section 2 belong to $\mathcal{B}_{\Sigma(X)}$. In fact, $\mathcal{B}_{\Sigma(X)}$ is sufficiently rich to permit us to count the numbers of accumulation points of configurations that are located in Borel sets of X. Furthermore, given a probability measure μ on X^∞, we obtain a probability measure $\hat{\mu}$ on Σ_X by defining, for all $A \in \mathcal{B}_{\Sigma(X)}$, $\hat{\mu}(A) = \mu(p^{-1}(A))$. Supposing μ to be quasi-invariant under $\mathit{Diff}(X)$, our construction ensures that $\hat{\mu}$ is also quasi-invariant as desired.

4. Measurable subsets of $S(X)$ and Σ_X with accumulation points

Here we demonstrate a series of lemmas showing that various interesting subsets of $S(X)$ are measurable. To simplify the presentation, we spell out the proofs for the relevant subsets of X^∞; the extensions to the corresponding subsets of $S(X)$ are obvious. In each case we display a measurable subset of X^∞ as obtained from cylinder sets by countable processes.

LEMMA 1. *The set $\check{X}^\infty$ is a measurable subset of X^∞.*

Let the diagonal $D_n = \{(\mathbf{x}_1, \ldots, \mathbf{x}_n) : \mathbf{x}_i = \mathbf{x}_j \text{ for some } i \neq j\}$. Then $D_n \subset X^n$ is Borel. Define $A_n = \{\omega \in X^\infty : (\mathbf{x}_1(\omega), \ldots, \mathbf{x}_n(\omega)) \in X^n \setminus D_n\}$. Since $\check{X}^\infty = \bigcap_{n=1}^\infty A_n$, it follows that $\check{X}^\infty$ is measurable. ♮

LEMMA 2. *Let $K \subset X$ be compact. Then the set of nonrepeating sequences having exactly n entries in K, defined by $\check{X}^\infty_{(n)}(K) = \{\omega \in \check{X}^\infty : |p(\omega) \cap K| = n\}$, is measurable.*

Note the presence of parentheses in the subscript of $\check{X}^\infty_{(n)}(K)$, indicating that n labels the number of actual sequence entries in K, not the number of accumulation points. First we observe that the set of nonrepeating sequences having no point in K is measurable, since $\check{X}^\infty_{(n=0)}(K) = \bigcap_{n=1}^\infty \{\omega \in \check{X}^\infty : \mathbf{x}_n(\omega) \in X \setminus K\}$, and $X \setminus K$ is Borel. For $n > 0$, let σ be an n-element subset of $\mathbb{N}$. Note that there are countably many such subsets. For $j \in \mathbb{N}$ such that $j \notin \sigma$, define $B_{\sigma,j}(K) = \{\omega \in \check{X}^\infty : \mathbf{x}_i(\omega) \in K \text{ for } i \in \sigma;\ \mathbf{x}_j(\omega) \in X \setminus K\}$. Then

$$\check{X}^\infty_{(n)}(K) = \bigcup_\sigma B_\sigma(K) = \bigcup_\sigma \left(\bigcap_{j \notin \sigma} B_{\sigma,j}(K) \right), \tag{18}$$

where σ ranges over all n-element subsets of $\mathbb{N}$; therefore $\check{X}^\infty_{(n)}(K)$ is measurable. ♮

LEMMA 3. *The set of all locally finite sequences $\check{X}^\infty_{N=0}$ is measurable.*

We have $\check{X}^\infty_0 = \{\omega \in \check{X}^\infty : |p(\omega) \cap K| < \infty, \text{ for all compact } K\}$. Since X is σ-compact we can choose a sequence of compact sets K_j, such that $K_j \subset K_{j+1}$, and $X = \bigcup_{j \in \mathbb{N}} K_j$. Then we can write $\check{X}^\infty_0 = \bigcap_{j=1}^\infty \left(\bigcup_{n=0}^\infty \check{X}^\infty_{(n)}(K_j) \right)$, whence $\check{X}^\infty_0$ is measurable. ♮

Now let $\mathcal{A}_N$ be the set of all nonrepeating sequences ω such that $p(\omega)$ has at least N accumulation points, and $\mathcal{A}_N(K)$ be the set of all nonrepeating sequences ω for which $p(\omega) \cap K$ contains at least N accumulation points of $p(\omega)$.

LEMMA 4. *The sets $\mathcal{A}_N(K)$ and $\mathcal{A}_N$ are measurable.*

First we demonstrate this for $N = 1$. Since $\mathcal{A}_1 = \check{X}^\infty \setminus \check{X}^\infty_0$, $\mathcal{A}_1$ is measurable. However, letting $\check{X}^\infty_{N=0}(K)$ be the set of nonrepeating sequences having no accumulation points in K, $\check{X}^\infty_{N=0}(K)$ is *not* simply equal to $\bigcup_{n=0}^\infty \check{X}^\infty_{(n)}(K)$. Indeed, a sequence ω of points in X may have accumulation points in a compact set K even if only finitely many entries belong to K. Thus we need a somewhat more sophisticated argument.

Let $\{\mathcal{U}_j : j \in \mathbb{N}\} \subset \mathcal{O}$ be a countable basis for the topology of X. Let $\bar{\mathcal{U}}_j$ denote the closure of $\mathcal{U}_j$. Now $\mathbf{x} \in K$ is an accumulation point of ω if and only if $|\,p(\omega) \cap \bar{\mathcal{U}}_j| = \aleph_0$ for all $\mathcal{U}_j$ containing $\mathbf{x}$. Thus ω has no accumulation points in K if and only if there exists a covering of K in the basis, written $\{\mathcal{U}_j : j \in \mathcal{I} \subset \mathbb{N}\}$, such that $|\,p(\omega) \cap \bar{\mathcal{U}}_j|$ is finite for all $j \in \mathcal{I}$. Since K is compact, such a covering has a finite subcovering with the same property. Thus

$$\breve{X}^{\infty}_{N=0}(K) = \bigcup_{\sigma} \left(\bigcap_{j \in \sigma} \left(\bigcup_{n=0}^{\infty} \breve{X}^{\infty}_{(n)}(\bar{\mathcal{U}}_j) \right) \right), \tag{19}$$

where σ ranges over all finite subsets of $\mathbb{N}$ such that $\{\mathcal{U}_j : j \in \sigma\}$ is a covering of K. Since there are countably many finite subsets of $\mathbb{N}$, $\breve{X}^{\infty}_{N=0}(K)$ is measurable. And as $\mathcal{A}_1(K) = \breve{X}^{\infty} \setminus \breve{X}^{\infty}_{N=0}(K)$, we now have that $\mathcal{A}_1(K)$ is measurable.

To complete the proof of the lemma, fix N and let τ range over the N-point subsets of $\mathbb{N}$ such that $\bar{\mathcal{U}}_j \cap \bar{\mathcal{U}}_k = \emptyset$ for $j, k \in \tau$. Then the set

$$\bigcap_{j \in \tau} \mathcal{A}_1(K \cap \bar{U}_j), \tag{20}$$

consisting of all the sequences with at least one accumulation point in each of the mutually disjoint $\bar{\mathcal{U}}_j$, $j \in \tau$, is measurable. This implies that

$$\mathcal{A}_N(K) = \bigcup_{\tau} \left(\bigcap_{j \in \tau} \mathcal{A}_1(K \cap \bar{O}_j) \right) \tag{21}$$

is also measurable. ♮

We can now state the main result of this section:

Proposition 1. *The set $\breve{X}^{\infty}_{N}(K)$ defined in Eq. (5) is measurable.*

The result follows directly, since $\breve{X}^{\infty}_{N}(K) = \mathcal{A}_N(K) \setminus \mathcal{A}_{N+1}(K)$. ♮

Having proved these results about the measurability of various sets of sequences, we can remark on the corresponding sets of configurations. The number of points from a nonrepeating sequence $\omega \in \breve{S}(X)$ that belong to some compact set $K \in X$ is determined by the configuration $\gamma = p(\omega)$, not ω itself. Likewise, we recall our earlier comment that $\mathbf{x} \in X$ is an accumulation point of a configuration γ if and only if it is an accumulation point of every sequence ω such that $p(\omega) = \gamma$. Thus, the sets of sequences we have discussed are actually the inverse images of corresponding sets of configurations, which are seen to belong to the σ-field $\mathcal{B}_{\Sigma(X)}$.

5. Conclusion

To sum up, suppose that we have constructed a probability measure μ on the sequence space $S(X)$, quasi-invariant under diffeomorphisms of X, for which (roughly speaking) sets of sequences with some distribution of accumulation points have positive measure, but the diagonal has zero measure. This is the situation for certain unitary representations of *Diff* (X), describing some quantum systems with infinitely many degrees of freedom. Then we have the machinery to obtain from μ a quasi-invariant measure $\hat{\mu}$, on the configuration space Σ_X. This means we can obtain from quantum theories of infinitely many labeled particles the corresponding theories of indistinguishable particles, described by means of diffeomorphism group representations.

Acknowledgements

We take pleasure in dedicating this paper to Sergio Albeverio. G.G. would like to acknowledge interesting discussions with Y. Kondratiev, T. Kuna, and M. Röckner, as well as S. Albeverio, and hospitality and support from the Università dell'Insubria in Como, the Universities of Bonn and of Bielefeld, and the Technical University of Clausthal. He thanks also the Alexander von Humboldt Foundation for its support.

References

[1] G. A. Goldin, J. Grodnik, R. T. Powers, and D. H. Sharp, *J. Math. Phys.* **21**, 88-100 (1974)

[2] A. M. Vershik, I. M. Gelfand, and M. I. Graev, *Dokl. Akad. Nauk. SSSR* **232**, 745 (1977).

[3] G. A. Goldin, R. Menikoff, and D. H. Sharp, *Phys. Rev. Lett.* **51**, 2246-2249 (1983).

[4] G. A. Goldin, "Diffeomorphism groups, semidirect products, and quantum theory," in *Fluids and Plasmas: Geometry and Dynamics, Contemp. Math.***28**, ed. by J. E. Marsden, Am. Math. Soc., Providence, RI, pp. 189-207 (1984).

[5] C. J. Isham, "Quantum geometry," in *Quantum Theory Of Gravity,* ed. by S. M. Christensen, Hilger, Bristol, UK (1984).

[6] S. Albeverio, Y. G. Kondratiev, M. Röckner, *J. Funct. An.* **154**, 444-500 (1998).

[7] S. Albeverio, Y. G. Kondratiev, M. Röckner, *J. Funct. An.* **157**, 242-291 (1998).

[8] S. Albeverio, Y. G. Kondratiev, M. Röckner, "Diffeomorphism groups and current algebras: Configuration space analysis in quantum theory," *Rev. Math. Phys.* (1998, in press).

[9] G. A. Goldin and U. Moschella, "Diffeomorphism groups, quasi-invariant measures, and infinite quantum systems," in *Symmetries in Science VIII,* ed. by B. Gruber, Plenum, New York, pp. 159-172 (1995).

[10] G. A. Goldin and U. Moschella, *J. Phys. A: Math. Gen.* **28**, L475-L481, (1995).

[11] G. A. Goldin and U. Moschella, "Diffeomorphism group representations and quantum phase transitions in one dimension," in *XIth International Congress on Mathematical Physics,* ed. by D. Iagolnitzer, International Press, Cambridge, MA, pp. 445-453 (1995).

[12] R. S. Ismagilov, *Functional. Anal. i Prilozhen* **5**, 45-53 (1971); also *Representations of Infinite Dimensional Groups,* AMS Translations of Math. Monographs **152** (Providence, RI: American Math. Society, 1996).

[13] G. A. Goldin, "Unitary representation of diffeomorphism groups from random fractal configurations," in *Lie Theory and its Applications,* ed. by H. D. Doebner, V. Dobrev and J. Hilgert, World Scientific, Singapore, pp. 82-88 (1996).

[14] Y. A. Neretin, *Sbornik: Mathematics* **187**, 857-868 (1996).

[15] G. A. Goldin and U. Moschella, "Random Cantor sets and measures quasi-invariant for diffeomorphism groups," *Rep. Math. Phys.,* to be published (1999).

DEPARTMENTS OF MATHEMATICS AND PHYSICS, RUTGERS - THE STATE UNIVERSITY OF NEW JERSEY, NEW BRUNSWICK, NJ 08903 USA
E-mail address: gagoldin@dimacs.rutgers.edu

DIPARTIMENTO DI SCIENZE CHIMICHE, FISICHE E MATEMATICHE, UNIVERSITÀ DELL'INSUBRIA, VIA LUCINI 3, 22100 COMO, ITALIA
E-mail address: moschella@fis.unico.it

Canadian Mathematical Society
Conference Proceedings
Volume **29**, 2000

A Composition Adjoint Lemma

Karl Gustafson

Dedicated to Sergio Albeverio on the occasion of his 60th birthday.

ABSTRACT. An interesting Lemma of Albeverio, Høegh–Krohn, and L. Streit, used in the regularization of singular forms, provides under certain conditions for self-adjoint operators A and B that $(AB)^* = B^*A^*$. Here I simplify and generalize that result.

1. Introduction

Sergio Albeverio is an expert in the use of quadratic forms, specifically Dirichlet forms, to treat singular stochastic and quantum processes, e.g., see [1, 2, 3]. When I visited ZIF in Bielefeld in 1980 (in connection with the conference of [4]), by chance while I was there I picked up the preprint of [1]. In this paper it is shown how to approximate and converge to such singular forms (e.g., representing processes with potentials more singular than distributions) by means of certain regularized forms. Suitably regularized, the convergence is then that of monotone form convergence or strong resolvent convergence, for these see e.g. [5].

While such approximation results are interesting of course for dealing with problems of singular physics, what caught my eye was a key lemma, namely

Lemma 2. [1] *Let A, B be self-adjoint operators in a Hilbert Space $\mathcal{H}$ and suppose that the domain $D(AB)$ is dense, that AB is closable, that $D(B)$ contains the range of A, that B^{-1} exists, and that one has $D(B^{-1}(AB)^*) \supset D(AB)^*)$. Then $(AB)^* = B^*A^*$.*

The deceptively clever proof of Lemma 2 [1] as given there went as follows. For densely defined operators A, B, and AB, one always has $(AB)^* \supset B^*A^*$, so it remains to show that $B^*A^* \supset (AB)^*$. "By the assumption that B^{-1} exists on a domain containing $D(A)$ we have

$$(AB)^{**}B^{-1} \supset A \qquad [1,(2.17)]$$

From this it follows easily, using the self-adjointness of B^{-1}, that

$$((AB)^{**}B^{-1})^* \subset A^* = A \qquad [1,(2.18)]$$

1991 *Mathematics Subject Classification.* 47A05, 47D45.

On the other hand

$$((AB)^{**}B^{-1})^* \supset B^{-1}(AB)^*; \qquad [1,(2.19)]$$

Hence, from (2.18),

$$Ax = B^{-1}(AB)^*x \qquad [1,(2.20)]$$

for all $x \in D(B^{-1}(AB)^*)$. Since by assumption the range of A is contained in $D(B)$, (2.20) yields

$$BAx = (AB)^*x \qquad [1,(2.21)]$$

for all $x \in D(B^{-1}(AB)^*)$; thus *a fortiori*, due to the assumption $D(B^{-1}(AB)^*) \supset D((AB)^*)$,

$$BA \supset (AB)^* \qquad [1,(2.22)]$$

But (2.22) yields then the Lemma."

At that time (1980) I jotted down a couple of minor improvements that one could make in the Lemma, but I never had any reason to spend further time on it. Here it seems appropriate to do so and in this note I shall simplify and generalize the Lemma and also give some accompanying commentary about the general question of when $(AB)^* = B^*A^*$.

2. A Simplification of the Lemma

First let me present a simplification of the Lemma. We need to show that $BA \supset (AB)^*$. We write

$$A = (AB)(B^{-1}). \qquad (1)$$

For this we need B^{-1} to exist on a domain containing $D(A)$, as was assumed in the proof above. Then $A^* \supset B^{-1}(AB)^*$. For this step we need to know that $D(AB)$ is dense. Thus $A \supset B^{-1}(AB)^*$ and applying B from the left, $BA \supset (AB)^*$. For this we need $D(B) \supset R(A)$. So we have shown the simplification

Lemma 1. *Let A and B be selfadjoint operators in a Hilbert space $\mathcal{H}$ and suppose that $D(AB)$ is dense, that $R(B) \supset D(A)$, and $D(B) \supset R(A)$. Then $(AB)^* = B^*A^*$.*

Note that the assumption $R(B) \supset D(A)$ is not explicitly stated in Lemma 2 [1] although it is used in the proof there and is hence tacitly assumed. Also it must be remembered (e.g., see [6] or [7], to which we will refer again below) that a dense range assumption for a selfadjoint operator B is equivalent to the existence of B^{-1}. Finally note that we have used the general operator composition associativity $(T_1T_2)T_3 = T_1(T_2T_3)$ rule in (1), e.g. see the discussions of such rules in [5, 8].

Is Lemma 1 more general than Lemma 2 [1]? As we just pointed out, the assumptions on $B^{-1}, R(B)$ and $D(B)$ vis a vis $D(A)$ and $R(A)$ are equivalent in the two versions. One need not explicitly assume AB to be closable because AB is closable iff $(AB)^*$ is densely defined, and the latter follows from $(AB)^* \supset B^*A^* = BA$, $D(BA)$ being dense because the assumption $D(B) \supset R(A)$ guarantees $D(BA) = D(A)$ dense. The assumption $D(B^{-1}(AB)^*) \supset D((AB)^*)$ is equivalent to $D(B^{-1}(AB)^*) = D((AB)^*)$ and implies that $R(B) \supset R((AB)^*) \supset R(BA)$. But because the other assumptions already imply the desired result $(AB)^* = BA$ and because one always has $R(BA) \subset R(B)$, this means that the assumption $D(B^{-1}(AB)^*) \supset D((AB)^*)$ is consistent but unnecessary.

Thus my Lemma 1 is equivalent to Lemma 2 [1] in content but carries much simplified assumptions whose effects are immediately clear from my proof. What really interests me about Lemma 1 is the power of domain and range assumptions when dealing with unbounded operators.

3. Earlier "Closed Range" Versions

Here I should mention that earlier [9] I had had some interests in the general question $(AB)^* = B^*A^*$, which is of course why Lemma 2 [1] caught my eye. So let me recall what is already known. To fix ideas I will just refer to these earlier results as "closed range" versions, as that was the essential ingredient in the thinking at that time. First however let us recall that it is well-known (e.g., [5, 8]) that when A is any bounded operator on $\mathcal{H}$, then $(AB)^* = B^*A^*$ for any operator B such that $D(B)$ is dense. Thus interest centers on the question for A unbounded.

For A and B Fredholm operators, we have $(AB)^* = B^*A^*$. See the early papers [10, 11]. Recall (e.g., [5]) that Fredholm operators are closed operators with closed range and finite index. In 1969–70 several papers independently appeared, all of which slightly generalized this result. In [9] I showed that one needs only that B be closed with $R(B)$ of finite codimension. Recall however that the finite codimension assumption implies that $R(B)$ was already closed. In [12] the same result as [9] was obtained assuming also that A be closed. In [13] these results were extended to Banach spaces. Namely, let B be a densely defined closed linear operator from X to Y, let A be a densely defined linear operator from Y to Z, X, Y, Z Banach spaces. If $R(B)$ is closed in Y and has finite codimension there, then $(AB)' = B'A'$, where $'$ denotes the dual (e.g., conjugate) operator. Some more "closed range" versions were given in [14]. Also some of the above known sufficient conditions for $(AB)^* = B^*A^*$ were shown in [14] to be sharp. For example, for B a closed densely defined operator in $\mathcal{H}$, if $(AB)^* = B^*A^*$ for all closed densely defined operators A in $\mathcal{H}$, then necessarily $R(B)$ is closed and of finite codimension.

Underlying these earlier "closed range" results is the fact that for abstract closed densely defined operators A and B in a Banach space, a finite codimension assumption on $R(B)$ guarantees not only that $R(B)$ be closed but also that $D(AB)$ is dense. Hence we may then speak abstractly of $(AB)^*$ and $(AB)'$,

Related to the question $(AB)^* = B^*A^*$, for Fredholm operators T we also have [15] that ascent $(T^*) = \text{descent}(T)$ and descent $(T^*) = \text{ascent}(T)$. For more details on the notions $\text{asc}(T)$ and $\text{des}(T)$ which are measures of the algebraic multiplicities of an operator and its adjoint, see [15] and the references therein.

On the other hand we may view the "domains" approach taken here as one in which we may know that $D(A)$, $D(B)$, and $D(AB)$ are already dense from the posing of the physical problem. For example the C_0^∞ functions are often dense in all domains and function spaces that we need consider. That fact underlies the whole theory of distributions and the earlier theory of weak solutions to partial differential equations [16]. In those settings we don't want to be restricted to Fredholm theory.

4. Some Generalizations of the Lemma

Here I do not want ultimate generality, rather I am interested in eventual utility to mathematical physics. Can we generalize Lemma 1 to A and B only assumed to be symmetric operators?

Lemma 2. *Let A and B be symmetric operators in a Hilbert space $\mathcal{H}$ and suppose that $D(AB)$ is dense, that $R(B) \supset D(A)$, and $D(B^*) \supset R(A^*)$. Then $(AB)^* = B^*A^*$.*

PROOF. The proof is a minor modification of my proof of Lemma 1. We know generally that $(AB)^* \supset B^*A^*$. We need to show $B^*A^* \supset (AB)^*$. We may write $A = (AB)(B^{-1})$ as in (1). Then

$$A^* = ((AB)(B^{-1}))^* \supset (B^{-1})^*(AB)^* = (B^*)^{-1}(AB)^*. \tag{2}$$

To obtain (2) we use the fact (e.g., see the state diagrams for symmetric operators in [6] or [7]) that for symmetric operators B, $R(B)$ dense guarantees that B^* is also 1–1. We also use the well-known fact [5, 8] that then $(B^{-1})^* = (B^*)^{-1}$. Given (2), from the assumption $D(B^*) \supset R(A^*)$ we may apply B^* on the left to conclude $B^*A^* \supset (AB)^*$. □

Let us note that although not explicitly stated therein, as in Lemma 1 the assumptions of Lemma 2 directly imply that AB is a closable operator. This may be verified as follows. $D(B^*) \supset R(A^*)$ implies $D(B^*A^*) = D(A^*)$ dense and hence $(AB)^* \supset B^*A^*$ is densely defined and hence AB was closable from the hypotheses. Similarly, from $(AB)^* \supset B^*A^* \supset BA$, both B^*A^* and BA are closable, although we cannot assure from the assumptions of Lemma 2 that BA need be densely defined. Of course from the conclusion of Lemma 2 we know that $B^*A^* = (AB)^*$ is closed.

The potential advantages of Lemma 2 over Lemma 1 hinge on the fact that self-adjointness is a delicate property with no leeway on domains, whereas symmetry has domain leeway and is almost immediately present from the posing of any problem in which it is physically inherent. On the other hand, to use Lemma 2 one will have to play with adjoints to arrange that the domain of B^* can be made large enough to contain the range of A^*.

Although we are not seeking ultimate generality, as a gesture toward that it is natural to ask if the method of proof of this paper does apply more generally.

Lemma 3. *Let A and B be arbitrary densely defined operators in a Hilbert space $\mathcal{H}$ and suppose that $D(AB)$ is dense, $R(B) \supset D(A)$, $D(B^*) \supset R(A^*)$ and that B is 1–1. Then $(AB)^* = B^*A^*$.*

PROOF. The proof is exactly that of Lemma 2. For general densely defined operators B in a Hilbert space a dense $R(B)$ guarantees [6, 7] that B^* is 1–1. With B also assumed 1–1, equations (1) and (2) and the Lemma follow. □

We did not assume A and B closed or even closable in Lemma 3. This means that the adjoint operators A^*, B^*, and $(AB)^*$ could usefully represent very singular objects, e.g. those whose Hilbert space domains could be arbitrarily small.

Finally, we should note a Banach space version of the lemma. One could do this for operators between different Banach spaces X, Y, Z as discussed above, but because we are emphasizing domain/range assumptions, for simplicity we stay within a single Banach space X.

Lemma 4. *Let A and B be arbitrary densely defined operators in a Banach space X and suppose that $D(AB)$ is dense, $R(B) \supset D(A)$, $D(B') \supset R(A')$ and B is 1–1. Then $(AB)' = B'A'$.*

PROOF. We know generally (e.g., [5]) that for densely defined B from X to Y, densely defined A from Y to Z, X, Y, Z Banach spaces, when AB is densely defined in X, then $(AB)' \supset B'A'$. This follows, by the way, directly from the definitions of the domains of dual operators of unbounded operators. So, as before, we need to show that $B'A' \supset (AB)'$. The relationship (1) above remains valid so long as $R(B) \supset D(A)$. Then (2) also holds with the * replaced by $'$ for the same reasons as enunciated above. For arbitrary densely defined linear B a dense range assumption guarantees (e.g., see [5, 6, 7] that B' is 1–1 and then because B is also assumed 1–1, the generally valid relation $(B^{-1})' = (B')^{-1}$ permits us to apply B' from the left on (2) to conclude $B'A' \supset (AB)'$. □

Because the duals of normed linear spaces are Banach spaces, one could also write down versions of Lemma 4 for X, Y, Z just normed linear spaces. For example the key relation $(B^{-1})' = (B')^{-1}$ for B of dense range holds there as well. Because the relation $(AB)' \supset B'A'$ extends essentially to topological vector spaces, one could also investigate versions of the lemma there.

Let me conclude by offering a clarifying delineation of the uses of the lemma in [1], the earlier "closed range" versions, and the "domain" versions obtained here. Although [1, 2, 3] permit more general uses of the lemma, one regularization procedure [1, Corollary to Lemma 2, p. 1638] assumes B and B^{-1} bounded selfadjoint. In that case $R(B) = \mathcal{H}$ and all versions coincide. In terms of operator state diagrams [6, 7], one is in range state I. However when $R(B)$ has finite codimension greater than zero, then generally $R(B)$ is not dense and hence cannot contain $D(A)$. In terms of operator state diagrams [6, 7], in those cases one is in range state III. In this paper we are treating the case of dense $R(B)$, that is, assuming that $R(B)$ is properly dense, we are treating operator state diagram [6, 7] range state II. As B here is also assumed to be 1–1, for closed B in the Hilbert space versions we are treating B in state II_2, which physically corresponds to the case of continuous spectrum.

References

1 S. Albeverio, R. Høegh–Krohn, and L. Streit, Regularization of Hamiltonians and processes, *J. Math. Physics* **21** (1980), 1636–1642.

2 S. Albeverio, T. Hida, J. Potthoff, M. Röckner, and L. Streit, Dirichlet forms in terms of white noise analysis I: Construction and QFT examples, *Reviews in Mathematical Physics* **1** (1990), 291–312.

3 S. Albeverio, T. Hida, J. Potthoff, M. Röckner, and L. Streit, Dirichlet forms in terms of white noise analysis II: Closability and diffusion processes, *Reviews in Mathematical Physics* **1** (1990), 313–323.

4 K. Gustafson, Irreversibility questions in chemistry, quantum-counting, and time delay, in *Energy Storage and Redistribution in Molecules* (J. Hinze, ed.), Plenum Press (1983), 516–526.

5 T. Kato, *Perturbation Theory for Linear Operators*, Springer, Berlin, 1980.

6 K. Gustafson, State diagrams for Hilbert space operators, *J. Math & Mech.* **18** (1968), 33–46.

7 K. Gustafson, Operator spectral states, *Computers Math. Applic.* **34** (1997), 467–508.

8 F. Riesz and B. Sz. Nagy, *Functional Analysis*, Ungar, N.Y., 1978.

9 K. Gustafson, On projections of self-adjoint operators and operator product adjoints, *Bull. Amer. Math. Soc.* **75** (1969), 739–741.
10 S. Kaniel and M. Schechter, Spectral theory for Fredholm operators, *Comm. Pure Appl. Math.* **16** (1963), 423–448.
11 H. Cordes and J. Labrousse, The invariance of the index in the metric space of closed operators, *J. Math. & Mech.* **12** (1963), 693–719.
12 S. Holland, On the adjoint of the product of operators, *J. Funct. Anal.* **3** (1969), 337–344.
13 M. Schechter, The conjugate of a product of operators, *J. Funct. Anal.* **6** (1970), 26–28.
14 J. Van Casteren and S. Goldberg, The conjugate of the product of operators, *Studia Math.* **38** (1970), 125–130.
15 K. Gustafson, On algebraic multiplicity, *Indiana U. Math. J.* **25** (1976), 769–781.
16 K. Gustafson, *Partial Differential Equations and Hilbert Space Methods*, 3rd Edition, Dover, 1999.

Department of Mathematics, University of Colorado, Boulder, CO 80309–0395 USA
E-mail address: gustafs@euclid.colorado.edu

Canadian Mathematical Society
Conference Proceedings
Volume **29**, 2000

An $Mp^{\mathbb{C}}$-spinorial Approach to a Geometric-type Quantization for Symplectic Manifolds

Katharina Habermann

ABSTRACT. Using $Mp^{\mathbb{C}}$-structures and symplectic spinor fields, concepts already applied in classical geometric quantization methods by Kostant and Souriau, a new natural scheme of a geometric-type quantization up to second order is given that is completely different from the classical one.

In honour of Sergio Albeverio's 60th birthday

Recently, the author has introduced symplectic Dirac operators. These Dirac operators are defined in a purely symplectic setting and act on symplectic spinor fields. See [**4, 5, 6, 7, 8**]. Dealing with symplectic spinors, it turns out that it is possible to extend the classical geometric quantization approach of Kostant and Souriau into a complete different direction than they did. As a consequence, one obtains a natural quantization scheme satisfying the weaker correspondence principle.

The purpose of this report is to present this concept.

1. Introduction

In the Hamilton formalism, the mathematical model of a classical mechanical system is given by the geometry of a symplectic manifold (M, ω). Observables of such a system are understood as functions on M which form the commutative associative algebra $C^\infty(M)$. Additionally, the Poisson bracket $\{\ ,\ \}$ induced by the symplectic structure ω gives $C^\infty(M)$ a Lie algebra structure.

In quantum mechanics, observables form a non-commutative operator algebra acting on a certain Hilbert space.

According to Dirac, the basic problem of quantization is to describe the way of passing from classical observables to the corresponding quantum observables, where one requires that Poisson brackets of classical observables yield the corresponding commutators of quantum observables. If $\mathfrak{q}$ denotes the quantization map, this reads as

$$\text{(QC)} \qquad [\mathfrak{q}(f), \mathfrak{q}(g)] = -i\hbar\mathfrak{q}(\{f, g\})$$

1991 *Mathematics Subject Classification.* 81S10 53C99.

which is known as Dirac's *quantization condition.* Here, $\hbar = h/2\pi$ is the positive Planck constant.

This has initiated the concept of geometric quantization which is to find Lie algebra morphisms between the Lie algebra $C^\infty(M)$ of classical observables and the Lie algebra of quantum observables, or on certain subalgebras. In this approach, however, a series of difficulties occurs, and therefore it seems reasonable to replace the above quantization condition by a weaker one. This weaker version of Dirac's quantization condition is the following so-called *correspondence principle*

$$[\mathfrak{q}(f), \mathfrak{q}(g)] = -i\hbar\mathfrak{q}(\{f, g\}) + \text{ higher order terms in } \hbar\,.$$

One way of producing a quantization satisfying this correspondence principle is given by formal deformation theory. See [**1, 2, 3, 9**]. The problem of deformation quantization is to define a non-commutative associative multiplication operation, the so-called $*$-product, on the linear space $C^\infty(M)[[\hbar]]$ of formal power series in $\hbar$ with coefficients in $C^\infty(M)$. That means, in the language of formal deformation theory, one understands quantization as deformation of the Poisson-algebra $C^\infty(M)$ with the formal deformation parameter $\hbar$. Furthermore, letting the formal parameter of deformation tend to zero should reproduce classical mechanics in the limit case.

It is well known that each symplectic manifold admits a deformation quantization, see [**2**]. There are different methods to prove the existence of and to construct such $*$-products. Mostly, the existence of a $*$-product is shown by constructing the coefficients in the product step by step using Hochschild cohomology. A simple and natural geometrical approach was presented by Fedosov [**3**]. Starting with a torsion-free symplectic connection of the underlying symplectic manifold, lifting it to a connection on the associated bundle of formal Weyl algebras, and modifying it, he produced a flat connection on the associated Weyl algebra bundle. Then this connection is used to produce a canonical $*$-product in $C^\infty(M)$.

In the present paper, we give a new construction of a natural geometric-type quantization scheme satisfying the correspondence principle which works for any symplectic manifold. The construction which we shall describe uses symplectic spinor fields which already played a key role as a tool for constructing half-densities in the context of geometric quantization of Kostant and Souriau. In difference to their classical quantization procedure, in the construction presented here no notions of polarizations and half-densities are needed.

As in the construction of Fedosov's quantization, we start with a fixed symplectic connection on the underlying symplectic manifold. However, in the case where the base manifold does not admit a metaplectic structure, additionally a $U(1)$-connection in the line bundle which is associated to each $Mp^{\mathbb{C}}$-structure has to be chosen. The quantization procedure results in a quantization map $\mathfrak{q}$ from $C^\infty(M)$ to a certain operator algebra acting on the Hilbert space of L^2-symplectic spinor fields, where the elements of $C^\infty(M)$ simply act as multiplication operators.

Our observation is that it is possible and quite natural to develop the idea of setting

$$\mathfrak{q}(f) = f + i\hbar\nabla_{X_f}\,,$$

which was the initial point in the classical scheme of geometric quantization of Kostant and Souriau, to another canonical geometric quantization procedure that is completely different from their.

The construction introduced here has some advantages. First, there are no topological obstructions and no need for cohomological considerations. Second, the construction itself is given within its own analytic context. Moreover, one does not need the notions of polarizations and half-densities. Finally, the expression $\hbar$ occures only up to second order. So, it is not necessary to ensure that the power series will converge, if one considers $\hbar$ not as a positive constant but as a formal parameter.

2. Symplectic Spinor Fields

In this section, we recall the construction of symplectic spinors and the symplectic spinor bundle. See also [**10, 11**] or [**4, 5**].

The metaplectic groups $Mp(2n, \mathbb{R})$ and $Mp^{\mathbb{C}}(2n, \mathbb{R})$. We consider the symplectic group $Sp(2n, \mathbb{R})$ as the group of real $2n \times 2n$-matrices leaving the standard symplectic structure ω_0 on $\mathbb{R}^{2n}$ invariant. Then $Sp(2n, \mathbb{R}) \cap O(2n, \mathbb{R}) \cong U(n)$ is a maximal compact subgroup of $Sp(2n, \mathbb{R})$. This implies $\pi_1(Sp(2n, \mathbb{R})) \cong \mathbb{Z}$ for the fundamental group.

The metaplectic group $Mp(2n, \mathbb{R})$ is defined to be the uniquely (up to isomorphism) determined two-fold covering group of $Sp(2n, \mathbb{R})$. Thus, we have the exact sequence

$$1 \longrightarrow \mathbb{Z}_2 \longrightarrow Mp(2n, \mathbb{R}) \overset{\rho}{\longrightarrow} Sp(2n, \mathbb{R}) \longrightarrow 1$$

with double covering map ρ. Furthermore, the group $Mp^{\mathbb{C}}(2n, \mathbb{R})$ is defined to be

$$Mp^{\mathbb{C}}(2n, \mathbb{R}) = Mp(2n, \mathbb{R}) \times_{\mathbb{Z}_2} U(1) ,$$

where the action of $\mathbb{Z}_2 = \{\pm 1\}$ on $Mp(2n, \mathbb{R}) \times U(1)$ is given by

$$(-1) \cdot (g, z) = (g(-\mathbf{1}), -z)$$

for $(g, z) \in Mp(2n, \mathbb{R}) \times U(1)$. Here $-\mathbf{1} \in Mp(2n, \mathbb{R})$ denotes that element of the inverse image of the unit element of $Sp(2n, \mathbb{R})$ with respect to ρ which is not the unit element of $Mp(2n, \mathbb{R})$.

Thus, we have the central exact sequence

$$1 \longrightarrow U(1) \longrightarrow Mp^{\mathbb{C}}(2n, \mathbb{R}) \overset{\rho^{\mathbb{C}}}{\longrightarrow} Sp(2n, \mathbb{R}) \longrightarrow 1$$

and the following group homomorphisms:

- the double covering $\rho : Mp(2n, \mathbb{R}) \to Sp(2n, \mathbb{R})$
- $\rho^{\mathbb{C}} : Mp^{\mathbb{C}}(2n, \mathbb{R}) \to Sp(2n, \mathbb{R})$ given by $\rho^{\mathbb{C}}[g, z] = \rho(g)$
- $\iota : Mp(2n, \mathbb{R}) \to Mp^{\mathbb{C}}(2n, \mathbb{R})$ given by $\iota(g) = [g, 1]$
- $\kappa : Mp^{\mathbb{C}}(2n, \mathbb{R}) \to U(1)$ given by $\kappa[g, z] = z^2$.

Furthermore, the following diagram commutes, where the rows as well as the columns are exact.

$$
\begin{array}{ccccccccc}
 & & 1 & & 1 & & & & \\
 & & \downarrow & & \downarrow & & & & \\
1 & \rightarrow & \mathbb{Z}_2 & \rightarrow & Mp(2n,\mathbb{R}) & \overset{\rho}{\rightarrow} & Sp(2n,\mathbb{R}) & \rightarrow & 1 \\
 & & \downarrow & & \downarrow \iota & & \| & & \\
1 & \rightarrow & U(1) & \rightarrow & Mp^{\mathbb{C}}(2n,\mathbb{R}) & \overset{\rho^{\mathbb{C}}}{\rightarrow} & Sp(2n,\mathbb{R}) & \rightarrow & 1 \\
 & & \downarrow .^2 & & \downarrow \kappa & & & & \\
 & & U(1) & = & U(1) & & & & \\
 & & \downarrow & & \downarrow & & & & \\
 & & 1 & & 1 & & & &
\end{array}
$$

The Metaplectic Representation. The group $Mp^{\mathbb{C}}(2n, \mathbb{R})$ has a natural representation acting on the Hilbert space $L^2(\mathbb{R}^n)$. This representation can be constructed by means of the Stone-von Neumann theorem for irreducible unitary representations of the $(2n+1)$-dimensional Heisenberg group $H = \mathbb{R}^{2n} \times \mathbb{R}$.

First, the Stone - von Neumann theorem yields the existence of a unique (up to equivalence) irreducible unitary representation

$$\sigma : H \rightarrow U(L^2(\mathbb{R}^n))$$

such that

$$\sigma(0,t) = \mathrm{e}^{\mathrm{i}t} \mathrm{id}_{L^2(\mathbb{R}^n)} .$$

Now the symplectic group $Sp(2n, \mathbb{R})$ acts on the the Heisenberg group by

$$g(w,t) = (gw, t) .$$

Setting $\sigma^g(w,t) = \sigma(gw,t)$ for any $g \in Sp(2n, \mathbb{R})$, this gives a new irreducible unitary representation σ^g of H satifying

$$\sigma^g(0,t) = \mathrm{e}^{\mathrm{i}t} \mathrm{id}_{L^2(\mathbb{R}^n)} .$$

Again by the Stone - von Neumann theorem, the representation σ^g is equivalent to σ. Therefore, there exists a unitary bijective operator $U(g) : L^2(\mathbb{R}^n) \rightarrow L^2(\mathbb{R}^n)$ (unique up to a constant multiple of modulus one) satisfying

$$\sigma^g = U(g) \circ \sigma \circ U(g)^{-1} .$$

Thus, the map $g \mapsto U(g)$ defines a projective unitary representation of $Sp(2n, \mathbb{R})$, i.e. for any $g, h \in Sp(2n, \mathbb{R})$, we have

$$U(gh) = c(g,h)\, U(g) \circ U(h)$$

with $c(g,h) \in U(1)$. This projective representation lifts uniquely to a unitary representation

$$L : Mp^{\mathbb{C}}(2n, \mathbb{R}) \rightarrow U(L^2(\mathbb{R}^n))$$

of $Mp^{\mathbb{C}}(2n, \mathbb{R})$.

Let $PU(L^2(\mathbb{R}^n))$ denote the projective unitary group $U(L^2(\mathbb{R}^n))/U(1)$ equipped with the quotient topology. Then one has the commutative diagram

$$\begin{array}{ccccccccc} 1 & \to & U(1) & \to & Mp^{\mathbb{C}}(2n,\mathbb{R}) & \overset{\rho^{\mathbb{C}}}{\to} & Sp(2n,\mathbb{R}) & \to & 1 \\ & & \| & & \downarrow L & & \downarrow & & \\ 1 & \to & U(1) & \to & U(L^2(\mathbb{R}^n)) & \to & PU(L^2(\mathbb{R}^n)) & \to & 1\,, \end{array}$$

where the rows are exact.

$Mp^{\mathbb{C}}$-Structures and associated line bundles. Let (M,ω) be a $2n$-dimensional symplectic manifold and let R be the $Sp(2n,\mathbb{R})$-principal fibre bundle of all symplectic frames over M. An $Mp^{\mathbb{C}}$-structure on (M,ω) is a principal fibre bundle P over M with structure group $Mp^{\mathbb{C}}(2n,\mathbb{R})$ together with a bundle morphism $f : P \to R$ which is equivariant with respect to the homomorphism $\rho^{\mathbb{C}} : Mp^{\mathbb{C}}(2n,\mathbb{R}) \to Sp(2n,\mathbb{R})$. That is, we have the following commutative diagram

$$\begin{array}{ccccc} P \times Mp^{\mathbb{C}}(2n,\mathbb{R}) & \to & P & & \\ & & & \searrow & \\ \downarrow f \times \rho^{\mathbb{C}} & & \downarrow f & & M\;. \\ & & & \nearrow & \\ R \times Sp(2n,\mathbb{R}) & \to & R & & \end{array}$$

Thus, an $Mp^{\mathbb{C}}$-structure is a lift of the symplectic frame bundle R with respect to the map $\rho^{\mathbb{C}}$.

Generally, one has a cohomological obstruction to the existence of such lifts. For instance, there are well known conditions for the existence of spin-structures, spin$^{\mathbb{C}}$-structures, and metaplectic structures. However, since each symplectic manifold admits an almost complex structure compatible with ω, there is no topological obstruction to the existence of $Mp^{\mathbb{C}}$-structures on (M,ω). The set of all inequivalent $Mp^{\mathbb{C}}$-structures is parametrized by the second cohomology group $H^2(M,\mathbb{Z})$. Moreover, every symplectic manifold (M,ω) carries a canonically $Mp^{\mathbb{C}}$-structure. Cf. [**11**].

Every $Mp^{\mathbb{C}}$-structure P defines a $U(1)$-principal bundle $P_{U(1)}$ by

$$P_{U(1)} = P/Mp(2n,\mathbb{R})\;.$$

Here $Mp(2n,\mathbb{R})$ acts on P by

$$\begin{array}{rcl} P \times Mp(2n,\mathbb{R}) & \to & P \\ (p,g) & \mapsto & p \cdot \iota(g)\;. \end{array}$$

Symplectic Spinor Bundle. If (M,ω) is a $2n$-dimensional symplectic manifold with fixed $Mp^{\mathbb{C}}$-structure P then the symplectic spinor bundle is defined to be the associated Hilbert bundle

$$\mathcal{Q} = P \times_L L^2(\mathbb{R}^n)\;.$$

Recall the Schrödinger quantization prescription

$$\begin{array}{lcll} 1 \in \mathbb{R} & \mapsto & \text{multiplication by } i/\hbar\;, & \\ a_j \in \mathbb{R}^{2n} & \mapsto & \text{multiplication by } ix_j/\hbar\;, & \text{and} \\ b_j \in \mathbb{R}^{2n} & \mapsto & \partial/\partial x_j & \text{for } \; j = 1,\dots,n, \end{array}$$

where $a_1, \ldots, a_n, b_1, \ldots, b_n$ denotes the canonical symplectic basis of $\mathbb{R}^{2n}$ with respect to the standard symplectic structure ω_0. In particular, this defines a multiplication of vectors in $\mathbb{R}^{2n}$ with functions on $\mathbb{R}^n$. As one easily checks, this multiplication is $Mp^{\mathbb{C}}(2n, \mathbb{R})$-equivariant. Hence, it induces a multiplication

$$\begin{array}{rcl} \mu : TM \otimes \mathcal{Q} & \rightarrow & \mathcal{Q} \\ X \otimes \varphi & \mapsto & \mu(X, \varphi) = X \cdot \varphi \end{array}$$

on the symplectic spinor bundle which is called symplectic Clifford multiplication. Obviously,

$$X \cdot Y \cdot \varphi - Y \cdot X \cdot \varphi = -\frac{\mathrm{i}}{\hbar}\omega(X, Y)\varphi$$

for vector fields X, Y and spinor fields φ.

The product of $L^2(\mathbb{R}^n)$ yields a canonical Hermitian metric $< \ , \ >$ on $\mathcal{Q}$. Then, for sections φ and ψ of $\mathcal{Q}$, their L^2-product is given by

$$(\varphi, \psi) = \int_M < \varphi, \psi > dM \,.$$

Now let $\nabla : \Gamma(TM) \rightarrow \Gamma(T^*M \otimes TM)$ be a covariant derivative on (M, ω) which is supposed to be symplectic, i.e. $\nabla\omega = 0$. Together with a connection A in the $U(1)$-principal bundle $P_{U(1)}$, it induces a covariant derivative on the symplectic spinor bundle $\mathcal{Q}$, again denoted by ∇, which is called spinor derivative.

Indeed, there is a one-to-one correspondence between symplectic covariant derivatives on the tangent bundle and connection 1-forms $Z : TP \rightarrow \mathfrak{sp}(2n, \mathbb{R})$ on the $Sp(2n, \mathbb{R})$-principal fibre bundle R of all symplectic frames of M. Furthermore, a connection A in the $U(1)$-principal bundle $P_{U(1)}$ is a certain 1-form $A : TP_{U(1)} \rightarrow \mathfrak{u}(1)$. Thus, lifting Z to the first component and A to the second component of $\mathfrak{mp}(2n, \mathbb{R}) \oplus \mathfrak{u}(1)$, one obtains a connection 1-form $\overline{Z} : TP \rightarrow \mathfrak{mp}^{\mathbb{C}}(2n, \mathbb{R}) \cong \mathfrak{mp}(2n, \mathbb{R}) \oplus \mathfrak{u}(1)$ in the $Mp^{\mathbb{C}}$-structure P which makes the following diagrams commutative.

$$\begin{array}{ccc} & \overline{Z} & \\ TP & \rightarrow & \mathfrak{mp}(2n, \mathbb{R}) \oplus \mathfrak{u}(1) \\ \downarrow & & \downarrow \\ TR & \rightarrow & \mathfrak{sp}(2n, \mathbb{R}) \cong \mathfrak{mp}(2n, \mathbb{R}) \\ & Z & \end{array} \qquad \begin{array}{ccc} & \overline{Z} & \\ TP & \rightarrow & \mathfrak{mp}(2n, \mathbb{R}) \oplus \mathfrak{u}(1) \\ \downarrow & & \downarrow \\ TP_{U(1)} & \rightarrow & \mathfrak{u}(1) \\ & A & \end{array}$$

Now, the spinor derivative

$$\nabla : \Gamma(\mathcal{Q}) \rightarrow \Gamma(T^*M \otimes \mathcal{Q})$$

is defined as the covariant derivative induced by $\overline{Z}$. Thus, if $\varphi = [\overline{s}, u]$ is a local section of the symplectic spinor bundle $\mathcal{Q}$ then

$$\nabla\varphi = [\overline{s}, du + L_*(\overline{Z} \circ \overline{s}_*)u] \,.$$

Finally, we recall that the Clifford multiplication, the spinor derivative, and the Hermitian scalar product are related by

$$\begin{array}{rcl} (X \cdot Y - Y \cdot X) \cdot \varphi & = & -\frac{\mathrm{i}}{\hbar}\omega(X, Y)\varphi \\ < X \cdot \varphi, \psi > & = & - < \varphi, X \cdot \psi > \\ \nabla_X(Y \cdot \varphi) & = & (\nabla_X Y) \cdot \varphi + Y \cdot \nabla_X \varphi \\ X < \varphi, \psi > & = & < \nabla_X \varphi, \psi > + < \varphi, \nabla_X \psi > \\ < \varphi, \psi > & = & \overline{< \psi, \varphi >} \end{array}$$

for vector fields X, Y on M and spinor fields $\varphi, \psi \in \Gamma(\mathcal{Q})$.

We point out that, in the case where M admits a metaplectic structure, the whole construction can be simplified by replacing the group $Mp^{\mathbb{C}}(2n, \mathbb{R})$ by the metaplectic group $Mp(2n, \mathbb{R})$. In particular, in this situation, one does not need the line bundle $P_{U(1)}$ with the $U(1)$-connection A.

3. Our Concept of Quantization

Consider a symplectic manifold (M, ω) and fix a symplectic connection ∇ and an $Mp^{\mathbb{C}}$-structure on it. Again, let $\mathcal{Q}$ denote the corresponding symplectic spinor bundle. Let $\mathfrak{H} = L^2(\mathcal{Q})$ be the Hilbert space of all L^2-sections of the symplectic spinor bundle.

We define a quantization map $\mathfrak{q}$ from the Poisson algebra $C^\infty(M)$ to the space of linear operators on $\mathfrak{H}$ by

$$\mathfrak{q}(f)(\varphi) = f\varphi + \mathrm{i}\hbar X_f \cdot \varphi + \mathrm{i}\hbar \nabla_{X_f}\varphi$$

for $f \in C^\infty(M)$ and $\varphi \in \Gamma(\mathcal{Q})$. Here, X_f for each function $f \in C^\infty(M)$ denotes the Hamiltonian vector field accociated to f, i.e. X_f is defined by

$$\omega(X_f, Y) = Y(f) = df(Y)$$

for any vector field Y on M.

We recall that the Poisson structure $\{\ ,\ \}$ on (M, ω) is given by

$$\{f, g\} = \omega(X_f, X_g)$$

for $f, g \in C^\infty(M)$. Furthermore, we recall that the torsion of the symplectic connection ∇ is the tensor field T^∇ given by

$$T^\nabla(X, Y) = \nabla_X Y - \nabla_Y X - [X, Y]$$

and that the curvature $\mathcal{R}^Q$ of the symplectic spinor derivative is defined by

$$\mathcal{R}^Q(X, Y)\varphi = \nabla_X \nabla_Y \varphi - \nabla_Y \nabla_X \varphi - \nabla_{[X,Y]}\varphi$$

for vector fields X, Y on M and $\varphi \in \Gamma(\mathcal{Q})$.

Now we can state

THEOREM 3.1. *The quantization map $\mathfrak{q}$ satisfies the following properties.*

1. $\mathfrak{q}(1) = id_{\mathfrak{H}}$.
2. *For $\hbar \to 0$ the quantization map $\mathfrak{q}$ tends to the identity in the sense that the elements of $C^\infty(M)$ act as multiplication operators on $\mathfrak{H}$.*
3. *For any $f, g \in C^\infty(M)$ and $\varphi \in \Gamma(\mathcal{Q})$ the correspondence principle*

$$[\mathfrak{q}(f), \mathfrak{q}(g)](\varphi) = -i\hbar\mathfrak{q}(\{f, g\})(\varphi) - \hbar^2(\mathcal{R}^Q(X_f, X_g)\varphi + T^\nabla(X_f, X_g) \cdot \varphi)$$

holds.

Proof: The first two assertions are obvious. We have to prove the third assertion. Since $X_{\{f,g\}} = -[X_f, X_g]$, we have

$$\begin{aligned} \mathrm{i}\hbar\mathfrak{q}(\{f,g\})(\varphi) &= \mathrm{i}\hbar\{f,g\}(\varphi) - \hbar^2 X_{\{f,g\}} \cdot \varphi - \hbar^2 \nabla_{X_{\{f,g\}}}\varphi \\ &= \mathrm{i}\hbar\{f,g\}(\varphi) + \hbar^2 [X_f, X_g] \cdot \varphi + \hbar^2 \nabla_{[X_f, X_g]}\varphi \,. \end{aligned}$$

On the other hand, using

$$\{f, g\} = \omega(X_f, X_g) = -X_f(g) = X_g(f) \,,$$

we deduce

$$
\begin{aligned}
&[\mathfrak{q}(f), \mathfrak{q}(g)](\varphi) = \\
&= \mathfrak{q}(f)(g\varphi + \mathrm{i}\hbar X_g \cdot \varphi + \mathrm{i}\hbar\nabla_{X_g}\varphi) - \mathfrak{q}(g)(f\varphi + \mathrm{i}\hbar X_f \cdot \varphi + \mathrm{i}\hbar\nabla_{X_f}\varphi) \\
&= fg\varphi + \mathrm{i}\hbar f X_g \cdot \varphi + \mathrm{i}\hbar f\nabla_{X_g}\varphi + \mathrm{i}\hbar g X_f \cdot \varphi - \hbar^2 X_f \cdot X_g \cdot \varphi \\
&\quad -\hbar^2 X_f \cdot \nabla_{X_g}\varphi + \mathrm{i}\hbar\nabla_{X_f}(g\varphi) - \hbar^2\nabla_{X_f}(X_g \cdot \varphi) - \hbar^2\nabla_{X_f}\nabla_{X_g}\varphi \\
&\quad -fg\varphi - \mathrm{i}\hbar g X_f \cdot \varphi - \mathrm{i}\hbar g\nabla_{X_f}\varphi - \mathrm{i}\hbar f X_g \cdot \varphi + \hbar^2 X_g \cdot X_f \cdot \varphi \\
&\quad +\hbar^2 X_g \cdot \nabla_{X_f}\varphi - \mathrm{i}\hbar\nabla_{X_g}(f\varphi) + \hbar^2\nabla_{X_g}(X_f \cdot \varphi) + \hbar^2\nabla_{X_g}\nabla_{X_f}\varphi \\
&= -\hbar^2(X_f \cdot X_g - X_g \cdot X_f) \cdot \varphi - \mathrm{i}\hbar X_g(f) + \mathrm{i}\hbar X_f(g) \\
&\quad -\hbar^2\nabla_{X_f}X_g \cdot \varphi + \hbar^2\nabla_{X_g}X_f \cdot \varphi - \hbar^2\nabla_{[X_f,X_g]}\varphi \\
&\quad -\hbar^2(\nabla_{X_f}\nabla_{X_g}\varphi - \nabla_{X_g}\nabla_{X_f}\varphi - \nabla_{[X_f,X_g]}\varphi) \\
&= \mathrm{i}\hbar\omega(X_f, X_g)\varphi - 2\mathrm{i}\hbar\{f,g\}\varphi - \hbar^2\nabla_{[X_f,X_g]}\varphi \\
&\quad -\hbar^2(\nabla_{X_f}X_g - \nabla_{X_g}X_f - [X_f, X_g]) \cdot \varphi - \hbar^2[X_f, X_g] \cdot \varphi \\
&\quad -\hbar^2\mathcal{R}^Q(X_f, X_g)\varphi \\
&= -\mathrm{i}\hbar\{f,g\}\varphi - \hbar^2[X_f, X_g] \cdot \varphi - \hbar^2\nabla_{[X_f,X_g]}\varphi \\
&\quad -\hbar^2(\mathcal{R}^Q(X_f, X_g)\varphi + T^\nabla(X_f, X_g) \cdot \varphi) \\
&= -\mathrm{i}\hbar\mathfrak{q}(\{f,g\})(\varphi) - \hbar^2(\mathcal{R}^Q(X_f, X_g)\varphi + T^\nabla(X_f, X_g) \cdot \varphi) .
\end{aligned}
$$

This proves the theorem. □

We remark that

$$
\begin{aligned}
\mathfrak{q}(f) \circ \mathfrak{q}(g) &= fg\varphi - \frac{\mathrm{i}}{2}\hbar\{f,g\}\varphi + \mathrm{i}\hbar(fX_g \cdot \varphi + gX_f \cdot \varphi + f\nabla_{X_g}\varphi + g\nabla_{X_g}\varphi) \\
&\quad -\hbar^2(\frac{1}{2}X_f \cdot X_g \cdot \varphi + \frac{1}{2}X_g \cdot X_f \cdot \varphi + X_f \cdot \nabla_{X_g}\varphi + X_g \cdot \nabla_{X_f}\varphi) \\
&\quad -\hbar^2(\nabla_{X_f}\nabla_{X_g}\varphi + \nabla_{X_f}X_g \cdot \varphi) .
\end{aligned}
$$

Thus, contrary to other constructions, the first order term of this composition does not agree with the Poisson structure $\{\ ,\ \}$ on (M, ω). However, the expression $[\mathfrak{q}(f), \mathfrak{q}(g)]$ for $f, g \in C^\infty(M)$ is related to $\{\ ,\ \}$ by

$$\left.\frac{[\mathfrak{q}(f), \mathfrak{q}(g)]}{\hbar}\right|_{\hbar=0} = -\mathrm{i}\mathfrak{q}(\{f,g\}) .$$

Moreover, the first order term in the above product forms a structure $\{\!\{\ ,\ \}\!\}$ which fulfills the Jacobi identity. Namely, for functions $f, g \in C^\infty(M)$, define $\{\!\{f,g\}\!\}$ acting on symplectic spinor fields by

$$\{\!\{f,g\}\!\}(\varphi) = \{f,g\}\varphi - 2\,s(f,g)(\varphi)$$

with

$$s(f,g)(\varphi) = fX_g \cdot \varphi + gX_f \cdot \varphi + f\nabla_{X_g}\varphi + g\nabla_{X_f}\varphi.$$

Then

LEMMA 3.2. *For any* $f, g, h \in C^\infty(M)$,

$$\{\!\{f, \{\!\{g,h\}\!\}\,\}\!\} + \{\!\{g, \{\!\{h,f\}\!\}\,\}\!\} + \{\!\{h, \{\!\{f,g\}\!\}\,\}\!\} = 0\,.$$

Proof: From

$$\begin{aligned}\{\!\{f, s(g,h)\}\!\}(\varphi) &= f \circ s(g,h)(\varphi) - s(g,h) \circ f(\varphi) \\ &= f\, s(g,h)(\varphi) - s(g,h)(f\varphi)\end{aligned}$$

we see that

$$\begin{aligned}&\{\!\{f, \{\!\{g,h\}\!\}\,\}\!\}(\varphi) = \\ &= \{f, \{g,h\}\,\}\varphi - 2(f\, s(g,h)(\varphi) - s(g,h)(f\varphi)) \\ &= \{f, \{g,h\}\,\}\varphi - 2(fgX_h \cdot \varphi + fhX_g \cdot \varphi + fg\nabla_{X_h}\varphi + fh\nabla_{X_g}\varphi \\ &\quad -gX_h \cdot (f\varphi) - hX_g \cdot (f\varphi) - g\nabla_{X_h}(f\varphi) - h\nabla_{X_g}(f\varphi)) \\ &= \{f, \{g,h\}\,\}\varphi + 2(gX_h(f) + hX_g(f))\varphi\,.\end{aligned}$$

Hence,

$$\begin{aligned}&\{\!\{f, \{\!\{g,h\}\!\}\,\}\!\} + \{\!\{g, \{\!\{h,f\}\!\}\,\}\!\} + \{\!\{h, \{\!\{f,g\}\!\}\,\}\!\} = \\ &= \{f, \{g,h\}\,\} + \{g, \{h,f\}\,\} + \{h, \{f,g\}\,\} \\ &\quad +2(gX_h(f) + hX_g(f) + hX_f(g) + fX_h(g) + fX_g(h) + gX_f(h)) \\ &= 2(g\{h,f\} + h\{g,f\} + h\{f,g\} + f\{h,g\} + f\{g,h\} + g\{f,h\}) \\ &= 0\,,\end{aligned}$$

which gives the desired result. □

We note that $\{\!\{\ ,\ \}\!\}$ is not a Poisson structure, since the expression $s(f,g)$ is symmetric in f and g.

4. Concluding Remarks

Usually, a deformation quantization is a formal deformation of the Poisson algebra which needs an analytical setting to work. First, one has to choose the Hilbert space on which the deformed algebra that gives the quantum observables should act appropriately. Second, one has to ensure that the power series which occur converge. Clearly, we do not have these problems in our construction.

Bayen et al. already noted in [**1**] that the term of order $\hbar^2$ in the commutator $[f,g] = f * g - g * f$ is strongly related to a symplectic connection on (M,ω). Theorem 3.1 implies that in the case of a flat torsion-free symplectic connection our quantization map $\mathfrak{q}$ even satisfies Dirac's strong *quantization condition* (QC).

In view of concrete examples, one might object that the space of smooth sections of the Hilbert space bundle $\mathcal{Q}$ is too large. Actually, if necessary, our construction has to be restricted to certain subspaces of suitable sections in $\mathcal{Q}$. For example, let us consider the harmonic oscillator on $\mathbb{R}^n$. The Hamiltonian H on the corresponding phase space $\mathbb{R}^{2n}$ is given by

$$H(x,y) = \frac{1}{2m}||y||^2 + \frac{k}{2}||x||^2\,.$$

Now consider the symplectic spinor bundle associated to the trivial metaplectic structure via the restricted metaplectic representation $L_{|Mp(2n,\mathbb{R})}$ and choose the flat Levi-Civita connection as symplectic connection on $\mathbb{R}^{2n}$. Then a symplectic spinor field is parallel if and only if it is constant, i.e.

$$\varphi(x) \equiv f \quad \text{for all } x \in \mathbb{R}^{2n} \text{ and some } f \in L^2(\mathbb{R}^n)\ .$$

Thus the space of all parallel symplectic spinor fields can be identified with $L^2(\mathbb{R}^n)$. Now let us single out this space as the space on which our quantized observables act. In order to preserve the space of constant sections in $\mathcal{Q}$, one has to modify the quantization map in such a way that only the symplectic Clifford multiplication by the corresponding Hamiltonian vector field is involved. Then, by the definition of the symplectic Clifford multiplication, the coordinate functions x_j are mapped to the multiplication by $-t_j$ and the momentum functions y_j are mapped to the differential operator $i\hbar\partial/\partial t_j$. Hence, the quantization of the harmonic oscillator H is

$$\sum_{j=1}^{n}\{-\frac{\hbar^2}{2m}\frac{\partial^2}{\partial t_j^2}+\frac{k}{2}t_j^2\} = \frac{\hbar^2}{2m}\Delta + \frac{k}{2}||t||^2$$

which is the known energy operator of H.

Acknowledgements

The major part of this paper was written during the authors stay at the Max-Planck Institute for Mathematics in the Sciences, Leipzig. I wish to thank for the kind hospitality and for the excellent working conditions.

Furthermore, I am grateful to Andreas Knauf for many useful remarks.

References

[1] F. Bayen, M. Flato, C. Fronsdal, A. Lichnerovicz & D. Sternheimer: *Deformation theory and quantization.* Annales of Physics **111**, 61-151 (1978)

[2] M. DeWilde, P. B. A. Lecomte: *Existence of star-products and of formal deformations of the Poisson Lie algebra of arbitrary symplectic manifolds.* Letters in Mathematical Physics **7**, 487-496 (1983)

[3] B. Fedosov: *A simple geometrical construction of deformation quantization.* Journal of Differential Geometry **40**, 213-238 (1994)

[4] K. Habermann: *The Dirac Operator on Symplectic Spinors.* Annals of Global Analysis and Geometry **13**, 155-168 (1995)

[5] K. Habermann: *Basic Properties of Symplectic Dirac Operators.* Communications in Mathematical Physics **184**, 629-652 (1997)

[6] K. Habermann: *Harmonic Symplectic Spinors on Riemann Surfaces.* manuscripta math. **94**, 465-484 (1997)

[7] K. Habermann: *Symplectic Dirac Operators on Kähler Manifolds.* MPI-MIS-Preprint Nr. 17 (1997) to appear in Math. Nach.

[8] K. Habermann: *Symplectic Dirac operators and a splitting property for the symplectic spinor bundle.* MPI-MIS-Preprint Nr. 19 (1999)

[9] M. Kontsevich: *Deformation Quantization of Poisson Manifolds, I.* Preprint q-alg/9709040

[10] B. Kostant: *Symplectic Spinors.* Symposia Mathematica. vol XIV (1974)

[11] P. L. Robinson, J. H. Rawnsley: *The metaplectic representation, $Mp^{\mathbb{C}}$-structures and geometric quantization.* Memoirs of the AMS **410** (1989)

Katharina Habermann, Institut für Mathematik und Informatik, Ernst-Moritz-Arndt-Universität Greifswald, Jahnstr. 15a, 17487 Greifswald

E-mail address: khaberma@mail.uni-greifswald.de

Canadian Mathematical Society
Conference Proceedings
Volume **29**, 2000

Nagel Stein Wainger Estimates for Balls Associated with the Bismut Condition

Astrid Hilbert and Remi Léandre

Dedicated to Professor Sergio Albeverio on occasion of his sixtieth birthday.

ABSTRACT. Nagel, Stein, and Wainger have given a uniform estimate for the volume of sub-Riemannian balls associated with an operator of the form $\frac{1}{2}\sum_{i=1}^{m} X_i^2$ satisfying a uniform strong Hörmander hypothesis. Léandre has given nonuniform estimates for the volume of balls associated with an operator $\mathcal{L} = \frac{1}{2}\sum_{i=1}^{m} X_i^2 + X_0$, and has shown that the Bismut condition is important to establish the relation between hypoelliptic kernels and volumes of balls. In this paper we settle the uniform case for balls associated with distances where the Bismut condition is involved.

1. Introduction

The properties of balls associated with the flow of diffusions for small times have been studied under different conditions with various techniques, [**PP81**], [**SC84**]. Nagel, Stein, and Wainger [**NSW85**] consider the same type of partial differential operators given by sums of squares of vector fields and a vector field X_0.

In a sufficiently small neighbourhood of any starting point certain exponential mappings approximating the flow corresponding to the operator $\mathcal{L}$ define diffeomorphisms on sufficiently small cubes independent of changes of bases. The algebraic properties of submodules constructed from the given vector fields are crucial since they appear when splitting off the evolution corresponding to the drift and changing bases. The Jacobian of an approximating exponential mapping is estimated which reveals the uniform bounds for the volumes of the balls corresponding to the cubes in a diffeomorphic way. Our main concern is to show the inclusions between the balls given as the image of the small cubes mentioned above and those corresponding to the subriemannian distances d_ε and $d_{\varepsilon,R}$ which contains the Bismut condition. When constructing the submodules the vector fields associated with the variation in time and space are treated differently.

1991 *Mathematics Subject Classification.* 47D06, 34E05, 60H07, 58J65.

R.Léandre and A. Hilbert gratefully acknowledge financial support by the Alexander von Humbold Stiftung and Svenska Institutet grant no. 202/10/1.7/97, respectively.

In this paper we generalize techniques in [**NSW85**] and [**Léa89**] in order to give uniform bounds on the volume of balls which are defined using the stochastic flow corresponding to a non elliptic partial differential operator of uniformly weak Hörmander type, deviating in this respect from [**NSW85**]. We adopt further techniques from stochastic variational calculus see e.g.[**Bis84**], [**Tan83**].

The estimates of non Euclidean balls have been used to derive uniform bounds for the corresponding kernel of various types of diffusion, see [**Léa94**] and [**NSW85**]. Alternative proofs for similar bounds were given e.g. in [**PP81**] and [**JSC86**].

In the sequel we formulate the main result of this paper. We begin with the main definitions and notations. For $i \in \{1, \dots, m\}$ we define an infinite arrangement $\mathcal{F}$ of multi indices $[\Gamma]$ by the following iteration: $[\Gamma] \in \{0, \dots, m\}$ is a multi index. If $[\Gamma_1], \dots, [\Gamma_n]$ are multi indices then $[\Gamma] = [[\Gamma_1], \dots, [\Gamma_n]]$ is a multi index. For simplicity we omit brackets around numbers, e.g. $\alpha \in \{0, 1, \dots, m\}$ and $[[1, 7, 4], 8, [5, 3]]$ are multi indices.

Let X_i, $i \in \{0, 1, \dots, m\}$, be real vector fields on $\mathbb{R}^d$ with bounded derivatives of all orders. Then $X_{[i,j]} := [X_i, X_j]$, $i, j \in \{0, \dots, m\}$, denotes the Lie bracket of the vector fields X_i and X_j. Moreover, let $X_{[\Gamma]}$ be the vector field corresponding to the multi index $[\Gamma] = [\Gamma_1, \dots \Gamma_n] \in \mathcal{F}$ for an arbitrary natural number n then we define by induction

$$X_{[[\Gamma_1], \dots, [\Gamma_n], i]} = X_{[[\Gamma], i]} := [X_{[\Gamma]},\ X_i] \tag{1.1}$$

where X_i is any of the given vector fields. This notation will play an important role when constructing an approximating solution cf. (2.11). All vector fields constructed from $X_0, \dots X_m$ according to (1.1) constitute the set $\mathcal{V} := \{X_{[\Gamma]} \mid [\Gamma] \in \mathcal{F}\}$. More frequently we will use the subset $\mathcal{V}_0 := \{X_{[\Gamma]} \in \mathcal{V} \mid [\Gamma] \neq 0]\}$. On the other hand we call

$$X_{(\alpha)} := X_{i_1} \cdots X_{i_n} \tag{1.2}$$

the m'th order differential operator, which arises by applying X_{i_j}, with $0 \le i_j \le m$ and $1 \le j \le n$, successively.

For a vector field $X_{[\Gamma]} \in \mathcal{V}$ with multi index $[\Gamma] := [i_1, \dots, i_n]$ where $0 \le i_j \le m$ for $1 \le j \le n$ the *length* is given by $|\Gamma| := \left|X_{[\Gamma]}\right| := n$. Moreover, the *weight* of $X_{[\Gamma]}$ is defined by $\|\Gamma\| := \left\|X_{[\Gamma]}\right\| := \sum_{k=1}^{n} \|\Gamma_k\|$ where $\|\Gamma_k\| = 2$ for $[\Gamma_k] = 0$ and $\|\Gamma_k\| = 1$ for $[\Gamma_k] \in \{1, \dots, m\}$. For the vector fields associated with $\mathcal{V}_0$ we assume the uniform weak Hörmander hypothesis WH: There exist $c > 0$ and $r \in \mathbb{N}$ such that

$$\inf_{x \in \mathbb{R}^d, |\zeta|=1} \sum_{\substack{\|\Gamma\| \le r, \\ [\Gamma] \neq 0}} \left\langle X_{[\Gamma]}(x)\ , \zeta \right\rangle^2 > c > 0 \quad . \tag{1.3}$$

For the vector fields X_i, $i \in \{0, \dots, m\}$, we introduce the following control differential equation

$$dy_{\varepsilon,s}(h) = \varepsilon^2 X_0(y_{\varepsilon,s}(h))ds + \sum_{i=1}^{m} X_i(y_{\varepsilon,s}(h))dh_s^i \tag{1.4}$$

with initial condition $y_{\varepsilon,0}(h) = x$ and control function $h \in \mathcal{H}$, the Cameron Martin space of $\mathbb{R}^d$ valued functions on [0,1] with $\|\cdot\|_{\mathcal{H}} < \infty$ where $\|h\|_{\mathcal{H}}^2 = \sum_{i=1}^{m} \int_0^1 \left|\dot{h}_s^i\right|^2 ds$. This solution defines a flow ϕ_ε on the Cameron Martin space, where we are assuming $s = 1$. For a proof see e.g. [**Bis84**] or works by Elworthy or Kunita to mention just a few names. Subsequently we define a pseudo metric

$$d_\varepsilon(x,y) := \inf_{\substack{h\in\mathcal{H},\\ \phi(h)=y}} \|h\|_{\mathcal{H}} \quad . \tag{1.5}$$

Moreover, following [**Bis84**], there exists another pseudo metric on $\mathbb{R}^d$ given by

$$d_{\varepsilon,R}(x,y) := \inf_{\phi_\varepsilon \in R} \|h\|_{\mathcal{H}} \tag{1.6}$$

for $R = \{h \in \mathcal{H} | \phi_\varepsilon$ is a submersion in h and $\phi_\varepsilon(h) = y\}$. The open balls of radius δ associated with the initial point $x \in \mathbb{R}^d$ corresponding to d_ε and $d_{\varepsilon,R}$ will be denoted by $B_\varepsilon(x,\delta)$ and $B_{\varepsilon,R}(x,\delta)$, respectively.

The balls $B_{\varepsilon,R}(x,\delta)$ are de facto centered at $x_\varepsilon := \exp[\varepsilon^2 X_0](x)$ which evolves from x according to the flow given by the vector field X_0, i.e. $x_\varepsilon := x_{\varepsilon,1}$ solves the deterministic equation $dx_{\varepsilon,s}/ds = \varepsilon^2 X_0(x_{\varepsilon,s})$. The pseudo metric d_ε fails to be a metric since for $X_0 \neq 0$ there exist $x \in \mathbb{R}^d$ and $\varepsilon > 0$ such that $d_\varepsilon(x,x) \neq 0$ whereas in any case $d_\varepsilon(x,x_\varepsilon) = 0$. For the pseudo metric $d_{\varepsilon,R}$ not only the inequality $d_{\varepsilon,R}(x,x) \neq 0$ may occur but $d_\varepsilon(x,x_\varepsilon)$ may be infinite. It may happen that the initial point x is not contained in $B_{\varepsilon,R}(x,\delta)$. This phenomenon already arises for some strongly hypoelliptic generators with drift cf. [**Léa89**], i.e. if the vector fields given by finite order Lie brackets of $X_1,\ldots,X_m$ span all tangent spaces. This example also shows that $d_t(x,y) < \infty$, and $d_{t,R}(x,y) = \infty$ occurs. In order to avoid problems of the type mentioned above we introduce the Bismut condition specified in the definition of the set R. Since in the definition of $d_{\varepsilon,R}$ following Bismut [**Bis84**] we assume ϕ_ε is a submersion in h we call $d_{\varepsilon,R}$ the Bismut metric. Let us stress that we have $B_{\varepsilon,R}(x,\delta) = B_\varepsilon(x,\delta)$ in case the generator of the Markov semi group associated with the stochastic differential equation corresponding to (1.4) is a sum of squares of vector fields and satisfies the strong Hörmander condition cf. [**Léa89**].

For a set $I = \{[\alpha_1],\ldots,[\alpha_d]\}$ of multi indices corresponding to vector fields $X_{[\alpha_i]} \in \mathcal{V}_0, 1 \le i \le d$, we denote by

$$\lambda_I(x) := \det\left(X_{[\alpha_1]},\ldots,X_{[\alpha_d]}\right)(x) \tag{1.7}$$

the determinant of the vector fields at $x \in \mathbb{R}^d$ (with $[\alpha_i] \neq 0$!), and $\|I\|$ as above. Our purpose is to prove that the volume of the balls given by $d_{\varepsilon,R}$ at time ε^2 is bounded from above and below up to constant factors by a product of λ_I and the polynomial $\varepsilon^{\|I\|}$ which is a generalization of Theorem I.3 in [**Léa89**], however, (under the condition WH in (1.3) above) restricting ourselves to the non uniform case and of Theorem 1 in [**NSW85**] where there is no X_0. In the following Vol(B) denotes the volume of a Borel set B.

THEOREM 1.1. *There exists a finite set of basis for all tangent spaces having cardinality smaller or equal to* $\binom{r(m+1)-1}{d}$ *corresponding to a finite set of index sets* I *and positive constants* c_1 *and* c_2 *such that for all* $x \in \mathbb{R}^d$ *and for* $0 < \varepsilon \le 1$ *the subsequent chain of inequalities holds*

$$c_1 \sum_I |\lambda_I(x)|\varepsilon^{\|I\|} \le \; Vol(B_{\varepsilon,R}(x,\varepsilon)) \le \; Vol(B_\varepsilon(x,\varepsilon)) \le c_2 \sum_I |\lambda_I(x)|\varepsilon^{\|I\|} \quad .$$

We point out that for an arbitrary given point x some of the index sets I in the summation might not correspond to a basis at this point such that $\lambda_I(x)$ vanishes. We furthermore remark that as a consequence of Lemma 3.2 and Proposition 3.1

the bound in the statement of the theorem is independent of the fact whether we evaluate λ_I at x or at $x_\varepsilon := \exp \varepsilon^2 X_0(x)$ which means that we split off the flow induced by the drift X_0. So the "apparent center" of these balls is x_ε. Moreover, there need not exist a basis of the tangent space at point x consisting of elements from $\mathcal{A}_r$ as in (2.3) revealing the infimum in the uniform weak Hörmander condition (1.3). However, the uniform weak Hörmander condition reveals the representation (2.4) which coincides with the central assumption in [**NSW85**]. Due to this representation it is possible to find the basis of elements from $\mathcal{A}_r$ satisfying the condition (3.2) of Proposition 3.1 (finite dimensional version of the chain of inequalities) in a forth comming step. Theorem 1.1 implies the doubling property for the balls $B_{\varepsilon,R}(x,\varepsilon)$ and $B_\varepsilon(x_\varepsilon)$ due to the fact that λ_I (as a determinant) scales with the matrix, i.e.

$$\mathrm{Vol}\,(B_{2\varepsilon}(x,2\varepsilon)) \leq c\ \mathrm{Vol}\,((B_{\varepsilon,R}(x,\varepsilon)))$$

where the constant c does not depend on x.

Our proof is more along the lines of [**NSW85**] instead of using the methods in [**Léa89**], the former clearly distinguishes the problems of estimating the volume of balls from the one of estimating the kernel. In [**Léa94**] Léandre solves the problem of deriving bounds for the associated heat kernel or transition probability uniformly in the initial point. The new aspect in comparison to [**NSW85**] is the Bismut condition which is taken from [**Léa89**] where we lack the uniform bounds.

In the subsequent section we recover the results for submodules of C^∞ functions involved in changes of basis given in [**NSW85**]. We repeatedly use the Campbell Hausdorff formula in order to approximate the solution of (1.4) using an exponential function given by a generating set for the tangent space at all points, to split off the exponential function of $\varepsilon^2 X_0$, and to incorporate the rest term of the approximation in the exponent of the first. In the next part we fix the starting point and the flow corresponding to the drift. We choose a particular basis and, forgetting about the Cameron Martin space, we restrict ourselves to the corresponding finite dimensional problem. In this setting we prove the chain of inequalities given in Theorem 1.1.

In the last section we show that the Bismut condition allows to relate the problem of estimating the volume of the hypoelliptic balls $B_\varepsilon, B_{\varepsilon,R}$ to the finite dimensional problem introduced earlier.

2. Algebraic Properties of Vector fields

REMARK 2.1. In case a generating system for all tangent spaces contains more vector fields than the dimension d we are facing the problem of choosing an appropriate basis for the tangent space at various points. We shall be using the Campbell Hausdorff formula in order to realize the change of basis for an exponential function approximating the solution to (1.4). Moreover, we want to determine the power of the parameter ε, which arises by exploiting the scaling property of the corresponding Jacobian.

In this chapter we therefore recall some results on the Lie algebra generated by vector fields $X_1, ..., X_m$. We introduce submodules of $C^\infty(\mathbb{R}^d)$ which are designed to classify the coefficient functions involved in a change of basis and their determinants. Exploiting the algebraic structure of these submodules we deduce an adequate approximation of the solution $\phi_\varepsilon(h)$ in (1.4) different from the one given in [**Aro89**].

We rewrite the elements the set $\mathcal{V}$ as linear combinations of standard type Lie brackets having the form $[\dots[X_{i_1}, X_{i_2}], \dots, X_{i_k}]$ where $X_{i_\ell} \in \{X_0, \dots, X_m\}$ for $1 \le \ell \le k$ using

$$[X,Y] + [Y,X] = 0 \quad \text{and} \quad [[X,Y],Z] + [[Y,Z],X] + [[Z,X],Y] = 0 \tag{2.1}$$

which hold for arbitrary vector fields X, Y, Z. The equation on the right is called Jacobi identity. Corresponding to the set of vector fields given by standard type Lie brackets we define sets of multi indices $\mathcal{A}_n := \{[\alpha] = [i_1, \dots, i_k] \mid 0 \le i_j \le m,\ 1 \le j \le k \le n,\ \|\alpha\| \le n\}$. We could also use the length of the standard vector fields for the definition of $\mathcal{A}_n$ since there exists a stochastic Taylor formula with rest term given in terms of length and weight, respectively. Since we shall see that the sums appearing in the Taylor formula can always be rearranged we seek the analogy to the line of arguments in [**NSW85**]. These subsets of multi indices constitute an increasing family $\mathcal{A}_1 \subset \dots \subset \mathcal{A}_n \subset \dots \subset \mathcal{A}_\infty$.

A well known result for the vector fields corresponding to these index sets is

$$X_{[\alpha']}(x) = \sum_{[\alpha] \in \mathcal{A}_{|\alpha'|},\ |\alpha| = |\alpha'|} c_{[\alpha']}^{[\alpha]} \cdot X_{[\alpha]}(x) \tag{2.2}$$

where $c_{[\alpha']}^{[\alpha]}$ are real constants. This can be seen directly by induction exploiting (2.1) or found in [**NSt82**].

Due to the uniform weak Hörmander condition (1.3) in combination with Remark 2.1 in [**KS87**], using identity (2.1), we deduce the following representation for $X_{[\Gamma]} \in \mathcal{V}_0$, i.e. $[\Gamma] \neq 0$, at $x \in I\!R^d$: there exist smooth functions $C_{[\Gamma]}^{[J]}(x)$ with bounded derivatives of all orders and such that

$$X_{[\Gamma]}(x) = \sum_{[J] \neq 0} C_{[\Gamma]}^{[J]}(x) \cdot X_{[J]}(x) \tag{2.3}$$

where $[J] \in \mathcal{A}_r$ and $r \in I\!N$ can be chosen uniformly for all $x \in I\!R$. Exploiting this representation, the one given in (2.2), and the fact that the vector field X_0 does not appear in this expansion we recover assumption 1 in [**NSW85**]

$$\left[X_{[\Gamma]}, X_{[\Gamma']}\right](x) = \sum_{\|\Lambda\| \le \|\Gamma\| + \|\Gamma'\| \wedge r} \bar{C}_{([\Gamma],[\Gamma'])}^{[\Lambda]}(x) \cdot X_{[\Lambda]}(x) \tag{2.4}$$

where $\bar{C}_{([\Gamma],[\Gamma'])}^{[\alpha]}$ are smooth functions on $I\!R^d$.

In particular, there exists a finite generating set $\bar{\mathcal{A}} := \{Y_1, \dots, Y_N\} \subset \mathcal{A}_r$, for the tangent space uniformly for all $x \in I\!R$. This is possible since $|\Gamma| \le \|\Gamma\| \le 2|\Gamma|$. In general we shall have $d \le N$. Let us fix $x \in I\!R^d$ and let $X_{[\alpha]}(x)$, $[\alpha] \in I := I(x)$, be such that the determinant $\lambda_I(x)$ cf. (1.4) does not vanish, i.e. $\lambda_I(x) \neq 0$ in a small neighbourhood of x. Moreover these vector fields may be chosen from the generating set $\bar{\mathcal{A}}$. Then the vector fields $X_{[\alpha]}(x)$, $[\alpha] \in I$, form a basis of the tangent space at x of dimension d (being equivalent to $I\!R^d$) and for any $X_{[\Gamma]}$ equation (2.3) becomes

$$\begin{aligned} X_{[\Gamma]}(x) &= \sum_{[\alpha] \in I} a_{[\Gamma]}^{[\alpha]}(x) X_{[\alpha]}(x) \\ &= \sum_{[\alpha] \in I} \left(\sum_{[J] \neq 0} C_{[\Gamma]}^{[J]}(x) \cdot a_{[J]}^{[\alpha]}(x) \right) X_{[\alpha]}(x) \end{aligned} \tag{2.5}$$

for $[\alpha] \in I \subset \bar{\mathcal{A}}$ and $[J]$ as in (2.3). In analogy to (2.4) we find

$$[X_{[\Gamma]}, X_{[\Gamma']}] = \sum_{[\alpha]\in I, \|\alpha\|\leq r} a^{[\alpha]}_{([\Gamma],[\Gamma'])}(x) X_{[\alpha]}(x), \quad [\alpha] \in I$$

where $a^{[\alpha]}_{[\Gamma],[\Gamma']}(x)$ are constants depending on x since the bases for the tangent spaces vary with $x \in \mathbb{R}^d$ may, however, stay constant in a sufficiently small neighbourhood of x. In fact, it does not matter how the vector fields involved were derived. If we replace the element $X_{[\bar{\alpha}]}$ of the basis corresponding to I by the vector field $X_{[\hat{\alpha}]}$ in such a way that $X_{[\alpha']}, [\alpha'] \in I'(x) = (I(x) \setminus [\bar{\alpha}]) \cup [\hat{\alpha}]$, again form a basis of the tangent space at x then the representation of $X_{[\hat{\alpha}]}$ with respect to the old basis introduced in (2.5) reveals the following coefficients which are functions depending on $x \in \mathbb{R}^d$,

$$a^{[\hat{\alpha}]}_{[\bar{\alpha}]}(x) = \frac{\lambda_{I'}(x)}{\lambda_I(x)} \,. \tag{2.6}$$

where the determinants $\lambda_{I'}$ and λ_I are given in (1.7). In order to keep track of the degree of the scaling parameter ε^2 in a suitable approximation of the solution of (1.4) and hence of the associated Jacobian we define submodules $\mathcal{A}^p_s := \mathcal{A}^p_{s,I}$, s, p positive integers, of $C^\infty(\mathbb{R}^d)$ generated by functions of the form $a^{\ell_1}_{j_1} \cdot \ldots \cdot a^{\ell_k}_{j_k}$, $a^{\ell_i}_{j_i}$, $1 \leq i \leq k$, as in (2.6) for $k \leq s$ and $p \leq (\|\ell_1\| + \ldots + \|\ell_k\|) - (\|j_1\| + \ldots + \|j_k\|)$. Here $\|\ell_n\|$, $1 \leq n \leq d$, are the weights of the elements of I and we may adopt the results of Lemmas 2.3-2.8 in [**NSW85**] with weight $\|\cdot\|$ replacing the degree $d(\cdot)$, see (2) in [**NSW85**]. Here the representation given in (2.5) is exploited. Since $\|\cdot\|$ and d are linear in the degree of the Lie brackets and the order of differentiation, the proofs carry over without further changes at this point and we don't repeat the line of arguments but recall the results.

LEMMA 2.2. (1) *For a vector field $Y_{[J]} \in \mathcal{A}_k$ and $J = (j_1, \ldots, j_k)$ any k-tuple of positive integers and $f \in \mathcal{A}^p_s$, we have $Y_{[J]}f \in \mathcal{A}^{p-\|J\|}_{s+k}$.*

(2) *Let $a^\ell_{[J]}$ as in (2.6) then it is $a^\ell_{[J]} \in \mathcal{A}^{\|i_\ell\|-\|J\|}_1$.*

(3) *For any k-tuple $J = (j_1 \ldots, j_k)$ there exists $f_{[J]} \in \mathcal{A}^{-\|J\|}_k$ so that for all $x \in \mathbb{R}^d$ for which I constitutes a basis we have*

$$Y_{[J]}\lambda_I(x) = f_{[J]}(x)\lambda_I(x) \quad .$$

(4) *For any multi index $[J] = ([j_1], \ldots, [j_N])$, every s-tuple $K = (k_1, \ldots, k_s)$, and x such that I corresponds to a basis there exists a real function $f_{[K],[J]} \in \mathcal{A}^{\|I\|-\|J\|-\|K\|}_{N+s}$ such that*

$$Y_{[K]}\lambda_J(x) = f_{[K],[J]}(x)\lambda_I(x) \qquad , \; x \in \mathbb{R}^d \quad .$$

For small times $t \geq 0$ the solution of a differential equation e.g. (1.4) can locally be described by an exponential mapping. A tool to split off a part of the evolution is the Campbell Hausdorff formula see [**NSW85**] and references therein. For C^∞-vector fields X and Y we have $\exp X \exp Y = \exp Z$ where

$$Z = X + Y + \sum_{m=2}^{\infty} \sum (-1)^{m-1} r(q,p) (\operatorname{ad} Y)^{q_m} (\operatorname{ad} X)^{p_m} \cdot \ldots \cdot (\operatorname{ad} Y)^{q_1} (\operatorname{ad} X)^{p_1}$$

with $r(q,p) := (m\,(\sum_{i=1}^m (p_i+q_i))\prod_{i=1}^m (p_i!q_i!))^{-1}$ and the inner sum is over all m-tuples of pairs of nonnegative integers (p_i, q_i) such that $p_i + q_i > 0$ and $\operatorname{ad} X(Y) := [X,Y]$ $(\operatorname{ad} X := X)$.

We proceed in studying the behaviour of the balls given by the pseudo metric $d_{\varepsilon,R}(\cdot,\cdot)$, i.e.

$$B_{\varepsilon,R}(x,\varepsilon) = \{y \in I\!R^d | d_{\varepsilon,R}(x,y) < \varepsilon\} \quad . \tag{2.7}$$

Changing variables, i.e. $h_s \longrightarrow \varepsilon h_s$, in (1.4) we obtain:

$$dy_{\varepsilon,s}(\varepsilon h) = \varepsilon^2 X_0(y_{\varepsilon,s}(\varepsilon h))ds + \varepsilon \sum_{l=1}^{m} X_i(y_{\varepsilon,s}(\varepsilon h))dh_s^i \ , \tag{2.8}$$

with initial condition $y_{\varepsilon,0}(\varepsilon h) = x$. Since the vector fields have bounded derivatives of all orders the solution $y_{\varepsilon,s}(\varepsilon h)$ defines a C^∞ flow in ε and x, see e.g. [**Aro89**] Thm. 4 and Thm. 10. Moreover, Ben Arous generalized a Taylor formula given in [**Aze81**] Thm. 1.6 where the terms of the expansion are terms of a chaos decomposition, i.e. iterated integrals

$$\int \cdots \int_{0<s_1<\ldots<s_m<t} d(h_1)^{j_1} \ldots d(h_m)^{j_m} \tag{2.9}$$

for natural numbers j_i, $1 \le i \le m$, combined with rational factors and derivatives of the coefficient functions of the given vector fields evaluated at the point x. For this expansion there exists an alternative representation in terms of corresponding Lie brackets. Ben Arous gives a convergence result for length as well as for weight. This allows for a comprehensive formula for the solution $y_{\varepsilon,s}$ in (2.8), i.e., changing variables $h \to \varepsilon h$ for $s = 1$ and ε sufficiently small. Since the vector fields have bounded derivatives of all orders we obtain uniformly in $x \in I\!R$ for natural numbers $r'' \ge r' \ge r$, r as in (2.3) (see also the definition of the set $\bar{\mathcal{A}}$ above)

$$y_{\varepsilon,1}(\varepsilon h)(x) = \exp\left[\varepsilon^2 X_0 + \sum_{[\alpha]\in\mathcal{A}_{r''}} \varepsilon^{\|\alpha\|} X_{[\alpha]} F_{[\alpha]}(h)\right](x) + o(\varepsilon^{r'}) \tag{2.10}$$

where the linear combinations $F_{[\alpha]}$ of the iterated integrals (2.9) are of the form mentioned above see [**Aze81**] or [**Aro89**]. Moreover r' and r'' correspond to the length and the weight of the vector fields of the exponential mapping, respectively. Given the Taylor formula admitting multi indices of maximal length r' in the summation then choosing r'' so that $|\alpha| < r'$ implies $\|\alpha\| < r''$ (e.g. $r'' = 2r'$) and including all terms associated with $\mathcal{A}_{r''}$ simply improves the approximation. The scaling property of the iterated integrals reveals (2.10). In a further step we apply Campbell-Hausdorff formula in order to split off the evolution described by the exponential function $\exp(\varepsilon^2 X_0)$, using the same order of approximation as in (2.10). This distinguishes our representation from previous approaches and leads to polynomials of $F_{[\alpha]}(h)$ of higher order than one, which complicates proving the existence of diffeomorphisms between different types of balls substantially. From the Campbell Hausdorff formula we get the sum over $[\alpha] \in \mathcal{A}_{r''}$ as well as Lie brackets of this sum with $\varepsilon^2 X_0$ and itself of various orders. Applying (2.4) we order these last terms to become a sum with respect to $\|\Gamma\| \le r$, $|\Gamma| > 1$. Here the number of X_0 in $[\Gamma]$ is strictly positive. The smooth bounded coefficient functions arising in this context are being evaluated at the point x_ε. Hence $y_{\varepsilon,1}(\varepsilon h)(x)$ in (2.10) becomes

$$(2.11)\quad \exp\left[\sum_{[\alpha]\in\mathcal{A}_{r''}} \varepsilon^{\|\alpha\|} X_{[\alpha]} F_{[\alpha]}(h) + \sum_{\substack{\|\Gamma\|\le r \\ |\Gamma|>1, |\Gamma|_0>0}} n_\Gamma X_{[\Gamma]} \prod_{[\alpha]\in\Gamma} \varepsilon^{\|\alpha\|} F_{[\alpha]}(h)\right](x_\varepsilon) + o(\varepsilon^{r'})$$

where $|\Gamma|_0 = 2\#\{\alpha_i \in \Gamma \mid \alpha_i = 0\}$. We still have unnecessarily many terms in the sum corresponding to the index set $\mathcal{A}_{r''}$. We recall that there exists a generating set for all tangent spaces corresponding to the index set $\bar{\mathcal{A}}$. Moreover, for any choice of initial point x we find a basis $X_{[\alpha]}$, $[\alpha] \in I = I(x_\varepsilon)$, $I \in \bar{\mathcal{A}}$ for the tangent space at x_ε. Using (2.1), and neglecting $o(\varepsilon^{r'})$ we get the following approximation for the solution (2.11):

(2.12)

$$\varphi_{\varepsilon,v}(u)(x_\varepsilon) := \exp\left[\sum_{[\alpha]\in I} X_{[\alpha]} u_{[\alpha]} + \sum_{[\beta]\in\bar{\mathcal{A}}\setminus I} X_{[\beta]} v_{[\beta]} + \sum_{\substack{\|\Gamma\|\le r \\ |\Gamma|>1, |\Gamma|_0>0}} X_{[\Gamma]} P_{[\Gamma]}(u)\right](x_\varepsilon)$$

where we also included the terms corresponding to $\mathcal{A}_{r'} \setminus \bar{\mathcal{A}}$ into the whole rest sum of the exponent. We are setting $P_\Gamma(u) := n'_\Gamma \prod_{[\alpha]\in\Gamma} u_{[\alpha]} \prod_{[\beta]\in\Gamma} v_{[\beta]}$, where $u_{[\alpha]} := \varepsilon^{\|\alpha\|} F_{[\alpha]}(h), [\alpha] \in I$, and $v_{[\beta]} := \varepsilon^{\|\beta\|} F_{[\beta]}(h)$, $\beta \in \mathcal{A}_{r'} \setminus I$. We stress that the coefficients $u_{[\alpha]}$ heavily depend on the choice of the function h in the Cameron Martin space.

The main difference between (2.12) and the expression for $\phi_v(u)$ in [**NSW85**] (chapter II(23)) is that we examine expressions which include polynomials in $u_{[\alpha]}$, $[\alpha] \in [\Gamma]$, as coefficient functions instead of linear ones.

3. Properties of the Exponential Function

The aim of this chapter is to study the behaviour of $\varphi_{\varepsilon,v}(u)$ for small ε and u, v small without taking care of the fact, that u, v depend on εh. We shall choose an auxiliary starting point y_0. Since all given vector fields are smooth for fixed v the mapping $u \to \varphi_{\varepsilon,v}(u)(y_0)$ depends in a smooth way on the parameter $u \in \mathbb{R}^d$ (cf. also the appendix of [**NSW85**] incase v varies in a sufficiently small compact set). Its image of a box in $\mathbb{R}^d$ constitutes a ball B_I. The estimates given for the Jacobian of the function $\varphi_{\varepsilon,v}(u)(y_0)$ reveal as upper and lower bounds for the volume of the ball B_I exactly bounds of the type given in Theorem 1.1 stated in the introduction. In the sequel we shall see why this also holds for all v in a sufficiently small box. The discussion on the correct dependence of y_0 on ε will be postponed to Section 4.

For each $y_0 \in \mathbb{R}^d$ we define the following balls induced by the flow $u \mapsto \varphi_{\varepsilon,v}(u)(y_0)$, cf. (2.12), approximating the flow corresponding to (1.4)

$$(3.1)\qquad B_I(y_0, v, \eta_2\varepsilon) = \{y \in \mathbb{R}^d | y = \varphi_{\eta_2\varepsilon,v}(u)(y_0) \text{ with } |u_j| < (\eta_2\varepsilon)^{\|\alpha_j\|}\} \quad .$$

Thus the ball $B_I(y_0, v, \eta_2\varepsilon)$ is exactly the image under the map $\varphi_{\eta_2\varepsilon,v}(u)(y_0)$ of the box $\{u \in \mathbb{R}^d | |u_j| < (\eta_2\varepsilon)^{\|\alpha_j\|}\}$. For $X_0 \equiv 0$ the balls B_I coincide with the balls of the same notation in [**NSW85**]. Now we introduce the variable v in order to account for the translation imposed by the vector fields $X_{[\beta]}, [\beta] \in \bar{\mathcal{A}} \setminus I$, with coefficients satisfying $|v_i| < (\eta_2\varepsilon)^{\|\beta_j\|}$.

PROPOSITION 3.1. *For fixed $y_0 \in \mathbb{R}^d$ let $\lambda_I(y_0)$ and $\lambda_{I'}(y_0)$, $I = (\alpha_1, \ldots, \alpha_d)$ for $\|\alpha_i\| \le r'$ as in (2.5) and $I' = (\alpha'_1, \ldots, \alpha'_d)$ for $\|\alpha'_i\| \le r'$, respectively, correspond to bases at y_0, be as defined in (1.7). Assume that there exists a real number $\eta_1 \in (0,1)$ such that:*

$$(3.2) \qquad |\lambda_I(y_0)|\varepsilon^{\|I\|} \ge \eta_1 \max_{I'} |\lambda_{I'}(y_0)|\varepsilon^{\|I'\|} \quad .$$

Then there exists $\eta_2 \in (0,1)$ such that for

$$(3.3) \qquad \left|v_{[\beta]}\right| \le (\eta_2\varepsilon)^{\|\beta\|}$$

the map $u \mapsto \varphi_{\eta_2\varepsilon,v}(u)(y_0)$ is one to one in the box

$$(3.4) \qquad \left|u_{[\alpha]}\right| \le (\eta_2\varepsilon)^{\|\alpha\|}$$

and for its Jacobian $J\varphi_{\eta_2\varepsilon,v}(u)(y_0)$ we have

$$(3.5) \qquad \frac{1}{4}|\lambda_I(y_0)| \le |J\varphi_{\eta_2\varepsilon,v}(u)(y_0)| \le 4|\lambda_I(y_0)| \quad .$$

PROOF. In analogy to the proof of Theorem 7 [**NSW85**] we start off by showing the line of inequalities (3.5). For this part of the proof we proceed along the line of arguments in [**NSW85**] replacing degree by weight as defined in the introduction. We have taken into account, that the coefficient of X_0 is ε^2 when choosing the boxes (3.3) and (3.4). For $a^{[\alpha]}_{[\alpha']}(y_0)$ being an iterated version of (2.6) we have, moreover, that for all $[\alpha']$ the estimate

$$(3.6) \qquad \left|a^{[\alpha]}_{[\alpha']}(y_0)\right| \le C\varepsilon^{\|\alpha\|-\|\alpha'\|}$$

is satisfied where $[\alpha] \in I \subset \bar{\mathcal{A}}$. For $[\alpha]$ and $[\alpha']$ as in (2.6) and I as in (3.2) the equation is a reformulation of (3.2). The general result (3.6) follows by iteration due to Lemma 2.2[2]. Inserting the definition of $a^{[\ell]}_{[\alpha']}$ given in (2.6) and adopting the notation of Lemma 2.2 then by combining Lemma 2.2[2] respectively Lemma 2.2[3] with (3.6) we get

$$(3.7) \qquad \left|X_\Gamma a^{[\ell]}_{[\alpha']}(y_0)\right| \le C\varepsilon^{\|\ell\|-\|\Gamma\|-\|\alpha'\|} \quad \text{and} \quad |X_\Gamma \lambda_{I},(y_0)| \le C\varepsilon^{\|I\|-\|J\|-\|K\|}$$

at a fixed point y_0 and for fixed parameter v. In order to show that these bounds hold for all $y \in B_I(y_0, v, \eta_2\varepsilon)$ uniformly in v satisfying (3.3), we need to recall that $\varphi_{\eta_2\varepsilon,v}(u)(y_0)$ depends in a smooth way on the variable u, such that for u with $|u_j| < \eta_2\varepsilon^{\|\alpha_j\|}$ there is a constant $C_n > 0$ and

$$(3.8) \qquad \left|\lambda_I(\varphi_{\eta_2\varepsilon,v}(u))(y_0) - \sum_{|k|\le n} \frac{1}{k!} \sum_{[\alpha]\in I} \lambda_I(\varphi_{\eta_2\varepsilon,v}(0)(y_0))(u_{[\alpha]}X_{[\alpha]})^k(y_0)\right| \le C_n|u|^{n+1} \quad .$$

Recalling the assumption (3.2) we can choose n uniformly in $B_I(y_0, v, \eta_2\varepsilon)$ and use (3.6) in such a way that $\frac{1}{4}|\lambda_I(y_0)|$ is an upper bound for $C_n|u|^{n+1}$ as well as for the approximating expansion in (3.8) if η_2 is sufficiently small. Introducing the bound in (3.8), for all $y \in B_I(y_0, v, \eta_2\varepsilon)$ we get

$$(3.9) \qquad |\lambda_I(y_0) - \lambda_I(y)| < \frac{1}{2}|\lambda_I(y_0)| \quad .$$

An analogous result holds for $\lambda_{I'}(y)$, I' another basis. This leads to the inequalities

$$(3.10) \qquad |\lambda_{I'}(y)| \leq C\varepsilon^{\|I\|-\|I'\|}|\lambda_I(y)| \qquad \text{and} \qquad \left|a_{[\alpha]}^{[\alpha']}(y)\right| \leq C\varepsilon^{\|\alpha\|-\|\alpha'\|}$$

where we inserted the definition (2.4). In analogy to the argument which lead to the inequalities (3.7) at point y_0 above, we furthermore find that for $[K]$, $[L]$, and $[J]$ being s-, r, and n-tuples with $s, r \leq n$, then

$$\left|X_{[K]}a_{[L]}^{\ell}(y)\right| \leq C\varepsilon^{\|i_\ell\|-\|K\|-\|L\|}, \text{ and } \left|X_{[K]}\lambda_{[J]}(y)\right| \leq C\varepsilon^{\|I\|-\|J\|-\|K\|}|\lambda_I(y)|$$

which is comparable to Theorem 6 in [**NSW85**]. Recalling the definition of the Jacobian and inserting the above estimates reveals the bounds for the Jacobian $J\varphi_{\eta_2\varepsilon,v}(u)(y_0)$ in terms of $\lambda_I(y)$ claimed in (3.5) uniformly in $y \in B_I(y_0, v, \eta_2\varepsilon)$, which can be seen by the following argument: For elements $y \in B_I(y_0, v, \eta_2\varepsilon)$, since the multiple integrals of functions of the Cameron Martin space are elements of a Banach space, we may simplify $\varphi_{\eta_2\varepsilon,v}(u)$ cf. (2.12) with $\eta_2\varepsilon$ replacing ε once more by fixing $s_\alpha := u_{[\alpha]}$, $s_\Gamma := P_\Gamma(u)$ and examine the function

$$(3.11) \qquad \psi(s) := \varphi_{\eta_2\varepsilon,v}(u)\ .$$

This allows to estimate the outer partial derivatives. For the sake of a simple notation for the exponent in (3.11) we introduce $s \diamond X := \sum_{j=1}^M s_j X_j$ for some enumeration of the multi indices $[\alpha]$ and $[\Gamma]$ in (2.12). We recall the following results from [**NSW85**] in Lemma 2.12

$$(3.12) \qquad \frac{\partial}{\partial s_j}\psi(s) := X_j + \sum_{k=2}^{n} \gamma_k[s \diamond X, \ldots, [s \diamond X, X_j]\ldots] + o(|s|^{n+1})\ .$$

In analogy to (2.12) the finite sum in (3.12) can be expanded with respect to the basis $\mathcal{B} := \{X_{[\alpha]} | [\alpha] \in I\}$, I as in (3.2), using the linearity of the Lie brackets. In the sequel we estimate the coefficients $b_j^{[\alpha]}$ with respect to the basis vectors $X_{[\alpha]}$, $\alpha \in I$. Inserting (3.10), (3.2) and (3.3) for each $k \geq 1$ and M as in (3.12) we have

$$\begin{aligned} &\sum_{j_1=1}^{M} \cdots \sum_{j_k=1}^{M} \prod_{l=1}^{k} s_{j_l}[X_{j_1}, \ldots, [X_{j_k}, X_j]\ldots] \\ &= \sum_{[\alpha]\in I}\sum_{j_1=1}^{M} \cdots \sum_{j_k=1}^{M} \prod_{l=1}^{k} s_{j_l} a_{[j_1,\ldots,j_k,j]}^{[\alpha]} X_{[\alpha]} \quad =: \quad \sum_{[\alpha\in I]} b_{k,j}^{[\alpha]} X_{[\alpha]} \end{aligned}$$

whence for the coefficient $b_j^{[\alpha]}$ we find by pulling the norm into the sum

$$\begin{aligned} \left|b_j^{[\alpha]}\right| &\leq \sum_{k=1}^{n}\sum_{j_1=1}^{M} \cdots \sum_{j_k=1}^{M} \left|\prod_{l=1}^{k} s_{j_l}\right| \left|a_{[j_1,\ldots,j_k,j]}^{[\alpha]}\right| \\ &\leq \sum_{j_1=1}^{M} \cdots \sum_{j_k=1}^{M} \sum_{k=1}^{n} (\eta_2\varepsilon)^{\sum_{l=1}^{k}\|j_l\|} C_k \varepsilon^{\|\alpha\|-\|j\|-\sum_{l=1}^{k}\|j_l\|} \\ (3.13) \qquad &= K_k(\eta_2\varepsilon)^{\|\alpha\|-\|j\|} \end{aligned}$$

where $K_k = nM^2\mathcal{C}_k$ does not depend on y_0 since the $a^{[\alpha]}_{[j_1,\dots,j_k,j]}$ are bounded functions. On the other hand by definition s depends on u, which for the inner derivatives means

$$\frac{\partial s_j}{\partial u_{[\alpha']}} = \begin{cases} \delta_j^{[\alpha']} & \text{for } j \in \mathcal{A}_r \ , \\ C_{[\Gamma]} \prod\limits_{[\alpha]\in\Gamma,\|\alpha\|\le r}{}' u_{[\alpha]} \prod\limits_{[\beta]\in\Gamma} v_{[\beta]} & \text{for all other } j \end{cases} \tag{3.14}$$

where $\delta_j^{[\alpha']}$ is the Dirac measure in $I\!N$, and $\prod'$ emerges from $\prod$ by reducing the power of $u_{[\alpha']}$ in $\prod$ by one. Since u and v are in the boxes given by (3.3) and (3.4), combining (3.13) and the corresponding estimates for (3.14) as well as neglecting $o\left((|u_\alpha| + |P_\Gamma(u)|)^{n+1}\right)$ in (2.12) reveals

$$\frac{\partial}{\partial u_{[\alpha']}} \varphi_{\eta_2\varepsilon,v}(u)(y_0) = \sum_{[\alpha]\in I} \left(\delta^{[\alpha]}_{[\alpha']} + \mathcal{C}^\alpha_{\alpha'}\right) X_{[\alpha]}(y_0)$$

with $|\mathcal{C}^\alpha_{\alpha'}| \le \mathcal{C}(\eta_2\varepsilon)^{\|\alpha\|-\|\alpha'\|}$. Recalling that $J\varphi_{\varepsilon,v} = \det\left(d\varphi_{\varepsilon,v}\frac{\partial}{\partial u_1}, \dots, d\varphi_{\varepsilon,v}\frac{\partial}{\partial u_d}\right)$ and using that for ε sufficiently small $\frac{1}{2}|\lambda_I(y_0)| \le |\lambda_I(\varphi_{\varepsilon,v}(u)(y_0))| \le 2|\lambda_I(y_0)|$ finally reveals the bounds (3.5).

Let us now show that $u \mapsto \varphi_{\eta_2\varepsilon,v}(u)(y_0)$ cf. (2.12) is one to one in the box $|u_{[\alpha]}| < (\eta_2\varepsilon)^{\|\alpha\|}$ for (3.3). We replace the method in [**NSW85**] by a more involved combinatoric argument. Let us suppose that

$$\varphi_{\eta_2\varepsilon,v}(u)(y_0) = \varphi_{\eta_2\varepsilon,v}(\bar{u})(y_0) \quad . \tag{3.15}$$

By Corollary 4.2, Appendix [**NSW85**], we know that $u \mapsto \varphi_{\eta_2\varepsilon,v}(u)(y)$ is a diffeomorphism on a subset of a neighbourhood of $\exp\sum_{[\Gamma]} s_{[\Gamma]}X_{[\Gamma]}(y_0)$. Hence $\varphi^{-1}_{\eta_2\varepsilon,v}(u)\cdot \varphi_{\eta_2\varepsilon,v}(u)(y_0) = y_0$ is defined. Since $X_{[J]}$, $[J] \in \mathcal{A}_r$, $r \ge d$, are not independent, we use Campbell Hausdorff formula when applying $\varphi^{-1}_{\eta_2\varepsilon,v}(u)$ to the right hand side of (3.15). Using the comprehensive description of $\varphi_{\eta_2\varepsilon,v}$ given in (3.12) for $s = s(u)$ and $\bar{s} = s(\bar{u})$, we may rewrite (3.15) using the Campbell Hausdorff formula and the smoothness of the vector fields

$$\begin{aligned} y_0 &= \exp(s \diamond X + (\bar{s} - s) \diamond X) \exp(-s \diamond X)(y_0) \\ &= \exp(H_n(s, \bar{s}, x) + \dots)(y_0) + R \end{aligned} \tag{3.16}$$

with $H_n(s,\bar{s},x) := (\bar{s}-s)\diamond X + [(\bar{s}-s)\diamond X, (-s)\diamond X]$ and $s\diamond X$ as in (3.12). Equation (3.16) includes only Lie brackets up to weight n cf. Appendix in [**NSW85**] and R corresponds to $o\left((|u_\alpha| + |P_\Gamma(u)|)^{n+1}\right) + o(\varepsilon^{r''})$. Since the terms, constructed from an arbitrary order of Lie brackets of $\pm s \diamond X$ with itself only, vanish due to the commutativity of the vector fields involved, each term in the truncated Campbell Hausdorff expansion (3.16) includes the vector field $(\bar{s} - s) \diamond X$ at least once. Rewriting (3.16) with respect to the basis $\mathcal{B}$ we deduce the following bound for the remainder

$$|R| \le C\,\varepsilon^{N_1} \sum_{[\alpha]\in I} \left|\bar{u}_{[\alpha]} - u_{[\alpha]}\right| \quad \text{with } N_1 := n - sup\,\|\alpha\| \tag{3.17}$$

cf. Lemma 2.2[2]. In the sequel we use the fact that for arbitrary vector fields Y_α having bounded derivatives and spanning $I\!R^d$ and for λ_α sufficiently small the equation $\exp[\sum \lambda_\alpha Y_\alpha](y_0) = 0$ implies $\sum \lambda_\alpha Y_\alpha(y_0) = 0$. In order to see this we use the fact that $\exp[\sum \lambda_\alpha Y_\alpha](y_0) = I + \sum \lambda_\alpha Y_\alpha(y_0) + \mathcal{R}$ where the remainder consists of higher order Lie brackets. On the linear submanifold $\sum \lambda_\alpha Y_\alpha = 0$ the rest $\mathcal{R}$ is equal to zero. Moreover, $\mathcal{R} = \mathcal{R}(y_0)$ defines a smooth function of y_0 vanishing on this submanifold. Therefore the following general result applies: For a smooth function f vanishing on a linear submanifold given by $H(\lambda) = \langle H, \lambda \rangle = 0$, H surjective and $\|H\| = 1$, then there exists $c > 0$ such that $|f(\lambda)| \leq c|\langle H, \lambda\rangle|$. This induces that there exist $C > 0$ such that $\mathcal{R} \leq C|\sum \lambda_\alpha Y_\alpha|$ which in particular tends to zero for $\lambda_\alpha \to 0$.

For n sufficiently large, and for $y_0 \in I\!R^d$ fixed due to (3.2) there exists a set $\tilde{I} := \tilde{I}(y_0)$ of multiindices $[\tilde{\alpha}_i] \neq 0$, $\|\tilde{\alpha}_i\| \leq r$, $1 \leq i \leq d$, such that the mappings $t_{[\alpha_i]} \mapsto \exp(\sum_{[\tilde{\alpha}] \in \tilde{I}} t_{[\tilde{\alpha}]} X_{[\tilde{\alpha}]})(y_0)$ are uniformly local diffeomorphisms cf. [**NSW85**, Appendix].

This holds because the set of vector fields $X_{[\tilde{\alpha}]}(y_0)$, $[\tilde{\alpha}] \in \tilde{I}(y_0)$, is a basis of $I\!R^d$ uniformly in y_0 with a determinant which has a strictly positive uniform lower value in norm due to assumption (3.2) and because all the derivatives of the vector fields given in the beginning are bounded. Moreover, smoothness is invariant with respect to a small perturbation in form of the rest R. Hence there exists $g_{[\tilde{\alpha}]} \in I\!R$, $[\tilde{\alpha}] \in \tilde{I}$, such that the rest R in (3.16) can be realized by an additional smooth vector field in the exponent in (3.16), i.e.

$$
\begin{aligned}
y_0 &= \exp\left(H_n(s, \bar{s}, X) + \sum_{[\tilde{\alpha}] \in \tilde{I}} g_{[\tilde{\alpha}]} X_{[\tilde{\alpha}]}\right)(y_0) \\
\left|g_{[\tilde{\alpha}]}\right| &\leq C\,\varepsilon^{N_1} \sum_{[\alpha'] \in \tilde{I}} \left|\bar{u}_{[\alpha']} - u_{[\alpha']}\right|
\end{aligned}
\tag{3.18}
$$

for all $[\tilde{\alpha}] \in \tilde{I}$ and the exponent will be a local diffeomorphism on the box (3.4) as well. Moreover for ε sufficiently small the vector fields in (3.18) and their derivatives are small since the given vector fields are bounded with bounded derivatives of all orders, and the coefficients are small according to (3.5). Hence, in y_0 we have

$$
H_n(s, \tilde{s}, X) + \sum_{[\tilde{\alpha}] \in \tilde{I}} g_{[\tilde{\alpha}]} X_{[\tilde{\alpha}]} = 0 \quad . \tag{3.19}
$$

We now examine the behaviour of the left hand side of (3.19). We recall the definition of $s, \bar{s}$ in (3.11), and represent it with respect to the basis in (3.2). For $\Gamma \notin I$ we deduce the following estimate for $\bar{s}_{[\Gamma]} - s_{[\Gamma]}$ since (3.10) holds (due to Lemma 2.2[2]) meeting the assumptions (3.3),

$$
|P_\Gamma(\bar{u}) - P_\Gamma(u)| \leq C\eta_2 \varepsilon^{\|\Gamma\|} \left(\sum_{\alpha \in I} \frac{\left|\bar{u}_{[\alpha]} - u_{[\alpha]}\right|}{\varepsilon^{\|\alpha\|}} \right) . \tag{3.20}
$$

Recalling the definition of H_n cf. (3.16) equation (3.18) can be written

$$
\begin{aligned}
y_0 &= \exp\left[H_n(s, \tilde{s}, X) + \sum_{[\tilde{\alpha}] \in \tilde{I}} g_{[\tilde{\alpha}]} X_{[\tilde{\alpha}]}\right](y_0) \\
&= \exp\left[\sum_{[\alpha] \in \tilde{I}} g_{[\alpha]}(s, \tilde{s}, x) X_{[\alpha]}\right](y_0).
\end{aligned}
$$

Following the general argument given above we deduce that $g_{[\alpha]}(s,\tilde{s},x)=0$ because $g_{[\alpha]}(s,\tilde{s},x)$ are small. This can be seen as follows. By proceeding as in (3.13) we get a representation of the Campbell Hausdorff approximation (3.16) with respect to the basis I where the coefficients $g_{[\alpha]}$, $[\alpha]\in I$, are bounded due to the estimates given in (3.18) and (3.20). From (3.19) we find

$$g_{[\alpha]} \le C\eta_2\varepsilon^{\|\alpha\|} \sum_{[\alpha']\in I} \frac{\left|\bar{u}_{[\alpha']}-u_{[\alpha']}\right|}{\varepsilon^{\|\alpha'\|}} \quad . \tag{3.21}$$

Analogously we prove the same behaviour for the vector fields realizing the rest term R in (3.16) for N_1 as above. Inserting into (3.19), and using (3.21) there follows for C sufficiently small that

$$\left|\bar{u}_{[\alpha]}-u_{[\alpha]}\right| \le C\eta_2\varepsilon^{\|\alpha\|} \sum_{[\alpha']\in I} \frac{\left|\bar{u}_{[\alpha']}-u_{[\alpha']}\right|}{\varepsilon^{\|\alpha'\|}} \, .$$

Choosing α_0 such that $\frac{|\bar{u}_{[\alpha]}-u_{[\alpha]}|}{\varepsilon^{\|\alpha\|}}$ has its maximum in α_0, yields a contradiction for $\bar{u}_{\alpha_0}-u_{\alpha_0}\neq 0$. This proves the injectivity of the map $u\mapsto \varphi_{\eta_2\varepsilon,v}(u)(y_0)$ for all $|v_\Gamma|\le \eta_2\varepsilon^{\|\Gamma\|}$ and finishes the proof of the proposition. □

Inserting the definition (3.1) of B_I the chain of inequalities (3.5) reveals the correct estimates for its volume $\mathrm{Vol}(B_I)$ at a given point incase the assumptions of Proposition 3.1 are satisfied. We recall that due to condition (1.3) we deduce (2.4), i.e. a finite number of basis generates all tangent spaces. Let $I=I(x)$ be a d-tuple such that $\|I\|$ is minimal among all d-tuples I' with $\lambda_{I'}\neq 0$ and such that

$$|\lambda_I(x)| = \max_{\|I'\|=\|I\|} |\lambda_{I'}(x)| \quad .$$

Then there exists ε_0 depending on y_0 such that

$$|\lambda_I(x)|\varepsilon^{\|I\|} \ge |\lambda_{I'}(x)|\varepsilon^{\|I'\|}$$

for all ε, $0<\varepsilon<\varepsilon_0$ and all d-tuples I'. Furthermore the Jacobian of the exponential map is the identity at the origin. Since the given vector fields are supposed to have bounded derivatives of all orders the rest terms in the approximations (2.11), (3.8), and (3.16) hold uniformly in $x\in \mathbb{R}^d$. Therefore we can replace N_1 and C in (3.17) as well as rewrite (3.21) in a uniform way. Hence we can choose ε_0 and η_2 sufficiently small such that for v satisfying (3.3) the mapping $\varphi_{\eta_2\varepsilon,v}(\cdot)(x)$ is one to one uniformly in x. The preceding results are summarized in the following lemma.

LEMMA 3.2. *For any $x\in\mathbb{R}^d$ there exist $0<\varepsilon<\varepsilon_0<1$ and an d-tuple of multi indices satisfying*

$$|\lambda_I(x)|\varepsilon^{\|I\|} \ge \frac{1}{2}\max_{I'}|\lambda_{I'}(x)|\varepsilon^{\|I'\|}$$

with $\lambda_{I'}\neq 0$. As a consequence for $\left|u_{[\beta]}\right|<(\eta_2\varepsilon)^{\|\beta\|}$, $[\beta]\in\bar{\mathcal{A}}\setminus I$ the mapping $u\mapsto\varphi_{\eta_2\varepsilon,v}(u)(x)$ with $u\in\mathbb{R}^d$ is globally one to one in $x\in\mathbb{R}^d$ for $\left|u_{[\alpha]}\right|<(\eta_1\varepsilon)^{\|\alpha\|}$, $[\alpha]\in I$

At this point we have collected all results necessary to prove inclusions for the various balls and give estimates for the volume of the balls B_I and may hence finish this paragraph.

4. Proof of the Theorem

Comparing the definitions we find that the following inclusion holds

$$B_{\delta,R}(x,\delta) \subseteq B_\delta(x,\delta) \quad . \tag{4.1}$$

In order to prove the theorem, we have to meet the assumption (3.2) and study the inclusions $B_{\eta_1\varepsilon}(x,\eta_1\varepsilon) \supseteq B_I(x,v,\eta\varepsilon)$ and $B_I(x,v,\varepsilon) \supseteq B_{\eta_2\varepsilon,\mathbb{R}}(x,\eta_2\varepsilon)$ for some $0 < \eta,\ \eta_1,\ \eta_2 < 1$. As will be explained below the first inclusion holds since for η_2 sufficiently small there exists a diffeomorphism between $B_{\eta_2\varepsilon}(x,\eta_2\varepsilon)$ and $B_I(x,v,\eta\varepsilon)$ such that for any $y \in B_I(x,v,\eta\varepsilon)$, v satisfying (3.3), there exists an element z of $B_{\eta_2\varepsilon}(x,\eta_2\varepsilon)$ so that the vectors corresponding to y and z in the boxes (3.4) are lying arbitrarily close. We therefore return to the infinite dimensional case, i.e. we reintroduce the functions h from Cameron Martin space. Let $y \in B_\varepsilon(x,\delta)$. Then there exists a function $h \in \mathcal{H}$ with $h(0) = x$ and $\phi_{\delta,1}(h) = y$ such that $\|h\| \leq \delta$ cf. (1.5). We shall extensively use the condition of Bismut given in the definition of R cf. [**Léa89**] and hence deviate from the procedure in [**NSW85**]. At first we turn to proving the lower bound of Theorem 1.1 formulated in the introduction which corresponds to the lower bound in (3.5).

Refering to the ball B_I we recall that there exists a basis of the tangent space at x_ε corresponding to the set $\tilde I := \tilde I(x_\varepsilon)$ of multiindices $[\alpha_i] \in \mathcal{A}_r$, $[\alpha_i] \neq 0$, $1 \leq i \leq d$, such that $\lambda_{\tilde I}(x_\varepsilon) > 0$. Moreover, according to [**NSW85**] (Appendix) for $[\alpha] \in \tilde I$ the mapping $u_{[\alpha]} \mapsto \varphi_{\eta_2\varepsilon,v}(u_\alpha)(x)$ being defined in (2.12), with I substituted by $\tilde I$ and ε replaced by $\eta_2\varepsilon$ for η_2 as in Proposition 3.1 is a diffeomorphism in u uniformly in x satisfying (3.3). Therefore introducing a small perturbation cf. (3.18) does not destroy the property of $\varphi_{\eta_2\varepsilon,v}(u)$ being a diffeomorphism in u. In particular, we can find $F_{[\tilde\alpha],\eta_2\varepsilon}(h)$, $[\tilde\alpha] \in \tilde I(x_\varepsilon)$, such that independent of x for $\|h\| \leq 1$, $s_{[\tilde\alpha],\varepsilon} := \varepsilon^{\|\tilde\alpha\|} F_{[\tilde\alpha],\eta_2\varepsilon}(h) + o(\varepsilon^{r'})$, ε sufficiently small, and $r' > r \geq \|\alpha\|$, we have

$$\phi_{\eta_2\varepsilon,1}(h)(x) = \exp\left[\sum_{[\tilde\alpha]\in\bar{\mathcal{A}}} s_{[\tilde\alpha],\varepsilon} X_{[\tilde\alpha]} + \sum_{\substack{\|\Gamma\|\leq r\\ |\Gamma|_0>0, |\Gamma|>1}} s_{[\Gamma],\varepsilon} X_{[\Gamma]}\right](x_\varepsilon) \tag{4.2}$$

where $s_{[\tilde\alpha],\varepsilon} := \varepsilon^{\|\tilde\alpha\|} F_{[\tilde\alpha],\varepsilon}(h)$ and $s_{[\Gamma],\varepsilon} := \varepsilon^{\|\Gamma\|} n_\Gamma \Pi_{[\tilde\alpha],[\tilde\beta]\in[\Gamma]} F_{[\tilde\alpha],\varepsilon}(h) F_{[\tilde\beta],\varepsilon}(h)$ for $[\tilde\alpha] \in \tilde I$ and $[\tilde\beta] \in \bar{\mathcal{A}} \backslash \tilde I$. Moreover, for $\varepsilon \longrightarrow 0$ the function $F_{[\tilde\alpha],\varepsilon}(h)$ tends to $F_{[\tilde\alpha]}(h)$ uniformly for $\|h\| \leq 1$, $x \in \mathbb{R}^d$.

Due to the Hörmander hypotheses (1.3) we recover representation (2.4) which allows following the line of arguments in [**NSW85**] and henceforth to choose I such that assumption (3.2) of Proposition 3.1 is satisfied, in particular $\lambda_I(x_\varepsilon) \geq C > 0$: Inserting into (4.2) we get

$$\phi_{\kappa,1}(h)(x) = \varphi_{\kappa,(\kappa)^{\|\beta\|}F_{[\beta],\kappa}(h)}\left((\kappa)^{\|\alpha\|} F_{[\alpha],\kappa}(h)\right)(x_\varepsilon) \tag{4.3}$$

with $\kappa := \eta_2\varepsilon$ for η_2 as in Proposition 3.1 and φ as in (2.12). Let us recall that it is equivalent that a functional which is defined on the Cameron-Martin space $\mathcal{H}$ is a submersion cf. (1.6) and has a deterministic Malliavin matrix which is invertible [**Bis84**][Chap 1]. As was pointed out in Theorem 1 of [**Léa87**], there exists h_0 in the Cameron Martin space, such that

$$\left|F_{[\alpha]}(h_0)\right| < \eta_2 \quad \text{and} \quad \left|F_{[\beta]}(h_0)\right| < \eta_2 \tag{4.4}$$

and such that $h \mapsto (F_{[\alpha]}(h), F_{[\beta]}(h))$, $[\alpha] \in I$ and $[\beta] \in \bar{A} \setminus I$ is almost surely invertible at $h = h_0(\omega)$, ω a path of d-dimensional Brownian Motion. This follows from the line of an argument given by Bismut. Namely, $[F_{[\alpha]}(h), F_{[\beta]}(h)]$ has a stochastic Malliavin matrix which is almost surely invertible if we replace dh by $g(\omega_s)ds$ where ω_s is a flat Brownian motion and g a small smooth function of t which vanishes at $t = 0$ but with derivatives of all orders different to zero, and $g'(0) = 1$ (see [**Léa87**]). This comes from traditional techniques of Malliavin calculus by looking at the differential equation satisfied by $[F_{[\alpha]}(h), F_{[\beta]}(h)]$ and enlarging the space $I\!R^m$ to $I\!R^m \times N(m;r)$ where $N(m;r)$ is the simply connected nilpotent Lie group of order r with m generators. We choose the canonical balls, and the canonical vector fields $\tilde{X}_i$, $i = 0, \ldots, m$. We consider the stochastic differential equation on $I\!R^m \times N(m+1;r)$ for $1 \leq i \leq m$ given by

$$\begin{array}{rcl} dx_t^i & = & dw_t^i \\ dF_{\cdot}(h_t) & = & \tilde{X}_0(F_{\cdot}(h_t))dt + \sum_{i=1}^m \tilde{X}_i(F_{\cdot}(h_t))g(x_t^i)dt \quad \equiv \quad \tilde{Y}_0(F_{\cdot}(h_t), x_t)dt \end{array}$$

in particular we have $[\tilde{Y}_0,\ e_i](0) = \tilde{X}_i(0)$. The system $(x_t,\ F_t)$ satisfies the weak Hörmander hypothesis, except for the component $F_{0,t} = t$ and the solution $(w_t^i, F_{[\alpha]}(h_t^i))$ where $h_t^i = g(w_t^i)$ ensures an invertible Malliavin matrix. Let G be the space spanned by $\tilde{X}_i$, $i \neq 0$, or the Lie brackets of the vector fields e_i, $\tilde{Y}_0$ at 0, respectively. Using the projection theorem of Taniguchi [**Tan83**] there follows that the Malliavin matrix of $\Pi_G(w_t^i, F_{[\alpha]}(h_t^i))$ is almost surely invertible at h_0 if $[\alpha] \neq 0$ where Π_G is the projection from $I\!R^m \times N(m;r)$ into G. Then $\Pi_G(x_1,\ F_1)$ has a smooth density. This, however, is nothing but $F_{[\alpha]}(h_1)$ for $[\alpha] \neq 0$. Therefore the deterministic Malliavin matrix of $F_{[\alpha]}(h_1)$, $[\alpha] \neq 0$ is invertible if $h^i = g(w_t^i)$ almost surely. Therefore we can choose $g(w_t^i)$ small in the Cameron Martin space as stated in (4.4). This implies that there exists ε_0 independent of x_ε such that for ε sufficiently small, the map $h \mapsto F_{[\alpha],\eta_2\varepsilon}(h)$ is a submersion at h_0, uniformly in ε, where (4.4) continues to hold [**Léa87**]. Moreover, by applying the implicit function theorem, we may choose $F_{[\beta],\eta_2\varepsilon}(h) = v_{[\beta],\varepsilon}$ in a neighbourhood V of h_0 and still prove that the map $h \mapsto (F_{[\alpha],\eta_2\varepsilon}(h),\ F_{[\beta],\eta_2\varepsilon}(h))$ is a submersion uniformly in V. Using [**Bis84**, Thm 1.3] and (4.3) there exist $v_{[\beta],\varepsilon}$ with $\left|v_{[\beta],\varepsilon}\right| \leq \eta_2 \varepsilon^{\|\beta\|}$, and a box $B = \prod_{[\alpha]\in I(x_\varepsilon)} [\eta_\alpha \varepsilon^{[\|\alpha\|}, \eta'_\alpha \varepsilon^{\|\alpha\|}]$ with $|\eta_\alpha| \leq \eta_2$, and $|\eta'_\alpha| \leq \eta_2$ where $|\eta_a - \eta'_\alpha| > c > 0$ uniformly in $\varepsilon < \varepsilon_0$ and in x such that $B_{\eta_2\varepsilon}(x, \eta_2\varepsilon)$ contains the image of $\varphi_{\eta_2\varepsilon,v}\ (\ \cdot\ )\ (x_\varepsilon)$ of the box B "up to an infinitesimally small error" in $I\!R^d$.

Then change of variables for integrals, assumption (3.2), and (3.5) reveal that $\mathrm{Vol}(B_I(x, v, \varepsilon)) \simeq |\lambda_I(x_\varepsilon)| \varepsilon^{\|I\|} \simeq \sum_I |\lambda_I(x)| \varepsilon^{\|I\|}$ where the single multi index I corresponds to the basis satisfying (3.2) in Proposition 3.1 and the summation is over all $I \subset \bar{A}$ corresponding to basis for arbitrary tangent spaces. Enlarging the subriemannian ball $B_{\eta_2\varepsilon}(x, \eta_2\varepsilon)$ due to monotonicity of integrals we get for $1 > \eta_1 > \eta_2 > 0$

$$c_1 \sum_I |\lambda_I(x_\varepsilon)| \varepsilon^{\|I\|} \leq c \left| \lambda_I(x_{\eta_1\varepsilon}) \varepsilon^{\|I\|} \right| \leq \mathrm{Vol}\ B_{\eta_1\varepsilon}(x, \eta_1\varepsilon)\ .$$

This finishes the proof of the lower bound.

Moreover, following the line of arguments above due to Theorem 1.3 in [**Bis84**] the function $\phi_{\kappa,1}(\cdot)(x)$ is a submersion. Hence this point is de facto an element of $B_{\eta_2\varepsilon,R}$ except for an infinitesimally small error as a consequence of the bounds in (4.4). The proof of the upper bound or more precisely of the chain of inequalities in Theorem 1.1 is a direct consequence of the above proof. It follows from (4.1) and the fact that for any point in the image of the box $|u_{[\alpha]}| \leq (\eta_2\varepsilon)^{\|\alpha\|}$, $[\alpha] \in I$, under the map $\varphi_{\eta_2\varepsilon,v}(\cdot)(x)$, v satisfying (3.3), equations (4.3) and (4.4) hold. This completes the proof due to monotonicity of the volume. We emphasize that all arguments in this chapter hold uniformly in x.

Acknowledgements

The authors thank Professors S. Albeverio and G. Ben Arous for interesting and stimulating comments. We are also deeply indebted to Y. Ishikawa and J.L. Wu for carefully reading the manuscript and many helpful discussions.

References

[Aro89] G. Ben Arous, *Flots et series de Taylor stochastiques*, Prob. Th. rel. Fields **81** (1989), 29–77.

[Aze81] R. Azencott, *Formule de Taylor stochastique et développement asymptotique d'inté grales de feynman — séminaire de probabilités XVI 1980/81*, Lecture Notes in Mathematics 921, eds J. Azéma, M. Yor, Springer Verlag (1981), 237–284.

[Bis84] J.M. Bismut, *Large deviations and Malliavin calculus*, Progress in Mathematics **45** (1984).

[JSC86] D. Jerison and A. Sánchez-Calle, *Estimates of the heat kernel for a sum of squares of vector fields*, Indiana Math. J. **35** (1986), 835–854.

[KS87] S. Kusuoka and D. Stroock, *Applications of the Malliavin calculus part III*, J. Fac. Sciences Univ. Tokyo **34** (1987), 391–442.

[Léa87] R. Léandre, *Minoration en temps petit de la densité d' une diffusion degenérée*, J. F. A. **74** (1987), 399–415.

[Léa89] R. Léandre, *Volume de boules sous-Riemannienes et explosion du noyau de la chaleur au sens de Stein*, Seminaire de Probabilités XXIII, Lect. Notes Math. (J.Azéma and P.A.Meyer, eds.), Springer Berlin, 1989.

[Léa94] R. Léandre, *Uniform upper bounds for hypoelliptic kernels with drift*, J. Math. Kyoto University **34** (1994).

[NSt82] M. Najmark and A. Stern, *Theory of Group Representation*, Grundlehren der Mathematischen Wissenschaften in Einzeldarstellungen **246**, Springer Verlag, New York, 1982.

[NSW85] A. Nagel, E. Stein, and St. Wainger, *Balls and metrics defined by vector fields I*, Basic Properties, Acta. Math. **155** (1985), 103–147.

[PP81] C. Pfefferman and D. Phong, *Subelliptic eigenvalue problems*, Proceedings of the Conference on Harmonic Analysis in Honor of A. Zygmund, eds. W. Becker et al, Wadsworth Belmont, 1981, pp. 590–600.

[SC84] A. Sánchez-Calle, *Fundamental solutions and geometry of the sums of squares of vector fields*, Inv. math. **78** (1984), 143–160.

[Tan83] S. Taniguchi, *Malliavin's stochastic calculus of variations for manifold-valued Wiener functionals and its application*, ZWVG **63** (1983), 269–290.

Department of Mathematics, University of Luleå, Sweden
E-mail address: `astrid@sm.luth.se`

Laboratoire Mathématiques, Université Henry Poincaré, Nancy, France
E-mail address: `leandre@iecn.u-nancy.fr`

Canadian Mathematical Society
Conference Proceedings
Volume **29**, 2000

Euclidean Gell-Mann-Low formula and double stochastic integrals

Fumio Hiroshima

ABSTRACT. Euclidean Gell-Mann-Low formula for a system of a quantum particle minimally coupled to a quantum field is established. We define *double stochastic integrals* in terms of a quadratic form q_0. As an application of the formula we prove the existence of asymptotic limits of the double stochastic integral.

1. Introduction

Let $\mathbf{B}(\mathbf{R}^d)$ be the family of Borel sets of $\mathbf{R}^d$. Fix $A, B \in \mathbf{B}(\mathbf{R}^d)$. Let $b(t) = \{b_\mu(t)\}_{\mu=1}^d$ denote the d-dimensional Brownian motion on a probability measure space (P, b) with $b_\mu(0) = 0$, $\mu = 1, ..., d$. Let $X_t := x + b(t)$, $x \in \mathbf{R}^d$, $W := P \times \mathbf{R}^d$ and $dX := db \otimes dx$. Define

$$D_{\mu\nu}(x,T) := \int_{\mathbf{R}^d} d_{\mu\nu}(k)\frac{|\hat{\rho}(k)|^2}{\omega(k)}e^{-|T|\omega(k)}e^{ikx}dk,$$

$$x \in \mathbf{R}^d, \quad T \in \mathbf{R}, \quad \mu,\nu = 1,...,d,$$

where $\hat{\rho}$ is the Fourier transform of ρ, $\omega(k) := \sqrt{|k|^2+\kappa^2}$, $\kappa \geq 0$, and $d_{\mu\nu}(k) := \delta_{\mu\nu} - k_\mu k_\nu/|k|^2$. Let $A_1, ..., A_m \in \mathbf{B}(\mathbf{R}^d)$ and $t_1 \leq t_2 \leq \cdots \leq t_m$. Let

$$Z_T(A_1,...,A_m,t_1,...,t_m) := \int_W \chi_A(X_0)\chi_{A_1}(X_{T+t_1})\cdots\chi_{A_m}(X_{T+t_m})\chi_B(X_{2T}) \times e^{-\int_0^{2T}V(X_s)ds}e^{-\alpha^2 W(2T)}dX,$$

where χ_K is the characteristic function of $K \subset \mathbf{R}^d$, V a real-valued function on $\mathbf{R}^d$, α a real parameter,

$$W(T) := \frac{1}{4}\sum_{\mu,\nu=1}^d \int_0^T db_\mu(t)\int_0^T db_\nu(s)D_{\mu\nu}(b(t)-b(s),t-s). \tag{1.1}$$

Note that *double stochastic integral* $\int_0^T db_\mu(t)\int_0^T db_\nu(s)$ in (1.1) has only formal meaning. Double stochastic integral (1.1) is mentioned in **[5]**,**[8]**,**[25]**. In this paper

1991 *Mathematics Subject Classification.* 81Q10,81U20,47B15,46G10.

This work is partially supported by Japan Society for the Promotion of Science (JSPS).

we consider asymptotic limit:

$$\mu^{t_1,\ldots,t_m}_{A_1,\ldots,A_m} := \lim_{T\to\infty} \frac{Z_T(A_1,\ldots,A_m,t_1,\ldots,t_m)}{Z_T}, \tag{1.2}$$

where Z_T a normalized constant:

$$Z_T := \int_W \chi_A(X_0)\chi_B(X_{2T})e^{-\int_0^{2T} V(X_s)ds} e^{\alpha^2 W(2T)} dX.$$

In the case where $\alpha = 0$, $\mu^{t_1,\ldots,t_m}_{A_1,\ldots,A_m}$ is a limit of a $P(\phi)_1$-process [**22**] and it is proved that $\mu^{t_1,\ldots,t_m}_{A_1,\ldots,A_m}$ is Gibbs measure in [**2**]. In such a case it plays an important role for $Z_T(A_1,\ldots,A_m,t_1,\ldots,t_m)/Z_T$ to have expression

$$\begin{aligned} &\frac{Z_T(A_1,\ldots,A_m,t_1,\ldots,t_m)}{Z_T} \\ &= \frac{\left(\chi_A, e^{-(T+t_1)K}\chi_{A_1}e^{-(t_2-t_1)K}\cdots\chi_{A_m}e^{-(2T-t_m)K}\chi_B\right)_{L^2(\mathbf{R}^d)}}{(\chi_A, e^{-2TK}\chi_B)_{L^2(\mathbf{R}^d)}} \end{aligned} \tag{1.3}$$

with self-adjoint operator $K = -\Delta/2 + V$ in $L^2(\mathbf{R}^d)$, where Δ is the Laplacian in $L^2(\mathbf{R}^d)$. Then the discussion of asymptotic limits on T in (1.2) with $\alpha = 0$ reduces to that of the analysis of the bottom of the spectrum of K. In [**3**] authors consider $\nu^{t_1,\ldots,t_m}_{A_1,\ldots,A_m}$ which is defined by $\mu^{t_1,\ldots,t_m}_{A_1,\ldots,A_m}$ with $W(T)$ replaced by $W_0(T)$:

$$W_0(T) := \frac{1}{4}\int_0^T dt \int_0^T ds \int_{\mathbf{R}^d} \frac{|\hat{\rho}(k)|^2}{\omega(k)} e^{-|t-s|\omega(k)} e^{ik(b(t)-b(s))} dk.$$

This model is related to the polaron model, which is investigated, for example, in [**4**],[**6**],[**20**],[**24**],[**25**],[**26**] by means of functional integrals and in [**17**],[**18**] by means of functional analysis. Actually the polaron model is defined by putting $d = 3$, $\hat{\rho}(k) = 1/|k|$ and $\omega(k) = 1$:

$$W_{\mathrm{polaron}}(T) := \frac{\pi^2}{8}\int_0^T dt \int_0^T ds \frac{e^{-|t-s|}}{|b(t)-b(s)|}.$$

For $Z_T(A_1,\ldots,A_m,t_1,\ldots,t_m)/Z_T$ with $W(T)$ replaced by $W_0(T)$, it is, however, impossible to give expression such as (1.3) on $L^2(\mathbf{R}^d)$. Then the authors in [**3**] take a large Hilbert space including $L^2(\mathbf{R}^d)$ and give an expression such as (1.3) with a self-adjoint operator on the large Hilbert space [**19**]. Thus the existence of $\nu^{t_1,\ldots,t_m}_{A_1,\ldots,A_m}$ is shown in [**3**] for arbitrary $\alpha \in \mathbf{R}$ through investigation of spectrum of the self-adjoint operator. The purpose of this paper is to extend results in [**3**] to the version including double stochastic integrals; we shall give the definition of (1.1) (see Definition 4.1) and show the existence of $\mu^{t_1,\ldots,t_m}_{A_1,\ldots,A_m}$ under conditions on $\hat{\rho}$, α and V. Strategy is as follow: First we construct a quadratic form q_0 ((2.1)) and define $W(T)$ by use of the quadratic form (Definition 4.1). Secondly we take Hilbert space $\mathbf{H} := L^2(\mathbf{R}^d) \otimes L^2(Q)$, where $L^2(Q)$ is the space of L^2-functions on a probability measure space (Q,μ) and construct a self-adjoint operator H in $\mathbf{H}$, which describes the Hamiltonian of a system of a quantum particle minimally coupled to a quantum field [**1**],[**7**],[**11**],[**12**],[**13**],[**14**],[**15**]. Finally we get

$$\begin{aligned} &\frac{Z_T(A_1,\ldots,A_m,t_1,\ldots,t_m)}{Z_T} \\ &= \frac{\left(\chi_A, e^{-(T+t_1)H}\chi_{A_1}e^{-(t_2-t_1)H}\cdots\chi_{A_m}e^{-(2T-t_m)H}\chi_B\right)_{\mathbf{H}}}{(\chi_A, e^{-2TH}\chi_B)_{\mathbf{H}}}. \end{aligned} \tag{1.4}$$

Hence the problem of asymptotic limits in (1.2) is reduces to that of the spectral analysis of H, which is well observed in [**11**],[**12**],[**13**].

This paper is organized as follows: In section 2 we give Euclidean Gell-Mann-Low formula ([**23**], Theorems 2.4 and 4.3) and see that $Z_T(A_1, ..., A_m, t_1, ..., t_m)/Z_T$ has expression (1.4). In Section 3 we mention the spectral properties of a Hamiltonian. In Section 4 we give a definition of (1.1) in terms of the quadratic form q_0 and show that $\mu_{A_1,...,A_m}^{t_1,...,t_m}$ exists for sufficiently small α.

2. Euclidean Gell-Mann-Low formula

Let $G(k) : \mathbf{C}^d \longrightarrow \mathbf{C}^{d-1}$ be defined by $G(k)\vec{v} := \begin{pmatrix} \vec{e}_1(k) \cdot \vec{v} \\ \vdots \\ \vec{e}_{d-1}(k) \cdot \vec{v} \end{pmatrix}$, $\vec{v} \in \mathbf{C}^d$, for each $k \in \mathbf{R}^d$, where $\vec{e}_r(k) \in \mathbf{R}^d$, $r = 1, ..., d-1$, satisfies that $\vec{e}_r(k) \cdot \vec{e}_s(k) = \delta_{rs}$ and $\vec{e}_r(k) \cdot k = 0$ for almost every $k \in \mathbf{R}^d$. Define $G := \int_{\mathbf{R}^d}^{\oplus} G(k)dk : \oplus^d L^2(\mathbf{R}^d) \to W := \oplus^{d-1} L^2(\mathbf{R}^d)$. Let q be a quadratic form on $\oplus^d L^2(\mathbf{R}^d)$ defined by $q(f,g) := (G\hat{f}, G\hat{g})_W$, where $\hat{h}$ denotes the Fourier transform of h and $(\cdot, *)_{\mathbf{K}}$ the scalar product of Hilbert space $\mathbf{K}$, which is antilinear in $\cdot$ and linear in $*$. Unless confusion arising we omit $\mathbf{K}$ in $(\cdot,\cdot)_{\mathbf{K}}$. Moreover we denote by $\|\cdot\|_{\mathbf{K}}$ the norm of $\mathbf{K}$ and abbreviate $\|\cdot\|_{\mathbf{K}}$ by $\|\cdot\|$. Let $\phi(f)$ be a Gaussian process indexed by $f \in \oplus^d L^2_{\mathrm{real}}(\mathbf{R}^d)$ on a probability measure space (Q, μ) with $\int_Q \phi(f)\phi(g)d\mu(\phi) = q(f,g)/2$, where $L^2_{\mathrm{real}}(\mathbf{R}^\nu)$ denotes the space of real-valued L^2-functions on $\mathbf{R}^\nu$. We define self-adjoint operator H_{f} in $L^2(Q)$ by

$$H_{\mathrm{f}} :\phi(f_1)\cdots\phi(f_n):= \sum_{j=1}^{n} :\phi(f_1)\cdots\phi(\langle\omega\rangle f_j)\cdots\phi(f_n):, \quad H_{\mathrm{f}}\Omega = 0,$$

$$f_1, ..., f_n \in D(\langle\omega\rangle), \quad n \geq 1,$$

where $\Omega \equiv 1$ in $L^2(Q)$, $:X:$ is the Wick product of X [**21**, p.226] and $\langle T\rangle := \oplus^d T : W \to W$ for $T : L^2(\mathbf{R}^d) \to L^2(\mathbf{R}^d)$. Let $\mathbf{H} := L^2(\mathbf{R}^d) \otimes L^2(Q)$ and

$$H := \frac{1}{2}\left(\vec{p}\otimes I - \alpha\vec{\phi}(\rho(\cdot - x))\right)^2 + I \otimes H_{\mathrm{f}} + V \otimes I,$$

where $\vec{p} := (-i\partial/\partial x_1, ..., -i\partial/\partial x_d)$, α is a real constant, V a function, and $\vec{\phi}(\lambda) := (\phi_1(\lambda), ..., \phi_d(\lambda))$, $\phi_\mu(\lambda) := \phi(\oplus_{\nu=1}^d \delta_{\mu\nu}\lambda)$, $\lambda \in L^2(\mathbf{R}^d)$. In what follows we abbreviate $X \otimes I$ and $I \otimes X$ by X, respectively.

> **Hypothesis I:** We assume that $\rho \in L^2_{\mathrm{real}}(\mathbf{R}^d)$, $\hat{\rho}/\omega \in L^2(\mathbf{R}^d)$, $\omega\hat{\rho} \in L^2(\mathbf{R}^d)$ and that V is relatively bounded with respect to the Laplacian Δ in $L^2(\mathbf{R}^d)$ with sufficiently small relative bound: there exist $a \ll 1$ and b such that $\|Vf\| \leq a\|\Delta f\| + b\|f\|$ for $f \in D(\Delta)$.

REMARK 2.1. For V in Hypothesis I, we see that $V_+ \in L^1_{\mathrm{loc}}(\mathbf{R}^d)$ and that $C_0^\infty(\mathbf{R}^d)$ is a core of $-\Delta + V$, where V_+ is the nonnegative part of V.

PROPOSITION 2.2 ([**14**, Theorem 2.9]). *We assume Hypothesis I. Then for arbitrary $\alpha \in \mathbf{R}$, H is essentially self-adjoint on $D(\Delta) \cap D(H_{\mathrm{f}})$ and bounded below. Moreover there exists α_0 such that, for $|\alpha| < \alpha_0$, H is self-adjoint on $D(\Delta) \cap D(H_{\mathrm{f}})$ and essentially self-adjoint on any core of $-\Delta + H_{\mathrm{f}}$.*

Let $W_0 := \oplus^{d-1} L^2(\mathbf{R}^{d+1})$. The family of linear isometries $\xi_s : L^2(\mathbf{R}^d) \to L^2(\mathbf{R}^{d+1})$ is defined by $\widehat{\xi_s f}(k, k_0) := \sqrt{\omega(k)/(\omega(k)^2 + k_0^2)} e^{-isk_0} \hat{f}(k)$, $(k, k_0) \in \mathbf{R}^d \times \mathbf{R}$, $s \in \mathbf{R}$. Let $\phi_0(F)$ be a Gaussian process indexed by $F \in W_0$ on a probability measure space (Q_0, μ_0) with $\int_{Q_0} \phi_0(F)\phi_0(K) d\mu_0(\phi_0) = q_0(F, K)/2$, where we define quadratic form q_0 on $W_0 \times W_0$ by

(2.1)

$$q_0(F, K) := ((I \otimes G)\hat{F}, (I \otimes G)\hat{K})_{W_0} = \sum_{\mu,\nu=1}^{d} \int_{\mathbf{R}^{d+1}} d_{\mu\nu}(k) \overline{\hat{F}(k, k_0)} \hat{K}(k, k_0) dk dk_0$$

for $F = (F_1, ..., F_d) \in W_0$ and $K = (K_1, ..., K_d) \in W_0$ under the identification $W_0 \cong L^2(\mathbf{R}) \otimes W$. Let $\vec{\phi_0}(f)$, $f \in L^2(\mathbf{R}^{d+1})$, be defined in the same way as $\vec{\phi}(\lambda)$, $\lambda \in L^2(\mathbf{R}^d)$. The second quantization [**21**] of $\langle \xi_s \rangle$ is denoted by Ξ_s so that Ξ_s is the family of isometries of $L^2(Q)$ to $L^2(Q_0)$. We define self-adjoint operator $\widehat{H}_\mathrm{f}$ in $L^2(Q_0)$ by

$$\widehat{H}_\mathrm{f} :\phi_0(f_1) \cdots \phi_0(f_n): = \sum_{j=1}^{n} :\phi_0(f_1) \cdots \phi_0(\langle \omega_0 \rangle f_j) \cdots \phi_0(f_n):, \quad \widehat{H}_\mathrm{f} \Omega_0 = 0,$$

$$f_1, ..., f_n \in D(\langle \omega \rangle), \quad n \geq 1,$$

where $\Omega_0 \equiv 1$ in $L^2(Q_0)$ and $\omega_0(k, k_0) := \sqrt{|k|^2 + k_0^2 + \kappa^2}$.

PROPOSITION 2.3 ([**14**, Theorem 2.10]). *Let* $H_0 := (1/2)\left(\vec{p} - \alpha\vec{\phi}(\rho(\cdot - x))\right)^2$ *and* $H_0(s) := (1/2)\left(\vec{p} - \alpha\vec{\phi_0}(\xi_s\rho(\cdot - x))\right)^2$ *for* $s \in \mathbf{R}$. *We assume Hypothesis I. Then for arbitrary* $\alpha \in \mathbf{R}$, H_0 *is essentially self-adjoint on* $D(\Delta) \cap D(H_\mathrm{f})$ *and* $H_0(s)$ *on* $D(\Delta) \cap D(\widehat{H}_\mathrm{f})$. *Moreover both of them are bounded below.*

We give the main theorem of this paper.

THEOREM 2.4. *Let* $\mathbf{H}_\infty := \left\{F \in \mathbf{H} \,|\sup_{x \in \mathbf{R}^d} \|F(x)\|_{L^2(Q)} < \infty\right\}$. *We assume Hypothesis I. Let* $t_1 \leq \cdots \leq t_m = t$, *and* $F_0, F_m \in \mathbf{H}$, $F_1, ..., F_m \in \mathbf{H}_\infty$. *Then*

$$\int_{W \times Q_0} \widehat{F}_0(X_0)\widehat{F}_1(X_{t_1})\widehat{F}_2(X_{t_2}) ... \widehat{F}_m(X_t) e^{-\int_0^t V(X_s)ds} e^{i\alpha\phi_0(\mathbf{W}_t)} dX d\mu$$

$$= (\overline{F_0}, e^{-t_1 H} F_1 e^{-(t_2 - t_1)H} F_2 \cdots e^{-(t - t_{m-1})H} F_m)_{\mathbf{H}}, \tag{2.2}$$

where $\widehat{F}_j := \Xi_{t_j} F_j$, $j = 0, ..., m$, $\mathbf{W}_t := \bigoplus_{\mu=1}^d \int_0^t \xi_s \rho(\cdot - \omega(s)) db_\mu(s) \in W_0$ *and* $\int db_\mu(s)$ *denotes* $L^2(\mathbf{R}^{d+1})$*-valued stochastic integral* [**9**],[**10**].

To prove Theorem 2.4, we show some lemmas.

LEMMA 2.5. *We assume Hypothesis I. Then for arbitrary* $\alpha \in \mathbf{R}$,

$$e^{-tH_0(T)} = s - \lim_{n \to \infty} \left(Q_{t/[n]}\right)^{[n]}, \tag{2.3}$$

where $[n] := 2^n$ *and* Q_s *is a contraction operator on* $\mathbf{H}$ *defined by*

$$Q_s F(x) := \int_{\mathbf{R}^d} p_s(x - y) \exp\left\{\frac{i}{2}\vec{\phi_0}\left(\xi_T(\rho(\cdot - x) + \rho(\cdot - y))\right)(x - y)\right\} F(y) dy,$$

$$Q_0 F(x) := F(x),$$

for $F = F(\cdot) \in \mathbf{H}$, *where* $p_s(x) := (2\pi s)^{-d/2} e^{-|x|^2/(2s)}$.

PROOF. It is known that the right-hand side of (2.3) is strongly continuous one-parameter symmetric semi-group in $t \geq 0$ and that its infinitesimal generator K is such that $H_0(T)\lceil_{D(\Delta)\cap D(\widehat{H}_f)} \subset K$ (the proof is similar as that of [**14**, Theorem 4.8]). Hence Proposition 2.3 yields (2.3). $\square$

We define projection E_s by $E_s := \Xi_s \Xi_s^*$, $s \in \mathbf{R}$.

LEMMA 2.6 ([**10**, Lemma 4.9]). *We assume Hypothesis I. Then for arbitrary* $\alpha \in \mathbf{R}$, $t \geq 0$, $s \in \mathbf{R}$, *it holds that* $\Xi_s e^{-tH_0} \Xi_s^* = E_s e^{-tH_0(s)} E_s$.

Proof of Theorem 2.4

Assume that V is continuous and that $\|V\|_\infty := \sup_{x\in\mathbf{R}^d} |V(x)| < \infty$. Let $s_j := t_j - t_{j-1}$, $j = 1, ..., m$, $t_0 := 0$, and $t_{jl} := s_j l/[n] + t_{j-1}$, $l = 0, ..., [n]$. Let

$$P_j := e^{-(s_j/[n])H_0} e^{-(s_j/[n])V}, \quad P_j^l := e^{-(s_j/[n])H_0(t_{jl})} e^{-(s_j/[n])V},$$

$$I_j^l := \Xi_{t_{jl}}, \quad J_j^l := (I_j^l)^*, \quad E_j^l := E_{t_{jl}}, \quad U_j := e^{-(s_j/[n])V}.$$

We obtain $e^{-tH} = s - \lim_{n\to\infty} \left(e^{-(t/[n])H_0} e^{-(t/[n])V} e^{-(t/[n])H_f}\right)^{[n]}$ by the Trotter product formula. Since $\|V\|_\infty < \infty$, we have

$$\begin{aligned}
&(F_0, e^{-s_1 H} F_1 e^{-s_2 H} F_2 \cdots e^{-s_n H} F_n) \\
&= \lim_{n\to\infty} \Big(F_0, J_0^* J_0 (P_1 J_1^0 I_1^1)(P_1 J_1^1 I_1^2) \cdots (P_1 J_1^{[n]-1} I_1^{[n]}) F_1 (P_2 J_2^0 I_2^1)(P_2 J_2^1 I_2^2) \\
&\cdots (P_2 J_2^{[n]-1} I_2^{[n]}) F_2 \cdots F_{m-1} (P_m J_m^0 I_m^1)(P_m J_m^1 I_m^2) \cdots (P_m J_m^{[n]-1} I_m^{[n]}) F_m\Big) \\
&= \lim_{n\to\infty} \Big(\widehat{F}_0, (E_1^0 P_1^0 E_1^0)(E_1^1 P_1^1 E_1^1) \cdots (E_1^{[n]-1} P_1^{[n]-1} E_1^{[n]-1})(E_1^{[n]} \widehat{F}_1 E_1^{[n]}) \times \\
&(E_2^0 P_2^0 E_2^0) \cdots (E_2^{[n]-1} P_2^{[n]-1} E_2^{[n]-1})(E_2^{[n]} \widehat{F}_2 E_2^{[n]}) \cdots (E_m^{[n]-1} P_m^{[n]-1} E_m^{[n]-1}) \widehat{F}_m\Big) \\
&= \lim_{n\to\infty} \Big(\widehat{F}_0, \left(P_1^0 P_1^1 \cdots P_1^{[n]-1}\right) \widehat{F}_1 \left(P_2^0 P_2^1 \cdots P_2^{[n]-1}\right) \widehat{F}_2 \cdots \\
&\qquad \cdots \left(P_{m-1}^0 P_{m-1}^1 \cdots P_{m-1}^{[n]-1}\right) F_{m-1} \left(P_m^0 P_m^1 \cdots P_m^{[n]-1}\right) \widehat{F}_m\Big) \\
&= \lim_{n\to\infty} \lim_{u\to\infty} \Big(\widehat{F}_0, (Q_1^0)^{[u]} U_1 (Q_1^1)^{[u]} U_1 \cdots \left(Q_1^{[n]-1}\right)^{[u]} U_1 \widehat{F}_1 (Q_2^0)^{[u]} U_2 (Q_2^1)^{[u]} \\
&\times U_2 \cdots \left(Q_2^{[n]-1}\right)^{[u]} U_2 \widehat{F}_2 \cdots (Q_m^0)^{[u]} U_m (Q_m^1)^{[u]} U_m \cdots \left(Q_m^{[n]-1}\right)^{[u]} U_m \widehat{F}_m\Big),
\end{aligned}$$

where the first equality is due to the fact that $\Xi_s^* \Xi_t = e^{-|t-s|H_f}$, the second is Lemma 2.6, the third is the Markov property of E_s ([**10**, Proposition 3.3]), the forth is Lemma 2.5, and $Q_j^l = Q_j^l(u, n)$ is defined by

$$Q_j^l F(x) := \int_{\mathbf{R}^d} p_{s_j/[n]}(x-y) \exp\left\{\frac{i}{2}\vec{\phi}_0\left(\xi_{t_{jl}}(\rho(\cdot - x) + \rho(\cdot - y))\right)(x-y)\right\} F(y)dy.$$

Thus we have

$$\begin{aligned}
&(F_0, e^{-s_1 H} F_1 e^{-s_2 H} F_2 \cdots e^{-s_m H} F_m) \\
&= \lim_{n\to\infty} \lim_{u\to\infty} \int_{\mathbf{R}^d \times \mathbf{R}^{m[u][n]d}} dx d\vec{x} P(\vec{x}) e^{-V_{n,u}} \left(\widehat{F}_0(x), e^{-i\alpha\chi_{n,u}} \Pi_{j=1}^m \widehat{F}_j\left(x_{j,[u]}^{[n]}\right)\right),
\end{aligned}$$

where $\vec{x}=(x_1^1,x_1^2,...,x_1^{[n]},x_2^1,x_2^2,...x_2^{[n]},......,x_m^{[n]})$, $x_j^l=(x_{j,1}^l,...,x_{j,[u]}^l)\in\mathbf{R}^{[u]d}$,

$$\chi_{n,u}:=\sum_{j=1}^m\sum_{l=1}^{[n]}\sum_{k=1}^{[u]}\vec{\phi_0}\left(\xi_{t_{jl}}\left(\rho(\cdot-x_{j,k-1}^l)+\rho(\cdot-x_{j,k}^l)\right)\right)(x_{j,k-1}^l-x_{j,k}^{i}),$$

$$V_{n,u}:=\sum_{j=1}^m\sum_{l=1}^{[n]}\frac{s_j}{[n]}V\left(x_{j,[u]}^l\right),\quad P(\vec{x}):=\Pi_{j=1}^m\Pi_{l=1}^{[n]}P_{s_j/n}(x_j^l)$$

where $P_s(x_j^l):=p_s\left(x_{j-1,[u]}^l-x_{j,1}^l\right)p_s\left(x_{j,1}^1-x_{j,2}^l\right)...p_s\left(x_{j,[u]-1}^l-x_{j,[u]}^l\right)$, $x_{j,0}^l:=x_{j,[u]}^{l-1}$, $x_{j,0}^0:=x_{j-1,[u]}^{[n]}$, $x_{0,0}^0:=x$, $j=1,....,m$. Then we get

$$(F_0,e^{-s_1H}F_1e^{-s_2H}F_2\cdots e^{-s_mH}F_m)=\lim_{n\to\infty}\lim_{u\to\infty}\int_W dXe^{-V_n}\left(\Phi,e^{-i\alpha\widehat{\chi}_{n,u}}\Psi\right),$$

where

$$\widehat{\chi}_{n,u}:=\sum_{j=1}^m\sum_{l=0}^{[n]-1}\sum_{k=1}^{[u]}$$
$$\times\vec{\phi_0}\left(\xi_{t_{jl}}\left(\rho\left(\cdot-X_{ks_j/([u][n])+t_{jl}}\right)+\rho\left(\cdot-X_{(k-1)s_j/([u][n])+t_{jl}}\right)\right)\right)$$
$$\times\left\{b\left(\frac{k}{[u]}\frac{s_j}{[n]}+t_{jl}\right)-b\left(\frac{(k-1)}{[u]}\frac{s_j}{[n]}+t_{jl}\right)\right\},$$

$$V_n:=\sum_{j=1}^m\sum_{l=0}^{[n]-1}\frac{s_j}{[n]}V\left(X_{t_{jl}}\right),\quad\Phi:=\widehat{F}_0(X_0),\quad\Psi:=\Pi_{j=1}^m\widehat{F}_j(X_{t_j}).$$

It follows that

$$(F_0,e^{-s_1H}F_1e^{-s_2H}F_2\cdots e^{-s_mH}F_m)=\lim_{n\to\infty}\int_W dXe^{-V_n}\left(\Phi,e^{-i\alpha\chi_n}\Psi\right),$$

where

$$\chi_n:=\sum_{j=1}^m\sum_{l=0}^{[n]-1}\phi_0\left(\bigoplus_{\mu=1}^d\int_{t_{jl}}^{t_{jl}+s_j/[n]}\xi_{t_{jl}}\rho(\cdot-X_s)db_\mu(s)\right).$$

Since

$$\sum_{l=0}^{[n]-1}\int_{t_{jl}}^{t_{jl}+s_j/[n]}\xi_{t_{jl}}\rho(\cdot-X_s)db_\mu(s)\to\int_{t_{j-1}}^{t_j}\xi_s\rho(\cdot-X_s)db_\mu(s)$$

strongly in $L^2(P;L^2(\mathbf{R}^d))$ as $n\to\infty$ [**10**, Theorem 2.5], we get (2.2). For V in Hypothesis 1, by a limiting argument we get (2.2). □

3. Spectral property

Here we mention the spectral properties of self-adjoint operator H without proofs.

PROPOSITION 3.1. [**11**] *We assume Hypothesis I. Then there exists $\alpha_*\le\alpha_0$ such that for $|\alpha|\le\alpha_*$, the ground states of H exist.*

PROPOSITION 3.2. [**12**] *We assume Hypothesis I. Then the ground state $\Psi_{\rm g}$ of H (if it exists) is unique up to multiple constants and has the form $\Psi_{\rm g}=e^{(i\pi/2)N}\Psi_0$, where N denotes the number operator in $L^2(Q)$ and Ψ_0 is a strictly positive vector in $\mathbf{H}$.*

REMARK 3.3. (1) In [**16**] examples of V such that, for sufficiently large α, the ground states of H exists but H with $\alpha = 0$ has no ground state are given. (2) In [**1**] condition $\lambda/\omega \in L^2(\mathbf{R}^d)$ is removed for proving the existence of the ground states of H. (3) In [**15**] examples of V such that H has degenerate ground states is given.

4. Existence of asymptotic limits of double stochastic integrals

DEFINITION 4.1. We define $W(t)$ by $W(t) := q_0(\mathbf{W}_t, \mathbf{W}_t)$.

REMARK 4.2. Note that taking subsequences $\{n'\}$ and $\{m'\}$ in $\mathbf{N}$ we have [**12**]

$$q_0(\mathbf{W}_t, \mathbf{W}_t)$$
$$= \lim_{n'\to\infty} \lim_{m'\to\infty} \sum_{\mu,\nu=1}^{d} \sum_{k=1}^{[n']} \sum_{l=1}^{[m']} D_{\mu\nu}\left(b\left(\frac{l-1}{[m']}t\right) - b\left(\frac{k-1}{[n']}t\right), \frac{l-1}{[m']}t - \frac{k-1}{[n']}t\right)$$
$$\times \left\{b_\mu\left(\frac{l}{[m']}t\right) - b\left(\frac{l-1}{[m']}t\right)\right\}\left\{b\left(\frac{k}{[n']}t\right) - b\left(\frac{k-1}{[n']}t\right)\right\}, \quad a.e. b \in P.$$

In this sense it is reasonable to define $W(t)$ by $q_0(\mathbf{W}_t, \mathbf{W}_t)$.

For $f, g \in L^2(\mathbf{R}^d)$ and $f, g \geq 0, f \not\equiv 0, g \not\equiv 0$, we define

$$dX^{fg}_{[0,2T]} := \frac{1}{N_T} f(X_0) g(X_{2T}) e^{-\int_0^{2T} V(X_s)ds} e^{(-\alpha^2/4)W(2T)} dX,$$

where N_T is the normalized constant such that $\int_W dX^{fg}_{[0,2T]} = 1$.

THEOREM 4.3. *We assume Hypothesis I. Let $t_1 \leq t_2 \leq ... \leq t_m$ and $h_j \in L^\infty(\mathbf{R}^d)$, $j = 1, ..., m$. Let $|\alpha| \leq \alpha_*$. Then*

$$\lim_{T\to\infty} \int_W h_1(X_{T+t_1}) h_2(X_{T+t_2}) ... h_m(X_{T+t_m}) dX^{fg}_{[0,2T]} \tag{4.1}$$
$$= (\Psi_g, h_1 e^{-(t_2-t_1)H} h_2 e^{-(t_3-t_2)H} h_3 .... e^{-(t_m - t_{m-1})H} h_m \Psi_g). \tag{4.2}$$

PROOF. From Propositions 3.1 and 3.2 it follows that $(f \otimes \Omega, \Psi_g) \neq 0$ and $(g \otimes \Omega, \Psi_g) \neq 0$ since $e^{i(\pi/2)N}\Omega = \Omega$. In particular we have

$$s - \lim_{t\to\infty} e^{-tH} f \otimes \Omega = (\Psi_g, f \otimes \Omega)\Psi_g, \quad s - \lim_{t\to\infty} e^{-tH} g \otimes \Omega = (\Psi_g, g \otimes \Omega)\Psi_g.$$

Note that

$$(\Omega, e^{-i\alpha\phi_0(\mathbf{W}_t)}\Omega)_{L^2(Q_0)} = e^{(-\alpha^2/4)q_0(\mathbf{W}_t, \mathbf{W}_t)} = e^{(-\alpha^2/4)W(t)}.$$

Let $\overline{H} := H - \inf \sigma(H)$. Then Theorem 2.4 yields that

$$(4.2) = \lim_{T\to\infty} \frac{\left(f \otimes \Omega, e^{-(T+t_1)\overline{H}}(h_1 \otimes I) e^{-(t_2-t_1)\overline{H}} \cdots (h_m \otimes I) e^{-(T-t_m)\overline{H}} g \otimes \Omega\right)}{\left(f \otimes \Omega, e^{-2T\overline{H}} g \otimes \Omega\right)}$$
$$= (4.1).$$

Thus theorem follows. □

We immediately have the following corollary:

COROLLARY 4.4. *Let $A, B, A_1, ..., A_m \in \mathbf{B}(\mathbf{R}^d)$. We assume Hypothesis I. Let $|\alpha| \leq \alpha_*$. Assume that the Lebesgue measure of A and B are finite, respectively. Then*

$$\mu^{t_1,...,t_m}_{A_1,...,A_m} = (\Psi_g, \chi_{A_1} e^{-(t_2-t_1)H} \chi_{A_2} e^{-(t_3-t_2)H} \chi_{A_3} .... e^{-(t_m-t_{m-1})H} \chi_{A_m} \Psi_g).$$

References

[1] V.Bach, J.Fröhlich, I.M.Sigal, Spectral Analysis for systems of atoms and molecules coupled to the quantized radiation field, 1998, preprint.

[2] V.Betz and J.Lorintzi, Gibbsian description of $P(\phi)_1$-process, 1999, preprint.

[3] V.Betz, F.Hiroshima, J.Lorintzi, Gibbs measure associated with a particle-field system, in preparation.

[4] M.D.Donsker and S.R.S.Varadhan, Asymptotics for the polaron, *Comm. Pure. Appl. Math.* **29** (1989), 505-528.

[5] R.P.Feynman, Mathematical formulation of the quantum theory of electromagnetic interaction, *Phys.Rev.* **80** (1950), 440-457,

[6] R.P.Feynman, Slow electrons in a polar crystal, *Phys.Rev.* **97** (1954), 660-665.

[7] L.Gross, The free euclidean Proca and electromagnetic fields, *Functional integral and its application*, Proceedings of the international conference held at the Cumberland Lodge, Windsor Great Park, London in 1974. A.M.Arthurs (editor), Clarendon Press, Oxford, 1975

[8] Z.Haba, Feynman integral in regularized nonrelativistic quantum electrodynamics, *J. Math. Phys.* **39** (1998), 1766-1787.

[9] F.Hiroshima, Functional integral representation of the Pauli-Fierz model in QED, *Proceedings of the 6th international colloquium on differential equations, Provdiv, Bulgaria*, **1** (1995), 183-192.

[10] F.Hiroshima, Functional integral representation of a model in quantum electrodynamics, *Rev. Math. Phys.* **9** (1997), 489-530.

[11] F.Hiroshima, Ground states of a model in nonrelativistic quantum electrodynamics I, to appear in *J.Math.Phys.*

[12] F.Hiroshima, Ground states of a model in nonrelativistic quantum electrodynamics II, to appear in *J.Math.Phys.*

[13] F.Hiroshima, Ground states and spectrum of quantum electrodynamics of non-relativistic particles, 1998, preprint.

[14] F.Hiroshima, Essential self-adjointness of translation-invariant quantum field models for arbitrary coupling constants, 1999, preprint.

[15] F.Hiroshima, Application of functional integral to models minimally coupled to massless quantum fields, 1999, preprint.

[16] F.Hiroshima and H.Spohn, Binding through a coupling to a field, in preparation.

[17] E.Lieb and L.E.Thomas, Exact ground state energy of the strong-coupling polaron, *Comm. Math. Phys.* **183** (1997), 511-519.

[18] E.Lieb and K.Yamazaki, Ground-state energy and effective mass of the polaron, *Phys. Rev.* **111** (1958), 728-733.

[19] E.Nelson, Interaction of nonrelativistic particles with a quantized scalar field, *J. Math. Phys.***5** (1964), 1190–1197.

[20] G.Roesendorf, *Path integral approach to quantum physics*, Berlin-Heidelberg-New York: Springer, 1994.

[21] M.Reed and B.Simon, *Method of Modern Mathematical Physics Vol.II*, Academic press, New York, 1975.

[22] B.Simon, Functional integral and quantum physics, Academic Press, 1979.

[23] B.Simon, *The $P(\phi)_2$ Euclidean (Quantum) Field Theory*, Princeton Univ. Press, 1974.

[24] H.Spohn, Roughening and pinning transitions for the polaron, *J.Phys. A: Math.Gen.* **19** (1986), 533-545.

[25] H.Spohn, Effective mass of polaron: A functional integral approach, *Ann.Phys.* **175** (1987), 278-318.

[26] H.Spohn, The polaron at large total momentum, *J.Phys. A: Math.Gen.* **21** (1988), 1199-1211.

DEPARTMENT OF MATHEMATICS, FACULTY OF SCIENCE, HOKKAIDO UNIVERSITY, SAPPORO 060 JAPAN

Current address: Technische Universität München, Zentrum Mathematik, Arcis str.21, D-80290 München, Germany

E-mail address: `hiro@mathematik.tu-muenchen.de, f-hirosh@math.sci.hokudai.ac.jp`

Canadian Mathematical Society
Conference Proceedings
Volume **29**, 2000

Operator Splitting Methods for Degenerate Convection–Diffusion Equations I: Convergence and Entropy Estimates

Helge Holden, Kenneth Hvistendahl Karlsen, and Knut–Andreas Lie

Dedicated to Sergio Albeverio on the occasion of his 60th birthday.

ABSTRACT. We present and analyze a numerical method for the solution of a class of scalar, multi-dimensional, nonlinear degenerate convection-diffusion equations. The method is based on operator splitting to separate the convective and the diffusive terms in the governing equation. The nonlinear convective part is solved using front tracking and dimensional splitting, while the nonlinear diffusion equation is solved by a suitable difference scheme. We verify L^1 compactness of the corresponding set of approximate solutions and derive precise entropy estimates. In particular, these results allow us to pass to the limit in our approximations and recover an entropy solution of the problem in question. The theory presented covers a large class of equations. Important subclasses are hyperbolic conservation laws, porous medium type equations, two-phase reservoir flow equations, and strongly degenerate equations coming from the recent theory of sedimentation-consolidation processes. A thorough numerical investigation of the method analyzed in this paper (and similar methods) is presented in a companion paper.

1. Introduction

We address the important issue of constructing and analyzing numerical methods for a class of scalar, nonlinear, degenerate convection–diffusion problems of the form

$$u_t + \sum_{j=1}^{m} V_j(x) f_j(u)_{x_j} = \varepsilon \sum_{j=1}^{m} \big(K_j(x) A_j(u)_{x_j}\big)_{x_j}, \quad u(x,0) = u_0(x), \tag{1}$$

for $(x,t) \in Q_T = \mathbb{R}^m \times (0,T)$. Here $u = u(x,t)$ is the unknown function; f_j, K_j, A_j, and u_0 are given functions; and $\varepsilon > 0$ is a (small) scaling constant. In applications related to fluid flow (see, e.g, [**27**]), $f = (f_1, \dots, f_m)$ is often referred to as the flux function, $V = (V_1, \dots, V_m)$ as the velocity field, $K = (K_1, \dots, K_m)$ as the permeability field, and $A = (A_1, \dots, A_m)$ as the function of diffusivity.

2000 *Mathematics Subject Classification.* Primary 35K65; Secondary 35L65, 65M06.

Key words and phrases. Degenerate convection–diffusion equations, entropy solutions, operator splitting, front tracking, convergence, entropy estimates.

Throughout this paper, we will assume that the data of our problem satisfy the following conditions:

$$
(2) \qquad \begin{cases} \text{(i) } f, A \in C^2(\mathbb{R};\mathbb{R}^m) \text{ and } A' \geq 0. \\ \text{(ii) } V \in C^2(\mathbb{R};\mathbb{R}^m) \text{ and } \sum_{j=1}^m V(x)_{x_j} = 0 \text{ (i.e., } V \text{ is divergence free).} \\ \text{(iii) } K \in C^2(\mathbb{R};\mathbb{R}^m) \text{ and } K > 0. \\ \text{(iv) } u_0 \in L^1(\mathbb{R}^m) \cap L^\infty(\mathbb{R}^m) \cap BV(\mathbb{R}^m). \end{cases}
$$

We mention that the "maximal regularity" conditions (i)–(iii) can be significantly weakened and they are assumed here only to make our presentation as short and as transparent as possible. On the other hand, condition (iv) concerning the initial function u_0 is fairly weak. Notice that the convective part of (1) is written on transport, or non-conservative, form, which is reasonable since V is assumed to be divergence free.

We only require that $A_j' \geq 0$ for each j and hence the parabolic term is allowed to degenerate. In fact, each A_j' may be zero on a set of positive measure, in which case we call the equation *strongly degenerate*, and thus the well known hyperbolic conservation law

$$
(3) \qquad u_t + \sum_{j=1}^m V_j(x) f_j(u)_{x_j} = 0
$$

is a special case of (1). Other subclasses of (1) include one-point degenerate porous medium type equations [**42**]; two-point degenerate two-phase reservoir flow equations [**17**]; and strongly degenerate equations coming from the recent theory of sedimentation-consolidation processes [**5, 6, 15**].

A characteristic feature of nonlinear partial differential equations of hyperbolic-parabolic type such as (1) is that solutions may exhibit quite complex behavior, like singularities and sharp transitions, in a small region of space (and time) and this makes them particularly hard to solve numerically. The aim of the paper is to construct and analyze a numerical method for nonlinear convection-diffusion equations of degenerate parabolic type (1) that works "uniformly" in the diffusion coefficient $A_j' \geq 0$, and as such is able to resolve the issues concerning singularities and sharp transitions in the solutions of (1); more details are given towards the end of this section.

If (1) is allowed to degenerate at certain points, that is, $A_j'(u) = 0$ for some values of u, solutions are not necessarily smooth, but typically continuous, and *weak solutions* must be sought. On the other hand, if $A_j(u)$ is zero on a set of positive measure, weak solutions may be discontinuous and are not necessarily uniquely determined by their initial data, as can be easily deduced from the maximum principle and what is known about the hyperbolic conservation law. Consequently, additional admissibility criteria — so-called *entropy conditions* — must be imposed to single out the physically correct solution. We use the following definition of an entropy or generalized solution:

DEFINITION 1.1 (Entropy Solution). *We call a function* $u \in L^1(Q_T) \cap L^\infty(Q_T)$ *an entropy solution of the Cauchy problem* (1) *provided*

$$
(4) \qquad \sqrt{\varepsilon K_j(x)} A(u)_{x_j} \in L^2_{loc}(Q_T), \quad j = 1, \dots, m,
$$

and the following entropy inequality holds

$$
\begin{aligned}
(5)\qquad &\iint_{Q_T} \Big(S(u,k)\phi_t + \sum_{j=1}^{m} F_j(u,k)(\phi V_j)_{x_j} + \varepsilon \sum_{j=1}^{m} Q_j(u,k)\big(K_j \phi_{x_j}\big)_{x_j} \Big)\, dt\, dx \\
&\qquad + \int |u_0 - k|\phi(x,0)\, dx - \int |u(x,T) - k|\phi(x,T)\, dx \geq 0,
\end{aligned}
$$

for every constant $k \in \mathbb{R}$ *and test function* $\phi \in C_0^\infty(\mathbb{R}^m \times [0,\infty))$, $\phi \geq 0$. *Here we have used the short-hand notations:* $S(u,k) = |u-k|$ *for the entropy,* $F_j(u,k) = \operatorname{sgn}(u-k)(f_j(u) - f_j(k))$ *for the convective entropy flux, and* $Q_j(u,k) = |A_j(u) - A_j(k)|$ *for the diffusive entropy flux.*

We denote the left-hand side of (5) by $\mathcal{L}(u,\phi,k; f, V, A, u_0, T)$ and sometimes $\mathcal{L}(u,\phi,k)$ or even simply $\mathcal{L}_\phi(u)$. Furthermore, $\mathcal{L}(u,\phi,k; f, V, 0, u_0, T)$ denotes the left-hand side of the entropy inequality corresponding to the purely hyperbolic case. Observe that (5) implies that (1) holds in the sense distributions, i.e., the equality

$$
\begin{aligned}
(6)\qquad &\iint_{Q_T} \Big(u\phi_t + \sum_{j=1}^{m} V_j(x) f_j(u)\phi_{x_j} + \varepsilon \sum_{j=1}^{m} A_j(u)\big(K_j(x)\phi_{x_j}\big)_{x_j} \Big)\, dt\, dx \\
&\qquad + \int u_0(x)\phi(x,0)\, dx - \int u(x,T)\phi(x,T)\, dx = 0,
\end{aligned}
$$

holds for each test function $\phi \in C_0^\infty(\mathbb{R}^m \times [0,\infty))$.

When $A_j' \equiv 0$ for all j, the entropy condition (5) reduces to the entropy condition introduced independently by Vol'pert [**44**] and Kružkov [**37**] for hyperbolic problems and thus Definition 1.1 contains the hyperbolic problem as a special case [**3**]. Definition 1.1 goes back to Vol'pert and Hudjaev [**45**], who were the first to consider strongly degenerate parabolic equations of nonlinear (or quasilinear) type. They also showed the existence of a BV entropy solution (i.e., an entropy solution whose first order partial derivatives are finite measures on Q_T), provided that u_0 is sufficiently smooth, by passing to the limit in a parabolic regularization and obtained some partial uniqueness results for BV entropy solutions. In the one-dimensional case, Wu and Yin [**46**] later provided the complete uniqueness proof for BV entropy solutions. See Bénilan and Touré [**1, 2**] for further results — via nonlinear semigroup theory — on entropy solutions in the one-dimensional case (without variable coefficients). In this context, let us also mention that Cockburn and Gripenberg [**13**] have established continuous dependence on the nonlinearities of semigroup solutions of the Cauchy problem for multi-dimensional equations of the type (with $f_j = f$ for all j)

$$
(7)\qquad u_t + \sum_{j=1}^{m} f_j(u)_{x_j} = \Delta_x A(u).
$$

In the multi-dimensional case, Brézis and Crandall [**4**] established uniqueness of weak solutions in $L^1(Q_T) \cap L^\infty(Q_T)$ of the Cauchy problem for (7) with no lower order terms ($f_j = 0$ for all j). Later, under the assumption that $A(u)$ is strictly increasing, which does not rule out the possibility of $A'(u)$ having infinite number of zero points, Yin [**47**] showed uniqueness of weak solutions in $L^\infty(Q_T) \cap BV(Q_T)$ of the Cauchy problem for (7). Note that (7) is a special case of (1). The assumption that u_t should be a finite measure was removed in [**49**]. The initial-boundary value problem for (7) with variable coefficients was treated in [**48**].

An important step forward in the general case of $A(u)$ being merely non-decreasing (i.e., the strongly degenerate case) was made recently by Carrillo [**12**], who showed uniqueness of entropy solutions for a particular boundary value problem for (7). His method of proof, which is an elegant extension of Kružkov's "doubling of the variables" device [**37**], also applies to the Cauchy problem for (7) as well weakly coupled systems of equations such as (7), see [**28**]. Moreover, the proof also applies to (7) with a more general flux function $f = f(x, t, u)$ which does not necessarily depend Lipschitz continuously on the spatial variable x, see [**35**]. In [**7**], the uniqueness proof of Carrillo was adopted to several initial-boundary value problems arising in the theory of sedimentation-consolidation processes [**5, 6, 15**], which in some cases call for the notion of entropy boundary condition (see also [**10, 11**] for the BV approach).

Later on we will introduce our numerical approximations and prove that they converge, based on a suitable compactness argument. We recall here that such convergence proofs merely ensure convergence along subsequences. However, when we are equipped with a uniqueness result, as in the cases mentioned above, we automatically get convergence of the whole sequence in question and not just some subsequence. To the best of our knowledge, uniqueness of entropy solutions for the general problem (1) that we consider here is still open when $A(u)$ is merely non-decreasing, so in the general case we have to resort to convergence along subsequences.

Although there seems to be an increasing interest in the (analysis of) numerical approximation of entropy solutions of degenerate convection-diffusion equations, the amount of literature on the subject is at the moment extremely modest, at least compared with the purely hyperbolic case which has long traditions. Recent literature include papers by Evje and Karlsen [**23, 19, 21, 22, 18**], Karlsen and Risebro [**36**], Cockburn and Shu [**14**], Kurganov and Tadmor [**38**], and Bouchut, Guarguaglini, and Natalini [**3**].

In [**23**], an operator splitting method for one-dimensional degenerate equations is proposed and analyzed. In particular, convergence results and precise entropy estimates are given for the proposed approximations which allow the authors to pass to the limit and recover the unique entropy solution of the equation in question. Related papers include [**20, 24, 8, 9, 17**].

In [**19, 36**], the authors build a convergence theory for explicit monotone difference approximations of degenerate convection-diffusion equations. This theory parallels (and includes) the classical theory of Harten, Hyman, and Lax [**25**] and Crandall and Majda [**16**] for conservation laws. Implicit monotone difference approximations are treated via nonlinear semigroup theory in [**21, 18**]. High order MUSCL type methods are analyzed in [**22**].

In [**14**], the authors present the so-called local discontinuous Galerkin method for (degenerate) convection-diffusion problems. This method is an extension of the Runge-Kutta discontinuous Galerkin method for hyperbolic problems. A convergence analysis within the entropy solution framework is, however, not provided.

In [**38**], the authors introduce and test a new family of central schemes for conservation laws and (degenerate) convection-diffusion equations. The main feature of the schemes is that they possess a much smaller numerical viscosity than the original central schemes [**41**] and as such apply to convection-diffusion equations. In particular, they admit a semi-discrete formulation. A convergence analysis of their convection-diffusion schemes within the entropy solution framework is not presented.

In [**3**], the authors introduce and analyze a class of discrete velocity BGK type methods for degenerate convection-diffusion equations, which in turn are extensions of similar methods for hyperbolic conservation laws. Convergence results and entropy estimates are proved, which imply that the BGK approximations converge to the unique entropy solution of the governing equation.

We conclude this section by giving more details about the contents of the present paper, which deals with a class of numerical methods called operator splitting methods. The underlying design principle behind any splitting method for an evolutionary partial differential equation is to split the time evolution into several partial steps in order to separate out the different (physical) effects present in the governing equation. In our context, the idea is to split the original problem (1) into a first order convection (hyperbolic) problem and a second order degenerate diffusion (parabolic) problem. The split problems are then solved sequentially to approximate the exact solution of the original problem. The main attraction of splitting methods lies, of course, in the fact that one can employ optimal existing methods for each split problem. A detailed description of the operator splitting method analyzed in this paper is given in §2. For an introduction to operator splitting methods in general, we refer to Espedal and Karlsen [**17**].

The main purpose of the present paper is to extend the analysis (and the splitting method) of Evje and Karlsen [**23**] (which in turn borrowed from Karlsen and Risebro [**34**]) to more general problems arising in applications. To be more precise, we extend the analysis of [**23**] as follows:

- In [**23**], the authors proved convergence to a limit satisfying (4) and (5) only for the semi-discrete method. For their fully discrete method, however, they did not establish (4). In the present paper, we establish at least that the limit of *one* of our fully discrete methods also satisfies (4).
- In [**23**], the authors proposed and analyzed operator splitting method for one-dimensional equations without variable coefficients. Here we propose and analyze an operator splitting method that works for a general class of multi-dimensional equations with variable coefficients, thereby containing [**23**] as a special case.

The convergence (or compactness) part of the analysis, together with the proof that the limit of one of the methods satisfies (4), is presented in §3. Precise entropy estimates are established in §4. These estimates imply in particular that the limit of a converging sequence of numerical approximations satisfies the entropy condition (5). Finally, we mention that a variety of (convincing) numerical applications are presented in our companion paper [**27**], including applications to oil reservoir simulation and sedimentation-consolidation processes.

Acknowledgement. KHK would like to thank Raimund Bürger and Steinar Evje for discussions and collaboration on the subject of degenerate parabolic equations and their applications to sedimentation-consolidation processes.

2. Operator Splitting Methods

To construct approximate solutions of (1) it is often favourable to use a splitting method. There are several motivations and possibilities for splitting the equation.

First, one could use dimensional splitting directly on (1), i.e., split it into m one-dimensional equations of the form

$$v_t + V(x)f(v)_x = \varepsilon\big(K(x)A(v)_x\big)_x, \tag{8}$$

and use ones favourite one-dimensional method to approximate the solution operator $\mathcal{P}_t$ of (8). In this case the splitting reads

$$u(x, n\Delta t) \approx \big[\mathcal{P}^m_{\Delta t} \circ \cdots \circ \mathcal{P}^1_{\Delta t}\big]^n u_0(x), \tag{9}$$

where $\Delta t > 0$ denotes the splitting step. To approximate $\mathcal{P}_j t$, one could for instance use a monotone difference method [**19, 21**] or a higher-order method based on MUSCL techniques [**22**]. Another alternative is the *corrected operator splitting method* (COS) introduced by Karlsen and Risebro [**33**], which is a large-time-step method. The method splits (8) into a hyperbolic step $v_t + V(x)f(v)_x = 0$ and a degenerate parabolic step $w_t = \varepsilon(K(x)A(u)_x)_x$. Once a shock is formed in the hyperbolic step, there will be an entropy loss due to Oleinik's convexification. The source of the entropy loss can be identified as a residual flux (defined locally as the flux f minus its convexification) which can be taken into account either in a separate correction step or included in the parabolic step. This way, COS resolves the correct balance between convection and diffusion in regions with sharp gradients for an (almost) arbitrarily large time step and can be used in combination with (9) to yield a very efficient numerical method (if the dynamics of (1) allows for large splitting steps). This method has been investigated in other studies [**24, 31, 30, 32**], see also the companion paper [**27**].

As an alternative to dimensional splitting, one can split with respect to physical mechanisms, which is what we will pursue in this paper. To this end, split (1) into a hyperbolic part

$$v_t + \sum_{j=1}^m V_j(x) f_j(v)_{x_j} = 0, \qquad v(x,0) = v_0(x) \tag{10}$$

and a degenerate parabolic part

$$w_t = \varepsilon \sum_{j=1}^m \big(K_j(x)A_j(w)_{x_j}\big)_{x_j}, \qquad w(x,0) = w_0(x). \tag{11}$$

In what follows, we choose a time step $\Delta t > 0$ and an integer N such that $N\Delta t = T$. We also use the notation $t^n = n\Delta t$ for $n = 0, \dots, N$. The corresponding (semi-discrete) splitting method then reads

$$u_{\Delta t}(t) = \big[\mathcal{H}_{\Delta t} \circ \mathcal{S}_{\Delta t}\big]^n u_0, \qquad \text{for } t \in (t^{n-1}, t^n], \quad n = 1, \dots, N, \tag{12}$$

where $\mathcal{S}_t$ and $\mathcal{H}_t$ denote the solution operators of (10) and (11), respectively.

A slightly different method is obtained if we apply dimensional splitting to the hyperbolic part, i.e., (10) is further split into one-dimensional equations

$$v_t + V_j(x) f_j(v)_{x_j} = 0, \qquad v(x,0) = v_0(x). \tag{13}$$

The corresponding semi-discrete splitting method takes the form

$$u_{\Delta t}(t) = \big[\mathcal{H}_{\Delta t} \circ \mathcal{S}^m_{\Delta t} \circ \cdots \circ \mathcal{S}^1_{\Delta t}\big]^n u_0, \qquad \text{for } t \in (t^{n-1}, t^n], \quad n = 1, \dots, N, \tag{14}$$

where $\mathcal{S}^j_t$ is the exact solution operator associated with (13).

Concerning the semi-discrete methods (12) and (14), we have the following main theorem:

THEOREM 2.1 (Semi-discrete methods). *Suppose that conditions* (i)-(iv) *in* (2) *hold. Let* $\{u_{\Delta t}\}$ *be the semi-discrete splitting sequence given by* (12) *or* (14). *Then* $\{u_{\Delta t}\}$ *converges along a subsequence in* $L^1_{loc}(Q_T)$ *to a limit* u *that satisfies* $u(\cdot,t) \in L^1(\mathbb{R}^m) \cap L^\infty(\mathbb{R}^m) \cap BV(\mathbb{R}^m)$ *uniformly in* t *and* $u(\cdot,t)$ *is uniformly* L^1 *Hölder continuous in time with exponent* $1/2$. *Finally, the limit* u *satisfies the regularity requirement* (4) *and the entropy inequality* (5).

This theorem shows that the semi-discrete splitting sequences given by (12) and (14) both converge along subsequences to entropy solutions of (1) in the sense of Definition 1.1. Moreover, when it is known that the entropy solution is unique (see §1), the whole sequence $\{u_{\Delta t}\}$ converges to this solution.

From a computational point of view, we need to replace the exact solution operators $\mathcal{S}_t$ and $\mathcal{H}_t$ in (12) by suitable numerical methods. In this paper we use an explicit-implicit finite difference method to approximate $\mathcal{H}_t$ and a large-step, front tracking method to approximate $\mathcal{S}_t$. The large-step method is based on dimensional splitting, which means that we shall here use (14) as the basis for constructing the fully discrete splitting method. We let $\mathcal{H}_{\Delta x,t}$ denote the finite difference solution operator associated with (11) at time t and $\mathcal{S}^j_{\delta,\Delta x,t}$ the front tracking solution operator associated with (13) at time t. To simplify the dimensional splitting process, each one-dimensional hyperbolic solution is projected onto a Cartesian grid with mesh parameter Δx by a projection operator

$$\pi v(x) = \frac{1}{\Delta x^m}\int_{\Omega_j} v(y)\,dy, \quad \text{for } x \in \Omega_j,$$

where Ω_j denotes

$$\Omega_j = [j_1\Delta x, (j_1+1)\Delta x) \times \cdots \times [j_m\Delta x, (j_m+1)\Delta x), \qquad j = (j_1,\dots,j_m) \in \mathbb{Z}^m.$$

Then our fully discrete splitting method reads

$$u_\eta(t) = \left[\mathcal{H}_{\Delta x,\Delta t} \circ \pi \circ \mathcal{S}^m_{\delta,\Delta x,\Delta t} \circ \cdots \circ \pi \circ \mathcal{S}^1_{\delta,\Delta x,\Delta t}\right]^n u^0, \tag{15}$$

for $t \in (t^{n-1}, t^n]$ and $n = 1,\dots,N$. Here η signifies the discretization parameters $(\Delta x, \Delta t, \delta)$ and $u^0 = \pi u_0$. For simplicity of notation, we sometimes use the shorthand

$$\mathcal{S}_{\delta,\Delta x,t} = \pi \circ \mathcal{S}^m_{\delta,\Delta x,t} \circ \cdots \circ \pi \circ \mathcal{S}^1_{\delta,\Delta x,t}, \qquad t > 0, \tag{16}$$

i.e., $\mathcal{S}_{\delta,\Delta x,t}$ denotes the front tracking-dimensional splitting solution operator associated with (10), see Holden and Risebro [**29**] and Lie [**39**] for more details about this approximate solution operator.

Concerning the fully discrete method (15), we have the following similar main theorem:

THEOREM 2.2 (Fully discrete method). *Suppose that conditions* (i)-(iv) *in* (2) *hold. Let* $\{u_\eta\}$ *be the fully discrete splitting sequence given by* (15). *Then* $\{u_\eta\}$ *converges along a subsequence in* $L^1_{loc}(Q_T)$ *to a limit* u *that satisfies* $u(\cdot,t) \in L^1(\mathbb{R}^m) \cap L^\infty(\mathbb{R}^m) \cap BV(\mathbb{R}^m)$ *uniformly in* t *and* $u(\cdot,t)$ *is uniformly* L^1 *Hölder continuous in time with exponent* $1/2$. *The precise entropy estimate for* u_η *reads* $\mathcal{L}_\phi(u_\eta) \geq -C\big(\sqrt{\Delta t} + \Delta x + \delta^2\big)$, *for some constant* $C > 0$ *independent of* η. *In particular, this implies that the limit* u *satisfies* (5). *In the case of a fully implicit diffusion solver, the limit* u *also satisfies the regularity requirement* (4).

Note that the case of an explicit-implicit (or fully explicit) diffusion solver $\mathcal{H}_{\Delta x,\Delta t}$ is not covered by Theorem 2.2. The difficulty in the explicit-implicit (or explicit) case is to establish the regularity requirement (4), which requires a substantially different argument than the one used herein (see [**36, 28**]).

Theorem 2.2 follows easily from the results stated and proved in §3 and §4. On the other hand, the proof of Theorem 2.1 will not be written out explicitly in this paper. It is however easy to modify the proof of Theorem 2.2 so that it applies to the semi-discrete methods (12) and (14). We leave the details to the interested reader as an exercise.

Next, we describe the front tracking and the finite difference method in more detail.

The Hyperbolic Step. In each hyperbolic step we solve an equation of the form

$$v_t + V(x)f(v)_x = 0, \qquad v(x,0) = v_0(x), \qquad x \in \mathbb{R},\ t > 0. \tag{17}$$

To this end we will use front tracking, which is an unconditionally stable method without an intrinsic time step. The method seeks approximations within the class of step functions. First, the initial data v_0 is approximated by a step function $v_{0,\Delta x}$. This is achieved by the projection in (15), i.e., $v_{0,\Delta x} = \pi v_0$. Thus the initial value problem is decomposed into a series of simple Riemann problems that can be solved analytically one by one. If the flux function is approximated by a continuous, piecewise linear function $f^\delta(v)$, the solution of each Riemann problem will again be a step function. The global solution, until the first wave interaction occurs, thus consists of constant states separated by discontinuities propagating along paths $x(t)$. Each path $x(t)$ is given by the differential equation $\dot{x} = V(x)s$, where s is the Rankine–Hugoniot speed $(f^\delta(v_L) - f^\delta(v_R))/(v_L - v_R)$. By approximating the velocity V by either a piecewise linear or a piecewise constant function $V_{\Delta x}$, the differential equations can be solved explicitly and the paths $x(t)$ are given in closed form. Since every wave interaction leads to a new Riemann problem, this construction can be continued up to any desired time. In fact, it can be shown [**26, 40**] that there is a finite number of wave interactions, even in infinite time, if the initial data is bounded and has finite total variation. We stress that we replace the flux functions f by f^δ in most of the following convergence analysis and derivation of entropy estimates. Only at the final step we consider the limit as $f^\delta \to f$. One can prove that the limit (as $\delta \to 0$) is indeed the entropy solution of (17), see Lie [**40**].

By construction, front tracking solutions are not increasing in L^∞ norm and have bounded total variation. Since all waves have finite speed of propagation, the solution is Lipschitz continuous in time with respect to the L^1 norm. Each solution satisfies an entropy condition for the perturbed equation with f^δ and $V_{\Delta x}$ and is thus an entropy solution. However, the solution operators are not L^1 contractions because of the velocity field and the non-conservative form. These claims can be verified by using a Kružkov type analysis, see Lie [**40**]. We summarize properties of solutions of (17) in the following lemma:

LEMMA 2.1. *Let $v(x,t)$ be a solution of* (17). *Then v satisfies the following estimates*

$$\|v\|_\infty \le \|v_0\|_\infty, \quad |v(\,\cdot\,,t)|_{BV} \le |v_0|_{BV}, \quad \|v(\,\cdot\,,t) - v_0\|_1 \le Ct,$$

where C is a constant depending on the data. Moreover, let $u(x,t)$ be a solution of (17) with flux function g, velocity field U, and initial data u_0. Then we have the stability estimate

$$\begin{aligned}\|v(\,\cdot\,,t)-u(\,\cdot\,,t)\|_1 \le e^{Dt}\big[&\|v_0-u_0\|_1\\ &+t\big(E_1\|U-V\|_\infty+E_2\|f-g\|_{Lip}\big)\min(|u_0|_{BV},|v_0|_{BV})\big],\end{aligned}$$

where the constants D, E_1, and E_2 depend on the data (f,g,V,U,v_0,u_0).

Note that each front tracking solution is an exact solution of (17) for special functions f, V, and v_0, so that Lemma 2.1 also gives the properties of the front tracking solutions used for the hyperbolic steps.

The Parabolic Step. In each parabolic step we solve a possibly degenerate equation of the form

$$w_t=\varepsilon\sum_{j=1}^{m}\big(K_j(x)A_j(w)_{x_j}\big)_{x_j},\qquad w(x,0)=w_0(x), \tag{18}$$

where we assume, without loss of generality, that w_0 has compact support. We will use an implicit–explicit finite difference method to solve (18). Difference methods for general degenerate parabolic equations are treated in [**19, 21, 22, 18**]. We assume a mesh with a uniform spacing Δx in each spatial direction and a time step τ. Let $\alpha=(\alpha_1,\dots,\alpha_m)\in\mathbb{Z}^m$ and e_j the index with the jth component equal unity and all other components equal zero. Let W_α^n denote the approximate solution in grid cell α at time $t=n\tau$. Furthermore, let $\beta_\alpha^j=\varepsilon\tau K^j_{\alpha+e_j/2}/\Delta x^2$ and $A_\alpha^{j,n}=A_j(W_\alpha^n)$. Then the scheme reads

$$\begin{aligned}W_\alpha^{n+1}=&W_\alpha^n+\theta\sum_{j=1}^{m}\Big[\beta_\alpha^j\big(A_{\alpha+e_j}^{j,n}-A_\alpha^{j,n}\big)-\beta_{\alpha-e_j}^j\big(A_\alpha^{j,n}-A_{\alpha-e_j}^{j,n}\big)\Big]\\ &+(1-\theta)\sum_{j=1}^{m}\Big[\beta_\alpha^j\big(A_{\alpha+e_j}^{j,n+1}-A_\alpha^{j,n+1}\big)-\beta_{\alpha-e_j}^j\big(A_\alpha^{j,n+1}-A_{\alpha-e_j}^{j,n+1}\big)\Big],\end{aligned} \tag{19}$$

where the parameter θ is in the interval $[0,1]$; $\theta=0$ giving a fully implicit scheme and $\theta=1$ a fully explicit scheme. We make the following simplifying assumption in this paper.

ASSUMPTION 2.1. *The scheme* (19) *admits a unique solution* $\{W_\alpha^{n+1}\}$.

The existence of a unique solution of (19) can be established by suitable fixed point argument. For $n=0,1\dots,$ we let $W^n=W^n(x)$ denote the piecewise constant function

$$W^n(x)=W_\alpha^n,\qquad\text{for } x\in\Omega_\alpha \tag{20}$$

and some $\alpha\in\mathbb{Z}^m$. Moreover, we let $\mathcal{H}_{\Delta x,t}$ denote the (finite difference) solution operator defined by

$$\mathcal{H}_{\Delta x,t}w_0(x)=W^n(x),\qquad x\in\mathbb{R}^m,\ t\in((n-1)\tau,n\tau],\ n=1,2,\dots, \tag{21}$$

for any initial function $w_0=w_0(x)$ that is piecewise constant with respect to the grid.

For later use, recall that the L^∞ norm and the L^1 norm of a grid function $W = \{W_\alpha\}$ are defined respectively as follows:

$$\|W\|_\infty = \sup_{\alpha\in\mathbb{Z}^m} \|W_\alpha\|, \quad \|W\|_1 = \sum_{\alpha\in\mathbb{Z}^m} |W_\alpha|\Delta x^m,$$

and the BV semi-norm as

$$TV(W) = \sum_{j,\alpha} |W_\alpha^n - W_{\alpha-e_j}^n|\Delta x^{m-1}.$$

For the scheme (19) we have the following properties.

LEMMA 2.2. *Let $W^n = \{W_\alpha^n\}$ and $V^n = \{V_\alpha^n\}$ be approximate solutions of* (18) *generated by* (19) *with a time step τ satisfying the stability condition*

$$\theta \max_{\alpha,j,n} \beta_\alpha^j (A_j')_\alpha^n \le \theta \frac{\varepsilon\tau}{\Delta x^2} \max_j \Big(\sup_x K_j(x) \sup_u A_j'(u)\Big) \le \frac{1}{2m}. \tag{22}$$

Then

$$\|W^n\|_\infty \le \|W^0\|_\infty, \quad TV(W^n) \le TV(W^0), \quad \|W^n - V^n\|_1 \le \|W^0 - V^0\|_1.$$

PROOF. The estimates are straightforward and we present the details only for $m = 1$ (with $K_1 = K$, $A_1 = A$, and $\alpha = i \in \mathbb{Z}$). First, we consider the L^∞ estimate for the approximation $\{W_i^n\}$. Using Taylor expansions of $A(W)$ we can rewrite the scheme as

$$\begin{aligned}\big[1 + (1-\theta)(\beta_i A'(\xi_i) + \beta_{i-1}A'(\xi_{i-1}))\big] W_i^{n+1} &= \\ &\big[1 - \theta(\beta_i A'(\zeta_i) + \beta_{i-1}A'(\zeta_{i-1}))\big] W_i^n \\ &\quad + \theta\big[\beta_i A'(\zeta_i) W_{i+1}^n + \beta_{i-1}A'(\zeta_{i-1}) W_{i-1}^n\big] \\ &\qquad + (1-\theta)\big[\beta_i A'(\xi_i) W_{i+1}^{n+1} + \beta_{i-1}A'(\xi_{i-1}) W_{i-1}^{n+1}\big],\end{aligned}$$

for suitable ξ_i, ζ_i. If we now choose $i = \ell$ such that $W_\ell^{n+1} = \max_i W_i^{n+1}$ and use the stability condition (22) we derive

$$\begin{aligned}\big[1 + (1-\theta)(\beta_\ell A'(\xi_\ell) + \beta_{\ell-1}A'(\xi_{\ell-1}))\big] W_\ell^{n+1} &\le \\ &\big[1 - \theta(\beta_i A'(\zeta_i) + \beta_{i-1}A'(\zeta_{i-1}))\big] \max_i W_i^n \\ &\quad + \theta\big[\beta_i A'(\zeta_i) \max_i W_{i+1}^n + \beta_{i-1}A'(\zeta_{i-1}) \max_i W_{i-1}^n\big] \\ &\qquad + (1-\theta)\big[\beta_\ell A'(\xi_\ell) W_{\ell+1}^{n+1} + \beta_{i-1}A'(\xi_{\ell-1}) W_{\ell-1}^{n+1}\big].\end{aligned}$$

Since $W_{\ell\pm1}^{n+1} \le W_\ell^{n+1}$, the inequality simplifies to $\max_i W_i^{n+1} \le \max_i W_i^n$. Similarly, by picking $i = k$ such that $W_k^{n+1} = \min_i W_i^{n+1}$ we can bound W_i^{n+1} from below. Hence, $\max_i |W_i^{n+1}| \le \max_i |W_i^n|$ and the L^∞ estimate follows by induction on n.

To prove estimate on the total variation, we use almost the same argument. Introduce $Z_i^n = W_{i+1}^n - W_i^n$. From the difference scheme (19) evaluated at i and $i+1$, we then get

$$\begin{aligned}Z_i^{n+1} = &Z_i^n + \theta\big[\beta_{i+1}A'(\zeta_{i+1}) Z_{i+1}^n - 2\beta_i A'(\zeta_i) Z_i^n + \beta_{i-1}A'(\zeta_{i-1}) Z_{i-1}^n\big] \\ &+ (1-\theta)\big[\beta_{i+1}A'(\xi_{i+1}) Z_{i+1}^{n+1} - 2\beta_i A'(\xi_i) Z_i^{n+1} + \beta_{i-1}A'(\xi_{i-1}) Z_{i-1}^{n+1}\big],\end{aligned}$$

for suitable ξ_i, ζ_i. Using once more the stability condition, we get

$$\begin{aligned}\sum_i \big(1 + 2(1-\theta)\beta_i A'(\xi_i)\big)|Z_i^{n+1}| &\le \sum_i (1-2\theta)\beta_i A'(\zeta_i)|Z_i^n| \\ &+ \theta\Big(\sum_i \beta_{i+1}A'(\zeta_{i+1})|Z_{i+1}^n| + \sum_i \beta_{i-1}A'(\zeta_{i-1})|Z_{i-1}^n|\Big) \\ &+ (1-\theta)\Big(\sum_i \beta_{i+1}A'(\xi_{i+1})|Z_{i+1}^{n+1}| + \sum_i \beta_{i-1}A'(\xi_{i-1})|Z_{i-1}^{n+1}|\Big).\end{aligned}$$

Hence $\sum_i |Z_i^{n+1}| \le \sum_i |Z_i^n|$, from which the *TV* estimate follows. The L^1 stability estimate is derived by a similar argument. □

3. Convergence Results

In this section we will prove that the fully discrete splitting method (15) converges to a limit $u(x,t)$. The proof is based on a standard L^1 compactness argument, where we first use Lemmas 2.1 and 2.2 to bound the approximate solutions u_η in suitable norms

LEMMA 3.1. *Let u_η denote the approximate solution defined by* (15). *Then u_η satisfies, for $n, \ell = 1, 2, \ldots$,*

$$\|u_\eta(\cdot, n\Delta t)\|_\infty \le \|u_0\|_\infty, \qquad |u_\eta(\cdot, n\Delta t)|_{BV} \le e^{C_1 n\Delta t}|u_0|_{BV},$$
$$\|u_\eta(\cdot, \ell\Delta t) - u_\eta(\cdot, n\Delta t)\|_1 \le C_2\sqrt{|n-\ell|\,\Delta t},$$

where C_1 and C_2 are constants independent of $\delta, \Delta x, \Delta t$.

PROOF. The first inequality follows since neither of the operators $\mathcal{H}_{\Delta x,\Delta t}$, $\mathcal{S}^j_{\delta,\Delta x,\Delta t}$, or π introduce new extrema into the solution (see Lemmas 2.1 and 2.2).

To prove the second inequality, we need to establish a bound on the total variation of the composite hyperbolic operator $\mathcal{S}_{\delta,\Delta x,\Delta t}$ defined by (16). This result is well-known and relies on the total variation non-increasing property and the L^1 stability of each one-dimensional hyperbolic operator given in Lemma 2.1, see Lie [**39**] and Holden and Risebro [**29**]. Combining this result with the fact that the total variation is non-increasing in the parabolic step (see Lemma 2.2), the second inequality follows.

It remains to prove the third inequality, which will be done by a technique used first in the context of splitting methods by Karlsen and Risebro [**34**]. For the product of the hyperbolic operators, i.e., for the composite operator $\mathcal{S}_{\delta,\Delta x,\Delta t}$ defined in (16), the result is valid with exponent 1 (i.e., Lipschitz continuity in time), see e.g., Lie [**39**]. Thus we focus on the parabolic operator here. The key point is to establish weak Lipschitz continuity in time for the finite difference approximation W^n defined in (20). To this end, let D_-^j and D_+^j denote the backward and forward differences in direction j. Then:

$$\begin{aligned}
&\Big|\int \phi(W^{n+1} - W^n)\,dx\Big| \\
&= \varepsilon\tau\Big|\sum_{j,\alpha}\big[\theta D^j_-\big(K^j_{\alpha+e_j/2} D^j_+ A^{j,n}_\alpha\big) + (1-\theta) D^j_-\big(K^j_{\alpha+e_j/2} D^j_+ A^{j,n+1}_\alpha\big)\big]\phi_\alpha \Delta x^m\Big| \\
&= \varepsilon\tau\Big|\sum_{j,\alpha}\Big(\theta K^j_{\alpha-e_j/2} A'_j(\zeta_\alpha) D^j_- W^n_\alpha + (1-\theta) K^j_{\alpha-e_j/2} A'_j(\xi_\alpha) D^j_- W^{n+1}_\alpha\Big) D^j_-\phi_\alpha \Delta x^m\Big| \\
&\le \varepsilon\tau \max_j\{\|K_j\|_\infty \|A'_j\|_\infty\}\,\|\nabla\phi\|_\infty \sum_{j,\alpha}\Big(\theta|D^j_- W^n_\alpha| + (1-\theta)|D^j_- W^{n+1}_\alpha|\Big)\Delta x^{m-1} \\
&\le \operatorname{Const}\tau\,\|\nabla\phi\|_\infty,
\end{aligned}$$

where summation by parts (once in space) was used to derive the second equality from the first one. Moreover, we have used the averaged test function $\phi_\alpha = \pi\phi(\alpha\Delta x)$, and the Taylor expansions $A^{j,n}_{\alpha+e_j} - A^{j,n}_\alpha = A'_j(\theta_\alpha)[W^n_{\alpha+e_j} - W^n_\alpha]$ for suitable θ_α and $|\phi_{\alpha+e_j} - \phi_\alpha| \le \|\partial_{x_j}\phi\|_\infty \Delta x$. By repeating the above argument, we easily derive

$$\Big|\int \phi(W^\ell - W^n)\,dx\Big| \le \operatorname{Const}\|\nabla\phi\|_\infty\,|\ell - n|\,\tau, \tag{23}$$

which implies that $|\int \phi(\mathcal{H}_{\Delta x,\Delta t} w_0 - w_0)\,dx| \le \operatorname{Const}\|\nabla\phi\|_\infty \Delta t$. Combining this with the strong Lipschitz continuity for the hyperbolic operator $\mathcal{S}_{\delta,\Delta x,\Delta t}$ we derive

$$\Big|\int \phi(u_\eta(x,\ell\Delta t) - u_\eta(x,n\Delta t))\,dx\Big| \le \operatorname{Const}\big(\|\phi\|_\infty + \|\nabla\phi\|_\infty\big)\,|\ell - n|\,\Delta t. \tag{24}$$

Let now ω_h be a smooth mollifier with support in $[-h,h]^m$ and define ψ to be equal $\operatorname{sgn}(u^\ell_\eta - u^n_\eta)$ for $|x| \le r - h$ (r is some fixed positive number) and zero otherwise, and finally $\psi_h = \omega_h * \psi$. Standard approximating arguments (see, e.g., [**34**]) show that

$$\int_{[-r,r]^m} |u_\eta(x,\ell\Delta t) - u_\eta(x,n\Delta t)|\,dx \le C_1 h + C_2\,|\ell - n|\,\Delta t/h.$$

By choosing $h = \sqrt{|\ell - n|\,\Delta t}$, and finally letting $r \to \infty$ we conclude that

$$\int |u_\eta(x,\ell\Delta t) - u_\eta(x,n\Delta t)|\,dx \le \operatorname{Const}\sqrt{|\ell - n|\,\Delta t}.$$

□

Using essentially the above lemma, Helly's theorem, and several diagonal arguments, we can prove that the sequence $\{u_\eta\}$ converges to a function u.

LEMMA 3.2. *Fix $T > 0$. Let u_η denote the approximate solution defined by* (15). *Then for any sequences $\Delta t_i \to 0$ and $\Delta x_i \to 0$ with $\Delta x_i/\Delta t_i$ fixed, there exists a subsequence of $\{u_\eta\}$, denoted $\{u_{\eta_i}\}$, and a function u such that*

$$u_{\eta_i} \to u \text{ in } L^1_{loc}(\mathbb{R}^m \times [0,T]).$$

Moreover, the limit $u(\cdot,t)$ belongs to $BV(\mathbb{R}^m)$ uniformly in t and is uniformly L^1 Hölder continuous in time with exponent $1/2$.

PROOF. The convergence proof is fairly standard, see, e.g., Smoller [**43**] or Karlsen and Risebro [**34**]. From Lemma 3.1 it follows that the sequence $\{u_\eta\}$ is uniformly bounded and has uniformly bounded total variation. Therefore, Helly's theorem ensures convergences of a subsequence $\{u_{\eta_i}\}$ in L^1 on bounded boxes $[-r,r]^m$

for each fixed t. Since r is arbitrary, the argument can be applied a countable number of times to form a further subsequence, still denoted by $\{u_{\eta_i}\}$, such that $\{u_{\eta_i}(\cdot,t)\}$ converges in $L^1_{loc}(\mathbb{R}^m)$ for each fixed t. Yet another diagonalization argument gives convergence for a dense countable subset $\{t_\ell\}$ in $[0,T]$. For $t \notin \{t_\ell\}$, there exists a sequence $\{t_k\} \subset \{t_\ell\}$ such that $t_k \to t$. By a triangle inequality

$$\int_{[-r,r]^m} |u_{\eta_i}(x,t) - u_{\eta_j}(x,t)|\,dx \le \int_{[-r,r]^m} |u_{\eta_i}(x,t) - u_{\eta_i}(x,t_k)|\,dx$$
$$+ \int_{[-r,r]^m} |u_{\eta_i}(x,t_k) - u_{\eta_j}(x,t_k)|\,dx + \int_{[-r,r]^m} |u_{\eta_j}(x,t_k) - u_{\eta_j}(x,t)|\,dx.$$

The first and third term can be made arbitrary small by making k large and using the Hölder continuity in time (see Lemma 3.1). The second term can be made arbitrary small by making i,j large and using that $\{u_{\eta_i}(\cdot,t_k)\}$ is a Cauchy sequence for each t_k. Hence, $\{u_{\eta_i}(\cdot,t)\}$ is a Cauchy sequence in $L^1_{loc}(\mathbb{R}^m)$ for all $t \in [0,T]$. We denote the limit by u. Finally, in view of Lemma 3.1, it is clear that the limit u possesses the regularity claimed in the lemma. □

It remains to prove that u is an entropy solution, i.e., that it satisfies (4) and (5). We start with (4). However, as already mentioned, we will establish this estimate only for the implicit diffusion solver (and, of course, the case of exact solution operators). In the next section we derive an upper bound on how far each approximate solution u_η is from satisfying (5).

LEMMA 3.3. *Suppose $\theta = 0$ in (19). Then the limit u constructed in Lemma 3.2 satisfies (4).*

PROOF. The proof is an adaption of the argument used in [**22**] (see also [**18**]). For simplicity of presentation, we only treat the two-dimensional equation

$$w_t = \varepsilon(K(x,y)A(w)_x)_x + \varepsilon(K(x,y)A(w)_y)_y \tag{25}$$

and the corresponding implicit difference scheme

$$\frac{W^{l+1}_{i,j} - W^l_{i,j}}{\tau} = \varepsilon D^x_-\left(K_{i+1/2,j} D^x_+ A(W^{l+1}_{i,j})\right) + \varepsilon D^y_-\left(K_{i,j+1/2} D^y_+ A(W^{l+1}_{i,j})\right), \tag{26}$$

where $W^l_{i,j}$ denotes the approximate solution of (25) at $(i\Delta x, j\Delta x, l\tau)$ and D^ℓ_-, D^ℓ_+ are the backward and forward differences in direction $\ell = x,y$. Then extension of the proof to the more general problem (11) is straightforward and is left to the reader. Let us introduce the function

$$A_\eta(x,y,t) = \begin{cases} A(W^{l+1}_{i,j}) + D^x_+ A(W^{l+1}_{i,j})(x-x_i) + D^y_+ A(W^{l+1}_{i+1,j})(y-y_j), \\ \quad \text{for } (x,y) \in T^L_{i,j},\ t \in (t^{n-1}+l\tau, t^{n-1}+(l+1)\tau], \\ A(W^{l+1}_{i,j}) + D^x_+ A(W^{l+1}_{i,j+1})(x-x_i) + D^y_+ A(W^{l+1}_{i,j})(y-y_j), \\ \quad \text{for } (x,y) \in T^U_{i,j},\ t \in (t^{n-1}+l\tau, t^{n-1}+(l+1)\tau], \end{cases}$$

for some $i,j \in \mathbb{Z}$, $l = 0,\dots,N_\tau - 1$, and $n = 1,\dots,N$ with $N_\tau \tau = \Delta t$ and $N\Delta t = T$. Here $T^L_{i,j}$ denotes the triangle with vertices (x_i,y_j), (x_{i+1},y_j), and (x_{i+1},y_{j+1}), while $T^U_{i,j}$ denotes the triangle with vertices (x_i,y_j), (x_i,y_{j+1}), and (x_{i+1},y_{j+1}). Let

$$R_{i,j} = [x_i, x_{i+1}] \times [y_i, y_{i+1}]$$

and note that $R_{i,j} = T^L_{i,j} \cup T^U_{i,j}$. For later use, observe that

$$\partial_x A_\eta \equiv D^x_+ A(W^{l+1}_{i,j}) \text{ on } P_{i,j} = T^L_{i,j-1} \cup T^U_{i,j},$$
$$\partial_y A_\eta \equiv D^y_+ A(W^{l+1}_{i,j}) \text{ on } Q_{i,j} = T^L_{i-1,j} \cup T^U_{i,j}.$$

Observe next that, independently of η,

$$\|\sqrt{\varepsilon K} A_\eta\|_{L^2(Q_T)} \leq \text{Const}(T),$$

which is due to the fact that K, $\{W^l_i\}$, and $\sum_i |W^l_i|$ are uniformly bounded. We next claim that, independently of η,

$$\sqrt{\varepsilon K(x,y)}\partial_z A_\eta \in L^2(Q_T), \quad z = x, y. \tag{27}$$

To see this, multiply the difference equation (26) by $\Delta x^2 A(W^{l+1}_{i,j})$, sum over i, j, and then do summation by parts. The result reads

$$\Delta x^2 \sum_{i,j} \frac{W^{l+1}_{i,j} - W^l_{i,j}}{\tau} A(W^{l+1}_{i,j}) + \Delta x^2 \sum_{i,j} \varepsilon K_{i+1/2,j} \big(D^x_+ A(W^{l+1}_{i,j})\big)^2 + \Delta x^2 \sum_{i,j} \varepsilon K_{i,j+1/2} \big(D^y_+ A(W^{l+1}_{i,j})\big)^2 = 0.$$

Since $\mathcal{A}(s) = \int_0^s A(\xi)\, d\xi$ is convex, it follows that

$$\mathcal{A}(W^{l+1}_{i,j}) - \mathcal{A}(W^l_{i,j}) \leq (W^{l+1}_{i,j} - W^l_{i,j}) A(W^{l+1}_{i,j}).$$

We thus get

$$\Delta x^2 \sum_{i,j} \frac{\mathcal{A}(W^{l+1}_{i,j}) - \mathcal{A}(W^l_{i,j})}{\tau} + \Delta x^2 \sum_{i,j} \varepsilon K_{i+1/2,j} \big(D^x_+ A(W^{l+1}_{i,j})\big)^2 + \Delta x^2 \sum_{i,j} \varepsilon K_{i,j+1/2} \big(D^y_+ A(W^{l+1}_{i,j})\big)^2 \leq 0.$$

Multiplying this inequality by τ and summing over l, we get

$$\Delta x^2 \sum_{i,j} \big(\mathcal{A}(W^{N_\tau}_{i,j}) - \mathcal{A}(W^0_{i,j})\big) + \Delta x^2 \tau \sum_{i,j} \sum_{l} \varepsilon K_{i+1/2,j} \big(D^x_+ A(W^{l+1}_{i,j})\big)^2 + \Delta x^2 \tau \sum_{i,j} \sum_{l} \varepsilon K_{i,j+1/2} \big(D^x_+ A(W^{l+1}_{i,j})\big)^2 \leq 0. \tag{28}$$

Introduce the splitting solution

$$\tilde{u}_\eta(t) = \big[\mathcal{H}_{\Delta x, t-t^{n-1}} \circ \mathcal{S}_{\delta,\Delta x,\Delta t}\big] u^{n-1}_\eta, \quad \text{for } t \in (t^{n-1}, t^n],$$

where $n = 1, \dots, N$ and $u^{n-1}_\eta = u_\eta(t^{n-1})$. Note that

$$\tilde{u}_\eta(x,y,t) = W^{l+1}_{i,j}, \quad \text{for } (x,y) \in R_{i,j},\ t \in (t^{n-1} + l\tau, t^{n-1} + (l+1)\tau],$$

where $l = 0, \dots, N_\tau - 1$. Moreover, we have that,

$$\begin{cases} \tilde{u}_\eta(t^n) = u_\eta(t^n) \text{ for all } n = 1, \dots, N, \\ \|\tilde{u}_\eta(t) - u_\eta(t)\|_1 = \mathcal{O}(\sqrt{\Delta t}) \text{ whenever } t \neq t^n \text{ for some } n = 1, \dots, N, \end{cases}$$

uniformly in $t \in [0,T]$, so that $\tilde{u}_\eta \to u$, with u being the limit constructed in Lemma 3.2. We can thus replace (28) by

$$\int \left(\mathcal{A}(\tilde{u}_\eta)\big|_{t=t^n} - \mathcal{A}(\tilde{u}_\eta)\big|_{t=t^{n-1}+}\right) dx\,dy + \iint_{t^{n-1}}^{t^n} \varepsilon K(x,y)\left(\partial_x A_\eta\right)^2 dt\,dx\,dy$$
$$+ \iint_{t^{n-1}}^{t^n} \varepsilon K(x,y)\left(\partial_y A_\eta\right)^2 dt\,dx\,dy = \mathcal{O}(\Delta x).$$

Observe that

$$\left|\int \left(\mathcal{A}(\tilde{u}_\eta)\big|_{t=t^{n-1}+} - \mathcal{A}(\tilde{u}_\eta)\big|_{t=t^{n-1}}\right) dx\,dy\right| = \mathcal{O}(\Delta x) + \mathcal{O}(\Delta t),$$

which yields

$$\int \left(\mathcal{A}(\tilde{u}_\eta)\big|_{t=t^n} - \mathcal{A}(\tilde{u}_\eta)\big|_{t=t^{n-1}}\right) dx\,dy + \iint_{t^{n-1}}^{t^n} \varepsilon K(x,y)\left(\partial_x A_\eta\right)^2 dt\,dx\,dy$$
$$+ \iint_{t^{n-1}}^{t^n} \varepsilon K(x,y)\left(\partial_y A_\eta\right)^2 dt\,dx\,dy = \mathcal{O}(\Delta x) + \mathcal{O}(\Delta t).$$

Summing over n, we get

$$\int \left(\mathcal{A}(\tilde{u}_\eta)\big|_{t=T} - \mathcal{A}(\tilde{u}_\eta)\big|_{t=0}\right) dx\,dy + \iint_0^T \varepsilon K(x,y)\left(\partial_x A_\eta\right)^2 dt\,dx\,dy$$
$$+ \iint_0^T \varepsilon K(x,y)\left(\partial_y A_\eta\right)^2 dt\,dx\,dy = \mathcal{O}(T).$$

Since, independently of η, $\mathcal{A}(\tilde{u}_\eta)\big|_{t=0,T} \in L^1(\mathbb{R}) \cap L^\infty(\mathbb{R})$, (27) follows. Therefore, passing if necessary to a subsequence,

$$\partial_z A_\eta \to \partial_z \overline{A} \text{ weakly in } L^2_{loc}(Q_T), \qquad z = x, y.$$

Since obviously

$$\iint_0^T \left|A_\eta(x,y,t) - A(\tilde{u}_\eta(x,y,t))\right| dt\,dx = \mathcal{O}(\Delta x)$$

and, thanks to Lemma 3.2, $A(\tilde{u}_\eta) \to A(u)$ a.e., we conclude that $\overline{A} = A(u)$ a.e., and thus (27) holds. This concludes the proof of the lemma. □

4. Entropy Estimates

In this section we derive a precise entropy estimate for u_η, i.e., an estimate saying how far the approximate solution u_η is from fulfilling the entropy condition (5). However, to accomplish this we need to introduce a new time interpolant $\hat{u}_\eta$ and first establish an entropy estimate for $\hat{u}_\eta$. To this end, we define intermediate solutions

$$u_\eta^{n,j} = \pi \circ \mathcal{S}^j_{\delta,\Delta x,\Delta t} \circ \cdots \circ \pi \circ \mathcal{S}^1_{\delta,\Delta x,\Delta t} u_\eta^{n-1}, \qquad u_\eta^{n,0} = u_\eta^{n-1} = u_\eta(t^{n-1}),$$

for $j = 1, \ldots, m$, and $n = 1, \ldots, N$. Next we split each time interval $[t^{n-1}, t^n]$ into subintervals $[t^n_j, t^n_{j+1}]$, $j = 0, \ldots, m$, where

$$t^n_j = \left((n-1) + \frac{j}{m+1}\right)\Delta t, \quad j = 0, \ldots, m+1,$$

so that $t_0^n = t^{n-1}$ and $t_{m+1}^n = t^n$. We then define (see [**34, 23**])

$$\hat{u}_\eta(t) = \begin{cases} \mathcal{S}^j_{\delta,\Delta x,(m+1)(t-t^n_{j-1})}\, u_\eta^{n,j-1}, & t \in (t^n_{j-1}, t^n_j), \\ u_\eta^{n,j}, & t = t^n_j, \\ \mathcal{H}_{\Delta x,(m+1)(t-t^n_m)}\, u_\eta^{n,m}, & t \in (t^n_m, t^n_{m+1}], \end{cases} \tag{29}$$

for $j = 1, \ldots, m$ and $n = 1, \ldots, N$. Observe that the sequence $\{\hat{u}_\eta\}$ converges to the limit function u constructed in Lemma 3.2, since obviously

$$\begin{cases} \hat{u}_\eta(t^n) = u_\eta(t^n) \text{ for all } n = 1, \ldots, N, \\ \|\hat{u}_\eta(t) - u_\eta(t)\|_1 = \mathcal{O}(\sqrt{\Delta t}) \text{ whenever } t \neq t^n \text{ for some } n = 1, \ldots, N. \end{cases} \tag{30}$$

4.1. Hyperbolic step. Since each one-dimensional hyperbolic step in (14) is the exact solution of a perturbed version of (13), the corresponding solution operator satisfies the entropy inequality

$$\mathcal{L}(v_{\Delta x,\delta}, \varphi, k; f^\delta, V^{\Delta x}, 0, v_{0,\Delta x}, T) \geq 0 \text{ for } m = 1,$$

where $v_{\Delta x,\delta}$ denotes the front tracking solution of (17). For sufficiently smooth V, f and initial data v_0 of bounded variation, a straightforward calculation gives (see, e.g., [**40, 23, 33**])

$$\begin{aligned} &\left|\mathcal{L}(v_{\Delta x,\delta}, \phi, k; f, V, 0, v_0, T) - \mathcal{L}(v_{\Delta x,\delta}, \phi, k; f^\delta, V^{\Delta x}, 0, v_{0,\Delta x}, T)\right| \\ &\qquad \leq \text{Const}\, T(\Delta x + \delta^2), \end{aligned}$$

In particular, it follows from this that the one-dimensional front tracking solution $v_{\Delta x,\delta}$ satisfies

$$\begin{aligned} &\int_{\mathbb{R}}\int_0^{\Delta t} \left\{|v_{\Delta x,\delta} - k|\phi_t + F(v_{\Delta x,\delta}, k)\big(\phi V_j(x,t)\big)_x\right\} dt dx \\ &\geq \int |v_{\Delta x,\delta}(x, \Delta t) - k|\phi(x, \Delta t)\, dx - \int |v_{\Delta x,\delta}(x,0) - k|\phi(x,0)\, dx \\ &\quad - \text{Const}\, \Delta t(\Delta x + \delta^2), \end{aligned} \tag{31}$$

for all test functions $\phi \geq 0$. In (31), $v_{\Delta x,\delta}(x,0)$ coincides, of course, with the initial function $v_{0,\Delta x}$. Exploiting the particular structure of $\hat{u}_\eta$ together with a suitable change of (time) variable, it is fairly easy to translate (31) into (see [**34, 23**] for details)

$$\begin{aligned} &\int_{\mathbb{R}^m}\int_{t^n_{j-1}}^{t^n_j} \left\{|\hat{u}_\eta - k|\phi_t + (m+1)F_j(\hat{u}_\eta, k)\big(\phi V_j(x,t)\big)_{x_j}\right\} dt dx \\ &\geq \int |\hat{u}_\eta(t^n_j-) - k|\phi(x, t^n_j)\, dx - \int |\hat{u}_\eta(t^n_{j-1}) - k|\phi(x, t^n_{j-1})\, dx \\ &\qquad - \text{Const}\, \Delta t(\Delta x + \delta^2), \end{aligned} \tag{32}$$

for $j = 1, \ldots, m$ and $n = 1, \ldots, N$. One should notice the factor $m+1$ coming from the particular time scaling in (29). In (32), $\hat{u}_\eta(x, t^n_j-)$ denotes the function before the operator π is applied.

4.2. Parabolic step. We now derive a corresponding entropy inequality for the (degenerate) parabolic step in (14). This will be established by first proving a discrete cell entropy inequality for the difference scheme (19). Similar (discrete) entropy inequalities have been derived in Evje and Karlsen [**23, 19, 21, 18**].

We derive the cell entropy inequality only for $m = 1$ (with $K_1 = K$, $A_1 = A$, and $\alpha = i \in \mathbb{Z}$) to avoid complicated notation. Let

$$S_k(u) = S(u,k) = |u-k|, \qquad Q_k(u) = Q(u,k) = |A(u) - A(k)|,$$

define the discrete difference operator

$$D_0 W_i^n = \frac{W_{i+1/2}^n - W_{i-1/2}^n}{\Delta x},$$

and let $a \vee k = \max(a,k)$ and $a \wedge k = \min(a,k)$.

We now set out to derive the following cell entropy condition

$$S_k(W_i^{n+1}) - S_k(W_i^n) - \varepsilon\tau D_0\Big[K_i D_0\Big(\theta Q_k(W_i^n) + (1-\theta)Q_k(W_i^{n+1})\Big)\Big] \le 0. \tag{33}$$

Using the scheme and the fact that A is an increasing function, we can write (here A_i^n denotes $A(W_i^n)$, etc.)

$$\begin{aligned} S_k(W_i^n) + \varepsilon\tau\theta D_0\big(K_i D_0 Q_k(W_i^n)\big) &= W_i^n \vee k - W_i^n \wedge k \\ &\quad + \theta\beta_i\Big[\big(A_{i+1}^n \vee A(k) - A_{i+1}^n \wedge A(k)\big) - \big(A_i^n \vee A(k) - A_i^n \wedge A(k)\big)\Big] \\ &\quad + \theta\beta_{i-1}\Big[\big(A_i^n \vee A(k) - A_i^n \wedge A(k)\big) - \big(A_{i-1}^n \vee A(k) - A_{i-1}^n \wedge A(k)\big)\Big] \\ &= H_i\big(W_{i+1}^n \vee k, W_i^n \vee k, W_{i-1}^n \vee k\big) - H_i\big(W_{i+1}^n \wedge k, W_i^n \wedge k, W_{i-1}^n \wedge k\big), \end{aligned}$$

where we have introduced the function

$$H_i(u,v,w) = v + \theta\beta_i\big(A(u) - A(v)\big) - \theta\beta_{i-1}\big(A(v) - A(w)\big).$$

The function H_i is increasing in all three arguments due to the stability condition (22), so that

$$\begin{aligned} & S_k(W_i^n) + \varepsilon\tau\theta D_0\big(K_i D_0 Q_k(W_i^n)\big) \\ &\ge H_i\big(W_{i+1}^n, W_i^n, W_{i-1}^n\big) \vee H_i(k,k,k) - H_i\big(W_{i+1}^n, W_i^n, W_{i-1}^n\big) \wedge H_i(k,k,k) \\ &= \big|H_i\big(W_{i+1}^n, W_i^n, W_{i-1}^n\big) - H_i(k,k,k)\big| \\ &= \big|W_i^{n+1} - k - (1-\theta)\varepsilon\tau D_0\big(K_i D_0 Q_k(W_i^{n+1})\big)\big|, \end{aligned} \tag{34}$$

where we have used that $H_i(k,k,k) = k$. Since A is a nondecreasing function, a case analysis will reveal that

$$|A(v) - A(k)| - |A(w) - A(k)| \le \operatorname{sgn}(v-k)\big(A(v) - A(w)\big).$$

Now, we estimate as follows

$$\begin{aligned} & S_k(W_i^{n+1}) - \varepsilon\tau(1-\theta)D_0\big(K_i D_0 Q_k(W_i^{n+1})\big) \\ &= W_i^{n+1} \vee k - W_i^{n+1} \wedge k + (1-\theta)\beta_i\Big[|A_i^{n+1} - A(k)| - |A_{i+1}^{n+1} - A(k)|\Big] \\ &\quad + (1-\theta)\beta_{i-1}\Big[|A_i^{n+1} - A(k)| - |A_{i-1}^{n+1} - A(k)|\Big] \\ &\le \operatorname{sgn}(W_i^{n+1} - k)(W_i^{n+1} - k) \\ &\quad + (1-\theta)\operatorname{sgn}(W_i^{n+1} - k)\big(\beta_i\big(A_i^{n+1} - A_{i+1}^{n+1}\big) + \beta_{i-1}\big(A_i^{n+1} - A_{i-1}^{n+1}\big)\big) \\ &\le \big|W_i^{n+1} - k - (1-\theta)\varepsilon\tau D_0\big(K_i D_0 Q_k(W_i^{n+1})\big)\big|. \end{aligned}$$

Combining this inequality with (34), we arrive at the cell entropy condition (33). To derive a global entropy condition, let $\phi \geq 0$ be a test function, multiply (33) with $\tau\Delta x\phi_i^n$ ($\phi_i^n = \phi(i\Delta x, n\tau)$) and sum over all $i \in \mathbb{Z}$ and $n = 0, \dots, N_\tau - 1$, where $N_\tau\tau = \Delta t$. Then, using summation by parts twice in space, we get

$$\begin{aligned}
0 \geq & \sum_i \sum_{n=0}^{N_\tau-1} \Big(S_k(W_i^{n+1}) - S_k(W_i^n)\Big)\phi_i^n \Delta x \\
& - \tau\varepsilon \sum_i \sum_{n=0}^{N_\tau-1} \Big[D_0\Big(K_i D_0\big[\theta Q_k(W_i^n) + (1-\theta)Q_k(W_i^{n+1})\big]\Big)\Big]\phi_i^n\Delta x \\
= & \sum_i S_k(W_i^{N_\tau})\phi_i^{N_\tau-1}\Delta x - \sum_i S_k(W_i^0)\phi_i^0\Delta x \\
& - \sum_i \sum_{n=0}^{N_\tau-1} \Big[S_k(W_i^{n+1})\frac{\phi_i^{n+1}-\phi_i^n}{\tau}\Big]\tau\Delta x \\
& + \varepsilon \sum_i \sum_{n=0}^{N_\tau-1} \Big[\big(\theta Q_k(W_i^n) + (1-\theta)Q_k(W_i^{n+1})\big)D_0(K_i D_0\phi_i^n)\Big]\tau\Delta x.
\end{aligned}$$

Using the L^1 Hölder continuity in time of $\{W_i^n\}$ (see Lemma 3.1 and its proof), we can write

$$\begin{aligned}
\sum_i \sum_{n=0}^{N_\tau-1} & \Big(\theta Q_k(W_i^n) + (1-\theta)Q_k(W_i^{n+1})\Big)\tau\Delta x \\
& = \sum_i \sum_{n=0}^{N_\tau-1} Q_k(W_i^{n+1})\tau\Delta x + \mathcal{O}\Big(\Delta t\sqrt{\tau}\Big),
\end{aligned}$$

so that we end up with

$$\begin{aligned}
\sum_i \sum_{n=0}^{N_\tau-1} & \Big[S_k(W_i^{n+1})\frac{\phi_i^{n+1}-\phi_i^n}{\tau} + \varepsilon Q_k(W_i^{n+1})D_0(K_i D_0\phi_i^n)\Big]\tau\Delta x \\
& \geq \sum_i S_k(W_i^{N_\tau})\phi_i^{N_\tau}\Delta x - \sum_i S_k(W_i^0)\phi_i^0\Delta x - \text{Const}\,\Delta t\sqrt{\tau}.
\end{aligned} \tag{35}$$

Letting $w(t) = \mathcal{H}_{\Delta x,t}w_0$ denote the finite difference solution (as defined in (21)), we can obviously replace the discrete inequality (35) by its continuous counterpart

$$\begin{aligned}
\int_{\mathbb{R}^m}\int_0^{\Delta t} & \Big\{|w-k|\phi_t + \varepsilon Q_k(w)\big(K(x)\phi_x\big)_x\Big\}\,dtdx \\
\geq & \int |w(x,\Delta t) - k|\phi(x,\Delta t)\,dx - \int |w(x,0)-k|\phi(x,0)\,dx \\
& - \text{Const}\,\Delta t(\sqrt{\Delta t} + \Delta x),
\end{aligned}$$

where $w(x,0)$ coincides with w_0. A similar inequality can be obtained (with the same argument) in the multi-dimensional case. Similar to the hyperbolic step, we can translate this multi-dimensional inequality into the following entropy inequality for the degenerate parabolic step (see [**34, 23**] for further details)

$$
\begin{aligned}
&\int_{\mathbb{R}^m}\int_{t_m^n}^{t_{m+1}^n} \Big\{|\hat{u}_\eta - k|\phi_t + (m+1)\varepsilon \sum_{j=1}^{m} |Q_j(\hat{u}_\eta, k)| \big(K_j(x)\phi_{x_j}\big)_{x_j}\Big\}\, dtdx \\
(36)\qquad &\geq \int |\hat{u}_\eta(t_{m+1}^n) - k|\phi(x, t_{m+1}^n)\, dx - \int |\hat{u}_\eta(t_m^n) - k|\phi(x, t_m^n)\, dx \\
&\qquad - \operatorname{Const} \Delta t(\sqrt{\Delta t} + \Delta x).
\end{aligned}
$$

4.3. The splitting method. For $j = 1, \dots, m+1$ and $n = 1, \dots, N$, let $\chi_j^n(t)$ be the characteristic function of the subinterval $[t_{j-1}^n, t_j^n]$. To derive an entropy estimate for the splitting method, we add (32) and (36), sum over $j = 1, \dots, m$ and $n = 1, \dots, N$ and rearrange terms, yielding

$$
(37)\qquad \mathcal{L}(\hat{u}_\eta, \phi, k; f, V, A, u_0, T) \geq E^{\mathcal{S}} + E^{\pi} + E^{\mathcal{H}} - \operatorname{Const} T\big(\sqrt{\Delta t} + \Delta x + \delta\big),
$$

where

$$
\begin{aligned}
E^{\mathcal{S}} &= \sum_{j=1}^{m} \iint_{Q_T} \Big[1 - (m+1)\sum_{n=1}^{N} \chi_j^n(t)\Big] F_j(\hat{u}_\eta, k)\partial_{x_j}\big(\phi V_j(x,t)\big)\, dt\, dx, \\
E^{\pi} &= \sum_{n=1}^{N}\sum_{j=1}^{m} \int_{\mathbb{R}^m} \Big(S_k(\hat{u}_\eta(x, t_j^n)) - S_k(\pi\hat{u}_\eta(x, t_j^n))\Big)\phi(x, t_j^n)\, dx, \\
E^{\mathcal{H}} &= \varepsilon \iint_{Q_T} \Big[1 - (m+1)\sum_{n=1}^{N} \chi_{m+1}^n(t)\Big] \sum_{j=1}^{m} Q_j(\hat{u}_\eta, k)\partial_{x_j}\big(K_j\partial_{x_j}\phi\big)\, dt\, dx.
\end{aligned}
$$

The three terms can be identified with the following sources for entropy production: $E^{\mathcal{S}}$ comes from the hyperbolic steps, E^{π} from the projections in the hyperbolic steps, and $E^{\mathcal{H}}$ from the (degenerate) parabolic steps. We shall next bound the three terms $E^{\mathcal{S}}$, E^{π}, and $E^{\mathcal{H}}$, starting with $E^{\mathcal{H}}$. By first splitting the integration over $[0, T]$ into a sum over intervals $[t^{n-1}, t^n]$, we see that the sum over n can be taken outside the double integral. Then writing

$$
Q_j(\hat{u}_\eta(t), k) = Q_j(\hat{u}_\eta(t^{n-1}), k) + \Big[Q_j(\hat{u}_\eta(t), k) - Q_j(\hat{u}_\eta(t^{n-1}), k)\Big],
$$

$E^{\mathcal{H}}$ becomes $E_1^{\mathcal{H}} + E_2^{\mathcal{H}}$, where

$$
\begin{aligned}
E_1^{\mathcal{H}} &= \varepsilon \sum_{j,n} \int_{\mathbb{R}^m}\int_{t^{n-1}}^{t^n} \Big[1 - (m+1)\chi_{m+1}^n(t)\Big] Q_j(\hat{u}_\eta(t^{n-1}), k)\partial_{x_j}\big(K_j\partial_{x_j}\phi\big)\, dt\, dx, \\
E_2^{\mathcal{H}} &= \varepsilon \sum_{j,n} \int_{\mathbb{R}^m}\int_{t^{n-1}}^{t^n} \Big[1 - (m+1)\chi_{m+1}^n(t)\Big] \\
&\qquad\qquad \times \Big[Q_j(\hat{u}_\eta(t), k) - Q_j(\hat{u}_\eta(t^{n-1}), k)\Big]\partial_{x_j}\big(K_j\partial_{x_j}\phi\big)\, dt\, dx.
\end{aligned}
$$

In $E_1^{\mathcal{H}}$ we can use the smoothness of $\phi(\,\cdot\,, t)$ to write

$$
\big(\partial_{x_j}(K_j\partial_{x_j}\phi)\big)\big|_{(x,t)} = \big(\partial_{x_j}(K_j\partial_{x_j}\phi)\big)\big|_{(x,t^{n-1})} + C_{j,n}(x,t)(t - t^{n-1}),
$$

for $t \in (t^{n-1}, t^n)$ and some suitable $C_{j,n}(x,t)$ uniformly bounded (in j, n, x, t, η) by a constant $C > 0$. Inserting this into $E_1^{\mathcal{H}}$ gives (where $\Omega = \operatorname{supp}(\phi)$)

$$|E_1^{\mathcal{H}}| \le \varepsilon \sum_{j,n} \int_\Omega \Bigg(\int_{t^{n-1}}^{t^n} Q_j(\hat{u}_\eta(t^{n-1})) \left(\partial_{x_j}(K_j \partial_{x_j} \phi)\right)\big|_{(x,t^{n-1})} \left[1 - (m+1)\chi_{m+1}^n(t)\right] dt$$
$$+ \int_{t^{n-1}}^{t^n} C(t - t^{n-1}) \Big[1 - (m+1)\chi_{m+1}^n(t)\Big]\, dt \Bigg)\, dx$$
$$\equiv \sum_{j,n} \int_\Omega \int_{t^{n-1}}^{t^n} C(t - t^{n-1}) \Big[1 - (m+1)\chi_{m+1}^n(t)\Big]\, dt\, dx = \mathcal{O}(\Delta t).$$

For the second term, we have

$$|E_2^{\mathcal{H}}| \le \sup_{|u| \le \|u_0\|_\infty} |A'(u)|\, m\varepsilon \sum_{j,n} \sup_{Q_T} \left|\partial_{x_j}(K_j \partial_{x_j} \phi)\right|$$
$$\times \int_{\mathbb{R}^m} \int_{t^{n-1}}^{t^n} |\hat{u}_\eta(x,t) - \hat{u}_\eta(x, t^{n-1})|\, dt dx$$
$$\le \text{Const} \sum_{j,n} \int_0^{\Delta t} \sqrt{t}\, dt = \text{Const} \sum_n \Delta t^{3/2} = \mathcal{O}(\sqrt{\Delta t}).$$

For $E^{\mathcal{S}}$ we use a similar argument (to obtain a similar estimate), so that

$$E^{\mathcal{S}} + E^{\mathcal{H}} \ge -\text{Const}\, T \sqrt{\Delta t}.$$

Now it remains to determine the entropy production due to the projection operator π. For simplicity we present the argument for $m = 1$; recall that each hyperbolic operator is one-dimensional, see also [**23, 33**]. Introduce the grid boundaries $x_i = i\Delta x$, let $v = v(x)$ be some function of bounded variation and let $\phi = \phi(x) \ge 0$ be a test function. Then

$$\int_{\mathbb{R}} \Big(S_k(v) - S_k(\pi v)\Big) \phi(x)\, dx = I_1 + I_2$$
$$:= \sum_i \int_{x_i}^{x_{i+1}} \Big(S_k(v) - S_k(\pi v)\Big) \phi(x_i)\, dx$$
$$+ \sum_i \int_{x_i}^{x_{i+1}} \Big(S_k(v) - S_k(\pi v)\Big) \Big(\phi(x) - \phi(x_i)\Big)\, dx.$$

The second term I_2 is straightforward estimated as follows:

$$|I_2| \le |\phi'| \Delta x \sum_i \int_{x_i}^{x_{i+1}} |v - \pi v|\, dx$$
$$\le |\phi'| \Delta x \sum_i \int_{x_i}^{x_{i+1}} \frac{1}{\Delta x} \left| \int_{x_i}^{x_{i+1}} v(y) - v(x)\, dy \right| dx \le \text{Const}\, \Delta x\, TV(v).$$

To estimate the first term, let us first assume that $S_k(u)$ is a C^2 approximation of $|u - k|$. We can then obviously write

$$S_k(v) = S_k(\pi v) + S_k'(\pi v)(v - \pi v) + C(x)(v - \pi v)^2 \text{ for some } C(x) \ge 0,$$

and therefore

$$I_1 = \sum_i \int_{x_i}^{x_{i+1}} \Big[S_k'(\pi v)(v - \pi v) + C(x)(v - \pi v)^2\Big]\, dx.$$

The first term integrates to zero and the second is always positive, so that $I_1 \geq 0$. This latter inequality can also be made rigorous for $S_k(u) = |u - k|$ by suitably approximating $|u - k|$ with C^2 functions, see [**23**]. Hence, $I_1 + I_2 \geq -C\Delta x$. From this we conclude that also

$$E^{\pi} \geq -\text{Const}\,\Delta x.$$

Summing up, we have derived the following entropy estimate for $\hat{u}_\eta$.

$$\mathcal{L}(\hat{u}_\eta, \phi, k; f, V, A, u_0, T) \geq -\text{Const}\,T\big(\sqrt{\Delta t} + \Delta x + \delta^2\big).$$

In view of (30), this estimate translates into the following similar entropy estimate for u_η:

LEMMA 4.1. *The entropy production of the splitting method* (15) *is of order* $\mathcal{O}(\sqrt{\Delta t} + \Delta x + \delta)$, *or more precisely*

$$\mathcal{L}(u_\eta, \phi, k; f, V, A, u_0, T) \geq -\text{Const}\,T\big(\sqrt{\Delta t} + \Delta x + \delta^2\big), \tag{38}$$

for some constant C > 0 *independent of the discretization parameters* $\eta = (\Delta x, \Delta t, \delta)$.

From this lemma we can conclude that if the approximating sequence $\{u_\eta\}$ generated by (15) converges, then it does so to an entropy solution of (1).

5. Summary

In this paper we have presented operator splitting methods for degenerate convection-diffusion equations and shown that these approximations converge to entropy solutions. In a companion paper [**27**] we apply these methods to examples from the simulation of two-phase flow in porous media and certain models for sedimentation-consolidation processes. We demonstrate that although the operator splittings may have certain potential pitfalls, such as splitting errors, lack of mass conservation and grid orientation effects, they perform very well on most cases and deliver more than the standard resolution with surprisingly high efficiency.

References

[1] P. Bénilan and H. Touré. Sur l'équation générale $u_t = a(\cdot, u, \phi(\cdot, u)_x)_x + v$ dans L^1. I. Étude du problème stationnaire. In *Evolution equations (Baton Rouge, LA, 1992)*, pages 35–62. Dekker, New York, 1995.

[2] P. Bénilan and H. Touré. Sur l'équation générale $u_t = a(\cdot, u, \phi(\cdot, u)_x)_x + v$ dans L^1. II. Le problème d'évolution. *Ann. Inst. H. Poincaré Anal. Non Linéaire*, 12(6):727–761, 1995.

[3] F. Bouchut, F. R. Guarguaglini, and R. Natalini. Diffusive BGK approximations for nonlinear multidimensional parabolic equations. *Indiana Univ. Math. J.* To appear.

[4] H. Brézis and M. G. Crandall. Uniqueness of solutions of the initial-value problem for $u_t - \Delta\varphi(u) = 0$. *J. Math. Pures Appl. (9)*, 58(2):153–163, 1979.

[5] R. Bürger, M. C. Bustos, and F. Concha. Settling velocities of particulate systems: Phenomenological theory of sedimentation: 9. Numerical simulation of the transient behavior of flocculated suspensions in an ideal batch or continuous thickener. *Int. J. Mineral Process.*, 55:267–282, 1999.

[6] R. Bürger and F. Concha. Mathematical model and numerical simulation of the settling of flocculated suspensions. *Int. J. Multiphase Flow*, 24:1005–1023, 1998.

[7] R. Bürger, S. Evje, and K. H. Karlsen. On strongly degenerate convection-diffusion problems modeling sedimentation-consolidation processes. J. Math. Anal. Appl. To appear.

[8] R. Bürger, S. Evje, and K. H. Karlsen. Viscous splitting methods for degenerate convection-diffusion equations with boundary conditions. Preprint (in preparation).

[9] R. Bürger, S. Evje, K. H. Karlsen, and K.-A. Lie. Numerical methods for the simulation of the settling of flocculated suspensions. *Separ. Purif. Technol.* To appear in a special issue.

[10] R. Bürger and W. L. Wendland. Entropy boundary and jump conditions in the theory of sedimentation with compression. *Math. Methods Appl. Sci.*, 21(9):865–882, 1998.
[11] R. Bürger and W. L. Wendland. Existence, uniqueness, and stability of generalized solutions of an initial-boundary value problem for a degenerating quasilinear parabolic equation. *J. Math. Anal. Appl.*, 218(1):207–239, 1998.
[12] J. Carrillo. Entropy solutions for nonlinear degenerate problems. *Arch. Rational Mech. Anal.*, 147(4):269–361, 1999.
[13] B. Cockburn and G. Gripenberg. Continuous dependence on the nonlinearities of solutions of degenerate parabolic equations. *J. Differential Equations*, 151(2):231–251, 1999.
[14] B. Cockburn and C.-W. Shu. The local discontinuous Galerkin method for time-dependent convection-diffusion systems. *SIAM J. Numer. Anal.*, 35(6):2440–2463 (electronic), 1998.
[15] F. Concha and R. Bürger. Wave propagation phenomena in the theory of sedimentation. In E. F. Toro and J. F. Clarke, editors, *Numerical Methods for Wave Propagation*, pages 173–196. Kluwer Academic Publishers, Dordrecht, The Netherlands, 1998.
[16] M. G. Crandall and A. Majda. Monotone difference approximations for scalar conservation laws. *Math. Comp.*, 34(149):1–21, 1980.
[17] M. S. Espedal and K. H. Karlsen. Numerical solution of reservoir flow models based on large time step operator splitting algorithms. In A. Fasano and H. van Duijn, editors, *Filtration in Porous Media and Industrial Applications*, Lecture Notes in Mathematics. Springer. To appear.
[18] S. Evje and K. H. Karlsen. Discrete approximations of BV solutions to doubly nonlinear degenerate parabolic equations. *Numer. Math.* To appear.
[19] S. Evje and K. H. Karlsen. Monotone difference approximations of BV solutions to degenerate convection-diffusion equations. *SIAM J. Numer. Anal.* To appear.
[20] S. Evje and K. H. Karlsen. A note on viscous splitting of degenerate convection–diffusion equations. Preprint, Institute of Mathematics and its Applications (IMA), University of Minnestota, Minneapolis, Minnesota, 1997.
[21] S. Evje and K. H. Karlsen. Degenerate convection-diffusion equations and implicit monotone difference schemes. In M. Fey and R. Jeltsch, editors, *Hyperbolic problems: theory, numerics, applications (seventh international conference in Zürich, 1998)*, volume 129 of *Int. Series of Numerical Mathematics*, pages 285–294. Birkhäuser Verlag, 1999.
[22] S. Evje and K. H. Karlsen. Second order difference schemes for degenerate convection-diffusion equations. Preprint, 2000.
[23] S. Evje and K. H. Karlsen. Viscous splitting approximation of mixed hyperbolic-parabolic convection-diffusion equations. *Numer. Math.*, 83(1):107–137, 1999.
[24] S. Evje, K. H. Karlsen, K.-A. Lie, and N. H. Risebro. Front tracking and operator splitting for nonlinear degenerate convection–diffusion equations. In P. Bjørstad and M. Luskin, editors, *Parallel Solution of Partial Differential Equations.* IMA Vol. Math. Appl., Vol. 120, Springer, New York, 2000.
[25] A. Harten, J. M. Hyman, and P. D. Lax. On finite-difference approximations and entropy conditions for shocks. *Comm. Pure Appl. Math.*, XXIX:297–322, 1976.
[26] H. Holden and L. Holden. On scalar conservation laws in one-dimension. In S. Albeverio, J. E. Fenstad, H. Holden, and T. Lindstrøm, editors, *Ideas and Methods in Mathematics and Physics*, pages 480–509. Cambridge University Press, Cambridge, 1988.
[27] H. Holden, K. H. Karlsen, and K.-A. Lie. Operator splitting methods for degenerate convection-diffusion equations II: numerical examples with emphasis on reservoir simulation. Preprint, 1999.
[28] H. Holden, K. H. Karlsen, K.-A. Lie, and N. H. Risebro. Operator splitting for nonlinear partial differential equations: An L^1 convergence theory. Preprint (in preparation).
[29] H. Holden and N. H. Risebro. A method of fractional steps for scalar conservation laws without the CFL condition. *Math. Comp.*, 60(201):221–232, 1993.
[30] K. H. Karlsen, K. Brusdal, H. K. Dahle, S. Evje, and K.-A. Lie. The corrected operator splitting approach applied to a nonlinear advection–diffusion problem. *Comput. Methods Appl. Mech. Engrg.*, 167:239–260, 1998.
[31] K. H. Karlsen and K.-A. Lie. An unconditionally stable splitting scheme for a class of nonlinear parabolic equations. *IMA J. Numer. Anal.* 19: 609–635, 1999.

[32] K. H. Karlsen, K.-A. Lie, N. H. Risebro, and J. Frøyen. A front-tracking approach to a two-phase fluid-flow model with capillary forces. *In Situ (Special Issue on Reservoir Simulation)*, 22(1):59–89, 1998.

[33] K. H. Karlsen and N. H. Risebro. Corrected operator splitting for nonlinear parabolic equations. *SIAM J. Numer. Anal.* 37(3):980 – 1003, 2000.

[34] K. H. Karlsen and N. H. Risebro. An operator splitting method for convection-diffusion equations. *Numer. Math.*, 77(3):365–382, 1997.

[35] K. H. Karlsen and N. H. Risebro. On the uniqueness and stability of entropy solutions of nonlinear degenerate parabolic equations with rough coefficients. Preprint, 2000.

[36] K. H. Karlsen and N. H. Risebro. Convergence of finite difference schemes for viscous and inviscid conservation laws with rough coefficients. Preprint, 2000.

[37] S. N. Kružkov. First order quasi-linear equations in several independent variables. *Math. USSR Sbornik*, 10(2):217–243, 1970.

[38] A. Kurganov and E. Tadmor. New high-resolution central schemes for nonlinear conservation laws and convection-diffusion equations. Preprint, 1999.

[39] K.-A. Lie. A dimensional splitting method for quasilinear hyperbolic equations with variable coefficients. *BIT*, 39(4): 683-700, 1999.

[40] K.-A. Lie. Front tracking for one-dimensional quasilinear hyperbolic equations with variable coefficients. Manuscript: `http://www.ifi.uio.no/~kalie/`

[41] H. Nessyahu and E. Tadmor. Nonoscillatory central differencing for hyperbolic conservation laws. *J. Comput. Phys.*, 87(2):408–463, 1990.

[42] A. A. Samarskii, V. A. Galaktionov, S. P. Kurdyumov, and A. P. Mikhailov. *Blow-up in quasilinear parabolic equations.* Walter de Gruyter & Co., Berlin, 1995. Translated from the 1987 Russian original by Michael Grinfeld and revised by the authors.

[43] J. Smoller. *Shock waves and reaction-diffusion equations.* Springer-Verlag, New York, second edition, 1994.

[44] A. I. Vol'pert. The spaces BV and quasi-linear equations. *Math. USSR Sbornik*, 2(2):225–267, 1967.

[45] A. I. Vol'pert and S. I. Hudjaev. Cauchy's problem for degenerate second order quasilinear parabolic equations. *Math. USSR Sbornik*, 7(3):365–387, 1969.

[46] Z. Wu and J. Yin. Some properties of functions in BV_x and their applications to the uniqueness of solutions for degenerate quasilinear parabolic equations. *Northeastern Math. J.*, 5(4):395–422, 1989.

[47] J. Yin. On the uniqueness and stability of BV solutions for nonlinear diffusion equations. *Comm. in Partial Differential Equations*, 15(12):1671–1683, 1990.

[48] J. Zhao. Uniqueness of solutions for higher dimensional quasilinear degenerate parabolic equation. *Chinese Ann. Math*, 13B(2):129–136, 1992.

[49] J. Zhao and P. Lei. Uniqueness and stability of solutions for Cauchy problem of nonlinear diffusion equations. *Sci. China Ser. A*, 40(9):917–925, 1997.

(H. Holden)
Department of Mathematical Sciences
Norwegian University of Science and Technology
N–7491 Trondheim, Norway
E-mail address: holden@math.ntnu.no
URL: http://www.math.ntnu.no/~holden/

(K. Hvistendahl Karlsen)
Department of Mathematics
University of Bergen
Johs. Brunsgt. 12
N–5008 Bergen, Norway
E-mail address: kennethk@mi.uib.no
URL: http://www.mi.uib.no/~kennethk/

(K.–A. Lie)
Department of Informatics
University of Oslo
P.O. Box 1080 Blindern
N–0316 Oslo, Norway
E-mail address: kalie@ifi.uio.no
URL: http://www.ifi.uio.no/~kalie/

Canadian Mathematical Society
Conference Proceedings
Volume **29**, 2000

A Class of SPDE Driven by Fractional White Noise

Yaozhong Hu

ABSTRACT. The following stochastic partial differential equation

$$\frac{\partial}{\partial t}u(t,x) = \frac{1}{2}\Delta u(t,x) + \mathcal{A}_t u(t,x) \diamond w^H(t,x)\,, \quad t \in \mathbb{R}_+\,, \quad x \in \mathbb{R}^d\,,$$

is studied, where $\mathcal{A}_t$ is a first order partial differential operator and $w^H(x)$ is a fractional white noise with parameter $H = (h_0, h_1, \cdots, h_d)$. Denote $|H| = h_1 + \cdots + h_d$. It is shown that when $4h_0 + |H| > d + 3$, the solution is in $\mathcal{L}_2$ and for any $\kappa > (4h_0 - 2)/(4h_0 + |H| - d - 3)$,

$$\limsup_{t\to\infty} \frac{\log \sup_x \mathbb{E}\,|u_t(x)|^2}{t^\kappa} < \infty\,.$$

1. Introduction

Fractional Brownian motion has been studied intensively in recent years (see for instance [**1**] and the references therein). There has also been study of fractional Brownian fields (abbreviated as FBF). However, the study of FBF seems relatively less intensive. (See [**12**], [**8**] and the references therein).

In this paper we shall use FBF as defined in [**6**]. Let $(\Omega, \mathcal{F}, P)$ be a probability space. The expectation on $(\Omega, \mathcal{F}, P)$ is denoted by $\mathbb{E}$, and $\mathcal{L}_2$ is the space of square integrable random variables $L^2(\Omega, \mathcal{F}, P)$. An FBF with parameter $H = (h_1, \cdots, h_d)$ is defined as a centered Gaussian random field $B^H(x)$ on $(\Omega, \mathcal{F}, P)$ satisfying

$$\text{(1.1)} \quad \mathbb{E}\left(B^H(x)B^H(y)\right) = \frac{1}{2^d}\prod_{i=1}^d \left(|x_i|^{2h_i} + |y_i|^{2h_i} - |x_i - y_i|^{2h_i}\right)\,, \quad x, y \in \mathbb{R}^d\,.$$

The corresponding fractional white noises with parameter H can be defined formally as the derivative of FBF: $w^H(x) = \frac{\partial^d}{\partial x_1 \partial x_2 \cdots \partial x_d} B^H(x)$. Formally the white noise (random field) can be thought as the fractional white noise with $H = (1/2, 1/2, \cdots, 1/2)$.

The stochastic partial differential equations (abbreviated as SPDE) involving white noise have been studied by many authors and have become an important topic of stochastic analysis. When the noise is space-time white or space white, it is known that the classical solutions ($\mathcal{L}_2$ solutions, say) of certain SPDE may not

---

2000 *Mathematics Subject Classification.* Primary 60H15; Secondary 60H05, 60G60.

Partly supported by a General Research Fund of the University of Kansas.

exist. See e.g. [**5**], [**11**], [**10**], [**13**], [**4**], and the references therein. The solutions are sought in generalized random variable spaces. However, when the white noise is replaced by a constant function, the classical solutions exist and possess nice properties. Let us replace the white noise in those equations by the fractional white noise w^H. The above observation states that when $H = (1/2, 1/2, \cdots, 1/2)$, the classical solution usually does not exist. However, when $H = (1, 1, \cdots, 1)$ (this corresponds to $W^H \equiv 1$), the classical solutions do exist. It is then an interesting problem to find out for what values of H the classical solutions do exist. In [**6**], this problem was studied for the stochastic heat equations, *i.e.*

$$\frac{\partial}{\partial t}u(t,x) = \frac{1}{2}\Delta u(t,x) + w^H(t,x) \diamond u(t,x)\,, \tag{1.2}$$

where $w^H(t,x)$ is fractional white noise with parameter $H = (h_0, h_1, \cdots, h_d)$. Let $1/2 < h_i < 1$ for all $0 \le i \le d$. Denote $|H| = h_1 + h_2 + \cdots + h_d$. It is shown in [**6**] that if

$$\rho_0 = \frac{2}{2h_0 - 1} + |H| - d > 0\,, \tag{1.3}$$

then the solution of (1.2) is in $\mathcal{L}_2$.

It is natural to consider the following SPDE

$$\frac{\partial}{\partial t}u(t,x) = \frac{1}{2}\Delta u(t,x) + \sum_{i=1}^{d} w_i^H(t,x) \diamond \frac{\partial}{\partial x_i}u(t,x)\,, \tag{1.4}$$

where $w_i^H(t,x)\,, i = 1, 2, \cdots, d$, are fractional white noises. When $w_i^H(t,x)\,, i = 1, 2, \cdots$, are white noises, this equation is studied in [**13**]. It is known that the solution of (1.4) is more singular than that of (1.2). Our approach can be applied to a slightly more general class of equations. Let $a_i(t,x)$, $t \in \mathbb{R}_+$, $x \in \mathbb{R}^d$, $i = 1, 2, \cdots, d$, be bounded continuously differentiable functions. Denote

$$\mathcal{A}_t f(x) = \sum_{i=1}^{d} a_i(t,x)\frac{\partial f}{\partial x_i}\,, \tag{1.5}$$

Consider the following SPDE

$$\frac{\partial}{\partial t}u(t,x) = \frac{1}{2}\Delta u(t,x) + \mathcal{A}_t u(t,x) \diamond w^H(t,x)\,, \tag{1.6}$$

where the initial condition $u_0(x)$ is a deterministic continuously differentiable function with bounded derivatives. Since $w^H(t,x)$ is generalized function, the product $\mathcal{A}_t u(t,x) w^H(t,x)$ does not exist in the classical sense. We use then the Wick product ([**3**], [**4**], [**9**]) $\mathcal{A}_t u(t,x) \diamond w^H(t,x)$. Its precise meaning shall be given in the next section.

The main result of this paper is to show that the solution of (1.6) is in $\mathcal{L}_2$ and determine its asymptotic behavior as $t \to \infty$, when $4h_0 + |H| > 3 + d$.

In Section 2, we give precise meaning for the solution of (1.6) and present the main results. In Section 3, we provide the proof. This paper uses some results from [**6**].

2. Preliminary and Main Results

Let us briefly recall some results about FBF. We refer to [**6**] for details. Denote $\phi_h(s,t) = h(2h-1)|s-t|^{2h-2}$ and

$$\Phi_H(u,v) = \prod_{i=1}^{d} h_i(2h_i-1)|u_i - v_i|^{2h_i-2}, \quad u,\ v \in \mathbb{R}^d. \tag{2.1}$$

Introduce the following Hilbert space

$$\mathcal{L}_H(\mathbb{R}^{dn}) = \left\{ f : \mathbb{R}^{dn} \to \mathbb{R}; \quad \|f\|_H^2 := \int_{\mathbb{R}^{2d}} \Phi_H(u,v) f(u) f(v) du dv < \infty \right\},$$

where $f(u_1, \cdots, u_n)$ is a symmetric function of $u_1, \cdots, u_n \in \mathbb{R}^d$. For any $f \in \mathcal{L}_H(\mathbb{R}^{dn})$, the following multiple stochastic integral

$$I_n(f) = \int_{\mathbb{R}^{dn}} f_n(u_1, \cdots, u_n) dB^H(u_1) \cdots dB^H(u_n)$$

is well defined and the following isometric equality holds

$$\mathbb{E}\ (I_n(f) I_n(g)) = n! \langle f, g \rangle_H. \tag{2.2}$$

Formally letting $H = (1, 1, \cdots, 1)$, we have $\Phi_H(u,v) = 1$. From this identity it follows that $\mathbb{E}\ (I_1(f) I_1(g)) = \int_{\mathbb{R}^d} f(x)dx \int_{\mathbb{R}^d} g(x)dx$. This suggests that when $H = (1, 1, \cdots, 1)$, $w^H \equiv 1$. The following theorem is stated in [**6**].

THEOREM 2.1. *Let $H = (h_1, \cdots, h_d)$ with $1/2 < h_i < 1$. Then any square integrable random variable F on $(\Omega, \mathcal{F}, P)$, i.e. $F \in \mathcal{L}_2$, admits the following chaos expansion*

$$F = \sum_{n=0}^{\infty} I_n(f_n), \tag{2.3}$$

where f_n is in $\mathcal{L}_H(\mathbb{R}^{nd})$, $n = 0, 1, 2, \cdots$. The above series is convergent in $\mathcal{L}_2$. Moreover,

$$\mathbb{E}\ F^2 = \sum_{n=0}^{\infty} \|f_n\|_H^2. \tag{2.4}$$

In [**6**], we introduced the class of "flat" Hilbert spaces

$$S_\rho := \{F = \sum_{n=0}^{\infty} F_n\ ; \quad \sum_{n=0}^{\infty} (n!)^\rho \mathbb{E}\ |F_n|^2 < \infty\},$$

where $\rho \in \mathbb{R}$ and F_n is the n-th chaos term of F. Note that an element in S_ρ has the following nice property: each of its chaos terms is in $\mathcal{L}_2$. For a random field $f(x)$ such that $f(x) \in S_\rho$, the stochastic integral $\int_{\mathbb{R}^d} f(x) dB^H(x)$ is well defined. In particular, the following lemma is stated in [**6**]

LEMMA 2.2. *Let $f(x)\,, x \in \mathbb{R}^d$, be a random field on $(\Omega, \mathcal{F}, P)$ such that*

$$\int_{\mathbb{R}^{2d}} \Phi_H(x,y) \langle f(x)\,, f(y) \rangle_{S_\rho} dx dy < \infty.$$

Then $\int_{\mathbb{R}^d} f(x) dB^H(x)$ exists as an element of S_ρ.

From now on, we assume that $B^H(t,x)$, $t\in\mathbb{R}_+$, $x\in\mathbb{R}^d$, is an FBF with parameter $H=(h_0,h_1,\cdots,h_d)$ such that $1/2<h_i<1$ for $i=0,1,\cdots,d$. Let $w^H(t,x)$ denote the corresponding fractional white noise.

DEFINITION 2.3. A random field $u:\mathbb{R}_+\times\mathbb{R}^d\times\Omega\to\mathbb{R}$ is called a solution of Eq (1.6) if
(i) $u:\mathbb{R}_+\times\mathbb{R}^d\times\Omega\to\mathbb{R}$ is jointly measurable;
(ii) $\int_{\mathbb{R}^d}P_{t-s}u_s(x)dB^H(x)$ is well-defined for all $t\in\mathbb{R}_+$ as an element of S_γ for certain $\gamma\in\mathbb{R}$; and
(iii) the following equation holds in S_γ:

$$u_t(x)=P_tu_0(x)+\int_0^t\int_{\mathbb{R}^d}P_{t-s}(x-z)\mathcal{A}_su_s(z)dB^H(s,z)\,. \tag{2.5}$$

Let

$$T_n=\{(s_1,s_2,\cdots,s_n);\quad 0<s_1<s_2<\cdots<s_n<t\}\,.$$

The formal chaos expansion of the solution to equation (2.5) is

$$u_t(x)=P_tu_0(x)+\sum_{n=1}^\infty I_n(f_n(t,x))\,, \tag{2.6}$$

where

$$\begin{aligned} &f_n(t,x;s_1,x_1,s_2,x_2,\cdots,s_n,x_n) \\ &= \operatorname{Sym}\left\{\int_{\mathbb{R}^d}P_{t-s_n}(x-x_n)\cdots\mathcal{A}_{s_2}P_{s_2-s_1}(x_2-x_1)\mathcal{A}_{s_1}P_{s_1}(x_1-y)u_0(y)dy\right\}, \end{aligned} \tag{2.7}$$

where Sym is the symmetrization with respect to n ($d+1$ dimensional) variables (s_1,x_1), $\cdots$, (s_n,x_n) and

$$\begin{aligned} I_n(f_n(t,x)) &= \int_{T_n\times\mathbb{R}^{nd}}f_n(t,x;s_1,x_1,s_2,x_2,\cdots,s_n,x_n) \\ &\qquad dB^H(s_1,x_1)dB^H(s_2,x_2)\cdots dB^H(s_n,x_n)\,. \end{aligned} \tag{2.8}$$

We only discuss (2.6). The fact that it is a solution to (2.5) can be shown in a similar way as in [**6**]. The main result of this paper is

THEOREM 2.4. *Let $B^H(t,x)$, $t\in\mathbb{R}_+$, $x\in\mathbb{R}^d$, be an FBF with parameter $H=(h_0,h_1,\cdots,h_d)$ such that $1/2<h_i<1$ for $i=0,1,\cdots,d$, and let $w^H(t,x)$ be the corresponding fractional white noise. Let $u_0(x)$ is a (deterministic) continuously differentiable function with bounded partial derivatives. Let $a_i(t,x)$ be bounded continuous function of t and x. Define $\mathcal{A}_t$ by (1.5). If*

$$4h_0+|H|>3+d\,, \tag{2.9}$$

then the solution $u_t(x)$ defined by (2.6) is in $\mathcal{L}_2$. Moreover, if

$$\kappa_0=\frac{2h_0-1}{4h_0+|H|-d-3}\,, \tag{2.10}$$

then for any $\kappa>\kappa_0$,

$$\limsup_{t\to\infty}\frac{\log\sup_{x\in\mathbb{R}^d}\mathbb{E}\,|u_t(x)|^2}{t^\kappa}<\infty\,. \tag{2.11}$$

3. Proof

Let us compute the $\mathcal{L}_2$ norm of each chaos term. Define

$$\Theta_n(t,x) := \mathbb{E}\ \left(I_n(f_n(t,x))\right)^2 .$$

In the remaining part of the paper C denotes a constant whose value may differ in different places. However, C is independent of t and n.

LEMMA 3.1. *Let $P_t(x) = (2\pi t)^{-d/2}e^{-\frac{x^2}{2t}}$. Then*

$$\sup_{0\le s<\infty} |\mathcal{A}_s P_t(x)| \le \frac{C}{\sqrt{t}} P_{2t}(x) \tag{3.1}$$

Proof This lemma is well known. Instead of sending readers to the literature, let us give a simple proof. It is easy to see that

$$\begin{aligned}\sup_{0\le s<\infty} |\mathcal{A}_s P_t(x)| &\le C \sup_{1\le i\le d} |\frac{\partial}{\partial x_i} P_t(x)| \\ &\le C \sup_{1\le i\le d} \frac{|x|}{t} P_t(x) \\ &= C \sup_{1\le i\le d} \frac{1}{\sqrt{t}} \left\{ \frac{|x|}{\sqrt{t}} e^{-\frac{x^2}{4t}} \right\} P_{2t}(x) .\end{aligned}$$

The proof is completed since $ye^{-\frac{y^2}{4}}$ is bounded.

The following lemma is obvious.

LEMMA 3.2. *If $\sup_{1\le i\le d} |\frac{\partial}{\partial x_i} u_0(x)| \le C < \infty$, then*

$$\sup_{0\le s<\infty,\ x\in\mathbb{R}^d} |\mathcal{A}_s P_t(x)| < \infty .$$

Assume that $\sup_{1\le i\le d} |\frac{\partial}{\partial x_i} u_0(x)| \le C < \infty$. Using (2.2) and (3.1), we obtain

$$\begin{aligned}&\Theta_n(t,x) \\ &= \int_{T_n^2}\int_{\mathbb{R}^{2(n+1)d}} \prod_{i=1}^n \phi_{h_0}(s_i-r_i)\Phi_H(\xi_i-\eta_i)P_{t-s_n}(x-\xi_n)\mathcal{A}_{s_n}P_{s_n-s_{n-1}}(\xi_n-\xi_{n-1})\cdots \\ &\quad \mathcal{A}_{s_2}P_{s_2-s_1}(\xi_2-\xi_1)\mathcal{A}_{s_1}P_{s_1}(\xi_1-y)u_0(y)P_{t-r_n}(x-\eta_n)\mathcal{A}_{r_n}P_{r_n-r_{n-1}}(\eta_n-\eta_{n-1})\cdots \\ &\quad \mathcal{A}_{r_2}P_{r_2-r_1}(\eta_2-\eta_1)\mathcal{A}_{r_1}P_{r_1}(\eta_1-\tilde{y})u_0(\tilde{y})dyd\tilde{y}d\xi d\eta dsdr \\ &\le C\int_{T_n^2}\int_{\mathbb{R}^{2nd}} \left\{\prod_{i=1}^n \phi_{h_0}(s_i-r_i)\Phi_H(\xi_i-\eta_i)\prod_{i=1}^{n-1}(s_{i+1}-s_i)^{-1/2}(r_{i+1}-r_i)^{-1/2}\right\} \\ &\quad P_{2(t-s_n)}(x-\xi_n)\cdots P_{2(s_2-s_1)}(\xi_2-\xi_1)P_{2(t-r_n)}(x-\eta_n)\cdots \\ &\quad P_{2(r_2-r_1)}(\eta_2-\eta_1)d\xi d\eta dsdr ,\end{aligned}$$

where, by convention, $s_{n+1} = t_{n+1} = t$. Denote

$$\begin{aligned}\Theta_{i,n}(t,x_i,s,r) &= \int_{\mathbb{R}^{2n}} \prod_{k=1}^n \phi_{h_k}(\rho_k-\tau_k)P_{2(t-s_n)}(x_i-\rho_n)\cdots P_{2(s_2-s_1)}(\rho_2-\rho_1) \\ &\qquad P_{2(t-r_n)}(x_i-\tau_n)\cdots P_{2(r_2-r_1)}(\tau_2-\tau_1)d\rho d\tau .\end{aligned} \tag{3.2}$$

Then

$$\Theta_n(t,x) = \int_{T_n^2} \left\{ \prod_{k=1}^n \phi_{h_0}(s_k - r_k) \prod_{k=1}^{n-1} (s_{k+1}-s_k)^{-1/2} (r_{k+1}-r_k)^{-1/2} \right\}$$

$$\prod_{i=1}^d \Theta_{i,n}(t,x_i,s,r) dsdr \,. \tag{3.3}$$

From (6.5) of [**6**], it follows that

LEMMA 3.3.

$$\Theta_{i,n}(t,x_i,s,r) \le C^n \prod_{k=1}^n (1 \wedge (s_{k+1}-s_k+r_{k+1}-r_k))^{-\frac{\gamma_i}{2}} \,,$$

for all $\gamma_i > 2 - 2h_i$ (the constant C may depend on γ_i, but not on t and n).

Thus by (3.3), for all $\gamma > 2d - 2|H|$, there is a constant C (which may depend on γ) such that

$$\Theta_n \le C^n \int_{T_n^2} \prod_{k=1}^n |s_k - r_k|^{2h_0-2} (1 \wedge (s_{k+1}-s_k+r_{k+1}-r_k))^{-\frac{\gamma}{2}}$$

$$\prod_{k=1}^{n-1} (s_{k+1}-s_k)^{-1/2}(r_{k+1}-r_k)^{-1/2} dsdr$$

$$\le C^n I_{n,p}^{1/p} \left\{ \int_{T_n^2} \prod_{k=1}^n |s_k - r_k|^{(2h_0-2)q} dsdr \right\}^{1/q} \,, \tag{3.4}$$

where

$$I_{n,p} = \int_{T_n^2} \prod_{k=1}^{n-1} (s_{k+1}-s_k)^{-p/2}(r_{k+1}-r_k)^{-p/2}$$

$$\prod_{k=1}^n (1 \wedge (s_{k+1}-s_k+r_{k+1}-r_k))^{-\frac{p\gamma}{2}} dsdr \,. \tag{3.5}$$

Denote

$$\mathcal{V}_n = \{(v_1, v_2, \cdots, v_n);\ \ v_1 > 0 \,, \cdots, v_n > 0 \,,\ \ v_1 + v_2 + \cdots + v_n < 1\} \,.$$

Making the substitution $s_k = tu_k$ and $r_k = tv_k$, we obtain that for any value of q such that $1 \le q < 1/(2 - 2h_0)$,

$$\left\{ \int_{T_n^2} \prod_{k=1}^n |s_k - r_k|^{(2h_0-2)q} dsdr \right\}^{1/q}$$

$$= \left\{ t^{(2+(2h_0-2)q)n} \int_{\mathcal{V}_n^2} \prod_{k=1}^n |u_k - v_k|^{(2h_0-2)q} dudv \right\}^{1/q}$$

$$\le C^n t^{\frac{2+2(h_0-1)q}{q} n} \,. \tag{3.6}$$

Now we estimate $I_{n,p}$.

$$\begin{aligned} I_{n,p} &\le \int_{T_n^2} \prod_{k=1}^{n-1} (s_{k+1}-s_k)^{-p/2}(r_{k+1}-r_k)^{-p/2} \\ &\quad \prod_{k=1}^{n} (1\wedge(s_{k+1}-s_k))^{-\frac{p\gamma}{4}} (1\wedge(r_{k+1}-r_k))^{-\frac{p\gamma}{4}}\, dsdr \\ &= J_{n,p}^2\,, \end{aligned} \tag{3.7}$$

where

$$J_{n,p} = \int_{T_n} \prod_{k=1}^{n-1} (s_{k+1}-s_k)^{-p/2} \prod_{k=1}^{n} (1\wedge(s_{k+1}-s_k))^{-\frac{p\gamma}{4}}\, ds\,.$$

Making the substitution $tv_1 = s_1$, $tv_2 = s_2 - s_2$, $\cdots$, $tv_n = s_n - s_{n-1}$, we obtain that when $p\gamma + 2p < 4$,

$$\begin{aligned} J_{n,p} &= t^{n-pn/2} \int_{\mathcal{V}_n} \prod_{k=2}^{n} v_k^{-p/2} (1\wedge(tv_k))^{-\frac{p\gamma}{4}} (1\wedge[t(1-v_1-\cdots-v_n)])^{-\frac{p\gamma}{4}}\, dv \\ &\le t^{n-pn/2} \int_{\mathcal{V}_n} \prod_{k=2}^{n} v_k^{-\frac{p\gamma+2p}{4}} (1-v_1-\cdots-v_n)^{-\frac{pr}{4}} dv \\ &= \frac{t^{n-pn/2}\Gamma\left(1-\frac{p\gamma+2p}{4}\right)^{n-1}\Gamma(1-pr/4)}{\Gamma\left[n\left(1-\frac{p\gamma+2p}{4}\right)+1-p\gamma/4\right]}\,, \end{aligned}$$

where $\Gamma(\alpha) = \int_0^\infty t^{\alpha-1}e^{-t}dt$ is the gamma function. From (3.7), it follows that

$$I_{n,p} \le \frac{C^n t^{2n-pn}}{\Gamma\left(2\left(1-\frac{p\gamma+2p}{4}\right)n+1\right)}\,. \tag{3.8}$$

Denote

$$\lambda = \frac{4-\gamma p-2p}{2p}\,.$$

When $\gamma p + 2p < 4$, we have

$$(I_{n,p})^{1/p} \le C^n t^{2n/p-n}/\Gamma(\lambda n+1)\,. \tag{3.9}$$

From (3.4), (3.6), and (3.9), it follows that

$$\Theta_n \le C^n t^{(2h_0-1)n}/\Gamma(\lambda n+1)\,,$$

when $1 \le q < 1/(2-2h_0)$, since the latter inequalities imply that

$$p = \frac{q}{q-1} > \frac{1}{2h_0-1}\,.$$

LEMMA 3.4. *Let $\lambda_0 = (4h_0+|H|-d-3)$. If (2.9) holds and $\lambda < \lambda_0$, then there are $\gamma > 0$, $p > 1$, and $q > 1$ such that $\gamma > 2d-2|H|$, $1/p+1/q = 1$, $q < 1/(2-2h_0)$, and $\lambda = (4-\gamma p-2p)/(2p)$.*

Then for any $\lambda < \lambda_0$, there is a constant $0 < C < \infty$ such that

$$\Theta_n \le \frac{C^n t^{(2h_0-1)n}}{\Gamma(\lambda n+1)}\,. \tag{3.10}$$

Consequently,

$$\sup_{x\in\mathbb{R}^d} \mathbb{E}\ |u_t(x)|^2 \le \sum_{n=0}^{\infty} \frac{C^n t^{(2h_0-1)n}}{\Gamma(\lambda n+1)} = E_\lambda(Ct^{(2h_0-1)}),$$

where E_λ is the Mittag-Leffler function [**2**]. Thus we have shown that

$$\limsup_{t\to\infty} \frac{\log \sup_{x\in\mathbb{R}^d} \mathbb{E}\ |u_t(x)|^2}{t^{(2h_0-1)/\lambda}} < \infty.$$

Let $\kappa = (2h_0 - 1)/\lambda$. Since $\lambda < \lambda_0$ is equivalent to $\kappa > \kappa_0$, this completes the proof of Theorem 2.4.

REMARK 3.5. From the estimate (3.10), we can also show that $u_t(x)$ is in S_ρ for some $\rho > 0$. One can also obtain the asymptotic behavior of $u_t(x)$ in these spaces.

REMARK 3.6. When the fractional white noise is homogeneous in the sense that $H = (h, h, \cdots, h)$, then the condition (2.9) becomes

$$h > \frac{d+3}{d+4}. \tag{3.11}$$

In this case $\lambda_0 = (4h + dh - d - 3)/2$ and

$$(2h_0 - 1)/\lambda > \kappa_0 = (4h - 2)/(4h + dh - d - 3).$$

Roughly speaking, this means that when $w^H(t, x)$ is a fractional white noise with parameter $H = (h, h, \cdots, h)$, then when $h > (d + 3)/(d + 4)$, the solution $u_t(x)$ of (1.6) is in $\mathcal{L}_2$ and for any $\kappa > \kappa_0$,

$$\limsup_{t\to\infty} \frac{\log \sup_{x\in\mathbb{R}^d} \mathbb{E}\ |u_t(x)|^2}{t^\kappa} < \infty. \tag{3.12}$$

Acknowledgment. The author thanks Professor Charles Himmelberg and the referee for careful reading of the original paper.

References

[1] Duncan, T.E.; Hu, Y.Z.; Pasik-Duncan, B. Stochastic calculus for fractional Brownian motion, I. Theory, Accepted by SIAM Journal on Control and Optimization, 1999.

[2] Erdélyi, A.; Magnus, W.; Oberhettinger F.; Tricomi, F.G. High Transcendental Functions, Vol. 3, McGraw-Hill Book Company, INC. 1955.

[3] Hida, T.; Kuo, H.-H.; Potthoff, J. and Streit, L. White noise. An infinite-dimensional calculus. Mathematics and its Applications, 253. Kluwer Academic Publishers Group, Dordrecht, 1993.

[4] Holden, H.; Øksendal, B.; Ubøe J. and Zhang, T.S. Stochastic Partial Differential Equations, a Modeling, White Noise Functional Analysis, Birkhäuser, 1996.

[5] Hu, Y.Z. Asymptotics of heat equation with white noise potentials, Preprint 1998.

[6] Hu, Y.Z. Heat equation with fractional white noise potentials, Preprint 1998.

[7] Hu, Y.Z.; Kallianpur, G. Exponential integrability and application to stochastic quantization. Appl. Math. Optim. 37 (1998), no. 3, 295–353.

[8] Lindstrøm, T. Fractional Brownian fields as integrals of white noise. Bull. London Math. Soc. 25 (1993), no. 1, 83–88.

[9] Meyer, P.A. Quantum probability for probabilist, Lect. Notes in Math. 1538, Springer, 1993.

[10] Nualart, D.; Rozovskii B. Weighted stochastic Sobolev spaces and bilinear SPDEs driven by space-time white noise, Journal of Functional Analysis, 149 (1997), 200-225,

[11] Nualart, D.; Zakai, M. Generalized Brownian functionals and the solution to a stochastic partial differential equation. J. Funct. Anal. 84 (1989), no. 2, 279- 296.

[12] Samorodnitsky, G. and Taqqu, M.S. Stable non-Gaussian random processes, Chapman & Hall, 1994.

[13] Potthoff, J.; Våge, G.; Watanabe, H. Generalized solutions of linear parabolic stochastic partial differential equations. Appl. Math. Optim. 38 (1998), 95–107.

Department of Mathematics, University of Kansas, 405 Snow Hall, Lawrence, KS 66045-2142

E-mail address: hu@math.ukans.edu

Canadian Mathematical Society
Conference Proceedings
Volume **29**, 2000

Stochastic partial differential equations driven by multiparameter fractional white noise

Yaozhong Hu, Bernt Øksendal, and Tusheng Zhang

ABSTRACT. We develop a multiparameter white noise theory for fractional Brownian motion with Hurst multiparameter $H = (H_1, \dots, H_d) \in (\frac{1}{2}, 1)^d$. The theory is used to solve the Poisson equation driven by multiparameter fractional white noise. The results are compared to the standard white noise case ($H_i = 1/2$ for all i).

1. Introduction

Recall that if $0 < H < 1$ then the *(1-parameter) fractional Brownian motion* with Hurst parameter H is the Gaussian process $B_H(t) = B_H(t, \omega)$; $t \in \mathbf{R}$, $\omega \in \Omega$ satisfying

$$B_H(0) = E[B_H(t)] = 0 \qquad \text{for all } t \in \mathbf{R} \tag{1.1}$$

and

$$E[B_H(s)B_H(t)] = \tfrac{1}{2}\{|s|^{2H} + |t|^{2H} - |s-t|^{2H}\} \qquad \text{for all } s, t \in \mathbf{R}\,. \tag{1.2}$$

Here E denotes the expectation with respect to the probability law P for $\{B_H(t,\omega)\}_{t\in\mathbf{R},\omega\in\Omega}$, where $(\Omega, \mathcal{F})$ is a measurable space.

If $H = \frac{1}{2}$ then $B_H(t)$ coincides with the standard Brownian motion $B(t)$. Much of the recent interest in fractional Brownian motion stems from its property that if $H > \frac{1}{2}$ then $B_H(t)$ has a *long range dependence*, in the sense that

$$\sum_{n=1}^{\infty} E[B_H(1)(B_H(n+1) - B_H(n))] = \infty\,.$$

Moreover, for any $H \in (0,1)$ and $\alpha > 0$ the law of $\{B_H(\alpha t)\}_{t\in\mathbf{R}}$ is the same as the law of $\{\alpha^H B_H(t)\}_{t\in\mathbf{R}}$, i.e. $B_H(t)$ is H-self-similar.

For more information on 1-parameter fractional Brownian motion see e.g. [**MV**], [**NVV**] and the references therein.

Recently a stochastic calculus based on Itô-type of integration with respect to $B_H(t)$ has been constructed for $H > \frac{1}{2}$ [**DHP**]. Subsequently a corresponding fractional white noise theory has been developed [**HØ**], and this has been used to study the corresponding fractional models in mathematical finance [**HØ**], [**HØS**].

2000 *Mathematics Subject Classification.* Primary 60H15, 35R60, 60G15.

Dedicated to Professor Sergio Albeverio on the occasion of his 60th birthday.

As in [**H1**], [**H2**] we define *d-parameter fractional Brownian motion* $B_H(x)$; $x = (x_1, \dots, x_d) \in \mathbf{R}^d$ with *Hurst parameter* $H = (H_1, \dots, H_d) \in (0,1)^d$ as a Gaussian process on $\mathbf{R}^d$ with mean

$$E[B_H(x)] = 0 \qquad \text{for all } x \in \mathbf{R}^d \tag{1.3}$$

and covariance

$$E[B_H(x)B_H(y)] = (\tfrac{1}{2})^d \prod_{i=1}^d (|x_i|^{2H_i} + |y_i|^{2H_i} - |x_i - y_i|^{2H_i}) \tag{1.4}$$

We also assume that

$$B_H(0) = 0 \quad \text{a.s.} \tag{1.5}$$

From now on we will assume that

$$\tfrac{1}{2} < H_i < 1 \qquad \text{for } i = 1, \dots, d\,. \tag{1.6}$$

The purpose of this paper is to extend the fractional white noise theory to the multiparameter case and use this theory to study the Poisson equation with a fractional white noise force.

2. Multiparameter fractional white noise

In this section we outline how the multiparameter white noise theory for standard Brownian motion (see e.g. [**HKPS**], [**HØUZ**] or [**K**]) can be extended to fractional Brownian motion. In the 1-parameter case such an extension was presented in [**HØ**].

Fix a parameter dimension $d \in \mathbf{N}$ and a *Hurst parameter*

$$H = (H_1, \dots, H_d) \in (\tfrac{1}{2}, 1)^d\,. \tag{2.1}$$

Define

$$\varphi(x,y) = \varphi_H(x,y) = \prod_{i=1}^d H_i(2H_i - 1)|x_i - y_i|^{2H_i - 2} \tag{2.2}$$

for $x = (x_1, \dots, x_d) \in \mathbf{R}^d$, $y = (y_1, \dots, y_d) \in \mathbf{R}^d$.

Let $L^2_\varphi(\mathbf{R}^d)$ be the space of measurable functions $f \colon \mathbf{R}^d \to \mathbf{R}$ satisfying

$$|f|^2_\varphi :=\int_{\mathbf{R}^d}\int_{\mathbf{R}^d} f(x)f(y)\varphi(x,y)dx\,dy < \infty \tag{2.3}$$

where $dx = dx_1 \dots dx_d$ and $dy = dy_1 \dots dy_d$ denotes Lebesgue measure.

Then $L^2_\varphi(\mathbf{R}^d)$ is a separable Hilbert space with the inner product

$$(f,g)_\varphi = \int_{\mathbf{R}^d}\int_{\mathbf{R}^d} f(x)g(y)\varphi(x,y)dx\,dy\,; \qquad f,g \in L^2_\varphi(\mathbf{R}^d)\,. \tag{2.4}$$

In fact, we have (see [**HØ**, Lemma 2.1] for the case $d = 1$):

LEMMA 2.1. *For $f \in L^2_\varphi(\mathbf{R}^d)$ and $u = (u_1, \dots, u_d) \in \mathbf{R}^d$ define*

$$\Gamma_\varphi f(u) = \int_{u_1}^\infty \cdots \int_{u_d}^\infty f(x_1, \dots, x_d) \prod_{i=1}^d c_{H_i}(x_i - u_i)^{H_i - 3/2} dx_1 \dots dx_d\,, \tag{2.5}$$

where

$$(2.6) \qquad c_{H_i} = \sqrt{\frac{H_i(2H_i-1)\cdot\Gamma(\frac{3}{2}-H_i)}{\Gamma(H_i-\frac{1}{2})\cdot\Gamma(2-2H_i)}}\ ; \qquad i=1,\dots,d\,.$$

Then Γ_φ *is an isometry from* $L^2_\varphi(\mathbf{R}^d)$ *into* $L^2(\mathbf{R}^d)$.

PROOF. For $f,g\in L^2_\varphi(\mathbf{R}^d)$ we have

$$\begin{aligned}
&(\Gamma_\varphi(f),\Gamma_\varphi(g))_{L^2(\mathbf{R}^d)}\\
&=\int_{\mathbf{R}^d}\Big(\int_{u_i}^{\infty}\cdots\int_{u_d}^{\infty} f(x)\prod_{i=1}^{d} c_{H_i}(x_i-u_i)^{H_i-3/2}dx\Big)\\
&\qquad\cdot\Big(\int_{u_1}^{\infty}\cdots\int_{u_d}^{\infty} g(y)\prod_{i=1}^{d} c_{H_i}(y_i-u_i)^{H_i-3/2}dy\Big)du_1\dots du_d\\
&=\int_{\mathbf{R}^d}\int_{\mathbf{R}^d} f(x)g(y)\Big(\prod_{i=1}^{d}\int_{-\infty}^{x_i\wedge y_i} c^2_{H_i}(x_i-u_i)^{H_i-3/2}(y_i-u_i)^{H_i-3/2}du_i\Big)dx\,dy\\
&=\int_{\mathbf{R}^d}\int_{\mathbf{R}^d} f(x)g(y)\varphi(x,y)dx\,dy\,,
\end{aligned}$$

where we have used the fact that (see e.g. [**GN**, p. 404])

$$(2.7) \qquad \int_{-\infty}^{x_i\wedge y_i} c^2_{H_i}(x_i-u_i)^{H_i-3/2}(y_i-u_i)^{H_i-3/2}du_i = H_i(2H_i-1)|x_i-y_i|^{2H_i-2}\,.$$

□

Let $\mathcal{S}(\mathbf{R}^d)$ be the Schwartz space of rapidly decreasing smooth functions on $\mathbf{R}^d$. The dual of $\mathcal{S}(\mathbf{R}^d)$, the space of tempered distributions, is denoted by $\mathcal{S}'(\mathbf{R}^d)$. The functional

$$f\to\exp(-\tfrac{1}{2}|f|^2_\varphi)\ ; \qquad f\in\mathcal{S}(\mathbf{R}^d)$$

is positive definite on $\mathcal{S}(\mathbf{R}^d)$, so by the Bochner-Minlos theorem there exists a probability measure μ_φ on $\mathcal{S}'(\mathbf{R}^d)$ such that

$$(2.8) \qquad \int_{\mathcal{S}'(\mathbf{R}^d)} e^{i\langle\omega,f\rangle}d\mu_\varphi(\omega)=e^{-\frac{1}{2}|f|^2_\varphi}\ ; \qquad f\in\mathcal{S}(\mathbf{R}^d)$$

where $\langle\omega,f\rangle$ denotes the action of $\omega\in\Omega:=\mathcal{S}'(\mathbf{R}^d)$ on $f\in\mathcal{S}(\mathbf{R}^d)$. From (2.8) one can deduce that if $f_n\in\mathcal{S}(\mathbf{R}^d)$ and $f_n\to f$ in $L^2_\varphi(\mathbf{R}^d)$ then

$$(2.9) \qquad \langle\omega,f\rangle:=\lim_{n\to\infty}\langle\omega,f_n\rangle \qquad \text{exists in } L^2(\mu_\varphi)$$

and defines a Gaussian random variable. Moreover,

$$(2.10) \qquad E[\langle\cdot,f\rangle]=0$$

and

$$(2.11) \qquad E[\langle\cdot,f\rangle\langle\cdot,g\rangle]=(f,g)_\varphi \qquad \text{for } f,g\in L^2_\varphi(\mathbf{R}^d)\,.$$

Here, and in the following, $E[\cdot] = E_{\mu_\varphi}[\cdot]$ denotes the expectation with respect to μ_φ.

In particular, we may define

$$\widetilde{B}_H(x) = \langle \omega, \mathcal{X}_{[0,x]}(\cdot) \rangle ; \qquad x = (x_1, \dots, x_d) \in \mathbf{R}^d \tag{2.12}$$

where

$$\mathcal{X}_{[0,x]}(y) = \prod_{i=1}^{d} \mathcal{X}_{[0,x_i]}(y_i) \qquad \text{for } y = (y_1, \dots, y_d) \in \mathbf{R}^d$$

and

$$\mathcal{X}_{[0,x_i]}(y_i) = \begin{cases} 1 & \text{if } 0 \le y_i \le x_i \\ -1 & \text{if } x_i \le y_i \le 0, \text{ except } x_i = y_i = 0 \\ 0 & \text{otherwise} \end{cases}$$

Using (2.10)–(2.11) and Kolmogorov's criterion, we see that $\widetilde{B}_H(x); x \in \mathbf{R}^d$ is a Gaussian process and it has a continuous version. Furthermore, we see that

$$E[B_H(x)] = 0$$

and

$$E[B_H(x)B_H(y)] = (\tfrac{1}{2})^d \prod_{i=1}^{d} (|x_i|^{2H_i} + |y|^{2H_i} - |x_i - y_i|^{2H_i}) . \tag{2.13}$$

Therefore $B_H(x); x \in \mathbf{R}^d$ is a d-parameter fractional Brownian motion with Hurst parameter $H = (H_1, \dots, H_d) \in (\frac{1}{2}, 1)^d$ (see (1.3)–(1.5)). It is this version of $B_H(x)$ we will use from now on.

Let $f \in L^2_\varphi(\mathbf{R}^d)$. The *stochastic integral* of f with respect to the fractional Brownian motion $B_H(x)$ is the Gaussian random variable on Ω defined by

$$\int_{\mathbf{R}^d} f(x) dB_H(x) = \int_{\mathbf{R}^d} f(x) dB_H(x, \omega) = \langle \omega, f \rangle . \tag{2.14}$$

Note that this is a natural definition from the point of view of Riemann sums:

If f_n is a simple integrand of the form

$$f_n(x) = \sum_{j=1}^{N_n} a_j^{(n)} \mathcal{X}_{(-\infty, y_j)}(x)$$

then (2.13) gives

$$\int_{\mathbf{R}^d} f_n(x) dB_H(x) = \langle \omega, f_n \rangle = \sum_{j=1}^{N_n} a_j^{(n)} B_H(y_j)$$

and if $f_n \to f$ in $L^2_\varphi(\mathbf{R}^d)$ then by (2.9) we have, as desired, that

$$\int_{\mathbf{R}^d} f_n(x) dB_H(x) = \langle \omega, f_n \rangle \to \langle \omega, f \rangle = \int_{\mathbf{R}^d} f(x) dB_H(x) .$$

Now we proceed in analogy with [**HØUZ**] (as done in [**HØ**] in the 1-parameter case) to obtain a multiparameter fractional chaos expansion:

Let

$$h_n(t) = (-1)^n e^{t^2/2} \frac{d^n}{dt^n}(e^{-t^2/2}) ; \qquad t \in \mathbf{R} , \quad n = 0,1,2,\dots \tag{2.15}$$

be the standard Hermite polynomials and let

$$\widetilde{h}_n(t) = \pi^{-1/4}((n-1)!)^{-1/2} h_{n-1}(\sqrt{2}\,t) e^{-t^2/2} ; \qquad n = 1,2,\dots \tag{2.16}$$

be the *Hermite functions.* Let $\mathbf{N} = \{1,2,\dots\}$. For $\alpha \in \mathbf{N}^d$ let $\eta_\alpha(x) = \prod_{i=1}^{d} \widetilde{h}_{\alpha_i}(x_i)$. Then $\{\eta_\alpha\}_{\alpha \in \mathbf{N}^d}$ constitutes an orthonormal basis of $L^2(\mathbf{R}^d)$. Therefore

$$e_\alpha(x) : = \Gamma_\varphi^{-1}(\eta_\alpha)(x) ; \qquad \alpha \in \mathbf{N}^d , \quad x \in \mathbf{R}^d$$

constitutes an orthonormal basis of $L^2_\varphi(\mathbf{R}^d)$. From now on we let $\{\alpha^{(i)}\}_{i=1}^\infty$ be a fixed ordering of $\mathbf{N}^d$ with the property that

$$i < j \Rightarrow |\alpha^{(i)}| \le |\alpha^{(j)}|$$

and we write

$$e_n(x) : = e_{\alpha^{(n)}}(x) . \tag{2.17}$$

Then just as in [**HØ**, Lemma 3.1] we can prove

LEMMA 2.2. *There exists a locally bounded function $C(x)$ on $\mathbf{R}^d$ such that*

$$\left| \int_{\mathbf{R}^d} e_n(y) \varphi(x,y) dy \right| \le C(x) n^{1/6}$$

Let $\mathcal{J} = (\mathbf{N}_0^{\mathbf{N}})_c$ denote the set of all (finite) multi-indices $\alpha = (\alpha_1, \dots, \alpha_m)$ with $\alpha_i \in \mathbf{N}_0 : = \mathbf{N} \cup \{0\}$. Then if $\alpha = (\alpha_1, \dots, \alpha_m) \in \mathcal{J}$ we define

$$\mathcal{H}_\alpha(\omega) = h_{\alpha_1}(\langle \omega, e_1 \rangle) \cdots h_{\alpha_m}(\langle \omega, e_m \rangle) . \tag{2.18}$$

In particular, if we put

$$\varepsilon^{(i)} = (0,0,\dots,1) \qquad \text{(the } i\text{'th unit vector)}$$

then by (2.14) we get

$$\mathcal{H}_{\varepsilon^{(i)}}(\omega) = h_1(\langle \omega, e_i \rangle) = \langle \omega, e_i \rangle = \int_{\mathbf{R}^d} e_i(x) dB_H(x) . \tag{2.19}$$

As is well-known in a more general context (see e.g. [**J**, Theorem 2.6]) we have the following Wiener-Itô chaos expansion theorem (see also [**DHP**] and [**HØ**]):

THEOREM 2.3. *Let $F \in L^2(\mu_\varphi)$. Then there exist constants $c_\alpha \in \mathbf{R}$ for $\alpha \in \mathcal{J}$, such that*

$$F(\omega) = \sum_{\alpha \in \mathcal{J}} c_\alpha \mathcal{H}_\alpha(\omega) \qquad \textit{(convergence in } L^2(\mu_\varphi)) . \tag{2.20}$$

Moreover, we have the isometry,

$$\|F\|^2_{L^2(\mu_\varphi)} = \sum_{\alpha \in \mathcal{J}} \alpha! c_\alpha^2 \tag{2.21}$$

where $\alpha! = \alpha_1! \alpha_2! \dots \alpha_m!$ if $\alpha = (\alpha_1, \dots, \alpha_m) \in \mathcal{J}$.

EXAMPLE 2.4. If $F(\omega) = \langle \omega, f \rangle$ for some $f \in L^2_\varphi(\mathbf{R}^d)$, then F has the expansion

$$F(\omega) = \left\langle \omega, \sum_{i=1}^{\infty} (f, e_i)_\varphi e_i \right\rangle = \sum_{i=1}^{\infty} (f, e_i)_\varphi \mathcal{H}_{\varepsilon^{(i)}}(\omega) \, . \tag{2.22}$$

In particular, for d-parameter fractional Brownian motion we get, by (2.12),

$$\begin{aligned} B_H(x) &= \langle \omega, \mathcal{X}_{[0,x]}(\cdot) \rangle = \sum_{i=1}^{\infty} (\mathcal{X}_{[0,x]}, e_i)_\varphi \mathcal{H}_{\varepsilon^{(i)}}(\omega) \\ &= \sum_{i=1}^{\infty} \Big[\int_0^x \Big(\int_{\mathbf{R}^d} e_i(v)\varphi(u,v)dv \Big) du \Big] \mathcal{H}_{\varepsilon^{(i)}}(\omega) \, , \end{aligned} \tag{2.23}$$

where $\int_0^x = \int_0^{x_d} \cdots \int_0^{x_1}$ and $\int_0^{x_i} = -\int_{x_i}^0$ if $x_i < 0$.

Next we proceed as in [**HØUZ**] to define the *multiparameter fractional Hida test function space* $(\mathcal{S})_H$ and *distribution space* $(\mathcal{S})^*_H$:

DEFINITION 2.5. **a)** (The multidimensional fractional Hida test function spaces) For $k \in \mathbf{N}$ define $(\mathcal{S})_{H,k}$ to be the space of all

$$\psi(\omega) = \sum_{\alpha \in \mathcal{J}} a_\alpha \mathcal{H}_\alpha(\omega) \in L^2(\mu_\varphi) \tag{2.24}$$

such that

$$\|\psi\|^2_{H,k} : = \sum_{\alpha \in \mathcal{J}} \alpha! a^2_\alpha (2\mathbf{N})^{k\alpha} < \infty \tag{2.25}$$

where

$$(2\mathbf{N})^\gamma = \prod_j (2j)^{\gamma_j} \qquad \text{if} \quad \gamma = (\gamma_1, \ldots, \gamma_m) \in \mathcal{J} \, .$$

Define $(\mathcal{S})_H = \bigcap_{k=1}^{\infty} (\mathcal{S})_{H,k}$ with the projective topology.

b) (The multiparameter fractional Hida distribution spaces)
For $q \in \mathbf{N}$ let $(\mathcal{S})^*_{H,-q}$ be the space of all formal expansions

$$G(\omega) = \sum_{\beta \in \mathcal{J}} b_\beta \mathcal{H}_\alpha(\omega) \tag{2.26}$$

such that

$$\|G\|^2_{H,-q} : = \sum_{\beta \in \mathcal{J}} \beta! b^2_\beta (2\mathbf{N})^{-q\beta} < \infty \, . \tag{2.27}$$

Define

$$(\mathcal{S})^*_H = \bigcup_{q=1}^{\infty} (\mathcal{S})^*_{H,-q}$$

with the inductive topology. Then $(\mathcal{S})^*_H$ becomes the dual of $(\mathcal{S})_H$ when the action of $G \in (\mathcal{S})^*_H$ given by (2.26) on $\psi \in (\mathcal{S})_H$ given by (2.24) is defined by

$$\langle\langle G, \psi \rangle\rangle = \sum_{\alpha \in \mathcal{J}} \alpha! a_\alpha b_\alpha \,. \tag{2.28}$$

EXAMPLE 2.6 (Multiparameter fractional white noise).
Define, for $y \in \mathbf{R}^d$

$$W_H(y) = \sum_{i=1}^{\infty} \Big[\int_{\mathbf{R}^d} e_i(v)\varphi(y,v)dv \Big] \mathcal{H}_{\varepsilon^{(i)}}(\omega) \tag{2.29}$$

Then as in [**HØ**, Example 3.6] we obtain that $W_H(y) \in (\mathcal{S})^*_H$ for all y. Moreover, $W_H(y)$ is integrable in $(\mathcal{S})^*_H$ for $0 \le y_i \le x_i$; $i = 1, \dots, d$, and

$$\int_0^x W_H(y)dy = \sum_{i=1}^{\infty} \Big[\int_0^x \Big(\int_{\mathbf{R}^d} e_i(v)\varphi(y,v)dv \Big] dy \Big] \mathcal{H}_{\varepsilon^{(i)}}(\omega) = B_H(x) \,, \tag{2.30}$$

by (2.23). Therefore $B_H(x)$ is differentiable with respect to x in $(\mathcal{S})^*_H$ and we have

$$\frac{\partial^d}{\partial x_1 \dots \partial x_d} B_H(x) = W_H(x) \qquad \text{in} \quad (\mathcal{S})^*_H \,. \tag{2.31}$$

This justifies the name (multiparameter) *fractional white noise* for $W_H(x)$.

The Wick product is defined just as in [**HØUZ**] and [**HØ**]:

DEFINITION 2.7. Suppose $F(\omega) = \sum_{\alpha \in \mathcal{J}} a_\alpha \mathcal{H}_\alpha(\omega)$ and $G(\omega) = \sum_{\beta \in \mathcal{J}} b_\beta \mathcal{H}_\beta(\omega)$ both belong to $(\mathcal{S})^*_H$. Then we define their *Wick product* $(F \diamond G)(\omega)$ by

$$(F \diamond G)(\omega) = \sum_{\alpha,\beta \in \mathcal{J}} a_\alpha b_\beta \mathcal{H}_{\alpha+\beta}(\omega) = \sum_{\gamma \in \mathcal{J}} \Big(\sum_{\alpha+\beta=\gamma} a_\alpha b_\beta \Big) \mathcal{H}_\gamma(\omega) \,. \tag{2.32}$$

EXAMPLE 2.8. **a)** ([**HØ**, Example 3.9]) If $f, g \in L^2_\varphi(\mathbf{R}^d)$ then

$$\Big(\int_{\mathbf{R}^d} f\, dB_H \Big) \diamond \Big(\int_{\mathbf{R}^d} g\, dB_H \Big) = \Big(\int_{\mathbf{R}^d} f\, dB_H \Big) \cdot \Big(\int_{\mathbf{R}^d} g\, dB_H \Big) - (f,g)_\varphi \,. \tag{2.33}$$

b) ([**HØ**, Example 3.10]) If $f \in L^2_\varphi(\mathbf{R}^d)$ then

$$\exp^\diamond(\langle \omega, f \rangle) \colon = \sum_{n=1}^{\infty} \frac{1}{n!} \langle \omega, f \rangle^{\diamond n}$$

converges in $(\mathcal{S})^*_H$ and is given by

$$\exp^\diamond(\langle \omega, f \rangle) = \exp(\langle \omega, f \rangle - \tfrac{1}{2}|f|^2_\varphi) \,. \tag{2.34}$$

We now use multiparameter fractional white noise to define integration with respect to multiparameter fractional Brownian motion, just as in [**HØ**, Definition 3.11] for the 1-parameter case:

DEFINITION 2.9. Suppose $Y : \mathbf{R}^d \to (\mathcal{S})^*_H$ is a given function such that $Y(x) \diamond W_H(x)$ is integrable in $(\mathcal{S})^*_H$ for $x \in \mathbf{R}^d$. Then we define the multiparameter fractional stochastic integral (of Itô type) of $Y(x)$ by

$$\int_{\mathbf{R}^d} Y(x) dB_H(x) = \int_{\mathbf{R}^d} Y(x) \diamond W_H(x) dx \,. \tag{2.35}$$

REMARK 2.10. If $H = \frac{1}{2}$ this definition gives an extension of the Itô-Skorohod integral. See [**HØUZ**, Section 2.5] for more details.

3. The Poisson equation with a fractional white noise force

To illustrate the theory we apply it to solve the stochastic partial differential equation

$$\Delta U(x) = -W_H(x)\,; \qquad x \in D \subset \mathbf{R}^d \tag{3.1}$$

$$U(x) = 0 \qquad \text{for} \quad x \in \partial D\,. \tag{3.2}$$

Here $\Delta = \sum_{i=1}^{d} \frac{\partial^2}{\partial x_i^2}$ is the Laplacian operator, D is a bounded open set in $\mathbf{R}^d$ with smooth boundary and we are looking for a solution $U : \bar{D} \to (\mathcal{S})_H^*$ which is continuous on the closure $\bar{D}$ of D and twice continuously differentiable (in $(\mathcal{S})_H^*$) on D.

In view of the corresponding situation in the classical deterministic case it is natural to guess that the solution must be (see (2.35))

$$U(x) = \int_D G(x,y)W_H(y)dy = \int_D G(x,y)dB_H(y)\,, \tag{3.3}$$

where $G(x,y)$ is the classical Green function for the deterministic Poisson problem.

First we verify that $U(x) \in (\mathcal{S})_H^*$ for all $x \in \bar{D}$: By (2.27) the expansion of $U(x)$ is

$$\begin{aligned} U(x) &= \int_D G(x,y) \sum_{k=1}^{\infty} \left[\int_{\mathbf{R}^d} e_k(v)\varphi(y,v)dv \right] \mathcal{H}_{\varepsilon^{(k)}}(\omega)dy \\ &= \sum_{k=1}^{\infty} a_k(x)\mathcal{H}_{\varepsilon^{(k)}}(\omega)\,, \end{aligned} \tag{3.4}$$

where

$$a_k(x) = a_{\varepsilon^{(k)}}(x) = \int_D G(x,y)\left(\int_{\mathbf{R}^d} e_k(v)\varphi(y,v)dv \right) dy\,. \tag{3.5}$$

By Lemma 2.2 we have

$$|a_k(x)| \le C_1 k^{1/6} \int_D G(x,y)dy \le C_2 k^{1/6}\,. \tag{3.6}$$

(Here, and in the following, $C_1, C_2, C_3, \ldots$ denote constants.) Therefore

$$\sum_{k=1}^{\infty} a_k^2(x)(2\mathbf{N})^{-q\varepsilon^{(k)}} \le C_2^2 \sum_{k=1}^{\infty} k^{1/3}(2k)^{-q} < \infty \qquad \text{for} \quad q > \tfrac{4}{3}$$

and we conclude that $U(x) \in (\mathcal{S})_{H,-q}^*$ for all $q > \frac{4}{3}$, for all x. This estimate also shows that $U(x) : \bar{D} \to (\mathcal{S})_H^*$ is uniformly continuous.

Next we verify that U satisfies (3.1). To this end, put

$$U_n(x) = \sum_{k=1}^{n} a_k(x)\mathcal{H}_{\varepsilon^{(k)}}(\omega)\,. \tag{3.7}$$

Then by the classical theory from the deterministic Poisson equation we know that

$$\Delta U_n(x) = -\sum_{k=1}^{n} \Big(\int_{\mathbf{R}^d} e_k(v)\varphi(x,v)dv \Big) \mathcal{H}_{\varepsilon^{(k)}}(\omega) \qquad \text{for} \quad x \in D\,.$$

Fix $q > \frac{4}{3}$.
Moreover, by Lemma 2.2 we have

$$\sum_{k \geq n} \Big(\int_{\mathbf{R}^d} e_k(v)\varphi(x,v)dv \Big)^2 (2k)^{-q} \leq C_3 \sum_{k \geq n} k^{\frac{1}{3}-q} \to 0 \qquad \text{as} \quad n \to \infty\,.$$

Therefore,

$$\Delta U_n(x) \to -W_H(x) \qquad \text{in } (\mathcal{S})^*_{H,-q} \text{ as } n \to \infty\,, \tag{3.8}$$

uniformly for $x \in D$.

From (3.6) we see that

$$U_n(x) \to U(x) \qquad \text{in } (\mathcal{S})^*_{H,-q} \text{ as } n \to \infty\,, \tag{3.9}$$

uniformly for $x \in \bar{D}$.

Combining (3.8) and (3.9) we see that (3.1) holds. So $U(x)$ given by (3.3) is indeed a solution of (3.1)–(3.2).

In the standard white noise case ($H_i = \frac{1}{2}$ for all i) we know that the solution of (3.1)–(3.2) exists as an element of $L^2(\mu)$ if and only if $d < 4$ (see e.g. [**HØUZ**, p. 4–5]). It is natural to ask what can be said for $H \in (\frac{1}{2}, 1)^d$. To investigate this we apply (2.11) to our solution $U(x)$ in (3.3): If we assume $d \geq 3$ then the singularity of $G(x,y)$ at $x = y$ is of the form $|x-y|^{2-d}$ and therefore we get

$$\begin{aligned} E[U^2(x)] &= \int_D \int_D G(x,y)G(x,z)\varphi(y,z)dy\,dz \\ &\leq C_4 \int_D \int_D |x-y|^{2-d}|x-z|^{2-d} \prod_{i=1}^{d} |y_i - z_i|^{2H_i-2} dy\,dz\,. \end{aligned} \tag{3.10}$$

To estimate (3.10), we shall use the following inequality

$$a_1 + a_2 + \cdots + a_d \geq a_1^{\beta_1} a_2^{\beta_2} \cdots a_d^{\beta_d}\,, \quad a_1, \cdots, a_d \geq 0\,, \tag{3.11}$$

where $\beta_1, \cdots, \beta_d$ are positive numbers such that

$$\beta_1 + \beta_2 + \cdots + \beta_d = 1\,. \tag{3.12}$$

Therefore

$$|x-y|^{2-d} = \Big(\sum_{i=1}^{d} |x_i - y_i|^2 \Big)^{\frac{2-d}{2}} \leq \prod_{i=1}^{d} |x_i - y_i|^{(2-d)\beta_i} \tag{3.13}$$

for any positive $\beta_1, \cdots, \beta_d$ satisfying (3.12). From (3.10) and (3.13) it follows that

(3.14)

$$E\left(U^2(x)\right) \leq C_5 \prod_{i=1}^{d} \int_{-R}^{R} \int_{-R}^{R} |x_i - y_i|^{(2-d)\beta_i} |x_i - z_i|^{(2-d)\beta_i} \, |z_i - y_i|^{2H_i-2} \, dy_i z_i\,,$$

where $R = \max\{|x|, x \in D\}$. It is obvious that the above integral is finite for $i = 1, 2, \cdots, d$ if the following inequalities hold:

$$(2-d)\beta_i > -1; \tag{3.15}$$

$$2H_i - 2 > -1; \tag{3.16}$$

$$2(2-d)\beta_i + 2H_i > 0\,. \tag{3.17}$$

Our assumption $H_i > 1/2$ implies (3.16). Thus $E\left(U^2(x)\right)$ is finite if

$$H_i > \beta_i(d-2)\,, \quad \beta_i > 0 \tag{3.18}$$

for $i = 1, 2, \cdots, d$ and $\beta_1 + \beta_2 + \cdots + \beta_d = 1$ since (3.17) implies (3.15). It is easy to see that if the following inequalities

$$\sum_{i=1}^{d} H_i > d-2\,, \quad 1/2 < H_i < 1 \tag{3.19}$$

hold, then we can find positive numbers $\beta_1, \cdots, \beta_d$ with $\beta_1 + \beta_2 + \cdots + \beta_d = 1$ such that (3.18) holds. We summarize what we have obtained as follows:

THEOREM 3.1. **a)** *For any dimension d there is a unique strong solution $U(x) : \bar{D} \to (\mathcal{S})^*_H$ satisfying the fractional Poisson equation*

$$\Delta U(x) = -W_H(x)\,; \qquad x \in D \subset \mathbf{R}^d \tag{3.20}$$

$$U(x) = 0 \qquad \text{for} \quad x \in \partial D\,. \tag{3.21}$$

The solution is given by

$$U(x) = \int_D G(x,y) W_H(y) dy = \int_D G(x,y) dB_H(y)\,. \tag{3.22}$$

The solution belongs to $C(\bar{D}) \cap C^2(D)$.

b) *If*

$$d - \sum_{i=1}^{d} H_i < 2\,, \tag{3.23}$$

then

$$U(x) \in L^2(\mu_\varphi) \qquad \text{for all} \quad x \in \bar{D}\,,$$

so the solution exists as an ordinary stochastic field.

REMARK 3.2. Note that if $H_i = \frac{1}{2}$ for all i then the condition (3.23) coincides with the known condition $d<4$ mentioned above. But if $H_i > \frac{1}{2}$ for all i we may get $L^2(\mu_\varphi)$ solutions also in higher dimensions. Thus it is easier to obtain $L^2(\mu_\varphi)$ solutions the higher the values of H_i are. This seems reasonable from the point of view that the paths of $B_H(x)$ get more regular as H increases.

References

[DHP] T. E. Duncan, Y. Hu and B. Pasik-Duncan: Stochastic calculus for fractional Brownian motion. I. Theory. To appear in SIAM Journal of Control and Optimization.

[GN] G. Gripenberg and I. Norros: On the prediction of fractional Brownian motion. *J. Appl. Prob.* **33** (1996), 400–410.

[H1] Y. Hu: Heat equation with fractional white noise potentials. Preprint 1998.

[H2] Y. Hu: A class of stochastic partial differential equations driven by fractional white noise. Preprint 1999.

[HKPS] T. Hida, H.-H. Kuo, J. Potthoff and L. Streit: White Noise Analysis. Kluwer 1993.

[HØ] Y. Hu and B. Øksendal: Fractional white noise calculus and applications to finance. Preprint, Dept. of Mathematics, University of Oslo 1999.

[HØUZ] H. Holden, B. Øksendal, J. Ubøe and T. Zhang: Stochastic Partial Differential Equations. Birkhäuser 1996.

[HØS] Y. Hu, B. Øksendal and A. Sulem: Optimal portfolio in a fractional Black & Scholes market. Preprint, Dept. of Mathematics, University of Oslo 1999.

[J] S. Janson: Gaussian Hilbert Spaces. Cambridge University Press 1997.

[K] H.-H. Kuo: White Noise Distribution Theory. CRC Press 1996.

[MV] B.B. Mandelbrot and J.W. Van Ness: Fractional Brownian motions, fractional noises and applications. *SIAM Rev.* **10** (1968), 422–437.

[NVV] I. Norros, E. Valkeila and J. Virtamo: An elementary approach to a Girsanov formula and other analytic results on fractional Brownian motions. *Bernoulli* **5** (1999), 571–587.

DEPT. OF MATHEMATICS, UNIVERSITY OF KANSAS, 405 SNOW HALL, LAWRENCE, KANSAS 66045–2142, USA.

E-mail address: hu@math.ukans.edu

DEPT. OF MATHEMATICS, UNIVERSITY OF OSLO, P. O. BOX 1053 BLINDERN, N–0316 OSLO, NORWAY.

Alternative address: NORWEGIAN SCHOOL OF ECONOMICS AND BUSINESS ADMINISTRATION, HELLEVEIEN 30, N–5035 BERGEN-SANDVIKEN, NORWAY.

E-mail address: oksendal@math.uio.no

DEPT. OF MATHEMATICS, AGDER COLLEGE, N–4684 KRISTIANSAND, NORWAY.

Current address: Dept. of Mathematics, Cornell University, Ithaca, NY 14853–4201, USA.

E-mail address: tusheng@math.cornell.edu

Canadian Mathematical Society
Conference Proceedings
Volume **29**, 2000

Some support properties of the laws of invariant spaces of stochastic differential equations

Peter Imkeller

ABSTRACT. We refine the smoothness results obtained in an earlier paper for the laws of the invariant flag and Oseledets spaces of random dynamical systems generated by linear(ized) stochastic differential equations. We prove connectedness of the support of the laws, show that zeroes of their densities are "flat" points, and that under restrictive Hörmander type conditions they are positive everywhere and therefore the supports of their laws are full.

Introduction

Consider a random dynamical system generated by a linear(ized) stochastic differential equation. The exponential growth rates of its trajectories take only a finite number of possible values, the Lyapunov exponents. They are characterized for large positive times by a nested family of random invariant linear subspaces of the state space, the *forward flag*: starting in a flag space means that the exponential growth rate at ∞ is at most equal to the Lyapunov exponent corresponding to the space. For small negative times a corresponding backward flag of invariant spaces takes over. Intersecting spaces of the two flags with corresponding Lyapunov exponents yields the *Oseledets spaces* of the system which, together with the Lyapunov exponents, form a complete random substitute of the spectral data of a real matrix generating a deterministic linear flow.

By identifying the various invariant linear spaces with the invariant states of flows of diffeomorphisms on the corresponding Grassmannian manifolds, and using Malliavin's calculus, we obtained in [**14**] smoothness properties of their laws. As it turns out, the knowledge of smoothness of the random invariant spaces in the Malliavin sense obtained as a by-product, produces much more subtle results about the laws. Indeed, the precise studies of small time asymptotics of heat kernels for diffusion processes done in a series of papers by Ben Arous [**3**], Léandre [**16**], [**17**], Ben Arous and Léandre [**5**], [**6**], in which ideas of Bismut [**7**] on the use of Malliavin's calculus for large deviations are employed, provide a complete understanding of the relationship between the positivity of transition densities and the regularity of the restriction of the sde to the Stroock-Varadhan support (or *skeleton*) of the diffusion:

1991 *Mathematics Subject Classification.* primary: 60 H 07, 58 F 11; secondary: 60 H 10, 58 G 32.

the restriction induces deterministic flows on the state space and thus creates a *control distance* intrinsic to the diffusions. To make the density of transition from x to y positive, y has to be reached from x via a deterministic flow and the latter has to vary regularly near y. For the control theoretic aspect see also Sussmann [**20**], Sussmann and Jurdjevic [**21**], Kunita [**15**], Ichihara and Kunita [**12**], and Arnold, Kliemann, Oeljeklaus [**2**]. These results are nicely explained in the recent Lecture Notes by Nualart [**19**], to which we also refer for a complete bibliography.

In this paper, we use them to establish some fine properties of the laws of the random invariant spaces described above. We prove that their supports are connected, that zeroes of the densities are *flat* points, and give Hörmander type conditions on the matrices $A_0, \cdots, A_m$ generating the linear system under which the densities are positive everywhere, and thus the support is full. If the drift vanishes, clearly ellipticity is sufficient for this. But if not, these conditions have to be considerably stronger than Hörmander's hypoellipticity. By an example we show that even if the drift can be described by Lie brackets of order 1 of the diffusion vector fields (on a Grassmannian), the densities of invariant spaces may have flat points.

Preliminaries and notation

Our basic probability space is the $m-$dimensional canonical Wiener space $(\Omega, \mathbf{F}, \mathbf{P})$, enlarged such as to carry an $m-$dimensional "Brownian motion" indexed by $\mathbf{R}$. More precisely, $\Omega = C(\mathbf{R}, \mathbf{R}^m)$ is the set of continuous functions on $\mathbf{R}$ with values in $\mathbf{R}^m$, $\mathbf{F}$ the $\sigma-$algebra of Borel sets with respect to uniform convergence on compacts of $\mathbf{R}$, $\mathbf{P}$ the probability measure on $\mathbf{F}$ for which the *canonical Wiener process* $W_t = (W_t^1, ..., W_t^m), t \in \mathbf{R}$, satisfies that both $(W_t)_{t\geq 0}$ and $(W_{-t})_{t\geq 0}$ are usual $m-$dimensional Brownian motions. The natural filtration $\{\mathbf{F}_s^t = \sigma(W_u - W_v : s \leq u, v \leq t) : \mathbf{R} \ni s \leq t \in \mathbf{R}\}$ of W is assumed to be completed by the sets of $\mathbf{P}-$ measure 0 in $\mathbf{F}$. For $t \in \mathbf{R}$, let $\theta_t : \Omega \to \Omega, \omega \mapsto \omega(t + \cdot) - \omega(t)$, the "shift" on Ω by t. It is well known that θ_t preserves Wiener measure $\mathbf{P}$ for any $t \in \mathbf{R}$ and is even ergodic. Hence $(\Omega, \mathbf{F}, \mathbf{P}, (\theta_t)_{t\in\mathbf{R}})$ is an ergodic metric dynamical system (see Arnold [**1**]). As usual, we use a "$\circ$" to denote Stratonovich integrals with respect to Wiener process. Since our parameter space is $\mathbf{R}$ and not just $\mathbf{R}_+$ let us briefly recall the conventions for the definition of the Stratonovitch integral. The forward Stratonovitch integral of a continuous $(\mathbf{F}_0^t)_{t\geq 0}$-semimartingale $(f_t)_{t\geq 0}$ w. r. t. a one dimensional Wiener process $(W_t)_{t\geq 0}$ is defined as usual and denoted by

$$\int_0^t f_s \circ dW_s, \quad t \geq 0,$$

the backward Stratonovitch integral of a continuous backward $(\mathbf{F}_t^0)_{t\leq 0}$-semimartingale $(g_t)_{t\leq 0}$ w. r. t. a backward one dimensional Wiener process $(W_t)_{t\leq 0}$ by

$$\int_t^0 g_s \circ dW_s, \quad t \leq 0.$$

For a process $(f_t)_{t\in\mathbf{R}}$ such that $(f_t)_{t\geq 0}$ is a continuous forward $(\mathbf{F}_0^t)_{t\geq 0}$-semimartingale and $(f_t)_{t\leq 0}$ a continuous backward $(\mathbf{F}_t^0)_{t\leq 0}$-semimartingale we introduce the convention

$$\int_0^t f_s \circ dW_s = \begin{cases} \int_0^t f_s \circ dW_s, & t \geq 0, \\ -\int_t^0 f_s \circ dW_s, & t \leq 0. \end{cases}$$

We shall be concerned with random invariant subspaces of a linear system of stochastic differential equations, objects described in the multiplicative ergodic theorem of Oseledets. We shall briefly recall its main assertions in the framework of linear sde. Let $A_0, A_1, ..., A_m$ be $d \times d$ matrices, and consider the linear flow $(\Phi_t)_{t\in\mathbf{R}}$ on $Gl(d;\mathbf{R})$ generated by

$$(0.1) \qquad \begin{aligned} d\Phi_t &= A_0\,\Phi_t\,dt + \sum_{i=1}^m A_i\,\Phi_t \circ dW_t^i, \\ \Phi_0 &= I. \end{aligned}$$

Then the *Ruelle matrix* $\Phi^\pm$ of the system given by $\Phi^\pm = \lim_{t\to\infty}[\Phi_t^*\,\Phi_t]^{\frac{1}{2|t|}}$ has a spectral decomposition

$$\Phi^\pm = \sum_{i=1}^r e^{\pm\lambda_i} S_i^\pm$$

with some r, $1 \le r \le d$, random orthogonal projectors $S_i^\pm$ of deterministic rank d_i such that $d_1 + \cdots + d_r = d$. The *Lyapunov exponents* λ_i describe the asymptotic exponential growth rate of the trajectories of the solutions of the linear equation: the orthogonal projectors $Q_i = S_i^+ + \cdots + S_r^+$ and $P_i = S_1^- + \cdots + S_i^-$ determine those subspaces of $\mathbf{R}^d$ on which trajectories have exponential growth rates $\le \lambda_i$ as $t \to \infty$ resp. $\ge \lambda_i$ as $t \to \infty$. More formally,

$$x \in Q_i(\mathbf{R}^d) \quad \text{iff} \quad \lim_{t\to\infty} \frac{1}{t} \ln ||\Phi_t\, x|| \le \lambda_i,$$

and

$$x \in P_i(\mathbf{R}^d) \quad \text{iff} \quad \lim_{t\to\infty} \frac{1}{t} \ln ||\Phi_t\, x|| \ge \lambda_i.$$

The intersections $E_i = Q_i(\mathbf{R}^d) \cap P_i(\mathbf{R}^d)$ of these two families, the *Oseledets spaces* correspondingly characterize those subspaces on which trajectories have exponential growth rate λ_i both at ∞ and at $-\infty$, thus generalizing the concept of eigenspaces of linear algebra. We equally associate orthogonal projectors with them which will be denoted R_i and have (P-a.s.) rank d_i, $1 \le i \le r$. The *forward flag* of orthogonal projectors associated with our system is the family $(Q_1, \cdots, Q_r)$, the *backward flag* the family $(P_1, \cdots, P_r)$.

So the random elements the laws of which we study can be supposed to take their values in some Grassmannian manifold of linear subspaces of $\mathbf{R}^d$, if we identify the latter with the space of orthogonal projectors in $\mathbf{R}^d$ of a certain fixed rank. Thus for dimension k, $1 \le k \le d$, we let

$$G_k(d) = \{p : p \in \mathbf{R}^{d\times d}, \operatorname{rk} p = k\}.$$

This simple identification provides an isometric embedding of the Grassmannian manifold, usually described as

$$O(d)/O(k) \times O(d-k)$$

of $k-$dimensional subspaces of $\mathbf{R}^d$ into the space $\mathbf{R}^{d\times d}$ of $d \times d$ matrices. The tangent space $T_pG_k(d)$ can then be seen to be the space of linear operators given by

$$T_pG_k(d) = \{(I-p)\,A\,p + p\,A^*\,(I-p) : \; A \in \mathbf{R}^{d\times d}\},$$

for $p \in G_k(d)$ (see **[14]**).

Hence the compact manifolds M we shall be working with are embedded in Euclidean spaces and inherit their Riemannian metrics from the usual Euclidean scalar product which shall be denoted by $\langle .,. \rangle$ irrespective of the dimension of the underlying space. The (co-) tangent bundle of M will be denoted by $TM(T^*M)$. Lie brackets are written $[.,.]$.

We finally give a brief reminder of the basic concepts of Malliavin's calculus we need. See Nualart [**18**] for a more detailed treatment. For $1 \leq j \leq m$ we shall denote by D^j the derivative operator which, for a smooth random variable of the form

$$F = f(W_{t_1}, ..., W_{t_n}), \quad f \in C_b^\infty((\mathbf{R}^m)^n), \quad t_1, ..., t_n \in \mathbf{R}_+,$$

takes the form

$$D_s^j F = \sum_{i=1}^n \frac{\partial}{\partial x_{i,j}} f(W_{t_1}, ..., W_{t_n}) 1_{[0,t_i]}(s).$$

Here the j in the partial derivative $\frac{\partial}{\partial x_{i,j}}$ denotes the component of the $m-$dimensional Wiener process, whereas i refers to the index of $t_1, ..., t_n$. For each $p \geq 1, \mathbf{D}_{p,1}$ will denote the Banach space of random variables on the Wiener space, defined as the closure of the set of smooth random variables with respect to the norm

$$||F||_{p,1} = ||F||_p + \sum_{1 \leq j \leq m} E([\int_0^\infty |D_s^j F|^2 \, ds]^{\frac{p}{2}})^{\frac{1}{p}},$$

to which D^j extends in a natural way.

For $k \in \mathbf{N}$, we define the $H-$derivative of order k by $k-$fold iteration of the above derivative on the space of smooth random variables. For such a random variable F, the expression $D_{s_1 \dots s_k}^{j_1 \dots j_k} F$ for $1 \leq j_i \leq m, 1 \leq i \leq k, s_1, ..., s_k \in \mathbf{R}_+$ is self explanatory. For $p \geq 1, k \in \mathbf{N}$, we denote by $\mathbf{D}_{p,k}$ the Banach space given by the completion of the set of smooth random variables with respect to the norm

$$\| F \|_{p,k} = \| F \|_p + \sum_{1 \leq j \leq k} \sum_{1 \leq l_i \leq m, 1 \leq i \leq j} [E([\int_0^\infty (D_{s_1 \dots s_j}^{l_1 \dots l_j} F)^2 \, ds_1 ... ds_j]^{\frac{p}{2}})]^{\frac{1}{p}}.$$

Finally, let $\mathbf{D}_\infty$ be the intersection of the preceding Sobolev spaces over all $p \geq 1, k \in \mathbf{N}$.

1. Some properties of the densities of invariant spaces

Let Q be a random orthogonal projector belonging to the forward flag of the linear system (0.1) and assume it takes its values in $M = G_k(d)$ for some $k, 1 \leq k \leq d$. We shall establish some fine structure properties of the density of the law of Q, which will then be extended to the Oseledets spaces of the system. These properties can easily be deduced from the smoothness results for Q in the Malliavin sense established in [**14**] and general relationships between the laws and the vector fields generating the flows investigated in Ben Arous and Léandre [**5**], [**6**], and Léandre [**16**], [**17**]. For a treatment of these relations in the flavour of potential theory see Hirsch and Song [**11**], for a recent complete and transparent presentation see Nualart [**19**].

The connectedness of the support of the law of Q is a consequence of a general result obtained by Fang [**8**].

THEOREM 1.1. *The support of the law of Q on M is connected.*

Proof:
[**14**] gives $Q \in \mathbf{D}_\infty$. Hence proposition 4.1.1 in Nualart [**19**] implies that the law of Q is connected in $\mathbf{R}^{d \times d}$, the space into which M is embedded, while concentrated

on M. Hence it is connected in M. □

A general result by Ben Arous and Léandre relates Malliavin smoothness of a random variable on Wiener space to "flatness" in zeros of its density. To be able to speak about densities of the law of Q, we shall have to assume some nondegeneracy.

For this purpose, recall (see [**14**]) that Q is the stationary state of a flow of diffeomorphisms on M which is generated by the"projections" $h_{A_0}, \cdots, h_{A_m}$ of the vector fields $A_0, \cdots, A_m$ on the tangent bundle of M, where

$$h_A(p) = (I-p)\,A\,p + p\,A^*\,(I-p) \quad \text{for} \quad A \in \mathbf{R}^{d\times d},$$

through the stochastic differential equations

$$dP_t = h_{A_0}(P_t)\,dt + \sum_{i=1}^m h_{A_i}(P_t) \circ dW_t^i, \tag{1.1}$$

$t \in \mathbf{R}$. We denote this flow by $(P(t,.))_{t\in\mathbf{R}}$, so that $P(t,.)\,p$ is just the solution at time t of (1.1) starting in p, for $p \in M$. According to [**14**], the invariance properties of Q can be expressed this way: the law, say ρ, of Q, equals the $\mathbf{P}\otimes\rho-$law of the mapping $(\omega, p) \mapsto P(t,.)\,p$. To formulate nondegeneracy conditions, let $\mathcal{L}^k$ be the Lie algebra generated by the vector fields $h_{A_0}, \cdots, h_{A_m}$, $\mathcal{I}^k$ the ideal in $\mathcal{L}^k$ generated by the vector fields $h_{A_1}, \cdots, h_{A_m}$, i.e.

$$\begin{aligned} \mathcal{I}^k &= \operatorname{span}\{h_{A_1}, ..., h_{A_m}, [h_{A_{i_p}}, [h_{A_{i_{p-1}}}, [...[h_{A_{i_1}}, h_{A_{i_0}}]...] : \\ &\quad 0 \le i_j \le m, 1 \le j \le p, p \in \mathbf{N}\}, \end{aligned}$$

and $\mathcal{M}^k$ the Lie algebra generated by $h_{A_1}, \cdots, h_{A_m}$. The following conditions will play a role in the sequel:

$$(WH)_k \qquad \mathcal{L}_p^k = T_pM, \quad p \in M,$$

$$(H)_k \qquad \mathcal{I}_p^k = T_pM, \quad p \in M,$$

$$(SH)_k \qquad \mathcal{M}_p^k = T_pM, \quad p \in M.$$

We note first that except in the trivial case $d = 2$ and $h_{A_1} = \cdots = h_{A_m} = 0$, $(WH)_k$ and $(H)_k$ are equivalent due to special properties of the Grassmann manifolds (see [**14**]). According to Hörmander's theorem, $(H)_k$ implies the existence of a smooth density of Q (see [**14**]) as well as for $P(t,.)\,p$ for $t > 0, p \in M$. So in the sequel, as long as we speak about densities of Q or the flow generating it, we shall tacitly assume $(H)_k$ to be fulfilled. We denote by ϕ the density of the law of Q, by $\phi_t(p,.)$ the density of the law of $P(t,.)\,p$.

THEOREM 1.2. *Let $p \in M$ be such that $\phi(p) = 0$. Then for any smooth curve $\gamma : I \to M$ such that $0 \in I, \gamma(0) = p$, we have*

$$\left(\frac{d}{dt}\right)^n (\phi\circ\gamma)|_{t=0} = 0 \quad \textit{for} \quad n \in \mathbf{N}.$$

Proof:
Again, we rely upon $Q \in \mathbf{D}_\infty$. In addition, we apply a version of the "flatness" theorem of Ben Arous and Léandre (see proposition 4.1.3 of Nualart [**19**]) for a local chart of M which contains p. This yields the desired result. □

Improving the support theorem of Stroock and Varadhan which characterizes the "skeleton" of a smooth diffusion process, Ben Arous and Léandre [**6**] obtain a necessary and sufficient condition for the positivity of the transition density of the solution of a stochastic differential equation. It amounts to regularity conditions for the restriction of the Malliavin matrix to the skeleton which were first introduced by Bismut [**7**]. For a complete treatment of these topics see the lecture notes by Nualart [**19**]. We shall next use these conditions to deduce positivity properties for the density of Q.

Note first that the invariance property of Q discussed above is clearly translated into a property of the densities via

$$\phi(p) = \int_M \phi_t(q,p)\, \phi(q)\, d\mu(q) \tag{1.2}$$

for $p \in M, t > 0$, where μ is the Riemannian volume on M. (1.2) implies that positivity properties of ϕ and ϕ_t for $t > 0$ are closely related. For this reason, following Bismut [**7**] and Ben Arous and Léandre [**6**], we have to consider the restriction of (1.1) to the skeleton. This leads to the following differential equation creating the "control distances" which play a fundamental role for the asymptotics of the corresponding heat kernels for small times. For $h = (h^1, \cdots, h^m) \in L^2(\mathbf{R}_+^m)$, consider

$$dP(h)_t = h_{A_0}(P(h)_t)\, dt + \sum_{i=1}^m h_{A_i}(P(h)_t)\, h^i(t)\, dt. \tag{1.3}$$

Let us consistently denote by $(P(h)(t))_{t\in\mathbf{R}}$ the flow on M created by (1.3), so that $P(h)(t)\,p$ is the solution of (1.3) at time t starting at p for some $p \in M$. In these terms, the positivity of $\phi_t(p,q)$ is related to the existence of h such that $P(h)(t)\,p = q$ and such that the mapping $h \mapsto P(h)(t)$ is a submersion into M. According to Bismut [7], the submersion property can be checked as nonsingularity of the corresponding "deterministic" Malliavin bilinear form. In our setting, this form has been calculated for the flow related to (1.1) in [**14**]. On the skeleton, it is given in the following way. If $(\Phi(h)_t)_{t\in\mathbf{R}}$ is the solution of the linear equation

$$\begin{aligned} d\Phi(h)_t &= A_0\, \Phi(h)_t\, dt + \sum_{i=1}^m A_i\, \Phi(h)_t\, h^i(t)\, dt, \\ \Phi(h)_0 &= I, \end{aligned}$$

with the $d \times d-$unit matrix I, setting

$$A_j^{s,t} = \Phi(h)_t\, \Phi(h)_s^{-1}\, A_j\, \Phi(h)_s\, \Phi(h)_t^{-1},$$

we have for $0 \le s \le t, 1 \le j \le m$,

$$\mathbf{D}_s^j P(h)_t = h_{A_j^{s,t}}(P(h)_t) \tag{1.4}$$

(see [**14**], proposition 2.1.). We then consider the bilinear form on the cotangent bundle of M given by

$$\begin{aligned} \langle\langle DP(h)_t, DP(h)_t\rangle\rangle_0^t : T^*_{P(h)_t}M \times T^*_{P(h)_t}M &\to \mathbf{R}, \\ (u,v) &\mapsto \sum_{j=1}^m \int_0^t \langle D_r^j P(h)_t, u\rangle \langle D_r^j P(h)_t, v\rangle\, dr, \end{aligned}$$

We recall that in these formulas $\langle .,. \rangle$ is just the usual scalar product in $\mathbf{R}^{d\times d}$. In these terms we have the following result.

THEOREM 1.3. *Let $p \in M$. Suppose that there exist $t > 0, q \in supp(\phi)$, and $h = (h^1, \cdots, h^m) \in L^2(\mathbf{R}_+^m)$ such that $P(h)(t)\, q = p$ and such that $\langle\langle DP(h)(t)\, p, DP(h)(t)\, p\rangle\rangle_0^t$ is nonsingular. Then $\phi(p) > 0$.*

Proof:
Under the given hypotheses, Ben Arous and Léandre [**6**], corollary II.2. (see also Nualart [**19**], proposition 4.3.1) yields $\phi_t(q,p) > 0$. Since, moreover, $\phi_t(.,p)$ is continuous, (1.2) clearly implies $\phi(p) > 0$. □

To describe the condition of theorem 1.3 in terms of the vector fields generating $P(h)(t)$ alone, we may develop the main part of (1.4) in the same way as in the stochastic proof of Hörmander's theorem, and use an argument which can be compared with Bismut [**7**], pp. 29, 30. For $\lambda = (\lambda^1, \cdots, \lambda^m) \in \mathbf{R}^m$ denote

$$A^\lambda = A_0 + \lambda^1 A_1 + \cdots \lambda^m A_m.$$

For $1 \le i \le m, s \ge 0, h \in L^2(\mathbf{R}_+^m)$, developing recursively $\Phi(h)_s^{-1}$ and $\Phi(h)_s$, we obtain

$$\begin{aligned}
(1.5)\quad & \Phi(h)_s^{-1} A_i\, \Phi(h)_s \\
&= A_i + \int_0^s \Phi(h)_{s_1}^{-1} [A_i, A^{h_{s_1}}]\, \Phi(h)_{s_1}\, ds_1 \\
& A_i + \int_0^s [A_i, A^{h_{s_1}}] ds_1 + \int_0^s \int_0^{s_1} \Phi(h)_{s_2}^{-1} [[A_i, A^{h_{s_1}}], A^{h_{s_2}}]\, \Phi(h)_{s_2}\, ds_2\, ds_1 \\
&= A_i + \sum_{j=1}^n \int_0^s \int_0^{s_1} \cdots \int_0^{s_{j-1}} [\cdots [A_i, A^{h_{s_1}}] \cdots], A^{h_{s_j}}]\, ds_j \cdots ds_1 \\
& \int_0^s \int_0^{s_1} \cdots \int_0^{s_n} \Phi(h)_u^{-1} [\cdots [A_i, A^{h_{s_1}}] \cdots], A^{h_u}]\, \Phi(h)_u\, du \cdots ds_1.
\end{aligned}$$

(1.5) motivates the following complicated condition on the vector fields $h_{A_i}, 0 \le i \le m$. For $\lambda = (\lambda_i)_{i \in \mathbf{N}} \subset \mathbf{R}^m$ let

$$\mathcal{N}^{k,\lambda} = \text{span}\{h_{A_1}, \cdots, h_{A_m}, [\cdots [h_{A_i}, h_{A^{\lambda_p}}], h_{A^{\lambda_{p-1}}}] \cdots], h_{A^{\lambda_1}}] : \ 1 \le i \le m, \ p \in \mathbf{N}\}.$$

Then we denote

$$(SSH)_k \qquad \mathcal{N}_p^{k,\lambda} = T_pM, \quad \text{for} \quad p \in M \quad \text{and} \quad \lambda \in (\mathbf{R}^m)^{\mathbf{N}}.$$

Obviously, $(SSH)_k$ implies $(H)_k$, and it includes the condition

(1.6)
$$\text{span}\{h_{A_1}(p), \cdots, h_{A_m}(p), [\cdots [h_{A_i}, h_{A_0}], h_{A_0}] \cdots], h_{A_0}](p) : \ 1 \le i \le m\} = T_pM$$

for $p \in M$. $(SSH)_k$ is fulfilled for example, if

$$\text{span}\{h_{A_1}(p), \cdots, h_{A_m}(p)\} = T_pM$$

for all $p \in M$.

THEOREM 1.4. *Suppose that $(SSH)_k$ is fulfilled. Then $\phi(p) > 0$ for all $p \in M$.*

Proof:
Due to theorem 1.1 and the connectedness of M, the proof reduces to showing that $\text{supp}(\phi)$ is open. Let $p \in \text{supp}(\phi)$. It is enough to prove $\phi(p) > 0$. Now by the remarks made above, $(H)_k$ is satisfied. In this case Chow's theorem gives that the Lie group corresponding to $\mathcal{L}^k$ is transitive on M (see Sussmann [**20**]). Hence there exist $t > 0, h \in L^2(\mathbf{R}_+^m)$ such that $P(h)(t)\,q = p$ for any $q \in M$. Moreover, the classical arguments of the stochastic proof of the theorem of Hörmander exposed with (1.5) show that $(SSH)_k$ imply the hypothesis of theorem 1.3. The latter applies and yields $\phi_t(p,p) > 0$, hence $\phi(p) > 0$ by continuity. □

The situation in which the drift vanishes is special. The conditions simplify drastically.

THEOREM 1.5. *Suppose that $A_0 = 0$. Then $\phi(p) > 0$ for any $p \in M$.*

Proof:
We emphasize that due to our conventions and $A_0 = 0$, $(SH)_k$ is satisfied. In this case, the "principe de la petite boucle" (see Léandre [**16**], théorème 1.1) yields in particular $\phi_t(p,p) > 0$ for any $p \in M$. This argument replaces the more complicated one in the preceding proof. □

As we shall see in an example below, in case the drift is non-zero, positivity of the density can rather quickly be violated.

Remark: Conditions on the level of the embedding space imply the validity of $(SSH)_k$ for any $k, 1 \le k \le d$. Indeed, if $\mathcal{L}, \mathcal{I}, \mathcal{M}$, and $\mathcal{N}^\lambda$ are defined by replacing h_{A_i} with A_i in the above definitions of $\mathcal{L}^k, \mathcal{I}^k, \mathcal{M}^k$, and $\mathcal{N}^{k,\lambda}$, and $(WH), (H), (SH)$, resp. (SSH) replace the above conditions with k, we have that (H) implies $(H)_k$ for any $k, 1 \le k \le d$, etc. This is an immediate consequence of the fact that the "projection" of vector fields is a Lie algebra epimorphism.

The preceding results can finally be knitted together to yield positivity results for the densities of the Oseledets spaces.

Recall for this purpose that the forward flag related to our linear system $(Q_r, \cdots, Q_1)$ takes its values in a flag manifold $F_{k_r,\cdots,k_1}(d)$ where $k_i = d_i + \cdots + d_r, 1 \le i \le r$. The condition corresponding to $(WH)_k$ for the flag manifold is given by

$$\text{(WHF)} \qquad \mathcal{L}_{p_r}^{k_r} \times \cdots \times \mathcal{L}_{p_1}^{k_1} = T_{(p_r,\cdots,p_1)} F_{k_r,\cdots,k_1}(d),$$

for $(p_r, \cdots, p_1) \in F_{k_r,\cdots,k_1}(d)$. In an analogous way, the other conditions are generalized to $(HF), (SHF), (SSHF)$. To be able to discuss properties of smooth densities for flag manifolds, we shall tacitly assume that (HF) is satisfied. Note that as before $(SSHF)$ implies (HF). Then we have the following result.

THEOREM 1.6. *Suppose that $(SSHF)$ or (HF) and $A_0 = 0$ is satisfied for the flag manifold $F_{k_r,\cdots,k_1}(d)$. Then the density ϕ_Q of $(Q_r, \cdots, Q_1)$ is strictly positive on $F_{k_r,\cdots,k_1}(d)$.*

Proof:
Theorem 1.4 applies componentwise. □

To obtain a statement about the density of the Oseledets spaces, recall that the backward flag $(P_1, \cdots, P_r)$ takes its values in $F_{l_1,\cdots,l_r}(d)$ where $l_i = d_1 + \cdots + d_i, 1 \le$

$i \leq r$. Then the projector R_i on the ith Oseledets space is the orthogonal projector onto $Q_i(\mathbf{R}^d) \cap P_i(\mathbf{R}^d)$ for $1 \leq i \leq r$.

THEOREM 1.7. *Suppose that $(SSHF)$ or (HF) and $A_0 = 0$ is satisfied both for the flag manifolds $F_{k_r,\cdots,k_1}(d)$ and $F_{l_1,\cdots,l_r}(d)$. Then the density ϕ_R of $(R_1, \cdots, R_r)$ is strictly positive on $G_{d_1}(d) \times \cdots \times G_{d_r}(d)$.*

Proof:
Apply theorem 1.6 to the forward and the backward flag separately first. Then apply the formula for the density of the vector $(R_1, \cdots, R_r)$ obtained from Federer's co-area formula (see [**14**]) from the densities for the forward and backward flags. Hence, the positivity of ϕ_Q and ϕ_P is inherited by ϕ_R. □

We finally exhibit an example to show that in general we cannot go far beyond the condition $(SSH)_k$.

Example :

Let $A_0 = \begin{bmatrix} 1 & 0 & 0 \\ 0 & 0 & 0 \\ 0 & 0 & 0 \end{bmatrix}, A_1 = \begin{bmatrix} 0 & 1 & 0 \\ -1 & 0 & 0 \\ 0 & 0 & 0 \end{bmatrix}, A_2 = \begin{bmatrix} 0 & 0 & 0 \\ 0 & 1 & 0 \\ 0 & 0 & 0 \end{bmatrix}, A_3 = \begin{bmatrix} 0 & 0 & 0 \\ 0 & 0 & 0 \\ 0 & 0 & 1 \end{bmatrix},$

$A_4 = \begin{bmatrix} 0 & 0 & 0 \\ 0 & 0 & 1 \\ 0 & 0 & 0 \end{bmatrix}, A_5 = A_4^*.$

Then direct calculations give

$$\begin{aligned} [A_0, A_1] &= -A_1, \\ [A_0, A_j] &= 0, \quad \text{for} \quad 2 \leq j \leq 5, \\ [A_1, A_2] &= \begin{bmatrix} 0 & 1 & 0 \\ 1 & 0 & 0 \\ 0 & 0 & 0 \end{bmatrix}, \\ [A_1, A_3] &= 0, \\ [A_1, A_4] &= \begin{bmatrix} 0 & 0 & 1 \\ 0 & 0 & 0 \\ 0 & 0 & 0 \end{bmatrix}, \\ [A_1, A_5] &= [A_1, A_4]^*, \\ [[A_1, A_2], A_1] &= 2(A_2 - A_0). \end{aligned}$$

Therefore we obtain

$$\mathcal{L} = \mathcal{I} = \mathcal{M} = \mathbf{R}^{3\times 3},$$

and hence $(SH)_k$ is satisfied for any $1 \leq k \leq 3$. More precisely, A_0 is in the Lie algebra generated by $A_1, \cdots, A_5$, but does not belong to their Lie brackets of order ≤ 2. However, for $1 \leq k \leq 3$, we have trivially $h_{A_0+A_2+A_3} = 0$ and hence h_{A_0} belongs to the Lie brackets of order ≤ 1 of $h_{A_1}, \cdots, h_{A_5}$. For $h = 0$ we consider the differential equation

$$d\Phi(0)_t = A_0 \, \Phi(0)_t \, dt, \quad \Phi(0)_0 = I.$$

Its solution is given by

$$\Phi(0)_t = \begin{bmatrix} e^t & 0 & 0 \\ 0 & 1 & 0 \\ 0 & 0 & 1 \end{bmatrix}, \quad t \in \mathbf{R}.$$

To calculate Malliavin bilinear forms, we need for $u \in \mathbf{R}$

$$\Phi(0)_u\, A_1\, \Phi(0)_{-u} = \begin{bmatrix} 0 & e^t & 0 \\ -e^{-t} & 0 & 0 \\ 0 & 0 & 0 \end{bmatrix}, \tag{1.7}$$

and

$$\Phi(0)_u\, A_i\, \Phi(0)_{-u} = A_i, \quad 2 \le i \le 5. \tag{1.8}$$

Let $k = 2$, and consider the differential equation on $G_2(3)$

$$dP(0)_t = h_{A_0}\, P(0)_t\, dt, \quad P(0)_0 = p = A_0 + A_2.$$

Since $h_{A_0}(p) = 0$, we obviously have $P(0)_t = p$ for all $t \in \mathbf{R}$. For $s, t \in \mathbf{R}, 0 \le s \le t$, set $A_j^{s,t} = \Phi(0)_{t-s}\, A_j\, \Phi(0)_{s-t}, 0 \le j \le 5$. Then (1.7) and (1.8) imply for $t \in \mathbf{R}$

$$h_{A_j^{s,t}}(P(0)_t) = h_{A_j^{s,t}}(p) = 0, \quad 1 \le j \le 4, \tag{1.9}$$

$$h_{A_5^{s,t}}(P(0)_t) = h_{A_5^{s,t}}(p) = A_4 + A_5. \tag{1.10}$$

We prove that for all $t \in \mathbf{R}$ the Malliavin matrix is not submersive at $h = 0$. For this purpose, take $v = [A_1, A_4] + [A_1, A_4]^*$. Then $v \in T_p G_2(3)$, and $||v||^2 = \frac{1}{2}\mathrm{tr}(vv) = 1$. But (1.9) and (1.10) imply

$$\langle h_{A_j^{s,t}}(P(0)_t), v\rangle = 0, \quad 1 \le j \le 4,$$

and

$$\langle h_{A_5^{s,t}}(P(0)_t), v\rangle = \frac{1}{2}\mathrm{tr} \begin{bmatrix} 0 & 0 & 0 \\ 1 & 0 & 0 \\ 0 & 0 & 0 \end{bmatrix} = 0.$$

Hence the Malliavin bilinear form vanishes on $v \in T_{P(0)_t} G_2(3)$, for all $t > 0$. This implies that $P(h)_t$ is not a submersion at $h = 0$. This again, according to Ben Arous and Leandre [**6**] (see also Nualart [**19**], proposition 4.3.1) implies that $\phi_t(p,p) = 0$ for any $t > 0$. But according to Ichihara, Kunita [**12**], corollary 5.2, this finally says that $\phi(p) = 0$.

References

[1] Arnold, L. *Random dynamical systems.* Preliminary version of book. Institut für Dynamische Systeme, Universität Bremen (1996).

[2] Arnold, L., Kliemann, W., Oeljeklaus, E. *Lyapunov exponents of linear stochastic systems.* in: Lyapunov Exponents. L. Arnold, V. Wihstutz (eds.) LNM 1186, 85-125. Springer: Berlin (1986).

[3] Ben Arous, G. *Flots et séries de Taylor stochastiques.* Probab. Th. Rel. Fields 81 (1989), 29-77.

[4] Ben Arous, G., Léandre, R. *Annulation plate du noyau de la chaleur.* CRAS 312 (1991), 463-464.

[5] Ben Arous, G., Léandre, R. *Decroissance exponentielle du noyau de la chaleur sur la diagonale I.* Probab. Th. Rel. Fields 90 (1991), 175-202.

[6] Ben Arous, G., Léandre, R. *Decroissance exponentielle du noyau de la chaleur sur la diagonale II.* Probab. Th. Rel. Fields 90 (1991), 377-402.

[7] Bismut, J. M. *Large deviations and the Malliavin calculus.* In: Progress in Mathematics, vol. 45. Birkhäuser: Boston, Basel 1984.

[8] Fang, S. *Une inégalité isopérimétrique sur l'espace de Wiener.* Bull. Sciences Math. 112 (1988), 345-355.

[9] Federer, H. *Geometric measure theory.* Springer: Berlin 1969.

[10] Helgason, S. *Differential geometry, Lie groups, and symmetric spaces.* Academic Press: New York 1978.

[11] Hirsch, F., Song, S. *Criteria of positivity for the density of the law of a Wiener functional.* Preprint, Univ. Evry 1995.

[12] Ichihara, K., Kunita, H. *A classification of the second order degenerate elliptic operators and its probabilistic characterization.* Z. Wahrscheinlichkeitstheorie verw. Gebiete 30 (1974), 235-254, and 39 (1977), 81-84 (Supplements).

[13] Ikeda, N., Watanabe, S. *Stochastic differential equations and diffusion processes.* North Holland (2 nd edition): 1989.

[14] Imkeller, P. *The smoothness of the laws of random flags and Oseledets spaces of linear stochastic differential equations.* Preprint, HU Berlin (1996).

[15] Kunita, H. *Supports of diffusion processes and controllability problems.* Proc. Intern. Symp. Stoch. Diff. Equs., Kyoto 1976, 163-185. New York: Wiley 1978.

[16] Léandre, R. *Intégration dans la fibre associée à une diffusion dégénérée.* Probab. Th. Rel. Fields 76 (1987), 341-358.

[17] Léandre, R. *Minoration en temps petit de la densité de diffusion dégénérée.* J. Funct. Anal. 74 (1987), 399-415.

[18] Nualart, D. *The Malliavin calculus and related topics.* Springer: Berlin 1995.

[19] Nualart, D. *Malliavin's calculus and anticipative calculus.* St Flour Lecture Notes (1996).

[20] Sussmann, H. *Orbits of families of vector fields and integrability of distributions.* TAMS 180 (1973), 171-188.

[21] Sussmann, H., Jurdjevic, V. *Controllability of nonlinear systems.* J. Diff. Eqns. 12 (1972), 95-116.

Institut für Mathematik, Humboldt-Universität zu Berlin, Unter den Linden 6, 10099 Berlin

E-mail address: `imkeller@mathematik.hu-berlin.de`

Canadian Mathematical Society
Conference Proceedings
Volume **29**, 2000

A Stability Theorem for Feynman's Operational Calculus

G.W. Johnson and Lance Nielsen

Dedicated to Sergio Albeverio on the occasion of his 60th birthday.

ABSTRACT. A variety of operational calculi in the spirit of Feynman's operational calculus for noncommuting operators can be studied via measures on the time interval of interest. We show that if a sequence of n-tuples of measures converges to another n-tuple of measures, then the corresponding sequence of operational calculi converges to the operational calculus determined by the limiting n-tuple of measures.

1. Introduction

In this paper we investigate some aspects of the formation of functions of noncommuting operators. Because of noncommutativity there are ambiguities involved in the formation of even simple functions of, say, two noncommuting operators A and B. Suppose, for example, that we wish to form the function $f(A, B)$ where $f(x, y) = xy$. Depending on the situation, any one of AB, BA, $\frac{1}{2}(AB + BA)$, etc. may be the most desirable. The problem of forming functions of noncommuting operators was of concern to Richard Feynman [7] because of his interest in quantum theory where the observables are noncommuting self-adjoint operators.

Feynman formulated some 'rules' to deal with the problem. Briefly, the 'rules' are:

(R1) Attach time indices to operators to indicate the order in which they act in products. (Note: Operators sometimes come with time indices naturally attached.) Feynman's time ordering convention was that an operator with a earlier time index should act before one with a later index no matter how the operators are ordered on the page.

(R2) With time indices attached, form functions of these operators as if they were commuting. (If one stops naively at this point, the 'equality' involved is usually false. For example, it might say that $e^{A+B} = e^A e^B$ even though A and B do not commute.)

(R3) After (R2) is completed, 'disentangle' the resulting expression; that is, restore the conventional ordering of the operators. In practice, this means to manipulate the expression (if possible) until the ordering on the page coincides with the time ordering.

2000 *Mathematics Subject Classification.* Primary 47A60, 60B10.

In Johnson and Lapidus [11] an extension of 'rules' (1) - (3) above was introduced. These ideas were pursued in basically the same framework in [12, 14-17] and in more general settings in [4, 5, 9, 10]. These extensions are accomplished using measures to attach time indices to the operators involved. For example, if the operator A is known to act earlier than B, the measure μ associated with A would be taken so that its support is to the left on the time interval of the support of the measure ν associated with B.

A much more extensive introduction to Feynman's operational calculus can be found in the book of Johnson and Lapidus [13], see especially Chapter 14.

The recent papers of Jefferies and Johnson [9], and of Jefferies, Johnson, and Nielsen [10] are the most closely related to the present paper. In the former a commutative Banach algebra, called the disentangling algebra, is constructed in which (R2) and (R3) above can be rigorously applied. A closely related Banach algebra is constructed in the latter paper. The primary difference between the two is that the former considers the disentangling operation (R3), i.e. the disentangling map, applied to fixed bounded operators and the latter constructs the disentangling formalism for operator-valued functions of time.

In this paper, we will work in the time independent setting. The time-ordering measures that we will consider will be, as in much of [9], probability measures. The restriction to probability measures can be avoided as was seen in [9] and especially in [10] thereby making possible applications to evolution problems. (The present work is relevant to evolution problems, but they will not be considered here.)

The main result, Theorem 3.1, comes from chapter two of the second author's dissertation [18] and is concerned with the stability of the operational calculus given by the disentangling map with respect to the time-ordering measures. It might be helpful at this point to clarify what we mean by stability in the present context. Let $A_1, \ldots, A_p$ be bounded linear operators on a Banach space X and let $\mu_1, \ldots, \mu_p$, respectively, be associated probability measures on a time interval which we take to be $[0,1]$ for convenience. These measures determine an operational calculus for the noncommuting (actually, not necessarily commuting) operators $A_1, \ldots, A_p$ as we will see in Section 2 where the basic framework of [9] will be reviewed. Now we take p sequences of probability measures $\{\mu_{in}\}_{n=1}^{\infty}$ for $i = 1, \ldots, p$, and suppose that $\mu_{in} \to \mu_i$ in some appropriate sense, i.e. weakly or in total variation norm. We then show in Theorem 3.1 and its corollaries that the functional calculus determined by the p-tuple $(\mu_{1n}, \ldots, \mu_{pn})$ converges in an appropriate sense to the functional calculus determined by $(\mu_1, \ldots, \mu_p)$. The convergence just mentioned is convergence of the 'disentangling maps'. (Definition 2.3 below describes these maps.)

We give two applications of Theorem 3.1 both of which have been done after the thesis [18] of the second author. The first, Example 3.1, involves a sequence of measures each of which is absolutely continuous with respect to Lebesgue measure on $[0,1]$ and which converges weakly to the continuous but singular Lebesgue-Stieltjes measure associated with the Cantor function. This example and others involving continuous, singular measures may be worthy of further study.

Section 4 of this paper provides a positive answer to a question asked of Nielsen by Brian Jefferies after learning about the theorem in [18] which corresponds to Theorem 3.1 below. We discuss here briefly why the result in Section 4 is of interest for the theory developed in [9]. The second part of [9] uses joint spectral properties which depend on the measures as well as on the noncommuting operators involved to extend the operational calculus beyond the level of analytic functions as developed in the first part of [9]. In order to apply this spectral theory, inequality (10.1) of [9] must be verified. The result in Section 4 below allows that key inequality to be easily proved in the setting of Weyl's functional calculus for any finite number of noncommuting, bounded, self-adjoint operators on a Hilbert space as studied for example in [1, 2, 20]. Thus the Weyl calculus can be viewed from the perspective of the more general theory developed in [9]. Although this fact has already been established in [9] (see the discussion following Example 10.2), the proof there depends on an application of the Trotter product formula found in [20]. In contrast, the proof found in Section 4 below uses an approach which is intrinsic to the present theory and which is potentially applicable to operational calculi other than the Weyl calculus. Such extensions will be pursued in later work of Brian Jefferies (and possibly of the present authors).

2. The Disentangling Map

We start this section with definitions of the commutative Banach algebras that will be used in the definition of the disentangling map. These definitions are much the same as used in the paper [9] in which the disentangling map was originally defined. (See also [10] for the discussion of this map in the time dependent setting.) After these algebras are defined, we turn to the definition of the disentangling map.

DEFINITION 2.1. Given a positive integer p and p positive numbers $r_1, \ldots, r_p$, we let $\mathbb{A} = \mathbb{A}(r_1, \ldots, r_p)$ be the space of complex-valued functions of n complex variables which are analytic at the origin and are such that their power series expansion

$$f(z_1, \ldots, z_p) = \sum_{m_1, \ldots, m_p = 0}^{\infty} c_{m_1, \ldots, m_p} z_1^{m_1} \cdots z_p^{m_p} \tag{2.1}$$

converges absolutely at least on the closed polydisk $|z_1| \leqslant r_1, \ldots, |z_p| \leqslant r_p$. Such functions are analytic, at least in the corresponding open polydisk.

REMARK 2.1. Fix $r_1, \ldots, r_p$. For $f \in \mathbb{A}(r_1, \ldots, r_p)$ given by (2.1), let

$$\|f\|_{\mathbb{A}} = \sum_{m_1, \ldots, m_p = 0}^{\infty} |c_{m_1, \ldots, m_p}| r_1^{m_1} \cdots r_p^{m_p}. \tag{2.2}$$

The function defined by (2.2) makes $\mathbb{A}$ into a Banach space. In fact, $\mathbb{A}$ with the norm defined in (2.2) is a commutative Banach algebra via pointwise multiplication. The verification of these facts is straightforward and can be found in [9].

Let X be a Banach space and let $A_1, \ldots, A_p \in \mathcal{L}(X)$. We create commuting formal objects $A_1^{\sim}, \ldots, A_p^{\sim}$ and look at power series in these objects. To this end, we will define a commutative Banach algebra $\mathbb{D}(A_1, \ldots, A_p)$ consisting of power series in these objects. Only the norms of the operators we have chosen will play a role in the definition of this Banach algebra. It is worth remarking that we do not require that all of the p operators are distinct, but we will regard the commuting objects coming from them as distinct.

DEFINITION 2.2. Consider the collection $\mathbb{D}(A_1, \dots, A_p)$ of all expressions of the form $f(A_1^{\sim}, \dots, A_p^{\sim})$ where

$$f(A_1^{\sim}, \dots, A_p^{\sim}) = \sum_{m_1, \dots, m_p = 0}^{\infty} c_{m_1, \dots, m_p} (A_1^{\sim})^{m_1} \cdots (A_p^{\sim})^{m_p} \tag{2.3}$$

where the coefficients $c_{m_1, \dots, m_p}$ are complex numbers and

$$\|f\|_{\mathbb{D}} = \sum_{m_1, \dots, m_p = 0}^{\infty} |c_{m_1, \dots, m_p}| \, \|A_1\|^{m_1} \cdots \|A_p\|^{m_p} \tag{2.4}$$

is finite.

Adding such expressions coordinate-wise produces another element of $\mathbb{D}$. Also, scalar multiples of elements of $\mathbb{D}$ are in $\mathbb{D}$ and pointwise multiplication of elements of $\mathbb{D}$ are in $\mathbb{D}$. Equation (2.4) defines a norm on $\mathbb{D}$ and $\mathbb{D}$ is a commutative Banach algebra with this norm. The proof of these assertions is the same as the proof of the previous statements about $\mathbb{A}$ and can be found in [9] as well. We will call $\mathbb{D}$ the *disentangling algebra* associated with the operators $A_1, \dots, A_p$.

With these initial definitions out of the way, we can now move on to the definition of the disentangling map. The first task is to establish some notation. Given nonnegative integers $m_1, \dots, m_p$, we let

$$P^{m_1, \dots, m_p}(z_1, \dots, z_p) = z_1^{m_1} \cdots z_p^{m_p} \tag{2.5}$$

so that

$$P^{m_1, \dots, m_p}(A_1^{\sim}, \dots, A_p^{\sim}) = (A_1^{\sim})^{m_1} \cdots (A_p^{\sim})^{m_p}. \tag{2.6}$$

It is convenient to let $m = m_1 + \cdots + m_p$. Let S_m be the set of all permutations of the first m positive integers. Given $\pi \in S_m$, let

$$\Delta_m(\pi) = \{(s_1, \dots, s_m) \in [0,1]^m : 0 < s_{\pi(1)} < \cdots < s_{\pi(m)} < 1\}. \tag{2.7}$$

For $j = 1, \dots, p$ and all $s \in [0,1]$, we let

$$A_j^{\sim}(s) = A_j^{\sim}. \tag{2.8}$$

For $\pi \in S_m$, we define

$$\tilde{C}_{\pi(i)}(s_{\pi(i)}) = \begin{cases} A_1^{\sim}(s_{\pi(i)}), & \text{if } \pi(i) \in \{1, \dots, m_1\} \\ A_2^{\sim}(s_{\pi(i)}), & \text{if } \pi(i) \in \{m_1, \dots, m_1 + m_2\} \\ \vdots & \\ A_p^{\sim}(s_{\pi(i)}), & \text{if } \pi(i) \in \{m_1 + \cdots + m_{p-1} + 1, \dots, m\}. \end{cases} \tag{2.9}$$

Let $\mu_1, \dots, \mu_p$ be continuous Borel probability measures on $[0,1]$. The following result is Proposition 2.2 of [9]. We include the short proof since it may help the reader to understand the motivation behind the definition of the disentangling map.

PROPOSITION 2.1. *Given nonnegative integers $m_1, \dots, m_p$ we can write*

$$\begin{aligned} &P^{m_1, \dots, m_p}(A_1^{\sim}, \dots, A_p^{\sim}) \\ &\quad = \sum_{\pi \in S_m} \int_{\Delta_m(\pi)} \tilde{C}_{\pi(m)} \cdots \tilde{C}_{\pi(1)} \left(\mu_1^{m_1} \times \cdots \times \mu_p^{m_p}\right)(ds_1, \dots, ds_m). \end{aligned} \tag{2.10}$$

PROOF. Since all of our measures are probability measures on $[0,1]$, it follows from (2.8) that

(2.11)

$$P^{m_1,\dots,m_p}\left(\tilde{A}_1,\dots,\tilde{A}_p\right)=\left(\int_{[0,1]}\tilde{A}_1(s)\mu_1(ds)\right)^{m_1}\cdots\left(\int_{[0,1]}\tilde{A}_p(s)\mu_p(ds)\right)^{m_p}$$

$$=\int_{[0,1]^m}\tilde{A}_1(s_1)\cdots\tilde{A}_1(s_{m_1})\tilde{A}_2(s_{m_1+1})\cdots\tilde{A}_2(s_{m_1+m_2})\cdots$$
$$\tilde{A}_p(s_{m_1+\dots+m_{p-1}+1})\cdots\tilde{A}_p(s_m)(\mu_1^{m_1}\times\cdots\times\mu_p^{m_p})(ds_1,\dots,ds_m).$$

By the continuity of the measures, we can restrict this integral to the union of the disjoint sets $\Delta_m(\pi)$, $\pi\in S_m$. Using the notation in (2.9), we obtain our result. □

We see from the definition of $\Delta_m(\pi)$ that the right-hand side of equation (2.11) is the sum of time-ordered expressions. Following Feynman's ideas, we now define the disentangling map $\mathcal{T}_{\mu_1,\dots,\mu_p}$ which will carry us from the commutative environment provided by our disentangling algebra to the noncommutative setting of $\mathcal{L}(X)$. We need notation as in (2.8) and (2.9), but involving our operators instead of the associated commuting objects. This is accomplished by omitting the tildes from the aforementioned equations. Having said this, we define the disentangling map.

DEFINITION 2.3.

$$\mathcal{T}_{\mu_1,\dots,\mu_p}P^{m_1,\dots,m_p}(\tilde{A}_1,\dots,\tilde{A}_p)=$$
$$\sum_{\pi\in S_m}\int_{\Delta_m(\pi)}C_{\pi(m)}\cdots C_{\pi(1)}\left(\mu_1^{m_1}\times\cdots\times\mu_p^{m_p}\right)(ds_1,\dots,ds_m). \tag{2.12}$$

Then for $f(\tilde{A}_1,\dots,\tilde{A}_p)\in\mathbb{D}(A_1,\dots,A_p)$ given by

$$f\left(\tilde{A}_1,\dots,\tilde{A}_p\right)=\sum_{m_1,\dots,m_p=0}^{\infty}d_{m_1,\dots,m_p}(\tilde{A}_1)^{m_1}\cdots(\tilde{A}_p)^{m_p} \tag{2.13}$$

we let

(2.14)

$$\mathcal{T}_{\mu_1,\dots,\mu_p}f(\tilde{A}_1,\dots,\tilde{A}_p)=\sum_{m_1,\dots,m_p=0}^{\infty}d_{m_1,\dots,m_p}\mathcal{T}_{\mu_1,\dots,\mu_p}P^{m_1,\dots,m_p}(\tilde{A}_1,\dots,\tilde{A}_p).$$

Of course, in order to know that this makes sense, it must be shown that the above sum converges absolutely in $\mathcal{L}(X)$ for any $f\in\mathbb{D}(A_1,\dots,A_p)$. This is the content of the following proposition whose proof we omit. (See Proposition 2.4 of [9].)

PROPOSITION 2.2. *(i) The series on the right-hand side of* (2.14) *converges absolutely in $\mathcal{L}(X)$ for all elements of the disentangling algebra $\mathbb{D}$.*

(ii) The disentangling map $\mathcal{T}_{\mu_1,\dots,\mu_p}$ is linear.

(iii) The map $\mathcal{T}_{\mu_1,\dots,\mu_p}$ is of norm one in $\mathcal{L}(\mathbb{D}(A_1,\dots,A_p),\mathcal{L}(X))$.

For the sake of brevity and in order to emphasize that $\mathcal{T}_{\mu_1,\dots,\mu_p}f(\tilde{A}_1,\dots,\tilde{A}_p)$ should be thought of as a function of the not necessarily commuting operators $A_1,\dots,A_p$ we will often use the alternate notation indicated below:

$$f_{\mu_1,\dots,\mu_p}(A_1,\dots,A_p)=\mathcal{T}_{\mu_1,\dots,\mu_p}(f(\tilde{A}_1,\dots,\tilde{A}_p)), \tag{2.15}$$

(2.16) $$P^{m_1,\dots,m_p}_{\mu_1,\dots,\mu_p}(A_1,\dots,A_p) = \mathcal{T}_{\mu_1,\dots,\mu_p} P^{m_1,\dots,m_p}(\tilde{A_1},\dots,\tilde{A_p}).$$

Jefferies and Johnson go on in the first part of [9] to extend and make rigorous many of the heuristic ideas in or suggested by Feynman's paper [7]. The paper of Jefferies, Johnson and Nielsen [10] focuses on the case of time-dependent operators and on the application of Feynman's operational calculus to evolution equations. Nielsen's thesis [18] is not limited to Theorem 3.1 below but emphasizes stability properties of various kinds for Feynman's operational calculus.

3. Stability of the Disentangling Map

In this section we obtain stability results for the disentangling map $\mathcal{T}_{\mu_1,\dots,\mu_p}$ that was introduced in the previous section. The main result will be realized by considering weakly convergent sequences of continuous Borel probability measures on the unit interval.

Before we state our theorem we remind the reader of the definition of weak convergence of a sequence of probability measures.

DEFINITION 3.1. Let $\{\nu_n\}_{n=1}^{\infty}$ be a sequence of Borel probability measures on $[0,1]$. We say that this sequence of measures converges weakly to the Borel probability measure ν on $[0,1]$ and write $\nu_n \rightharpoonup \nu$ provided that $\int_{[0,1]} f(s)\,\nu_n(ds) \to \int_{[0,1]} f(s)\,\nu(ds)$ for every bounded continuous function f on $[0,1]$. (Since $[0,1]$ is compact every continuous function on this interval is, of course, bounded.) We remark that the convergence $\nu_n \rightharpoonup \nu$ is weak-* convergence in the terminology of functional analysis.

THEOREM 3.1. *Let $A_1,\dots,A_p \in \mathcal{L}(X)$ where X is a Banach space. Let $\{\mu_{in}\}_{n=1}^{\infty}$ for $i=1,\dots,p$ be sequences of continuous Borel probability measures on $[0,1]$ such that $\mu_{in} \rightharpoonup \mu_i$ for each i. Let $\mathcal{T}_{\mu_{1n},\dots,\mu_{pn}}$ denote the disentangling map corresponding to the nth term of each of our sequences of measures. It follows that $s-\lim_{n\to\infty} \mathcal{T}_{\mu_{1n},\dots,\mu_{pn}} = \mathcal{T}_{\mu_1,\dots,\mu_p}$ as operators in $\mathcal{L}(\mathbb{D}(A_1,\dots,A_p),\mathcal{L}(X))$; that is $\mathcal{T}_{\mu_{1n},\dots,\mu_{pn}} f(\tilde{A_1},\dots,\tilde{A_p})$ converges in norm to $\mathcal{T}_{\mu_1,\dots,\mu_p} f(\tilde{A_1},\dots,\tilde{A_p})$ for every element $f(\tilde{A_1},\dots,\tilde{A_p})$ of $\mathbb{D}(A_1,\dots,A_n)$.*

PROOF. Let $A_i(s) := A_i$ for all i and all s in $[0,1]$. First we consider what happens when our sequence of disentangling maps is applied to $P^{m_1,\dots,m_p}(\tilde{A}_1,\dots,\tilde{A}_p)$. We see that

(3.1)
$$\begin{aligned}
&\|\mathcal{T}_{\mu_{1n},\dots,\mu_{pn}} P^{m_1,\dots,m_p}(\tilde{A_1},\dots,\tilde{A_p}) - \mathcal{T}_{\mu_1,\dots,\mu_p} P^{m_1,\dots,m_p}(\tilde{A_1},\dots,\tilde{A_p})\|_{\mathcal{L}(X)} \\
&= \| \sum_{\pi\in S_m} \int_{\Delta_m(\pi)} C_{\pi(m)}(s_{\pi(m)})\cdots C_{\pi(1)}(s_{\pi(1)})(\mu_{1n}^{m_1}\times\cdots\times\mu_{pn}^{m_p})(ds_1,\dots,ds_m) - \\
&\sum_{\pi\in S_m} \int_{\Delta_m(\pi)} C_{\pi(m)}(s_{\pi(m)})\cdots C_{\pi(1)}(s_{\pi(1)})(\mu_1^{m_1}\times\cdots\times\mu_p^{m_p})(ds_1,\dots,ds_m)\|_{\mathcal{L}(X)} \\
&\le \sum_{\pi\in S_m} \| \int_{\Delta_m(\pi)} C_{\pi(m)}(s_{\pi(m)})\cdots C_{\pi(1)}(s_{\pi(1)})(\mu_{1n}^{m_1}\times\cdots\times\mu_{pn}^{m_p})(ds_1,\dots,ds_m) - \\
&\int_{\Delta_m(\pi)} C_{\pi(m)}(s_{\pi(m)})\cdots C_{\pi(1)}(s_{\pi(1)})(\mu_1^{m_1}\times\cdots\times\mu_p^{m_p})(ds_1,\dots,ds_m)\|_{\mathcal{L}(X)}.
\end{aligned}$$

Since $A_i(s)$ is a constant-valued map for each i, given any permutation $\pi\in S_m$ the integrand in each of the integrals in (3.1) is then a product of fixed operators, and so for any of these permutations we can remove the $C_{\pi(k)}$'s from the integrals. Also, $\{\mu_{1n}^{m_1} \times \cdots \times \mu_{pn}^{m_p}\}_{n=1}^{\infty}$ is clearly a sequence of continuous probability measures on $[0,1]^m$ since each term in the product is a continuous probability measure. Moreover it is well known that $[0,1]^m$ is separable for any integer m (in the product topology) and so, since $\mu_{in} \rightharpoonup \mu_i$ for each i, $\mu_{1n}^{m_1} \times \cdots \times \mu_{pn}^{m_p} \rightharpoonup \mu_1^{m_1} \times \cdots \times \mu_p^{m_p}$ using Theorem 3.2 on page 21 of [3]. Putting this weak convergence together with the above norm estimate and using the fact that our integrands are constant-valued, we can apply the continuity of the measures $\mu_1, \ldots, \mu_p$ and hence of $\mu_1^{m_1} \times \cdots \times \mu_p^{m_p}$ along with (v) of Theorem 2.1 on page 11 of [3] to conclude that

(3.2)

$$\sum_{\pi \in S_m} \| \int_{\Delta_m(\pi)} C_{\pi(m)}(s_{\pi(m)}) \cdots C_{\pi(1)}(s_{\pi(1)})(\mu_{1n}^{m_1} \times \cdots \times \mu_{pn}^{m_p})(ds_1, \ldots, ds_m) -$$
$$\int_{\Delta_m(\pi)} C_{\pi(m)}(s_{\pi(m)}) \cdots C_{\pi(1)}(s_{\pi(1)})(\mu_1^{m_1} \times \cdots \times \mu_p^{m_p})(ds_1, \ldots, ds_m)\|_{\mathcal{L}(X)}$$
$$= \sum_{\pi \in S_m} \|C_{\pi(m)} \cdots C_{\pi(1)}\|_{\mathcal{L}(X)} |(\mu_{1n}^{m_1} \times \cdots \times \mu_{pn}^{m_p})(\Delta_m(\pi)) -$$
$$(\mu_{1n}^{m_1} \times \cdots \times \mu_{pn}^{m_p})(\Delta_m(\pi))| \longrightarrow 0$$

as $n \to \infty$. We therefore conclude, using (3.1) and (3.2), that

(3.3)

$$\lim_{n\to\infty} \|\mathcal{T}_{\mu_{1n},\ldots,\mu_{pn}} P^{m_1,\ldots,m_p}(\tilde{A_1}, \ldots, \tilde{A_p}) - \mathcal{T}_{\mu_1,\ldots,\mu_p} P^{m_1,\ldots,m_p}(\tilde{A_1}, \ldots, \tilde{A_p})\|_{\mathcal{L}(X)}$$
$$= 0.$$

We now turn to $\mathcal{T}_{\mu_{1n},\ldots,\mu_{pn}} f(\tilde{A_1}, \ldots, \tilde{A_p}) \in \mathbb{D}(\tilde{A_1}, \ldots, \tilde{A_p})$. Write this element of the disentangling algebra as in (2.3). Using the definition of the disentangling map we have, provided that the inequality below can be established and provided that we can justify interchanging the limit and the sum,

$$\|\mathcal{T}_{\mu_{1n},\ldots,\mu_{pn}} f(\tilde{A_1}, \ldots, \tilde{A_p}) - \mathcal{T}_{\mu_1,\ldots,\mu_p} f(\tilde{A_1}, \ldots, \tilde{A_p})\|_{\mathcal{L}(X)} =$$
$$\| \sum_{m_1,\ldots,m_p=0}^{\infty} c_{m_1,\ldots,m_p} \mathcal{T}_{\mu_{1n},\ldots,\mu_{pn}} P^{m_1,\ldots,m_p}(\tilde{A_1}, \ldots, \tilde{A_p}) -$$
$$\sum_{m_1,\ldots,m_p=0}^{\infty} c_{m_1,\ldots,m_p} \mathcal{T}_{\mu_1,\ldots,\mu_p} P^{m_1,\ldots,m_p}(\tilde{A_1}, \ldots, \tilde{A_p})\|_{\mathcal{L}(X)} \tag{3.4}$$
$$\leq \sum_{m_1,\ldots,m_p=0}^{\infty} |c_{m_1,\ldots,m_p}| \, \|\mathcal{T}_{\mu_{1n},\ldots,\mu_{pn}} P^{m_1,\ldots,m_p}(\tilde{A_1}, \ldots, \tilde{A_p}) -$$
$$\mathcal{T}_{\mu_1,\ldots,\mu_p} P^{m_1,\ldots,m_p}(\tilde{A_1}, \ldots, \tilde{A_p})\|_{\mathcal{L}(X)} \longrightarrow 0$$

as $n \to \infty$. Now, to verify that this inequality is valid, we note that

$$\|\mathcal{T}_{\mu_{1n},\ldots,\mu_{pn}} f(\tilde{A_1}, \ldots, \tilde{A_p}) - \mathcal{T}_{\mu_1,\ldots,\mu_p} f(\tilde{A_1}, \ldots, \tilde{A_p})\|_{\mathcal{L}(X)}$$
$$\leq \sum_{m_1,\ldots,m_p=0}^{\infty} |c_{m_1,\ldots,m_p}| \, \|\mathcal{T}_{\mu_{1n},\ldots,\mu_{pn}} (\tilde{A_1})^{m_1} \cdots (\tilde{A_p})^{m_p} -$$
$$\mathcal{T}_{\mu_1,\ldots,\mu_p} (\tilde{A_1})^{m_1} \cdots (\tilde{A_p})^{m_p}\|_{\mathcal{L}(X)} \tag{3.5}$$
$$\leq 2 \sum_{m_1,\ldots,m_p=0}^{\infty} |c_{m_1,\ldots,m_p}| \, \|A_1\|_{\mathcal{L}(X)}^{m_1} \cdots \|A_p\|_{\mathcal{L}(X)}^{m_p} = 2\|f\|_{\mathbb{D}} < \infty.$$

It follows that the series in (3.4) converges and that the function

$$(m_1, \dots, m_p) \mapsto 2|c_{m_1,\dots,m_p}| \, \|A_1\|^{m_1}_{\mathcal{L}(X)} \cdots \|A\|^{m_p}_{\mathcal{L}(X)} \tag{3.6}$$

is a dominating function for the last sum in (3.4). Lebesgue's dominated convergence theorem enables us to pass the limit on n through the summation in (3.4) and we obtain the stated limit. This finishes the proof. □

Now that our theorem is proved, we can state three corollaries. These results simply state that, under the hypotheses of the theorem, we can obtain strong convergence with other topologies on $\mathcal{L}(X)$. More precisely we state the following.

COROLLARY 3.1. *Assume the same hypotheses as in Theorem 3.1. Then the sequence of disentangling maps converges in the strong operator topology on* $\mathcal{L}(\mathbb{D}(A_1, \dots, A_p), \mathcal{L}_s(X))$ *where* $\mathcal{L}_s(X))$ *denotes the space* $\mathcal{L}(X)$ *with the strong operator topology.*

PROOF. Let $f \in \mathbb{D}(A_1, \dots, A_p)$. Then for $x \in X$ we have

$$\begin{aligned} &\|(\mathcal{T}_{\mu_{1n},\dots,\mu_{pn}} f - \mathcal{T}_{\mu_1,\dots,\mu_p} f)(x)\|_X \leq \\ &\|\mathcal{T}_{\mu_{1n},\dots,\mu_{pn}} f - \mathcal{T}_{\mu_1,\dots,\mu_p} f\|_{\mathcal{L}(X)} \|x\|_X \longrightarrow 0 \end{aligned} \tag{3.7}$$

as $n \to \infty$ by Theorem 3.1. This finishes the proof. □

COROLLARY 3.2. *Assume the same hypotheses as in Theorem 3.1. Then the sequence of disentangling maps converges in the strong operator topology on* $\mathcal{L}(\mathbb{D}(A_1, \dots, A_p), \mathcal{L}_w(X))$ *where* $\mathcal{L}_w(X))$ *denotes the space* $\mathcal{L}(X)$ *with the weak operator topology.*

PROOF. Let $f \in \mathbb{D}(A_1, \dots, A_p)$. Then given $x^* \in X^*$ and $x \in X$ it follows that

$$\begin{aligned} &|x^*\{(\mathcal{T}_{\mu_{1n},\dots,\mu_{pn}} f)(x)\} - x^*\{(\mathcal{T}_{\mu_1,\dots,\mu_p} f)(x)\}| \leq \\ &\qquad \|x^*\|_{X^*} \|(\mathcal{T}_{\mu_{1n},\dots,\mu_{pn}} f - \mathcal{T}_{\mu_1,\dots,\mu_p} f\|_{\mathcal{L}(X)} \|x\|_X \longrightarrow 0 \end{aligned} \tag{3.8}$$

as $n \to \infty$ by Theorem 3.1. □

COROLLARY 3.3. *Assume the same hypotheses as in Theorem 3.1 except for the assumption that the sequences of measures converge weakly. Instead, assume that these sequences of measures converge in total variation norm. Then the sequence of disentangling maps converges in the operator topologies cited in the three previous results.*

PROOF. The proof is immediate since the convergence of the measures in total variation norm implies that the sequences converge weakly. □

EXAMPLE 3.1. Here we consider an example of Theorem 3.1. Let C denote the Cantor set. Also, let f_C denote the Cantor function. We briefly recall the definition of this function. If $x \in C$ we may write $x = \sum_j a_j 3^{-j}$ where $a_j = 0$ or 2 for all j. Let $b_j = \frac{a_j}{2}$ and define

$$f_C(x) = \sum_j b_j 2^{-j}.$$

Via this series it is seen that f_C is continuous, nondecreasing, and is a surjection onto $[0, 1]$. Also, as is well known, $f'_C(x) = 0$ almost everywhere in $[0, 1]$ with respect to Lebesgue measure. Using standard results of measure theory (see for instance, section 19 of [8]) there is a unique continuous Borel measure μ_C such that

$$\mu_C([x,y]) = f_C(y) - f_C(x) \tag{3.9}$$

for $0 \leq x < y \leq 1$. (Recall that this measure is singular with respect to Lebesgue measure on $[0,1]$ by virtue of the fact that $f'_C(x) = 0$ almost everywhere in $[0,1]$ with respect to Lebesgue measure.) We wish to construct a sequence of functions that converge uniformly to f_C. At the first step, denote by $I_{1,1}$ the open middle-third interval of $[0,1]$. Define $\psi_1(x)$ to be equal to $\frac{1}{2}$ on $I_{1,1}$, $\psi_1(0) = 0$, $\psi_1(1) = 1$, and define $\psi_1(x)$ to be linear on the closed intervals $J_{1,1}, J_{1,2}$ that remain after $I_{1,1}$ is removed. For the second step, denote by $I_{2,j}$, $j = 1,2$ the two open intervals of length $\frac{1}{9}$ removed from the centers of $J_{1,k}$, $k = 1,2$. Define $\psi_2(x)$ to be equal to $\frac{1}{4}$ on $I_{2,1}$, equal to $\frac{3}{4}$ on $I_{2,2}$, equal to $\frac{1}{2}$ on $I_{1,1}$, $\psi_2(0) = 0$, $\psi_2(1) = 1$, and define $\psi_2(x)$ to be linear on the four closed intervals $J_{2,k}$, $k = 1,2,3,4$ that remain after the removal of the $I_{2,k}$. Continue this process. At the nth step we remove open intervals of length $\frac{1}{3^n}$ from the centers of the closed intervals $J_{n-1,k}$, $k = 1, \ldots, 2^{n-1}$, left over in the previous step. Define $\psi_n(x)$ by $\psi_n(0) = 0$, $\psi_n(1) = 1$, $\psi_n(x) = \frac{2k-1}{2^j}$ for $x \in I_{j,k}$, $j = 1, \ldots, n$ and $k = 1, \ldots, 2^{j-1}$, and define $\psi_n(x)$ to be linear on the closed intervals $J_{n,k}$ for $k = 1, \ldots, 2^n$ that remain after the removal of the $I_{n,k}$. It is well known that the sequence $\{\psi_n(x)\}_{n=1}^{\infty}$ converges uniformly to $f_C(x)$. As with f_C, we obtain unique continuous Borel probability measures μ_n on $[0,1]$ such that

$$\mu_n([x,y]) = \psi_n(y) - \psi_n(x) \tag{3.10}$$

for $0 \leq x < y \leq 1$. In fact, each of the measures μ_n, $n \in \mathbb{N}$, is absolutely continuous with respect to Lebesgue measure l on $[0,1]$. We wish to show that the sequence $\{\mu_n\}_{n=1}^{\infty}$ converges weakly to μ_C.

To establish this claim, we let A be any open set in $[0,1]$. Write $A = \bigcup_j V_j$ where $\{V_j\}$ is a disjoint countable family of open intervals. For any $n \in \mathbb{N}$ we have

$$\mu_n(A) = \sum_j \mu_n(V_j). \tag{3.11}$$

Let us consider therefore $\mu_n((x,y))$ for $0 \leq x < y \leq 1$. We see at once that $\mu_n((x,y)) = \psi_n(y) - \psi_n(x) \to f_C(y) - f_C(x) = \mu_C((x,y))$. Appealing to a theorem of E.H. Moore on iterated limits ([6], page 28) we see that the limit on n and the sum on j can be interchanged and we obtain

$$\lim_{n\to\infty} \mu_n(A) = \sum_j \mu_C(V_j) = \mu_C(A). \tag{3.12}$$

Hence we certainly have $\lim_{n\to\infty} \mu_n(A) = \liminf_{n\to\infty} \mu_n(A) \geq \mu_C(A)$ for any open set A in $[0,1]$. Thus by the portmanteau theorem for weak convergence of measures (see p. 12 of [3]), we have $\mu_n \rightharpoonup \mu_C$.

We can now apply Theorem 3.1 to obtain,

$$f_{l,\mu_n}(A,B) \to f_{l,\mu_C}(A,B) \tag{3.13}$$

for any bounded linear operators A, B on a Banach space X, and any $f \in \mathbb{D}(A,B)$, where l=Lebesgue measure.

The operators $f_{l,\mu_n}(A,B)$ can be explicitly calculated for various functions f and these calculations provide some insight into the operators $f_{l,\mu_C}(A,B)$. However we forgo these calculations here due to lack of space.

Next, we briefly sketch a second example relating to Theorem 3.1. As the reader will note, this second example amounts to a proof of the theorem in a particular setting. The argument used in this simpler setting much shorter than the the proof of Theorem 3.1 as might be expected.

EXAMPLE 3.2. Suppose that p_j, $j = 1, \dots, n$ is a collection of probability densities on the interval $[0,1]$. We then obtain probability measures μ_j, $j = 1, \dots, n$, on the interval $[0,1]$ via $\mu_j(E) = \int_E p_j(x)dl(x)$ where l, as usual, denotes Lebesgue measure and E is a Borel set in $[0,1]$. Now let $\{p_{jk}(x)\}_{k=1}^{\infty}$, $j = 1, \dots, n$, be sequences of probability densities which are such that $p_{jk}(x) \to p_j(x)$ almost everywhere with respect to Lebesgue measure for each j. It follows from Scheffe's Theorem ([3], page 224) that the sequences $\{\mu_{jk}\}_{k=1}^{\infty}$, $j = 1, \dots, n$, converge weakly to μ_j for each j. Now, Theorem 3.1 could be applied at this point to obtain the limit of the sequence of disentangling maps corresponding to the sequences of measures, but it is interesting to see how these particular probability measures come into play in this theorem. Using our sequences and bounded linear operators $A_1, \dots, A_n$ on a Banach space X, we obtain, for $f \in \mathbb{D}(A_1, \dots, A_n)$,

(3.14)
$$\mathcal{T}_{\mu_{1k}, \dots, \mu_{nk}} f(\tilde{A_1}, \dots, \tilde{A_n}) = \sum_{m_1, \dots, m_n = 0}^{\infty} d_{m_1, \dots, m_n} \sum_{\pi \in S_m} C_{\pi(m)} \cdots C_{\pi(1)} \int_{\Delta_m(\pi)} p_{1k}(s_{1,1}) \cdots p_{1k}(s_{1,m_1}) \cdots p_{nk}(s_{n,1}) \cdots p_{nk}(s_{n,m_n}) l^m(ds_{1,1}, \dots, ds_{n,m_n}).$$

Now, because the product of the probability densities that appears in the integral in the last expression above is another probability density, we obtain from this product a sequence of probability densities which converge to the corresponding product of the limiting probability densities. Therefore we can apply Scheffe's Theorem once again to obtain the required limit. A more detailed exposition can be found in [10]. (The difference in that paper is that the disentangled exponential is considered, but the main points are the same.)

4. Application of Theorem 3.1

Here we will consider an application of Theorem 3.1 which was discussed in the last paragraph of Section 1.

We will make use of Proposition 9.2 of [9] which we now state.

PROPOSITION 4.1. *Suppose that* $\mu_1, \mu_1', \nu_1, \nu_1', \nu_j, j = 2, \dots, l$ *are finite, continuous measures on* $\mathcal{B}([0,1])$, *and there exists* $a, b \in (0,1)$ *such that* $a < b$ *and* $S(\mu_1') \subsetneqq [0,a]$, $S(\nu_1') \subsetneqq [a,b]$, $S(\mu_1) \subsetneqq [b,1]$, *and* $S(\nu_j) \subsetneqq [b,1]$ *for all* $j = 1, \dots, l$. *Let* $A, B_1, \dots, B_l$ *be elements of* $\mathcal{L}(X)$. *Then for nonnegative integers* $m, n_1, \dots, n_l$, *the equality*

$$P^{m,n_1,\dots,n_l}_{\mu_1+\mu_1',\nu_1+\nu_1',\nu_2,\dots,\nu_l}(A, B_1, \dots, B_l) = \sum_{k_0+k_1=n_1} \sum_{j_0+j_1=m} \frac{m!n_1!}{j_0!j_1!k_0!k_1!} \times P^{j_1,k_1,n_2,\dots,n_l}_{\mu_1,\nu_1,\dots,\nu_l}(A, B_1, \dots, B_l)(\|\nu_1'\|B_1)^{k_0}(\|\mu_1'\|A)^{j_0} \tag{4.1}$$

is valid. The sum is over nonnegative integers j_0, j_1, k_0, k_1.

Now suppose that $\mu_j, j = 0, \dots, k$, and $\nu_j, j = 1, \dots, k$ are finite, continuous measures on $\mathcal{B}([0,1])$ and $\mu = \sum_{j=0}^k \mu_j$ and $\nu = \sum_{j=1}^k \nu_j$. Furthermore, suppose that

$$0 = a_0 < b_0 < a_1 < b_1 < \cdots < a_k < b_k = 1$$

and $S(\mu_j) \subseteqq [a_j, b_j]$ for all $j = 0, \dots, k$ and $S(\nu_j) \subseteqq [b_{j-1}, a_j]$ for all $j = 1, \dots, k$.

In the case $l = 1$ in which we are interested, it follows from equation (4.1) and induction that

$$\mathcal{T}_{\mu,\nu} e^{(A^{\sim}+B^{\sim})} = e^{\|\mu_k\|A} e^{\|\nu_k\|B} e^{\|\mu_{k-1}\|A} \cdots e^{\|\nu_1\|B} e^{\|\mu_1\|A}; \tag{4.2}$$

a Trotter - type product.

In particular, if X is a Hilbert space and $A, B \in \mathcal{L}(X)$ are self-adjoint, then for all $\xi \in \mathbb{R}^2$ the operator

$$\mathcal{T}_{\mu,\nu} e^{i(\xi_1 A^{\sim}+\xi_2 B^{\sim})} = e^{i\xi_1\|\mu_k\|A} e^{i\xi_2\|\nu_k\|B} e^{i\xi_1\|\mu_{k-1}\|A} \cdots e^{i\xi_2\|\nu_1\|B} e^{i\xi_1\|\mu_1\|A} \tag{4.3}$$

has norm one.

We are now ready to state the question posed by Brian Jefferies: Can you find two sequences of continuous Borel probability measures $\{\lambda_k\}$ and $\{\sigma_k\}$ on $[0,1]$ such that both sequences converge weakly to Lebesgue measure and an equality like equation (4.3) holds for each k and for any pair of bounded, self-adjoint operators A and B? If so, then for every $\xi_1, \xi_2 \in \mathbb{R}$,

$$\mathcal{T}_{\lambda_k,\sigma_k} e^{i(\xi_1 A^{\sim}+\xi_2 B^{\sim})} \to \mathcal{T}_{l,l} e^{i(\xi_1 A^{\sim}+\xi_2 B^{\sim})}$$

in operator norm by Theorem 3.1. But also every operator in the sequence just above has norm one (as in (4.3)), and so the norm of the limiting operator is also one. With this information it is trivial to conclude that inequality (10.1) from [9] holds. We turn now to producing the sequences $\{\lambda_k\}$ and $\{\sigma_k\}$. Let the partition

$$0 = a_0 < b_0 < a_1 < b_1 < \cdots < a_k < b_k = 1$$

be defined by

$$\begin{aligned} a_j &= \frac{2j}{2k+1}, \\ b_j &= \frac{2j+1}{2k+1} \end{aligned} \tag{4.4}$$

for $j = 0, \dots, k$. Next define continuous Borel measures $\mu_0, \dots, \mu_k$ on $[0,1]$ by

$$\mu_j = l|_{[a_j,b_j]} \tag{4.5}$$

where l is Lebesgue measure and then define

$$\lambda_k = \frac{2k+1}{k+1} \sum_{j=0}^{k} \mu_j. \tag{4.6}$$

Clearly λ_k is a continuous probability measure on $[0,1]$ for each $k \in \mathbb{N}$. We wish to show that the sequence $\{\lambda_k\}_{k=1}^{\infty}$ converges weakly to Lebesgue measure l on $[0,1]$. We will establish this claim in two steps. The first step will be to show that, if $f(x) = x^q$, $q \in \mathbb{N}$, then

$$\lim_{k\to\infty} \int_{[0,1]} f(x) d\lambda_k(x) = \int_{[0,1]} f(x) dx. \tag{4.7}$$

Then it will immediately follow that

$$\lim_{k\to\infty}\int_{[0,1]} f(x)d\lambda_k(x) = \int_{[0,1]} f(x)dx \tag{4.8}$$

for all polynomials $f(x)$ on $[0,1]$. For the second step, we will apply a theorem of E.H. Moore on iterated limits ([6], page 28) mentioned above to obtain

$$\lim_{k\to\infty}\int_{[0,1]} f(x)d\lambda_k(x) = \int_{[0,1]} f(x)dx \tag{4.9}$$

for any $f \in C_b([0,1])$.

To establish the limit in equation (4.7) for $f(x) = x^q$ we first do some calculations:

(4.10)

$$\begin{aligned}
&\int_{[0,1]} f(x)d\lambda_k(x)\\
&= \frac{2k+1}{k+1}\sum_{j=0}^{k}\int_{[a_j,b_j]} f(x)dx\\
&= \frac{2k+1}{k+1}\sum_{j=0}^{k}\frac{1}{q+1}\left(b_j^{q+1} - a_j^{q+1}\right)\\
&= \frac{2k+1}{k+1}\sum_{j=0}^{k}\frac{1}{(q+1)(2k+1)^{q+1}}\left((2j+1)^{q+1} - (2j)^{q+1}\right)\\
&= \frac{2k+1}{k+1}\sum_{j=0}^{k}\frac{2^{q+1}}{(q+1)(2k+1)^{q+1}}\left((j+\frac{1}{2})^{q+1} - j^{q+1}\right)\\
&= \left(\frac{2k+1}{k+1}\right)\left(\frac{2^{q+1}}{(q+1)(2k+1)^{q+1}}\right)\cdot\\
&\qquad\sum_{j=0}^{k}\frac{1}{2}\left((j+\frac{1}{2})^{q} + (j+\frac{1}{2})^{q-1}j + \cdots + (j+\frac{1}{2})j^{q-1} + j^q\right)\\
&= \left(\frac{2k+1}{k+1}\right)\left(\frac{2^{q}}{(q+1)(2k+1)^{q+1}}\right)\left\{\frac{(k+1)^{q+1}}{q+1} + (\text{other lower degree terms}) +\right.\\
&+ \frac{(k+1)^{q+1}}{q+1} + (\text{other lower degree terms}) + \cdots\\
&+ \left.\frac{(k+1)^{q+1}}{q+1} + (\text{other lower degree terms})\right\}\\
&= \left(\frac{2k+1}{k+1}\right)\left(\frac{2^{q}}{(q+1)(2k+1)^{q+1}}\right)\left\{(k+1)^{q+1} + (\text{other lower degree terms})\right\}\\
&\to 2\cdot\frac{2^q}{2^{q+1}(q+1)} = \frac{1}{q+1}
\end{aligned}$$

as $k\to\infty$. Hence it follows that

$$\lim_{k\to\infty}\int_{[0,1]} x^q d\lambda_k(x) = \lim_{k\to\infty}\frac{2k+1}{k+1}\sum_{j=0}^{k}\int_{[a_j,b_j]} x^q dx = \frac{1}{q+1} = \int_{[0,1]} x^q dx. \tag{4.11}$$

Consequently, if $f(x)$ is any polynomial on $[0,1]$ we have

$$\lim_{k\to\infty}\int_{[0,1]} f(x)d\lambda_k(x) = \int_{[0,1]} f(x)dx. \tag{4.12}$$

Now let $g(x)$ be any continuous function on $[0,1]$. Since polynomials are dense in $C([0,1])$ with respect to the uniform norm there is a sequence $\{p_n(x)\}_{n=1}^{\infty}$ of polynomials on $[0,1]$ which converge uniformly to $g(x)$ on $[0,1]$. We may write, using the uniform convergence of the sequence of polynomials,

$$\lim_{k\to\infty}\int_{[0,1]} g(x)d\lambda_k(x) = \lim_{k\to\infty}\lim_{n\to\infty}\int_{[0,1]} p_n(x)d\lambda_k(x). \tag{4.13}$$

To use the previously mentioned theorem of Moore, first let $n \in \mathbb{N}$ be fixed. We consider

$$\lim_{k\to\infty}\int_{[0,1]} p_n(x)d\lambda_k(x). \tag{4.14}$$

Using the calculation above and the remarks that follow that calculation we see that

$$\lim_{k\to\infty}\int_{[0,1]} p_n(x)d\lambda_k(x) = \int_{[0,1]} p_n(x)dx. \tag{4.15}$$

The second ingredient needed in order to apply Moore's theorem is to show that the limit on n is uniform in k. Once again applying the uniform convergence of our sequence of polynomials we get

$$\lim_{n\to\infty}\int_{[0,1]} p_n(x)d\lambda_k(x) = \int_{[0,1]} g(x)d\lambda_k(x) \tag{4.16}$$

and this limit clearly does not depend on k; i.e. is uniform in k. Now we can apply the theorem of E.H. Moore to obtain

$$\begin{aligned}\lim_{k,n\to\infty}&\int_{[0,1]} p_n(x)d\lambda_k(x)\\ &= \lim_{k\to\infty}\lim_{n\to\infty}\int_{[0,1]} p_n(x)d\lambda_k(x)\\ &= \lim_{n\to\infty}\lim_{k\to\infty}\int_{[0,1]} p_n(x)d\lambda_k(x)\\ &= \lim_{n\to\infty}\int_{[0,1]} p_n(x)dx\\ &= \int_{[0,1]} g(x)dx.\end{aligned} \tag{4.17}$$

Therefore we have established the weak convergence $\lambda_k \rightharpoonup l$ as claimed.

To continue, we now use the partition given in (4.4) to define a different, although very similar, sequence of continuous Borel probability measures. Specifically, we define measures $\nu_1, \ldots, \nu_k$ by

$$\nu_j = l|_{[b_{j-1},a_j]}. \tag{4.18}$$

Using these k measures, define the measures σ_k by

$$\sigma_k = \frac{2k+1}{k}\sum_{j=1}^{k}\nu_j. \tag{4.19}$$

Virtually the same argument as above shows that $\sigma_k \rightharpoonup l$.

With the sequences of measures $\{\lambda_k\}_{k=0}^{\infty}$ and $\{\sigma_k\}_{k=1}^{\infty}$ and bounded self adjoint operators A, B on a Hilbert space $\mathcal{H}$ we have, using equation (4.3),

$$\begin{aligned} &\mathcal{T}_{\lambda_k,\sigma_k} e^{i(\xi_1 A^{\sim} + \xi_2 B^{\sim})} \\ &\quad = e^{i\xi_1 \|\mu_k\| A} e^{i\xi_2 \|\nu_k\| B} e^{i\xi_1 \|\mu_{k-1}\| A} e^{i\xi_2 \|\nu_{k-1}\| B} \cdots e^{i\xi_2 \|\nu_1\| B} e^{i\xi_1 \|\mu_0\| A} \qquad (4.20) \\ &\quad = e^{(\frac{i\xi_1}{k+1})A} e^{(\frac{i\xi_2}{k})B} e^{(\frac{i\xi_1}{k+1})A} \cdots e^{(\frac{i\xi_2}{k})B} e^{(\frac{i\xi_1}{k+1})A} \end{aligned}$$

Using Theorem 3.1 and then Proposition 8.5 of [9] we obtain

$$\lim_{k\to\infty} \mathcal{T}_{\lambda_k,\sigma_k} e^{i(\xi_1 A^{\sim} + \xi_2 B^{\sim})} = \mathcal{T}_{l,l} e^{i(\xi_1 A^{\sim} + \xi_2 B^{\sim})} = e^{i(\xi_1 A + \xi_2 B)} \qquad (4.21)$$

as desired. It follows that we obtain the limit of the Trotter product on the right hand side of equation (4.20) as $k \to \infty$ and it is given on the right hand side of equation (4.21).

As a result of the limits given in equations (4.20) and (4.21) it is seen that Weyl's functional calculus for two bounded self-adjoint linear operators on a Hilbert space fits into the framework of the disentangling map and particularly with the formalism developed in sections 9 and 10 of [9].

REMARK 4.1. We could carry out the above analysis using p measures and any p-tuple of bounded self-adjoint linear operators on a Hilbert space. The main change that would need to be made in the preceding argument consists of changing the partition of $[0, 1]$ that provides the prescription for the definition of the weakly convergent sequences of measures. The rest of the argument would be very similar to the argument above except that we would have p sequences of weakly convergent measures converging to Lebesgue measure. Theorem 3.1 would still immediately apply. Hence it would follow that Weyl's functional calculus for p bounded self-adjoint linear operators would also fit into the formalism developed in Part II of [9].

References

[1] Anderson, R.F.V.: The Weyl functional calculus, J. Funct. Analy., **4** (1969), 240-267.

[2] ______: On the Weyl functional calculus, J. Funct. Analy., **6** (1970), 110-115.

[3] Billingsley, P.: *Convergence of Probability Measures*, Wiley, New York, 1968.

[4] DeFacio, B., Johnson, G.W., and Lapidus, M.L.: Feynman's operational calculus and evolution equations, Acta. Applic. Math., **47** (1997), 155-211.

[5] ______: Feynman's operational calculus as a generalized path integral, in: S. Cambanis *et al.* (eds), *Stochastic Processes: A Festschrift in Honour of Gopinath Kallianpur*, Springer-Verlag, New York, 1993, pp. 51-60.

[6] Dunford, N. and Schwartz, J. T.: *Linear Operators*, Part I, Wiley Interscience, New York, 1958.

[7] Feynman, R.: An operator calculus having applications in quantum electrodynmanics, Phys. Rev., **84** (1951), 108-128.

[8] Hewitt, E. and Stromberg, K.: *Real and Abstract Analysis*, Springer-Verlag, 1965.

[9] Jefferies, B. and Johnson, G. W.: Feynman's operational calculi for noncommuting systems of operators, submitted for publication.

[10] Jefferies, B., Johnson, G. W., and Nielsen, L.: Feynman's operational calculi for time dependent noncommuting operators, submitted for publication.

[11] Johnson, G.W. and Lapidus, M.L.:*Generalized Dyson series, generalized Feynman diagrams, the Feynman integral and Feynman's operational calculus*, Mem. Amer. Math. Soc., **62** (1986), 1-78.

[12] ______:*Noncommutative operations on Wiener functionals and Feynman's operational calculus*, J. Funct. Anal., **81** (1988), 74-99.

[13] ———: *The Feynman integral and Feynman's operational calculus*, Oxford Mathematical Monographs, Oxford Univ. Press (anticipated publication date, January 2000).

[14] Johnson, G.W. and Park, Y.H.: Decomposing the disentangling algebra on Wiener space, submitted for publication.

[15] Lapidus, M.L.: *The differential equation for the Feynman-Kac formula with Lebesgue-Stieltjes measure*, Lett. Math. Phys., **11** (1986), 1-13.

[16] ———: *The Feynman-Kac formula with Lebesgue-Stieltjes measure and Feynman's operational calculus*, Stud. Appl. Math., **76** (1987), 93-132.

[17] ———: *The Feynman-Kac formula with Lebesgue-Stieltjes measure: An integral equation in the general case*, Int. Eq. Op. Theory, **12** (1989), 163-210.

[18] Nielsen, L.: Stability properties for Feynman's operational calculus, Thesis, Univ. of Nebraska, Lincoln, May 1999.

[19] Reyes, J.T.: Exponentials of noncommuting operators via Feynman's operational calculus, and associated evolution equations, Thesis, Univ. of Nebraska, Lincoln, Dec. 1995.

[20] Taylor, M.E.: Functions of several self-adjoint operators, Proc. Amer. Math. Soc., **19** (1968), 91-98.

DEPARTMENT OF MATHEMATICS AND STATISTICS, UNIVERSITY OF NEBRASKA, LINCOLN, NE
E-mail address: gjohnson@math.unl.edu

DEPARTMENT OF MATHEMATICS, VALLEY CITY STATE UNIVERSITY, VALLEY CITY, ND
E-mail address: Lance_Nielsen@mail.vcsu.nodak.edu

Canadian Mathematical Society
Conference Proceedings
Volume **29**, 2000

On a biorthogonal approach to the construction of non–Gaussian analysis

N.A. Kachanovsky

ABSTRACT. In this paper we describe a framework of a biorthogonal approach to a non–Gaussian generalization of White Noise Analysis in the most general (at the moment) situation. In particular, we describe the test and generalized functions spaces of non–Gaussian analysis, inner description of these spaces, integral S–transform, Wick product etc. The applications to the Poisson analysis and to symmetric random walks are also considered.

1. Introduction

At the moment there are two main approaches to construction of a non–Gaussian generalization of White Noise Analysis: a spectral one and a biorthogonal one.

A *spectral approach* (see e.g. [1,2] and references therein) considers a so–called Jacobi field, which generates a family of polynomials and a probability measure μ such that these polynomials are orthogonal with respect to the scalar product in $L^2(\mu)$. Then one can use the constructed polynomials as orthogonal bases in spaces of test and generalized functions so as Hermite polynomials in Gaussian analysis. A merit of such approach consists in coincidence of orthogonal bases in the test and generalized functions spaces and in *orthogonality* of these bases with respect to the scalar product in $L^2(\mu)$, which is very convenient for e.g. 'stochastic applications'. The main lacks of the spectral approach are: 1) it is difficult to construct orthogonal polynomials and a Jacobi field for a given probability measure; 2) in many cases it is very hard to give the inner description of test functions spaces; 3) in the majority of cases it is impossible to write explicit formulas for integral transforms; 4) a scale of test and generalized functions spaces is not flexible.

A *biorthogonal approach* was proposed in [3] and developed in [4–11]. The main idea of this approach is to construct the so–called generalized Appell–like polynomials with generating function $\gamma(\theta)h(\langle x, \alpha(\theta)\rangle)$ (where θ belongs to some Hilbert or nuclear space, x is from the corresponding dual space, $\langle \cdot, \cdot \rangle$ denotes the corresponding pairing); or, more generally, the generalized Appell characters with generating function $\gamma(\theta)h(x; \alpha(\theta))$ (here x can belong to a sufficiently general metric space X). These polynomials or characters form the orthogonal bases in test functions spaces. Then one has to consider a probability measure μ on X. Under certain additional conditions test functions spaces are embedded in $L^2(\mu)$ and the embeddings are

2000 *Mathematics Subject Classification.* Primary 46A13, 46A20, 46F05, 46F12, 46F25.

dense. So one can construct generalized functions spaces that are dual to test functions ones with respect to $L^2(\mu)$. Unfortunately, under this approach in the most of cases generalized Appell–like polynomials (or characters) are *not orthogonal in* $L^2(\mu)$ and, therefore, the orthogonal bases in test and generalized functions spaces do not coincide (but they are biorthogonal, see Section 3). Moreover, there is no natural orthogonal basis in $L^2(\mu)$. The main merit of biorthogonal approach is absence of the lacks of type 1)–4).

In this paper we describe a framework of the biorthogonal approach to a non–Gaussian generalization of Gaussian White Noise Analysis in the most general (at the moment) situation. The presentation is based on the paper of Yu.M. Berezansky [10] and on the author's papers [8,11].

The paper is organized as follows. In Section 2 we deal with generalized Appell characters and test functions spaces, Section 3 is devoted to generalized functions spaces and examples. Note that because of the limited volume of this paper we do not deal with some objects of biorthogonal analysis, such as generalized translation operator, integral C–transform, 'stochastic applications' etc. By the same reason we can not present here full proofs of our statements.

2. The generalized Appell characters and the test functions spaces

Let $H_2 \subseteq H_1 \subseteq H_0$ be real separable Hilbert spaces and the embeddings $H_1 \hookrightarrow H_0$, $H_2 \hookrightarrow H_1$ be dense and of Hilbert–Schmidt type. Let H_{-1}, H_{-2} be dual spaces of the chain

$$H_2 \subseteq H_1 \subseteq H_0 \subseteq H_{-1} \subseteq H_{-2}$$

(see e.g.[12]). We denote the complexifications of spaces by subindex $\mathbb{C}$, the norms in H_p $(p = -2,-1,0,1,2)$ and its tensor powers and complexifications by $|\cdot|_p$. (Note that the spaces H_1, H_{-1} are necessary for some technical estimates only, see e.g. [11] for details.) Let X be some separable metric space. We consider a function $h : X \times H_{1,\mathbb{C}} \to \mathbb{C}$ such that

1) the function $X \ni x \mapsto h(x;\theta) \in \mathbb{C}$ is continuous for θ from some neighbourhood of $0 \in H_{1,\mathbb{C}}$;

2) for all $x \in X$ $h(x;\cdot) \in Hol_0(H_{1,\mathbb{C}})$, where $Hol_0(H_{1,\mathbb{C}})$ denotes the algebra of germs of holomorphic at $0 \in H_{1,\mathbb{C}}$ functions with values in $\mathbb{C}$;

3) $h(\cdot;\theta)$ is locally bounded uniformly with respect to θ *or* (if $X \subseteq H_{-2}$) there exists nondecreasing function $\Psi : \mathbb{R} \to \mathbb{R}_+$ such that $|h(x;\theta)| \le \Psi(|x|_{-1}|\theta|_1)$;

4) there exists *independent on* $x \in X$ neighbourhood of zero $\widetilde{\mathcal{U}}_0 \subset H_{1,\mathbb{C}}$ such that

$$h(x;\theta) = \sum_{n=0}^{\infty} \frac{1}{n!}\langle \chi_n(x), \theta^{\otimes n}\rangle,$$

for all $\theta \in \widetilde{\mathcal{U}}_0 \cap H_{2,\mathbb{C}}$, where $\chi_n(x) \in H_{-2,\mathbb{C}}^{\widehat{\otimes} n}$, $H_{-2,\mathbb{C}}^{\widehat{\otimes} 0} := \mathbb{C}$, $\langle\cdot,\cdot\rangle$ denotes the (real) dual pairing in the tensor powers of spaces, generated by the scalar product in the tensor powers of H_0, $\widehat{\otimes}$ denotes the symmetric tensor product;

5) there exists a 'zero element' $e \in X$ such that $h(e;\cdot) \equiv 1$;

6) $h(\cdot;0) \equiv 1$.

Let us assume also that

7) there exists a sequence $\varphi_n \in H_{2,\mathbb{C}}^{\widehat{\otimes} n}$, $n \in \mathbb{Z}_+ := \mathbb{N}\cup\{0\}$ such that $\big\{\langle \chi_n(x), \varphi_n\rangle : n = 0,\dots,N\big\}$ are lineary independent for all $N \in \mathbb{Z}_+$.

Let $Hol_0(H_{1,\mathbb{C}}, H_{1,\mathbb{C}})$ be the algebra of germs of holomorphic at $0 \in H_{1,\mathbb{C}}$ functions with values in $H_{1,\mathbb{C}}$. We consider the functions γ and α such that

(1) $\gamma \in Hol_0(H_{1,\mathbb{C}})$, $\gamma(0) \neq 0$;

(2) $\alpha \in Hol_0(H_{1,\mathbb{C}}, H_{1,\mathbb{C}})$, $\alpha(H_2) \subseteq H_2$, α is invertible on some neighbourhood of zero $\mathcal{U}_0 \subseteq H_{1,\mathbb{C}}$ in the sense that there exists α^{-1} such that $\alpha^{-1}(\alpha(\theta)) = \theta$, $\forall \theta \in \mathcal{U}_0$, and $\alpha(0) = 0$.

EXAMPLE 2.1. One can consider the following example of introduced objects: let $H_p = W_2^p(\mathbb{R}, (1+|y|)^p dy)$, $p = 1, 2$ be Sobolev spaces, $H_0 = L^2(\mathbb{R})$, $X = H_{-2}$, $h(x;\theta) = e^{\langle x,\theta\rangle}$, $\gamma \equiv 1$, $\alpha = id$ (i.e. $\alpha(\theta) = \theta$ for all θ). It is not hard to verify that all demands above are satisfied now (in particular, $\Psi(u) = e^u$ in property 3) and $e = 0 \in H_{-2}$ in property 5)). The considered objects can be used for construction of the 'classical' Gaussian analysis (see Example 3.4 below).

Another (more complicated) example can be constructed if we put $h(x;\theta) = \Lambda_s(\langle x,\theta\rangle)$, $s \geq -1/2$, where

$$\Lambda_s(u) := 1 + \sum_{m=1}^{\infty} u^{2m}\Big(-\frac{1}{4}\Big)^m \frac{1}{m! \prod_{k=1}^{m}(s+k)}. \tag{2.1}$$

This function connected with the so-called symmetric Wiener process, see more detailed description below.

Now let us consider a function $\gamma(\theta)h(x;\alpha(\theta))$. It is not hard to prove (using Cauchy inequalities and kernel theorem, see e.g. [6,10]) that under the conditions above

$$\gamma(\theta)h(x;\alpha(\theta)) = \sum_{n=0}^{\infty} \frac{1}{n!}\langle P_n^{\gamma,\alpha}(x), \theta^{\otimes n}\rangle,$$

where $\theta \in H_{2,\mathbb{C}}$ is from some neighbourhood of zero, $P_n^{\gamma,\alpha}(x) \in H_{-2,\mathbb{C}}^{\widehat{\otimes} n}$ and for any $x \in X$ the series converges uniformly on some neighbourhood of $0 \in H_{1,\mathbb{C}}$.

DEFINITION 2.2. The system $\{\langle P_n^{\gamma,\alpha}(x), \varphi^{(n)}\rangle,\ \varphi^{(n)} \in H_{2,\mathbb{C}}^{\widehat{\otimes} n},\ n \in \mathbb{Z}_+\}$ is called the system of generalized Appell characters (we use the terminology of [10]).

EXAMPLE 2.3. Note that for $h(x;\theta) = e^{\langle x,\theta\rangle}$, $\gamma \equiv 1$, $\alpha = id$ $P_n^{1,id}(x) = x^{\otimes n}$.

Of course, the generalized kernels $P_n^{\gamma,\alpha}(x)$ depend on h. But we do not stress this dependence by the corresponding index for simplification of notations.

REMARK 2.4. In particular case, when $X \subseteq H_{-2}$ and $h(x;\theta) = \chi(\langle x,\theta\rangle)$, where $\chi : \mathbb{C} \to \mathbb{C}$ is entire function, $\{\langle P_n^{\gamma,\alpha}(x), \varphi^{(n)}\rangle\}$ are called *the generalized Appell-like polynomials*. An analysis connected with these polynomials is studied in details in [8,11]. Note that in this case the assumption of 'linear independence' 7) above is fulfilled automatically.

Let us consider a linear set $\mathcal{P} := \{\varphi(x) = \sum_{n=0}^{N_\varphi}\langle P_n^{\gamma,\alpha}(x), \varphi^{(n)}\rangle \colon N_\varphi \in \mathbb{Z}_+\}$ (it is not difficult to see that it does not depend on γ and α). We define on this set the scalar product $(\cdot,\cdot)_{q,\gamma,\alpha}$, $q \in \mathbb{N}$, putting for $\varphi_j(\cdot) = \sum_{n=0}^{N_{\varphi_j}}\langle P_n^{\gamma,\alpha}(\cdot), \varphi_j^{(n)}\rangle \in \mathcal{P}$, $j = 1, 2$

$$(\varphi_1, \varphi_2)_{q,\gamma,\alpha} := \sum_{n=0}^{\min(N_{\varphi_1}, N_{\varphi_2})} (n!)^2 K^{qn} \langle \varphi_1^{(n)}, \varphi_2^{(n)}\rangle_{H_{2,\mathbb{C}}^{\widehat{\otimes} n}},$$

where $K > 1$ is a fixed constant. Let $\|\varphi\|_{q,\gamma,\alpha} := \sqrt{(\varphi,\varphi)_{q,\gamma,\alpha}}$ be the corresponding norm.

DEFINITION 2.5. By $(H_2)_{q,\gamma,\alpha}$ we denote a Hilbert space which is the completion of $\mathcal{P}$ with respect to the norm $\|\cdot\|_{q,\gamma,\alpha}$.

REMARK 2.6. Note that the elements of $(H_2)_{q,\gamma,\alpha}$ are the series

$$\varphi_j(x) = \sum_{n=0}^{\infty} \langle P_n^{\gamma,\alpha}(x), \varphi_j^{(n)} \rangle,\ j = 1, 2, \tag{2.2}$$

and $(\varphi_1, \varphi_2)_{(H_2)_{q,\gamma,\alpha}} = \sum_{n=0}^{\infty} (n!)^2 K^{qn} \langle \varphi_1^{(n)}, \varphi_2^{(n)} \rangle_{H_{2,\mathbb{C}}^{\hat{\otimes} n}}$. So, $\{\langle P_n^{\gamma,\alpha}(\cdot), \varphi^{(n)} \rangle\}_{n=0}^{\infty}$ is the orthogonal basis in $(H_2)_{q,\gamma,\alpha}$.

Now the natural question arises: are $(H_2)_{q,\gamma,\alpha}$ functional spaces? In other words, if $\varphi \in (H_2)_{q,\gamma,\alpha}$, can we guarantee that the series (2.2) converges at least pointwisely and, if 'yes', can we interpret φ as a usual function? I.e. if $\varphi \equiv 0$, can we guarantee that in decomposition (2.2) $\varphi^{(n)} = 0$ for all $n \in \mathbb{Z}_+$? The answer is given by the following

THEOREM 2.7. *1. Let $\varphi \in (H_2)_{q,\gamma,\alpha}$ for sufficiently large $q \in \mathbb{N}$. Then the series (2.2) converges uniformly on every neighbourhood of $e \in X$.*

2. Assume that for some γ and α under the conditions above there exists $q \in \mathbb{N}$ such that the space $(H_2)_{q,\gamma,\alpha}$ is a functional one. Then for any γ and α under the conditions above there exists $\widetilde{q} := \widetilde{q}(\gamma,\alpha) \in \mathbb{N}$ such that for any $q \geq \widetilde{q}$ the space $(H_2)_{q,\gamma,\alpha}$ is a functional one.

The proof is a simple modification of the proof of the corresponding theorem in [13]. Note that it comes out by estimation of $|\varphi(x)|$ and presentation φ in the form (2.2) with different γ and α.

REMARK 2.8. If $X \subseteq H_{-2}$ and $h(x;\theta) = \chi(\langle x, \theta \rangle)$, then $P_n^{1,id}(x) = \chi_n x^{\otimes n}$, $n \in \mathbb{N}$, $P_0^{1,id}(x) = 1$ (here χ_n are from the decomposition $\chi(u) = \sum_{n=0}^{\infty} \frac{1}{n!} \chi_n u^n$). So for $\gamma \equiv 1$ and $\alpha = id$ (2.2) is the Taylor series and, therefore, $(H_2)_{q,\gamma,\alpha}$ are functional spaces for all γ, α under the conditions above and sufficiently large $q \in \mathbb{N}$.

Starting from this place we assume that *all* assumptions of Theorem 2.7 are fulfilled, so $(H_2)_{q,\gamma,\alpha}$ are functional spaces.

REMARK 2.9. Note that an analysis connected with generalized Appell characters for special γ and $\alpha = id$ is studied in [10]. But functionality of test functions spaces is not discussed therein.

Let us consider a space $(H_2)_{\gamma,\alpha} := \operatorname{pr} \lim_{q \in \mathbb{N}} (H_2)_{q,\gamma,\alpha}$. It was proved in [11] for $h(x;\theta) = \chi(\langle x, \theta \rangle)$ and in [10] for $\alpha = id$ and special γ that such a space does not depend on γ and α as topological one. Now we have the following

THEOREM 2.10. *The space $(H_2)_{\gamma,\alpha} = \operatorname{pr} \lim_{q \in \mathbb{N}} (H_2)_{q,\gamma,\alpha}$ does not depend on γ and α as topological one. (So we denote it by (H_2).)*

The proof is completely analogous to the proofs of the corresponding statements in [10,11] and can be obtained by estimation of the norms $\|\cdot\|_{q,\gamma,\alpha}$ with different γ and α with each other.

Now let us study inner description of (H_2).

THEOREM 2.11. *All elements of* (H_2) *are continuous locally bounded functions on* X. *Moreover, if* $X \subseteq H_{-2}$ *and* $h(x;\theta) = \chi(\langle x, \theta\rangle)$, *where* $\chi : \mathbb{C} \to \mathbb{C}$ *is an entire function and in decomposition* $\chi(u) = \sum_{n=0}^{\infty} \chi_n u^n$ $\chi_n \neq 0$ *for all* $n \in \mathbb{Z}_+$ *then* (H_2) *consists of restrictions on* H_2 *of entire on* $H_{1,\mathbb{C}}$ *functions.*

The first statement follows from Theorem 2.10 and the corresponding theorem in [10]. The second one follows from properties of *polynomial* $\langle P_n^{\gamma,\alpha}(x), \varphi^{(n)}\rangle$ and uniform convergence of the corresponding series (2.2) (Theorem 2.7).

EXAMPLE 2.12. If $h(x;\theta) = e^{\langle x,\theta\rangle}$ then the corresponding space (H_2) coincides with the space of entire functions of first order and minimal type ([6]).

Finally, let us introduce a generalized differential operator $\langle \Phi^{(m)}, : D :^{\otimes m}\rangle$ (with constant coefficient $\Phi^{(m)} \in H_{-2,\mathbb{C}}^{\widehat{\otimes} m}$) on $\mathcal{P}$, which will be used in the next section.

DEFINITION 2.13. We put on 'monomials' $\langle P_n^{\gamma,\alpha}(x), \varphi^{(n)}\rangle$, $\varphi^{(n)} \in H_{2,\mathbb{C}}^{\widehat{\otimes} n}$

$$\langle \Phi^{(m)}, : D :^{\otimes m}\rangle \langle P_n^{\gamma,\alpha}(x), \varphi^{(n)}\rangle := 1_{\{n\geq m\}} \frac{n!}{(n-m)!} \langle P_{n-m}^{\gamma,\alpha}(x), \langle \Phi^{(m)}, \varphi^{(n)}\rangle\rangle,$$

where $1_{\{n\geq m\}}$ denotes the indicator of $\{n \geq m\}$, $\langle \Phi^{(m)}, \varphi^{(n)}\rangle \in H_{2,\mathbb{C}}^{\widehat{\otimes}(n-m)}$ is defined for $\varphi^{(n)} = \varphi^{\otimes n}$ by $\langle \Phi^{(m)}, \varphi^{\otimes n}\rangle := \varphi^{\otimes(n-m)}\langle \Phi^{(m)}, \varphi^{\otimes m}\rangle$ and for general $\varphi^{(n)} \in H_{2,\mathbb{C}}^{\widehat{\otimes} n}$ by well known polarization identity.

Of course, the operator $\langle \Phi^{(m)}, : D :^{\otimes m}\rangle$ depends on h and α (it can be shown that it does not depend on γ), but we do not mark it for simplicity of notations.

EXAMPLE 2.14. Let $h(x;\theta) = e^{\langle x,\theta\rangle}$, $\alpha = id$. Then

$$\langle \Phi^{(m)}, : D :^{\otimes m}\rangle \langle x^{\otimes n}, \varphi^{(n)}\rangle = 1_{\{n\geq m\}} \frac{n!}{(n-m)!} \langle x^{\otimes(n-m)}, \langle \Phi^{(m)}, \varphi^{(n)}\rangle\rangle,$$

i.e. $\langle \Phi^{(m)}, : D :^{\otimes m}\rangle$ is the usual differential operator with constant coefficient, which is used in the 'classical' Gaussian analysis.

Now it is possible to study the operator $\langle \Phi^{(m)}, : D :^{\otimes m}\rangle$ in detail. In particular, it is not hard to describe the dependence of this operator on α; to continue it as a linear continuous operator on $(H_2)_{q,\gamma,\alpha}$; to study the operators $\nu(: D :) := \sum_{m=0}^{\infty} \langle \nu_m, : D :^{\otimes m}\rangle$, where $\nu(\theta) := \sum_{m=0}^{\infty} \langle \nu_m, \theta^{\otimes m}\rangle$ ($\nu_m \in H_{-2,\mathbb{C}}^{\widehat{\otimes} m}$) is holomorphic function; and to introduce a generalized translation operator as a particular case of $\nu(: D :)$ etc. But we will not to do it in this paper. Note that investigations of such a type were made for the case $h(x;\theta) = \chi(\langle x,\theta\rangle)$ in [14].

REMARK 2.15. Sometimes it is convenient to introduce a chain

$$H_2 \subseteq H_1 \subseteq H_0^\sigma \subseteq H_{-1}^\sigma \subseteq H_{-2}^\sigma, \tag{2.3}$$

where $H_0^\sigma (\neq H_0)$ is a separable real Hilbert space and the embedding $H_1 \hookrightarrow H_0^\sigma$ is dense and of Hilbert–Schmidt type. Then we can introduce on $\mathcal{P}$ the operator $\langle \Phi_\sigma^{(m)}, : D :^{\otimes m}\rangle_\sigma$, $\Phi_\sigma^{(m)} \in H_{-2,\mathbb{C}}^{\sigma\widehat{\otimes} m}$ puting on 'monomials'

$$\langle \Phi_\sigma^{(m)}, : D :^{\otimes m}\rangle_\sigma \langle P_n^{\gamma,\alpha}(x), \varphi^{(n)}\rangle := 1_{\{n\geq m\}} \frac{n!}{(n-m)!} \langle P_{n-m}^{\gamma,\alpha}(x), \langle \Phi_\sigma^{(m)}, \varphi^{(n)}\rangle_\sigma\rangle,$$

where $\langle \cdot, \cdot\rangle_\sigma$ denotes the (real) pairing in tensor powers of spaces, generated by the scalar product in the tensor powers of H_0^σ, and defined for $m \neq n$ by the analogy with $\langle\Phi^{(m)},\varphi^{(n)}\rangle$, see Definition 2.13. Note that $\langle\Phi_\sigma^{(m)},:D:^{\otimes m}\rangle_\sigma=\langle\Phi^{(m)},:D:^{\otimes m}\rangle$ with some $\Phi^{(m)}\in H_{-2,\mathbb{C}}^{\widehat{\otimes}m}$; it easily follows from the fact that $\nu(\theta):=\langle\Phi_\sigma^{(m)},\theta^{\otimes m}\rangle_\sigma$, $\Phi_\sigma^{(m)}\in H_{-2,\mathbb{C}}^{\sigma\widehat{\otimes}m}$ can be presented in the form $\nu(\theta)=\langle\Phi^{(m)},\theta^{\otimes m}\rangle$, $\Phi^{(m)}\in H_{-2,\mathbb{C}}^{\widehat{\otimes}m}$.

3. The generalized functions spaces and examples

Let us consider a space $L^2(X,\mathcal{B}(X),\mu)$ ($\mathcal{B}$ denotes a Borel σ–algebra) of square integrable with respect to some probability measure μ functions $\varphi:X\to\mathbb{C}$. We assume that μ satisfies the following conditions:

1) For $\gamma\equiv 1$ and $\alpha=id$

$$\||P_n^{1,id}(\cdot)|_{-2}\|_{L^2(X,\mu)}\le n!C^nM,$$

where $C>1$, $M>0$ are some constants. (This implies that for any γ and α under the conditions above the analogous estimate with other constants holds.)

2) μ is positive on each nonempty open subset of X.

3) The set $\mathcal{P}$ is dense in $L^2(X,\mu)$.

REMARK 3.1. 1. For the *polynomials* $\langle P_n^{\gamma,\alpha}(x),\varphi^{(n)}\rangle$ (see Remark 2.4) there exists a simple sufficient condition for density of $\mathcal{P}$ in $L^2(X,\mu)$ ([15]): if for the measure μ its Laplace transform $l_\mu(\theta):=\int_X e^{\langle x,\theta\rangle}\mu(dx)$ is holomorphic at $0\in H_{2,\mathbb{C}}$ function (i.e. μ is an analytic measure in terms of [5,6,8,11]) then $\mathcal{P}$ is dense in $L^2(X,\mu)$.

2. Note also that for the polynomials $\langle P_n^{\gamma,\alpha}(x),\varphi^{(n)}\rangle$ it is possible to relax the condition 2) on μ: it is sufficient to specify that *there exists* a nonempty open set $O\subseteq X$ such that μ is positive on each nonempty open subset of O (see [13] for details).

3. If the condition 3) on μ does not hold we can consider the space $\widetilde{L}^2(X,\mu)$ which is the completion of $\mathcal{P}$ with respect to the norm in $L^2(X,\mu)$, and to continue all our considerations using $\widetilde{L}^2(X,\mu)$ instead of $L^2(X,\mu)$ (cf. [8]).

It follows from [10] and Theorem 2.10 that for any γ and α under the conditions above there exists $\widetilde{q}=\widetilde{q}(\gamma,\alpha)\in\mathbb{N}$ such that for any $q\ge\widetilde{q}$ we have $(H_2)_{q,\gamma,\alpha}\subseteq L^2(X,\mu)$, and this embedding is dense and of Hilbert–Schmidt type. So one can consider $(H_{-2})'_{-q,\gamma,\alpha}$ ($q\ge\widetilde{q}$)—the dual spaces to $(H_2)_{q,\gamma,\alpha}$ with respect to $L^2(X,\mu)$, and $(H_{-2})'=\text{ind}\lim_{q\in\mathbb{N}}(H_{-2})'_{-q,\gamma,\alpha}$—the dual space to (H_2).

Let us construct the orthogonal bases in $(H_{-2})'_{-q,\gamma,\alpha}$. At first, we define δ–function $\delta_x\in(H_{-2})'$, $x\in X$ by the formula $\langle\langle\delta_x,\varphi\rangle\rangle:=\varphi(x)$, $\forall\varphi\in(H_2)$. Here $\langle\langle\cdot,\cdot\rangle\rangle$ denotes the dual pairing generated by the scalar product in $L^2(X,\mu)$. It is not difficult to prove that this definition is correct (see e.g. [10]). Furtheron, let $\langle\Phi^{(m)},:D:^{\otimes m}\rangle^*$ ($\Phi^{(m)}\in H_{-2,\mathbb{C}}^{\widehat{\otimes}m}$) be a linear continuous operator on $(H_{-2})'_{-q,\gamma,\alpha}$ (q is sufficiently large) adjoint to $\langle\Phi^{(m)},:D:^{\otimes m}\rangle$ with respect to $L^2(X,\mu)$, i.e. $\langle\langle\langle\Phi^{(m)},:D:^{\otimes m}\rangle^*\Phi,\varphi\rangle\rangle\equiv\langle\langle\Phi,\langle\Phi^{(m)},:D:^{\otimes m}\rangle\varphi\rangle\rangle$, $\forall\Phi\in(H_{-2})'_{-q,\gamma,\alpha}$, $\forall\varphi\in(H_2)_{q,\gamma,\alpha}$. We put

$$Q_m^{\gamma,\alpha}(\Phi^{(m)};\cdot):=\sum_{k=m}^{\infty}\frac{1}{(k-m)!}\langle\Phi^{(m)}\widehat{\otimes}\widetilde{\gamma}_{k-m},:D:^{\otimes k}\rangle^*\delta_e,\ m\in\mathbb{Z}_+,$$

where $\widetilde{\gamma}_{k-m}\in H_{-2,\mathbb{C}}^{\widehat{\otimes}(k-m)}$ are defined by $\frac{1}{\gamma(\theta)}=\sum_{n=0}^{\infty}\frac{1}{n!}\langle\widetilde{\gamma}_n,\theta^{\otimes n}\rangle$.

REMARK 3.2. It is easy to see that for $\gamma\equiv 1$ we have $Q_m^{1,\alpha}(\Phi^{(m)};\cdot)=\langle\Phi^{(m)},:D:^{\otimes m}\rangle^*\delta_e$, which corresponds to the corresponding definitions in [10,11].

THEOREM 3.3. *1. The system of generalized functions $Q_m^{\gamma,\alpha}(\Phi^{(m)};\cdot)$ is biorthogonal to the system of generalized Appell characters in the sense that*

$$\langle\langle Q_m^{\gamma,\alpha}(\Phi^{(m)};\cdot), \langle P_n^{\gamma,\alpha}(\cdot), \varphi^{(n)}\rangle\rangle\rangle = \delta_{mn} n! \langle \Phi^{(n)}, \varphi^{(n)}\rangle. \tag{3.1}$$

2. A generalized function $\Phi \in (H_{-2})'_{-q,\gamma,\alpha}$ if and only if there exists a sequence $\{\Phi^{(m)}\}_{m=0}^{\infty}$, $\Phi^{(m)} \in H_{-2,\mathbb{C}}^{\widehat{\otimes} m}$ such that $\Phi(\cdot) = \sum_{m=0}^{\infty} Q_m^{\gamma,\alpha}(\Phi^{(m)};\cdot)$ and $\|\Phi\|^2_{-q,\gamma,\alpha} := \sum_{m=0}^{\infty} K^{-qm} |\Phi^{(m)}|^2_{-2} < \infty$.

The proof of this Theorem consists in direct calculations, one can find analogous proofs (in particular cases) in [8,10]. Note that the norm $\|\cdot\|_{-q,\gamma,\alpha}$ introduced in the Theorem is Hilbert one and the system of generalized functions $Q_m^{\gamma,\alpha}(\Phi^{(m)};\cdot)$ plays a role of orthogonal bases in $(H_{-2})'_{-q,\gamma,\alpha}$.

EXAMPLE 3.4. (*'Classical' Gaussian analysis*) Let μ be the Gaussian measure on H_{-2}, i.e. the measure with the Laplace transform $l_\mu(\theta) = \int_{H_{-2}} e^{\langle x,\theta\rangle}\mu(dx) = e^{(1/2)\langle\theta,\theta\rangle}$, $h(x;\theta) = e^{\langle x,\theta\rangle}$, $\alpha = id$, $\gamma(\theta) = 1/l_\mu(\theta)$. Then $\langle P_n^{\gamma,id}(x), \varphi^{(n)}\rangle$ are the generalized Hermite polynomials, $Q_m^{\gamma,id}(\Phi^{(m)};\cdot)$ are the generalizations of the generalized Hermite polynomials (in the sense that for 'smooth' $\Phi^{(m)} \in H_{2,\mathbb{C}}^{\widehat{\otimes} m}$ $Q_m^{\gamma,id}(\Phi^{(m)};x) = \langle P_m^{\gamma,id}(x), \Phi^{(m)}\rangle$ is the usual generalized Hermite polynomial), and the biorthogonality relation (3.1) is the natural generalization of the well known orthogonality relation for the Hermite polynomials. The spaces $(H_2)_{q,\gamma,id}$ and $(H_{-2})'_{-q,\gamma,id}$ now are known as Kondratiev test and generalized functions spaces, see e.g. [5,6].

Now it is possible to construct the so–called *integral S–transform* on $(H_{-2})'$ putting for $\Phi \in (H_{-2})'$ $(S\Phi)(\theta) := \langle\langle \Phi, \gamma(\theta)h(\cdot;\alpha(\theta))\rangle\rangle$. Then one can introduce the so–called *Wick product* on $(H_{-2})'$, putting for $\Phi, \Psi \in (H_{-2})'$ $\Phi\diamond\Psi := S^{-1}(S\Phi S\Psi) \in (H_{-2})'$; and *Wick version of holomorphic at* $(S\Phi)(0)$, $\Phi \in (H_{-2})'$ *function* $F : \mathbb{C} \to \mathbb{C}$ putting $F^{\diamond}(\Phi) := S^{-1}F(S\Phi) \in (H_{-2})'$. These objects one can apply, for example, to study a wide class of generalized stochastic equations, see e.g. [16] for details.

REMARK 3.5. For fixed α let us consider a function $\widetilde{\gamma}(\theta) := \int_X h(x;\alpha(\theta))\mu(dx)$, $\theta \in H_{1,\mathbb{C}}$. It is not hard to prove (by the full analogy with [10]) that this function is holomorphic at $0 \in H_{1,\mathbb{C}}$. It is obvious that $\widetilde{\gamma}(0) = 1 \neq 0$. So we can use $\widetilde{\gamma}$ as γ. Note that $\widetilde{\gamma}$ plays a very important role in the biorthogonal analysis. For example, for $\gamma = \widetilde{\gamma}$ we have $(S1)(\cdot) \equiv 1$ (and $Q_0^{\widetilde{\gamma},\alpha} \equiv 1$). This implies that for $\Phi, \Psi \in \mathbb{C}$ we have $\Phi \diamond \Psi = \Phi\Psi$, i.e. 'from stochastic point of view' the Wick product of nonrandom elements coincides with its usual product. Note also that for $\Phi^{(m)} \in H_{-2,\mathbb{C}}^{\widehat{\otimes} m}$ we have $Q_m^{\widetilde{\gamma},\alpha}(\Phi^{(m)};\cdot) = \langle \Phi^{(m)}, :D:^{\otimes m}\rangle^* 1$ (cf. [5,6,8,10]).

Now we will construct orthogonal bases in $(H_{-2})'_{-q,\gamma,\alpha}$ in a different way. Let us consider a chain of the type (2.3)

$$H_2 \subseteq H_1 \subseteq H_0^{\sigma} \subseteq H_{-1}^{\sigma} \subseteq H_{-2}^{\sigma}$$

and the operator $\langle \Phi_\sigma^{(m)}, :D:^{\otimes m}\rangle_\sigma$, $\Phi_\sigma^{(m)} \in H_{-2,\mathbb{C}}^{\sigma\widehat{\otimes} m}$ introduced in Remark 2.15. Using this operator instead of $\langle \Phi^{(m)}, :D:^{\otimes m}\rangle$, we put

$$Q_m^{\sigma,\gamma,\alpha}(\Phi_\sigma^{(m)};\cdot) := \sum_{k=m}^{\infty} \frac{1}{(k-m)!} \langle \Phi_\sigma^{(m)} \widehat{\otimes} \widetilde{\gamma}_{k-m}^{\sigma}, :D:^{\otimes k}\rangle_\sigma^* \delta_e, \ m \in \mathbb{Z}_+,$$

where $\widetilde{\gamma}^{\sigma}_{k-m} \in H^{\sigma\widehat{\otimes}(k-m)}_{-2,\mathbb{C}}$ is defined by $\frac{1}{\gamma(\theta)} = \sum_{n=0}^{\infty} \frac{1}{n!}\langle \widetilde{\gamma}^{\sigma}_{n}, \theta^{\otimes n}\rangle_{\sigma}$.

For generalized functions $Q^{\sigma,\gamma,\alpha}_{m}(\Phi^{(m)}_{\sigma};\cdot)$ with kernels $\Phi^{(m)}_{\sigma} \in H^{\sigma\widehat{\otimes}m}_{-2,\mathbb{C}}$ the full analog of Theorem 3.3 holds. For example, a biorthogonality relation has now the form

$$\langle\!\langle Q^{\sigma,\gamma,\alpha}_{m}(\Phi^{(m)}_{\sigma};\cdot), \langle P^{\gamma,\alpha}_{n}(\cdot), \varphi^{(n)}\rangle\rangle\!\rangle = \delta_{mn} n! \langle \Phi^{(n)}_{\sigma}, \varphi^{(n)}\rangle_{\sigma}.$$

REMARK 3.6. Let N be a real separable nuclear Fréchet space, densely embedded in a real separable Hilbert space H_0 by nuclear operator. Then, as it is well–known (see e.g. [12]), there exists a family of real separable Hilbert spaces $H_1, H_2, \ldots$ such that $N = \mathrm{pr}\lim_{p\in\mathbb{N}} H_p$ and $H_{p+1} \hookrightarrow H_p$ $(p \in \mathbb{Z}_+)$ by the Hilbert–Schmidt type operator. So one can consider a chain

$$N \subseteq H_p \subseteq H_0 \subseteq H_{-p} \subseteq N',$$

where H_{-p} is dual to H_p space, $N' = \mathrm{ind}\lim_{p\in\mathbb{N}} H_{-p}$ is dual to N. Now one can construct the biorthogonal analysis using N instead of H_2, see e.g. [6,8].

Finally, some examples will be considered.

EXAMPLE 3.7. (*Poisson analysis on Schwartz space*) Let $\mathcal{S}$ be a Schwartz space of quickly decreasing functions on $\mathbb{R}^d$ $(d \in \mathbb{N})$, $\mathcal{S}'$ be dual to $\mathcal{S}$ with respect to $L^2(\mathbb{R}^d)$ space. One can consider a Poisson measure π on $\mathcal{S}'$, which can be defined via Minlos theorem (e.g. [15]) by its Laplace transform

$$l_\pi(\theta) = \int_{\mathcal{S}'} e^{\langle x,\theta\rangle}\pi(dx) = \exp\{\langle 1, e^{\theta} - 1\rangle\},\ \theta \in \mathcal{S}_{\mathbb{C}}.$$

We consider the so–called *Charlier polynomials* $\langle C_n(x), \varphi^{(n)}\rangle$ $(\varphi^{(n)} \in \mathcal{S}^{\widehat{\otimes}n}_{\mathbb{C}})$ with generating function $e^{\langle x,\alpha(\theta)\rangle}/l_\pi(\alpha(\theta))$ (cf. Remark 3.5), where $\alpha(\theta) = \log(1+\theta)$, i.e.

$$e^{\langle x,\log(1+\theta)\rangle - \langle 1,\theta\rangle} = \sum_{n=0}^{\infty} \frac{1}{n!}\langle C_n(x), \theta^{\otimes n}\rangle.$$

It is well-known (see e.g. [7] and references therein) that Charlier polynomials are orthogonal in $L^2(\mathcal{S}', \pi)$, i.e.

$$\int_{\mathcal{S}'} \langle C_n(x), \varphi^{(n)}\rangle \langle C_m(x), \psi^{(m)}\rangle \pi(dx) = \delta_{mn} n! \langle \varphi^{(n)}, \psi^{(n)}\rangle,$$

where $\varphi^{(n)} \in \mathcal{S}^{\widehat{\otimes}n}_{\mathbb{C}}$, $\psi^{(m)} \in \mathcal{S}^{\widehat{\otimes}m}_{\mathbb{C}}$. So one can use them as orthogonal bases for test and generalized functions spaces of Poisson analysis. Note that if we change γ and α then the corresponding polynomials will not be orthogonal in $L^2(\mathcal{S}', \pi)$. But, nevertheless, one can use a biorthogonal approach and 'limit spaces' $(\mathcal{S})$ and $(\mathcal{S}')'$ of test and generalized functions will be invariant with respect to γ and α.

EXAMPLE 3.8. (*Poisson analysis on configuration space*) Let σ be a non–atomic σ–finite measure on $\mathcal{B}(\mathbb{R}^d)$. We denote by $\mathcal{D} := \mathcal{D}(\mathbb{R}^d)$ and by $\mathcal{D}' := \mathcal{D}(\mathbb{R}^d)'$ the classical Schwartz space and its dual one respectively. Following [17], we introduce a set $\widetilde{\Gamma}$ of locally finite configurations on $\mathbb{R}^d$:

$$\widetilde{\Gamma} := \{\widetilde{x} \subset \mathbb{R}^d :\ |\widetilde{x} \cap K| < \infty \text{ for any compact } K \subset \mathbb{R}^d\},$$

where $|A|$ denotes the cardinality of A. One can identify $\widetilde{\Gamma}$ with a set Γ of measures on $\mathcal{B}(\mathbb{R}^d)$:

$$\widetilde{\Gamma} \ni \widetilde{x} \leftrightarrow \sum_{u\in\widetilde{x}} \delta_u =: x \in \Gamma,$$

where δ_u is the Dirac measure. Let us endow Γ with vague topology ([17]). Namely, we say that a sequence of measures $\mu_n \in \Gamma$, $n \in \mathbb{N}$ converges to $\mu \in \Gamma$ if and only if for any $f \in C_0(\mathbb{R}^d)$ (the set of all continuous functions on $\mathbb{R}^d$ with compact support) we have

$$\int_{\mathbb{R}^d} f(u)\mu_n(du) \to \int_{\mathbb{R}^d} f(u)\mu(du) \text{ as } n \to \infty.$$

So for any $f \in C_0(\mathbb{R}^d)$ one can consider a continuous functional

$$\Gamma \ni x \mapsto \langle x, f\rangle := \int_{\mathbb{R}^d} f(u)x(du) = \sum_{u\in\widetilde{x}} f(u) \in \mathbb{R}. \tag{3.2}$$

Conversely, functionals (3.2) generate topology on Γ. Therefore we can say that $\Gamma \subset \mathcal{D}'$ and dual pairing between elements of $\mathcal{D}$ and Γ is given by (3.2).

We define a Poisson measure π_σ (with intensity measure σ) on $(\Gamma, \mathcal{B}(\Gamma))$ by its Laplace transform

$$l_{\pi_\sigma}(\theta) = \int_\Gamma e^{\langle x,\theta\rangle}\pi_\sigma(dx) = \exp\{\langle 1, e^\theta - 1\rangle_\sigma\},\ \theta \in \mathcal{D},$$

where $\langle\cdot,\cdot\rangle_\sigma$ denotes dual pairing between test and generalized functions in a chain $\mathcal{D}^{\widehat{\otimes}n} \subset L^2(\mathbb{R}^d,\sigma)^{\widehat{\otimes}n} \subset \mathcal{D}_\sigma^{'\widehat{\otimes}n}$, $n \in \mathbb{Z}_+$ generated by the scalar product in $L^2(\mathbb{R}^d,\sigma)^{\widehat{\otimes}n}$. In particular, $\langle 1, e^\theta - 1\rangle_\sigma = \int_{\mathbb{R}^d}(e^{\theta(u)} - 1)\sigma(du)$.

Introducing a generating function of Charlier polynomials by the analogy with the previous example we have

$$e^{\langle x,\log(1+\theta)\rangle - \langle 1,\theta\rangle_\sigma} = \sum_{n=0}^\infty \frac{1}{n!}\langle C_n^\sigma(x), \theta^{\otimes n}\rangle,\ \theta \in \mathcal{D},\ x \in \Gamma.$$

An orthogonality relation now has the form ([17])

$$\int_\Gamma \langle C_n^\sigma(x), \varphi^{(n)}\rangle\langle C_m^\sigma(x), \psi^{(m)}\rangle\pi_\sigma(dx) = \delta_{mn}n!\langle\varphi^{(n)},\psi^{(n)}\rangle_\sigma, \tag{3.3}$$

$\varphi^{(n)} \in \mathcal{D}^{\widehat{\otimes}n}$, $\psi^{(m)} \in \mathcal{D}^{\widehat{\otimes}m}$. Using $\langle C_n^\sigma(\cdot), \varphi^{(n)}\rangle$ as the orthogonal bases in the spaces of test and generalized functions one can construct a Poisson analysis on configuration space Γ. Note that due to (3.3) it is convenient now to construct the orthogonal bases in the generalized functions spaces for general γ and α by the operators $\langle\Phi_\sigma^{(m)}, : D :^{\otimes m}\rangle_\sigma^*$ (see Remark 2.15, $H_0^\sigma = L^2(\mathbb{R}^d,\sigma)$) in order to obtain the coincidence of the bases in the test and generalized functions spaces for special γ and α, when $P_n^{\gamma,\alpha}(x) = C_n^\sigma(x)$.

EXAMPLE 3.9. (*The biorthogonal analysis connected with symmetric Wiener process*) Let us consider on Schwartz space $\mathcal{S}'(\mathbb{R}) \equiv \mathcal{S}'$ the measure μ which can be characterized by its Hankel–type transform

$$l_\mu^{\Lambda_s}(\theta) = \int_{\mathcal{S}'} \Lambda_s(\langle x,\theta\rangle)\mu(dx) = e^{-a\langle\theta,\theta\rangle},\ s \geq -\frac{1}{2},\ a > 0,\ \theta \in \mathcal{S},$$

where Λ_s was defined by (2.1). One can prove that such a measure is correctly defined and satisfies all demands formulated above (now $h(x;\theta) = \Lambda_s(\langle x,\theta\rangle)$). Let us consider μ as a probability measure on the probability space $(\mathcal{S}', \mathcal{B}(\mathcal{S}'), \mu)$. Then the random process $Z_t(\cdot) := \langle\cdot, 1_{[0,t)}\rangle$ (defined as $L^2(\mathcal{S}',\mu)$–limit of $\langle\cdot,\varphi_n\rangle$, $\mathcal{S}_\mathbb{C}^{\widehat{\otimes}n} \ni \varphi_n \to 1_{[0,t)}$ as $n \to \infty$ in $L^2(\mathbb{R})$) is the so–called *symmetric Wiener process* ([18]). Such a process plays the important role in the theory of random walks with spherical symmetry (see [18]). So, it is natural to construct the analysis connected with μ (which is the 'white noise measure' for Z_t, so as the Gaussian measure is the white noise measure for the Wiener process). Since it is impossible, unfortunately, to obtain the explicit expression for the Laplace transform of μ, our considerations (which allow us to use Hankel–type transform instead of Laplace one) are useful and convenient now.

References

1. E. W. Lytvynov, *Multiple Wiener integrals and non–Gaussian white noise: a Jacobi field approach*, Methods Funct. Anal. Topol., Vol. 1, No. 1 (1995), 61–85.
2. Yu. M. Berezansky, *Commutative Jacobi fields in Fock space*, Integr. Equ. Oper. Theory 30 (1998), 163–190.
3. Yu. L. Daletsky, *A Biorthogonal Analog of the Hermite Polynomials and the Inversion of the Fourier Transform with respect to a Non-Gaussian Measure*, Funct. Anal. and Appl., Vol. 25, No. 2 (1991), 68–70 (Russian).
4. S. Albeverio, Yu. G. Kondratiev and L. Streit, *How to generalize White Noise Analysis to non–Gaussian spaces*, In: 'Dynamics of Complex and Irregular Systems', Eds.: Ph. Blanchard et al., World Scientific (1993).
5. S. Albeverio, Yu. L. Daletsky, Yu. G. Kondratiev and L. Streit, *Non-Gaussian infinite dimensional analysis*, J. Funct. Anal. Vol. 138, No. 2 (1996), 311–350.
6. Yu. G. Kondratiev, L. Streit, W. Westercamp and J. Yan, *Generalized functions in infinite dimensional analysis*, IIAS Reports No. 1995–002, Kyoto (1995).
7. G. F. Us, *Dual Appell systems in Poissonian analysis*, Methods Funct. Anal. Topol., Vol. 1, No. 1 (1995), 93–108.
8. N. A. Kachanovsky, *Biorthogonal Appell–like systems in a Hilbert space*, Methods Funct. Anal. Topol., Vol. 2, No. 3–4 (1996), 36–52.
9. Yu. M. Berezansky and Yu. G. Kondratiev, *Biorthogonal systems in hypergroups: an extension of non–Gaussian analysis*, Methods Funct. Anal. Topol., Vol. 2, No. 2 (1996), 1–50.
10. Yu. M. Berezansky, *Infinite dimensional analysis, related to generalized translation operator*, Ukr. Math. J., Vol. 49, No. 3 (1997), 364–409 (Russian).
11. N. A. Kachanovsky, *Dual Appell–like systems and finite order spaces in non–Gaussian infinite–dimensional analysis*, Methods Funct. Anal. Topol., Vol. 4, No. 2 (1998), 41–52.
12. I. M. Gel'fand and N. Ya. Vilenkin, *Generalized Functions, Vol. IV*, Academic Press, New York and London (1968).
13. N. A. Kachanovsky and S. V. Koshkin, *Minimality of Appell-like systems and embeddings of test function spaces in a generalization of White Noise Analysis*, Methods Funct. Anal. Topol., Vol. 5, No. 3 (1999), 13–25.
14. N. A. Kachanovsky, *Pseudodifferential equations and generalized translation operator in non–Gaussian infinite dimensional analysis*, Ukr. Math. J., Vol. 51, No. 10 (1999), 1334–1341 (Russian).
15. A. V. Skorokhod, Integration in Hilbert Space, Springer, Berlin (1974).
16. H. Holden, B. Oksendal, J. Uboe and T. S. Zhang, *Stochastic Partial Differential Equations*, Birkhäuser (1996).
17. Yu. G. Kondratiev, J. L. Silva, L. Streit and G. F. Us, *Analysis on Poisson and Gamma spaces*, Probab. and Related Topics, Vol. 1 (1998), 91–117.
18. J. F. C. Kingman, *Random walks with spherical symmetry*, Acta Mathematica zeitschrift, Vol. 109, No. 1–2 (1963), 11–53.

Institute of Mathematics,
3, Tereshchenkivs'ka Street, 252601, GSP, Kiev – 4, UKRAINE
E-mail address: nkachano@scientist.com

Canadian Mathematical Society
Conference Proceedings
Volume **29**, 2000

Twist Maps, Kicked Rotors, and Quantum Chaos

Gunther KARNER

ABSTRACT. This article describes a rigorous method of investigating formal objects such as

$$\mathcal{K}_{\kappa V} = -i\partial/\partial t - \partial^2/\partial x^2 + \kappa V(x)\sum_j \delta(t - j\Gamma),\ \Gamma, \kappa \geqslant 0,\ x \in S^1$$

which represent "kicked quantum rotors". Corresponding self-adjoint realizations $K_{\kappa V}$ on an extended Hilbert space are determined and related to the Floquet operators $U_{\kappa V}(0,\Gamma)$ of these quantum systems. The main findings are : (i) $\sigma(K_{\kappa V}) = \overline{\sigma_{pp}(K_{\kappa V})}$ for $V \in \mathcal{C}^2(S^1)$, $\kappa \geqslant 0$ and Γ typically diophantine. (ii) The form of the trajectories $U_{\kappa V}(0, k\Gamma)\psi_0$, $\psi_0 \in \mathcal{L}^2(S^1)$, $k \in \mathsf{N}$, depends on the smoothness of the eigenfunctions to $K_{\kappa V}$ and number theoretic properties of the coupling κ. Based on that information, quantum and classical dynamics (KAM, Aubry-Mather theorems) are compared and common features are highlighted.

1. Quantum Kicked Rotors

The stability problem for Hamiltonian systems has been in the focus of enormous research activities since the groundbreaking work of Poincaré on the restricted three-body problem. Subsequent milestones were Birkhoff's contribution to the understanding of periodic orbits, the celebrated KAM-theorem, which clarified the meaning of "small perturbations" of an integrable Hamiltonian system, and finally the studies of Aubry and Mather on the breakdown of stability for "larger perturbations". (Among the innumerable amount of literature, see [**1**] for a review.)

Based on this knowledge, the following problem emerged naturally : Replace the phase space variables by "corresponding" operators on some Hilbert space to generate a quantum dynamical system. Does the latter possess the same dynamical characteristics as the original classical system or is the quantum dynamics qualitatively different ? Expressed in a different phrase, the question is : *Does "quantum chaos" exist ?*

Among the various scenarios studied are so-called "kicked" quantum systems, which consist of a stationary ($\equiv$ integrable) Hamiltonian under time-periodic external perturbations acting in form of "δ−kicks". (See below for more details.) Although numerical results for theses problems were available early on, for example [**2**], the number of analytic discussions is still rather limited [**3, 4, 5, 6**], for

1991 *Mathematics Subject Classification.* Primary 35P05; Secondary 81Q50.

instance. The present article is an outline of some recent work of the author and intended as a contribution to the rigorous understanding of kicked operator systems and the relationship to their classical analogons. Most of the statements, however, are presented without proofs due to the space limitations. Yet, all the details can be found in [**6, 7, 8**].

The subject of the present study are one-dimensional quantum systems under time-periodic external perturbations, which act in form of $\delta(t - n\Gamma)$-pulses, where δ is the Dirac delta distribution, $n \in \mathsf{Z}$ and the period $\Gamma > 0$. These systems are described in a *formal* manner by periodic families $\{\eta(t), t \in \mathsf{R}\}$ of Hamiltonians, where

$$\eta(t) = H_0 + W \cdot \sum_{j \in \mathsf{Z}} \delta(t - j\Gamma) \tag{1.1}$$

with a (discrete) self-adjoint and time-constant part H_0 and a suitable real-valued multiplication W acting on $\mathcal{L}^2(\Omega)$. Integration of the corresponding Schrödinger equation over a full period Γ provides the Floquet operator to the family $\{\eta(t), t \in \mathsf{R}\}$ as

$$U_W(0, \Gamma) = \exp(-i\Gamma H_0) \exp(-iW). \tag{1.2}$$

Hence, the physical effect of a δ-pulse at $t = j\Gamma$ is a discontinuity in the otherwise free time evolution of the system represented by H_0. Remark that (1.1) and therefore the Schrödinger equation *cannot* be furnished with a rigorous meaning. Nevertheless, the form of $U_W(0, \Gamma)$ is appealing from a physical point of view and therefore often used as the defining property of kicked quantum systems [**9**], which leads to the notion of quantum twist maps as discussed in [**6**].

In the opinion of the author, however, the most suitable method for a sound discussion of the system (1.1), (1.2) is an adapted version of the Howland-Yajima formalism, [**10**] for the latter. It has been first applied to kicked systems in [**6**] and the strategy for the implementation of the $\delta-$ kicks is derived from the work of *S. Albeverio et al* [**11**] on spatial point interactions :

Restrict the time variable to the period $\pi_\Gamma = [0, \Gamma]$, introduce an extended Hilbert space $\mathcal{H} = \mathcal{L}^2(\pi_\Gamma, ds) \otimes \mathcal{L}^2(\Omega, dx)$ and on it a *formal* kicked Floquet Hamiltonian $k_W = (-i\partial/\partial s + W \cdot \delta(s - \Gamma)) + 1 \otimes H_0$ as well as a "free" Floquet Hamiltonian

$$K_0 = \overline{-i\partial_s \otimes 1 + 1 \otimes H_0}. \tag{1.3}$$

Here $-i\partial_s$ is defined as the derivative on suitable Γ-periodic functions, which renders K_0 self-adjoint. Thus, the δ-kicks enter the formalism as a perturbation of the time derivative rather than the spatial part H_0, which provides a set-up just as for spatial point interactions [**11**]. In analogy to the latter, the rigorous accomodation of the $\delta-$ kicks is in form of a self-adjoint *singular* time derivative $-i\partial_{s,W}$

[6]:

$$\mathcal{D}(-i\partial_{s,W}) = \Big\{ f \in \mathcal{H} : f(\cdot, x) \in \mathcal{D}\left(-i\partial_{s,W(x)}\right) a.e.,$$

$$\mathcal{D}\left(-i\partial_{s,W(x)}\right) = \{ g \in \mathcal{L}^2(\pi_\Gamma) : \partial g/\partial s \in \mathcal{L}^2(\pi_\Gamma),$$

$$g(0) = \exp(-iW(x))\, g(\Gamma)\} \Big\},$$

$$-i\partial_{s,W} f := -i\frac{\partial f}{\partial s}. \tag{1.4}$$

That provides the Floquet Hamiltonian K_W for the kicked system (1.1) as the closure of

$$\overset{\circ}{K}_W = -i\partial_{s,W} + 1 \otimes H_0, \tag{1.5}$$

defined on $\mathcal{D}(-i\partial_{s,W}) \cap \mathcal{D}\left(\overline{1 \otimes H_0}\right)$. As shown in [6], K_W is self-adjoint under the condition $W \in \mathcal{D}(H_0)$. However, since the main interest in this paper is focused on the principle rather than the ultimate technical generality, the operators H_0, W and Ω are chosen as simple as possible, namely

$$H_0 = -\triangle_x,\ \Omega = S^1, W \in \mathcal{C}^2(S^1)$$

$$\mathcal{D}(H_0) = \Big\{ g \in \mathcal{H}^{2,1}(S^1) \cap \mathcal{H}^{2,2}(S^1 \backslash \{0\}) :$$

$$g(0_+) = g(2\pi_-), g_x(0_+) = g_x(2\pi_-) \Big\}. \tag{1.6}$$

Then $(K_0 - z)^{-1}$, defect functions, etc. are explicitly known and therefore a more detailed analysis of the nature of $\sigma(K_W)$ is possible. The main feature, which makes K_W worth studying, is its relation to the Floquet operator $U_W(0, \Gamma)$ in form of:

$$\exp[-i(k\Gamma) K_W] \Psi(\Gamma) = U_W(0,\Gamma)^k \Psi(\Gamma) =: U_W(0, k\Gamma) \Psi(\Gamma) \tag{1.7}$$

for all $\Psi \in \mathcal{D}(K_W)$ and $k \in \mathsf{Z}$, see [6] for details. Therefore, the spectrum of $U_W(0,\Gamma)$, which is characteristic for the long-term dynamics of the system (1.1), see for example the RAGE-methods developed in [6], can be determined by the computation of $\sigma(K_W)$. To achieve that goal, another item extensively employed for spatial point interactions turns out to be very helpful : *Krein's formula* for the resolvent difference of two self-adjoint realizations of a given symmetric operator [11].

In the present case, Krein's resolvent formula for the operators K_W and K_0 reads in strong sense

$$(K_W - z)^{-1} - (K_0 - z)^{-1} = \sum_{m \in \mathsf{Z}} \left\langle \sum_{n \in \mathsf{Z}} \overline{\lambda_{mn}(z, W)} \Phi_n(\overline{z}, \Gamma), \cdot \right\rangle_{\mathcal{H}} \Phi_m(z, \Gamma) \tag{1.8}$$

for all $z \in \mathsf{C} \backslash \mathsf{R}$. The numbers $\lambda_{mn}(z, W)$ are uniquely determined by the domain properties of K_W, respectively K_0 and the choice of the orthonormal set $\mathcal{S}(z)$, here given by

$$\mathcal{S}(z) = \left\{ \Phi_n(z, \Gamma; s, x) = ic(z, \Gamma) \exp\left[i\left(z - n^2\right) s\right] \exp(inx), n \in \mathsf{Z} \right\} \tag{1.9}$$

with normalization

$$c(z,\Gamma) = \left[\Im z\left(\pi - \pi \exp\left(-2\Gamma \Im z\right)\right)^{-1}\right]^{1/2}.$$

For the determination of the spectral type of K_W in *Theorem 1.1* below the "standard version" (1.8) of Krein's formula suffices. To obtain a more detailed information on $\sigma_{pp}(K_W)$, however, a modified version of (1.8) is used. It has been first introduced in [**12**], see [**13**] as well for a more recent contribution, and its derivation is a "variation of constant" in Krein's formula for $K_{W=\kappa}$. In particular, for the constant kick $W = \kappa$ the $\lambda_{mn}(z,\kappa)$ can be explicitly computed [**12**]. The replacement

$$\kappa \mapsto \kappa V = W$$

$$\lambda_{mn}(z,\kappa) \mapsto \begin{cases} \Lambda_{nn}(z,W) := \lambda_{nn}(z,W) \\ \Lambda_{mn}(z,W) = 0, \ m \neq n \end{cases} \tag{1.10}$$

of course destroys the resolvent equality in first place. Yet it turns out that an explicit correction term can be computed to recover $(K_W - z)^{-1}$. The form of the equation used is

$$(K_W - z)^{-1}\left[1 + D^2\Lambda(z,W)\right] = (K_0 - z)^{-1} + \Lambda(z,W) \tag{1.11}$$

with

$$\Lambda(z,W) = \sum_{n\in Z}\left(1 \otimes \Lambda_{nn}(z,W)\right)\langle\Phi_n(\overline{z},\Gamma),\cdot\rangle_{\mathcal{H}}\,\Phi_n(z,\Gamma),$$

$$D^2\Lambda(z,W) = \sum_{n\in \mathsf{Z}}\Big(1\otimes\left(-\partial^2/\partial x^2 - 2in\partial/\partial x\right)\Lambda_{nn}(z,W)\Big)\langle\Phi_n(\overline{z},\Gamma),\cdot\rangle_{\mathcal{H}}\,\Phi_n(z,\Gamma),$$

$$\Lambda_{nn}(z,W) = i\Gamma\left(1 - \exp(iW)\right) \times \tag{1.12}$$

$$\times\left[\left(\exp[-i\left(z-n^2\right)\Gamma] - 1\right)\left(\exp[i\left(z-n^2\right)\Gamma] - \exp(iW)\right)\right]^{-1}.$$

(Details are in the upcoming article [**8**] and [**12, 13**].)

Together with the limiting absorption, i.e. $\lim\left\langle\Psi,(K_W - z)^{-1}\Psi\right\rangle_{\mathcal{H}}$ for $\Im z \to 0$, and de la Vallèe Poussin's theorem, [**14**] for example, the above tools are sufficient to shed light on $\sigma(K_W)$ and $\sigma_{pp}(K_W)$.

In this regard, the main results available so far are the following:

THEOREM 1.1. *Let $H_0 = -\Delta_x$ and $W = \kappa V$ a real-valued multiplication on $\mathcal{L}^2(S^1)$ in the model (1.1), with $\kappa \geq 0$ and $V \in \mathcal{C}^2(S^1)$ such that $V_x \neq 0$ a.e. Then $U_W(0,\Gamma)$ as defined in (1.2) is*

(i) pure absolutely continuous for all $\kappa > 0$ and $\Gamma = 2\pi p/q$, $p, q \in \mathsf{N}$,

(ii) dense pure point for all $\kappa \geq 0$ and $\nu = \Gamma/2\pi$ a typically diophantine number.

The proof of *Theorem* 1.1 is a direct application of the corresponding statements in [**6, 7**] and some number theoretics involving the diophantine estimate, [**15**] for the latter. The details are repeated in [**8**].

Hence, from a number theoretic point of view, the "generic" situation is a pure point Floquet operator whereas the "exceptions", though dense, are pure quantum mechanical resonances. Intermediate cases, i.e. ν a Liouville irrational, presumably lead to singular continuous $U_W(0,\Gamma)$, see also [**5**].

It turns out that the smoothness of the eigenfunctions to K_W and the nature of $\sigma_{pp}(K_W)$ are intimately related :

THEOREM 1.2. *Let* $H_0 = -\Delta_x$, $W = \kappa V$, $V \in \mathcal{C}^2(S^1)$ *with* $V_x \neq 0$ *a.e. and* $\nu = \Gamma/2\pi$ *typically diophantine. Furthermore, assume that the eigenfunctions* $\Psi_\alpha(\kappa)$ *to* K_W *fulfill* $\Psi_\alpha(\kappa) \in \mathcal{C}^1(\pi_\Gamma) \times \mathcal{C}^{2+\varepsilon}(S^1)$ *for all* α, *some* $\varepsilon > 0$ *and* $\kappa \in [0, \widehat{\kappa}]$, $\widehat{\kappa}$ *sufficiently small. Then, for all* $0 \leq \kappa < \widehat{\kappa}$ *it follows that*

$$\sigma_{pp}(K_W) \subset \left\{2\pi j/\Gamma + n^2 + \kappa\widetilde{V_m}/\Gamma,\ m \in \mathsf{M}, (j,n) \in \mathsf{Z}^2\right\}.$$

Here the $\widetilde{V}_m$ *denote the local extrema of* V.

The proof of *Theorem* 1.2 relies on the modified Krein's formula (1.11) and the assumed smoothness of the eigenfunctions $\Psi_\alpha(\kappa)$. In fact, this "assumption" is a true statement as shown in [**8**], however, for didactical reasons the above formulation has been chosen. The smoothness of the $\Psi_\alpha(\kappa)$ follows from $\Psi_\alpha(\kappa = 0) \in \mathcal{C}^\infty(\pi_\Gamma) \times \mathcal{C}^\infty(S^1)$ and $(K_{\kappa V} - z)^{-1} \overset{s}{\to} (K_0 - z)^{-1}$, which is a consequence of (1.11) and (1.12).

An interesting implication, which relates as well to the classical model (as explained below), is the following property of "trajectories" $\{U_W(0, k\Gamma)\Phi_0, k \in \mathsf{N}\}$:

COROLLARY 1.3. *Define the kicked system as in Theorem 1.2 and assume that* $\kappa\widetilde{V_m} = \left(\frac{p}{q}\right)_m \Gamma$ *for all* $m \in \mathsf{M}$. *Then the sets* $T_W(\Phi_0) := \{U_W(0, j\Gamma)\Phi_0, j \in \mathsf{N}\} \subset \mathcal{L}^2(S^1)$ *are closable and the* $\overline{T_W(\Phi_0)}$ *are invariant under* $U_W(0,\Gamma)$ *for all* $\Phi_0 \in \mathcal{L}^2(S^1)$.

PROOF. As $\nu \in \mathsf{R}\backslash\mathsf{Q}$, $\{\exp[2\pi i k\nu], k \in \mathsf{N}\}$ is dense on S^1, for instance [**15**]. Thus, for proper subsequences $\{k_j; k, j \in \mathsf{N}\}$ it follows that

$$\left|\left[\exp\left[2\pi i\nu\left(k_{j,r} - k_{j,s}\right)\right]\right]^{n^2 + \left(\frac{p}{q}\right)_m} - 1\right| \underset{r,s\to\infty}{\longrightarrow} 0\ \ \forall n \in \mathsf{Z}, m \in \mathsf{M}$$

i.e. together with the statements in *Theorem* 1.2 and uniform convergence that $\left\{U_W(0,\Gamma)^{k_j}\Phi_0\right\}$ is Cauchy for all $\Phi_0 \in \mathcal{L}^2(S^1)$. Denote its limit by $f(\{k_j\}, \Phi_0)$. Then $U_W(0,\Gamma) f(\{k_j\}, \Phi_0) = \lim_{j\to\infty} U_W(0,\Gamma)^{k_j+1}\Phi_0$ is contained in $\overline{T_W(\Phi_0)}$, since $\left\{U_W(0,\Gamma)^{k_j+1}\Phi_0\right\}$ is Cauchy by the above arguments as well. □

REMARK 1.4. In case of the "original rotor", where $V = \cos x$, the above assumption means $\kappa \in \left\{\frac{p}{q}\Gamma; p, q \in \mathsf{N}\right\}$, which excludes $\kappa = \frac{m}{n}\pi$, for instance. Note that irrational multiples of $\kappa\widetilde{V_k}\Gamma^{-1}$ might lead to certain (strongly) divergent sequences $\left\{U_W(0,\Gamma)^{k_j}\Phi_0\right\}$. The existence of number theoretical restrictions on the coupling κ and the period Γ are important for the comparison to the classical model as well, see below.

Corollary 1.3 is a "typical stability" result for small couplings since the evolution of an initial state Φ_0 is trapped in $\overline{T_W(\Phi_0)}$ for all times. Thus, the next question naturally imposing itself is : What about increasing the kick strength ?

The "Maryland model", which basically relates the Fourier transform of the eigenvalue equation for $U_W(0,\Gamma)$ to the Anderson localization in the theory of random Schrödinger operators, indicates that a significant change occurs in the decay rate of the Fourier coefficients of the eigenfunctions to $U_W(0,\Gamma)$, at least for $W = \kappa \cos x$ and $\kappa > \pi$. (See [**3**] fore more information.) For the time being, the author is not aware of any proof that the smoothness properties of the eigenfunctions indeed change for increasing kick strength. Assuming the latter, however, has exciting consequences:

THEOREM 1.5. *Define the kicked system as in Theorem 1.1, (ii) and assume that there are eigenfunctions $\Psi_\beta(\kappa)$ to K_W such that $\Psi_{\beta,xx}(\kappa;\cdot) \notin \mathcal{C}^2(S^1)$ for some $s \in \pi_\Gamma$. Then there exist $E_\beta(\kappa) \in \sigma_{pp}(K_W)$ such that*

$$E_\beta(\kappa) \notin \left\{2\pi j/\Gamma + n^2 + \kappa\widetilde{V_m}/\Gamma,\ m \in \mathsf{M}, (j,n) \in \mathsf{Z}^2\right\}$$

and the corresponding sets $\{U_W(0,k\Gamma)\,\Phi_0, k \in \mathsf{N}\}$ are not closable.

Again, the proof employs the modified Krein's formula and is part of [**8**]. The question, whether or not $\Psi_{\beta,xx}(\kappa;\cdot) \notin \mathcal{C}^2(S^1)$ for κ sufficiently large indeed occurs, is still under consideration.

2. Quantum and Classical Dynamics Compared

The discussion about the (non-)existence of chaotic effects in (kicked) quantum systems has been focused so far almost exclusively on the spectral type of the corresponding Floquet operators. In particular, point spectrum has been regarded as a "non-chaotic" feature with the implication that quantum and classical dynamics are qualitatively different in the classically chaotic regime. However, the spectral type of the quantum evolution operator is too crude a measure in the comparison of quantum and classical dynamics. As classical chaos refers to the fate of individual trajectories, the same "fine structure" has to be examined in the quantum models as well. In this sense, the results presented in *Section* 1 demonstrate a global similarity between quantum and classical dynamics, see [**6**] as well for a simplified model with $H_0 = -d/dx$. To underline that claim, a short review of the classical situation:

"Small" perturbations away from the integrable H_0 are described by the *KAM-theorem* [**1**]. Generically, elliptic fixed points of a twist map are surrounded by invariant curves, so-called KAM-tori, which are smooth and densely filled by the iterates (p_n, q_n) of the map. That again is a "typical stability" result since the orbits are trapped by the invariant curves. For increasing perturbation strength, however, the fixed points become unstable and the continuous curves disintegrate into Cantor sets (CAN-tori) through loss of smoothness. These so-called *Mather sets* are no longer invariant under the map, i.e. orbits may leak "through the gaps". Now stability has been lost and the motion has become chaotic.

According to *Theorems 1.2, 1.5* and *Corollary 1.3* the quantum rotor behaves similar: For *small coupling* κ the original, i.e. $\kappa = 0$, eigenvalues bifurcate and the local extrema of the kick potentials determine the new values. The same is true for the period-one fixed points of the classical maps, as a linearization of the quantum problem shows, see [**16**] for more. The quantum orbit structure is characterized by stability as the invariant sets $\overline{T_W(\Phi_0)}$, which play the role of the KAM-tori, prohibit large excursions in Hilbert space in the sense of the RAGE-theorem, [**6, 17**]. Also, number theoretic restrictions on the period Γ and the coupling κ, which are similiar to the classical conditions on the rotation numbers of the orbits [**1**], apply. Hence, generically, a KAM-type dynamics dominates for small couplings κ.

If the increase of the kick strength indeed goes along with the loss of the invariant sets $\overline{T_W(\Phi_0)}$, then the quantum scenario resembles the classical break-up of the KAM-tori into CAN-tori. Thus, further investigations have to decide about the (non-)occurrence of such a quantum transition, in particular, whether the possible loss of smoothness of the eigenfunctions is individual or uniform at a certain coupling strength. If such a quantum transition indeed exists, it would, at least in the opinion of the author, well deserve the name "route to quantum chaos".

References

[1] Moser, J.: *Recent developments in the theory of Hamiltonian systems,* SIAM Review **28**, 459-485 (1986)

[2] Casati, G., Guarneri, I.: *Chaos and special features of quantum systems under external perturbations,* Phys. Rev. Lett. **50**, 640-643 (1983)

[3] Bellissard, J.: *Stability and instability in quantum mechanics,* in *Trends and developments in the eighties*, S. Albeverio and Ph. Blanchard eds., Singapore, World Scientific 1985

[4] De Bièvre, S., Forni, G.: *Transport properties of kicked and quasi-periodic Hamiltonians,* J. Stat. Phys., to appear

[5] Geisler III, L., Howland, J.S.: *Spectra of quasienergies*, J. Math. Anal. Appl. **207**, 397-408 (1997)

[6] Karner, G.: *On quantum twist maps and spectral properties of Floquet Hamiltonians*, Ann. Inst. H. Poincaré A **68**, 139-157 (1998)

[7] Karner, G.: *The Schrödinger equation on non-stationary domains*, Elec. J. Diff. Eqns. **1998-20**, 1-20 (1998)

[8] Karner, G.: *On dynamical similiarities in quantum and classical twist maps,* preprint

[9] Combescure, M.: *Recurrent versus diffusive dynamics for a kicked quantum system,* Ann. Inst. H. Poincaré A **57**, 67-79 (1992)

[10] Cycon, H.L., Froese, R.G., Kirsch, W., Simon, B.: *Schrödinger operators*, New York, Springer 1987

[11] Albeverio, S., Gesztesy, F., Høegh-Krohn, R., Holden, H.: *Solvable models in quantum mechanics*, Heidelberg, Springer 1988

[12] Karner, G.: *De l'application standard et l'application Fermi en mécaniques quantique et classique*, Habilitation, Université de Toulon et du Var (1994)

[13] Duclos, P., Števiček, P., Váňa, O.: *About a resolvent formula,* in Proceedings of the conference "Rigorous Results in Quantum Mechanics", Dittrich, Exner and Tater eds., Zürich, Birkäuser 1999

[14] Reed, M., Simon, B.: *Methods of modern mathematical physics, vol 4*, New York, Academic Press 1978

[15] Kuipers, L., Niedereiter, H.: *Uniform distribution of sequences,* New York, Wiley 1974

[16] Karner, G.: *A note on the quantum standard map and its linearization*, Nonlinearity **7**, 623-631 (1994)

[17] Enss, V., Veselic, K.: *Bound states and propagating states for time-dependent Hamiltonians,* Ann. Inst. H. Poincaré A **39**, 159-191 (1983)

Institut für Kerntechnik und Reaktorsicherheit, Universität Karlsruhe (TH), Postfach 3640, D - 76021 KARLSRUHE, Germany

E-mail address: karner@irs.fzk.de

Canadian Mathematical Society
Conference Proceedings
Volume **29**, 2000

The Generalized Laplace Operator in $L_2(\mathbb{R}^n)$

W. Karwowski and V. Koshmanenko

Dedicated to Sergio Albeverio on the occasion of his 60th birthday.

ABSTRACT. By definition the generalized Laplace operator, $-\tilde{\Delta}$, is a self-adjoint operator coinciding with the usual Laplace operator $-\Delta$ on a dense subset in the Hilbert space. We show that each $-\tilde{\Delta}$ with the domain in the Sobolev space W_2^1 under natural conditions admits the representation in the form of a generalized sum, $-\tilde{\Delta} = -\Delta \tilde{+} V$, where V is a singular "potential" acting in the Sobolev scale, from W_2^1 to W_2^{-1}.

1. Introduction

Let $-\Delta = -(\frac{d^2}{dx_1^2} + \frac{d^2}{dx_2^2} + \cdots + \frac{d^2}{dx_n^2})$, $n \geq 1$ be the usual positive self-adjoint Laplace operator in the Hilbert space $L_2(\mathbb{R}^n, dx) \equiv L_2$. It is well known that the domain $\mathcal{D}(-\Delta)$ of the Laplace operator coincides with the Sobolev space $W_2^2(\mathbb{R}^n) \equiv W_2^2$. Recall that $W_2^k(\mathbb{R}^n)$, $k \in \mathbb{Z}$ can be defined as the completion of $C_0^\infty(\mathbb{R}^n)$ in the norm

$$\|f\|_{W_2^k} := \left(\int_{\mathbb{R}^n} |(I-\Delta)^{k/2} f(x)|^2 \, dx\right)^{1/2} \tag{1.1}$$

where I stands for the identity operator.

By definition, the generalized Laplace operator, $-\tilde{\Delta}$, in L_2 is a self-adjoint operator which coincides with $-\Delta$ on a linear subset $\mathcal{D} \subset W_2^2$ dense in L_2. In other words a self-adjoint operator $-\tilde{\Delta}$ in L_2 is said to be the generalized Laplace operator if the set

$$\mathcal{D} := \{f \in W_2^2 \mid -\Delta f = -\tilde{\Delta} f\} \tag{1.2}$$

is dense in L_2. In the singular perturbation theory as studied in references [**Ko1**] – [**Ko3**] and [**AK1**] – [**AKM**] $-\tilde{\Delta}$ is called a singular perturbation of $-\Delta$ and the class of all singular perturbations of $-\Delta$ is denoted by $\mathcal{A}_s(-\Delta)$.

1991 *Mathematics Subject Classification.* Primary 47A10, 47A55, ; Secondary 46E35.

Key words and phrases. Laplace operator, Sobolev scale of spaces, singular perturbations, generalized sum.

The second author was supported in part by DAAD and SFB-256 Grants.

It is not hard to see that for any generalized Laplace operator $-\tilde{\Delta} \neq -\Delta$, the corresponding set $\mathcal{D}$ is a proper closed subspace in W_2^2. Thus every generalized Laplace operator $-\tilde{\Delta}$ generates an orthogonal decomposition:

$$W_2^2 = \mathcal{M}_2 \oplus \mathcal{N}_2 \tag{1.3}$$

with $\mathcal{N}_2 \neq 0$ and $\mathcal{M}_2 = \mathcal{D}$. We have

$$\mathcal{M}_2 = \mathcal{D} \text{ is dense in } L_2. \tag{1.4}$$

Conversely every fixed decomposition (1.3) with the condition (1.4) leads to some subfamily of generalized Laplace operators. Indeed for given (1.3) with (1.4) we can introduce in L_2 the operator

$$-\dot{\Delta} := -\Delta \restriction \mathcal{M}_2. \tag{1.5}$$

Obviously $-\dot{\Delta}$ is a densely defined closed symmetric operator in L_2. Therefore each self-adjoint extension $-\tilde{\Delta}$ of the operator $-\dot{\Delta}$ belongs to $\mathcal{A}_s(-\Delta)$.

Example. Let a point $y \in \mathbb{R}^n$ be fixed. Define in L_2 the domain

$$\mathcal{D}_y = \{f \in W_2^2 \mid f(y) = 0\}.$$

Under condition $n \leq 3$ we have the decomposition (1.3) with $\mathcal{M}_2 = \mathcal{D}_y$, which obviously is dense in L_2, and $\mathcal{N}_2 = W_2^2 \ominus \mathcal{M}_2$, is a one-dimensional subspace. Then the family of generalized Laplace operators obeying (1.2) with $\mathcal{D} = \mathcal{D}_y$ can be described by one parameter $b \in \mathbb{R} \cup \{\infty\}$ and we denote it by $\{\tilde{\Delta}_{y,b}\}$ (see [**AGHKH, AK1, GS, HaSn, KS**]). Let $-\tilde{\Delta} \in \{-\tilde{\Delta}_{y,b}\}$. Assume that $-1 \in \rho(-\tilde{\Delta})$, i.e., -1 is in the resolvent set of $-\tilde{\Delta}$. Then operators $(-\Delta + I)^{-1}$ and $(-\tilde{\Delta} + I)^{-1}$ coincide on subspace $(-\Delta + I)\mathcal{D}_y$. Therefore their difference is a rank one operator:

$$(-\tilde{\Delta} + I)^{-1} - (-\Delta + I)^{-1} = \frac{1}{b}(\cdot, \eta)\eta, \ b \neq 0,$$

where $\eta \in L_2$ is the function given by $\eta(x) = G_y(x)$, and where $G_y(x) = G(x - y)$ is the integral kernel of the operator $(-\Delta + I)^{-1}$. Thus we have $-\tilde{\Delta} = -\tilde{\Delta}_{y,b}$, where

$$\mathcal{D}(-\tilde{\Delta}_{y,b}) = \{g \in L_2 \mid g = f + \frac{1}{b} f(y)\eta, \ f \in W_2^2\}, \quad b \neq 0$$

$$-\tilde{\Delta}_{y,b}\, g = -\Delta f - \frac{1}{b} f(y)\eta, \quad g \in \mathcal{D}(-\tilde{\Delta}_{y,b}).$$

If $-1 \notin \rho(-\tilde{\Delta})$, then $(-\tilde{\Delta} + I)\eta = 0$. In this case we set $b = 0$, and the corresponding operator $-\tilde{\Delta}_{y,0}$ has η as eigenvector, $-\tilde{\Delta}_{y,0}\eta = -\eta$. For $n = 2, 3$ the operator $-\tilde{\Delta}_{y,\infty}$ is the Friedrichs extension of $-\dot{\Delta}_y = -\Delta \restriction \mathcal{D}_y$, which coincides with $-\Delta$. In the case $n = 1$, $-\tilde{\Delta}_{y,\infty} = \Delta$ does not coincide with the Friedrichs extension of $-\dot{\Delta}_y$ (for details see [**AK1, GS, KS, Ko6**]). Thus all operators $-\tilde{\Delta}_{y,b} \in \mathcal{A}_s(-\Delta)$. In fact the family $\{-\tilde{\Delta}_{y,b}\}_{b \in \mathbb{R} \cup \{\infty\}}$ coincides with the set of all self-adjoint extensions of $-\dot{\Delta}_y$.

Let us decompose the set of $\mathcal{A}_s(-\Delta)$ into two disjoint subfamilies $\mathcal{A}_{s_i}(-\Delta)$, $i = 1, 2$. It goes as follows. Consider any $-\tilde{\Delta} \in \mathcal{A}_s(-\Delta)$. Then the set $\mathcal{D}$ in (1.2) is uniquely determined and so is $-\overset{\bullet}{\Delta}$ (see (1.5)). Let $-\Delta_\infty$ denote the Friedrichs extension of $-\overset{\bullet}{\Delta}$. Then either:

1) $-\Delta_\infty \neq -\Delta$,

or

2) $-\Delta_\infty = -\Delta$.

Respectively in the case 1) we say that

$$-\tilde{\Delta} \in \mathcal{A}_{s_1}(-\Delta),$$

and in the case 2) we write

$$-\tilde{\Delta} \in \mathcal{A}_{s_2}(-\Delta).$$

In this paper we shall study the operators $-\tilde{\Delta} \in \mathcal{A}_{s_1}(-\Delta)$ under the additional assumption that $-\tilde{\Delta}$ and $-\Delta_\infty$ are relatively prime with respect to $-\overset{\bullet}{\Delta}$. The last requirement means that $-\overset{\bullet}{\Delta}$ is the maximal common part for the operators $-\tilde{\Delta}$ and $-\Delta_\infty$. This in turn is equivalent to the condition:

$$\mathcal{M}_1 \cap W_2^2 = \mathcal{D}, \quad \mathcal{M}_1 := \mathcal{D}^{cl,1} \tag{1.6}$$

where $cl, 1$ stands for the closure in W_2^1.

2. Generalized operator sum

In this section we introduce the notion of the generalized operator sum $\tilde{+}$ and show that the wide class of the generalized Laplace operators may be presented in the form $-\tilde{\Delta} = -\Delta \tilde{+} V$ with a suitable V.

Let $V : W_2^1 \to W_2^{-1}$ be a closed symmetric operator $V \subseteq V^*$. Note that the adjoint operator V^* is defined with respect to the duality $\langle \cdot, \cdot \rangle$ between the spaces W_2^1 and W_2^{-1}. So

$$\langle V\varphi, \psi \rangle = \overline{\langle V\psi, \varphi \rangle}, \quad \varphi, \psi \in \mathcal{D}(V) \subseteq \mathcal{D}(V^*) \subseteq W_2^1.$$

Often but not in general we assume that V is self-adjoint. Operator V will be called *singular* with respect to L_2 if the set

$$\text{Ker}V \ \text{ is dense in } \ L_2. \tag{2.1}$$

Then it is singular in the sense of references [**KK, AKK, Ko1, Ko2**].

Denoting Ker$V$ by $\mathcal{M}_1$ (see (1.6)) we note that it is a closed subspace in W_2^1 since V is a closed operator. Thus we have the orthogonal decomposition:

$$W_2^1 = \mathcal{M}_1 \oplus \mathcal{N}_1, \quad \mathcal{N}_1 \neq 0. \tag{2.2}$$

We shall say that an operator $V : W_2^1 \to W_2^{-1}$ singular with respect to L_2 belongs the $\mathcal{H}_{-1}$-class,and write

$$V \in \mathcal{H}_{-1} - \text{class}, \tag{2.3}$$

if

$$\text{Ker}V \cap \mathcal{D}(-\Delta) \ \text{ is dense in } \ \mathcal{M}_1. \tag{2.4}$$

Usually we interpret operators V from $\mathcal{H}_{-1}$-class as objects carrying the weak singular perturbations of $-\Delta$. Here we show that in our construction such operators lead to the generalized Laplace operators $-\tilde{\Delta} \in \mathcal{A}_{s_1}(-\Delta)$. For the theory of the strong singular perturbations, which in literature are denoted by $\mathcal{H}_{-2}$-class and which lead in general to the perturbed operators $-\tilde{\Delta} \in \mathcal{A}_{s_2}(-\Delta)$ the corresponding abstract approach has been developed in [**AKK, KKO, Ko1, Ko3**] (see also [**KN**]).

Put $\mathcal{M}_2 = \mathrm{Ker}V \cap \mathcal{D}(-\Delta)$. Obviously it is a proper closed subspace in W_2^2. Thus similarly to (1.3) we again have, $W_2^2 = \mathcal{M}_2 \oplus \mathcal{N}_2$, where $\mathcal{N}_2 \neq 0$. Now the conditions (2.1) and (2.4) might be rewritten as follows:

$$\mathcal{M}_1^{cl,0} = L_2, \quad \mathcal{M}_2^{cl,1} = \mathcal{M}_1, \tag{2.5}$$

where $cl, 0$ stands for the closure in L_2. Of course, (2.5) implies that $\mathcal{M}_2^{cl,0} = L_2$. Thus each $V \in \mathcal{H}_{-1}$-class defines a symmetric restriction of $-\Delta$:

$$-\overset{\bullet}{\Delta} := -\Delta \upharpoonright \mathcal{M}_2, \tag{2.6}$$

which has the non-trivial deficiency indices:

$$\mathbf{n}^{\pm}(-\overset{\bullet}{\Delta}) = \dim \mathcal{N}_2.$$

It follows from (2.4) that $-\Delta$ and $-\Delta_\infty$ are relatively prime with respect to $-\overset{\bullet}{\Delta}$.

Let $V : W_2^1 \to W_2^{-1}$ be symmetric. Following [**Be, KrYa, Ko5, KarKo**] we introduce a notion of sum for $-\Delta$ and V. Let

$$-\boldsymbol{\Delta} : W_2^1 \to W_2^{-1}$$

denote the closure of $-\Delta$ as the operator from W_2^1 to W_2^{-1}.

We define the *generalized sum*, $-\tilde{\Delta} = -\Delta \tilde{+} V$, as the "restriction" onto L_2 of the operator sum $-\boldsymbol{\Delta} + V : W_2^1 \to W_2^{-1}$. Precisely

$$\mathcal{D}(-\tilde{\Delta}) = \{\varphi \in W_2^1 \cap \mathcal{D}(V) \mid -\boldsymbol{\Delta}\varphi + V\varphi \in L_2\}, \quad -\tilde{\Delta}\varphi = -\boldsymbol{\Delta}\varphi + V\varphi. \tag{2.7}$$

Of course $-\Delta \tilde{+} V$ coincides with usual sum $-\Delta + V$ if V acts in L_2 with $\mathcal{D}(V) \subseteq \mathcal{D}(-\Delta)$. It has been shown [**AKM, KarKo, Ko4**] that the sum in the sense of quadratic forms is a particular case of the generalized sum.

The important problem arises, to find conditions when the operator $-\tilde{\Delta}$ defined by (2.7) is self-adjoint. We present below some simple results.

THEOREM 2.1. *Let a symmetric operator $V : W_2^1 \to W_2^{-1}$ satisfies the following condition. The range of the operator $(-\boldsymbol{\Delta} + I)^{-1}V + I$ in W_2^1 contains the domain of the operator $-\Delta$, i.e.*

$$\mathcal{R}((-\boldsymbol{\Delta} + I)^{-1}V + I) \supset \mathcal{D}(-\Delta). \tag{2.8}$$

Then $-\tilde{\Delta} = -\Delta \tilde{+} V$ is self-adjoint in L_2.

PROOF. We will prove that the range of the operator $-\tilde{\Delta} + I$ coincides with whole space L_2, i.e. $\mathcal{R}(-\tilde{\Delta} + I) = L_2$. Since -1 is a regular point for $-\Delta$, for each $h \in L_2$ there exists $f \in W_2^2$ such that $h = (-\Delta + I)f$. By (2.8) there exists $g \in \mathcal{D}(V)$ such that $f = ((-\boldsymbol{\Delta} + I)^{-1}V + I)g$. Thus we have $(-\boldsymbol{\Delta} + I + V)g = (-\Delta + I)((-\boldsymbol{\Delta} + I)^{-1}V + I)g = (-\Delta + I)f = h$. When g runs over $\mathcal{D}(V)$ the vectors h fill whole of the space L_2. Therefore $(-\Delta + I)\tilde{+}V$ is self-adjoint since it is symmetric and its range is L_2. So the operator $-\Delta \tilde{+} V = (-\Delta + I)\tilde{+}V - I$ is also self-adjoint. □

It easily follows from the above theorem that if V is self-adjoint and the point -1 is regular for the operator

$$V_1 = (-\mathbf{\Delta} + I)^{-1}V, \tag{2.9}$$

which is self-adjoint in W_2^1;

$$-1 \in \rho(V_1), \tag{2.10}$$

then $-\Delta\tilde{+}V$ is self-adjoint in L_2. It is so because obviously (2.10) implies (2.8).

We remark that the above simple theorem contains as a particular case the well-known KLMN-theorem [**ReS**]. Indeed if a quadratic form β satisfies the condition:

$$|\beta[\cdot]| \le b\|\cdot\|_{W_2^1} + a\|\cdot\|_{L_2}^2, \quad 0 \le b < 1, \; 0 \le a$$

then it is closable in W_2^1 and for the corresponding associated operator $V_\beta : W_2^1 \to W_2^{-1}$ the condition (2.8) if fulfiled (for details see [**AKM, Ko4, KarKo**]). In fact slightly stronger result is true (cf. with Theorem 3.2 in [**AK4**]).

THEOREM 2.2. *Let the operator V_1 defined by (2.9) has the point spectrum in a $\varepsilon-$neighborhood of the point -1. Then $-\tilde{\Delta} = -\Delta\tilde{+}V$ is self-adjoint in L_2.*

Now we observe that under suitable condition on V the operator $-\tilde{\Delta} = -\Delta\tilde{+}V$ belongs to the family $\mathcal{A}_{s_1}(-\Delta)$.

THEOREM 2.3. *If $V \in \mathcal{H}_{-1}$ –* class *and the generalized sum $-\tilde{\Delta} = -\Delta\tilde{+}V$ is self-adjoint in L_2, then $-\tilde{\Delta} \in \mathcal{A}_{s_1}(-\Delta)$.*

PROOF. Let us consider for the pair $-\tilde{\Delta}$ and $-\Delta$ the set $\mathcal{D}$ defined by (1.2). Obviously $-\tilde{\Delta}\varphi = -\Delta\varphi$ implies that $\varphi \in \mathrm{Ker}V \cap \mathcal{D}(-\Delta)$, i.e. $\mathcal{D} \subset \mathrm{Ker}V \cap W_2^2 = \mathcal{M}_1 \cap W_2^2 = \mathcal{M}_2$. By conditions (2.4) and (2.3) we conclude that $-\tilde{\Delta} \in \mathcal{A}_s(-\Delta)$. Since $V \neq 0$, $\mathcal{M}_1$ is a proper subspace in W_2^1. Therefore $\Delta_\infty \neq \Delta$. This means that in fact $-\tilde{\Delta} \in \mathcal{A}_{s_1}(-\Delta)$. □

3. Construction of $-\tilde{\Delta} \in \mathcal{A}_{s_1}(-\Delta)$ by the method of generalized sum

Here we solve the inverse problem. Given $-\tilde{\Delta} \in \mathcal{A}_{s_1}(-\Delta)$ find $V \in \mathcal{H}_{-1}$ – class such that $-\tilde{\Delta} = -\Delta\tilde{+}V$. We show that under some natural condition this problem has the positive solution. First we need some preparations.

Let $-\tilde{\Delta} \in \mathcal{A}_{s_1}(-\Delta)$ be fixed. Since $-\tilde{\Delta}$ is one of the self-adjoint extensions of the symmetric operator $-\overset{\bullet}{\Delta}$, the resolvents of $-\tilde{\Delta}$ and $-\Delta$ are connected by the Krein formula [**Bi**]:

$$\tilde{R}_z = R_z + \tilde{B}_z^{-1}, \quad z \in \rho(\Delta) \cap \rho(-\tilde{\Delta}), \tag{3.1}$$

where $\tilde{R}_z = (zI - \tilde{\Delta})^{-1}$, $R_z = (zI - \Delta)^{-1}$ and $\tilde{B}_z^{-1} = B_z^{-1}P_{\mathfrak{N}_z}$, with $B_z : \mathfrak{N}_{\bar{z}} \to \mathfrak{N}_z$, where $\mathfrak{N}_z$ denote the deficiency subspace of $-\overset{\bullet}{\Delta}$ and $P_{\mathfrak{N}_z}$ stands for the orthogonal projection.

If we assume that $-1 \in \rho(-\tilde{\Delta})$, then (3.1) implies

$$(I - \tilde{\Delta})^{-1} = (I - \Delta)^{-1} + \tilde{B}^{-1}, \tag{3.2}$$

where $\tilde{B}^{-1} = B_z^{-1}P_{\mathcal{N}_0}$ with $B := B_{z=-1}$ and where $P_{\mathcal{N}_0}$ stands for the orthogonal projection onto the subspace $\mathcal{N}_0 := \mathfrak{N}_{-1}$. We note that B is an operator-valued parameter acting in $\mathcal{N}_0$. A choice of B fixes the extension $-\tilde{\Delta}$.

In particular from (3.1) for the Friedrichs extension $-\Delta_\infty$ we have,

$$(I-\Delta_\infty)^{-1} = (I-\Delta)^{-1} + \tilde{B}_\infty^{-1}, \tag{3.3}$$

where the operator-valued parameter B_∞ admits the following description [**Ko4, Ko6**]:

$$B_\infty \eta = (I-\Delta)\eta \in \mathcal{N}_0, \quad \eta \in \mathcal{D}(B_\infty), \tag{3.4}$$

where

$$\mathcal{D}(B_\infty) := \{\eta \in \mathcal{N}_1 | \eta = P_{\mathcal{N}_1}\varphi,\ \varphi \in \mathcal{N}_2 \subset \mathcal{D}(-\Delta)\}, \tag{3.5}$$

where $P_{\mathcal{N}_1}$ is the orthogonal projection onto subspace $\mathcal{N}_1 = W_2^1 \ominus \mathcal{M}_1$.

The next theorem gives our main result (cf. with Theorem 7 in [**Ko4**]).

THEOREM 3.1. *Suppose $-\tilde{\Delta} \in \mathcal{A}_{s_1}(-\Delta)$ is such that*

$$\mathcal{D}(-\tilde{\Delta}) \subset W_2^1, \tag{3.6}$$

and the operators $-\tilde{\Delta}$ and $-\Delta_\infty$ are relatively prime with respect to $-\dot{\Delta}$. Then $-\tilde{\Delta}$ admits the representation by the generalized sum: $-\tilde{\Delta} = -\Delta \tilde{+} V$, where the operator V belongs to $\mathcal{H}_{-1}$ – class *and has the representation,*

$$V = -(I-\boldsymbol{\Delta})(B_\infty^{-1}B + I_{\mathcal{N}_1})^{-1}P_{\mathcal{N}_1}, \tag{3.7}$$

in which operators B and B_∞ are defined by (3.2) and (3.4) resp.

PROOF. Let us show first that the domain of the operator B belongs to the subspace $\mathcal{N}_1$ in W_2^1,

$$\mathcal{D}(B) \subseteq \mathcal{N}_1. \tag{3.8}$$

Recall that $\mathcal{N}_1 = (\mathcal{D}^{cl,1})^\perp$ in W_2^1. Indeed, due to $\mathcal{D}(-\tilde{\Delta}) \subset W_2^1$ we have $\mathcal{D}(B) \subset W_2^1$. Therefore $\mathcal{D}(B) \subset W_2^1 \cap \mathcal{N}_0$ since B is an operator in $\mathcal{N}_0$. In fact $\mathcal{D}(B) \subset \mathcal{N}_1$ due to $(\mathcal{D}(B), \mathcal{M}_1)_{W_2^1} = 0$, where $\mathcal{M}_1 = W_2^1 \ominus \mathcal{N}_1$. Indeed

$$(\mathcal{D}(B), \mathcal{M}_2)_{W_2^1} = (\mathcal{D}(B), (I-\Delta)^{-1}\mathcal{M}_2) = (\mathcal{D}(B), \mathcal{M}_0) = 0,$$

since $\mathcal{D}(B) \subset \mathcal{N}_0 \perp \mathcal{M}_0$, where $\mathcal{M}_0 := L^2 \ominus \mathcal{N}_0$. Therefore $\mathcal{D}(B) \perp \mathcal{M}_1$ since $\mathcal{M}_2$ is dense in $\mathcal{M}_1$. This proves (3.8).

Let us introduce the operator

$$W := B_\infty^{-1}B.$$

Since both B_∞^{-1}, B^{-1} are bounded, we have: $\mathcal{R}(B_\infty) = \mathcal{R}(B) = \mathcal{N}_0$. This implies that the domain $\mathcal{D}(W) = \mathcal{D}(B) \subseteq \mathcal{N}_1$ and the range $\mathcal{R}(W) = \mathcal{D}(B_\infty) \subseteq \mathcal{N}_1$. So we have

$$W : \mathcal{N}_1 \supseteq \mathcal{D}(B) = \mathcal{D}(W) \to \mathcal{R}(W) = \mathcal{D}(B_\infty) \subseteq \mathcal{N}_1. \tag{3.9}$$

Further since B is self-adjoint in $\mathcal{N}_0$ the operator $W = B_\infty^{-1}\ B$ is symmetric in $\mathcal{N}_1$. Indeed for any $\eta_1, \eta_2 \in \mathcal{D}(\ B)$ we can write

$$(B_\infty^{-1}\ B\eta_1, \eta_2)_{W_2^1} = (P_{\mathcal{N}_1}(I-\Delta)^{-1}B\eta_1, \eta_2)_{W_2^1} = ((I-\Delta)^{-1}\ B\eta_1, \eta_2)_{W_2^1} = (\ B\eta_1, \eta_2)$$
$$= (\eta_1,\ B\eta_2) = \langle (I-\boldsymbol{\Delta})\eta_1, (I-\Delta)^{-1}B\eta_2\rangle = (\eta_1, P_{\mathcal{N}_1}(I-\Delta)^{-1}\ B\eta_2)_{W_2^1}$$
$$= (\eta_1, B_\infty^{-1}\ B\eta_2)_{W_2^1},$$

where we used (3.4). We remark that the inverse operator $W^{-1} = B^{-1}B_\infty$ exists due to $\mathcal{R}(B_\infty) = \mathcal{R}(B) = \mathcal{N}_0$. We also used the fact that

$$\mathcal{N}_1 \subset \mathcal{N}_0.$$

This is so because for each $\eta_1 \in \mathcal{N}_1$ and $\mu_0 = (I - \Delta)\mu_2 \in \mathcal{M}_0$, $\mu_2 \in \mathcal{M}_2$ we have: $(\eta_1, \mu_0) = (\eta_1, (I - \Delta)\mu_2) = (\eta_1, \mu_2)_{W_2^1} = 0$.

Let us prove now that the operator

$$W + I_{\mathcal{N}_1} = B_\infty^{-1} B + I_{\mathcal{N}_1}$$

where $I_{\mathcal{N}_1}$ denotes the identity in $\mathcal{N}_1$ is invertible in $\mathcal{N}_1$. For this aim we will prove that

$$\operatorname{Ker}(B_\infty^{-1} B + I_{\mathcal{N}_1}) = \operatorname{Ker}(W + I_{\mathcal{N}_1}) = \{0\}.$$

Assume that $(W+I_{\mathcal{N}_1})\eta_1 = 0$ for some vector $0 \neq \eta_1 \in \mathcal{D}(B) = \mathcal{D}(W) \subset \mathcal{N}_1 \subset \mathcal{N}_0$. Then $W\eta_1 = -\eta_1$ and equivalently $B\eta_1 = -B_\infty\eta_1$, where we used the fact that $W = B_\infty^{-1} B$ and $\mathcal{R}(W) = \mathcal{N}_1 = \mathcal{D}(B_\infty)$. Further put $\phi = B\eta_1 = -B_\infty\eta_1$. Since $\mathcal{R}(B) = \mathcal{R}(B_\infty) = \mathcal{N}_0$ we have $B^{-1}\phi = \eta_1$, $B_\infty^{-1}\phi = -\eta_1$. Using (3.2) and (3.3) we have:

$$\begin{aligned}(I - \tilde{\Delta})^{-1}\phi &= (I - \Delta)^{-1}\phi + B^{-1}\phi = (I - \Delta_\infty)^{-1}\phi + B_\infty^{-1}\phi + B^{-1}\phi \\ &= (I - \Delta_\infty)^{-1}\phi - \eta_1 + \eta_1 = (I - \Delta_\infty)^{-1}\phi.\end{aligned}$$

Thus for $0 \neq \phi \in \mathcal{N}_0$ we obtain,

$$(I - \Delta_\infty)^{-1}\phi = (I - \tilde{\Delta})^{-1}\phi = \xi \in \mathcal{D}(-\Delta_\infty) \cap \mathcal{D}(-\tilde{\Delta}).$$

This gives the contradiction with the condition that $-\Delta_\infty$ and $-\tilde{\Delta}$ are relatively prime with respect to $-\dot{\Delta}$. Indeed on one hand a vector ξ can not belong to the set $\mathcal{D}$ if $\phi \in \mathcal{N}_0$, (recall that $\mathcal{N}_0 = ((I - \dot{\Delta})\mathcal{D})^\perp)$. On the other hand if $0 \neq \xi \in \mathcal{D}(-\Delta_\infty) \cap \mathcal{D}(-\tilde{\Delta})$, $\xi \notin \mathcal{D}$ and $\tilde{\Delta}\xi = -\Delta_\infty\xi$, then $-\dot{\Delta}$ is not a maximal common part for a pair $-\Delta_\infty$, $-\tilde{\Delta}$, contrary to the assumption.

Thus for any $\eta_1 \in \mathcal{D}(B) = \mathcal{D}(W)$ we have $B_\infty^{-1} B\eta_1 \neq -\eta_1$. This means that $(W + I_{\mathcal{N}_1})\eta_1 = 0$ implies $\eta_1 = 0$ and therefore the operator $W + I_{\mathcal{N}_1}$ is invertible. Consequently the operator

$$\check{V}_1 := -(B_\infty^{-1} B + I_{\mathcal{N}_1})^{-1} = -(W + I_{\mathcal{N}_1})^{-1} \tag{3.10}$$

is symmetric in $\mathcal{N}_1$. Moreover the operator

$$\begin{aligned}V &:= -(I - \mathbf{\Delta})\check{V}_1 P_{\mathcal{N}_1} = -(I - \mathbf{\Delta})(B_\infty^{-1}B + I_{\mathcal{N}_1})^{-1}P_{\mathcal{N}_1} \\ &= -(I - \mathbf{\Delta})(W + I_{\mathcal{N}_1})^{-1}P_{\mathcal{N}_1} : W_2^1 \to W_2^{-1} \end{aligned} \tag{3.11}$$

is symmetric in the sense of a dual pair of spaces.

Evidently V belongs to the $\mathcal{H}_{-1}$-class since the set $\mathcal{D} = \operatorname{Ker}V \cap \mathcal{D}(-\Delta) \equiv \mathcal{M}_2$ is dense inL_2 and (1.6) is fulfilled. By construction the condition (2.8) is also fulfilled.

Therefore the above constructed operator V satisfies conditions of Theorem 2.1. Thus the generalized sum $\tilde{\Delta}' = -\Delta \tilde{+} V$ defines the self-adjoint operator in L_2 with the range $\mathcal{R}(\tilde{\Delta}') = L_2$ and with the domain $\mathcal{D}(\tilde{\Delta}') \subset W_2^1$. In other words the operator $-\tilde{\Delta}' \in \mathcal{A}_{s_1}(-\Delta)$. In fact operator $-\tilde{\Delta}'$ and the starting operator $-\tilde{\Delta}$ coincide. To see that $-\tilde{\Delta}' = -\tilde{\Delta}$ define

$$B' = -B_\infty(\check{V}^{-1}(I - \mathbf{\Delta})P_{\mathcal{N}_1} + I_{\mathcal{N}_1}), \tag{3.12}$$

where $\check{V}$ is the restriction of V onto $\mathcal{N}_1$. From (3.10), (3.4), (3.2), and (3.12) we find that $B' = B$. Therefore $-\tilde{\Delta}' = -\tilde{\Delta}$. □

COROLLARY. *Each operator $\tilde{\Delta} \in \mathcal{A}_{s_1}(A)$ with $\mathcal{D}(\tilde{\Delta}) \subset \overset{\bullet}{W}{}_2^1$, under condition that a pair $-\tilde{\Delta}$, $-\Delta_\infty$ is relatively prime with respect to $-\overset{\bullet}{\Delta}$, admits two representations:*

$$(I - \tilde{\Delta})^{-1} = (I - \Delta_\infty)^{-1} + B^{-1}P_{\mathcal{N}_0}$$

and

$$-\tilde{\Delta} = -\Delta \tilde{+} V.$$

These formulae establish the one-to-one correspondence between operators $-\tilde{\Delta} \in \mathcal{A}_{s_1}(-\Delta)$ with $\mathcal{D}(-\tilde{\Delta}) \subset W_2^1$ and singular perturbations $V \in \mathcal{H}_{-1}$-class.

References

[AGHKH] S. Albeverio, F. Gesztesy, R. Høegh- Krohn, H. Holden, *Solvable models in quantum mechanics*, Springer, Berlin,1988.

[AK1] S. Albeverio, V. Koshmanenko, *Singular rank one perturbations of self-adjoint operators and Krein theory of self-adjoint extensions*, to appear Potential Anal., **11**, No 3 (1999).

[AK2] S. Albeverio, V. Koshmanenko, *Form-sum approximations of singular perturbations of self-adjoint operators*, to appear in J. Funct. Anal., **198** (1999).

[AK3] S. Albeverio, V. Koshmanenko, *On the problem of the right Hamiltonian under singular form-sum perturbations*, SFB 237 Preprint, Nr. 375, Institute für Mathematik Ruhr-Universität-Bochum, 30 p. (1997) to appear in Rev. Math. Phys.

[AK4] S. Albeverio, V. Koshmanenko, *On Schrödinger operators perturbed by fractal potentials*, Preprint of Bonn University, no. 593, to appear in Reports on Math. Phys.

[AKK] S. Albeverio, V. Koshmanenko, W. Karwowski *Square Power of Singularly Perturbed Operators*, Math. Nachr., **173** (1995), 5–24 .

[AKM] S. Albeverio, V. Koshmanenko, K. Makarov *Generalized eigenfunctions under singular perturbations*, Methods of Functional Analysis and Topology, **5**, No. 1 (1999) 13-27.

[Be] Yu.M. Berezanskij *The bilinear forms and Hilbert equipments*, In "Spectral analysis of differential operators", Institute of Mathemetics, Kiev (1980), 83-106 .

[Bi] M.Sh. Birman, *To self-adjoint extensions theory of positive definite operators*, Math. Sbornik, **38** (80), N 4 (1956), 431-450.

[GS] F. Gesztesy, B. Simon, *Rank-One Perturbations at Infinite Coupling*, J. Funct. Anal., **128** (1995), 245-252.

[HaSn] S. Hassi, H. de Snoo, *On rank one perturbations of selfadjoint operators, Integral equations and operator theory*, **29**, 288-300 (1997).

[KarKo] T.V. Karataeva, V. Koshmanenko, *Generalized sum for operators*, to appear in Math. Notes, **66**, No. 5, (1999).

[KKO] W. Karwowski, V. Koshmanenko, S. Ôta, *Schrödinger operator perturbed by operators related to null-sets*, Positivity, **2**, No.1, 77-99 (1998).

[KK] W. Karwowski, V. Koshmanenko, *On definition of the singular bilinear form and singular operator*, Ukrainian Math. J., **45**, No. 8, (1993), 1084-1089.

[KS] A. Kiselev, B. Simon, *Rank One Perturbations with Infinitesimal Coupling*, J.Funct.Anal., **130** (1995), 345-356.

[Ko1] V. Koshmanenko, *Singular Quadratic Forms in Perturbation Theory*, Kluwer Academic Publishers, (1999).

[Ko2] V. D. Koshmanenko, *Towards the rank-one singular perturbations of self-adjoint operators*, Ukrainian Math. J., **43**, No. 11, (1991), 1559-1566.

[Ko3] V. D. Koshmanenko, *Perturbations of self-adjoint operators by singular bilinear forms*, Ukrainian Math.J., **41** (1989), 1- 14.

[Ko4] V. Koshmanenko, *Singular operator as a parameter of self-adjoint extensions*, Proceedings of M. Krein Conference, Odessa, (1997), to appear in Operator Theory. Advances and Applications.

[Ko5] V. D. Koshmanenko, *Regular approximations of singular perturbations of $\mathcal{H}_{-2}$-class*, to appear in Ukrainian Math. J., (1999).

[Ko6] V.D. Koshmanenko, *Singular perturbations at infinite coupling*, Funct. Anal. Appl., **33**, No. 2, (1999).

[KrYa] M.G. Krein, V.A. Yavrian *Spectral shift functions arising in perturbations of a positive operator*, J. Operator Theory, **6** (1981), 155-191.

[KN] S.T. Kuroda and H. Nakagami, $\mathcal{H}_{-2}$ – *construction and some applications*, in: Mathematical Results in Quantum Mechanics, Operator Theory Advances and Applications **108** (1998), 99-105.

[ReS] M. Reed, B. Simon, *Methods of Modern Mathematical Physics. II. Fourier Analysis, Self-adjointness*, Academic Press, New York - San Francisco - London, 1975.

Institute of Theoretical Physics, University of Wroclaw, Wroclaw, Poland
E-mail address: `Wkar@ift.uni.wroc.pl`

Institute of Mathematics, Kyiv, Ukraine
E-mail address: `kosh@imath.kiev.ua`

Canadian Mathematical Society
Conference Proceedings
Volume **29**, 2000

Learning of p-adic neural networks

Andrei Khrennikov[1] and Brunello Tirozzi[2]

ABSTRACT. A p-adic model which describes a large class of neural networks is presented. In this model the states of neurons are described by digits in the canonical expansion of a p-adic number. Thus each p-adic number represents a configuration of firing and non firing neurons. We present the algorithm of learning for p-adic neural networks based on the minimization of the error-functional (here we use a random search procedure in the space of p-adic weights). This algorithm (or its more advanced versions) could be applied for image recognition.

1. Introduction

In recent years there has been an enormous development of the applications of neural networks for solving practical problems (see, for example, [1]-[4]). Neural networks take their name and structure from the neuronal system of animals and humans. They can be described as a unit with an input and corresponding output. The input is taken from any time series composed by vectors of n variables containing the results of some experiment at time t and the output might be some state variable which summarizes the conclusion of the experiment at time $t+1$ represented by a vector of n variables. The transformation which associates the output to a given input is defined in analogy with the behaviour of the real neurons. For example, the simplest neural structure contains the input described by N neurons and the output by M neurons. The input neurons are connected by *synaptic weights* w_{jk}, $j=0,\ldots,M-1$, $k=0,\ldots,N-1$, to the neurons of the output layer The input vector $x=(x_0,...,x_{N-1})$ is transformed into a vector of outputs $y=(y_0,...,y_{M-1})$ by the law defined by the weights w_{jk}:

(1) $$z_j=\sum_{k=0}^{N-1} w_{jk}x_k, j=0,...,M-1;$$

(2) $$y_j=f(z_j-\theta_j)$$

[1]This investigation was supported by the grant "Strategical investigations" of Växjö University and visiting professor fellowship at University of Rome, "La Sapienza.".

[2]This investigation was supported by the grant "Strategical investigations" of Växjö University.

for $f(x)$ = signx (or some other function with a similar behaviour). Here $\theta_j \geq 0$ are threshold constants.

The general problem solved by neural networks can be state as follows. Given a sequence of pairs $(x^{(\alpha)}, y^{(\alpha)}), \alpha = 1, ..., m$, obtained by experiments, measurements or whatever, we want to find what is the law which transforms the input $x^{(\alpha)}$ into the output $y^{(\alpha)}$. The *learning process* for the neural networks is obtained by applying a minimization procedure to the vector of all the synaptic weights $W \equiv w_{jk}$, in order to get the minima of the *network prediction error*:

$$E(W) = \frac{1}{m} \sum_{\alpha} \|y^{(\alpha)} - f(x^{(\alpha)} - \theta)\|^2. \tag{3}$$

We use p-adic numbers (see [5] and section 2 for an introduction to p-adic numbers, see, for example, [6]- [8] and [8]-[11] for applications of p-adic numbers to quantum physics and cognitive sciences, respectively) to describe a large class of neural networks having a kind of hierarchic structure [9]. In this model every neuron can have p different states, $\beta = 0, ..., p-1$, (here $p > 1$ is a fixed prime number) which are described by digits in the canonical expansion of a p-adic number. If $p = 2$, then every neuron can only be in one of the two states: $\beta = 1$, firing, and $\beta = 0$, non firing. Since a p-adic number may have an infinite number of nonzero digits, we study configurations where, in general, an infinite number of neurons can be in an active state.

In this note we propose an algorithm of learning for p-adic neural networks based on the minimization procedure of a p-adic analogue of the functional (3).

2. A model of p-adic dynamical neural network

We start this section with a brief introduction to p-adic numbers.

The field of real numbers $\mathbf{R}$ is constructed as the completion of the field of rational numbers Q with respect to the metric $\rho(x, y) = |x - y|$, where $|\cdot|$ is the usual valuation given by the absolute value. The fields of p-adic numbers $\mathbf{Q}_p$ are constructed in a corresponding way, by using other valuations. For any prime number the p-adic valuation $|\cdot|_p$ is defined in the following way. First we define it for natural numbers. Every natural number n can be represented as the product of prime numbers : $n = 2^{r_2} 3^{r_3} \cdots p^{r_p} \cdots$. Then we define $|n|_p = p^{-r_p}$, we set in addition $|0|_p = 0$ and $|-n|_p = |n|_p$. We extend the definition of the p-adic valuation $|\cdot|_p$ to all rational numbers by setting for $m \neq 0 : |n/m|_p = |n|_p/|m|_p$. The completion of Q with respect to the metric $\rho_p(x, y) = |x - y|_p$ is the locally compact field of p-adic numbers $\mathbf{Q}_p$. It is well known (Ostrovsky's theorem), see [5], that $|\cdot|$ and $|\cdot|_p$ are the only possible valuations on Q. The p-adic valuation satisfies so-called strong triangle inequality: $|x + y|_p \leq \max[|x|_p, |y|_p]$, which makes ρ_p into an ultrametric. Set $U_r(a) = \{x \in \mathbf{Q}_p : |x - a|_p \leq r\}$ and $U_r^-(a) = \{x \in \mathbf{Q}_p : |x - a|_p < r\}, r = p^n, n = 0, \pm 1, \pm 2, ...$. These are ("closed" and "open") balls in $\mathbf{Q}_p$. Set $S_r(a) = \{x \in K : |x - a|_p = r\}$. These are spheres in $\mathbf{Q}_p$. Any p-adic ball $U_r(0)$ is an additive subgroup of $\mathbf{Q}_p$. The ball $U_1(0)$ is also a ring. It is called the *ring of p-adic integers* and denoted by $\mathbf{Z}_p$. For any $x \in \mathbf{Q}_p$ we have a unique canonical expansion (converging in the $|\cdot|_p$-norm) of the form

$$x = \alpha_{-n}/p^n + \cdots \alpha_0 + \cdots + \alpha_k p^k + \cdots , \tag{4}$$

where $\alpha_j = 0, 1, ..., p-1$, are the "digits" of the p-adic expansion. The elements $x \in \mathbf{Z}_p$ have the expansion: $x = \alpha_0 + \alpha_1 p + \cdots + a_k p^k + \cdots$, i.e., they can be identified with sequences of digits $x = (\alpha_0, ..., \alpha_k, ...), \alpha_j = 0, 1, ..., p-1$.

We set $\mathbf{Q}_p^N = \mathbf{Q}_p \times \cdots \times \mathbf{Q}_p$($N$-times); $\mathbf{Z}_p^N = \mathbf{Z}_p \times \cdots \times \mathbf{Z}_p$. We define the norm $\|x\|_p = \max_{0 \le j \le N-1} |x_j|_p, x = (x_0, ..., x_{N-1}) \in \mathbf{Q}_p^N$, and define the balls $U_r^N(a) = \{x \in \mathbf{Q}_p^N : \|x - a\|_p \le r\}$, and spheres $S_r^N(a) = \{x \in \mathbf{Q}_p^N : \|x - a\|_p = r\}, r = p^n (n = 0, \pm 1, \pm 2, ...), a \in \mathbf{Q}_p^N$. We have $U_r^N(a) = U_r(a_0) \times \cdots \times U_r(a_{N-1}), a = (a_0, ..., a_{N-1})$. We remark that balls, as well as spheres are open and closed at the same time, 'clopen'.

We consider a model of p-adic neural network based on the weight matrix $w = (w_{ij})_{i,j=0}^{N-1,M-1}, w_{ij} \in \mathbf{Z}_p$, where N is the length of the input layer $x = (x_0, ..., x_{N-1})$ and M is the length of the output layers $y = (y_0, ..., y_{M-1})$; here coordinates x_j, y_j yield the values $0, 1, ..., p-1$. As usual we start with linear transform (1). However, further we might not develop the model in the same way as in the standard real framework based on thresholds $\theta_j, j = 0, ..., M-1$, and nonlinear transform (2) for $f(x) = \text{signx}$ (or some other function with a similar behaviour). The real framework is based on the order structure on $\mathbf{R}$: if a value of z_j is larger than the threshold θ_j then y_j becomes firing; if not then y_j stays nonfiring. In the p-adic case (as there is no order structure on $\mathbf{Q}_p$) a model cannot be based on the similar arguments.

We propose a new model. The 2-adic case will be considered. Let $x = (x_0, ..., x_{N-1}), x_j = 0, 1$, be an input data and $y = (y_0, ..., y_{M-1}), y_j = 0, 1$, be an output data (both are coded in the 2-adic system). The input signal x is transformed in a multilayer massive $z = (z_0, ..., z_{M-1}), z_j \in \mathbf{Z}_p$, by the linear transform (1). The output of the neural network is determined by computing the 2-adic distance among the z_j and the thresholds θ_j, and that this system is able to recognize and classify the input signals $x^{(\alpha)}, \alpha = 1, ..., m$ using the weights that are constructed using the algorithm developed below in the paper.

We use the scheme which is similar to the usual back-propagation scheme for the real networks. In the ideal case the equality $z_j = \theta_j$ must imply that $y_j = 0$. However, it is natural to assume that the cognitive system contains a comparator which compares the 2-adic distance between layers z_j and θ_j. If these layers are close in the 2-adic distance then the coordinate y_j of the response y yields the value 0; in the opposite case y_j yields the value 1. Of course, there is a precision $\epsilon_k = 1/2^k, k = 1, 2, ...$, which determines the regime of functioning of the comparator:

$$\text{(5)} \qquad \text{if } |z_j - \theta_j|_2 \le \epsilon_k \text{ then } y_j = 0;$$

$$\text{(6)} \qquad \text{if } |z_j - \theta_j|_2 > \epsilon_k \text{ then } y_j = 1.$$

These conditions can be rewritten in the standard form (2) with the function

$$\text{(7)} \qquad f(x) = 1 - I_{U_{\epsilon_k}(0)}(x)$$

where $I_{U_{\epsilon_k}(0)}$ is the characteristic function of the ball $U_{\epsilon_k}(0)$. Thus our model is similar to the standard one[1].

We going to study the process of learning for the 2-adic neural network. As usual there is a set $x^{(\alpha)} = (x_0^{(\alpha)}, ..., x_{N-1}^{(\alpha)})$ of inputs datas and the set of $y^{(\alpha)} =$

[1]We underline again that the analogy in mathematical expressions is only a formal one. The law (2) with function (7) is not connected with the absolute increasing or decreasing of z_j. This law describes nearness between z_j and θ_j in the information space with 2-adic metric.

$(y_0^{(\alpha)}, ..., y_{M-1}^{(\alpha)})$ of output datas for $\alpha = 1, ..., m$ (there $x_j^{(\alpha)}, y_j^{(\alpha)} = 0, 1$). We have to find the coefficients w_{jk} which give the minimum of the functional:

$$\mathcal{L}(w; x^{(\alpha)}, y^{(\alpha)}) = \frac{1}{m} \sum_{\alpha=1}^{m} \sum_{j=0}^{M-1} |y_j^{(\alpha)} - f(z_j^{(\alpha)})|. \tag{8}$$

We can reduce the process of learning to the case $M = 1$, i.e., we consider the set $x^{(\alpha)} = (x_0^{(\alpha)}, ..., x_{N-1}^{(\alpha)})$ of input datas and the set of $y^{(\alpha)} = 0, 1$ of output datas for $\alpha = 1, ..., m$. Thus we consider the numbers $z^{(\alpha)} = \sum_{l=0}^{N-1} w_l x_l^{(\alpha)}, \alpha = 1, ..., m, w = (w_0, ..., w_{N-1}) \in \mathbf{Z}_2^N$.

We shall study the model with "generalized information vectors" $x^{(\alpha)} \in \mathbf{Z}_2^N$ and restrict our consideration to the case of zero thresholds. As usual, we want to minimize the functional

$$\mathcal{L}(w; x^{(\alpha)}, y^{(\alpha)}) = \frac{1}{m} \sum_{\alpha=1}^{m} |y^{(\alpha)} - f(z^{(\alpha)})|, \ w \in \mathbf{Z}_2^N, \tag{9}$$

where $f(x) = 1 - I_{U_{\epsilon_k}}(0)(x)$.

Every input data $x^{(\alpha)}$ determines a linear functional on $\mathbf{Z}_2^N, X_\alpha(w) = \sum_{l=0}^{N-1} w_l x_l^{(\alpha)}$. We set $V_\alpha = U_{\epsilon_k}(0)$ if $y^{(\alpha)} = 0$ and $V_\alpha = \mathbf{Z}_2 \setminus U_{\epsilon_k}(0)$ if $y^{(\alpha)} = 1$. We set also $W_\alpha = X_\alpha^{-1}(V_\alpha)$. The set W_α is said to be a *domain of* $[x^\alpha/y^\alpha]$*-learning.*[2] As V_α is open, then W_α is also open. Thus, for each $w \in W_\alpha$, there exists a ball $U_r^N(w) \subset W_\alpha$. Moreover, we have

Lemma 1. *Let x^α and y^α be input and output and let w belong to the domain of $[x^\alpha/y^\alpha]$-learning W_α. Then the ball $U_{\epsilon_k}^N(w) \subset W_\alpha$, where ϵ_k is the precision of the 2-adic comparator.*

Proof. 1). Let $y^{(\alpha)} = 0$. Then $w \in W_\alpha$ implies that $|X_\alpha(w)|_2 \le \epsilon_k$. It is also evident that, for every $u \in \mathbf{Z}_2^N$, $|X_\alpha(u)|_2 \le \|u\|_2$. Thus if $w' = w + u, \|u\|_2 \le \epsilon_k$, then

$$|X_\alpha(w')|_2 \le \max[|X_\alpha(w)|_2, |X_\alpha(u)|_2] \le \epsilon_k.$$

2). Let $y^{(\alpha)} = 1$. Then $w \in W_\alpha$ implies that $|X_\alpha(w)|_2 > \epsilon_k$. Thus if $w' = w + u, \|u\|_2 \le \epsilon_k$, then

$$|X_\alpha(w')|_2 = \max[|X_\alpha(w)|_2, |X_\alpha(u)|_2] > \epsilon_k.$$

(here we have used the following property of the p-adic valuation: if $|a|_p \neq |b|_p$ then $|a + b|_p = \max[|a|_p, |b|_p]$).

Sets V_α are closed. Hence sets W_α are also closed. The set $\mathbf{Z}_2$ is compact. Hence sets W_α are compact. As a consequence of this lemma, we obtain that the domain of $y^{(\alpha)}$-learning W_α can be determined by the choice of a finite number of points $w^j, j = 1, ..., s_\alpha$:

$$W_\alpha = \cup_{j=1}^{s_\alpha} U_{\epsilon_k}^N(w^j), \ U_{\epsilon_k}^N(w^j) \cap U_{\epsilon_k}^N(w^i) = \emptyset, i \neq j.$$

Therefore, to find the domain of $y^{(\alpha)}$-learning, it is sufficient to find the points $w^j = (w_0^j, ..., w_{N-1}^j) \in \mathbf{Z}_2^N, j = 1, ..., s_\alpha$.

We expand the coordinates of these points in 2-adic fractions:

$$w_l^j = \sum_{n=0}^{\infty} \beta_{ln}^j 2^n, \beta_{ln}^j = 0, 1.$$

[2]This is the domain of all weight vectors $w = (w_0, ..., w_{N-1})$ which give the right uotput $y^{(\alpha)}$ for the input $x^{(\alpha)}$.

Then we set $[w_l^j]_k = \sum_{n=0}^{k-1} \beta_{ln}^j 2^n$. It is evident that $U_{\epsilon_k}(w_l^j) = U_{\epsilon_k}([w_l^j]_k)$ and $U_{\epsilon_k}^N(w^j) = U_{\epsilon_k}^N([w^j]_k)$, where $[w^j]_k = ([w_0^j]_k, ..., [w_{N-1}^j]_k)$. Thus we can always determine the domain of $[x^{(\alpha)}/y^{(\alpha)}]$-learning by points $d^j = (d_0^j, ..., d_{N-1}^j), j = 1, ..., s_\alpha$, with coordinates belonging to the set

$$\mathbf{N}_k = \{n = \sum_{l=0}^{k-1} \beta_l 2^l, \beta_l = 0, 1\},$$

i.e., $d^j \in \mathbf{N}_k^N$. The set $\mathbf{N}_k^N$ contains $H_k = 2^{kN}$ points. If H_k is not very large (comparing with our computing abilities) then we can find W_α exactly by checking H_k times the condition $X_\alpha(d) \in V_\alpha, d \in \mathbf{N}_k^N$. If H_k is rather large then we can find an approximation of W_α by using different algorithms of random search.

We propose an algorithm of a *random extension* of 2-adic balls. Our algorithm is based on the following heuristic considerations. If we have found a point $w \in W_\alpha$ then by Lemma 1 the ball $U_{\epsilon_k}^N(w) \subset W_\alpha$ and it is natural to extend this ball in W_α as much as possible

We shall use the following technical result.

Lemma 2. *Let $v, d \in \mathbf{N}_k^N$ and let $v \in S_{2\epsilon_k}^N(d)$. Then*

$$(\{d\} \cup S_{2\epsilon_k}^N(d)) \cap \mathbf{N}_k^N = (\{v\} \cup S_{2\epsilon_k}^N(v)) \cap \mathbf{N}_k^N. \tag{10}$$

Proof. Let $v = (v_j), d = (d_j), x = (x_j), j = 0, ..., N-1, v_j, d_j, x_j \in \mathbf{N}_k^N$ and let

$$v_j = \sum_{n=0}^{k-2} \beta_{jn} 2^n + \lambda_j 2^{k-1}, \beta_{jn}, \lambda_j = 0, 1. \tag{11}$$

Then, as $v \in S_{2\epsilon_k}^N(d)$, we have

$$d_j = \sum_{n=0}^{k-2} \beta_{jn} 2^n + \gamma_j 2^{k-1}, \gamma_j = 0, 1, \tag{12}$$

where, for some j, $\gamma_j \neq \lambda_j$. If $x = (x_j) \in S_{2\epsilon_k}^N(d) \cap \mathbf{N}_k^N$ then $x_j = \sum_{n=0}^{k-2} \beta_{jn} 2^n + \delta_j 2^{k-1}, \delta_j = 0, 1$. If $\delta_j = \lambda_j$ for all j then $x = v$; if there exists $j = q$ such that $\delta_q \neq \lambda_q$ then $x \in S_{2\epsilon_k}^N(v)$.

By this lemma if we take a point $v \in S_{2\epsilon_k}^N(d) \cap \mathbf{N}_k^N$ and consider the sphere of the same radius with center at v (in $\mathbf{N}_k^N$), $S_{2\epsilon_k}^N(v) \cap \mathbf{N}_k^N$, then we shall not obtain new points. This result will be used in the following algorithm of learning. It implies that we have to use spheres of larger radii for extending the sphere $S_{2\epsilon_k}^N(d)$.

3. Algorithm of random learning

(A1) We choose randomly a point $w^0 \in \mathbf{N}_k^N$ and test the condition

$$X_\alpha(v) \in V_\alpha \tag{13}$$

for $v = w^0$.

(A2) If (13) is not satisfied we choose randomly a new point $w^1 \in \mathbf{N}_k^N, w^1 \neq w^0$ and check (13). Suppose that after l steps we have found the point $w^l \in \mathbf{N}_k^N$, which satisfies (13). We set $v^0 = w^l$.

Remark. In the latter case (by Lemma 1) the ball $U_{\epsilon_k}^N(v^0) \subset W_\alpha$. Thus to extend the domain of learning, we have to check the points belonging to the sphere $S_{2\epsilon_k}(v^0) \cap \mathbf{N}_k^N$. We note that the set $S_{2\epsilon_k}(v^0) \cap \mathbf{N}_k^N$ contains 2^N points. If the coordinates of v^0 have expansions (11), then elements of $S_{2\epsilon_k}(v^0) \cap \mathbf{N}_k^N$ have the coordinates with expansions (12). Thus we can essentially reduce a volume of computations on the sphere $S_{2\epsilon_k}(v^0)$ if we use the "local coordinates" γ_j.

(A3) We choose randomly $w^0 \in S_{2\epsilon_k}(v^0) \cap \mathbf{N}_k^N$ and check the condition (13). If (13) is not satisfied we choose randomly a new point $w^1 \in S_{2\epsilon_k}(v^0) \cap \mathbf{N}_k^N, w^1 \neq w^0$ and check (13). Suppose that after l steps we have found the point $w^l \in S_{2\epsilon_k}(v^0) \cap \mathbf{N}_k^N$ which satisfies (13). We set $v^1 = w^l$.

By repeating (A3)[3] we shall find a sequence of points $v^j \in W_\alpha \cap S_{2\epsilon_k}(v^0) \cap \mathbf{N}_k^N, j = 1, \dots$.

If the number 2^N is rather large for our computer system we can stop this process after some time interval, $T_{\lim}$, and find the approximation

$$W'_\alpha = \cup_{j=0}^s U_{\epsilon_k}^N(v^j), \tag{14}$$

where

$$v^1, \dots, v^s \in W_\alpha \cap S_{2\epsilon_k}(v^0) \cap \mathbf{N}_k^N, \tag{15}$$

are the points which have been found in the process of algorithm's work. If 2^N is not so large we test all points of the set $S_{2\epsilon_k}(v^0) \cap \mathbf{N}_k^N$ and find all points which satisfies (15). These points give approximation (14) of the domain of $y^{(\alpha)}$-learning.

We can now extend domain (14). We have to study condition (13) on the larger sphere $S_{4\epsilon_k}(v^j)$ (for some $j = 1, \dots, s$), because by Lemma 2 we have $S_{2\epsilon_k}(v^j) \cap \mathbf{N}_k^N \subset (\{v\} \cup S_{2\epsilon_k}(v^0)) \cap \mathbf{N}_k^N$ and we cannot extend domain (14) by considering points which belong $S_{2\epsilon_k}(v^j)$.

(A4) We choose randomly $v^j \in S_{2\epsilon_k}(v^0) \cap \mathbf{N}_k^N$ (for some $j = 1, \dots, s$) and realize the previous scheme for points of the set $S_{4\epsilon_k}(v^j) \cap \mathbf{N}_k^N$.

(A5) We can continue this procedure by considering spheres of radii $2^m \epsilon_k, m = 1, \dots, k$. When m will approaches k, the procedure will be terminated.

(A6) If there is no time restriction we shall repeat the process by starting with (A1) and considering the points of $\mathbf{N}_k^N$ which have not been considered on the previous steps.

4. Algorithm of minimization

We will minimize the functional (9). If there $\cap_{\alpha=1}^m W_\alpha \neq \emptyset$, then (at least theoretically) we can find a point w which gives the minimum $\mathcal{L} = 0$. In general case we have to find a point $w \in \cap_{\alpha_j=1}^s W_{\alpha_j}$ with the maximal s. If m is rather large then it is also natural to use an algorithm of the random search of such an s.

(B0) We apply first the algorithm of $[x^{(\alpha)}/y^{(\alpha)}]$-learning for every $\alpha = 1, \dots, m$. Thus we shall obtain approximations $W'_\alpha \subset \mathbf{N}_k^N$ of the domains of $[x^{(\alpha)}/y^{(\alpha)}]$-learning W_α.

(B1) We choose randomly α_1 and a point $w^1 \in W'_{\alpha_1}$. Then choose randomly $\alpha_2 \neq \alpha_1$ and test the condition $w^1 \in W'_{\alpha_2}$ and put $s_1(w^1) = 2$ if w^1 satisfies this condition and $s_1(w^1) = 1$ in the opposite case; then we choose randomly $\alpha_3 \neq \alpha_1, \alpha_2$ and put $s_2(w^1) = s_2(w^1) + 1$ if $w^1 \in W'_{\alpha_3}$ and $s_2(w^1) = s_1(w^1)$ in the opposite case. After L random choices (where L is some number determined by our computing ability) we terminate the process and save the number $s_L(w^1)$.

---

[3]One has to make sure that in the random search we do not consider points that have been already examined so there should be a list of counter given to the points such that if counter=1 the point will not be visited by the algorithm again and if counter=0 it can be visited.

(B2) We repeat (B1) T times (where T is some number determined by the our computing ability) and obtain a sequence of numbers $s_L(w^1), s_L(w^2), ..., s_L(w^T)$.

(B3) We find the number $s_L(w^j) = \max_{1\leq i\leq T} s_L(w^i)$ and choose the point w^j as an approximation for the solution of the minimization problem.

We have that $\mathcal{L}(w^j; x^{(\alpha)}, y^{(\alpha)}) \leq (1 - \frac{s_L(w^j)}{m})$ (because w^j belongs the intersection of (at least) $s_L(w^j)$ domains of y^α-learning). The weight vector $w^j = (w_0^j, ..., w_{N-1}^j)$ gives the approximate solution for the learning problem.

References

1. Amit D., *Modeling Brain Function*, Cambridge University Press (1989).

2. Geszti T., *Physical Model of Neural Networks*, World Scientific (1990)

3. Hopfield J.J., Neural networks and physical systems with emergent collective computational abilities, *Proc. Natl. Acad. Sci. USA*, **79**, 1554-2558(1982).

4. Albeverio S., Tirozzi B., An introduction to the mathematical theory of neural networks, Proceedings of "Fourth Granada Lectures in Computational Physics", Ed. P. Garrido, J. Marro, *Lecture Notes on Physics*, **493**, Springer Verlag, 197-222 (1997).

5. Schikhov W., *Ultrametric calculus*, Cambridge Univ. Press, Cambridge (1984).

6. Vladimirov V. S. , Volovich I. V. , and Zelenov E. I. , *p-adic Analysis and Mathematical Physics*, World Scientific Publ., Singapore (1994).

7. Khrennikov A.Yu. , *p-adic valued distributions in mathematical physics.* Kluwer Academic Publishers, Dordrecht (1994).

8. Khrennikov A.Yu. , *Non-Archimedean analysis: quantum paradoxes, dynamical systems and biological models.* Kluwer Academic Publishers, Dordrecht (1997).

9. Albeverio S., Khrennikov A.Yu., Tirozzi B., p-adic dynamical systems and neural networks. *Math. Models and Methods in Appl. Sc.,* (1998).

10. Albeverio S., Khrennikov A.Yu., P. Kloeden, Memory retrieval as a p-adic dynamical system. *Biosystems,* **49,** 105-115 (1999).

11. Dubischar D., Gundlach V.M., Steinkamp O., Khrennikov A. Yu., A p-adic model for the process of thinking disturbed by physiological and information noise. *J. Theor. Biology,* **197,** 451-467 (1999).

DEPARTMENT OF MATHEMATICS, STATISTICS AND COMPUTER SCIENCES, UNIVERSITY OF VÄXJÖ, 35195, VÄXJÖ, SWEDEN

DIPARTIMENTO DI FISICA, UNIVERSITA DI ROMA "LA SAPIENZA", PIAZZALE ALDO MORO, I-00185 ROMA, ITALY

Canadian Mathematical Society
Conference Proceedings
Volume **29**, 2000

Quantum Effects in Lattice Models of Vector Anharmonic Oscillators

Yuri Kozitsky

ABSTRACT. A lattice model of d-dimensional quantum anharmonic oscillators with a polynomial anharmonicity and a ferroelectric interaction is considered. For arbitrary $d \in I\!N$, it is shown that the critical fluctuations of the position operator, peculiar to the critical point, are suppressed at all temperatures provided the particles are "strongly quantum". The latter means that for the corresponding one-dimensional model, the smallest distance between the energy levels of the isolated oscillator is large enough or that the reduced mass of the oscillator is less than some threshold value depending on the anharmonicity parameters. These results are the extension to the vector case of the statements obtained recently in [**2**] and [**3**] regarding such effects in similar lattice models.

1. Posing of the problem and results.

The results described in this note were obtained in the framework of the approach in quantum statistical mechanics initiated by now classical paper [**1**] published by Sergio Albeverio and Raphael Høegh–Krohn almost 25 years ago.

We consider a countable system of interacting quantum particles (oscillators) with the reduced mass $m = \mu/\hbar^2$, performing d-dimensional oscillations around their equilibrium positions, the set of which is a lattice $I\!L$, e.g. $I\!L = Z\!\!Z^\nu$ with certain $\nu \in I\!N$. The dynamics of the oscillations around the position l is described by means of the momentum and displacement operators $\{p_l, q_l\}$ densely defined on the complex Hilbert space $\mathcal{H}_l = L^2(I\!R_l^d, dx_l)$. The whole system is described by the formal Hamiltonian

$$H = \frac{1}{2} \sum_{l,l' \in I\!L} d_{ll'}(q_l, q_{l'}) + \sum_{l \in I\!L} H_l; \tag{1.1}$$

$$H_l = \frac{p_l^2}{2m} + V(q_l), \quad p_l^2 \stackrel{\text{def}}{=} (p_l, p_l), \tag{1.2}$$

1991 *Mathematics Subject Classification.* 82B10; 82B27.

Dedicated to Sergio Albeverio on the occasion of his 60th birthday.

This research was supported in part by the Polish Scientific Research Committee, Grant KBN 2 PO3A 02915.

where $(\, . \, , \, . \,)$ stands for the scalar product in $\mathbb{R}^d$ and $d_{ll'}$ form the dynamical matrix. The one-particle potential V has the form

$$V(x) = v((x,x)), \quad v(\xi) = \frac{1}{2}a\xi + \sum_{s=2}^{r} b_s \xi^s, \quad a \in \mathbb{R}, \; b_s \geq 0, \; b_r > 0. \tag{1.3}$$

The formal Hamiltonian cannot be defined directly and used is "represented" by its local versions – indexed by finite subsets $\Lambda \subset \mathbb{L}$ local Hamiltonians H_Λ. For such Λ, we set $\mathcal{H}_\Lambda = \bigotimes_{l\in\Lambda} \mathcal{H}_l$. Then each H_Λ is defined as an essentially self-adjoint lower bounded operator acting in the Hilbert space $\mathcal{H}_\Lambda$. For our purposes, it is enough to consider these Hamiltonians indexed by boxes $\Lambda = \{l = (l_1, \ldots, l_\nu) \mid l_j^0 \leq l_j \leq l_j^1, \; j = 1, \ldots, \nu; \; l_j^0 < l_j^1, \; l_j^0, l_j^1 \in \mathbb{Z} \}$ with the periodic conditions on the boundaries of Λ. Let $\mathcal{L}$ stand for the family of all such boxes. For a box Λ, let $\mathcal{L}(\Lambda)$ denote the partition of $\mathbb{L}$ by the boxes which are obtained as the translations of Λ. We also denote by $\mathcal{G}$ the group of all translations of $\mathbb{L}$. Let $\mathcal{G}(\Lambda) \subset \mathcal{G}$ be the subgroup consisting of the translations which generate the partition $\mathcal{L}(\Lambda)$. This means $\mathcal{L}(\Lambda) = \{t(\Lambda), \; t \in \mathcal{G}(\Lambda)\}$, where $t(\Lambda) = \{t(l), \; l \in \Lambda\}$. Then for a box Λ, the dynamical matrix $(d_{ll'}^\Lambda)_{l,l'\in\Lambda}$ corresponding to the periodic conditions on the boundary of Λ is defined as follows

$$d_{ll'}^\Lambda = \min\{d_{lt(l')} \; : \; t \in \mathcal{G}(\Lambda)\}. \tag{1.4}$$

Thereafter, the dynamical matrices $(d_{ll'})_{l,l'\in\mathbb{L}}$ and $(d_{ll'}^\Lambda)_{l,l'\in\Lambda}$ are supposed to have the following properties

(D1): $d_{ll'}$ is invariant under translations on $\mathbb{L}$;
(D2): $d_{ll'}^\Lambda \leq 0$, $d_{ll'} \leq 0$ (ferroelectricity), $\quad d_{ll}^\Lambda = d_{ll} = 0$;
(D3): the series $\sum_{l'\in\mathbb{L}} d_{ll'}$ converges for all $l \in \mathbb{L}$.

The mentioned local Hamiltonian H_Λ corresponding to the periodic boundary conditions is

$$H_\Lambda = \frac{1}{2} \sum_{l,l'\in\Lambda} d_{ll'}^\Lambda (q_l, q_{l'}) + \sum_{l\in\Lambda} H_l. \tag{1.5}$$

For every $\Lambda \in \mathcal{L}$ and the inverse temperature $\beta = T^{-1}$, the periodic Gibbs state in Λ is defined as a positive functional on appropriate C^*-algebra of observables as follows:

$$\gamma_{\beta,\Lambda}(A) = \frac{\text{trace}(Ae^{-\beta H_\Lambda})}{\text{trace}(e^{-\beta H_\Lambda})}. \tag{1.6}$$

By means of this state one can define temperature Green functions. For $\{\tau_0, \tau_1, \ldots, \tau_n\}$, $\tau_i \leq \tau_{i+1}$, $\tau_i \in I_\beta := [0, \beta]$, and appropriate operators $\{A_0, A_1, \ldots, A_n\}$, the temperature Green function is

$$\begin{aligned} &\Gamma^{\beta,\Lambda}_{A_0;A_1;\ldots;A_n}(\tau_0, \tau_1, \ldots, \tau_n) \\ = \; &\gamma_{\beta,\Lambda}\left(A_0 e^{-(\tau_1-\tau_0)H_\Lambda} A_1 e^{-(\tau_2-\tau_1)H_\Lambda} \ldots A_n e^{-(-\tau_n+\tau_0)H_\Lambda}\right). \end{aligned} \tag{1.7}$$

These functions can be obtained by using the analytic continuation of the usual Green functions corresponding to the unitary time evolution, see [**1**], [**8**]. Considering the limits $\Lambda \nearrow \mathbb{L}$ we will use collections of boxes $\mathcal{C}$, which are countable and completely ordered by inclusion. Our intention is to show that the large fluctuations of the displacements of particles in the model considered are absent (suppressed) at any temperature if these particles are "strongly quantum". The precise meaning of this will be clarified below. The first results in this direction (on the "physical level" of strictness) were published in [**9**]. Since that time the suppression of critical phenomena in lattice systems by quantum effects has been studied theoretically and experimentally in various physical publications (see [**7**], [**11**], [**12**]. The first mathematically strict results for the models with unbounded operators were obtained in [**2**], [**3**]. The research described in this note is a direct extension to the vector case of the statements from the latter paper.

Thus we restrict ourselves to the consideration of the operators which describe the mentioned fluctuations, namely

$$A^{(k)}(\Lambda) = \frac{1}{\sqrt{|\Lambda|}} \sum_{l\in\Lambda} q_l^{(k)}, \quad k = 1, 2, \ldots, d. \tag{1.8}$$

We set

$$\Gamma_\Lambda^{(k)}(\tau, \tau') = \Gamma^{\beta,\Lambda}_{A^{(k)}(\Lambda);A^{(k)}(\Lambda)}(\tau, \tau'). \tag{1.9}$$

The $O(d)$-symmetry of the model implies that such functions with different k coincide with each other. Further, we set

$$U_\Lambda = \sum_{k=1}^{d} U_\Lambda^{(k)}, \quad U_\Lambda^{(k)} = \frac{1}{\beta}\int\int_{I_\beta^2} \Gamma_\Lambda^{(k)}(\tau, \tau')d\tau d\tau'. \tag{1.10}$$

This parameter U_Λ describes, in some sense, variance of the fluctuations. It depends on β and is known as local (for $|\Lambda| < \infty$) static ferroelectric susceptibility (see [**7**], [**11**]).

DEFINITION 1.1. For given inverse temperature β and arbitrary choice of the sequence of boxes $\mathcal{C}$, let the sequence $\{U_\Lambda, \Lambda \in \mathcal{C}\}$ be bounded. Then the critical fluctuations of the displacements of particles are absent at this temperature.

In case this sequence is convergent, one may establish analyticity properties for thermodynamic potentials describing this model. Now we are in a position to state our main result.

THEOREM 1.2. *For any positive integer d, there exists a positive threshold value of the reduced mass m_* such that for all $m < m_*$, the critical fluctuations of the displacement of particles in the d-dimensional model described above are absent at all temperatures.*

The proof of this statement is based upon the comparison of our vector model with the model of one–dimensional quantum oscillators described by the following local Hamiltonians

$$\tilde{H}_\Lambda = \frac{1}{2}\sum_{l,l'\in\Lambda} d_{ll'}^\Lambda q_l^{(1)} q_{l'}^{(1)} + \sum_{l\in\Lambda} \tilde{H}_l, \tag{1.11}$$

$$\tilde{H}_l = \frac{1}{2m}\left(p_l^{(1)}\right)^2 + v\left((q_l^{(1)})^2\right), \tag{1.12}$$

where the canonical operators $p_l^{(1)}, q_l^{(1)}$ act in the space $L^2(\mathbb{R}_l, dx_l)$. They can be viewed as the components of the momentum and displacement operators introduced above. In the same sense we consider the operators $A^{(1)}(\Lambda)$. The other parameters of the Hamiltonians $\tilde{H}_\Lambda$, $\tilde{H}_l$ are the same as in (1.5), (1.2), and (1.3). Let the state $\tilde{\gamma}_{\beta,\Lambda}$ be defined by (1.6) with $\tilde{H}_\Lambda$ instead of H_Λ. For $0 \leq \tau \leq \tau' \leq \beta$, we set

$$\tilde{\Gamma}_\Lambda^{(1)}(\tau, \tau') = \tilde{\gamma}_{\beta,\Lambda}\left(A^{(1)}(\Lambda)e^{-(\tau'-\tau)\tilde{H}_\Lambda}A^{(1)}(\Lambda)e^{-(\tau-\tau')\tilde{H}_\Lambda}\right), \tag{1.13}$$

$$\tilde{U}_\Lambda^{(1)} = \frac{1}{\beta}\int\int_{I_\beta^2}\tilde{\Gamma}_\Lambda^{(k)}(\tau, \tau')d\tau d\tau'. \tag{1.14}$$

LEMMA 1.3. *For every β, finite Λ, and any $k = 1, 2, \ldots, d$*

$$U_\Lambda^{(k)} \leq \tilde{U}_\Lambda^{(1)}. \tag{1.15}$$

In what follows, in order to have the critical d-dimensional fluctuations suppressed it is enough to suppress these fluctuations at the same temperature in the corresponding one–dimensional model. This conclusion is evident from the physical point of view but its mathematical justification was impossible so far, mainly due of the fact that the tool of proving here – Griffiths and Lebowitz inequalities – works only in one dimension (although the Griffiths inequality holds also for $d = 2$, it is not enough to obtain such effective estimates like that described below in Proposition 1.4, which was proved in [**3**]). Having the bound (1.15) we may use the results obtained in [**3**] for the one–dimensional quantum oscillators described by the local Hamiltonians (1.11).

The dynamical matrix $(d_{ll'})$ can be defined as a linear bounded operator D on the Hilbert space $l^2(\mathbb{L})$. Since it is translation invariant (D1) and ferroelectric (D2), the series mentioned in (D3) does not depend on l. Thus

$$\|D\| = -\sum_{l' \in \mathbb{L}} d_{ll'}. \tag{1.16}$$

Define

$$T = \tilde{H}_l + \|D\|\left(q_l^{(1)}\right)^2, \quad l \in \mathbb{L}, \tag{1.17}$$

where $\tilde{H}_l$ is given by (1.12). The operator T has purely discrete nondegenerate spectrum. We set

$$T\psi_n = E_n\psi_n, \quad n \in \mathbb{Z}_+, \tag{1.18}$$

where the eigenvalues E_n are numbered in increasing order. Denote

$$\Delta = \min\{E_{n+1} - E_n \mid n \in \mathbb{Z}_+\}. \tag{1.19}$$

PROPOSITION 1.4. [**3**] Let the mass m, the spectral parameter Δ, and the interaction parameter $\|D\|$ satisfy the condition

$$m\Delta^2 > 2\|D\|. \tag{1.20}$$

Then for every $\beta > 0$ and arbitrary $\Lambda \in \mathcal{L}$, the following estimate holds

$$\tilde{U}_\Lambda^{(1)} \leq \frac{1}{m\Delta^2 - 2\|D\|}. \tag{1.21}$$

Thus in order to have the critical one-dimensional fluctuations suppressed for all temperatures it is enough to satisfy (1.20). In this case both one-dimensional and d-dimensional models may be called "strongly quantum" and the parameter appeared in the left hand side of (1.20) may be used to characterize this property quantitatively. It can be set large by means of strong hydrostatic pressure (see e.g. [**11**] or [**12**]). For the harmonic oscillator (i.e. if in (1.3) $a > 0$ and all $b_s = 0$), this parameter does not depend on m (one has $m\Delta^2 = a$ which is growing with pressure), but for the anharmonic oscillator described by the Hamiltonian (1.12), (1.3) with $r \geq 2$, it does depend.

PROPOSITION 1.5. [**3**] For the parameter $m\Delta^2$ defined by (1.19), (1.18), (1.17), (1.12), and (1.3), there exists $K > 0$ such that

$$\lim_{m \to 0+} m^{(r-1)/(r+1)} \left(m\Delta^2\right) = K. \tag{1.22}$$

Therefore, there exists $m_* > 0$ such that for $m < m_*$, the condition (1.20) is satisfied.

The proof of Theorem 1.2 easily follows from the latter three statements. The above results may be extended to the models with dynamical matrices, which are not translation invariant. For such matrices, we write $\hat{D} \leq D$ if for their elements one has $\hat{d}_{ll'} \leq d_{ll'}$ for every pair $l, l' \in \mathbb{L}$.

COROLLARY 1.6. For the model described by the Hamiltonian (1.1) - (1.3), let conditions (D2), (D3) be satisfied. Let also there exist a dynamical matrix $\hat{D}$ obeying (D1) - (D3) and such that $\hat{D} \leq D$. Then Theorem 1.2 holds for this model as well.

The proof of this statement follows from the Griffiths inequality, that will be explained in the next section. It is also worth to mention the following property of the one-dimensional models, relevant to the problems discussed here (see [**4**], [**6**]).

PROPOSITION 1.7. The local susceptibility $\tilde{U}_\Lambda^{(1)}$ defined by (1.14) is a monotone increasing continuous function of the reduced mass m.

2. The Proofs

For given inverse temperature β, let

$$C_\beta = \{\omega \in C(I_\beta \to \mathbb{R}^d) | \omega(0) = \omega(\beta)\},$$

and

$$\Omega_{\beta,\Lambda} = \{\omega_\Lambda = (\omega_l)_{l \in \Lambda} | \omega_l \in C_\beta\}, \quad \Lambda \in \mathcal{L}.$$

Since Λ is finite, $\Omega_{\beta,\Lambda}$ may be equipped with the usual Banach space structure. Let $\mathcal{B}_{\beta,\Lambda}$ stand for the Borel σ–algebra of its subsets. The space $\Omega_{\beta,\Lambda}$ may be embedded densely into the following real Hilbert space

$$\mathcal{X}_{\beta,\Lambda} = \bigotimes_{l \in \Lambda} \mathcal{X}_l, \quad \mathcal{X}_l \cong \mathcal{X} = L^2(I_\beta \to \mathbb{R}^d).$$

The scalar product in the space $\mathcal{X}$ is

$$< \omega, \omega' >= \int_{I_\beta} (\omega(\tau), \omega'(\tau)) d\tau. \tag{2.1}$$

Consider the following strictly positive trace class operator on $\mathcal{X}$

$$S = (-m\Delta_\beta + 1)^{-1} \mathbf{1}, \tag{2.2}$$

where Δ_β is the Laplace operator in $L^2(I_\beta)$, m is the reduced mass, and $\mathbf{1}$ is the identity operator in $I\!R^d$. The operator S determines on $\mathcal{X}$ an isotropic Gaussian measure χ_β having the following Laplace transform

$$\int_{\mathcal{X}} \exp\left(< \varphi, \omega >\right) \chi_\beta(d\omega) = \exp\left(< S\varphi, \varphi >\right).$$

One has $\chi_\beta(C_\beta) = 1$ (see e.g. [**8**]), thus we can define on $\mathcal{X}$ the measure σ_β which is supported on C_β and has there the following representation

$$\sigma_\beta(d\omega) = \frac{1}{Y_\beta} \exp\left(-\int_{I_\beta} V(\omega(\tau))d\tau\right) \chi_\beta(d\omega), \tag{2.3}$$

where V is given by (1.3) and Y_β is the normalization constant.

In the framework of the approach mentioned at the beginning of this note, the states in the boxes Λ are defined by means of Euclidean Gibbs measures. The periodic Euclidean Gibbs measure which corresponds to the d–dimensional model we consider in this work is the following probability measure on the Hilbert space $\mathcal{X}_{\beta,\Lambda}$

$$\mu_{\beta,\Lambda}(d\omega_\Lambda) = \frac{1}{Z_{\beta,\Lambda}} \exp\left\{\frac{1}{2}\sum_{l,l' \in \Lambda} d^\Lambda_{ll'} < \omega_l, \omega_{l'} >\right\} \bigotimes_{l\in\Lambda} \sigma_\beta(d\omega_l), \tag{2.4}$$

where $Z_{\beta,\Lambda}$ is the normalization constant and the periodic dynamical matrix $(d^\Lambda_{ll'})_{l,l'\in\Lambda}$ is defined by (1.4). Just in the same way one introduces the objects which describe the one-dimensional version of the model considered here. We denote them by $\tilde{C}_\beta$, $\tilde{\mathcal{X}}$, $\tilde{\Omega}_{\beta,\Lambda}$, $\tilde{\mathcal{X}}_{\beta,\Lambda}$, $\tilde{S}_\beta$, $\tilde{\chi}_\beta$. Obviously

$$C_\beta = \tilde{C}^d_\beta, \quad \mathcal{X} = \tilde{\mathcal{X}}^d, \quad \Omega_{\beta,\Lambda} = \tilde{\Omega}^d_{\beta,\Lambda}, \tag{2.5}$$

and

$$\gamma_\beta(d\omega_l) = \bigotimes_{k=1}^{d} \tilde{\gamma}_\beta(d\omega_l^{(k)}). \tag{2.6}$$

Then we set

$$\tilde{\sigma}_\beta(d\omega^{(1)}) = \frac{1}{\tilde{Y}_\beta} \exp\left\{-\int_{I_\beta} v\left((\omega^{(1)}(\tau))^2\right)d\tau\right\} \tilde{\chi}_\beta(d\omega^{(1)}), \tag{2.7}$$

where $\tilde{Y}_\beta$ is the normalization constant and v is given by (1.3). Thus the periodic Euclidean Gibbs measure of the one-dimensional model is

$$\tilde{\mu}_{\beta,\Lambda}(d\omega_\Lambda^{(1)}) = \frac{1}{\tilde{Z}_{\beta,\Lambda}} \exp\left\{\frac{1}{2}\sum_{l,l'\in\Lambda} d^\Lambda_{ll'} \int_{I_\beta} \omega_l^{(1)}(\tau)\omega_{l'}^{(1)}(\tau)d\tau\right\} \bigotimes_{l\in\Lambda} \tilde{\sigma}_\beta(d\omega_l^{(1)}). \tag{2.8}$$

In our approach the Green functions (1.7), (1.9), (1.13) are written as expectations with respect to the above introduced Euclidean Gibbs measures. That is

$$\Gamma^{(k)}_\Lambda(\tau,\tau') = \frac{1}{|\Lambda|}\sum_{l,l'\in\Lambda} I\!E_{\mu_{\beta,\Lambda}}\left\{\omega_l^{(k)}(\tau)\omega_{l'}^{(k)}(\tau')\right\}, \quad k = 1, 2, \dots, d, \tag{2.9}$$

$$\tilde{\Gamma}^{(1)}_\Lambda(\tau,\tau') = \frac{1}{|\Lambda|}\sum_{l,l'\in\Lambda} I\!E_{\tilde{\mu}_{\beta,\Lambda}}\left\{\omega_l^{(1)}(\tau)\omega_{l'}^{(1)}(\tau')\right\}. \tag{2.10}$$

Proof of Lemma 1.3. The proof is based on an infinite dimensional version of the Griffiths-Kelly-Sherman (GKS) inequality (see e.g. [**10**] p..119-124). A detailed study of this and other correlation inequalities for Euclidean Gibbs measures will be given in our forthcoming paper [**6**]. Let us consider

$$\mathbb{E}_{\mu_{\beta,\Lambda}}\left\{\omega_l^{(k)}(\tau)\omega_{l'}^{(k)}(\tau')\right\} = \int_{\mathcal{X}_{\beta,\Lambda}} \omega_l^{(k)}(\tau)\omega_{l'}^{(k)}(\tau')\mu_{\beta,\Lambda}(d\omega_\Lambda) =$$

$$\frac{1}{Z_{\beta,\Lambda}Y_\beta^{|\Lambda|}}\int_{\Omega_{\beta,\Lambda}} \omega_l^{(k)}(\tau)\omega_{l'}^{(k)}(\tau')\exp\{-\frac{1}{2}\sum_{\lambda,\lambda'\in\Lambda} d^\Lambda_{\lambda\lambda'}\sum_{k=1}^d\int_{I_\beta}\omega_l^{(k)}(t)\omega_{\lambda'}^{(k)}(t)dt-$$

$$\frac{a}{2}\sum_{\lambda\in\Lambda}\sum_{k=1}^d\int_{I_\beta}\left(\omega_\lambda^{(k)}(t)\right)^2 dt - \sum_{\lambda\in\Lambda}\sum_{s=1}^r b_s\int_{I_\beta}\left(\sum_{k=1}^d(\omega_\lambda^{(k)}(t))^2\right)^s dt\}\bigotimes_{\lambda\in\Lambda}\chi(d\omega_\lambda).$$

For $\eta\in[0,1]$, we set

$$\Xi_{l,l'}^{(k)}(\eta|\tau,\tau') = \mathbb{E}_{\rho_\eta}\left\{\omega_l^{(k)}(\tau)\omega_{l'}^{(k)}(\tau')\right\}, \tag{2.11}$$

where the expectation is taken with respect to such measure on $\tilde{\Omega}_{\beta,\Lambda}$

$$\begin{aligned}\rho_\eta(d\omega_\Lambda^{(k)}) &= \frac{1}{\Upsilon^{(k)}(\eta)}\exp\{-\frac{1}{2}\sum_{\lambda,\lambda'\in\Lambda} d^\Lambda_{\lambda\lambda'}\int_{I_\beta}\omega_\lambda^{(k)}(t)\omega_{\lambda'}^{(k)}(t)dt \\ &- \frac{a}{2}\sum_{\lambda\in\Lambda}\int_{I_\beta}\left(\omega_\lambda^{(k)}(t)\right)^2 dt - \sum_{\lambda\in\Lambda}\sum_{s=1}^r b_s\int_{I_\beta}\left(\omega_\lambda^{(k)}(t)\right)^{2s} dt \\ &- \eta\sum_{\lambda\in\Lambda}\sum_{n=1}^{r-1}\int_{I_\beta}\left[\left(\omega_\lambda^{(k)}(t)\right)^{2n}P_{\lambda,n}^{(k)}(t)\right]dt\}\bigotimes_{\lambda\in\Lambda}\tilde{\gamma}(d\omega_\lambda^{(k)}).\end{aligned} \tag{2.12}$$

Here

$$P_{\lambda,n}^{(k)} := \sum_{s=n+1}^r b_s\left(\sum_{k'\neq k}(\omega_\lambda^{(k')}(t))^2\right)^{s-n} \geq 0, \tag{2.13}$$

$$\begin{aligned}\Upsilon^{(k)}(\eta) &:= \int_{\tilde{\Omega}_{\beta,\Lambda}}\exp\{-\frac{1}{2}\sum_{\lambda,\lambda'\in\Lambda} d^\Lambda_{\lambda\lambda'}\int_{I_\beta}\omega_\lambda^{(k)}(t)\omega_{\lambda'}^{(k)}(t)dt \\ &- \frac{a}{2}\sum_{\lambda\in\Lambda}\int_{I_\beta}\left(\omega_\lambda^{(k)}(t)\right)^2 dt - \sum_{\lambda\in\Lambda}\sum_{s=1}^r b_s\int_{I_\beta}\left(\omega_\lambda^{(k)}(t)\right)^{2s} dt \\ &- \eta\sum_{\lambda\in\Lambda}\sum_{n=1}^{r-1}\int_{I_\beta}\left\{\left(\omega_\lambda^{(k)}(t)\right)^{2n}P_{\lambda,n}^{(k)}(t)\right\}dt\}\bigotimes_{\lambda\in\Lambda}\tilde{\gamma}(d\omega_\lambda^{(k)}).\end{aligned}$$

Taking into account (2.5) and (2.6) one obtains from (2.11)–(2.13)

$$\mathbb{E}_{\mu_{\beta,\Lambda}}\left\{\omega_l^{(k)}(\tau)\omega_{l'}^{(k)}(\tau')\right\} = \frac{1}{Z_{\beta,\Lambda}Y_\beta^{|\Lambda|}}\int_{\tilde{\Omega}_{\beta,\Lambda}^{d-1}}\Xi_{ll'}^{(k)}(1|\tau,\tau')\Upsilon^{(k)}(1) \tag{2.14}$$

$$\exp\{-\frac{1}{2}\sum_{\lambda,\lambda'\in\Lambda} d^\Lambda_{\lambda\lambda'}\sum_{k'\neq k}\int_{I_\beta}\omega_\lambda^{(k')}(t)\omega_{\lambda'}^{(k')}(t)dt - \frac{a}{2}\sum_{\lambda\in\Lambda}\sum_{k'\neq k}\int_{I_\beta}\left(\omega_\lambda^{(k')}(t)\right)^2 dt$$

$$-\sum_{\lambda\in\Lambda}\sum_{s=1}^r b_s\int_{I_\beta}\left(\sum_{k'\neq k}(\omega_\lambda^{(k')}(t))^2\right)^s dt\}\bigotimes_{\lambda\in\Lambda}\bigotimes_{k'\neq k}\tilde{\chi}(d\omega_\lambda^{(k')}).$$

It is not difficult to check that for any choice of k, l, l', and τ, τ', $\Xi^{(k)}_{l,l'}(\eta|\tau,\tau')$ given by (2.11) is a continuous function of η on $[0,1]$, which is differentiable on $(0,1)$, where its derivative is

$$\frac{\partial}{\partial\eta}\Xi^{(k)}_{l,l'}(\eta|\tau,\tau') = -\sum_{\lambda\in\Lambda}\sum_{n=1}^{r-1}\int_{I_\beta} P^{(k)}_{\lambda,n}(t) \tag{2.15}$$

$$\{\mathbb{E}_{\rho_\eta}\left[\left(\omega^{(k)}_\lambda(t)\right)^{2n}\omega^{(k)}_l(\tau)\omega^{(k)}_{l'}(\tau')\right]$$

$$-\quad \mathbb{E}_{\rho_\eta}\left[\left(\omega^{(k)}_\lambda(t)\right)^{2n}\right]\mathbb{E}_{\rho_\eta}\left[\omega^{(k)}_l(\tau)\omega^{(k)}_{l'}(\tau')\right]\}dt.$$

The measure ρ_η is such that, for every $\eta\in[0,1]$ and for any value of $P^{(k)}_{\lambda,n}(t)\geq 0$ (2.13), its moments obey the GKS-inequality

$$\mathbb{E}_{\rho_\eta}\left[\left(\omega^{(k)}_\lambda(t)\right)^{2n}\omega^{(k)}_l(\tau)\omega^{(k)}_{l'}(\tau')\right] \geq \mathbb{E}_{\rho_\eta}\left[\left(\omega^{(k)}_\lambda(t)\right)^{2n}\right]\mathbb{E}_{\rho_\eta}\left[\omega^{(k)}_l(\tau)\omega^{(k)}_{l'}(\tau')\right],$$

for every τ, τ', and t. Taking into account that $P^{(k)}_{\lambda,n}(t)\geq 0$ one deduces from (2.15)

$$\Xi^{(k)}_{l,l'}(1|\tau,\tau') \leq \Xi^{(k)}_{l,l'}(0|\tau,\tau') = \mathbb{E}_{\tilde{\mu}_{\beta,\Lambda}}\left\{\omega^{(k)}_l(\tau)\omega^{(k)}_{l'}(\tau')\right\}.$$

We use the latter estimate in (2.14) and obtain for every $\tau, \tau'\in[0,\beta]$

$$\mathbb{E}_{\mu_{\beta,\Lambda}}\left\{\omega^{(1)}_l(\tau)\omega^{(1)}_{l'}(\tau')\right\} \leq \mathbb{E}_{\tilde{\mu}_{\beta,\Lambda}}\left\{\omega^{(1)}_l(\tau)\omega^{(1)}_{l'}(\tau')\right\}\int_{\tilde{\Omega}^{d-1}_{\beta,\Lambda}}\frac{\Upsilon^{(1)}(1)}{Z_{\beta,\Lambda}Y^{|\Lambda|}_\beta}$$

$$\exp\{-\frac{1}{2}\sum_{\lambda,\lambda'\in\Lambda}d^\Lambda_{\lambda\lambda'}\sum_{k=2}^{d}\int_{I_\beta}\omega^{(k)}_\lambda(t)\omega^{(k)}_{\lambda'}(t)dt - \frac{a}{2}\sum_{\lambda\in\Lambda}\sum_{k=2}^{d}\int_{I_\beta}\left(\omega^{(k)}_\lambda(t)\right)^2dt -$$

$$\sum_{\lambda\in\Lambda}\sum_{s=1}^{r}b_s\int_{I_\beta}\left(\sum_{k=2}^{d}(\omega^{(k)}_\lambda(t))^2\right)^s dt\}\bigotimes_{\lambda\in\Lambda}\bigotimes_{k=2}^{d}\tilde{\chi}(d\omega^{(k)}_\lambda) = \mathbb{E}_{\tilde{\mu}_{\beta,\Lambda}}\left\{\omega^{(1)}_l(\tau)\omega^{(1)}_{l'}(\tau')\right\}.$$

Now we use the representation (2.9), (2.10) and the definitions (1.10), (1.14) and obtain the estimate (1.15). □

3. Acknowledgments

It is a honor and pleasure to collaborate with Sergio Albeverio that is gratefully acknowledged by the author.

References

[1] S. Albeverio, R. Høegh–Krohn: Homogeneous Random Fields and Quantum Statistical Mechanics, J. Funct. Anal., **19** 242–279 (1975)

[2] S. Albeverio, Yu. Kondratiev, Yu Kozitsky: Absence of Critical Points for a Class of Quantum Hierarchical Models, Commun. Math. Phys., **187**, 1–18 (1997)

[3] S. Albeverio, Yu. Kondratiev, Yu Kozitsky: Suppression of Critical Fluctuations by Strong Quantum Effects in Quantum Lattice Systems, Commun. Math. Phys., **194**, 493-512 (1998)

[4] S. Albeverio, Yu. Kondratiev, Yu Kozitsky: Classical Limits of Euclidean Gibbs States for Quantum Lattice Models, to appear in Letters in. Math. Phys., (1999)

[5] S. Albeverio, Yu. Kondratiev, Yu Kozitsky, M.Röckner: Uniqueness for Gibbs States of Quantum Lattices in Small Mass Regime, Preprint SFB 343, Bielefeld, 1999

[6] S. Albeverio, Yu. Kondratiev, Yu Kozitsky, M.Röckner: Correlation Inequalities for Temperature Loop Measures, (in preparation)

[7] A.D. Bruce, R.A. Cowley:Structural Phase Transitions, Taylor and Francis Ltd., 1981
[8] S.A. Globa, Yu.G. Kondratiev: The Construction of Gibbs States of Quantum Lattice Systems, Selecta Math. Sov., **9**, 297–307 (1990)
[9] T. Schneider, H. Beck, E. Stoll: Quantum Effects in an n-component Vector Model for Structural Phase Transitions, Phys. Rev. **B13**, 1123–1130 (1976)
[10] B. Simon: Functional Integration and Quantum Physics, Academic Press, New York San Francisco London, 1979
[11] S.Stamenkovic: Unified Model description of Order-Disorder and Displacive Structural Phase Transitions. Condensed Matter Physics, **2(14)**, 257-309 (1998)
[12] J.E. Tibballs, R.J. Nelmes, G.J. McIntyre: The Crystal Structure of Tetragonal KH_2 PO_4 and KD_2 PO_4 as a Function of Temperature and Pressure, J. Phys. C: Solid State Phys., **15**, 37–58 (1982)

Institute of Mathematics, Marie Curie-Sklodowska University, Lublin PL 20-31, Poland

Institute for Condensed Matter Physics. Lviv, Ukraine
E-mail address: jkozi@golem.umcs.lublin.pl

Canadian Mathematical Society
Conference Proceedings
Volume **29**, 2000

On Rank One H_{-3}-Perturbations of Positive Self–adjoint Operators

P. Kurasov and K. Watanabe

Dedicated with great admiration to S. Albeverio on the occasion of his 60th birthday

ABSTRACT. Rank one H_{-3} perturbations of positive self–adjoint operators are constructed using a certain extended Hilbert space and regularization procedures. Applications to Schrödinger operators with point interactions are discussed.

1. Introduction

Self–adjoint operators with complicated spectral structure can be obtained using finite rank perturbations of well studied operators with rather simple structure of the spectrum. Nevertheless these perturbed operators are exactly solvable in the sense that all eigenfunctions, spectral decomposition and scattering matrix can be written explicitly, sometimes using analytic functions. Suppose that A is a certain self–adjoint operator acting in the Hilbert space H. Then a finite rank perturbation of it is formally defined by the formula

$$A_V = A + V, \tag{1.1}$$

where V is a finite dimensional operator. The domains of the perturbed and unperturbed operators coincide if the perturbation is a bounded operator in the Hilbert space $V \in \mathcal{B}(H)$. But it is possible to consider more general perturbations determined by operators V acting in the scale of Hilbert spaces associated with the original operator A

$$V : H_2(A) \to H_{-2}(A). \tag{1.2}$$

In the latter case the perturbed operator can be defined using the form perturbation technique and the extension theory for symmetric operators. Really the operators A and A_V restricted to the domain

$$\mathrm{Dom}(A^0) = \{\psi \in \mathrm{Dom}(A) : \psi \in \mathrm{Ker}(V)\}$$

coincide. If the operator V is symmetric and finite dimensional then the restricted operator has finite deficiency indices and the resolvents of the perturbed and original operators are related via Krein's formula [**19, 20, 21, 24**]. This approach has been

1991 *Mathematics Subject Classification.* Primary 47B25, 47A20, 81Q10; Secondary 35P05.
Supported by The Royal Swedish Academy of Sciences, Sweden-Japan Sasakawa Foundation and Gakushuin Univ. Foundation.

used to construct the so-called operators with point interactions [**2, 7, 9, 25, 26**]. Perturbations with the support on a discrete set of points or on low dimensional manifolds have been studied in detail, especially in applications to Schrödinger operators. More general finite rank perturbations of self–adjoint operators have been investigated recently [**3, 4, 14, 18, 22, 23, 32**]. It has been shown that standard perturbation theory allows one to define only perturbations determined by operators acting as $H_1(A) \to H_{-1}(A)$, the so-called H_{-1} perturbations. It is possible to generalize this approach to include more general perturbations (1.2). The perturbed operator cannot be determined uniquely in this case only a finite parameter family corresponding to the formal expression (1.1) can be established. To determine a unique operator one needs to use extra assumptions, for example scaling properties of the original operator and the perturbation [**3, 4, 5**]. This approach does not work for H_{-3} perturbations determined by the operators acting as

$$V : H_3(A) \to H_{-3}(A). \tag{1.3}$$

Really consider the one dimensional perturbation formally determined by

$$A_\alpha = A + \alpha\langle\varphi, \cdot\rangle\varphi, \tag{1.4}$$

where $\varphi \in H_{-3}(A) \setminus H_{-2}(A)$. The restriction of the operator A to the domain $\mathrm{Dom}(A^0) = \{\psi \in H_3(A) : \langle\varphi, \psi\rangle = 0\}$ is essentially self–adjoint. Every self–adjoint extension of this symmetric operator coincides with the original operator A. Therefore the operator corresponding to (1.4) cannot be defined in the original Hilbert space using standard methods of extension theory. The aim of the current paper is to define the operator A_α as a self–adjoint operator in a certain extended Hilbert space.

Operators formally defined by (1.4) are important in different applications. For example, consider the Laplace operator with one point interaction acting in $L_2(\mathbf{R}^3)$. The deficiency indices of Laplace operator restricted to the set of functions vanishing at one point are equal to $(1, 1)$. All self–adjoint extensions of this symmetric operator are rotationally invariant. Therefore no non-spherically symmetric point interaction for the Laplace operator in $\mathbf{R}^3$ can be constructed. Different modifications of this approach have been suggested. The first mathematically rigorous approach uses perturbations in Pontryagin spaces, where the perturbed operator has been defined in a certain extension of the original Hilbert space with non positive definite scalar product [**12, 17, 27, 28, 29, 30, 31**].[1] Another approach has been used by physicists who suggested considering certain abstract boundary conditions connecting coefficients in the asymptotics of wave functions [**10, 11**] without paying any attention to the fact that the wave functions corresponding to the p-state have non-normalizable singularities of order r^{-2}. Therefore no scalar product space has been taken into consideration. Another constructive approach has been suggested by Yu.E. Karpeshina [**16**]. In the first step a set of functions having the form of p-scattering states is constructed. Then the self–adjoint operator and the Hilbert space are defined using this set of scattering waves. Recently I. Andronov suggested using extended Hilbert spaces to model non-spherically symmetric scatterers in $\mathbf{R}^2$ [**6**]. This model is very similar to the generalized point interactions introduced by B. Pavlov [**5, 25**]. We are going to use some of the ideas suggested by

[1] After finishing this manuscript the authors learned about the recent paper [**13**] where the idea of using Pontryagin spaces to define singular perturbations is developed.

I. Andronov to construct rank one H_{-3} perturbations. Nonspherically symmetric point interactions appear naturally in the studies of the Aharonov-Bohm effect but no modification of the Hilbert space is needed [**1, 8**].

In [**3, 4**] it was proven that H_{-2} perturbations can be defined uniquely using a certain regularization procedure. It appears that a similar approach works for H_{-3}-perturbations. In our paper we are going to show how to define arbitrary rank one H_{-3}-perturbations of self–adjoint operators restricting our consideration to the case of semibounded (positive) operators. The approach suggested can easily be modified to include finite rank and not necessarily semibounded operators. We are not going to discuss how to construct regularizations and relations to symmetry properties of the problem. We are planning to present those results in one of our forthcoming publications.

2. The Hilbert Space

Our aim is to define the self–adjoint operator corresponding to (1.4), where A is a certain positive self–adjoint operator acting in the Hilbert space H and φ is an element from the space H_{-3} from the scale of Hilbert spaces associated with the operator A. In what follows we are going to consider the case where

$$\varphi \in H_{-3} \setminus H_{-2}, \tag{2.1}$$

since rank one perturbations from the class H_{-2} have already been studied in detail in [**3, 4, 22, 23, 32**].

It is natural to determine the operator A_α using the restriction-extension method. The operator A_α coincides with one of the extensions of the operator A^0, which is the restriction of the operator A to the set of functions u from the domain $\mathrm{Dom}(A)$ of the operator A satisfying the condition

$$\langle \varphi, u \rangle = 0. \tag{2.2}$$

Suppose that the operator A is considered as an operator in the original Hilbert space H. Then the domain of the operator coincides with the space H_2, $\mathrm{Dom}(A) = H_2$ and the operator A^0 is essentially self–adjoint if φ satisfies (2.1). The operator A^0 is not essentially self–adjoint for such φ only if the domain of the unperturbed operator A is a subset of H_3. Therefore let us consider the operator A as a self–adjoint operator acting in the Hilbert space H_1 equipped with the scalar product

$$\langle u, v \rangle_1 = \langle u, (1 + bA)v \rangle, \tag{2.3}$$

where b is a positive real number,

$$b > 0. \tag{2.4}$$

The norm determined by the latter scalar product is equivalent to the standard norm in the space H_1 and is given by

$$\| u \|_1^2 = \langle u, (1 + bA)u \rangle.$$

Then the domain of the operator A coincides with the space H_3 and the operator is self–adjoint on this domain. The operator A^0 being the restriction of the operator A to the domain

$$\mathrm{Dom}(A_0) = \{u \in H_3 : \langle \varphi, u \rangle = 0\} \tag{2.5}$$

is a densely defined symmetric operator, since φ satisfies (2.1). The domain of the self–adjoint operator corresponding to the formal expression (1.4) necessarily contains the element $g_1 = \frac{1}{A+a_1}\varphi \in H_{-1}$. Therefore the extension of the operator A^0 corresponding to formal expression (1.4) cannot be constructed in the Hilbert space H_1. Let us consider the one dimensional extension $\mathcal{H}$ of this space,

$$\mathcal{H} = \mathrm{Dom}(A^0)\dot{+}\mathbf{C} \ni \mathcal{U} = (u, u_1). \tag{2.6}$$

Note that $\mathcal{H} \subset H_3\dot{+}\mathbf{C}$. We define the following natural embedding ρ of the space $\mathcal{H}$ into the space H_{-1}:

$$\begin{array}{rcl} \rho & : & \mathcal{H} \to H_{-1} \\ & & (u, u_1) \mapsto u + u_1 g_1. \end{array} \tag{2.7}$$

Then the scalar product in the space $\mathcal{H}$ can be introduced using the following formal calculations where b is a certain positive constant:

$$\begin{array}{rcl} \ll \mathcal{U}, \mathcal{V} \gg_{\mathcal{H}} & = & \langle \rho\mathcal{U}, \rho\mathcal{V}\rangle + b\langle \rho\mathcal{U}, A\rho\mathcal{V}\rangle \\ & = & \langle u + u_1 g_1, v + v_1 g_1\rangle + b\langle u + u_1 g_1, A(v + v_1 g_1)\rangle \\ & = & \langle u, v\rangle + b\langle u, Av\rangle + \bar{u}_1 v_1 \left(\| g_1 \|^2 + b\langle g_1, Ag_1\rangle\right) \\ & & + \bar{u}_1 \left(\langle g_1, v\rangle + b\langle Ag_1, v\rangle\right) + v_1 \left(\langle u, g_1\rangle + b\langle u, Ag_1\rangle\right). \end{array}$$

The last two terms can be simplified taking into account that

$$Ag_1 = -a_1 g_1 + \varphi$$

and the fact that the functions $u, v \in H_3$ satisfy (2.2). Then the scalar product is given by the expression

$$\begin{array}{rcl} \ll \mathcal{U}, \mathcal{V} \gg_{\mathcal{H}} & = & \langle u, v\rangle + b\langle u, Av\rangle + \bar{u}_1 v_1 \left(\| g_1 \|^2 + b\langle g_1, Ag_1\rangle\right) \\ & & + (1 - ba_1)\left(u_1\langle g_1, v\rangle + v_1\langle u, g_1\rangle\right), \end{array}$$

which can be considered only formally, since the scalar product $\langle g_1, Ag_1\rangle$ and the norm $\| g_1 \|^2$ are not defined (since φ is an element from $H_{-3} \setminus H_{-2}$). To define the scalar product we extend φ as a bounded linear functional using the equalities

$$\langle g_1, g_1\rangle = c_1, \quad \langle g_1, Ag_1\rangle = c_2, \tag{2.8}$$

where c_1 and c_2 are arbitrary positive real constants.[2] In what follows we are going to use thenotation

$$d = c_1 + bc_2 \in \mathbf{R}_+. \tag{2.9}$$

The scalar product determined by the following expression will also be considered:

$$\ll \mathcal{U}, \mathcal{V} \gg_{\mathcal{H}} = \langle u, v\rangle + b\langle u, Av\rangle + d\bar{u}_1 v_1 + (1 - ba_1)\left\{\bar{u}_1\langle g_1, v\rangle + v_1\langle u, g_1\rangle\right\}. \tag{2.10}$$

This formula defines a sesquilinear form on the domain $\mathrm{Dom}(A^0)\dot{+}\mathbf{C}$. This form defines a scalar product only if it is positive definite.

Let us denote by $\| \mathcal{U} \|^2_{\mathcal{H}} = \ll \mathcal{U}, \mathcal{U} \gg_{\mathcal{H}}$ the norm associated with the previously introduced scalar product. The space $\mathcal{H}$ with this norm is not complete, and the following lemma describes its completion with respect to this norm.

LEMMA 2.1. *Let the following inequality be satisfied*

$$d > |1 - ba_1|^2 \| g_1 \|^2_{-1}. \tag{2.11}$$

Then the norm $\| \cdot \|_{\mathcal{H}}$ *is equivalent to the standard norm in the Hilbert space* $H_1 \oplus \mathbf{C}$

$$\| \mathcal{U} \|^2 \equiv \| (u, u_1) \|^2 = \langle u, (1 + bA)u\rangle + |u_1|^2.$$

---

[2] One can use some regularization procedure similar to those introduced for H_{-2}-pertubations in [3, 4].

The completion of the space $\mathcal{H} = \mathrm{Dom}(A^0)\dot{+}\mathbf{C}$ *with respect to the norm* $\| \cdot \|_{\mathcal{H}}$ *coincides with the space* $H_1\dot{+}\mathbf{C}$.

Proof. We prove first that the norm $\| \cdot \|_{\mathcal{H}}$ can be estimated from above by the standard norm. The last term in (2.10) for $\mathcal{U} = \mathcal{V}$ can be estimated as follows

(2.12)
$$\begin{array}{rcl} |(1-ba_1)\left\{\bar{u}_1\langle g_1, u\rangle + u_1\langle u, g_1\rangle\right\}| & \leq & 2|1-ba_1||u_1| \parallel u \parallel_1 ||g_1||_{-1} \\ & \leq & |1-ba_1| \parallel g_1 \parallel_{-1} \{|u_1|^2 + \parallel u \parallel_1^2\}. \end{array}$$

It follows that

$$\begin{array}{rcl} \ll \mathcal{U},\mathcal{U} \gg_{\mathcal{H}} & \leq & \parallel u \parallel_1^2 + d|u_1|^2 + |1-ba_1|\{\parallel u \parallel_1^2 + |u_1|^2\} \\ & \leq & (\max\{1,d\} + |1-ba_1|) \parallel \mathcal{U} \parallel^2 . \end{array} \tag{2.13}$$

This upper estimate is proven without using any assumption on the parameters d, b, a_1.

Let us prove the lower estimate provided that condition (2.11) is satisfied. Using estimate (2.12) we can estimate the norm as follows

$$\begin{array}{rcl} \ll \mathcal{U},\mathcal{U} \gg & \geq & \|u\|_1^2 + d|u_1|^2 - 2|1-ba_1| \parallel g_1 \parallel_{-1} |u_1| \parallel u \parallel_1 \\ & = & (\parallel u \parallel - |1-ba_1| \parallel g_1 \parallel_{-1})^2 + \left(d - |1-ba_1|^2 \parallel g_1 \parallel_{-1}^2\right)|u_1|^2 \\ & \geq & \epsilon \parallel \mathcal{U} \parallel^2 \end{array}$$

where $\epsilon = \min\{\frac{d-|1-ba_1|^2\|g_1\|_{-1}^2}{2}, \frac{d-|1-ba_1|^2\|g_1\|_{-1}^2}{d+|1-ba_1|^2\|g_1\|_{-1}^2}\} > 0$. This completes the proof of the Lemma.

Note that the scalar product in the space $\overline{\mathcal{H}}$ calculated on the vectors with the component u_1 equal to zero is equivalent to the scalar product in the space H_1. But the decomposition of the space $\overline{\mathcal{H}} = H_1\dot{+}\mathbf{C}$ is not orthogonal. Only if $ba_1 = 1$ does the decomposition become orthogonal. We are going to use the same notation $\mathcal{H}$ for the completed space.

3. The Operator

We define the operator $\mathcal{A}$ on the set of regular elements $\mathrm{Dom}_r \subset \mathcal{H}$ which possess the representation

$$\mathcal{U} = (u, u_1) = (u_r + u_2 g_2, u_1),$$

where $u_r \in H_3, u_2 \in \mathbf{C}$. The vector

$$g_2 = \frac{1}{A+a_2} g_1 = \frac{1}{A+a_2}\frac{1}{A+a_1}\varphi \in H_1$$

is defined using another one positive parameter, $a_2 > 0$. The embedding operator ρ maps every such element to a vector from H_{-1} as follows:

$$\rho(u_r + u_2 g_2, u_1) = u_r + u_2 g_2 + u_1 g_1.$$

Then the operator $\mathcal{A}$ in $\mathcal{H}$ is defined on Dom_r in such a way that the following equality holds:

$$A\rho\mathcal{U} \equiv \rho\mathcal{A}\mathcal{U}(\mathrm{mod}\varphi), \tag{3.1}$$

where this equality in H_{-3} holds if and only if the difference between the left and right hand sides is proportional to $\varphi \in H_{-3}$. In other words, the operator $\mathcal{A}$ acts as formal adjoint operator. There exists a unique operator $\mathcal{A}$ in $\mathcal{H}$ satisfying (3.1)

$$\mathcal{A}\mathcal{U} = \mathcal{A}\begin{pmatrix} u_r + u_2 g_2 \\ u_1 \end{pmatrix} = \begin{pmatrix} Au_r - a_2 u_2 g_2 \\ u_2 - a_1 u_1 \end{pmatrix}. \tag{3.2}$$

The operator $\mathcal{A}$ is not self–adjoint. In fact it is not even symmetric. The boundary form of the operator can be calculated explicitly. We present here the result of these tedious but otherwise straightforward calculations

$$\begin{aligned} \ll \mathcal{A}\mathcal{U}, \mathcal{V} \gg_{\mathcal{H}} - \ll \mathcal{U}, \mathcal{A}\mathcal{V} \gg_{\mathcal{H}} = &< u_r, \varphi > (bv_2 + (1 - ba_1)v_1) \\ &- (b\bar{u}_2 + (1 - ba_1)\bar{u}_1) < \varphi, v_r > + a\{\bar{u}_2 v_1 - \bar{u}_1 v_2\}, \end{aligned} \tag{3.3}$$

where we used the notation

$$a = d + (ba_1 - 1) < g_1, g_2 > (a_2 - a_1).$$

This formula defines the following sesquilinear form in $\mathbf{C}^3$:

$$\ll \mathcal{U}, \mathcal{A}\mathcal{V} \gg_{\mathcal{H}} - \ll \mathcal{A}\mathcal{U}, \mathcal{V} \gg_{\mathcal{H}} = \tag{3.4}$$

$$= \left\langle \begin{pmatrix} 0 & -c & b \\ c & 0 & -a \\ -b & a & 0 \end{pmatrix} \begin{pmatrix} < \varphi, u_r > \\ u_1 \\ u_2 \end{pmatrix}, \begin{pmatrix} < \varphi, v_r > \\ v_1 \\ v_2 \end{pmatrix} \right\rangle,$$

where $c = ba_1 - 1 \in \mathbf{R}$. The rank of the 3×3 matrix appeared in this formula is equal to 2 if at least one of the parameters a, b, c is different from zero, since the characteristic polynomial for the matrix is given by $-\lambda(\lambda^2 + a^2 + b^2 + c^2)$. Then all symmetric restrictions of the operator $\mathcal{A}$ can be defined by certain boundary conditions imposed on the functions from the domain of the operator. The problem of defining a symmetric restriction of $\mathcal{A}$ is equivalent to the problem of finding a Lagrangian plane of the boundary form.

Suppose that the boundary conditions are written in the form

$$\alpha < \varphi, u_r > + \beta u_1 + \gamma u_2 = 0, \tag{3.5}$$

where $\alpha, \beta, \gamma \in \mathbf{C}$ are certain complex parameters, not all equal to zero simultaneously. Suppose that the parameter α is different from zero. Then the boundary form of the operator restricted to the linear set of functions satisfying the boundary conditions is given by

(3.6)

$$\begin{aligned} \ll \mathcal{A}\mathcal{U}, \mathcal{V} \gg_{\mathcal{H}} - \ll \mathcal{U}, \mathcal{A}\mathcal{V} \gg_{\mathcal{H}} &= \bar{u}_1 v_1 c\left(\tfrac{\beta}{\alpha} - \tfrac{\bar{\beta}}{\bar{\alpha}}\right) + \bar{u}_2 v_2 b\left(-\tfrac{\gamma}{\alpha} + \tfrac{\bar{\gamma}}{\bar{\alpha}}\right) \\ &+ \bar{u}_1 v_2 \left(a + b\tfrac{\bar{\beta}}{\bar{\alpha}} + c\tfrac{\gamma}{\alpha}\right) - \bar{u}_2 v_1 \overline{\left(a + b\tfrac{\bar{\beta}}{\bar{\alpha}} + c\tfrac{\gamma}{\alpha}\right)}. \end{aligned}$$

This expression vanishes for arbitrary $u_{1,2}, v_{1,2}$ if and only if the following three conditions are satisfied:

$$\frac{\beta}{\alpha} \in \mathbf{R}; \quad \frac{\gamma}{\alpha} \in \mathbf{R}; \quad a + b\frac{\bar{\beta}}{\bar{\alpha}} + c\frac{\gamma}{\alpha} = 0. \tag{3.7}$$

The first two conditions imply that the complex parameters α, β and γ have equal phase. Hence, without loss of generality, we can restrict our consideration to the case of real parameters, since the boundary condition (3.5) is linear. Then the first two conditions are fulfilled automatically. The third condition can be written as

$$\alpha a + \beta b + \gamma c = 0 \Leftrightarrow (\alpha, \beta, \gamma) \perp (a, b, c). \tag{3.8}$$

The symmetric restrictions of $\mathcal{A}$ have been described by three real parameters $(\alpha, \beta, \gamma) \in \mathbf{R}^3$ satisfying (3.8). Since the length of the vector (α, β, γ) does not play any role, all of the Lagrangian planes can be parameterized by one real parameter $\theta \in [0, 2\pi)$ as follows:

$$(\alpha, \beta, \gamma) = (b \sin \theta, -a \sin \theta - c \cos \theta, b \cos \theta), \tag{3.9}$$

where we have taken into account that b is not equal to zero due to (2.4). We are going to use the following definition in what follows

DEFINITION 3.1. The operator $\mathcal{A}_\theta$, $\theta \in [0, 2\pi)$, is the restriction of the operator $\mathcal{A}$ defined by (3.2) to the domain of functions $\mathcal{U} = (u, u_1) \in \mathcal{H}$ possessing the representation

$$(u, u_1) = (u_r + u_2 g_2, u_1), \ u_r \in H_3, u_{1,2} \in \mathbf{C}$$

and satisfying the boundary condition

$$b \sin \theta \langle \varphi, u_r \rangle - (a \sin \theta + c \cos \theta) u_1 + b \cos \theta u_2 = 0. \tag{3.10}$$

4. Self–adjointness

THEOREM 4.1. *The operator $\mathcal{A}_\theta$ is a self adjoint operator in $\mathcal{H}$ with the scalar product $\ll \cdot, \cdot \gg_{\mathcal{H}}$.*

Proof. It has already been proven that the operator $\mathcal{A}_\theta$ is symmetric. We are going to prove that it is self–adjoint by calculating its resolvent for large negative λ, $\lambda \ll 0$.

We prove that the range of the operator $\mathcal{A}_\theta - \lambda$ coincides with the Hilbert space $\mathcal{H}$

$$\mathcal{R}(\mathcal{A}_\theta - \lambda) = \mathcal{H},$$

i.e. that for for any $\mathcal{V} = (v, v_1) \in \mathcal{H}$ there exits an element $\mathcal{U} = (u_r + u_2 g_2, u_1) \in \mathrm{Dom}(\mathcal{A}_\theta)$ such that

$$(\mathcal{A}_\theta - \lambda)\mathcal{U} = \mathcal{V}.$$

The last equation can be written as

$$\begin{cases} (A - \lambda) u_r - (a_2 + \lambda) u_2 g_2 & = & v; \\ u_2 - (a_1 + \lambda) u_1 & = & v_1. \end{cases}$$

The first of these equations can be rewritten as

$$u_r - (a_2 + \lambda) \frac{1}{A - \lambda} g_2 u_2 = \frac{1}{A - \lambda} v,$$

which implies

$$\langle \varphi, u_r \rangle - (a_2 + \lambda) \langle \varphi, \frac{1}{A - \lambda} g_2 \rangle u_2 = \langle \varphi, \frac{1}{A - \lambda} v \rangle.$$

The vector $\mathcal{U} = (u_r + u_2 g_2, u_1)$ should satisfy the boundary condition (3.5). Hence the vector $(\langle \varphi, u_r \rangle, u_1, u_2) \in \mathbf{C}^3$ solves the system of linear equations

$$\begin{pmatrix} 1 & 0 & -(a_2 + \lambda)\langle \varphi, \frac{1}{A-\lambda} g_2 \rangle \\ 0 & -(a_1 + \lambda) & 1 \\ b \sin \theta & -a \sin \theta - c \cos \theta & b \cos \theta \end{pmatrix} \begin{pmatrix} \langle \varphi, u_r \rangle \\ u_1 \\ u_2 \end{pmatrix} = \begin{pmatrix} \langle \varphi, \frac{1}{A-\lambda} v \rangle \\ v_1 \\ 0 \end{pmatrix}. \tag{4.1}$$

The determinant of this system is given by

$$-(a_1 + \lambda) b \cos \theta - b \sin \theta (a_1 + \lambda)(a_2 + \lambda) \langle \varphi, \frac{1}{A - \lambda} g_2 \rangle + a \sin \theta + c \cos \theta. \tag{4.2}$$

To prove the theorem we have to show that the determinant is not equal to zero for some λ.

Suppose first that $\theta \neq 0$. Then the second term is dominant for very small negative λ, $\lambda < 0$. Let us denote by σ_φ the spectral measure associated with the positive self–adjoint operator A and the element $\varphi \in H_{-3} \setminus H_{-2}$. Then we have:

$$\int_0^\infty \frac{1}{(\mu+1)^3} d\sigma_\varphi(\mu) = \langle \varphi, \frac{1}{(A+1)^3}\varphi\rangle; \quad \lim_{N\to\infty} \int_0^N \frac{1}{(\mu+1)^2} d\sigma_\varphi(\mu) = \infty. \tag{4.3}$$

Actually the estimate

$$\begin{aligned} \left|\left\langle \varphi, \tfrac{\lambda}{A-\lambda}\tfrac{1}{A+a_1}\tfrac{1}{A+a_2}\varphi\right\rangle\right| &= \int_0^\infty \tfrac{|\lambda|}{\mu+|\lambda|}\tfrac{1}{\mu+a_1}\tfrac{1}{\mu+a_2} d\sigma_\varphi(\mu) \\ &\geq \int_0^N \tfrac{|\lambda|}{\mu+|\lambda|}\tfrac{1}{\mu+a_1}\tfrac{1}{\mu+a_2} d\sigma_\varphi(\mu) \\ &\geq \tfrac{|\lambda|}{N+|\lambda|}\int_0^N \tfrac{1}{\mu+a_1}\tfrac{1}{\mu+a_2} d\sigma_\varphi(\mu), \end{aligned}$$

which is uniform in $N > 0$, implies that

$$\lim_{\lambda\to-\infty} \left|\left\langle \varphi, \frac{\lambda}{A-\lambda}\frac{1}{A+a_1}\frac{1}{A+a_2}\varphi\right\rangle\right| = \infty,$$

if one takes into account (4.3). It follows that the asymptotics of the determinant for small negative λ is given by

$$-b\sin\theta\lambda\langle\varphi, \frac{\lambda}{A-\lambda}g_2\rangle$$

and it follows that the determinant is different from zero in a neighborhood of $-\infty$.

Consider now the case $\theta = 0$. The determinant is then given by

$$-a_1 b + c - b\lambda$$

and is different from zero for small negative λ, since $b \neq 0$ according to (2.4).

We have proven that the linear system (4.1) has a unique solution. Then the element $\mathcal{U}$ can be calculated to be

$$\mathcal{U} = \left(\frac{1}{A-\lambda}v + \frac{1}{A-\lambda}g_1 u_2, u_1\right).$$

The theorem is proven.

We have actually proven that the operator $\mathcal{A}_\theta$ is not only self–adjoint, but semibounded from below. Moreover the resolvent of the operator can easily be calculated. The solution of the linear system (4.1) is given by

$$\begin{aligned} \langle\varphi, u_r\rangle &= \frac{(a\sin\theta+(c-b(a_1+\lambda))\cos\theta)\langle\varphi,\frac{1}{A-\lambda}v\rangle+(a_2+\lambda)\langle\varphi,\frac{1}{A-\lambda}g_2\rangle(a\sin\theta+c\cos\theta)v_1}{\cos\theta(c-b(a_1+\lambda))+\sin\theta\big(a-b(a_1+\lambda)(a_2+\lambda)\langle\varphi,\frac{1}{A-\lambda}g_2\rangle\big)}; \\ u_1 &= \frac{b\sin\theta\langle\varphi,\frac{1}{A-\lambda}v\rangle+b\big(\cos\theta+(a_2+\lambda)\langle\varphi,\frac{1}{A-\lambda}g_2\rangle\sin\theta\big)v_1}{\cos\theta(c-b(a_1+\lambda))+\sin\theta\big(a-b(a_1+\lambda)(a_2+\lambda)\langle\varphi,\frac{1}{A-\lambda}g_2\rangle\big)}; \\ u_2 &= \frac{b\sin\theta(a_1+\lambda)\langle\varphi,\frac{1}{A-\lambda}v\rangle+(a\sin\theta+c\cos\theta)v_1}{\cos\theta(c-b(a_1+\lambda))+\sin\theta\big(a-b(a_1+\lambda)(a_2+\lambda)\langle\varphi,\frac{1}{A-\lambda}g_2\rangle\big)}. \end{aligned}$$

Then the resolvent can be calculated as

$$\frac{1}{\mathcal{A}_\theta-\lambda}(v,v_1) = \left(\frac{1}{A-\lambda}v + \left(\frac{1}{A-\lambda}g_1\right)u_2, u_1\right). \tag{4.4}$$

The resolvent restricted to the subspace $H_1 \subset \mathcal{H}$ of functions $\mathcal{V} = (v, v_1) \in \mathcal{H}$ with zero component $v_1 = 0$ is given by

$$\rho \frac{1}{\mathcal{A}_\theta - \lambda}|_{H_1} v = \frac{1}{A-\lambda} v + \tag{4.5}$$

$$+ \frac{b \sin\theta}{\cos\theta(c - b(a_1+\lambda)) + \sin\theta\left(a - b(a_1+\lambda)(a_2+\lambda)\langle\varphi, \frac{1}{A-\lambda} g_2\rangle\right)} \left(\frac{1}{A-\lambda}\varphi\right) \left\langle \varphi, \frac{1}{A-\lambda} v \right\rangle.$$

Consider the special case $\theta = 0$. In this case the resolvent and the restricted resolvent are given by

$$\frac{1}{\mathcal{A}_0 - \lambda}\mathcal{V} = \left(\frac{1}{A-\lambda} v + \left(\frac{1}{A-\lambda} g_1\right) \frac{1}{\frac{c}{b} - a_1 - \lambda} v_1, \frac{1}{\frac{c}{b} - a_1 - \lambda} v_1\right) \tag{4.6}$$

and

$$\frac{1}{A-\lambda}|_{H_1} = \frac{1}{\mathcal{A}_0 - \lambda}|_{H_1}, \tag{4.7}$$

respectively. The range of the restricted resolvent in this case is a subset of H_1 again. Moreover the restricted resolvent coincides with the resolvent of the original operator A, and this property is characteristic of the operator $\mathcal{A}_0$. In other words, the domain of the operator $\mathcal{A}_0$ contains the domain of the original operator A, and the action of the operators $\mathcal{A}_0$ and A restricted to this domain coincide,

$$\mathcal{A}_0|_{\mathrm{Dom}(A)} = A.$$

Therefore the operator $\mathcal{A}_0$ should be considered as an **unperturbed operator**, since this is the unique operator possessing the properties described above. All of the other operators $\mathcal{A}_\theta$ corresponding to $\theta \neq 0$ are perturbations of $\mathcal{A}_0$.

References

[1] R.Adami and A.Teta, On the Aharonov–Bohm Hamiltonian, *Lett. Math. Phys.*, **43**, 1998, 43–53.

[2] S.Albeverio, F.Gesztesy, R.Høegh-Krohn and H.Holden, Solvable models in quantum mechanics, Springer, 1988.

[3] S.Albeverio and P.Kurasov, Rank one perturbations, approximations, and selfadjoint extensions, *J. Funct. Anal.*, **148**, 1997, 152–169.

[4] S.Albeverio and P.Kurasov, Rank one perturbations of not semibounded operators, *Int. Eq. Oper. Theory*, **27**, 1997, 379–400.

[5] S.Albeverio and P.Kurasov, Singular perturbations of differential operators, Cambridge Univ. Press, to appear.

[6] I.Andronov, Zero-range potential model of a protruding stiffener, *J.Phys.A*, **32** , 1999, L231–238.

[7] F.A.Berezin and L.D.Faddeev, Remark on the Schrödinger equation with singular potential, *Dokl. Akad. Nauk SSSR*, **137**, 1011–1014, 1961.

[8] L.Dąbrowski and P.Šťovíček, Aharonov–Bohm effect with δ–type interaction, *J. Math. Phys.*, **39**, 1998, 47–62.

[9] Yu.N.Demkov and V.N.Ostrovsky, Zero-range potentials and their applications in atomic physics, Plenum Press, New York and London, 1988.

[10] Yu.N.Demkov and G.F.Drukarev, Loosely bound particle with nonzero orbital angular momentum in an electric or magnetic field, Sov. Phys. JETP, **54** (4), 1981, 650–656.

[11] Yu.N.Demkov and G.F.Drukarev, Potential curves for the p-state of the system AB^- in the approximation of small radius potentials, in: Problems of atomic collisions, vyp. 3, Leningrad Univ. Press, Leningrad, 1986, 184–195.

[12] J.F.van Diejen and A.Tip, Scattering from generalized point interactions using self–adjoint extensions in Pontryagin spaces, *J. Math. Phys.*, **32**, 1991, 630–641.

[13] A.Dijksma, H.Langer, Yu.Shondin and C.Zeinstra, Self–adjoint operators with inner singularities and Pontryagin spaces, to be published in Proceedings of Krein conference.

[14] F.Gesztesy and B.Simon, Rank-one perturbations at infinite coupling, *J. Funct. Anal.*, **128**, 1995, 245–252.

[15] S.Hassi and H. de Snoo, One-dimensional graph perturbations of self–adjoint relations, *Ann. Acad. Sci. Fenn. Math.*, **22**, 1997, 123–164.

[16] Yu.E.Karpeshina, Zero-Range Model of p-scattering by a Potential Well, Preprint Forschungsinstitut für Mathematik, ETH, Zürich, 1992.

[17] A.Kiselev and I.Popov, An indefinite metric and scattering by regions with a small aperture, *Mat. Zametki*, **58**, 1995, 837–850, 959.

[18] A.Kiselev and B.Simon, Rank one perturbations with infinitesimal coupling, *J. Funct. Anal.*, **130**, 1995, 345–356.

[19] M.Krein, On Hermitian operators whose deficiency indices are 1, *Comptes Rendus (Doklady) Acad. Sci. URSS (N.S.)*, **43**, 131-134, 1944.

[20] M.Krein, On Hermitian operators whose deficiency indices are equal to one. II, *Comptes Rendus (Doklady) Acad. Sci. URSS (N.S.)*, **44**, 131-134, 1944.

[21] M.Krein, On a remarkable class of Hermitian operators, *Comptes Rendus (Doklady) Acad. Sci. URSS (N.S.)*, **44**, 175-179, 1944.

[22] P.Kurasov and S.T.Kuroda, Krein's formula in perturbation theory, in preparation.

[23] S.T.Kuroda, H.Nagatani, $\mathcal{H}_{-2}$-construction and some applications, in: Mathematical Results in Quantum Mechanics (QMath 7, Prague, 1998), J.Dittrich, P.Exner, M.Tartar (eds.), *Operator Theory: Advances and Applications*, **108** Birkhäuser Verlag, 1999.

[24] M.Naimark, On spectral functions of a symmetric operator, *Bull. Acad. Sciences URSS*, **7**, 285-296, 1943.

[25] B.Pavlov, The theory of extensions and explicitly solvable models, *Uspekhi Mat. Nauk*, **42**, 99-131, 1987.

[26] B.Pavlov, Boundary conditions on thin manifolds and the semiboundedness of the three-body Schrödinger operator with point potential, *Mat. Sb. (N.S.)*, **136**, 163-177, 1988.

[27] B.S.Pavlov and I.Popov, An acoustic model of zero-width slits and the hydrodynamic stability of a boundary layer, *Teoret. Mat. Fiz.*, **86**, 1991, 391–401.

[28] B.S.Pavlov and I.Popov, Scattering by resonators with small and point holes, *Vestnik Leningrad. Univ. Mat. Mekh. Astronom.*, vyp. 3, 1984, 116–118.

[29] I.Popov, The Helmholtz resonator and operator extension theory in a space with indefinite metric, *Mat. Sb.*, **183**, 1992, 3–27.

[30] Yu.Shondin, Perturbation of differential operators on high-codimension manifold and the extension theory for symmetric linear relations in an indefinite metric space, *Teoret. Mat. Fiz.*, **92**, 1992, 466–472.

[31] Yu.Shondin, Quantum mechanical models in $\mathbf{R}^n$ connected with extensions of the energy operator in a Pontryagin space, *Teoret. Mat. Fiz.*, **74**, 1988, 331–344.

[32] B.Simon, Spectral analysis of rank one perturbations and applications, in: Mathematical quantum theory. II. Schrödinger operators (Vancouver, BC, 1993), 109–149, *CRM Proc. Lecture Notes*, **8**, AMS, Providence, RI, 1995.

DEPT. OF MATH., STOCKHOLM UNIVERSITY, 10691 STOCKHOLM, SWEDEN
E-mail address: `pak@matematik.su.se`

DEPT. OF MATHEMATICS, GAKUSHUIN UNIV., 1-5-1 MEJIRO, TOSHIMA-KU, TOKYO, 171-8588 JAPAN
E-mail address: `kazuo.watanabe@gakushuin.ac.jp`

Canadian Mathematical Society
Conference Proceedings
Volume **29**, 2000

On a strong maximum principle for Dirichlet forms

Kazuhiro Kuwae

Dedicated to Professor Sergio Albeverio on the occasion of his sixtieth birthday.

ABSTRACT. We prove a strong maximum principle for subharmonic functions on relatively compact domains in the framework of strongly local regular symmetric Dirichlet forms whose part spaces are associated with strong Feller processes.

1. Introduction

The maximum principle for subharmonic functions on a bounded domain has been treated in various fields of analysis. Classically, subharmonic functions on the disk D of complex plane $\mathbb{C}$ are defined to be (upper semi) continuous functions which satisfy the inequalities related to the mean value property of harmonic functions on D. In the framework of smooth Riemannian manifolds and its analysis, there is a notion of subharmonic functions which strongly depends on the smooth (or C^2-)structure of the underlying manifold (see V. Definition 3.6 in [**10**]). The maximum principle for smooth Riemannian manifolds plays a fundamental role in spectral geomtery (cf. Appendix 4 in [**10**]). On the other hand, in recent progress of differential geometry, non-smooth manifolds and singular geometric objects have been treated, for example, Lipschitz Riemannian manifolds, Alexandrov spaces (see [**3**],[**9**],[**7, 8**]), the limit space of measured Gromov-Hausdorff convergence of closed Riemannian manifolds with Ricci curvature bounded below and diameter bounded above, and so on. Such singular spaces do not admit the C^2-structure even if we delete some singular points, hence the subharmonics as in smooth Riemannian manifolds can not be defined in the same way. Fortunately, the notion of subharmonics for strongly local regular Dirichlet forms was done by Biroli-Mosco [**1**] and Sturm [**11**]. In [**1**], a weak maximum principle is proved. However, it is insufficient to analyse the above singular spaces.

The purpose of this paper is to prove the strong maximum principle for subharmonic functions in a relatively compact open domain G for strongly local regular symmetric Dirichlet forms, whose part spaces on the domain G are associated with

---

1991 *Mathematics Subject Classification.* Primary 31C25, 60J45; Secondary 53C23, 58G25.

The author was partially supported by a Grant-in-Aid for Scientific Research No. 11740112 from the Japanese Ministry of Education, Science, Sports and Culture.

strong Feller processes. This framework includes the case of Alexandrov spaces (cf. [**7**, **8**]).

2. Framework

Let $(\mathcal{E},\mathcal{F})$ be a strongly local regular Dirichlet form on $L^2(X;m)$ with a special standard core $\mathcal{C}$, where X is a locally compact separable metric space with the one point compactification $X_\Delta := X\cup\Delta$ of X and m is a positive Radon measure on X with full topological support (cf. [**5**]). The *strong local property* of $(\mathcal{E},\mathcal{F})$ means that $\mathcal{E}(u,v)=0$ if $u,v\in\mathcal{F}$ have compact supports and v is m-a.e. constant on a neighbourhood of supp$[u]$. According to the basic theory of Dirichlet forms, every strongly local regular Dirichlet form $(\mathcal{E},\mathcal{F})$ admits the following expression:

$$\mathcal{E}(u,v)=\frac{1}{2}\int_X \mu_{\langle u,v\rangle}(dx),\qquad u,v\in\mathcal{F},$$

where $\mu_{\langle u,v\rangle}$ is a signed finite Borel measure on X, called the *energy measure* of $(\mathcal{E},\mathcal{F})$. We simply set $\mu_{\langle u\rangle}:=\mu_{\langle u,u\rangle}$ for $u\in\mathcal{F}$. The strong local property of $(\mathcal{E},\mathcal{F})$ can be extended to $\mu_{\langle u,v\rangle}$ as follows: $\mu_{\langle u,v\rangle}=0$ if $u,v\in\mathcal{F}$ have compact supports and v is m-a.e. constant on a neighbourhood of supp$[u]$.

Let $\mathbf{M}:=(\Omega,X_t,\{P_x\}_{x\in X_\Delta})$ be an m-symmetric diffusion process associated with $(\mathcal{E},\mathcal{F})$:

$$T_tu(x)=p_tu(x):=E_x[u(X_t)]\ \ m\text{-a.e. } x\in X$$

for any Borel function $u\in L^2(X;m)$. Here T_t is the $L^2(X;m)$-semigroup associated with $(\mathcal{E},\mathcal{F})$. Throughout this paper, we fix a non-empty open subset G of X and denote by $(\mathcal{E}_G,\mathcal{F}_G)$ the part space on G:

$$\begin{cases}\mathcal{F}_G:=\{u\in\mathcal{F}\mid \tilde{u}=0 \text{ q.e. on } X\setminus G\}\\ \mathcal{E}_G(u,v):=\mathcal{E}(u,v) \text{ for } u,v\in\mathcal{F}_G.\end{cases}$$

Here $\tilde{u}$ is a quasi continuous m-version of u with respect to 1-capacity of $(\mathcal{E},\mathcal{F})$ (cf. Chapter 2 in [**5**]). Then $(\mathcal{E}_G,\mathcal{F}_G)$ is a regular Dirichlet form on $L^2(G;m)$ having $\mathcal{C}_G:=\{u\in\mathcal{C}\mid \text{supp}[u]\subset G\}$ as its special standard core (see Theorem 4.4.3 in [**5**]). Let $(L^G,\mathcal{D}(L^G))$ be the $L^2(G;m)$-generator associated with $(\mathcal{E}_G,\mathcal{F}_G)$. We denote by $\mathbf{M}^G:=(\Omega,X_t^G,\{P_x\}_{x\in G_\Delta})$ the part process on G:

$$X_t^G(\omega):=\begin{cases}X_t(\omega) & 0\le t<\dot{\sigma}_{X\setminus G}(\omega)\\ \Delta & \dot{\sigma}_{X\setminus G}(\omega)\le t.\end{cases}$$

Here $\dot{\sigma}_{X\setminus G}(\omega):=\inf\{t\ge 0\mid X_t\in X\setminus G\}$ is the first entry time to $X\setminus G$ (see A. 2 in [**5**]). The transition kernel of $\mathbf{M}^G$ is given by

$$p_t^G(x,dy)=P_x(X_t\in dy,\, t<\sigma_{X\setminus G}),\ x\in G,$$

where $\sigma_{X\setminus G}:=\inf\{t>0\mid X_t\in X\setminus G\}$ is the first hitting time to $X\setminus G$. Then $\mathbf{M}^G$ is an m-symmetric diffusion process which is associated with $(\mathcal{E}_G,\mathcal{F}_G)$ (see Theorem 4.4.3 in [**5**]): The $L^2(G;m)$-semigroup T_t^G associated with $(\mathcal{E}_G,\mathcal{F}_G)$ satisfies

$$T_t^Gf(x)=p_t^Gf(x):=E_x[f(X_t):t<\sigma_{X\setminus G}]\ \ m\text{-a.e. } x\in G$$

for Borel function $f\in L^2(G;m)$. $\mathbf{M}^G$ is said to be a *strong Feller process* if for each $t>0$ and any bounded Borel functions f on G, $p_t^Gf\in C_b(G)$, where $C_b(G)$

is the set of bounded continuous functions on G. We say that $(\mathcal{E}_G, \mathcal{F}_G)$ is *transient* (cf. §1.5 in [**5**]) if there exists an m-a.e. strictly positive $g \in L^1(G;m)$ such that

$$\int_G |u| g dm \le \sqrt{\mathcal{E}(u,u)} \text{ for any } u \in \mathcal{F}_G,$$

which is equivalent to that there exists an m-a.e. strictly positive $f \in L^1(G;m)$ with $\int_0^\infty T_t^G f dt < \infty$ m-a.e. on G. Here T_t^G is the $L^1(G;m)$-contraction semigroup for $(\mathcal{E}_G, \mathcal{F}_G)$. Under the transience of $(\mathcal{E}_G, \mathcal{F}_G)$, $\mathcal{E}(u,u) = 0$ implies $u = 0$ m-a.e. for $u \in \mathcal{F}_G$. A sufficient condition for the transience of $(\mathcal{E}_G, \mathcal{F}_G)$ is discussed in Theorem 8.1 of [**7**]. We say that $\mathbf{M}^G$ is *transient* if and only if so is $(\mathcal{E}_G, \mathcal{F}_G)$. We define the *space of functions locally in* $\mathcal{F}$ (resp. $\mathcal{F}_G$), say *local space of* $\mathcal{F}$ (resp. *local space of* $\mathcal{F}_G$), denoted by $\mathcal{F}_{loc}$ (resp. $(\mathcal{F}_G)_{loc}$):

$$\begin{aligned}\mathcal{F}_{loc} &:= \{u : X \to \mathbb{R} \mid \text{ for any relatively compact open set } O \text{ there} \\ &\qquad \text{exists } u_O \in \mathcal{F} \text{ such that } u = u_O \ m\text{-a.e. on } O\}. \\ (\mathcal{F}_G)_{loc} &:= \{u : G \to \mathbb{R} \mid \text{ for any relatively compact open set } O \text{ with} \\ &\qquad \overline{O} \subset G, \text{ there exists } u_O \in \mathcal{F}_G \text{ such that } u = u_O \ m\text{-a.e. on } O\}.\end{aligned}$$

The energy measures $\mu_{\langle\cdot,\cdot\rangle}$ for the elements of $\mathcal{F}$, $\mathcal{F}_G$ can be extended for those of $\mathcal{F}_{loc}$, $(\mathcal{F}_G)_{loc}$ as signed Radon measures by the strong locality, respectively. We have

$$\mathcal{E}(u,v) = \frac{1}{2}\int_G \mu_{\langle u,v\rangle}(dx) \text{ for } u \in \mathcal{F},\ v \in \mathcal{F}_G$$

(see Lemma 5.1 and 5.2(Γ6) in [**6**]). Hence we may set $\mathcal{E}(u,v) := \frac{1}{2}\int_G \mu_{\langle u,v\rangle}(dx)$ for $u \in \mathcal{F}_{loc}$ and $v \in \mathcal{F}_G$ with compact support $\text{supp}[v] \subset G$.

3. Subharmonic functions

Let X, m, $(\mathcal{E}, \mathcal{F})$, $\mathcal{C}$, $(\mathcal{E}_G, \mathcal{F}_G)$, $(L^G, \mathcal{D}(L^G))$, $\mathcal{C}_G$, $\mathcal{F}_{loc}$, $\mathbf{M}^G$ be as in Section 2.

DEFINITION 1 (Subharmonic functions). Take an open subset U of X with $G \subset U$.

(1) A function $u : X \to \mathbb{R}$ is said to be *subharmonic in* X if $u \in \mathcal{F}_{loc}$ and $\mathcal{E}(u,\varphi) \le 0$ for any $\varphi \in \mathcal{C}$ with $\varphi \ge 0$.
(2) A function $u : G \to \mathbb{R}$ is said to be *subharmonic in* G if $u \in (\mathcal{F}_G)_{loc}$ and $\mathcal{E}(u,\varphi) \le 0$ for any $\varphi \in \mathcal{C}_G$ with $\varphi \ge 0$.
(3) A function $u : U \to \mathbb{R}$ is said to be *subharmonic in* G if $u \in (\mathcal{F}_U)_{loc}$ and $\mathcal{E}(u,\varphi) \le 0$ for any $\varphi \in \mathcal{C}_G$ with $\varphi \ge 0$.

Denote by $sub\mathcal{H}(G,U)$ the family of subharmonic functions in G on U and set $sub\mathcal{H}(G) := sub\mathcal{H}(G,G)$

DEFINITION 2 (Harmonic functions). Take an open subset U of X with $G \subset U$.

(1) A function $u : X \to \mathbb{R}$ is said to be *superharmonic in* X if $-u$ is subharmonic in X, and is said to be *harmonic in* X if u and $-u$ are subharmonic in X.
(2) A function $u : G \to \mathbb{R}$ is said to be *superharmonic in* G if $-u$ is subharmonic in G, and is said to be *harmonic in* G if u and $-u$ are subharmonic in G.
(3) A function $u : U \to \mathbb{R}$ is said to be *superharmonic in* G if $-u$ is subharmonic in G, and is said to be *harmonic in* G if u and $-u$ are subharmonic in G.

Denote by $\mathcal{H}(G,U)$ the family of harmonic functions in G on U and set $\mathcal{H}(G) := \mathcal{H}(G,G)$

LEMMA 3. *Let U be an open set with $G \subset U$. The following are equivalent.*

(1) $u \in sub\mathcal{H}(G,U)$.

(2) $u \in (\mathcal{F}_U)_{loc}$ *and* $\mathcal{E}(u,v) \le 0$ *for any* $v \in \mathcal{F} \cap C_0(G)$ *with* $v \ge 0$.

(3) u *is a subsolution of* $L^U w = 0$ *on* G *in the sense of* [**11**]*(see (2.1) in* [**11**]*):* $u \in (\mathcal{F}_U)_{loc}$ *and* $\mathcal{E}(u,v) \le 0$ *for any* $v \in \mathcal{F}_G$ *with compact support* $\mathrm{supp}[v] \subset G$ *and* $v \ge 0$ *m-a.e.*

PROOF. The implications (3)$\Longrightarrow$(2) $\Longrightarrow$(1) are trivial. We only show (1)$\Longrightarrow$(3). Suppose $u \in (\mathcal{F}_U)_{loc}$. Take $v \in \mathcal{F}_G$ with $K :=\mathrm{supp}[v]$ is compact and $\mathrm{supp}[v] \subset G$. We may assume $0 \le v \le 1$ by truncating v without loss of generality. Then there exists a sequence $\{v_n\}$ in $\mathcal{C}_G$ such that $v_n \to v$ in $\mathcal{E}_1^{1/2}$-norm as $n \to \infty$. We take a relatively compact open set O with $K \subset O \subset \overline{O} \subset G$ and nonnegative $\varphi \in \mathcal{C}_G$ with $\varphi = 1$ on K and $= 0$ on $X \setminus O$. We further set $\varphi_n := (\frac{1}{n} + \phi_{\frac{1}{n}}(v_n))\varphi$. Here $\phi_\varepsilon, \varepsilon > 0$ is the function specified in $(\mathcal{E}, 4)$ of [**5**]. Then $\varphi_n \in \mathcal{C}_G$ and $0 \le \varphi_n \le \|\varphi\|_\infty(1 + \frac{2}{n})$ and

$$\begin{aligned}\mu_{\langle\varphi_n\rangle}(X) &\le 2\int_X (1+\frac{2}{n})^2 \mu_{\langle\varphi\rangle}(dx) + 2\int_X \varphi(x)^2 \mu_{\langle\phi_{\frac{1}{n}}(v_n)\rangle} \\ &\le 4(1+\frac{2}{n})^2 \mathcal{E}(\varphi,\varphi) + 4\|\varphi\|_\infty^2 \mathcal{E}(v_n,v_n),\end{aligned}$$

which implies that $\{\varphi_n\}_{n\in\mathbb{N}}$ is $\mathcal{E}_1^{1/2}$-bounded and $\varphi_n \to v\varphi = v$ in $L^2(G;m)$ as $n \to \infty$. The Banach-Saks theorem tells us that there exists a subsequence $\{n_k\}$ of $\{n\}$ such that $w_N := (1/N)\sum_{k=1}^N \varphi_{n_k} \in \mathcal{C}_G$ is $\mathcal{E}_1^{1/2}$-convergent to v. Since $w_N \ge 0$ on X, we have $\mathcal{E}(u, w_N) \le 0$, hence $\mathcal{E}(u,v) \le 0$. This completes the proof. □

COROLLARY 4. *The following are equivalent.*

(1) $u \in sub\mathcal{H}(G)$.

(2) $u \in (\mathcal{F}_G)_{loc}$ *and* $\mathcal{E}(u,v) \le 0$ *for any* $v \in \mathcal{F} \cap C_0(G)$ *with* $v \ge 0$.

(3) u *is a subsolution of* $L^G w = 0$ *on* G *in the sense of* [**11**]*(see (2.1) in* [**11**]*):* $u \in (\mathcal{F}_G)_{loc}$ *and* $\mathcal{E}(u,v) \le 0$ *for any* $v \in \mathcal{F}_G$ *with compact support* $\mathrm{supp}[v] \subset G$ *and* $v \ge 0$ *m-a.e.*

Here after, we assume $\partial G := \overline{G} \setminus G \neq \emptyset$ except Proposition 8. For a function f on G, we write $f \in C(\overline{G})$ if f is continuous on G and f has a continuous extension $\bar{f}$ to $\overline{G}$. For simplicity, we use the same symbol f for such extension $\bar{f}$.

LEMMA 5. *Suppose that G is relatively compact. Let U be an open set with $G \subset U$. Assume $u \in sub\mathcal{H}(G,U)$ and $u|_G \in C(\overline{G})$. Then $u|_G \in sub\mathcal{H}(G)$.*

PROOF. It suffices to show $u|_G \in (\mathcal{F}_G)_{loc}$. Set $M_G := \sup_{x\in\overline{G}} |u(x)|$. Then $u = (-M_G) \vee u \wedge M_G$ m-a.e. on G. Take a relatively compact open set O with $\overline{O} \subset G \subset U$. We can take a $\varphi_O \in \mathcal{C}_G$ with $\varphi_O = 1$ on $\overline{O}$ and a $u_O \in \mathcal{F}_U$ with $u = u_O$ m-a.e. on O. Then we have $u = ((-M_G) \vee u_O \wedge M_G)\varphi_O$ m-a.e. on O, which implies $u|_G \in (\mathcal{F}_G)_{loc}$. This completes the proof. □

4. Maximum principle

Let X, m, $(\mathcal{E},\mathcal{F})$, $\mathcal{C}$, $(\mathcal{E}_G,\mathcal{F}_G)$, $(L^G,\mathcal{D}(L^G))$, $\mathcal{C}_G$, $\mathcal{F}_{loc}$, $\mathbf{M}^G$ be as in Section 2. The following weak maximum principle due to Ôkura is a version of what was proved in [**1**].

PROPOSITION 6 (Weak maximum principle, cf. (*i*) in [**1**]). *Let U be an open set with $G \subset U$. Suppose that G is relatively compact and $(\mathcal{E}_G, \mathcal{F}_G)$ is a transient Dirichlet form on $L^2(G;m)$. Then, for $u \in sub\mathcal{H}(G,U)$ with $u|_G \in C(\overline{G})$, we have*

$$u \leq \max_{x \in \partial G} u(x) \text{ on } G.$$

PROOF. The proof is the same as in (*i*) of [**1**]. We only note that Lemma 3 is used in this proof. We omit it □

THEOREM 4.1 (Strong maximum principle). *Let U be an open set with $G \subset U$. Suppose that G is a connected relatively compact open set and $\mathbf{M}^G$ is a strong Feller process on G. For $u \in sub\mathcal{H}(G,U)$ with $u|_G \in C(\overline{G})$, if there exists an $x_0 \in G$ with $u(x_0) = \max_{x \in \overline{G}} u(x)$, then u is constant on G.*

COROLLARY 7. *Under the same condition in Proposition 6 or Theorem 4.1, if $u, v \in (\mathcal{F}_U)_{loc}$ satisfies $u|_G, v|_G \in C(\overline{G})$, $\mathcal{E}(u,\varphi) \leq \mathcal{E}(v,\varphi)$ for any $\varphi \in \mathcal{C}_G$ with $\varphi \geq 0$, and $u \leq v$ on ∂G, then $u \leq v$ on G.*

PROOF. Let $w := u - v$. Then $w \in sub\mathcal{H}(G,U)$. The weak maximum principle implies the assertion. □

PROPOSITION 8. *Suppose that $\mathbf{M}^G$ is a strong Feller process. Then G is connected if and only if G is finely connected with respect to $\mathbf{M}^G$ in the sense that for any finely open and finely closed Borel subset Y of G, $Y = \emptyset$ or $Y = G$.*

PROOF. The "if"part is trivial. We show the "only if"part. Let Y_1 and Y_2 be finely open Borel subset of G with $G = Y_1 \cup Y_2$ and $Y_1 \cap Y_2 = \emptyset$. Set $G_i := \{x \in G \mid P_x^G(\sigma_{Y_i} < \infty) > 0\}$ for $i = 1, 2$. Here $\{P_x^G\}_{x \in G \cup \Delta}$ is the probability law of $\mathbf{M}^G$ and $\sigma_{Y_i} := \inf\{t > 0 \mid X_t \in Y_i\}$ is the first hitting time to Y_i with respect to $\mathbf{M}^G$. Then the Borel measurability and the finely openness of Y_i imply $Y_i \subset G_i$ for each $i = 1, 2$. Note that $G \ni x \mapsto h_i(x) := P_x^G(\sigma_{Y_i} < \infty)$ is excessive with respect to $\mathbf{M}^G$, $i = 1, 2$. The strong Feller property of $\mathbf{M}^G$ tells us that each h_i is Borel measurable by the absolute continuity: $p_t^G(x, dy) \ll m(dy)$, $t > 0$, $x \in G$. Then we have that each h_i is lower semi continuous on G, hence G_i is open in G. Since each Y_i is quasi-open and quasi-closed with respect to the 1-capacity of $(\mathcal{E}_G, \mathcal{F}_G)$, there exists a properly exceptional set N of $\mathbf{M}^G$ such that $Y_1 \setminus N$ and $Y_2 \setminus N$ are $\mathbf{M}^G$-invariant by Lemma 4.6.3 in [**5**]. This implies $G_i \subset Y_i \cup N$, $i = 1, 2$. Then we have $G_1 \cap G_2 \subset N$, hence $m(G_1 \cap G_2) = 0$, consequently $G_1 \cap G_2 = \emptyset$. In particular, $G_i = Y_i$, $i = 1, 2$. The connectivity of G implies the assertion. This completes the proof. □

REMARK 9. Under the same assumption of Proposition 8, Takeda[**12**] proved the irreducibility of $(\mathcal{E}_G, \mathcal{F}_G)$ without using the strong locality.

PROOF OF THEOREM 4.1. Put $M := \max_{x \in \overline{G}} u(x)$. Then $u \leq M$ on $\overline{G}$. Owing to Lemma 3, $v := M - u$ is a nonnegative bounded (on $\overline{G}$) supersolution of the equation $L^U w = 0$ on G in the sense of [**11**]. In view of Lemma 5, $v|_G$ is a nonnegative bounded supersolution of the equation $L^G w = 0$ on G in the sense of [**11**]. Then v is an excessive function on G in the sense of [**11**](see pp. 177 in [**11**]) by Lemma 3 of [**11**]. Then, we have that the sets $Y_1 := \{v > 0\} := \{x \in G \mid u(x) < M\}$ and $Y_2 := G \setminus Y_1 := \{x \in G \mid u(x) = M\}$ are finely open with respect to $\mathbf{M}^G$. In fact, v is nonnegative bounded on G and excessive in the sense that $p_t^G v \leq v$ m-a.e. on G. The strong Feller property of $\mathbf{M}^G$ and the continuity of v on $\overline{G}$ imply

$p_t^G v \le v$, $\forall t > 0$ on G and $\lim_{t\to 0} p_t^G v(x) = v(x)$ for $x \in G$. We see $p_t^G v(x) \uparrow v(x)$ and $p_t^G(nv \wedge 1)(x) \uparrow (nv \wedge 1)(x)$ as $t \downarrow 0$ for each $x \in G$ and $n \in \mathbb{N}$. Then we have, for $x \in G$

$$\begin{aligned} p_t^G I_{Y_1}(x) &= \uparrow \lim_{n\to\infty} p_t^G(nv \wedge 1)(x) \\ &\le \uparrow \lim_{n\to\infty} nv(x) \wedge 1 = I_{Y_1}(x) \end{aligned}$$

and

$$\begin{aligned} \uparrow \lim_{t\downarrow 0} p_t^G I_{Y_1}(x) &= \uparrow \lim_{t\downarrow 0} \uparrow \lim_{n\to\infty} p_t^G(nv \wedge 1)(x) \\ &= \uparrow \lim_{n\to\infty} \uparrow \lim_{t\downarrow 0} p_t^G(nv \wedge 1)(x) \\ &= \uparrow \lim_{n\to\infty} nv(x) \wedge 1 = I_{Y_1}(x). \end{aligned}$$

Thus I_{Y_1} is a Borel measurable excessive function with respect to $\mathbf{M}^G$ and lower semi continuous on G by the strong Feller property of $\mathbf{M}^G$. Note that every excessive function of $\mathbf{M}^G$ is finely continuous with respect to the fine topology of $\mathbf{M}^G$ (see [**2**]). Therefore we can conclude that Y_1 and Y_2 are finely open and finely closed with respect to $\mathbf{M}^G$. On the other hand, G is finely connected by the preceding proposition and $Y_2 \neq \emptyset$ by the assumption, hence $Y_2 = G$. This completes the proof. □

REMARK 10. (1) In [**12**], it is proved that the irreducibility and the strong Feller property of $\mathbf{M}^G$ hold under the connectivity of G and the doubly-Feller property of $\mathbf{M}$. Here $\mathbf{M}$ is said to be *doubly Feller* if $\mathbf{M}$ possesses
  - (a) (Feller property). For each $t > 0$, $p_t u \in C_\infty(X)$ for $u \in C_\infty(X)$ and $\lim_{t\to 0} p_t u(x) = u(x)$ for each $u \in C_\infty(X)$ and $x \in X$
  - (b) (Strong Feller property). For each $t > 0$ and bounded Borel function u on X, $p_t u$ is a bounded continuous function on X.

  Here $C_\infty(X)$ is the set of continuous functions u on X_Δ with $u(\Delta) = 0$ (see [**4**],[**12**]).

(2) In [**7**], we construct a jointly continuous heat kernel $p_t^G(x,y)$ associated with $(\mathcal{E}_G, \mathcal{F}_G)$ for the canonical strongly local regular Dirichlet form $(\mathcal{E}, \mathcal{F})$ on Alexandrov spaces. In this case, strong Feller property of $\mathbf{M}^G$ is satisfied. Hence Theorem 4.1 is applicable to Alexandrov spaces.

(3) In our proof of Theorem 4.1, the result of weak maximum principle is not used.

Acknowledgement. The author would like to thank Professors Y. Machigashira, T. Shioya and M. Takeda for valuable discussions.

References

[1] M. Biroli and U. Mosco, A Saint-Venant type principle for Dirichlet forms on discontinuous media. *Ann. Mat. Pura Appl.* (4) **169** (1995), 125–181.

[2] R. M. Blumenthal and R. K. Getoor, "Markov processes and potential theory,"Pure and Applied Mathematics, Vol. **29** Academic Press, New York-London 1968.

[3] Yu. Burago, M. Gromov and G. Perel'man, A. D. Aleksandrov spaces with curvatures bounded below, (Russian) *Uspekhi Mat. Nauk* **47** (1992), 3–51, 222; *translation in Russian Math. Surveys* **47** (1992), 1–58.

[4] K. L. Chung, Doubly-Feller process with multiplicative functional, "Seminar on stochastic processes,"1985. Edited by E. Cinlar, K. L. Chung, R. K. Getoor and J. Glover. Progress in Probability and Statistics, **12** Birkhäuser Boston, Inc., Boston, Mass., 63–78, 1986.

[5] M. Fukushima, Y. Oshima and M. Takeda, "Dirichlet Forms and Symmetric Markov Processes,"de Gruyter Studies in Mathematics, **19** Walter de Gruyter & Co., Berlin, 1994.

[6] K. Kuwae, Functional calculus for Dirichlet forms, *Osaka J. Math.* **35** (1998), 683–715.

[7] K. Kuwae, Y. Machigashira and T. Shioya, Sobolev spaces, Laplacian, and heat kernel on Alexandrov spaces, preprint, 1998.

[8] K. Kuwae, Y. Machigashira and T. Shioya, Beginning of analysis on Alexandrov spaces, preprint, 1999.

[9] Y. Otsu and T. Shioya, The Riemannian structure of Alexandrov spaces, *J. Differential Geom.* **39** (1994), 629–658.

[10] T. Sakai, "Riemannian geometry,"Translated from the 1992 Japanese original by the author. Translations of Mathematical Monographs, **149**. American Mathematical Society, Providence, RI, 1996.

[11] K.-Th. Sturm, Analysis on local Dirichlet spaces. I. Recurrence, conservativeness and L^p-Liouville properties, *J. Reine Angew. Math.* **456** (1994), 173–196.

[12] M. Takeda, On exit times of symmetric Levy processes from connected open sets, *Probability theory and mathematical statistics (Tokyo, 1995)*, 478–484, World Sci. Publishing, River Edge, NJ, 1996.

Department of Mathematical Sciences, Yokohama City University, Yokohama, 236-0027 Japan

E-mail address: `kuwae@math.yokohama-cu.ac.jp`

Canadian Mathematical Society
Conference Proceedings
Volume **29**, 2000

On the relation between the model product and the harmonic product of distributions

Bang-He Li and Ya-Qing Li

Dedicated to Prof. S. Albeverio on the occasion of his sixtieth birthday.

ABSTRACT. In the first one of the twelve open problems of Oberguggenberger in his book [O$_1$], it stated that the existence of the model product $[uv]$ implies the existence of the product $u \circ v$ by analytic regularization, for distributions u, v on R^1, and suggested to prove the analogous statment in higher dimensions for the product $u \circ v$ obtained by harmonic regularization. Here, we prove that the analogous statment is indeed true in any dimension.

1. Introduction

Let S and T be distributions on R^n, i.e. $S, T \in \mathcal{D}'(R^n)$, the product of S and T were defined in several ways by several authors. Here, we review only some definitions needed in this paper.

Following Oberguggenberger's paper [O$_2$], we say that:
A net $\{\phi^\varepsilon\}_{0<\varepsilon\leq 1} \subset \mathcal{D}(R^n)$ with

$$\phi^\varepsilon \geq 0 \text{ and } \int \phi^\varepsilon(x)dx = 1 \text{ for all } \varepsilon$$

is called a C_i-delta-net $(i = 2, 4)$, provided it satisfies condition (C_i) as follows:

(C_2). support $(\phi^\varepsilon) \longrightarrow \{0\}$ as $\varepsilon \longrightarrow 0$ and

$$\forall \alpha \in N_0^n \ \exists A_\alpha > 0 \text{ such that } \int |x|^{|\alpha|} |\partial^\alpha \phi^\varepsilon(x)| dx \leq A_\alpha \text{ for all } \varepsilon;$$

(C_4). $\phi^\varepsilon(x) = \varepsilon^{-n}\phi(\frac{x}{\varepsilon})$ for some $\phi \in \mathcal{D}(R^n)$.

For $S, T \in \mathcal{D}'(R^n)$, we say that the M_i-product $(i = 2, 4)$ of S and T exists if

$$\lim_{\varepsilon\longrightarrow 0} (S * \phi^\varepsilon)(T * \phi^\varepsilon) = M_i(S,\ T)$$

exists in $\mathcal{D}'(R^n)$ for all C_i-delta-nets $\{\phi^\varepsilon\}_{0<\varepsilon\leq 1}$ and is independent of the particular C_i-delta-net chosen.

C_2-nets were introduced by Shiraishi [S], called "restricted delta-nets" there. The M_2-product was studied by Itano [I$_1$]. C_4-nets appear at many places [C$_1$,C$_2$,F,I$_2$,K] and are called model nets by Kaminski in [K]. The M_4-product is

Key words and phrases. distribution, model product, harmonic representation.

called model product then by Oberguggenberger in [O_1]. It was proved by Wawak that M_2- and M_4- product are equivalent.

To state harmonic product, we review a few facts about harmonic representations of distributions on R^n first. A harmonic function $u(x,t)$ on $H_{n+1} = R^n \times R_+$ is called a harmonic representation of $S \in \mathcal{D}'(R^n)$, if

$$\lim_{t \longrightarrow 0} u(x,t) = S \quad \text{in } \mathcal{D}'(R^n)$$

Such u exists always and if for an open set V in R^n with $S|_V = 0$ in $\mathcal{D}'(V)$, then u extends to a harmonic function on $H_{n+1} \cup (-H_{n+1}) \cup V$ such that $u(x,0) = 0$ for $x \in V$, where $-H_{n+1} = R^n \times R_-$ (cf [LL_1], Theorem 1). Thus, especially, the difference of any two harmonic representations u_1 and u_2 of a distribution is a harmonic function $u_1 - u_2$ on R^{n+1} skew-symmetric with respect to the last variable, which is equivalent to the fact that

$$(u_1 - u_2)(x,0) = 0, \ x \in R^n$$

For $S, T \in \mathcal{D}'(R^n)$, let u and v be harmonic representations of S and T respectively, if

$$\lim_{t \longrightarrow 0} u(\cdot,t)v(\cdot,t) = S \circ T$$

exists in $\mathcal{D}'(R^n)$, we call $S \circ T$ the harmonic product of S and T ([LL_2,LL_3]). This product is independent of the choice of harmonic representations.

For $n = 1$, the harmonic product is equivalent to the product defined by analytic representations, called Tillmann product in [O_2]. It was proved in [LL_4] that the existence of the product defined by symmetric C_2-nets implies the existence of Tillmann product. The same result was independently proved later in [O_2]. Due to the equivalence of C_2- and C_4- product, the result of the authors and Oberguggenberger implies then that the existence of the model product implies the existence of Tillmann product.

In his book [O_1], §27, Oberguggenberger presented twelve open problems, the first one asked to generalize the above statement to higher dimensions. In this paper, we prove that the same statement is indeed true in any dimension.

Recall that $\phi \in \mathcal{D}(R^n)$ is rotationally symmetric, if $x, y \in R^n$ and $|x| = |y|$ imply $\phi(x) = \phi(y)$. Our result is the following

Theorem. Let $S, T \in \mathcal{D}'(R^n)$. If $\lim\limits_{\varepsilon \longrightarrow 0} (S * \phi^\varepsilon)(T * \phi^\varepsilon) = S \cdot T$ exists in $\mathcal{D}'(R^n)$ for all rotationally symmetric C_2-nets $\{\phi^\varepsilon\}_{0<\varepsilon\leq 1}$,then the harmonic product $S \circ T$ exists also and coincides with $S \cdot T$.

As a consequence of the theorem, the existence of the model product implies the existence of the harmonic product, and the two products coincide. An interesting example of Colombeau and Oberguggenberger [CO] is : $(H \otimes \delta) \cdot (\delta \otimes H) = \frac{1}{4}(\delta \otimes \delta)$ (hence, $(H \otimes \delta) \circ (\delta \otimes H) = \frac{1}{4}(\delta \otimes \delta)$ by the theorem), while the model product of $H \otimes \delta$ and $\delta \otimes H$ does not exsist.

2. Some Lemmas

Recall that

$$P_y(x) = c_n \frac{y}{(|x|^2 + y^2)^{\frac{n+1}{2}}}, \quad x \in R^n, y \in R_+$$

is the Poisson kernel, where $c_n = \pi^{-\frac{n+1}{2}}\Gamma(\frac{n+1}{2})$. And if $S \in \mathcal{D}'(R^n)$ is compactly supported, then the Poisson integral

$$u(x,y) = (S * P_y)(x), x \in R^n, y \in R_+$$

is a harmonic representation of S.

To compare $T \cdot S$ and $T \circ S$, we need to construct rotationally symmetric restricted delts-nets relavant to P_ε.

Let $a \in \mathcal{D}(R)$ satisfying: 1). $0 \le a(x) \le 1$, 2). $a(x) = 0$, if $|x| \ge 1$; $a(x) = 1$, if $|x| \le \frac{1}{2}$. For any integer $k \ge 2$, and $0 < \varepsilon \le 1$, let

$$h_\varepsilon(x) = a(\varepsilon^{-\frac{1}{k}}|x|)P_\varepsilon(x)$$

It is obvious that $h_\varepsilon \ge 0, h_\varepsilon \in \mathcal{D}(R^n)$ and is rotationally symmetric. We have

Lemma 1. $\lim_{\varepsilon \longrightarrow 0} \int_{R^n} h_\varepsilon(x)dx = 1$

Proof.

$$1 = \int_{R^n} P_\varepsilon(x)dx \ge \int_{R^n} h_\varepsilon(x)dx \ge \int_{|x|\le\frac{1}{2}\varepsilon^{\frac{1}{k}}} P_\varepsilon(x)dx$$

$$= \int_{|y|\le\frac{1}{2}\varepsilon^{\frac{1}{k}-1}} P_1(y)dy \qquad \text{(obtained by } x = \varepsilon y)$$

$$= 1 - \int_{|y|>\frac{1}{2}\varepsilon^{\frac{1}{k}-1}} P_1(y)dy \qquad \longrightarrow 1$$

since $\varepsilon^{\frac{1}{k}-1} \longrightarrow \infty$, as $\varepsilon \longrightarrow 0$. The lemma is proved.

Lemma 2. Let $\alpha \in N_0^n$, then

$$\frac{\partial^{|\alpha|}}{\partial x_1^{\alpha_1}\cdots\partial x_n^{\alpha_n}} P_\varepsilon(x) = \frac{\varepsilon c_n P_\alpha(x,\varepsilon)}{(|x|^2+\varepsilon^2)^{\frac{n+1}{2}+|\alpha|}}$$

where $P_\alpha(x,\varepsilon)$ is a homogenous polynomial of $x_1,\cdots,x_n$ and ε with degree $|\alpha|$.

Proof. It is obviously true for $|\alpha| = 0$. Now suppose it is true for $|\alpha| = m$, we have

$$\frac{\partial^{m+1}}{\partial x_1^{\alpha_1+1}\partial x_2^{\alpha_2}\cdots\partial x_n^{\alpha_n}} P_\varepsilon(x) = \varepsilon c_n(|x|^2+\varepsilon^2)^{-n-2m-1}\times$$

$$((|x|^2+\varepsilon^2)^{\frac{n+1}{2}+m}\tfrac{\partial P_\alpha}{\partial x_1}(x,\varepsilon) - (\tfrac{n+1}{2}+m)(|x|^2+\varepsilon^2)^{\frac{n+1}{2}+m-1}2x_1P_\alpha(x,\varepsilon))$$

$$= \varepsilon c_n(|x|^2+\varepsilon^2)^{-\frac{n+1}{2}-m-1}((|x|^2+\varepsilon^2)\tfrac{\partial P_\alpha}{\partial x_1}(x,\varepsilon) - (\tfrac{n+1}{2}+m)2x_1P_\alpha(x,\varepsilon))$$

Thus it is true for $|\alpha| = m+1$, hence the lemma.

Lemma 3. For any $\alpha \in N_0^n$, there is an A_α independent of ε such that

$$\int_{R^n} |x|^{|\alpha|}\,|\frac{\partial^\alpha}{\partial x_1^{\alpha_1}\cdots\partial x_n^{\alpha_n}}h_\varepsilon(x)|\,dx \le A_\alpha$$

for $0 < \varepsilon \le 1$.

Proof. If $|\alpha| = 0$, it is obvious by Lemma 1. In the following, we assume $|\alpha| > 0$. For $x \in R^n$, let $b(x) = a(|x|)$, then $b \in \mathcal{D}(R^n)$. And by Lemma 2,

$$|\frac{\partial^{|\alpha|}}{\partial x_1^{\alpha_1}\cdots\partial x_n^{\alpha_n}}h_\varepsilon(x)|$$

$$= |\sum_{\beta_1=0}^{\alpha_1}\sum_{\beta_2=0}^{\alpha_2}\cdots\sum_{\beta_n=0}^{\alpha_n}\prod_{q=1}^{n}\binom{\alpha_q}{\beta_q}\frac{\varepsilon c_n P_\beta(x,\varepsilon)}{(|x|^2+\varepsilon^2)^{\frac{n+1}{2}+|\beta|}}\varepsilon^{-\frac{|\alpha-\beta|}{k}}\times$$

$$\times\frac{\partial^{|\alpha-\beta|}b}{\partial x_1^{\alpha_1-\beta_1}\ldots\partial x_n^{\alpha_n-\beta_n}}(\varepsilon^{-\frac{1}{k}}x)|$$

$$\le B_\alpha\sum_{j=0}^{|\alpha|}\varepsilon^{1-\frac{1}{k}(|\alpha|-j)}\frac{\sum_{|\beta|=j}|P_\beta(x,\varepsilon)|}{(|x|^2+\varepsilon^2)^{\frac{n+1}{2}+j}} \qquad \text{(for some constant } B_\alpha\text{)}.$$

Thus,

$$\int_{R^n}|x|^{|\alpha|}|\frac{\partial^\alpha}{\partial x_1^{\alpha_1}\cdots\partial x_n^{\alpha_n}}h_\varepsilon(x)|\,dx \le B_\alpha\sum_{j=0}^{|\alpha|}\varepsilon^{1-\frac{1}{k}(|\alpha|-j)}\int_{|x|\le\varepsilon^{\frac{1}{k}}}\frac{|x|^m\sum_{|\beta|=j}|P_\beta(x,\varepsilon)|}{(|x|^2+\varepsilon^2)^{\frac{n+1}{2}+j}}dx$$

$$= B_\alpha\sum_{j=0}^{|\alpha|}\varepsilon^{1-\frac{1}{k}(|\alpha|-j)}\varepsilon^{|\alpha|+n+j-(n+1-2j)}\int_{|t|\le\varepsilon^{\frac{1}{k}-1}}\frac{|t|^{|\alpha|}\sum_{|\beta|=j}|Q_\beta(t)|}{(|t|^2+1)^{\frac{n+1}{2}+j}}dt$$

(by $x_i=\varepsilon t_i$, where Q_β is a homogenous polynomial of $t_1,\cdots,t_n$ with degree $|\beta|$)

$$\le C_\alpha\sum_{j=0}^{|\alpha|}\varepsilon^{(1-\frac{1}{k})(|\alpha|-j)}\int_{(|t|\le\varepsilon^{\frac{1}{k}-1}}\frac{|t|^{|\alpha|+j}}{(|t|^2+1)^{\frac{n+1}{2}+j}}dt \qquad \text{(for some constant } C_\alpha\text{)}$$

$$= C_\alpha\Omega_{n-1}\sum_{j=0}^{|\alpha|}\varepsilon^{(1-\frac{1}{k})(|\alpha|-j)}\int_0^{\varepsilon^{\frac{1}{k}-1}}\frac{r^{|\alpha|+j+n-1}}{(|r|^2+1)^{\frac{n+1}{2}+j}}dr \qquad \text{(by polar coordinates)}$$

$$\le C_\alpha\Omega_{n-1}\sum_{j=0}^{|\alpha|}\varepsilon^{(1-\frac{1}{k})(|\alpha|-j)}(1+\int_1^{\varepsilon^{\frac{1}{k}-1}}r^{|\alpha|-j-2}\,dr)$$

$$= C_\alpha\Omega_{n-1}(\sum_{j=0}^{|\alpha|}\varepsilon^{(1-\frac{1}{k})(|\alpha|-j)}$$

$$+\sum_{|\alpha|-j\ne1}\varepsilon^{(1-\frac{1}{k})(|\alpha|-j)}\frac{(\varepsilon^{(\frac{1}{k}-1)(|\alpha|-j-1)}-1)}{|\alpha|-j-1}+\varepsilon^{1-\frac{1}{k}}\ln\varepsilon^{\frac{1}{k}-1})$$

$$\le A_\alpha \qquad \text{(for some constant } A_\alpha\text{)}.$$

The lemma is proved.

Lemma 4. For any integer $m\ge0$, there is a constant M_m such that

$$\max_{0\le|\alpha|\le m}\sup_{x\in R^n}|\frac{\partial^{|\alpha|}}{\partial x_1^{\alpha_1}\cdots\partial x_n^{\alpha_n}}(P_\varepsilon(x)-h_\varepsilon(x))|\le M_m\varepsilon^{1-\frac{m+n+1}{k}}$$

Proof. Let $c(x) = 1 - b(x)$ for $x \in R^n$, then for any $\alpha \in N_0^n$.

$$\frac{\partial^{|\alpha|}}{\partial x_1^{\alpha_1} \cdots \partial x_n^{\alpha_n}} c(x)$$

is bounded on R^n.

Notice that $P_\varepsilon(x) - h_\varepsilon(x) = 0$ if $|x| \leq \frac{1}{2}\varepsilon^{\frac{1}{k}}$. Now, by Lemma 2, we have, for $|x| \geq \frac{1}{2}\varepsilon^{\frac{1}{k}}$,

$$|\frac{\partial^{|\alpha|}}{\partial x_1^{\alpha_1} \cdots \partial x_n^{\alpha_n}}(P_\varepsilon(x) - h_\varepsilon(x))| = |\frac{\partial^{|\alpha|}}{\partial x_1^{\alpha_1} \cdots \partial x_n^{\alpha_n}}(c(\varepsilon^{-\frac{1}{k}}x)P_\varepsilon(x))|$$

$$= |\sum_{\beta_1=0}^{\alpha_1}\sum_{\beta_2=0}^{\alpha_2}\cdots\sum_{\beta_n=0}^{\alpha_n}\prod_{q=1}^{n}\begin{pmatrix}\alpha_q\\ \beta_q\end{pmatrix}\frac{\varepsilon c_n P_\beta(x,\varepsilon)}{(|x|^2+\varepsilon^2)^{\frac{n+1}{2}+|\beta|}}\varepsilon^{-\frac{|\alpha-\beta|}{k}} \times$$

$$\times \frac{\partial^{|\alpha-\beta|}c}{\partial x_1^{\alpha_1-\beta_1}\cdots\partial x_n^{\alpha_n-\beta_n}}(\varepsilon^{-\frac{1}{k}}x)|$$

$$\leq B_\alpha \sum_{j=0}^{|\alpha|}\varepsilon^{1-\frac{1}{k}(|\alpha|-j)}\frac{\sum_{|\beta|=j}|P_\beta(x,\varepsilon)|}{(|x|^2+\varepsilon^2)^{\frac{n+1}{2}+j}} \qquad (\text{ for some constant } B_\alpha)$$

$$\leq A_\alpha \sum_{j=0}^{|\alpha|}\varepsilon^{1-\frac{1}{k}(|\alpha|-j)}|x|^{j-(n+1+2j)} \qquad (\text{ for some constant } A_\alpha)$$

$$\leq A_\alpha \sum_{j=0}^{|\alpha|}\varepsilon^{1-\frac{1}{k}(|\alpha|-j)}(\frac{1}{2}\varepsilon^{\frac{1}{k}})^{-(n+1+j)} = A_\alpha\sum_{j=0}^{|\alpha|}2^{n+1+j}\varepsilon^{1-\frac{|\alpha|+n+1}{k}}$$

Therefore, for some M_m,

$$\max_{0\leq|\alpha|\leq m}\sup_{x\in R^n}|\frac{\partial^{|\alpha|}}{\partial x_1^{\alpha_1}\cdots\partial x_n^{\alpha_n}}(P_\varepsilon(x)-h_\varepsilon(x))| \leq M_m\varepsilon^{1-\frac{m+n+1}{k}}$$

3. Proof of the Theorem

Let $\widetilde{h_\varepsilon}(x) = h_\varepsilon(x)/\int_{R^n} h_\varepsilon(x)\,dx$. Then by Lemma 1 and 3, $\{\widetilde{h}_\varepsilon\}_{0<\varepsilon\leq 1}$ is a rotationally symmetric C_2-delta-net. If $S \cdot T$ exists, then for any $\phi \in \mathcal{D}(R^n)$,

$$\begin{aligned} < S\cdot T, \phi > &= \lim_{\varepsilon\longrightarrow 0} < (S*\widetilde{h_\varepsilon})(T*\widetilde{h_\varepsilon}), \phi > \\ &= \lim_{\varepsilon\longrightarrow 0} < (S*h_\varepsilon)(T*h_\varepsilon), \phi > \end{aligned}$$

Take $\beta \in \mathcal{D}(R^n)$, such that $\beta(x) \equiv 1$ on some neighborhood U of the support of ϕ.

Let

$$\widetilde{S} = \beta S, \;\; \widetilde{T} = \beta T$$

Then by the localizability of the harmonic product ([LL$_2$],p.719), for any harmonic representations u and v of S and T respectively, we have

$$\lim_{\varepsilon\longrightarrow 0} < u(\cdot,\varepsilon)v(\cdot,\varepsilon) - (\widetilde{S}*P_\varepsilon)(\widetilde{T}*P_\varepsilon), \psi >= 0 \;\; \text{for } \psi \in \mathcal{D}(U)$$

It is easy to see that

$$\lim_{\varepsilon \longrightarrow 0} < (S * h_\varepsilon)(T * h_\varepsilon),\, \phi >= \lim_{\varepsilon \longrightarrow 0} < (\widetilde{S} * h_\varepsilon)(\widetilde{T} * h_\varepsilon),\, \phi >$$

Thus, to show that $S \circ T$ exists and $S \circ T = S \cdot T$, we need only prove that

$$\lim_{\varepsilon \longrightarrow 0} < (\widetilde{S} * P_\varepsilon)(\widetilde{T} * P_\varepsilon),\, \phi >= \lim_{\varepsilon \longrightarrow 0} < (\widetilde{S} * h_\varepsilon)(\widetilde{T} * h_\varepsilon),\, \phi >$$

For $f \in C^\infty(R^n)$, and $j \in N_0$, define the norm

$$||f||_j = \max_{0 \le |\alpha| \le j} \max_{|x| \le j} |\frac{\partial^\alpha}{\partial x_1^{\alpha_1} \cdots \partial x_n^{\alpha_n}} f(x)|$$

Since $\widetilde{S}$ and $\widetilde{T}$ are compactly supported, there exist $m \in N_0$ and a constant M such that for $f \in C^\infty$

$$| < \widetilde{S}, f > | \le M||f||_m, \quad | < \widetilde{T}, f > | \le M||f||_m \tag{1}$$

Hereafter, we take $k > 2m + n + 1$ in the definition of h_ε. Then by Lemma 4,

$$\lim_{\varepsilon \longrightarrow 0} \max_{0 \le |\alpha| \le 2m} \sup_{x \in R^n} |\frac{\partial^{|\alpha|}}{\partial x_1^{\alpha_1} \cdots \partial x_n^{\alpha_n}} (P_\varepsilon(x) - h_\varepsilon(x))| = 0$$

Since

$$(\frac{\partial^{|\alpha|}}{\partial x_1^{\alpha_1} \cdots \partial x_n^{\alpha_n}} \widetilde{S} * (P_\varepsilon - h_\varepsilon))(x) =< \widetilde{S}(t), (\frac{\partial^{|\alpha|}}{\partial x_1^{\alpha_1} \cdots \partial x_n^{\alpha_n}} (P_\varepsilon - h_\varepsilon))(x - t) >$$

we see from (1) that if $0 \le |\alpha| \le m$,

$$\lim_{\varepsilon \longrightarrow 0} \sup_{x \in R^n} |\frac{\partial^{|\alpha|}}{\partial x_1^{\alpha_1} \cdots \partial x_n^{\alpha_n}} (\widetilde{S} * (P_\varepsilon - h_\varepsilon))(x)| = 0 \tag{2}$$

Similarly, if $0 \le |\alpha| \le m$,

$$\lim_{\varepsilon \longrightarrow 0} \sup_{x \in R^n} |\frac{\partial^{|\alpha|}}{\partial x_1^{\alpha_1} \cdots \partial x_n^{\alpha_n}} (\widetilde{T} * (P_\varepsilon - h_\varepsilon))(x)| = 0 \tag{3}$$

Now

$$\begin{aligned} &< (\widetilde{S} * P_\varepsilon)(\widetilde{T} * P_\varepsilon) - (\widetilde{S} * h_\varepsilon)(\widetilde{T} * h_\varepsilon),,\, \phi > \\ =&< (\widetilde{S} * (P_\varepsilon - h_\varepsilon))(\widetilde{T} * h_\varepsilon), \phi > + < (\widetilde{S} * h_\varepsilon)(\widetilde{T} * (P_\varepsilon - h_\varepsilon)), \phi > \\ &+ < (\widetilde{S} * (P_\varepsilon - h_\varepsilon))(\widetilde{T} * (P_\varepsilon - h_\varepsilon)), \phi > \equiv I(\varepsilon) + II(\varepsilon) + III(\varepsilon) \end{aligned}$$

We see from (2) and (3) that

$$\lim_{\varepsilon \longrightarrow 0} III(\varepsilon) = 0$$

Take an open subset V of U such that $supp.\phi \subset V$ and $\bar{V}$ (the closure of V) $\subset U$. Let $\mathcal{D}^m(\bar{V})$ consist of elements of $\mathcal{D}^m(U)$ with supports in $\bar{V}$. Then according to the norm

$$||\psi|| = \max_{0 \le |\alpha|| \le m} \max_{x \in \bar{V}} |\frac{\partial^{|\alpha|}}{\partial x_1^{\alpha_1} \cdots \partial x_n^{\alpha_n}} \psi(x)|, \qquad \text{for } \psi \in \mathcal{D}^m(\bar{V})$$

$\mathcal{D}^m(\bar{V})$ is a Banach space.

From Theorem 2 in [LL$_2$], we have

$$\lim_{\varepsilon \longrightarrow 0} < \widetilde{S} * P_\varepsilon,\, \psi >=< \widetilde{S}, \psi >, \qquad \psi \in \mathcal{D}^m(\bar{V})$$

And from (2), we have

$$\lim_{\varepsilon \longrightarrow 0} < \widetilde{S} * (P_\varepsilon - h_\varepsilon), \psi >= 0, \qquad \psi \in \mathcal{D}^m(\bar{V})$$

Hence

$$\lim_{\varepsilon \longrightarrow 0} < \widetilde{S} * h_\varepsilon, \psi >=< \widetilde{S}, \psi >, \qquad \psi \in \mathcal{D}^m(\bar{V})$$

Thus $\{\widetilde{S} * h_\varepsilon\}_{0<\varepsilon\leq 1}$ is a family of weakly bounded continuous functionals on $\mathcal{D}^m(\bar{V})$, which is thus also strongly bounded by the famous Resonance Theorem, i.e. there is a constant A such that

$$| < \widetilde{S} * h_\varepsilon, \psi > | \leq A||\psi||, \qquad \text{for } 0 < \varepsilon \leq 1 \tag{4}$$

Now

$$(\widetilde{T} * (P_\varepsilon - h_\varepsilon))\phi \in \mathcal{D}^m(\bar{V})$$

and by (3)

$$\lim_{\varepsilon \longrightarrow 0} ||(\widetilde{T} * (P_\varepsilon - h_\varepsilon)\phi|| = 0$$

Hence, in virtue of (4),

$$\lim_{\varepsilon \longrightarrow 0} < \widetilde{S} * h_\varepsilon, (\widetilde{T} * (P_\varepsilon - h_\varepsilon)\phi >= 0$$

which is equivalent to

$$\lim_{\varepsilon \longrightarrow 0} II(\varepsilon) = 0$$

Similarly,

$$\lim_{\varepsilon \longrightarrow 0} I(\varepsilon) = 0$$

Therefore $S \circ T$ exists and coincides with $S \cdot T$. The proof is complete.

Acknowledgement. The authors are grateful to the referee for pointing out that the main result here was obtained independently by Boie [B], using a quite different method.

References

: [B] V.Boie, Multiplication of distributions, Comm.Math.Univ. Carolinae, 39 (1998),309-321.

: [C_1] J.F.Colombeau,New generalized functions and multiplication of distributions, Amsterdam-New York-Oxford:North Holland Publ.1984.

: [C_2] J.F.Colombeau,Elementary introduction to new generalized functions, Amsterdam-New York-Oxford:North Holland Publ.1985.

: [CO] J.F.Colombeau & M. Oberguggenberger, On a hyperbolic system with a compatible quadratic term: generalized solutions, delta waves and multiplication of distributions, Comm.Part.Diff.Eq. 15 (1990) 905-938.

: [F] B.Fisher, The Product of distributions, Quart.J.Math.Oxford 22 (1971) 291-298.

: [I_1] M. Itano, Remark on the multiplicative products of distributions, J. Sci. Hiroshima Univ., 6 (1976) 365-375.

: [I_2] M. Itano, On the theory of multiplicative products of distributions, J. Sci. Hiroshima Univ., A-I, 30 (1966) 151-181

: [K] A.Kaminski, Convolution, product and Fourier transform of distributions, Studia math. 74 (1982) 83-96.

: [LL$_1$] Bang-He Li & Ya-Qing Li, On the harmonic and analytic representations of distributions, Scientia Sinica, XXVIII (1985) 716-729.

: [LL$_2$] Bang-He Li & Ya-Qing Li, Nonstandard analysis and multiplication of distribution in any dimension, Scientia Sinica, XXVIII (1985) 923-937.

: [LL$_3$] Bang-He Li & Ya-Qing Li, Multiplication of distributions via harminoc representations, Functional Analysis in China, Kluwer Academic Publishers, 1996 90-106.

: [LL$_4$] Bang-He Li & Ya-Qing Li, On the multiplication of distributions, Acta scientiarum naturalium universitatis Jilinenesis, (1981) 13-30.

: [O$_1$] M. Oberguggenberger, Multiplication of distributions and applications to partial diferential equations, 1992, Longman Scientific & Technical, Copublished in the United states with John Wiley & Sons. Inc., New York.

: [O$_2$] M. Oberguggenberger, Products of distributions, nonstandard methods, Zeitschr Anal. Anw. 7 (1988) 347-365. Corrections to this article: Z. A. A., 10 (1991) 263-264.

: [S] R. Shiraishi, On the value of distributions at a point and the multiplicative products, J. Sci. Hiroshima Univ., A-I, 31 (1967) 89-104.

: [W] R. Wawak, On the Colombeau product of distributions, In P. Antosik, A. Kaminski, Generalized functions and convergence, World Scientific Publ., Singapore 1990.

Authors' addresses:

Bang-He Li & Ya-Qing Li
Institute of Systems Science
Academia Sinica
Beijing 100080
P. R. China
Libh@iss06.iss.ac.cn
yli@iss06.iss.ac.cn

Canadian Mathematical Society
Conference Proceedings
Volume **29**, 2000

Sobolev spaces and capacity theory on path spaces

Xiang Dong Li

Dedicated to Prof. S. Albeverio on the occasion of his 60th birthday.

ABSTRACT. We study Sobolev spaces and capacity theory on the path space over a compact Riemannian manifold M. We prove a comparison theorem for the Sobolev norms and the related capacities between the Wiener space and the path space. As a corollary, we prove the tightness of the (r,p)-capacity on the path space.

Introduction

Sobolev spaces and capacity theory have been intensively studied on the classical or an abstract Wiener space by many authors (see [AM], [AFHMR], [BH], [FL], [FOT], [M2], [M3], [MN], [MR], [N], [Sh1], [Su]). The tightness of the (r,p)-capacities is an *a priori* condition to obtain a co-area formula or Sugita's finite measure representation theorem of the positive generalized functional on the Wiener space. This allows us to use *the quasi-sure analysis* to study the finite co-dimensional submanifold of the Wiener space. In [Sh2], Shigekawa studied the Sobolev spaces and the related potential theory on the path group over a compact Lie group equipped with the left Cartan connection. He proved that the Itô map defined by a stochastic differential equation on the Lie group has an ∞-quasicontinuous modification which is a quasihomeomorphism between the path group and the Wiener space, implying the tightness of the any (r,p)-capacity on the path group, $r \in N, p > 1$. But in the general case of path space of a compact Riemannian manifold, Driver [D] has shown that the quasihomeomorphism property of the Itô map doesn't hold. The purpose of this paper is to study the relation between Sobolev spaces and capacities on the path space over a compact Riemannian manifold and those on the Wiener space and to prove the tightness of all (r,p)-capacities on the path space.

1991 *Mathematics Subject Classification.* 58B20, 60D05, 60H07, 60J45.

Partially supported by the National Science Foundation of China and the Fundação para a Ciência e Tecnologia of Portugal.

1. Notation and Definition

We denote by (X, H, μ_0) the classical Wiener space, i.e., $X = \{x \in C([0,1], R^m) : x(0) = 0\}$, $H = \{h \in X : ||h||_H^2 = \int_0^1 |\dot{h}(t)|^2 dt < \infty\}$, μ_0 is the Wiener measure on X. We denote by $W^{r,p}(X)$ and $C_{r,p}$ the (r,p)-Sobolev space and its related (r,p)-capacity on (X, H, μ_0). It has been proved independently by Airault and Malliavin [AM] and by Sugita [Su] that $C_{r,p}$ is tight.

Let M be a compact Riemannian manifold equipped with the Levi-Civita connection, $m = dimM$, $m \geq 2$. For a fixed point $m_0 \in M$, the path space based on m_0 is defined as $P_{m_0}(M) := \{\gamma \in C([0,1], M) \mid \gamma(0) = m_0\}$. For any $\gamma_1, \gamma_2 \in P_{m_0}(M)$, let

$$d_\infty(\gamma_1, \gamma_2) := \max_{\tau \in [0,1]} d(\gamma_1(\tau), \gamma_2(\tau)).$$

We denote the uniform convergence topology on $P_{m_0}(M)$ by $(P_{m_0}(M), d_\infty)$.

We denote by $O(M)$ the orthonormal frame principal bundle over M, $\pi : O(M) \to M$ the canonical projection. Let $O(M)^* = \{r^{-1} : r \in O(M)\}$. Then $O(M)$ and $O(M)^*$ are compact Riemannian manifolds whose Riemannian metrics can be intrinsically defined by using the canonical Levi-Civita parallelism on $O(M)$.

Recall the definition of Brownian motion on M (cf. [FM]). Let $\{A_1, A_2, \ldots, A_m\}$ be the basic horizontal vector fields on $O(M)$, i.e., A_i is the horizontal vector on $O(M)$ such that

$$d\pi_r(A_i) = r \circ e_i, \quad \forall i = 1, \ldots, m,$$

where $(e_1, \ldots, e_m)$ is the standard basis of R^m.

We fix a point $r_0 \in O(M)$ such that $\pi(r_0) = m_0$. Let $r_x(t)$ be the solution to the following Stratonovich SDE on $O(M)$:

$$\begin{cases} dr_x(t) = \sum_{i=1}^m A_i(r_x(t)) \circ dx^i(t), & x \in X; \\ r_x(0) = r_0. \end{cases} \tag{1.1}$$

Then $\gamma_x(\tau) = \pi(r_x(\tau))$ is a diffusion on M with infinitesimal generator Δ, i.e., the Laplace-Beltrami operator on M. The Itô map is defined as

$$I : X \to P_{m_0}(M), \quad x \mapsto \gamma_x = \pi(r_x). \tag{1.2}$$

The Wiener measure μ on $P_{m_0}(M)$ is defined as the law of the Riemannian Brownian motion $\{\gamma_x(\tau)\}$, i.e.,

$$\mu = I_* \mu_0. \tag{1.3}$$

Let $D^{\alpha_1 \ldots \alpha_n}_{s_1 \ldots s_n} r_x(t)$ be the Malliavin derivatives of $r_x(t)$, $s_1, \ldots, s_n \in [0,1]$, $1 \leq \alpha_1, \ldots, \alpha_n \leq m$. Then

PROPOSITION 1.1. For any $p \geq 1$, we have

$$\sup_{s_1, \ldots, s_n \in [0,1]} E[\sup_{s_1 \vee \ldots \vee s_n \leq s \leq 1} ||D^{\alpha_1 \ldots \alpha_n}_{s_1 \ldots s_n} r_x(s)||^p_{T_{r_x(s)} O(M)}] < +\infty. \tag{1.4}$$

Proof. Similar to Theorem 2.2 in [N]. □

Let E be a real separable Banach space, we denote by $\mathcal{F}C^\infty(P_{m_0}(M), E)$ the space of all E-valued cylinder functionals on $P_{m_0}(M)$. When $E = R$, we use $\mathcal{F}C^\infty(P_{m_0}(M))$ instead of $\mathcal{F}C^\infty(P_{m_0}(M), R)$.

We denote by $\mathcal{L}(H,E)$ the space of all bounded linear operators from H into E, which can be identified with $H \otimes E$. We define the gradient operator ∇ on $P_{m_0}(M)$ by a densely defined operator $\nabla : L^2(P_{m_0}(M), E) \to L^2(P_{m_0}(M), H \otimes E)$ with domain $Dom(\nabla) = \mathcal{F}C^\infty(P_{m_0}(M), E)$ such that for any $h \in H$, we have

$$\nabla F(\gamma)(h) = \nabla_h F(\gamma), \tag{1.5}$$

where $\nabla_h F(\gamma)$ is the derivative of the cylinder functional F along the direction $h \in H$, namely,

$$\nabla_h F(\gamma) := \sum_{i=1}^k d_i f|_{T_{\gamma_{s_i}} M} \left(t^\gamma_{s_i \leftarrow 0} h_{s_i} \right),$$

$t^\gamma_{s_i \leftarrow 0} = r_x(s_i) r_0^{-1}$ is the Itô's stochastic parallel translation from $T_{m_0}M$ onto $T_{\gamma_x(s_i)}M$ along the Brownian path $\gamma, x = I^{-1}(\gamma)$, and $d_i f$ denotes the exterior differentiation (in the sense of Fréchet-Gâteaux) of the E-valued smooth application $(m_1, \dots, m_k) \to f(m_1, \dots, m_k)$ with respect to the i-th variable.

Taking $h \in H$ such that $\dot{h} = \mathbf{1}_{[\tau,1]} e_\alpha$, $\tau \in [0,1]$, $\alpha = 1, \dots, m$, we obtain

$$\nabla_{\tau,\alpha} F(\gamma) = \sum_{i=1}^k \mathbf{1}_{\tau < s_i} d_i f|_{T_{\gamma_{s_i}} M} \left(t^\gamma_{s_i \leftarrow 0} e_\alpha \right). \tag{1.6}$$

Using the fact that the stochastic parallel transport is infinite differentiable along H-directions, we can define the high order derivatives of F along H-directions: for μ-a.s $\gamma \in P_{m_0}(M)$, for any $n \in N$, $\nabla^n F(\gamma)$ is defined as the unique element in $H^{\otimes n} \otimes E$ such that for any $h_1, \dots, h \in H$, we have

$$\nabla^n F(h_1 \otimes \dots \otimes h_n) := \nabla_{h_n} \dots \nabla_{h_1} F. \tag{1.7}$$

Using the usual method for real valued cylinder functional, we can prove that an integration by parts formula holds on $P_{m_0}(M)$ for all $F \in \mathcal{F}C^\infty(P_{m_0}(M), E)$; moreover the gradient operator ∇ is closable from $L^p(P_{m_0}(M), E)$ into $L^p(P_{m_0}(M), H \otimes E)$ for all $p > 1$.

Let Γ be the Christoffel symbols of the Levi-Civita connection on M, which is considered as a bilinear $TO(M)$-valued application on $TM \times TO(M)$. We define the horizontal lift U of γ from $P_{m_0}(M)$ into $P_{r_0}(O(M))$ as follows

$$\begin{aligned} &dU_s + \Gamma_{\gamma_s}(\circ d\gamma_s, U_s) = 0, \\ &U_o = r_0. \end{aligned} \tag{1.8}$$

We denote by $U(\gamma)$ the solution to the $O(M)$-valued SDE (1.8) driven by the Riemannian Brownian motion γ on M.

We denote by R the Riemannian curvature tensor with respect to the Levi-Civita connection on M. The following formulas are due to Bismut (cf. [B]):

$$\nabla_h U_s(\gamma) = U_s \int_0^s U_u^{-1} R(\circ d\gamma_u, U_u h(u)) U_u, \tag{1.9}$$

$$\nabla_h U_s^{-1}(\gamma) = -\int_0^s U_u^{-1} R(\circ d\gamma_u, U_u h(u)) U_u U_s^{-1}. \tag{1.10}$$

PROPOSITION 1.2. For all $n \in N, p \geq 1$, we have

$$\sup_{\tau_1, \dots, \tau_n \in [0,1]} E[\sup_{\tau_1 \vee \dots \vee \tau_n \leq s \leq 1} ||\nabla^{\alpha_1 \dots \alpha_n}_{\tau_1 \dots \tau_n} U_s(\gamma)||^p] < +\infty, \tag{1.11}$$

$$\sup_{\tau_1, \dots, \tau_n \in [0,1]} E[\sup_{\tau_1 \vee \dots \vee \tau_n \leq s \leq 1} ||\nabla^{\alpha_1 \dots \alpha_n}_{\tau_1 \dots \tau_n} U_s^{-1}(\gamma)||^p] < +\infty, \tag{1.12}$$

where $||\nabla^{\alpha_1,\cdots,\alpha_n}_{\tau_1,\cdots,\tau_n}U_s(\gamma)||$ (resp. $||\nabla^{\alpha_1,\cdots,\alpha_n}_{\tau_1,\cdots,\tau_n}U_s^{-1}(\gamma)||$) denotes the Riemannian norm of the Malliavin derivatives of $\nabla^n U_s(\gamma)$ in $T_{U_s(\gamma)}O(M)$ (resp. the Malliavin derivative of $\nabla^n U_s^{-1}(\gamma)$ in $T_{U_s^{-1}(\gamma)}O(M)^*$).

Proof. By induction, use (1.9) and (1.10) and Buckholder-Davis-Gundy inequality. □

As a corollary, we can prove that for any $p>1$ and for any $F\in\mathcal{F}C^\infty(P_{m_0}(M),E)$, we have

$$E\left[\int_{[0,1]^n}||\nabla^{\alpha_1\ldots\alpha_n}_{\tau_1\ldots\tau_n}F(\gamma)||_E^2 d\tau_1\ldots d\tau_n\right]^{\frac{p}{2}}<+\infty, \tag{1.13}$$

where $\nabla^{\alpha_1,\cdots,\alpha_n}_{\tau_1,\cdots,\tau_n}F\in E$ denotes the E-valued Malliavin derivatives of $\nabla^n F\in H^{\otimes n}\otimes E$.

DEFINITION 1.3. We define the Sobolev norm $||F||_{D^{r,p}(P_{m_0}(M),E)}$ of an E-valued cylinder functional $F\in\mathcal{F}C^\infty(P_{m_0}(M),E)$ by

$$||F||^p_{D^{r,p}(P_{m_0}(M),E)}:=\sum_{k=0}^r\sum_{j=1}^k\sum_{\alpha_j=1}^m E\left[\int_{[0,1]^k}||\nabla^{\alpha_1\ldots\alpha_k}_{\tau_1\ldots\tau_k}F||_E^2 d\tau_1\ldots d\tau_k\right]^{\frac{p}{2}}. \tag{1.14}$$

We define the E-valued Sobolev space $D^{r,p}(P_{m_0}(M),E)$ as the completion of $\mathcal{F}C^\infty(P_{m_0}(M),E)$ with respect to the norm $||\cdot||_{D^{r,p}(P_{m_0}(M),E)}$ defined in (1.14). We denote by $D^{\infty,\infty}(P_{m_0}(M),E)=\cap_{r,p}D^{r,p}(P_{m_0}(M),E)$, and we call its elements E-valued smooth functionals on $P_{m_0}(M)$ in the sense of Malliavin calculus. In the case of $E=R$, we denote $D^{r,p}(P_{m_0}(M),E)$ by $D^{r,p}(P_{m_0}(M))$ for $r\in N, p>1$, etc.

REMARK 1.4. The Banach space-valued Sobolev spaces on the Wiener space have been defined and studied in [FL], [MN], [Sh1]. The real-valued Sobolev spaces on the path or the loop space have been defined and studied in [CM], [FM], [JL], [Le1], [Le2].

Now we define the (r,p)-capacity $\widehat{C}_{r,p}$ on $P_{m_0}(M)$ and the notion of slim set on $P_{m_0}(M)$.

DEFINITION 1.5. The (r,p)-capacity $\widehat{C}_{r,p}$ on $P_{m_0}(M)$ is defined as follows: for an open subset $O\subset P_{m_0}(M)$,

$$\widehat{C}_{r,p}(O):=inf\{||f||^p_{D^{r,p}(P_{m_0}(M))}:f\in D^{r,p}(P_{m_0}(M)), f\geq 1\ \ \mu-\text{a.e. on } O\}; \tag{1.15}$$

for any subset $A\subset P_{m_0}(M)$,

$$\widehat{C}_{r,p}(A):=inf\{C_{r,p}(O):A\subset O,\ \ O\text{ is an open subset of } P_{m_0}(M)\}. \tag{1.16}$$

A subset S of $P_{m_0}(M)$ is called a *slim* set if $\widehat{C}_{r,p}(S)=0,\forall r\in\mathbf{N}, p>1$.

The notion of the ∞-quasicontinuous modification or the redefinition of a functional on $P_{m_0}(M)$ is similar to the one on the Wiener space. If E is a Banach space with a smooth norm, by using Chebyshev type inequality and Borel-Cantelli lemma for capacity (cf. [MN], [Su]), we can prove that any smooth functional $f\in D^\infty(P_{m_0}(M),E)$ admits an ∞-quasicontinuous modification.

We need to introduce the Hölder topology $(X,||\cdot||_{2m,\alpha})$ and the Hölder topology $(P_{m_0}(M),d_{2m,\alpha})$.

DEFINITION 1.6 (CF. [AM], [MN], [SU]). For any $0 < \alpha < \frac{1}{2}$, let

$$||x||_{2m,\alpha}^{2m} := \int_0^1 \int_0^1 \frac{|x(t) - x(s)|^{2m}}{|t - s|^{1+2m\alpha}} dtds. \tag{1.17}$$

For any $\gamma_1, \gamma_2 \in P_{m_0}(M)$, let

$$d_{2m,\alpha}^{2m}(\gamma_1, \gamma_2) := \int_0^1 \int_0^1 \frac{|d(\gamma_1(t), \gamma_2(t)) - d(\gamma_1(s), \gamma_2(s))|^{2m}}{|t - s|^{1+2m\alpha}} dtds. \tag{1.18}$$

We denote by $(X, ||\cdot||_{2m,\alpha})$ (resp. $(P_{m_0}(M), d_{2m,\alpha})$) the Hölder topology on X (resp. $P_{m_0}(M)$) induced by the Hölder norm $||\cdot||_{2m,\alpha}$ (resp. the Hölder metric $d_{2m,\alpha}$).

Note that $||\cdot||_{2m,\alpha}$ is a smooth norm. If $\alpha \in (\frac{1}{2m}, \frac{1}{2})$, by Lemma $A.3$ of [AM], there exists a constant $C(m, \alpha) > 0$ such that $\forall x \in X$,

$$||x||_\infty \leq C(m, \alpha) ||x||_{2m,\alpha}^{2m}, \tag{1.19}$$

where

$$||x||_\infty := \max_{\tau \in [0,1]} ||x(\tau)||.$$

THEOREM 1.7. For any $0 < \alpha < \frac{1}{2}$, we have

$$I \in D^{\infty,\infty}(X, (P_{m_0}(M), d_{2m,\alpha})), \tag{1.20}$$

$$I^{-1} \in D^{\infty,\infty}(P_{m_0}(M), (X, ||\cdot||_{2m,\alpha})). \tag{1.21}$$

Proof. By Nash embedding and Theoreom 3.4 of [MN] (or Example 4.2 of [Su]), we can prove (1.20). For the proof of (1.21), we use Bismut's formula (1.10) and using Proposition 1.2. See also Shigekawa [Sh1] if the topology on the target space is the uniform convergence topology. □

COROLLARY 1.8. The Itô map $I : X \to (P_{m_0}(M), d_{2m,\alpha})$ has an ∞-quasi-continuous modification $\widetilde{I} : X \to (P_{m_0}(M), d_{2m,\alpha})$.

Proof. By Chebyshev type inequality and Borel-Cantelli lemma for capacity. □

By (1.19), we conclude that the uniform convergence topology $(P_{m_0}(M), d_\infty)$ is weaker than the Hölder topology $(P_{m_0}(M), d_{2m,\alpha})$ if $\alpha \in (\frac{1}{2m}, \frac{1}{2})$. Hence $\widetilde{I} : X \to (P_{m_0}(M), d_\infty)$ is also ∞-quasi-continuous.

2. Comparison Theorem of Sobolev norms and Tightness of Capacities

The follwoing result is a comparison theorem of Sobolev norms between the Wiener space and the path space.

THEOREM 2.1. For any $r \in N$, $p \geq 2$, $\epsilon > 0$, there exists a constant $C = C(r, p, \epsilon) > 0$ such that for any cylinder functional F on $P_{m_0}(M)$, we have

$$||F||_{D^{r,p}(P_{m_0}(M))} \leq C ||F \circ I||_{W^{2r,p+\epsilon}(X)}. \tag{2.1}$$

Sketch of Proof. The complete proof can be found in [Li]. Using the intertwining formula for the differential of the Itô map (Theorem 2.6 of [CM]), and using Theoreom 2.3.8 of [CM], we obtain an explicit representation formula of $\nabla_{\tau,\alpha} F$ for

$F \in \mathcal{F}C^\infty(P_{m_0}(M))$ by means of a Skorohod integration of $D\widehat{F}(x)$, $\widehat{F} = F \circ I$. More precisely, for any $\alpha = 1, \dots, m$, and any $\tau \in [0,1]$, we have

$$\nabla_\tau^\alpha F(\gamma) = D_\tau^\alpha \widehat{F}(x) + \frac{1}{2}\int_\tau^1 < Ric_{r_x(s)}(D_s\widehat{F}(x)), e_\alpha > ds + \int_0^1 q_\alpha(\tau, s) D_s \widehat{F}(x) dx_s, \tag{2.2}$$

where $x = I^{-1}(\gamma)$, Ric_r is the Ricci curvature read in $O(M)$, i.e.,

$$Ric_r(a) = \sum_{i=1}^d r^{-1} \circ R(re_i, ra) \circ re_i, \quad \forall a \in R^d, \quad \forall r \in O(M);$$

$$q_\alpha(\tau, s) = \mathbf{1}_{[\tau, \mathbf{1}]}(s) \int_\tau^s \Omega_{r_x(u)}(e_\alpha, \circ dx(u)),$$

where $\Omega_r(a,b)c = r^{-1}R(ra, rb)rc$, for any $a, b, c \in R^d$ and any $r \in O(M)$.

By induction we can obtain the general representation for $\nabla_{\tau_1, \dots, \tau_r}^{\alpha_1, \dots, \alpha_r} F$ by means of a Skorohod integration of $D^i\widehat{F}, i = 1, \dots, r$. Applying the L^p-estimate for Skorohod integration (cf. Proposition 3.2.1 in [N]) and Proposition 1.1, we can prove (2.1) for all $r \in N, p \geq 2$. □

In the following theorem, we obtain the tightness of all (r,p)-capacities on the path space, which generalize the results of Airault and Malliavin [AM], Sugita [Su] on the Wiener space and the result of Shigekawa [Sh2] on the path group over a compact Lie group.

THEOREM 2.2. For any $r \in N, p > 1$, $\widehat{C}_{r,p}$ is tight, i.e., there exists a sequence of increasing compact subsets $\{K_n\}$ in $(P_{m_0}(M), d_\infty)$ such that

$$\lim_{n\to\infty} \widehat{C}_{r,p}(K_n^c) \to 0. \tag{2.3}$$

Proof. For any $r \in N, p \geq 2$, using the tightness of (r,p)-capacity on X, by a slight modification of Shigekawa's argument in [Sh2], we prove that there is a sequence of increasing compact subsets $\{F_n\}$ of X, such that

$$\lim_{n\to\infty} C_{r,p}(F_n^c) = 0, \quad \forall (r,p) \in N \times (1, \infty),$$

and

$$\widetilde{I}|_{F_n} : F_n \to P_{m_0}(M) \quad \text{is continous.}$$

Set $K_n = \widetilde{I}(F_n)$. By the compactness of F_n and the continuty of the map $\widetilde{I}|_{F_n} : F_n \to (P_{m_0}(M), d_\infty)$, we conclude $\{K_n\}$ is a sequence of increasing compact subsets of $(P_{m_0}(M), d_\infty)$. Moreover, using a similar result of (2.1), we can prove

$$\left(\widehat{C}_{r,p}(K_n^c)\right)^{\frac{p+1}{p}} \leq \alpha C_{2r, p+1}(F_n^c) \tag{2.4}$$

This implies the tightness of $\widehat{C}_{r,p}$ for any $r \in N, p \geq 2$. Furthermore, the tightness of $\widehat{C}_{r,p}$ for $r \in N, p \in (1,2)$ can be driven by the tightness of $\widehat{C}_{r,2}$. □

COROLLARY 2.3. $\widehat{C}_{r,p}$ is inner-regular, i.e., for any $\epsilon > 0$ and any Borel subset $A \subset (P_{m_0}(M), d_\infty)$, there exists a compact subset $K \subset A$ such that

$$\widehat{C}_{r,p}(A \setminus K) < \epsilon. \tag{2.5}$$

Proof. By Theorem 2.2 and the standard argument as in [M3]. □

Similar to Theorem 2.1, we can obtain a comparison theorem for capacities between the Wiener space and the path space.

THEOREM 2.4. For any $r \in N$, $p \geq 2$, $\epsilon > 0$, there exists a constant $\alpha = \alpha(r,p,\epsilon) > 0$ for any $A \subset P_{m_0}(M)$, we have

$$\widehat{C}_{r,p}(A) \leq \alpha C_{2r,p+\epsilon}((\widetilde{I})^{-1}(A))^{\frac{p}{p+\epsilon}}. \tag{2.6}$$

Proof. Use Theorem 2.1 and some modification of Shigekawa's method in [Sh2]. □

REMARK 2.5. In fact, the result in Theorem 2.1 has been implicitly obtained in the case of $r = 1$ by Cruzeiro and Malliavin [CM]. The comparison theorem for Sobolev norms for forms with assumption of regularity about their kernels has been stated in Léandre [Le2].

REMARK 2.6. The tightness of the $(1,2)$-capacity on path space is an important condition for using the quasi-regular Dirichlet form theory to construct the Ornstein-Uhlenbeck process on path space. In [DR], Driver and Röckner have proved the $(1,2)$-tightness by using the method of [RS]. Their method can also be used to prove the $(1,p)$-tightness for any $p > 1$. But it seems that we cannot use their methods to prove the tightness of all (r,p)-capacities ($r \in N$, $p > 1$) on $P_{m_0}(M)$.

REMARK 2.7. The main results in this paper can be generalized to the path space over a compact Riemannian manifold equipped with a Driver connection, i.e., a connection which preserves the Riemannian metric and has skew-torsion. We can also prove the tightness of all (r,p)-capacities on the path space $(P_{m_0}(M), d_{2m,\alpha})$ on which we consider the Hölder topology. We will discuss these problems in another paper.

Acknowledgement

This work is a part of my PhD thesis [Li] under the joint supervision of Professors Z. M. Ma, P. Malliavin and A. B. Cruzeiro. I would like to express my profound thanks to them for their supervision and encouragement. For valuable discussions, I also thank Professors S. Fang, D. Feyel, F. Z. Gong, F. Hirsch, R. Léandre, A. de La Pradelle, J. Ren and S. Song. I am grateful to Prof. R. Léandre for his kindful invitation to the international conference "Probability and Geometry" held in Nancy, 20–25 September 1999, in which I had the opportunity to report this work. Finally, I want to express my gratitude to editors and referees for the publication of this paper in this Conference Proceeding in honor of Prof. S. Albeverio for his 60th birthday.

References

[AM] H. Airault, and P. Malliavin, Intégration géométrique sur l'espace de Wiener. Bull. Sci. Math. (2) **112** (1988), 3-52.

[AFHMR] S. Albeverio, M. Fukushima, W. Hansen, Z. M. Ma, and M. Röckner. An invariance result for capacities on Wiener space. J. Funct. Anal. **106** (1992), 35-49.

[B] J. M. Bismut. *Large Deviations and the Malliavin calculus.* Progress in Mathematics. Birkhäuser 1984.

[BH] N. Bouleau, and F. Hirsch, *Dirichlet forms and analysis on the Wiener space.* de Gruyter 1991.

[CM] A. B. Cruzeiro, and P. Malliavin, Renormalized differential geometry on path space: Structure equation, curvature. J. Funct. Anal. **139** (1996), 119-181.

[D] B. K. Driver, The non-equivalence of the Dirichlet forms on path spaces. Proc. U.S.-Japan Bilateral Seminar 1994.

[DR] B. K. Driver, and M. Röckner, Construction of diffusions on path and loop spaces of compact Riemannian manifolds. C. R. Acad. Sci. Paris, Sér. I. **315** (1992), 603-608.

[FL] D. Feyel, and A. de La Pradelle, Capacités guassiennes. Ann. Inst. Fourier, Grenoble, **41**, 1 (1991), 49-76.

[FOT] M. Fukushima, Y. Oshima, and M. Takeda, *Dirichlet forms ans symmetric Markov process.* de Gruyter 1994.

[FM] S. Fang, and P. Malliavin, Stochastic analysis on the path space over a compact Riemannian manifold. J. Funct. Anal. **131** (1995) 228-252.

[JL] J. D. S Jones, and R. Léandre, L^p-Chen forms on loop spaces. In *Stochastic Analysis.* M.T. Barlow, N. Bingham eds., Cambridge University Press, 104-162 (1991).

[Le1] R. Léandre, Integration by parts formulas and rotationally invariant Sobolev calculus on the loop space. In "XXVII School of Mathematical Physics of Karpacz". A. Borowiec, R. Gielerak, eds., J. Geometry and Physics **11** (1993), 517-528.

[Le2] R. Léandre, Brownian cohomology of an homogeneous manifold. In "New Trends in Stochastic Analysis", K. D. Elworthy, S. Kusuoka, I. Shigekawa, eds., World Scientific, Singapore, 305–348 (1997).

[Li] X. D. Li, Stochastic Analysis and Geometry on Path and Loop Space, PhD thesis in the Institute of Applied Mathematics of Academia Sinica and in the University of Lisbon, 1999.

[M1] P. Malliavin, Formule de la moyenne, calcul de perturbation et théorème d'annulation pour les formes harmoniques, J. Funct. Anal. **17** (1974), 274-291.

[M2] P. Malliavin, Implicit functions in finite corank on the Wiener space. In: Proc. of the Taniguchi Symp. Stoch. Anal., Katata/Kyoto 1982 (K. Itô, ed.). North-Holland (1984) 369-386.

[M3] P. Malliavin, *Stochastic analysis.* Springer-Verlag 1997.

[MN] P. Malliavin, and D. Nualart, Quasi sure analysis of stochastic flows and Banach space valued smooth functionals on the Wiener space, J. Funct. Anal. **112** (1993), 287-317.

[MR] Z. M. Ma, and M. Röckner, *An introduction to the theory of (non-symmetric) Dirichlet forms.* Springer-Verlag 1992.

[N] D. Nualart, *The Malliavin calculus and related topics.* Springer-Verlag 1996.

[RS] M. Röckner, and B. Schmuland, The tightness of general $C_{1,p}$ -capacities on Banach space. J. Funct. Anal. **108** (1992), 1-12.

[Sh1] I. Shigekawa, Sobolev spaces of Banach valued functions associated to a Markov process, Probab. Theory Relat. Fields **99** (1992) 425-443.

[Sh2] I. Shigekawa, A quasihomeomorphism of the Wiener space. In: Proc. Sump. Pure and Appl. Math. **57** (M. Cranston and M. Pinsky eds.) Cornell, 1993, 473-487.

[Su] H. Sugita, Positive generalized functions and potential theory over an abstract Wiener space. Osaka J. Math. **25** (1988) 665-696.

Institute of Applied Mathematics, Academia Sinica, P. O. Box 2734, Beijing 100080, P. R. China, and Grupo de Física-Matemática, Universidade de Lisboa Av. Prof. Gama Pinto 2, 1649-003 Lisboa, Portugal

Canadian Mathematical Society
Conference Proceedings
Volume **29**, 2000

Integration by Parts Formulae for Point Processes

Volkmar Liebscher

ABSTRACT. On the configuration space over a locally compact Riemannian manifold Gibbsian measures can be characterized by means of an integration by parts formula with respect to the canonical gradient. The rôle of the logarithmic energy is taken by the canonical gradient of the conditional energy. We prove that generally, such an integration by parts measure fulfils the condition Σ_m^c of [**15**]. Our approach, based on the concept of the (reduced) Campbell measure, sharpens the results of [**1, 2**].

1. Introduction

Beginning with [**1, 2**] there were studied Dirichlet forms on so-called configuration spaces. It was shown how tempered Gibbs measures, especially mixed Poisson processes, give rise to a class of canonical Dirichlet forms associated with a stationary reversible Markov process on such configuration spaces. One (side-) result were Integration by Parts formulae for (mixed) Poisson processes and more generally, for tempered Gibbs measures.

This note is part of a more complex work [**9**], which is still in progress and aims to find a natural setting for the differential and Riemannian structure on configuration space. These structures were declared in [**1, 2**] just in analogy with spaces of distributions.

The present approach is based on the observation that the gradient defined in [**1, 2**] is the composition of the gradient on the base manifold and the Malliavin gradient $\mathbb{D}$ in configuration space language. This gradient and its counterpart, the Skorohod integral $\mathbb{S}$ are well-connected to the so-called reduced Campbell measure $C_\mu^!$ associated to any point process μ [**7, 10**]. Gibbsian point processes [**16**] are characterized by an absolute continuity property of this measure, see also [**11**]. An extension of this absolute continuity property is property Σ_m^c as defined below [**15**]. We show that all point processes with an integration by parts formula have this property, which constitutes a point process version of the Malliavin lemma [**13**].

1991 *Mathematics Subject Classification.* Primary 58-99, Secondary 60-99.

This work was inspired by an invitation to BiBoS Institute and supported by the Deutsche Forschungsgemeinschaft.

2. Preliminary Definitions and Notations

For the natural numbers,real, and positive real numbers we use the symbols $\mathbb{N} = \{0, 1, 2, \dots\}$, $\mathbb{R}$, and $\mathbb{R}_+ = [0, \infty)$ respectively. In our framework we follow [**1, 2**], but we alter some notations. From the point process point of view we have to allow for multiple points in configurations too [**10**]. So let X be a locally compact Riemannian manifold with Levi-Civita connection ∇^X and Lebesgue measure m. Further, let $\mathfrak{X}$ denote the Borel sets of X, $\mathfrak{X}^{\mathsf{b}}$ the ring of relatively compact Borel sets on X and $\mathfrak{O}$ the set of relatively compact open sets. For the Dirac measure concentrated in $x \in X$ we use the abbreviation δ_x, the zero measure is denoted by $\mathfrak{o}$ ($\mathfrak{o}(K) = 0\ \forall K \in \mathfrak{X}$). The complement of $K \in \mathfrak{X}$ we denote by K^{C}, the indicator function of K is χ_K.

Let Γ be the set of all locally finite counting measures on X, i.e.

$$\Gamma = \left\{\gamma : \gamma \text{ is measure on } X,\ \gamma(K) \in \mathbb{N} \text{ for all } K \in \mathfrak{X}^{\mathsf{b}}\right\}.$$

Further, $\operatorname{supp}\gamma = \{x \in X : \gamma(\{x\}) > 0\}$ is the support of $\gamma \in \Gamma$.

REMARK 2.1. For each $\gamma \in \Gamma$ the set $\operatorname{supp}\gamma$ is enumerable and we have the representation $\gamma = \sum_{x \in \operatorname{supp}\gamma} \gamma(\{x\})\delta_x$ (cf. [**10**]). Thus we can interprete γ as locally finite point configuration in X, $\mathfrak{o}$ represents the void configuration. Γ, but also

$$\Gamma^{\leq 1} = \{\gamma \in \Gamma : \gamma(\{x\}) \leq 1 \text{ for all } x \in X\}.$$

is called *configuration space*. Any $\gamma \in \Gamma^{\leq 1}$ may be identified with $\operatorname{supp}\gamma$, which allows to translate all measure-theoretic notions below into a set-theoretic one. For the following considerations it makes not much difference which of the two spaces we consider as all measures of interest are concentrated on $\Gamma^{\leq 1}$ (see remark 4.6).

On Γ we introduce the σ-field $\mathfrak{B}(\Gamma)$ which is the smallest σ-field making the map $\gamma \mapsto \gamma(K')$ measurable for all $K' \in \mathfrak{X}^{\mathsf{b}}$. In the case that K' is variing only over the bounded Borel sets inside $K \in \mathfrak{X}$, the corresponding σ-field is denoted by $\mathfrak{B}(\Gamma)_K$. A probability measure on $(\Gamma, \mathfrak{B}(\Gamma))$ is called *point process on G*.

The best explored class of point processes constitute *Poisson processes* [**4, 8, 10**]. For a fixed locally finite intensity measure σ this is the point process π_σ which fulfils for all $r, n_1, n_2, \dots, n_r \in \mathbb{N}$ and all disjoint sets $K_1, K_2, \dots, K_r \in \mathfrak{X}^{\mathsf{b}}$

$$\pi_\sigma\left(\{\gamma \in \Gamma : \gamma(K_i) = n_i\, \forall i, 1 \leq i \leq r\}\right) = \prod_{i=1}^{r} \mathrm{e}^{-\sigma(K_i)} \frac{\sigma(K_i)^{n_i}}{n_i!}.$$

Note that π_σ is concentrated on $\Gamma^{\leq 1}$ if and only if $\sigma(\{x\}) = 0\, \forall x \in X$.

3. A differentiable structure on Γ

How to introduce a concept of differentiability on a certain space like Γ for example? The most natural thing to start with is the concept of a local diffeomorphism. This can be done, for a broad workout we refer to [**9**] or to the slightly different ansatz of [**1**]. We just summarize, that a group of such local diffeomorphisms is generated by a map $\mathcal{V} : \Gamma \times X \longmapsto \mathrm{T}X$ for which

$$\mathcal{V}(\gamma, x) \in \mathrm{T}_x X, \qquad (\gamma \in \Gamma, x \in X) \tag{3.1}$$

and further, there are $K \in \mathfrak{X}^{\mathsf{b}}$ and $n \in \mathbb{N}$ such that

$$\mathcal{V}(\gamma, x) = \mathcal{V}(\gamma_K, x), \qquad (\gamma \in \Gamma, x \in X) \tag{3.2}$$

and $\mathcal{V}(\gamma, x) = 0$ unless $x \in K$ and $\gamma(K) \leq n$. Smoothness is assured by

$$K^k \ni (x_1, \ldots, x_{k-1}, x) \mapsto \mathcal{V}(\delta_{x_1} + \cdots + \delta_{x_{k-1}}, x)$$

being smooth and bounded. Such a map we call *local smooth vector field.*

EXAMPLE 3.1. One example, extensively used in [**1, 2**] is the *independent vector field*, defined for some C_0^∞ vector field v on X by

$$\mathcal{V}_v(\gamma, x) = v(x), \qquad (\gamma \in \Gamma, x \in X). \tag{3.3}$$

For getting a local smooth vector field, we have truncate it w.r.t. the particle number. Say, v has support in K and set

$$\mathcal{V}_{v,n}(\gamma, x) = \begin{cases} v(x) & \text{if } \gamma(K) \leq n \\ 0 & \text{otherwise} \end{cases}, \qquad (\gamma \in \Gamma, x \in X).$$

For $F : \Gamma \longmapsto \mathbb{R}$ we introduce $\mathbb{D}F : \Gamma \times X \longmapsto \mathbb{R}$ by

$$\mathbb{D}F(\gamma, x) = F(\gamma + \delta_x), \qquad (\gamma \in \Gamma, x \in X).$$

REMARK 3.2. $\mathbb{D}$ is exactly the Malliavin gradient operator if we represent Wiener space as L^2 space over the configuration space equipped with an exponential measure [**6**].

Then, $F : \Gamma \longmapsto \mathbb{R}$ is differentiable if

$$x \mapsto \mathbb{D}F(\gamma, x) = F(\gamma + \delta_x)$$

is differentiable for all $\gamma \in \Gamma$. The derivative of F along the local smooth vector field $\mathcal{V}$ is given by

$$\nabla_{\mathcal{V}}^{\Gamma} F(\gamma) = \int \gamma(\mathrm{d}x) \left\langle \mathcal{V}(\gamma - \delta_x, x), 1 \otimes \nabla^X \mathbb{D}F(\gamma - \delta_x, x) \right\rangle_{\mathrm{T}_x X} \tag{3.4}$$

Similarly, we have higher orders of differentiability. In [**1, 2**] e.g., there was extensively used the class $\mathcal{F}C_{\mathrm{b}}^\infty(\Gamma)$ of all functions $F : \Gamma \longmapsto \mathbb{R}$ of the form

$$F(\gamma) = f(\langle \varphi_1, \gamma \rangle, \langle \varphi_2, \gamma \rangle), \ldots, \langle \varphi_n, \gamma \rangle)$$

where $f \in C_{\mathrm{b}}^\infty(\mathbb{R}^n)$, $\varphi_1, \ldots, \varphi_n \in C_0^\infty(X)$ are differentiable functions with compact support in X and $\langle \varphi, \gamma \rangle = \int \varphi \mathrm{d}\gamma$. Clearly, functions from $\mathcal{F}C_{\mathrm{b}}^\infty(\Gamma)$ are smooth and for a local smooth vector fields $\mathcal{V}$ the map $\gamma \mapsto \nabla_{\mathcal{V}}^{\Gamma} F(\gamma)$ is bounded.

Formula (3.4) suggests to introduce the operator $\mathbb{S}$ (the *Skorohod integral*) acting as

$$\mathbb{S}F(\gamma) = \int \gamma(\mathrm{d}x) F(\gamma - \delta_x, x), \qquad (\gamma \in \Gamma).$$

Thus we obtain

$$\nabla_{\mathcal{V}}^{\Gamma} F = \mathbb{S} \langle \mathcal{V}, \cdot \rangle 1 \otimes \nabla^X \mathbb{D}F.$$

We introduce the tangent space $\mathrm{T}_\gamma \Gamma$ as direct integral: $\mathrm{T}_\gamma \Gamma = \int_\oplus \gamma(\mathrm{d}x) \mathrm{T}_x X$. Correspondingly, we equip Γ with the Riemannian structure

$$\langle \mathcal{V}, \mathcal{W} \rangle_{\mathrm{T}_\gamma \Gamma} = \int \gamma(\mathrm{d}x) \langle \mathcal{V}(\gamma - \delta_x, x), \mathcal{W}(\gamma - \delta_x, x) \rangle_{\mathrm{T}_x X} \tag{3.5}$$

REMARK 3.3. In this scheme, the tangent space $\mathrm{T}_\gamma\Gamma$ is the space of γ-square integrable sections $v : X \longmapsto \mathrm{T}X$. One should not take this definition too seriously, as any local smooth vector field $\mathcal{V}$ induces for all $\gamma \in \Gamma$ a linear form on all sections $v : X \longmapsto \mathrm{T}X$ by

$$\langle \mathcal{V}, v\rangle_\gamma = \int \gamma(\mathrm{d}x)\, \langle \mathcal{V}(\gamma - \delta_x, x), v(x)\rangle_{\mathrm{T}_x X}\,.$$

Then, for smooth F, we obtain $\nabla^\Gamma_\mathcal{V} F(\gamma) = \left\langle \mathcal{V}, \nabla^\Gamma F\right\rangle_{\mathrm{T}_\gamma\Gamma}$, where the *gradient* $\nabla^\Gamma F$ is defined via

$$\nabla^\Gamma F(\gamma, x) = 1 \otimes \nabla^X \mathbb{D} F(\gamma, x), \qquad (\gamma \in \Gamma, x \in X).$$

4. Integration by Parts Formulae and the Property Σ^{c}_m

Remember that a measure σ on $[X, \mathfrak{X}]$ is an "integration by parts" (ibp-) measure, if there exists a locally integrable section $\beta^\sigma : X \longmapsto \mathrm{T}X$, the *logarithmic derivative of* σ such that for all local smooth vector fields v and all $f \in C^\infty_0(X)$

$$\int \nabla^X_v f \mathrm{d}\sigma = -\int \sigma(\mathrm{d}x) f(x) (\langle v(x), \beta^\sigma(x)\rangle_{\mathrm{T}_x X} + \mathrm{div}^X v(x))$$

where $\mathrm{div}^X v$ is the divergence of v.

On Γ, we define for a vector field $\mathcal{V} : \Gamma \longmapsto \mathrm{T}\Gamma$ its *divergence*

$$\mathrm{div}^\Gamma \mathcal{V}(\gamma) = \mathbb{S} 1 \otimes \mathrm{div}^X \mathcal{V}(\gamma) = \int \gamma(\mathrm{d}x) 1 \otimes \mathrm{div}^X \mathcal{V}(\gamma - \delta_x, x), \qquad (\gamma \in \Gamma).$$

DEFINITION 4.1. We say that a point process μ is an *ibp-measure* if there is a measurable section $B^\mu : \Gamma \times X \longmapsto \mathrm{T}\Gamma$ (called *logarithmic derivative of* μ) such that

(LD1) For all local smooth vector fields $\mathcal{V}$ the function $\gamma \mapsto \langle \mathcal{V}, B^\mu\rangle_{\mathrm{T}_\gamma\Gamma}$ is integrable. Define the directional logarithmic derivative

$$B^\mu_\mathcal{V}(\gamma) = \langle \mathcal{V}, B^\mu\rangle_{\mathrm{T}_\gamma\Gamma} + \mathrm{div}^\Gamma \mathcal{V}(\gamma).$$

(LD2) For all $F, G \in \mathcal{F}C^\infty_{\mathrm{b}}(\Gamma)$ and all local smooth vector fields $\mathcal{V}$ it holds

$$\int \nabla^\Gamma_\mathcal{V} F G \mathrm{d}\mu = -\int F \nabla^\Gamma_\mathcal{V} G \mathrm{d}\mu - \int F G B^\mu_\mathcal{V} \mathrm{d}\mu \tag{4.1}$$

We recall some definitions from [**15**] connected to the question to characterize ibp-measures.

DEFINITION 4.2. Let μ be a point process. The *reduced Campbell measure* $C^!_\mu$ is defined by

$$C^!_\mu(Z) = \int \mu(\mathrm{d}\gamma) \mathbb{S} \chi_Z(\gamma) = \int \mu(\mathrm{d}\gamma) \int \gamma(\mathrm{d}x) \chi_Z(\varphi - \delta_x, x), \qquad (Z \in \mathfrak{B}(\Gamma) \otimes \mathfrak{X}).$$

Another point process μ' is called *supporting measure* for μ if

$$C^!_\mu(\,\cdot\, \times X) \sim \mu'.$$

A *balancing kernel* $\tilde{\mu}$ fulfils for such a supporting measure μ'

$$C^!_\mu(Z) = \int \mu'(\mathrm{d}\gamma) \int \tilde{\mu}(\gamma, \mathrm{d}x) \chi_Z(\gamma, x), \qquad (Z \in \mathfrak{B}(\Gamma) \otimes \mathfrak{X}).$$

We want to prove

THEOREM 4.3. *A point process μ is an ibp-measure iff the measures $\tilde{\mu}(\gamma,\,\cdot\,)$ are ibp-measures for μ'–a.a. γ and*

$$B^{\mu}(\gamma, x) = \beta^{\tilde{\mu}(\gamma,\cdot)}(x), \qquad (\gamma \in \Gamma, x \in X) \tag{4.2}$$

fulfils (LD1). *Then B^{μ} is the logarithmic derivative of μ.*

Especially, any ibp-measure μ has property Σ^{c}_{m}, *i.e.*

$$C^{!}_{\mu}(\Gamma \times \,\cdot\,) \ll m.$$

Moreover, the set ibp(B) *of all ibp-measures with logarithmic derivative B fixed is the convex set of all point processes μ for which there is a balancing kernel $\tilde{\mu}$ with*

$$\beta^{\tilde{\mu}(\gamma,\cdot)}(x) = B(\gamma, x), \qquad (C^{!}_{\mu}\text{–a.a.}[\gamma, x] \in \Gamma \times X). \tag{4.3}$$

REMARK 4.4. First note [**15**] that two balancing kernels $\tilde{\mu}$, $\tilde{\mu}'$ fulfil essentially

$$\tilde{\mu}(\gamma,\,\cdot\,) = g(\gamma)\tilde{\mu}'(\gamma,\,\cdot\,)$$

for a strictly positive function $g : \Gamma \longmapsto \mathbb{R}_{+}$. Thus the formula for B^{μ} does not depend on the choice of μ', as the logarithmic derivative β^{σ} does not depend on the scaling of σ.

REMARK 4.5. Especially, Gibbs measures associated to smooth potentials are ibp-measures, which is a special case of a result proved in [**2**]. Then even $C^{!}_{\mu} \ll \mu \otimes m$, a property named Σ'_{m} [**11**]. Explicitely, $\tilde{\mu}(\gamma, \mathrm{d}x) = \mathrm{e}^{-H(\delta_x|\gamma)}m(\mathrm{d}x)$ where the conditional energy $H(\,\cdot\,|\,\cdot\,)$ is formally given by

$$H(\gamma'|\gamma) = H(\gamma + \gamma') - H(\gamma),$$

with a (formal) global energy function H, see [**11, 16**]. If $X = \mathbb{R}^d$ and there is only a pair potential ϕ present one finds

$$H(\delta_x|\gamma) = \int \gamma(\mathrm{d}y)\phi(x - y).$$

Then

$$B^{\mu}_{V_v}(\gamma) = \frac{1}{2}\int \gamma(\mathrm{d}x) \int (\gamma - \delta_x)(\mathrm{d}y)\,\langle v(x), \nabla^{X}\phi(x - y)\rangle_{\mathrm{T}_x X}$$

which is the same expression as used in [**2**, Lemma 4.1].

The quantity $e_x(\gamma) = \beta^{\tilde{\mu}(\gamma,\cdot)}(x)$ was called canonical gradient in [**5, 15**]. In the Gibbsian case, it holds $e_x(\gamma) = -\nabla^{X}_{x} H(\delta_x|\gamma)$. Thus, integration of the canonical gradient leads to conditional energy functions [**12**].

REMARK 4.6. For measures μ with property Σ^{c}_{m} it is easy to see that $\mu(\Gamma^{\leq 1}) = 1$. Thus the result translates easily to $\Gamma^{\leq 1}$.

PROOF. It is standard to show (using the product formula for the gradient ∇^{Γ}_{V}) that (4.1) is equivalent to

$$\int \nabla^{\Gamma}_{V} F \mathrm{d}\mu = -\int F B^{\mu}_{V} \mathrm{d}\mu. \tag{4.4}$$

But

$$\begin{aligned}\int \mu(\mathrm{d}\gamma)\nabla^\Gamma_{\mathcal{V}}F &= \int \mu(\mathrm{d}\gamma)\int \gamma(\mathrm{d}x)\left\langle \mathcal{V}(\gamma-\delta_x,x), 1\otimes\nabla^X\mathbb{D}F(\gamma-\delta_x,x)\right\rangle_{\mathrm{T}_xX}\\ &= \int C^!_\mu(\mathrm{d}\gamma,\mathrm{d}x)\left\langle \mathcal{V}(\gamma,x), 1\otimes\nabla^X\mathbb{D}F(\gamma,x)\right\rangle_{\mathrm{T}_xX}\\ &= \int \mu'(\mathrm{d}\gamma)\int \tilde\mu(\gamma,\mathrm{d}x)\left\langle \mathcal{V}(\gamma,x), 1\otimes\nabla^X\mathbb{D}F(\gamma,x)\right\rangle_{\mathrm{T}_xX}.\end{aligned}$$

Similarly, we obtain

$$\begin{aligned}\int FB^\mu_{\mathcal{V}}\mathrm{d}\mu &= \int \mu(\mathrm{d}\gamma)F(\gamma)\int \gamma(\mathrm{d}x)\left\langle \mathcal{V}(\gamma-\delta_x,x), B^\mu(\gamma-\delta_x,x)\right\rangle_{\mathrm{T}_xX}\\ &= \int \mu'(\mathrm{d}\gamma)\int \tilde\mu(\gamma,\mathrm{d}x)\mathbb{D}F(\gamma,x)\left\langle \mathcal{V}(\gamma,x), B^\mu(\gamma,x)\right\rangle_{\mathrm{T}_xX}.\end{aligned}$$

Suppose that μ is an ibp-measure. Inserting vector fields $\mathcal{V}$ like $\mathcal{V}'(\gamma,x)=G(\gamma)\mathcal{V}(\gamma,x)$ for suitably smooth G we find for all $\mathcal{V}$ and F for μ'–a.a. $\gamma\in\Gamma$

$$\int \tilde\mu(\gamma,\mathrm{d}x)\left\langle \mathcal{V}(\gamma,x), 1\otimes\nabla^X\mathbb{D}F(\gamma,x)\right\rangle_{\mathrm{T}_xX} = \int \tilde\mu(\gamma,\mathrm{d}x)\mathbb{D}F(\gamma,x)\left\langle \mathcal{V}(\gamma,x), B^\mu(\gamma,x)\right\rangle_{\mathrm{T}_xX}.$$

We know that we can select an enumerable set of $F\in\mathcal{F}C^\infty_{\mathrm{b}}(\Gamma)$ which separates points in Γ. Thus, we find an enumerable set of such F and vector fields $\mathcal{V}$, for which the above equation is true for μ'–a.a. $\gamma\in\Gamma$ and for all $\gamma\in\Gamma$ the set of functions $\mathbb{D}F(\gamma,\,\cdot\,)$ is dense in $C_0(X)$. Thus the Malliavin lemma, see [**14**, Satz2.8], shows that $\tilde\mu(\gamma,\,\cdot\,)$ is an ibp-measure for μ'–a.a. $\gamma\in\Gamma$. Moreover, we find (4.2) $C^!_\mu$–a.e. The converse direction is clear from the same arguments.

The properties of ibp-measures imply $\tilde\mu(\gamma,\,\cdot\,)\ll m$ for μ'–a.a.$\gamma\in\Gamma$. As Γ is Polish, this is equivalent to Σ^{c}_m.

Clearly, for a function B the set ibp(B) is convex. The above derivation shows for any $\mu\in$ ibp(B) that μ'–a.a. $\gamma\in\Gamma$ fulfil (4.3) for $\tilde\mu(\gamma,\,\cdot\,)$-a.a. $x\in X$. This completes the proof. □

REMARK 4.7. The use of another class of vector fields, compared to [**1, 2**], proves useful in two directions: Firstly, we can avoid moment conditions like $\int\mu(\mathrm{d}\gamma)\gamma(K)<\infty$ for $K\in\mathfrak{X}^{\mathrm{b}}$. Secondly, the proof becomes much shorter and more clear, it ties together the operators $\mathbb{D}$, $\mathbb{S}$ well-related to point process theory and ∇^X, crucial for the differential structure of X.

The case of free particles, which was the subject of [**1**], is interesting in its own. Recall [**10**] that a point process μ possesses the *property* (g_σ) if for all $K\in\mathfrak{X}^{\mathrm{b}}$

$$\mu(Y|\{\gamma\in\Gamma:\gamma(K)=n\})=\pi_\sigma(Y|\{\gamma\in\Gamma:\gamma(K)=n\}),\qquad (Y\in\mathfrak{B}(\Gamma)_K).$$

Moreover, let Ξ_n denote the map $(x_1,\dots,x_n)\mapsto\delta_{x_1}+\cdots+\delta_{x_n}$.

THEOREM 4.8. *Suppose σ is a σ-finite measure. Then π_σ is an ibp-measure if and only if σ is an ibp-measure.*

If X is connected, define B^σ by

$$B^\sigma(\gamma,x)=\beta^\sigma(x),\qquad (\gamma\in\Gamma,x\in X).$$

Then ibp(B^σ) *consists of all measures with property (g_σ).*

Especially, if $\sigma(X) = \infty$, there exists for all $\mu \in \mathrm{ibp}(B^\sigma)$ a measure ν on $\mathbb{R}_+$ such that

$$\mu = \int \nu(\mathrm{d}z)\pi_{z\sigma}.$$

If $\sigma(X) < \infty$ then there is a sequence $(\lambda_n)_{n\in\mathbb{N}} \subset \mathbb{R}_+$, $\sum_{n\in\mathbb{N}} \lambda_n \sigma(X)^n = 1$, with

$$\mu = \sum_{n\in\mathbb{N}} \lambda_n \sigma^{\otimes n} \circ \Xi_n.$$

PROOF. We may set $\mu' = \pi_\sigma$, $\tilde{\mu}(\gamma, \cdot\,) = \sigma(\,\cdot\,)$. This proves the first assertion.

If X is connected, all measures on X with the same logarithmic derivative are multiples of each other, which follows easily from [**13**, Theorem 3.3]. Consequently,

$$\tilde{\mu}(\gamma, \cdot\,) = c(\gamma)\sigma(\,\cdot\,),, \qquad (\gamma \in \Gamma)$$

or $C^!_\mu = \hat{\mu} \otimes \sigma$ for some measure $\hat{\mu}$. This relation characterizes the property (g_σ), see [**10, 15**]. The former reference establishes also the representations of μ. □

REMARK 4.9. In [**2**], the above result was obtained for $\sigma(X) = \infty$ and the condition that $(z\sigma)_{z\in\mathbb{R}_+}$ are the only Radon measures ν on X such that div^X_σ is the dual operator on $\mathcal{L}^2(\nu)$ of ∇^X consider with domains $V_0(K)$ (smooth vector fields with support in K), $C_0^\infty(K)$ for all $K \in \mathfrak{D}$. Certainly, the latter condition implies that X is connected and our result covers this case.

REMARK 4.10. If X is connected, one may define a *Lebesgue measure* as an ibp-measure M with $B^M = 0$. Then, by the above theorem, any finite Lebesgue measure is of the form

$$M = \int \nu(\mathrm{d}z)\pi_{zm}$$

for some finite measure ν on $\mathbb{R}$ if X is not compact or

$$M = \lambda_0 \delta_{\mathfrak{o}} + \sum_{n\in\mathbb{N}^*} \lambda_n m^{\otimes n} \circ \Xi_n^{-1}$$

with nonegative $(\lambda_n)_{n\in\mathbb{N}} \in \mathbb{R}_+$, $\sum_{n\in\mathbb{N}} \lambda_n < \infty$ if X is compact and m normalized.

If we leave the class of finite measures, there are more candidates. One candidate, e.g., is the exponential measure [**3**]

$$F^m(Y) = \chi_Y(\mathfrak{o}) + \sum_{n=1}^{\infty} \frac{1}{n!} \int \chi_Y(\Xi_n([x_1, \dots, x_n]))m^{\otimes n}(\mathrm{d}x_1, \dots, \mathrm{d}x_n), \qquad (Y \in \mathfrak{B}(\Gamma)).$$

We close with one example which shows that Theorem 4.3 becomes false (on Γ) if we consider, like in [**1, 2**] only independent vector fields.

EXAMPLE 4.11. Suppose a random variable ξ on X is distributed according to the ibp-measure σ and let μ be the distribution of $2\delta_\xi$. For a smooth symmetric function F on X^2, set $f(x) = F(x, x)$. Then, as μ is concentrated on Γ_2, we find for any local smooth vector field v on X

$$\begin{aligned}\int \nabla^{\Gamma}_{\mathcal{V}_v} F \mathrm{d}\mu &= 2\int \sigma(\mathrm{d}x)\nabla^X_v f(x)\\ &= -2\int \sigma(\mathrm{d}x)(\langle v(x),\beta^\sigma(x)\rangle_{\mathrm{T}_xX} + \mathrm{div}^X v(x))f(x)\\ &= -\int \mu(\mathrm{d}\gamma)(\langle \mathcal{V}_v, B^\mu\rangle_{\mathrm{T}_\gamma\Gamma} + \mathrm{div}^\Gamma \mathcal{V}_v(\gamma))F(\gamma)\end{aligned}$$

with $B^\mu(\gamma,x) = \beta^\sigma(x)$. On the other hand, it is clear that μ has not property Σ^{c}_m because it is not absolutely continuous with respect to $m^2 \circ \Xi_2^{-1}$ [**15**, Theorem 2.13].

References

[1] S. Albeverio, Yu. G. Kondratiev, and M. Röckner **Analysis and Geometry on Configuration Spaces** J. Funct. Anal. **154**(2):440–500 1998

[2] S. Albeverio, Yu. G. Kondratiev, and M. Röckner **Analysis and Geometry on Configuration Spaces: The Gibbsian Case** J. Funct. Anal. **157**(1):242–291 1998

[3] D.S. Carter and P.M. Prenter **Exponential spaces and counting processes** Z. Wahrscheinlichkeitstheor. Verw. Geb. **21**:1–19 1972

[4] D.J. Daley and D. Vere-Jones **An Introduction to the Theory of Point Processes** Springer New York, Berlin, Heidelberg 1988

[5] K.-H. Fichtner and W. Freudenberg **On a probabilistic model of infinite quantum mechanical particle systems** Math. Nachr. **121**:171 – 210 1985

[6] K.-H. Fichtner and G. Winkler **Generalized Brownian Motion, Point Processes and Stochastic Calculus for Random Fields** Math. Nachr. **161**:291 – 307 1993

[7] O. Kallenberg **Random Measures** fourth edition Akademie-Verlag and Academic Press Berlin, London, New York 1986

[8] J.F.C. Kingman **Poisson processes** Clarendon Press Oxford 1993

[9] V. Liebscher **Analysis and Geometry on Configuration Spaces: A Pedestrian Approach** in preparation

[10] K. Matthes, J. Kerstan, and J. Mecke **Infinitely Divisible Point Processes** Wiley Chichester 1978

[11] K. Matthes, W. Warmuth, and J. Mecke **Bemerkungen zu einer Arbeit von Nguyen Xuan Xanh und Hans Zessin** Math. Nachr. **88**:117 – 127 1978

[12] B. Rauchenschwandtner **Gibbsprozesse und Papangeloukerne** VWGÖ Wien 1980

[13] O. G. Smolyanov and H. v. Weizscker **Differentiable families of measures** J. Funct. Anal. **118**:454–476 1993

[14] H. von Weizäcker **Differenzierbare Maße und Stochastische Analysis** 6 lectures atthe graduate college 'Stochastische Prozesse und probabilistische Analysis' Berlin 1998

[15] A. Wakolbinger and G. Eder **A condition $\Sigma^{\mathrm{c}}_\lambda$ for point processes** Math. Nachr. **116**:209 – 232 1984

[16] N. X. Xanh and H. Zessin **Integral and differential characterizations of the Gibbs process** Math. Nachr. **88**:105–115 1979

GSF-National Research Center for Environment and Health, Institute for Biomathematics and Biometry, Ingolstädter Landstr.1, D–85758 Neuherberg, Germany

E-mail address: liebsche@gsf.de

Canadian Mathematical Society
Conference Proceedings
Volume **29**, 2000

Fock Space Markovian Cocycles: Their Representation, Generation, and Dilation

J Martin Lindsay and Stephen J Wills

Happy birthday Sergio

ABSTRACT. Current work on Markovian cocycles on operator algebras that are adapted to a Fock space filtration is reviewed. We focus on the generation and characterisation of cocycles as solutions of quantum stochastic differential equations, and the recognition of properties of the cocycle, such as positivity, from its stochastic generator. Our framework goes beyond von Neumann algebras to the topological context of C^*-algebras. Some stochastic dilation results, in which completely positive cocycles are realised in terms of *-homomorphic cocycles, are given.

0. Introduction

If a one-parameter unitary group U on a Hilbert space $\mathfrak{h}$ is randomised using a Brownian motion: $u_t = U_{B_t} (t \geq 0)$, the result is a family of unitary operators on the Hilbert space of vector-valued functions $L^2(\Omega; \mathfrak{h})$ that satisfy the cocycle identity $u_{s+t} = u_s \theta_s(u_t)$, in which $(\theta_s)_{s \geq 0}$ is the right shift on Brownian paths. Equally, randomisation via a Poisson process yields such a cocycle. Similarly, starting with a one-parameter group of automorphisms of a von Neumann algebra $\mathcal{M}$, one obtains cocycles of automorphisms of the von Neumann algebra $L^\infty(\Omega; \mathcal{M})$ ([**LiS**]). Fock space provides a common home for each of these. Quantum probability allows noncommutative randomisations, and cocycles may be explicitly constructed with the use of quantum stochastic calculus ([**Bia**],[**Mey**],[**Par**]).

Let $(\sigma_t)_{t \geq 0}$ be the one-parameter semigroup of right shifts on $\mathcal{F}$, the symmetric Fock space over a Hilbert space of the form $L^2(\mathbb{R}_+; \mathsf{k})$, and let $\mathfrak{h}$ be an additional Hilbert space. In this paper we consider the two classes of equation:

$$V_{s+t} = V_s \sigma_s(V_t), \quad V_0 = 1 \tag{0.1}$$

$$k_{s+t} = \widehat{k}_s \circ \sigma_s \circ k_t, \quad k_0(a) = a \otimes 1_{\mathcal{F}} \tag{0.2}$$

where $s, t \geq 0$. A solution of (0.1) consists of a family of bounded operators on $\mathfrak{h} \otimes \mathcal{F}$. The equation (0.2) is studied with reference to a C^*-algebra $\mathcal{A}$ acting on $\mathfrak{h}$; a solution consists of a family of bounded maps from $\mathcal{A}$ to the tensor product

1991 *Mathematics Subject Classification.* Primary 47-02, 46L53, 81S25; Secondary 60H20.

von Neumann algebra $\mathcal{A}'' \otimes \mathcal{B}(\mathcal{F})$, and here $\widehat{k}_s$ is an extension of k_s whose domain includes $\bigcup_{t\geq 0} \sigma_s(k_t(\mathcal{A}))$. In each case solutions are required to satisfy an adaptedness condition with respect to the natural filtration of $\mathfrak{h} \otimes \mathcal{F}$ (which induces a filtration of $\mathcal{A}'' \otimes \mathcal{B}(\mathcal{F})$) arising from the continuous tensor product structure of Fock space; they are called *Markovian cocycles.*

The equations are related in several ways. If V satisfies (0.1) then $k_t(a) = V_t a$ defines a Markovian cocycle on the von Neumann algebra $\mathcal{A} = \mathcal{B}(\mathfrak{h})$; k is completely bounded and normal, and V is recoverable from k by evaluation at the identity. The basic analysis of (0.1) is thereby subsumed, to some degree, by that of (0.2), and so our focus will be on the second equation. Alternatively, setting $k_t(a) = V_t a V_t^*$ defines a Markovian cocycle on $\mathcal{B}(\mathfrak{h})$, which is completely positive and normal. The filtration of $\mathcal{A}''\otimes\mathcal{B}(\mathcal{F})$ comes with conditional expectation maps $(\mathbb{E}_s)_{s\geq 0}$, compatible with the vacuum of Fock space. Their basic properties quickly reveal that $(\mathbb{E}_0 \circ k_t)$ and $(\mathbb{E}_0[V_t])$ are one-parameter semigroups on $\mathcal{A}$ and $\mathfrak{h}$ respectively — the *Markov semigroups* associated to these cocycles. Indeed the cocycle equations generalise the defining equations of a one parameter semigroup on $\mathfrak{h}$ and $\mathcal{A}$ respectively (taking $\mathsf{k} = \{0\}$), and so this work may be viewed as a stochastic extension of the generation and dilation of such semigroups.

In previous work the cocycles consisted of normal injective unital *-homomorphisms on a von Neumann algebra ([**Bra**],[**AcM**]), and unitary operators on a Hilbert space ([**HuL**],[**Jou**]) respectively. The generator of the Markov semigroup of the cocycle is used to obtain an associated martingale whose quantum stochastic integral representation ([**PaS**]) leads to the existence of a *stochastic generator* for the cocycle. This entails realising the cocycle as the solution of a quantum stochastic differential equation ([**HuP**],[**Par**]). In current work many previous constraints are set aside. Most importantly we use an equivalent definition of cocycles which also makes good sense in the C^*-algebra context. A second aspect of current work is the characterisation of the stochastic generators of certain classes of Markovian cocycle: completely positive, contractive, and unital cocycles on an algebra; contractive, isometric and coisometric cocycles on a Hilbert space. This extends the original work of Hudson and Parthasarathy on unitary processes ([**HuP**]), and M P Evans on *-homomorphic flows ([**Eva**]).

An axiomatic framework for quantum stochastic processes was formulated in [**AFL**], which incorporated Accardi's earlier ideas on Markovian cocycles on both Hilbert spaces and operator algebras ([**Acc**]). Here such cocycles provided a means of perturbing a Markovian semigroup, giving rise to the notion of a quantum Feynman-Kac formula. These ideas have been developed further in [**HKK**], where an axiomatic description of quantum white noise over a von Neumann algebra is given. Unitary cocycles are defined with respect to such a noise (which plays the role of our shift semigroup σ) and are shown to satisfy a noncommutative stochastic differential equation in which the integrator process is constructed from the particular cocycle under consideration.

Unitary cocycles also appear in the work of Arveson ([**Ar1**]) on classifying one-parameter semigroups of unital *-endomorphisms of $\mathcal{B}(\mathcal{H})$ (for a separable Hilbert space $\mathcal{H}$), called E_0-semigroups. Given such a semigroup α, if $(U_t)_{t\geq 0}$ is a strongly continuous family of unitaries on $\mathcal{H}$ satisfying $U_{s+t} = U_s \alpha_s(U_t)$, then $X \mapsto U_t \alpha_t(X) U_t^*$ is again an E_0-semigroup. The type and index of an E_0-semigroup are invariant under such perturbations. These ideas are now being extended to the classification of quantum dynamical semigroups (QDS's) on $\mathcal{B}(\mathcal{H})$, that is, semigroups of unital and completely positive maps on this algebra. Bhat showed that given any QDS τ one can construct a *minimal* dilation of τ to an E_0-semigroup θ acting on $\mathcal{B}(\mathcal{H}')$, where $\mathcal{H}'$ is a Hilbert space containing $\mathcal{H}$ as a subspace. That is

$$\tau_t(X) = E^*\theta_t(EXE^*)E \quad \forall X \in \mathcal{B}(\mathcal{H}), \tag{0.3}$$

where E is the inclusion map $\mathcal{H} \to \mathcal{H}'$, and minimality means that $\mathcal{H}'$ is the closed linear span of $\{\theta_t(EXE^*)u : t \geq 0, X \in \mathcal{B}(\mathcal{H}), u \in E\mathcal{H}\}$. The index of τ defined in [**Bh1**] is the index of its minimal dilation. An equivalent more intrinsic definition appears in [**Ar2**]. The methods of quantum stochastic calculus are now being introduced into this study ([**Bh2**]) as a result of the following observation: given any QDS τ with bounded generator it is possible to construct a Markovian cocycle j satisfying (0.2), for some choice of the Fock space $\mathcal{F}$, in which the maps j_t are unital and *-homomorphic, and whose Markov semigroup is τ ([**HuS**],[**GoS**]). Composing the cocycle j with the shift σ then gives an E_0-semigroup $(\widehat{\jmath}_t \circ \sigma_t)_{t\geq 0}$ that dilates τ in the sense that it satisfies (0.3). For a particular class of dilations realised in this manner a "deficiency index" is computed in [**Bh2**] which measures how far this dilation is from being minimal. This is achieved through a characterisation of all positive contraction operator cocycles in the case when $\mathfrak{h} = \mathbb{C}$.

The plan of this paper is as follows. After setting up the necessary notation in Section 1 we describe a new approach to Markovian cocycles on Fock space, which extends the scope of previous definitions to C^*-algebras (and unbounded operators) in Section 2. We give a brief review of quantum stochastic calculus in Section 3, including existence and uniqueness results for quantum stochastic differential equations (QSDE's). These lead naturally to our first result on the stochastic generation of cocycles, Theorem 3.4, in which we give sufficient conditions for a cocycle to weakly satisfy a QSDE. In Section 4 we specialise to completely positive and completely positive contraction cocycles, first characterising the generators of these cocycles, and then showing how this *a priori* assumption on the cocycle enables us to strengthen the results of Section 3. Finally in Section 5 we give a number of stochastic dilation results, showing first that completely positive contraction cocycles can be realised as filtered *-homomorphic cocycles, and then, more generally, that (unbounded) completely positive cocycles arise by conjugating a *-homomorphic cocycle by an operator-valued process. The former result is a stochastic extension of the Goswami-Sinha dilation theorem, while the latter result constitutes a continuous time stochastic Stinespring theorem.

1. Preliminaries

First some notations and identifications to be used throughout the paper. We shall use the symbol $\odot$ to denote algebraic tensor products, reserving the symbol $\otimes$ for tensor products of Hilbert spaces, bounded Hilbert space operators, von Neumann algebras and normal completely bounded linear maps between von Neumann algebras.

Fix a Hilbert space $\mathfrak{h}$, called the *initial* space, a separable Hilbert space k, called the *noise dimension* space, and an orthonormal basis $\eta = (e_i)_{i\geq 1}$ for k which is extended to a basis $(e_\alpha)_{\alpha\geq 0}$ of $\widehat{\mathsf{k}} := \mathbb{C} \oplus \mathsf{k}$ by setting $e_0 = 1 \in \mathbb{C}$. Let $D = \dim \widehat{\mathsf{k}}$, and let $c(D)$ denote $\mathbb{C}^{D-1}$ if $D < \infty$, and $c_{00}(\mathbb{N})$, the space of eventually zero sequences in $\mathbb{C}$, otherwise. The basis η allows us to identify the Hilbert spaces

$\mathfrak{h} \otimes \widehat{\mathsf{k}}$ and $\oplus_{\alpha \geq 0} \mathfrak{h}$ through the isometric embeddings $E_\alpha : \mathfrak{h} \to \mathfrak{h} \otimes \widehat{\mathsf{k}}$ defined by $E_\alpha u = u \otimes e_\alpha$.

For any complex vector space V, $\mathrm{M}_D(V)$ denotes the collection of $(D \times D)$ matrices $v = [v^\alpha_\beta]$ with entries in V, the indices ranging over the set $\{\gamma : 0 \leq \gamma < D\}$. $\mathrm{M}_D(V)$ is endowed with linear structure by pointwise defined operations, and when V has an involution $\dagger$, this gives rise to an involution $v \mapsto v^\dagger$ on $\mathrm{M}_D(V)$ defined by $(v^\dagger)^\alpha_\beta = (v^\beta_\alpha)^\dagger$. Throughout the paper Einstein's summation convention for repeated indices will apply, with greek indices running over the set $\{\gamma : 0 \leq \gamma < D\}$ and roman indices running over the set $\{i : 1 \leq i < D\}$ indexing the basis η (k, and so also D, will be evident from the context). For $v \in \mathrm{M}_D(V)$ and $w, z \in c(D)$, we use the notation

$$v_{w,z} := w_\alpha v^\alpha_\beta z^\beta \tag{1.1}$$

where $w_\alpha = \overline{w^\alpha}$, and $w^0 = z^0 = 1$.

Elements of $\mathrm{M}_D(\mathcal{B}(\mathcal{A}))$ are called *mapping matrices*, and elements of $\mathrm{M}_D(\mathcal{A})$ are called *algebra matrices*, where $\mathcal{A} \subset \mathcal{B}(\mathfrak{h})$ is a C^*-algebra. Two particularly useful mapping matrices are

$$\Delta^\alpha_\beta = \begin{cases} \mathrm{id} & \text{if } \alpha = \beta \neq 0 \\ 0 & \text{otherwise} \end{cases}, \quad \iota^\alpha_\beta = \begin{cases} \mathrm{id} & \text{if } \alpha = \beta \\ 0 & \text{otherwise} \end{cases}. \tag{1.2}$$

The first is used to define transforms on $\mathrm{M}_D(\mathcal{B}(\mathcal{A}))$, and on $\mathrm{M}_D(\mathcal{A})$, as follows:

$$\widehat{\theta} = \theta + \Delta, \quad \widehat{l} = l + \Delta(1). \tag{1.3}$$

An algebra matrix l is *bounded* if there is an operator in $\mathcal{B}(\mathfrak{h} \otimes \widehat{\mathsf{k}})$, also denoted l, such that $l^\alpha_\beta = (E_\alpha)^* l E_\beta$ for all α, β. Note that all algebra matrices are bounded if $D < \infty$. A mapping matrix θ is *bounded* if each algebra matrix $[\theta^\alpha_\beta(a)]$ is bounded. In this case, by the Banach-Steinhaus Theorem, $\theta \in \mathcal{B}(\mathcal{A}; \mathcal{B}(\mathfrak{h} \otimes \widehat{\mathsf{k}}))$, and it also follows that $\theta(\mathcal{A}) \subset \mathcal{A}'' \otimes \mathcal{B}(\widehat{\mathsf{k}})$.

Fock space. For a Hilbert space $\mathcal{K}$ containing a vector f, $\Gamma(\mathcal{K})$ denotes the symmetric Fock space over $\mathcal{K}$, and $\varepsilon(f) \in \Gamma(\mathcal{K})$ the exponential vector associated to f. When $\mathcal{K} = L^2(I; \mathsf{k})$ for some measurable subset I of $\mathbb{R}_+$ we write $\mathcal{F}_{I,\mathsf{k}}$ for $\Gamma(\mathcal{K})$, $\mathcal{B}_{I,\mathsf{k}}$ for the algebra of all bounded operators on $\mathcal{F}_{I,\mathsf{k}}$, and $\Omega_{I,\mathsf{k}}$ for the vacuum vector $\varepsilon(0)$, dropping the subscript I when $I = \mathbb{R}_+$. The exponential property of Γ, that is $\Gamma(\mathcal{K}_1 \oplus \mathcal{K}_2) = \Gamma(\mathcal{K}) \otimes \Gamma(\mathcal{K}_2)$, is best viewed via exponential vectors: $\varepsilon(f_1, f_2) \leftrightarrow \varepsilon(f_1) \otimes \varepsilon(f_2)$. It is the source of a number of natural identifications, in particular $\mathcal{F}_{[0,t],\mathsf{k}}$ is identified as a closed subspace of $\mathcal{F}_\mathsf{k}$ through the isometric embedding $E_t : \xi \mapsto \xi \otimes \Omega_{[t,\infty[,\mathsf{k}}$. This gives rise to a conditional expectation of $\mathcal{B}(\mathfrak{h}) \otimes \mathcal{B}_\mathsf{k}$ onto $\mathcal{B}(\mathfrak{h}) \otimes \mathcal{B}_{[0,t],\mathsf{k}}$ defined by $\mathbb{E}_t[c] = E_t^* c E_t \otimes 1_{[t,\infty[,\mathsf{k}}$, where we identify $\mathcal{B}_{I,\mathsf{k}}$ with $\mathcal{B}_{I,\mathsf{k}} \otimes 1_{I^c,\mathsf{k}} \subset \mathcal{B}_\mathsf{k}$. Product vectors of the form $u \otimes \varepsilon(f)$ in $\mathfrak{h} \otimes \mathcal{F}_\mathsf{k}$ will usually be abbreviated to $u\varepsilon(f)$.

For each $t \geq 0$ let s_t be the isometric right shift on $L^2(\mathbb{R}_+; \mathsf{k})$ defined by

$$(s_t f)(u) = \begin{cases} f(u-t), & u \geq t \\ 0, & u < t \end{cases}, \tag{1.4}$$

and let S_t denote the isometry on $\mathfrak{h} \otimes \mathcal{F}_\mathsf{k}$ obtained through second quantisation and ampliation, which satisfies $S_t u\varepsilon(f) = u\varepsilon(s_t f)$. In fact S_t defines an isometry of $\mathfrak{h} \otimes \mathcal{F}_\mathsf{k}$ onto $\mathfrak{h} \otimes \mathcal{F}_{[t,\infty[,\mathsf{k}}$, and so given $X \in \mathcal{B}(\mathfrak{h}) \otimes \mathcal{B}_\mathsf{k}$, it follows that $S_t X S_t^*$

can be considered as an operator on $\mathfrak{h} \otimes \mathcal{F}_{[t,\infty[,\mathsf{k}}$. Ampliating this with the identity operator on $\mathcal{F}_{[0,t],\mathsf{k}}$ produces an operator on $\mathfrak{h} \otimes \mathcal{F}$, which we shall denote $\sigma_t(X)$. The family of maps $\sigma = (\sigma_t)_{t \geq 0}$ is an ultraweakly continuous semigroup of normal unital *-homomorphisms of $\mathcal{B}(\mathfrak{h}) \otimes \mathcal{B}_{\mathsf{k}}$.

Given $f \in L^2(\mathbb{R}_+; \mathsf{k})$ define the sequence of functions (f^α) by

$$f^0(t) = 1; \quad f^i(t) = \langle e_i, f(t) \rangle, \ i \geq 1,$$

for each e_i in the basis η of k and set $f_\alpha = \overline{f^\alpha}$. Let $\mathbb{S}[\eta]$ be the dense subspace of $L^2(\mathbb{R}_+; \mathsf{k}) = L^2(\mathbb{R}_+) \otimes \mathsf{k}$ defined by

$$\mathbb{S}[\eta] = \operatorname{Lin}\{\mathbf{1}_I \otimes e_i : I \subset \mathbb{R}_+ \text{ a bounded interval}, i \geq 1\}.$$

Thus for $f \in \mathbb{S}[\eta]$ only finitely many terms in the sequence (f^α) are nonzero. Here $\mathbf{1}_I$ denotes the indicator function of I; we shall also abbreviate $f\mathbf{1}_I$ to f_I in what follows.

The dense subspace $\operatorname{Lin}\{\varepsilon(f) : f \in \mathbb{S}[\eta]\}$ of $\mathcal{F}_{\mathsf{k}}$ is denoted $\mathcal{E}[\eta]$. We will also make use of the set

$$\mathbb{S}_{\text{loc}}[\eta] = \{f : \mathbb{R}_+ \to \mathsf{k} \mid f_I \in \mathbb{S}[\eta] \text{ for all bounded subintervals } I \subset \mathbb{R}_+\}.$$

For each $f \in L^2_{\text{loc}}(\mathbb{R}_+; \mathsf{k})$ and bounded interval $I \subset \mathbb{R}_+$, let D^f_I be the operator $\mathfrak{h} \to \mathfrak{h} \otimes \mathcal{F}_{I,\mathsf{k}} \subset \mathfrak{h} \otimes \mathcal{F}_{\mathsf{k}}$ defined by

$$D^f_I u = u\varepsilon(f_I). \tag{1.5}$$

The shift semigroup σ extends naturally to $L^2_{\text{loc}}(\mathbb{R}_+; \mathsf{k})$.

Quantum stochastic processes. An $\mathcal{E}[\eta]$-*process on* $\mathfrak{h}$, the initial space, is a time-indexed family $(X_t)_{t \geq 0}$ of operators on $\mathfrak{h} \otimes \mathcal{F}_{\mathsf{k}}$ satisfying

(a) $\operatorname{Dom} X_t = \mathfrak{h} \odot \mathcal{E}[\eta]$,
(b) $t \mapsto X_t u\varepsilon(f)$ is weakly measurable, and
(c) $X_t u\varepsilon(f_{[0,t]}) \in \mathfrak{h} \otimes \mathcal{F}_{[0,t],\mathsf{k}}$ and $X_t u\varepsilon(f) = [X_t u\varepsilon(f_{[0,t]})] \otimes \varepsilon(f_{[t,\infty[})$.

An $\mathcal{E}[\eta]$-process X is *measurable/continuous* if it satisfies

(b)′ $t \mapsto X_t u\varepsilon(f)$ is strongly measurable/continuous.

The linear space of $\mathcal{E}[\eta]$-processes on $\mathfrak{h}$ is denoted $\mathbb{P}(\mathfrak{h}, \mathcal{E}[\eta])$.

Fix now a unital C^*-algebra $\mathcal{A}$, called the *initial algebra*, acting on $\mathfrak{h}$. An $\mathcal{E}[\eta]$-*process on* $\mathcal{A}$ is a linear map $k : \mathcal{A} \to \mathbb{P}(\mathfrak{h}, \mathcal{E}[\eta])$, satisfying the affiliation condition

$$k_t(a)a'u\varepsilon(f) = (a' \otimes 1)k_t(a)u\varepsilon(f) \quad \forall a' \in \mathcal{A}'. \tag{1.6}$$

The linear space of $\mathcal{E}[\eta]$-processes on $\mathcal{A}$ is denoted $\mathbb{P}(\mathcal{A}, \mathcal{E}[\eta])$. Such a process is called *bounded, completely bounded, contractive, completely positive*, or, when $\mathcal{A}$ is a von Neumann algebra, *normal*, if each k_t has that property. Since bounded operators are uniquely determined by their action on any dense subspace, we write $\mathbb{P}_{\text{b}}(\mathcal{A}, \mathsf{k})$ for the subspace of bounded processes. The process $k \in \mathbb{P}(\mathcal{A}, \mathcal{E}[\eta])$ has an adjoint if there is some $j \in \mathbb{P}(\mathcal{A}, \mathcal{E}[\eta])$ such that $\langle \xi, k_t(a)\zeta \rangle = \langle j_t(a^*)\xi, \zeta \rangle$ for all ξ, ζ and a. This j is unique, and is denoted $k^\dagger$. Clearly all bounded processes have adjoints: $k^\dagger_t(a) = k_t(a^*)^*$.

Given $k \in \mathbb{P}(\mathcal{A}, \mathcal{E}[\eta])$, and recalling the notation (1.5), define linear maps $k^{f,g}_t$ from $\mathcal{A}$ into the space of linear maps on $\mathfrak{h}$ by

$$k^{f,g}_t(a) = (D^f_{[0,t]})^* k_t(a) D^g_{[0,t]}$$

for all $t > 0$ and $f, g \in \mathbb{S}_{\mathrm{loc}}[\eta]$. An $\mathcal{E}[\eta]$-process is *weakly initial algebra bounded* (weakly i.a.b.) if $k_t^{f,g} \in \mathcal{B}(\mathcal{A}; \mathcal{B}(\mathfrak{h}))$ for all f, g and t. It follows from (1.6) that if k is weakly i.a.b. then $k_t^{f,g} \in \mathcal{B}(\mathcal{A}; \mathcal{A}'')$. If, furthermore, the maps $\{k_t^{f,g}\}_{t \geq 0}$ are locally uniformly bounded for each choice of f, g, then k is said to be *weakly regular*. Thus k is weakly regular if and only if

(1.7) $$\sup\{|\langle u\varepsilon(f), k_t(a)v\varepsilon(g)\rangle| : \|u\| = \|a\| = \|v\| = 1, t \in [0, T]\} < \infty$$

for all $f, g \in \mathbb{S}[\eta]$ and $T > 0$.

2. Markovian cocycles on Fock space

In the category of completely bounded normal operators from a von Neumann algebra acting on a Hilbert space into the algebra of bounded operators on another Hilbert space we may tensorise two such operators to obtain a third. This is one consequence of Haagerup's extension of the Wittstock-Pauslen Theorem ([**Haa**]) which gives a representation $S(a) = V_1^*(a \otimes 1_{\mathcal{L}})V_2$ for such operators, in which $\mathcal{L}$ is a supplementary Hilbert space and V_1 and V_2 are bounded operators, generalising the better known representation of completely positive operators due to Stinespring ([**Sti**]). In particular, for a completely bounded normal process k on a von Neumann algebra $\mathcal{M}$, we have well-defined operators $\widehat{k}_s := k_s \otimes \mathrm{id}_{\mathcal{B}(\mathcal{F}_{[s})}$ in which $\mathcal{F}_{[s} = S_s(\mathcal{F}_{\mathsf{k}})$, that is Fock space over $L^2([s, \infty[; \mathsf{k})$. This provides a precise meaning of the defining equation (0.2). The key to extending the definition further, from von Neumann algebras to C^*-algebras, and to possibly unbounded processes, is the following result in which s_s is the isometric shift defined in (1.4):

PROPOSITION 2.1 ([**LW2**]). *Let $k \in \mathbb{P}_{\mathrm{b}}(\mathcal{M}, \mathsf{k})$ be a completely bounded normal process on the von Neumann algebra $\mathcal{M}$. The following are equivalent*:

(i) *k satisfies the identity* (0.2).

(ii) $k_0^{f,g} = \mathrm{id}_{\mathcal{M}}$, *and* $k_{s+t}^{f,g} = k_s^{f,g} \circ k_t^{s_s^* f, s_s^* g}$, $\quad \forall f, g \in L^2(\mathbb{R}_+; \mathsf{k}), s, t \geq 0$.

This proposition is the basis for our new definition of cocycle, allowing us to bypass topological questions concerning the completion/closure of certain algebraic tensor products.

DEFINITION. A Markovian cocycle on the C^*-algebra $\mathcal{A}$, with dimension space k having fixed basis η, is a weakly i.a.b. process k in $\mathbb{P}(\mathcal{A}, \mathcal{E}[\eta])$ such that, for all $f, g \in \mathbb{S}_{\mathrm{loc}}[\eta]$ and $s, t \geq 0$,

(a) $k_0(a) = a \otimes 1$

(b) $k_t^{f,g}(\mathcal{A}) \subset \mathcal{A}$

(c) $k_{s+t}^{f,g} = k_s^{f,g} \circ k_t^{s_s^* f, s_s^* g}$.

Note that if $\mathcal{A}$ is a von Neumann algebra then condition (b) above is automatically satisfied as a consequence of the affiliation condition (1.6) on processes and the assumption of weak initial algebra boundedness. If $\mathcal{A}$ is not a von Neumann algebra then (b) should be considered as a *generalised Feller condition*. When k is a bounded process the maps $k_t^{f,g} \in \mathcal{B}(\mathcal{A}; \mathcal{A}'')$ are defined for all $f, g \in L^2([0, t]; \mathsf{k})$, and moreover $(f, g) \mapsto k_t^{f,g}$ is continuous, hence if the conditions (b) and (c) hold for all $f, g \in \mathbb{S}_{\mathrm{loc}}[\eta]$, then they must hold for all $f, g \in L^2_{\mathrm{loc}}(\mathbb{R}_+; \mathsf{k})$. Thus whether or not a bounded process is a Markovian cocycle does not depend on the basis of the noise dimension space used in the definition. Furthermore, for completely bounded processes there is a very satisfactory meaning that we may attach to $\widehat{k}_s$ under which cocycles are again characterised by (0.2). In brief, $k_t(\mathcal{A}) \subset \mathrm{M}_{\mathcal{F}_{t]}}(\mathcal{A})_{\mathrm{b}}$, a *matrix space* over $\mathcal{A}$, and $\widehat{k}_s = k_s^{(\mathcal{F}_{s]})}$, a natural lifting of k_s to a completely bounded operator $\mathrm{M}_{\mathcal{F}_{[s}}(\mathcal{A})_{\mathrm{b}} \to \mathrm{M}_{\mathcal{F}}(\mathcal{A})_{\mathrm{b}} \subset \mathcal{A}'' \otimes \mathcal{B}(\mathcal{F})$ (see [**LW3**]).

Throughout this article we will view $c(D)$ as the subspace of $\mathbb{S}_{\mathrm{loc}}[\eta]$ consisting of constant functions, identifying $(z^i) \in c(D)$ with $z^i e_i \in \mathbb{S}_{\mathrm{loc}}[\eta]$. The following result is then an immediate consequence of our definition since the elements of $c(D)$ are left unaltered by the action of the left shifts, $(s_t^*)_{t\geq 0}$.

THEOREM 2.2 ([**LW2**]). *Let $k \in \mathbb{P}(\mathcal{A}, \mathcal{E}[\eta])$. The following are equivalent:*

(i) *k is a Markovian cocycle on $\mathcal{A}$.*

(ii) *For all $w, z \in c(D)$, $(\mathcal{P}_t^{w,z} := k_t^{w,z})_{t\geq 0}$ is a one-parameter semigroup on $\mathcal{A}$, and for all $t > 0$ and $f, g \in \mathbb{S}_{\mathrm{loc}}[\eta]$,*

$$k_t^{f,g} = \mathcal{P}_{t_1 - t_0}^{[0]} \circ \cdots \circ \mathcal{P}_{t-t_n}^{[n]} \tag{2.1}$$

where $\mathcal{P}^{[i]} = \mathcal{P}^{w,z}$ for $w = f(t_i), z = g(t_i)$ whenever $\{0 = t_0 \leq \cdots \leq t_n \leq t_n\}$ contains all of the discontinuities of $f_{[0,t]}$ and $g_{[0,t]}$.

Thus our definition of Markovian cocycle gives a simple algebraic characterisation of those processes that are cocycles. It also resonates perfectly with earlier work on the solutions of QSDE's which we discuss in the following section. Cocycles whose Markov semigroup $\mathcal{P}^{0,0}$ is norm continuous are called *regular Markovian cocycles.*

Recall (0.1), the defining equation for a Markovian cocycle on a Hilbert space. Interpreting this for bounded operator-valued cocycles is obviously unproblematic and, by an analogous result to Proposition 2.1, we find that a bounded operator process X is a cocycle if and only if

$$X_0 = 1, \quad X_{s+t}^{f,g} = X_s^{f,g} X_t^{s_s^* f, s_s^* g} \tag{2.2}$$

for all $s, t \geq 0$. We may therefore extend the definition of Markovian cocycle on a Hilbert space to be any (possibly unbounded) process for which $X_t^{f,g} \in \mathcal{B}(\mathfrak{h})$ for all $f, g \in \mathbb{S}_{\mathrm{loc}}[\eta]$ and $t \geq 0$, and which satisfies (2.2). This is again useful since unbounded cocycles arise naturally in Belavkin's approach to quantum filtering ([**Be1**]). One of the ideas running through [**LiP**], [**LW1**] and [**LW2**] is that basic properties of such cocycles should be deduced from an understanding of cocycles on an algebra. Indeed, let $X \in \mathbb{P}(\mathfrak{h}, \mathcal{E}[\eta])$, let $\mathcal{A} = \mathcal{B}(\mathfrak{h})$, and define $k \in \mathbb{P}(\mathcal{A}, \mathcal{E}[\eta])$ by $k_t(a) = X_t a$. Then k is a Markovian cocycle on $\mathcal{A}$ if X is a Markovian cocycle on $\mathfrak{h}$. This illustrates why it is useful to be able to deal with map-valued cocycles that are not even completely positive.

3. Quantum stochastic differential equations

In a seminal paper Hudson and Parthasarathy developed a quantum stochastic calculus for operator-valued processes with the creation, annihilation and number processes on Fock space as integrators ([**HuP**]). This was extended to infinite dimensions of quantum noise by Mohari and Sinha ([**MoS**],[**Mey**],[**Par**]). The matrix of integrator processes $\Lambda = [\Lambda_\beta^\alpha]_{\alpha,\beta\geq 0}$ is defined in terms of the orthonormal basis η by:

Creation:	$\Lambda_j^0(t) := a^*(e_j\mathbf{1}_{[0,t]})$;
Conservation:	$\Lambda_j^i(t) := d\Gamma(\|e_j\rangle\langle e_i\| \otimes \mathrm{M}_{[0,t]})$;
Annihilation:	$\Lambda_0^i(t) := a(e_i\mathbf{1}_{[0,t]})$;
Time:	$\Lambda_0^0(t) := t1$;

where $a^*, d\Gamma$ and a denote, respectively, Fock space creation, differential second quantisation and annihilation, and $\mathrm{M}_{[0,t]}$ is the operator of multiplication by $\mathbf{1}_{[0,t]}$ on $L^2(\mathbb{R}_+)$. The integral of a process X on $\mathfrak{h}$ against any of these integrators, denoted $\int_0^t X_s \, d\Lambda_\beta^\alpha(s)$, is defined whenever X is measurable and $s \mapsto X_s\xi$ is locally square integrable for all $\xi \in \mathfrak{h} \odot \mathcal{E}[\eta]$. A matrix $Z = [Z_\beta^\alpha]$ of measurable processes is *stochastically integrable* if

$$\int_0^t \sum_{\alpha \geq 0} \|Z_\beta^\alpha u\varepsilon(f)\|^2 \, ds < \infty$$

for all t, β, u and f, and for such families the process $\int_0^t Z_\beta^\alpha(s) \, d\Lambda_\alpha^\beta(s)$ is defined as the strong limit (on $\mathfrak{h} \odot \mathcal{E}[\eta]$) of partial sums over α, β. They are characterised by the identity

$$\langle u\varepsilon(f), \int_0^t Z_\beta^\alpha(s) \, d\Lambda_\alpha^\beta(s) v\varepsilon(g)\rangle = \int_0^t f_\alpha(s) g^\beta(s) \langle u\varepsilon(f), Z_\beta^\alpha(s) v\varepsilon(g)\rangle \, ds. \tag{3.1}$$

Central to our study is the following QSDE:

$$dk_t = k_t \circ \theta_\beta^\alpha \, d\Lambda_\alpha^\beta(t), \quad k_0(a) = a \otimes 1, \tag{3.2}$$

where $\theta = [\theta_\beta^\alpha] \in \mathrm{M}_D(\mathcal{B}(\mathcal{A}))$. A *weak solution* of (3.2) is any process k on $\mathcal{A}$ such that

$$t \mapsto f_\alpha(s) g^\beta(s) \langle u\varepsilon(f), k_t(\theta_\beta^\alpha(a)) v\varepsilon(g)\rangle \text{ is locally integrable,}$$

and

$$\langle u\varepsilon(f), (k_t(a) - a) v\varepsilon(g)\rangle = \int_0^t f_\alpha(s) g^\beta(s) \langle u\varepsilon(f), k_s(\theta_\beta^\alpha(a)) v\varepsilon(g)\rangle \, ds.$$

This is the weakest notion of solution that we can define in $\mathbb{P}(\mathcal{A}, \mathcal{E}[\eta])$, in the light of (3.1). A *strong solution* is any process k such that

$$[k.(\theta_\beta^\alpha(a))] \text{ is stochastically integrable,}$$

and

$$k_t(a) = a \otimes 1 + \int_0^t k_s(\theta_\beta^\alpha(a)) \, d\Lambda_\alpha^\beta(s).$$

Recall the transformation $\theta \mapsto \widehat{\theta}$ defined on $\mathrm{M}_D(\mathcal{B}(\mathcal{A}))$ by (1.3), the notation (1.1), and the definition of weak regularity (1.7).

THEOREM 3.1 ([**LiP**],[**LW1**]). *Let $\theta \in \mathrm{M}_D(\mathcal{B}(\mathcal{A}))$ and let $k \in \mathbb{P}(\mathcal{A}, \mathcal{E}[\eta])$ be a weakly regular process on $\mathcal{A}$. The following are equivalent*:

(i) *k is a weak solution of* (3.2) *for this θ.*

(ii) *For all* $f, g \in \mathbb{S}_{\rm loc}[\eta]$ *and* $t \geq 0$,

$$k_t^{f,g} = \mathcal{P}^{[0]}_{t_1 - t_0} \circ \cdots \circ \mathcal{P}^{[n]}_{t - t_n} \tag{3.3}$$

where $\mathcal{P}^{[i]}$ *is the one-parameter semigroup on* $\mathcal{A}$ *with generator* $\widehat{\theta}_{w,z}$ *with* $w = f(t_i)$ *and* $z = g(t_i)$, *whenever* $\{0 = t_0 \leq \ldots \leq t_{n+1} = t\}$ *contains all the points of discontinuity of* $f_{[0,t]}$ *and* $g_{[0,t]}$.

Since we assume that our processes are adapted, it follows that k is determined by the family of maps $\{k_t^{f,g}\}$. In particular the above representation implies that to any given mapping matrix θ there is at most one weakly regular weak solution. Comparing (3.3) with the decomposition formula (2.1) for a Markovian cocycle also yields the following:

COROLLARY 3.2 ([**LW2**]). *Let* $k \in \mathbb{P}(\mathcal{A}, \mathcal{E}[\eta])$. *If* k *is a weakly regular weak solution to* (3.2) *for some* $\theta \in \mathrm{M}_D(\mathcal{B}(\mathcal{A}))$ *then* k *is a Markovian cocycle on* $\mathcal{A}$.

Thus, as in the case of the injective * homomorphic flows considered in [**Bra**], Markovian cocycles (according to our definition) arise whenever there is a (weak) solution to an Evans-Hudson type equation. Thus, perhaps the next question that should be tackled is for which θ is it possible to construct a solution to (3.2)? When k is finite dimensional (3.2) may be solved for *all* $\theta \in \mathrm{M}_D(\mathcal{B}(\mathcal{A}))$ via a Picard iteration scheme ([**Eva**]). When $D = \infty$ the situation becomes more complicated, but for most purposes (c.f. Proposition 4.4) the following is sufficient:

THEOREM 3.3 ([**LW3**]). *Let* $\theta \in \mathrm{M}_D(\mathcal{B}(\mathcal{A}))$. *If* θ *is completely bounded as a map* $\mathcal{A} \to \mathcal{A}'' \otimes \mathcal{B}(\widehat{\mathsf{k}})$ *then there is a strong solution* k *to* (3.2) *satisfying*

$$\sup\{\|k_t(a)u\varepsilon(f)\| : \|u\| = \|a\| = 1, t \in [0,T]\} < \infty \tag{3.4}$$

for all $f \in \mathbb{S}[\eta], T > 0$.

As in the case when $D < \infty$, the solution is constructed by an iteration scheme, and in fact can be achieved under far weaker conditions than those given in the theorem above. Indeed a sufficient condition is that for each a, u and d the sequence

$$\left\{ \sum_{\alpha_1, \ldots, \alpha_n \geq 0} \sum_{\beta_1, \ldots, \beta_n \leq d} \|\theta^{\alpha_1}_{\beta_1} \circ \cdots \theta^{\alpha_n}_{\beta_n}(a)u\|^2 : n \geq 0 \right\}$$

satisfies a certain growth condition ([**Mey**]). If this condition is satisfied uniformly for all $\|u\| = \|a\| = 1$ then the resulting solution satisfies the strong regularity condition (3.4) ([**LW1**]). The main result of [**LW3**] is to show that completely bounded mapping matrices uniformly satisfy the Meyer regularity condition. See [**Be2**] for work on the construction of completely positive solutions to (3.2) when the maps θ^α_β are unbounded.

The other obvious question arising from Corollary 3.2 is the following: what conditions on a Markovian cocycle imply that it satisfies a QSDE of the form (3.2)? This is the main question tackled in [**LW2**], and the most general answer obtained there, in the absence of any special assumptions on k such as complete positivity or boundedness, is the following:

THEOREM 3.4 ([**LW2**]). *Let* $k \in \mathbb{P}(\mathcal{A}, \mathcal{E}[\eta])$ *be a Markovian cocycle on* $\mathcal{A}$ *such that the following hold:*

(a) *The adjoint process* $k^\dagger$ *exists, and*

$$\lim_{t\to 0} k_t(a)\xi = \lim_{t\to 0} k_t^\dagger(a)\xi = a\xi \quad \forall a, \xi.$$

(b) *The semigroups* $\{(k_t^{f,g})_{t\geq 0}\}$ *are norm continuous for all choices of* $f, g \in \{0, e_1, e_2, \dots\} \subset c(D) \subset \mathbb{S}_{\text{loc}}[\eta]$.

Then k *is the unique weakly regular weak solution to the* QSDE (3.2) *for some* $\theta \in \mathrm{M}_D(\mathcal{B}(\mathcal{A}))$.

NOTE. If $D < \infty$ then, by the existence theorem in [**Eva**], there is a strong solution to (3.2) satisfying (3.4) for this θ, and so the cocycle in the above theorem is in fact a *strong* solution of the QSDE (3.2) by the uniqueness result contained in Theorem 3.1.

The essence of the proof of Theorem 3.4 is very simple. The conditions (a) and (b) above imply that the semigroup $\mathcal{P}^{w,z}$ is norm continuous for all choices of w and z, and that its generator is related to the generators of $\mathcal{P}^{0,0}$, $\mathcal{P}^{e_i,0}$, $\mathcal{P}^{0,e_j}$ and $\mathcal{P}^{e_i,e_j}$ through appropriate linear combinations determined by w and z. The equivalence in Theorem 3.1 then does the rest.

As we shall see in the next section when we focus our attention on CP contraction cocycles not only do conditions (a) and (b) of Theorem 3.4 simplify greatly, but also the conclusion of the theorem is strengthened.

4. Completely positive cocycles

Having extended the definition of Markovian cocycle to unbounded processes on C^*-algebras, and found conditions under which a general cocycle can be represented as the solution of a QSDE, we now impose certain desirable conditions on the cocycle, in particular that it be completely positive (CP). We characterise the generators of CP cocycles, and show how matters simplify in the case of CP contraction cocycles.

THEOREM 4.1 ([**LiP**],[**LW1**]). *Let* $\theta \in \mathrm{M}_D(\mathcal{B}(\mathcal{A}))$, *let* $k \in \mathbb{P}(\mathcal{A}, \mathcal{E}[\eta])$, *and suppose that* k *is a weakly regular weak solution to the* QSDE (3.2) *for this* θ. *The following are equivalent*:

(i) k *is completely positive.*

(ii) θ_0^0 *is real and there is a quadruple* $\mathcal{Q} = (\pi, \mathcal{K}, \delta, \{W_i\})$ *consisting of a representation* $(\pi, \mathcal{K})$ *of* $\mathcal{A}$, *a* π*-derivation* $\delta : \mathcal{A} \to \mathcal{B}(\mathfrak{h}; \mathcal{K})$ *and a sequence* $\{W_i\}$ *in* $\mathcal{B}(\mathfrak{h}; \mathcal{K})$ *such that*

$$\left.\begin{aligned} \partial\theta_0^0(a_1, a_2) &= \delta(a_1)^*\pi(1)\delta(a_2) \\ \theta_0^i(a) = \theta_i^0(a^*)^* &= W_i^*\pi(1)\delta(a) + \theta_0^i(1)a \\ \theta_j^i(a) &= W_i^*\pi(a)W_j - \delta_j^i a. \end{aligned}\right\} \tag{4.1}$$

(iii) *There is a quadruple* $\mathcal{P} = (\pi, \mathcal{K}, \{K_\gamma\}, \{W_\gamma\})$ *consisting of a representation* $(\pi, \mathcal{K})$ *of* $\mathcal{A}$ *and sequences of operators* $\{K_\gamma\}$ *in* $\mathcal{B}(\mathfrak{h})$ *and* $\{W_\gamma\}$ *in* $\mathcal{B}(\mathfrak{h}; \mathcal{K})$ *such that*

$$\widehat{\theta}_\beta^\alpha(a) = W_\alpha^*\pi(a)W_\beta + \delta_0^\alpha a K_\beta + \delta_\beta^0 K_\alpha^* a.$$

where δ *here is the Kronecker delta.*

Moreover the quadruple $\mathcal{Q}$ in (ii) *may be chosen so that $(\pi, \mathcal{K})$ is unital and*

$$\mathcal{K} = \overline{\operatorname{Lin}}\left\{\delta(a)u^0 + \textstyle\sum_{i=1}^n \pi(a)W_i u^i : a \in \mathcal{A}, n \in \mathbb{N}, u^\alpha \in \mathfrak{h}\right\}. \tag{4.2}$$

NOTE. The coboundary-like operator ∂ acts on linear maps ψ on the algebra $\mathcal{A}$ to give sesquilinear maps $\partial\psi : \mathcal{A} \times \mathcal{A} \to \mathcal{A}$ in the following manner:

$$\partial\psi(a_1, a_2) = \psi(a_1^* a_2) - a_1^*\psi(a_2) - \psi(a_1^*)a_2 + a_1^*\psi(1)a_2.$$

The map $\partial\psi$ determines ψ up to the sum of a derivation and an anticommutator on $\mathcal{A}$.

The quadruple $\mathcal{Q}$ for which π is unital and the minimality condition (4.2) is satisfied, is unique up to isomorphism. The above theorem simultaneously generalises M P Evans' characterisation of the generators of *-homomorphic cocycles ([**Eva**]), as well as the characterisation of the generators of norm continuous CP semigroups on C^*-algebras due to Linblad, Christensen and Evans, et al. ([**ChE**] and references therein). A similar characterisation has been obtained independently by Belavkin (see [**Be2**] and references therein). Further refinement is obtained in the following result. Belavkin refers to a CP cocycle as *filtering* (respectively *subfiltering*) if the positive operator-valued process $(k_t(1))_{t\geq 0}$ is a martingale (resp. a supermartingale):

$$\mathbb{E}_s[k_t(1)] = k_s(1) \quad (\text{resp. } \mathbb{E}_s[k_t(1)] \leq k_s(1)) \quad \forall s \leq t.$$

In view of the identity $\mathbb{E}_s \circ k_{s+t} = k_s \circ \mathbb{E}_0 \circ k_t$, these conditions are respectively equivalent to unitality (or conservativity) and contractivity of the associated Markov semigroup. If k is unital and CP then it is necessarily bounded ([**LiP**], Proposition 2.1); the (sub-)filtering condition is a tool for the study of unbounded processes.

Unbounded Markovian cocycles are easily constructed. A simple example is obtained by putting $k_t(a) = V_t^* a V_t$ where $V_t u\varepsilon(f) = \exp\{l \int_0^t f(s)\, ds\} u\varepsilon(f)$ for $l \in \mathcal{A}$ and $\mathsf{k} = \mathbb{C}$, so that formally $V_t = e^{l \otimes A_t}$. Our next result, whose idea comes from [**Moh**], gives a sufficient condition for unboundedness.

PROPOSITION 4.2. *Let $k \in \mathbb{P}(\mathcal{A}, \mathcal{E}[\eta])$ be a completely positive process weakly satisfying* (3.2). *If there is some $i \geq 1$ such that $\theta_i^i(1)$ is uniformly positive, that is $\theta_i^i(1) \geq \epsilon 1$ for some $\epsilon > 0$, then k is unbounded.*

PROOF. We may assume without loss that $M := \begin{bmatrix} \theta_0^0(1) & \theta_i^0(1) \\ \theta_0^i(1) & \theta_i^i(1) \end{bmatrix} \geq \epsilon 1$ by replacing k_t by $e^{t\lambda}k_t$ for sufficiently large λ. Put $\xi = \|u\varepsilon(f)\|^{-1}u\varepsilon(f)$ where $u \neq 0$ and $f = ne_i \mathbf{1}_{[0,T]}$. Then, for $t \leq T$,

$$\begin{aligned}
\langle \xi, k_t(1)\xi\rangle &= 1 + \int_0^t f_\alpha(s) f^\beta(s) \langle \xi, k_s(\theta_\beta^\alpha(1))\xi\rangle\, ds \\
&= 1 + \int_0^t \left\langle \begin{bmatrix} \xi \\ n\xi \end{bmatrix}, k_s^{(2)}(M) \begin{bmatrix} \xi \\ n\xi \end{bmatrix} \right\rangle ds \\
&\geq 1 + \epsilon \int_0^t \left\langle \begin{bmatrix} \xi \\ n\xi \end{bmatrix}, k_s^{(2)}(1) \begin{bmatrix} \xi \\ n\xi \end{bmatrix} \right\rangle ds \\
&= 1 + \epsilon(1 + n^2) \int_0^t \langle \xi, k_s(1)\xi\rangle\, ds.
\end{aligned}$$

In particular, $\langle \xi, k_s(1)\xi\rangle \geq 1$ so that

$$\langle \xi, k_t(1)\xi\rangle \geq 1 + \epsilon(1 + n^2)t.$$

Hence $k_t(1)$ is unbounded, and so k is unbounded. □

PROPOSITION 4.3 ([**LiP**],[**LW1**]). *Let $\theta \in \mathrm{M}_D(\mathcal{B}(\mathcal{A}))$ and let $k \in \mathbb{P}(\mathcal{A}, \mathcal{E}[\eta])$ be a weakly regular weak solution of* (3.2)*. Then*

(a) *k is unital if and only if $\theta(1) = 0$;*
(b) *if k is completely positive, then k is a contraction process if and only if $\theta^{[d]}(1) \leq 0$ for all $1 \leq d < D$.*

Here $\theta^{[d]}(1)$ is the element of $\mathrm{M}_{d+1}(\mathcal{A})$ taken from the top-left hand corner of the (possibly infinite) matrix $\theta(1)$. This necessary and sufficient condition on the sequence $(\theta^{[d]}(1))_{d\geq 1}$ for k to be a contraction process entails global boundedness of the generator:

PROPOSITION 4.4 ([**LW1**]). *Let $\theta \in \mathrm{M}_D(\mathcal{B}(\mathcal{A}))$ satisfy the CP structure equations* (4.1)*, and suppose that $\theta^{[d]}(1) \leq 0$ for all d. Then θ defines a bounded map $\mathcal{A} \to \mathcal{A}'' \otimes \mathcal{B}(\widehat{\mathsf{k}})$ with components θ^α_β. Moreover, the transformed matrix $\widehat{\theta}$ can be written in the form*

$$\widehat{\theta}(a) = \psi(a) + K^* a E_0^* + E_0 a K \tag{4.3}$$

where $\psi \in \mathcal{B}(\mathcal{A}; \mathcal{B}(\mathfrak{h} \otimes \widehat{\mathsf{k}}))$ is a CP map, $K \in \mathcal{B}(\mathfrak{h} \otimes \widehat{\mathsf{k}}; \mathfrak{h})$, E_0 is the map $u \mapsto u \otimes e_0 \in \mathfrak{h} \otimes \widehat{\mathsf{k}}$, and

$$\psi(1) + K^* E_0^* + E_0 K \leq \Delta(1).$$

The significance of this result is not merely that the generator θ of a CP contraction flow is bounded, but that it is *completely* bounded. This leads to the following refinement of Theorem 3.4:

THEOREM 4.5 ([**LW2**]). *Let k be a regular Markovian CP contraction cocycle on $\mathcal{A}$. Then k is a strong solution to the* QSDE (3.2) *for some completely bounded $\theta \in \mathrm{M}_D(\mathcal{B}(\mathcal{A}))$ of the form* (4.3)*.*

This result follows by combining Theorems 3.1, 3.3, 3.4 and 4.1, and Propositions 4.3 and 4.4. First of all some careful estimates show that since the Markov semigroup $\mathcal{P}^{0,0}$ is norm continuous then so are all of the associated semigroups $(\mathcal{P}^{w,z})$, and so part (b) of the hypothesis of Theorem 3.4 holds. Now k is its own adjoint process, and a combination of the norm continuity of $\mathcal{P}^{0,0}$, the decomposition (2.1) for cocycles, and the operator Schwarz inequality shows that k is pointwise strongly continuous at 0. Thus it must satisfy (3.2) weakly for some $\theta \in \mathrm{M}_D(\mathcal{B}(\mathcal{A}))$. But now by Theorem 4.1 and Propositions 4.3 and 4.4 we know that θ is completely bounded, and Theorem 3.3 states that there is a strongly regular strong solution to (3.2) for this θ. The result is then clinched by the uniqueness of weakly regular weak solutions that follows from Theorem 3.1.

The first characterisations of Markovian cocycles were for cocycles on a Hilbert space. All regular unitary Markovian cocycles were shown to satisfy a QSDE ([**HuL**]), via a proof that made use of the quantum martingale representation theorem ([**PaS**]). These results can now be extended by applying results above in the case $\mathcal{A} = \mathcal{B}(\mathfrak{h})$. This yields more transparent proofs, and also allows us to drop the separability assumption on $\mathfrak{h}$ of earlier approaches. Here is a sample:

THEOREM 4.6 ([**LW2**]). *Let $X \in \mathbb{P}_{\mathrm{b}}(\mathfrak{h}, \mathsf{k})$ be a regular Markovian contraction cocycle. Then there is an operator matrix $F \in \mathrm{M}_D(\mathcal{B}(\mathfrak{h}))$ such that X strongly satisfies the equation $dX_t = X_t F^\alpha_\beta \, d\Lambda^\beta_\alpha(t)$.*

In fact the operator matrix F governing the cocycle is bounded and satisfies $F + F^* + F^*\Delta(1)F \leq 0$, by virtue of X being a contraction process ([**LW1**]).

5. Stochastic dilations

We conclude this article with a selection of dilation theorems, showing how CP cocycles can be realised in terms of *-homomorphic cocycles. Throughout this section we will assume that the initial space $\mathfrak{h}$ is separable. This assumption allows us to conclude that the supplementary noise spaces that appear in what follows are separable, and so we can continue to use the calculus of Mohari and Sinha. The key to all of the results in this section is the following refinement given in [**GLSW**] of Theorem 4.1. It is always possible to choose the quadruple $(\pi, \mathcal{K}, \delta, \{W_i\})$ satisfying (4.1) in such a way that $\mathcal{K} = \mathfrak{h} \otimes \mathsf{k}_1$ for some separable Hilbert space k_1 and such that $\pi(\mathcal{A}) \subset \mathcal{A}''\otimes\mathcal{B}(\mathsf{k}_1), \delta(\mathcal{A}) \subset \mathcal{A}''\otimes\mathsf{k}_1$ and $W_i \in \mathcal{A}''\otimes\mathsf{k}_1$. Here the Hilbert W^*-module $\mathcal{A}'' \otimes \mathsf{k}_1$ is defined as follows $\{T \in \mathcal{B}(\mathfrak{h}; \mathfrak{h} \otimes \mathsf{k}_1) : (a'\otimes 1)T = Ta' \ \forall a' \in \mathcal{A}'\}$. In achieving this refinement we lose the minimality (4.2), however this is inconsequential for what follows.

Given Hilbert spaces k_0 and k, with k_0 a subspace of k, the conditional expectation $\mathbb{E}^{(0)} : \mathcal{B}(\mathfrak{h}) \otimes \mathcal{B}_{\mathsf{k}} \to \mathcal{B}(\mathfrak{h}) \otimes \mathcal{B}_{\mathsf{k}_0}$ is defined by $\mathbb{E}^{(0)}[X] = E^{(0)*}XE^{(0)}$, where $E^{(0)}$ is the isometric embedding $\mathfrak{h} \otimes \mathcal{F}_{\mathsf{k}_0} \to \mathfrak{h} \otimes \mathcal{F}_{\mathsf{k}}$ given by $\xi \mapsto \xi \otimes \Omega_{\mathsf{k}\ominus\mathsf{k}_0}$, where $\Omega_{\mathsf{k}\ominus\mathsf{k}_0}$ is the vacuum vector in $\mathcal{F}_{\mathsf{k}\ominus\mathsf{k}_0}$. Using this notation we have the following:

THEOREM 5.1 ([**GLSW**]). *Let $k \in \mathbb{P}_{\mathrm{b}}(\mathcal{M}, \mathsf{k}_0)$ be a regular Markovian CP contraction cocycle on the von Neumann algebra $\mathcal{M}$ such that each k_t is normal. Then there exists a separable Hilbert space k containing k_0 as a subspace, and a *-homomorphic Markovian cocycle $j \in \mathbb{P}_{\mathrm{b}}(\mathcal{M}, \mathsf{k})$ such that*

$$k_t = \mathbb{E}^{(0)} \circ j_t \quad \forall t \geq 0.$$

The proof consists of a careful analysis of the generator θ of k. The Hilbert space k is taken to be $\mathsf{k}_0 \oplus \mathsf{k}_1 \oplus \mathbb{C}$ where k_1 is the Hilbert space that features in the representation π above, and it is shown that there is a bounded linear map $\phi : \mathcal{M} \to \mathcal{M} \otimes \mathcal{B}(\mathbb{C} \oplus \mathsf{k}_0 \oplus \mathsf{k}_1 \oplus \mathbb{C})$ of the form

$$\phi = \begin{bmatrix} \theta & \chi \\ \chi^\dagger & \psi \end{bmatrix},$$

with $\chi : \mathcal{M} \to \mathcal{M} \otimes \mathcal{B}(\mathsf{k}_1 \oplus \mathbb{C}; \widehat{\mathsf{k}_0})$ and $\psi : \mathcal{M} \to \mathcal{M} \otimes \mathcal{B}(\mathsf{k}_1 \oplus \mathbb{C})$, such that

$$\phi(a^*a) = \iota(a)^*\phi(a) + \phi(a)^*\iota(a) + \phi(a)^*\Delta(1)\phi(a).$$

where ι is the identity matrix (1.2). From this it follows that ϕ generates a *-homomorphic cocycle j ([**LW1**], Theorem 6.5), and that j dilates k, since θ forms the top-left hand corner of ϕ. By taking the original noise space k_0 to be $\{0\}$ we see that this result includes a dilation theorem for normal CP contraction semigroups on the von Neumann algebra $\mathcal{M}$; the main result of [**GoS**] is thereby a corollary.

At the Hilbert space level the analogous result for Markovian contraction cocycles also holds:

THEOREM 5.2 ([**GLSW**]). *Let $X \in \mathbb{P}_{\mathrm{b}}(\mathfrak{h}, \mathsf{k}_0)$ be a regular Markovian contraction cocycle. Then there exists a unitary Markovian cocycle $U \in \mathbb{P}_{\mathrm{b}}(\mathfrak{h}, \mathsf{k})$, where $\mathsf{k} = \mathsf{k}_0 \oplus \mathsf{k}_0 \oplus \mathbb{C}$, such that*

$$X_t = \mathbb{E}^{(0)}[U_t] \quad \forall t \geq 0.$$

Since the conditional expectation $\mathbb{E}^{(0)}$ is a contraction it is clear that a dilation of the form contained in Theorem 5.1 can only be achieved for contraction cocycles. However we have successfully extended the definition of cocycle to unbounded processes, so what can be said for these cocycles? Our answer provides a continuous time stochastic Stinespring theorem, and once more it is proved for cocycles over a von Neumann algebra $\mathcal{M}$.

Let η_0 be an orthonormal basis for the noise dimension space k_0.

THEOREM 5.3 ([**GLW**]). *Let $\theta \in \mathrm{M}_D(\mathcal{B}(\mathcal{M}))$ with each θ^α_β ultraweakly continuous, and suppose that $k \in \mathbb{P}(\mathcal{M}, \mathcal{E}[\eta_0])$ is a completely positive weakly regular process on the von Neumann algebra $\mathcal{M}$ that weakly satisfies the* QSDE (3.2). *Then there is a separable Hilbert space k_1, a normal regular *-homomorphic Markovian cocycle $j \in \mathbb{P}_\mathrm{b}(\mathcal{M}, \mathsf{k}_0 \oplus \mathsf{k}_1)$ and a process $W \in \mathbb{P}(\mathfrak{h}, \mathcal{E}[\eta])$ such that*

$$k_t(a) \odot P^{(1)}_t = W^*_t j_t(a) W_t \quad \forall t \geq 0$$

where $P^{(1)}_t$ is the orthogonal projection $\mathcal{F}_{\mathsf{k}_1} \to \mathcal{F}_{[t,\infty[,\mathsf{k}_1}$, and η is an orthonormal basis of $\mathsf{k}_0 \oplus \mathsf{k}_1$ containing η_0. Moreover (k_1, j, W) may be chosen so that W is affiliated to the von Neumann algebra $\mathcal{M} \otimes \mathcal{B}_{\mathsf{k}_0 \oplus \mathsf{k}_1}$ in the sense that $W_t(a' \otimes 1) = (a' \otimes 1)W_t$ $\forall a' \in \mathcal{M}'$. If, furthermore, k is a contraction process, then W may be chosen to be a contraction process.

Again the proof of this decomposition proceeds via an analysis of the generator of k. The generator θ of the *-homomorphic cocycle j is constructed from the components of θ, and W is realised as the solution of the perturbed HP equation $dW_t = j_t(l^\alpha_\beta) W_t \, d\Lambda^\beta_\alpha(t)$, where $l = [l^\alpha_\beta] \in \mathrm{M}_D(\mathcal{M})$ is an algebra matrix also constructed from the components of θ. Thus, in the language of [**EvH**], k is a perturbation of the *-homomorphic process j.

Since we are working throughout this section with ultraweakly continuous maps on von Neumann algebras it is pertinent to ask if this stochastic Stinespring theorem can be recast in a form where the *-homomorphic cocycle j is the trivial cocycle, that is $j_t(a) = a \otimes 1$, thereby mirroring the usual Stinespring Theorem for normal completely positive maps. The following answers this in the affirmative:

THEOREM 5.4 ([**GLW**]). *Let $\theta \in \mathrm{M}_D(\mathcal{B}(\mathcal{M}))$ with each θ^α_β ultraweakly continuous, and suppose that $k \in \mathbb{P}(\mathcal{M}, \mathcal{E}[\eta_0])$ is a completely positive weakly regular process that weakly satisfies the equation* (3.2). *If k_0 is infinite dimensional then there is a process $W \in \mathbb{P}(\mathfrak{h}, \mathcal{E}[\eta_0])$ such that*

$$k_t(a) = W^*_t (a \otimes 1) W_t \quad \forall t \geq 0. \tag{5.1}$$

REMARKS. (i) The assumption that k_0 be infinite dimensional is not an essential limitation — we can always embed k_0 into a separable infinite dimensional Hilbert space k'_0, and extend θ to an infinite matrix by making all of the new entries zero. It then follows that $k_t(a) \odot 1 \in \mathbb{P}(\mathcal{M}, \mathcal{E}[\eta'_0])$ satisfies the equation (3.2) on this larger space.

(ii) The cost paid for achieving a spatial representation of the form (5.1) for k is that the conjugating process W need no longer be affiliated to $\mathcal{M} \otimes \mathcal{B}_{\mathsf{k}_0}$.

(iii) Clearly k is a contraction (respectively unital) process if and only if W is a contraction (resp. isometry process).

(iv) Finally if k is a homomorphic process then W is a partial isometry process.

Acknowledgement. SJW gratefully acknowledges receipt of a grant from the European Union TMR programme, provided through Université de Paris VI.

References

[Acc] L Accardi, On the quantum Feynman-Kac formula, *Rend Sem Mat Fis Milano* **48** (1978), 135–180.

[AFL] L Accardi, A Frigerio and J L Lewis, Quantum stochastic processes, *Publ Res Inst Math Sci* **18** (1982), 97–133.

[AcM] L Accardi and A Mohari, On the structure of classical and quantum flows, *J Funct Anal* **135** (1996), 421–455.

[Ar1] W Arveson, Continuous analogues of Fock space, *Mem Amer Math Soc* **80** (1989) no. 409.

[Ar2] W Arveson, The index of a quantum dynamical semigroup, *J Funct Anal* **146** (1997), 557–588.

[Be1] V P Belavkin, Quantum stochastic calculus and quantum nonlinear filtering, *J Multivariate Anal* **42** (1992), 171–201.

[Be2] V P Belavkin, Quantum stochastic positive evolutions: characterization, construction, dilation, *Comm Math Phys* **184** (1997), 533-566.

[Bh1] B V Rajarama Bhat, An index theory for quantum dynamicl semigroups, *Trans Amer Math Soc* **348** (1996), 561–583.

[Bh2] B V Rajarama Bhat, Cocycles of CCR flows, *Mem Amer Math Soc*, to apppear.

[Bia] Ph Biane, Calcul stochastique non-commutatif, *in*, "Lectures on Probability Theory: Ecole d'Eté de Probabilités de Saint-Flour XXIII — 1993," *ed. P Bernard*, Springer Lecture Notes in Mathematics **1608**, Heidelberg (1995).

[Bra] W S Bradshaw, Stochastic cocycles as a characterisation of quantum flows, *Bull Sci Math (2)* **116** (1992), 1–34.

[ChE] E Christensen and D E Evans, Cohomology of operator algebras and quantum dynamical semigroups, *J London Math Soc* **20** (1979), 358–368.

[Eva] M P Evans, Existence of quantum diffusions, *Probab Theory Related Fields* **81** (1989), 473–483.

[EvH] M P Evans and R L Hudson, Perturbations of quantum diffusions, *J London Math Soc* **41** (1990), 373–384.

[GLSW] D Goswami, J M Lindsay, K B Sinha and S J Wills, Dilation of Markovian cocycles on a von Neumann algebra, *Preprint* 1999.

[GLW] D Goswami, J M Lindsay and S J Wills, A stochastic Stinespring theorem, *Preprint* 1999.

[GoS] D Goswami and K B Sinha, Hilbert modules and stochastic dilation of a quantum dynamical semigroup on a von Neumann algebra, *Comm Math Phys* **205** (1999), 377–405.

[Haa] U Haagerup, The Grothendieck inequality for bilinear forms on C^*-algebras, *Adv in Math* **56** (1985), 93–116.

[HKK] J Hellmich, C Köstler and B Kümmerer, Stationary quantum Markov processes as solutions of stochastic differential equations, *in*, "Quantum Probability (Gdańsk, 1997)," *eds. R Alicki, M Bożejko and W A Majewski*, Banach Center Publications **43**, Warsaw (1998), 217–229.

[HuL] R L Hudson and J M Lindsay, On characterizing quantum stochastic evolutions, *Math Proc Cambridge Philos Soc* **102** (1987), 363–369.

[HuP] R L Hudson and K R Parthasarathy, Quantum Itô's formula and stochastic evolutions, *Comm Math Phys* **93** (1984), 301–323.

[HuS] R L Hudson and P Shepperson, Stochastic dilation of quantum dynamical semigroups using one-dimensional quantum stochastic calculus, *in*, "Quantum Probability and Applications V," *eds. L Accarid and W von Waldenfels*, Springer Lecture Notes in Mathematics **1442** (1990), 216–218.

[Jou] J-L Journé, Structure des cocycles markoviens sur l'espace de Fock, *Probab Theory Related Fields* **75** (1987), 291–316.

[LiP] J M Lindsay and K R Parthasarathy, On the generators of quantum stochastic flows, *J Funct Anal* **158** (1998), 521–549.

[LiS] J M Lindsay and K B Sinha, Feynman-Kac representation of some noncommutative elliptic operators, *J Funct Anal* **147** (1997), 400–419.

[LW1] J M Lindsay and S J Wills, Existence, positivity, and contractivity for quantum stochastic flows with infinite dimensional noise, *Probab Theory Related Fields*, to appear.

[LW2] J M Lindsay and S J Wills, Markovian cocycles on operator algebras, adapted to a Fock filtration, *Preprint* 1998.

[LW3] J M Lindsay and S J Wills, Existence of Feller cocycles on a C^*-alegbra, *Preprint* 1999.

[Mey] P-A Meyer, "Quantum Probability for Probabilists," 2nd Edition, Springer Lecture Notes in Mathematics **1538** Heidelberg (1993).

[Moh] A Mohari, "Quantum Stochastic Calculus with Infinite Degrees of Freedom," PhD thesis, Indian Statistical Institute, Delhi (1992).

[MoS] A Mohari and K B Sinha, Quantum stochastic flows with infinite degrees of freedom and countable state Markov processes, *Sankhyā Ser A* **52** (1990), 43–57.

[Par] K R Parthasarathy, "An Introduction to Quantum Stochastic Calculus," Birkhäuser, Basel (1992).

[PaS] K R Parthasarathy and K B Sinha, Stochastic integral representation of bounded quantum martingales in Fock space, *J Funct Anal* **67** (1986), 126–151.

[Sti] W F Stinespring, Positive functions on C^*-algebras, *Proc Amer Math Soc* **6** (1955), 211–216.

School of Mathematical Sciences, University of Nottingham, Nottingham NG7 2RD
E-mail address: `jml@maths.nott.ac.uk`
E-mail address: `sjw@maths.nott.ac.uk`

Canadian Mathematical Society
Conference Proceedings
Volume **29**, 2000

Wiener space approach to a perturbative Chern-Simons Integral

Itaru Mitoma

ABSTRACT. We discuss a stationary phase method for a heuristic Chern-Simons integral.

1. Introduction and a heuristic definition

Following Witten's work[**10**], various papers have been published concerning the relation between Chern-Simons integrals and topological invariants of 3-manifolds, among them the linking number [**3**],[**4**]. Albeverio and his colleagues have been studying the integral for the abelian case by the Fresnel integral [**1**] and the non-commutative case (in the setting of Fröhlich-King [**5**]) by using a white noise distribution [**2**].

Inspired by [**1**] and based on Bar-Natan-Witten[4] and the idea of Itô [**7**] for the Feynman integral, we will work with a definition of the Chern-Simons integral in a Wiener space setting.

In this paper, we examine the stationary phase method(infinite dimensional) for the integral, where the S-operator method of [**7**] plays a crucial role.

Let M be a compact oriented 3-manifold and $\mathcal{G}$ a skew symmetric matrix algebra.

Let Ω^p be the space of $\mathcal{G}$-valued p-forms and let $(\omega,\eta)_p = \int_M -Tr\omega \bigwedge *\eta$. Let $L^2(\Omega_+) = L^2(\Omega^1 \oplus \Omega^3)$ and $L^2(\Omega_-) = L^2(\Omega^0 \oplus \Omega^2)$ be the Hilbert spaces with the inner products $(\cdot,\cdot)_+$ and $(\cdot,\cdot)_-$ induced from the above inner products $(\cdot,\cdot)_p, p = 1,3$ and $(\cdot,\cdot)_p, p = 0,2$ respectively. Let $J\phi = -\phi$ if ϕ is a zero or a three form and $J\phi = \phi$ if ϕ is a one or a two form. Set $x = (A,\phi), Q = \frac{1}{4\pi}(*D^0 + D^0*)J$, where $D^0 = d + [A_0,]$, with A_0 being the 1-form specified below and $[A,B] = A\bigwedge B + B\bigwedge A$.

Using the method of super fields[**4**], we arrive at the interesting heuristic form of the normalized Chern-Simons integral[**8**].

$$\frac{\int_{L^2(\Omega_+)} F(A_0 + x_A)e^{ik(x,Qx)_+ - \frac{ik}{4\pi}\int_M Tr\frac{2}{3}x_A \wedge x_A \wedge x_A}\mathcal{D}(x)}{\int_{L^2(\Omega_+)} e^{ik(x,Qx)_+}\mathcal{D}(x)}, \tag{1.1}$$

where k is assumed to be a natural number in this paper, x_A = the projection of x on $L^2(\Omega^1)$ and A_0 is a critical point of the Chern-Simons Lagrangian

2000 *Mathematics Subject Classification.* Primary 60H07, 81T13, 81T45.

$$L(A) = -\frac{ik}{4\pi} Tr \int_M A \bigwedge dA + \frac{2}{3} A \bigwedge A \bigwedge A.$$

Let $\widehat{\Omega}_p$ be the set of all real valued p-forms, $(\alpha, \beta)_{\check{p}} = \int_M \alpha \bigwedge *\beta$ and $L^2(\widehat{\Omega}_p)$ the Hilbert space with the inner product $(\cdot, \cdot)_{\check{p}}$. Let $E_j, j = 1, 2, ..., d$ be the orthonormal basis of $\mathcal{G}$ and $x_A = \sum_{t=1}^d x_A^t \cdot E_t$ and $x_\phi = \sum_{t=1}^d x_\phi^t \cdot E_t$, where d is the dimension of the skew symmetric complex matrix algebra $\mathcal{G}$. Then corresponding to (x_A^t, x_ϕ^t), we conveniently denote by H^t the Hilbert space $L^2(\widehat{\Omega}_1 \oplus \widehat{\Omega}_3)$ with the inner product $(\cdot, \cdot)_{H^t}$ induced from inner products $(\cdot, \cdot)_{\check{p}}, p = 1, 3$. Set $x^t = (x_A^t, x_\phi^t)$. Since Q is self-adjoint and Q^2 is elliptic, so that Q has pure point spectrum [**6**]. Then

$$(x, Qx)_+ = CS(x) = \sum_{i=1}^{\infty} \mu_i [\sum_{t=1}^{d} (x^t, (e_i^{t,A}, e_i^{t,\phi}))_{H^t}]^2$$

where $\mu_i, (e_i^A = \sum_{t=1}^d e_i^{t,A} \cdot E_t, e_i^\phi = \sum_{t=1}^d e_i^{t,\phi} \cdot E_t), i = 1, 2, \cdots$ are the eigenvalues of Q and the corresponding CONS(Complete Orthonormal System) of $L^2(\Omega_+)$.

Noticing that

$$x_A = \sum_{i=1}^{\infty} [\sum_{t=1}^{d} (x^t, (e_i^{t,A}, e_i^{t,\phi}))_{H^t}] e_i^A$$

and using the notation

$$\langle x, (e_i^A, e_i^\phi) \rangle = \sum_{t=1}^{d} (x^t, (e_i^{t,A}, e_i^{t,\phi}))_{H^t},$$

we have

$$\int_M Tr \frac{2}{3} x_A \bigwedge x_A \bigwedge x_A$$

$$= \sum_{t,u,v=1}^{d} \sum_{i,j,\ell=1}^{\infty} \langle x, (e_i^A, e_i^\phi) \rangle \langle x, (e_j^A, e_j^\phi) \rangle \langle x, (e_\ell^A, e_\ell^\phi) \rangle$$

$$\times Tr E_t E_u E_v \int_M \frac{2}{3} e_i^{t,A} \bigwedge e_j^{u,A} \bigwedge e_\ell^{v,A}.$$

Henceforth, we assume the condition that

(A.1) $Tr E_t E_u E_v$ is real.

In fact this holds for $\mathfrak{S}\,\mathfrak{U}(2)$.

Set

$$\beta_{tuv}^{ij\ell} = -Tr E_t E_u E_v \int_M \frac{2}{12\pi} e_i^{t,A} \bigwedge e_j^{u,A} \bigwedge e_\ell^{v,A}.$$

Using techniques of [**7**], we will find a realization of the expression (1.1) in the setting of Wiener space.

Let $H^t = L^2(\widehat{\Omega}_1 \oplus \widehat{\Omega}_3)$. Let (B^t, H^t, μ_t) be the abstract Wiener space such that a separable Hilbert space H^t is continuously embedded into B^t equipped with the norm $\| \cdot \|_{B^t}$ and μ_t is a Gaussian measure satisfying

$$\int_{B^t} e^{i\langle x^t, \xi \rangle} d\mu_t(x) = e^{-\frac{\|\xi\|^2_{H^t}}{2}},$$

where $\langle x^t, \xi \rangle$ again denotes the canonical bilinear form on $B^t \times (B^t)^*$.

Let $H = H^1 \times H^2 \times \cdots \times H^d$ be a separable Hilbert space equipped with the norm $\|x\|_H^2 = \sum_{t=1}^d \|x^t\|^2_{H^t}, B = B^1 \times B^2 \times \cdots \times B^d$ and $\mu = \otimes_{t=1}^d \mu_t$. We denote $(B^1)^* \times (B^2)^* \times \cdots \times (B^d)^*$ by B^* and for $x \in B$, $\xi \in B^*$, $\sum_{t=1}^d \langle x^t, \xi^t \rangle$ again by $\langle x, \xi \rangle$. Then we have

$$\int_B e^{i\langle x, \xi \rangle} d\mu(x) = e^{-\frac{\|\xi\|^2_H}{2}}.$$

Since

$$((e_i^{1,A}, e_i^{1,\phi}), (e_i^{2,A}, e_i^{2,\phi}), \cdots, (e_i^{d,A}, e_i^{d,\phi})), i = 1, 2, \cdots$$

forms a complete orthonormal system of the Hilbert space H, Q induces naturally the linear operator defined in H and having the same spectrum that Q has. We denote it again by the same notation Q. Using the above CONS of H, we can construct a positive definite and self-adjoint operator S on the Hilbert space H such that

$$\sum_{j=1}^{\infty} \lambda_j \mid \mu_j \mid < +\infty,$$

where $\{\lambda_i > 0, i = 1, 2, \cdots\}$ are the eigenvalues of S([**9**],[**11**]).

A Wiener space realization of (1.1) can be defined by

$$\overline{\lim_{n\to\infty}} \frac{1}{\widehat{Z}_n} \int_B F(A_0 + \sum_{i=1}^{\infty} \langle \sqrt{nS}x, (e_i^A, e_i^\phi) \rangle e_i^A) e^{ikCS(\sqrt{nS}x)} \tag{1.2}$$

$$\times e^{ik\{\sum_{t,u,v=1}^d \sum_{i,j,\ell=1}^\infty \langle \sqrt{nS}x, (e_i^A, e_i^\phi)\rangle \langle \sqrt{nS}x, (e_j^A, e_j^\phi)\rangle \langle \sqrt{nS}x, (e_\ell^A, e_\ell^\phi)\rangle \beta^{ij\ell}_{tuv}\}} \mu(dx),$$

where

$$\langle x, (e_i^A, e_i^\phi) \rangle = \sum_{t=1}^d \langle x^t, (e_i^{t,A}, e_i^{t,\phi}) \rangle,$$

$$\widehat{Z}_n = \int_B e^{ikCS(\sqrt{nS}x)} \mu(dx) = \frac{1}{\sqrt{det(1 - 2\sqrt{-1}k\sqrt{nS}Q\sqrt{nS})}},$$

and for real x_n, y_n,

$$\overline{\lim_{n\to\infty}} (x_n + iy_n) = \overline{\lim_{n\to\infty}} x_n + i \overline{\lim_{n\to\infty}} y_n.$$

Now we recall the growth order of the integrand which was obtained in [**8**] for $F(x) = \prod_{j=1}^s TrW^\epsilon_{\gamma_j}(x)$, where $TrW^\epsilon_\gamma(x)$ is the Wiener space version of the

ϵ-regularization of the Wilson line $Tr_R Pe^{i\int_\gamma A}$, (see [**1**]) and $\gamma_j, j = 1, 2, ..., s$ are the closed oriented loops ;

$$[\int_B |F(A_0 + \sum_{i=1}^{\infty} \langle \sqrt{n}Sx, (e_i^A, e_i^\phi)\rangle e_i^A)|^2 \mu(dx)]^{\frac{1}{2}} \leq c_0(s)(\sum_{j=0}^{\infty} \frac{(c_1\sqrt{n})^j}{\sqrt{j!}})^s$$

where $c_0(s), c_1$ are positive constants. In the sequel we denote by $c_p(\cdot)$ the positive constants depending on $(\cdot)$.

Since

$$E_\mu[e^{\sum_{i=1}^{\infty} |\langle \sqrt{S}x, (e_i^A, e_i^\phi)\rangle|}] < +\infty,$$

by the Chebyshev inequality, we have

$$\mu[(x \in B \mid \sum_{i=1}^{\infty} |\langle \sqrt{S}x, (e_i^A, e_i^\phi)\rangle| > (s+2)c_1^2 kn^2)] \leq c_2 e^{-(s+2)c_1^2 kn^2}.$$

Set $B_{k,n} = \{x \in B \mid \sum_{i=1}^{\infty} |\langle \sqrt{S}x, (e_i^A, e_i^\phi)\rangle| \leq (s+2)c_1^2 kn^2\}$. Then since for $n \geq 1$,

$$\sum_{j=0}^{\infty} \frac{(c_1\sqrt{n})^j}{\sqrt{j!}} exp[-c_1^2 kn^2]$$

$$\leq \sum_{j=0}^{\infty} \frac{(c_1\sqrt{n})^{2j}}{\sqrt{(2j)!}} exp[-c_1^2 kn^2] + \sum_{j=0}^{\infty} \frac{(c_1\sqrt{n})^{2j+1}}{\sqrt{(2j+1)!}} exp[-c_1^2 kn^2]$$

$$\leq \sum_{j=0}^{\infty} \frac{j!}{\sqrt{(2j)!}} + \sum_{j=0}^{\infty} \frac{(j+1)!}{c_1\sqrt{(2j+1)!}}$$

$$\leq \sum_{j=0}^{\infty} \frac{1}{\sqrt{2^j \frac{1\cdot 3\cdots(2j-1)}{j!}}} + \sum_{j=0}^{\infty} \frac{1}{c_1\sqrt{2^j \frac{1\cdot 3\cdots(2j-1)(2j+1)}{(j+1)!(j+1)}}} < +\infty,$$

we get

$$\lim_{n\to\infty} \sqrt{\det(1 - 2nk\sqrt{-1}\sqrt{S}Q\sqrt{S})} \int_{B_{k,M}^c} F(A_0 + \sum_{i=1}^{\infty} \langle \sqrt{n}Sx, (e_i^A, e_i^\phi)\rangle e_i^A)$$

$$\times e^{ik\{CS(\sqrt{n}Sx) + \sum_{t,u,v=1}^{d} \sum_{i,j,\ell=1}^{\infty} \langle \sqrt{n}Sx, (e_i^A, e_i^\phi)\rangle \langle \sqrt{n}Sx, (e_j^A, e_j^\phi)\rangle \langle \sqrt{n}Sx, (e_\ell^A, e_\ell^\phi)\rangle \beta_{tuv}^{ij\ell}\}} \mu(dx) = 0.$$

Here we assume the basic condition

(A.2) $\quad | \beta_{tuv}^{ij\ell} | \leq c_3.$

Then , together with (A.2), the integral (1.2) is localized on $B_{k,n}$, so that we obtain an expansion form of (1.2) such that

$$(1.3) \qquad \overline{\lim_{n\to\infty}} \frac{1}{\widehat{Z}_n} \sum_{m=0}^{\infty} \int_{B_{k,n}} F(A_0 + \sum_{i=1}^{\infty} \langle \sqrt{n}Sx, (e_i^A, e_i^\phi)\rangle e_i^A) e^{ikCS(\sqrt{n}Sx)}$$

$$\times(ik\sum_{t,u,v=1}^{d}\sum_{i,j,\ell=1}^{\infty}\langle\sqrt{n}Sx,(e_i^A,e_i^\phi)\rangle\langle\sqrt{n}Sx,(e_j^A,e_j^\phi)\rangle\langle\sqrt{n}Sx,(e_\ell^A,e_\ell^\phi)\rangle$$

$$\times\beta_{tuv}^{ij\ell})^m/m!\mu(dx).$$

However, at present we cannot calculate the above (1.3) without any change. Now we settle what change we need. First we explain ε-cut-off of Q. Let $\mu_j, j = 1, 2, \cdots$ be the eigenvalues of Q. We replace μ_j by $\mu_j exp(\varepsilon|\mu_j|)$ for $CS(x)$ and Q and denote them by $CS_\varepsilon(x)$ and Q_ϵ.

Next we explain a modification(δ-cut-off) of the Borel summability. Namely we replace $\sum_{m=0}^{\infty}$ in (1.3) by $\int_0^\infty \sum_{m=0}^{\infty}(\cdots)/m!(te^{-\delta t})^m e^{-t}dt$. We denote this by $B\sum_{m=0}^{\infty}$.

Using these cut-offs, instead of (1.3) we consider

$$\overline{\lim_{n\to\infty}}\frac{1}{\widehat{Z}_n}\int_0^\infty\sum_{m=0}^{\infty}\int_{B_{k,n}}F(A_0+\sum_{i=1}^{\infty}\langle\sqrt{n}Sx,(e_i^A,e_i^\phi)\rangle e_i^A)e^{ikCS(\sqrt{n}Sx)} \tag{1.4}$$

$$\times(ik\sum_{t,u,v=1}^{d}\sum_{i,j,\ell=1}^{\infty}\langle\sqrt{n}Sx,(e_i^A,e_i^\phi)\rangle\langle\sqrt{n}Sx,(e_j^A,e_j^\phi)\rangle\langle\sqrt{n}Sx,(e_\ell^A,e_\ell^\phi)\rangle$$

$$\times\beta_{tuv}^{ij\ell})^m(te^{-\delta t})^m/(m!)^2\mu(dx)e^{-t}dt.$$

Then we have

THEOREM. Let $F(x) = \prod_{j=1}^{s} TrW_{\gamma_j}^{\epsilon}(x)$. Then under the conditions (A.1)~(A.4), the integral (1.4) $\longrightarrow$ $F(A_0)$ as $k \longrightarrow \infty$.

2. Proof of Theorem

First we recall the ϵ-regularization of the Wilson line $Tr_R Pe^{i\int_\gamma A}$;

$$W_\gamma^\epsilon(x) = I + \sum_{k=1}^{\infty} i^k \int_0^1\int_0^{t_1}\cdots\int_0^{t_{k-1}} d\overline{x}(t_1)d\overline{x}(t_2)\cdots d\overline{x}(t_k). \qquad \text{(see [8])}$$

where $\overline{x}(t) = \sum_{j=1}^{2\hat{d}^2}\overline{x}_j(t)\widehat{E}_j$,

$$\overline{x}_j(t) = \sum_{u=1}^{d}\langle(x^1,\cdots,x^u,\cdots,x^d),((0,0),\cdots,(*C_t^{j,\varepsilon}(\gamma),0),\cdots,(0,0))\rangle(E_u)_j,$$

$(E_u)_j$ is the j-unit entry of E_u and $C_t^{j,\varepsilon}(\gamma)$ is a compact real 2-form such that

$$\lim_{\varepsilon\to 0}\sup_{0\le s\le 1}|\int_{\gamma[0,s]}A_j - (A_j, *C_s^{j,\varepsilon}(\gamma))_{\mathbb{1}}| = 0.$$

Here the $\mathcal{G}$-valued 1-form A is represented as a complex $\hat{d}\times\hat{d}$-matrix-valued 1-form such that

$$A = \sum_{j=1}^{2\hat{d}^2} A_j\cdot\widehat{E}_j$$

where $\widehat{E}_j$ is a complex unit matrix.

Setting

$$W^J_{\gamma_j}(x) = I + \sum_{k=1}^{J} i^k \int_0^1 \int_0^{t_1} \cdots \int_0^{t_{k-1}} d\overline{x}(t_1) d\overline{x}(t_2) \cdots d\overline{x}(t_k),$$

$$F_J(x) = \prod_{j=1}^{s} TrW^J_{\gamma_j}(x),$$

$$G_{N,J}(n,X) = F_J(A_0 + \sum_{i=1}^{N} \sqrt{n\lambda_i} x_i e_i^A)$$

$$\times (ik \sum_{t,u,v=1}^{d} \sum_{i,j,\ell=1}^{N} \sqrt{n\lambda_i} x_i \sqrt{n\lambda_j} x_j \sqrt{n\lambda_\ell} x_\ell \beta^{ij\ell}_{tuv})^m / m!.$$

Then we have the integral (1.4)

$$= \overline{\lim_{n\to\infty}} \frac{1}{\widehat{Z}_n} B \sum_{m=0}^{\infty} \lim_{J\to\infty} \lim_{N\to\infty} \int \cdots \int_{\{\sum_{i=1}^N \sqrt{\lambda_i}|x_i| \le Mkn^2\}} G_{N,J}(n,X)$$

$$\times \prod_{i=1}^{N} \frac{1}{\sqrt{2\pi}} e^{-\frac{1}{2}(1-2ikn\lambda_i\mu_i e^{\epsilon|\mu_i|})x_i^2} dx_i = \overline{\lim_{n\to\infty}} \frac{1}{\widehat{Z}_n} B \sum_{m=0}^{\infty} \lim_{N,J\to\infty} *.$$

From now on, for the brevity we set $(s+2)c_1^2 = M$ and often use μ_i instead of $\mu_i e^{\epsilon|\mu_i|}$. Setting

$$z_j = 1 - 2ikn\lambda_j\mu_j = |z_j| e^{-i\alpha_j}, 0 < (\text{sign}\mu_j)\alpha_j < \frac{\pi}{2},$$

we get by the Cauchy integral formula,

$$\int_{|x_N| \le \frac{1}{\sqrt{\lambda_N}}(Mkn^2 - \sum_{i=1}^{N-1} \sqrt{\lambda_i}|x_i|)} G_{N,J}(n,X) \frac{1}{\sqrt{2\pi}} e^{-\frac{1}{2}(1-2ikn\lambda_N\mu_N)x_N^2} dx_N$$

$$= \int_{|x_N| \le \frac{1}{\sqrt{\lambda_N}}(Mkn^2 - \sum_{i=1}^{N-1} \sqrt{\lambda_i}|x_i|)} e^{i\frac{\alpha_N}{2}} G_{N,J}(n, x_1, \cdots, e^{i\frac{\alpha_N}{2}} x_N) \frac{1}{\sqrt{2\pi}} e^{-\frac{|z_N|}{2} x_N^2} dx_N$$

$$+ \Big(\int_0^{\frac{\alpha_N}{2}} + \int_{\pi}^{\pi + \frac{\alpha_N}{2}} \Big) \frac{1}{\sqrt{2\pi}} e^{-\frac{z_N}{2}[\frac{1}{\sqrt{\lambda_N}}(Mkn^2 - \sum_{i=1}^{N-1}\sqrt{\lambda_i}|x_i|)e^{i\theta}]^2} G_{N,J}(n, x_1, \cdots,$$

$$\frac{1}{\sqrt{\lambda_N}}(Mkn^2 - \sum_{i=1}^{N-1} \sqrt{\lambda_i}|x_i|)e^{i\theta}) \frac{1}{\sqrt{\lambda_N}}(Mkn^2 - \sum_{i=1}^{N-1} \sqrt{\lambda_i}|x_i|) i e^{i\theta} d\theta.$$

Hence we have the above $*$

$$= \int \int \cdots \int_{\{\sum_{i=1}^N \sqrt{\lambda_i}|x_i| \le Mkn^2\}} e^{i \sum_{i=1}^{N} \frac{\alpha_i}{2}} G_{N,J}(n, e^{i\frac{\alpha_1}{2}} x_1, e^{i\frac{\alpha_2}{2}} x_2, \cdots, e^{i\frac{\alpha_N}{2}} x_N)$$

$$\times \prod_{i=1}^{N} \frac{1}{\sqrt{2\pi}} e^{-\frac{|z_i| x_i^2}{2}} dx_i + R_{m,N},$$

where

$$|R_{m,N}| \le \sum_{r=1}^{N} \int_{|x_1| \le \frac{Mkn^2}{\sqrt{\lambda_1}}} \int_{|x_2| \le \frac{1}{\sqrt{\lambda_2}}(Mkn^2 - \sqrt{\lambda_1}|x_1|)} \cdots \tag{2.1}$$

$$\cdots \int_0^{\frac{|\alpha_r|}{2}} \frac{1}{\sqrt{2\pi}} e^{-\frac{Re(z_r e^{2i\theta})}{2\lambda_r}(Mkn^2 - \sum_{i=1}^{r-1}\sqrt{\lambda_i}|x_i|)^2} \frac{1}{\sqrt{\lambda_r}} Mkn^2 d\theta$$

$$\cdots \int_{|x_N| \le \frac{1}{\sqrt{\lambda_N}}(Mkn^2 - \sum_{i=1}^{N-1}\sqrt{\lambda_i}|x_i|)} \{\sum_{\nu=0}^{1}$$

$$|G_{N,J}(n, x_1, \cdots, (-1)^\nu \frac{e^{i\theta}(Mkn^2 - \sum_{i=1}^{r-1}\sqrt{\lambda_i}|x_i|)}{\sqrt{\lambda_r}}, e^{i\frac{\alpha_{r+1}}{2}} x_{r+1}, \cdots, e^{i\frac{\alpha_N}{2}} x_N)|\}$$

$$\prod_{i=1}^{r-1} \frac{1}{\sqrt{2\pi}} e^{-\frac{x_i^2}{2}} dx_i \prod_{i=r+1}^{N} \frac{1}{\sqrt{2\pi}} e^{-\frac{|z_i| x_i^2}{2}} dx_i.$$

Note that

$$|G_{N,J}(\cdots, (-1)^\nu \frac{e^{i\theta}(Mkn^2 - \sum_{i=1}^{r-1}\sqrt{\lambda_i}|x_i|)}{\sqrt{\lambda_r}}, e^{i\frac{\alpha_{r+1}}{2}} x_{r+1}, \cdots, e^{i\frac{\alpha_N}{2}} x_N)|$$

$$\le e^{c_1 s Mkn^2 \sqrt{n}} (Mkn^2\sqrt{n})^{3m}.$$

Now first we consider the case where for small $\delta > 0$,

$$Mkn^2 - \sum_{j=1}^{r-1} \sqrt{\lambda_j}|x_j| > \sqrt{\lambda_r}\delta k n^2.$$

Noticing that $\alpha_r \le 2k\lambda_r|\mu_r|$, $|z_i| = |1 - 2\sqrt{-1}k\lambda_i\mu_i| \ge 1$ and $Re(z_r e^{2i\theta}) \ge 1$, we have the righthand side of (2.1), where the r-th integral is restricted on $\{Mkn^2 - \sum_{j=1}^{r-1}\sqrt{\lambda_j}|x_j| > \sqrt{\lambda_r}\delta kn^2\}$

$$\le \sum_{r=1}^{N} \frac{2e^{c_1 s Mkn^2\sqrt{n}}(Mkn^2\sqrt{n})^{3m}}{\sqrt{2\pi}} \sqrt{\lambda_r}|\mu_r| Mk^2n^2 e^{-\frac{\delta^2 k^2 n^4}{2}}. \tag{2.2}$$

Here we need the condition such that

$$\sum_{r=1}^{\infty} \sqrt{\lambda_r}|\mu_r| < +\infty. \tag{A.3}$$

Secondly assume

$$Mkn^2 - \sum_{j=1}^{r-1} \sqrt{\lambda_j}|x_j| \le \sqrt{\lambda_r}\delta k n^2.$$

In this case, dominating

$$e^{-\frac{Re(z_r e^{2i\theta})}{2\sqrt{\lambda_r}}(Mkn^2 - \sum_{i=1}^{r-1}\sqrt{\lambda_i}|x_i|)^2}$$

by 1 in the above estimation, we have the estimation of the righthand side of (2.1) for this case is reduced to

$$\mu_{r-1}(Mkn^2 - \sqrt{\lambda_r}\delta kn^2 \leq \sum_{j=1}^{r-1} \sqrt{\lambda_j}|x_j| \leq Mkn^2), \tag{2.3}$$

where

$$\mu_{r-1}(dx) = \prod_{i=1}^{r-1} \frac{1}{\sqrt{2\pi}} e^{-\frac{x_i^2}{2}} dx_i.$$

Suppose

$$\sum_{i=1}^{\infty} \sqrt{\lambda_i} = \rho < \frac{1}{2}. \tag{A.4}$$

By the inequality that

$$|1-y| \geq e^{-2|y|}, \text{ if } |y| < \frac{1}{2},$$

we have

$$E_{\mu_{r-1}}[e^{(\sum_{j=1}^{r-1} \sqrt{\lambda_j}|x_j|)^2}] \leq E_{\mu_{r-1}}[e^{\rho(\sum_{j=1}^{r-1} \sqrt{\lambda_j}|x_j|^2)}]$$

$$= \frac{1}{\sqrt{\prod_{i=1}^{r-1}(1-2\rho\sqrt{\lambda_i})}} \leq e^{2\rho\sum_{i=1}^{\infty}\sqrt{\lambda_i}} < +\infty.$$

Hence under the condition (A.4), we have the above (2.3)

$$\leq \mu_{r-1}((Mkn^2 - \sqrt{\lambda_r}\delta kn^2)^2 \leq (\sum_{j=1}^{r-1} \sqrt{\lambda_j}|x_j|)^2)$$

$$\leq c_4 e^{-(Mkn^2 - \sqrt{\lambda_r}\delta kn^2)^2}.$$

Therefore we get in the latter case, the righthand side of (2.1), where the r-th integral is restricted on $\{Mkn^2 - \sum_{j=1}^{r-1} \sqrt{\lambda_j}|x_j| \leq \sqrt{\lambda_r}\delta kn^2\}$

$$\leq c_4 \sum_{r=1}^{N} \frac{2e^{c_1 sMkn^2\sqrt{n}}(Mkn^2\sqrt{n})^{3m}}{\sqrt{2\pi}} \sqrt{\lambda_r}|\mu_r|Mk^2n^2 e^{-(Mkn^2 - \sqrt{\lambda_r}\delta kn^2)^2}. \tag{2.4}$$

By (2.2),(2.4) and $\int_0^\infty (te^{-\delta t})^m e^{-t} dt \leq c_\delta^m$,
we have

$$B\sum_{m=0}^{\infty} sup_N |R_{m,N}|$$

$$\leq c_5 \int_0^\infty \sum_{m=0}^{\infty} e^{-3c_6 k^2 n^4} e^{c_1 sMkn^2\sqrt{n}} (Mkn^2\sqrt{n})^{3m} (te^{-\delta|t|})^m/(m!)^2 e^{-t} dt$$

$$\leq c_7 \sum_{m=0}^{\infty} e^{-2c_6 k^2 n^4} (Mkn^2\sqrt{n})^{3m} c_\delta^m/(m!)^2$$

$$= c_7 \sum_{m=0}^{\infty} \sum_{q=0}^{7} e^{-2c_6 k^2 n^4} (Mk^3 \sqrt{n}^{15})^{8m+q} c_\delta^{8m+q} / ((8m+q)!)^2$$

$$\leq c_7 e^{-c_6 k^2 n^4} \sum_{m=0}^{\infty} \sum_{q=0}^{7} (Mk^3 \sqrt{n}^{15})^{8m+q} c_\delta^{8m+q}$$

$$\times (15m+2q)!/((8m+q)!)^2 (c_6 k^2 \sqrt{n}^8)^{15m+2q}$$

$$\leq c_8 e^{-c_6 k^2 n^4} \sum_{m=0}^{\infty} \sum_{q=0}^{7} \frac{(Mk^3 \sqrt{n}^{15} c_\delta)^{8m+q}}{(c_6 k^2 \sqrt{n}^8)^{15m+2q}}$$

$$\times \frac{(8m+1)\cdots(9m+1)\cdots(14m+1)\cdots(15m+1)\cdots(15m+2q)}{m!(m+1)\cdots(2m+1)\cdots(7m+1)\cdots\{(8m+1)\cdots(8m+q)\}^2}$$

$$\leq c_9 e^{-c_6 k^2 n^4}.$$

Therefore letting $N, J, n \to \infty$, we have (1.4)

$$= \overline{\lim_{n\to\infty}} B \sum_{m=0}^{\infty} \int_{D_{n,k}} F(A_0 + \sum_{i=1}^{\infty} \langle R^\varepsilon_{n,k} x, (e_i^A, e_i^\phi)\rangle e_i^A)$$

$$\times (ik \sum_{t,u,v=1}^{d} \sum_{i,j,\ell=1}^{\infty} \langle R^\varepsilon_{n,k} x, (e_i^A, e_i^\phi)\rangle \langle R^\varepsilon_{n,k} x, (e_j^A, e_j^\phi)\rangle \langle R^\varepsilon_{n,k} x, (e_\ell^A, e_\ell^\phi)\rangle$$

$$\times \beta^{ij\ell}_{tuv})^m / m! \mu(dx).$$

where

$$D_{n,k} = \{x \in B \mid \sum_{i=1}^{\infty} |(1 - 2\sqrt{-1}k\sqrt{nS}Q_\varepsilon \sqrt{nS})|^{\frac{-1}{2}} |\langle \sqrt{S}x, (e_i^A, e_i^\phi)\rangle| \leq Mkn^2\}$$

and

$$R^\varepsilon_{n,k} = (1 - 2\sqrt{-1}k\sqrt{nS}Q_\varepsilon \sqrt{nS})^{\frac{-1}{2}} \sqrt{nS}.$$

Since

$$\left|\frac{\sqrt{n\lambda_i}}{\sqrt{1 - 2\sqrt{-1}kn\lambda_i\mu_i e^{\varepsilon|\mu_i|}}}\right| \leq \frac{e^{-\frac{\varepsilon|\mu_i|}{2}}}{\sqrt{2k|\mu_i|}} = T_i,$$

using the independence of $\langle x, (e_i^A, e_i^\phi)\rangle, \langle x, (e_j^A, e_j^\phi)\rangle$ and $\langle x, (e_\ell^A, e_\ell^\phi)\rangle$ and (A.2), we have

$$\frac{1}{(m!)^2} E[|(ik \sum_{t,u,v=1}^{d} \sum_{i,j,\ell=1}^{\infty} \langle R^\varepsilon_{n,k} x, (e_i^A, e_i^\phi)\rangle \langle R^\varepsilon_{n,k} x, (e_j^A, e_j^\phi)\rangle \langle R^\varepsilon_{n,k} x, (e_\ell^A, e_\ell^\phi)\rangle$$

$$\times \beta^{ij\ell}_{tuv})^m|^2]^{\frac{1}{2}}$$

$$\leq \frac{(d^3kc_3)^m}{(m!)^2}\{(\sum_{i,j,\ell=1}^{\infty} E[\langle x, R^{\varepsilon}_{n,k}(e^A_i, e^{\phi}_i)\rangle^{2m}]^{\frac{1}{2m}}$$

$$\times E[\langle x, R^{\varepsilon}_{n,k}(e^A_j, e^{\phi}_j)\rangle^{2m}]^{\frac{1}{2m}} E[\langle x, R^{\varepsilon}_{n,k}(e^A_\ell, e^{\phi}_\ell)\rangle^{2m}]^{\frac{1}{2m}})^{2m}\}^{\frac{1}{2}}$$

$$\leq \frac{(d^3kc_3)^m}{\sqrt{m!}}\sqrt{(\frac{(2m)!}{2^m(m!)^2}(\sum_{i=1}^{\infty} T_i)^{2m})^3}.$$

Since $F(A_0 + \sum_{i=1}^{\infty}\langle R^{\varepsilon}_{n,k}x, (e^A_i, e^{\phi}_i)\rangle e^A_i)$ is very well integrable [**8**] and

$$\sum_{m=0}^{\infty} \frac{(d^3kc_3)^m}{\sqrt{m!}}\sqrt{(\frac{(2m)!}{2^m(m!)^2}(\sum_{i=1}^{\infty} T_i)^{2m})^3} < +\infty,$$

the Lebesgue convergence theorem, together with the one loop approximation argument [**8**] such that

$$\lim_{k\to\infty}\overline{\lim_{n\to\infty}}\int_B F(A_0 + \sum_{i=1}^{\infty}\langle R^{\varepsilon}_{n,k}x, (e^A_i, e^{\phi}_i)\rangle e^A_i)\mu(dx) = F(A_0),$$

yields the completion of the proof .

Acknowledgement. The author would like to express his sincere thanks to the referee for many valuable suggestions for improving the original manuscript.

References

[1] S. Albeverio, J. Schäfer, *Abelian Chern-Simons theory and link invariants*, J. Math. Phys. **36**, (1995), 2157-2169.

[2] S. Albeverio, A. Sengupta, *The Chern-Simons functional integral as an infinite dimensional distribution*, Proc. 2nd world congress of nonlinear analysis, to appear.

[3] S. Axelrod, I.M. Singer, *Chern-Simons perturbation theory*, Proc. XXth DGM Conf. World Scientific, (1992), 1-36.

[4] D. Bar-Natan, E. Witten, *Perturbation expansion of Chern-Simons theory with non-compact Gauge group*, Commun.Math.Phys. **141**, (1991), 423-440.

[5] J. Frölich, C. King, *The Chern-Simons theory and knot polynomials*, Commun.Math.Phys. **126**, (1989), 167-199.

[6] P.B. Gilkey, *Invariance Theory, The Heat Equation, And the Atiyah-Singer Index Theorem*, Publish or Perish, Inc.(1984).

[7] K. Itô, *Generalized uniform complex measures in the Hilbertian metric space with their application to the Feynman integral* , Proc. Fifth Berkeley Symp. Math. Stat. Probab., **11**, (1965), 145-161.

[8] I. Mitoma, *One loop approximation of the Chern-Simons integral*, to appear in the volume in honor of 70th birthday of T.Hida, World Scientific, (1999).

[9] M. Reed, B. Simon , *Method of modern mathematical physics Vol 1*, Academic press,(1980).

[10] E. Witten, *Quantum field theory and the Jones polynomial*, Commun.Math.Phys. **121**, (1989), 351-399.

[11] K. Yosida, *Functional Analysis*, Springer, (1971).

Department of Mathematics, Saga University, Saga 840-8502, Japan
E-mail address: `mitoma@ms.saga-u.ac.jp`

Canadian Mathematical Society
Conference Proceedings
Volume **29**, 2000

From Nelson's kinematics to symmetric and non symmetric ground–state transformations

L.M. Morato

ABSTRACT. Nelson's stochastic kinematics is revisited in the spirit of [**AHKS77**] and extended to the case of diffusions associated to symmetric and non symmetric Dirichlet Forms. This allows to generalize the classical ground state transformation to non symmetric cases and open the possibility of studying the asymptotic behaviour of a non symmetric diffusion, with assigned invariant density, exploiting the properties of the spectrum of a suitable selfadjoint operator on a complex Hilbert space.

1. Introduction

In 1966 E. Nelson introduced a quantization procedure (named Stochastic Mechanics) where the lagrangian paths of a physical system with a finite number of degrees of freedom was substituted by a markovian diffusion with coefficient proportional to the Planck's constant [**Nel67**].

As in any true "mechanics" the first step consisted in introducing kinematics, namely suitable definitions of velocity and acceleration.

Dynamics then was imposed by substituting the classical acceleration with the new stochastic one in the expression of 2nd Newton law.

Nelson's kinematics actually represents an original way of considering a diffusion process [**RW94**], which originates a number of interesting probabilistic problems. In addition, after the basic paper [**AHKS77**] Stochastic Mechanics is known to have a non trivial connection with Dirichlet Forms theory.

(All these mathematical aspects do not share, of course, the alternate luck and misfortune Stochastic Mechanics has experienced as a physical theory, as well as the unending sequence of controversies ...).

In the spirit of [**AHKS77**] I revisit here Nelson's kinematics for diffusion processes associated with Dirichlet Forms.

For such diffusions one can write formally the expression of Nelson's stochastic acceleration field and observe that it is proportional either to the gradient of a scalar potential or to a suitable "Lorentz force" (see [**Mor82**] for the smooth general time dependent case), with a vector potential defined up to a gauge transformation.

The first case corresponds to symmetric diffusions with given invariant density so that, under weak conditions, the quantum hamiltonian operator for the related scalar potential is well defined in the sense of bilinear forms and it turns out to be

2000 *Mathematics Subject Classification.* Primary 60J60; Secondary 47D07.

unitary equivalent, through a ground–state transformation, to the generator of the symmetric diffusion [**AHKS77**].

The second case corresponds to the non symmetric diffusions with given invariant density. The quantum hamiltonian associated with the vector potential is unitary equivalent, under mild conditions, to a linear complex combination of the generator and its adjoint in the Hilbert space of complex square integrable functions with respect to the assigned invariant measure. Under additional assumptions such an unitary transformation can be seen as a generalization of a ground state transformation. In analogy with the symmetric case the spectrum of the possible self–adjoint extension of the hamiltonian operator determines the asymptotic behaviour of the associated non symmetric diffusion.

The reader is referred to [**AMU98**] for a probabilistic interpretation of the Hamiltonian operator in the case of non symmetric diffusions which are solutions of a Stochastic Differential Equation with constant diffusion coefficient.

Most results reported here can be virtually extended to infinite dimensional cases following the method introduced in [**AR90**] and described for example in [**MR92**].

2. Nelson's kinematics of a diffusive motion revisited

Let us consider firstly the case of a $\mathbb{R}^d$–valued process $\{\xi_t\}_{t\in[0,1]}$ which satisfies the S.D.E.

$$\xi_t - \xi_0 = \int_0^t b_s^+(\xi_s)ds + \int_0^t \sigma(\xi_s)dW_s, \quad t \in [0,1] \tag{2.1}$$

where $\{W_t\}_{t\in[0,1]}$ is a standard Wiener process independent of ξ_0. b_t^+ and σ are assigned vector and, respectively, matrix fields on $\mathbb{R}^d$.

Let P denote the corresponding measure on $(C([0,1]), B(C([0,1])))$ and assume that the process admits a time dependent density $\{\rho_t\}_{t\in[0,1]}$.

Introducing the time–reversal map

$$\begin{aligned} T: \qquad C\left([0,1]\to\mathbb{R}^d\right) &\mapsto C\left([0,1]\to\mathbb{R}^d\right) \\ \xi_t &\mapsto \hat{\xi}_t := \xi_{1-t} \\ P &\mapsto \hat{P} \end{aligned}$$

we assume that there exists a Wiener process $\left\{\hat{W}_t\right\}_{t\in[0,1]}$ and a time dependent vector field $\left\{\hat{b}_t^+\right\}_{t\in[0,1]}$ such that

$$\hat{\xi}_t - \hat{\xi}_0 = \int_0^t \hat{b}_s^+(\hat{\xi}_s)ds + \int_0^t \sigma(\hat{\xi}_s)d\hat{W}_s, \quad t \in [0,1] \tag{2.2}$$

Putting, for all $s, s' \in [0,1]$,

$$\begin{aligned} 1 - s' &= s \\ W_{1-s}^* &= \hat{W}_s \\ b_s^- &= -\hat{b}_{1-s}^+ \end{aligned}$$

(2.2) reads, recalling $\hat{\xi}_t := \xi_{1-t}$,

$$\xi_t - \xi_0 = \int_0^t b^-_{s'}(\xi_{s'})ds' + \int_0^t \sigma(\xi_{s'})dW^*_{s'}, \quad t \in [0,1] \tag{2.3}$$

Notice that $\{W^*_t\}_{t\in[0,1]}$ is a reversed Wiener process with $W^*_0 = \hat{W}_1$ and $W^*_1 = \hat{W}_0 = 0$.

The problem of the existence of $\hat{b}$ and $\hat{W}$ in order for (2.2) to be satisfied by the time reversal of the process $\{\xi_t\}_{t\in[0,1]}$ has been widely investigated [**Föl86**] [**HP86**] [**MNS89**] . If σ is equal to the identity matrix a sufficient condition is that P has finite relative entropy with respect to the Wiener measure [**Föl86**]. This result can be extended to infinite dimensional cases [**FW86**].

Once (2.2), or equivalently (2.3), is assumed, one can prove under mild assumptions, the "duality relation"

$$(b^+_t - b^-_t) = \frac{1}{2}\frac{1}{\rho_t}\sum \frac{\partial}{\partial x_j}(\sigma\sigma^T)_{ij}\rho_t$$

An important rôle is played by the so called "current velocity field", namely

$$b_t := \frac{b^+_t + b^-_t}{2}$$

In fact from the Fokker–Planck equation

$$\frac{\partial}{\partial t}\rho_t = -\mathrm{div}(\rho b^+_t) + \frac{1}{2}\frac{\partial}{\partial x_i}\frac{\partial}{\partial x_j}(\sigma\sigma^T)\rho_t$$

one gets the continuity equation

$$\frac{\partial}{\partial t}\rho_t = -\mathrm{div}(\rho b_t)$$

Nelson's stochastic kinematics is introduced as follows : firstly we define the mean forward and backward derivatives of any smooth real function f

$$D_+ f(\xi_t) := \lim_{\triangle\downarrow 0}\frac{1}{\triangle}E_t\left[f(\xi_{t+\triangle}) - f(\xi_t)\right] \tag{2.4}$$

$$D_- f(\xi_t) := \lim_{\triangle\downarrow 0}\frac{1}{\triangle}E_t\left[f(\xi_t) - f(\xi_{t-\triangle})\right] \tag{2.5}$$

E_t denoting the conditional expectation given the σ–algebra generated by ξ_t up to time t. Assuming the existence of the limits in (2.4) (2.5) the explicit expressions are

$$D_+ f = \left[\frac{\partial}{\partial t} + \sum(b^+_t)_i\frac{\partial}{\partial x_i} + \frac{1}{2}\sum(\sigma\sigma^T)_{ij}\frac{\partial}{\partial x_i}\frac{\partial}{\partial x_j}\right] f \tag{2.6}$$

$$D_- f = \left[\frac{\partial}{\partial t} + \sum(b^-_t)_i\frac{\partial}{\partial x_i} - \frac{1}{2}\sum(\sigma\sigma^T)_{ij}\frac{\partial}{\partial x_i}\frac{\partial}{\partial x_j}\right] f \tag{2.7}$$

Then one introduce the forward and backward velocities by

$$D_+\xi_t = b^+(\xi_t)$$
$$D_-\xi_t = b^-(\xi_t)$$

and, up to smoothness conditions on $b^\pm$, the Nelson's acceleration

$$(a_N)(\xi_t, t) := \left(\frac{D_+ b_- + D_- b_+}{2}\right)(\xi_t, t)$$

The problem of the existence of diffusions (with possibly singular drift) arising in Stochastic Mechanics was first solved by Carlen [**Car85**](see also [**AHKS77**]).

Let us now consider the case of a diffusion $\{\xi_t\}_{t>0}$ on $\mathbb{R}^d$ with assigned invariant density ρ, "current velocity" b and diffusion matrix a, such that, in the sense of generalized functions

$$\operatorname{div}\rho b = 0$$

Assume also that such a diffusion is associated to the elliptic operator

$$L_+ f = \sum \left[\frac{1}{2} a_{ij} \frac{\partial}{\partial x_i}\frac{\partial}{\partial x_j} + b_i^+ \frac{\partial}{\partial x_i}\right] f, \quad f \in C_0^\infty(\mathbb{R}^d \to \mathbb{R}) \tag{2.8}$$

where

$$b_i^+ = b_i + \frac{1}{2\rho}\sum \frac{\partial}{\partial x_j} a_{ij}\rho$$

in the following sense: there exists an extension of L^+ which is the generator of a strongly continous Markov transition semigroup $\{T_t\}_{t>0}$ in $L^2(\mathbb{R}^d \to \mathbb{R}, \rho dx)$ such that the transition semigroup of $\{\xi_t\}_{t>0}$ is a ρdx version of $\{T_t\}_{t>0}$ on a suitable domain.

In this case, recalling (2.6) (2.7), we can introduce a weaker version of Nelson's kinematics, by putting, for all $f \in C^\infty$

$$[D_+ f](\xi_t) := [L_+ f](\xi_t)$$
$$[D_- f](\xi_t) := [L_- f](\xi_t)$$

where

$$L_- f := \sum \left[-\frac{1}{2} a_{ij} \frac{\partial}{\partial x_i}\frac{\partial}{\partial x_j} + b_i^- \frac{\partial}{\partial x_i}\right] f, \quad f \in C_0^\infty(\mathbb{R}^d \to \mathbb{R})$$
$$b_i^- := b_i - \frac{1}{2\rho}\sum \frac{\partial}{\partial x_j} a_{ij}\rho$$

so that Nelson's kinematics reads, formally,

$$D_+\xi_t := b_+(\xi_t)$$
$$D_-\xi_t := b_-(\xi_t)$$
$$a_N(\xi_t) = \left(\frac{L_+ b_- + L_- b_+}{2}\right)(\xi_t)$$

3. Symmetric and non symmetric ground–state transformations

We consider in this section diffusion processes with generator L_+ given by

$$L_+ f = \sum \left[\frac{1}{2} a_{ij} \frac{\partial}{\partial x_i}\frac{\partial}{\partial x_j} + b_i^+ \frac{\partial}{\partial x_i}\right] f, \quad f \in C_0^\infty(\mathbb{R}^d \to \mathbb{R})$$

where

$$b_i^+ := b_i + \delta b_i$$

and, in the sense of generalized functions

$$\delta b_i := \frac{1}{2\rho}\sum \frac{\partial}{\partial x_j} a_{ij}\rho \tag{3.1}$$

$$\operatorname{div}\rho b = 0 \tag{3.2}$$

a being a given (symmetric and satisfying a strict ellipticity condition) matrix field in $\mathbb{R}^d$, b a vector field and ρ a non–negative density. (Most part of what is reported in the following can be extendend to the case where $\mathbb{R}^d$ is replaced by a separable real Banach space following the "component–wise approach" as reported for example in [**MR92**]).

Assuming $a_{ij}, b_i \in L^1_{loc}(\mathbb{R}^d \to \mathbb{R}, \rho dx), i, j = 1, \ldots, d$ we can construct the following non negative densely defined bilinear form on $L^2(\mathbb{R}^d \to \mathbb{R}; \rho dx)$

$$\mathcal{E}_{(a,b,\rho)}(u,v) = \frac{1}{2}\int \sum \frac{\partial u}{\partial x_i}\frac{\partial v}{\partial x_j} a_{ij}\rho dx + \\ + \int \sum \left(b_i \frac{\partial}{\partial x_i} u\right) v\rho dx, \quad u, v \in C_0^\infty(\mathbb{R}^d \to \mathbb{R})$$

Introducing

$$b_i^- := b_i - \delta b_i$$

$$L_- f := \sum \left[-\frac{1}{2} a_{ij}\frac{\partial}{\partial x_i}\frac{\partial}{\partial x_j} + b_i^- \frac{\partial}{\partial x_i}\right] f, \quad f \in C_0^\infty(\mathbb{R}^d \to \mathbb{R})$$

and assuming $\delta b_i \in L^1_{loc}(\mathbb{R}^d \to \mathbb{R}, \rho dx)$ one can easily see, integrating by parts, that

$$\mathcal{E}_{(a,b,\rho)}(u,v) = (L_- u, v)_{\rho dx} = (u, -L_+ v)_{\rho dx}, \quad \forall u, v \in C_0^\infty(\mathbb{R}^d \to \mathbb{R})$$

$(,)_{\rho dx}$ denoting the inner product in $L^2(\mathbb{R}^d \to \mathbb{R}, \rho dx)$.

We consider firstly the case where $b_i \equiv 0$. Thus $\mathcal{E}_{(a,b,\rho)}$ is identical to its symmetric part, namely, for all $u, v \in C_0^\infty(\mathbb{R}^d \to \mathbb{R})$

$$\tilde{\mathcal{E}}_{a,\rho}(u,v) := \frac{1}{2}\int \sum \frac{\partial u}{\partial x_i}\frac{\partial v}{\partial x_j} a_{ij}\rho dx = (-Lu, v)_{\rho dx} = (u, -Lv)_{\rho dx}$$

where

$$L_0 f = \sum \left[\frac{1}{2} a_{ij}\frac{\partial}{\partial x_i}\frac{\partial}{\partial x_j} + \delta b_i \frac{\partial}{\partial x_i}\right] f, \quad f \in C_0^\infty(\mathbb{R}^d \to \mathbb{R})$$

Sufficient conditions for closability on $L^2(\mathbb{R}^d \to \mathbb{R}, \rho dx)$ of this form are very well known [**Fuk80**] [**MR92**].

For example it is sufficient to assume

$$a_{i,j}, \delta b_i \in L^2_{loc}(\mathbb{R}^d \to \mathbb{R}, \rho dx) \tag{3.3}$$

Thus its smallest closed extension is a (regular and local) Dirichlet Form on $L^2(\mathbb{R}^d \to \mathbb{R}, \rho dx)$ to which is associated a strongly continous Markov semigroup with densely defined generator L_0. In case a is the identity matrix the associated diffusion process is a distorted Brownian Motion [**AHKS77**]. For more general results and extension to infinite dimensional cases see [**AR91**] and references quoted therein.

Nelson's kinematics in this case reads, in the sense of generalized functions

$$b^+ = -b^- := \delta b$$

and

$$a_N := L_0 \delta b \tag{3.4}$$

Expliciting the right side of (3.4) in terms of the assigned a and ρ we get

$$a_N = -\nabla \Phi_0$$

where the "scalar potential" Φ_0 is defined by

$$\Phi_0 = \frac{1}{2} \frac{\sum \frac{\partial}{\partial x_i} a_{ij} \frac{\partial}{\partial x_j} \rho^{\frac{1}{2}}}{\rho^{\frac{1}{2}}}$$

Introducing the Hamiltonian operator

$$Hf := \left[-\frac{1}{2} \sum \frac{\partial}{\partial x_i} a_{ij} \frac{\partial}{\partial x_j} + \Phi_0 \right] f, \quad f \in C_0^\infty(\mathbb{R}^d \to \mathbb{R})$$

we have, putting $\phi := \rho^{\frac{1}{2}}$

$$H\phi = 0$$

and

$$U_\phi H U_\phi^{-1} = L_0$$

where U_ϕ denotes the unitary transformation

$$U_\phi : \quad L^2(\mathbb{R}^d \to \mathbb{R}, dx) \to L^2(\mathbb{R}^d \to \mathbb{R}, \rho dx)$$
$$f \mapsto \frac{f}{\phi}$$

so that, in particular

$$\mathcal{E}_{a,b,\rho}(u,v) = (-L_0 u, v)_{\rho dx} = (Hu, v)_{dx}, \quad u, v \in C_0^\infty(\mathbb{R}^d \to \mathbb{R})$$

Such a connection between Dirichlet Forms and Stochastic Mechanics was firstly observed by Albeverio, Høegh–Krohn and Streit in the celebrated paper [**AHKS77**].

Let now consider the non symmetric case. We will still assume (3.3) to ensure that the symmetric part is closable. Then, if b satisfies a sufficient condition in order for $\mathcal{E}_{a,b,\rho}$ to be sectorial, which means that it can be extended to a (regular and local) coercive closed form $(\mathcal{E}, D(\mathcal{E}))$ in the sense of [**MR92**] we know from the basic theorem in non–symmetric Dirichlet Forms Theory [**MR92**] that there exists a couple of strongly continous Markov semigroups $\{T_t\}_{t>0}$ and $\{\hat{T}_t\}_{t>0}$ associated to $(\mathcal{E}, D(\mathcal{E}))$ to which corresponds a couple of properly associated diffusions.

It is easy to see that if $b \in L^2_{loc}(\mathbb{R}^d \to \mathbb{R}, \rho dx)$ then the two generators coincide with $-L_-$ and L_+, respectively, on $C_0^\infty(\mathbb{R}^d \to \mathbb{R})$ [**Mor98**].

Nelson's kinematics reads, in the sense of generalized functions

$$b_+ := b + \delta b$$
$$b_- := b - \delta b$$

(3.5) $$a_N := \frac{L_+ b_- + L_- b_+}{2}$$

To explicit the right side of (3.5) we introduce the "vector potential"

$$\mathcal{A}_i^0 := -(a)^{-1}_{ij} b_j$$

In case a is the identity matrix it is very easy to see that the right side of (3.5) takes the form of a "Lorentz force"

$$(a_N)_i = -\frac{\partial}{\partial x_i} \Phi_0 + b_j \left(\frac{\partial}{\partial x_j} \mathcal{A}_i^0 - \frac{\partial}{\partial x_i} \mathcal{A}_j^0 \right)$$

For a general strictly elliptic diffusion matrix an analogous compact expression can be written by considering $\mathbb{R}^n$ as the coordinate space of a n–dimensional Riemannian manifold with metric tensor a^{-1} [**Mor82**] .

The basic fact here is that, under mild assumptions, the Hamiltonian operator for the $(n+1)$–vector potential $(\Phi^0, \mathcal{A}^0)$ is unitary equivalent to the "complex generator" $L_{\mathbb{C}}$ define by

$$L_{\mathbb{C}} f = L_0 f + \imath \sum b_i \frac{\partial}{\partial x_i} f, \quad f \in C_0^\infty(\mathbb{R}^d \to \mathbb{R})$$

The precise statement is given by the following theorem

THEOREM 3.1. *Let $a : \mathbb{R}^d \to \mathbb{R}^d \times \mathbb{R}^d$, $b : \mathbb{R}^d \to \mathbb{R}^d$, be given (symmetric and positive definite) matrix and vector fields on $\mathbb{R}^d$ and let ρ be a non–negative density on $\mathbb{R}^d$.*

Define formally for any $S \in C^\infty(\mathbb{R}^d \to \mathbb{R})$ and $E \in \mathbb{R}$ the vector field $\mathcal{A}^S$ and the scalar field Φ_E on $\mathbb{R}^d$ as

$$\mathcal{A}_i^S := \frac{\partial}{\partial x_i} S - \sum a_{ij}^{-1} b_j, \quad i = 1, \dots, d \tag{3.6}$$

$$\Phi_E := \frac{1}{2} \sum \left[\frac{\frac{\partial}{\partial x_i} a_{ij} \frac{\partial}{\partial x_j} \rho^{\frac{1}{2}}}{\rho^{\frac{1}{2}}} - a_{ij}^{-1} b_i b_j \right] + E \tag{3.7}$$

$a_{ij}^{-1}(x)$ denoting the i, j element of the inverse of the matrix $a_{ij}(x)$ $\quad x \in \mathbb{R}^d$.

Let us assume (3.2) *holds and*

$$a_{ij}, \frac{\partial}{\partial x_i} a_{ij}, b_i, \frac{\partial}{\partial x_i} b_i \in L^2_{loc}(\mathbb{R}^d \to \mathbb{R}, dx) \tag{3.8}$$

$$\sum \frac{\frac{\partial}{\partial x_i} a_{ij} \frac{\partial}{\partial x_j} \rho^{\frac{1}{2}}}{\rho^{\frac{1}{2}}} \in L^2_{loc}(\mathbb{R}^d \to \mathbb{R}, dx) \tag{3.9}$$

with $i, j = 1, \dots, d$. Then the equality

(3.10)

$$H_E^S f := \left[\frac{1}{2} \sum (i \frac{\partial}{\partial x_i} + \mathcal{A}_i^{(S)}) a_{ij} (i \frac{\partial}{\partial x_j} + \mathcal{A}_j^{(S)}) + \Phi_{(E)} \right] f, \quad f \in C_0^\infty(\mathbb{R}^d \to \mathbb{C})$$

defines $(H_E^S, C_0^\infty(\mathbb{R}^d \to \mathbb{C}))$ as a linear symmetric operator on $L^2(\mathbb{R}^d \to \mathbb{R}, dx)$. In addition, if

$$a_{ij}, b_i, \delta b_i \in L^2_{loc}(\mathbb{R}^d \to \mathbb{R}, \rho dx)$$

δb_i been defined as in (3.1)*, then the operator $(L_{\mathbb{C}}, C_0^\infty(\mathbb{R}^d \to \mathbb{C}))$ where*

$$L_{\mathbb{C}} g := \sum \left\{ \frac{1}{2} a_{ij} \frac{\partial}{\partial x_i} \frac{\partial}{\partial x_j} + (\delta b_i + i b_i) \frac{\partial}{\partial x_i} \right\} g, \quad g \in C_0^\infty(\mathbb{R}^d \to \mathbb{C})$$

is a symmetric operator on $L^2(\mathbb{R}^d \to \mathbb{C}, \rho dx)$ and, introducing the unitary transformation

$$\begin{aligned} U_{\phi^{(S)}} :=& L^2(\mathbb{R}^d \to \mathbb{C}; dx) \to L^2(\mathbb{R}^d \to \mathbb{C}, \rho dx) \\ f \mapsto& \frac{f}{\phi^{(S)}} \\ \phi^{(S)} :=& \rho^{\frac{1}{2}} e^{iS}, \quad S \in C^\infty \end{aligned} \tag{3.11}$$

we have for all $S \in C^\infty$ and $E \in \mathbb{R}$,

$$(3.12) \qquad (-L_{\mathbb{C}} + E)g = (U_{\phi(S)} H_E^S U_{\phi(S)}^{-1})g, \quad g \in C_0^\infty(\mathbb{R}^d \to \mathbb{C})$$

in the sense of distributions.

PROOF. Inserting (3.6) and (3.7) into (3.10) one finds that the terms involving $a_{ij}^{-1} b_i b_j$ cancel each other and that , under the regularity conditions (3.8) and (3.9), $H_{(E)}^{(S)} f, f \in C_o^\infty$, is an element of $L^2(\mathbb{R}^d \to \mathbb{C}; dx)$ for all $S \in C^\infty$ and $E \in \mathbb{R}$.

The symmetry of these operators (which are familiar Hamiltonians in Quantum Mechanics) can be easily verified.

A tedious but straightforward computation gives, for any $S \in C^\infty$,

$$\left(\frac{1}{\phi^S} H_{(O)}^{(S)} \phi^S\right) g =$$

$$= -\left\{\frac{1}{2}\sum a_{ij} \frac{\partial}{\partial x_i}\frac{\partial}{\partial x_j} + (\delta b_i + i b_i)\frac{\partial}{\partial x_i}\right\} g + i\frac{1}{2}\sum g \frac{\partial}{\partial x_i}(\rho b_i),$$

$$g \in C_0^\infty(\mathbb{R}^d \to \mathbb{C})$$

which, by (3.2), proves (3.12).

Finally an integration by parts shows that $(L_{\mathbb{C}}, C_0^\infty(\mathbb{R}^d \to \mathbb{C}))$ is symmetric on $L^2(\mathbb{R}^d \to \mathbb{C}, \rho dx)$. □

Notice that, in addition, with $\phi := \rho^{\frac{1}{2}}$, we have

$$H_0^0 \phi = 0$$

If H_0^0 is bounded from below, H_0^0 has a selfadjoint extension and (3.11) is a generalization of a "ground–state transformation".

A condition is given in the following

COROLLARY 3.1. Let a, b be given matrix and vector fields on $\mathbb{R}^d$ as in Theorem 3.1 and let ρ be a non-negative density on $\mathbb{R}^d$. Let (3.2) hold and let δb and Φ_o be defined as in (3.1) and (3.7) respectively.

Assume

$$a_{ij}, b_i, \delta b_i \in L_{loc}^1(\mathbb{R}^d \to \mathbb{R}, \rho dx), \quad i, j = 1, \ldots, d$$

then

$$\mathcal{E}_{L_{\mathbb{C}}}^E(f, g) := ((-L_{\mathbb{C}} + E)f, g)_{\rho dx}, \quad f, g \in C_0^\infty(\mathbb{R}^d \to \mathbb{C}), \quad E \in \mathbb{R}$$

is a well defined symmetric sesquilinear form on $C_0^\infty(\mathbb{R}^d \to \mathbb{C})$.

If in addition

(3.13)

$$|(\Phi_o \phi, \phi)| \le \left(-\frac{1}{2}\sum \frac{\partial}{\partial x_i} a_{ij} \frac{\partial}{\partial x_j}\phi, \phi\right) + b(\phi, \phi), \quad \phi \in \{\phi : \phi = \rho^{\frac{1}{2}} |f|, f \in C_0^\infty\}$$

for some $b \in \mathbb{R}$, then there exists $E \in \mathbb{R}$ such that $(\mathcal{E}_{L_{\mathbb{C}}}^E, C_0^\infty(\mathbb{R}^d \to \mathbb{C}))$ is positive.

PROOF. The first part of the corollary is obvious. To prove the positivity under condition (3.13) we observe that, by the definition of $L_{\mathbb{C}}$ we can write

$$\mathcal{E}_{L_{\mathbb{C}}}^E(f, g) = \left(H_{(E)}^{(0)} \rho^{\frac{1}{2}} f, \rho^{\frac{1}{2}} g\right)_{dx} \quad f, g \in C_0^\infty(\mathbb{R}^d \to \mathbb{C})$$

Introducing for all $f \in C_0^\infty(\mathbb{R}^d \to \mathbb{C})$

$$\hat{\rho} := \rho\,|f|$$
$$\hat{S} := \Im\,(\ln(\rho f))$$

so that $\rho^{\frac{1}{2}} f = \hat{\rho}^{\frac{1}{2}} e^{i\hat{S}}$, we get, for all $f \in C_0^\infty(\mathbb{R}^d \to \mathbb{C})$

$$\mathcal{E}^E_{L_\mathbb{C}}(f,f) = \int \left\{ \frac{1}{2} \sum a_{ij} \frac{\partial}{\partial x_i}(\frac{1}{2}\ln\hat{\rho}) \frac{\partial}{\partial x_j}(\frac{1}{2}\ln\hat{\rho}) + \frac{1}{2}\sum a_{ij}^{-1}\hat{b}_i\hat{b}_j + \Phi_o + E \right\} \hat{\rho}dx$$

where $\hat{b}_i := a_{ij}(\mathcal{A}^o_j - \frac{\partial}{\partial x_j}\hat{S})$ and since $\hat{\rho}^{\frac{1}{2}} = \rho^{\frac{1}{2}}\,|f|$ and

$$(-\frac{1}{2}\sum \frac{\partial}{\partial x_i} a_{ij} \frac{\partial}{\partial x_j} \hat{\rho}^{\frac{1}{2}}, \hat{\rho}^{\frac{1}{2}}) = \int \frac{1}{2}\sum a_{ij} \frac{\partial}{\partial x_i}(\frac{1}{2}\ln\hat{\rho}) \frac{\partial}{\partial x_j}(\frac{1}{2}\ln\hat{\rho})\hat{\rho}dx$$

there exists $E \in \mathbb{R}$ such that $\mathcal{E}^E_{L_\mathbb{C}}$ is positive on $C_0^\infty(\mathbb{R}^d \to \mathbb{C})$. □

In particular it is possible to exploit the Cauchy–Schwarz inequality and the identity

$$L_\mathbb{C}\,(f + \imath f) = -L_- f + \imath L_+ f, \quad f \in C_0^\infty(\mathbb{R}^d \to \mathbb{R})$$

to show easily [**Mor82**] that, under the assumptions of Corollary 3.1, $\mathcal{E}_{a,b,\rho}$ is sectorial.

Actually condition (3.13) on Φ_0 is close to others given in the literature to ensure coercivity of $\mathcal{E}_{a,b,\rho}$ on $L^2(\mathbb{R}^d \to \mathbb{R}, dx)$ (see [**MR92**] Ch. II).

Weaker sufficient conditions for the existence of diffusion processes associated to extensions of $\mathcal{E}_{a,b,\rho}$ to generalized Dirichlet Forms were recently given by Stannat [**Sta96**] [**Sta97**] both in finite and in infinite dimensional cases.

Nelson's kinematics in the sense of Section 2 can be formally introduced in all cases mentioned above, but the interpretation of (3.11) as a generalization of a "ground state transformation" does not hold in general.

The ground state transformation is interesting in the symmetric case also as far as the asymptotic properties of the associated diffusion is concerned, since they are related to the spectrum of a Schroedinger operator with scalar potential.In the non symmetric case we are considering if H_0^0 has a unique selfadjoint extension on $L^2(\mathbb{R}^d \to \mathbb{C}, dx)$ one could try to relate its spectrum to the asymptotic behaviour of a diffusion with generator L_+ associated to the closed extension $(\mathcal{E}, D(\mathcal{E}))$ of $\mathcal{E}_{a,b,\rho}$.

This fact can give a probabilistic interpretation of $L_\mathbb{C}$ (see [**AMU98**] for a detailed discussion in the case of a being equal to the identity matrix).

Indeed, following [**AMU98**] we can observe that $\left\{\tilde{T}_t\right\}_{t>0}$ defined by

$$\tilde{T}_t(f + \imath g) = T_t f + \imath\hat{T}_t g, \quad f, g \in C_0^\infty(\mathbb{R}^d \to \mathbb{R})$$

is, in our case, a strongly continous contraction semigroup on the complex Hilbert space $L^2(\mathbb{R}^d \to \mathbb{C}, \rho dx)$.

Denoting by $\tilde{L}$ its infinitesimal generator we have

$$\tilde{L}(f + \imath g) = -L_- f + \imath L_+ g, \quad f, g \in C_0^\infty(\mathbb{R}^d \to \mathbb{R}) \tag{3.14}$$

Putting

$$\text{diag}\, C_0^\infty(\mathbb{R}^d \to \mathbb{C}) := \{f \in C_0^\infty(\mathbb{R}^d \to \mathbb{C}) : \Im f = \Re f\}$$
$$\text{diag}\, L^2(\mathbb{R}^d \to \mathbb{C}, \rho dx) := \{f \in L^2(\mathbb{R}^d \to \mathbb{C}, \rho dx) : \Im f = \Re f\}$$

We can observe that for $f \equiv g$ (3.14) reduces to

$$\tilde{L}(f + \imath f) = -L_- f + \imath L_+ f = L_{\mathbb{C}}(f + \imath f), \quad f \in C_0^\infty(\mathbb{R}^d \to \mathbb{R})$$

and the closures in diag $L^2(\mathbb{R}^d \to \mathbb{C}, \rho dx)$ of $L_{\mathbb{C}}$ and $\tilde{L}$ coincide.

Acknowledgements

The autor is grateful to S. Albeverio, A. Posilicano and M. Röckner for very useful comments.

References

[AHKS77] S. Albeverio, R. Høegh-Krohn, and L. Streit, *Energy Forms, Hamiltonians and distorted Brownian paths*, J.Math.Phys **18** (1977), 907–917.

[AMU98] S. Albeverio, L.M. Morato, and S. Ugolini, *Non symmetric diffusions and Related Hamiltonians*, Potential Analysis **8** (1998), 195–204.

[AR90] S. Albeverio and M. Röckner, *Classical Dirichlet Forms on topological vector spaces. Closability and Cameron–Martin formula*, J. Funct. Anal. **88** (1990), 395–436.

[AR91] S. Albeverio and M. Röckner, *Stochastic Differential Equations in infinite dimensions: a solution via Dirichlet Forms*, Probab.Th. Rel. Fields (1991), 347–386.

[Car85] E. Carlen, *Conservative Diffusions*, Comm.Math.Phys **94** (1985), 293–315.

[Föl86] H. Föllmer, *Time reversal on Wiener space*, Stochastic Process in Mathematics and Physics. Lecture Notes in Math. Vol. 1158, 119–129, Springer, New York, 1986.

[Fuk80] M. Fukushima, *Dirichlet Forms and Markov Processes*, North Holland and Kodansha LTD Japan, 1980.

[FW86] H. Föllmer and A. Wakolbinger, *Time reversal of infinite dimensional diffusions*, Stoch. Processes and Appl. **22** (1986), 56–77.

[HP86] U.G. Haussman and E. Pardoux, *Time reversal of diffusions*, The Annals of Probability **14** (1986), no. 4, 1188–1206.

[MNS89] A. Millet, D. Nualart, and M. Sanz, *Integration by parts and time reversal for diffusion processes*, The Annals of Probability **17** (1989), no. 1, 206–238.

[Mor82] L.M. Morato, *On the dynamics of diffusions and the related electromagnetic potentials*, J. Math. Phys. **23** (1982), no. 6, 1020–1024.

[Mor98] L.M. Morato, *A note on a class of Non Symmetric Dirichlet Forms associated with diffusions*, (preprint), 1998.

[MR92] Z.M. Ma and M. Röckner, *Introduction to the Theory of (Non–Symmetric) Dirichlet Forms*, Springer–Verlag, Berlin Heidelberg, 1992.

[Nel67] E. Nelson, *Dynamical Theories of Brownian Motion*, Princeton U.P., 1967.

[RW94] L.C.G. Rogers and D. Williams, *Diffusions, Markov processes and martingales.*, vol. II, John Wiley & Sons Ltd., 1994.

[Sta96] W. Stannat, *The theory of generalized Dirichlet forms and its application in analysis and stochastics*, Doctor Thesis, SFB 343, 1996, (to appear in Memoirs of the AMS).

[Sta97] W. Stannat, *(Nonsymmetric) Dirichlet Operators on L^1: existence, uniqueness and associated Markov processes*, Preprint, SFB 343, 1997.

Facoltà di Scienze, Università di Verona, Strada le Grazie, I 37134 Verona, Italy
E-mail address: morato@sci.univr.it

Canadian Mathematical Society
Conference Proceedings
Volume **29**, 2000

Representations of *-algebras in indefinite inner product spaces

G. Morchio and F. Strocchi

ABSTRACT. Weak extensions of indefinite inner product spaces are discussed in connection with representations of *-algebras, relevant for quantum field theory and Lie group representations. Regularity of representations in Krein extensions allows for a classification of the representations of compact Lie groups and of the Heisenberg algebra in indefinite inner product spaces.

1. Introduction

Representations of (unital) *-algebras often occur in theoretical and mathematical physics and their mathematical study is therefore very relevant. The analysis of the representations π of a *-algebra $\mathcal{A}$ can be done in terms of the normalized linear functionals ω on $\mathcal{A}$, satisfying the hermiticity condition $\omega(A^*) = \overline{\omega(A)}$, since by a generalized GNS (or Wightman) reconstruction theorem, given ω one obtains an inner product vector space $\mathcal{D}_\omega$ and a distinguished vector $x_\omega \in \mathcal{D}_\omega$ such that $\mathcal{D}_\omega$ is the carrier of a *-representation π_ω of $\mathcal{A}$ and $\forall A \in \mathcal{A}$

$$\omega(A) = < x_\omega,\, \pi_\omega(A)\, x_\omega >; \tag{1.1}$$

conversely, any vector of a *-representation of $\mathcal{A}$ defines a hermitean linear functional on $\mathcal{A}$.

The class of representations defined by *positive* functionals ω (or states), i.e. such that $\omega(A^*\, A) \geq 0, \forall A \in \mathcal{A}$, is rather well understood and one gets *-representations π_ω of $\mathcal{A}$ in a Hilbert space $\mathcal{H}_\omega$ with a common dense domain $\mathcal{D}_\omega = \{\pi_\omega(\mathcal{A})\, x_\omega\}$. This is the case of the standard representation theory of C^*-algebras (GNS theory) [**1**] as well the Wightman theory of relativistic field algebras [**2**]. As emphasized by Gelfand and Naimark, this also provides a general approach to the Hilbert space representations of Lie algebras.

However, for the solution of interesting theoretical physics problems (like gauge quantum field theory [**3**], indefinite metric quantum field theory of the Albeverio-Gottschalk-Wu project [**4**], damped oscillation problems [**5**] etc.)) one has to deal with representations of in general not normed *-algebras defined by a non-positive functional ω, which only yields an indefinite inner product space (IIPS) $\mathcal{D}_\omega$, with no intrinsic or natural pre-Hilbert structure associated to it.

1991 *Mathematics Subject Classification.* 22E46, 46C20, 46K10.

On the other hand, the full physical interpretation of the theory as well the mathematical control of the steps required for the solution of physical problems require to consider some extension, in particular a Hilbert closure, of $\mathcal{D}_\omega$, with the same rôle of the Hilbert space $\mathcal{H}_\omega$ in the positive case. In general, $\mathcal{D}_\omega$ is not even a complete topological space and the completion of $\mathcal{D}_\omega$ in the weak topology τ_w is a too large space, on which even the extension of the inner product is problematic (this is true also in the positive case). Thus, one does not have an intrinsic topological solution and actually, in general, there are more than one relevant extension of $\mathcal{D}_\omega$. Since the solution of physical problems or more generally the mathematical analysis of a given representation a priori depends on the choice of such an extension, one is facing the problem of analyzing the extensions of $\mathcal{D}_\omega$ in some generality.

In gauge quantum field theory, the representation of the charged local field algebra defined by the vacuum functional ω_0 cannot satisfy positivity [**3**] and the construction of the physical charged states requires the solution of the Gupta-Bleuler (GB) or BRST subsidiary condition in a suitable closure of $\mathcal{D}_{\omega_0}$ [**3, 6**]. It is in fact well known [**3**] that in quantum electrodynamics (QED) the solutions of the subsidiary condition in $\mathcal{D}_{\omega_0}$ have all vanishing electric charge, as a consequence of Gauss' law, and physical charged states have been constructed as suitable limits of local states [**6**]. In two dimensional QFT, the bosonization in terms of a massless scalar field requires to construct the exponential of the field in terms of its non-positive Wightman functions and again some closure of $\mathcal{D}_{\omega_0}$ is involved [**3, 7**].

Indefinite metric quantum field theories (IMQFT) also yield interesting examples of representations of Lie groups and of Lie algebras in IIPS. The (physically motivated) requirement that such Lie groups G are represented by $< .,. >$ isometric operators T_g, $g \in G$, i.e. such that $< T_g\, x,\, T_g\, y >=< x,\, y >$, $\forall x, y \in \mathcal{D}_\omega$, does not imply that T_g are bounded operators in Hilbert space closures of $\mathcal{D}_\omega$, (the Lorentz group in the GB formulation of free QED is an example of unbounded representations). Thus, the analysis of $< .,. >$ isometric representations of Lie groups in general faces the mathematical problems of representations of unbounded operator algebras [**8**]. As far as we know, not much is known about $< .,. >$ isometric representations of Lie groups or of Lie algebras in IIPS. A case which has recently been controlled is that of the Heisenberg algebra, where a general classification of irreducible representations has been obtained by exploiting a regularity condition in suitable majorant topologies (see below).

The lesson emerging from the examples mentioned above is that, in the analysis of the representations of *-algebras in IIPS, one is given a (common) domain $\mathcal{D}_0$, with an indefinite inner product with no intrinsic topology or closure associated to it. The systematic analysis of IIPS presented in Bognar's book [**9**] is therefore not directly applicable, since concepts like admissible majorant topology and Hilbert topology, which require some completeness property of $\mathcal{D}_0$, almost never apply. From Bognar's analysis, one may even get the impression that the route from $\mathcal{D}_0$ to a Krein-Hilbert space, if it exists, is unique, whereas this is almost never the case; in fact, in general there are infinitely many Krein extensions of $\mathcal{D}_0$, as well as infinitely many relevant weak extensions.

In Sect.2 we discuss the possible weak extensions of $\mathcal{D}_0$ by relating them to the concept of set of weakly or u-weakly compatible sequences, namely in terms of the structure which directly defines the constructions employed for the solution of physical and related mathematical problems. In Sect.3 we use the analysis of Sect.2 to discuss the $< .,. >$ isometric representations of compact Lie groups and of the Heisenberg group. In Sect.4, the relevance of the previous analysis for QFT is discussed.

2. Weak extensions of indefinite inner product spaces

The general structure common to the various representation problems discussed in the Introduction is based on a *vector space* $\mathcal{D}_0$ equipped with an *indefinite inner product* $< \cdot, \cdot >$.

Without loss of generality, we can assume that $\mathcal{D}_0$ is *non–degenerate* , i.e. that $< x, y >= 0, \ \ \forall y \in \mathcal{D}_0$, implies $x = 0$. In fact, if $\mathcal{D}_0$ is the carrier of a representation of a $*$ *algebra* $\mathcal{F}$ (in quantum field theory $\mathcal{F}$ is the polynomial algebra generated by the Wightman fields, in the mathematical problem of representations of Lie algebras it is the polynomial algebra generated by the Lie algebra, etc.), the subspace

$$\mathcal{I}_0 \equiv \{x \in \mathcal{D}_0 \ : < x, y > = 0, \ \ \forall y \in \mathcal{D}_0\}$$

is left invariant by $\mathcal{F}$, since $\forall A \in \mathcal{F}$

$$< Ax, y >=< x, A^* y >= 0, \ \forall y \in \mathcal{D}_0.$$

We can therefore take the quotient $\mathcal{D}_0/\mathcal{I}_0$, and $\mathcal{F}$ is still well defined on such a quotient space.

We denote by τ_w the weak topology on $\mathcal{D}_0$ defined by the inner product, i.e. by the family of seminorms

$$p_y(x) \equiv | < y, x > | \ \ , \quad y \in \mathcal{D}_0,$$

(see **[9]** p.60). In the following we shall conside extensions by sequences of $\mathcal{D}_0$, which is natural if $\mathcal{D}_0$ is τ_w separable, i.e. it contains a countable τ_w dense subset. In the QFT case this follows from $\mathcal{D}_0$ being defined by the application of the field distributions, smeared with test functions in the separable spaces $\mathcal{D}$ or $\mathcal{S}$, to the vacuum.

Since, in the applications, one is only given a vector space $\mathcal{D}_0$ which (even in the positive definite case) is not a Krein (Hilbert) space and, in general, not even a weakly complete inner product space, the analysis of the representations of $\mathcal{F}$ requires in general to extend $\mathcal{D}_0$ by means of weak limits.

To analyze such extensions we introduce the following notion:

DEFINITION 2.1. *A non–degenerate vector space $\mathcal{D}$ is called a* **weak extension** *of a non–degenerate (indefinite) inner product space $\mathcal{D}_0$ if*
i) $\mathcal{D} \supset \mathcal{D}_0$
ii) the inner product $< \cdot, \cdot >$ of $\mathcal{D}_0$ has an extension to $\mathcal{D}$
iii) $\mathcal{D}_0$ is sequentially weakly dense in $\mathcal{D}$, i.e. $\forall v \in \mathcal{D}, \ \ \exists v_n \in \mathcal{D}_0, \ n \in \mathbf{N}$, such that $\forall z \in \mathcal{D}$

$$\lim_n < v_n, z > \ = \ < v, z > \ . \tag{2.1}$$

In general, given a non–degenerate IIPS $\mathcal{D}_0$ there are more than one weak extensions, even under the requirement of maximality (see below). To classify the weak extensions of $\mathcal{D}_0$ we note that, if $\mathcal{D}$ is one of them, then the set of sequences $\{x_n\}, x_n \in \mathcal{D}_0$ which converge weakly in $\mathcal{D}$ have the following properties: if $x_n \to x \in \mathcal{D}, \ \ y_n \to y \in \mathcal{D}$, then

$$\lim_n \lim_m < x_n, y_m > = \lim_n \ < x_n, y > \ = \ < x, y > \ = \lim_m \lim_n < x_n, y_m > . \tag{2.2}$$

DEFINITION 2.2. *Given a non–degenerate IIPS* $\mathcal{D}_0$, *a set* T *of sequences* $\underline{x} = \{x_n\}$, $x_n \in \mathcal{D}_0$, *which includes the constant sequences* $x_n \equiv x \in \mathcal{D}_0$ *and such that the limits of their inner products exist and satisfy*

$$\lim_n \lim_m < x_n, y_m > = \lim_m \lim_n < x_n, y_m >, \tag{2.3}$$

will be called a **set of weakly compatible sequences** .

A set S *of weakly compatible sequences* $\{x_n\}$, $x_n \in \mathcal{D}_0$, *with the property that the limit in eq. (2) is uniform, i.e.* $\forall \underline{x} \equiv \{x_n\},\ \underline{y} \equiv \{y_n\} \in S_\alpha,\ \forall \varepsilon > 0,\ \ \exists M$ *such that*

$$|< x_n, y_m > - \lim_n \lim_m < x_n, y_m >| \ < \ \varepsilon, \quad \forall n, m > M, \tag{2.4}$$

is called a **set of uniformly weakly** *(briefly* **u-weakly***)* **compatible sequences**.

Thus, each extension of $\mathcal{D}_0$ determines a set of weakly compatible sequences; conversely, each set T_α (α running on some index set) of weakly compatible sequences generates a vector space, again denoted by T_α, with inner product defined by

$$< \underline{x}, \underline{y} > \equiv \lim_n \lim_m < x_n, y_m > = \lim_m \lim_n < x_n, y_m > . \tag{2.5}$$

$\mathcal{D}_0$ is naturally identified with the subspace of constant sequences, which is weakly dense in T_α. Similarly, each S_α uniquely defines a vector space, still denoted by S_α, with an inner product defined by eq.(2.5), and also given by

$$< \underline{x}, \underline{y} > = \lim_n \ < x_n, y_n > . \tag{2.6}$$

DEFINITION 2.3. *Given a set* T_α, *the following equivalence relation is defined:* $\underline{x} \sim \underline{y}$ *if* $\forall \underline{z} \in T_\alpha$

$$< \underline{x} - \underline{y}, \underline{z} > \equiv \lim_n \lim_m \ < x_n - y_n, z_m > = 0 \ . \tag{2.7}$$

One can therefore consider the quotient space $[T_\alpha] = T_\alpha/\!\sim\, = T_\alpha/\mathcal{I}_0^\alpha$, *with*

$$\mathcal{I}_0^\alpha \equiv \{\underline{x} \in T_\alpha : < \underline{x}, \underline{y} > = 0, \quad \forall \underline{y} \in T_\alpha\} = \{\underline{x} \in T_\alpha : < \underline{x}, z > = 0, \quad \forall z \in \mathcal{D}_0\}. \tag{2.8}$$

The above equation follows from the fact that, if $\lim_n \ < x_n, z > = 0,\ \ \forall z \in \mathcal{D}_0$, then $< \underline{x}, \underline{y} > = \lim_m \lim_n \ < x_n, y_m > = 0$, $\forall \underline{y} \in T_\alpha$. Clearly, the quotient space $[T_\alpha]$ is a non–degenerate IIPS and one easily has

PROPOSITION 2.1. *A vector space* $[T_\alpha]$ *of weakly compatible sequences of a non–degenerate IIPS* $\mathcal{D}_0$, *modulo sequences converging to zero, is a weak extension of* $\mathcal{D}_0$. *Conversely, every weak extension* $\mathcal{D}_\alpha$ *of* $\mathcal{D}_0$ *can be identified with the vector space* $[T_\alpha]$ *of weakly compatible sequences of* $\mathcal{D}_0$ *which converge weakly as elements of* $\mathcal{D}_\alpha$.

The important advantage of the sets S_α is that, thanks to eq.(2.6), if the inner product is (semi-)definite on $\mathcal{D}_0$, so is its extension to $[S_\alpha]$, whereas this is not the case in general for sets T_α, as the following example shows.

Example 1. Let $\mathcal{D}_0 \equiv C^0([0,1])$, with inner product $< g, f > \equiv \int \bar{g}(x) f(x)\, dx$, and $\underline{f} \equiv \{f_n\}$ a fixed sequence of continuous function which converges, in the dual of C^0, to $\delta(x - 1/2)$, with $f_n(1/2) = 0,\ \forall n$. Let T be obtained by adjoining such a sequence to the set of constant sequences. Then $< \underline{f}, \underline{f} > = \lim_n \lim_m \ < f_n, f_m >$

$= 0$ and $< \underline{f}, g > = g(1/2)$, $\forall g \in C^0$, which is incompatible with semi-definiteness of the inner product.

The weak extensions of an indefinite IIPS $\mathcal{D}_0$ are partially ordered by inclusion, and one can consider *maximal weak extensions* .

By the above identification, such maximal extensions are given by maximal vector spaces of weakly compatible sequences modulo sequences converging to zero in $\mathcal{D}_0$ (see eq.(2.8)). One should notice that such an ordering on vector spaces of weakly compatible sequences modulo sequences converging to zero in $\mathcal{D}_0$, is weaker (thanks to eq.(2.8)) than the ordering by inclusion, and in fact one can construct vector spaces of weakly compatible sequences which are maximal with respect to the ordering by inclusion, but which do not define maximal weak extensions of $\mathcal{D}_0$. The two notions coincide if the inner product in $\mathcal{D}_0$ is definite.

PROPOSITION 2.2. *Weak extensions of a non–degenerate IIPS which are sequentially weakly complete are maximal.*

Proof. If $\mathcal{D}$ is such a weak extension, the set $T_{\mathcal{D}}$ of weakly compatible sequences of $\mathcal{D}_0$ which converge to elements of $\mathcal{D}$ define a maximal space $[T_{\mathcal{D}}]$; in fact, if $[T'] \supset [T_{\mathcal{D}}]$ is a larger weak extension of $\mathcal{D}_0$, then by density of $\mathcal{D}_0$ each $x' \in [T']$ is the weak limit, in $[T']$, of a sequence $x_n \in \mathcal{D}_0$. By weak sequential completeness of $[T_{\mathcal{D}}]$, x_n is also convergent in $[T_{\mathcal{D}}]$ to $x \in [T_{\mathcal{D}}]$. It then follows that

$$< x - x', z > = 0, \quad \forall z \in \mathcal{D}_0,$$

which implies, by definitions of quotients and by eq.(2.8), $x' = x \in [T_{\mathcal{D}}]$.

A special class of weak extensions of $\mathcal{D}_0$ are those defined by sets S_α of u-weakly compatible sequences; they will be called *u-weak extensions.* Such extension are partially ordered by inclusion, and one may consider *maximal u-weak extensions.* The definite IIPS have a unique maximal u-weak estension, since in this case sequences $x_n \in \mathcal{D}_0$ satisfying eq.(2.5) with $y_n = x_n$ are automatically uniformly compatible by Schwarz' inequality; in general, however, such spaces have weak extensions which are not contained in the maximal u-extension and are not even positive (see Example 1).

DEFINITION 2.4. *A weak extension $\mathcal{D}$ of $\mathcal{D}_0$ is called a Krein extension if $\mathcal{D}$ is a Krein space, i.e. it is a Hilbert space, and $< x, y > = (x, \eta y)$ with $\eta^2 = 1$, $(\cdot, \cdot)$ denoting the Hilbert scalar product.*

PROPOSITION 2.3. *Krein extensions are maximal u-weak extensions; they are also maximal as weak extensions.*

Proof. If $\mathcal{D}$ is a Krein extension of $\mathcal{D}_0$, (sequential) weak density of $\mathcal{D}_0$ implies weak and therefore strong density in the Hilbert topology, and strong convergent sequences are u–weakly compatible by Schwarz' inequality. Maximality as a weak extension follows by Proposition 2.2, since Krein spaces are weakly complete.

A set $[T_\alpha]$ of weakly compatible sequences of $\mathcal{D}_0$, equivalently (Proposition 2.1), a weak extension of $\mathcal{D}_0$ defines a locally convex topology τ_α on $\mathcal{D}_0$, through the seminorms (labeled by the elements $[\underline{y}]$ of $[T_\alpha]$)

$$p_{[\underline{y}]}(x) = \inf_{\underline{y} \in [\underline{y}]} \sup_{y_n \in \underline{y}} | < y_n, x > | = \lim_n | < y_n, x > |. \tag{2.9}$$

PROPOSITION 2.4. *The above defined topology τ_α is a partial majorant of the inner product, i.e. the latter is separately τ_α continuous on $\mathcal{D}_0$.*

Conversely, every weak extension $\mathcal{D}_\alpha$ of $\mathcal{D}_0$ is identified with the space of sequences $x_n \in \mathcal{D}_0$ which are Cauchy sequences with respect to all the seminorms p_y, $y \in \mathcal{D}_\alpha$ (see eq.(2.6)) and such that the limit $n \to \infty$ of $< x_n, y >$ defines a τ_α continuous functional on $\mathcal{D}_0$.

Proof. In fact, since in general a linear functional continuous with respect to a set of seminorms defined by a family of linear functionals is linear combinations of them, τ continuity of $\lim_n < x_n, y >$ implies that it is of the form $< x, y >$, with $x \in \mathcal{D}_\alpha$. Moreover, if x_n, y_n are Cauchy sequences in the above seminorms and converge as functionals on $\mathcal{D}_0$ to $x, y \in \mathcal{D}_\alpha$, then

$$\lim_m \lim_n \; < y_m, x_n > = \lim_m \; < y_m, x > = < y, x > \quad ,$$

the last equality following from τ convergence and the identification of the null subspace of $\mathcal{D}_\alpha$ with the space of null functionals on $\mathcal{D}_0$, eq.(2.8). The same limit is reached in the other order, so that the sequences x_n and y_n are compatible.

One should notice that the Cauchy property with respect to a family of seminorms, even defined by a weak extension $\mathcal{D}_\alpha$, is not in general sufficient for compatibility, so that τ_α continuity of the limit functionals is essential both for convergence in $\mathcal{D}_\alpha$ and for compatibility.

Similarly, a locally convex topology τ which is *a majorant* of the inner product, i.e. the inner product is *jointly* τ–continuous ([**9**] p.77), defines a set S_τ of u-weakly compatible sequences of $\mathcal{D}_0$. In fact, is x_n,y_n are τ convergent sequences, then $\lim_{n,m} \; < x_n, y_m >$ exists, uniformly in n, m as a consequence of the joint continuity of the scalar product. In general, neither the set S_τ nor the corresponding $u - weak$ extension of $\mathcal{D}_0$ are maximal. The joint τ-continuity of the inner product is essential for u-weak compatibility. It is also crucial for the possibility of extending the inner product from $\mathcal{D}_0 \times \mathcal{D}_0$ to $\mathcal{D}_\tau \times \mathcal{D}_\tau$, where D_τ is the τ-closure of $\mathcal{D}_0$. For a partial majorant topology in general one gets only the extension of the inner product to $\mathcal{D}_0 \times \mathcal{D}_\tau$.

In the applications, the existence of at least one majorant topology is guaranteed by a distributional regularity of $\mathcal{D}_0$ which implies the Hilbert space condition [**3, 11**].

In general, u-weak extensions of $\mathcal{D}_0$ are not defined by majorant topologies. This follows from the existence of maximal u-weak extensions (see Example 2 below) which are not Banach spaces and from Lemma 1.2 at p. 78 of Bognár's book [**9**], by which for every majorant on $\mathcal{D}_0$ there is a weaker majorant defined by a single (Banach) seminorm, so that every u-weak extension defined by a majorant topology is contained in a u-weak extension defined by a Banach majorant.

Example 2 Let V_0 be a reflexive topological space on the reals, $\tilde{V}_0$ a sequentially dense proper subspace of the dual V_0' of V_0, and V_0, V_0' be weakly complete. Let $\mathcal{D}_0$ be the vector space of the pairs (x, y'), $x \in V_0, y' \in \tilde{V}_0$, briefly $\mathcal{D}_0 = \{(V_0, \tilde{V}_0)\}$ and the inner product on $\mathcal{D}_0$ be

$$< (x, y'), (z, w') >= w'(x) + y'(z).$$

let S_α be the set of sequences (x, y_n') with y_n' weakly convergent. They are u-weakly compatible and define a vector space $\mathcal{D}_\alpha = \{(V_0, V_0')\}$ which is maximal even as a weak extension; in fact, if $< (x_n, y'_m), (z, w') >$ converge $\forall (z, w') \in \mathcal{D}_\alpha$, then

$$\lim_{n\to\infty} x_n \in V_0, \quad \lim_{m\to\infty} y'_m \in V'_0,$$

by weak completeness of V_0 and V'_0. However, in general, $\mathcal{D}_\alpha$ is not a Banach (nor a Hilbert) space (e.g. one can take $V_0 = \mathcal{S}(\mathbf{R}^n)$, $V'_0 = \mathcal{S}'(\mathbf{R}^n)$). This example can be turned into an example which fits in free QED, by taking

$$V_0 = \{k_\mu f(k),\, f(k) \in \mathcal{S}(\mathbf{R}^4)\}, \quad \tilde{V}_0 = \{\bar{k}_\mu g(k),\, g(k) \in \mathcal{S}(\mathbf{R}^4),\, \bar{k}_\mu = (k_0, -\mathbf{k})\},$$

$$V'_0 = \{\bar{k}_\mu g(k),\, (k_0^2 + \mathbf{k}^2) g(k) \in \mathcal{S}'(\mathbf{R}^4)\}.$$

For a related example see Example 5.6 at p.90 of Bognár book.

In general, given a non–degenerate IIPS $\mathcal{D}_0$ and a τ majorant there exists a weaker majorant τ_B which is Banach and this implies that there is a minimal majorant τ_{min}, equivalently a Banach self-polar majorant (see Lemma 4.1 at p.85 and Theorem 4.2 at p.86 of Ref. [**9**]. In the applications, in general one is given an IIPS $\mathcal{D}_0$ which is not weakly complete and if τ is a (partial) majorant, then $\mathcal{D}_0 \subseteq \mathcal{D}'_0 \equiv$ the τ dual of $\mathcal{D}_0$; the property that τ is *admissible* ([**9**] Chap.III, Sect.5), namely $\mathcal{D}_0 \supseteq \mathcal{D}'_0$ is almost never satisfied. It is therefore relevant to introduce the following notion, which for simplicity we give only for majorant topologies.

DEFINITION 2.5. *A majorant topology τ on $\mathcal{D}_0$ is said* **pre-admissible** *if the weak extension $\mathcal{D}_\tau$ of $\mathcal{D}_0$ has the property that for any τ continuous functional φ on $\mathcal{D}_\tau$ there exists a $y \in \mathcal{D}_\tau$ such that, $\forall x \in \mathcal{D}_\tau$, $\varphi(x) =< y, x >$, i.e. if the natural extension of τ to $\mathcal{D}_\tau$ is admissible in the sense of Ref.* [**9**].

DEFINITION 2.6. *A* **pre-Hilbert topology** *τ_h on $\mathcal{D}_0$ is a topology defined by a seminorm $p(x) = (x, x)$, with $(.,.)$ a Hilbert product on $\mathcal{D}_0$.*

Any pre-Hilbert majorant is pre-admissible, since $\mathcal{D}_0$ is a pre-Hilbert space and its weak extension $\mathcal{D}_\tau$ is a Hilbert space. If τ_h is a pre-Hilbert majorant and it is minimal, then $\mathcal{D}_\tau$ is a Krein space.

3. Representations of compact Lie groups in indefinite inner product spaces

The framework discussed in Sect.2 appears useful for the analysis of the representations of Lie Groups in IIPS.

We recall that a representation of *-algebra $\mathcal{A}$ acting on an IIPS $\mathcal{D}$ is a *-morphism ρ of $\mathcal{A}$ into the algebra of operators acting on $\mathcal{D}$, namely $\rho(\mathcal{A})\,\mathcal{D} \subseteq \mathcal{D}$ and $\forall\, x, y \in \mathcal{D}$, $\forall A \in \mathcal{A}$,

$$< x,\, \rho(A)\, y >=< \rho(A^*)\, x,\, y > .$$

DEFINITION 3.1. *Two representations ρ_1, ρ_2 of $\mathcal{A}$ in the IIPS $\mathcal{D}_1$, $\mathcal{D}_2$, respectively, are said to be* **strictly equivalent** *if there is an invertible $< .,. >$ isometry $U : \mathcal{D}_1 \to \mathcal{D}_2$ such that*

$$\rho_2(A) = U\, \rho_1(A)\, U^{-1}. \tag{3.1}$$

Such strict equivalence coincides with the standard notion of equivalence if $\mathcal{D}_1$, $\mathcal{D}_2$ are Hilbert spaces, but it is in general too strict, as one can easily see also in the positive definite case. In fact, in this case the representations ρ_1, ρ_2 of *bounded* operator algebras on different invariant dense domains are in general not strictly equivalent, as the following Example 3 shows, even if they arise as restrictions of equivalent Hilbert space representations.

Example 3. Consider the representations of the translation group on the domains $\mathcal{D}(\mathbf{R}^s)$, $\mathcal{S}(\mathbf{R}^s)$, $H^k(\mathbf{R}^s)$, with the standard L^2 inner product; they are all not strictly equivalent, since all such spaces are dense in $L^2(\mathbf{R}^s)$ and any possible isometry relating any two of them (eq.(3.1)), by continuity extends to a unitary operator U in L^2 which commutes with the translations and therefore corresponds to a multiplication of the Fourier transform by a phase. Thus, it cannot interpolate between two above spaces. On the other hand, the extensions by continuity of all such representations to $L^2(\mathbf{R}^s)$ are all equivalent as well as strictly equivalent.

This problem becomes more serious in the case of unbounded operator algebras. In general, in Hilbert spaces, if one has a unitary representation of a Lie group, the corresponding representations of the Lie algebra on different invariant domains are in general not strictly equivalent, even in the case of domains of analytic vectors. A more flexible notion of equivalence can be given in terms of the closures of the unbounded operators involved. In fact, to each representation ρ of a *-algebra $\mathcal{A}$ is associated an extension by closure (not necessarily a representation) defined on the common domain $\mathcal{D}_c$ of the closures $\overline{\rho(A)}$, $A \in \mathcal{A}$. One can then introduce the following notion of equivalence.

DEFINITION 3.2. *Two representations ρ_1, ρ_2 of a *-algebra $\mathcal{A}$, defined on the positive-definite inner product spaces $\mathcal{D}_1, \mathcal{D}_2$ are said to be* **equivalent** *if, up to Hilbert space isometries, each of them is contained in the extension by closure of the other.*

In the case of unitary Hilbert space representations of a Lie group, every pair of *-representations ρ_1, ρ_2 of the corresponding Lie algebra, defined on dense domains contained in the Nelson domain [**10**] of analytic vectors are equivalent in this sense.

In the case of bounded operator algebras two representations are equivalent in the above sense iff they are restrictions to dense subdomains of equivalent Hilbert space representations.

The adaptation of the above concept of equivalence to the case of *-representations in IIPS $\mathcal{D}$ requires to make reference to extensions and topologies of $\mathcal{D}$. The simplest case is when the carriers $\mathcal{D}_1$, $\mathcal{D}_2$ of the two representations admits Krein extensions $\mathcal{K}_1$, $\mathcal{K}_2$, in terms of which the equivalence can be defined.

DEFINITION 3.3. *Given two Krein spaces $\mathcal{K}_1$, $\mathcal{K}_2$ a* **Krein isometry** *is an invertible isometry between the Hilbert spaces $\mathcal{K}_1$, $\mathcal{K}_2$, which is also a $< .,. >$ isometry.*

DEFINITION 3.4. *Two *-representations ρ_1, ρ_2 of a *-algebra $\mathcal{A}$, defined on the dense domains $\mathcal{D}_1$, $\mathcal{D}_2$ of the Krein spaces $\mathcal{K}_1$, $\mathcal{K}_2$, respectively, are called* **Krein equivalent** *if, up to Krein space isometries, each one is contained in the extension by closure of the other.*

This is the close analog of the equivalence relation given above for representations in positive-definite inner product spaces. On the other hand, since IIPS may in general have more than one Krein closure it is worthwhile to have a weaker form of equivalence.

DEFINITION 3.5. *Two *-representations ρ_1, ρ_2 of a *-algebra $\mathcal{A}$ defined in the IIPS $\mathcal{D}_1$, $\mathcal{D}_2$, respectively, are said to be* **Krein isometrically related** *if there exist Krein closures $\mathcal{K}_1$, $\mathcal{K}_2$ of $\mathcal{D}_1$, $\mathcal{D}_2$, respectively, such that ρ_1, ρ_2, as representations in $\mathcal{K}_1$, $\mathcal{K}_2$, are Krein equivalent.*

The above notions also apply to representations of groups.

DEFINITION 3.6. *A $< .,. >$* **isometric** *or briefly a* **pseudounitary representation** *of a group $\mathcal{G}$ in an IIPS $\mathcal{D}_0$ is a homomorphism U of $\mathcal{G}$ into the group of linear $< .,. >$ isometric operators in $\mathcal{D}_0$:*

$$< U(g)\,x, U(h)\,y >=< x, y >, \quad \forall\, x, y \in \mathcal{D}_0, \;\; g, h \in \mathcal{G}. \tag{3.2}$$

A pseudounitary representation of a group $\mathcal{G}$ will be called **non-degenerate** *if it has no null subrepresentation, i.e. for any non trivial subspace V, $U(g)V \subset V$, $\forall g \in \mathcal{G}$ implies that the inner product does not vanish identically in V.*

Non-degeneracy of U implies non-degeneracy of $\mathcal{D}_0$ since, by eq. (3.1), the subspace $\mathcal{I}_0 \equiv \{x \in \mathcal{D}_0 :< x, y >= 0, \;\; \forall y \in \mathcal{D}_0\}$ is automatically invariant.

Given a group $\mathcal{G}$, one may consider the abstract *-algebra $\mathcal{A}_{\mathcal{G}}$ consisting of complex linear combinations of elements of $\mathcal{G}$, with the * operation defined by $g^* \equiv g^{-1}$, $\forall g \in \mathcal{G}$ and extended to $\mathcal{A}_{\mathcal{G}}$ by antilinearity.

A pseudounitary representation of a group $\mathcal{G}$ is a representation of the associated *-algebra $\mathcal{A}_{\mathcal{G}}$ and viceversa.

We shall consider the case of Lie groups. As mentioned in the Introduction, the control of their representations in IIPS requires, as in the Hilbert space case, continuity properties, which can be formulated on the basis of partial majorant or majorant topologies. To this purpose, we introduce the following notions.

DEFINITION 3.7. *A representation U of a group $\mathcal{G}$ in an IIPS $\mathcal{D}_0$ is called* **unitarizable** *if there exists a Krein extension of $\mathcal{D}_0$ in which U is a unitary Hilbert space representation.*

DEFINITION 3.8. *A non–degenerate representation of a topological group in an IIPS is called $\tau_{\mathcal{K}}$* **regular** *if*

i) $U(g)x$ is continuous in g, $\forall x \in \mathcal{D}_0$, in the Krein topology $\tau_{\mathcal{K}}$

ii) no null representations is contained in the common domain of the extension by closure of the operators $U(g)$.

By the group relations, τ_K continuity, i.e. strong continuity in the Hilbert space $\mathcal{K} = \overline{\mathcal{D}_0}^{\tau_K}$ is equivalent to weak continuity in $\mathcal{K}$.

The following Theorem shows that, in the case of compact Lie groups, the analysis of regular pseudounitary representations can be reduced to the analysis of unitary representations in Hilbert spaces.

THEOREM 3.1. *Any non–degenerate pseudounitary representation of a compact Lie group in an IIPS $\mathcal{D}_0$, which is $\tau_{\mathcal{K}}$ regular for some Krein topology $\tau_{\mathcal{K}}$, is Krein isometrically related to a unitarizable representation.*

Proof. We construct a Krein isometrically related unitarizable representation $\mathcal{U}$ in the space $\mathcal{D}$ of finite linear combinations of elements of spaces V^l, $\mathcal{D} = \sum_l V^l$, where l runs in a subset of the index set of the (finite dimensional) irreducible representations R^l of $\mathcal{G}$. The inner product is definite in each V^l, and such spaces are orthogonal for different indexes,

$$< V^l, V^k >= 0 \quad \forall l \neq k$$

Each V^l is invariant under $\mathcal{U}$ and each V^l consists of a sum of finite dimensional unitary representations all equivalent to R^l. The spaces V^l are constructed as subspaces of $\mathcal{K}$ as follows.

Let dg denote the invariant Haar measure on $\mathcal{G}$ and R^l_{mn} the matrix elements of $\mathcal{R}^l$. $\forall x \in \mathcal{D}_0$, the integral

$$x^l_{mn} \equiv \int dg\ R^{*l}_{mn}(g)\, U(g)x \tag{3.3}$$

is well defined by weak $\tau_{\mathcal{K}}$ continuity and it defines, by weak completeness, a vector in $\mathcal{K}$.

Let V^l denote the subspace of $\mathcal{K}$ of vectors of the form (2.3). The spaces V^l are in the domain of the closure of the operators $U(g)$ and the orthogonality relations of the functions R^l_{mn} in $L^2(\mathcal{G}, dg)$ [**12**] imply, as usual, that all vectors in V^l transform under (the extension of) $U(g)$ as vectors of $\mathcal{R}^l$,

$$U(g)x^l_{mn} = \sum_k R^l_{mk}(g)x^l_{kn}. \tag{3.4}$$

The (extension by continuity of the) inner product is semidefinite on each space V^l; in fact, each $x \in V^l$ defines an irreducible representation R^l_x, and $< x, x >\neq 0$ implies that the inner product is definite in R^l_x. Now, if $< x, x >> 0$, $< y, y >< 0$ one could easily construct a null representation in the sum $R^l_x + R^l_y$, which is excluded by regularity. Semidefiniteness implies definiteness again by absence of null sub-representations. Orthogonality of V^l for different indexes follows from the fact that each vector in such spaces belongs to irreducible representations, which are inequivalent for different indexes.

The representation on $\mathcal{D}$ becomes unitary with the Hilbert scalar product defined by changing the sign of the inner product in the spaces V^l in which it is negative.

The representation defined on $\sum_l V^l$ is contained in the common domain of the closure of the operators $U(g)$, $g \in \mathcal{G}$ in (any) Krein space yielding the regularity condition. Conversely, each vector in $x \in \mathcal{D}_0$ is obtained as a strong limit

$$x = \lim_n \int dg\ U(g)x\, f_n(g)$$

with f_n converging to $\delta(g)$ in the dual of $C^0(\mathcal{G})$, and therefore, by density of the set $R^l_{mn}(g)$ in $L^2(\mathcal{G})$, as a strong limit of vectors $x_i \in \mathcal{D}$, with $U(g)x_i$ converging to $U(g)x$. The representation in $\mathcal{D}_0$ is therefore contained in the closure of the operators $U(g)$ defined on $\mathcal{D}$, and the two representations are Krein isometrically related, as dense domains of the same Krein space $\mathcal{K}$.

A very similar analysis can be done with a weaker condition of regularity, defined by replacing, in Definition 3.8, the Krein topology by the weak topology defined by a weakly complete weak extension of $\mathcal{D}_0$. The result is that all representations which are regular in this sense are related, by an isometry of weakly complete spaces, to a unitarizable representation, with the same construction as in Proposition 3.1.

The above result is also relevant for the control of representations of non-compact groups and of *-algebras in IIPS in terms of representations of compact subgroups. An interesting and physically motivated example is the classification [**13**] of the representations of the Heisenberg algebra; in the following we will derive such a classification as a direct consequence of Proposition 3.1.

DEFINITION 3.9. *The* **Heisenberg algebra** $\mathcal{A}_H$ *is the complex *-algebra generated by elements* p, q, *with the relations*

$$[p, q] = -i, \quad p^* = p, \quad q^* = q$$

In the following, we will use the standard notation

$$a \equiv 1/\sqrt{2}\,(q + ip), \qquad N = a^* a.$$

The Heisenberg relations become

$$[a, a^*] = 1.$$

The analysis of Hilbert space representations of the Heisenberg algebra has been done by requiring regularity conditions which assure the exponentiability of N to a compact $U(1)$ group [**14**]. A similar strategy can be used for IIPS representations, by using Theorem 3.1 for the $U(1)$ group generated by N and by an induced representation construction, which exploits irreducibility.

To this purpose, we consider representations of the Heisenberg algebra which extend to representations of the (abstract) *-algebra $\mathcal{A}_{ext}$ generated by $\mathcal{A}_H$ and the group $U(s)$, $s \in \mathbf{R}$, satisfying

$$U(s)\, a\, U(-s) = e^{-is}\, a, \qquad U(s)^* = U(-s). \tag{3.5}$$

DEFINITION 3.10. *A representation* ρ *of* $\mathcal{A}_{ext}$ *on a domain* $\mathcal{D}$ *will be called* $\tau_{\mathcal{K}}$ **regular** *if* $\rho(N)$ *is the weak derivative of* $U(s)$ *on* $\mathcal{D}$, *in the* $\tau_{\mathcal{K}}$ *topology.*

DEFINITION 3.11. *A representation* ρ *of* $\mathcal{A}_{ext}$ *in an IIPS* $\mathcal{D}$ *is said to be* **commutant irreducible** *if any operator on* $\mathcal{D}$ *commuting with* ρ *is a multiple of the identity. A* $\tau_{\mathcal{K}}$ *regular representation is called* $\tau_{\mathcal{K}}$ **commutant irreducible** *if so are all the representations contained in the extension of* ρ *by closure in* $\mathcal{K}$.

THEOREM 3.2. *All* $\tau_{\mathcal{K}}$ *regular and* $\tau_{\mathcal{K}}$ *commutant irreducible representations* ρ *of* $\mathcal{A}_{ext}$ *in an IIPS* $\mathcal{D}$ *are Krein isometrically related to representations uniquely induced by unitarizable representations of the group* $U(s)$, $s \in \mathbf{R}$ *with*

$$\rho(U(2\pi)) = e^{2\pi i \theta}, \quad \theta \in [0, 1),$$

defined on the linear span of the eigenvectors of $U(s)$ *and characterized by the following simple spectrum* $\sigma(N)$ *of the generator* N*:*

1) π_+: $\sigma(N) = \mathbf{N}$. *The inner product is definite and the representation is equivalent to the ordinary Fock representation.*

2) π_-: $\sigma(N) = \{-1, -2, \dots\}$. *The inner product of the eigenvectors* ψ_n *is*

$$< \psi_l, \psi_m > = \eta_1\, (-1)^m\, \delta_{l,m}\ \eta_1 = \pm 1.$$

3) π_θ: $\sigma(N) = \mathbf{Z} + \theta$, $\theta \in (0, 1)$. *The inner product of the eigenvectors is*

$$< \psi_l, \psi_m > = \eta_\theta\, \delta_{l,m}, \quad m \geq 0, \quad < \psi_l, \psi_m > = \eta_\theta\, \delta_{l,m}\, (-1)^{m+1}, \quad m < 0, \quad \eta_\theta = \pm 1.$$

The representations of $\mathcal{A}_{ext}$ *are uniquely determined by the representations of the group* $U(s)$ *as a consequence of the simplicity of the spectrum.*

The proof is essentially the same as in Ref. [**13**]. One should remark that the Krein metric operators appearing in the formulation of the theorem given in [**13**] make the representations of $U(s)$ unitary and do not necessarily coincides with the Krein structures which provide the Krein equivalence.

4. Representations of relativistic field algebras and construction of physical states

As mentioned in the Introduction, the vacuum representation of a relativistic gauge (local) field algebra $\mathcal{F}$ cannot satisfy positivity and one has an indefinite IIPS $\mathcal{D}_0$, of "local vectors", with no intrinsic Krein topology [**3**]. The subsidiary condition selects a semidefinite subspace $\mathcal{D}_{0\,(phys)}$ of $\mathcal{D}_0$, the closure of which describes the vacuum sector of the observable algebra $\mathcal{A}_{obs}$. The charged state representations have to be obtained as suitable limits of the local vectors, stable under the application of $\mathcal{A}_{obs}$. Thus, e.g. in QED, one is lead to consider a set $\mathcal{D}_{phys}$ of sequences of the form $\{\mathcal{A}_{obs}\, x_n\}$, $x_n \in \mathcal{D}_0$ with the following properties

i) $\lim_{n,m} < x_n,\, \mathcal{A}_{obs}\, y_m >$ is uniform in n, m and defines a semidefinite inner product on $\mathcal{D}_{phys}$

ii) the Gauss' law constraint holds on $\mathcal{D}_{phys}$ i.e.

$$\lim_{n,m} < x_n,\, (j - \partial F)\, \mathcal{A}_{obs}\, y_m >= 0$$

iii) the electric charge Q^{el}, defined as a limit of local charges Q_R, with a suitable smearing in time [**15, 16**], takes integer values on $\mathcal{D}_{phys}$.

From a technical point of view, it is natural and convenient to obtain $\mathcal{D}_{phys}$ as a subspace of a space $\mathcal{D}$ of u-weakly compatible sequences . As discussed in Sect.2, the most convenient way of constructing $\mathcal{D}$ is to consider a Krein topology on $\mathcal{D}_0$. In this case, $\mathcal{D}_{phys}$ is identified as the solution of the subsidiary condition in such a Krein extension of $\mathcal{D}_0$.

One should remark that not every Krein topology on $\mathcal{D}_0$ identifies u-weakly compatible sequences fulfilling the above requirements ii), iii), and therefore it may not define a physically relevant closure of $\mathcal{D}_0$.

Similar constraints appear in two dimensional QFT, e.g. in the bosonization problem, where one has to construct fermionic states as suitable limits of bosonic states, which involve the construction of the Wick exponential of the massless scalar field. Such construction can be done in the Krein closure of $\mathcal{D}_0$ defined by the unique Krein topology of the one-particle space [**7**].The construction would not go through in the closure defined by a Sobolev majorant topology.

Finally, also in the Albeverio-Gottschalk-Wu project, the physical interpretation of the theory in terms of the asymptotic states requires taking suitable limits in the IIPS which carries the representation of the field algebra and again the structures discussed in Sect.2 may be relevant.

References

[1] M.A. Naimark, *Normed Rings*, Noordhoff 1972

[2] A.S. Wightman and L. Garding, Arkiv for Fysik, t.**28**, 129 (1964)

[3] F. Strocchi, *Selected Topics on the General Properties of Quantum Field Theory*, World Scientific 1993

[4] S. Albeverio, H. Gottschalk and J.-L. Wu, Comm. Math. Phys. **184**, 509 (1997); Rep. Math. Phys. **40**, 243 (1997)

[5] T.Ya. Azizov and I.S. Iokhvidov, *Linear Operators in Spaces with an Indefinite Metric*, J. Wiley 1989
[6] G. Morchio and F. Strocchi, Nucl. Phys. **B211**, 471 (1984): **B232**, 547 (1984)
[7] G. Morchio, D. Pierotti and F. Strocchi, Jour. Math. Phys. **31**, 1467 (1990); **33**, 777 (1992)
[8] K. Schmüdgen, *Unbounded Operator Algebras and Representation Theory*, Akademie-Verlag 1990
[9] J. Bognàr, *Indefinite Inner Product Spaces*, Springer 1974
[10] E. Nelson, Ann. Math. **70**, 572 (1959); E. Nelson and W.F. Stinespring, Am. J. Math. **81**, 547 (1959)
[11] G. Morchio and F. Strocchi, Ann. Inst. H. Poincaré **33**, 251 (1980)
[12] A.O. Barut and R. Raczka, *Theory of Group Representations and Applications*, World Scientific 1986
[13] M. Mnatsakanova, G. Morchio, F. Strocchi and Yu. Vernov, Jour. Math. Phys. **39**, 2969 (1998)
[14] C.R. Putnam, *Commutation Properties of Hilbert Space Operators and Related Topics*, Springer 1967
[15] M. Requardt, Comm. Math. Phys. **50**, 259 (1976)
[16] D. Buchholz, Phys. Lett. **B174**, 331 (1986)

Dipartimento di Fisica, Universita' di Pisa and INFN, Pisa

Scuola Normale Superiore and INFN, Pisa

Canadian Mathematical Society
Conference Proceedings
Volume **29**, 2000

Delta Interactions and Electrodynamics of Point Particles

Diego Noja and Andrea Posilicano

ABSTRACT. We report on some recent work of the authors showing the relations between singular (point) perturbation of the Laplacian and the dynamical system describing a charged point particle interacting with the self–generated radiation field (the Maxwell–Lorentz system) in the dipole approximation. We show that in the limit of a point particle, the dynamics of the system is described by an abstract wave equation containing a selfadjoint operator H_m of the class of point interactions; the classical Abraham–Lorentz–Dirac third order equation, or better its integrated second order version, emerges as the evolution equation of the singular part of the field and is related to the boundary conditions entering in the definition of the operator domain of H_m. We also give the Hamiltonian structure of the limit model and, in the case of no external force, we study the reduced dynamics on the linear stable manifold.

Dedicated to Sergio Albeverio

1. Introduction

Over a century after the discover of the electron by Thompson and the first theoretical studies by Lorentz, a satisfying mathematical description of the interaction of point charged particles and electromagnetic field is still lacking. As it is well known, the Maxwell–Lorentz system, which correctly describes the bulk matter, looses its meaning in the case of a point particle, due to the elementary fact that the solutions of the field equations are not regular enough to be evaluated at a single point, as required by the particle equation. This classical singularity yields directly to the need of infinite mass renormalization and to the difficulties which plague the classical theory of the electron. This theory, as emerges in the work of Lorentz, Abraham and Dirac among the others, leads to a reduced equation for the particle alone, the so called Abraham Lorentz Dirac equation (ALD for short), a third order differential equation which embodies the interaction between particle and field in an inertial term (mass renormalization) and a third order term, describing radiation reaction. The mathematically dubious procedures involved in its deduction from Maxwell–Lorentz system, and the presence of unphysical solutions, the so called runaway solutions, surround the entire subject of a legitimate suspicion, and the usual way out is to invoke quantum electrodynamics for the solution of the problem. But, in turn, mathematical foundations of quantum electrodynamics

2000 *Mathematics Subject Classification.* Primary 34Gxx, 37Kxx, 47B25, 70Sxx, 78A35.

are far from being clear, and in the opinion of the present authors and many others a solution to the classical problem is a first necessary step toward a clarification of the quantum one. In the following (see [2,3] for complete proofs, more details and references) we describe the main results obtained in the last years by us in the so called dipole approximation of classical electrodynamics, a well known model which in essence amounts to linearize the interaction (see the Hamiltonian (3.1)) in the Maxwell–Lorentz system. For this model we show that a limit dynamical system describing the coupled dynamics of both particle and field indeed exists (theorem 3.3), and is an abstract wave equation in which the operator part is a point interaction, the elements of the domain of which have a well precise relation (see the system (3.5)) with the particle velocity. To simplify the presentation we introduce first the special case of a free particle and zero total momentum, where the standard delta potential appears, and then we give the general case in which an external force is present. To treat this second case a slight generalization of the point interaction is needed, where the linear boundary condition of standard delta interactions are replaced by an affine one in which the particle momentum enters (see definition 3.2). Such a boundary condition allows to interpret the solutions of the limit system as the solutions of the system in which a standard wave equation with a delta source is coupled with an ordinary second order equation. This equation is a low order version of the ALD equation.

Our description allows also an immediate Hamiltonian formulation of the limit dynamics, a result which solves an old problem in classical electron theory.

In the last paragraph we study, in the free case, the reduced dynamics on the stable linear manifold, also called, in the classical literature, the "non runaway dynamics". In particular we show that the flow of the reduced Hamiltonian system is correlated, through a canonical transformation, to another Hamiltonian system which lives on the standard Hilbert space of the free wave equation, the symplectic form not being however the usual one.

2. Notations

- $L^2_*(\mathbb{R}^3)$ is the Hilbert space of square integrable, divergence–free, vector fields on $\mathbb{R}^3$.
- M is be the projection from $L^2_3(\mathbb{R}^3)$, the Hilbert space of square integrable vector fields on $\mathbb{R}^3$, onto $L^2_*(\mathbb{R}^3)$.
- $\langle\cdot,\cdot\rangle$ ($\|\cdot\|_2$ being the corresponding Hilbert norm) denotes the scalar products in $L^2(\mathbb{R}^3)$, $L^2_3(\mathbb{R}^3)$, $L^2_*(\mathbb{R}^3)$ and also the obvious pairing between an element of $L^2_3(\mathbb{R}^3)$ and one of $L^2(\mathbb{R}^3)$ (the result being a vector in $\mathbb{R}^3$).
- Given two functions f and g in $L^2(\mathbb{R}^3)$, $f\otimes g$ is the operator in $L^2_3(\mathbb{R}^3)$ defined by $f\otimes g(A) := f\langle g, A\rangle$.
- $H^s(\mathbb{R}^3)$, $s\in\mathbb{R}$, is the usual scale of Sobolev–Hilbert spaces, and $H^s_3(\mathbb{R}^3)$ and $H^s_*(\mathbb{R}^3)$ are defined correspondingly.
- θ denotes the Heaviside function.
- $I(T)$ denotes the compact time interval $[-T,T]$.
- $\text{Lip}(\mathbb{R}^3;\mathbb{R}^3)$ is the space of Lipschitz vector fields.
- Given a measurable non negative function ρ, its energy $E(\rho)$ is defined as

$$E(\rho) := \frac{1}{4\pi}\int_{\mathbb{R}^3}\int_{\mathbb{R}^3}\frac{\rho(x)\rho(y)}{|x-y|}\,dx\,dy\ .$$

3. The Point Limit of the Maxwell–Lorentz System

Let us consider, on the symplectic vector space $(H^1_*(\mathbb{R}^3)\times L^2_*(\mathbb{R}^3)\times\mathbb{R}^6,\Omega)$,

$$\Omega((A_1,E_1,q_1,p_1),(A_2,E_2,q_2,p_2)) := \langle A_1,E_2\rangle - \langle A_2,E_1\rangle + q_1\cdot p_2 - q_2\cdot p_1\ ,$$

the Hamiltonian associated to the regularized Maxwell–Lorentz system in the dipole approximation with an external force $F=-\nabla V$, i.e.

$$\mathcal{H}_r(A,E,q,p) = 2\pi c^2\|E\|_2^2 + \frac{1}{8\pi}\,\|\nabla A\|_2^2 + \frac{1}{2m_r}\left|p-\frac{e}{c}\,\langle\rho_r,A\rangle\right|^2 + V(q)\ , \tag{3.1}$$

where c is the velocity of light and the square integrable density ρ_r describes an extended particle with electric charge e and radius r. The corresponding Hamilton equations give rise to the Cauchy problem

$$\begin{cases} \dot{A}_r = 4\pi c^2 E_r \\ \dot{E}_r = \frac{1}{4\pi}\,\Delta A_r - \frac{e^2}{m_r c^2}\,M\langle\rho_r,A_r\rangle\rho_r + \frac{e}{m_r c}\,Mp_r\rho_r \\ \dot{q}_r = \frac{1}{m_r}\,p_r - \frac{e}{m_r c}\,\langle\rho_r,A_r\rangle \\ \dot{p}_r = -\nabla V(q) \\ A_r(0) = A_0^r\in H^1_*(\mathbb{R}^3),\quad E_r(0)=E_0\in L^2_*(\mathbb{R}^3), \\ q_r(0)=q_0,\quad p_r(0)=p_0\ . \end{cases} \tag{3.2}$$

Let us begin with the simplest situation, i.e. $V=\text{const.}$ and $p_0=0$. In this case the fields equations decouple from the particle ones and one is led to study the convergence as $r\downarrow 0$ of the self–adjoint operator

$$H_r := -\Delta + \frac{4\pi e^2}{m_r c^2}\,M\cdot\rho_r\otimes\rho_r\ .$$

This is a first rank perturbation of the Laplacian; its resolvent is readily calculated and it is given by

$$(H_r+z)^{-1} = (-\Delta+z)^{-1} + \Gamma_r(z)^{-1}M\cdot(-\Delta+z)^{-1}\rho_r\otimes(-\Delta+z)^{-1}\rho_r\ ,$$

where $\operatorname{Im} z\neq 0$,

$$\Gamma_r(z) = -\frac{m_r c^2}{4\pi e^2} - \frac{2}{3}\langle(-\Delta+z)^{-1}\rho_r,\rho_r\rangle\ ,$$

and one immediately obtains that, in order to obtain a non trivial (i.e. different from $-\Delta$) limit, the mass must be renormalized according to the classical prescription

$$m_r := m - \frac{8\pi e^2}{3c^2}\,E(\rho_r)\ ,$$

where m is the phenomenological mass. With this definition of m_r, $(H_r+z)^{-1}$ converges in norm to

$$R_m(z) := (-\Delta+z)^{-1} + \Gamma_m(z)^{-1}M\cdot G_z\otimes G_z\ ,$$

where

$$\Gamma_m(\lambda) = -\frac{mc^2}{4\pi e^2} + \frac{\sqrt{\lambda}}{6\pi}\ ,$$

and

$$G_z(x) := \frac{1}{4\pi}\,\frac{e^{-\sqrt{z}\,|x|}}{|x|}\ ,\qquad \operatorname{Re}\sqrt{z}>0\,.$$

It is not difficult to show then that $R_m(z)$ is the resolvent of a self–adjoint operator. In more detail one has the following result, which is no more than an adaptation to our situation of [1, §II.1.] as regards the operator aspects and of [4] as regards the form ones.

THEOREM 3.1. *As $r \downarrow 0$, i.e as $\rho_r(x) := r^{-3}\rho(r^{-1}x)$, ρ a spherically symmetric probability density with bounded support, weakly converges to δ_0, the self–adjoint operator H_r converges in norm resolvent sense in $L^2_*(\mathbb{R}^3)$ to the self–adjoint operator H_m so defined:*
1. $A \in D(H_m)$ if and only if

$$\exists\, Q_A \in \mathbb{R}^3 \;:\; A_\lambda := A - \frac{4\pi e}{c}\, MQ_A G_\lambda \in H^2_*(\mathbb{R}^3)\,, \qquad -\lambda \in \rho(H_m), \quad \lambda > 0\,,$$

and the following boundary condition holds:

$$\lim_{r\downarrow 0} \frac{1}{4\pi r^2} \int_{S_r} \left(A - \frac{4\pi e}{c}\, MQ_A G_0 \right) d\mu_r = -\frac{mc}{e}\, Q_A\,,$$

where S_r denotes the sphere of radius r and μ_r is the corresponding surface measure.
2.

$$(H_m + \lambda)A := (-\Delta + \lambda)A_\lambda\,.$$

Moreover

$$\sigma_{ess}(H_m) = \sigma_{ac}(H_m) = [0, +\infty), \quad \sigma_{sc}(H_m) = \emptyset\,,$$

and

$$\sigma_p(H_m) = \left\{ -\left(\frac{3mc^2}{2e^2} \right)^2 \right\} \equiv \{-\lambda_0\}\,,$$

where $-\lambda_0$ has a threefold degeneration and

$$X^0_j = 2\sqrt{2\pi m}\; \frac{c}{e} M e_j G_{\lambda_0}\,,$$

are the corresponding normalized eigenvectors, where $\{e_j\}_1^3$ is an orthonormal basis.
If F_m is the quadratic form corresponding to H_m then

$$F_m(A, A) + \lambda \|A\|_2^2 = \|(-\Delta + \lambda)^{\frac{1}{2}} A_\lambda\|_2^2 + \left(\frac{4\pi e}{c} \right)^2 \Gamma_m(\lambda) |Q_A|^2\,, \qquad \lambda > 0\,,$$

where the vector A is in the form domain $D(F_m)$ if and only if

$$\exists\, Q_A \in \mathbb{R}^3 \;:\; A_\lambda := A - \frac{4\pi e}{c}\, MQ_A G_\lambda \in H^1_*(\mathbb{R}^3)\,.$$

Finally, given $A \in D(F_m)$, Q_A can be explicitly computed by the formula

$$Q_A = \frac{3c}{2e} \lim_{r\downarrow 0} r\, \frac{1}{4\pi r^2} \int_{S_r} A(x)\, d\mu_r\,.$$

The norm resolvent convergence of H_r to H_m implies, by using the explicit solution for abstract linear wave equations in terms of sine and cosine operator functions, that if

$$\lim_{r\downarrow 0} \|(H_r + \lambda)^{\frac{1}{2}} A_0^r - (H_m + \lambda)^{\frac{1}{2}} A_0\|_2 = 0\,, \qquad \lambda > \lambda_0$$

(we used the natural distance in the energy norm between elements in the form domain of H_r and H_m), then (see [2, corollary 2.9])

$$\lim_{r\downarrow 0} \sup_{|t|\le T} \|(H_r + \lambda)^{\frac{1}{2}} A_r(t) - (H_m + \lambda)^{\frac{1}{2}} A(t)\|_2 = 0\,. \tag{3.3}$$

Here $A(t)$ is the solution of the Cauchy problem

$$\begin{cases} \frac{1}{c^2}\ddot{A} = -H_m A_m \\ A(0) = A_0 \in D(F_m), \quad \dot{A}(0) = 4\pi c^2 E_0 \in L^2_*(\mathbb{R}^3) \end{cases}$$

This gives the limit field dynamics.

As regards the behaviour of the particle dynamics in the limit $r \downarrow 0$, by [3, lemma 2.4], relation (3.3) implies

$$\lim_{r\downarrow 0} \sup_{|t|\leq T} \left| -\frac{e}{m_r c}\langle \rho_r, A_r(t)\rangle - Q_{A(t)} \right| = 0 \,.$$

This follows from (here $X \in L^2_*(\mathbb{R}^3)$)

$$-\frac{e}{m_r c}\langle \rho_r, (H_r+\lambda)^{-1/2}X\rangle = -\frac{e}{c}\frac{\langle \rho_r, (-\Delta+\lambda)^{-\frac{1}{2}}X\rangle}{m_r}$$
$$-\frac{2e}{3\pi c}\int_0^\infty \Gamma_r(x+\lambda)^{-1}\langle(-\Delta+x+\lambda)^{-1}\rho_r, X\rangle \frac{\langle(-\Delta+\lambda)^{-1}\rho_r,\rho_r\rangle}{m_r}\frac{dx}{\sqrt{x}}\,,$$

$$Q_{(H_m+\lambda)^{-1/2}X} = \left(\frac{4\pi e}{c}\right)^{-1}\frac{1}{\pi}\int_0^\infty \Gamma_m(x+\lambda)^{-1}\langle G_{x+\lambda}, X\rangle\frac{dx}{\sqrt{x}}$$

and

$$\lim_{r\downarrow 0}\Gamma_r(\lambda) = \Gamma_m(\lambda)\,,$$

$$\lim_{r\downarrow 0}\left|\frac{\langle \rho_r, (-\Delta+\lambda)^{-\frac{1}{2}}X\rangle}{m_r}\right| = 0\,,$$

$$\lim_{r\downarrow 0}\frac{\langle(-\Delta+\lambda)^{-1}\rho_r,\rho_r\rangle}{m_r} = -\frac{3c^2}{8\pi e^2}\,.$$

Therefore

$$\lim_{r\downarrow 0}\dot{q}_r = \lim_{r\downarrow 0} -\frac{e}{m_r c}\langle \rho_r, A_r\rangle = Q_A$$

uniformly in time over compact intervals.

So, to summarize, the dynamics of the system corresponding to the limit of (3.2) is completely specified (in the case $V=$ const., $p_0=0$) by solving the abstract wave equation $c^{-2}\ddot{A} = -H_m A$ and then recovering the time evolution of the particle position from the relation $\dot{q} = Q_A$. Since the system is linear this can be done explicitly (see [2, §3]).

Let us now consider the general situation $V \neq$ const., $p_0 \neq 0$. In this case the fields equations and the particle ones no more decouple. However, considering the field equation with an assigned path $t \mapsto p(t)$ not depending on the dynamics, one is led to study the Cauchy problem

$$\begin{cases} \frac{1}{c^2}\ddot{A}_r = -H_r A_r + \frac{4\pi e}{m_r c}Mp\rho_r \\ A(0) = A_0^r \in H^1_*(\mathbb{R}^3), \quad \dot{A}(0) = \dot{A}_0 \in L^2_*(\mathbb{R}^3)\,. \end{cases}$$

As in the case $p=0$ one would like to show that the solution of the above system converges to the solution of the Cauchy problem

$$\begin{cases} \frac{1}{c^2}\ddot{A} = -H^p_m A \\ A(0) = A_0 \in D(F_m), \quad \dot{A}(0) = \dot{A}_0 \in L^2_*(\mathbb{R}^3)\,, \end{cases}$$

where H_m^p is a (not necessarily linear) operator to be identified. To this end let us consider the system

$$\begin{cases} \frac{1}{c^2}\ddot{A}_r = -(H_r+\lambda)A_r + \frac{4\pi e}{m_r c}Mp\rho_r \\ A(0) = A_0^r \in H^1_*(\mathbb{R}^3), \quad \dot{A}(0) = \dot{A}_0 \in L^2_*(\mathbb{R}^3) , \end{cases}$$

where the parameter $\lambda > \lambda_0$ is inserted only to make $H_r + \lambda$, and successively $H_m + \lambda$, invertible. Its (mild) solution is readily found and, after an integration by parts, can be rewritten as

$$\begin{aligned} A_r(t) =& \cos(ct(H_r+\lambda)^{\frac12})A_0^r + \sin(ct(H_r+\lambda)^{\frac12})(H_r+\lambda)^{-\frac12}\dot{A}_0 \\ &-\frac{4\pi e}{m_r c}\int_0^t \cos(c(t-s)(H_r+\lambda)^{\frac12})(H_r+\lambda)^{-1}M\dot{p}(s)\rho_r ds \\ &+\frac{4\pi e}{m_r c}(H_r+\lambda)^{-1}Mp(t)\rho_r - \frac{4\pi e}{m_r c}\cos(ct(H_r+\lambda)^{\frac12})(H_r+\lambda)^{-1}Mp(0)\rho_r \, . \end{aligned}$$

By [3, lemma 2.5]

$$\lim_{r\downarrow 0}\left\| \frac{4\pi e}{m_r c}(H_r+\lambda)^{-\frac12}Mp\rho_r + \frac{c}{e}\Gamma_m(\lambda)^{-1}(H_m+\lambda)^{\frac12}MpG_\lambda \right\|_2 = 0$$

and so, by thm. 3.1, if

$$\lim_{r\downarrow 0} \|(H_r+\lambda)^{\frac12}A_0^r - (H_m+\lambda)^{\frac12}A_0\|_2 = 0 \, ,$$

then

$$\lim_{r\downarrow 0}\ \sup_{|t|\le T} \|(H_r+\lambda)^{\frac12}A_r(t) - (H_m+\lambda)^{\frac12}A(t)\|_2 = 0 \, ,$$

where

$$\begin{aligned} A(t) =& \cos(ct(H_m+\lambda)^{\frac12})A_0 + \sin(ct(H_m+\lambda)^{\frac12})(H_m+\lambda)^{-\frac12}\dot{A}_0 \\ &+\frac{c}{e}\Gamma_m(\lambda)^{-1}\int_0^t \cos(c(t-s)(H_m+\lambda)^{\frac12})M\dot{p}(s)G_\lambda \, ds \\ &-\frac{c}{e}\Gamma_m(\lambda)^{-1}Mp(t)G_\lambda + \frac{c}{e}\Gamma_m(\lambda)^{-1}\cos(ct(H_m+\lambda)^{\frac12})Mp(0)G_\lambda \, . \end{aligned}$$

Finally, again integrating by parts, it is easily seen that

$$A_p(t) := A(t) + \frac{c}{e}\Gamma_m(\lambda)^{-1}Mp(t)G_\lambda$$

solves the Cauchy problem

$$\begin{cases} \frac{1}{c^2}\ddot{A}_p = -(H_m+\lambda)A_p + \frac{c}{e}\Gamma_m(\lambda)^{-1}M\ddot{p}G_\lambda \\ A_p(0) = A_0 + \frac{c}{e}\Gamma_m(\lambda)^{-1}Mp(0)G_\lambda, \quad \dot{A}_p(0) = \dot{A}_0 + \frac{c}{e}\Gamma_m(\lambda)^{-1}M\dot{p}(0)G_\lambda \, , \end{cases}$$

and so $A(t)$ solves the Cauchy problem

$$\begin{cases} \frac{1}{c^2}\ddot{A} = -(H_m+\lambda)\left(A + \frac{c}{e}\Gamma_m(\lambda)^{-1}MpG_\lambda\right) \\ A(0) = A_0 \in D(F_m), \quad \dot{A}(0) = \dot{A}_0 \in L^2_*(\mathbb{R}^3) \, . \end{cases}$$

This induces us to define, on the domain

$$D(H_m^p) := D(H_m) - \frac{c}{e}\Gamma_m(\lambda)^{-1}MpG_\lambda \, ,$$

the affine operator H_m^p according to

$$(H_m^p+\lambda)A := (H_m+\lambda)\left(A + \frac{c}{e}\Gamma_m(\lambda)^{-1}MpG_\lambda\right) .$$

Alternatively (see [3, lemma 4.1]) H^p_m can be defined in the following way:

DEFINITION 3.2. 1. $A \in D(H^p_m)$ if and only if

$$\exists Q_A \in \mathbb{R}^3 \ : \ A_\lambda := A - \frac{4\pi e}{c} MQ_A G_\lambda \in H^2_*(\mathbb{R}^3)\,, \qquad -\lambda \in \rho(H_m), \quad \lambda > 0\,,$$

and the following boundary condition holds:

$$\lim_{r\downarrow 0} \frac{1}{4\pi r^2} \int_{S_r} \left(A - \frac{4\pi e}{c} MQ_A G_0\right) d\mu_r = -\frac{mc}{e} Q_A + \frac{c}{e} p\,.$$

2.

$$(H^p_m + \lambda)A := (-\Delta + \lambda)A_\lambda\,.$$

Note that, as it is evident, the affine operator H^p_m reduces, for $p = 0$, to the linear operator H_m describing the standard point interaction.

The same considerations leading to the definition of H^p_m give, coupled with a fixed point argument and estimates uniform in r, the following

THEOREM 3.3. (see [3, thm. 3.4]) *Let V such that $\nabla V \in \text{Lip}(\mathbb{R}^3;\mathbb{R}^3)$, $|\nabla V(x)| \le K\,(1+|x|)$, $\lambda > \lambda_0$, and $E_0 \in L^2_*(\mathbb{R}^3)$. Let $A^r_0 \in H^1_*(\mathbb{R}^3)$, $A_0 \in D(F_m)$, such that*

$$\lim_{r\downarrow 0} \|(H_r + \lambda)^{\frac{1}{2}} A^r_0 - (H_m + \lambda)^{\frac{1}{2}} A_0\|_2 = 0\,. \tag{3.4}$$

Then there exists $T > 0$, not depending on r, such that, denoting by

$$(A_r, E_r, q_r, p_r) \in C(I(T); H^1_*(\mathbb{R}^3)) \times C(I(T); L^2_*(\mathbb{R}^3)) \times C^2(I(T);\mathbb{R}^3) \times C^1(I(T);\mathbb{R}^3)$$

the unique mild solution of the Cauchy problem (3.2), one has

$$\lim_{r\downarrow 0} \sup_{|t|\le T} \|(H_r + \lambda)^{\frac{1}{2}} A_r(t) - (H_m + \lambda)^{\frac{1}{2}} A(t)\|_2 = 0,$$
$$\lim_{r\downarrow 0} \sup_{|t|\le T} \|E_r(t) - E(t)\|_2 = 0,$$
$$\lim_{r\downarrow 0} \sup_{|t|\le T} |\dot q_r(t) - \dot q(t)| + \sup_{|t|\le T} |q_r(t) - q(t)| = 0,$$
$$\lim_{r\downarrow 0} \sup_{|t|\le T} |\dot p_r(t) - \dot p(t)| + \sup_{|t|\le T} |p_r(t) - p(t)| = 0\,,$$

where

$$(A, E, q, p) \in C(I(T); D(F_m)) \times C(I(T); L^2_*(\mathbb{R}^3)) \times C^1(I(T);\mathbb{R}^3) \times C^1(I(T);\mathbb{R}^3)$$

denotes the unique mild solution of the Cauchy problem

$$\begin{cases} \dot A = 4\pi c^2 E \\ \dot E = -\frac{1}{4\pi}\, H^p_m A \\ \dot q = Q_A \\ \dot p = -\nabla V(q) \\ A(0) = A_0 \in D(F_m), \quad E(0) = E_0 \in L^2_*(\mathbb{R}^3), \\ q(0) = q_0, \quad p(0) = p_0\,. \end{cases} \tag{3.5}$$

An alternative description of the limit dynamics defined by the previous system is provided by the following

THEOREM 3.4. (see [3, thm. 4.2]) *Given V such that $\nabla V \in \mathrm{Lip}(\mathbb{R}^3;\mathbb{R}^3)$, let $(A,E,q,p)\in C^1([0,T];D(F_m))\times C^1([0,T];L^2_*(\mathbb{R}^3))\times C^2([0,T];\mathbb{R}^3)\times C^1([0,T];\mathbb{R}^3)$, $A_0\in D(H^{p_0}_m)$, $E_0\in D(F_m)$, be the unique strict solution of the Cauchy problem (3.5). Then*

$$A(t)=A_f(t)+\frac{4\pi e}{c}\,MA_\delta(t)\ ,$$

where $A_f(t)$ is the solution of the free wave equation with initial data A_0, E_0 and A_δ is the retarded potential of the source $Q_A\delta_0$, i.e.

$$A_\delta(t,x)=\frac{1}{4\pi}\,\frac{\theta(ct-|x|)}{|x|}Q_{A(t-|x|/c)}\ .$$

Moreover Q_A satisfies the equation

$$\dot{Q}_{A(t)}=c\sqrt{\lambda_0}Q_{A(t)}+\frac{3c^2}{2e}\,A_f(t,0)-\frac{3c^3}{2e^2}\,p(t)\ . \tag{3.6}$$

Let us sketch the proof of the above theorem. Given an arbitrary function $Q(t)$, consider the function

$$A(t,x):=A_f(t,x)+\frac{1}{4\pi}\,\frac{\theta(ct-|x|)}{|x|}Q(t-|x|/c)\ .$$

It solves the distributional equation

$$\frac{1}{c^2}\ddot{A}=\Delta A+\frac{4\pi e}{c}MQ\delta_0\ .$$

Kirchhoff formula shows that A_f gives no contribution to Q_A (see [1, §3]) and so $Q_{A(t)}=Q(t)$. Moreover, by [3, lemma 4.1] and [2, thm. 3.3] there follows

$$A-\frac{4\pi e}{c}MQG_\lambda\in H^2_*(\mathbb{R}^3)\ .$$

Therefore if $A(t)\in D\left(H^{p(t)}_m\right)$, since

$$\frac{1}{c^2}\ddot{A}=\Delta A+\frac{4\pi e}{c}MQ\delta_0=\Delta\left(A-\frac{4\pi e}{c}MQG_\lambda\right)+\lambda\frac{4\pi e}{c}MQG_\lambda=-H^p_mA\ ,$$

then the thesis will follow from unicity of the solution of (3.5). The conditions on $Q(t)$ leading to $A(t)\in D\left(H^{p(t)}_m\right)$ are found as follows. By an elementary integration

$$\begin{aligned}&\lim_{r\downarrow 0}\frac{1}{4\pi r^2}\int_{S_r}\left(A-\frac{4\pi e}{c}MQG_0\right)d\mu_r(x)\\ =&A_f(t,0)+\frac{2}{3}\frac{e}{c^2}\lim_{r\downarrow 0}\left(\frac{Q(t-r/c)-Q(t)}{r/c}\right)\\ =&A_f(t,0)-\frac{2}{3}\frac{e}{c^2}\dot{Q}(t)\end{aligned}$$

and so A satisfies the boundary condition in Definition 3.2 if and only if $Q(t)$ solves (3.6).

Let us remark that the above theorem gives the connection with the traditional description: indeed the field variable satisfies a standard wave equation with a point source and the equation satisfied by Q_A (i.e. $\dot{q}$) is the integrated version of the ALD equation

$$m\ddot{q}(t)=\frac{2e^2}{3c^3}\dddot{q}(t)-\frac{e}{c}\dot{A}_f(t,0)+F(q(t))\ .$$

Moreover the above argument also gives the converse statement:
the solution of the distributional Cauchy problem

$$\begin{cases} \frac{1}{c^2}\ddot{A} = \Delta A + \frac{4\pi e}{c} M\dot{q}\delta_0 \\ \ddot{q}(t) = c\sqrt{\lambda_0}\dot{q}(t) + \frac{3c^2}{2e} A_f(t,0) - \frac{3c^3}{2e^2}\, p(t) \\ \dot{p} = -\nabla V(q) \\ A(0) = A_0 \in D(H_m^{p_0}), \quad \dot{A}(0) = 4\pi c^2 E_0 \in D(F_m), \\ q(0) = q_0, \quad \dot{q}(0) = Q_{A_0}, \quad p(0) = p_0 \end{cases}$$

solves (3.5). Also note that the equivalence between the two descriptions holds true if and only if the initial data for the field are chosen coherently with the particle's ones (i.e. $\dot{q}_0 = Q_{A_0}$ and $\ddot{q}_0 = Q_{\dot{A}_0}$).

Let us now come to the Hamiltonian character of system (3.5). By definition 3.2, and by $D(H_m^p) \subset D(F_m)$, one can check that

$$\langle H_m^p A_1, A_2\rangle = F_m(A_1, A_2) + 4\pi p \cdot Q_{A_2} \ .$$

Therefore equations (3.5) are nothing but the Hamilton equations corresponding to the (degenerate) Hamiltonian

$$\mathcal{H}_m(A, E, q, p) := 2\pi c^2 \|E\|_2^2 + \frac{1}{8\pi}\, F_m(A, A) + p \cdot Q_A + V(q) \ ;$$

this is defined on the symplectic vector space $(D(F_m) \times L_*^2(\mathbb{R}^3) \times \mathbb{R}^6, \Omega)$. Moreover one has the following convergence result:

THEOREM 3.5. (see [3, thm.4.5]) *Let*

$$\mathcal{H}_r(A, E, q, p) = 2\pi c^2 \|E\|_2^2 + \frac{1}{8\pi}\, \|\nabla A\|_2^2 + \frac{1}{2m_r}\left|p - \frac{e}{c}\langle \rho_r, A\rangle\right|^2 + V(q)$$

be the Hamiltonian giving the equations (3.2), let $E \in L_*^2(\mathbb{R}^3)$, $(q,p) \in \mathbb{R}^6$, and let $A_r \in H_*^1(\mathbb{R}^3)$, $A \in D(F_m)$ satisfy the condition (3.4). Then

$$\lim_{r\downarrow 0}\, \mathcal{H}_r(A_r, E, q, p) = \mathcal{H}_m(A, E, q, p) \ .$$

4. The Non Runaway Dynamics

The negative eigenvalue in the spectrum of the operator H_m gives rise to unstable behaviour which corresponds, in classical electron theory, to the presence of the so called "runaway solutions". In this section we briefly describe, in the free case and, for simplicity of presentation, vanishing particle momentum, the reduced dynamics on the stable manifold.

Given any vector subspace $\mathcal{V} \subseteq L_*^2(\mathbb{R}^3)$, we define the corresponding "non runaway" subspace $[\mathcal{V}]_{\rm nr}$ by

$$[\mathcal{V}]_{\rm nr} := \{\ A \in \mathcal{V}\ :\ \langle A, X_j^0\rangle = 0,\ j = 1,2,3\ \}\ ,$$

X_j^0 being the eigenvectors corresponding to $-\lambda_0$ (see thm. 3.1). Observe that, if $A = A_\lambda + \frac{4\pi e}{c} MQ_A G_\lambda \in D(F_m)$, then one has

$$\begin{aligned} \langle A, X_j^0\rangle &= \langle A_\lambda, X_j^0\rangle + \frac{4\pi e}{c}\, \langle MQ_A G_\lambda, X_j^0\rangle \\ &= 2\sqrt{2\pi m}\frac{c}{e}\left(\langle A_\lambda^j, G_{\lambda_0}\rangle + \frac{8\pi e}{3c}\, Q_A^j \langle G_\lambda, G_{\lambda_0}\rangle\right) . \end{aligned}$$

Therefore $A \in [D(F_m)]_{\rm nr}$ if and only if

$$Q_A = -\frac{3c}{8\pi e}\frac{\langle G_{\lambda_0}, A_\lambda\rangle}{\langle G_\lambda, G_{\lambda_0}\rangle}\ ,$$

so that any $A \in [D(F_m)]_{\rm nr}$ is univocally determined by its regular part A_λ. This implies, since $\langle G_{\lambda_0}, G_{\lambda_0}\rangle^{-1} = 8\pi\sqrt{\lambda_0}$, that the map

$$\Phi : H^1_*(\mathbb{R}^3) \to [D(F_m)]_{\rm nr}, \quad \Phi A := A + MP_A G_{\lambda_0}, \quad P_A := -12\pi\sqrt{\lambda_0}\langle G_{\lambda_0}, A\rangle\ ,$$

is bijective, with inverse given by

$$\Phi^{-1} : [D(F_m)]_{\rm nr} \to H^1_*(\mathbb{R}^3), \quad \Phi^{-1}A := A - \frac{4\pi e}{c} MQ_A G_{\lambda_0} \equiv A_{\lambda_0}\ .$$

LEMMA 4.1. *Φ is a continuous bijection between $H^1_*(\mathbb{R}^3)$ and $[D(F_m)]_{\rm nr}$. Moreover*

$$\Phi^{-1}([D(H_m)]_{\rm nr}) = \{A \in H^2_*(\mathbb{R}^3)\ :\ A(0) = 0\}\ . \tag{4.1}$$

Proof. If $\lambda > \lambda_0$ we can use $\|A\|_\lambda := F^\lambda_m(A, A)$ as a norm on $[D(F_m)]_{\rm nr}$. Then

$$\begin{aligned}
\|\Phi A\|^2_\lambda &= \|A + MP_A(G_{\lambda_0} - G_\lambda) + MP_A G_\lambda\|_\lambda \\
&=\|(-\Delta+\lambda)^{\frac12}(A + MP_A(G_{\lambda_0} - G_\lambda))\|^2_2 + \frac{2}{3}\Gamma_m(\lambda)|P_A|^2 \\
&\le 2(\|(-\Delta+\lambda)^{\frac12}A\|^2_2 + \|(-\Delta+\lambda)^{\frac12}MP_A(G_{\lambda_0} - G_\lambda)\|^2_2) + \frac{2}{3}\Gamma_m(\lambda)|P_A|^2 \\
&\le 2\|(-\Delta+\lambda)^{\frac12}A\|^2_2 + c_1|P_A|^2 \le 2\|(-\Delta+\lambda)^{\frac12}A\|^2_2 + c_2\|A\|^2_2 \\
&\le c_3\|(-\Delta+\lambda)^{\frac12}A\|^2_2\ .
\end{aligned}$$

Since, for any $\lambda \neq \lambda_0$, $A = A_\lambda + \Gamma_m(\lambda)^{-1}MA_\lambda(0)G_\lambda \in [D(H_m)]_{\rm nr}$ if and only if $A_\lambda(0) = \Gamma_m(\lambda)P_{A_\lambda}$, one has

$$\begin{aligned}
(\Phi^{-1}A)(0) &= A_\lambda(0) + (MP_A(G_\lambda - G_{\lambda_0}))(0) \\
&= A_\lambda(0) + \frac{2}{3}\frac{\sqrt{\lambda_0} - \sqrt{\lambda}}{4\pi}P_A \\
&= A_\lambda(0) - \Gamma_m(\lambda)P_{A_\lambda} = 0\ .
\end{aligned}$$

Then Φ^{-1} is continuous by bijectivity and by the open mapping theorem. □

Obviously the map Φ can be extended to the whole $L^2_*(\mathbb{R}^3)$, giving rise the orthogonal projection onto $[L^2_*(\mathbb{R}^3)]_{\rm nr}$.

Let us now consider the symplectic space $([D(F_m)]_{\rm nr} \times [L^2_*(\mathbb{R}^3)]_{\rm nr}, \Omega_0)$, where Ω_0 denotes the canonical symplectic form induced by $\langle\cdot,\cdot\rangle$, i.e.

$$\Omega_0((A_1, E_1), (A_2, E_2)) = \langle A_1, E_2\rangle - \langle A_2, E_1\rangle\ .$$

On $([D(F_m)]_{\rm nr} \times [L^2_*(\mathbb{R}^3)]_{\rm nr}, \Omega_0)$ we have the *non negative* Hamiltonian

$$\begin{aligned}
\mathcal{H}^+_m(A, E) &= 2\pi c^2\,\|E\|^2_2 + \frac{1}{8\pi}\,F_m(A, A) \\
&= 2\pi c^2\,\|E\|^2_2 + \frac{1}{8\pi}\left(\|\nabla A_{\lambda_0}\|^2_2 + \frac{\sqrt{\lambda_0}}{12\pi}\left(\frac{4\pi e}{c}\right)^2|Q_A|^2\right)\ ,
\end{aligned}$$

with corresponding Hamiltonian vector field

$$X_{\mathcal{H}^+_m}(A, E) = \left(4\pi c^2 E, -\frac{1}{4\pi}H_m A\right)\ ,$$

defined on the domain $[D(H_m)]_{\rm nr} \times [D(F_m)]_{\rm nr}$. If we pull–back Ω_0 and $\mathcal{H}_m^+$ to $H^1_*(\mathbb{R}^3) \times L^2_*(\mathbb{R}^3)$ by using the map

$$\Psi := \Phi \times \Phi : H^1_*(\mathbb{R}^3) \times L^2_*(\mathbb{R}^3) \to [D(F_m)]_{\rm nr} \times [L^2_*(\mathbb{R}^3)]_{\rm nr} ,$$

we obtain the following

THEOREM 4.2. *1.* $(H^1_*(\mathbb{R}^3) \times L^2_*(\mathbb{R}^3), \Omega_{\rm nr})$ *is a symplectic space, where the (weakly) nondegenerate symplectic form* $\Omega_{\rm nr}$ *is given by*

$$\begin{aligned}&\Omega_{\rm nr}((A_1, E_1), (A_2, E_2)) := \Psi^*\Omega_0((A_1, E_1), (A_2, E_2))\\ =&\Omega_0((A_1, E_1), (A_2, E_2)) - \frac{1}{12\pi\sqrt{\lambda_0}} (P_{A_1} \cdot P_{E_2} - P_{A_2} \cdot P_{E_1}) .\end{aligned}$$

2. Defining

$$X_{\mathcal{H}_{\rm nr}} : H^2_*(\mathbb{R}^3) \times H^1_*(\mathbb{R}^3) \to H^1_*(\mathbb{R}^3) \times L^2_*(\mathbb{R}^3)$$

$$X_{\mathcal{H}_{\rm nr}}(A, E) := \left(4\pi c^2 E, \frac{1}{4\pi} \Delta A + \frac{3}{2} \sqrt{\lambda_0} M A(0) G_{\lambda_0}\right) ,$$

one has, for any $(A, E) \in H^2_*(\mathbb{R}^3) \times H^1_*(\mathbb{R}^3)$,

$$\frac{1}{2} \Omega_{\rm nr}(X_{\mathcal{H}_{\rm nr}}(A, E), (A, E)) = \mathcal{H}_{\rm nr}(A, E) ,$$

where

$$\mathcal{H}_{\rm nr} : H^1_*(\mathbb{R}^3) \times L^2_*(\mathbb{R}^3) \to \mathbb{R}$$

$$\begin{aligned}&\mathcal{H}_{\rm nr}(A, E) := \Psi^*\mathcal{H}_m^+(A, E)\\ =&2\pi c^2 \|E\|_2^2 + \frac{1}{8\pi} \|\nabla A\|_2^2 - \frac{c^2}{6\sqrt{\lambda_0}} |P_E|^2 + \frac{\sqrt{\lambda_0}}{96\pi^2} |P_A|^2 .\end{aligned}$$

i.e. $X_{\mathcal{H}_{\rm nr}}$ *is the (unique) Hamiltonian vector field corresponding to* $\mathcal{H}_{\rm nr}$.
3. The Hamiltonian vector fields $X_{\mathcal{H}_m^+}$ *and* $X_{\mathcal{H}_{\rm nr}}$ *are* Ψ*-correlated, i.e.*

$$X_{\mathcal{H}_m^+} \circ \Psi = \Psi \circ X_{\mathcal{H}_{\rm nr}} , \tag{4.2}$$

and

$$U_m(t) \circ \Psi = \Psi \circ U_{\rm nr}(t) , \tag{4.3}$$

where $U_m(t)$ *and* $U_{\rm nr}(t)$, $t \in \mathbb{R}$, *denote the one parameter groups of canonical transformation of* $([D(F_m)]_{\rm nr} \times [L^2_*(\mathbb{R}^3)]_{\rm nr}, \Omega_0)$ *and* $(H^1_*(\mathbb{R}^3) \times L^2_*(\mathbb{R}^3), \Omega_{\rm nr})$ *given by the flows of* $X_{\mathcal{H}_m^+}$ *and* $X_{\mathcal{H}_{\rm nr}}$ *respectively. Moreover, if* $U_f(t)$, $t \in \mathbb{R}$, *denotes the flow, on* $H^1_*(\mathbb{R}^3) \times L^2_*(\mathbb{R}^3)$, *given by solving the free wave equation, then*

$$U_m(t)_{|\,[D(H_m)]_{\rm nr} \times [D(F_m)]_{\rm nr}} = \Psi \circ U_f(t) \circ \Psi^{-1} . \tag{4.4}$$

Proof. 1. A simple calculation shows that

$$\begin{aligned}&\Omega_0((\Phi A_1, \Phi E_1), (\Phi A_2, \Phi E_2)) = \langle \Phi A_1, \Phi E_2 \rangle - \langle \Phi A_2, \Phi E_1 \rangle\\ =&\Omega_0((A_1, E_1), (A_2, E_2)) - \frac{1}{12\pi\sqrt{\lambda_0}} (P_{A_1} \cdot P_{E_2} - P_{A_2} \cdot P_{E_1}) ,\end{aligned}$$

2. Since

$$\begin{aligned}&F_m(\Phi A, \Phi A) = \|(-\Delta + \lambda_0)^{\frac{1}{2}} A\|_2^2 - \lambda_0 \|\Phi A\|_2^2\\ =&\|(-\Delta + \lambda_0)^{\frac{1}{2}} A\|_2^2 - \lambda_0 \langle A, \Phi A \rangle\\ =&\|\nabla A\|_2^2 + \frac{\sqrt{\lambda_0}}{12\pi} |P_A|^2 ,\end{aligned}$$

$$\|\Phi E\|_2^2 = \langle E, \Phi E\rangle = \|E\|_2^2 - \frac{1}{12\pi\sqrt{\lambda_0}}\,|P_E|^2\ ,$$

posing $4\pi H_{\rm nr}A = -\Delta A - 6\pi\sqrt{\lambda_0}MA(0)G_{\lambda_0}$, we need to verify the relation

$$-\langle \Delta A, A\rangle + \frac{\sqrt{\lambda_0}}{12\pi}\,|P_A|^2 = 4\pi\langle H_{\rm nr}A, A\rangle - \frac{1}{12\pi\sqrt{\lambda_0}}\,P_{4\pi H_{\rm nr}A}\cdot P_A\ .$$

Since

$$\begin{aligned}P_{\Delta A} =& -\,12\pi\sqrt{\lambda_0}(-\langle G_{\lambda_0}, (-\Delta+\lambda_0)A\rangle + \lambda_0\langle G_{\lambda_0}, A\rangle)\\ =&12\pi\sqrt{\lambda_0}A(0) + \lambda_0 P_A\ ,\end{aligned}$$

and

$$P_{MA(0)G_{\lambda_0}} = -12\pi\sqrt{\lambda_0}\frac{2}{3}\,\frac{1}{8\pi\sqrt{\lambda_0}}\,A(0) = -A(0)\ ,$$

one has

$$\begin{aligned}&4\pi\langle H_{\rm nr}A, A\rangle - \frac{1}{12\pi\sqrt{\lambda_0}}\,P_{4\pi H_{\rm nr}A}\cdot P_A\\ =&-\,\langle \Delta A, A\rangle - \frac{1}{2}\,A(0)\cdot P_A\\ &+\frac{1}{12\pi\sqrt{\lambda_0}}\,\left(12\pi\sqrt{\lambda_0}A(0) + \lambda_0 P_A\right)\cdot P_A - \frac{1}{2}\,A(0)\cdot P_A\\ =&-\,\langle \Delta A, A\rangle + \frac{\sqrt{\lambda_0}}{12\pi}\,|P_A|^2\ .\end{aligned}$$

3. $X_{\mathcal{H}_{\rm nr}}$ generates a group of continuous (w.r.t. the Hilbert norm on $H^1_*(\mathbb{R}^3)\oplus L^2_*(\mathbb{R}^3)$) linear transformations since we can write

$$4\pi H_{\rm nr} + \lambda_0 = (1 - 6\pi\sqrt{\lambda_0}G_{\lambda_0}\otimes G_{\lambda_0})\circ(-\Delta+\lambda_0)\ .$$

Moreover such transformations are symplectic, since $X_{\mathcal{H}_{\rm nr}}$ is Hamiltonian. Formula (4.2) follows from the definitions of $\Omega_{\rm nr}$ and $\mathcal{H}_{nr}$, and (4.3) follows from (4.2) by the unicity of generators. Finally, (4.4) follows from (4.1) and the definition of $X_{\mathcal{H}_{nr}}$. □

References

[1] Albeverio S., Gesztesy F., Høegh–Krohn R., Holden H.: *Solvable Models in Quantum Mechanics.* New York: Springer–Verlag 1988

[2] Noja D., Posilicano A.: The Wave Equation with One Point Interaction and the (Linearized) Classical Electrodynamics of a Point Particle. *Ann. Inst. Henri Poincaré* **68** (1998), pp. 351–377

[3] Noja D., Posilicano A.: On the Point Limit of the Pauli–Fierz Model. *Ann. Inst. Henri Poincaré* **71** (1999), pp. 425–457

[4] Teta A.: Quadratic Forms for Singular Perturbations of the Laplacian. *Publ. RIMS Kyoto Univ.* **26** (1990), pp. 803–819

Dipartimento di Matematica dell'Università, Via Saldini 50, I–20133 Milano, Italy
E-mail address: noja@berlioz.mat.unimi.it

Dipartimento di Scienze, Università dell'Insubria, Via Lucini 3, I–22100 Como, Italy
E-mail address: posilicano@mat.unimi.it

Canadian Mathematical Society
Conference Proceedings
Volume **29**, 2000

Coherent State Representations in White Noise Calculus

Nobuaki Obata

Dedicated to Professor Sergio Albeverio on the occasion of his 60th birthday.

ABSTRACT. An infinite dimensional analogue of the coherent state representation in quantum mechanics is discussed on the basis of the recent framework of white noise distribution theory on a complex Gaussian space. A resolution of the identity and the diagonal coherent state representation of an arbitrary white noise operator are proved. Examples include a differential operator, a translation operator, the number operator, the Gross Laplacians, etc. As an application, inversion formulas for the S-transform and the operator symbol are derived.

1. Introduction

In contemporary literatures the appellation "coherent states" appears in various contexts with wide applications (see e.g., [**7**], [**15**]), however, in this paper we focus on the so-called "canonical" coherent states generated by annihilation and creation operators of Boson field. We first recall the most basic formula in the simplest case. Let a and a^* denote respectively the annihilation and creation operators satisfying the canonical commutation relation $[a, a^*] = I$. Then the Boson Fock space $\Gamma(\mathbf{C})$ is constructed from the Fock vacuum $|0\rangle$ with the action of the creation operator. Then, for any $z \in \mathbf{C}$ we define

$$|z\rangle = e^{za^* - \bar{z}a}\,|0\rangle = e^{-|z|^2/2} \sum_{n=0}^{\infty} \frac{z^n}{\sqrt{n!}}\,|n\rangle,$$

where $|n\rangle \in \Gamma(\mathbf{C})$ are the number vectors. The above $|z\rangle \in \Gamma(\mathbf{C})$ is called a (normalized) *coherent state*. A coherent state is also characterized as an eigenvector of a with eigenvalue z, i.e., $a|z\rangle = z|z\rangle$. It is well known that $\{|z\rangle\,;\, z \in \mathbf{C}\}$ is linearly independent and is overcomplete. Moreover, the resolution of the identity

$$I = \frac{1}{\pi} \int_{\mathbf{C}} |z\rangle\langle z|\, d^2 z \tag{1.1}$$

1991 *Mathematics Subject Classification.* 46F25, 60G15, 81R30.

Work supported in part by Grant-in-Aid for Scientific Research No. 09440057, Ministry of Education, Japan.

holds, where $|z\rangle\langle z|$ stands for the one-dimensional projection onto the subspace spanned by $|z\rangle$ and d^2z is the Lebesgue measure on $\mathbf{C}$. The above formula (1.1), due to Klauder [**14**], is an essential property of the coherent states and has extensive applications in quantum physics, e.g., in quantum optics [**8**], [**16**], [**29**].

In this paper we discuss an infinite dimensional generalization of the formula (1.1), of which derivation is intuitively straightforward. Namely, the Lebesgue measure d^2z is replaced with the complex Gaussian white noise measure on the infinite dimensional complex vector space $E^*_{\mathbf{C}} = \mathcal{S}'(\mathbf{R}) + i\mathcal{S}'(\mathbf{R})$, and the normalized coherent states $|z\rangle$ with non-normalized coherent states or exponential vectors ϕ_z in the Boson Fock space (for the precise statement see Theorem 6.3). This procedure is rigorously performed by means of white noise distribution theory [**11**], [**21**], [**24**], or more precisely on the basis of the most recent framework given by Cochran, Kuo and Sengupta [**6**] and on the associated operator theory [**4**]. We then derive the diagonal coherent state representation of an arbitrary white noise operator with some examples including a differential operator D_y, a translation operator T_y, the number operator N, the Gross Laplacian Δ_G, and so on. We also show some properties pertaining to the diagonal coherent state representation, in particular, the inversion formulas for the S-transform and for the operator symbol, both of which are expected to have many applications, e.g., in a study of classical or quantum stochastic differential equations.

The complex white noise was already introduced at an early stage of white noise calculus [**9**], [**10**] and has been discussed in connection with complex Hermite polynomials [**13**], with Segal–Bargmann transform [**20**], [**30**], and so forth. Our results in this paper (see also [**25**], [**26**]) shed light on a different role of complex white noise and will exploit some applications, e.g., a direct approach to some questions of coherent states for infinitely many degrees of freedom [**16**, Chapter 7]. Moreover, it is our hope that non-Gaussian generalization [**1**], [**2**] will be related to our approach. Further investigation is now in progress.

2. Complex Gaussian Space

Although more general formulation is possible, our consideration will be based on a particular real Gelfand triple:

$$E \equiv \mathcal{S}(\mathbf{R}) \subset H \equiv L^2(\mathbf{R}, dt) \subset E^* \equiv \mathcal{S}'(\mathbf{R}), \tag{2.1}$$

where $\mathcal{S}(\mathbf{R})$ is the space of rapidly decreasing functions and $\mathcal{S}'(\mathbf{R})$ the dual space, i.e., the space of tempered distributions. We denote by $\langle \cdot, \cdot \rangle$ the canonical bilinear form on $E^* \times E$ and by $|\cdot|_0$ the norm of H. For notational convenience, the $\mathbf{C}$-bilinear extension on $E^*_{\mathbf{C}} \times E_{\mathbf{C}}$ is denoted by the same symbol so that $|\xi|_0^2 = \langle \bar{\xi}, \xi \rangle$ holds for $\xi \in H_{\mathbf{C}}$.

Let μ' be the Gaussian measure on E^* with variance $\sigma^2 = 1/2$, namely, a probability measure on E^* determined uniquely by the characteristic function:

$$e^{-|\xi|_0^2/4} = \int_{E^*} e^{i\langle x, \xi \rangle} \mu'(dx), \qquad \xi \in E.$$

Let $\Gamma(H_{\mathbf{C}})$ be the Boson Fock space over $H_{\mathbf{C}}$. Then, through the celebrated Wiener–Itô–Segal isomorphism we have

$$L^2(E^*, \mu') \cong \Gamma(H_{\mathbf{C}}), \tag{2.2}$$

where the isometric isomorphism is uniquely determined by the correspondence:

$$\psi_\xi(x) \equiv e^{\sqrt{2}\langle x,\,\xi\rangle - \langle \xi,\,\xi\rangle/2} \leftrightarrow \phi_\xi \equiv \left(1, \xi, \frac{\xi^{\otimes 2}}{2!}, \cdots, \frac{\xi^{\otimes n}}{n!}, \cdots\right), \qquad \xi \in E_{\mathbf{C}}. \tag{2.3}$$

The above ϕ_ξ is called an *exponential vector* or a *coherent state* (without normalization). Note that

$$\langle\langle \phi_\xi,\, \phi_\eta \rangle\rangle = e^{\langle \xi,\,\eta\rangle}, \qquad \xi, \eta \in E_{\mathbf{C}},$$

where the left hand side stands for the canonical $\mathbf{C}$-bilinear form on $\Gamma(H_{\mathbf{C}})$.

In view of the topological isomorphism $E_{\mathbf{C}}^* \cong E^* \times E^*$, we define a probability measure $\nu = \mu' \times \mu'$ on $E_{\mathbf{C}}^*$ by

$$\nu(dz) = \mu'(dx)\mu'(dy), \qquad z = x + iy \in E_{\mathbf{C}}^*.$$

Following Hida [**10**, Chapter 6] the probability space $(E_{\mathbf{C}}^*, \nu)$ is called the *complex Gaussian space.* We note the following formula:

$$\int_{E_{\mathbf{C}}^*} e^{\langle \bar{z},\,\xi\rangle + \langle z,\,\eta\rangle}\, \nu(dz) = e^{\langle \xi,\,\eta\rangle}, \qquad \xi, \eta \in E_{\mathbf{C}}, \tag{2.4}$$

where $\bar{z} = x - iy$ for $z = x + iy \in E^* + iE^*$. Recall that $\langle \cdot,\, \cdot \rangle$ is the canonical $\mathbf{C}$-bilinear form on $E_{\mathbf{C}}^* \times E_{\mathbf{C}}$.

3. A Review of CKS-Space

First a general notation is needed. For a Hilbert space H and a sequence $\alpha = \{\alpha(n)\}_{n=0}^\infty$ of positive numbers we put

$$\Gamma_\alpha(H) = \left\{ \phi = (f_n)_{n=0}^\infty \,;\, f_n \in H^{\widehat{\otimes} n},\ \|\phi\|^2 \equiv \sum_{n=0}^\infty n!\,\alpha(n)|f_n|^2 < \infty \right\}.$$

In an obvious manner $\Gamma_\alpha(H)$ becomes a Hilbert space and is called a *weighted Fock space.* The Boson Fock space is by definition the special case of $\alpha(n) \equiv 1$ and is denoted by $\Gamma(H)$. In the sequel we assume that the weight sequence $\alpha = \{\alpha(n)\}_{n=0}^\infty$ satisfies the following conditions:

(A1) $\alpha(0) = 1$ and $\inf_n \alpha(n) > 0$;

(A2) $G_\alpha(t) \equiv \displaystyle\sum_{n=0}^\infty \frac{\alpha(n)}{n!} t^n$ has an infinite radius of convergence;

(A3) $\widetilde{G}_\alpha(t) \equiv \displaystyle\sum_{n=0}^\infty t^n \frac{n^{2n}}{n!\alpha(n)} \left\{ \inf_{s>0} \frac{G_\alpha(s)}{s^n} \right\}$ has a positive radius of convergence;

(A4) there exists a constant $C_1 > 0$ such that $\alpha(n)\alpha(m) \le C_1^{n+m}\alpha(n+m)$ for any n, m;

(A5) there exists a constant $C_2 > 0$ such that $\alpha(n+m) \le C_2^{n+m}\alpha(n)\alpha(m)$ for any n, m;

(A6) $\{\alpha(n)\}$ is non-decreasing.

Conditions (A1)–(A3) are needed for the characterization for S-transform (see [**6**]) and (A4)–(A6) for white noise operator theory (see [**4**]). An almost ultimate description of conditions for $\{\alpha(n)\}$ has been investigated very recently, however, in this paper we use the above conditions for simplicity. (The labels (A1)–(A6) are kept the same as in the previous works [**4**], [**25**], [**26**] for convenience.)

For $p \geq 0$ let $E_{\pm p}$ be the Hilbert space obtained by completing $E = \mathcal{S}(\mathbf{R})$ with respect to the norm $|\xi|_{\pm p} = |A^{\pm p}\xi|_H$, where $A = 1 + t^2 - d^2/dt^2$. We then have

$$E = \operatorname*{proj\,lim}_{p\to\infty} E_p, \qquad E^* = \operatorname*{ind\,lim}_{p\to\infty} E_{-p}.$$

Given a sequence $\alpha = \{\alpha(n)\}_{n=0}^{\infty}$ satisfying the above conditions, we put

$$\Gamma_\alpha(E) = \operatorname*{proj\,lim}_{p\to\infty} \Gamma_\alpha(E_p). \tag{3.1}$$

Then $\Gamma_\alpha(E)$ becomes a nuclear space, the topology of which is given by the family of norms:

$$\|\phi\|_{p,+}^2 = \sum_{n=0}^{\infty} n!\,\alpha(n)|f_n|_p^2, \qquad \phi = (f_n), \quad p \geq 0.$$

By a standard argument we see that

$$\Gamma_\alpha(E)^* \cong \operatorname*{ind\,lim}_{p\to\infty} \Gamma_{\alpha^{-1}}(E_{-p}), \tag{3.2}$$

where $\Gamma_\alpha(E)^*$ carries the strong dual topology and $\cong$ stands for a topological linear isomorphism.

In what follows, we write $\mathcal{W}$ for $\Gamma_\alpha(E)$ for simplicity Then, in view of (3.1) and (3.2) we come to a Gelfand triple:

$$\mathcal{W} \subset \Gamma(H_{\mathbf{C}}) \subset \mathcal{W}^*, \tag{3.3}$$

which is called the *Cochran–Kuo–Sengupta space* or the *CKS-space* for short [**6**]. The canonical $\mathbf{C}$-bilinear form on $\mathcal{W}^* \times \mathcal{W}$ is denoted also by $\langle\!\langle \cdot, \cdot \rangle\!\rangle$. Then

$$\langle\!\langle \Phi, \phi \rangle\!\rangle = \sum_{n=0}^{\infty} n!\,\langle F_n, f_n\rangle, \qquad \Phi = (F_n) \in \mathcal{W}^*, \quad \phi = (f_n) \in \mathcal{W}, \tag{3.4}$$

and it holds that

$$|\langle\!\langle \Phi, \phi \rangle\!\rangle| \leq \|\Phi\|_{-p,-}\|\phi\|_{p,+},$$

where

$$\|\Phi\|_{-p,-}^2 = \sum_{n=0}^{\infty} \frac{n!}{\alpha(n)}|F_n|_{-p}^2, \qquad \Phi = (F_n) \in \mathcal{W}^*.$$

Concrete examples of the weight sequence $\{\alpha(n)\}$ are $\alpha(n) \equiv 1$ (Kubo and Takenaka [**19**]), $\alpha(n) = (n!)^\beta$ for $0 \leq \beta < 1$ (Kondratiev and Streit [**18**]), the k-th order Bell numbers $B_k(n)$ (Cochran, Kuo and Sengupta [**6**]), and so on.

4. CKS-Space over Complex Gaussian Space

By duplicating the Gelfand triple (3.3) we obtain

$$\mathcal{W} \otimes \mathcal{W} \subset \Gamma(H_{\mathbf{C}}) \otimes \Gamma(H_{\mathbf{C}}) \subset (\mathcal{W} \otimes \mathcal{W})^*. \tag{4.1}$$

On the other hand, identifying a function on $E_{\mathbf{C}}^*$ with one on $E^* \times E^*$ in such a way that

$$\phi \otimes \psi(x + iy) = \phi(x)\psi(y), \qquad x, y \in E^*, \quad \phi, \psi \in L^2(E^*, \mu'),$$

we come to the isomorphisms:

$$\Gamma(H_{\mathbf{C}}) \otimes \Gamma(H_{\mathbf{C}}) \cong L^2(E^*, \mu') \otimes L^2(E^*, \mu') \cong L^2(E_{\mathbf{C}}^*, \nu). \tag{4.2}$$

Combination of (4.1) and (4.2) gives rise to a Gelfand triple which we shall denote by

$$\mathcal{D} \subset L^2(E^*_{\mathbf{C}}, \nu) \subset \mathcal{D}^*. \tag{4.3}$$

Since each equivalence class contains a unique continuous function on $E^*_{\mathbf{C}}$, we may regard $\mathcal{D}$ as a space of continuous functions on $E^*_{\mathbf{C}}$. More detailed properties of functions in $\mathcal{D}$ are examined in a similar manner as in [**22**], [**28**].

5. White Noise Operators on CKS-Space

In general, a continuous operator from $\mathcal{W}$ into $\mathcal{W}^*$ is called a *white noise operator* and is a main subject to study in this paper. Note that both $\mathcal{L}(\mathcal{W},\mathcal{W})$ and $\mathcal{L}(\mathcal{W},\Gamma(H_{\mathbf{C}}))$ are subspaces of $\mathcal{L}(\mathcal{W},\mathcal{W}^*)$. Moreover, so is the space of all bounded operators $\mathcal{L}(\Gamma(H_{\mathbf{C}}),\Gamma(H_{\mathbf{C}}))$. Note also that $\mathcal{L}(\mathcal{W}^*,\mathcal{W}^*)$ is isomorphic to $\mathcal{L}(\mathcal{W},\mathcal{W})$ by duality.

We start with the most basic white noise operators. An integral kernel operator with kernel distribution $\kappa_{l,m} \in (E_{\mathbf{C}}^{\otimes(l+m)})^*$ admits a formal integral expression:

$$\Xi_{l,m}(\kappa_{l,m}) = \tag{5.1}$$
$$= \int_{\mathbf{R}^{l+m}} \kappa_{l,m}(s_1,\cdots,s_l,t_1,\cdots,t_m)a^*_{s_1}\cdots a^*_{s_l}a_{t_1}\cdots a_{t_m}ds_1\cdots ds_l dt_1\cdots dt_m.$$

The precise definition is as follows, for more details see [**24**, Chapter 4]. Let $\phi = (f_n) \in \mathcal{W}$. Then $\Xi_{l,m}(\kappa_{l,m})\phi = (g_n)$ is defined by

$$g_n = 0, \quad 0 \le n < l; \qquad g_n = \frac{(n-l+m)!}{(n-l)!}\,\kappa \otimes_m f_{n-l+m}, \quad n \ge l,$$

where $\otimes_m$ denotes the right contraction of tensor products. It is shown that an integral kernel operator $\Xi_{l,m}(\kappa_{l,m})$ always belongs to $\mathcal{L}(\mathcal{W},\mathcal{W}^*)$ for any $\kappa_{l,m} \in (E_{\mathbf{C}}^{\otimes(l+m)})^*$. Moreover, it is proved [**4**] that any $\Xi \in \mathcal{L}(\mathcal{W},\mathcal{W}^*)$ admits an infinite series expansion:

$$\Xi = \sum_{l,m=0}^{\infty} \Xi_{l,m}(\kappa_{l,m}), \tag{5.2}$$

where the right hand side converges in $\mathcal{L}(\mathcal{W},\mathcal{W}^*)$. Notice that the class $\mathcal{L}(\mathcal{W},\mathcal{W}^*)$ contains all bounded operators on $\Gamma(H_{\mathbf{C}})$. In this connection we only mention the following

PROPOSITION 5.1. [**5**], [**24**] *Let* $\kappa \in (E_{\mathbf{C}}^{\otimes(l+m)})^*$.
(1) $\Xi_{l,m}(\kappa) \in \mathcal{L}(\mathcal{W},\mathcal{W})$ *if and only if* $\kappa \in E_{\mathbf{C}}^{\otimes l} \otimes (E_{\mathbf{C}}^{\otimes m})^*$.
(2) $\Xi_{l,m}(\kappa) \in \mathcal{L}(\mathcal{W},\Gamma(H_{\mathbf{C}}))$ *if and only if* $\kappa \in H_{\mathbf{C}}^{\otimes l} \otimes (E_{\mathbf{C}}^{\otimes m})^*$.

The *symbol* of $\Xi \in \mathcal{L}(\mathcal{W},\mathcal{W}^*)$ is by definition a $\mathbf{C}$-valued function on $E_{\mathbf{C}} \times E_{\mathbf{C}}$ defined by

$$\widehat{\Xi}(\xi,\eta) = \langle\langle \Xi\phi_\xi,\, \phi_\eta \rangle\rangle, \qquad \xi,\eta \in E_{\mathbf{C}}.$$

Every operator in $\mathcal{L}(\mathcal{W},\mathcal{W}^*)$ is uniquely determined by its symbols since the set of exponential vectors $\{\phi_\xi\,;\,\xi \in E_{\mathbf{C}}\}$ spans a dense subspace of $\mathcal{W}$. For an integral kernel operator $\Xi_{l,m}(\kappa)$ we have

$$\widehat{\Xi_{l,m}(\kappa)}(\xi,\eta) = \langle\langle \Xi_{l,m}(\kappa)\phi_\xi,\, \phi_\eta \rangle\rangle = \left\langle \kappa,\, \eta^{\otimes l} \otimes \xi^{\otimes m} \right\rangle e^{\langle \xi,\, \eta \rangle}, \qquad \xi,\eta \in E_{\mathbf{C}}. \tag{5.3}$$

For $\Xi \in \mathcal{L}(\mathcal{W},\mathcal{W}^*)$, the symbol $\Theta = \widehat{\Xi}$ obviously satisfies:

(O1) for any $\xi, \xi_1, \eta, \eta_1 \in E_{\mathbf{C}}$ the function $(z, w) \mapsto \Theta(z\xi + \xi_1, w\eta + \eta_1)$ is entire holomorphic on $\mathbf{C} \times \mathbf{C}$;

(O2) there exist constant numbers $C \ge 0$ and $p \ge 0$ such that

$$|\Theta(\xi, \eta)|^2 \le C G_\alpha(|\xi|_p^2) G_\alpha(|\eta|_p^2), \qquad \xi, \eta \in E_{\mathbf{C}}.$$

More important, the converse is also true. The following assertion, known as analytic characterization of operator symbols, shows one of the most significant features of white noise calculus.

THEOREM 5.2. [**4**] *A function $\Theta : E_{\mathbf{C}} \times E_{\mathbf{C}} \to \mathbf{C}$ is the symbol of an operator $\Xi \in \mathcal{L}(\mathcal{W}, \mathcal{W}^*)$ if and only if (O1) and (O2) are satisfied. In that case*

$$\|\Xi\phi\|_{-(p+q),-}^2 \le C \widetilde{G}_\alpha^2(\|A^{-q}\|_{\mathrm{HS}}^2) \|\phi\|_{p+q,+}^2, \qquad \phi \in \mathcal{W}, \tag{5.4}$$

where $q > 1/2$ is taken as $\widetilde{G}_\alpha(\|A^{-q}\|_{\mathrm{HS}}^2) < \infty$. (This choice is always possible since $\|A^{-q}\|_{\mathrm{HS}} \to 0$ as $q \to \infty$.)

We note that $\Xi \in \mathcal{L}(\mathcal{W}, \mathcal{W}^*)$ is not uniquely specified by the diagonal elements $\{\widehat{\Xi}(\xi, \xi) \,;\, \xi \in E_{\mathbf{C}}\}$ nor $\{\widehat{\Xi}(\bar{\xi}, \xi) \,;\, \xi \in E_{\mathbf{C}}\}$. (The description in [**16**, Chapter 7.2.C] is misleading.) In fact, fix $\zeta \in E_{\mathbf{C}}$ and consider an operator

$$\Xi\phi = (\langle \zeta, f_1 \rangle, -f_0\zeta, 0, \cdots), \qquad \phi = (f_0, f_1, \cdots) \in \mathcal{W}.$$

Then, $\Xi \in \mathcal{L}(\mathcal{W}, \mathcal{W})$ and

$$\widehat{\Xi}(\xi, \eta) = \langle \xi, \zeta \rangle - \langle \eta, \zeta \rangle, \qquad \xi, \eta \in E_{\mathbf{C}}.$$

Consequently, $\widehat{\Xi}(\xi, \xi) = 0$ for all $\xi \in E_{\mathbf{C}}$ but $\Xi \ne 0$.

6. Diagonal Coherent State Representations

For any $z \in E_{\mathbf{C}}^*$ the exponential vector $\phi_z \in \mathcal{W}^*$ is defined by the same formula as in (2.3). We then put

$$Q_z\phi = \langle\!\langle \phi_{\bar{z}}, \phi \rangle\!\rangle \, \phi_z, \qquad \phi \in \mathcal{W}.$$

As is easily verified, $Q_z \in \mathcal{L}(\mathcal{W}, \mathcal{W}^*)$, and both maps $z \mapsto \phi_z \in \mathcal{W}^*$ and $z \mapsto Q_z \in \mathcal{L}(\mathcal{W}, \mathcal{W}^*)$ are continuous. Similarly, for $z_1, z_2 \in E_{\mathbf{C}}^*$ we put

$$R_{z_2,z_1}\phi = \langle\!\langle \phi_{\bar{z}_1}, \phi \rangle\!\rangle \, \phi_{z_2}, \qquad \phi \in \mathcal{W}.$$

Then $R_{z_2,z_1} \in \mathcal{L}(\mathcal{W}, \mathcal{W}^*)$. Here we record some useful formulas, the proofs of which are straightforward from definition:

$$R_{z,z} = Q_z, \qquad R_{z_2,z_1}^* = R_{\bar{z}_1,\bar{z}_2}, \qquad Q_z^* = Q_{\bar{z}},$$

$$Q_{z_2}Q_{z_1} = e^{\langle \bar{z}_2, z_1 \rangle} R_{z_2,z_1}.$$

The symbols are also easily computed:

$$\widehat{R}_{z_2,z_1}(\xi, \eta) = \langle\!\langle \phi_{\bar{z}_1}, \phi_\xi \rangle\!\rangle \langle\!\langle \phi_{z_2}, \phi_\eta \rangle\!\rangle = e^{\langle \bar{z}_1, \xi \rangle + \langle z_2, \eta \rangle}, \tag{6.1}$$

$$\widehat{Q}_z(\xi, \eta) = e^{\langle \bar{z}, \xi \rangle + \langle z, \eta \rangle}. \tag{6.2}$$

We are now in a position to discuss the coherent state representations for operators. For simplicity we put

$$q_{\xi,\eta}(z) = \widehat{Q}_z(\xi, \eta) = e^{\langle \bar{z}, \xi \rangle + \langle z, \eta \rangle}, \qquad z \in E_{\mathbf{C}}^*, \quad \xi, \eta \in E_{\mathbf{C}}. \tag{6.3}$$

Then, for $z = x + iy$ we obtain

$$q_{\xi,\eta}(x+iy) = e^{\langle x,\, \xi+\eta\rangle} e^{\langle y,\, i(-\xi+\eta)\rangle} = e^{\langle \xi,\, \eta\rangle} \psi_{(\xi+\eta)/\sqrt{2}}(x)\psi_{i(-\xi+\eta)/\sqrt{2}}(y), \tag{6.4}$$

where ψ_ξ is defined as in (2.3). In other words, $q_{\xi,\eta} = e^{\langle \xi,\, \eta\rangle}\psi_{(\xi+\eta)/\sqrt{2}} \otimes \psi_{i(-\xi+\eta)/\sqrt{2}}$ and hence $q_{\xi,\eta} \in \mathcal{W} \otimes \mathcal{W} \cong \mathcal{D}$.

LEMMA 6.1. *For $w \in \mathcal{D}^*$ there is a unique operator $\Xi \in \mathcal{L}(\mathcal{W}, \mathcal{W}^*)$ such that*

$$\langle\!\langle \Xi\phi_\xi,\, \phi_\eta \rangle\!\rangle = \langle\!\langle w,\, q_{\xi,\eta} \rangle\!\rangle, \qquad \xi, \eta \in E_{\mathbf{C}}. \tag{6.5}$$

PROOF. By definition we have

$$\begin{aligned} |\langle\!\langle w,\, q_{\xi,\eta}\rangle\!\rangle|^2 &= \left| e^{\langle \xi,\, \eta\rangle} \langle\!\langle w,\, \psi_{(\xi+\eta)/\sqrt{2}} \otimes \psi_{i(-\xi+\eta)/\sqrt{2}} \rangle\!\rangle \right|^2 \\ &\le e^{2|\langle \xi,\, \eta\rangle|} \|w\|_{-p}^2 \|\psi_{(\xi+\eta)/\sqrt{2}}\|_p^2 \, \|\psi_{i(-\xi+\eta)/\sqrt{2}}\|_p^2 \\ &= e^{2|\langle \xi,\, \eta\rangle|} \|w\|_{-p}^2 G_\alpha\Big(|(\xi+\eta)/\sqrt{2}|_p^2\Big)\, G_\alpha\Big(|i(-\xi+\eta)/\sqrt{2}|_p^2\Big), \end{aligned} \tag{6.6}$$

where $p \ge 0$ is chosen as $\|w\|_{-p} < \infty$. It follows from conditions (A4) and (A5) mentioned in §2 that

$$G_\alpha(s+t) \le G_\alpha(C_2 s) G_\alpha(C_2 t), \quad G_\alpha(s) G_\alpha(t) \le G_\alpha(C_1(s+t)), \quad s, t \ge 0,$$

and from (A6) that

$$e^s G_\alpha(t) \le G_\alpha(s+t), \qquad s, t \ge 0.$$

Moreover, in view of $|\xi|_p \le 2^{-q} |\xi|_{p+q}$ for $\xi \in E_{\mathbf{C}}$ and $q \ge 0$, we see that (6.6) yields an estimate of the form:

$$|\langle\!\langle w,\, q_{\xi,\eta}\rangle\!\rangle|^2 \le \|w\|_{-p}^2 G_\alpha(|\xi|_{p+q}^2) G_\alpha(|\eta|_{p+q}^2).$$

Then the assertion is a consequence of the characterization theorem for operator symbols (Theorem 5.2). □

The operator Ξ defined as in (6.5) is written in a formal integral:

$$\Xi = \int_{E_{\mathbf{C}}^*} w(z) Q_z \, \nu(dz) \tag{6.7}$$

and is called the *diagonal coherent state representation* of Ξ.

THEOREM 6.2. *Every operator in $\mathcal{L}(\mathcal{W}, \mathcal{W}^*)$ admits a unique diagonal coherent state representation.*

PROOF. Given $\Xi \in \mathcal{L}(\mathcal{W}, \mathcal{W}^*)$ we consider

$$\Theta(\xi, \eta) = \langle\!\langle \Xi\phi_{(\xi+i\eta)/\sqrt{2}},\, \phi_{(\xi-i\eta)/\sqrt{2}} \rangle\!\rangle e^{-\langle \xi+i\eta,\, \xi-i\eta\rangle/2}, \qquad \xi, \eta \in E_{\mathbf{C}}. \tag{6.8}$$

Modelled after the proof of Lemma 6.1, we see that there exists $W \in \mathcal{L}(\mathcal{W}, \mathcal{W}^*)$ such that

$$\Theta(\xi, \eta) = \langle\!\langle W\phi_\xi,\, \phi_\eta \rangle\!\rangle, \qquad \xi, \eta \in E_{\mathbf{C}}. \tag{6.9}$$

It then follows by changing the parameters in (6.8) and (6.9) that

$$\langle\!\langle \Xi\phi_\xi,\, \phi_\eta \rangle\!\rangle = \langle\!\langle W\phi_{(\xi+\eta)/\sqrt{2}},\, \phi_{i(-\xi+\eta)/\sqrt{2}} \rangle\!\rangle\, e^{\langle \xi,\, \eta\rangle}, \qquad \xi, \eta \in E_{\mathbf{C}}.$$

Choose $w \in (\mathcal{W} \otimes \mathcal{W})^*$ corresponding to $W \in \mathcal{L}(\mathcal{W}, \mathcal{W}^*)$ in accordance with the canonical isomorphism $(\mathcal{W} \otimes \mathcal{W})^* \cong \mathcal{L}(\mathcal{W}, \mathcal{W}^*)$. Then,

$$\langle\!\langle \Xi\phi_\xi,\, \phi_\eta \rangle\!\rangle = \langle\!\langle w,\, \phi_{(\xi+\eta)/\sqrt{2}} \otimes \phi_{i(-\xi+\eta)/\sqrt{2}} \rangle\!\rangle\, e^{\langle \xi,\, \eta\rangle}. \tag{6.10}$$

Now, taking (2.3) into account, we consider the functional realization of the right hand side of (6.10). As is seen already in (6.3) and (6.4), we have for $z = x + iy$,

$$\psi_{(\xi+\eta)/\sqrt{2}}(x)\psi_{i(-\xi+\eta)/\sqrt{2}}(y)e^{\langle \xi, \eta \rangle} = q_{\xi,\eta}(x+iy) = \langle\!\langle Q_z\phi_\xi, \phi_\eta \rangle\!\rangle .$$

Hence (6.10) is written in a formal integral

$$\langle\!\langle \Xi\phi_\xi, \phi_\eta \rangle\!\rangle = \int_{E^*_{\mathbf{C}}} w(z) \langle\!\langle Q_z\phi_\xi, \phi_\eta \rangle\!\rangle \, \nu(dz), \tag{6.11}$$

which means that Ξ admits a diagonal coherent state representation as in (6.7).

The uniqueness is already shown since w is uniquely specified essentially by the symbol of Ξ. A direct proof is also possible. From (6.3) and (6.5) we see that for $w \in \mathcal{D}^*$,

$$\int_{E^*_{\mathbf{C}}} w(z)Q_z \, \nu(dz) = 0$$

holds if and only if

$$\int_{E^*_{\mathbf{C}}} w(z)e^{\langle \bar{z}, \xi \rangle + \langle z, \eta \rangle} \nu(dz) = 0, \qquad \xi, \eta \in E_{\mathbf{C}}.$$

In that case, by differentiation we obtain

$$\int_{E^*_{\mathbf{C}}} w(z) \langle \bar{z}, \xi \rangle^l \langle z, \eta \rangle^m \nu(dz) = 0$$

for any $l, m \geq 0$ and $\xi, \eta \in E_{\mathbf{C}}$. Hence $w = 0$ which proves the uniqueness of the diagonal coherent state representation. □

As a particular case of Theorem 6.2, we obtain the resolution of the identity.

THEOREM 6.3. *It holds that*

$$I = \int_{E^*_{\mathbf{C}}} Q_z \, \nu(dz). \tag{6.12}$$

PROOF. We write Ξ for the right hand side. By means of formulas (6.3) and (2.4) the symbol of Ξ is computed as follows:

$$\begin{aligned} \widehat{\Xi}(\xi, \eta) &= \langle\!\langle \Xi\phi_\xi, \phi_\eta \rangle\!\rangle = \int_{E^*_{\mathbf{C}}} \langle\!\langle Q_z\phi_\xi, \phi_\eta \rangle\!\rangle \, \nu(dz) \\ &= \int_{E^*_{\mathbf{C}}} e^{\langle \bar{z}, \xi \rangle + \langle z, \eta \rangle} \, \nu(dz) = e^{\langle \xi, \eta \rangle} = \langle\!\langle \phi_\xi, \phi_\eta \rangle\!\rangle . \end{aligned}$$

Hence $\Xi = I$ as desired. □

7. Examples And Comments

For $y \in E^*_{\mathbf{C}}$ consider the integral kernel operator

$$\Xi_{0,1}(y) = \int_{\mathbf{R}} y(t)a_t \, dt. \tag{7.1}$$

Recall that $\Xi_{0,1}(y) \in \mathcal{L}(\mathcal{W}, \mathcal{W})$. For a real $y \in E^*$ we also write $D_y = \Xi_{0,1}(y)$. It is known that D_y is the differential operator:

$$D_y\phi(x) = \lim_{\theta \to 0} \frac{\phi(x + \theta y) - \phi(x)}{\theta}, \qquad \phi \in \mathcal{W}, \quad x \in E^*.$$

(The limit always exists.) In particular, $a_t = D_{\delta_t}$ for $t \in \mathbf{R}$. We now show that

$$\Xi_{0,1}(y) = \int_{E^*_{\mathbf{C}}} \langle z, y \rangle \, Q_z \, \nu(dz). \tag{7.2}$$

Since $w(z) = \langle z, y \rangle$ belongs to $\mathcal{D}^*$ by definition, the right hand side of (7.2), denoted by Ξ, is an operator in $\mathcal{L}(\mathcal{W}, \mathcal{W}^*)$ characterized by

$$\widehat{\Xi}(\xi, \eta) = \int_{E^*_{\mathbf{C}}} \langle z, y \rangle \, e^{\langle \bar{z}, \xi \rangle + \langle z, \eta \rangle} \nu(dz), \qquad \xi, \eta \in E_{\mathbf{C}}, \tag{7.3}$$

where the integral is understood through the canonical pairing on $\mathcal{D}^* \times \mathcal{D}$. Using the formula:

$$\int_{E^*} \langle x, \xi \rangle \, e^{\langle x, \eta \rangle} \mu'(dx) = \frac{1}{2} \, \langle \xi, \eta \rangle \, e^{\langle \eta, \eta \rangle / 4}, \qquad \xi, \eta \in E_{\mathbf{C}},$$

the derivation of which is straightforward, one can compute (7.3) easily to obtain

$$\widehat{\Xi}(\xi, \eta) = \langle y, \xi \rangle \, e^{\langle \xi, \eta \rangle} = \Xi_{0,1}(y)\widehat{\ }(\xi, \eta), \qquad \xi, \eta \in E_{\mathbf{C}},$$

which proves (7.2). At a somewhat formal level one may obtain (7.2) easily by observing

$$\Xi_{0,1}(y) Q_z = \langle z, y \rangle \, Q_z$$

and by the resolution of the identity. (If $y \in E_{\mathbf{C}}$ the above formula is valid in a usual sense.)

In a similar manner, we have

$$\Xi_{0,m}(\kappa) = \int_{E^*_{\mathbf{C}}} \left\langle z^{\otimes m}, \kappa \right\rangle Q_z \nu(dz), \qquad \kappa \in (E_{\mathbf{C}}^{\otimes m})^*. \tag{7.4}$$

By duality and the invariance $\nu(dz) = \nu(d\bar{z})$ we also obtain

$$\Xi_{l,0}(\kappa) = \int_{E^*_{\mathbf{C}}} \left\langle \bar{z}^{\otimes l}, \kappa \right\rangle Q_z \nu(dz), \qquad \kappa \in (E_{\mathbf{C}}^{\otimes l})^*.$$

In particular,

$$a_t = \int_{E^*_{\mathbf{C}}} z(t) \, Q_z \nu(dz), \qquad a_t^* = \int_{E^*_{\mathbf{C}}} \overline{z(t)} \, Q_z \nu(dz). \tag{7.5}$$

Here $z(t) = \langle z, \delta_t \rangle$ is the complex white noise.

For $y \in E^*$ let T_y be the translation operator defined by

$$T_y \phi(x) = \phi(x + y), \qquad \phi \in \mathcal{W}.$$

It is known that $T_y \in \mathcal{L}(\mathcal{W}, \mathcal{W})$ and $T_y = \exp D_y$, see e.g., [**24**]. Then we see from (7.4) that

$$T_y = \int_{E^*_{\mathbf{C}}} e^{\langle z, y \rangle} Q_z \nu(dz), \qquad T_y^* = \int_{E^*_{\mathbf{C}}} e^{\langle \bar{z}, y \rangle} Q_z \nu(dz).$$

The adjoint operator $T_y^* \in \mathcal{L}(\mathcal{W}^*, \mathcal{W}^*)$ is related to the Wick product and is used to express a solution of a stochastic differential equation [**12**]. Quantum stochastic differential equations are also discussed from this aspect [**5**], [**27**].

The uniqueness property mentioned in Theorem 6.2 is in contrast to the case of white noise functions. First recall that the pointwise multiplication gives rise to a continuous bilinear map from $\mathcal{W} \times \mathcal{W}$ into $\mathcal{W}$. Hence, by duality the pointwise multiplication $\Phi\phi$ is defined for $\Phi \in \mathcal{W}^*$ and $\phi \in \mathcal{W}$. Moreover, it can be checked without difficulty that $\phi \mapsto \Phi\phi \in \mathcal{W}^*$, $\phi \in \mathcal{W}$, is a continuous linear map. Thus, every $\Phi \in \mathcal{W}^*$ is regarded as a multiplication operator, denoted by the same symbol, and this identification gives rise to a continuous injection: $\mathcal{W}^* \to \mathcal{L}(\mathcal{W}, \mathcal{W}^*)$. Now for $\Phi \in \mathcal{W}^*$ let

$$\Phi = \int_{E_{\mathbf{C}}^*} w(z) Q_z \, \nu(dz) \tag{7.6}$$

be the diagonal coherent state representation. Note that the kernel $w(z)$ is uniquely determined. Then, in view of $\Phi\phi_0 = \Phi$ and $Q_z\phi_0 = \phi_z$, where ϕ_0 is the vacuum, we have

$$\Phi = \int_{E_{\mathbf{C}}^*} w(z) \phi_z \, \nu(dz) \tag{7.7}$$

as generalized functions in $\mathcal{W}^*$. On the other hand, there are many different coherent state representations of white noise functions due to the famous overcompleteness of the exponential vectors $\{\phi_z\}$. For example,

$$\int_{E_{\mathbf{C}}^*} e^{\langle \bar{z}, \xi \rangle} \phi_z \nu(dz) = \phi_\xi,$$

which is verified by the S-transform. Thus the representation (7.7) is "canonical" in the sense that the kernel $w(z)$ is determined uniquely in such a way that the diagonal coherent state representation of the associated multiplication operator is given by (7.6) with the same $w(z)$. For example,

$$\phi_\xi = \int_{E_{\mathbf{C}}^*} e^{\langle z, \xi \rangle + \langle \bar{z}, \xi \rangle - \langle \xi, \xi \rangle} \phi_z \, \nu(dz), \qquad \xi \in E_{\mathbf{C}},$$

is the canonical coherent state representation.

The classical white noise $W_t(x) = \langle x, \delta_t \rangle \in \mathcal{W}^*$ and the quantum white noise $\{a_t, a_t^*\}$ are related as $W_t = a_t + a_t^*$, where W_t is regarded as a multiplication operator. Hence by (7.5),

$$W_t = \int_{E_{\mathbf{C}}^*} (z(t) + \overline{z(t)}) \phi_z \, \nu(dz)$$

is the canonical coherent state representation of the classical white noise.

8. Some Applications

For $\Phi \in \mathcal{W}^*$ the S-transform is defined by

$$S\Phi(\xi) = \langle\!\langle \Phi, \phi_\xi \rangle\!\rangle, \qquad \xi \in E_{\mathbf{C}}.$$

Then, in general $S\Phi$ is defined merely on $E_{\mathbf{C}}$; however, the S-transform of $\phi \in \mathcal{W}$ becomes naturally a $\mathbf{C}$-valued function defined on $E_{\mathbf{C}}^*$ by

$$S\phi(z) = \langle\!\langle \phi_z, \phi \rangle\!\rangle, \qquad z \in E_{\mathbf{C}}^*.$$

In that case the inversion formula for the S-transfrom is written down by means of the complex white noise measure.

THEOREM 8.1. *For $\phi \in \mathcal{W}$ it holds that*

$$\phi = \int_{E_{\mathbf{C}}^*} S\phi(\bar{z}) \phi_z \, \nu(dz). \tag{8.1}$$

PROOF. Write ψ for the right hand side of (8.1). It is easy to see that $\psi \in \mathcal{W}$ and the S-transform is computed as

$$\begin{aligned} S\psi(\xi) &= \int_{E^*_{\mathbf{C}}} S\phi(\bar{z})S\phi_z(\xi)\,\nu(dz) \\ &= \int_{E^*_{\mathbf{C}}} \langle\langle \phi_{\bar{z}},\, \phi\rangle\rangle\,\langle\langle \phi_z,\, \phi_\xi\rangle\rangle\,\nu(dz) = \int_{E^*_{\mathbf{C}}} \langle\langle Q_z\phi,\, \phi_\xi\rangle\rangle\,\nu(dz). \end{aligned}$$

It then follows from Theorem 6.3 that the last expression is equal to $\langle\langle \phi,\, \phi_\xi\rangle\rangle = S\phi(\xi)$, and hence $\phi = \psi$ by the uniqueness of the S-transform. □

The inversion formula for the S-transform was discussed also by Kondratiev [**17**] in a different context, see also [**3**, Chapter 2.5]. A rather different type of inversion formula is also known, see e.g., [**20**].

In harmonic analysis on Gaussian space rotation-invariant operators are interesting to investigate. Let $O(E;H)$ denote the group of all linear topological isomorphisms g of E such that $|g\xi|_0 = |\xi|_0$ for all $\xi \in E$. In other words, $O(E;H)$ is the automorphism group of the Gelfand triple: $E \subset H \subset E^*$. An operator $\Xi \in \mathcal{L}(\mathcal{W}, \mathcal{W}^*)$ is called *rotation-invariant* if

$$\Gamma(g)\Xi = \Xi\Gamma(g), \qquad g \in O(E;H),$$

where $\Gamma(g)$ is the second quantization of g. It is known that $\Gamma(g) \in \mathcal{L}(\mathcal{W}, \mathcal{W}) \cap \mathcal{L}(\mathcal{W}^*, \mathcal{W}^*)$. In this connection we only mention the following

PROPOSITION 8.2. *Let $w \in \mathcal{D}^*$. Then*

$$\Xi = \int_{E^*_{\mathbf{C}}} w(z)Q_z\,\nu(dz)$$

is rotation-invariant if and only if $\Gamma(g)^*w = w$ for any $g \in O(E;H)$.

PROOF. First note that

$$\Gamma(g)Q_z = Q_{gz}\Gamma(g), \qquad g \in O(E;H), \quad z \in E^*_{\mathbf{C}}.$$

In fact, since $\Gamma(g)\phi_{\bar{z}} = \phi_{g\bar{z}} = \phi_{\overline{gz}}$ we have

$$\Gamma(g)Q_z\phi = \langle\langle \phi_{\bar{z}},\, \phi\rangle\rangle\,\phi_{gz} = \langle\langle \phi_{\overline{gz}},\, \Gamma(g)\phi\rangle\rangle\,\phi_{gz} = Q_{gz}\Gamma(g)\phi.$$

Then the assertion follows by the rotation-invariance of $\nu = \mu' \times \mu'$ (see [**10**]). □

The rotation-invariant operators are characterized completely in [**24**]. In fact, two "Laplacian operators"

$$N = \Xi_{1,1}(\tau) = \int_{\mathbf{R}} a_t^* a_t\,dt, \qquad \Delta_G = \Xi_{0,2}(\tau) = \int_{\mathbf{R}} a_t^2\,dt,$$

the former being called the *number operator* and the latter the *Gross Laplacian*, generate all rotation-invariant operators and thereby an infinite dimensional analogue of Weyl's invariance theory is deemed to be within our scope. Recall that the "trace" admits the Fourier series expansion:

$$\tau = \sum_{n=0}^{\infty} e_n \otimes e_n \tag{8.2}$$

in $(E \otimes E)^*$, where $\{e_n\}_{n=0}^{\infty} \subset E$ is the usual orthonormal basis of H, see [**24**]. We now define

$$\langle z \otimes \bar{z}, \tau \rangle_{\mathrm{ren}} = \sum_{n=0}^{\infty} \{\langle z, e_n \rangle \langle \bar{z}, e_n \rangle - 1\}.$$

Then $w(z) = \langle z \otimes \bar{z}, \tau \rangle_{\mathrm{ren}} \in \mathcal{D}^*$ and looks like an infinite dimensional Euclidean norm (with renormalization).

PROPOSITION 8.3. *The diagonal coherent state representation of the number operator is given by*

$$N = \int_{E_{\mathbf{C}}^*} \langle z \otimes \bar{z}, \tau \rangle_{\mathrm{ren}} Q_z \, \nu(dz). \tag{8.3}$$

For the Gross Laplacian we have

$$\Delta_G = \int_{E_{\mathbf{C}}^*} \langle z \otimes z, \tau \rangle Q_z \, \nu(dz). \tag{8.4}$$

PROOF. The proof of (8.4) is immediate from (7.4). We shall prove (8.3). Put

$$w_n(z) = \langle z, e_n \rangle \langle \bar{z}, e_n \rangle = \langle x, e_n \rangle^2 + \langle y, e_n \rangle^2, \qquad z = x + iy,$$

and consider

$$\Xi_n = \int_{E_{\mathbf{C}}^*} w_n(z) Q_z \, \nu(dz).$$

Then, $\Xi_n \in \mathcal{L}(\mathcal{W}, \mathcal{W}^*)$ and is characterized by the symbol:

$$\widehat{\Xi}_n(\xi, \eta) = \int_{E_{\mathbf{C}}^*} \langle z, e_n \rangle \langle \bar{z}, e_n \rangle e^{\langle \bar{z}, \xi \rangle + \langle z, \eta \rangle} \nu(dz).$$

With the help of the formula:

$$\int_{E^*} \langle x, e_n \rangle^2 e^{\langle x, \zeta \rangle} \mu'(dx) = \left(\frac{1}{2} + \frac{1}{4} \langle e_n, \zeta \rangle^2 \right) e^{\langle \zeta, \zeta \rangle / 4}, \qquad \zeta \in E_{\mathbf{C}},$$

which can be verified easily, we come to

$$\widehat{\Xi}_n(\xi, \eta) = (\langle e_n, \xi \rangle \langle e_n, \eta \rangle + 1) e^{\langle \xi, \eta \rangle}. \tag{8.5}$$

On the other hand, by (5.3) we have

$$\Xi_{1,1}(e_n \otimes e_n)\widehat{\ }(\xi, \eta) = \langle e_n \otimes e_n, \eta \otimes \xi \rangle e^{\langle \xi, \eta \rangle}. \tag{8.6}$$

It then follows from (8.5) and (8.6) that

$$\Xi_n = \Xi_{1,1}(e_n \otimes e_n) + I.$$

Finally, in view of (8.2) we conclude that

$$\begin{aligned} N &= \Xi_{1,1}(\tau) = \sum_{n=0}^{\infty} \Xi_{1,1}(e_n \otimes e_n) = \sum_{n=0}^{\infty} (\Xi_n - I) \\ &= \sum_{n=0}^{\infty} \int_{E_{\mathbf{C}}^*} (w_n(z) - 1) Q_z \, \nu(dz) = \int_{E_{\mathbf{C}}^*} \langle z \otimes \bar{z}, \tau \rangle_{\mathrm{ren}} Q_z \, \nu(dz), \end{aligned}$$

which completes the proof. □

The diagonal coherent state representation seems useful in computing the trace of an operator Ξ. Consider a one-dimensional operator

$$\Xi\phi = \langle\!\langle \psi_1, \phi \rangle\!\rangle\, \psi_2, \tag{8.7}$$

where $\psi_1, \psi_2 \in \mathcal{W}$. Then, $\Xi \in \mathcal{L}(\mathcal{W}^*, \mathcal{W})$, which is regarded as a subspace of $\mathcal{L}(\mathcal{W}, \mathcal{W}^*)$ in a natural manner, and

$$\operatorname{Tr} \Xi = \langle\!\langle \psi_1, \psi_2 \rangle\!\rangle .$$

On the other hand,

$$\widehat{\Xi}(\bar{z}, z) = \langle\!\langle \Xi\phi_{\bar{z}}, \phi_z \rangle\!\rangle = \langle\!\langle \psi_1, \phi_{\bar{z}} \rangle\!\rangle \langle\!\langle \psi_2, \phi_z \rangle\!\rangle = \langle\!\langle \psi_2, Q_z\psi_1 \rangle\!\rangle .$$

Therefore,

$$\operatorname{Tr} \Xi = \int_{E_{\mathbf{C}}^*} \langle\!\langle Q_z\psi_1, \psi_2 \rangle\!\rangle\, \nu(dz) = \int_{E_{\mathbf{C}}^*} \widehat{\Xi}(\bar{z}, z)\nu(dz). \tag{8.8}$$

In particular,

$$\operatorname{Tr} R_{\zeta_1,\zeta_2} = e^{\langle \bar{\zeta}_1, \zeta_2 \rangle} = \int_{E_{\mathbf{C}}^*} \widehat{R}_{\zeta_1,\zeta_2}(\bar{z}, z)\, \nu(dz), \qquad \zeta_1, \zeta_2 \in E_{\mathbf{C}}.$$

Obviously, the trace formula (8.8) is valid for a finite dimensional operator $\Xi \in \mathcal{L}(\mathcal{W}^*, \mathcal{W})$. It is an interesting question to extend (8.8) to a larger class of operators. In this context one may hope that trace class operators on $\Gamma(H_{\mathbf{C}})$ are characterized.

Finally, we show the inversion formula for operator symbols. If an operator Ξ belongs to $\mathcal{L}(\mathcal{W}^*, \mathcal{W})$ the operator symbol $\widehat{\Xi}$ becomes a holomorphic function on $E_{\mathbf{C}}^* \times E_{\mathbf{C}}^*$ by

$$\widehat{\Xi}(z_1, z_2) = \langle\!\langle \Xi\phi_{z_1}, \phi_{z_2} \rangle\!\rangle , \qquad z_1, z_2 \in E_{\mathbf{C}}^*.$$

THEOREM 8.4. *For any $\Xi \in \mathcal{L}(\mathcal{W}^*, \mathcal{W})$ it holds that*

$$\Xi = \int_{E_{\mathbf{C}}^*} \int_{E_{\mathbf{C}}^*} \widehat{\Xi}(z_1, \bar{z}_2) R_{z_2,z_1} \nu(dz_1)\nu(dz_2). \tag{8.9}$$

PROOF. We denote the right hand side of (8.9) by X. Then, in view of (6.1) we have

$$\begin{aligned}
\widehat{X}(\xi, \eta) &= \int_{E_{\mathbf{C}}^*} \int_{E_{\mathbf{C}}^*} \widehat{\Xi}(z_1, \bar{z}_2)\, \langle\!\langle \phi_{\bar{z}_1}, \phi_\xi \rangle\!\rangle \langle\!\langle \phi_{z_2}, \phi_\eta \rangle\!\rangle\, \nu(dz_1)\nu(dz_2) \\
&= \int_{E_{\mathbf{C}}^*} \int_{E_{\mathbf{C}}^*} \langle\!\langle \Xi\phi_{z_1}, \phi_{\bar{z}_2} \rangle\!\rangle \langle\!\langle \phi_{\bar{z}_1}, \phi_\xi \rangle\!\rangle \langle\!\langle \phi_{z_2}, \phi_\eta \rangle\!\rangle\, \nu(dz_1)\nu(dz_2) \\
&= \int_{E_{\mathbf{C}}^*} \int_{E_{\mathbf{C}}^*} \langle\!\langle \Xi \langle\!\langle \phi_{\bar{z}_1}, \phi_\xi \rangle\!\rangle\, \phi_{z_1}, \langle\!\langle \phi_{z_2}, \phi_\eta \rangle\!\rangle\, \phi_{\bar{z}_2} \rangle\!\rangle\, \nu(dz_1)\nu(dz_2) \\
&= \int_{E_{\mathbf{C}}^*} \int_{E_{\mathbf{C}}^*} \langle\!\langle \Xi Q_{z_1}\phi_\xi, Q_{\bar{z}_2}\phi_\eta \rangle\!\rangle\, \nu(dz_1)\nu(dz_2).
\end{aligned}$$

Then (8.9) follows by the resolution of the identity. □

References

[1] S. Albeverio, Yu. G. Kondratiev and L. Streit: *How to generalize white noise analysis to non-Gaussian measures*, in "Dynamics of Complex and Irregular Systems (Ph. Blanchard, L. Streit, M. Sirugue-Collin and D. Testard, Eds.)," pp. 120–130, World Scientific, 1993.

[2] S. Albeverio, Yu. L. Daletsky, Yu. G. Kondratiev and L. Streit: *Non-Gaussian infinite dimensional analysis*, J. Funct. Anal. **138** (1996), 311–350.

[3] Y. M. Berezansky and Y. G. Kondratiev: "Spectral Methods in Infinite-Dimensional Analysis," Kluwer Academic, 1995.

[4] D. M. Chung, U. C. Ji and N. Obata: *Higher powers of quantum white noises in terms of integral kernel operators*, Infinite Dimen. Anal. Quantum Prob. **1** (1998), 533–559.

[5] D. M. Chung, U. C. Ji and N. Obata: *Normal-ordered white noise differential equations II: Regularity properties of solutions*, to appear in "Prob. Theory and Math. Stat. (B. Grigolionis et al. Eds.)," VSP/TEV, 1999.

[6] W. G. Cochran, H.-H. Kuo and A. Sengupta: *A new class of white noise generalized functions*, Infinite Dimen. Anal. Quantum Prob. **1** (1998), 43–67.

[7] C. G. Emch, G. C. Hergerfeldt and L. Streit: "On Klauder's Path: A Field Trip," World Scientific, 1994.

[8] R. J. Glauber: *Coherent and incoherent states of the radiation field*, Phys. Rev. **131** (1963), 2766–2788.

[9] T. Hida: "Analysis of Brownian Functionals," Carleton Math. Lect. Notes, no. 13, Carleton University, Ottawa, 1975.

[10] T. Hida: "Brownian Motion," Springer–Verlag, 1980.

[11] T. Hida, H.-H. Kuo, J. Potthoff and L. Streit: "White Noise," Kluwer Academic, 1993.

[12] H. Holden, B. Øksendal, J. Ubøe and T. Zhang: "Stochastic Partial Differential Equations," Birkhäuser, 1996.

[13] F. Jondral: *Some remarks about generalized functionals of complex white noise*, Nagoya Math. J. **81** (1981), 113–122.

[14] J. R. Klauder: *The action option and a Feynman quantization of spinor fields in terms of ordinary C-numbers*, Ann. Phys. **11** (1960), 123–168.

[15] J. R. Klauder and B.-S. Skagerstam: "Coherent States," World Scientific, 1985.

[16] J. R. Klauder and E. C. G. Sudarshan: "Fundamentals of Quantum Optics," W. A. Benjamin, Inc., 1968.

[17] Ju. G. Kondratiev: *Nuclear spaces of entire functions in problems of infinite-dimensional analysis*, Soviet Math. Dokl. **22** (1980), 588–592.

[18] Yu. G. Kondratiev and L. Streit: *Spaces of white noise distributions: Constructions, descriptions, applications I*, Rep. Math. Phys. **33** (1993), 341–366.

[19] I. Kubo and S. Takenaka: *Calculus on Gaussian white noise I*, Proc. Japan Acad. **56A** (1980), 376–380.

[20] I. Kubo and Y. Yokoi: *Generalized functions and functionals in fluctuation analysis*, in "Mathematical Approach to Fluctuations II (T. Hida, Ed.)," pp. 203–230, World Scientific, 1995.

[21] H.-H. Kuo: "White Noise Distribution Theory," CRC Press, 1996.

[22] Y.-J. Lee: *Analytic version of test functionals, Fourier transform and a characterization of measures in white noise calculus*, J. Funct. Anal. **100** (1991), 359–380.

[23] Y.-J. Lee: *Integral representation of second quantization and its application to white noise analysis*, J. Funct. Anal. **133** (1995), 253–276.

[24] N. Obata: "White Noise Calculus and Fock Space," Lect. Notes in Math. Vol. 1577, Springer–Verlag, 1994.

[25] N. Obata: *A note on Hida's whiskers and complex white noise*, in "Analysis on Infinite-Dimensional Lie Groups and Algebras (H. Heyer and J. Marion, Eds.)," pp. 321–336, World Scientific, 1998.

[26] N. Obata: *Complex white noise and coherent state representations*, to appear in the commemorative volume of T. Hida, 1999.

[27] N. Obata: *Complex white noise and normal-ordered white noise equations*, preprint, 1999.

[28] H. Ouerdiane: *Fonctionnelles analytiques avec condition de croissance, et application à l'analyse gaussienne*, Japan. J. Math. **20** (1994), 187–198.

[29] E. C. G. Sudarshan: *Equivalence of semiclassical and quantum mechanical descriptions of statistical light beams*, Phys. Rev. Lett. **10** (1963), 277–279.

[30] Y. Yokoi: *Simple setting for white noise calculus using Bargmann space and Gauss transform*, Hiroshima Math. J. **25** (1995), 97–121.

Graduate School of Mathematics, Nagoya University, Nagoya, 464-8602 Japan
E-mail address: obata@math.nagoya-u.ac.jp

Canadian Mathematical Society
Conference Proceedings
Volume **29**, 2000

Certain ratio limit theorem for time inhomogeneous Markov chains

Yoichi Oshima

ABSTRACT. For a family of transition functions $\{p_i(x,dy)\}_{i\geq 0}$, put $p_i^j(x,dy) = p_i p_{i+1}\cdots p_{j-1}(x,dy)$ and $G_n\varphi(i,x) = \sum_{j=i}^{i+n} p_i^j\varphi(x)$. In this paper, under certain conditions, we show the existence of the ratio limit $F(i,x) = \lim_{n\to\infty} G_n\varphi(i,x)/G_n\psi(i,x)$. Further, under an additional condition on the oscillation of the transition function, we give an explicit expression of $F(i,x)$.

1. Introduction

Let X be a locally compact separable metric space and $\{m_i\}_{i=1,2,\cdots}$ a mutually equivalent family of positive Radon mesures with full support. We assume that there exist equivalent positive Radon measures μ_1 and μ_2 such that

$$\mu_1 \leq m_i \leq \mu_2, \quad \forall i \geq 0. \tag{1.1}$$

Let $\{p_i(x,dy)\}_{i=0,1,2,\cdots}$ be a Markov family of transition functions and $\{p_i^j\}_{0\leq i\leq j}$ be its compositions defined by

$$\begin{aligned} p_i^i(x,dy) &= \delta_{\{x\}}(dy) \\ p_i^j(x,dy) &= p_i p_{i+1}\cdots p_{j-1}(x,dy). \end{aligned}$$

We assume that $\{m_i\}$ is an invariant family of measures relative to $\{p_i\}$ in the sense

$$m_i p_i = m_{i+1}, \quad \forall i \geq 0. \tag{1.2}$$

Let $Z = \{0,1,2,\cdots\}\times X$ and P be the space-time transition function corresponding to $\{p_i\}$, that is, the transition function on Z determined by

$$Pf(i,x) = \int_X p_i(x,dy)f(i+1,y).$$

Then its n-step transition function P^n is given by

$$P^n f(i,x) = \int_X p_i^{i+n}(x,dy)f(i+n,y).$$

2000 *Mathematics Subject Classification.* Primary 60J45; Secondary 60J10.

This research was partially supported by Grand-in-Aid Scientific Research, Ministry of Education, Science and Culture, Japanese Government, No.10640127.

p_i can be naturally considered as an operator from $L^1(X;m_{i+1})$ to $L^1(X;m_i)$. Let $\hat{p}_{i+1}$ be its dual operators from $L^\infty(X;m_i)$ to $L^\infty(X;m_{i+1})$ given by

$$\int_X p_i\varphi(x)\psi(x)dm_i(x) = \int_X \hat{p}_{i+1}\psi(y)\varphi(y)dm_{i+1}(y), \quad \forall i \geq 0, \tag{1.3}$$

for all $\varphi \in L^1(X;m_{i+1})$ and $\psi \in L^\infty(X;m_i)$. For any $f(i,x)$ such that $f(i,\cdot) \in L^\infty(X;m_i)$, let

$$\hat{P}f(i,y) = \hat{p}_i(f(i-1,\cdot))(y).$$

Then for any non-negative functions f, g such that $f(i,\cdot) \in L^1(X;m_i)$ and $g(i,\cdot) \in L^\infty(X;m_i)$,

$$\sum_{i\geq 0}\int_X \hat{P}g(i+1,y)f(i+1,y)dm_{i+1}(y) = \sum_{i\geq 0}\int_X g(i,x)Pf(i,x)dm_i(x).$$

Define the kernel G_n on Z by

$$G_n g(i,x) = \sum_{k=0}^{n} P^k g(i,x). \tag{1.4}$$

Then the purpose of this paper is to get a ratio limit theorem related to $\{p_i\}$, that is, to show the existence of the limit

$$F(i,x) = \lim_{n\to\infty}\frac{G_n\varphi(i,x)}{G_n\psi(i,x)}$$

for any non-negative μ_2-integrable functions φ, ψ on X and give a characterization of $F(i,x)$.

Introducing the assumptions (A.1) and (A.2), the existence of the limit is shown in §2. In the case of time homogeneous recurrent Markov chain, the limit $F(i,x)$ is independent of (i,x) and given explicitly by $F(i,x) = \langle\mu,\psi\rangle/\langle\mu,\varphi\rangle$, where μ is the unique invariant measure of the Markov chain (see e.g.[**1**], [**2**], [**3**], [**4**], [**5**]). In the case of $m_0(X) < \infty$, under an additional condition on the oscilation of the transition functions, we show in §3 an explicit representation of $F(i,x)$ given by

$$\lim_{n\to\infty}\frac{1}{n}G_n\varphi(i,x) = \frac{1}{m_0(X)}\lim_{n\to\infty}\frac{1}{n}\sum_{j=0}^{n}\int_X \varphi(y)dm_j(y). \tag{1.5}$$

Related result can be found in [**7**]. For the proof, we use an ergodic maximal inequality given by Meyer [**4**] (see also [**2**], [**3**]).

2. A ratio limit theorem

For any $i \geq 1$ and $\varphi \in L^1(X;\mu_2)$, let $\Gamma_i(\varphi)$ be a subset of X defined by

$$\Gamma_i(\varphi) = \bigcap_{k\geq 0}\left\{x \in X : \sup_{n\geq k}\sum_{j=k}^{n} P^j\varphi(i,x) > 0\right\}. \tag{2.1}$$

For any measurable subset A of Γ_i of with compact closure, put $A_i = \{i\}\times A$. Denote by $\hat{e}_{A_i}$ the equilibrium potential of A_i relative to $\hat{P}$, that is the smallest non-negative measurable function $h(j,x)$ satisfying $\hat{P}h(j,x) \leq h(j,x)$ and $h(i,x) \leq 1$ a.e.$x \in A$.

Now we introduce the first assumption (A.1).
(A.1) For any i, compact subset K of X and bounded μ_2-integrable function φ on X,

$$\sum_{j=i}^{\infty} \int_X \hat{e}_{A_i}(j,x)\varphi(x)dm_j(x) < \infty. \tag{2.2}$$

Since

$$\sum_{j=i}^{\infty} \int_X \hat{e}_{A_i}(j,x)\varphi(x)dm_j(x) = \sum_{j=i}^{\infty} \int_A p_i \ {}^A p_{i+1}^{j-1}\varphi(x)dm_i(x),$$

Assumption (A.1) is satisfied if $\sum_{j=i}^{\infty} {}^A p_{i+1}^{j-1}\varphi(x)$ is bounded relative to x, where ${}^A p_i(x,dy) = I_{X-A}(x)p_i(x,dy)I_{X-A}(y)$ and ${}^A p_i^j(x,dy) = {}^A p_i {}^A p_{i+1} \cdots {}^A p_{j-1}(x,dy)$.

For any fixed i and $\varphi \in L^1(X;\mu_2)$, define $\{K_p\}$ inductively by

$$K_1 = \inf\{k : \sum_{j=0}^{k} P^j\varphi(i,x) \geq 0\}, \quad K_{p+1} = \inf\{k : \sum_{j=K_p+1}^{k} P^j\varphi(i,x) \geq 0\}.$$

Taking a compact subset A of $\Gamma_i(\varphi)$, put $B^k = A \cap \{x : k \leq K_p(i,x) \leq n\}$ and $B_i^k = \{i\} \times B^k$. For a sequence $(g_0, g_1, \cdots)$ of functions g_k on Z such that $g_k(j,\cdot) \in L^\infty(X;m_i)$, let $S(g_0,g_1,g_2,\cdots) = (\hat{P}g_1, \hat{P}g_2, \cdots)$ and $e'_B = (b_0, b_1, \cdots)$ be the equilibrium potential of the set $B = (B_i^0, B_i^1, \cdots, B_i^n, \emptyset, \cdots)$ with respect to S. Under the Assumption (A.1), we can then apply the proof of Meyer [**4**] to show that $b_0 \uparrow \hat{e}_{A_i}$ as $n \uparrow \infty$, $p \uparrow \infty$ and get the following result.

LEMMA 2.1. *Let $i \geq 0$ and $\varphi \in L^\infty(X;\mu_2) \cap L^1(X;\mu_2)$. Under Assumption (A.1), for any compact set $A \subset \Gamma_i(\varphi)$,*

$$\sum_{j \geq i} \int_X \varphi(x) \cdot \hat{e}_{A_i}(j,x)dm_j(x) \geq 0. \tag{2.3}$$

We introduce further assumption (A.2).
(A.2) For any i and compact set A with $\mu_1(A) > 0$,

$$\sum_{j=i}^{\infty} \hat{e}_{A_i}(j,x) > 0 \ \ a.e.,$$

where $A_i = \{i\} \times A$.

Clearly, (A.2) is satisfied if $p_j(x,\cdot)$ has positive density for any i and x.

THEOREM 2.2. *Under Assumptions (A.1) and (A.2), if there exists an μ_2-integrable function $\varphi > 0$ such that $\sum_{j=i}^{\infty} p_i^j\varphi < \infty$ a.e., then $\sum_{j=i}^{\infty} p_i^j\psi < \infty$ a.e. for any $\psi \in L_+^1(\mu_2)$.*

Proof. Let φ be a function satisfying the condition of the theorem. If there exists $\psi \in L_+^1(\mu_2)$ such that $\sum_j p_i^j\psi = \infty$ a.e. on a compact A of X of positive measure, then $A \subset \Gamma_i(\psi - a\varphi)$ for any positive number a. Lemma 2.1 then implies that

$$\sum_{j=i}^{\infty} \int_X \hat{e}_{A_i}(j,x)(\psi - a\varphi)(x)dm_j(x) \geq 0.$$

Since a is arbitrary, this gives a contradiction.

Let G_n be the kernel defined by (1.4). Then we have the following theorem.

THEOREM 2.3. *Under Assumptions (A.1) and (A.2), for any i and non-negative non-trivial μ_2-integrable functions φ, ψ on X,*

$$F(i,x) = \lim_{n\to\infty} \frac{G_n\varphi(i,x)}{G_n\psi(i,x)} \tag{2.4}$$

exists μ_1-a.e. $x \in X$.

Proof. Suppose that there exists i, $a < b$ and a compact set A of positive μ_1-measure such that

$$\varliminf_{n\to\infty} \frac{G_n\varphi(i,x)}{G_n\psi(i,x)} < a < b < \varlimsup_{n\to\infty} \frac{G_n\varphi(i,x)}{G_n\psi(i,x)}$$

for all $x \in A$. Then $A \subset \Gamma_i(a\psi - \varphi) \cap \Gamma_i(\varphi - b\psi)$. Hence, by putting $A_i = \{i\} \times A$, we have

$$\sum_{j=i}^{\infty} \int_X \hat{e}_{A_i}(j,x)(a\psi - \varphi)dm_j \geq 0, \text{ and } \sum_{j=i}^{\infty} \int_X \hat{e}_{A_i}(j,x)(\varphi - b\psi)dm_j \geq 0,$$

that is,

$$\begin{aligned} b\sum_{j=i}^{\infty} \int_X \hat{e}_{A_i}(j,x)\psi(x)dm_j(x) &\leq \sum_{j=i}^{\infty} \int_X \hat{e}_{A_i}(j,x)\varphi(x)dm_j(x) \\ &\leq a\sum_{j=i}^{\infty} \int_X \hat{e}_{A_i}(j,x)\psi(x)dm_j(x) \end{aligned}$$

which is a contradiction.

3. A characterization of $F(i,x)$

In this section, we assume that $m_0(X) < \infty$. For any transition function $p(x,dy)$ put

$$\delta(p) = \frac{1}{2} \sup_{x,y} ||p(x,\cdot) - p(y,\cdot)||$$

with the total variation norm $||\cdot||$. Then $\delta(p_i^k p_{k+1}^\ell) \leq \delta(p_i^k)\delta(p_{k+1}^\ell)$ for any $i \leq k < \ell$ (see [**7**]).

THEOREM 3.1. *Assume that (A.1) and (A.2) hold and $\mu_2(X) < \infty$. Then, for any non-negative bounded function φ on X, the limit*

$$F(i,x) = \lim_{n\to\infty} \frac{1}{n} G_n\varphi(i,x) \tag{3.1}$$

exists a.e. Further, if there exist $\delta \in (0,1)$ and an increasing sequence $n_k \uparrow \infty$ such that

$$\delta(p_{n_k}^{n_{k+1}}) \leq \delta \ \text{ and } \sum_{k=1}^{\infty}(n_{k+1} - n_k)\delta^k < \infty, \tag{3.2}$$

then $F(i,x)$ is independent of i and x and given by

$$F(i,x) = \frac{1}{m_0(X)} \lim_{n\to\infty} \frac{1}{n} \sum_{j=1}^{n} \int_X \varphi(x) dm_j(x). \tag{3.3}$$

Proof. Let $n_k \uparrow \infty$ be the sequence stated above. Then, for any $i \in [n_\ell, n_{\ell+1})$ and $n \in [n_k, n_{k+1})$, $\delta(p_i^n) \le \delta^{k-\ell-1}$. Therefore,

$$\begin{aligned} |G_n\varphi(i,x) - G_n\varphi(i,y)| &= \sum_{k=j}^{i+n} \int_X |p_i^k(x,dz) - p_i^k(y,dz)| \varphi(z) \\ &\le ||\varphi||_\infty \sum_{k=\ell}^{\infty} (n_{k+1} - n_k)\delta^{k-\ell-1} < \infty. \end{aligned}$$

In particular, $F(i,x)$ is independent of x. Then, integrating both sides of (3.1) by $m_i(\cdot)/m_i(X)$ and noting that $m_i(X) = m_0(X)$ for all $i \ge 1$, we have

$$\begin{aligned} F(i,x) &= \frac{1}{m_i(X)} \int_X F(i,y) dm_i(y) = \frac{1}{m_i(X)} \lim_{n\to\infty} \frac{1}{n} \int_X \sum_{j=i}^{i+n} p_i^j \varphi(y) dm_i(y) \\ &= \frac{1}{m_i(X)} \lim_{n\to\infty} \frac{1}{n} \sum_{j=i}^{i+n} \int_X \varphi(y) dm_j(y) \\ &= \frac{1}{m_0(X)} \lim_{n\to\infty} \frac{1}{n} \sum_{j=0}^{n} \int_X \varphi(y) dm_j(y). \end{aligned}$$

References

[1] J. Azèma, M. Duflo et D. Revuz, *Mesure invariante sur les classes récurrentes des processus de Markov*, Z. Wahrscheinlichkeitstheorie und verw. Gebiete, **8** (1967), 157-181.

[2] P.J. Fitzsimmons, *The quasi-sure ratio ergodic theorem*, Ann. Inst. Henri Poincaré, **34** (1998), 385-405.

[3] M. Fukushima, *Almost polar sets and an ergodic theorem*, J. Math. Soc. Japan, **26** (1974), 17-32.

[4] P.A. Meyer, *Théorie ergodique et potentiels*, Ann. Inst. Fourier, Grenoble, **15** (1965), 89-102.

[5] J. Neveu, *Mathematical foundations of the Calculus of Probability*, Holden-Day Inc., 1965.

[6] Y. Oshima, *On the recurrence of some time inhomogeneous Markov processes*, Forum Math., **10** (1998), 119-134.

[7] T. Ueno, Some limit theorems for temporally discrete Markov processes, I, II, J. Fac. Sci. Univ. of Tokyo, **7** (1958), 449-462, 557-565.

DEPARTMENT OF COMPUTER SCIENCE, FACULTY OF ENGINEERING, KUMAMOTO UNIVERSITY, KUMAMOTO 860-8555, JAPAN

E-mail address: oshima@gpo.kumamoto-u.ac.jp

Canadian Mathematical Society
Conference Proceedings
Volume **29**, 2000

Infinitesimals in abstract Wiener spaces

Horst Osswald

Dedicated to Sergio Albeverio

ABSTRACT. We fix a Loeb probability measure over an internal Gaussian measure and we fix an internal hyperfinite dimensional Brownian motion B. For any abstract Wiener space $(\mathbb{B}, \mathbb{H})$ we construct a Brownian motion b using the standard part of B with respect to the measurable norm on $\mathbb{H}$. The set of continuous functions on $[0,1]$ with values in $\mathbb{B}$ is identical with the the sample paths of b, whenever $\mathbb{B}$ has a Schauder basis.

1. Introduction

In the classical work of Kuo [**11**] or Duncan and Varaiya [**5**] Brownian motion in abstract Wiener spaces $(\mathbb{H}, \mathbb{B})$ is constructed as a sequence of evaluation maps on infinite products of $\mathbb{B}$ endowed with suitable Gaussian measures. (Here $\mathbb{B}$ is the completion of a separable Hilbertspace $\mathbb{H}$ with respect to a measurable norm.) Duncan and Varaiya constructed Brownian motion on the whole $\mathbb{B}^{[0,1]}$ and used results of Fernique [**6**] and Nelson [**17**] to obtain a continuous modification. Kuo used the smaller space of continuous $\mathbb{B}$-valued functions starting in 0. In their recent work [**19**] Üstünel and Zakai followed Kuo's way in an introduction to abstract Wiener spaces. In our approach Brownian motion is constructed pathwise as a continuous process on a probability space not depending on the abstract Wiener space.

If we assume that $\mathbb{B}$ has a basis, then each continuous $\mathbb{B}$-valued functions is a path of the constructed Brownian motion. Those continuous functions also appear which are not 0 at 0, although b starts at 0 almost surely.

Let $\hat{\Gamma}$ be the Loeb measure over the internal Gaussian measure Γ on the internal Borel sets of $\mathbb{F}^H$ with variance $\frac{1}{H}$, where $\mathbb{F}$ is an ω-dimensional Euclidean space and ω and H are unlimited positive integers. We may assume that $\mathbb{F}$ contains any separable Hilbert as an (external) subspace. In analogy to Cutland's construction of finite dimensional Brownian motion [**3**] we set $B(X, n) := \sum_{i \leq n} X_i$ for each $X \in \mathbb{F}^H$ and for each $n \in T := \{1, ..., H\}$. For any abstract Wiener space $(\mathbb{B}, \mathbb{H})$ we construct a Brownian motion b using the standard part of B with respect to the measurable norm on $\mathbb{H}$.

2000 *Mathematics Subject Classification.* Primary 60J65; Secondary 26E35.

Gross [**7**] introduced the notion "measurable norm" to establish Gaussian measures in infinite dimensional spaces. Recall that a norm $|\ \ |$ on $\mathbb{H}$ is called measurable if for each $\varepsilon > 0$ there exists a finite dimensional subspace $E_\varepsilon \subset \mathbb{H}$ such that for each finite dimensional subspace $E \subset \mathbb{H}$ with $E \perp E_\varepsilon$

$$\mu_1^E \{x \in E \mid |x| \geq \varepsilon\} < \varepsilon,$$

where μ_1^E is the centered Gaussian measure on the Borel sets of E with variance 1. This equality is equivalent to

$$\widehat{\Gamma}\left\{X \in \mathbb{F}^H \mid |\pi_E(B(X,H))| \geq \varepsilon\right\} < \varepsilon,$$

where π_E denotes the orthogonal projection from $\mathbb{F}$ onto E. Using Lévy's [**13**] celebrated inequality for symmetric sequences of random variables, one can extend the previous equality to

$$\widehat{\Gamma}\left\{X \in \mathbb{F}^H \mid \max_{n \in T} |\pi_E(B(X,n))| \geq \varepsilon\right\} < \varepsilon$$

for special finite dimensional subspaces E of $\mathbb{H}$ (see Lemma 4.1). This result is used to show that the standard part b of B exists pathwise with respect to $|\ \ |$ and is a continuous $\mathbb{B}$-valued Brownian motion on $[0,1]$.

An earlier construction of infinite dimensional diffusion using infinitesimals is due to Lindstrøm [**16**] (see [**1**]). He also constructs Brownian motion as the standard part of an internal process. However, the underlying probability space is a hyperfinite Loeb space with uniform counting measure and he didn't use the uniformity on the internal time line T mentioned previously. In our approach we have smoothness of B in the state space, which can be applied to differential operators, see Cutland and Ng [**4**] for the classical Wiener space and in more details in [**18**]).

We use Nonstandard Analysis in the framework of the book of Albeverio, Høegh-Krohn, Fenstad, Lindstrøm [**1**]. Undefined notions and notations can be found there.

I would like to thank my colleagues Josef Berger, Thomas Kriecherbauer and Georg Schlüchtermann for helpful discussions.

2. The measure space $(\mathbb{F}^H, L_\Gamma(\mathcal{B}), \widehat{\Gamma})$

Fix a separable real Hilbert space $\mathbb{H}$ inside of the standard model of mathematics with scalar product $\langle\ ,\ \rangle$ and norm $\|\ \ \|$. By saturation, there exists a *finite dimensional space $\mathbb{F}$ with $\mathbb{H} \subseteq \mathbb{F} \subseteq {}^*\mathbb{H}$. Let ω be the internal dimension of $\mathbb{F}$. Then ω is an unlimited positive integer. Again by saturation, each orthonormal basis of $\mathbb{H}$ can be extended to an internal orthonormal basis of $\mathbb{F}$. Notice that $\mathbb{F}$ does not essentially depend on $\mathbb{H}$.

Let $\mathbb{B}$ be a real separable Banach space inside of the standard model with basis $(\mathfrak{b}_i)_{i \in \mathbb{N}}$ and norm $|\ \ |$. Then for each function $f : [0,1] \to \mathbb{B}$ there exists a function $\alpha_f : [0,1] \times \mathbb{N} \to \mathbb{R}$ such that $f(t) = \sum_{i=1}^\infty \alpha_f(t,i)\mathfrak{b}_i$ for each $t \in [0,1]$. Therefore, there exists a function $g : \mathbb{B}^{[0,1]} \times \mathbb{N} \to \mathbb{N}$ such that for each $f : [0,1] \to \mathbb{B}$ and for each $n \in \mathbb{N}$ and for all $k, m, l \in \mathbb{N}$ with $k \leq m \leq n$ and $g(f,m) \leq l \leq n$

$$(+) \quad \left| f(\frac{k}{m}) - \sum_{i=1}^{l} \alpha_f(\frac{k}{m}, i)\mathfrak{b}_i \right| < \frac{1}{m}.$$

By saturation, for each $f : [0,1] \to \mathbb{B}$ there is an unlimited $H_f \in {}^*\mathbb{N}$ such that $(+)$ is true for all $k, m, l \in {}^*\mathbb{N}$ with $k \leq m \leq H_f$ and $g(f,m) \leq l \leq H_f$. Again by saturation, there exists an unlimited $H \in {}^*\mathbb{N}$ with $H \leq H_f$ for each $f : [0,1] \to \mathbb{B}$. Again by a saturation argument, it is possible to choose H independent of each Banach space $\mathbb{B}$ with basis. Set $T := \{1, ..., H\}$. Thus we have proved

PROPOSITION 2.1. *For each standard Banach space $\mathbb{B}$ with basis say $(\mathfrak{b}_i)_{i\in\mathbb{N}}$ and for all $f : [0,1] \to \mathbb{B}$ and for all unlimited positive integers $\omega \leq H$*

$$(++)\ {}^*f(\frac{k}{H}) \approx_{\|\ \|} \sum_{i=1}^{\omega} {}^*\alpha_f(\frac{k}{H}, i)\mathfrak{b}_i \text{ for all } k \in T.$$

In the following we assume that $\omega \leq H$. For each $n \in T$ denote by $\mathcal{B}(\mathbb{F}^n)$ the internal Borel ${}^*\sigma$-algebra on $\mathbb{F}^n$. Let μ be the internal centered Gaussian measure with variance $\frac{1}{H}$ on the set $\mathcal{B}(\mathbb{F})$ of internal Borel sets in $\mathbb{F}$. Define the internal Gaussian measure Γ on $\mathcal{B}(\mathbb{F}^H)$ to be the H-fold product of μ. In order to apply elementary martingale theory in the internal setting, we define on $\mathbb{F}^H$ an internal filtration $(\mathcal{B}_n)_{n\in T}$ by setting $\mathcal{B}_n := \{B \times \mathbb{F}^{H-n} \mid B \in \mathcal{B}(\mathbb{F}^n)\}$. Notice the following simple lemma:

LEMMA 2.2. *An internal function $F : \mathbb{F}^H \to {}^*\mathbb{R}$ is $\mathcal{B}_n$-measurable if and only if F is $\mathcal{B}(\mathbb{F}^H)$-measurable and $F(X) = F(Y)$ for each $X, Y \in \mathbb{F}^H$ with $X_s = Y_s$ for all $s \leq n$.*

Set $\mathcal{B} := \mathcal{B}(\mathbb{F}^H)$. Denote by $(\mathbb{F}^H, L_\Gamma(\mathcal{B}), \widehat{\Gamma})$ the Loeb space over $(\mathbb{F}^H, \mathcal{B}, \Gamma)$ (see [**14**]).

3. The internal Brownian motion

First we introduce some notations: By $\mathcal{E}(\mathbb{H})$ we denote the finite dimensional subspaces of $\mathbb{H}$, and for each $E \in \mathcal{E}(\mathbb{F})$ let π_E be the orthogonal projection from $\mathbb{F}$ onto $\mathbb{E}$.

For each $E \in \mathcal{E}(\mathbb{H})$ we denote the centered Gaussian measure on the Borel sets on E with variance 1 by μ_1^E. By μ_1 we denote the finitely additive measure on the cylindersets of $\mathbb{H}$ given by the projective system $(\mu_1^E)_{E\in\mathcal{E}(\mathbb{H})}$ of measures.

We shall often use the *extension of entities to the nonstandard model, but we omit "*", if there is no risk of confusion.

In analogy to Cutland's construction of finite dimensional Brownian motion [**3**] we define an internal process $B : \mathbb{F}^H \times T \to \mathbb{F}$ setting $B(X, n) := \sum_{s=1}^{n} X_s$. It can be easily proved (see [**18**] for the one-dimensional case) that B is an internal Brownian motion, indeed, we obtain the following result:

PROPOSITION 3.1. *Fix $a, b \in \mathbb{F}$ with $\|a\| = \|b\| = 1$:*

(1) $\left(e^{\lambda\langle a, B(\cdot,n)\rangle - \frac{1}{2}\lambda^2\frac{n}{H}}\right)_{n\in T}$ is an internal $(\mathcal{B}_n)_{n\in T}$-martingale for each $\lambda \in {}^*\mathbb{R}$. It follows that $\langle a, B(\cdot, n)\rangle$ is normally distributed with variance $\frac{n}{H}$ and $\langle a, B(\cdot, n) - B(\cdot, m)\rangle$ is independent of $\mathcal{B}_m$ for $m < n$ in T.

(2) $\mathbb{E}(\langle a, B(\cdot, n)\rangle \cdot \langle b, B(\cdot, n)\rangle) = \langle a, b\rangle \cdot \frac{n}{H}$.

(3) For each $E \in \mathcal{E}(\mathbb{H})$ and each Borel set D in E

$$\mu_1^E(D) = \Gamma\left\{\pi_E(B(\cdot, H)) \in {}^*D\right\}.$$

Of course, $B(X, n) = \sum_{s\leq n} X_s = \sum_{i\leq\omega}\sum_{s\leq n}\langle X_s, \mathfrak{e}_i\rangle \cdot \mathfrak{e}_i$ for any orthonormal basis $\mathfrak{E} := (\mathfrak{e}_i)_{i\leq\omega}$ of $\mathbb{F}$. Moreover, we define for each $i \leq \omega$

$$B^{\mathfrak{E}}(i) : \mathbb{F}^H \times T \to \mathbb{R},\ (X, n) \mapsto \sum_{s\leq n}\langle X_s, \mathfrak{e}_i\rangle.$$

Then $B(X,n) = \sum_{i\leq\omega} B^{\mathfrak{E}}(i)\mathfrak{e}_i$. We need several lemmata:

LEMMA 3.2. *For each $i \leq \omega$, $B^{\mathfrak{E}}(i)$ is an internal $(\mathcal{B}_n)_{n\leq H}$-martingale such that $B^{\mathfrak{E}}(i)^2(\cdot,n)$ is Γ-integrable for each $n \in T$.*

PROOF. Using Lemma 2.2, it is easy to see that $B^{\mathfrak{E}}(i)(\cdot,n)$ is $\mathcal{B}_n$-measurable. Now $\int_{\mathbb{F}^H} B^{\mathfrak{E}}(i)^2(\cdot,n)d\Gamma = \alpha+\beta$ with

$$\alpha = \sum_{s\leq n}\int_{\mathbb{F}^H}\langle X_s,\mathfrak{e}_i\rangle^2 d\Gamma(X) = n\cdot\int_{\mathbb{F}}\langle X,\mathfrak{e}_i\rangle^2 d\mu(X) \leq$$

$$H\cdot\int_{^*\mathbb{R}^\omega} X_i^2 e^{-\frac{H}{2}\sum_{j=1}^{\omega}X_j^2}\sqrt{\frac{H}{2\pi}}^{\omega} d(X_j)_{j\leq\omega} = H\cdot\int_{^*\mathbb{R}} X^2 e^{-\frac{H}{2}X^2}\sqrt{\frac{H}{2\pi}}dX = 1,$$

and

$$\beta = 2\sum_{s<t\leq n}\int_{\mathbb{F}^H}\langle X_s,\mathfrak{e}_i\rangle\langle X_t,\mathfrak{e}_i\rangle d\Gamma(X) =$$

$$2\sum_{s<t\leq n}\int_{^*\mathbb{R}^2} XYe^{-\frac{H}{2}(X^2+Y^2)}\sqrt{\frac{H}{2\pi}}^{2} dXY = 0.$$

Finally, in order to prove the martingale property, fix $n,m \leq H$ with $m<n$ and let $E \in \mathcal{B}_m$. Then $E = C\times\mathbb{F}^{H-m}$ with $C \in \mathcal{B}(\mathbb{F}^m)$. It suffices to show that $\int_{C\times\mathbb{F}^{H-m}}\langle X_s,\mathfrak{e}_i\rangle d\Gamma(X) = 0$ for each $s>m$, thus, by Fubini's theorem, it suffices to prove that $\int_{\mathbb{F}}\langle X,\mathfrak{e}_i\rangle d\mu(X) = 0$. But

$$\int_{\mathbb{F}}\langle X,\mathfrak{e}_i\rangle d\mu(X) = \int_{^*\mathbb{R}} Xe^{-\frac{H}{2}X^2}\sqrt{\frac{H}{2\pi}}dX = 0. \square$$

LEMMA 3.3. *For each $i \leq \omega$, $\int_{\mathbb{F}^H}\max_{n\in T} B^{\mathfrak{E}}(i)^2(\cdot,n)d\Gamma$ is limited.*

PROOF. Using Doob's inequality and the martingale properties of $B^{\mathfrak{E}}(i)$, we obtain similar to the proof of Lemma 3.2

$$\mathbb{E}\max_{n\in T} B^{\mathfrak{E}}(i)^2(\cdot,n) \leq 4\mathbb{E}B^{\mathfrak{E}}(i)^2(\cdot,H) \text{ is limited. } \square$$

COROLLARY 3.4. *For $\widehat{\Gamma}$-almost all X, $\max_{n\in T} B^{\mathfrak{E}}(i)^2(X,n)$ is limited for all $i \in \mathbb{N}$.*

Recall that an internal function $F : T \to {}^*\mathbb{T}$, where $\mathbb{T}$ is a uniform space, is called S-**continuous** in $\mathbb{T}$, if $F(n)$ is nearstandard in $\mathbb{T}$ for each $n \in T$, and if $F(n) \approx_{\mathbb{T}} F(m)$ for each $n,m \in T$ with $\frac{n}{H}\approx\frac{m}{H}$.

Moreover, recall that the quadratic variation $\left[B^{\mathfrak{E}}(i)\right]$ of $B^{\mathfrak{E}}(i)$ is defined by setting

$$\left[B^{\mathfrak{E}}(i)\right] : \mathbb{F}^H\times T \to {}^*\mathbb{R},\ (X,n)\mapsto\sum_{s\leq n}\left(B^{\mathfrak{E}}(i)(X,s) - B^{\mathfrak{E}}(i)(X,s^-)\right)^2,$$

where $\alpha(s^-) := \alpha(s-1)$, if $1<s$, and $\alpha(1^-) := 0$ for each internal function α defined on T. Notice that $\left[B^{\mathfrak{E}}(i)\right](X,s) = \sum_{s\leq n}\langle X_s,\mathfrak{e}_i\rangle^2$.

The next lemma presents a result of Cutland [**3**]. The nice proof is due to Josef Berger

LEMMA 3.5. *For each $i \in \mathbb{N}$, the process $B^{\mathfrak{E}}(i)$ is S-continuous in $\mathbb{R}$ $\widehat{\Gamma}$-a.s.*

PROOF. Define

$$G : \mathbb{F}^H \times T \to {}^*\mathbb{R},\ (X,s) \mapsto H \cdot \langle X_s, \mathfrak{e}_i \rangle^2 .$$

Let ν be the internal uniform counting measure on T, i. e., $\nu(A) := \frac{1}{H}|A|$ for each internal subset $A \subseteq T$, where $|A|$ denotes the internal number of elements in A. Since

$$\int_{\mathbb{F}^H \times T} G^2 d\Gamma \otimes \nu =$$

$$\sum_{t\in T} H^2 \int_{\mathbb{F}^H} \langle X_t, \mathfrak{e}_i \rangle^4 d\Gamma(X) \frac{1}{H} = H^2 \int_{{}^*\mathbb{R}} X^4 e^{-\frac{H}{2}X^2} \sqrt{\frac{H}{2\pi}} dX = 3,$$

it follows that G is $S_{\Gamma\otimes\nu}$-integrable. By Keisler's Fubini theorem (see [**10**] or [**1**]), $G(X,\cdot)$ is S_ν-integrable for $\widehat{\Gamma}$-almost all $X \in \mathbb{F}^H$. Choose such an X. Then we obtain for each $n, m \subset T$ with $\frac{n}{H} \approx \frac{m}{H}$ and $n < m$

$$[B^{\mathfrak{E}}(i)](X,m) - [B^{\mathfrak{E}}(i)](X,n) = \int_{]n,m]} G(X,\cdot) d\nu \approx 0,$$

because $]n,m] := \{s \in \omega \mid n < s \le m\}$ is an internal $\widehat{\nu}$-nullset. Moreover

$$[B^{\mathfrak{E}}(i)](X,m) \le \int_T G(X,\cdot) d\nu \text{ is nearstandard in } \mathbb{R}.$$

This proves that $[B^{\mathfrak{E}}(i)]$ is S-continuous in $\mathbb{R}$ $\widehat{\Gamma}$-a.s. Using Lemma 3.3, we can apply a result due to Lindstrøm [**15**] (see also [**1**]) and Hoover and Perkins [**8**], which says that the S-continuity of $[B^{\mathfrak{E}}(i)]$ $\widehat{\Gamma}$-a.s. implies the S-continuity of $B^{\mathfrak{E}}(i)(X,\cdot)$ for $\widehat{\Gamma}$-almost all $X \in \mathbb{F}^H$. □

The result of Lindstrøm has been proved in the case of hyperfinite Loeb probability spaces and can be easily converted to our setting (see [**18**]).

4. The Brownian motion b on abstract Wiener spaces

Fix a measurable norm $|\ \ |$ on $\mathbb{H}$. Let $(\mathbb{B}, |\ \ |)$ be the completion of $(\mathbb{H}, |\ \ |)$. By $\mathbb{B}'$ we denote the topological dual of $\mathbb{B}$. Then $\mathbb{H}$ is dense in $\mathbb{B}$ and $\mathbb{B}'$ is dense in $\mathbb{H}' = \mathbb{H}$. Thus, we may assume that $\mathbb{B}' \subseteq \mathbb{H}$.

Gross [**7**] proved that the Gaussian measure μ_1 on the cylinder sets of $\mathbb{H}$ can also be defined for the cylinder sets in $\mathbb{B}$ and has an extension to a σ-additive Borel measure on $\mathbb{B}$.

The following lemma is the crucial point to see that the standard part of B exists and results in a continuous process. Here we use the characterization of measurable norms by the Loeb measure $\widehat{\Gamma}$ in the introduction and in Proposition 3.1.

LEMMA 4.1. *There exists an orthonormal basis $(\mathfrak{c}_i)_{i\in\mathbb{N}}$ of $\mathbb{H}$ such that for each $m \in \mathbb{N}$ there exists a $k_m \in \mathbb{N}$ such that for all $k, l \in \mathbb{N}$ with $k_m \le k < l$*

$$\widehat{\Gamma}\left\{ X \in \mathbb{F}^H \mid \max_{n\in T} \left| \sum_{i=k+1}^{l} \langle B(X,n), \mathfrak{c}_i \rangle \mathfrak{c}_i \right| \ge \frac{1}{m} \right\} < \frac{1}{m}.$$

PROOF. Since $\mathbb{B}'$ is dense in $\mathbb{H}$, there exists an orthonormal basis $(\mathfrak{c}_i)_{i\in\mathbb{N}}$ of $\mathbb{H}$ in $\mathbb{B}'$ such that the assertion of the lemma is true, if we replace $\max_{n\in T}\left|\sum_{i=k+1}^{l}\langle B(X,n),\mathfrak{c}_i\rangle\,\mathfrak{c}_i\right|$ with $\left|\sum_{i=k+1}^{l}\langle B(X,H),\mathfrak{c}_i\rangle\,\mathfrak{c}_i\right|$ and $\frac{1}{m}$ with $\frac{1}{3m}$. Set $F_s(X):=\sum_{i=k+1}^{l}\langle X_s,\mathfrak{c}_i\rangle\,\mathfrak{c}_i$ for each $s\in T$ and $X\in\mathbb{F}^H$. Since Γ is a Gaussian measure, we obtain, using Fourier transformation, that for all internal H-tuples $(\sigma_1,...,\sigma_H)$ in $\{1,-1\}$ the sequence $(F_s)_{s\in T}$ has the same distribution as $(\sigma_s F_s)_{s\in T}$. From Levy's inequality (see Ledoux and Talagrand [**12**], Proposition 2.3) it follows that

$$\Gamma\left\{\max_{n\in T}\left|\sum_{s=1}^{n}F_s\right|>\frac{1}{3m}\right\}\leq 2\Gamma\left\{\left|\sum_{s=1}^{H}F_s\right|>\frac{1}{3m}\right\}\leq\frac{2}{3m},$$

thus

$$\widehat{\Gamma}\left\{\max_{n\in T}\left|\sum_{i=k+1}^{l}\langle B(\cdot,n),\mathfrak{c}_i\rangle\,\mathfrak{c}_i\right|\geq\frac{1}{m}\right\}<\frac{1}{m}.\ \square$$

For each $\gamma\leq\omega$ we set $B^{\gamma}:=\pi_0^{\gamma}(B)$, that is,

$$B^{\gamma}(X,n):=\sum_{i=1}^{\gamma}\langle B(X,n),\mathfrak{c}_i\rangle\,\mathfrak{c}_i.$$

By a straighforward saturation argument we conclude

COROLLARY 4.2. *There exists an unlimited $\gamma\leq\omega$ such that for each $m\in\mathbb{N}$ and for all δ,ρ with $k_m<\delta<\rho\leq\gamma$*

$$\Gamma\left\{X\in\mathbb{F}^H\mid\max_{n\in T}\left|\sum_{i=\delta+1}^{\rho}\langle B(X,n),\mathfrak{c}_i\rangle\,\mathfrak{c}_i\right|\geq\frac{1}{m}\right\}<\frac{1}{m},$$

in particular, for all unlimited δ,ρ with $\delta<\rho\leq\gamma$

$$\widehat{\Gamma}\left\{X\in\mathbb{F}^H\mid\max_{n\in T}\left|\sum_{i=\delta+1}^{\rho}\langle B(X,n),\mathfrak{c}_i\rangle\,\mathfrak{c}_i\right|\not\approx 0\right\}=0.$$

For the rest of the paper fix an unlimited positive integer $\gamma\leq\omega$ such that the preceding equalities are true. We are now able to convert the internal process B into a continuous process $b:\mathbb{F}^H\times[0,1]\to\mathbb{B}$.

LEMMA 4.3. *There exists a set $U\subseteq\mathbb{F}^H$ with $\widehat{\Gamma}(U)=1$ and a strictly increasing function $g:U\times\mathbb{N}\to\mathbb{N}$ such that for all $X\in U$, $n\in T$,*

(1) $B^{\gamma}(X,n)$ is nearstandard with respect to $|\ |$ and

$${}^{\circ}B^{\gamma}(X,n)=\lim_{m\to\infty}\sum_{i=1}^{g(X,m)}{}^{\circ}\langle B^{\gamma}(X,n),\mathfrak{c}_i\rangle\,\mathfrak{c}_i.$$

(2) $B^{\gamma}(X,\cdot)$ is S-continuous in $\mathbb{B}$ with respect to $|\ |$.

PROOF. In order to prove (1), it suffices to prove that for $\widehat{\Gamma}$-almost all $X\in\mathbb{F}^H$ and for all $m\in\mathbb{N}$ there exists a $k\in\mathbb{N}$ (the smallest of those k is $g(X,m)$) such that for each $n\in T$

$$\left|B^{\gamma}(X,n)-\sum_{i=1}^{k}\langle B^{\gamma}(X,n),\mathfrak{c}_i\rangle\,\mathfrak{c}_i\right|<\frac{1}{m},\ \text{i. e.,}\ \left|\sum_{i=k+1}^{\gamma}\langle B^{\gamma}(X,n),\mathfrak{c}_i\rangle\,\mathfrak{c}_i\right|<\frac{1}{m}.$$

But this is true because of the construction of $(\mathfrak{c}_i)_{i\in\mathbb{N}}$. This proves (1).

Assertion (2) follows from (1), because, by Lemma 3.5, the function $n \mapsto \langle B^\gamma(\cdot,n), \mathfrak{c}_i\rangle$ on T is S-continuous $\widehat{\Gamma}$-a.s. for all $i \in \mathbb{N}$. □

Now we are able to define the process b. For each $X \in \mathbb{F}^H$ and each $n \in T$ set

$$b(X, {}^\circ\frac{n}{H}) := \begin{cases} {}^\circ B^\gamma(X,n) & \text{if } B^\gamma(X,\cdot) \text{ is } S\text{-continuous in the norm } |\ | \\ 0, & \text{otherwise,} \end{cases}$$

where ${}^\circ B^\gamma(X,n)$ denotes the standard part of $B^\gamma(X,n)$ with repect to the norm $|\ |$. We obtain

THEOREM 4.4.
(A) For $\hat{\Gamma}$-almost all X,

$$b(X, {}^\circ\frac{n}{H}) = {}^\circ B^\gamma(X,n) \text{ for all } n \in T.$$

The process b has continuous paths. If $\mathbb{B}$ has a basis, then each continuous function from $[0,1]$ into $\mathbb{B}$ is a path of b.
(B) b is a Brownian motion.

PROOF. (A) The first part of (A) follows from Lemma 4.3. Moreover, $b(X,\cdot)$ is continuous. Let $(\mathfrak{e}_i)$ be a basis of $\mathbb{B}$. Fix $f : [0,1] \to \mathbb{B}$. By Lemma 2.1, ${}^*f(\frac{t}{H}) \approx \sum_{i=1}^{\gamma} {}^*\alpha_f(\frac{t}{H}, i)\mathfrak{e}_i$ for each $t \in T$. We may assume that $span(\mathfrak{e}_i)_{i\le\gamma}$ is an internal subspace of $\mathbb{F}$. We define $X := (X_s)_{s\in T} \in \mathbb{F}^H$ by setting

$$X_s := \sum_{i\le\gamma} \left({}^*\alpha(\frac{s}{H}, i) - {}^*\alpha(\frac{s^-}{H}, i)\right)\mathfrak{e}_i,$$

where ${}^*\alpha(\frac{s^-}{H}, i) := 0$ for $s=1$ and ${}^*\alpha(\frac{s^-}{H}, i) := {}^*\alpha(\frac{s-1}{H}, i)$ for $s > 1$. Now we assume that f is continuous. Then $B^\gamma(X,\cdot)$ is S-continuous. Therefore, for each $n \in T$

$$b(X, {}^\circ\frac{n}{H}) \approx B^\gamma(X,n) = \sum_{i\le\gamma} {}^*\alpha(\frac{n}{H}, i)\mathfrak{e}_i \approx {}^*f(\frac{n}{H}) \approx f({}^\circ\frac{n}{H}).$$

This proves that $f = b(X,\cdot)$.

(B) Fix $\mathfrak{b}_1 \in \mathbb{B}'$ with $\|\mathfrak{b}_1\| = 1$ and recall that $\mathbb{B}' \subseteq \mathbb{H}$. Let $\mathfrak{F} := (\mathfrak{b}_i)_{i\le\gamma}$ be an extension of $\mathfrak{b}_1$ to an internal orthonormal basis of $\mathbb{F}$. Then for each $X \in \mathbb{F}^H$ such that $B^\gamma(X,\cdot)$ is S-continuous and for each $n \in T$

$$\mathfrak{b}_1(b(X, {}^\circ\frac{n}{H})) \approx {}^*\mathfrak{b}_1(B^\gamma(X,n)) = \sum_{s\le n}\sum_{i\le\gamma} \langle X_s, \mathfrak{b}_i\rangle {}^*\mathfrak{b}_1(\mathfrak{b}_i) = \sum_{s\le n} \langle X_s, \mathfrak{b}_1\rangle.$$

N. Cutland [**3**] has shown that $\mathfrak{b}_1 \circ b$ is a one-dimensional Brownian motion.

Fix $\mathfrak{b}_1, \mathfrak{b}_2' \in \mathbb{B}'$ with $\|\mathfrak{b}_1\| = \|\mathfrak{b}_2'\| = 1$. Set $r := {}^\circ\frac{n}{H}$. If $\mathfrak{b}_1 = \mathfrak{b}_2'$, then

$$\mathbb{E}\left(\mathfrak{b}_1 \circ b(\cdot,r)\mathfrak{b}_1 \circ b(\cdot,r)\right) \approx \mathbb{E}\sum_{s\le n} \langle \cdot_s, \mathfrak{b}_1\rangle \sum_{s\le n} \langle \cdot_s, \mathfrak{b}_1\rangle = \mathbb{E}\sum_{s\le n} \langle \cdot_s, \mathfrak{b}_1\rangle^2 =$$

$$\sum_{s\le n} \frac{1}{H} = \frac{n}{H}\langle \mathfrak{b}_1, \mathfrak{b}_1\rangle \approx r\langle \mathfrak{b}_1, \mathfrak{b}_1\rangle.$$

Let $\mathfrak{b}_1 \neq \mathfrak{b}_2'$. Set $\mathfrak{b}_2 := \frac{1}{1-\langle \mathfrak{b}_1, \mathfrak{b}_2'\rangle^2}\left(\mathfrak{b}_2' - \langle \mathfrak{b}_1, \mathfrak{b}_2'\rangle \mathfrak{b}_1\right)$. Then $\mathfrak{b}_2$ is orthogonal to $\mathfrak{b}_1$ and $\|\mathfrak{b}_2\| = 1$. Since $\mathbb{E}\sum_{s\le n} \langle \cdot_s, \mathfrak{b}_1\rangle \sum_{s\le n} \langle \cdot_s, \mathfrak{b}_2\rangle = 0$, we obtain

$$\mathbb{E}\left(\mathfrak{b}_1 \circ b(\cdot,r)\mathfrak{b}_2' \circ b(\cdot,r)\right) \approx \mathbb{E}\sum_{s\le n} \langle \cdot_s, \mathfrak{b}_1\rangle \sum_{s\le n} \langle \cdot_s, \mathfrak{b}_2'\rangle =$$

$$\mathbb{E}\langle\mathfrak{b}_1,\mathfrak{b}_2'\rangle\sum_{s\leq n}\langle\cdot_s,\mathfrak{b}_1\rangle\sum_{s\leq n}\langle\cdot_s,\mathfrak{b}_1\rangle=\langle\mathfrak{b}_1,\mathfrak{b}_2'\rangle\sum_{s\leq n}\frac{1}{H}\approx r\langle\mathfrak{b}_1,\mathfrak{b}_2'\rangle\,.\,\square$$

References

[1] S. Albeverio, J. E. Fenstad, R. Høegh Krohn, T. Lindstrøm, *Nonstandard methods in stochastic analysis and mathematical physics*, Academic Press, 1986.

[2] R. M. Anderson, A nonstandard representation of Brownian motion and Itô integration, *Israel J. Math.* **25**(1976), 15-46.

[3] N. Cutland, Infinitesimals in Action, *J. Lond. Math. Soc.* (2), **35**(1987), 202-216.

[4] N. Cutland, S.-A. Ng, A nonstandard approach to the Malliavin calculus, *Proc. of the Blaubeuren Conf.*, Univ. Tübingen, Walter de Gruyter, 1995.

[5] T. Duncan, P. Varaiya, On the solutions of a stochastic control system, *SIAM J. Control 13 Nr. 5 (1975)* **13** Nr. 5 (1975) 1077-1092.

[6] M. X. Fernique, Intégrabilité des Vecteurs Gaussian, *Academie des Sciences, Paris, Compte Rendus* **270** *Séries A* (1970) 1698-1699.

[7] L. Gross, Measurable functions on Hilbert space, *Trans. Amer. Math. Soc.* **105** (1962) 372-390.

[8] D. L. Hoover, E. A. Perkins, Nonstandard construction of the stochastic integral and applications to stochastic differential equations I and II, *Trans. Amer. Math. Soc.* **275**(1983), 1-58.

[9] A. E. Hurd, P. A. Loeb, *An introduction to Nonstandard Real Analysis*, Academic Press, 1985.

[10] H. J. Keisler, An infinitesimal approach to stochastic analysis, *Memoirs Amer. Math. Soc.*, **48**(1984).

[11] H. H. Kuo, Gaussian measures on Banach spaces, *Lecture notes in Mathematics* **463** (1975).

[12] M. Ledoux, M. Talagrand, Probability in Banach spaces, *Ergebnisse der Mathematik und ihrer Grenzgebiete*, Springer Verlag **23** (1991).

[13] P. Lévy, Téorie de l'addition des variables alétoires, Gauthier-Villars, Paris 1937.

[14] P. A. Loeb, Conversion from nonstandard to standard measure spaces and applications in probability theory, *Trans. Amer. Math. Soc.*, **211**(1975), 113-122.

[15] T. Lindstrøm, Hyperfinite stochastic integration I, II, III, and Addendum, *Math. Scand.* **46** (1980) 265-333.

[16] T. Lindstrøm, A Loeb measure approach to theorems of Prohorov, Sazonov and Gross, *Trans. Amer. Math. Soc.* **269** (1982) 521-534.

[17] E. Nelson, An existence theorem for second order parabolic equations, *Trans.Amer. Math. Soc.* **88** (1958) 414-429.

[18] H. Osswald, Probability Theory, to appear in *Nonstandard Analysis for the working mathematician*, P. Loeb, M. Wolff eds., Kluwer Scientific.

[19] A. S. Üstünel, M. Zakai, *Transformation of measure on Wiener space*, preprint.

Horst Osswald

Mathematisches Institut der Universität München

Theresienstr. 39

D-80333 München

E-mail: osswald@rz.mathematik.uni-muenchen.de

Canadian Mathematical Society
Conference Proceedings
Volume **29**, 2000

The Flux-Across-Surfaces Theorem for a Point Interaction Hamiltonian

G. Panati and A. Teta

ABSTRACT. The flux-across-surfaces theorem establishes a fundamental relation in quantum scattering theory between the asymptotic outgoing state and a quantity which is directly measured in experiments. We prove it for a hamiltonian with a point interaction, using the explicit expression for the propagator. The proof requires only assuptions on the initial state and it covers also the case of zero-energy resonance. We also outline a different approach based on generalized eigenfunctions, in view of a possible extension of the result.

Dedicated to Sergio Albeverio

1. Introduction

In quantum scattering theory one is concerned with the derivation of the experimentally measurable quantities from the large time asymptotics of the quantum state. In particular it is reasonable to expect that the probability that a particle crosses the active surface Σ of a very far detector equals the probability that for large times the particle has a momentum in the cone $C(\Sigma) = \{\lambda\mathbf{x} \in \mathbb{R}^3 : \mathbf{x} \in \Sigma, \lambda \geq 0\}$ generated by the surface Σ. The precise mathematical formulation of this flux-across-surfaces (FAS) conjecture is ([**1**], see also the discussion in [**4**])

$$\lim_{R\to\infty} \int_T^\infty dt \int_{\Sigma_R} \mathbf{j}^{\Psi_t} \cdot \mathbf{n}\, d\sigma = \int_{C(\Sigma)} |\widehat{\Psi}_{\text{out}}(\mathbf{k})|^2\, d^3k \tag{1}$$

for any $T \in \mathbb{R}$, where $\mathbf{j}^{\Psi_t} := 2\text{Im}(\Psi_t^* \nabla \Psi_t)$ is the density of probability current associated to $\Psi_t = e^{-iHt}\Psi_0$, $\Sigma_R = \{\mathbf{x} \in C(\Sigma) : |\mathbf{x}| = R\}$, $\Psi_{\text{out}} = \Omega_+^{-1}\Psi_0$ is the asymptotic outgoing state and $\widehat{\Psi}$ denotes the Fourier transform of Ψ. We have chosen units in which $\hbar = 1$ and $m = \frac{1}{2}$.

The proof of (1) in the free case was given in [**2**], exploiting the explicit form of the free unitary group (see also [**3**] or [**4**] for a more direct proof). The interacting case has been studied in [**5**] for short range and in [**6**] for long range potentials

1991 *Mathematics Subject Classification.* Primary 81-06, 81U05.

respectively. The proof relies on the basic assumption that the asymptotic outgoing state Ψ_{out} has a Fourier transform with compact support not containing the origin. More recently, a different proof has been given in [**7**] for sufficiently smooth potentials, assuming the absence of zero-energy resonances and requiring Ψ_{out} in the Schwartz space $\mathcal{S}(\mathbb{R}^3)$.

All these results are obtained under assumptions that avoid the difficulty due to zero-energy resonances which produce a slower decay of the wave function for large times and then make problematic the convergence of the left hand side of (1).

Here we consider a specific model hamiltonian, i.e. hamiltonian with a point interaction, and we prove the FAS theorem using the explicit expression for the propagator, derived in [**9**].

This paper has, in a sense, a pedagogical purpose. It indicates the possibility to extend the FAS theorem to more general situations. In particular we show that (1) holds true in the case of point interaction even with a zero-energy resonance. The analysis of zero-energy resonances in the general case of potential scattering will be approached in a further work ([**10**]). We also stress that the result is proved assuming only some regularity on the initial state.

In order to formulate the result we denote by $H_{\alpha,\mathbf{y}}$ the Schrödinger operator in $L^2(\mathbb{R}^3)$ which corresponds to one point interaction placed at $\mathbf{y} \in \mathbb{R}^3$ whose strength is parametrized by $\alpha \in \mathbb{R}$.

It is well known that $H_{\alpha,\mathbf{y}}$ can be constructed as self-adjoint operator using the standard extension theory ([**8**]). Here we only recall that the continuous spectrum is purely absolutely continuous and $\sigma_{\text{ac}}(H_{\alpha,\mathbf{y}}) = [0, +\infty)$. The point spectrum is empty if $\alpha \leq 0$ and $\sigma_{\text{p}}(H_{\alpha,\mathbf{y}}) = \left\{ -(4\pi\alpha)^2 \right\}$ if $\alpha < 0$. For $\alpha = 0$ the hamiltonian exhibits a zero-energy resonance.

The scattering wave functions are

$$\psi_{\alpha,\mathbf{y}}(\mathbf{x}, \mathbf{k}) = e^{i\mathbf{k}\cdot\mathbf{x}} + \frac{e^{i\mathbf{k}\cdot\mathbf{y}}}{(4\pi\alpha - ik)} \frac{e^{ik|\mathbf{x}-\mathbf{y}|}}{|\mathbf{x} - \mathbf{y}|} \tag{2}$$

Using the generalized eigenfunctions $\Phi_-(\mathbf{x}, \mathbf{k}) = \psi_{\alpha,\mathbf{y}}(\mathbf{x}, \mathbf{k})$ and $\Phi_+(\mathbf{x}, \mathbf{k}) = \psi^*_{\alpha,\mathbf{y}}(\mathbf{x}, -\mathbf{k})$ one can define two unitary maps $\mathcal{F}_\pm : \mathcal{H}_{ac}(H_{\alpha,\mathbf{y}}) \to L^2(\mathbb{R}^3)$ by

$$(\mathcal{F}_\pm f)(\mathbf{k}) = \operatorname*{s-lim}_{R\to+\infty} \frac{1}{(2\pi)^{3/2}} \int_{|\mathbf{x}|<R} \Phi^*_\pm(\mathbf{x}, \mathbf{k}) f(\mathbf{x}) \, d^3x. \tag{3}$$

which spectralise the operator $H_{\alpha,\mathbf{y}}$ restricted to $\mathcal{H}_{ac}$, in the sense that $\mathcal{F}_\pm H_{\alpha,\mathbf{y}} \mathcal{F}_\pm^{-1}$ is a multiplication operator in $L^2(\mathbb{R}^3)$.

The well known relation with the wave operators $\Omega_\pm = \text{s-lim}_{t\to\pm\infty} e^{iH_{\alpha,y}t} e^{-iH_0 t}$ (where $H_0 = -\Delta$) is expressed by the intertwining properties

$$\Omega_\pm^{-1} = \mathcal{F}^{-1}\mathcal{F}_\pm \qquad \text{and} \qquad \Omega_\pm = \mathcal{F}_\pm^{-1}\mathcal{F}. \tag{4}$$

With the above notation, our result is the following.

THEOREM 1. *Let us fix $\Psi_0 \in \mathcal{S}(\mathbb{R}^3) \cap \mathcal{H}_{ac}(H_{\alpha,\mathbf{y}})$. Then $\Psi_t := e^{-iH_{\alpha,y}t}\Psi_0$ is continuously differentiable in $\mathbb{R}^3 \backslash \{0\}$ and relation (1) holds true, for every $T \in \mathbb{R}$.*

In Section 2 we shall give the details of the proof for the more interesting case $\alpha = 0$ and then we outline the procedure for the other cases. It will also be clear from the proof that the strong assumption $\Psi_0 \in \mathcal{S}(\mathbb{R}^3)$ has been considered only for the

sake of simplicity; it can be replaced by the assumption that Ψ_0 is sufficiently many times differentiable and decays rapidly at infinity.
In Section 3 we briefly sketch a different proof based on the generalized eigenfunctions of $H_{\alpha,\mathbf{y}}$. In the case of one point interaction such method is less satisfactory than the method of the propagator, since it requires regularity of Ψ_{out}. Nevertheless it may be suitable for the extension of the result to general potential scattering in presence of zero-energy resonances.

2. Proof of the theorem

We shall denote by w the modulus of the vector $\mathbf{w} \in \mathbb{R}^3$ and with $\omega_{\mathbf{w}} := \frac{\mathbf{w}}{|\mathbf{w}|}$ the unit vector in the direction defined by $\mathbf{w} \neq \mathbf{0}$.

Without loss of generality we fix $\mathbf{y} = 0$ and, since the interaction is not trivial only in the s-wave ([**8**]), we choose a spherically symmetric initial state Ψ_0.
Using the explicit propagator for $\alpha = 0$ ([**9**]) we have

$$
\begin{aligned}
\Psi_t(\mathbf{x}) &= \int_{\mathbb{R}^3} \frac{e^{i\frac{|\mathbf{x}-\mathbf{y}|^2}{4t}}}{(4\pi i t)^{3/2}} \Psi_0(y)\, d^3y + \frac{2it}{x} \int_{\mathbb{R}^3} \frac{\Psi_0(y)}{y} \frac{e^{i\frac{(x+y)^2}{4t}}}{(4\pi i t)^{3/2}}\, d^3y \\
&= \frac{e^{i\frac{x^2}{4t}}}{(2it)^{3/2}} \left[\frac{1}{(2\pi)^{3/2}} \int_{\mathbb{R}^3} \Psi_0(y) \left(e^{-i\frac{\mathbf{x}\cdot\mathbf{y}}{2t}} + \frac{2it}{x} \frac{e^{i\frac{x}{2t}y}}{y} \right) d^3y \right] + R(\mathbf{x},t) \\
&= \frac{e^{i\frac{x^2}{4t}}}{(2it)^{3/2}} \widehat{\Psi}_{\text{out}}\left(\frac{\mathbf{x}}{2t}\right) + R(\mathbf{x},t) \equiv P(\mathbf{x},t) + R(\mathbf{x},t) \qquad (5)
\end{aligned}
$$

where we used (4) and we have denoted

$$
\begin{aligned}
R(\mathbf{x},t) &= \frac{e^{i\frac{x^2}{4t}}}{(4\pi i t)^{3/2}} \int_{\mathbb{R}^3} e^{-i\frac{\mathbf{x}\cdot\mathbf{y}}{2t}} \left(e^{i\frac{y^2}{2t}} - 1 \right) \Psi_0(y)\, d^3y + \\
&+ \frac{e^{i\frac{x^2}{4t}}}{(2it)^{1/2} x} \frac{1}{(2\pi)^{3/2}} \int_{\mathbb{R}^3} e^{i\frac{xy}{2t}} \left(e^{i\frac{y^2}{2t}} - 1 \right) \frac{\Psi_0(y)}{y}\, d^3y \\
&=: R_1(\mathbf{x},t) + R_2(\mathbf{x},t) \qquad (6)
\end{aligned}
$$

An explicit computation gives

$$
\nabla P(\mathbf{x},t) = \frac{i}{2}\frac{\mathbf{x}}{t} P(\mathbf{x},t) + \frac{e^{i\frac{x^2}{4t}}}{(2it)^{3/2}} \nabla_{\mathbf{x}} \widehat{\Psi}_{\text{out}}\left(\frac{\mathbf{x}}{2t}\right) \equiv \frac{i}{2}\frac{\mathbf{x}}{t} P(\mathbf{x},t) + \mathbf{Q}(\mathbf{x},t) \qquad (7)
$$

Taking (7) into account one obtains

$$
\mathbf{j}^{\Psi_t}(\mathbf{x},t) := 2\mathrm{Im}\left(\Psi_t^*(\mathbf{x},t)\nabla\Psi_t(\mathbf{x},t)\right) = \frac{\mathbf{x}}{t}|P(\mathbf{x},t)|^2 + \mathbf{N}(\mathbf{x},t)
$$

where

$$
\mathbf{N} = 2\mathrm{Im}\left(P^*\mathbf{Q} + P^*\nabla R + R^*\nabla P + R^*\nabla R\right) \qquad (8)
$$

A change of the integration variable yields

$$
\begin{aligned}
\lim_{R\to\infty} \int_T^\infty dt \int_{\Sigma_R} |P(\mathbf{x},t)|^2 \frac{\mathbf{x}}{t} \cdot \mathbf{n}\, d\sigma &= \lim_{R\to\infty} \int_T^\infty dt \int_{\Sigma_R} \frac{1}{(2t)^3} \left| \widehat{\Psi}_0\left(\frac{\mathbf{x}}{2t}\right) \right|^2 \frac{\mathbf{x}}{t} \cdot \mathbf{n}\, d\sigma \\
&= \int_{C(\Sigma)} |\widehat{\Psi}_{\text{out}}(\mathbf{k})|^2\, d^3k.
\end{aligned}
$$

We are then reduced to prove that

$$\lim_{R\to\infty}\int_T^\infty dt\int_{\Sigma_R}\mathbf{N}(\mathbf{x},t)\cdot\mathbf{n}\,d\sigma=0 \tag{9}$$

It is convenient to introduce the following definition.

DEFINITION 1. We say that $F:\mathbb{R}^3\times\mathbb{R}\to C^r$ ($r\in N$) satisfies the hypothesis $\mathcal{O}(n)$ (for $n\in N$), and we write $F=\mathcal{O}(n)$, if there exist $T_0>0$, $R_0>0$ such that

$$\sup_{x\geq R_0,t\geq T_0}\left(\frac{x}{t}\right)^q|F(\mathbf{x},t)|\leq C$$

for each $q\leq n$. If $F=\mathcal{O}(n)$ for every $n\in N$ we will write $F=\mathcal{O}(\infty)$.

It is easy to check that for every $f\in\mathcal{S}(\mathbb{R}^3)$ we have

$$F(\mathbf{x},t):=\int_{\mathbb{R}^3}e^{-i\frac{\mathbf{x}\cdot\mathbf{y}}{2t}}f(\mathbf{y})\,d^3y=\mathcal{O}(\infty). \tag{10}$$

Likewise, by iterated integration by parts, one can prove that for every $f\in\mathcal{S}(\mathbb{R}^3)$

$$F_0(\mathbf{x},t):=\int_{\mathbb{R}^3}e^{i\frac{x}{2t}y}\,f(\mathbf{y})\,d^3y=\mathcal{O}(2) \tag{11}$$

$$F_{-1}(\mathbf{x},t):=\int_{\mathbb{R}^3}e^{i\frac{x}{2t}y}\frac{f(\mathbf{y})}{y}\,d^3y=\mathcal{O}(1) \tag{12}$$

Using (10), (11) and (12) we can write

$$\begin{aligned}P(\mathbf{x},t)&=\frac{e^{i\frac{x^2}{4t}}}{(2it)^{3/2}}\widehat{\Psi}_0\left(\frac{\mathbf{x}}{2t}\right)+\frac{e^{i\frac{x^2}{4t}}}{(2it)^{1/2}x}\frac{1}{(2\pi)^{3/2}}\int_{\mathbb{R}^3}\Psi_0(y)\,\frac{e^{i\frac{x}{2t}y}}{y}\,d^3y\\&=\frac{\mathcal{O}(\infty)}{t^{3/2}}+\frac{\mathcal{O}(1)}{xt^{1/2}}.\end{aligned} \tag{13}$$

By a direct computation we obtain

$$\begin{aligned}\mathbf{Q}(\mathbf{x},t)&=\frac{e^{i\frac{x^2}{4t}}}{(2it)^{3/2}}\frac{1}{(2\pi)^{3/2}}\left\{-\frac{i}{2t}\int_{\mathbb{R}^3}e^{-i\frac{\mathbf{x}\cdot\mathbf{y}}{2t}}\mathbf{y}\Psi_0(y)\,d^3y+\right.\\&\quad-\left.i\frac{2t}{x^2}\omega_{\mathbf{x}}\int_{\mathbb{R}^3}e^{i\frac{x}{2t}y}\,\frac{\Psi_0(y)}{y}\,d^3y-\frac{1}{x}\omega_{\mathbf{x}}\int_{\mathbb{R}^3}e^{i\frac{x}{2t}y}\,\Psi_0(y)\,d^3y\right\}\\&=\frac{1}{t^{5/2}}\mathcal{O}(\infty)+\frac{1}{x^2t^{3/2}}\mathcal{O}(1)+\frac{1}{xt^{3/2}}\mathcal{O}(2).\end{aligned} \tag{14}$$

where we have used (10) and (12) and the inequality

$$\left|e^{iw}-1\right|\leq|w|\,. \tag{15}$$

Now we analyse $R_1(\mathbf{x},t)$ in (6) following the line of [**4**]. For any $q\in N$ one has

$$\left(\frac{x}{2t}\right)^q e^{-i\frac{\mathbf{x}\cdot\mathbf{y}}{2t}}=i^q\,(\omega_{\mathbf{x}}\cdot\nabla_y)^q\,e^{-i\frac{\mathbf{x}\cdot\mathbf{y}}{2t}}$$

and, by iterated integration by parts, we obtain

$$\begin{aligned}\left(\frac{x}{t}\right)^q|R_1(\mathbf{x},t)|&=\frac{1}{(4\pi it)^{3/2}}\left|\int_{\mathbb{R}^3}\left[(\omega_{\mathbf{x}}\cdot\nabla_y)^q e^{-i\frac{\mathbf{x}\cdot\mathbf{y}}{2t}}\right]\left(e^{i\frac{y^2}{2t}}-1\right)\Psi_0(y)\,d^3y\right|\\&\leq\frac{1}{(4\pi it)^{3/2}}\left|\int_{\mathbb{R}^3}(\omega_{\mathbf{x}}\cdot\nabla_y)^q\left(\left(e^{i\frac{y^2}{2t}}-1\right)\Psi_0(y)\right)\,d^3y\right|\leq\frac{C_q}{t^{5/2}}\end{aligned}$$

where the last inequality follows from an explicit differentiation and from (15). We conclude that

$$R_1(\mathbf{x},t) = \frac{1}{t^{5/2}}\mathcal{O}(\infty) \tag{16}$$

An analogous computation gives

$$\begin{aligned} \nabla R_1(\mathbf{x},t) &= \frac{\mathbf{x}}{t}R_1(\mathbf{x},t) + \frac{e^{i\frac{x^2}{4t}}}{(4\pi i t)^{3/2}}\frac{-i}{2t}\int_{\mathbb{R}^3} e^{-i\frac{x}{2t}y}\left(e^{i\frac{y^2}{2t}}-1\right)\mathbf{y}\Psi_0(y)\,d^3y \\ &= \frac{\mathbf{x}}{t^{7/2}}\mathcal{O}(\infty) + \frac{1}{t^{7/2}}\mathcal{O}(\infty). \end{aligned} \tag{17}$$

We analyse now $R_2(\mathbf{x},t)$. Using the inequality (15) one easily gets

$$|R_2(\mathbf{x},t)| \leq \frac{C}{xt^{3/2}}. \tag{18}$$

Moreover, using the radial symmetry of Ψ_0, one has (for $q=1,2$)

$$\begin{aligned} i^q\left(\frac{x}{2t}\right)^q R_2(\mathbf{x},t) &= \frac{e^{i\frac{x^2}{4t}}}{(2\pi)^{3/2}(2it)^{1/2}x}\int_{\mathbb{R}^3}\left(\frac{d^q}{dy^q}e^{i\frac{x}{2t}y}\right)\left(e^{i\frac{y^2}{2t}}-1\right)\frac{\Psi_0(y)}{y}\,d^3y \\ &= \frac{e^{i\frac{x^2}{4t}}}{(\pi i t)^{1/2}x}\int_0^{+\infty}\left(\frac{d^q}{dy^q}e^{i\frac{x}{2t}y}\right)\left(e^{i\frac{y^2}{2t}}-1\right)\Psi_0(y)y\,dy. \end{aligned}$$

Integrating by parts one obtains

$$\left|\left(\frac{x}{2t}\right)^q R_2(\mathbf{x},t)\right| \leq \frac{1}{(\pi t)^{1/2}x}\int_0^{+\infty}\left|\frac{d^q}{dy^q}\left(\left(e^{i\frac{y^2}{2t}}-1\right)\Psi_0(y)y\right)\right|\,dy \leq \frac{C_q}{xt^{3/2}} \tag{19}$$

By inequalities (18) and (19) one has

$$R_2(\mathbf{x},t) = \frac{1}{xt^{3/2}}\mathcal{O}(2) \tag{20}$$

A direct computation yields

$$\begin{aligned} \nabla R_2(\mathbf{x},t) &= \frac{\mathbf{x}}{t}R_2(\mathbf{x},t) + \frac{e^{i\frac{x^2}{4t}}}{(2it)^{1/2}}\frac{1}{(2\pi)^{3/2}}\left\{-\frac{\omega_{\mathbf{x}}}{x^2}\int_{\mathbb{R}^3} e^{i\frac{x}{2t}y}\left(e^{i\frac{y^2}{2t}}-1\right)\frac{\Psi_0(y)}{y}\,d^3y\right. \\ &+ \left.\frac{1}{x}\frac{i}{2t}\int_{\mathbb{R}^3} e^{i\frac{x}{2t}y}\left(e^{i\frac{y^2}{2t}}-1\right)\mathbf{y}\Psi_0(y)\,d^3y\right\}. \end{aligned}$$

and, by an argument analogous to the previous one, we obtain

$$\nabla R_2(\mathbf{x},t) = \frac{1}{t^{5/2}}\mathcal{O}(2) + \frac{1}{x^2t^{3/2}}\mathcal{O}(1) + \frac{1}{xt^{5/2}}\mathcal{O}(2). \tag{21}$$

In order to prove (9) we first observe that

$$\begin{aligned} &P^*(\mathbf{x},t)\mathbf{Q}(\mathbf{x},t) = \\ &= \mathbf{A}_0(\mathbf{x},t) + \frac{\mathcal{O}(\infty)}{t^4} + \frac{\mathcal{O}(\infty)\mathcal{O}(1)}{x^2t^3} + \\ &+ \frac{\mathcal{O}(\infty)\mathcal{O}(2)}{xt^3} + \frac{\mathcal{O}(1)\mathcal{O}(\infty)}{xt^3} + \frac{\mathcal{O}(1)\mathcal{O}(2)}{x^2t^2}. \end{aligned}$$

where

$$\mathbf{A}_0(\mathbf{x},t) := \frac{1}{(2\pi)^3}\frac{\mathbf{x}}{2tx^4}\left|\int_{\mathbb{R}^3} e^{i\frac{x}{2t}y}\frac{\Psi_0(y)}{y}\,d^3y\right|^2. \tag{22}$$

Since we are interested in the imaginary part of (22) we can neglet the term $\mathbf{A}_0(\mathbf{x},t)$. For the other terms in (22) a direct application of the dominated convergence theorem yields

$$\lim_{R\to\infty}\int_T^{+\infty}\int_{\Sigma_R}|\mathrm{Im}\,(P(\mathbf{x},t)\mathbf{Q}(\mathbf{x},t))|\,R^2 d\Omega dt=0.$$

Using the estimates (13), (14), (16), (17), (20) and (21) we similarly prove that

$$\lim_{R\to\infty}\int_T^{+\infty}\int_{\Sigma_R}|\mathrm{Im}\,(P^*\nabla R+R^*\nabla P+R^*\nabla R)|\,R^2 d\Omega dt=0,$$

proving the claim for an appropriate $T>0$. Using the invariance by finite time translations like in [**7**] we obtain the thesis for every $T\in\mathbb{R}$, in the case $\alpha=0$.

Now we sketch the proof in the case $\alpha\neq 0$. For $\alpha>0$, using the explicit form of the propagator, one obtains

$$\begin{aligned}\Psi_t(\mathbf{x}) &= \frac{e^{i\frac{x^2}{4t}}}{(4\pi it)^{3/2}}\int_{\mathbb{R}^3}\Psi_0(y)\left(e^{-i\frac{\mathbf{x}\cdot\mathbf{y}}{2t}}+\frac{1}{4\pi\alpha-i\frac{x}{2t}}\frac{e^{i\frac{x}{2t}y}}{y}\right)d^3y+\sum_{j=1}^3 R_j(\mathbf{x},t)\\ &= \frac{e^{i\frac{x^2}{4t}}}{(2it)^{3/2}}\widehat{\Psi}_{\mathrm{out}}\left(\frac{\mathbf{x}}{2t}\right)+\sum_{j=1}^3 R_j(\mathbf{x},t)\end{aligned}$$

where R_1 and R_2 are given by (6) and

$$R_3(\mathbf{x},t):=\frac{-2\alpha e^{i\frac{x^2}{4t}}}{(4\pi it)^{1/2}x}\int_{\mathbb{R}^3}d^3y\;e^{i\frac{x}{2t}y}\frac{\Psi_0(y)}{y}\int_0^{+\infty}du\;e^{-4\pi\alpha u}\left(e^{\frac{i}{4t}(u^2+y^2+2uy)}-1\right)e^{\frac{i}{2t}ux}.$$

All the estimates proved for $\alpha=0$ hold true in the present case. Moreover, concerning the leading term it is easy to see that

$$\left|\frac{e^{i\frac{x^2}{4t}}}{(2it)^{3/2}}\widehat{\Psi}_{\mathrm{out}}\left(\frac{\mathbf{x}}{2t}\right)\right|\leq\frac{C}{t^{3/2}}. \tag{23}$$

We stress that the estimate (23) holds true only for $\alpha\neq 0$. Using the radial symmetry of Ψ_0 and applying Fubini's theorem we can write

$$\begin{aligned}\widetilde{R}_3(\mathbf{x},t) &:= \int_{\mathbb{R}^3}d^3y\;e^{i\frac{x}{2t}y}\frac{\Psi_0(y)}{y}\int_0^{+\infty}du\;e^{-4\pi\alpha u}\left(e^{\frac{i}{4t}(u^2+y^2+2uy)}-1\right)e^{\frac{i}{4t}2ux}\\ &= 4\pi\int_0^{+\infty}du\int_0^{+\infty}dy\;e^{i\frac{x}{2t}(y+u)}y\Psi_0(y)e^{-4\pi\alpha u}\left(e^{\frac{i}{4t}(u+y)^2}-1\right)\\ &= 4\pi\int_0^{+\infty}du\int_u^{+\infty}dw\;e^{i\frac{x}{2t}w}(w-u)\Psi_0(w-u)e^{-4\pi\alpha u}\left(e^{\frac{i}{4t}w^2}-1\right)\end{aligned}$$

where we used the change of integration variable $w=y+u$. The integration domain is

$$D=\left\{(u,w)\in\mathbb{R}^2_+: u\in[0,+\infty), w>u\right\}=\left\{(u,w)\in\mathbb{R}^2_+: w\in[0,+\infty), u<w\right\}$$

so

$$\begin{aligned}\widetilde{R}_3(x,t) &= 4\pi\int_0^{+\infty}dw\int_0^w du\;e^{i\frac{x}{2t}w}\left(e^{\frac{i}{4t}w^2}-1\right)(w-u)\Psi_0(w-u)e^{-4\pi\alpha u}\\ &= 4\pi\int_0^{+\infty}dw\;e^{i\frac{x}{2t}w}\left(e^{\frac{i}{4t}w^2}-1\right)e^{-4\pi\alpha w}\int_0^w ds\;s\Psi_0(s)e^{4\pi\alpha s}.\end{aligned}$$

Using the fact that the function

$$\varphi(w) := e^{-4\pi\alpha w}\int_0^w ds\ s\Psi_0(s)e^{4\pi\alpha s}$$

satisfies

(24) $$\lim_{w\to+\infty} w^n\varphi(w) = 0 \qquad (n\in N)$$

we can integrate by parts showing that $\widetilde{R}_3(x,t) = \frac{1}{t}\mathcal{O}(2)$. Then we conclude that

(25) $$R_3(\mathbf{x},t) = \frac{\mathcal{O}(2)}{xt^{3/2}}.$$

Following the same line we obtain the corresponding estimate for $\nabla R_3(\mathbf{x},t)$

(26) $$\nabla R_3(\mathbf{x},t) = \frac{\mathcal{O}(2)}{t^{5/2}} + \frac{\mathcal{O}(2)}{x^2t^{3/2}} + \frac{\mathcal{O}(2)}{xt^{5/2}}.$$

The previous estimates (23), (25) and (26) allow us to use dominated convergence and then to obtain the result in the case $\alpha > 0$.

Finally, in the case $\alpha < 0$ the propagator can be written in the form

$$\begin{aligned}\Psi_t(\mathbf{x}) &= \left(e^{-iH_0t}\Psi_0\right)(\mathbf{x}) - e^{i(4\pi\alpha)^2t}\Psi_\alpha(\mathbf{x})\int_{\mathbb{R}^3}\Psi_\alpha^*(y)\Psi_0(y)\ d^3y + \\ &+ \int_{\mathbb{R}^3}\frac{\Psi_0(y)}{y}\frac{e^{i\frac{(x+y)^2}{4t}}}{(2it)^{1/2}x}d^3y - 4\pi\alpha\frac{2it}{x}\int_{\mathbb{R}^3}d^3y\ \frac{\Psi_0(y)}{y}\int_0^{+\infty}du\ e^{4\pi\alpha u}\frac{e^{i\frac{(u+x+y)^2}{4t}}}{(4\pi it)^{3/2}}\end{aligned}$$

where Ψ_α is the eigenfunction relative to the eigenvalue $\lambda_\alpha = -(4\pi\alpha)^2$. The second term is identically zero due to the assumption $\Psi_0 \in \mathcal{H}_{\text{ac}}(H_{\alpha,\mathbf{y}}) = \mathcal{H}_{\text{p}}(H_{\alpha,\mathbf{y}})^\perp$ and all the remaining terms can be treated exactly as in the case $\alpha > 0$, so we omit the details.

Remark 1. In the proof of Theorem 1 for $\alpha = 0$ we have estimated the *absolute value* of the terms in parenthesis in (8), except for $\mathbf{A}_0$ (see (22)), which is real and then it doesn't contribute to $\mathbf{N}$.
This is a crucial point since, in general, one has

(27) $$\lim_{R\to\infty}\int_T^\infty dt\int_{\Sigma_R}|\mathbf{A}_0\cdot\mathbf{n}|\ d\sigma = +\infty$$

(choose e.g. $\Psi_0(y) = e^{-y}$).
The divergent limit (27) is a consequence of the slower decay of the wave function in presence of a zero-energy resonance. This means that, in such case, one cannot hope to prove the theorem by simply estimating the absolute value of $\mathbf{N}$, unless one assumes the pseudo-orthogonality condition $\Psi_0 \in \mathcal{W}$ (see Remark 2). On the other hand this difficulty doesn't arise for $\alpha \neq 0$. In fact the term $\mathbf{A}_0$ is now replaced by

(28) $$\mathbf{A}_\alpha(\mathbf{x},t) = \frac{1}{t^3}\left|\int_{\mathbb{R}^3}e^{i\frac{x}{t}y}\ \frac{\Psi_0(y)}{y}\ d^3y\right|^2\frac{1}{\alpha+i\frac{x}{t}}\left(\frac{d}{dx}\frac{1}{\alpha-i\frac{x}{t}}\right)\omega_{\mathbf{x}}$$

which is easily estimated taking the absolute value.

3. The method of generalized eigenfunctions

In this section we outline a proof of the FAS conjecture based on the generalized eigenfunctions of $H_{\alpha,\mathbf{y}}$ and we compare it with the result obtained in the previous section. This technique was previosly used in [**7**]. Here we generalize the method in order to allow for the presence of zero-energy resonances.

THEOREM 2. *Let us fix $\Psi_{\rm out} \in \mathcal{S}(\mathbb{R}^3)$. Then $\Psi_t := e^{-iH_{\alpha,\mathbf{y}}t}\Omega_+\Psi_{\rm out}$ is continuously differentiable in $\mathbb{R}^3\backslash\{0\}$ and (1) holds true for every $T \in \mathbb{R}$.*

Proof. Let be $\Psi_0 = \Omega_+\Psi_{\rm out}$. Using the properties of $\mathcal{F}_+$ and (4) we obtain

$$\begin{aligned}\Psi_t(\mathbf{x}) &= \frac{1}{(2\pi)^{\frac{3}{2}}}\int_{\mathbb{R}^3} e^{-ik^2t}\widehat{\Psi}_{\rm out}(\mathbf{k})\Phi_+(\mathbf{x},\mathbf{k})\,d^3k \\ &= \frac{1}{(2\pi)^{\frac{3}{2}}}\int_{\mathbb{R}^3} e^{-ik^2t}\widehat{\Psi}_{\rm out}(\mathbf{k})e^{i\mathbf{k}\cdot\mathbf{x}}\,d^3k + \frac{1}{(2\pi)^{\frac{3}{2}}}\int_{\mathbb{R}^3} e^{-ik^2t}\frac{\widehat{\Psi}_{\rm out}(\mathbf{k})}{4\pi\alpha+ik}\frac{e^{-ikx}}{x}d^3k \\ &=: a(\mathbf{x},t) + b(\mathbf{x},t)\,.\end{aligned}$$

The density current is

$$\mathbf{j}^{\Psi_t} = \mathrm{Im}(a^*\nabla a + a^*\nabla b + b^*\nabla a + b^*\nabla a). \tag{29}$$

The first term $\mathbf{j}_0 = \mathrm{Im}(a^*\nabla a)$ corresponds to the free evolution of $\Psi_{\rm out}$, so using the free flux-across-surfaces theorem [**2**] one has

$$\lim_{R\to\infty}\int_T^{+\infty} dt\int_{\Sigma_R} \mathbf{j}_0(\mathbf{x},t)\cdot\mathbf{n}\,d\sigma = \int_{C(\Sigma)} |\widehat{\Psi}_{\rm out}(\mathbf{k})|^2\,d^3k\,.$$

It remains to show that

$$\lim_{R\to\infty}\int_T^{+\infty} dt\int_{\Sigma_R} |\mathbf{j}_1(\mathbf{x},t)\cdot\mathbf{n}|\,d\sigma = 0 \tag{30}$$

where $\mathbf{j}_1 := \mathrm{Im}(a^*\nabla b + b^*\nabla a + b^*\nabla b)$.
In order to prove (30) we need estimates on a, b and their gradients. In the notation of the previous section (see Definition 1), one has

$$a(\mathbf{x},t) = \frac{\mathcal{O}(\infty)}{t^{3/2}} \qquad \text{and} \qquad \nabla a(\mathbf{x},t) = \frac{\mathcal{O}(\infty)}{t^{3/2}}. \tag{31}$$

Concerning b,∇b we distinguish between the cases $\alpha \neq 0$ and $\alpha = 0$.
Case I. In the case $\alpha \neq 0$ we use a stationary phase technique, following [**7**]. Posing

$$\chi(k) \equiv \frac{k^2t+kx}{t+x} \qquad \text{and} \qquad \omega \equiv t+x \tag{32}$$

and denoting by $'$ the derivation with respect to k, one has

$$\begin{aligned}|b(\mathbf{x},t)| &= \frac{1}{(2\pi)^{\frac{3}{2}}}\left|\int_{\mathbb{R}^3} e^{i\omega\chi(k)}\widehat{\Psi}_{\rm out}(\mathbf{k})\frac{1}{4\pi\alpha+ik}\frac{1}{x}d^3k\right| \\ &= \frac{1}{(2\pi)^{\frac{3}{2}}}\frac{1}{x}\left|\int \frac{1}{\omega\chi'}\left[\frac{d}{dk}e^{-i\omega\chi}\right]\widehat{\Psi}_{\rm out}(\mathbf{k})\frac{1}{4\pi\alpha+ik}k^2\,dk\,d\Omega_{\mathbf{k}}\right|\end{aligned}$$

Integrating by parts and observing that the boundary term vanishes for $k=0$ since $\chi'(k) \geq \min(1,2k)$, we obtain

$$\begin{aligned}
\sup_{\mathbf{x}\in S_R} |b(\mathbf{x},t)| &\leq \frac{1}{R(R+t)} \int \left| \frac{d}{dk}\left[\frac{1}{\chi'}\widehat{\Psi}_{\text{out}}(\mathbf{k})\frac{1}{4\pi\alpha+ik}k^2\right]\right| dk\, d\Omega_{\mathbf{k}} \\
&\leq \frac{C}{R(R+t)} \qquad (33)
\end{aligned}$$

The function $b(x,t)$ is continuously differentiable in $\mathbb{R}^3\backslash\{0\}$ and for every $\mathbf{x}\neq 0$ one has

$$(\nabla b)(\mathbf{x},t) = -\frac{1}{(2\pi)^{\frac{3}{2}}}\frac{\omega_{\mathbf{x}}}{x}\int e^{-i(k^2t+kx)}\frac{\widehat{\Psi}_{\text{out}}(\mathbf{k})}{4\pi\alpha+ik}\left(\frac{1}{x}+ik\right) d^3k. \qquad (34)$$

The first term is given by $-\frac{\omega_{\mathbf{x}}}{x}b(\mathbf{x},t)$ and the second term is similar to $b(\mathbf{x},t)$ when we replace $\widehat{\Psi}_{\text{out}}(\mathbf{k})$ by $k\widehat{\Psi}_{\text{out}}(\mathbf{k})$. We obtain

$$\sup_{\mathbf{x}\in S_R} |\nabla b(\mathbf{x},t)| \leq \frac{c}{R(R+t)}. \qquad (35)$$

Using (31), (33) and (35) it is now easy to verify (30) and then to prove the thesis.
Case II. In the case $\alpha = 0$ we proceed in a slightly different way. Denoting $\Psi(k) := \int_{S^2}\widehat{\Psi}_{\text{out}}(\mathbf{k})d\Omega_{\mathbf{k}}$ we write

$$b(x,t) = \frac{c}{x}\int_0^{+\infty} e^{-ik^2t-ixk}\, k\Psi(k)\, dk.$$

In order to isolate the contribution arising from the value of $\widehat{\Psi}_{\text{out}}$ in zero, we write $b(x,t) = b_1(x,t) + b_2(x,t)$ where

$$\begin{aligned}
b_1(x,t) &:= \frac{c}{x}\int_0^{+\infty} e^{-ik^2t-ixk}\, k\left(\Psi(k)-\Psi(0)e^{-k^2}\right) dk \\
b_2(x,t) &:= \frac{c}{x}\Psi(0)\int_0^{+\infty} e^{-ik^2t-ixk}\, k\, e^{-k^2}\, dk
\end{aligned}$$

The term $b_1(x,t)$ can be dealt with by the same technique used in the case $\alpha\neq 0$, so we analyse the term $b_2(x,t)$. Using the formula (valid for $\text{Re}\xi > 0, \eta\in C$)

$$\int_0^{+\infty} w\exp\left(-\xi w^2-\eta w\right) dw = -\frac{1}{2}\sqrt{\frac{\pi}{\xi}}\frac{\partial}{\partial\eta}\left[\exp\left(\frac{\eta^2}{4\xi}\right)\text{erfc}\left(\frac{\eta}{2\sqrt{\xi}}\right)\right]$$

one obtains that (posing $\xi := 1-it$ and $R = |x|$)

$$b_2(x,t) = -\frac{1}{4}\frac{c}{R\,\xi^{3/2}}\left(iR\sqrt{\pi}e^{-\frac{R^2}{4\xi}} - 2\sqrt{\xi} - iR\sqrt{\pi}e^{-\frac{R^2}{4\xi}}\text{erf}\left(\frac{iR}{2\sqrt{\xi}}\right)\right).$$

Then we conclude that

$$|b(x,t)| \leq \frac{C_1}{R(R+t)} + \frac{C_2}{R\,t}. \qquad (36)$$

Concerning the gradient we have

$$\nabla b_2(x,t) = -\frac{c}{x^2}\,\omega_x \int_0^{+\infty} e^{-ik^2t-ixk}\, ke^{-k^2}\, dk + \tag{37}$$
$$- i\frac{c}{x}\,\omega_x \int_0^{+\infty} e^{-ik^2t-ixk}\, k^2e^{-k^2}\, dk$$

From (37) one easily deduces the estimate

$$|\nabla b(x,t)| \leq \frac{C_1}{R(R+t)} + \frac{C_2}{R^2t}. \tag{38}$$

Estimates (36) and (38) allow us to apply the dominated convergence theorem, and then to prove (30).

Remark 2. Using (4), (3) and (2) one obtains that

$$\begin{aligned}\widehat{\Psi}_{\mathrm{out}}(\mathbf{k}) &= (\mathcal{F}_+\Psi_0)(\mathbf{k}) = \int_{\mathbb{R}^3} \Phi_+(\mathbf{x},\mathbf{k})^*\Psi_0(\mathbf{x})\, d^3x \\ &= \int_{\mathbb{R}^3}\left(e^{-i\mathbf{k}\cdot\mathbf{x}} + \frac{1}{(4\pi\alpha - ik)}\frac{e^{ikx}}{x}\right)\Psi_0(\mathbf{x})\, d^3x \\ &= \widehat{\Psi}_0(\mathbf{k}) + \frac{1}{4\pi\alpha - ik}\int_{\mathbb{R}^3}\frac{e^{ikx}}{x}\Psi_0(\mathbf{x})\, d^3x.\end{aligned}$$

The above expression shows that, in presence of zero-energy resonances (i.e. for $\alpha = 0$), and if $\widehat{\Psi}_0$ is regular, the asymptotic outgoing state has a singularity in the origin of the momentum space, unless the initial state Ψ_0 belongs to the linear subspace

$$\mathcal{W} = \left\{\Psi \in L^2(\mathbb{R}^3) : \int_{\mathbb{R}^3}\frac{1}{x}\Psi(\mathbf{x})\, d^3x = 0\right\}$$

We underline that the set $\mathcal{W}$ is not closed in $L^2(\mathbb{R}^3)$. Note that, if $\Psi_0 \in \mathcal{W}$, then the problematic term $\mathbf{A}_0$ can be estimated taking the absolute value. The condition $\Psi_0 \in \mathcal{W}$ can be read as a condition of pseudo-orthogonality between Ψ_0 and the resonance function $\Psi_{\mathrm{res}}(\mathbf{x}) = \frac{1}{|\mathbf{x}|} \in L^2_{\mathrm{loc}}(\mathbb{R}^3)$. It is then clear that an assumption on the smoothness of $\widehat{\Psi}_{\mathrm{out}}$ is related to this condition of pseudo-orthogonality on the initial state. Since in Theorem 1 we don't need such restrictive assumption on Ψ_0 we conclude that the method of the propagator allows a better result. Nevertheless the method cannot be extended to more general situations while, on the other hand, this seems to be the case for the method of generalized eigenfunctions.

Acknowledgements. We are indebted to Prof. G. Dell'Antonio for many helpful discussions and comments. We also thank Prof. D. Dürr and Dr. S. Teufel for some stimulating discussions during the preparation of this paper.

References

[1] J. M. Combes, R.G. Newton, R. Shtokhamer, Scattering into cones and flux across surfaces, *Phys. Rev. D* **11**(1975), 366–372.

[2] M. Daumer, D. Dürr, S. Goldstein, N. Zanghì, On the Flux-Across-Surfaces Theorem, *Lett. Math. Phys.* **38** (1996), 103–116.

[3] S. Teufel, *The flux-across-surfaces theorem and its implications for scattering theory*, Ph.D. thesis at Ludwig-Maximilians-Universität München, March 1999.

[4] D. Dürr, S. Teufel, On the role of the flux in scattering theory, in these proceedings.

[5] W.O. Amrein, J.L. Zuleta, Flux and scattering into cones in potential scattering, *Helv. Phys. Acta* **70** (1997), 1-15.

[6] W.O. Amrein, D.B. Pearson, Flux and scattering into cones for long range and singular potentials, *Journal of Physics A*, **30** (1997), 5361-5379.

[7] S. Teufel, D. Dürr, K. Munch-Berndl, The flux-across-surfaces theorem for short range potentials and wave functions without energy cutoffs, *Journ. Math. Phys.* **40** (1999), 1901–1922.

[8] S. Albeverio, F. Gesztesy, R. Høegh-Krohn, H. Holden, *Solvable Models in Quantum Mechanics*, Springer Verlag, New York, 1988.

[9] S. Scarlatti, A.Teta, Derivation of the time-dependent propagator for the three-dimensional Schrödinger equation with one point interaction, *J. Phys. A* **23** (1990), L1033-L1035.

[10] G. Dell'Antonio, G. Panati, The Flux-Across-Surfaces Theorem with Zero-energy Resonances, in preparation.

S.I.S.S.A/I.S.A.S, International School for Advanced Studies, Via Beirut, 2 - 34014 Trieste, Italy.

E-mail address: panati@sissa.it

Dipartimento di Matematica, Università di Roma "La Sapienza", P.le Aldo Moro, 2 - 00185 Roma, Italy

E-mail address: teta@mat.uniroma1.it

Canadian Mathematical Society
Conference Proceedings
Volume **29**, 2000

Construction by subordination of processes with jumps on infinite dimensional state spaces and corresponding non local Dirichlet forms

Barbara Rüdiger and Jiang-Lun Wu

ABSTRACT. We show how symmetric non local , quasi-regular Dirichlet forms on infinite dimensional state spaces can be constructed by subordination. We start by considering any symmetric Markov semigroup $(T_t)_{t\geq 0}$ on $L^2(X,m)$, with (X,m) any measure space, and consider the subordinate Markov semigroup $(T_t^f)_{t\geq 0}$ with subordinator corresponding to any Bernstein function f. We construct the generator on its whole domain. We analyze properties of essentially self-adjointness of the subordinate generators, and properties of closability and quasi-regularity of the subordinate Dirichlet forms. We then show how these results can be applied to construct processes with jumps on infinite dimensional state spaces and apply them as a particular example to the subordinate of Ornstein-Uhlenbeck processes.

1. Introduction

The theory of Dirichlet forms made it possible to understand the relation between Markov processes, Markov semigroups, Dirichlet operators and Dirichlet forms on finite and also on infinite dimensional state spaces. With the theory of Dirichlet forms it was in particular possible to understand the relation between the "perturbed Laplace operator" $\Delta + \beta\nabla$, obtained by a (singular) h-transform from a Schroedinger operator (in finite as well as in infinite dimensions), and the distorted Brownian motion(see for example [**1**],[**2**],[**3**], [**6**], [**12**]); e.g. existence and uniqueness of weak solutions of the Cauchy problem for the distorted Brownian motion was proven , i.e. for the equation

$$dZ_t = dW_t + \beta(Z_t)dt \tag{1.1}$$

$$Z_0 = z\,, \tag{1.2}$$

where W_t is an X-valued Brownian motion, with X a separable Banach space, and for very singular drifts β [**6**]. Z_t are diffusion processes, and the corresponding symmetric Dirichlet forms $(\mathcal{E}, Dom(\mathcal{E}))$ on $L^2(X,m)$ (m being the symmetrizing measure of the process) are therefore local Dirichlet forms.
The theory of Dirichlet forms provides also the possibility to consider non local Dirichlet forms: the Beurling-Deny formula makes it possible to distinguish the

1991 *Mathematics Subject Classification.* 60H10, 60J75.

local part of the Dirichlet forms from the non local one, corresponding respectively to a diffusion process and a "nice" (m-special standard [**19**]) process with jumps [**13**],[**14**], [**19**], [**4**] , [**5**]. Nevertheless, the theory was not used for the concrete construction of processes with jumps on infinite dimensional state spaces, even not for the (restricted) case where X is a separable Banach space. It is on the other hand interesting to construct concrete examples of Dirichlet forms on infinite dimensional state spaces, which have also a non-local part. The analysis of Stochastic Differential Equations for processes with jumps on infinite dimensional state spaces is in general still very poor. Even Lévy processes on separable Banach spaces where not really analyzed. From statistical mechanics we know however that Lévy processes might be important for the description of critical phenomena (see for example the physical phenomena described in [**18**], [**23**], [**25**]). In the theory of quantum fields there is an interest in using symmetric infinite dimensional Lévy processes (random fields) as a constructive tool (see related work [**15**]).

Starting a longer research project in common with S.Albeverio we show that non local quasi-regular symmetric Dirichlet forms on separable Banach spaces $(X, \mathcal{B}(X))$ (in fact on the more general spaces where quasi-regular Dirichlet forms are constructed, see Section 3 below) and the corresponding processes with jumps can be constructed by subordination.

We start by considering any symmetric Markov semigroup $(T_t)_{t\geq 0}$ on $L^2(X, m)$, with (X, m) any given measure space, and consider the subordinate Markov semigroup $(T_t^f)_{t\geq 0}$ with subordinator corresponding to any Bernstein function f. In Section 2 we construct the corresponding generator L^f in its whole domain $Dom(L^f)$; see Theorem 2.4. Theorem 2.4 makes it possible to start in the following Sections the analysis of subordinate symmetric Dirichlet forms on infinite dimensional state spaces and the corresponding processes. Moreover , as remarked at the end of Section 2, Theorem 2.4 extends known results concerning the generator of subordinate symmetric Markov semigroups: in $L^2(\mathbf{R}^d)$ the subordinate generator was constructed in [**17**] only for very particular cases where the starting generator corresponding to a symmetric Markov semigroup is a particular pseudo-differential operator and only when the subordinater is given by a complete Bernstein function (see the comments in Section 2); these results were already covered by the much more general results in [**22**] where the subordinate generator is constructed starting by any generator corresponding to a Markov semigroup on a Banach space, but again only for a subordinator corresponding to a complete Bernstein function; for the case of symmetric Markov semigroups Theorem 2.4 is more general than in [**22**], as any Bernstein function is taken for subordination (see comments in Section 2); moreover we prove (Theorem 2.11) that if the generator L of $(T_t)_{t\geq 0}$ is essentially self-adjoint on $D \subset Dom(L)$, then the same holds for the subordinate generator L^f, i.e. corresponding to the subordinate Markov semigroup $(T_t^f)_{t\geq 0}$; we then prove in [**9**] the same property for $\sqrt{-L}$, see Theorem 2.12; this is equivalent to state that if the starting Dirichlet form $(\mathcal{E}, Dom(\mathcal{E}))$ is closable on D, $D \subset Dom(\sqrt{-L}) = Dom(\mathcal{E})$, then this also holds for the subordinate Dirichlet form $(\mathcal{E}^f, Dom(\mathcal{E}^f))$.

In Section 3 we use Theorem 2.12 to prove that if the Dirichlet form $(\mathcal{E}, Dom(\mathcal{E}))$ corresponding to the starting semigroup $(T_t)_{t\geq 0}$ is quasi-regular, this also holds for the subordinate Dirichlet form $(\mathcal{E}^f, Dom(\mathcal{E}^f))$, see Theorem 3.4.

In Section 4 we discuss how processes with jumps can be constructed on infinite dimensional state spaces by the previous results: It is known [**8**] that the essentially self-adjointness of the generator L of a Markov semigroup $(T_t)_{t\geq 0}$ on $D \subset Dom(L)$ and the quasi-regularity of the corresponding Dirichlet form imply the existence and uniqueness of the solution to the martingale problem (we call it $\mathcal{E}$-MP) for L on D (see [**8**], where the $\mathcal{E}$-MP for m-tight special standard processes is defined). It then follows by our results that the existence and uniqueness of the solution to the ($\mathcal{E}^f$-) martingale problem is also solved for the subordinate generator L^f on D. In Proposition 4.3 we give an argument to show that the process solving the ($\mathcal{E}^f$-) martingale problem for (L^f, D) is a process with jumps, according to our starting project.
In order to strength the power and generality of our results we apply them in Section 5 to the subordinate of the Ornstein-Uhlenbeck semigroups with drift given by strictly negative definte self-adjoint operators $-A$ on abstract Wiener spaces with covariance operator A and discuss uniqueness of the solutions to the corresponding martingale problems. In [**9**] we will also construct the corresponding SDEs.

2. The subordinate Dirichlet forms

We consider a symmetric Markov semigroup $(T_t)_{t\geq 0}$ on $L^2(X,m) := L^2(X,\mathcal{B},m)$, with $(X, \mathcal{B}, m)$ a given measure space. From the general theory [**13**], [**14**], [**19**] we know that the generator $(L, Dom(L))$ of $(T_t)_{t\geq 0}$ is a densly defined , self- adjoint , negative-definite Dirichlet-operator , the corresponding Dirichlet form beeing given by $(\mathcal{E}, Dom(\mathcal{E}))$ with

$$Dom(\mathcal{E}) = Dom(\sqrt{-L}) \tag{2.1}$$

$$\mathcal{E}(\phi, \psi) = (\sqrt{-L}\phi, \sqrt{-L}\psi) \quad \forall \phi, \psi \in Dom(\sqrt{-L}) \tag{2.2}$$

DEFINITION 2.1. A subordinator is a vaguely continuous convolution semigroup of probability measures $(\mu_t)_{t\geq 0}$ on $\mathbf{R}$ supported on $[0, \infty)$.

It is well known (see [**17**] and references there) that there is a one to one correspondence between Bernstein functions f and subordinators $(\mu_t^f)_{t\geq 0}$, the relation given by the Laplace transform:

$$\tilde{\mu}_t^f(x) := \int_0^\infty \exp(-sx)\mu_t^f(ds) = \exp(-tf(x)) \tag{2.3}$$

Moreover any Bernstein function is uniquely defined by a constant $b \geq 0$ and a non negative measure μ^f on $(0, \infty)$, which satisfies

$$\int_0^\infty \frac{s}{(1+s)}\mu^f(ds) < \infty\,, \tag{2.4}$$

in the sense that any Bernstein function f can be uniquely represented in the following way:

$$f(x) = bx + \int_0^\infty \quad (1 - \exp(-xs)) \quad \mu^f(ds) \tag{2.5}$$

DEFINITION 2.2. **Subordinate semigroup** Let f be a Bernstein function and $(\mu_t^f)_{t\geq 0}$ the corresponding convolution semigroup. Then the semigroup $(T_t^f)_{t\geq 0}$ defined by the Bochner integral

$$T_t^f(u) = \int_0^\infty T_s u \, \mu_t^f(ds) \quad u \in L^2(X,m), \, t \in \mathbf{R}_+ \tag{2.6}$$

is called the subordinate semigroup of $(T_t)_{t\geq 0}$ with respect to f

REMARK 2.3. $(T_t^f)_{t\geq 0}$ is a symmetric Markov semigroup [**21**], [**10**], [**26**].

We start computing the generator $(L^f, Dom(L^f))$ of the subordinate semigroup $(T_t^f)_{t\geq 0}$ in its whole domain, $Dom(L^f)$:

THEOREM 2.4.

$$(L^f, Dom(L^f)) = (-f(-L), Dom(f(-L))) \tag{2.7}$$

where

$$-f(-L) = -\int_0^\infty f(\lambda)\, dP_\lambda \,, \tag{2.8}$$

(recalling that $(L, Dom(L))$ is the generator of $(T_t)_{t\geq 0}$), P_λ the projection valued measure associated to the operator $-L$, i.e.

$$-L = \int_0^\infty \lambda \, dP_\lambda \,, \tag{2.9}$$

Here we avoid the detailed presentation of the easy parts of the proof, the complete proof can however be found in [**9**]. The problem of computing the domain $Dom(L^f)$ of L^f , generator of a given subordinated symmetric Markov semigroup on $L^2(\mathbf{R}^d)$ and to know how L^f acts on its whole domain was analyzed by N.Jacobs [**17**], but only for the case where the Bernstein function f is a complete Bernstein function , and when the starting generator is a pseudo- differential operator with its symbol satisfying a series of conditions (see Corollary 5.1 of [**17**]). In [**22**] however Schilling proved Theorem 2.4 but only for the case where f is a complete Bernstein function. (In [**22**] also the case where the semigroup $(T_t)_{t\geq 0}$ is defined on a Banach space is considered.) Differently from us , Schilling computes the resolvent of the subordinate semigroup, using the particular structure of the complete Bernstein functions. Similar to [**22**] , we use however the following Propositions 2.5, 2.6 [**21**] in our proof:

PROPOSITION 2.5. [**21**]
For any $u \in Dom(L)$

$$L^f u = bLu + \int_0^\infty (T_t u - u)\, \mu^f(dt) \tag{2.10}$$

PROPOSITION 2.6. [**21**] $(L^f, Dom(L))$ is an operator core for $(L^f, Dom(L^f))$

Proposition 2.6 is proven by showing that $T^f(Dom(L)) = Dom(L)$.

REMARK 2.7. As $(L^f, Dom(L^f))$ is by Remark 2.3 self-adjoint, Proposition 2.6 is equivalent to state that $(L^f, Dom(L))$ is essentially self-adjoint and (denoting the closability by -)

$$\overline{(L^f, Dom(L))} = (L^f, Dom(L^f)) \tag{2.11}$$

Remark also that from (2.10) we obtain heuristically $L^f\phi = -f(-L)\phi$, $\forall\phi \in Dom(L)$. We will start by making this statement rigorous; this will then be usefull to prove Theorem 2.4.

PROPOSITION 2.8.

$$(L^f, Dom(L)) = (-f(-L), Dom(L)) \tag{2.12}$$

To prove Proposition 2.8 we need the following Lemma.

LEMMA 2.9. *The following properties hold*
i) $Dom(L) \subseteq Dom(-f(-L))$
ii) there are two constants $c \geq 0$ *and* $K \geq 0$ *such that* $(f(-L)\phi, \phi) \leq c(-L\phi, \phi) + K\|\phi\|^2 \quad \forall\phi \in Dom(L)$
iii) there are two constants $c_2 \geq 0$ *and* $K_2 \geq 0$ *such that* $\| -f(-L)\phi\|^2 \leq c_2\|L\phi\|^2 + K_2\|\phi\|^2 \quad \forall\phi \in Dom(L)$

The proof [**9**] of Lemma 2.9 uses the representation

$$Dom(-f(-L)) = \{\phi \in L^2(X, m) : \int_0^\infty f^2(\lambda) \quad d(\phi, P_\lambda\phi) < \infty\} \tag{2.13}$$

(and the corresponding one for $f(x) = x$), see [**9**].

Proof of Proposition 2.8 As $Dom(L)$ is a closed linear subspace of $L^2(X, m)$ it is sufficient to prove that

$$(L^f\phi, \phi) = (-f(-L)\phi, \phi) \quad \forall\phi \in Dom(L) \tag{2.14}$$

From (2.5) it follows

$$(-f(-L)\phi, \phi) = -b\int_0^\infty \lambda d(\phi, P_\lambda\phi) \tag{2.15}$$

$$+\int_0^\infty \int_0^\infty (\exp(-t\lambda) - 1)\mu^f(dt)d(\phi, P_\lambda\phi) \quad \forall\phi \in Dom(L) \tag{2.16}$$

In [**9**] it is proven

$$\int_0^\infty \int_0^\infty |\exp(-t\lambda) - 1|\mu^f(dt)d(\phi, P_\lambda\phi) < \infty \quad \forall\phi \in Dom(L) \tag{2.17}$$

so that we can apply Fubini's theorem to the last term in (2.16), obtaining then

$$(-f(-L)\phi, \phi) = -b(-L\phi, \phi) \tag{2.18}$$

$$+\int_0^\infty \int_0^\infty (\exp(-t\lambda) - 1)d(\phi, P_\lambda\phi)\mu^f(dt) \quad \forall\phi \in Dom(L) \tag{2.19}$$

Applying Stone's theorem we obtain

$$(-f(-L)\phi,\phi) = b(L\phi,\phi) + \int_0^\infty ((T_t-1)\phi,\phi)\mu^f(dt) \quad \forall\phi\in Dom(L) \tag{2.20}$$

We will prove

$$\int_0^\infty ((T_t-1)\phi,\phi)\mu^f(dt) = \left(\int_0^\infty (T_t-1)\phi\mu^f(dt),\phi\right) \quad \forall\phi\in Dom(L) \tag{2.21}$$

so that (2.20) becomes (2.14). In fact (2.21) is a consequence of the following inequality proven in [**9**]

$$\int_0^\infty ((T_t-1)\phi,\phi)\mu^f(dt) < \infty \tag{2.22}$$

Q.E.D.

We will use the statement in proposition 2.8 to prove the following

LEMMA 2.10. *$(-f(-L), Dom(f(-L)))$ is a closed extension of $(L^f, Dom(L))$*

Proof of Lemma 2.10 $(-f(-L), Dom(f(-L)))$ is the unique closed extension of $(-f(-L), Dom(L))$. The statement follows then from Proposition 2.8. Q.E.D.

Theorem 2.4 is then easily proved by Lemma 2.10, as we can immediatly check

Proof of Theorem 2.4 There is a unique closed and also self- adjoint extension of $(L^f, Dom(L))$, which by Remark 2.7 is given by $(L^f, Dom(L^f))$. From Lemma 2.10 it follows that $(-f(-L), Dom(f(-L)))$ is a closed self-adjoint extension of $(L^f, Dom(L))$, so that equation (2.7) holds. Q.E.D.

Using the above results we will now prove the following

THEOREM 2.11. *If $(-L, D)$ is an operator core for $(-L, Dom(-L))$ then $(-f(-L), D)$ is an operator core for $(-f(-L), Dom(-f(-L))$*

Proof of Theorem 2.11 We will prove that given any $\phi\in Dom(-f(-L)))$ there is a sequence $(\phi_n)_{n\in\mathbf{N}}$ such that

$$\phi_n \in D \tag{2.23}$$

$$\lim_{n\to\infty}\|\phi_n-\phi\| = 0 \tag{2.24}$$

$$\lim_{n\to\infty}\|f(-L)\phi_n - f(-L)\phi\| = 0. \tag{2.25}$$

By hypothesis , given $\phi\in Dom(L)$, there is a sequence $(\phi_n)_{n\in\mathbf{N}}$ such that equations (2.23), (2.24) holds and

$$\lim_{n\to\infty}\|L\phi_n - L\phi\| = 0 \tag{2.26}$$

From property iii) of Lemma 2.9 it follows that (2.26) implies (2.25), so that for all $\phi\in Dom(L)$ (2.23), (2.24), (2.25) hold. The statement is then a consequence of Proposition 2.6 and Theorem 2.4. Q.E.D.

In order to prove that the subordinate Dirichlet form is quasi-regular , if this holds for the starting one, we need the following result

THEOREM 2.12. [**9**] *If $(\sqrt{-L}, D)$ is an operator core for $(\sqrt{-L}, Dom(\sqrt{-L})$ then $(\sqrt{-f(-L)}, D)$ is an operator core for $(\sqrt{-f(-L)}, Dom(\sqrt{-f(-L)})$*

The proof (see [**9**]) of Theorem 2.12 is rather long, we won't give it here. It uses inequality (3.7) in the next Section and the following inequality

$$\|\sqrt{-f(-L)}\phi\|^2 \leq \| - f(-L)\phi\|\|\phi\| \quad \forall\phi \in Dom(f(-L)) \tag{2.27}$$

3. Quasi-regularity of Dirichlet forms obtained via subordination

Starting from here we assume that X is a Hausdorff topological space, $\mathcal{B}(X) = \sigma(C(X))$, with $\mathcal{B}(X)$ the Borel σ-algebra, $C(X)$ the set of continuous functions on X, and m a σ- finite, positive measure on $(X, \mathcal{B}(X))$. Moreover, we assume that the Dirichlet form associated to the starting symmetric Markov semigroup $(T_t)_{t\geq 0}$ on $L^2(X, m)$ is quasi-regular.

REMARK 3.1. The property of quasi-regularity of a Dirichlet form $(\mathcal{E}, Dom(\mathcal{E}))$ on $L^2(X, m)$ is equivalent to the existence of a unique (up to m-equivalence) "nice" process (an m-tight special standard process) properly associated to $(\mathcal{E}, Dom(\mathcal{E}))$ [**19**].

We recall the definition of quasi-regular Dirichlet form. In order to do this we first give some notations and further definitions.

Notations We denote with $(\mathcal{E}_1, Dom(\mathcal{E})$, the symmetric form defined through

$$\mathcal{E}_1(\phi, \psi) = \mathcal{E}(\phi, \psi) + (\phi, \psi) \quad \forall \phi, \psi \in Dom(\mathcal{E}) \tag{3.1}$$

and correspondingly with $\mathcal{E}$ replaced by $\mathcal{E}^f$.
$\|\cdot\|_1$ ($\|\cdot\|_1^f$) is the norm induced on $Dom(\mathcal{E})$ ($Dom(\mathcal{E}^f)$) through $\mathcal{E}_1$ ($\mathcal{E}_1^f$).

DEFINITION 3.2. **Capacity** Given an open set $A \subset X$, the capacity $Cap^{\mathcal{E}}(A)$ of A with respect to the Dirichlet form $(\mathcal{E}, Dom(\mathcal{E})$ is given by

$$Cap^{\mathcal{E}}(A) = \inf_{y\in\mathcal{L}_A^{\mathcal{E}}} \mathcal{E}_1(\phi, \phi) \quad if \mathcal{L}_A^{\mathcal{E}} \neq \emptyset \tag{3.2}$$

$$\infty \quad if \mathcal{L}_A^{\mathcal{E}} = \emptyset \tag{3.3}$$

where

$$\mathcal{L}_A^{\mathcal{E}} := \{\phi \in Dom(\mathcal{E}) : \phi \geq 1 \quad m \quad a.e. \quad on\, A\} \tag{3.4}$$

If N is any subset of X then

$$Cap^{\mathcal{E}}(N) = \inf_{N\subset A\in\mathcal{O}} Cap^{\mathcal{E}}(A) \tag{3.5}$$

where $\mathcal{O}$ is the collection of open subsets of X.

DEFINITION 3.3. **Quasi-regular Dirichlet form** A symmetric Dirichlet form $(\mathcal{E}, Dom(\mathcal{E})$ on $L^2(X, m)$ is quasi-regular if
1) there exists an $\mathcal{E}$-nest $(F_k)_{k\in\mathbf{N}}$ consisting of compact sets
2) there exists a $\|\cdot\|_1$-dense subset D of $Dom(\mathcal{E})$ whose elements have an $\mathcal{E}$-quasi-continuous m -version.
3) there exists a sequence $(\phi_n)_{n\in\mathbf{N}}$ with $\phi_n \in Dom(\mathcal{E})$, having $\mathcal{E}$-quasi-continuous m- versions $\tilde{\phi}_n$, and an $\mathcal{E}$-exceptional set $N \subset X$ such that $\{\tilde{\phi}_n, n \in \mathbf{N}\}$ separates the points of $X\backslash N$

(We refer to [**13**], [**19**] for the definitions of $\mathcal{E}$-nest, $\mathcal{E}$-quasi-continuous m -version, $\mathcal{E}$-exceptional set.)

We will prove the following Theorem

THEOREM 3.4. *Let f be a Bernstein function. If the symmetric Dirichlet form $(\mathcal{E}, Dom(\mathcal{E})$ on $L^2(X,m)$ with generator $(L, Dom(L))$ is quasi-regular then the symmetric Dirichlet form $(\mathcal{E}^f, Dom(\mathcal{E}^f)$ with generator $(-f(-L), Dom(f(-L))$ is quasi-regular.*

In order to prove Theorem 3.4, we need first to prove the following Lemma

LEMMA 3.5. *There is a constant $C \geq 0$ such that for any subset $N \subset X$*

$$Cap^{\mathcal{E}^f}(N) \leq C Cap^{\mathcal{E}}(N) \tag{3.6}$$

Proof of Lemma 3.5 Using the spectral theorem it can be checked [**9**] that there are two constants $c \geq 0$, $K \geq 0$ such that

$$\|\sqrt{-f(-L)}\phi\|^2 \leq c\|\sqrt{-L}\phi\|^2 + K\|\phi\|^2 \quad \forall \phi \in Dom(\sqrt{-L}) \tag{3.7}$$

so that there is a constant $C \geq 0$ such that

$$\mathcal{E}_1^f(\phi,\phi) \leq C\mathcal{E}_1(\phi,\phi) \quad \forall \phi \in Dom(\mathcal{E}) \tag{3.8}$$

This implies the Lemma. Q.E.D.

REMARK 3.6. From Lemma 3.5 it follows in particular that
i) any $\mathcal{E}$-nest is also an $\mathcal{E}^f$-nest,
ii) any $\mathcal{E}$-exceptional set is also $\mathcal{E}^f$-exceptional,
iii) if $\phi \in L^2(X,m)$ is $\mathcal{E}$-quasi-continuous then it is also $\mathcal{E}^f$-quasi-continuous.

Proof of Theorem 3.4 Properties 1) and 3) in Definition 3.3 are a consequence of Remark 3.6.
Proof of property 2):
by hypothesis there exists an $\|\cdot\|_1$-dense subset D of $Dom(\mathcal{E})$ whose elements have an $\mathcal{E}$-quasi-continuous version. By Remark 3.6 , if $\phi \in D$ then ϕ has also an $\mathcal{E}^f$-quasi-continuous version. Moreover the statement that D is $\|\cdot\|_1$-dense in $Dom(\mathcal{E})$ is equivalent to the statement that $(\sqrt{-L}, D)$ is an operator core for $(\sqrt{-L}, Dom(\sqrt{-L}))$ and by Theorem 2.12 this implies that $(\sqrt{-f(-L)}, D)$ is an operator core for $(\sqrt{-f(-L)}, Dom(\sqrt{-f(-L)}))$, i.e. D is an $\|\cdot\|_1^f$-dense subset of $Dom(\mathcal{E}^f)$ Q.E.D.

4. Applications: construction of processes with jumps

In order to understand the power of the results of Section 2 and Section 3 we recall the following

THEOREM 4.1. [**8**] *Let $(\mathcal{E}, Dom(\mathcal{E}))$ be a symmetric quasi-regular Dirichlet form with generator $(L, Dom(L))$ on $L^2(X,m)$, and $D \subset Dom(L)$, (L, D) essentially self-adjoint, then up to m-equivalence there is a unique m-symmetric, m-tight special standard process $M = (\Omega, \mathcal{F}_\infty, (X_t)_{t\geq 0}, (P_z)_{z\in E_\Delta})$ satisfying the following martingale problem ($\mathcal{E}$-martingale problem) for (L, D):*
for any $u \in D$

$$M_t^u := \tilde{u}(X_t) - \tilde{u}(X_0) - \int_0^t Lu(X_s)ds \tag{4.1}$$

is an $\mathcal{F}_t$*-Martingale under* $(P_z)_{z\in E_\Delta}$ *for* $\mathcal{E}$*-q.e. starting point* $X_0 = z \in E$*, where* $\mathcal{F}_t$ *is the natural filtration of* M*, and* $\tilde{u}$ *is any* $\mathcal{E}^f$*-quasi continuous version of* u*.*

As a consequence of Theorem 4.1 and the results in the previous Sections it follows

THEOREM 4.2. *Let* f *be a Bernstein function and* $(\mathcal{E}, Dom(\mathcal{E}))$ *be a symmetric quasi-regular Dirichlet form with generator* $(L, Dom(L))$ *on* $L^2(X,m)$*,* $D \subset Dom(L)$*,* (L,D) *essentially self-adjoint, then up to* m*-equivalence there is a unique* m*-symmetric,* m*-tight special standard process*
$M^f = (\Omega, \mathcal{F}^f_\infty, (X_t)^f_{t\geq 0}, (P^f_z)_{z\in E_\Delta})$ *satisfying the following martingale problem (*$\mathcal{E}^f$*-martingale problem) for* $(f(-L), D)$*:*
for any $u \in D$

$$M_t^u := \tilde{u}(X_t) - \tilde{u}(X_0) - \int_0^t f(-L)u(X_s)ds \tag{4.2}$$

is an $\mathcal{F}^f_t$*-Martingale under* $(P^f_z)_{z\in E_\Delta}$ *for* $\mathcal{E}^f$*-q.e. starting point* $X_0 = z \in E$*, where* $\mathcal{F}^f_t$ *is the natural filtration of* M^f*, and* $\tilde{u}$ *is any* $\mathcal{E}^f$ *-quasi continuous version of* u*.*

We would like to stress now, that according to our main aim, we really construct in this way processes with jumps on infinite dimensional state spaces. In fact, as a consequence of the following Proposition, by starting from any diffusion process we obtain by subordination a process with jumps. We can prove the following result

PROPOSITION 4.3. Suppose that $(\mathcal{E}, Dom(\mathcal{E}))$ is a symmetric, quasi-regular, local Dirichlet form on $L^2(X,m)$ corresponding to the semigroup $(T_t)_{t\geq 0}$, let $\phi \geq 0$, $\phi \in Dom(\mathcal{E})$, and suppose that there is a function $\psi \geq 0$, $\psi \in Dom(L)$, s.t. $supp\,\phi \cap supp\,\psi = \emptyset$ then $\mathcal{E}^f(\phi,\psi) = 0$ implies

$$T_t\psi(x) = \psi(x) \quad for \quad m - a.e. \quad x \in supp\,\phi \tag{4.3}$$

Proof of Proposition 4.3 From Proposition 2.5 we know that

$$-\mathcal{E}^f(\phi,\psi) = b(\phi, L\psi) + (\phi, \int_0^\infty (T_t\psi - \psi)\mu^f(dt)) \tag{4.4}$$

and from equation (2.21) it follows

$$-\mathcal{E}^f(\phi,\psi) = b(\phi, L\psi) + \int_0^\infty (\phi, (T_t\psi - \psi))\mu^f(dt)) \tag{4.5}$$

By hypothesis however $(\phi, L\psi) = 0$, so that $\mathcal{E}^f(\phi,\psi) = 0$ implies

$$0 = (\phi, (T_t\psi - \psi)) \geq 0 \quad \mu^f - a.s. \tag{4.6}$$

so that

$$T_t\psi(x) = \psi(x) = 0 \quad for \quad m \times \mu^f a.e. \quad x \in supp\,\phi \tag{4.7}$$

and the Proposition follows then by the strong continuity of $(T_t)_{t\geq 0}$ Q.E.D.

REMARK 4.4. Similar to Theorem 3.4 it can be proven that if $(\mathcal{E}, Dom(\mathcal{E}))$ is a symmetric, regular Dirichlet form on $L^2(X, m)$, where X is a locally compact space (and further assumptions on (X, m) in [**13**]), then this holds also for $(\mathcal{E}^f, Dom(\mathcal{E}^f))$. Starting from this one could procede in a similar way as above to discuss the corresponding martingale problem ($\mathcal{E}^f$- MP). We do not go into details, since in the locally compact case such problems could also be discussed by other means ,e.g. proving the Feller property, (see e.g. [**16**] , [**17**] and references there), as on locally compact state spaces there is a correspondence between Feller semigroups, Hunt processes and martingale problems (MPs) (MPs in its classical (most known) meaning). As also mentioned in [**11**], in the case where the state space is not locally compact, it is not clear yet wether this correspondence holds. Our main aim in this paper was to discuss the infinite dimensional case, where the technique of Dirichlet forms is indeed the most established one for the analysis of general processes with infinite dimensional state space, including singular processes (e.g. singular perturbations of Ornstein-Uhlenbeck processes, see [**7**]). This theory provides the possibility to characterize m-tight special standard processes by proving the existence and uniqueness of the solution to the corresponding martingale problem (we call it here $\mathcal{E}-$ MP) as defined in Theorem 4.1

5. The subordinate of the Ornstein- Uhlenbeck semigroup

Let us see , as an example, how the the previous results apply to the semigroup of the Ornstein-Uhlenbeck process. Let (E, H, μ) be an abstract Wiener space , i.e. E a separable Banach space , H a separable real Hilbert space densely and continuously imbedded in E , and μ a mean zero Gaussian measure on $(E, \mathcal{B}(E))$, where $\mathcal{B}(E)$ is the Borel field on E. The Ornstein-Uhlenbeck semigroup $(T_t)_{t\geq 0}$ on $L^2(E, \mu)$ is a Markovian semigroup [**20**], [**24**]; it is defined by the "Mehler formula"

$$T_t F(\omega) = \int_E F(\exp(-t)\omega + \sqrt{1 - \exp(-2t)}u)\, \mu(du) \quad \omega \in E,\ F \in L^2(E, \mu) \tag{5.1}$$

(Another well known equivalent definition of this semigroup, its generator , and associated Dirichlet form is in terms of the Wiener chaos decomposition, which we do not give here, because of limited space)
We set

$$\mathcal{F}C_b^\infty := \{f(l_1, ...l_m) : m \in \mathbf{N},\ f \in C_b^\infty(\mathbf{R}^m),\ l_1...l_m \in E'\}, \tag{5.2}$$

and define $\mathcal{FP}$ correspondingly, but with the functions f being polynomials. Starting from now we set $D = \mathcal{F}C_b^\infty$ or $D = \mathcal{FP}$. The following three Theorems hold:

THEOREM 5.1. [**7**], [**5**] *The Dirichlet form $(\mathcal{E}, Dom(\mathcal{E}))$ corresponding to the Ornstein Uhlenbeck process is the closure of the symmetric bilinear form $(\mathcal{E}, D)$, with*

$$\mathcal{E}(u, v) = \int < \nabla u(\omega), \nabla v(\omega) >_H \mu(d\omega) = \sum_{i\in\mathbf{N}} \int \frac{\partial u}{\partial e_i}(\omega)\frac{\partial v}{\partial e_i}(\omega)\mu(d\omega)\,, \tag{5.3}$$

and with $e_1, ...e_i, ...$ an ONB of H, [**7**], [**19**] *(we refer to* [**19**] *for the definition of the gradient operator and directional derivative). $(\mathcal{E}, Dom(\mathcal{E}))$ is a quasi- regular local Dirichlet form* [**5**]. *The generator $(L, Dom(L))$ of the Ornstein-Uhlenbeck process is $(\nabla^\star\nabla, Dom(\nabla^\star\nabla))$*

REMARK 5.2. From Theorem 5.1 it follows that $(\sqrt{\nabla^\star\nabla}, D)$, is essentially self-adjoint. Its closure is the Dirichlet form of the Ornstein-Uhlenbeck process $(\mathcal{E}, Dom(\mathcal{E})) = (\sqrt{\nabla^\star\nabla}, Dom(\sqrt{\nabla^\star\nabla}))$.

THEOREM 5.3. **[7]** $(\nabla^\star\nabla, D)$, *is essentially self- adjoint*

We now consider the subordinate semigroup $(T_t^f)_{t\geq 0}$ on $L^2(E,\mu)$ of the Ornstein-Uhlenbeck semigroup $(T_t)_{t\geq 0}$ with respect to the Bernstein function f. By the theory of Sections 2 and 3 it follows

THEOREM 5.4. $(-f(-L), Dom(f(-L)) = (-f(-\nabla^\star\nabla), Dom(f(-\nabla^\star\nabla))$ *is the generator of* $(T_t^f)_{t\geq 0}$. *The corresponding Dirichlet form is quasi regular and is given by* $(\mathcal{E}^f, Dom(\mathcal{E}^f))$, *with*

$$Dom(\mathcal{E}^f) = Dom(\sqrt{f(-\nabla^\star\nabla)}) \tag{5.4}$$

and

$$\mathcal{E}^f(u,v) = (\sqrt{f(-\nabla^\star\nabla)}u, \sqrt{f(-\nabla^\star\nabla)}v) \quad \forall u,v \in Dom(\mathcal{E}^f) \tag{5.5}$$

Moreover, applying Proposition 4.3 and remarking that $D \subset Dom(L)$ it can easily be checked that the following result holds

THEOREM 5.5. $(\mathcal{E}^f, Dom(\mathcal{E}^f))$ *is a quasi-regular non local Dirichlet form*

Moreover from Theorem 2.11 and Theorem 5.3 it follows

THEOREM 5.6. $(-f(-\nabla^\star\nabla), D)$ *is essentially self- adjoint and its closure is* $(-f(-\nabla^\star\nabla), Dom(f(-\nabla^\star\nabla)))$

From Theorem 4.2 it follows:

THEOREM 5.7. *Up to m-equivalence there is a unique m-symmetric, m-tight special standard process* $M^f = (\Omega, \mathcal{F}^f_\infty, (X_t^f)_{t\geq 0}, (P_z^f)_{z\in E_\Delta})$ *satisfying the following martingale problem:*
for any $u \in D$

$$M_t^u := \tilde{u}(X_t) - \tilde{u}(X_0) + \int_0^t f(-\nabla^\star\nabla)u(X_s)ds \tag{5.6}$$

is an $\mathcal{F}_t^f$*-Martingale under* $(P_z^f)_{z\in E_\Delta}$ *for* $\mathcal{E}^f$*-q.e. starting point* $X_0 = z \in E$, *where* $\mathcal{F}_t^f$ *is the natural filtration, and* $\tilde{u}$ *is any* $\mathcal{E}^f$*-quasi continuous version of* u.

REMARK 5.8. We can also consider the Ornstein Uhlenbeck semigroup on an abstract Wiener space (E, H, μ_A), with μ_A a Gaussian measure with covariance operator given by a strictly positive self-adjoint operator A in H. In this case, the Dirichlet form $(\mathcal{E}, Dom(\mathcal{E}))$ is the closure of the symmetric bilinear form $(\mathcal{E}, D)$, with $D = \mathcal{F}C_b^\infty$, or $D = \mathcal{F}P$ and with

$$\mathcal{E}(u,v) = \int <\nabla u(\omega), \nabla v(\omega)>_H \mu_A(d\omega), \tag{5.7}$$

$(\mathcal{E}, Dom(\mathcal{E}))$ is a quasi- regular local Dirichlet form, and the generator $L = \nabla^\star\nabla$ acting on $D = \mathcal{F}C_b^\infty$, or $D = \mathcal{F}P$ is essentially self-adjoint. Applying again the results of the previous Section there exists a unique solution to the martingale problem for $-(f(-\nabla^\star\nabla), D)$ with $D = \mathcal{F}C_b^\infty$, or $D = \mathcal{F}P$.

REMARK 5.9. If, as in the above examples, there is a unique solution to the ($\mathcal{E}$-) martingale problem on $\mathcal{F}P(E) := \mathcal{F}P$ the corresponding Stochastic Differential Equation (SDE) is determined by the finite dimensional SDE satisfied by $< k, X_t >$, with $k \in H$. In fact by the Ito-formula the SDE satisfied by $u(X_t)$ is then uniquely given for any $u \in \mathcal{F}P$. In [**9**] we use this remark to construct the SDE corresponding to the $\mathcal{E}^f$- MPs.

Acknowlegements. We thank Sergio Albeverio for allowing us to use in this paper part of our joint results from [**9**], and for many fruitful discussions. We thank Prof. Kondratiev, Prof. Mandrekar, Prof. Röckner, Dipl.Math. Ivan Veselic and the "group of Bonn" for useful comments and fruitful discussions. The first author acknowledges the financial supports of the European Community (TMR-fellowship of stochastic analysis and its applications), and " Wissenschaftsministerium Nordrhein- Westfalen" for the "Lise-Meitner-Habilitationsstipendium". The second author acknowledges the financial support of DFG through the project "Stochastische Analysis und Systeme mit unendlich vielen Freiheitsgraden".

References

[1] Albeverio S. , Høegh-Krohn R., Quasi-invariant measures, symmetric diffusion processes and quantum field, Proc. Les methodes mathematiques de la theorie quantique des champes, Colloques Internationaux du Centre Nat. Rech. Sci. Marseille, 23-27 juin 75, CNRS 1976

[2] Albeverio S. , Høegh-Krohn R.,Dirichlet forms and diffusion processes on rigged Hilbert spaces, Z. Wahr. theo. verw. Geb. **40** (1977), 1-57

[3] Albeverio S. , Høegh-Krohn R.,Streit , L., Energy forms, Hamiltonians, distorted Brownian paths, Journal Mathematical Physics **18** (1997) 907-917

[4] Albeverio S. ,Ma Z.M., Röckner M., A Beurling-Deny type structure theorem for Dirichlet forms on general state space, Memorial Volume for R. Høegh-Krohn Vol I, Ideas and methods in math.an., stoch. and appl, Ed. Albeverio S., Holden H., Lindstrϕm T., Cambridge University Press 1992

[5] Albeverio S. ,Ma Z.M., Röckner M., Quasi-regular Dirichlet forms and Markov processes, J. Funct. Anal. **111** (1993), 118-154

[6] Albeverio S. , Röckner M., Stochastic differential equations in infinite dimension: solutions via Dirichlet forms. Prob. Prob. Th. Rel. Fiels 89 (1992), 347-386

[7] Albeverio S. , Röckner M., Classical Dirichlet forms on topological vector spaces- Closability and a Cameron-Martin Formula. Jour. Funct.An. **88** (1990) 395-436

[8] Albeverio S. , Röckner M., Dirichlet Form Methods for Uniqueness of Martingale Problems and Applications, Proceedings to the "1993 Summer Research Institute on Stochastic Analysis, Cornell, July 1993", Eds. R. Dürrett et al., Symp. AMS

[9] Albeverio S., Rüdiger B., Wu J.-L., Processes with jumps on infinite dimensional state spaces obtained via subordination. in preparation

[10] Berg Ch., Boyadzhiev Kh., deLaubenfels R., Generation of generators of holomorphic semigroups. J.Austral.Math.Soc. (Ser A) **55** (1993), 246-269

[11] van Casteren J.A., Some Problems in Stochastic Analysis and Semigroup Theory. preprint Universitaire Instelling Antwerpen, 99-10

[12] Fukushima M., On a stochasic calculus related to Dirichlet forms and distorted Brownian motion. Phys. Rep. 77 (1981); 255-262

[13] Fukushima M.,Dirichlet forms and Markov processes. Amsterdam: North Holland 1991

[14] Fukushima M.,Oshima Y., Takeda M. , Dirichlet forms and symmetric Markov processes. Walter de Gruyter, Berlin, 1994
[15] Gottschalk H., Wu J.-L., On the formulation of SPDE's leading to relativistic QFT with indefinite metric and non trivial S-matrix. Bonn SFB 256 preprint Nr. 608, 1999.
[16] Hoh W., The martingale problem for a class of pseudo -differential operators. Math. Ann. , 300: 121-147, 1994
[17] Jacob N., Pseudo-differential operators and Markov processes. Mathematical Research **Vol 94** Akademie Verlag,Berlin
[18] Klafter J., Shlesinger F., Zumofen G., Beyond Brownian motion. Physics today **February** (1996)
[19] Ma Z.M., Röckner M., An introduction on the theory of (non-symmetric) Dirichlet forms. Springer Verlag, Berlin 1992
[20] Nualart D., The Malliavin Calculus and related Topics. Probability and its Applications, Springer Verlag 1995
[21] Philipps, R.S. , On the generation of semigroups of linear operators. Pacific J. Math. 2 (1952), 343-369
[22] Schilling R. , On the domain of the generator of a subordinate semigroup. Potential Theory -ICTP94, Eds. Kraĺ , Lokes, Netuka, Vesely, Walter de Gruyter and Co., Berlin New York 1996
[23] Shlesinger F., Zalavski G.M., Frisch U., Lévy flights and related topics in Physics. Proceedings of the International Workshop held in Nice, France, 27-30 June 1994
[24] Watanabe S., Lectures on Stochastic Differential Equations and Malliavin Calculus. Tata Institute of Fundamental Research , Bombay 1984
[25] West J.B., An essay of the importance of being non linear. Lecture Notes in Biomathematics, Springer Verlag **62**
[26] Yosida, Fractional Powers of infinitesimal generators and the analyticity of the semigroup generated by them . Proc. Japan. Acad. **36** (1960)

Institut für Angewandte Mathematik der Universität Bonn, D-53013 Bonn, Germany

Institut für Angewandte Mathematik der Universität Bonn, D-53013 Bonn, Germany; and Institute of Applied Mathematics, Academia Sinica, Beijing 100080, China

Canadian Mathematical Society
Conference Proceedings
Volume **29**, 2000

Diff$_+(S^1)/S^1$ AS A SPACE OF COMPLEX STRUCTURES ON LOOP SPACES OF COMPACT LIE GROUPS

Armen Sergeev

To Sergio Albeverio on the occasion of his 60th birthday.

ABSTRACT. The homogeneous Fréchet space Diff$_+(S^1)/S^1$ is realized as a space of complex structures on the loop space of a compact Lie group, compatible with the symplectic structure. The Kähler geometry of this manifold is analyzed. We define a symplectic twistor bundle over the loop space, having Diff$_+(S^1)/S^1$ as a fibre, and present a scheme of twistor quantization of the loop space.

The study of the Kähler geometry of Diff$_+(S^1)/S^1$ was initiated in [**6**] and [**10**]. In [**10**] several models for this space were proposed. In particular, it was realized as an orbit of the coadjoint representation of the Virasoro group, provided with the canonical symplectic form. This realization is closely related to two other models for Diff$_+(S^1)/S^1$ as the moduli space of Hill's operators under the action of the diffeomorphism group Diff$_+(S^1)$ and as the space of projective structures (cf. [**8**]). Another realization, which is important for the Kähler geometry of Diff$_+(S^1)/S^1$, identifies Diff$_+(S^1)/S^1$ as the space of univalent holomorphic functions on the unit disc, smooth up to the boundary (cf. [**9**]). We briefly recall these two realizations of Diff$_+(S^1)/S^1$ in Sec.2.

In [**6**] still another model for Diff$_+(S^1)/S^1$ was proposed. Namely, this space was realized as a space of complex structures on the space of loops in Minkowski space. This realization was later extended to loop spaces of compact Lie groups in [**11, 12**] (cf. also [**7**]). We have collected basic results on the Kähler geometry of this model in Sec.2.

The realization of Diff$_+(S^1)/S^1$ as a space of complex structures, compatible with the symplectic structure, on loop spaces of compact Lie groups is closely related to the geometric quantization of these spaces. In Sec.3 we define a symplectic twistor bundle over the loop space of a compact Lie group, having Diff$_+(S^1)/S^1$ as a fibre. This bundle, as in the conventional twistor theory, has a natural almost complex structure, which makes it possible to reformulate a geometric quantization

2000 *Mathematics Subject Classification.* Primary 22E67, 58D05, 81T30.

Partly supported by Russian Foundation for Basic Research (grants 98-01-00640, 96-15-96131).

problem in twistor terms. In Sec.3 a scheme of twistor quantization of loop spaces of compact Lie groups is presented, the details may be found in [**13, 7**].

1. Loop spaces of compact Lie groups as Kähler Frechet manifolds

1.1. Loop spaces of compact Lie groups. Let G be a connected compact Lie group. Denote by $LG = \mathrm{Map}(S^1, G)$ the corresponding loop group, i.e. the space of smooth maps $S^1 \to G$. It is a Lie-Frechet group with the group law, given by the pointwise multiplication. The Frechet structure on $LG = \mathrm{Map}(S^1, G)$ is induced by the exponential map $\exp : \mathfrak{g} \to G$, where $\mathfrak{g}$ is the Lie algebra of G.

In more detail, the exponential exp defines a local diffeomorphism of some neighborhood $\mathfrak{u}$ of zero in the Lie algebra $\mathfrak{g}$ onto a neighborhood U of the identity e in the group G. This diffeomorphism generates a local chart of LG in a neighborhood of the identity in LG. Indeed, the identity in LG is given by the map $\mathbf{e} : S^1 \to e \in G$ and we take for a coordinate neighborhood of $\mathbf{e}$ the open subset $\mathbf{U} := \mathrm{Map}(S^1, U)$ in LG. The corresponding local chart is given by the bijection

$$\chi : \quad \mathfrak{U} = \mathrm{Map}(S^1, \mathfrak{u}) \longrightarrow \mathrm{Map}(S^1, U) = \mathbf{U} ,$$

induced by the exponential $\exp : \mathfrak{u} \to U$. Then the inverse map $\varphi_{\mathbf{e}} := \chi^{-1} : \mathbf{U} \to \mathfrak{U}$ is a bijection of $\mathbf{U}$ onto an open subset $\mathfrak{U}$ in the Frechet space $\mathrm{Map}(S^1, \mathfrak{g})$.

In the same way, a local chart in the neighborhood $\mathbf{U}_\gamma := \gamma \cdot \mathbf{U}$ of an arbitrary point $\gamma \in \mathrm{Map}(S^1, G)$ is given by the composition of the left shift $\gamma^{-1} : \mathbf{U}_\gamma \to \mathbf{U}$ and the map $\varphi_{\mathbf{e}} : \mathbf{U} \to \mathfrak{U} \subset \mathrm{Map}(S^1, \mathfrak{g})$:

$$\varphi_\gamma := \varphi_{\mathbf{e}} \circ \gamma^{-1} : \mathbf{U}_\gamma \to \mathfrak{U} .$$

The coordinate neighborhoods $\{\mathbf{U}_\gamma\}$ and local charts $\{\varphi_\gamma\}$, $\gamma \in \mathrm{Map}(S^1, G)$, define a Frechet manifold structure on LG with the local model, given by the Frechet space $\mathrm{Map}(S^1, \mathfrak{g})$.

We define the associated loop space ΩG to be the Frechet homogeneous space (of right conjugate classes) of the loop group LG of the form

$$\Omega G = LG/G ,$$

where G in the denominator is identified with the group of constant maps $S^1 \to g_0 \in G$.

The loop group LG acts on ΩG by left shifts. We denote by o the origin (neutral element) in ΩG, identified with the class of constant maps

$$o := [\mathbf{e}] = [G] .$$

The loop space ΩG (as a homogeneous space) can be identified with the subgroup $L_1 G$ of based maps in LG, sending $1 \in S^1$ to the identity of G: $\gamma(1) = e \in G$, i.e. with a closed Frechet submanifold (of codimension 1) of the Frechet manifold LG.

1.2. Symplectic structure. We shall define a symplectic structure on ΩG, which is invariant under the action of the loop group LG. Such a structure is uniquely determined by its value at the origin $o \in \Omega G$, i.e. by its restriction to the tangent space $T_o(\Omega G)$, which can be identified with the Frechet space $\Omega\mathfrak{g} := L\mathfrak{g}/\mathfrak{g}$. In terms of Fourier decompositions vectors from this space are given by the series

$$\xi(z) = \sum_{k \neq 0} \xi_k z^k \quad , \ z = e^{i\theta} \ , \quad \xi_k \in \mathfrak{g}^{\mathbb{C}} ,$$

where $\xi_{-k} = \bar{\xi}_k$ and "bar" stands for the complex conjugation in the complexified Lie algebra $\mathfrak{g}^{\mathbb{C}}$. By dropping the reality condition, we obtain a description of the complexified tangent space $T_o^{\mathbb{C}}(\Omega G) = \Omega\mathfrak{g}^{\mathbb{C}} :== L\mathfrak{g}^{\mathbb{C}}/\mathfrak{g}^{\mathbb{C}}$.

We fix now an inner product $<\cdot,\cdot>$ on the Lie algebra $\mathfrak{g}$, invariant under the adjoint action Ad of G on $\mathfrak{g}$, and define a 2-form ω_0 on $L\mathfrak{g} = \mathrm{Map}(S^1, \mathfrak{g})$ by the formula

$$\omega_0(\xi,\eta) = \frac{1}{2\pi}\int_0^{2\pi} < \xi(e^{i\theta}), \frac{d\eta(e^{i\theta})}{d\theta} > d\theta \tag{1}$$

for $\xi = \xi(e^{i\theta}), \eta = \eta(e^{i\theta}) \in L\mathfrak{g}$. It is a skew-symmetric bilinear form on $L\mathfrak{g}$, which vanishes if at least one of the maps ξ, η is constant. Because of the invariance of $<\cdot,\cdot>$, the form ω_0 is invariant under the adjoint action of the group G of constant loops on $L\mathfrak{g}$. By pushing ω_0 down to $\Omega\mathfrak{g}$, we get a non-degenerate form on $\Omega\mathfrak{g}$, which extends to an invariant non-degenerate 2-form ω on ΩG. The cocycle condition

$$\omega([\xi,\eta],\zeta) + \omega([\eta,\zeta],\xi) + \omega([\zeta,\xi],\eta) = 0 ,$$

which can be checked by a direct computation, implies that ω is closed on ΩG and thus determines an invariant symplectic structure on ΩG.

The choice of the formula (1) for the symplectic structure ω on ΩG looks ambiguous, however one can show that ω is uniquely determined by the invariant inner product $<\cdot,\cdot>$ on $\mathfrak{g}$ in the case of a semisimple group G. More precisely, we have the following

THEOREM 1.1 ([**14**]). *If the group G is semisimple, then any 2-form ω_0 on $L\mathfrak{g}$, satisfying the cocycle condition and invariant under the adjoint action of G on $L\mathfrak{g}$, is given by the formula* (1) *for some symmetric invariant bilinear form $<\cdot,\cdot>$ on $L\mathfrak{g}$.*

1.3. Complex structure. A complex structure on ΩG is provided by the "complex" representation of ΩG as a homogeneous space of a complex Lie-Frechet group. More precisely, we have the following equality

$$\Omega G = LG/G = LG^{\mathbb{C}}/\,L^+G^{\mathbb{C}} , \tag{2}$$

where $L^+G^{\mathbb{C}} = \mathrm{Hol}(\Delta, G^{\mathbb{C}})$ is the subgroup of maps from $LG^{\mathbb{C}} = \mathrm{Map}(S^1, G^{\mathbb{C}})$, which extend to holomorphic (smooth up to the boundary) maps of the unit disc $\Delta \subset \mathbb{C}$ into the complexified Lie group $G^{\mathbb{C}}$.

For the unitary group $G = \mathrm{U}(n)$ the equality (2) means, in other words, that any smooth non-degenerate (i.e. with values in $\mathrm{GL}(n,\mathbb{C})$) matrix function $T(z)$ on the circle may be written as the product

$$T(z) = U(z)\cdot H_+(z) , \quad z = e^{i\theta} , \tag{3}$$

where $U(z)$ is a smooth unitary (i.e. with values in $\mathrm{U}(n)$) matrix function and $H_+(z)$ extends to a holomorphic (smooth up to the boundary) non-degenerate matrix function in the disc Δ. It is a parametric analogue of the standard representation of a non-degenerate matrix as the product of a unitary and an upper-triangular matrices.

The representation (3) is unique, if one fixes the value of U at 1 by assuming that $U \in L_1\mathrm{U}(n)$. Moreover, the product map $(U, H_+) \mapsto U \cdot H_+$ is a diffeomorphism $\Omega\mathrm{U}(n) \times L^+\mathrm{GL}(n,\mathbb{C}) \to L\mathrm{GL}(n,\mathbb{C})$. In the same way one should understand the equality (2) for an arbitrary (simply connected compact) Lie group G. Namely, we have the following

THEOREM 1.2 ([**14, 7**]). *The product map*

$$\Omega G \times L^+G^{\mathbb{C}} \longrightarrow LG^{\mathbb{C}}$$

is a diffeomorphism of Frechet manifolds.

The "complex" representation (2) yields a complex structure J^0 on ΩG. This complex structure is LG-invariant and its restriction J^0_o to the tangent space $T^{\mathbb{C}}_o(\Omega G)$ may be described by an explicit formula. Namely, the value of J^0_o on a vector $\xi = \sum_{k\neq 0} \xi_k z^k \in \Omega\mathfrak{g}^{\mathbb{C}}$ is given by the formula

$$J^0_o\xi = -i\sum_{k>0} \xi_k z^k + i\sum_{k<0} \xi_k z^k \ .$$

The subspace $T^{1,0}_o(\Omega G)$ for this complex structure consists of vectors $\sum_{k<0}\xi_k z^k$ and $T^{0,1}_o(\Omega G)$ of vectors, having the form $\sum_{k>0}\xi_k z^k$.

In order to introduce local complex coordinates on ΩG, we use again the "complex" representation (2) and the Birkhoff theorem. Namely, denote by $L^-_1 G^{\mathbb{C}}$ the closed subgroup of $LG^{\mathbb{C}}$, consisting of maps $\gamma \in LG^{\mathbb{C}}$, which extend to holomorphic, smooth up to the boundary, maps of the complement Δ_- of the unit disc $\overline{\Delta}$ on the Riemann sphere $\mathbb{CP}^1$, taking the value $e \in G^{\mathbb{C}}$ at infinity:

$$L^-_1 G^{\mathbb{C}} = \{\gamma \in \mathrm{Hol}(\Delta_-, G^{\mathbb{C}}) : \gamma(\infty) = e\} \ .$$

It follows from the Birkhoff's theorem (cf. [**3, 4**]) that for all $\gamma \in LG^{\mathbb{C}}$ in a neighborhood of the identity $\mathbf{e} \in LG^{\mathbb{C}}$ the following representation holds

$$\gamma = \gamma_+ \cdot \gamma_- \ ,$$

where $\gamma_+ \in L^+G^{\mathbb{C}}$, $\gamma_- \in L^-_1 G^{\mathbb{C}}$. The maps $\gamma_\pm$ are uniquely determined by γ and the product map

$$L^+G^{\mathbb{C}} \times L^-_1 G^{\mathbb{C}} \longrightarrow LG^{\mathbb{C}}$$

is a local diffeomorphism. In particular, ΩG is locally diffeomorphic to the complex Lie–Frechet group $L^-_1 G^{\mathbb{C}}$.

We note that the symplectic structure ω and the complex structure J^0 on ΩG are compatible in the sense that $\omega(J^0_o\xi, J^0_o\eta) = \omega(\xi,\eta)$ for any $\xi,\eta \in T_o(\Omega G)$ and the symmetric form

$$g^0_o(\xi,\eta) := \omega(\xi, J^0_o\eta)$$

on $T_o(\Omega G) \times T_o(\Omega G)$ is positive definite. In other words, because of the invariance of ω and J^0, g^0_o extends to an invariant Riemannian metric g^0 on ΩG. These two properties together mean that ΩG is a Kähler Frechet manifold, provided with a Kähler metric g^0.

1.4. Flag manifolds. The existence of two kinds of representations of the loop space ΩG as a homogeneous space of a real and a complex Lie–Frechet groups

$$\Omega G = LG/G = LG^{\mathbb{C}}/L^+G^{\mathbb{C}}$$

is a characteristic property of ΩG. Finite-dimensional Kähler manifolds, having an analogous property, are called flag manifolds.

More precisely, a flag manifold F of a compact semisimple Lie group G can be represented in the form

$$F = G/G \cap P = G^{\mathbb{C}}/P$$

where $G^{\mathbb{C}}$ is the complexification of G and P is a parabolic (triangular) subgroup of $G^{\mathbb{C}}$. The complex representation $F = G^{\mathbb{C}}/P$ implies that F is a complex manifold, which has a G-invariant complex structure. The real representation implies that F is compact and Kähler. Note also that F is simply connected, if the group G has this property. One can show that the flag manifolds F exhaust all simply connected compact Kähler G-manifolds with a transitive action of a compact semisimple Lie group G (cf. [**5, 16**]).

An important example of flag manifolds of G is given by the manifold of full flags, having a homogeneous space representation of the form

$$F_T = G/T = G^{\mathbb{C}}/B_+ ,$$

where T is a maximal torus in G, B_+ is the standard Borel subgroup in $G^{\mathbb{C}}$. The importance of this flag manifold is due to the fact that, by the Borel–Weil construction (cf., e.g., [**2**]), all irreducible representations of G can be realized in the spaces of holomorphic sections of complex line bundles $L_\lambda \to F_T$, associated with antidominant lowest weights λ of $\mathfrak{g}^{\mathbb{C}}$.

The loop space ΩG may be considered as a universal flag manifold of the group G in the sense that all flag manifolds of G can be canonically embedded in ΩG.

To show that, we note, first, that the real representation $F = G/L$ of a flag manifold F can be interpreted, otherwise, as the realization of F as an orbit of the adjoint action Ad of the group G on its Lie algebra $\mathfrak{g}$ (or as an orbit of the coadjoint action Ad^* of G on the dual Lie algebra $\mathfrak{g}^*$). Namely, the orbit of an element $\xi \in \mathfrak{g}$ under the adjoint action has the form

$$G/G(\xi) ,$$

where the isotropy subgroup $G(\xi)$ at ξ coincides with the centralizer of ξ, i.e. with $G(\xi) = \{g \in G : \mathrm{Ad}\, g(\xi) = \xi\}$. All such orbits are flag manifolds and, conversely, every flag manifold of a compact semisimple Lie group may be realized in this way.

Consider now the natural S^1-action of the circle group on the loop space ΩG (identified with the subgroup L_1G of LG), given by the rotation of loops:

$$\lambda \cdot \gamma(z) = \gamma(\lambda)^{-1}\gamma(\lambda z) , \quad \lambda \in S^1 ,$$

where $\gamma \in \Omega G$. A loop γ is the fixed point of S^1-action if and only if

$$\gamma(\lambda z) = \gamma(\lambda)\gamma(z) \quad \text{for all}\, \lambda, z \in S^1 ,$$

i.e. if γ is a homomorphism $S^1 \to G$. If $\gamma : S^1 \to G$ is a homomorphism, then so are all loops $\gamma_g = g\gamma g^{-1}$, $g \in G$, conjugated to γ. Hence, the set of all such loops (conjugacy classes of a homomorphism γ) is parametrized by points of the homogeneous space

$$F_\gamma = G/G(\gamma) ,$$

where $G(\gamma)$ is the centralizer of the 1-parameter subgroup $\gamma(S^1)$ in G. The homogeneous space F_γ can be identified, as we have just pointed out, with a flag manifold of the group G.

Hence, the set of fixed points of the S^1-action on ΩG is the disjoint union

$$\mathrm{Fix}(S^1) = \bigcup_\gamma F_\gamma$$

of flag manifolds F_γ, where γ runs over the set of conjugacy classes of homomorphisms $S^1 \to G$, and flag manifolds F_γ are embedded into ΩG as finite-dimensional Kähler submanifolds. This relation between flag manifolds and fixed points of the S^1-action on ΩG was pointed out in [**1**].

Along with ΩG, one can also define a "loop-space" analogue of the full flags manifold F_T. Namely, we fix again a maximal torus T in G and set

$$\Omega_T G = LG/T \ .$$

It is also a Kähler Frechet manifold. In order to define a symplectic structure on $\Omega_T G$, note that the loop group LG, as a manifold, is diffeomorphic to the direct product $G \times \Omega G$. This diffeomorphism maps a loop $\gamma \in LG$ into the pair $(\gamma(1), \gamma(1)^{-1}\gamma) \in G \times \Omega G$, where ΩG is identified with the subgroup $L_1 G = \{\gamma \in LG : \gamma(1) = e\}$ in LG. Hence, as a Frechet manifold, $\Omega_T G = LG/T = G/T \times \Omega G$. A symplectic structure on $\Omega_T G$ is generated by the symplectic structure on ΩG, defined above, and Kirillov symplectic structure on the manifold of full flags $F_T = G/T$. (Recall that flag manifolds of G can be considered as orbits of the coadjoint representation of G on the dual Lie algebra $\mathfrak{g}^*$, which have Kirillov symplectic structure).

A complex structure on $\Omega_T G$ is provided, as in the case of ΩG, by the "complex" representation of $\Omega_T G$ as a homogeneous space of the complexified loop group of the form

$$\Omega_T G = LG^{\mathbb{C}} / B_+ G^{\mathbb{C}} \ ,$$

where $B_+ G^{\mathbb{C}}$ is the subgroup in $L_+ G^{\mathbb{C}} = \mathrm{Hol}(\Delta, G^{\mathbb{C}})$, consisting of maps $\gamma \in \mathrm{Map}(S^1, G^{\mathbb{C}})$, which can be extended to holomorphic (smooth up to the boundary) maps of the unit disc $\gamma : \Delta \to G^{\mathbb{C}}$, such that $\gamma(0) \in B_+$.

This complex structure on $\Omega_T G$ is compatible with the symplectic one and so defines the structure of a Kähler Frechet manifold on $\Omega_T G$.

2. $\mathrm{Diff}_+(S^1)/S^1$ as a Kähler Frechet manifold

We shall study now the action of the diffeomorphism group of the circle on loop spaces ΩG.

2.1. The diffeomorphism group of the circle. The group $\mathrm{Diff}(S^1)$ consists of diffeomorphisms of the circle S^1 with the group law, given by the composition of diffeomorphisms. This group is an open subset in the Frechet manifold $\mathrm{Map}(S^1, S^1)$ and is provided with the induced structure of a Frechet manifold, so that it is a Lie-Frechet group. $\mathrm{Diff}_+(S^1)$ has two connected components and we denote by $\mathrm{Diff}_+(S^1)$ the connected component of the identity, formed by the diffeomorphisms of S^1, preserving the orientation.

By pulling up to the universal covering $\mathbb{R} \to \mathbb{R}/2\pi\mathbb{Z} \approx S^1$, we can associate with a diffeomorphism $f \in \mathrm{Diff}_+(S^1)$ a smooth function $h : \mathbb{R} \to \mathbb{R}$, which is defined up to a multiplicative constant $2\pi n$, $n \in \mathbb{Z}$, and satisfies the relation $h(x + 2\pi) = h(x) + 2\pi$, $x \in \mathbb{R}$. It follows that the derivative $h'(x)$, which is already uniquely determined by f, is a smooth 2π-periodic function on $\mathbb{R}$. Moreover, $h'(x) > 0$ and the average of $h'(x)$ over S^1 is equal to 1. Hence, we have associated with $f \in \mathrm{Diff}_+(S^1)$ a smooth 2π-periodic function h' on $\mathbb{R}$, which is strictly positive and has the average 1 over the period. We denote the set of such functions by C. It is a convex open subset in a closed affine subspace of codimension 1 in the Frechet space of smooth 2π-periodic functions on $\mathbb{R}$. The function h is uniquely determined by $h' \in C$ up to an additive constant $h(0) \in \mathbb{R}$ and also h is defined by $f \in \mathrm{Diff}_+(S^1)$ up to an additive constant $2\pi n$, $n \in \mathbb{Z}$. So f is uniquely determined by $h' \in C$ up to an element of $\mathbb{R}/2\pi\mathbb{Z} \approx S^1$. Hence, $\mathrm{Diff}_+(S^1)$ is diffeomorphic to $S^1 \times C$ and, in particular, homotopy equivalent to S^1.

The Lie algebra of $\mathrm{Diff}(S^1)$ concides with the algebra $\mathrm{Vect}(S^1)$ of smooth vector fields on the circle S^1. In other words, $\mathrm{Vect}(S^1)$ concides with the Frechet space of smooth sections of the tangent bundle of the circle $TS^1 \to S^1$. Elements $v \in \mathrm{Vect}(S^1)$ are written down in the form $v = v(\theta)\frac{d}{d\theta}$, where $v(\theta)$ is a smooth 2π-periodic function of θ. The bracket of two vector fields $v_1, v_2 \in \mathrm{Vect}(S^1)$ is given by the standard formula

$$\left[v_1(\theta)\frac{d}{d\theta}, v_2(\theta)\frac{d}{d\theta}\right] = \{v_1(\theta)v_2'(\theta) - v_1'(\theta)v_2(\theta)\}\frac{d}{d\theta},$$

so that $\mathrm{Vect}(S^1)$ is a Lie–Frechet algebra.

In terms of Fourier decompositions, the coefficients $v(\theta)$ of vector fields $v = v(\theta)\frac{d}{d\theta}$ are given by the series

$$v(\theta) = \sum_{n=-\infty}^{\infty} v_n e^{in\theta}, \quad v_n \in \mathbb{C},$$

subject to the relation: $v_{-n} = \bar{v}_n$.

2.2. The action of $\mathrm{Diff}_+(S^1)$ on loop spaces ΩG. The diffeomorphism group $\mathrm{Diff}_+(S^1)$ acts naturally on ΩG by the reparametrization of loops:

$$f_*\gamma(\theta) := \gamma(f(\theta)) \quad \text{for } \gamma \in LG,\ f \in \mathrm{Diff}_+(S^1).$$

It follows from the definition of symplectic structure ω on ΩG, associated with the form

$$\omega_0(\xi, \eta) = \frac{1}{2\pi}\int_0^{2\pi} < \xi(e^{i\theta}), \eta'(e^{i\theta}) > d\theta$$

on $L\mathfrak{g}$, that this action preserves ω, i.e. diffeomorphisms $f \in \mathrm{Diff}_+(S^1)$ generate symplectomorphisms of ΩG.

Consider next the action of $\mathrm{Diff}_+(S^1)$ on the complex structure J^0 on ΩG. A diffeomorphism $f \in \mathrm{Diff}_+(S^1)$ transforms it into a complex structure

$$J := f_*^{-1} \circ J^0 \circ f_*, \tag{4}$$

where f_* is the differential of f.

THEOREM 2.1 (cf. [**7**]). *The complex structure J coincides with J^0 if and only if the diffeomorphism $f \in \mathrm{Diff}_+(S^1)$ is a rotation, i.e. $f \in S^1 \subset \mathrm{Diff}_+(S^1)$.*

Indeed, if $f \in \mathrm{Diff}_+(S^1)$ does not change the complex structure, it also preserves the "complex" representation (2) of ΩG. This implies that f leaves the subspaces $L^+G^{\mathbb{C}}$ and $L_1^-G^{\mathbb{C}}$ in $LG^{\mathbb{C}}$ invariant, which is possible only if $f \in S^1$.

Recall that the complex structure J^0 on ΩG is invariant under left LG-shifts on ΩG and compatible with the symplectic structure ω on ΩG. The invariance of ω with respect to the action of $\mathrm{Diff}_+(S^1)$ implies that the complex structures J, obtained from J^0 by formula (4), are also invariant under left LG-shifts and compatible with ω. In particular, any such J generates an invariant Kähler metric g on ΩG, given by the formula

$$g_\gamma(\xi, \eta) := \omega_\gamma(\xi, J_\gamma\eta), \quad \gamma \in \Omega G,$$

for any $\xi, \eta \in T_\gamma(\Omega G)$.

We shall call the complex structures J on ΩG, obtained from J^0 by the action of $\mathrm{Diff}_+(S^1)$ according to the formula (4), the admissible complex structures on ΩG.

According to Theorem 2.1, admissible complex structures on ΩG are parametrized by points of the manifold $\mathrm{Diff}_+(S^1)/S^1$, which we shall study next.

2.3. Symplectic structure of $\mathrm{Diff}_+(S^1)/S^1$. The space $\mathbf{S} := \mathrm{Diff}_+(S^1)/S^1$ is a homogeneous space of the group $\mathrm{Diff}_+(S^1)$, which acts on $\mathbf{S}$ by left shifts. It can be identified (as a homogeneous space) with the subgroup of $\mathrm{Diff}_+(S^1)$, consisting of all diffeomorphisms $f \in \mathrm{Diff}_+(S^1)$, fixing the point $1 \in S^1$: $f(1) = 1$. This provides $\mathbf{S}$ with a Frechet manifold structure.

As in the case of ΩG, invariant geometric structures on ΩG are uniquely determined by their values at the origin $o \in \mathrm{Diff}_+(S^1)/S^1$.

The tangent space $T_o\mathbf{S}$ at the origin is identified with the space $\mathrm{Vect}_0(S^1)$, consisting of smooth 2π-periodic functions $\xi(\theta)$ with zero average: $\frac{1}{2\pi}\int_0^{2\pi}\xi(\theta)d\theta = 0$. In terms of Fourier decompositions tangent vectors $\xi \in T_o\mathbf{S}$ are given by the series

$$\xi = \sum_{n\neq 0} \xi_n e_n\ , \quad \xi_n \in \mathbb{C}\ ,$$

subject to the condition $\bar{\xi}_n = -\xi_n$, where $e_n = ie^{in\theta}\frac{d}{d\theta}$, $n = 0, \pm 1, \pm 2, \ldots$, is the standard base of the Lie algebra $\mathrm{Vect}(S^1)$.

In particular, an invariant symplectic structure on $\mathbf{S}$ is uniquely determined by its restriction to $T_o\mathbf{S}$, i.e. by a 2-cocycle w on the Lie algebra $\mathrm{Vect}(S^1)$, invariant under rotations. It is well known (cf., e.g., [**14**]), that all such cocycles are parametrized by two real numbers α and β, their values on the base elements $\{e_n\}$ are given by the formula

$$w(e_n, e_m) = (\alpha n^3 + \beta n)\delta_{n,-m}\ .$$

The form $w_{\alpha,\beta}$, corresponding to parameters α, β, is non-degenerate on $\mathrm{Vect}_0(S^1)$ if and only if $\alpha n^3 + \beta n \neq 0$ for all natural n. The latter condition is satisfied if either $\alpha = 0$, $\beta \neq 0$ or $-\beta/\alpha$ is not the square of a natural number. In the first case the form $w_{\alpha,\beta}$, as it's easy to see, is exact, we take the second possibility.

In a more invariant way the above cocycle $w_{\alpha,\beta}$ may be written in the form

$$\omega_{\alpha,\beta}(\xi,\eta) = \frac{1}{2\pi}\int_0^{2\pi} \xi(\theta)\left(\beta\eta'(\theta) - \alpha\eta'''(\theta)\right)d\theta\ ,$$

where $\xi, \eta \in \mathrm{Vect}_0(S^1)$.

The family of symplectic structures $w_{\alpha,\beta}$, determined by these cocycles, has a natural interpretation in terms of coadjoint orbits of the group $\mathrm{Diff}_+(S^1)$ on the Virasoro algebra vir^*. Recall that the Virasoro algebra vir is the central extension of the Lie algebra $\mathrm{Vect}(S^1)$, determined by the cocycle $w_{1,0}$ on $\mathrm{Vect}(S^1)$, generating the cohomology group $H^2(\mathrm{Vect}(S^1), \mathbb{R})$. The dual algebra $\mathrm{Vect}^*(S^1)$ may be identified with the space of quadratic differentials with distributional coefficients on S^1, so the regular part of $\mathrm{Vect}^*(S^1)$ may be identified with the space $Q(S^1)$ of quadratic differentials on S^1 of the form $q = q(\theta)(d\theta)^2$, where $q(\theta)$ is a smooth 2π-periodic function of θ. The coadjoint action of $\mathrm{Diff}_+(S^1)$ on the regular part of the dual Virasoro algebra, identified with $Q(S^1) \oplus \mathbb{R}$, is given by the formula (cf. [**8**])

$$f \in \mathrm{Diff}_+(S^1) \longmapsto K(f)(q,t) = (g^*(q) + tS(g)d\theta^2, t)\ ,$$

where $g = f^{-1}$ and $g^*(q)$ denotes the natural action of a diffeomorphism g on a quadratic differential q. The functional $S(\varphi)$ coincides with the Schwarz derivative of a diffeomorphism φ, i.e.

$$S(\varphi) = \left(\frac{\varphi''}{\varphi'}\right)' - \frac{1}{2}\left(\frac{\varphi''}{\varphi'}\right)^2 .$$

One can describe explicitly the orbits (under the action of $\mathrm{Diff}_+(S^1)$) of the elements $(q,t) = (qd\theta^2, t)$, where q and t are real numbers. If $2q/t$ is not the square of a natural number, then the orbit of $(qd\theta^2, t)$ coincides with $\mathbf{S}$. If $2q/t = n^2$, then the orbit of $(qd\theta^2, t)$ coincides with the homogeneous space

$$\mathrm{Diff}_+(S^1)/\,\mathrm{PSL}^{(n)}(2,\mathbb{R}) ,$$

where $\mathrm{PSL}^{(n)}(2,\mathbb{R})$ is the n-fold covering of the Möbius group $\mathrm{PSL}(2,\mathbb{R})$ of fractional-linear automorphisms of the unit disc Δ.

Recall now that the orbits of the coadjoint representation have a canonical symplectic structure, given by the Kirillov form. Identifying the homogeneous space $\mathbf{S}$ with the orbits of elements $(qd\theta^2, t)$ for the different choice of (q,t) with $2q/t \neq n^2$ and providing these orbits with the corresponding Kirillov forms, we reproduce again the above two-parameter family of symplectic structures on $\mathbf{S}$.

2.4. Complex structure of $\mathrm{Diff}_+(S^1)/S^1$. We switch now to the invariant complex structure on $\mathbf{S}$. Its restriction to $T_o\mathbf{S} = \mathrm{Vect}_0(S^1)$ is given by the Hilbert transform, associating with $\xi \in \mathrm{Vect}_0(S^1)$ the element

$$(I_o\xi)(\theta) = \frac{1}{2\pi}\int_0^{2\pi} \mathrm{ctg}\,\frac{\theta - \psi}{2} \cdot [\xi(\theta) - \xi(\psi)]\, d\psi , \quad 0 \le \theta \le 2\pi .$$

In terms of Fourier decompositions $\xi = \sum_{n\neq 0} \xi_n e_n \in \mathrm{Vect}_0(S^1)$ we obtain

$$I_o\xi = -i\sum_{n>0} \xi_n e_n + i\sum_{n<0} \xi_n e_n ,$$

so that the subspace $T_o^{0,1}(\mathbf{S})$ consists of the vectors of the form $\sum_{n>0}\xi_n e_n$ and $T_o^{1,0}(\mathbf{S})$ — of the vectors $\sum_{n<0}\xi_n e_n$. The complex structure I_o on $T_o\mathbf{S}$ uniquely determines a $\mathrm{Diff}_+(S^1)$-invariant almost complex structure I on $\mathbf{S}$ (obtained from I_o by left translations). It's easy to show that I is integrable, i.e. the Lie bracket of two $(1,0)$-vector fields is again a $(1,0)$-vector field. Moreover, I is the only integrable $\mathrm{Diff}_+(S^1)$-invariant almost complex structure on $\mathbf{S}$ (cf. [**15**]).

Moreover, one can introduce local complex coordinates on $\mathbf{S}$ (i.e. realize $\mathbf{S}$ as a complex Frechet manifold), using a realization of $\mathbf{S}$ as a space of holomorphic univalent functions in the disc from [**9**].

Consider the Frechet space $\mathbf{A}$ of all smooth complex-valued functions on the closure of the unit disc Δ, which are holomorphic inside Δ and vanish at the origin. Let $\mathbf{A}_0$ be the open subset of $\mathbf{A}$, consisting of all $f \in \mathbf{A}$ which are univalent on Δ and map diffeomorphically $S^1 = \partial\Delta$ onto its image. We denote by $\mathfrak{S}$ the set of functions $f \in \mathbf{A}_0$, such that $f'(0) = 1$. It is an open subset of the affine subspace of $\mathbf{A}$, consisting of $f \in \mathbf{A}$, subject to $f'(0) = 1$. Functions $f \in \mathfrak{S}$ are holomorphic and univalent in Δ, they map diffeomorphically $\overline{\Delta} \to \overline{f(\Delta)}$ and satisfy the normalization conditions $f(0) = 0$, $f'(0) = 1$.

Functions $f \in \mathfrak{S}$ have the power series decompositions of the form

$$f(z) = z + c_2 z^2 + c_3 z^3 + \dots .$$

Hence, $\mathfrak{S}$ is a complex Frechet manifold with the complex coordinates, given by the coefficients $(c_2, c_3, \dots)$.

THEOREM 2.2 (cf. [**9**]). *The space $\mathfrak{S}$ of smooth univalent functions on Δ is diffeomorphic to the space* $\mathbf{S}$ *of admissible complex structures on* ΩG.

This provides $\mathbf{S}$ with the structure of a complex Frechet manifold.

2.5. Kähler structure of $\mathbf{Diff}_+(S^1)/S^1$. The complex structure I on $\mathbf{S}$ is compatible with all symplectic structures $w_{\alpha,\beta}$ and so generates a family of Kähler metrics $g_{\alpha,\beta}(\xi,\eta) := w_{\alpha,\beta}(\xi, I\eta)$ on $\mathbf{S}$, given at the origin by the formula

$$g_{\alpha,\beta,o}(\xi,\eta) = \sum_{n\geq 1}(\alpha n^3 + \beta n)\xi_n\bar{\eta}_n \ ,$$

where $\xi = \sum_{n\neq 0}\xi_n e_n$, $\eta = \sum_{n\neq 0}\eta_n e_n \in T_o\mathbf{S}$. Thus, $\mathbf{S}$ is a Kähler Frechet manifold with a two-parameter family of Kähler metrics $g_{\alpha,\beta}$.

The Kähler metrics $g_{\alpha,\beta}$, being $\mathrm{Diff}_+(S^1)$-invariant, are uniquely determined by their restriction to the tangent space $T_o\mathbf{S}$ at the origin. The Ricci curvature of $g_{\alpha,\beta}$ in the standard base $\{e_n\}$ of $T_o\mathbf{S}$ was computed in [**6**]. Surprisingly, it is finite and given by the formula

$$\mathrm{Ric}_{-n,m} := \mathrm{Ric}(e_{-n}, e_m) = \left(-\frac{13}{3}n^3 + \frac{1}{3}n\right)\delta_{n,m} \ .$$

The coefficient of n in this formula is not essential, since it can be changed by modifying the symplectic structure $w_{\alpha,\beta}$ inside its cohomology class in $H^2(\mathrm{Vect}(S^1),\mathbb{R})$ (or, in physical terms, by changing the normal ordering). The coefficient of n^3 does not alter inside the given cohomology class of $w_{\alpha,\beta}$ and so is more interesting. First of all, note that it does not depend on the values of parameters α and β (assuming, that $\alpha \neq 0$). Secondly, one can consider the Lie algebra $\mathrm{Vect}(S^1)$ as an infinite-dimensional analogue of the semisimple Lie algebra $\mathrm{sl}(2,\mathbb{R})$ and, accordingly, the space $\mathrm{Diff}_+(S^1)/S^1$ as a Frechet analogue of the unit disc $\mathrm{SL}(2,\mathbb{R})/S^1$. In this way, one should think of the $\mathrm{Diff}_+(S^1)$-invariant Kähler metrics $g_{\alpha,\beta}$ on $\mathrm{Diff}_+(S^1)/S^1$ as infinite-dimensional analogues of the $\mathrm{SL}(2,\mathbb{R})$-invariant Poincaré metric on the disc. They have the Ricci curvature, equal (essentially) to $-\frac{13}{3}n^3$, which should be compared to the curvature $-\frac{1}{4}$ of the Poincaré metric on the disc.

3. A symplectic twistor bundle over ΩG and the geometric quantization of loop spaces

3.1. A twistor bundle over ΩG. Consider the bundle $\pi : Z \to \Omega G$ of admissible complex structures on ΩG with the fibre $\pi^{-1}(\gamma)$ at $\gamma \in \Omega G$, given by the restrictions J_γ of admissible complex structures J on ΩG to the tangent space $T_\gamma\Omega G$. Points $z \in Z$ are the pairs $z = (\gamma, J_\gamma)$, where $\gamma \in \Omega G$, J_γ is an admissible complex structure on $T_\gamma\Omega G$.

Admissible complex structures are invariant under the LG-action on ΩG, so there is a natural action of LG on Z and the quotient of this action coincides with the manifold $\mathbf{S} = \mathrm{Diff}_+(S^1)/S^1$. Thus we have a double fibration

$$\begin{array}{ccc} Z & \xrightarrow{\ p\ } & \mathbf{S} \\ {\scriptstyle\pi}\big\downarrow & & \\ \Omega G & & \end{array} \tag{5}$$

where p is the natural projection of Z onto $\mathbf{S}$. The fibre $p^{-1}(J)$ at $J \in \mathbf{S}$ can be identified with the complex manifold $(\Omega G, J)$, i.e. with the space ΩG, provided with the complex structure, corresponding to $J \in \mathbf{S}$ (we denote a point $J \in \mathbf{S}$ and the corresponding complex structure on ΩG by the same letter J).

As in the conventional twistor theory, the space Z of admissible complex structures on ΩG has a natural almost complex structure. Namely, the tangent bundle TZ of Z can be split into the direct sum $TZ = V \oplus H$ of vertical and horizontal subbundles. The vertical subbundle $V \subset Z$ coincides with the kernel of the differential π_* of projection $\pi : Z \to \Omega G$. Respectively, the horizontal subbundle $H \subset Z$ is given by the kernel of the differential p_* of projection $p : Z \to \mathbf{S}$. The fibre V_z at $z \in Z$ is identified by p_* with the tangent space $T_J\mathbf{S}$ at $J = p(z) \in \mathbf{S}$, and so has a natural complex structure $\mathbf{J}^v$, induced by the Kähler complex structure of $\mathbf{S}$. We define now an almost complex structure $\mathbf{J}$ on Z in terms of the splitting $TZ = V \oplus H$ as

$$\mathbf{J} = \mathbf{J}^v \oplus \mathbf{J}^h \ ,$$

where the complex structure $\mathbf{J}^h$ at $z \in Z$ is equal to the complex structure J_γ on $T_zH \approx T_{\pi(z)}\Omega G$, determined by the point $z = (\gamma, J_\gamma)$. Note that the projection $p : Z \to \mathbf{S}$ is holomorphic with respect to $\mathbf{J}$ (while π is not!).

The double fibration (5) is somewhat similar to the twistor fibration over hyperkähler manifolds. Recall that these are the Riemannian manifolds M, provided with three covariantly constant complex structures I, J, K, satisfying the usual quaternion relations. The fibre of the twistor bundle $\pi\colon Z \to M$ of admissible (compatible with the Riemannian metric) complex structures on M is identified with the sphere $S^2 \cong \mathbb{CP}^1$, consisting of complex structures of the form $aI + bJ + cK$, where (a, b, c) is a point of the unit sphere in $\mathbb{R}^3$. The twistor space Z has a natural complex structure and the natural projection $p\colon Z \to \mathbb{CP}^1$ is a holomorphic mapping. Hence, we have the commutative diagram

$$\begin{array}{ccc} Z & \xrightarrow[p]{} & \mathbb{CP}^1 \\ {\scriptstyle\pi}\big\downarrow & & \\ M & & \end{array} \tag{6}$$

The main difference between this twistor diagram and the double diagram (5) in the case of loop spaces ΩG is that in (6) we consider complex structures, compatible with the Riemannian metric, while in (5) the involved complex structures are compatible with the symplectic structure. This fact implies, in particular, that the fibre of $\pi : Z \to M$ in (6) coincides with the Riemann sphere and so has the positive curvature and a non-trivial topology, while in (5) it is identified with the "Frechet disc" $\mathrm{Diff}_+(S^1)/S^1$ and so is contractible and, in the above sense, negatively curved.

3.2. Twistor quantization of the loop space of a vector space. The space $\mathbf{S}$ plays a central role in the twistor approach to the geometric quantization of loop spaces. We recall here briefly the twistor quantization construction, details may be found in [**7**]. In this Subsection we present the main ideas for the case of the loop space of the d-dimensional vector space $\mathbb{R}^d$, where the role of the group G is played by the group $\mathbb{R}^d$ of translations of $\mathbb{R}^d$, and in the next Subsection we generalize them to the loop spaces of compact Lie groups.

We consider again the twistor double fibration (5) with $G = \mathbb{R}^d$. The twistor quantization of $\Omega\mathbb{R}^d$ in terms of (5) means (cf. [**14, 7**]) the construction of a holomorphic (Fock) Hilbert bundle over the space $\mathbf{S}$ of admissible complex structures on $\Omega\mathbb{R}^d$ and a flat unitary connection on it.

We start from the construction of a Fock bundle over $\mathbf{S}$. For that we should enlarge first our Frechet loop space $\Omega\mathbb{R}^d$ to a Hilbert space. The natural choice is the Sobolev space $V^d := H_0^{1/2}(S^1, \mathbb{R}^d)$, consisting of functions $v \in L^2(S^1, \mathbb{R}^d)$, which have the zero mean value over the circle and generalized derivative of order $1/2$ in $L^2(S^1, \mathbb{R}^d)$. (In some sense, it is the largest Sobolev space, for which our construction is still valid). In other words, the elements of V^d are the functions $v \in L^2(S^1, \mathbb{R}^d)$, having Fourier series

$$v(\theta) := v(e^{i\theta}) = \sum_{k \neq 0} v_k e^{ik\theta} , \quad v_k = \bar{v}_{-k}$$

with finite Sobolev norm

$$\|v\|_{1/2}^2 = \sum_{k \neq 0} |k| \|v_k\|^2 .$$

(In physical applications $\mathbb{R}^d$ is provided with the Lorentz metric of signature $(d-1, 1)$).

The definitions of the complex structure J^0 and symplectic structure ω, given in Sec.1 for the space ΩG, are easily extended to the space V^d so that V^d becomes a Kähler Hilbert space. Its complexification

$$V_{\mathbb{C}}^d = H_0^{1/2}(S^1, \mathbb{C}^d)$$

is a complex Hilbert space with the Hermitian inner product, defined by the Kähler metric on V^d. The symplectic form ω and complex structure operator J^0 extend complex linearly to $V_{\mathbb{C}}^d$ and determine a decomposition of $V_{\mathbb{C}}^d$ into the direct orthogonal sum

$$V_{\mathbb{C}}^d = W_+^0 \oplus W_-^0$$

of $(\pm i)$-eigenspaces of J^0 on $V_{\mathbb{C}}^d$, which are isotropic with respect to the symplectic form ω, i.e. $\omega(x, y) = 0$ if $x, y \in W_+^0$ or $x, y \in W_-^0$.

The action of the diffeomorphism group $\mathrm{Diff}_+(S^1)$ on the loop spaces ΩG by the reparametrization, defined in Sec.2.2, naturally extends to its action on the space V^d by symplectic transformations. Under the action of $f \in \mathrm{Diff}_+(S^1)$ the original complex structure J^0 on V^d transforms into another complex structure J, which coincides with J^0 iff $f \in S^1$. This complex structure J is also compatible with the symplectic form ω and so generates a decomposition

$$V_{\mathbb{C}}^d = W_+ \oplus W_- \tag{7}$$

into the direct sum of isotropic eigenspaces $W_\pm$ of the operator J on $V_{\mathbb{C}}^d$. The complex structure J, being compatible with ω, defines a Kähler metric on V^d, which extends to a Hermitian inner product on $V_{\mathbb{C}}^d$ and the decomposition (7) is orthogonal with respect to this inner product. We call again the complex structures J on V^d, obtained from J^0 by the action of $\mathrm{Diff}_+(S^1)$, admissible.

With any admissible complex structure on V^d, generating a decomposition (7), we associate a Fock space F_J of holomorphic functions on $V_{\mathbb{C}}^d$ (with respect to J), defined in the following way. We consider the algebra $\mathfrak{S}(W_+)$ of symmetric polynomials in variables $z \in W_+$ and provide it with a natural inner product, generated by the inner product on $V_{\mathbb{C}}^d$. The completion of $\mathfrak{S}(W_+)$ with respect to this inner product is called the Fock space $F_J = F(V_{\mathbb{C}}^d, J)$ over $V_{\mathbb{C}}^d$ with respect to the admissible complex structure J.

We introduce next the Heisenberg Lie algebra $\mathrm{heis}(V^d)$, which is the central extension of the Abelian Lie algebra V^d. In other words, it coincides with the vector space

$$\mathrm{heis}(V^d) = V^d \oplus \mathbb{R} ,$$

provided with the Lie bracket

$$[(x,s),(y,t)] := (0, \omega(x,y)) \ , \quad x, y \in V^d, \ s, t, \in \mathbb{R} .$$

This algebra has a standard (Heisenberg) irreducible representation r_J in the Fock space F_J, acting on vectors $f \in F_J$ by the formula

$$v \in V^d_{\mathbb{C}} \longmapsto r_J(v)f(z) = -\partial_v f(z) + < z, v > f(z) ,$$

where ∂_v is the derivative in the direction of v, and $< z, v >$ is the multiplication by the function $< z, v >$.

According to a theorem of Shale, all representations r_J are unitary equivalent. In other words, for any admissible complex structure J on V^d there exists a unitary operator $T_J : F_0 \to F_J$, intertwining the representation r_J in F_J with the representation r_0 in F_0, corresponding to the original complex structure J^0. Moreover, the map $\mathrm{Diff}_+(S^1) \ni f \longmapsto T_J : F_0 \to F_J$, where $J = J_f := f_* J$, defines a projective unitary action of $\mathrm{Diff}_+(S^1)$ (or a unitary action of the Virasoro group) on the union F of Fock spaces F_J over all admissible J. The infinitesimal version of this action yields a projective representation ρ_J of the Lie algebra $\mathrm{Vect}(S^1)$ of $\mathrm{Diff}_+(S^1)$ (or a representation of the Virasoro algebra vir) in the Fock space F_J. It is given explicitly in terms of the Virasoro operators L_n (cf. [**6, 14, 7**]).

This Fock space construction admits a natural interpretation in terms of the twistor bundle $\pi : Z \to V^d$ of admissible complex structures on V^d with the fibre at $v \in V^d$, formed by the restrictions of admissible complex structures J to the tangent space $T_v V^d \cong V^d$. We have again the twistor diagram

$$\begin{array}{ccc} Z & \xrightarrow{p} & \mathbf{S} \\ {\scriptstyle\pi}\big\downarrow & & \\ V^d & & \end{array}$$

where the fibre $p^{-1}(J)$ over $J \in \mathbf{S}$ is identified with the space $V_J = (V, J)$, provided with the complex structure J. We define an almost complex structure $\mathbf{J}$ on Z, as in Sec.3.1. Then the Fock space F_J for an admissible complex structure $J \in \mathbf{S}$ is identified with the Fock space $F(p^{-1}(J))$ of holomorphic functions on the fibre $p^{-1}(J)$ (with respect to the complex structure $\mathbf{J}$). The union of Fock spaces

$$F = \bigcup_{J \in \mathbf{S}} F_J$$

is a holomorphic Hilbert bundle over $\mathbf{S}$, which is trivial on $\mathbf{S}$ (since $\mathbf{S}$ is contractible). The holomorphic trivialization of H is given explicitly by the above map

$$T_J : F_0 \longrightarrow F_J .$$

Note that this trivialization is equivariant with respect to the action of the Virasoro group.

Recall that the infinitesimalization of this $\mathrm{Diff}_+(S^1)$-action yields a projective representation ρ_0 of the Lie algebra $\mathrm{Vect}(S^1)$ in the Fock space F_0. Having such a representation, we can construct a connection A on the Fock bundle F, whose curvature is equal to the cocycle of ρ_0 (cf. [**7**]). More explicitly, we identify $\mathbf{S} =$

$\text{Diff}_+(S^1)/S^1$ with the subgroup $\text{Diff}_1(S^1)$, which consists of diffeomorphisms in $\text{Diff}_+(S^1)$ fixing the point $1 \in S^1$. Then the covariant derivative of A at $f \in \text{Diff}_1(S^1)$ in the direction $\xi \in T_f\mathbf{S}$ is equal to

$$\nabla_{A,\xi} v = \rho_0(L_{f^{-1},*}\xi)v , \quad v \in F_0 ,$$

where $L_{f^{-1}}$ is the left translation on $\text{Diff}_1(S^1)$ by the element f^{-1}. This connection is invariant under the Virasoro group action.

The curvature of A at the origin $o \in \mathbf{S}$ in the standard base $e_n = ie^{in\theta}\frac{d}{d\theta} \in \text{Vect}_0(S^1)$ of $T_o\mathbf{S}$ is given by (cf. [**6, 12**])

$$R_A(e_{-n}, e_m) = -\frac{d}{12}(n^3 - n)\delta_{n,m} .$$

As we see from this formula, this curvature never vanishes.

In order to get a flat connection, one should tensorially multiply F by, what physisists call, the "ghost bundle". This bundle in mathematical terms is equal to the square root $K^{-1/2} \to \mathbf{S}$ of the anticanonical bundle $K^{-1} \to \mathbf{S}$ of the space $\mathbf{S}$. The bundle $K^{-1/2} \to \mathbf{S}$ is a holomorphic line bundle over $\mathbf{S}$ with an action of the Virasoro group, covering the action of $\text{Diff}_+(S^1)$ on $\text{Diff}_+(S^1)/S^1$ by left shifts. It can be provided with a canonical (spinor) connection B, invariant under this action. The curvature of B is equal to $-\frac{1}{2}\text{Ric}$, where Ric is the Ricci curvature of $\mathbf{S}$, computed at the origin of $\mathbf{S}$ in Sec.2. In other words, the curvature of B in the standard base $\{e_n\}$ of $T_o\mathbf{S}$ is equal to

$$R_B(e_{-n}, e_m) = \left(\frac{13}{6}n^3 - \frac{1}{6}n\right)\delta_{n,m} .$$

Now we define the quantization bundle $\mathbf{H}$ as the tensor product of the Fock bundle $F \to \mathbf{S}$ and the ghost bundle $K^{-1/2} \to \mathbf{S}$:

$$\mathbf{H} := F \otimes K^{-1/2} \to \mathbf{S}$$

and provide it with the Vir-invariant tensor product connection

$$\mathbf{D} := A \otimes 1 + 1 \otimes B .$$

Then the curvature of $\mathbf{D}$ in the base $\{e_n\}$ of $T_o\mathbf{S}$ is equal to

$$R_{\mathbf{D}}(e_{-n}, e_m) = \left(\frac{26 - d}{12}n^3 + \frac{d-2}{12}n\right)\delta_{n,m} .$$

As we have remarked in Sec.2, the coefficient of n may be made arbitrary by changing the symplectic structure inside its cohomology class (or modifying the normal ordering), so we can assert that the curvature $R_{\mathbf{D}}$ vanishes for $d = 26$. Hence, the *twistor quantization of the loop space* $\Omega\mathbb{R}^d$, *i.e. the construction of a Vir-invariant flat (Hermitian) connection on the quantization bundle* $\mathbf{H} = F \otimes K^{-1/2} \to \mathbf{S}$ *is possible for* $d = 26$.

3.3. Twistor quantization of loop spaces of compact Lie groups. In the case of the loop space of a compact Lie group G, we take for a twistor bundle the bundle $\pi : Z \to \Omega_T G$ of admissible complex structures on $\Omega_T G$ with the fibre $\mathbf{S} = \text{Diff}_+(S^1)/S^1$ and associated double fibration

$$\begin{array}{ccc} Z & \xrightarrow{\ p\ } & \mathbf{S} \\ {\scriptstyle\pi}\big\downarrow & & \\ \Omega_T G & & \end{array} \tag{8}$$

We provide Z with the canonical almost complex structure as above, so that p becomes a holomorphic map. The group LG acts on Z horizontally, preserving the fibres of p without changing the complex structure, and the group $\mathrm{Diff}_+(S^1)$ acts on Z vertically, preserving the fibres of π, this action being induced by the action of $\mathrm{Diff}_+(S^1)$ on $\mathbf{S}$.

Let $\widetilde{L\mathfrak{g}}$ be the central extension of the loop algebra $L\mathfrak{g}$ and λ be an antidominant lowest weight of $\widetilde{L\mathfrak{g}}$. The algebra $\widetilde{L\mathfrak{g}}$ plays the role of the Heisenberg algebra (cf. Sec.3.2), whereas its lowest weight representation is an analogue of the Heisenberg representation.

The twistor quantization of $\Omega_T G$ is based on a construction, similar to that of Borel–Weil (cf. [**2**]). We consider the complex line bundle $L^\lambda \to \Omega_T G$ over the loop space $\Omega_T G$, associated with the lowest weight λ. The pull-back $\widetilde{L} = \pi^* L$ of this bundle to the twistor space Z has a natural almost complex structure.

Recall that any fibre $p^{-1}(J)$, $J \in \mathbf{S}$, of the projection $p : Z \to \mathbf{S}$ is identified with the loop space $\Omega_T G$, provided with the complex structure J. Accordingly, for any $J \in \mathbf{S}$ the irreducible lowest weight representation of $\widetilde{L\mathfrak{g}}$ with the weight λ may be realized in a Hilbert space H^λ_J of holomorphic sections of $\widetilde{L}_\lambda$ along the fibre $p^{-1}(J)$. The union H^λ of the Hilbert spaces H^λ_J over all $J \in \mathbf{S}$ forms a holomorphic Hilbert bundle $H^\lambda \to \mathbf{S}$ over $\mathbf{S}$, which is trivial since $\mathbf{S}$ is contractible. The action of the diffeomorphism group $\mathrm{Diff}_+(S^1)$ on $\mathbf{S}$, as in Sec.3.2, is covered by a projective action of $\mathrm{Diff}_+(S^1)$ (or a $\widetilde{\text{Virasoro}}$ group action) on H^λ, intertwining the lowest weight representations of $\widetilde{L\mathfrak{g}}$ in H^λ_J for different $J \in \mathbf{S}$.

Using this action, we define, as in Sec.3.2, a connection A on H^λ, invariant under the Virasoro group action. The curvature of this connection at the origin of $\mathbf{S}$, computed in the base $e_n = ie^{in\theta}\frac{d}{d\theta}$ of $T_o\mathbf{S}$, has the form (cf. [**11, 12**])

$$R_A(e_{-n}, e_m) = \frac{c(\mathfrak{g})}{12}(-n^3 + n)\delta_{n,m} \ .$$

Here the number $c(\mathfrak{g})$ is determined by the formula

$$c(\mathfrak{g}) = \frac{h \dim \mathfrak{g}}{h + \kappa(\mathfrak{g})} \ ,$$

where h is the level of the lowest weight representation and $\kappa(\mathfrak{g})$ is the dual Coxeter number of $\mathfrak{g}$.

Following the scheme of Sec.3.2, we take the tensor product $\mathbf{H}$ of the bundle $H^\lambda \to \mathbf{S}$ with the ghost bundle $K^{-1/2} \to \mathbf{S}$:

$$\mathbf{H} := H^\lambda \otimes K^{-1/2} \to \mathbf{S}$$

and provide it with the Vir-invariant tensor product connection

$$\mathbf{D} := A \otimes 1 + 1 \otimes B \ .$$

Then the curvature of $\mathbf{D}$ in the base $\{e_n\}$ of $T_o\mathbf{S}$ is equal to

$$R_{\mathbf{D}}(e_{-n}, e_m) = \left(\frac{26 - c(\mathfrak{g})}{12}n^3 + \frac{c(\mathfrak{g}) - 2}{12}n\right)\delta_{n,m} \ .$$

Having this formula, we can assert that *the twistor quantization of the loop space $\Omega_T G$ is possible if*

$$c(\mathfrak{g}) = \frac{h \dim \mathfrak{g}}{h + \kappa(\mathfrak{g})} = 26 \ .$$

References

[1] M.F.Atiyah, Instantons in two and four dimensions, Commun. Math. Phys., 1984, vol.93, pp.437–451

[2] R.J.Baston, M.G.Eastwood, The Penrose Transform: its Interaction with Representation Theory. Oxford: Clarendon Press, 1989

[3] G.D.Birkhoff, Singular points of ordinary differential equations, Trans. Amer. Math. Soc., 1909, vol.10, pp.436–470

[4] G.D.Birkhoff, Equivalent singular points of ordinary linear differential equations, Math. Ann., 1913, vol.74, pp.134–139

[5] A.Borel, Kählerian coset spaces of semi-simple Lie groups, Proc. Nat. Acad. Sci. USA, 1954, vol.40, pp.1147–1151

[6] M. J. Bowick, S. G. Rajeev, The holomorphic geometry of closed bosonic string theory and $\mathrm{Diff}(S^1)/S^1$, Nucl. Phys., 1987, vol.B293, pp.348–384

[7] J.Davidov, A.G.Sergeev, Twistor quantization of of loop spaces, Proc. Steklov Inst. Math., 1997, vol.217, pp.1–90

[8] A. A. Kirillov, Infinite dimensional Lie groups: their orbits, invariants and representations. The geometry of moments, Lect. Notes Math., 1982, vol.970, pp.101–123

[9] A.A.Kirillov, Kähler structure on K-orbits of the diffeomorphism group of the circle, Funct. Anal. Appl., 1987, vol.21, pp.122–125

[10] A. A. Kirillov, D. V. Juriev, Kähler geometry of the infinite-dimensional homogeneous space $M = \mathrm{Diff}(S^1)/\mathrm{Rot}(S^1)$, Func. Anal. Appl., 1987 vol.21, pp.284–294

[11] J. Mickelsson, String quantization on group manifolds and the holomorphic geometry of $\mathrm{Diff}(S^1)/S^1$, Commun. Math. Phys., 1987, vol.112, pp.653–661

[12] J. Mickelsson, Current Algebras and Groups. New York: Plenum Press, 1989

[13] A.D.Popov, A.G.Sergeev, Symplectic twistors and geometric quantization of strings, in: Algebraic Geometry and Appl. (eds. A.Tikhomirov, A.Tyurin). Wiesbaden: Vieweg Verlag, 1994; pp.137–157

[14] A. Pressley, G. Segal, Loop Groups. Oxford: Clarendon Press, 1986

[15] C.Vizman, Coadjoint orbits in infinite dimensions, preprint, 1995

[16] H.C.Wang, Closed manifolds with complex homogeneous structure, Amer. J. Math., 1954, vol.76, pp.1–32

Steklov Mathematical Institute, Gubkina 8, GSP-1, Moscow, 117966, RUSSIA
E-mail address: sergeev@mi.ras.ru

Canadian Mathematical Society
Conference Proceedings
Volume **29**, 2000

Brownian motion on a manifold as limit of stepwise conditioned standard Brownian motions

O. G. SMOLYANOV, H. v. WEIZSÄCKER, and O. WITTICH

1. Introduction

Let M be a smooth m-dimensional compact Riemannian manifold without boundary. Due to Nash's theorem ([**8**]) we can and shall assume that it is isometrically embedded into a Euclidean space $\mathbb{R}^n$ and we shall fix this embedding. This work is part of a general study of the relation between the law of Brownian motion on M and the flat Wiener measure on the space of $\mathbb{R}^n$-valued paths. In this paper we consider specifically the following scheme: Let $\mathcal{P} = \{0 = t_0 < \cdots < t_k = 1\}$ be a partition of the unit time interval and let $x \in M$. By $\mathbf{W}^x_{M,\mathcal{P}}$ denote the law of the continuous process which starts in x and then in each partition interval moves like a standard n-dimensional Brownian motion in the surrounding space which is conditioned to return to M at the end of the partition interval. We prove the following

THEOREM 1. *For each $x \in M$, as the mesh $|\mathcal{P}|$ of the partition $\mathcal{P}$ converges to 0, the probability measure $\mathbf{W}^x_{M,\mathcal{P}}$ converges in law to the distribution $\mathbf{W}^x_M$ of a Brownian motion on the manifold M starting in x.*

Here as usual Brownian motion on the manifold is the continuous Markov process whose transition semigroup is generated by $-\frac{1}{2}\Delta_M$ where Δ_M is the Laplace-Beltrami operator on M.
In this result it is a little surprising that in the limit the particular choice of the the isometric embedding does not matter. The proof follows traditional analytic lines. It is divided into several parts which may have independent interest. We start with a comparison of the Taylor expansions of the geodesic metric and the Euclidean distance induced by the embedding (Proposition 1) which appears in the associated Gaussian integrals. This gives explicit estimates for the short time asymptotics of the conditional integral operator (Proposition 2). These estimates allow to apply versions of a result of Chernoff in semigroup theory (Proposition 3) and of classical tightness arguments. We do not use stochastic development but we need the smoothing properties of the heat semigroup on M. The proof yields a slightly more general result (see Theorem 2 in section 7) which allows an arbitrary fourth order perturbation of the integral kernels. This includes an intrinsic approximation based on geodesic distance.
As an important application Theorem 1 allows to determine also the asymptotic law

of the standard Brownian motion under the condition that it visits M at *all* times in the partition $\mathcal{P}$. This leads to a new measure $\mathbf{W}^{x,s}_{M,\mathcal{P}}$. We prove convergence of these measures to a measure $\mathbf{W}^{x,s}_M$ which is equivalent to $\mathbf{W}^x_M$. This measure $\mathbf{W}^{x,s}_M$ can be considered as the surface measure which the n-dimensional Wiener measure induces on the infinite dimensional submanifold of M-valued paths. This second convergence result holds also in the more general perturbed setting mentioned above, and the corresponding Radon-Nikodym derivative $\frac{d\mathbf{W}^{x,s}_M}{d\mathbf{W}^x_M}$ can be computed in terms of the perturbation function. (Theorem 3, section 8). As a particular case we get a variant of a result of Andersson-Driver ([**1**]).
At the end of the paper we briefly mention a few possible extensions and implications of our approach.

2. Definition of the measure $\mathbf{W}^x_{M,\mathcal{P}}$.

For each $x \in \mathbb{R}^n$ let $(W^x_t)_{t\geq 0}$ be a standard n-dimensional Brownian motion starting in the point x and let $p(t,x,y)$ for $t > 0$ and $y \in \mathbb{R}^n$ denote its usual Gaussian density with respect to Lebesgue measure λ^n . For every $\varepsilon > 0$ and every subset A of $\mathbb{R}^n$ let $U_\varepsilon(A)$ denote the ε-neighbourhood of A in $\mathbb{R}^n$. For each $t \geq 0$ and points x, y in M let

$$q(t,x,y) = \frac{p(t,x,y)}{\int_M p(t,x,z)\,\mathrm{vol}_M(z)} \tag{1}$$

where vol_M is the Riemannian volume form on M . Then $q(t,x,\cdot)$ can be viewed as the vol_M-density of the conditional distribution of the random vector W^x_t conditioned to be in M. In fact we have the following straightforward

LEMMA 1. *For every Borel set $A \subset G$ we have*

$$\int_A q(t,x,z)\,vol_M(z) = \lim_{\varepsilon\to 0} \frac{\int_{U_\varepsilon(A)} p(t,x,y)\,\lambda^n(dy)}{\int_{U_\varepsilon(G)} p(t,x,y)\,\lambda^n(dy)} \tag{2}$$

Now let $\mathcal{P} = (0 = t_0, \cdots, t_k = 1)$ be a partition of the interval $[0,1]$. Using the transition functions q we can define a Markov process with state space M and time parameter set $\mathcal{P}$ whose joint distribution on the set $M^{\mathcal{P}}$ given the start in x is the probability measure $\mathbf{P}^x_{M,\mathcal{P}}$ where

$$\begin{aligned}(3)\qquad &\mathbf{P}^x_{M,\mathcal{P}}(A_1 \times \cdots \times A_k) = \\ &\int_{A_1}\cdots\int_{A_n} q(t_1,x,y_1)\cdots q(t_k - t_{k-1}, y_{k-1}, y_k)\,\mathrm{vol}_M(y_1)\cdots\mathrm{vol}_M(y_k).\end{aligned}$$

Let for each $x \in \mathbb{R}^n$ the symbol $\mathbf{W}^x_{M,s,t}$ denote the conditional law of the Brownian motion on the time interval $[s,t]$ conditioned to start at time s in x and to end up at time t in M. More precisely let

$$\int_{C([s,t],\mathbb{R}^n)} f\,d\mathbf{W}^x_{M,s,t} = \lim_{\varepsilon\to 0} \frac{\mathbb{E}(f\,\mathbf{1}_{W^x_{t-s}\in U_\varepsilon(M)})}{\mathcal{P}(W^x_{t-s} \in U_\varepsilon(M))}$$

where f is a bounded measurable function.
Now we piece these transition kernels together. Every $\omega \in C([0,1],\mathbb{R}^n)$ can be identified with a unique m-tuple

$$(\omega_1, \dots, \omega_m) \in C([0,t_1],\mathbb{R}^n) \times \cdots \times C([t_{m-1},1],\mathbb{R}^n)$$

which satisfies $\omega_j(t_j) = \omega_{j+1}(t_j)$ for all $j \in \{1, \dots, m-1\}$. Using this identification we define, in analogy to (3), the measure $\mathbf{W}^x_{M,\mathcal{P}}$ on $C([0,1], \mathbb{R}^n)$ by

$$\int_{C([0,1],\mathbb{R}^n)} f(\omega)\mathbf{W}^x_{M,\mathcal{P}}(d\omega) =$$

$$\int \mathbf{W}^{x,s}_{M,\mathcal{P}} 0t_1(d\omega_1) \int \mathbf{W}^{\omega_1(t_1)}_{M,t_1,t_2}(d\omega_2) \cdots \int \mathbf{W}^{\omega_{m-1}(t_{m-1})}_{M,t_{m-1},t_m}(d\omega_m) f(\omega_1, \dots, \omega_m) .$$

Another, equivalent, way to pass from $\mathbf{P}^x_{M,\mathcal{P}}$ to $\mathbf{W}^x_{M,\mathcal{P}}$ is to define $\mathbf{W}^x_{M,\mathcal{P}}$ as the image of $\mathbf{P}^x_{M,\mathcal{P}}$ under the transition kernel associating to each point $(x_{t_0}, \cdots, x_{t_m}) \in M^{\mathcal{P}}$ the law of the process which connects each pair $(x_{t_i}, x_{t_{i+1}})$ of succesive points by an independent n-dimensional Brownian bridge.
Conversely, evaluation at the partition points gives a canonical projection $\pi_{\mathcal{P}}$ of the path space $C([0,1], \mathbb{R})$ to $(\mathbb{R}^n)^{\mathcal{P}}$. One can easily verify that under this projection, the measure $\mathbf{W}^x_{M,\mathcal{P}}$ has the marginal measure $\mathbf{P}^x_{M,\mathcal{P}}$. A particular right inverse of this projection is provided by geodesic interpolation. The image of $\mathbf{P}^x_{M,\mathcal{P}}$ under this right inverse gives another measure on $C([0,1], \mathbb{R}^n)$ whose marginal is $\mathbf{P}^x_{M,\mathcal{P}}$ again. That measure is close to the measure $\nu^1_{\mathcal{P}}$ studied by Andersson–Driver ([**1**]).

3. Some geometric Taylor approximations.

We need the fact that for an isometric embedding $\iota : M \longrightarrow \mathbb{R}^n$ the square of the geodesic distance d in M and the square of the Euclidean distance induced by the embedding agree up to the third order: $d(y,z)^2 - ||\iota(y) - \iota(z)||^2 = O(d(y,z)^4)$. We indicate the proof since we did not find a reference. Moreover we need a Taylor approximation of the volume form.
We use the Einstein summation convention. Greek indices run from 1 to n, latin indices from 1 to m.

PROPOSITION 1. *Let $\iota : M \longrightarrow \mathbb{R}^n$ be an isometric embedding of the smooth m-dimensional Riemannian manifold M into $\mathbb{R}^n$. Let $y \in M$ and let V be a neighbourhood of $0 \in \mathbb{R}^m$ and let $\psi : V \longrightarrow U$ provide normal coordinates of a neighborhood U of y. Then for all $x \in V$*

$$|x|^2 = d(\psi(x), y)^2 = |\iota \circ \psi(x) - \iota(y)|^2 + O(|x|^4). \tag{4}$$

The volume form $vol_M(x) = \sqrt{det g_{ij}(x)}\ dx$ satisfies in these coordinates

$$vol_M(x) = \big(1 + \frac{1}{6} R_{iuvi} x^u x^v + O(|x|^3)\big) dx, \tag{5}$$

in particular

$$\Delta \sqrt{det g_{ij}(x)}_{|x=0} = -\frac{\mathrm{Scal}(0)}{3}. \tag{6}$$

Proof: The first equality in (4) follows from ([**6**] p. 18). Moreover in normal coordinates the metric tensor has the form (cf. [**10**] p. 8)

$$g_{ij}(x) = \delta_{ij} + \frac{1}{3} R_{iuvj} x^u x^v + O(|x|^4). \tag{7}$$

We write $\iota \circ \psi(x) = (i^1(x), \dots, i^n(x))$. Since the embedding ι is isometric the metric tensor on the manifold agrees with the pullback of the metric tensor δ of $\mathbb{R}^n$, i.e.

(cf. [**6**], p.29 f and p. 156)

$$g_{ij}(x) = ((\iota \circ \psi)^* \delta)_{ij}(x) = \sum_\alpha \frac{\partial i^\alpha}{\partial x^i} \frac{\partial i^\alpha}{\partial x^j}(x).$$

Comparing with (7) we conclude that

$$0 = \frac{\partial}{\partial x^k} g_{ij}(0) = \sum_\alpha \frac{\partial i^\alpha}{\partial x^j} \frac{\partial^2 i^\alpha}{\partial x^k \partial x^i}(0) + \frac{\partial i^\alpha}{\partial x^i} \frac{\partial^2 i^\alpha}{\partial x^k \partial x^j}(0).$$

We may assume $\iota(y) = 0$. The second order Taylor expansion of the coordinates $i^\alpha(x)$ gives

$$\begin{aligned} |\iota \circ \psi(x) - \iota(y)|^2 &= \sum_\alpha (i^\alpha(x))^2 \\ &= \sum_\alpha \frac{\partial i^\alpha}{\partial x^i} \frac{\partial i^\alpha}{\partial x^j}(0)\, x^i x^j + \frac{1}{2} \frac{\partial i^\alpha}{\partial x^i} \frac{\partial^2 i^\alpha}{\partial x^k \partial x^j}(0) x^k x^i x^j + O(|x|^4) \\ &= g_{ij}(0) x^i x^j + \frac{1}{2} \sum_\alpha \frac{\partial i^\alpha}{\partial x^i} \frac{\partial^2 i^\alpha}{\partial x^k x^j}(0)\, x^k x^i x^j + O(|x|^4) \\ &= \delta_{ij} x^i x^j + \frac{1}{4} \frac{\partial g_{ij}}{\partial x^k}(0) x^i x^j x^k + O(|x|^4) \\ &= |x|^2 + O(|x|^4). \end{aligned}$$

This proves (4). For the volume form we have according to ([**6**], p. 81) and (7)

$$\begin{aligned} \mathrm{vol}_M(x) &= \sqrt{\det g(x)}dx \\ &= \sqrt{\det\big(\delta_{ij} + \frac{1}{3} R_{iuvj}(0) x^u x^v + O(|x|^3)\big)dx} \\ &= \sqrt{e^{1/3\ \mathrm{tr}(R_{iuvj}(0)) x^u x^v} + O(|x|^3)dx} \\ &= (e^{1/6\ R_{iuvi}(0) x^u x^v} + O(|x|^3))\ dx \\ &= (1 + \frac{1}{6}\ R_{iuvi}(0) x^u x^v + O(|x|^3))\ dx. \end{aligned}$$

The scalar curvature is defined by Scal $= g^{ik} R_{ik} = g^{ik} g^{jl} R_{ijkl}$ (Jost, Def. 3.3.3, p.131) which in normal coordinates can be written as Scal $= R_{iuiu} = -R_{iuui}$ by the antisymmetry of the curvature tensor (Jost, Lemma 3.3.1, p.129). Thus (5) implies (6). ∎

4. Short time asymptotics for Gaussian integrals

LEMMA 2. *Let V be a neighbourhood of the origin in $\mathbb{R}^m$. Then there is a constant K and some $t_0 > 0$ such that for all functions $f \in C^5(V)$, $h \in C^3(V)$ whose Taylor expansion has the following form*

$$\begin{aligned} f(x) &= \frac{1}{4!} \frac{\partial^4 f}{\partial x^t \partial x^u \partial x^v \partial x^w}(0)\ x^t x^u x^v x^w + O(|x|^5) \\ h(x) &= h(0) + \frac{\partial h}{\partial x^k}(0) x^k + \frac{1}{2} \frac{\partial^2 h}{\partial x^l \partial x^m}(0)\ x^l x^m + O(|x|^3). \end{aligned}$$

and all $t \in (0, t_0)$ the following estimate holds:

$$\begin{aligned} &\frac{1}{\sqrt{2\pi t}^m} \left| \int_V \exp\left(-\frac{|x|^2 + f(x)}{2t} \right) h(x) dx \right. \\ &\qquad \left. - \left(h(0) + \frac{t}{2}\Delta h(0) - \frac{t}{16} h(0)\Delta\Delta f(0) \right) \right| \\ &\leq K\|h\|_3(1 + \|f\|_5)\, t^{3/2}. \end{aligned} \tag{8}$$

Proof: Let ν_t denote the standard Gaussian distribution on $\mathbb{R}^m$ with covariance tI. We choose t_0 such that for all $t \in (0, t_0)$ and all k with $0 \leq k \leq 8$ one has

$$\int_{\mathbb{R}^m - V} |x|^k \; \nu_t(dx) \leq t^{3/2}. \tag{9}$$

In the following the symbols θ_i denote real functions resp. with $|\theta_i| \leq 1$. Taylor expansion for f, g and the exponential function yields

$$\begin{aligned} &\frac{1}{\sqrt{2\pi t}^m} \int_V \exp\left(-\frac{|x|^2 + f(x)}{2t} \right) h(x) \; dx \\ = \; &\int_V \left(1 - \frac{1}{48t} \frac{\partial^4 f}{\partial x^t \partial x^u \partial x^v \partial x^w}(0) x^t x^u x^v x^w + \|f\|_5 \theta_1(x)|x|^5/t \right) \\ &\times \left(h(0) + \frac{\partial h}{\partial x^k}(0)x^k + \frac{1}{2}\frac{\partial^2 h}{\partial x^l \partial x^m}(0)x^l x^m + \|h\|_3 \theta_2(x)|x|^3 \right) \nu_t(dx). \end{aligned}$$

If we disregard in both brackets the highest order term and, moreover, integrate over the whole space we get the main term

$$\begin{aligned} &\int_{\mathbb{R}^m} \left(h(0) + \frac{\partial h}{\partial x^k} x^k + \frac{1}{2}\frac{\partial^2 h}{\partial x^l \partial x^m} x^l x^m \right. \\ &\qquad \left. - \frac{h(0)}{48t} \frac{\partial^4 f}{\partial x^t \partial x^u \partial x^v \partial x^w}(0) x^t x^u x^v x^w \right) \nu_t(dx) \\ = \; & h(0) + \frac{1}{2}\frac{\partial^2 h}{\partial x^l \partial x^m}(0) \int_{\mathbb{R}^m} x^l x^m \nu_t(dx) \\ &\qquad - \frac{h(0)}{48t} \frac{\partial^4 f}{\partial x^t \partial x^u \partial x^v \partial x^w}(0) \int_{\mathbb{R}^m} x^t x^u x^v x^w \nu_t(dx) \\ = \; & h(0) + \frac{t}{2}\Delta h(0) - \frac{h(0)}{48t} \frac{\partial^4 f}{\partial x^t \partial x^u \partial x^v \partial x^w}(0)(\delta^{tu}\delta^{vw} + \delta^{tv}\delta^{uw} + \delta^{tw}\delta^{uv})\; t^2 \\ = \; & h(0) + \frac{t}{2}\Delta h(0) - \frac{t}{16} h(0)\Delta\Delta f(0), \end{aligned}$$

where the remaining error integral satisfies

$$\begin{aligned} |R_t| \; \leq \; & (1 + \|f\|_5)\|h\|_3) \left(\int_{\mathbb{R}^m} \left(|x|^3 + \frac{|x|^5 + |x|^6 + |x|^7}{t} \right) \nu_t(dx) \right. \\ & \left. + \int_{\mathbb{R}^m - V} (1 + |x|^4 + |x|^5)(1 + |x| + |x|^2 + |x|^3)\nu_t(dx) \right) \end{aligned}$$

and hence because of (9) $|R_t| \leq K(1 + \|f\|_5)\|h\|_3)t^{3/2}$. This implies (8). ■

5. Short time asymptotics of the conditional kernel

We prove a similar estimate as in Lemma 2 for the integral kernels which govern our approximation scheme. The norms $\|\cdot\|_k$ on the spaces $C^k(M)$ can be defined in an invariant way as in [**7**]. However for our computations we use the fact that (as is carried out in the analogous case of Sobolev spaces in [**7**], Prop. 2.11, 2.14, p. 175-176) an equivalent norm can be defined with the help of an admissible atlas: If a covering of M by charts $\mathcal{V} := \{(V, \psi_y) : y \in M\}$ is such such that the differentials up to order k of the chart maps are uniformly bounded then

$$\|f\|_k^{\mathcal{V}} := \sum_{|\lambda|=0}^{k} \sup_y \sup_{x\in V} |\partial^\lambda f \circ \psi_y(x)|$$

defines an norm which is equivalent to $\|\cdot\|_k$. We can even find a covering with additional properties: Since the map

$$\rho : TM \to M \times M$$

given by $\rho(v) := (\pi(v), \exp_{\pi(v)}(v))$ is a diffeomorphism from a neighbourhood of the zero section $M \subset TM$ to a neighbourhood of the diagonal in $M \times M$ (see [**5**], Lemma 2, p.150) we can choose a covering such that the charts are given by the exponential maps $\exp_y$, in particular each ψ_y is a normal coordinate system of some neighbourhood of y whose diameter is bounded away from 0 as y varies. Since $|x| = d(y, \psi_y(x))$ for normal coordinates this implies that the domain V of these coordinate systems can be chosen independently of y . Moreover for any given point $y_0 \in M$ we can choose this covering in such a way such that for some neighbourhood U_0 of y_0 the map $(y, z) \mapsto \psi_y^{-1}(z)$ is smooth in U_0^2 ([**5**], Theorem 8.7 (2), p. 149).

PROPOSITION 2. *Let M be a compact m-dimensional Riemannian manifold. Let $\varphi \in C^5(M^2)$ such that $\varphi(y,y) = 0$ and $\big(\nabla^\lambda \varphi(y,\cdot)\big)(y) = 0$ for all $y \in M$ and $1 \le |\lambda| \le 3$. Let $p_\varphi(t,y,z) = \exp\Big(-\frac{d(y,z)^2+\varphi(y,z)}{2t}\Big)$. Then there are a function $W_\varphi \in C^1(M)$, depending linearly on φ, and constants $\kappa, \tilde{\kappa}$ such that for all $u \in C^3(M)$ and all points $y \in M$ the following estimates hold: (a)*

(10)
$$\left|\frac{1}{\sqrt{2\pi t}^m}\int_M p_\varphi(t,y,z)u(z)vol_M(z) - u(y)\big(1 - t(\frac{Scal(y)}{6} + W_\varphi(y))\big) + \frac{t}{2}\Delta_M u(y)\right| \le \kappa \|u\|_3\, t^{3/2}.$$

(b) If $Q(t)$ denotes the associated Markov integral operator $Q(t)u(y) = \int_M q_\varphi(t,y,z)u(z)vol_M(z)$ where

$$q_\varphi(t,y,z) = \frac{p_\varphi(t,y,z)}{\int_M p_\varphi(t,y,z)vol_M(z)}. \tag{11}$$

one has for every $u \in C^3(M)$

$$|Q(t)u(y) - u(y) + \frac{t}{2}\Delta_M u(y)| \le \tilde{\kappa}\|u\|_3\, t^{3/2}. \tag{12}$$

Proof: We choose a neighbourhood V of the origin in $\mathbb{R}^m$ and normal charts $\psi_y : V \to M$ described as above and for each y we apply Lemma 2 with $f_y(x) =$

$\varphi(y, \psi_y(x))$ and $h_y(x) = \sqrt{detg_{ij}(x)}u(\psi_y(x))$. This is possible since all partial derivatives up to the third order of the functions f_y at 0 vanish. Since the map $y \mapsto \varphi(y, \cdot) \in C^5(M)$ is continuous the set of functions $\{\varphi(y, \cdot) : y \in M\}$ is compact in $C^5(M)$ and hence the set $\{f_y : y \in M\}$ is uniformly bounded in $C^5(V)$. Moreover by the Taylor expansion of the volume form given in Proposition the gradient $\nabla\sqrt{detg_{ij}}$ vanishes at $x = 0$ and the passage from Δ to Δ_M according to the usual convention changes signs. Thus 1

$$\begin{aligned} & \Delta h_y(0) \\ = \quad & \Delta(u \circ \psi_y)(0)\sqrt{detg_{ij}}(0) + 2(\nabla u \circ \psi_y)(0)\nabla\sqrt{detg_{ij}}(0) + u \circ \psi_y(0)\Delta\sqrt{detg_{ij}}(0) \\ = \quad & -\Delta_M u(y) + 0 - u(y)\frac{1}{3}Scal(y). \end{aligned}$$

Finally by a suitable choice of the normal charts the set $\{g_{ij} : y \in M\}$ can be chosen to be uniformly bounded in $C^3(V)$ and hence there is a constant γ such that $\|h_y\|_3 \leq \gamma\|u\|_3^y$ for all $y \in M$. Thus we get from Lemma 2 the estimate (10) for all $t \in (0, t_0)$ where t_0 depends only on the set V and

$$W_\varphi(y) = \frac{1}{16}\Big(\Delta\Delta\varphi(y, \psi_y(\cdot))\Big)(0). \tag{13}$$

To see that this value does not depend on the choice of the normal charts note that from (10) with $u = 1$ we get the invariant representation

$$W_\varphi(y) = -\lim_{t\to 0}\frac{1}{t}\left(\frac{1}{\sqrt{2\pi t}^m}\int_M p_\varphi(t, y, z)vol_M(z) - 1\right) - \frac{Scal(y)}{6}. \tag{14}$$

To see that this function is in $C^1(M)$ we fix a point y_0 and choose the normal charts of the neighbouring points in a smooth way as indicated above. Then the representation (13) shows that W_φ is continuously differentiable at y_0 since in this neighbourhood it is a smooth function of the fourth order derivatives of φ. If we write down the same estimate for the constant function $u = 1$ we get

$$Q(t)u(y) = \frac{u(y)\big(1 - t(\frac{Scal(y)}{6} + W_\varphi(y))\big) + \frac{t}{2}\Delta_M u(y) + \theta_1(y)\kappa\|u\|_3\, t^{3/2}}{\big(1 - t(\frac{Scal(y)}{6} + W_\varphi(y))\big) + \theta_2(y)\kappa\|u\|_3\, t^{3/2}}$$

with two functions θ_i of absolute value ≤ 1. From this one easily verifies that there is some $t_0 > 0$ and some constant $\tilde{\kappa}$ such that the estimate (12) holds for all $u \in C^3(M)$ and all $t \in (0, t_0)$. Finally both estimates (10) and (12) remain also valid , with possibly different constants, for all positive t by the compactness of M. ■

A consequence is the following short time asymptotic for the normalization constants of these 'pseudo-Gaussian' kernels.

COROLLARY 1. *As $t \to 0$ one has*

(15)
$$\frac{1}{\sqrt{2\pi t}^m}\int_M \exp\Big(-\frac{d(y,z)^2 + \varphi(y,z)}{2t}\Big)vol_M(z) = 1 - t\Big(\frac{Scal(y)}{6} + W_\varphi(y)\Big) + O(t^{3/2})$$

uniformly in $y \in M$ where W_φ is given by (14) resp. in local normal coordinates by (13), in particular $W_\varphi = 0$ for $\varphi = 0$.

Proof: This is the special case $u = 1$ of (10). Note that from (13) we get $W_\varphi = 0$ if $\varphi = 0$. ■

This corollary can be considered as the basic reason of the appearance of the Radon-Nikodym density $e^{\frac{1}{6}\int_0^1 Scal(x(s))\,ds}$ in the main result of [**1**]. If we replace however the geodesic distance by the Euclidean distance induced by the embedding (which by Proposition 1 still fits into the frame of Proposition 2) the resulting correction term $tW_\varphi(y)$ will depend on the particular choice of the embedding. Fortunately this term cancels in the asymptotics of the operator $Q(t)$.

6. A version of Chernoff's Theorem

We need the following tool concerning convergence to contraction semigroups on Banach spaces. It is closely related to a similar result of P. Chernoff ([**2**]) which however deals only with equipartitions (i.e. $t_i^n = t_j^n$).

PROPOSITION 3. *Let A be the generator of a strongly continuous semigroup on a Banach space B. Let $Q(t)$, $t > 0$ be a one-parameter family of contractions on B such that*

$$\frac{Q(t) - I}{t} e^{aA} g \to A e^{aA} g \tag{16}$$

as $t \downarrow 0$, for all $a > 0$ and all g in the domain $D(A)$. Let $(t_i^n)_{1 \le i \le m_n}$, $n \in \mathbb{N}$ be positive numbers such that $\sum_{i=1}^{m_n} t_i^n \longrightarrow t \ge 0$ and $\max_i t_i^n \longrightarrow 0$. Then, for all $f \in B$, we have

$$Q(t_1^n) \dots Q(t_{m_n}^n)\, f \to e^{tA} f$$

as $n \to \infty$.

Proof: Fix $a > 0$ For all $t > 0$ the operators $(\frac{Q(t)-I}{t} - A)e^{aA} : D(A) \to B$ are bounded if $D(A)$ is equipped with the graph norm. By Banach-Steinhaus and our assumption these operators are even equicontinuous. Moreover for every $g \in D(A)$ the map $s \mapsto (e^{sA}g, Ae^{sA}g)$ into the graph is continuous and hence the set $\{e^{sA}g\}_{0 \le s \le b-a}$ is compact for the graph norm. Therefore the convergence in (16) is uniform on this set, i.e.

$$\lim_{t \to 0} \sup_{a \le s \le b} \|(\frac{Q(t) - I}{t} - A) e^{sA} g\| = 0. \tag{17}$$

Let $t_1, \cdots, t_m$ be positive numbers and let $s_j = \sum_{i=j+1}^{m} t_i$ for $0 \le j \le m$. Then

$$Q(t_1) \dots Q(t_m) - e^{s_0 A} = \sum_{j=1}^{m} Q(t_1) \dots Q(t_{j-1})(Q(t_j) - e^{t_j A})\, e^{s_j A}.$$

Since the $Q(t)$ are contractions we get

$$\begin{aligned} &\|(Q(t_1) \dots Q(t_m) - e^{s_0 A})\, f\| \\ \le\ & \sum_{j=1}^{n} t_j \|(\frac{Q(t_j) - I}{t_j} - \frac{e^{t_j A} - I}{t_j})\, e^{s_j A} f\| \\ \le\ & (\sum_{j=1}^{m} t_j) \sup\{\|(\frac{Q(\tau) - I}{\tau} - \frac{e^{\tau A} - I}{\tau})\, e^{sA} f\| : \tau \le \max t_i,\ s \le \sum_{j=1}^{m} t_j\} \end{aligned} \tag{18}$$

Now we apply this to a vector of the form $f = e^{aA}g$, $a > 0$, $g \in D(A)$. From (17) (and the fact that (e^{tA}) is a strongly continuous semigroup) it follows that the right-hand-side converges to 0 as $n \to \infty$ if the parameters t_i are of the form t_i^n as indicated in the assumption. Moreover in this case $e^{s_0^n A}f \to e^{tA}f$. Again the contraction property and the fact that the vectors f of the above form are dense in B imply the assertion for all $f \in B$. ∎

The same idea yields under extra conditions even some quantitative version of this convergence.

COROLLARY 2. *In the situation of Proposition 2 there are a finite constant K and some $t_0 > 0$ such that for $u \in C^4(M), y \in M$ and all positive numbers t_i with $\sum_{i=1}^m t_i = t < t_0$ one has*

$$\frac{|(Q(t_1)\cdots Q(t_m) - e^{-\frac{t}{2}\Delta_M})u(y)|}{t} \leq K\|u\|_4 \max_{1\leq i\leq m} \sqrt{t_i}. \tag{19}$$

and

$$\frac{|\big(Q(t_1)\cdots Q(t_m) - (I - \frac{t}{2}\Delta_M)\big)u(y)|}{t} \leq K\|u\|_4 \max_{1\leq i\leq m} \sqrt{t_i}. \tag{20}$$

Proof: Let $u \in C^4(M)$ be in the domain of the bilaplacian $\Delta_M\Delta_M$. Thus the map $t \mapsto e^{-\frac{t}{2}\Delta_M}u \in C(M)$ is twice continuously differentiable. Therefore by Taylor's formula $\lim_{t\downarrow 0} \frac{\|I-\frac{t}{2}\Delta_M - e^{-\frac{t}{2}\Delta_M})u\|}{t^2} < \infty$. Comparing with (12) we get some $t_0 > 0$ such that for $0 < t < t_0$

$$\sup_y \frac{|\big(Q(t) - e^{\frac{t}{2}\Delta_M}\big)u(y)|}{t} = O(t^{1/2}).$$

By the smoothing properties of the heat semigroup the set $\{e^{\frac{s}{2}\Delta_M}u :\ 0 \leq s \leq t_0\}$ is a compact subset of $C^4(M)$ for each u in this space (cf. [**10**]). Hence even

$$\sup_y \sup_{0\leq s\leq t_0} \frac{|\big(Q(t) - e^{-\frac{t}{2}\Delta_M}\big)e^{-\frac{s}{2}\Delta_M}u(y)|}{t} = O(t^{1/2}) \tag{21}$$

The estimate (18) in the previous proof now implies that the left hand side of (19) is $O(\max_{1\leq i\leq m}\sqrt{t_i})$ for every $u \in C^4(M)$. Again by the Banach-Steinhaus theorem there is a constant K such that (19) holds. The proof of (20) follows the same line. ∎

7. Tightness and end of the proof of Theorem 1

In this section we consider the following situation: Let $M \subset \mathbb{R}^n$ be isometrically imbedded and let $(Q(t))_{0\leq t\leq 1}$ be a family of integral Markov operators as in Proposition 2 part (b). For each partition $\mathcal{P}$ of $[0,1]$ and each $x \in M$ let $\mathbf{P}_{\mathcal{P}}^x$ be the law of the Markov process $(X_t)_{t\in\mathcal{P}}$ with state space M starting in y and $\mathbb{E}\big(f(X_{t_{k+1}})|X_{t_k}\big) = Q(t_{k+1} - t_k)f(X_{t_k})$. Moreover let $\mathbf{W}_{\mathcal{P}}^x$ be law of a continuous process which interpolates the successive steps of this Markov process either by independent n-dimensional Brownian bridges or by geodesics. With these notations we have the following theorem. According to Proposition 1 the kernels $q(t,\cdot,\cdot)$ defined in section 2 satisfy the assumption of Proposition 2. Therefore by the remark at the end of section 2 the measures $\mathbf{W}_{M,\mathcal{P}}^x$ in Theorem 1 are special examples of the present situation, i.e. Theorem 2 implies Theorem 1.

THEOREM 2. *Let $(\mathcal{P}_k)$ be a sequence of partitions of $[0,1]$ whose mesh $|\mathcal{P}_k|$ converges to 0. Let $x \in M$. Then the sequence of the probability measures $\mathbf{W}^x_{\mathcal{P}_k}$ is uniformly tight over the space $C([0,1],\mathbb{R}^n)$ and converges to the law of a Brownian motion on the manifold M starting in x.*

Proof: The tightness is provided by the three lemmas below. Proposition 3 implies the convergence of the finite dimensional marginals to those of the Brownian motion on M. This completes the proof. ∎

LEMMA 3. *Let $X_0, \dots, X_n$ be a (not necessarily time homogeneous) Markov process with values in a metric space (M,d). Then for every $\alpha > 0$,*

$$P(\max_{1\le k\le n} d(X_0,X_k) > 2\alpha) \le 2 \max_{0\le k\le n} \sup_{x\in M} P(d(X_k,X_n) > \alpha | X_k = x). \tag{22}$$

Proof: Let T be the first instant at which $d(X_0,X_k) > 2\alpha$ if there is such a time and let $T = n$ otherwise. Then T is a stopping time and $d(X_0,X_T) > 2\alpha$ on $\{T \le n-1\}$. Thus by the strong Markov property

$$\begin{aligned}
&P(\max_{1\le k\le n} d(X_0,X_k) > 2\alpha)\\
\le\ & P(\{T\le n-1\} \cup \{d(X_0,X_n) > \alpha\})\\
\le\ & P(\{T\le n-1\} \cap \{d(X_0,X_n) \le \alpha\}) + P(d(X_0,X_n) > \alpha)\\
\le\ & P(\{T\le n-1\} \cap \{d(X_T,X_n) > \alpha\}) + P(d(X_0,X_n) > \alpha)\\
\le\ & \max_{1\le k\le n} \sup_{x\in M} P(d(X_k,X_n) > \alpha | X_k = x) + P(d(X_0,X_n) > \alpha)\\
\le\ & 2 \max_{0\le k\le n} \sup_{x\in M} P(d(X_k,X_n) > \alpha | X_k = x).
\end{aligned}$$

∎

LEMMA 4. *For a finite subset E of $[0,1]$ and a function $f : E \to M$ and $\delta > 0$ let*

$$w(f,E,\delta) := \sup\{d(f(s),f(t)) : s,t \in E, |s-t| < \delta\}.$$

Then for every sequence of partitions $\mathcal{P}_k$ with $|\mathcal{P}_k| \to 0$ and every $\varepsilon > 0$ one has

$$\sup_{k\in N} \mathbf{P}^x_{\mathcal{P}_k}\{\omega \in M^{\mathcal{P}_k} : w(\omega,\mathcal{P}_k,\delta) > \varepsilon\} \to 0.$$

In particular the sequence of measures on $C([0,1],\mathbb{R}^n)$ obtained from the $\mathbf{P}^x_{\mathcal{P}_k}$ by linear interpolation or by geodesic interpolation on M is uniformly tight.

Proof: We would like to apply Proposition 2 to the functions $d^4(y,\cdot)$. In order to circumvent the loss of smoothness of d at the cut locus let $r > 0$ be the radius of injectivity on M. Then d^4 is smooth on the set $T = \{(x,y) \in M^2 : d(x,y) < r\}$. Let $\chi : \mathbb{R}_+ \to \mathbb{R}_+$ be smooth, increasing such that $\chi(\alpha) = \alpha$ on $[0,r/2)$ and $\chi = const$ on $[r,\infty)$. Then

$$p(t) := \sup_{y\in M} \sup\{(Q(t_1)\cdots Q(t_m)(\chi\circ d)^4(y,\cdot))(y) : \sum_{i=1}^m t_i < t\} = O(t^{3/2}). \tag{23}$$

In fact $\chi \circ d$ is smooth, the set $\{(\chi\circ d)^4(y,\cdot) : y \in M\}$ is even bounded in $C^4(M)$. Moreover

$$((I - \frac{t}{2}\Delta_M)((\chi\circ d)^4)(y,\cdot))(y) = 0$$

for each $y \in M$. Therefore (23) follows from (20).
Given $\eta > 0$ and $\varepsilon \in (0, r)$ we can find with the help of (23) a number $\delta > 0$ such that

$$p(2\delta) < (\varepsilon/4)^4 \frac{\delta}{4} \eta. \tag{24}$$

Let $k \in \mathbb{N}$ be such that $|\mathcal{P}_k| < \delta$. Then there is a subpartition $0 = t_0 < \ldots < t_m = 1$ such that $\delta/2 < t_{i+1} - t_i < \delta$ for each i. Clearly $m \leq 2/\delta$. Then by Markov's inequality and Lemma 1 applied to $\alpha = \varepsilon/2$ and to the restriction of the coordinate process to the points in $\mathcal{P}_k \cap [t_i, t_i + 2\delta]$ we get

$$\begin{aligned} & \mathbf{P}^y_{\mathcal{P}_k} \{\omega \in M^{\mathcal{P}_k} : w(\omega, \mathcal{P}_k, \delta) > \varepsilon\} \\ \leq\ & \mathbf{P}^y_{\mathcal{P}_k} \left(\cup_{i=0}^{m-1} \{d(\omega(t_i), \omega(t)) > \varepsilon/2 \text{ for some } t \in \mathcal{P}_k \cap [t_i, t_i + 2\delta]\}\right) \\ \leq\ & \sum_{i=0}^{m-1} P\left(\{d(\omega(t_i), \omega(t)) > \varepsilon/2 \text{ for some } t \in \mathcal{P}_k \cap [t_i, t_i + 2\delta]\}\right) \\ \leq\ & 2 \sum_{i=0}^{m-1} \frac{p(2\delta)}{(\varepsilon/4)^4} = m \frac{p(\delta)}{(\varepsilon/4)^4} < \eta \end{aligned}$$

For a fixed partition $\mathcal{P}$ one has $w(f, \mathcal{P}, \delta) = 0$ as soon as δ is smaller than the smallest distance of two different partition points. Together with the previous estimate this proves the lemma. ∎

LEMMA 5. *Let $(\mathcal{P}_k)$ be a sequence of partitions on $[0,1]$. Let (X^k) be a sequence of $\mathbb{R}^n$-valued continuous processes. Let (Y^k) be the sequence of processes which are formed by taking the values of the X_k at the partition points and interpolating with independent Brownian bridges. If the laws of the X^k are uniformly tight over the path space $C(\mathbb{R}_+, \mathbb{R}^n)$ then this holds also for the Y^k.*

Proof: Let W be a Brownian motion independent of the X^k. Let W^k be the process which we get from W by linear interpolation in the partition intervals and let $B^k = W - W^k$. The continuity of the paths of W imply that the laws of the B^k are uniformly tight. Each B^k is an 'iterated' Brownian bridge, i.e. a concatenation of independent Brownian bridges in the partition intervals. Without changing the hypothesis we may assume that the X^k are linear in the partition intervals. Then $X^k + B^k$ has the same law as Y^k. Since addition is continuous this implies the assertion. ∎

8. Conditioning the whole path

Fix again a finite partition $\mathcal{P}$ of the time interval. We modify the previous construction (cf. Proposition 2) in which we started from a unnormalized kernel $p_\varphi(t, y, z) = \exp\left(-\frac{d^2(y,z)+\varphi(y,z)}{2t}\right)$ and renormalized it in each partition interval separately. Now we normalize at all partition times simultaneously. Thus, we study the probability measure $\mathbf{P}^{x,s}_{\mathcal{P}}$ with

$$\begin{aligned} & \mathbf{P}^{x,s}_{\mathcal{P}}(A_1 \times \cdots \times A_m) \\ = & c_{\mathcal{P}}(x) \int_{A_1} \cdots \int_{A_m} p_\varphi(t_1, x, y_1) \cdots p_\varphi(t_m - t_{m-1}, y_{m-1}, y_m)\, vol_M(y_1) \cdots vol_M(y_m). \end{aligned} \tag{25}$$

where the constant is chosen so that this becomes a probability distribution. In the special case $p(t,y,z) = c\exp\big(-\frac{|y-z|^2}{2t}\big)$ where $|\cdot|$ denotes the Euclidean distance the resulting probability measure on $M^{\mathcal{P}}$ describes the 'marginal distribution of the free Brownian motion W^x under the condition that it visits M at all times t_i in the partition'. In this case the right hand side can also be expressed using the usual surface measure limiting procedure as

(26)
$$\lim_{\varepsilon\to 0}\frac{\int_{U_\varepsilon(A_1)\times\cdots\times U_\varepsilon(A_m)} p(t_1,x,y_1)\cdots p(t_m-t_{m-1},y_{m-1},y_m)\,\lambda^n(dy_1)\cdots\lambda^n(dy_m)}{\int_{U_\varepsilon(G)^m} p(t_1,x,y_1)\cdots p(t_m-t_{m-1},y_{m-1},y_m)\,\lambda^n(dy_1)\cdots\lambda^n(dy_m)}$$

Therefore in this case we can consider $\mathbf{P}^{x,s}_{\mathcal{P}}$ as a kind of surface measure induced by the n-dimensional Wiener measure on $M^{\mathcal{P}}$. Hence the upper index 's'.

In the other special case $\varphi = 0$ the measure $\mathbf{P}^{x,s}_{\mathcal{P}}$ is essentially the image of the measure $\nu^0_{\mathcal{P}}$ of [**1**] (which is defined over the set of piecewise geodesic paths) under the canonical restriction $\pi_{\mathcal{P}}$ to the partition points; see [**1**] Proposition 5.6: We say 'essentially' because this holds only on a set of the form $\{(y_{t_i}) \in M^{\mathcal{P}} : d(y_{t_i}, y_{t_{i-1}}) < \varepsilon\}$ whose mass however becomes close to 1 as the partition becomes finer.

Comparing (25) with (11), Corollary 1 and the definition of $\mathbf{P}^y_{\mathcal{P}}$ at the beginning of the previous section we see that $\mathbf{P}^{x,s}_{\mathcal{P}}$ is absolutely continuous with respect to $\mathbf{P}^x_{\mathcal{P}}$ and the Radon–Nikodym density can be written as

$$\begin{aligned}
&\frac{d\mathbf{P}^{x,s}_{\mathcal{P}}}{d\mathbf{P}^x_{\mathcal{P}}}(y_1,\dots,y_m)\\
&= c_{\mathcal{P}}(x)\frac{p_\varphi(t_1,x,y_1)\cdots p_\varphi(1-t_{m-1},y_{m-1},y_m)}{q_\varphi(t_1,x,y_1)\cdots q_\varphi(1-t_{m-1},y_{m-1},y_m)}\\
&= c_{\mathcal{P}}(x)\Big(\int_M p_\varphi(t_1,x,z)\,dz\Big)\cdots\Big(\int_M p_\varphi(1-t_{m-1},y_{m-1},z)\,dz\Big)\\
&= c_{\mathcal{P}}(x)\prod_{j=1}^m\big(\mathrm{e}^{-\left(\frac{\mathrm{Scal}(\mathrm{y_{j-1}})}{6}+\mathrm{W}_\varphi(\mathrm{y_{j-1}})\right)(\mathrm{t_j}-\mathrm{t_{j-1}})}+\mathrm{O}((\mathrm{t_j}-\mathrm{t_{j-1}})^{3/2}\big)\,.
\end{aligned}$$

Moreover by the compactness of M, the uniformity statement in Corollary 1 and the continuity of the scalar curvature these products are bounded away from zero and infinity uniformly in m and y and hence the same is true for the normalizing constants. From this we get for every path ω

(27)
$$\lim_{|\mathcal{P}|\to 0}\frac{d\mathbf{P}^{x,s}_{\mathcal{P}}}{\mathbf{P}^x_{\mathcal{P}}}(\pi_{\mathcal{P}}(\omega)) = c(x)\mathrm{e}^{-\int_0^1\frac{1}{6}\,\mathrm{Scal}(\omega(t))+W_\varphi(\omega(t))\,dt}\,.$$

Concluding these observations we now can state the following result. As in Proposition 2 let $\varphi \in C^5(M^2)$ be a function such that $\varphi(y,y) = 0$ and $(\nabla^\lambda\varphi(y,\cdot))(y) = 0$ for all $y \in M$ and $1 \le |\lambda| \le 3$ and consider the kernel $p_\varphi(t,y,z) = \exp\Big(-\frac{d(y,z)^2+\varphi(y,z)}{2t}\Big)$. Let W_φ be given by (13) resp (14). After we have chosen for a partition $\mathcal{P}$ the multivector $(y_t)_{t\in\mathcal{P}}$ according to the distribution $\mathbf{P}^{x,s}_{\mathcal{P}}$ defined in (25) we can, like in the preceding paragraph, interpolate in the partition intervals either with geodesics or with Brownian bridges which connect each $y_{t_{j-1}}$ with y_{t_j}. The distribution of the resulting continuous time process is denoted by $\mathbf{W}_{\mathcal{P}}^{x,s}$. Combining (27) and the uniform boundedness of that convergence with Theorem 2 we get

THEOREM 3. *As* $|\mathcal{P}| \to 0$, *the measures* $\mathbf{W}_{\mathcal{P}}^{x,s}$ *induced by the kernels* p_φ *via (25) converge in law to a probability measure* $\mathbf{W}_M^{x,s}$ *which is absolutely continuous with respect to the Wiener measure* $\mathbf{W}_M^x$ *with Radon–Nikodym density*

$$\frac{d\mathbf{W}_M^{x,s}}{d\mathbf{W}_M^x}(\omega) = c(x)\,\mathrm{e}^{-S(\omega)}$$

where $S(\omega) = \int_0^1 (\frac{1}{6}\,\mathrm{Scal} + W_\varphi)(\omega(t))\,dt$ *and* $c(x) = (\mathbb{E}_{\mathbf{W}_M^x}(\mathrm{e}^{-S}))^{-1}$.

8. Concluding remarks

First note that - as already remarked in the introduction - if $p_\varphi = p$ is the restriction of the n-dimensional Gauss kernel then the measure $\mathbf{W}_{s,M}^x$ of Theorem 3 can be considered, in analogy to (26) as the surface measure induced by the n-dimensional Wiener measure $\mathbf{W}^x$ on the infinite-dimensional submanifold $C([0,1], M)$ of $C([0,1], \mathbb{R}^n)$, or in vague probabilistic language as the law of a free Brownian motion under the condition that it never leaves the manifold M.
We expect that this point of view will allow us to show how some properties of the Wiener process on M are naturally 'inherited' from the corresponding properties of flat Wiener measures.
Among these are the 'generalized density' (which was introduced in different but equivalent ways in [**12**] and [**4**]; see also [**13**]) which can be considered as a rigorous version of the Stratonovich probability functional of a diffusion process (introduced in [**11**]); the logarithm of the generalized density coincides in most interesting cases with the so-called Onsager-Machlup functionals ([**9**], [**3**]). In particular we believe that this approach is also an alternative way to arrive at the results of ([**15**]).
Moreover the same idea can be applied if we replace the embedding space $\mathbb{R}^n$ by another manifold. This would give 'relative' versions of the concepts and results mentioned above.
Some first results in these directions, including a more explicit representation of the function W_φ, will be announced in [**14**].

Acknowledgement We are grateful to a referee who pointed out an error in our original proof of Theorem 1.

References

[1] L. Andersson and B.K. Driver. Finite dimensional approximations to Wiener measure and path integral formulas on manifolds. *Journ. Funct. Anal.*, 165:430–498, 1999.

[2] R. Chernoff. Product formulas, Nonlinear semigroups, and Addition of Unbounded operators. *Memoirs of American Math. Soc.*, 140, 1974.

[3] Graham, R. Path integral formulation of general diffusion process. Z. Physik, B26 (1979), 281–290.

[4] Kirillov, A.V. Infinite-dimensional analysis and quantum theory as semimartingale calculus. Russ. Math. Surv. 3 (1994), 41–95.

[5] Sh. Kobayashi and K. Nomizu. *Foundations of Differential Geometry I.* J. Wiley & Sons, New York, 1997.

[6] J. Jost. *Riemannian Geometry and Geometric Analysis, 2nd edition.* Springer, Heidelberg-New York, 1998.

[7] H.B. Lawson Jr., and M.-L. Michelson. *Spin Geometry*, Princeton University Press, Princeton NJ, 1989.

[8] J.F. Nash. The imbedding problem for Riemannian manifolds. *Ann. of Math.*, 63:20–63, 1956.

[9] Onsager, L., Machlup S. Fluctuations and irreversible processes, I, II. Phys. Rev. 91 (1953), 1505–1512, 1512–1515.

[10] J. Roe. *Elliptic Operators, topology and asymptotic methods.* Longman, London, 1988.

[11] Stratonovich, R.L. On the probability functional of diffusion processes. Select. Transl. in Math. Stat. Prob. 10 (1971), 273–286.

[12] Smolyanov, O.G., Weizsäcker, H.v. Change of measures and their logarithmic derivatives under smooth transformations. C.R. Acad. Sci., Paris, t. 321, Serie 1, p. 103–108, 1995.

[13] Smolyanov, O.G., Weizsäcker, H.v. Smooth probability measures and associated differential operators. Infinite dimensional analysis, quantum probability and related topics. V. 2, No 1, 1999, 51–78.

[14] Smolyanov, O.G., Weizsäcker, H.v., Wittich, O. Diffusions on compact Riemannian manifolds and surface measures (in Russian). Submitted.

[15] Takahashi Y., Watanabe S. The probability functionals (Onsager-Machlup functionals) of diffusion processes. Lecture Notes in Math, v. 851 (1980), 432–463.

Faculty of Mechanics and Mathematics,, Moscow State University, 119899, Moscow, Russia

Fachbereich Mathematik,, Universität Kaiserslautern, 67653 Kaiserslautern, Germany

GSF-National Research Center for Environment and Health,, 85764 Neuherberg, Germany

Canadian Mathematical Society
Conference Proceedings
Volume **29**, 2000

The Analytic Quantum Information Manifold

R. F. Streater

ABSTRACT. Let $H \geq I$ be a self-adjoint operator on a Hilbert space, such that $e^{-\beta H}$ is of trace-class for some $\beta < 1$. Let V be a symmetric operator such that $\|V\|_\omega := \|RV\| < \infty$, where $R = H^{-1}$. We show that the partition function $\mathrm{Tr}e^{-(H+\lambda V)}$ is analytic in λ in a hood of the origin in the sense of Fréchet, in the Banach space with norm $\|\bullet\|_\omega$. This is applied to the quantum information manifold defined by H. **Keywords**: Fisher metric, Bogoliubov Kubo Mori Green function, Orlicz space.

1. Introduction

The information manifold in non-parametric estimation theory has been introduced and studied by Pistone and Sempi [**19, 5**]. For a finite number of parameters, the theory is described in the books [**1, 8**]. The quantum version has been introduced by Hasagawa [**6, 7**] and by Nagaoka [**13**], and by Petz [**15, 16**]. This study is strictly valid only if the Hilbert space $\mathcal{H}$ of states is finite dimensional. It provides a geometrical picture, as well as rigour, to the theory of linear response [**9, 12, 4, 11, 3, 21**]. For $\dim\mathcal{H} = \infty$, but limited to bounded potentials, the theory has been developed by Araki [**2**] in the context of Tomita-Takesaki theory. Our more concrete Hilbert-space version of this goes as follows.

Denote by $\mathcal{C}_1$ the set of trace-class operators on $\mathcal{H}$ and let $\Sigma \subseteq \mathcal{C}_1$ denote the set of density operators. Given $\rho_0 \in \Sigma$ let $H_0 = -\log \rho_0 + cI \geq I$ be a self-adjoint operator with domain $\mathcal{D}(H_0)$ such that

$$\rho_0 = Z_0^{-1} e^{-H_0}. \tag{1.1}$$

We recognise H_0, which is unique up to the constant c, as the Hamiltonian, and Z_0 as the partition function, whose finiteness expresses thermodynamic stability. Let X be a symmetric bounded operator, so $X \in \mathcal{B}(\mathcal{H})$, and perturb H_0 to $H_X = H_0 + X$, with corresponding state $\rho_X = Z_X^{-1}\exp(-H_X)$ and free energy $\psi_X = \log Z_X$. Araki proves that ψ_X is an analytic functional with a convergent Taylor series, whose first

This continues work started at CNRS Luminy, Marseille; I thank P. Combe for arranging the visit, and CNRS for financial support. Thanks are due to P. Combe, G. Burdet and H. Nencka for useful discussions.

few terms are

$$\nabla^+_Y \psi_X = \rho_X.Y := \mathrm{Tr}(\rho_X Y) \tag{1.2}$$

$$\nabla^+_Y \nabla^+_Z \psi_X = \int_0^1 d\lambda \mathrm{Tr}\left\{\rho_X^\lambda \hat{Y} \rho_X^{\lambda'} \hat{Z}\right\} := g_X(Y,Z) \tag{1.3}$$

$$\nabla^+_Y \nabla^+_Z \nabla^+_W \psi_X = \int_0^1 \prod d\alpha_i \delta(\sum \alpha_i - 1) \mathrm{Tr}\left(\rho_X^{\alpha_1} Y \rho_X^{\alpha_2} Z \rho_X^{\alpha_3} W\right)_c$$

$$:= t_X(Y,Z,W). \tag{1.4}$$

Here, ∇^+_Y denotes the Gateaux derivative of ψ_X in the direction $Y \in \mathcal{B}(\mathcal{H})$, $\hat{Y} = Y - \rho_X.Y$ is the centred observable (the socalled "score"), $g_X(Y,Z)$ is the Bogoliubov-Kubo-Mori (*BKM*) metric, and $\mathrm{Tr}(\dots)_c$ denotes the cumulant of third order. We adopt the convention throughout this paper that if $\lambda \in [0,1]$ then $\lambda' = 1 - \lambda$. The torsion, t, measures the failure of g to be invariant under (+1)-parallel transport.

In [**22, 23, 14**] we began a study of the unbounded case. Let $\mathcal{C}_p$, $0 < p < 1$ denote the set of compact maps $A : \mathcal{H} \to \mathcal{H}$ such that $|A|^p \in \mathcal{C}_1$, and put

$$\mathcal{C}_{<1} := \bigcup_{0<p<1} \mathcal{C}_p. \tag{1.5}$$

It is known [**18**] that if $p < q$ then $\mathcal{C}_p \subset \mathcal{C}_q$ and that $\mathcal{C}_p$ is a complete linear space with quasinorm

$$\|A\|_p := [\mathrm{Tr}\,(A^* A)^{p/2}]^{1/p}. \tag{1.6}$$

In [**23**] we took the underlying set of the quantum information manifold to be

$$\mathcal{M} = \mathcal{C}_{<1} \cap \Sigma. \tag{1.7}$$

$\mathcal{M}$ has a natural affine structure, coming from the linear structure of each $\mathcal{C}_p$; it is called the (−1)-affine structure. Thus if $\rho_i \in \mathcal{C}_{p_i} \cap \Sigma, i = 1,2$ let $p = \max\{p_1, p_2\}$; then $\rho_i \in \mathcal{C}_p \cap \Sigma, i = 1,2$. We define $\lambda\rho_1 + \lambda'\rho_2, 0 \le \lambda \le 1$ as the usual sum of operators in $\mathcal{C}_p$. We see that $\mathcal{M}$ is (−1)-convex.

There is a quantum analogue of the classical Orlicz space $L \log L$: let

$$\mathcal{C}_1 \log \mathcal{C}_1 := \{\rho \in \mathcal{C}_1 : S(\rho) := -\sum_{i=1}^\infty \lambda_i \log \lambda_i < \infty\} \tag{1.8}$$

where $\{\lambda_i\}$ are the singular numbers of ρ. Thus the set of normal states of finite entropy is $\mathcal{C}_1 \log \mathcal{C}_1 \cap \Sigma$. It is easy to show that $\mathcal{C}_{<1} \subset \mathcal{C}_1 \log \mathcal{C}_1$.

The topology given to $\mathcal{M}$ in [**23**] is not that induced on it as a subset of $\mathcal{C}_1$. Following an idea in [**19**] we constructed a hood of $\rho_0 \in \mathcal{M}$ consisting of all $\rho \in \mathcal{M}$ which are 'connected to ρ_0 by a one-parameter exponential family of states'. In [**23**] this is the family

$$\rho_{\lambda X} = Z_\lambda^{-1} e^{-(H_0 + \lambda X + c(\lambda))} \in \mathcal{M}, \qquad 0 \le \lambda \le 1, \tag{1.9}$$

where X is a symmetric quadratic form defined on $\mathcal{D}(H_0^{1/2})$ and form-bounded relative to the energy form

$$q_0(\phi, \psi) := \langle H_0^{1/2}\phi, H_0^{1/2}\psi\rangle. \tag{1.10}$$

The state $\rho_{\lambda X}$ does not depend on the choice of $c(\lambda)$ and we do this so that $H := H_0 + \lambda X + c(\lambda) \ge I$. We write $R = R_X := H^{-1}$ for the inverse of H at $\lambda = 1$. If $\rho_0 \in \mathcal{C}_{\beta_0}$, to make sense we need that the q_0-bound of X be less than $1 - \beta_0$. We then showed that $\rho_X \in \mathcal{C}_{\beta_X}$ for some $\beta_X < 1$. The set of q_0-bounded quadratic forms X is a Banach space $\mathcal{T}_0$ with norm

$$\|X\|_0 := \|R_0^{1/2} X R_0^{1/2}\|, \tag{1.11}$$

where $\|\bullet\|$ is the operator norm for $\mathcal{H}$, and $R_0 = H_0^{-1}$. We showed that sufficient for $\rho_X \in \mathcal{M}$ is that $\|X\|_0 < 1 - \beta_0$, and that Z_X is a Lipschitz function of X; however, we could not show that the first moment $\mathrm{Tr}(\rho_0 X)$ is finite. We introduced the regularised mean of $X \in \mathcal{T}_0$,

$$\rho_0.X := \mathrm{Tr}[\rho_0^{\lambda} X \rho_0^{\lambda'}] \tag{1.12}$$

and showed that it is finite and independent of $\lambda \in (0,1)$.

The hood of ρ_0 defined by $\{\rho_X : \|X\|_0 < 1 - \beta_0\}$ can be mapped bijectively onto the open ball of radius $1 - \beta_0$ in the Banach space

$$\hat{\mathcal{T}}_0 := \{X \in \mathcal{T}_0 : \rho_0.X = 0\}; \tag{1.13}$$

thus, $\hat{\mathcal{T}}_0$ consists of score variables. This map makes the hood into a manifold modelled on $\hat{\mathcal{T}}_0$. The linear structure of $\hat{\mathcal{T}}_0$ induces a local affine structure on $\mathcal{M}$, called the $(+1)$-affine structure. To any point ρ_X, we may similarly define a hood of states of ρ_X to be

$$\left\{\rho_Y : \|Y\|_X := \|R_X^{1/2} Y R_X^{1/2}\| < a \leq 1 - \beta_X\right\}, \tag{1.14}$$

which is in bijection with an open ball in the Banach space

$$\hat{\mathcal{T}}_X := \{Y : \|Y\|_X < \infty \text{ and } \rho_X.Y = 0\}. \tag{1.15}$$

We showed that on the overlap region, the norms $\| \bullet \|_0$ and $\| \bullet \|_X$ are equivalent norms; the $(+1)$-affine structures on the region of overlap are also the same. Thus we have covered $\mathcal{M}$ by a Banach manifold with a local affine structure. We showed that the connected component containing ρ_0 is $(+1)$-convex.

In the present paper we consider the problem of furnishing $\mathcal{M}$ with a Riemannian metric. As shown by Petz [**16**], in the finite-dimensional case there are several candidates, but for geometric reasons we choose the *BKM* metric. Indeed, the studies of Nagaoka [**13**] show that $+1$ and -1-affine structures are not dual relative to the 'Uhlmann' metric, (the real part of $\mathrm{Tr}\,\rho X^* Y$)). They *are* dual, however, relative to the *BKM*-metric. Since this is the second derivative of ψ, in infinite dimensions we must impose more regularity than we have so far. Apart from having this 'Amari' duality, similar to the classical Fisher-Rao metric, the *BKM*-metric is the one entering Kubo's theory of linear response, as well as the one arising in quantum statistical dynamics.

In this paper we assume that X is an operator which is H_0-bounded. We then say that the tangent direction X is analytic. We show in §(2) that this leads to the Fréchet differentiability of ψ_0 in the analytic direction X, with Lipschitz continuous derivative

$$\nabla_X \psi_0 = \rho_0.X, \tag{1.16}$$

and that $\rho_0 X$ is of trace class. This means that the regularisation of the mean is not necessary. We show that a metric g can be defined by eq. (1.3).

In §(3) we show that the n-point Kubo cumulant is finite, and that for small $|\lambda|$ the Taylor series for $\psi_{\lambda X}$ converges if X is an analytic direction. We show that when we furnish $\mathcal{M}$ with the analytic topology, it becomes an analytic Banach manifold.

2. H_0-bounded perturbations

Let $H_0 \geq I$ be a selfadjoint operator with domain $\mathcal{D}(H_0)$ and quadratic form q_0. A necessary and sufficient condition for a symmetric operator $X : \mathcal{D}(H_0) \to \mathcal{H}$ to be H_0-bounded is that R_0X be bounded [**20**]. One can show that

$$\|X\|_0 := \|R_0^{1/2} X R_0^{1/2}\| \leq \|R_0 X\| \tag{2.1}$$

It follows that if X is H_0-bounded, then its form is q_0-bounded. Suppose now that $\rho_0 := Z_0^{-1}e^{-H_0} \in \mathcal{C}_{\beta_0}$, $\beta_0 < 1$, and that $\|R_0X\| < 1 - \beta_0$. Then by eq. (2.1), $\|X\|_0 < 1 - \beta_0$ and so by [**23**] the perturbed state

$$\rho_X = Z_X^{-1} e^{-H_X} \in \mathcal{C}_{\beta_X} \subset \mathcal{C}_{<1} \tag{2.2}$$

lies in a hood of ρ_0 in the topology induced by $\mathcal{T}_0$. Let us define a stronger topology than this, by taking hoods of ρ_0 to be

$$\{\rho_X : \|X\|_\omega := \|R_0 X\| < a \leq 1 - \beta_0\}. \tag{2.3}$$

We call this the analytic or ω-topology, and $\| \bullet \|_\omega$ the ω-norm. Similarly, if ρ_X is in the analytic hood of ρ_0, and Y is H_X-bounded with $\|R_X Y\| < 1 - \beta_X$, then $\rho_{X+Y} \in \mathcal{M}$ and the analytic hoods of ρ_X can be defined by

$$\{\rho_{X+Y} : \|R_X Y\| < a \leq 1 - \beta_X\}. \tag{2.4}$$

Let $\hat{\mathcal{T}}_\omega(\rho_X) := \{Y : \|YR_X\| < \infty, \quad \rho_X.Y = 0\}$. Then the construction of the ω-hoods covers $\mathcal{M}$ by sets homeomorphic to balls in the Banach spaces $\hat{\mathcal{T}}_\omega(\rho)$; these have mutually equivalent norms on the overlaps of the hoods. To see this, suppose that Y is in the hood of ρ_0 and also of ρ_X. The charts around ρ_0 and ρ_X take ρ_Y to balls in the Banach spaces $\hat{\mathcal{T}}_\omega(\rho_0)$ and $\hat{\mathcal{T}}_\omega(\rho_X)$, with norms $\|R_0Y\|$ and $\|R_XY\|$. These are equivalent, since

$$m\|YR_0\| \leq \|YR_X\| \leq M\|R_0Y\| \qquad \text{for all } Y, \tag{2.5}$$

where $m = \|(H_X + I)R_0\|^{-1}$ and $M = \|(H_0 + I)R_X\|$. Thus $\mathcal{M}$ is made into a Banach manifold, called the ω quantum information manifold. This name will be justified in the next section. It is easy to see that the local $(+1)$-affine structure coming from the linear structure of $\hat{\mathcal{T}}_\omega(\rho_X)$ is compatible with that given in [**23**].

The following lemma will be needed later. It is clear that R_{X+Y} maps $\mathcal{H}$ into $\mathcal{D}(H_0) = \mathcal{D}(H_X) = \mathcal{D}(H_{X+Y})$. It follows from lemma 6, [**23**] that $H_X R_{X+Y}$ is bounded. We find

$$(I + YR_X)(H_X R_{X+Y}) = H_X R_{X+Y} + YR_{X+Y} = I. \tag{2.6}$$

Hence $H_X R_{X+Y} = (I + YR_X)^{-1}$, and we get if $\|YR_X\| < 1$,

$$\|H_X R_{X+Y}\| \leq (1 - \|YR_X\|)^{-1}. \tag{2.7}$$

From this we see

LEMMA 2.1. *If $a < 1$ and X is H_0-small, then $\|H_X R_{X+Y}\|$ is bounded in the set $\{Y : \|YR_X\| \leq a\}$.* □

Now suppose that X perturbs H_0 and Y perturbs H_X, as above. Then

THEOREM 2.2. *If $1 - 2\delta > \beta_X$, then $\rho_X^{\delta'} Y$ has a trace-class extension.*

Proof. Choose $\delta > 0$ so that $\rho_X^{1-2\delta} \in \mathcal{M}$. Then

$$\begin{aligned}\|\rho_X^{1-\delta}Y\|_1 &= \|Z_X^{\delta-1}e^{-(1-\delta)H_X}Y\|_1 \\ &\leq C\|\rho_X^{1-2\delta}\|_1\|e^{-\delta H_X}H_X\|_\infty\|R_XY\|_\infty \\ &< \infty.\end{aligned}$$

□

Corollary. The regularised mean coincides with the true mean:

$$\mathrm{Tr}(\rho_X^\lambda Y\rho_X^{1-\lambda}) = \mathrm{Tr}(\rho_X Y), \qquad \text{for all } \lambda \in (0,1). \tag{2.8}$$

For, we can write $\rho_X Y = \rho_X^\delta \rho_X^{\delta'} Y$ and use the cyclicity of the trace.

THEOREM 2.3. *In the analytic manifold,* $\psi_X := \log Z_X$ *is* ∇_Y^+ *Fréchet differentiable, and*

$$\nabla_Y^+\psi_X = \rho_X.Y \tag{2.9}$$

Proof. By two applications of Duhamel's formula [**23**], Th. 9, we get for the difference quotient,

$$\lambda^{-1}\mathrm{Tr}\left[e^{-(H_X+\lambda Y)} - e^{-H_X}\right] - \mathrm{Tr}\left[e^{-H_X}Y\right] =$$

$$= \mathrm{Tr}\int_0^1 \alpha d\alpha \int_0^1 d\beta \left\{e^{-\alpha\beta(H_X+\lambda Y)}\lambda Y e^{-\alpha\beta' H_X} Y e^{-\alpha' H_X}\right\}.$$

We can put the trace inside the integral, by Fubini's theorem, if we can show that by doing so we get an absolutely convergent integral of the trace norm. The right hand side is then obviously $O(\lambda)$, with constant equal to the trace of an operator of the form

$$C = \int\int\int_0^1 \rho_1^{\alpha_1} Y\rho_2^{\alpha_2} Z\rho_3^{\alpha_3} \prod_{i=1}^3 d\alpha_i \delta\left(\sum_{i=1}^3 \alpha_i - 1\right), \tag{2.10}$$

which is the quantum version of the integral remainder in Taylor's theorem.

We divide the region of integration into three (overlapping) parts; in region i, $\alpha_i \geq 1/3, (i = 1,2,3)$. The estimate is similar in each case. The hardest case is when the variable larger than 1/3, say α_1, involves the state $\rho_1 = Z_\lambda^{-1}e^{-(H_X+\lambda X)}$, as this requires an estimate independent of λ. We therefore do this case in detail. So take $\alpha_1 \geq 1/3$, and $\alpha_2, \alpha_3 > 0$, and $H_i \geq I$. Put $R_i = H_i^{-1}$. We now show that for small $\delta > 0$, $\rho_1^{\alpha_1-\delta}Y\rho_2^{\alpha_2}Z\rho_3^{\alpha_3}$ is of trace class. Indeed, by Hölder,

$$\begin{aligned}\|\rho_1^{\alpha_1-\delta}Y\rho_2^{\alpha_2}Z\rho_3^{\alpha_3}\|_1 &\leq \|\rho_1^{\alpha_1-2\delta}\|_{1/\alpha_1}\|Y\rho_1^\delta\|_\infty\|\rho_2^{\alpha_2}\|_{1/\alpha_2} \\ &\quad \|Z\rho_3^\gamma\|_\infty\|\rho_3^{\alpha_3-\gamma}\|_{1/\alpha_3} \\ &\leq \|\rho_1^{1-2\delta/\alpha_1}\|_1^{\alpha_1}\|YR_1\|_\infty\|H_1\rho_1^\delta\|_\infty \\ &\quad \|ZR_3\|_\infty\|H_3\rho_3^\gamma\|_\infty\|\rho_3^{1-\gamma/\alpha_3}\|_1^{\alpha_3}.\end{aligned}$$

If $2\delta < \beta_1'\alpha_1$ and $\gamma < \alpha_3\beta_3'$, then all the norms are finite; here, the β_i are such that $\rho_i \in \mathcal{C}_{\beta_i}$. The size of γ depends on α_3, but in the region $\alpha_1 \geq 1/3$, δ can be chosen independent of α_i. Now take $2\delta < \beta_1'/3$. Then for each $\alpha_1 \geq 1/3$, $\alpha_2 > 0$, $\alpha_3 > 0$ we can take a bit, ρ_1^δ, of the dominant factor to the right end. We get, using

$\sum \alpha_i = 1$ and Hölder,

$$\begin{aligned}\|C\|_1 &= \|\rho_1^{\alpha_1-2\delta}(\rho_1^\delta Y)(\rho_2^{\alpha_2})(ZR_3)\rho_3^{\alpha_3}(H_3\rho_1^\delta)\|_1 \\ &\leq \|\rho_1^{\alpha_1-2\delta}\|_{1/\alpha_1}\|\rho_1^\delta Y\|_\infty \|\rho_2^{\alpha_2}\|_{1/\alpha_2}^{\alpha_2}\|ZR_3\|_\infty \\ &\quad \|\rho_3^{\alpha_3}\|_{1/\alpha_3}^{\alpha_3}\|H_3R_1\|_\infty \|H_1\rho_1^\delta\|_\infty.\end{aligned}$$

The first norm is bounded by

$$\|\rho_1^{1-2\delta/\alpha_1}\|_1^{\alpha_1} \leq \|\rho_1^{1-6\delta}\|_1 < \infty$$

as $1-6\delta > \beta_1$. The second factor,

$$\|\rho_1^\delta Y\|_\infty \leq \|\rho_1^\delta H_1\|_\infty \|R_1 Y\|_\infty$$

is finite, independent of λ, by the spectral theorem, and by lemma (2.1). The same can be said of the factors $\|H_1\rho_1^\delta\|_\infty$ and $\|H_3R_1\|_\infty$. The factors involving ρ_2 and ρ_3 are unity, as they are states. Hence the trace norm of C is bounded (in the region 1) independent of α and λ, and its integral is bounded. The other regions, 2 and 3, are treated similarly. □

Corollary. The *BKM* metric

$$g_X(Y,Z) = \int_0^1 d\alpha \mathrm{Tr}\left(\rho_X^\alpha Y \rho_X^{1-\alpha} Z\right) \tag{2.11}$$

is finite if Y, Z are ω-directions. Moreover, ψ_X is Fréchet differentiable in the Y direction in the analytic manifold, with derivative $\rho_X.Y$; for

$$|\rho_X.Y| = |\mathrm{Tr}[\rho_X^{1-\delta}(\rho_X H_X)(R_X Y)]| \leq C\|R_X Y\| \tag{2.12}$$

is continuous in the ω-norm

3. The analyticity of the free energy

In this section, we show that the free energy $\psi_{\lambda X}$ is an analytic function in a hood of $\lambda = 0$ if X is an ω-direction. Suppose $\rho = Z^{-1}\exp(-H) \in \mathcal{C}_\beta \in \mathcal{M}$, and V_j are H-bounded, $j = 1, \ldots, n$. As before, we assume that $H \geq 1$ and write $R = H^{-1}$.

We start with the n-point function

$$\mathrm{Tr}\int_0^1 d\alpha_1 \int_0^1 d\alpha_2 \ldots \int_0^1 d\alpha_{n-1}\left[\rho^{\alpha_1}V_1\rho^{\alpha_2}V_2\ldots\rho^{\alpha_n}V_n\right].$$

Here, $\alpha_n = 1 - \alpha_1 - \ldots - \alpha_{n-1}$. Write this as the trace of

$$\begin{aligned}&\left[\rho^{\alpha_1\beta}\right]\left[H^{1-\delta_n+\delta_1}\rho^{(1-\beta)\alpha_1}\right]\left[R^{\delta_1}V_1R^{1-\delta_1}\right]\left[\rho^{\alpha_2\beta}\right]\left[H^{1-\delta_1+\delta_2}\rho^{(1-\beta)\alpha_2}\right] \\ &\left[R^{\delta_2}V_2R^{(1-\delta_2)}\right]\ldots\left[\rho^{\alpha_n\beta}\right]\left[H^{1-\delta_{n-1}+\delta_n}\rho^{(1-\beta)\alpha_n}\right]\left[R^{\delta_n}V_nR^{1-\delta_n}\right],\end{aligned}$$

where $\delta_j \in (0,1)$ are yet to be chosen. This is the product of n factors of the form $\left[\rho^{\alpha_j\beta}\right]$, n factors of the form $\left[H^{1-\delta_{j-1}+\delta_j}\rho^{(1-\beta)\alpha_j}\right]$, where δ_0 means δ_n, and n factors of the form $\left[R^{\delta_j}V_jR^{1-\delta_j}\right]$. By interpolation, the last type is bounded in operator norm by [**20**]

$$\|R^{\delta_j}V_jR^{1-\delta_j}\| \leq \|RV_j\| = \|V_j\|_\omega.$$

This bound is independent of α. The factors $\rho^{\alpha_j\beta}$ are bounded in trace-norm by the Hölder inequality

$$\|A_1\ldots A_n\|_1 \leq \|A_1\|_{p_1}^{1/p_1}\ldots\|A_n\|_{p_n}^{1/p_n}.$$

Since $\sum \alpha_j = 1$ we may put $p_j = 1/\alpha_j$ to get

$$\begin{aligned} \| \left[\rho^{\alpha_1\beta}\right] \quad \ldots \quad \left[\rho^{\alpha_n\beta}\right] \|_1 &\leq \|\rho^\beta\|_1^{\alpha_1} \cdots \|\rho^\beta\|_1^{\alpha_n} \\ &= \|\rho^\beta\|_1 < \infty. \end{aligned}$$

This bound is also independent of the $\alpha's$. We bound the remaining factors in operator norm. By the spectral theorem we have the bound

$$\|\rho^{(1-\beta)\alpha_j} H^{1-\delta_{j-1}+\delta_j}\|_\infty = Z^{-\alpha_j(1-\beta)} \sup_{x\geq 1}\{e^{-(1-\beta)\alpha_j x} x^{1-\delta_{j-1}+\delta_j}\} \tag{3.1}$$

$$\leq Z^{-\alpha_j(1-\beta)} \left(\frac{1-\delta_{j-1}+\delta_j}{(1-\beta)\alpha_j}\right)^{1-\delta_{j-1}+\delta_j} e^{-(1-\delta_{j-1}+\delta_j)}. \tag{3.2}$$

We have to integrate $\alpha_j^{-(1-\delta_{j-1}+\delta_j)} d\alpha_j$; the region of integration is the union of n (overlapping) regions $S_j := \{\alpha : \alpha_j \geq 1/n\}$. We treat each in a similar manner. For S_n we can ensure integrability at $\alpha_j = 0$, $j = 1, \ldots, n-1$ if we choose δ_j so that $1 - \delta_{j-1} + \delta_j < 1$, that is, $\delta_j < \delta_{j-1}$. So we choose $\delta_n = \delta_0 > \delta_1 > \delta_2 > \ldots > \delta_{n-1}$. To ensure that $\delta_j \in [0,1]$ it is enough to choose $\delta_n = 1$, $\delta_1 = 1 - 1/n$, $\ldots$, $\delta_{n-1} = 1/n$. Then each of the $n-1$ integrals for $j = 1, \ldots n-1$,

$$\int_0^1 d\alpha_j \alpha_j^{-(1-\delta_{j-1}+\delta_j)} = (\delta_{j-1} - \delta_j)^{-1}$$

is equal to n, giving a factor n^{n-1}. The remaining α factor in the region S_n is $\alpha_n^{-(1-\delta_{n-1}+\delta_n)} \leq n^2$. We get the same bound in the other regions, S_j $j = 1, \ldots n-1$, so we get the bound $n^3 n^{n-1} = n^2 n^n$. The other factors in the estimate (3.2) can be bounded independent of α:

$$\begin{aligned} &\prod_{j=1}^n Z^{-\alpha_j(1-\beta)} (1-\delta_{j-1}+\delta_j)^{1-\delta_{j-1}+\delta_j} (1-\beta)^{-(1-\delta_{j-1}+\delta_j)} e^{-(1-\delta_{j-1}+\delta_j)} \\ &\leq 4Z^{-(1-\beta)}(1-\beta)^{-n} e^{-n} \end{aligned}$$

since $(1 - \delta_{j-1} + \delta_j) < 1$ except for one term, when it is less than 2. Thus we get for the n-point function the bound

$$4\|\rho^\beta\|_1 Z^{-\beta} n^2 n^n e^{-n} \prod_j \|V_j/(1-\beta)\|_\omega.$$

It follows that ψ_V is infinitely Fréchet differentiable in analytic directions, and that the Taylor series converges if $\|V\|_\omega < 1 - \beta$. We recall that this is the condition that ρ_V lies in $\mathcal{M}$ in an ω-hood of ρ.

4. Conclusion

Suppose that $\beta < 1$. We have seen that the condition that the perturbation of the Hamiltonian $H \geq I$ of a state $\rho \in \mathcal{C}_\beta$ by a small enough symmetric operator V leads to another state $\rho_V \in \mathcal{M}$. We defined a topology on $\mathcal{M}$ in a hood of ρ using the norm $\|V\|_\omega(\rho) = \|RV\|$, where $RH = I$; this hood $\mathcal{U}$ is set of states ρ_V coming from the ball of operators V with $\|V\|_\omega < 1 - \beta$. This is finer than the topology defined by the norm $\|V\| = \|R^{1/2}VR^{1/2}\|$ which is finite for all form-bounded perturbations. We saw that the Taylor series for the free energy converges in the same hood of ρ provided that the perturbation V is in an analytic direction. The ω-norms associated with different points of $\mathcal{M}$ are equivalent on the overlap of two hoods, so together we have a Banach manifold modelled on the Banach spaces of H-bounded symmetric operators. The question arises, is there an analytic structure of this manifold? In fact, an analytic structure is deemed to be provided if we specify the ring of germs of analytic functions at each point. Let us say that a map $\psi : \mathcal{U} \to \mathbf{C}$ is $+1$-analytic if in $\mathcal{U}$ it is infinitely often Fréchet differentiable and $\psi(\rho_{\lambda V})$ has a convergent Taylor expansion in λ, for all ω-directions V in the tangent space. Because of the equivalence of norms, this concept is coordinate-free. Thus an analytic structure for $\mathcal{M}$ has been specified. In particular, the mixture coordinates $\eta_X = \rho.X$ are analytic, as they are derivatives of the free energy. In his theory of expansionals, Araki [**2**] showed that the free energy ψ_X is an entire function in the Banach space $\mathcal{BH}$ if the perturbation X is bounded, in the more general context of Tomita-Takesaki theory. With unbounded perturbations we cannot hope for entire functions; for we need only take $V = -H_0$ to hit a singularity in ψ_X. However, it is likely that some analyticity remains, even in the thermodynamic limit, for relatively bounded perturbations.

Compared with [**22**], we have improved the result in several ways. We show analyticity instead of twice-differentiability; we have dropped the commutator condition altogether; the manifold is infinite dimensional instead of finite-dimensional; and we have enlarged the class of states to $\mathcal{C}_{<1}$.

References

[1] S.-i. Amari, *Differential Geometric Methods in Statistics,* **Lecture Notes in Statistics**, **28**, 1985. Springer-Verlag.
[2] Araki, H., Publ. R. I. M. S. (Kyoto), **9**, 165-209, 1968.
[3] Balian, R., Y. Alhassid, H. Reinhardt, *Dissipation in many-body systems: a geometrical approach based on information theory*, Physics Reports, **131**, 1-146, 1986. North Holland.
[4] Bogoliubov, N. N., Phys. Abh. Sov. Union, **1**, 229-, 1962.
[5] Gibilisco, P., and G. Pistone, *Connections on nonparametric statistical manifolds by Orlicz space geometry*, Infinite-dimensional Analysis, Quantum Probability and Related Topics, **1**, 325-347, 1998.
[6] Hasagawa, H. Reports on Math. Phys., **33**, 87, 1993.
[7] Hasagawa, H. *Noncommutative Extension of the Information Geometry*, in **Quantum Communication and Measurement**, Eds V. P. Balavkin, O. Hirota and R. L. Hudson, 1995, Plenum Press, pp 327-337,
[8] Kass, R. A. and P. W. Vos, "Geometric Foundations of Asymptotic Inference", Wiley, NY 1997.
[9] Kubo, R. Reports on Progress in Physics, **29**, 255-284, 1966.
[10] Lesniewski, A., and M. B. Ruskai, *Relative Entropy and Monotone Riemannian Metrics on Noncommutative Probability Spaces*, preprint.
[11] Matsubara, T., Prog. Theor. Phys.,**14**, 351-, 1955.
[12] Mori, H., Prog. Theor. Phys., **33**, 423-, 1965.
[13] Nagaoka, H., pp 449-452 in **Quantum Communication and Measurement**, Eds V. P. Belavkin, O. Hirota and R. L. Hudson, Plenum Press, New York, 1995.
[14] H. Nencka and R. F. Streater, *Information Geometry for Some Lie Algebras*, to appear in Infinite Dimensional Analysis and Quantum Probability, World Scientific.
[15] Petz, D., and Toth, G. Lett. Math. Phys., **27**, 205-216, 1993.
[16] Petz, D., *Monotone Metrics on Matrix Spaces*, Linear Alg. and Appl., **244**, 81-96, 1996.
[17] Petz, D. and C. Sudar, *Geometries of Quantum States*, Journ. Math. Phys., **37**, 2662-73, 1996.
[18] Pietsch, A. **Nuclear Locally Convex Spaces**, Springer-Verlag, 1972.
[19] Pistone, G. and C. Sempi, Annals of Statistics, **23**, 1543-1561, 1995.
[20] Reed, M. and B. Simon, **Methods of Modern Mathematical Physics**, Academic Press, Vol. 2, 1975.
[21] Roepstorff, G. Commun. Math. Phys.,**46**, 253-262 1976.

[22] R. F. Streater, *Information Geometry and Reduced Quantum Description*, Reports on Mathematical Physics, **38**, 419-436 1996.
[23] R. F. Streater, *The Information Manifold for Relatively Bounded Potentials*, to appear in the Bogoliubov 90th Birthday Memorial Volume. Ed A. A. Slavnov.

DEPT. OF MATHEMATICS, KING'S COLLEGE LONDON,, STRAND, WC2R 2LS
E-mail address: ray.streater@kcl.ac.uk

Canadian Mathematical Society
Conference Proceedings
Volume **29**, 2000

L^p-independence of the spectral radius of symmetric Markov semigroups

Masayoshi Takeda

ABSTRACT. We show L^p-independence of the spectral radius of symmetric Markov semigroups by using Donsker-Varadhan type large deviation principle. As its application, we obtain a necessary and sufficient condition for gaugeability.

1. Introduction

In this paper, we consider L^p-independence of the spectral radius of symmetric Markov semigroups. In [Si], he proved the L^p-independence of the spectral radius of Schrödinger semigroups. Our methed is completely different from [Si]; we use arguments of Donsker-Varadhan large deviation theory. More precisely, let $\mathbf{M} = (\Omega, X_t, P_x, \zeta)$ be an m-symmetric Markov process on a locally compact separable metric space X. Here ζ is the lifetime of $\mathbf{M}$ and m is a positive Radon measure with full support. In [T], we considered Donsker-Varadhan's type large deviations of symmetric Markov processes with finite lifetime and proved that the full large deviation principle holds if the Markov process explodes so fast that the 1-resolvent of the identity function 1, $R_1 1(x)(= 1 - E_x(\exp(-\zeta)))$, belongs to $C_\infty(X)$, the space of continuous functions vanishing at infinity. These results can be a little bit strengtened as in Theorem 2.1 below. As its corollary, the following variational formula on the lifetime is shown (Corollary 2.2):

$$\lim_{t\to\infty} \frac{1}{t} \log \sup_{x\in X} P_x(t < \zeta) = -\inf\left\{\mathcal{E}(u,u) : u \in \mathcal{F},\ \int_X u^2 dm = 1\right\}, \tag{1.1}$$

where $(\mathcal{E}, \mathcal{F})$ is the Dirichlet form on $L^2(X; m)$ of $\mathbf{M}$. Let us denote by λ_p the spectral radius of the semigroup $p_t f(x) = E_x(f(X_t))$ as an operator on $L^p(X; m)$:

$$\lambda_p = -\lim_{t\to\infty} \frac{1}{t} \log \|p_t\|_{p,p}, \quad 1 \le p \le \infty,$$

where $\|p_t\|_{p,p}$ is the operator norm from $L^p(X;m)$ to $L^p(X;m)$. Then the equation (1.1) implies that $\lambda_\infty = \lambda_2$, as a result, λ_p is independent of p.

In section 3, we obtain a necessary and sufficient condition for gaugeability, integrability of Feynman-Kac functionals, by applying the L^p-independence to time changed Brownian motions.

2000 *Mathematics Subject Classification.* Primary 47D07, 31C25, 60J55, 60J65.

2. Donsker-Varadhan type large deviation principle

Let X be a locally compact separable metric space and m a positive Radon measure on X with full support. Let $\mathbf{M} = (\Omega, X_t, P_x, \zeta)$ be an m-symmetric Borel right process on X. Here Ω is the space of all right continuous maps of $[0,\infty]$ into $X \cup \{\Delta\}$ (Δ is a extra point) and $X_t(\omega) = \omega(t)$ for $\omega \in \Omega$ ($X_\infty(\omega) = \Delta$). ζ is the lifetime of $\mathbf{M}$, $\zeta = \inf\{t > 0 : X_t = \Delta\}$. Let us denote by $\{p_t\}_{t\geq 0}$ and $\{R_\alpha\}_{\alpha>0}$ the semigroup and resolvent of $\mathbf{M}$, i.e., $p_t f(x) = E_x(f(X_t))$ and $R_\alpha f(x) = E_x(\int_0^\infty e^{-\alpha t} f(X_t)dt)$. Throughout this section, we make following assumptions:

I. (Irreducibility) If a Borel set A is p_t-invarient, i.e., $p_t(I_A f)(x) = I_A p_t f(x)$ m-a.e. for any $f \in L^2(X;m) \cap \mathcal{B}_b(X)$ and $t > 0$, then A satisfies either $m(A) = 0$ or $m(X \setminus A) = 0$. Here $\mathcal{B}_b(X)$ is the space of bounded Borel functions on X.

II. (Absolute continuity condition) The transition probability $p_t(x, dy)$ of $\mathbf{M}$ is absolutely continuous with respect to m for each $t > 0$ and $x \in X$.

III. $p_t(C_\infty(X)) \subset C_\infty(X)$, where $C_\infty(X)$ is the space of continuous functions vanishing at infinity.

IV. $R_1 1(x) \in C_\infty(X)$.

Define the generator A by

$$Au = \alpha u - f \quad \text{for } u = R_\alpha f,\ f \in \mathcal{B}_b(X).$$

For $\phi(x) = R_\alpha f(x) > 0$, $f \in \mathcal{B}_b(X)$, define the martingale additive functional M_t^ϕ by

$$M_t^\phi = \phi(X_t) - \phi(X_0) - \int_0^t A\phi(X_s)ds,$$

and let

$$M_t = \int_0^t \frac{1}{\phi(X_{s-}) + \epsilon} dM_s^\phi.$$

It is known that the unique solution of so called Doléans-Dade equation,

$$X_t = 1 + \int_0^t X_{s-} dM_s,$$

is given by

$$\exp(M_t - \frac{1}{2} < M^c >_t) \prod_{0<s\leq t} (1 + \Delta M_s) \exp(-\Delta M_s), \tag{2.1}$$

where $< M^c >_t$ is the quadratic variation of the continuous martingale part of M_t and $\Delta M_s = M_s - M_{s-}$ (cf. [RW]). Denoting by $N_t^{\phi,\epsilon}$ the solution, we see

$$E_x(N_t^{\phi,\epsilon}) \leq 1. \tag{2.2}$$

On the other hand, by applying of Itô formula to $\log(x + \epsilon)$ ($\epsilon > 0$), we have

$$\log(\phi(X_t) + \epsilon) = \log(\phi(X_0) + \epsilon) + M_t + \int_0^t \frac{A\phi(X_s)}{\phi(X_{s-}) + \epsilon} ds - \frac{1}{2} < M^c >_t$$
$$+ \sum_{0<s\leq t} \left(\log(\phi(X_s) + \epsilon) - \log(\phi(X_{s-}) + \epsilon) - \frac{1}{\phi(X_{s-}) + \epsilon}(\phi(X_s) - \phi(X_{s-})) \right).$$

Since

$$\Delta M_s = \frac{\phi(X_s) - \phi(X_{s-})}{\phi(X_{s-}) + \epsilon},$$

$N_t^{\phi,\epsilon}$ is expressed as

$$N_t^{\phi,\epsilon} = \frac{\phi(X_t)+\epsilon}{\phi(X_0)+\epsilon} \exp\left(- \int_0^t \frac{A\phi}{\phi+\epsilon}(X_s)ds \right). \tag{2.3}$$

Set

$$\mathcal{D}^+(A) = \{R_\alpha f : \ \alpha > 0, \ f \in L^2(X;m) \cap C_b^+(X) \text{ and } f \not\equiv 0\} \ (\subset \mathcal{F}).$$

Here $C_b^+(X)$ denotes the set of non-negative bounded continuous functions. Note that any function in $\mathcal{D}^+(A)$ is strictly positive. Indeed, given $\phi = R_\alpha^\mu f \in \mathcal{D}^+(A)$, put $O = \{x \in X : f(x) > 0\}$. Then, it is known in [FOT;Theorem 4.6.6] that $P_x(\sigma_O < \infty) > 0$ for q.e. x, so that $P_x(\sigma_O < \infty) > 0$ for any x because of II. Here $\sigma_O = \inf\{t > 0 : X_t \in O\}$.

Denote by $(\mathcal{E}, \mathcal{F})$ the Dirichlet space on $L^2(X;m)$ of $\mathbf{M}$ ([FOT; p. 141]). Let $\mathcal{P}$ be the set of probability measures on X equipped with the weak topology and define the function $I_\mathcal{E}$ on $\mathcal{P}$ by

$$I_\mathcal{E}(\mu) = \begin{cases} \mathcal{E}(\sqrt{f}, \sqrt{f}) & \text{if } \mu = f\cdot m, \ \sqrt{f} \in \mathcal{F} \\ \infty & \text{otherwise.} \end{cases}$$

For $\omega \in \Omega$ with $\zeta(\omega) > t$, we define the normalized occupation time distribution $L_t(\omega) \in \mathcal{P}$ by

$$L_t(\omega)(A) = \frac{1}{t}\int_0^t I_A(X_s(\omega))ds, \quad A \in \mathcal{B}(X).$$

We then have

THEOREM 2.1. (i) *For any open set G of $\mathcal{P}$*

$$\liminf_{t\to\infty} \frac{1}{t} \log P_x(L_t \in G, t < \zeta) \ge - \inf_{\mu\in G} I_\mathcal{E}(\mu) \quad \textit{for all } x \in X.$$

(ii) *For any closed set K of $\mathcal{P}$*

$$\limsup_{t\to\infty} \frac{1}{t} \log \sup_{x\in X} P_x(L_t \in K, t < \zeta) \le - \inf_{\mu\in K} I_\mathcal{E}(\mu).$$

PROOF. (i) See Theorem 3.1 in [T].
(ii) Denote

$$Q_{x,t}(C) = P_x(L_t \in C; t < \zeta) \quad \text{for } C \in \mathcal{B}(\mathcal{P}).$$

For any $u \in \mathcal{D}^+(A)$ and any $\epsilon > 0$

$$E_x\left(\exp\left(- \int_0^t \frac{Au}{u+\epsilon}(X_s)ds \right); t < \zeta \right) \le \frac{u(x)+\epsilon}{\epsilon},$$

by virtue of (2.2) and (2.3), and thus, for any Borel subset C of $\mathcal{P}$

$$\limsup_{t\to\infty} \frac{1}{t} \log \sup_{x\in X} Q_{x,t}(C) \le \inf_{\substack{u\in\mathcal{D}^+(A)\\ \epsilon>0}} \sup_{\mu\in C} \int_X \frac{Au}{u+\epsilon} d\mu. \tag{2.4}$$

Let K be a compact set of $\mathcal{P}$ and set

$$\ell = \sup_{\mu\in K} \inf_{\substack{u\in\mathcal{D}^+(A)\\ \epsilon>0}} \int_X \frac{Au}{u+\epsilon} d\mu.$$

Then, for any $\delta > 0$ and $\mu \in K$, there exist $u_\mu \in \mathcal{D}^+(A)$ and $\epsilon_\mu > 0$ such that

$$\int_X \frac{Au_\mu}{u_\mu+\epsilon_\mu} d\mu \le \ell + \delta.$$

Since the function $\dfrac{Au_\mu}{u_\mu + \epsilon_\mu}$ belongs to $C_\infty(X)$, there exists a neighbourhood $N(\mu)$ of μ such that

$$\int_X \frac{Au_\mu}{u_\mu + \epsilon_\mu} d\nu \le \ell + 2\delta \quad \text{for any } \nu \in N(\mu).$$

Since $\cup_{\mu\in K} N(\mu)$ is an open covering of K, there exist $\mu_1, \dots, \mu_k$ in K such that $K \subset \cup_{j=1}^k N(\mu_j)$. Put $K_j = N(\mu_j)$. We then have for $1 \le j \le k$

$$\sup_{\mu\in K_j} \int_X \frac{Au_{\mu_j}}{u_{\mu_j} + \epsilon_{\mu_j}} d\mu \le \ell + 2\delta,$$

and thus

$$\max_{1\le j\le k} \inf_{\substack{u\in\mathcal{D}^+(A)\\ \epsilon>0}} \sup_{\mu\in K_j} \int_X \frac{Au}{u+\epsilon} d\mu \le \ell + 2\delta.$$

Therefore, by (2.4)

$$\begin{aligned}
\limsup_{t\to\infty} \frac{1}{t} \log \sup_{x\in X} Q_{x,t}(K) &\le \max_{1\le j\le k} \limsup_{t\to\infty} \frac{1}{t} \log \sup_{x\in X} Q_{x,t}(K_j) \\
&\le \max_{1\le j\le k} \inf_{\substack{u\in\mathcal{D}^+(A)\\ \epsilon>0}} \sup_{\mu\in K_j} \int_X \frac{Au}{u+\epsilon} d\mu \\
&\le \ell + 2\delta. \qquad (2.5)
\end{aligned}$$

Since

$$-\inf_{\substack{u\in\mathcal{D}^+(A)\\ \epsilon>0}} \int_X \frac{Au}{u+\epsilon} d\mu = I_{\mathcal{E}}(\mu) \qquad (2.6)$$

by Proposition 4.1 in [T], the proof is completed for any compact set K.

Put $\phi = R_1 1$ and

$$V_\epsilon(x) = -\frac{A\phi}{\phi+\epsilon}(x) = \frac{1-\phi(x)}{\phi(x)+\epsilon} \ge 0.$$

Then, on account of (2.2)

$$\int_{\mathcal{P}} \exp\left(t \int_X V_\epsilon(x)\mu(dx)\right) dQ_{x,t} \le \frac{\phi(x)+\epsilon}{\epsilon} \le \frac{1+\epsilon}{\epsilon}. \qquad (2.7)$$

Put

$$K_\epsilon = \left\{x \in X : V_\epsilon(x) \le \frac{1}{2\epsilon}\right\} = \left\{x \in X : \phi(x) \ge \frac{\epsilon}{2\epsilon+1}\right\}.$$

Then, note that K_ϵ is a compact set by the assumption IV. Since $V_\epsilon > 1/2\epsilon$ on K_ϵ^c, it follows from (2.7) that

$$\int_{\mathcal{P}} \exp\left(\frac{t}{2\epsilon}\mu(K_\epsilon^c)\right) dQ_{x,t} \le \frac{1+\epsilon}{\epsilon},$$

and thus

$$Q_{x,t}(\mathcal{M}_\epsilon^\delta) \le \frac{1+\epsilon}{\epsilon} \exp\left(-\frac{t\delta}{2\epsilon}\right), \qquad \mathcal{M}_\epsilon^\delta = \{\mu \in \mathcal{P} : \mu(K_\epsilon^c) > \delta\}.$$

Hence, for any $\lambda > 0$

$$Q_{x,t}(J) \le \lambda\left(\frac{e^{-\lambda t} + e^{-\lambda t/2}}{(1-e^{-\lambda t/2})^3}\right) + \frac{e^{-\lambda t/2}}{1-e^{-\lambda t/2}}, \quad J = \cup_{n=1}^\infty \mathcal{M}_{\frac{1}{\lambda n^2}}^{1/n},$$

and thus

$$\limsup_{t\to\infty}\frac{1}{t}\log\sup_{x\in X}Q_{x,t}(J)\le-\frac{\lambda}{2}.$$

Noting J^c is compact, we have for any closed subset K of $\mathcal{P}$

$$\begin{aligned}&\limsup_{t\to\infty}\frac{1}{t}\log\sup_{x\in X}Q_{x,t}(K)\\ \le\ &\left(\limsup_{t\to\infty}\frac{1}{t}\log\sup_{x\in X}Q_{x,t}(K\cap J^c)\right)\vee\left(\limsup_{t\to\infty}\frac{1}{t}\log\sup_{x\in X}Q_{x,t}(K\cap J)\right)\\ \le\ &\left(-\inf_{\mu\in K\cap J^c}I_{\mathcal{E}}(\mu)\right)\vee\left(-\frac{\lambda}{2}\right)\le\left(-\inf_{\mu\in K}I_{\mathcal{E}}(\mu)\right)\vee\left(-\frac{\lambda}{2}\right).\end{aligned}$$

By letting λ to ∞, the proof is completed. □

Setting $G=K=\mathcal{P}$ in Theorem 2.1, we have

COROLLARY 2.2.

$$\lim_{t\to\infty}\frac{1}{t}\log\sup_{x\in X}P_x(t<\zeta)=-\inf\left\{\mathcal{E}(u,u):u\in\mathcal{F},\ \int_X u^2dm=1\right\}. \tag{2.8}$$

Let us denote by $\|p_t\|_{p,p}$ the operator norm of p_t from $L^p(X;m)$ to $L^p(X;m)$ and put

$$\lambda_p=-\lim_{t\to\infty}\frac{1}{t}\log\|p_t\|_{p,p},\quad 1\le p\le\infty.$$

Note that $\sup_{x\in X}P_x(t<\zeta)=\|p_t\|_{\infty,\infty}$ and the right hand side of (2.8) is equal to $-\lambda_2$ by the spectral theorem. We then see from Corollary 2.2 that

$$\lambda_\infty=\lambda_2. \tag{2.9}$$

Since $\|p_t\|_{2,2}\le\|p_t\|_{p,p}\le\|p_t\|_{\infty,\infty}$, $1<p<\infty$ by the symmetry of p_t and Riesz-Thorin interpolation theorem, we can obtain

THEOREM 2.3. *Under the assumptions $I\sim IV$, the spectral radius λ_p ($1\le p\le\infty$) is independent of p.*

For p-independence of λ_p, the condition $R_1 1(x)<1$ is not sufficient. In fact, let p_t^B be the semigroup of the Brownian motion on hyperbolic space of dimension $d+1$ and consider the semigroup $p_t=e^{-kt}p_t^B$, $k>0$. Then $\lambda_2=k+\frac{d^2}{8}$ by [D;Theorem 5.7.1], while $\lambda_\infty=k$. As for λ_∞, the next probabilistic interpretation is known:

THEOREM 2.4. ([Sa])

$$\sup_{x\in X}E_x(\exp(\lambda\zeta))<\infty \quad \textit{if and noly if} \quad \lambda<\lambda_\infty.$$

Therefore, we obtain

COROLLARY 2.5. *Assume $I\sim IV$. Then*

$$\sup_{x\in X}E_x(\exp(\zeta))<\infty \quad \textit{if and only if}\ \ \lambda_2>1.$$

3. A Necessary and Sufficient Condition for Gaugeability

Let $\mathbf{D}$ be the classical Dirichlet form:

$$\mathbf{D}(u,v) = \int_{R^d} \nabla u \cdot \nabla v dx \quad \text{for } u, v \in H^1(R^d), \tag{3.1}$$

where $H^1(R^d)$ is the Sobolev space of order 1. Let us denote by (B_t, P_x^W) the d-dimensional Brownian motion on R^d. Given a domain D of R^d, set

$$H_0^1(D) = \{u \in H^1(R^d) : \widetilde{u} = 0, \text{ q.e. on } R^d \setminus D\},$$

where $\widetilde{u}$ is a quasi continuous version of u. By identifying $L_D^2(R^d)(= \{u \in L^2(R^d) : u = 0 \text{ a.e. on } R^d \setminus D\})$ with $L^2(D)$, $(\frac{1}{2}\mathbf{D}, H_0^1(D))$ becomes a regular Dirichlet form on $L^2(D)$ (Theorem 4.4.3 in [FOT]). We call it the *part of the Dirichlet form* $\frac{1}{2}\mathbf{D}$ *on the domain* D. Set

$$B_t^D = \begin{cases} B_t & \text{on } t < \tau_D \\ \Delta & \text{on } t \geq \tau_D \end{cases} \qquad (B_\infty^D = \Delta).$$

Here τ_D is the exit time from D, $\tau_D = \inf\{t > 0 : B_t \in R^d \setminus D\}$, and Δ is an extra point. The process B_t^D is called the *absorbing Brownian motion on* D and its transition function is given by

$$p_t^D(x, E) = P_x^W(B_t \in E, t < \tau_D), \quad t > 0,\ x \in D,\ E \in \mathcal{B}(D).$$

The Dirichlet form of B_t^D is $(\frac{1}{2}\mathbf{D}, H_0^1(D))$ (Theorem 4.4.3 in [FOT]). Suppose that the domain D is regular, $P_x^W(\tau_D = 0) = 1$ for any $x \in \partial D$. We then see from Theorem 2.2 and Theorem 2.4 in [CZ] that B_t^D satisfies I, II and III in §2. Let us denote by $p_t^D(x,y)$ and $g^D(x,y)$ its transition density and Green densitiy of B_t^D.

A positive Radon measure μ on R^d is said to be in the *Kato class* K_d if

$$\lim_{\alpha \downarrow 0} \sup_{x \in R^d} \int_{|x-y|<\alpha} \frac{\mu(dy)}{|x-y|^{d-2}} = 0, \quad d \geq 3$$

$$\lim_{\alpha \downarrow 0} \sup_{x \in R^d} \int_{|x-y|<\alpha} (\log |x-y|^{-1}) \mu(dy) = 0, \quad d = 2$$

$$\sup_{x \in R^d} \int_{|x-y|\leq 1} \mu(dy) < \infty, \quad d = 1.$$

Set

$$K_{d,0} = \{\mu \in K_d : \mu(R^d) < \infty\}.$$

For $\mu \in K_d$, there exists a unique (up to equivalence) positive continuous additive functional A_t^μ which is in Revuz correspondence with μ (cf. [ABM]): for any γ-excessive function h ($\gamma \geq 0$) and positive Borel function f

$$\lim_{t \to 0} \frac{1}{t} \int_{R^d} E_x^W \left(\int_0^t f(B_s) dA_s^\mu \right) h(x) dx = \int_{R^d} f(x) h(x) d\mu.$$

If μ is absolutely continuous with respect to the Lebesgue measure, say $\mu = V(x)dx$, then A_t^μ is nothing but $\int_0^t V(B_s)ds$.

Let A_t^μ be the continuous additive functional associated with $\mu \in K_{d,0}$. Let $\{\tau_t\}_{t \geq 0}$ be the right continuous inverse of $A_{\tau_D \wedge t}^\mu$:

$$\tau_t = \inf\{s > 0 : A_{\tau_D \wedge s}^\mu > t\},$$

and denote by F the fine support of μ, $F = \{x \in D : P_x^W(\tau_0 = 0) = 1\}$. The time changed process Y_t^μ of B_t^D with respect to $A^\mu_{\tau_D \wedge t}$ is defined by

$$Y_t^\mu = B_{\tau_t}^D.$$

The process Y_t^μ is then a μ-symmetric Hunt process on F with lifetime $\zeta = A^\mu_{\tau_D}$ (Theorem 6.2.3 in [FOT]). We assume that F equals to the topological support of μ:

$$F = \mathrm{supp}[\mu]. \tag{3.2}$$

Set

$$H_F u(x) = E_x^W(u(B_{\sigma_F}); \sigma_F < \tau_D),$$

where $\sigma_F = \inf\{t > 0; B_t \in F\}$. Then, the Dirichlet form $(\check{\mathcal{E}}, \check{\mathcal{F}})$ on $L^2(F;\mu)$ of the time changed process Y_t^μ is as follows:

$$\begin{cases} \check{\mathcal{F}} = \{\varphi \in L^2(F;\mu) : \varphi = u \ \mu\text{-a.e. on } F \text{ for some } u \in H_e^1(D)\} \\ \check{\mathcal{E}}(\varphi,\varphi) = \frac{1}{2}\mathbf{D}(H_F u, H_F u), \ \varphi \in \check{\mathcal{F}}, \ \varphi = u \ \mu\text{-a.e. on } F \ u \in H_e^1(D). \end{cases} \tag{3.3}$$

Here $H_e^1(D)$ stands for the extended Dirichlet space of $(\frac{1}{2}\mathbf{D}, H_0^1(D))$ (Theorem 6.2.1 in [FOT]).

A domain D is said to be *Green-bounded* if

$$\sup_{x \in D} E_x^W(\tau_D) < \infty.$$

From now on, we assume, moreover, that for $d = 1$, D is bounded and for $d = 2$, D is Green-bounded.

LEMMA 3.1. *The time changed process Y_t^μ satisfies Assumption I $\sim$ IV.*

PROOF. Let $\{G_\alpha^\mu(x,dy)\}_{\alpha \geq 0}$ be the resolvent kernel of Y_t^μ. Note that

$$G_0^\mu(x,dy) = g^D(x,y)\mu(dy) \tag{3.4}$$

by the definition of the time changed process. Since $g^D(x,y) > 0$ for any $(x,y) \in D \times D$ (Theorem 2.5 in [CZ]),

$$G_0^\mu I_A(x) = \int_F g^D(x,y) I_A(y) d\mu(y) > 0$$

for any $A \in \mathcal{B}(F)$ with $\mu(A) > 0$. Thus the irreducibility of Y_t^μ follows. $G_\alpha^\mu(x,dy)$ is absolutely continuous with respect to μ because of (3.4), which implies the absolute continuity of the transition probability of Y_t^μ by Theorem 4.2.4 in [FOT]. For $f \in \mathcal{B}_b(F)$ and $\alpha > 0$

$$G_\alpha^\mu f(x) = G_0^\mu f(x) - \alpha G_0^\mu(G_\alpha^\mu f)(x) = \int_F g^D(x,y)(f - \alpha G_\alpha^\mu f)(y)\mu(dy). \tag{3.5}$$

Since $(f - \alpha G_\alpha^\mu f)(y)\mu(dy) \in K_{d,0}$, the right hand side of (3.5) belongs to $C_\infty(D)$ on account of Theorem 3.2 and Theorem 4.3 in [CZ]. Thus, $\{G_\alpha^\mu\}$ becomes a strongly continuous resolvent on $C_\infty(F)$, and its semigroup satisfies Assumption III by virtue of Hille-Yoshida theorem. □

THEOREM 3.2. *Suppose that $\mu \in K_{d,0}$ satisfies (3.2). Then*

$$\sup_{x \in D} E_x^W\left(\exp(A^\mu_{\tau_D})\right) < \infty \tag{3.6}$$

if and only if

$$\inf\left\{\frac{1}{2}\mathbf{D}(u,u) : u \in C_0^\infty(D),\ \int_D u^2 d\mu = 1\right\} > 1. \tag{3.7}$$

PROOF. As mentioned above, the time changed process Y_t^μ satisfies Assumption I $\sim$ IV. Since $A_{\tau_D}^\mu$ is the lifetime of Y_t^μ, Corollary 2.5 tells us that the equation (3. 6) holds if and only if

$$\inf\left\{\check{\mathcal{E}}(u,u) : u \in \check{\mathcal{F}},\ \int_F u^2 d\mu = 1\right\} > 1.$$

By Lemma 3.1 in [T2], the left hand side above is equal to the left hand side of (3.7). □

REMARK 3.3. (i) If $D \subset R^3 (d \geq 3)$ is of finite Lebesgue measure, Theorem 3.1 can be verified for any $\mu \in K_d$ on account of Theorem 4.3 (iii) in [CZ].
(ii) For $d \geq 3$ we introduce a subclass K_d^∞ of K_d by following [Z]:

$$K_d^\infty = \left\{\mu \in K_d : \lim_{A\to\infty}\left[\sup_{x\in R^d}\int_{|y|\geq A}\frac{d\mu(y)}{|x-y|^{d-2}}\right] = 0\right\}.$$

It is known in [Z;Proposition 1] that $K_{d,0} \subset K_d^\infty$. For $\mu \in K_d^\infty$,

$$\int_{R^d}\frac{d\mu(y)}{|x-y|^{d-2}} \in C_\infty(R^d). \tag{3.8}$$

Indeed, since $\int_{R^d} 1/|x-y|^{d-2} I_{\{|y|<A\}} d\mu(y) \in C_\infty(R^d)$, (3.8) follows from the definition of K_d^∞. As a result, Lemma 3.1 and Theorem 3.1 can be extended to $\mu \in K_d^\infty$.

Denote by Δ_D the Laplacian on D with the Dirichlet boundary condition and by λ_D the bottom of spectrum of $-\Delta_D$:

$$\lambda_D = \inf\left\{\frac{1}{2}\mathbf{D}(u,u) :\ u \in C_0^\infty(D),\ \int_D u^2 dx = 1\right\}.$$

By the exactly same argument as in [Si], we see that for any domain D

$$-\lambda_D = \lim_{t\to\infty}\frac{1}{t}\log\|p_t^D\|_{\infty,\infty}. \tag{3.9}$$

Hence, we see form (3.9) and [S;Corollary 3.2] that for a domain D

$$D \text{ is Green-bounded} \iff \lambda_D \text{ is positive.}$$

On account of this fact, the next lemma was verified in [T2].

LEMMA 3.4. *Let D be a Green-Bounded and $\mu \in K_d$. Then the following conditions are equivalent:*
(i) $\inf\left\{\frac{1}{2}\mathbf{D}(u,u) + \int_D u^2 d\mu : u \in C_0^\infty(D),\ \int_D u^2 dx = 1\right\} > 0.$
(ii) $\inf\left\{\frac{1}{2}\mathbf{D}(u,u) : u \in C_0^\infty(D),\ \int_D u^2 d\mu = 1\right\} > 1.$

The assertion (i) in Lemma 3.4 is known as one of necessary and sufficient conditions for gaugeability if the domain D has finite volume ([CZ;Theorem 4.19 (v)]).

EXAMPLE 3.5. Let $\partial B(0,r) = \{x \in R^d : |x| = r\}$ $(d \geq 3)$ and μ the surface measure σ of the sphere $\partial B(0,r)$. Then, it is known in [BEKS] that $\sigma \in K_d$. Denote by $\ell_r(t)$ the continuous additive functional corresponding to σ and by $(\check{\mathcal{E}}, \check{\mathcal{F}})$ the Dirichlet form of the time changed process by $\ell_r(t)$. In this case,

$$Hf(x) = H_{\partial B(0,r)}f(x) = \frac{1}{2\pi R}\int_{B(0,r)} \frac{|x|^2 - |r|^2}{|x-y|^d} f(y)d\sigma \quad |x| > r.$$

We see (cf. [K;Lemma 12.5]) that

$$\begin{aligned}\check{\mathcal{E}}\left(f, \frac{1}{r^{d-2}}\right) &= \mathcal{E}\left(Hf, H\left(\frac{1}{r^{d-2}}\right)\right) = \lim_{\rho\downarrow r} \frac{1}{2}\int_{|x|\geq\rho} \nabla Hf(x)\cdot\nabla\left(\frac{1}{|x|^{d-2}}\right)\\ &= \frac{d-2}{2r^{d-1}}\int_{|x|=r} f(x)d\sigma(x).\end{aligned}$$

On the other hand, the right hand side is written as

$$\frac{1}{r^{d-2}}\int_{|x|=r} f(x)k(dx),$$

where k is the killing measure of $(\check{\mathcal{E}}, \check{\mathcal{F}})$ in Beurling-Deny decomposition. Hence, the measure k equals to $\frac{d-2}{2r}\sigma$, and consequently

$$\inf\left\{\check{\mathcal{E}}(u,u); u \in \check{\mathcal{F}}, \int_{|x|=r} u^2 d\sigma = 1\right\} = \frac{d-2}{2r}.$$

Therefore

$$\sup_{x\in R^d} E_x^W(e^{\ell_r(\infty)}) < \infty$$

if and only if $r < \frac{d-2}{2}$.

We see from [T1] that

$$\lim_{t\to\infty}\frac{1}{t}\log E_x^W\left(e^{\ell_r(t)}\right) = -\inf_{\substack{u\in H^1(R^d)\\ \|u\|_2=1}}\left(\frac{1}{2}\mathbf{D}(u,u) - \int_{|x|=r} u(x)^2 d\sigma\right),$$

and from [AGS] that the right hand side is equal to 0 if $r \leq \frac{d-2}{2}$ and positive if $r > \frac{d-2}{2}$. Therefore $-\frac{1}{2}\Delta - \sigma$ is subcritical in $r < \frac{d-2}{2}$, critical in $r = \frac{d-2}{2}$ and supercritical in $r > \frac{d-2}{2}$ (for definition of criticality and supercriticality, see [Mu]).

EXAMPLE 3.6. Let $\mu \in K_d$. For any compact set $K \subset D$, define

$$\pi(K,D) = \begin{cases} \frac{\mu(K)}{\mathrm{Cap}(K,D)} & \text{for } \mathrm{Cap}(K,D) > 0,\\ 0 & \text{for } \mathrm{Cap}(K,D) = 0,\end{cases}$$

where the capacity $\mathrm{Cap}(K,D)$ defined by

$$\mathrm{Cap}(K,D) = \inf\{\mathbf{D}(u,u); u \geq 1 \text{ on } K,\ u \in C_0^\infty(D)\}.$$

It is known from Theorem 2.5.2/1 in [Ma] that if

$$\sup_{K\subset D} \pi(K,D) < \frac{1}{8},$$

then

$$\inf\left\{\frac{1}{2}\mathbf{D}(u,u) : u \in C_0^\infty(D), \int_D u^2 d\mu = 1\right\} > 1.$$

In particular, for $d \geq 3$ and a closed set F

$$\sup_{x \in R^d} E_x^W \left(\exp\left(\int_0^\infty I_F(B_t)dt \right) \right) < \infty$$

if

$$\sup_{K \subset R^d} \frac{|F \cap K|}{\mathrm{Cap}(K)} < \frac{1}{8}. \tag{3.10}$$

Here $|\cdot|$ means the Lebesgue measure and $\mathrm{Cap}(K) = \mathrm{Cap}(K, R^d)$. For a closed set F denote by B_F the ball with the same volume as F:

$$B_F = B(0, r_F), \qquad r_F = \frac{\left(|F|\Gamma(\frac{d}{2}+1)\right)^{1/d}}{\sqrt{\pi}}.$$

Since by [Ma;2.2.3, 2.2.4]

$$\begin{aligned} \frac{|F \cap K|}{\mathrm{Cap}(K)} &\leq \frac{|F \cap K|}{\mathrm{Cap}(F \cap K)} \leq \frac{|F \cap K|}{\mathrm{Cap}(B_{F \cap K})} \\ &= \frac{|B_{F \cap K}|}{\mathrm{Cap}(B_{F \cap K})} \leq \frac{|B_F|}{\mathrm{Cap}(B_F)} = \frac{{r_F}^2}{d(d-2)}, \end{aligned}$$

the equation (3.10) holds if

$$|F| < \frac{\left(\frac{1}{8}d(d-2)\pi\right)^{d/2}}{\Gamma\left(\frac{d}{2}+1\right)}.$$

References

[ABM] S. Albeverio, P. Blanchard, Z.M. Ma, *Feynman-Kac semigroups in terms of signed smooth measures*, "Random Partial Differential Equations" ed. U. Hornung et al., Birkhäuser, 1991.

[AGS] J.-P. Antoine, F. Gesztesy, J. Shabani, *Exactly solvable models of sphere interactions in quantum mechanics*, J. Phys. A **20**, (1987), 3687-3712.

[BEKS] J. Brasche, P. Exner, Yu. Kuperin, P. Seba, *Schrödinger operators with singular interactions*, J. Math. Anal. Appl. **184**, (1994), 112-139.

[CZ] K.L. Chung, Z. Zhao, *From Brownian Motion to Schrödinger's Equation*, Springer (1995).

[D] E.B. Davies *Heat kernals and spectral theory*, Cambridge University Press (1989).

[FOT] M. Fukushima, Y. Oshima, M. Takeda, *Dirichlet Forms and Symmetric Markov Processes*, de Gruyter (1994).

[K] M. Kishi, *Potential Theory*, Morikita (1974) (in Japanease).

[Ma] V.G. Maz'ja, *Sobolev Spaces*, Springer (1985).

[Mu] M. Murata, *Structure of positive solutions of* $(-\Delta + V)u = 0$ *in* R^d, Duke Math. J. **53**, (1986) 869-943.

[RW] L.C.G. Rogers, D. Williams, *Diffusions, Markov Processes, and Martingales*, Vol 2, John Wiley and Sons (1987).

[Sa] S. Sato, *An inequality for the spectral radius of Markov processes*, Kodai Math. J. **8**, (1985) 5-13.

[Si] B. Simon, *Schrödinger semigroups*, Bull. Am. Math. Soc. **7**, (1982) 447-536.

[T] M. Takeda, *A Large deviation for symmetric Markov processes with finite lifetime*, Stochastics and Stochastic Reports **59**, (1996) 143-167.

[T1] M. Takeda, *Asymptotic properties of generalized Feynman-Kac functionals*, Potential Analysis **9**, (1998) 261-291.

[T2] M. Takeda, *Exponential decay of lifetime and a Theorem of Kac on total occupation times*, to appear in Potential Analysis.

[Z] Z. Zhao, *Subcriticality and gaugeability of the Schrödinger operator*, Trans. Amer. Math. Soc. **334**, (1992) 75-96.

TOHOKU UNIVERSITY, MATHEMATICAL INSTITUTE, SENDAI 980-8578, JAPAN
E-mail address: takeda@math.tohoku.ac.jp

Canadian Mathematical Society
Conference Proceedings
Volume **29**, 2000

Measure theoretic Laplace operators on fractals

M. Zähle

1. Motivation

The treatment of (stochastic) differential equations on fractal sets is stimulated by physical applications to porous media.
In the last decade several attempts have been made in order to create a corresponding harmonic calculus. Classical notions and relationships like Dirichlet form, Laplace operator, Green function, Neumann derivative and Gauss–Green formulae have been extended to certain self–similar fractal sets in $\mathbb{R}^n$.
The following main approaches are now established in the literature:

(1) A discrete probabilistic approximation via *random walks*. The Laplace operator arises as *infinitesimal generator* of a limit *diffusion process* on nested fractals. (Since 1987: Kusuoka, Goldstein, Barlow, Bass, Perkins, Lindstrøm, Kumagai, Hambly a.o.)
(2) A discrete probabilistic approximation by means of *Markov chains*. The (special) self–similar sets are interpreted as *Martin boundaries* and the Laplacian is determined by *continuous extension* of a transion operator. (Since 1997: Denker and Sato.)
(3) Discrete analytical (or graph theoretical) approximation via harmonic structures for certain second order *difference operators* on post critically finite self–similar sets. The Laplace operator is introduced as *limit* in such difference schemes and some variational calculus is developed. (Since 1989: Kigami, Fukushima, Shima, Lapidus, Mosco a.o.)
(4) Application of the theory of function spaces. Δ^{-1} is defined on fractal d–sets in $\mathbb{R}^n$ as *trace of the euclidean* Δ^{-1} on the corresponding Besov spaces. (Since 1996: Triebel a.o.)

REMARKS. 1. In (1)–(3) the construction steps of the self–similar fractals play an essential role in the approximation schemes. The Laplacian is here always given as limit operator.
2. Results of Sabot and Metz imply that the approaches (1) and (3) are equivalent for nested fractals.
3. Denker ans Sato have obtained the equivalence of (2) and (3) for the Sierpinski gasket.
4. Comparison of the spectral asymptotics of the Laplace operator in (3) and (4) shows that (4) is different from the other models.

2000 *Mathematics Subject Classification.* Primary 28A80; Secondary 47B05, 35P20.

In all four definitions the Laplacian is determined implicitly and the first three methods are restricted to certain self–similar sets. The aim of this note is to present an explicit measure *theoretic approach* which does not depend on the underlying fractal constructions. As in (4) in this paper our restriction consists in considering compact d-sets in $\mathbb{R}^n$ with $d > n-2$.
We start with a version extending the classical euclidean notions in the context of geometric measure theory by means of random directional derivatives. In order to obtain better function spaces for the domain of definition of Dirichlet form and Laplace operator we then adopt the inverse approach starting from the definition of $(-\Delta)^{-1/2}$ in form of a Riesz potential with respect to Hausdorff measure on the given set.
Comparison with the distribution method (4) shows that in our approach $(-\Delta)^{-1/2}$ (or $(-\Delta)^{1/2}$) (but *not* Δ or Δ^{-1}) may be interpreted as trace of the euclidean boundary free variant. This is in a better correspondence with the case $n=1$.
Our variant may further be improved by taking *fractional Riesz potentials* in order to include all fractal dimensions and to get a scaling behavior of the Laplacian which is compatible with physical interpretations.

2. Classical euclidean notions

In this section we recall related euclidean results and present them in the light of "*random line sections*".
Let U be an open set in $\mathbb{R}^n$ with scalar product $\langle\cdot,\cdot\rangle$. For smooth functions $f: U \to \mathbb{R}$ we denote the *differential* by $Df(x) =: \nabla f(x)$, so that $Df(x)\,v =: \nabla_v f(x)$ is the *derivative in direction* $v \in S^{n-1}$. The *gradient* of f is dermined by $\langle \operatorname{grad} f, v\rangle = \nabla_v f$ and the *Laplace operator* by $\Delta f := \operatorname{div} \operatorname{grad} f = \sum_{i=1}^{n} \frac{\partial^2 f}{\partial x_i^2}$ in coordinates.
Let σ^{n-1} be the normalized surface area measure on S^{n-1} which may be interpreted as *uniform distribution of a random direction* in $\mathbb{R}^n$. Straightforward calculations (preferably in coordinates) show that for all $x \in U$,

$$\Delta f(x) = \int_{S^{n-1}} (\nabla_v)^2 f(x)\, \sigma^{n-1}(dv) \tag{5}$$

$$\langle \operatorname{grad} f(x),\ \operatorname{grad} g(x)\rangle = \int_{S^{n-1}} \nabla_v f(x)\, \nabla_v g(x)\, \sigma^{n-1}(dv). \tag{6}$$

These are basic interpretations leading to our fractal extensions. Similar relationships hold for the *Neumann derivatives*: Let $G \subset \bar{G} \subset U$ be a bounded domain with smooth boundary ∂G and $N(x)$ be the outer unit normal to G at $x \in \partial G$. Then the Neumann derivative of f at $x \in \partial G$ is given by $\frac{\partial f}{\partial N}(x) := \nabla_{N(x)} f(x)$. The following result (which I have not found in the literature) may be proved by means of Federer's co–area formula (cf. [**2**]): $\mathcal{H}^k$ denotes the k–dimensional Hausdorff measure in $\mathbb{R}^n$ with Lebesgue normalization. Then $\mathcal{H}^{n-1}|_{\partial G}$ agrees with the surface area measure. $L(v)$ is the linear hull of $v \in S^{n-1}$ and $L(v)^{\perp}$ its orthogonal complement in $\mathbb{R}^n$. $\mathcal{H}^{n-1}|_{L(v)^{\perp}}$ agrees with the Lebesgue measure on this subspace.

THEOREM.

$$\int_{\partial G} f\,\frac{\partial g}{\partial N}\,d\mathcal{H}^{n-1} = \int_{S^{n-1}} \int_{L(v)^\perp} \sum_{x\in\partial G\cap(z+L(v))} \operatorname{sgn}(v,x)\; f(x)\,\nabla_v g(x)\;\mathcal{H}^{n-1}(dz)\,\sigma^{n-1}(dv) \tag{7}$$

where $\operatorname{sgn}(v,x) := \operatorname{sgn}\langle v, N(x)\rangle$ *is positive iff* v *points into the exterior of* G.

The classical *Dirichlet form* on G is defined by

$$\mathcal{E}(f,g) = \int_G \langle \operatorname{grad} f,\ \operatorname{grad} g\rangle d\mathcal{L}^n \tag{8}$$

with Lebesgue measure $\mathcal{L}^n$, and the *Gauss–Green formulae* read:

$$\int_G f\,\Delta g\,d\mathcal{L}^n = -\mathcal{E}(f,g) + \int_{\partial G} f\,\frac{\partial g}{\partial N}\,d\mathcal{H}^{n-1} \tag{9}$$

$$\int_G f\,\Delta g\,d\mathcal{L}^n - \int g\,\Delta f\,d\mathcal{L}^n = \int_{\partial G} f\,\frac{\partial g}{\partial N}\,d\mathcal{H}^{n-1} - \int_{\partial G} g\,\frac{\partial f}{\partial N}\,d\mathcal{H}^{n-1}\,. \tag{10}$$

Substituting the expressions from (5)–(7) one obtains an integral geometric version of (9) and (10) in terms of ∇_v and $(\nabla_v)^2$ only. This is the key for measure theoretic extension to fractals. We start with the case $n = 1$ which also serves as an auxiliary tool for $\mathbb{R}^n$.

3. Harmonic calculus with respect to a measure on an interval

Let μ finite atomless Borel regular measure on the interval $I := [a,b]$. Without loss of generality we may assume that the endpoints a, b belong to the support K of μ. The complement set of K in I is representable as $K^c = \bigcup_{i=1}^{\infty}(a_i, b_i)$ with disjoint open intervals. We call the set $\partial K := \bigcup_{i=0}^{\infty}\{a_i, b_i\}$ with $a_0 := a, b_0 := b$ the *measure geometric boundary* of K. Denote $L_2 := L_2(K,\mu)$.

DEFINITION.

$$D_1^\mu := \Big\{ f \in L_2 : \exists\, f' \in L_2 \text{ with } f(x) = \int_a^x f'd\mu + \sum_{i=1}^{\infty} 1(b_i \le x)\big(f(b_i) - f(a_i)\big),\ x \in K\,,\quad \sum_{i=1}^{\infty}\big|f(b_i) - f(a_i)\big| < \infty \Big\}$$

PROPOSITION. f' is unique in L_2 and $D_1^\mu \subset C(K)$.

$\nabla^\mu f = \frac{df}{d\mu} := f'$ is called the μ*–derivative* of $f \in D_1^\mu$.
Recall that L_2 with the scalar product $\langle f, g\rangle_\mu := \int fg\,d\mu$ is a separable Hilbert space. Then we introduce the corresponding *Dirichlet form* on the linear subspace D_1^μ:

$$\mathcal{E}^\mu(f,g) := \langle \nabla^\mu f,\ \nabla^\mu g\rangle_\mu\,.$$

DEFINITION. $D_2^\mu := \{f \in D_1^\mu : \nabla^\mu f \in D_1^\mu\}$.

The *μ-Laplacian* on D_2^μ is defined by

$$\Delta^\mu := (\nabla^\mu)^2$$

and satisfies the following *Gauss–Green formulae*:

THEOREM.

$$\langle f, \Delta^\mu g\rangle_\mu = -\mathcal{E}(f,g) - \sum_{i=0}^{\infty} \left[f(b_i)\, \nabla^\mu g(b_i) - f(a_i)\, \nabla^\mu g(a_i) \right] \tag{11}$$

$$\begin{aligned} \langle f, \Delta^\mu g\rangle\mu - \langle \Delta^\mu f, g\rangle_\mu = -\sum_{i=0}^{\infty} \big[& f(b_i)\, \nabla^\mu g(b_i) - f(a_i)\, \nabla^\mu g(a_i) \\ & \nabla^\mu f(b_i)\, g(b_i) + \nabla^\mu f(a_i)\, g(a_i) \big] . \end{aligned} \tag{12}$$

This can be shown by Fubini's theorem (see [**3**]*).*

Note that $f(b_i)\, \nabla^\mu g(b_i) - f(a_i)\, \nabla^\mu g(a_i) = f(b_i)\, \big(\nabla^\mu g(b_i) - \nabla^\mu g(a_i)\big)$ $\nabla^\mu g(a_i)\, \big(f(b_i) - f(a_i)\big)$. Thus, for Dirichlet or Neumann boundary conditions with vanishing second summand on the right–hand side of (11) we may assume that we have only finitely many jumps of the functions and their derivations at ∂K and for the remaining boundary points the classical conditions. For simplicity, we consider only the following case:

DEFINITIONS.

$D_{2,0}^\mu := \big\{ f \in D_2^\mu : f(a_i) = f(b_i) \text{ and } \nabla^\mu f(a_i) = \nabla^\mu f(b_i),\ i \in \mathbb{N} \big\}$

(no jumps at ∂K, classical Gauss–Green formulae)

$D_{2,D}^\mu := \big\{ f \in D_{2,0}^\mu : f(a) = f(b) = 0 \big\}$

(*Dirichlet boundary condition*)

$D_{2,N}^\mu := \big\{ f \in D_{2,0}^\mu : \nabla^\mu f(a) = \nabla^\mu f(b) \big\}$

(*Neumann boundary condition.*)

THEOREM. *$D_{2,\cdot}^\mu$ is dense in $L_2(K,\mu)$ and Δ^μ on $D_{2,\cdot}^\mu$ is a nonnegative symmetric operator with discrete spectrum whose only accumulation point is $-\infty$.*

For the proof (see [**3**]) we represent the resolvent operator of Δ^μ by means of an explicitly determined Green function. Then we use well–known results from functional analysis in order to obtain compactness. self–adjointness and finite trace. The orthonormal basis of eigenfunctions of the resolvent operator is also one of the Laplacian and the eigenvalues are in the obvious correspondence. (Such an approach is known from generalized diffusion processes in a probabilistic context.) The latter conrrespondence is exploited in order to determine the spectral asymptotics for the special case of *self–similar measures*:

THEOREM. *(U. Freiberg, see* [**3**]*)*
If μ is self–similar and $\lambda_1 \geq \lambda_2 \geq \ldots$ are the eigenvalues of Δ^μ then λ_k is asymptotically equivalent to $-k^2$ as $k \to \infty$.

Here the local scaling properties of μ are essentially used. Note that one obtains the *Weyl exponent* 2. When interpreting the Lebesgue measure on $[a,b]$ as a special self–similar measure the classical model of Courant and Weyl for $n = 1$ is included.

4. Extension to $\mathbb{R}^n$

Let μ be a finite Borel regular measure in $\mathbb{R}^n$ with compact support K. Suppose first that the $n-1$–*energy* of μ is *finite*; i.e., $\iint |x-y|^{-(n-1)}\,\mu(dx)\,\mu(dy) < \infty$. Then for σ^{n-1}–almost all $v \in S^{n-1}$ the measure μ can be disintegrated with respect to Lebesgue measure on $L(v)^\perp$:

$$\mu = \int\limits_{L(v)^\perp} \mu_{v,z}\; d\mathcal{H}^{n-1}(dz)$$

where the section measure $\mu_{v,z}$ concentrated on the line l through the point z with direction v is determined for $\mathcal{H}^{n-1}$–almost all $z \in L(v)^\perp$ (cf. Mattila [**4**]). Note that

$$\kappa(dl) := \mathcal{H}^{n-1}|_{L(v)^\perp}(dz)\,\sigma^{n-1}(dv)$$

is the natural motion invariant measure on the space of oriented lines in $\mathbb{R}^n$ and we get

$$\mu = \int \mu_l\,\kappa(dl)$$

with $\mu_l := \mu_{v,z}$ in the parametrization $l = l(v,z)$. We next consider the analogues of the spaces D_1^μ and D_2^μ where the first, resp. second order derivatives w.r.t μ_l are determined in the sense of section 3 for κ–almost all lines l and satisfy the square integrabilities w.r.t. $\mu_l(dy)\,\kappa(dl)$. Then

$$\nabla_v^\mu f(x) := \nabla^{\mu_{v,\Pi_{L(v)^\perp}(x)}} f(x)$$

is called the *derivative* of $f \in D_1^\mu$ *in direction* v at point x w.r.t. the measure μ. In analogy to (6) we define for $f, g \in D_1^\mu$

$$\langle \mathrm{grad}^\mu f(x),\ \mathrm{grad}^\mu g(x)\rangle := \int\limits_{S^{n-1}} \nabla_v^\mu f(x)\,\nabla_v^\mu g(x)\,\sigma^{n-1}(dv)$$

and the corresponding "*Dirichlet form*" on D_1^μ is introduced by

$$\mathcal{E}^\mu(f,g) := \int \langle \mathrm{grad}^\mu f,\ \mathrm{grad}^\mu g\rangle\, d\mu\,.$$

Similarly,

$$\Delta^\mu f(x) := \int\limits_{S^{n-1}} (\nabla_v^\mu)^2 f(x)\,\sigma^{n-1}(dv)$$

is a candidate for the *Laplace operator* on D_2^μ and we get

$$\int f\,\Delta^\mu g\,d\mu = \iint f\,\Delta^\mu f\,d\mu_l\,d\kappa(l)\,.$$

Thus, formula (11), Fubini and the invariance properties of σ^{n-1} imply the following.

COROLLARY.

$$\langle f, \Delta^\mu g\rangle_\mu = -\mathcal{E}^\mu(f,g) + \int \sum_{i=0}^{\infty} \Big[f\big(b_i)(l)\big)\,\nabla^{\mu_l} g\big(b_i(l)\big) - f\big(a_i(l)\big)\,\nabla^{\mu_l} g\big(a_i(l)\big)\Big]\,d\kappa(l)\,, \tag{13}$$

$f, g \in D_2^\mu$, where the ordered boundary points $a_i(l)$, $b_i(l)$ of $K \cap l$ on the directed line l are chosen according to the case $n = 1$ (section 3).

Compared to the euclidean case (7) the integral on the right hand side may formally be denoted by $\int_{\partial K} f \frac{\partial^\mu g}{\partial N} \, d(\partial\mu)$, though the notions $\frac{\partial^\mu g}{\partial N}$ and $\partial\mu$ are not explicitly introduced. (Think of a measure on the boundary ∂K and of the mean one–sided directional derivatives.) Then (13), Fubini and the rotation invariance of σ^{n-1} imply the *Gauss–Green formula*:

$$\text{(14)} \qquad \langle f, \Delta^\mu g\rangle_\mu - \langle g, \Delta^\mu f\rangle_\mu = \int_{\partial K} f \frac{\partial^\mu g}{\partial N} \, d(\partial\mu) - \int_{\partial K} g \frac{\partial^\mu f}{\partial N} \, d(\partial\mu) \,.$$

The boundary conditions on ∂K, which is given by the boundary points of the intersections with almost all lines, may be arranged similarly as in the one–dimensional case.
If $\int_{\partial K} f \frac{\partial^\mu g}{\partial N} \, d(\partial\mu) = 0$ for all f and g under consideration then Δ^μ is symmetric.

A disadvantage of this approach is that the spaces D_1^μ or D_2^μ may be poor of "good" functions and the above boundary conditions for symmetry are rather artificial. In general, self–adjointness cannot be proved. However , if Δ^μ would be self–adjoint with discrete spectrum and μ self–similar with fractal dimension d, then some scaling arguments would provide the spectral dimension

$$d_s = \frac{d}{d-n+1} \,.$$

This has led to the following *improvement* of our definition. Note that in the euclidean case $\nabla := (-\Delta)^{1/2}$ is closer related to the metric than $-\Delta$ by the linear scaling property of ∇. If the domain is the whole $\mathbb{R}^n$, i.e., the problem is boundary free, then the operator ∇^{-1} may be realized as special Riesz potential:

$$\nabla^{-1} f(x) = \int_{\mathbb{R}^n} \frac{1}{|x-y|^{n-1}} \, f(x) \, dy \,, \; n > 1 \,.$$

Using polar coordinates this may be interpreted as integrating f along a random one–dimensional direction starting from the point x. We now will keep the boundary free point of view for fractal measures. This is consistent with the interpretation of certain self similar sets as Martin boundaries (cf. Denker and Sato [**1**]) and an analogue to differential geometry, where a boundary of a manifold or a current has no boundary. Summarizing all these arguments we now introduce in an inverse direction:

$$(\nabla^\mu)^{-1} f(x) := \int \frac{1}{|x-y|^{n-1}} \, f(y) \, \mu(dy) \,.$$

The operator $(\nabla^\mu)^{-1}$ may be interpreted as integrating f along a random one–dimensional direction according to the corresponding section measure of μ starting from the point x. Thus, it is related to the former version of $(-\Delta^\mu)^{-1/2}$ in a sense. Suppose now that the measure μ is the d-dimensional Hausdorff measure on a compact d-set. Then our operator agrees with the trace of the boundary free euclidean ∇^{-1} on the d-set in the sense of Triebel's approach [**5**] and the corresponding function spaces are available. The corresponding *Dirichlet form* is given by

$$\mathcal{E}^\mu(f,g) := \langle \nabla^\mu f, \; \nabla^\mu g\rangle_\mu$$

and the *Laplace operator* by

$$\Delta^\mu = -(\nabla^\mu)^2 \ .$$

The spectral dimension of Δ^μ is equal to $d_s = \frac{d}{d-n+1}$.
Our approach may be *modified* when taking fractional powers of the inverse of the boundary free euclidean Laplacian which are representable as *Riesz potentials of fractional order.* An appropriate choice of this order leads to fractal Laplacians whose spectral behaviour is closely related to physical interpretations.
All details will be presented in part II of the paper [**3**].

References

[1] Denker, M., Sato, H., Sierpinski gasket as a Martin boundary I: Martin kernels, Kyushu–MP5-1997-19 (to appear in : Potential Analysis).
[2] Federer, H., *Geometric Measure Theory*, Springer, 1969.
[3] Freiberg, U., Zähle, M., Harmonic calculus on fractals – A measure–geometric approach I (Preprint).
[4] Mattila, P., *Geometry of Sets and Measures in Euclidean Spaces*, Cambridge Univ. Press, 1995.
[5] Triebel, H., *Fractals and Spectra*, Birkhäuser 1997.

UNIVERSITY OF JENA, MATHEMATICAL INSTITUTE, 07740 JENA, GERMANY
E-mail address: zaehle@minet.uni-jena.de

Canadian Mathematical Society
Conference Proceedings
Volume **29**, 2000

Estimation of the Phase Transition Time in the Stefan Problem

Weian Zheng

ABSTRACT. Suppose a volume of water is surrounded by ice. How long it will take that the entire water will be frozen? We use a martingale model associated to the Stefan problem to give a pair of concrete estimates.

There have been many fruitful results [**2**] in the study of one dimensional classical Stefan problem. Because of the existence of "mushy region", which is a region with co-existence of two phases at the critical temperature, the study of the multi-dimensional case is much more difficult. We discovered in [**12**] a relationship between the phase transition speed and the shape changing of the free boundary. However, that estimation depends on the evolution of the boundary shape, which is not easy to decide in many applications. In this paper, we study more precisely the dependence of the solution to Stefan's problem on the initial distribution and therefore we obtain a pair of concrete bounds which depends only on the initial temperature distribution. This result is quite satisfactory. The author also would like to point out that the method of random time changing plays an important role in the proof, which has not been interpreted by the other non-probabilistic languages yet.

Let $b_1, b_2, ..., b_n$, $L_2, L_3, ..., L_n$ and $u_2 < ... < u_n$ be given finite positive numbers. Define a step function:

$$b(u) = b_1\chi_{[0,u_2)}(u) + \sum_{i=2}^{n-1} b_i\chi_{[u_i,u_{i+1})}(u) + b_n\chi_{[u_n,\infty)}(u).$$

Consider a weak solution (see p.63 [**1**], [**6**] and [**8**], explained also in Remark 1) of the following equation in R^d :

$$\frac{\partial}{\partial t}H(u(x,t)) = \frac{1}{2}\sum_j \frac{\partial^2}{\partial x_j^2}u(x,t) \tag{0.1}$$

---

[1]1991 Mathematical Subject Classification: Primary 60J65; secondary 60J60, 60J35, 58G32, 58G11.

where $H(u)$ is defined by

$$H(u) = \int_0^u b(v)dv + \sum_{\{i;u_i<u\}} L_i$$

and the initial condition is given by $u(x,0) = u_0(x)$. We will assume $u_0(.)$ is a bounded continuous function and $\int H(u_0(x))dx = 1$ in this paper to simplify our discussions.

For $i = 2, ..., n$, denote $\xi_i = \inf\{t, \ sup_x u(x,t) < u_i\}$. Then ξ_i is the first time that the values of $u(x,t)$ is uniformly less than u_i. The purpose of this paper is to give an upper bound and a lower bound estimates to the value of ξ_i.

REMARK 1. We say the couple $(u(x,t), h(x,t))$ gives a weak solution to (0.1) if

i) $h(x,t) = H(u(x,t))$ when $u(x,t) \notin \{u_2, ..., u_n\}$ and $H(u_i-) \leq h(x,t) \leq H(u_i)$ when $u(x,t) = u_i$;

ii) for all test functions ϕ,

$$\int[\int h(x,s)\frac{\partial}{\partial t}\phi(x,s) + \frac{1}{2}u(x,t)\sum_j \frac{\partial^2}{\partial x_j^2}\phi(x,s)]dx\, ds$$
$$= \int h(x,t)\phi(x,t)dx - \int h(x,0)\phi(x,0)dx.$$

To simplify our notation, we will still denote $h(x,t)$ by $H(u(x,t))$. However, we should bare in our mind that $H(u_i)$ may be multi-valued. We may also consider the case where the initial set carrying the phase transition. Then we should assume $H(u_0(x))$ may be also multi-valued. Our method will be independent of the multi-value hypothesis.

REMARK 2. If we denote by $u = \phi(v)$ the inverse function of $2H(u) = v$, then (0.1) becomes the porous media equation:

$$\frac{\partial}{\partial t}v(x,t) = \sum_j \frac{\partial^2}{\partial x_j^2}\phi(v(x,t)). \tag{0.2}$$

(0.1) is directly associated to the famous Stefan problem (see [**1**], [**11**], [**12**]) where $u(x,t)$ is the temperature and $H(u(x,t))$ is the enthalpy (see [**1**], p43). Here we assume that the product of the thermal conductivity and thermal diffusivity is equal to one without losing generality (see [**5**] for details). u_i is the fusion temperature between the $(i-1)$–th state and the i–th state, and L_i (> 0) is the latent heat of fusion at temperature u_i.

The physical meaning of our problem is given in the following example. Let us consider the case where $n = 2$. Suppose there is some water (where the absolute temperature $u > u_2 = 273.16K$) surrounded by ice (where $u < u_2 = 273.16K$, we are interested in estimating the time at which the water is entirely frozen (i.e. $u < 273.16K$ everywhere). (0.1) is just formulae (2.28) and (2.29) of [**1**]. We intend to establish an upper bound and a lower bound for this time of freezing.

In [**11**] [**12**], we have proved that there is a solution (X_t) of Ito stochastic differential equation

$$X_t = X_0 + \int_0^t \sqrt{uH^{-1}(u(X_s, s))}dW_s \tag{0.3}$$

which has the enthalpy $H(u(x,t))$ as its density function with respect to Lebesgue measure. In the above formula, (W_s) is a standard d-dimensional Brownian motion. This martingale representation is due to our hypothesis that the product of the thermal conductivity and thermal diffusivity is constant. Actually we regularized H in [**12**] to get (0.3) for the regularized H, which is a diffusion process. Then let the regularized H tends to the real H to get the desired representation by weak convergence theory of martingales.

The generator of X_t may be formally written as

$$\frac{1}{2}u(x,t)H^{-1}(u(x,t))\sum_{j-1}^{d}\frac{\partial^2}{\partial x_i^2}. \tag{0.4}$$

However, we will not need this formal generator in the following discussions as our proof only used the martingale property of (X_t).

This study has its own interests in probability theory as follows. Let (X_t) be a martingale of the form (0.3). Our question is how long it will take that its instant density function will be uniformly under certain given value?

Let us consider two properties of the solutions to (0.1) first. Without losing generality, we assume in the proof of the following two lemmas that $H(u)$ appeared in (0.1) is a smooth increasing function such that its first derivative is bounded below by a positive constant. As we have proved in [**11**] the general case is just the limit situation of the smooth cases, which enables us to make the simplification.

Consider first a result about the monotonicity relation between the solutions of (0.1) and their initial values. This result has its strong intuition in physics: suppose there are two initial temperature distributions such that one is higher or equal to the other everywhere. Then this relation will be kept at any other time later. A similar result for the porous media equation (0.2) can be found in [**7**].

LEMMA 0.1. *Suppose $u_0(x)$ and $\overline{u}_0(x)$ are two initial functions such that $0 \leq u_0(x) \leq \overline{u}_0(x)$ and $u(x,t)$ and $\overline{u}(x,t)$ are the corresponding solutions to (0.1). Then*

$$H(u(x,t)) \leq H(\overline{u}(x,t)). \qquad a.e.$$

Proof. Denote $Q_t = \{x,\ u(x,t) > \overline{u}(x,t)\}$. It is sufficient to show that

$$\int_{Q_t} H(u(x,t))dx = \int_{Q_t} H(\overline{u}(x,t))dx. \tag{0.5}$$

Now we have

$$\begin{aligned} &\int_{Q_t}\{H(u(x,t)) - H(\overline{u}(x,t))\}dx \\ = \ &\sum_{i=0}^{k-1}\{\int_{Q_{\frac{(i+1)t}{k}}}[H(u(x,\frac{(i+1)t}{k})) - H(\overline{u}(x,\frac{(i+1)t}{k}))]dx \\ &-\int_{Q_{\frac{it}{k}}}[H(u(x,\frac{it}{k}))dx - H(\overline{u}(x,\frac{it}{k}))]\} \end{aligned}$$

$$
\begin{aligned}
= & \sum_{i=0}^{k-1}\{\int_{Q_{\frac{(i+1)t}{k}}}[H(u(x,\frac{(i+1)t}{k}))-H(\overline{u}(x,\frac{(i+1)t}{k}))]dx \\
& -\int_{Q_{\frac{(i+1)t}{k}}}[H(u(x,\frac{it}{k}))-H(\overline{u}(x,\frac{it}{k}))]dx\} \\
+ & \sum_{i=0}^{k-1}\{\int_{Q_{\frac{(i+1)t}{k}}}[H(u(x,\frac{it}{k}))-H(\overline{u}(x,\frac{it}{k}))]dx \\
& -\int_{Q_{\frac{it}{k}}}[H(u(x,\frac{it}{k}))-H(\overline{u}(x,\frac{it}{k}))]dx\} \\
\leq & \sum_{i=0}^{k-1}\{\int_{Q_{\frac{(i+1)t}{k}}}[H(u(x,\frac{(i+1)t}{k}))-H(\overline{u}(x,\frac{(i+1)t}{k}))]dx \\
& -\int_{Q_{\frac{(i+1)t}{k}}}[H(u(x,\frac{it}{k}))-H(\overline{u}(x,\frac{it}{k}))]dx\} \\
+ & \sum_{i=0}^{k-1}\{\int_{Q_{\frac{(i+1)t}{k}}\cup Q_{\frac{it}{k}}}[H(u(x,\frac{it}{k}))-H(\overline{u}(x,\frac{it}{k}))]dx \\
& -\int_{Q_{\frac{it}{k}}}[H(u(x,\frac{it}{k}))-H(\overline{u}(x,\frac{it}{k}))]dx\}.
\end{aligned}
$$

The last inequality is true because $H(u(x,\frac{it}{k}))-H(\overline{u}(x,\frac{it}{k}))$ is non-negative on $Q_{\frac{it}{k}}$. So we get

$$
\begin{aligned}
& \int_{Q_t}\{H(u(x,t))-H(\overline{u}(x,t))\}dx \\
\leq & \sum_{i=0}^{k-1}\{\int_{Q_{\frac{(i+1)t}{k}}}[H(u(x,\frac{(i+1)t}{k}))-H(u(x,\frac{it}{k}))]dx \\
& -\int_{Q_{\frac{(i+1)t}{k}}}[H(\overline{u}(x,\frac{(i+1)t}{k}))-H(\overline{u}(x,\frac{it}{k}))]dx\} \\
+ & \sum_{i=0}^{k-1}\{\int_{Q_{\frac{(i+1)t}{k}}\setminus Q_{\frac{it}{k}}}[H(u(x,\frac{it}{k}))-H(\overline{u}(x,\frac{it}{k}))]dx \\
= & \int\{\sum_{i=0}^{k-1}\chi_{Q_{\frac{(i+1)t}{k}}}(x)[H(u(x,\frac{(i+1)t}{k}))-H(u(x,\frac{it}{k}))]dx \\
& -\sum_{i=0}^{k-1}\chi_{Q_{\frac{(i+1)t}{k}}}(x)[H(\overline{u}(x,\frac{(i+1)t}{k}))-H(\overline{u}(x,\frac{it}{k}))]dx\} \\
+ & \int\sum_{i=0}^{k-1}\chi_{Q_{\frac{(i+1)t}{k}}\setminus Q_{\frac{it}{k}}}(x)[H(u(x,\frac{it}{k}))-H(\overline{u}(x,\frac{it}{k}))]dx
\end{aligned}
$$

We have

$$\begin{aligned}
&\int \chi_{Q_{\frac{(i+1)t}{k}}}(x)[H(u(x,\frac{(i+1)t}{k}))-H(u(x,\frac{it}{k}))]dx \\
&-\int \chi_{Q_{\frac{(i+1)t}{k}}}(x)[H(\overline{u}(x,\frac{(i+1)t}{k}))-H(\overline{u}(x,\frac{it}{k}))]dx \\
= &\ \frac{1}{k}\int \chi_{Q_{\frac{(i+1)t}{k}}}(x)\frac{\partial}{\partial s}H(u(x,s))|_{s=\frac{(i+1)t}{k}}dx \\
&-\frac{1}{k}\int \chi_{Q_{\frac{(i+1)t}{k}}}(x)\frac{\partial}{\partial s}H(\overline{u}(x,s))|_{s=\frac{(i+1)t}{k}}dx+O(\frac{1}{k}) \\
= &\ \frac{t}{2k}\int \chi_{Q_{\frac{(i+1)t}{k}}}(x)[\sum_j \frac{\partial^2}{\partial x_j^2}u(x,\frac{(i+1)t}{k})-\frac{\partial^2}{\partial x_j^2}\overline{u}(x,\frac{(i+1)t}{k})]dx+O(\frac{1}{k}) \\
\geq &\ O(1)
\end{aligned}$$

The last inequality is from the Green's formula. The main part is the surface integral of the normal derivative of $u(.,\frac{(i+1)t}{k})-\overline{u}(.,\frac{(i+1)t}{k})$ on the boundary of $Q_{\frac{(i+1)t}{k}}$. According to the definition of $Q_{\frac{(i+1)t}{k}}$, its outer normal derivative on the boundary is non-negative. On the set $Q_{\frac{(i+1)t}{k}} \setminus Q_{\frac{it}{k}}$, $H(u(x,\frac{(i+1)t}{k}))-H(\overline{u}(x,\frac{(i+1)t}{k}))\geq 0$ but $H(u(x,\frac{it}{k}))-H(\overline{u}(x,\frac{it}{k}))\leq 0$. So from the continuity,

$$\sum_{i=0}^{k-1}\chi_{Q_{\frac{(i+1)t}{k}}\setminus Q_{\frac{it}{k}}}(x)[H(u(x,\frac{it}{k}))-H(\overline{u}(x,\frac{it}{k}))]\to 0 \qquad a.e.$$

Therefore we deduce

$$\int_{Q_t}\{H(u(x,t))-H(\overline{u}(x,t))\}dx\geq 0$$

The physical meaning of the following lemma is clear. Since the space variable in (0.1) affects the equation through the unknown u, if u is spherical symmetrical at the beginning, so will it be later.

LEMMA 0.2. *Suppose that $u_0(x)=u_0$ is spherical symmetrical with center x_0 and there is a decreasing function $q(r)$ on R^1 such that $u_0(x)=q(|x|)$. Then $H(u(x,t))$ is always spherical symmetric with respect to its maximum point x_0. Moreover, $H(u(x,t))$ for each fixed t is a decreasing function of $|x-x_0|$.*

Proof. The spherical symmetry of $u(x,t)$ is an easy consequence of the symmetry of the initial condition and the form of (0.1). So it is sufficient to show that the symmetric center is just the maximum point.

Let us prove the Lemma by disproving its opposite statement. Denote by I_1 the vector in R^d such that its first component is 1 and all the other components are 0. If the statement of the lemma is not true, then we may find two differentiable functions $f_1(t)$ and $f_2(t)$ defined on an interval $[T,T+\delta$ such that

1) $u(f_1(T)I_1,T)=u(f_2(T)I_1,T)$

2) $u(f_1(t)I_1,t)<u(f_2(t)I_1,t);|f_1(t)-x_0|<|f_2(t)-x_0|,\ \forall t\in(T,T+\delta]$;

3) each $f_1(t)I_1$ is a minimum point of $u(x,t)$ in x, and each $f_2(t)I_1$ is a maximum point of $u(x,t)$ in x.

Therefore, there must be a time t_0 in $(T, T+\delta]$ such that

$$\frac{\partial}{\partial t}H(u(f_2(t_0),t_0)) > \frac{\partial}{\partial t}H(u(f_1(t_0),t_0)). \tag{0.6}$$

However

$$\begin{aligned}
&\frac{\partial}{\partial t}H(u(f_1(t_0)I_1,t_0))\\
= \;& \frac{\partial}{\partial t}H(u(f_2(t_0)I_1,t))|_{t=t_0} + \frac{\partial}{\partial x_1}H(u(f_1(t_0)I_1,t_0))\frac{d}{dt}f_1(t)|_{t=t_0}\\
= \;& \frac{1}{2}\sum_j \frac{\partial^2}{\partial x_j^2}u(f_1(t_0),t_0) \geq 0
\end{aligned}$$

the last equality holds because $\frac{\partial}{\partial x_1}u(f_1(t_0)I_1,t_0) = 0$. Similarly,

$$\frac{\partial}{\partial t}H(u(f_2(t_0),t_0)) = \frac{1}{2}\sum_i \frac{\partial^2}{\partial x_i^2}u(f_2(t_0),t_0) \leq 0$$

which are contradict to (0.6). So we get the conclusion.

Let Q be the class of all decreasing function $q(r)$ on R^1. Denote

$$Q_1 = \{q(.) \in Q,\ \exists y \in R^d,\ s.t.\ q(|x-y|) \leq H(u_0(x))\ \ a.e.x\};$$

$$Q_2 = \{q(.) \in Q,\ \exists y \in R^d,\ s.t.\ q(|x-y|) \geq H(u_0(x))\ \ a.e.x\}.$$

We call y as the defining point of $q(.)$. For any $q(.) \in Q_1 \cup Q_2$, define a new function

$$\begin{aligned}
\phi(x,q(.),t) &= \int (2\pi t)^{-\frac{d}{2}} \exp\{-\frac{|x-z|^2}{2t}\}q(|z-y|)dz\\
&= \int (2\pi t)^{-\frac{d}{2}} \exp\{-\frac{|z|^2}{2t}\}q(|x-z-y|)dz
\end{aligned}$$

where y is the defining point of $q(.)$

Denote

$$\gamma(q(.),t) = \sup_x\{\phi(x,q(.),t)\} = \phi(0,q(.),t).$$

Then it is easy to see that $\gamma(t) \searrow 0$ when $t \nearrow \infty$.

Denote $c = \inf_u uH^{-1}(u)$ and $C = \sup_u uH^{-1}(u)$. Then $0 < c < C < \infty$. The following theorem gives an upper bound and a lower bound estimates for the time that the i-th phase disappears everywhere.

THEOREM 0.3. *We have the following inequalities:*

1)
$$\begin{aligned}
&\inf\{t;\ ess\sup_x[H(u(x,t))] \leq H(u_i-)\}\\
\geq\;& C^{-1}\sup\{s;\ \gamma(q(.),2s) > C(2c^{-1}-C^{-1})H(u_i-)\}, \qquad \forall q(.) \in Q_1
\end{aligned}$$

and

2)
$$\begin{aligned}
&\inf\{t;\ ess\sup_x[H(u(x,t))] \leq H(u_i-)\}\\
\leq\;& 2\sup\{s;\ \gamma(q(.),cs) \geq c(2C-c)^{-1}H(u_i-)\}, \qquad \forall q(.) \in Q_2.
\end{aligned}$$

Proof. By Lemma 1, we may assume that the initial temperature is given by $q(|x|)$. Since X_t is a continuous martingale, there is a time change T_t such that there is a Brownian motic ı W_t and $X_t = W_{T_t}$. It is easy to see that

$$c \le \frac{dT_t}{dt} \le C.$$

Denote $B(\delta) = \{y \in R^d;\ |y_i| \le \delta\}$. Then the volume of $B(\delta)$ is $(2\delta)^d$.

1) Denote

$$t_* = \sup\{s;\ \gamma(q(.), 2s) > C(2c^{-1} - C^{-1})H(u_i-)\}.$$

Then

$$\begin{aligned} &\int_{B(\delta)} \int_{t_*}^{2t_*} \phi(x, q(.), s) ds\, dy \\ = &\int_{t_*}^{2t_*} E\{\chi_{B(\delta)}(W_s)\} ds \\ = &E\{\int_{T^{-1}(t_*)}^{T^{-1}(2t_*)} \chi_{B(\delta)}(W_{T_s}) \frac{dT_s}{ds} ds\} \\ \le &CE\{\int_{C^{-1}t_*}^{2c^{-1}t_*} \chi_{B(\delta)}(W_{T_s}) ds\} \\ = &CE\{\int_{C^{-1}t_*}^{2c^{-1}t_*} \chi_{B(\delta)}(X_s) ds\} \\ = &C\int_{C^{-1}t_*}^{2c^{-1}t_*} \int \chi_{B(\delta)}(H(u(y,s))) dy ds \end{aligned}$$

If the desired inequality is not true, then

$$H(u(y,s)) \le H(u_i-) \qquad \forall s \ge C^{-1}t_*.$$

So when δ is sufficiently small,

$$\int_{B(\delta)} \int_{t_*}^{2t_*} \phi(x, q(.), s) ds\, dy \le Ct_*(2\delta)^d(2c^{-1} - C^{-1})H(u_i-).$$

On the other hand, by the definition of t_* and the monotonicity of $\phi(0, q(.), s)$,

$$\int_{B(\delta)} \int_{t_*}^{2t_*} \phi(x, q(.), s) ds\, dy > (2\delta)^d C(2c^{-1} - C^{-1})H(u_i-)t_*$$

for small enough δ. So we get a contradiction.

2) Denote $t^* = \sup\{s;\ \gamma(q(.), cs) \ge c(2C - c)^{-1}H(u_i-)\}$. We have

$$\begin{aligned} &\int_{B(\delta)} \int_{t*}^{2t^*} H(u(y,s)) ds\, dy \\ = &\int_{t*}^{2t^*} E\{\chi_{B(\delta)}(X_s)\} ds \\ = &\int_{t*}^{2t^*} E\{\chi_{B(\delta)}(W_{T_s})\} ds \\ = &E\{\int_{t*}^{2t^*} \chi_{B(\delta)}(W_{T_s}) ds\} \end{aligned}$$

$$\begin{aligned} &\leq c^{-1}E\{\int_{t^*}^{2t^*} \chi_{B(\delta)}(W_{T_s})\frac{dT_s}{ds}ds\} \\ &= c^{-1}E\{\int_{T_{t*}}^{T_{2t^*}} \chi_{B(\delta)}(W_s)ds\} \\ &\leq c^{-1}E\{\int_{ct^*}^{2Ct^*} \chi_{B(\delta)}(W_s)ds\} \end{aligned}$$

So we get for small enough δ :

$$\int_{B(\delta)}\int_{t*}^{2t^*} H(u(y,s))ds\,dy < H(u_i-)(2\delta)^d t^*$$

On the other hand, if the conclusion fails, then

$$\int_{B(\delta)}\int_{t*}^{2t^*} H(u(y,s))ds\,dy \geq H(u_i-)(2\delta)^d t^*$$

So we get the lemma by finding another contradiction.

Remark. We replaced the initial distribution by a symmetrical one by virtue of Lemma 1 in the proof. So it is easy to see that the $ess\sup_x$ in the statement of the theorem may be replaced by $\sup_x$.

References

[1] C.M.Elliott and H.R.Ockendon, Weak and Variational Methods for Moving Boundary Problems, Pitman Publishing Inc. (1982);

[2] A.Fasano, M.Primicerio and A.A.Lacey, "New results on some classical parabolic free boundary problems", Q. Appl. Math. 38 (1981), 439-460;

[3] A.Fasano, "On a mathematical model for phase change accounting for coexistence of phases", Free Boundary Problems: Applications and Theory, Vol II, Pitman Publishing Inc., (1990), 694-698;

[4] J.M.Hill and J.N.Dewynne, Heat Conduction, Blackwell Scientific Publications, (1987);

[5] O.A.Ladyzenskaya, V.A.Solonnikov and N.N.Uralceva, Linear and Quasilinear Equations of Parabolic Type, AMS, (1968);

[6] J.R.Ockendon, "Generalised phase-change problems", Free Boundary Problems: Applications and Theory, Vol I, Pitman Publishing Inc., (1983), 231-236;

[7] M.Pierre, "Uniqueness if the solutions of $u_t - \Delta\phi(u) = 0$ with initial datum a measure", Nonlinear Analysis. Theory, Methods & Applications, Vol 6, No.2, 175-187 (1982);

[8] M.Primicerio and M.Ughi, "Phase-change problems with mushy region", Free Boundary Problems: Applications and Theory, Vol III, Pitman Publishing Inc., (1985);

[9] L.Rubinstein, "The Stefan problem: comments on its present state", J. Inst. Maths Applics, 24, (1979), 259-277;

[10] W.Zheng, "Conditional propagation of chaos and a class of quasi-linear PDE", The Annals in Probability Theroy, Vol.23, No.3, (1995), 1389-1413;

[11] W.Zheng, "Meyer's topology and Brownian motion in a composite medium, (In honors of P.A.Meyer and J.Neveu), ", Séminair de Probabilités XXX, Lecture Notes in Mathematics 1626 (1996), 108-116;

[12] W.Zheng, "A phase transition speed estimate in Stefan problem", Proceedings of the Royal Society of Edinburgh, 126A (1996), 1341-1347.

Department of Mathematics, University of California, Irvine, CA92697, USA, and Department of Statistics, East China Normal University, Shanghai, China

E-mail address: `wzheng@uci.edu`

Canadian Mathematical Society
Conference Proceedings
Volume **29**, 2000

Cocycles of Loop Groups

WANG ZHENGDONG

ABSTRACT. Let G be a compact semisimple Lie group, and LG=$\{\gamma : S^1 \xrightarrow{C^\infty} G\}$ be the loop group. Two types of representation of LG are known. One of them has been discussed extensively, namely the highest weight representation or positive energy representation (see [1]). Another type of representation is the energy representation (see [2]). The energy representation of LG is highly reducible. Its reduction theory has been worked out and its components are mainly based on a cocycle of LG (see [3], [4]). A geometric construction of such a cocycle is presented in this paper. As an application, we will derive all components of the energy representation and the Murray's construction of the central extension of loop groups.

Let K be a topological group, which acts continuously on a topological space X. We will consider some interesting representation of K on $\mathbf{L}^2(X,\mu)$, where μ is a probability measure on X.

Let H be an abelian topological group and ρ: H$\to\mathbf{C}^*$ be a character of H. We can define a line bundle $\mathcal{L}^\rho$ over X such that the space of its continuous sections is given by

$$\Gamma^0(\mathcal{L}^\rho) = \{s : X \times H \xrightarrow{C^0} \mathbf{C} \mid s(x, h_1h_2) = \rho^{-1}(h_2)s(x, h_1), \forall x \in X, h_1, h_2 \in H\}$$

Using the measure μ, we can define a scalar product on $\Gamma^0(\mathcal{L}^\rho)$ by

$$< s_1, s_2 >= \int_X s_1(x,h)s_2^*(x,h)d\mu(x)$$

for all $s_1, s_2 \in \Gamma^0(\mathcal{L}^\rho)$. The completion space of $\Gamma^0(\mathcal{L}^\rho)$, with respect to this scalar product, will be denoted by $\Gamma(\mathcal{L}^\rho)$. In fact, $\Gamma(\mathcal{L}^\rho)$ is isomorphic to $\mathbf{L}^2(X,\mu)$ by identifying a section s of $\mathcal{L}^\rho$ with a function f_s on X defined by $f_s(x) = \rho(h)s(x,h)$, $\forall x \in$X, $h \in$H.

Definition 1. A continuous map c: K$\times$X$\to$H is called an H-valued cocycle of K on X if c satisfies the following cocycle condition

$$c(\gamma_1 \cdot \gamma_2, x) = c(\gamma_1, \gamma_2 \cdot x)c(\gamma_2, x), \qquad \forall \gamma_1, \gamma_2 \in K, x \in X.$$

2000 *Mathematics Subject Classification.* Primary 55P35, 55P47.

Using a cocycle c, an action of K on X×H can be defined as

$$\gamma\cdot(x,h)=(\gamma\cdot x, c(\gamma,x)h), \qquad \forall\gamma\in \mathrm{K}, x\in \mathrm{X},$$

Furthermore, if the measure μ is quasi invariant with respect to the action of K, a unitary representation $\mathbf{U}^\rho$ of K can be defined on $\Gamma(\mathcal{L}^\rho)$ as

$$\left(\mathbf{U}^\rho(\gamma)s\right)(x,h)=s(\gamma^{-1}\cdot(x,h))\left(\frac{d\mu(\gamma^{-1}\cdot x)}{d\mu(x)}\right)^{\frac{1}{2}}$$

for any $\gamma\in$K, $s\in\Gamma(\mathcal{L}^\rho)$. Identifying $\Gamma(\mathcal{L}^\rho)$ with $\mathbf{L}^2$(X,μ), we have the following

Theorem 1. Suppose that c is an H valued cocycle of K on X, and μ is a K-quasi invariant measure on X. Then the unitary representation $\mathbf{U}^\rho:\gamma\rightarrow\mathbf{U}^\rho(\gamma)$ ($\gamma\in$K) of K can be realized on $\mathbf{L}^2$(X,μ) and is given by

$$\left(\mathbf{U}^\rho(\gamma)f\right)(x)=\rho^{-1}\left(c(\gamma^{-1},x)\right)\left(\frac{d\mu(\gamma^{-1}\cdot x)}{d\mu(x)}\right)^{\frac{1}{2}}f(\gamma^{-1}\cdot x)$$

for all f in $\mathbf{L}^2$(X,μ).

Definition 2. Suppose that W is a manifold (it might be infinite dimensional)and denote by PW the path space of smooth maps from $[0,2\pi]$ into W. Let $\mathrm{P}_{w_0}\mathrm{W}$ be the subspace of all smooth paths with starting point w_0 , i.e., $\mathrm{P}_{w_0}\mathrm{W}=\{w\in\mathrm{PW}|\ w(0)=w_0\}$. The loop space $\{w:\mathrm{S}^1\xrightarrow{C^\infty}\mathrm{W}\}$ is denoted by LW and it can be regarded as a subspace of PW.

If Q is a Lie group, then the path space PQ and the loop space LQ are groups with respect to the pointwise multiplication. Let π: B⟶M be a principal bundle with an abelian structural group T and projection map π. It is called Q-principal bundle if the group Q acts both on B and M such that $\gamma\cdot\pi(b)=\pi(\gamma\cdot b)$ for all $\gamma\in$Q and $b\in$B. Now we present a construction of a cocycle of the path group PQ or the loop group LQ as follows.

Suppose that π: B⟶M is a Q-principal bundle. With respect to a given connection α of this bundle, any smooth path $m\in$PM can be horizontally lifted to a path $\tilde{m}\in$PB. For any path $\gamma\in$PQ, the horizontally lifted path $\widetilde{\gamma\cdot m}$ of $\gamma\cdot m$ with starting point

$$\widetilde{\gamma\cdot m}(0)=\gamma(0)\cdot\tilde{m}(0).$$

is uniquely defined. Let o and e be the unit elements of T and Q respectively, and m_0 be a given point in M. A continuous map c_α: PQ×PM→P_oT can be defined such that

$$\gamma\cdot\tilde{m}=(\widetilde{\gamma\cdot m})c_\alpha(\gamma,m)$$

holds for any $\gamma\in\mathrm{PQ}, m\in\mathrm{PM}$. In fact, we have the following

Theorem 2. (1). c_α is a P_oT valued cocycle of PQ on PM.

(2). The restriction of c_α on LQ×LM is a P_oT valued cocycle of LQ on LM.

(3). The restriction of c_α on $\mathrm{P}_e\mathrm{Q}\times\mathrm{P}_{m_0}\mathrm{M}$ is a $\mathrm{P}_o\mathrm{T}$ valued cocycle of $\mathrm{P}_e\mathrm{Q}$ on $\mathrm{P}_{m_0}\mathrm{M}$.

PROOF. . It is easy to see that (2) and (3) follows from (1). Thus we only need to prove (1).

By the definition of c_α, we see that both of $(\gamma_1\cdot\gamma_2)\cdot\tilde{m}\cdot(c_\alpha(\gamma_1\cdot\gamma_2,m))^{-1}$ and $(\gamma_1\cdot(\widetilde{\gamma_2\cdot m}))\cdot(c_\alpha(\gamma_1,\gamma_2\cdot m))^{-1}$ represent the horizontally lifted paths of $\gamma_1\cdot\gamma_2\cdot m$ with same starting point $\gamma_1(0)\cdot\gamma_2(0)\cdot\tilde{m}(0)$ for any $m\in PM$, and $\gamma_1,\gamma_2\in PQ$. Hence we have

$$(\gamma_1\cdot\gamma_2)\cdot\tilde{m}\cdot(c_\alpha(\gamma_1\cdot\gamma_2,m))^{-1}=(\gamma_1\cdot(\widetilde{\gamma_2\cdot m}))\cdot(c_\alpha(\gamma_1,\gamma_2\cdot m))^{-1}$$

Combining this with

$$\widetilde{\gamma_2\cdot m}=(\gamma_2\cdot\tilde{m})\cdot(c_\alpha(\gamma_2,m))^{-1}$$

yields the cocycle condition for c_α:

$$(c_\alpha(\gamma_1\cdot\gamma_2,m))^{-1}=(c_\alpha(\gamma_2,m))^{-1}\cdot(c_\alpha(\gamma_1,\gamma_2\cdot m))^{-1}.$$

This completes the proof. □

In the following we will present two applications of our theorems. We will derive all components of the energy representation and the Murray's construction of the central extension of loop groups.

Application 1. Let G be a semisimple compact Lie group. Fix a maximal torus T of G. Let M=G/T be the quotient manifold. Given a connection α of the G-principal bundle G→M, by Theorem 2, we have a $\mathrm{P}_o\mathrm{T}$ valued cocycle c_α of LG on LM,

$$c_\alpha(\gamma,m)=(\widetilde{\gamma\cdot m})^{-1}\cdot(\gamma\cdot\tilde{m})\qquad\forall\gamma\in\mathrm{LG},\quad m\in\mathrm{LM},$$

where $\widetilde{\gamma\cdot m}$ and $\tilde{m}$ are the horizontally lifted paths of $\gamma\cdot m$ and m respectively such that

$$\widetilde{\gamma\cdot m}(0)=\gamma(0)\cdot\tilde{m}(0).$$

In the case of α being the G left invariant connection α_0, we have

$$\mathcal{P}(\widetilde{\gamma\cdot m}^{-1}d\widetilde{\gamma\cdot m})=0,$$

where $\mathcal{P}$ denotes the projection map from the Lie algebra $\mathcal{G}$ of G into the Lie algebra $\mathcal{T}$ of T. Combining this fact with $\mathcal{P}\circ Ad_t=\mathcal{P}$, $\forall t\in$T, (see [5]), we derive

$$\mathcal{P}\Big(c_{\alpha_0}(\gamma,m)dc_{\alpha_0}(\gamma,m)^{-1}+Ad_{\tilde{m}^{-1}}(\gamma^{-1}d\gamma)\Big)=0.$$

This yields

$$c_{\alpha_0}(\gamma,m)(\theta)=\exp\Big[\int_0^\theta\mathcal{P}\Big(Ad_{g_m^{-1}}(\gamma^{-1}d\gamma)\Big)\Big],\qquad\theta\in[0,2\pi],$$

where g_m: $\mathrm{S}^1\to$G is a lifted loop of m.

Set $q=c_{\alpha_0}^{-1}dc_{\alpha_0}$, i.e.,

$$q(\gamma,m)=\mathcal{P}\Big(Ad_{g_m^{-1}}(\gamma^{-1}d\gamma)\Big).$$

q is a L$\mathcal{T}$ valued cocycle of LG on LM. In fact, q can be defined on LG×L^{c}M (L^{c}M being the set of continuous loops in M) without any difficulty and q is thus a L$^c\mathcal{T}$ (the set of continuous loops in $\mathcal{T}$) valued cocycle of LG on L^{c}M.

Let μ be the standard Brownian measure on L^{c}M and λ: L$^c\mathcal{T}\to$R^1 be a linear function. By Theorem 1, we have a unitary representation $\mathbf{U}^\lambda$ of LG. The operator $\mathbf{U}^\lambda(\gamma)$ ($\gamma\in$LG) is given by

$$\Big(\mathbf{U}^\lambda(\gamma)f\Big)(m)=e^{-i<\lambda,q(\gamma^{-1},m)>}\left(\frac{d\mu(\gamma^{-1}\cdot m)}{d\mu(m)}\right)^{\frac{1}{2}}f(\gamma^{-1}\cdot m)$$

for all f in $\mathbf{L}^2$(L^{c}M,μ). Clearly, the unitary representation $\mathbf{U}^\lambda$ appears as a component of the energy representation of LG studied in [3] and [4]. We remark here that a similar construction for the representation of the loop algebra is given in [6]. To give a unitary representation of LG intertwined with the natural action of S^1, we need a S^1 invariant cocycle of LG, which has been constructed in [7].

Application 2. Murray (1988) gives an interesting construction of the central extension U(1)→ $\widetilde{\text{LG}}$ →LG of the loop group LG (see [8]). Murray constructs the central extension P(LG)×U(1) of the group P(LG) of paths in LG with a cocycle (called the Murray's cocycle) given by the holonomy around a certain path in LG. Then he shows that there is a normal subgroup N of the central extension of P(LG) such that the quotient group is $\widetilde{\text{LG}}$, i.e.,

$$\widetilde{\text{LG}}=\text{P(LG)}\times\text{U(1)/N}.$$

A similar discussion can also be found in Mickelsson's paper [9].

As an application of Theorem 2, we will derive the Murray's construction of the central extension of loop groups.

It is known that the central extension $\widetilde{\text{LG}}$ of the loop group LG is determined by a LG invariant two form ω on LG (in fact ω is the curvature of the LG left invariant connection α_0 of the principal bundle U(1)→ $\widetilde{\text{LG}}$ →LG). By Theorem 2, we have a P$_o$(U(1)) valued cocycle c_{α_0} of P$_e$($\widetilde{\text{LG}}$) on P$_e$(LG) (the unit elements of $\widetilde{\text{LG}}$ and LG are denoted by same symbol e for simplicity) given by

$$c_{\alpha_0}(\gamma,m)=(\widetilde{\gamma\cdot m})^{-1}\cdot(\gamma\cdot\tilde{m})$$

for any $\gamma\in$P$_e$($\widetilde{\text{LG}}$), $m\in$P$_e$(LG). Using this cocycle, the action of P$_e$($\widetilde{\text{LG}}$) on P$_e$(LG)×P$_o$(U(1)) can be defined by

$$\gamma\cdot(m,h)=(\gamma\cdot m,c_{\alpha_0}(\gamma,m)h)$$

for all $\gamma\in$P$_e$($\widetilde{\text{LG}}$), $m\in$ P$_e$(LG) and $h\in$ P$_o$(U(1)).

Define a map

$$\phi : \mathrm{P}_e(\mathrm{LG}) \times \mathrm{P}_o(\mathrm{U}(1)) \longrightarrow \mathrm{P}_e(\widetilde{\mathrm{LG}})$$

by

$$\phi : (m, h) \to \tilde{m} \cdot h, \quad m \in \mathrm{P}_e(\mathrm{LG}), h \in \mathrm{P}_o(\mathrm{U}(1)),$$

where $\tilde{m}$ represents the horizontally lifted path of m with starting point $\tilde{m}(0) = e$. It is easy to see that ϕ is an isomorphic map. Furthermore, $\mathrm{P}_e(\mathrm{LG}) \times \mathrm{P}_o(\mathrm{U}(1))$ is a group with multiplication given by

$$(m_1, h_1) \cdot (m_2, h_2) = (m_1 \cdot m_2, h_1 h_2 c(m_1, m_2))$$

for all $m_1, m_2 \in \mathrm{P}_e(\mathrm{LG}), h_1, h_2 \in \mathrm{P}_o(\mathrm{U}(1))$, where

$$c(m_1, m_2) = c_{\alpha_0}(\tilde{m}_1, m_2),$$

$\tilde{m}_1$ being the horizontally lifted path of m_1 with starting point $\tilde{m}_1(0) = e$ and we have the following

Fact 1. $\phi : \mathrm{P}_e(\mathrm{LG}) \times \mathrm{P}_o(\mathrm{U}(1)) \longrightarrow \mathrm{P}_e(\widetilde{\mathrm{LG}})$ is a group isomorphic map.

Set $\mathrm{L}_e(\widetilde{\mathrm{LG}}) = \{\gamma \in \mathrm{P}_e(\widetilde{\mathrm{LG}}) \mid \gamma(0) = \gamma(2\pi) = e\}$. It is easy to see that $\mathrm{L}_e(\widetilde{\mathrm{LG}})$ is a normal subgroup of $\mathrm{P}_e(\widetilde{\mathrm{LG}})$ and the quotient group is $\widetilde{\mathrm{LG}}$, i.e.,

Fact 2. $\widetilde{\mathrm{LG}} = \mathrm{P}_e(\widetilde{\mathrm{LG}})/\mathrm{L}_e(\widetilde{\mathrm{LG}})$.

Let $\mathrm{Z} = \phi^{-1}(\mathrm{L}_e(\widetilde{\mathrm{LG}}))$ be the preimage of ϕ. Clearly, Z is a normal subgroup of $\mathrm{P}_e(\mathrm{LG}) \times \mathrm{P}_o(\mathrm{U}(1))$ and is given by

$$\mathrm{Z} = \{(m, h) \mid m \in \mathrm{L}_e(\mathrm{LG}), h \in \mathrm{P}_o(\mathrm{U}(1)), h(2\pi) = hol^{-1}(m)\},$$

where $hol(m)$ denotes the holonomy around m, i.e., $hol(m) = \tilde{m}(0)^{-1}\tilde{m}(2\pi)$. By Fact 1 and Fact 2, we get the following

Fact 3. $\widetilde{\mathrm{LG}} \cong \mathrm{P}_e(\mathrm{LG}) \times \mathrm{P}_o(\mathrm{U}(1))/\mathrm{Z}$.

Now we define a group homomorphism

$$\psi : \mathrm{P}_e(\mathrm{LG}) \times \mathrm{P}_o(\mathrm{U}(1)) \longrightarrow \mathrm{P}_e(\mathrm{LG}) \times \mathrm{U}(1)$$

by

$$\psi(m, h) = (m, h(2\pi)), \quad \forall m \in \mathrm{P}_e(\mathrm{LG}), h \in \mathrm{P}_o(\mathrm{U}(1)).$$

Notice that $\mathrm{P}_e(\mathrm{LG}) \times \mathrm{U}(1)$ is a group with the multiplication given by

$$(m_1, u_1) \cdot (m_2, u_2) = (m_1 \cdot m_2, u_1 u_2 \kappa(m_1, m_2))$$

for any $m_1, m_2 \in \mathrm{P}_e(\mathrm{LG}), u_1, u_2 \in \mathrm{U}(1)$, where

$$\kappa(m_1, m_2) = c(m_1, m_2)(2\pi).$$

Hence $\mathrm{P}_e(\mathrm{LG}) \times \mathrm{U}(1)$ is a central extension of $\mathrm{P}_e(\mathrm{LG})$ with the Murray's cocycle κ.

Set $\mathrm{N} = \psi(\mathrm{Z})$. It is a normal subgroup of $\mathrm{P}_e(\mathrm{LG}) \times \mathrm{U}(1)$ and is given by

$$\mathrm{N} = \{(m, hol^{-1}(m)) \mid m \in \mathrm{L}_e(\mathrm{LG})\}.$$

Observe that $\mathrm{Z} = \psi^{-1}(\mathrm{N})$. Thus we have

$$\mathrm{P}_e(\mathrm{LG}) \times \mathrm{P}_o(\mathrm{U}(1))/\mathrm{Z} \cong \mathrm{P}_e(\mathrm{LG}) \times \mathrm{U}(1)/\mathrm{N}.$$

It follows from Fact 3 that

$$\widetilde{\mathrm{LG}} \cong \mathrm{P}_e(\mathrm{LG}) \times \mathrm{U}(1)/\mathrm{N}.$$

Now the Murray's construction of the central extension of loop groups is derived.

By the relation between the cocycles κ and c_{α_0}, we see that

$$\kappa(m_1, m_2) = (\widetilde{m_1 \cdot m_2})^{-1}(2\pi) \cdot \tilde{m}_1(2\pi) \cdot \tilde{m}_2(2\pi),$$

where $\tilde{m}_1, \tilde{m}_2$ and $\widetilde{m_1 \cdot m_2}$ represent the horizontally lifted paths of m_1, m_2 and $m_1 \cdot m_2$ with same starting point $\tilde{m}_1(0) = \tilde{m}_2(0) = \widetilde{m_1 \cdot m_2}(0) = e$. This yields the Murray's formula

$$\kappa(m_1, m_2) = hol^{-1}(m_1 \cdot m_2) hol(m_1) hol(m_2)$$

for any $m_1, m_2 \in \mathrm{P}_e(\mathrm{LG})$. Though the cocycle c_{α_0} depends on the multiplication of $\widetilde{\mathrm{LG}}$, the Murray's cocycle κ only depends the holonomy of loops in LG. Note that the holonomy of a loop m in LG is the exponential of the integral of the curvature ω over any disk D in LG whose boundary is the loop m, i.e.

$$hol(m) = \exp(i \int_D \omega).$$

Acknowledgement: I would like to express my thanks to the referee for his helpful remarks. This work is partially supported by Chinese National Science Foundation and China Education Committee Foundation.

References

[1] A. Pressley and G. Segal, Loop groups, Oxford: Clarendon Press, 1986.

[2] S. Albeverio and R. Høegh-Krohn, The energy representation of Sobolev-Lie groups, Compositio Math. 36 (1978), 37-52.

[3] S. Albeverio, R. Høegh-Krohn, D. Testard and A. Vershik, Factorial representations of path groups, J. Funct. Anal. 51 (1981), 115-131.

[4] S. Albeverio, R. Høegh-Krohn, J. Marion, D. Testard and B. Torrésani, Noncommutative Distributions, Marcel Dekker, New York, 1993.

[5] S. Helgason, Differential Geometry, Lie Group and Symmetric Spaces, 2nd ed., Academic Press, New York, 1978.

[6] Wang Zheng-dong, The construction of a representation of loop algebras, Lett. Math. Phys., 38 (1996), 377 - 388.

[7] Wang Zheng-dong, Non highest weight representation of loop groups, preprint.

[8] M. Murray, Another construction of the central extension of the loop group, Commun. Math. Phys., 116, 73-80 (1988).

[9] J. Mickelsson, Kac-Moody groups, topology of the Dirac determinant bundle, and Fermionzation, Commun. Math. Phys., 110, 173-183 (1987).

Department of Mathematics, Peking University, Beijing 100871,P. R. China
E-mail address: zdwang@pku.edu.cn